Marine Mammal Acoustics in a Noisy Ocean

Marine Mammal Acoustics in a Noisy Ocean

Christine Erbe • Dorian Houser
Ann Bowles • Michael B. Porter
Editors

Marine Mammal Acoustics in a Noisy Ocean

Editors
Christine Erbe
Centre for Marine Science
and Technology
Curtin University
Perth, WA, Australia

Dorian Houser
National Marine Mammal
Foundation
San Diego, CA, USA

Ann Bowles
Hubbs-Sea World Research Institute
San Diego, CA, USA

Michael B. Porter
HLS Research, Inc.
San Diego, CA, USA

ISBN 978-3-031-77021-0   ISBN 978-3-031-77022-7   (eBook)
https://doi.org/10.1007/978-3-031-77022-7

© The Editor(s) (if applicable) and The Author(s) 2025. This book is an open access publication.
Jointly published with ASA Press

Open Access This book is licensed under the terms of the Creative Commons Attribution 4.0 International License (http://creativecommons.org/licenses/by/4.0/), which permits use, sharing, adaptation, distribution and reproduction in any medium or format, as long as you give appropriate credit to the original author(s) and the source, provide a link to the Creative Commons license and indicate if changes were made.
The images or other third party material in this book are included in the book's Creative Commons license, unless indicated otherwise in a credit line to the material. If material is not included in the book's Creative Commons license and your intended use is not permitted by statutory regulation or exceeds the permitted use, you will need to obtain permission directly from the copyright holder.
The use of general descriptive names, registered names, trademarks, service marks, etc. in this publication does not imply, even in the absence of a specific statement, that such names are exempt from the relevant protective laws and regulations and therefore free for general use.
The publishers, the authors, and the editors are safe to assume that the advice and information in this book are believed to be true and accurate at the date of publication. Neither the publishers nor the authors or the editors give a warranty, express or implied, with respect to the material contained herein or for any errors or omissions that may have been made. The publishers remain neutral with regard to jurisdictional claims in published maps and institutional affiliations.

Cover photo: Southern resident killer whales (*Orcinus orca*) in British Columbia, Canada.
© *Katherine Gavrilchuk, Fisheries and Oceans Canada.*

This Springer imprint is published by the registered company Springer Nature Switzerland AG
The registered company address is: Gewerbestrasse 11, 6330 Cham, Switzerland

If disposing of this product, please recycle the paper.

The ASA Press

ASA Press, which represents a collaboration between the Acoustical Society of America and Springer Nature, is dedicated to encouraging the publication of important new books as well as the distribution of classic titles in acoustics. These titles, published under a dual ASA Press/Springer imprint, are intended to reflect the full range of research in acoustics. ASA Press titles can include all types of books that Springer publishes, and may appear in any appropriate Springer book series.

Editorial Board

Ning Xiang (Chair), University of Texas at Austin
Timothy F. Duda, Woods Hole Oceanographic Institution
Gary W. Elko, MH Acoustics LLC
Robin S. Glosemeyer-Petrone, Threshold Acoustics
Skyler G. Jennings, Skyler G. Jennings
Darlene R. Ketten, Boston University
Philip L. Marston, Washington State University
Andrew N. Norris, Rutgers University
Christine H. Shadle, Haskins Laboratories

Ex Officio

William M. Hartmann, Michigan State University
James F. Lynch, Woods Hole Oceanographic Institution
Arthur N. Popper, University of Maryland

Preface

It is done! It only took 12 years—during which I sometimes felt that it could not be done.

The idea for this book was planted and replanted by so many people who said to me that somebody should attempt to update Richardson et al. 1995 "Marine Mammals and Noise" (Academic Press)—a seminal book that strongly influenced research and policy.

In late 2011, I found my team of co-editors: Dorian, Ann, and Mike. We sent proposals to potential funders, who would cover some of our time and purchase of material to be reviewed (articles and reports). In 2013, we had raised sufficient support to get started—thinking it would only take us a couple of years.

Yet, the task soon felt Herculean. The literature expanded exponentially. Methodologies for assessing and managing underwater noise improved; for example, NOAA drastically changed its technical guidance for assessing noise effects while we were in the middle of writing this book. Standards on underwater acoustic terminology, as well as on the measurement of underwater noise from ships and pile driving, were published after we had already completed those chapters. While it is exciting to be part of this continually advancing field, writing this book became an emotional rollercoaster. Keeping all chapters up-to-date at the same time started to seem unattainable.

That's when I reached out to colleagues around the globe and within my own team. Not a single person declined; the response was overwhelming. I am not exaggerating when I say that everyone dropped what they were doing at the time, in order to revise the various chapters concurrently.

And here we are. And here it is.

I am extremely grateful to my co-editors and co-authors. This book would not be in our hands now if it weren't for your expert inputs and grand efforts.

I would like to thank Mike Weise at the Office of Naval Research, Anu Kumar of the Naval Facilities Engineering and Expeditionary Warfare Center, Jason Gedamke of the National Oceanic and Atmospheric Administration, and the various industry representatives on the International Association of Oil & Gas Producers' Sound and Marine Life Joint Industry Programme for supporting the early drafts of these chapters, and Joel Feldmeier of the Office of Naval Research Global for covering the open-access fees. I know you all feared we would never get this book done. I apologize it has taken so long.

I wish to thank Ross Chapman, Isabelle Charrier, Randall Davis, Alec Duncan, Christ de Jong, Heike Herata, John Hildebrand, Dorian Houser, Saana Isojunno, Ryan Jones, Ron Kastelein, Klaus Lucke, Laura May-Collado, Leslie New, Sofia Pastor, Colleen Reichmuth, Brandi Ruscher, Jillian Sills, John Terhune, Joy Tripovich, Peggy Walton, and Bernd Würsig for reviewing one or more chapters.

Lars Koerner from Springer Verlag, thank you for your unwavering support—editorial and emotional.

I know this book will soon be out-of-date (hopefully not already by the time it's printed). But I hope it might serve our marine-mammals-and-noise community—for a little while.

Perth, WA Australia
June 2024

Christine Erbe

Marine Mammal Taxonomic Species and their Common Names

Taxonomic Group and Species	Common Name
Infraorder Cetacea	***Whales and Dolphins***
Mysticeti	**Baleen Whales**
Family Balaenidae	Right whales
Balaena mysticetus Linnaeus, 1758	Bowhead whale, Greenland whale
Eubalaena australis Desmoulins, 1822	Southern right whale
Eubalaena glacialis Müller, 1776	North Atlantic right whale
Eubalaena japonica Lacépède, 1818	North Pacific right whale
Family Neobalaenidae	
Caperea marginata Gray, 1846	Pygmy right whale
Family Eschrichtiidae	
Eschrichtius robustus Lilljeborg, 1861	Gray whale
Family Balaenopteridae	Rorquals
Balaenoptera acutorostrata Lacépède, 1804	Common minke whale
B. a. acutorostrata Lacépède, 1804	North Atlantic minke whale
B. a. scammoni Deméré, 1986	North Pacific minke whale
Balaenoptera bonaerensis Burmeister, 1867	Antarctic minke whale
Balaenoptera borealis Lesson, 1828	Sei whale
B. b. borealis Lesson, 1828	Northern sei whale
B. b. schlegelii Flower, 1865	Southern sei whale
Balaenoptera edeni Anderson, 1879	Bryde's whale
B. e. brydei Olsen, 1913	Bryde's whale
B. e. edeni Anderson, 1879	Eden's whale
Balaenoptera musculus Linnaeus, 1758	Blue whale
B. m. brevicauda Ichihara, 1966	Pygmy blue whale
B. m. indica Blyth, 1859	Antarctic blue whale
B. m. intermedia Burmeister, 1871	Northern blue whale
B. m. musculus Linnaeus, 1758	Blue whale
Baleanoptera omurai Wada, Oishi and Yamada, 2003	Omura's whale
Balaenoptera physalus Linnaeus, 1758	Fin whale
B. p. physalus Linnaeus, 1758	North Atlantic fin whale
B. p. quoyi Fischer, 1829	Southern fin whale
B. p. velifera Cope in Scammon, 1869	North Pacific fin whale
Balaenoptera ricei Rosel, Wilcox, Yamada and Mullin, 2021	Rice's whale
Megaptera novaeangliae Borowski, 1781	Humpback whale
M. n. australis Lesson, 1828	Southern humpback whale

(continued)

Taxonomic Group and Species	Common Name
M. n. kuzira Gray, 1850	North Pacific humpback whale
M. n. novaeangliae Borowski, 1781	North Atlantic humpback whale
Odontoceti	**Toothed Whales**
Physeter macrocephalus Linnaeus, 1758	Sperm whale, cachalot
Family Kogiidae	
Kogia breviceps Blainville, 1838	Pygmy sperm whale
Kogia sima Owen, 1866	Dwarf sperm whale
Family Ziphiidae	Beaked whales
Berardius arnuxii Duvernoy, 1851	Arnoux's beaked whale
Berardius bairdii Stejneger, 1883	Baird's beaked whale
Berardius minimus Yamada, Kitamura and Matsuishi, 2019	Sato's beaked whale
Hyperoodon ampullatus Forster in Kalm, 1770	Northern bottlenose whale
Hyperoodon planifrons Flower, 1882	Southern bottlenose whale
Indopacetus pacificus Longman, 1926	Longman's beaked whale, tropical bottlenose whale
Mesoplodon bidens Sowerby, 1804	Sowerby's beaked whale
Mesoplodon bowdoini Andrews, 1908	Andrew's beaked whale
Mesoplodon carlhubbsi Moore, 1963	Hubbs' beaked whale
Mesoplodon densirostris Desmarest, 1817	Blainville's beaked whale
Mesoplodon eueu Carroll et al. 2021	Ramari's beaked whale
Mesoplodon europaeus Gervais, 1855	Gervais' beaked whale
Mesoplodon ginkgodens Nishiwaki and Kamiya, 1958	Ginkgo-toothed beaked whale
Mesoplodon grayi von Haast, 1876	Gray's beaked whale
Mesoplodon hectori Gray, 1871	Hector's beaked whale
Mesoplodon hotaula Deraniyagala, 1963	Deraniyagala's beaked whale
Mesoplodon layardii Gray, 1865	Strap-toothed beaked whale, Layard's beaked whale
Mesoplodon mirus True, 1913	True's beaked whale
Mesoplodon perrini Dalebout, Mead, Baker, Baker and van Helden, 2002	Perrin's beaked whale
Mesoplodon peruvianus Reyes, Mead and Van Waerebeek, 1991	Pygmy beaked whale
Mesoplodon stejnegeri True, 1885	Stejneger's beaked whale
Mesoplodon traversii Gray, 1874	Spade-toothed beaked whale
Tasmacetus shepherdi Oliver, 1937	Shepherd's beaked whale, Tasman beaked whale
Ziphius cavirostris G. Cuvier, 1823	Cuvier's beaked whale, goose-beaked whale
Family Platanistidae	
Platanista gangetica Lebeck, 1801	Ganges river dolphin, susu
Platanista minor Owen, 1853	Indus river dolphin, bhulan
Family Iniidae	
Inia geoffrensis Blainville, 1817	Amazon river dolphin
I. g. boliviensis d'Orbigny, 1834	Bolivian bufeo
I. g. geoffrensis Blainville, 1817	Common boto
Family Lipotidae	
Lipotes vexillifer Miller, 1918	Baiji, Yangtze river dolphin (possibly extinct)
Family Pontoporiidae	
Pontoporia blainvillei Gervais and d'Orbigny, 1844	Franciscana, toninha
Family Monodontidae	
Delphinapterus leucas Pallas, 1776	Beluga, white whale
Monodon monoceros Linnaeus, 1758	Narwhal

(continued)

Taxonomic Group and Species	Common Name
Family Delphinidae	
Cephalorhynchus commersonii Lacépède, 1804	Commerson's dolphin
C. c. commersonii Lacépède, 1804	Commerson's dolphin
C. c. kerguelenensis Robineau, Goodall, Pichler and C. S. Baker, 2007	Kerguelen dolphin
Cephalorhynchus eutropia Gray, 1846	Chilean dolphin
Cephalorhynchus heavisidii Gray, 1828	Heaviside's dolphin
Cephalorhynchus hectori Van Beneden, 1881	Hector's dolphin
C. h. hectori Van Beneden, 1881	South Island Hector's dolphin
C. h. maui A. Baker, Smith and Pichler, 2002	Māui dolphin, North Island Hector's dolphin
Delphinus delphis Linnaeus, 1758	Common dolphin
D. d. bairdii Dall, 1873	Eastern North Pacific long-beaked common dolphin
D. d. delphis Linnaeus, 1758	Common dolphin
D. d. ponticus Barabash, 1935	Black Sea common dolphin
D. d. tropicalis van Bree, 1971	Indo-Pacific common dolphin
Feresa attenuata Gray, 1874	Pygmy killer whale
Globicephala macrorhynchus Gray, 1846	Short-finned pilot whale
Globicephala melas Traill, 1809	Long-finned pilot whale
G. m. edwardii A. Smith 1834	Southern long-finned pilot whale
G. m. melas Traill, 1809	North Atlantic long-finned pilot whale
Grampus griseus G. Cuvier, 1812	Risso's dolphin, grampus
Lagenodelphis hosei Fraser, 1956	Fraser's dolphin
Lagenorhynchus acutus Gray, 1828	Atlantic white-sided dolphin
Lagenorhynchus albirostris Gray, 1846	White-beaked dolphin
Lagenorhynchus australis Peale, 1849	Peale's dolphin
Lagenorhynchus cruciger Quoy and Gaimard, 1824	Hourglass dolphin
Lagenorhynchus obliquidens Gill, 1865	Pacific white-sided dolphin
Lagenorhynchus obscurus Gray, 1828	Dusky dolphin
L. o. fitzroyi Waterhouse, 1838	Fitzroy's dolphin
L. o. obscurus Gray, 1828	African dusky dolphin
L. o. posidonia Philippi, 1893	Peruvian/Chilean dusky dolphin
Lissodelphis borealis Peale, 1849	Northern right-whale dolphin
Lissodelphis peronii Lacépède, 1804	Southern right-whale dolphin
Orcaella brevirostris Owen in Gray, 1866	Irrawaddy dolphin, pesut
Orcaella heinsohni Beasley, Robertson and Arnold, 2005	Australian snubfin dolphin
Orcinus orca Linnaeus, 1758	Killer whale, orca
Peponocephala electra Gray, 1846	Melon-headed whale
Pseudorca crassidens Owen, 1846	False killer whale
Sousa teuszii Kükenthal, 1892	Atlantic humpback dolphin
Sousa chinensis Osbeck, 1765	Indo-Pacific humpback dolphin
S. c. chinensis Osbeck, 1765	Chinese Humpback dolphin
S. c. taiwanensis Wang, Yang and Hung, 2015	Taiwanese humpback dolphin
Sousa plumbea G. Cuvier, 1829	Indian Ocean humpback dolphin
Sousa sahulensis Jefferson and Rosenbaum, 2014	Australian humpback dolphin
Sotalia fluviatilis Gervais and Deville in Gervais, 1853	Tucuxi
Sotalia guianensis P.J. Van Beneden, 1864	Guiana dolphin, costero
Stenella attenuata Gray, 1846	Pantropical spotted dolphin
S. a. attenuata Gray, 1846	Offshore pantropical spotted dolphin
S. a. graffmani Lönnberg, 1934	Coastal pantropical spotted dolphin

(continued)

Taxonomic Group and Species	Common Name
Stenella clymene Gray, 1850	Clymene dolphin
Stenella coeruleoalba Meyen, 1833	Striped dolphin
Stenella frontalis G. Cuvier, 1829	Atlantic spotted dolphin
Stenella longirostris Gray, 1828	Spinner dolphin
S. l. centroamericana Perrin, 1990	Central American spinner dolphin
S. l. longirostris Gray, 1828	Gray's spinner dolphin
S. l. orientalis Perrin, 1990	Eastern spinner dolphin
S. l. roseiventris Wagner, 1846	Dwarf spinner dolphin
Steno bredanensis Lesson, 1828	Rough-toothed dolphin
Tursiops aduncus Ehrenberg, 1832	Indo-Pacific bottlenose dolphin
Tursiops erebennus Cope 1865	Tamanend's bottlenose dolphin
Tursiops truncatus Montagu, 1821	Common bottlenose dolphin
T. t. gephyreus Lahille, 1908	Lahille's bottlenose dolphin
T. t. nuuanu Andrews, 1911	Eastern Tropical Pacific bottlenose dolphin
T. t. ponticus Barabash-Nikiforov, 1940	Black Sea bottlenose dolphin
T. t. truncatus Montagu, 1821	Common bottlenose dolphin
Family Phocoenidae	Porpoises
Neophocaena phocaenoides G. Cuvier, 1829	Indo-Pacific finless porpoise
Neophocaena asiaeorientalis Pilleri and Gihr, 1972	Narrow-ridged finless porpoise
N. a. asiaeorientalis Pilleri and Gihr, 1972	Yangtze finless porpoise
N. a. sunameri Pilleri and Gihr, 1975	East Asian finless porpoise
Phocoena dioptrica Lahille, 1912	Spectacled porpoise
Phocoena phocoena Linnaeus, 1758	Harbor porpoise
P. p. phocoena Linnaeus, 1758	Atlantic harbor porpoise
P. p. relicta Abel, 1905	Black Sea harbor porpoise
P. p. vomerina Gill, 1865	Pacific harbor porpoise
Phocoena sinus Norris and McFarland, 1958	Vaquita, Gulf of California harbor porpoise
Phocoena spinipinnis Burmeister, 1865	Burmeister's porpoise
Phocoenoides dalli True, 1885	Dall's porpoise
P. d. dalli True, 1885	*dalli*-type Dall's porpoise
P. d. truei Andrews, 1911	*truei*-type Dall's porpoise
Order Carnivora	***Carnivores***
Family Ursidae	
Ursus maritimus Linnaeus, 1758	Polar bear
Family Mustelidae	Otters (marine species)
Enhydra lutris Linnaeus, 1758	Sea otter
E. l. kenyoni Wilson, 1991	Eastern sea otter
E. l. lutris Linnaeus, 1758	Western sea otter
E. l. nereis Merriam, 1904	Southern sea otter
Lontra felina Molina, 1782	Chungungo, marine otter
Pinnipedia	**Seals and Sea Lions**
Family Otariidae	Eared seals and sea lions
Arctocephalus australis Zimmermann, 1783	South American fur seal
Arctocephalus forsteri Lesson, 1828	Long-nosed fur seal, New Zealand fur seal
Arctocephalus galapagoensis Heller, 1904	Galapagos fur seal
Arctocephalus gazella Peters, 1875	Antarctic fur seal
Arctocephalus philippii Peters, 1866	Juan Fernandez fur seal
A. p. philippii Peters, 1866	Juan Fernandez fur seal
A. p. townsendi Merriam, 1897	Guadalupe fur seal

(continued)

Marine Mammal Taxonomic Species and their Common Names

Taxonomic Group and Species	Common Name
Arctocephalus pusillus Schreber, 1775	Cape fur seal
A. p. pusillus Schreber, 1775	Cape fur seal
A. p. doriferus Wood Jones, 1925	Australian fur seal
Arctocephalus tropicalus Gray, 1872	Subantarctic fur seal
Callorhinus ursinus Linnaeus, 1758	Northern fur seal
Eumetopias jubatus Schreber, 1776	Steller sea lion
E. j. jubatus Schreber, 1776	Western Steller sea lion
E. j. monteriensis Gray, 1859	Loughlin's Steller sea lion
Neophoca cinerea Peron, 1816	Australian sea lion
Otaria byronia Blainville, 1820	South American sea lion
Phocarctos hookeri Gray, 1844	New Zealand sea lion, Hooker's sea lion,
Zalophus californianus Lesson, 1828	California sea lion
Zalophus wollebaeki Sivertsen, 1953	Galapagos sea lion
Family Odobenidae	Walruses
Odobenus rosmarus Linnaeus, 1758	Walrus
O. r. divergens Illiger, 1815	Pacific walrus
O. r. rosmarus Linnaeus, 1758	Atlantic walrus
Family Phocidae	Earless seals
Cystophora cristata Erxleben, 1777	Hooded seal
Erignathus barbatus Erxleben, 1777	Bearded seal
E. b. barbatus Erxleben, 1777	Atlantic bearded seal
E. b. nauticus Pallas, 1881	Pacific bearded seal
Halichoerus grypus Fabricius, 1791	Gray seal
H. g. grypus Fabricius, 1791	Baltic gray seal
H. g. atlantica Nehring, 1866	Atlantic gray seal
Histriophoca fasciata Zimmerman, 1783	Ribbon seal
Hydrurga leptonyx Blainville, 1820	Leopard seal
Leptonychotes weddellii Lesson, 1826	Weddell seal
Lobodon carcinophaga Hombron and Jacquinot, 1842	Crabeater seal
Mirounga leonina Linnaeus, 1758	Southern elephant seal
Mirounga angustirostris Gill, 1866	Northern elephant seal
Monachus monachus Hermann, 1779	Mediterranean monk seal
Neomanachus schauinslandi Matschie, 1905	Hawaiian monk seal
Ommatophoca rossii Gray, 1844	Ross seal
Pagophilus groenlandicus Erxleben, 1777	Harp seal
Phoca vitulina Linnaeus, 1758	Harbor seal, common seal
P. v. vitulina Linnaeus, 1758	Atlantic harbor seal
P. v. mellonae Doutt, 1942	Ungava harbor seal
P. v. richardii Gray, 1864	Pacific harbor seal
Phoca largha Pallas, 1811	Spotted seal, largha seal
Pusa hispida Schreber, 1775	Ringed seal
P. h. botnica Gmelin, 1788	Baltic ringed seal
P. h. hispida Schreber, 1775	Arctic ringed seal
P. h. ladogensis Nordquist, 1889	Lake Ladoga seal
P. h. ochotensis Pallas, 1811	Okhotsk ringed seal
P. h. saimensis Nordquist, 1889	Saima seal
Pusa caspica Gmelin, 1788	Caspian seal
Pusa sibirica Gmelin, 1788	Baikal seal

(continued)

Taxonomic Group and Species	Common Name
Order Sirenia	***Sea Cows***
Family Trichechidae	Manatees
Trichechus inunguis Natterer, 1883	Amazonian manatee
Trichechus manatus Linnaeus, 1758	West Indian manatee
T. m. latirostris Harlan, 1824	Florida manatee
T. m. manatus Linnaeus, 1758	Antillean manatee
Trichechus senegalensis Link, 1795	West African manatee, African manatee
Family Dugongidae	
Dugong dugon Müller, 1776	Dugong

Taxonomy based on https://marinemammalscience.org/science-and-publications/list-marine-mammal-species-subspecies/; accessed 14 June 2024

Contents

1 **Fundamentals of Ocean Acoustics** 1
   Michael B. Porter, Laurel J. Henderson, Dorian Houser,
   and Christine Erbe

2 **Sources of Underwater Noise** 85
   Christine Erbe, Alec J. Duncan, Alexander Gavrilov,
   Montserrat Landero, Robert D. McCauley, Iain Parnum,
   Chandra Salgado-Kent, and Evgeny Sidenko

3 **Mysticete Sounds** 179
   Christine Erbe, Anita Murray, Meghan Aulich, Ann Bowles,
   Ciara Browne, Brodie Elsdon, Emily K. Evans, Adam Frankel,
   Alexander Gavrilov, Corinna Gosby, Lauren Hawkins,
   Capri Jolliffe, Paul Nguyen Hong Duc, and Chong Wei

4 **Odontocete Sounds** 267
   Christine Erbe and Chong Wei

5 **Pinniped Sounds** 351
   Sylvia K. Parsons, Christine Erbe, Sarah A. Marley,
   and Miles J. Parsons

6 **Otter Sounds** 441
   Renata S. Sousa-Lima, Samara Almeida, Izabela Laurentino,
   and Caroline Leuchtenberger

7 **Sirenian Sounds** 459
   Renata S. Sousa-Lima, Juan Carlos Azofeifa-Solano,
   Vera M. F. da Silva, Giovanna A. Dantas, Isadora M. Carletti,
   Ann Bowles, Rodney Rountree, and Christine Erbe

8 **Marine Mammal Hearing** 491
   Dorian Houser

9 **Physiological Effects of Sound on Marine Mammals** 579
   Dorian Houser

10	**Behavioral Responses to Underwater Noise**	611
	Christine Erbe, Ann Bowles, Dorian Houser, Capri Jolliffe, Shyam Madhusudhana, Sarah A. Marley, Angela Recalde Salas, Chandra Salgado-Kent, Renee Schoeman, Valeria Senigaglia, Cristina Tollefsen, Leah Trigg, and Rebecca Wellard	
11	**Biological Significance of Responses to Noise**	699
	Valeria Senigaglia, Dorian Houser, Capri Jolliffe, and Christine Erbe	
12	**Management of Noise**	731
	Capri Jolliffe, Christine Erbe, Carina Juretzek, Jill Lewandowski, Nathan D. Merchant, Brian Miller, Valeria Senigaglia, and Sheila J. Thornton	

About the Editors

Christine Erbe holds an M.Sc. degree in Physics (University of Dortmund, Germany) and a Ph.D. in Geophysics (University of British Columbia, Canada). She worked as a Research Scientist at Fisheries and Oceans Canada, was Director of JASCO Applied Sciences Australia, and eventually returned to academia, where she is now Director of the Centre for Marine Science and Technology at Curtin University (Perth, WA, Australia) and Director of the Centre of Ocean and Earth Science and Technology at Curtin University (Moka, Mauritius). Christine's interests are underwater sound (biotic, abiotic, and anthropogenic), sound propagation, signal processing, and noise effects on marine fauna. She's a John Curtin Distinguished Professor, Fellow of the Acoustical Society of America, Board member of the International Commission for Acoustics, former member of the ISO Technical Subcommittee on Underwater Acoustics, former Chair of the Animal Bioacoustics Technical Committee of the Acoustical Society of America, and former Chair of the international conference series on The Effects of Noise on Aquatic Life.

Dorian Houser holds a Ph.D. in Biology from the University of California—Santa Cruz (USA). After completing a National Research Council Post-Doctoral Associateship at the US Navy Marine Mammal Program, he started the research consultation company, Biomimetica, which he operated until 2010. In 2009, he joined the National Marine Mammal Foundation where he currently serves as the Director of Conservation Biology. Dorian's scientific interests include marine mammal diving, fasting and stress physiology, bioacoustics, and behavior. Dorian is a Fellow of the Acoustical Society of America, from which he received the R. Bruce Lindsay Award (2007). He served multiple terms as the Chair and Vice-Chair of ANSI/ASA S3/SC1 Animal Bioacoustics Accredited Standards Committee and as a member of the ASA Animal Bioacoustics Technical Committee. He has also served as a subject matter expert to the US Navy and multiple governments concerned with the impact of ocean noise on marine life.

Ann Bowles received her Ph.D. in Marine Biology from the Scripps Institution of Oceanography and now leads the Animal Behavior and Senses Program at Hubbs-SeaWorld Research Institute (HSWRI). She has 45 years of experience in the study of animal behavior, sound production, perception of sound, and effects of human-made noise on marine and terrestrial species. Research on behavioral responses has included responses of a range of terrestrial birds and mammals to diverse human-made sound sources, ranging from intense geoacoustic sources and low-flying military aircraft to low-amplitude net alarms and ultrasonic-coded transmitters. Ann has studied a wide range of marine mammal species, including large whales, toothed whales and dolphins, seals and sea lions, polar bears, and manatees. She has served on a wide range of advisory panels for agencies and organizations such as the Office of Naval Research, International Union for the Conservation of Nature, National Park Service, and National Oceanographic and Atmospheric Administration. Ann is a Fellow of the Acoustical Society of America. She helped develop current science-based criteria for protecting marine mammal hearing from noise and is involved in ongoing efforts to establish criteria for behavioral responses. She feels strongly about fostering young investigators in her field and has formal affiliations with the Department of Environmental and Ocean Sciences at the University of San Diego and the Department of Biology at the University of California at San Diego.

Michael B. Porter has obtained his Ph.D. in 1984 from Northwestern University (Applied Math.). He was a scientist at the then Naval Ocean Systems Center in San Diego, then took a position as a research physicist at the Naval Research Laboratory in Washington, D.C. He spent 4 years at the NATO Undersea Research Centre in La Spezia, Italy, in the Environmental Acoustics group, when he co-authored the standard text on Computational Ocean Acoustics (with Jensen, Kuperman, and Schmidt). He subsequently joined the Mathematical Sciences Department at the New Jersey Institute of Technology as a Professor for 10 years. He moved to Science Applications International Corp. in San Diego as Assistant Vice President and Chief Scientist. In 2004, he founded Heat, Light, and Sound Research, Inc., with his old group from SAIC where he remains to this day. Michael held visiting positions at the University of Algarve (1996) and the Scripps Institution of Oceanography (1997). Michael's work has been recognized with the A.B. Wood Medal of the Institute of Acoustics, the Innovators Award for the Naval Research Laboratory, the Rabbi Award from SAIC, the Pioneers Medal of the Acoustical Society of America. He served as the Chief Scientist for two large at-sea experiments and established the Ocean Acoustics Library, which continues as the primary internet source for ocean acoustics models.

Fundamentals of Ocean Acoustics

Michael B. Porter, Laurel J. Henderson, Dorian Houser, and Christine Erbe

Contents

1.1	**Introduction**	2
1.2	**Terminology and Units**	2
1.2.1	The Decibel (dB)	6
1.3	**Basics of Sound Propagation**	7
1.4	**Spherical Spreading (Point Source in Free Space in the Frequency Domain)**	9
1.5	**Seawater Attenuation**	10
1.6	**Lloyd's Mirror Pattern**	11
1.7	**Deep-Water Propagation**	13
1.8	**Environmental Information and Databases**	17
1.8.1	Bathymetry	18
1.8.2	Geo-acoustic Databases	19
1.8.3	Oceanographic Databases and Models	20
1.9	**Numerical Models**	22
1.9.1	Rays and Beams	23
1.9.2	Wavenumber Integration	25
1.9.3	Normal Modes	26
1.9.4	Parabolic Equation	32

M. B. Porter (✉) · L. J. Henderson
Heat, Light, and Sound Research, La Jolla, CA, USA
e-mail: porter@HLSResearch.com; laurel@HLSresearch.com

D. Houser
National Marine Mammal Foundation, San Diego, CA, USA
e-mail: dorian.houser@nmmpfoundation.org

C. Erbe
Centre for Marine Science and Technology, Curtin University, Perth, WA, Australia
e-mail: c.erbe@curtin.edu.au

© The Author(s) 2025
C. Erbe et al. (eds.), *Marine Mammal Acoustics in a Noisy Ocean*,
https://doi.org/10.1007/978-3-031-77022-7_1

1.10	**Model Selection**	35
1.10.1	Range Independent or Range Dependent	35
1.10.2	Low Frequency or High Frequency (Shallow Water or Deep Water)	36
1.10.3	Broadband or Narrowband	37
1.11	**3D Effects**	37
1.12	**Time-Domain Calculations**	42
1.13	**Aero-Acoustics and Air-to-Water Propagation**	47
1.14	**Background on Noise (Soundscape) Modeling**	49
1.15	**Noise in a Homogeneous Halfspace with a Sheet of Monopoles**	52
1.16	**Stratified (Range-Independent) Media with a Sheet of Monopoles**	54
1.16.1	Normal Mode Representation	54
1.16.2	Wavenumber Integral Representation	55
1.16.3	Field Representation	55
1.16.4	Examples	56
1.17	**Modeling Specific Noise Sources**	60
1.17.1	Wind Noise	60
1.17.2	Shipping	62
1.17.3	Quick Approximate Solutions for Range-Dependent Problems	63
1.17.4	Estimating Shipping Noise Levels from Track Data	65
1.17.5	Pile Driving	67
1.17.6	Explosives	69
1.17.7	Seismic Exploration and Directional Sources	73
1.18	**Other Applications of Propagation Models**	74
1.18.1	Animats	74
1.18.2	Tracking Marine Mammals	76
1.19	**Summary**	80
1.20	**Abbreviations and Symbols**	80
	References	81

1.1 Introduction

A cubic meter of water weighs roughly 1000 kg, that is, a metric ton. Swimming in waves at the beach or stepping into a moving stream, one is reminded about the powerful forces associated with modest volumes of water. Indeed, rogue waves in deep water can break the hulls of large vessels. If you imagine yourself sitting at the bottom of an ocean 5000 m deep, you would be supporting a column of water of enormous weight. The resulting pressure would be about 500 times what we are used to, with 1 atmosphere of pressure at sea level.

Sound pressure is simply a small perturbation to the background pressure field described above. It is akin to the ripples on a pond that can be generated by tossing a rock. Thus, when we talk about the acoustic pressure, it is understood that these massive quasi-static pressures due to the huge volume of the ocean have been removed. It is remarkable that such sound waves in the ocean can travel thousands of kilometers across an ocean basin and be processed as a coherent signal.

1.2 Terminology and Units

Sound under water can be generated by a marine animal, such as a whale, or by a subsea volcano, or by a ship. There are many sources of sound in the ocean. By definition, sound is a small perturbation to the ambient, hydrostatic pressure at any location. Such a perturbation might be ignited by the vibration of the swim bladder of a fish. Some fishes have swim bladders and sonic muscles that surround it so that they can actively compress and relax the gas-filled swim bladder sending a pressure perturbation into the surrounding water. A pressure perturbation might also be caused by the sudden release of gas into the water from a seismic air gun.

Such pressure perturbations travel through the water by oscillation of the water particles. Each particle moves back and forth, in line with (or longitudinal to) the direction of propagation of the sound wave. Individual particles do not move with the sound wave away from the source. Rather, it is the ensemble of water particles and their coupled oscillations that effect the propagation of the sound wave. Such longitudinal oscillations result in an alternating pattern of compressions and rarefactions. At the locations of the compressions, the sound pressure is high; at the locations of the rarefactions, it is low (Fig. 1.1). At a fixed point in space, the pressure cycles from high to low, and again to high. The time it takes for one full cycle is called the period. The inverse of the period is the frequency of the pressure wave. The distance between two successive pressure peaks is the wavelength of sound.

The velocity at which the water particles move is an oscillatory quantity, too. At the midpoint of a particle's periodic displacement, the magnitude of the velocity is highest, and the amplitude is either maximally positive or maximally negative. At the two turning points, the velocity changes direction and the velocity amplitude passes through zero. The velocity of the water particles is different from the speed of sound. The speed at which the sound wave travels does not oscillate; rather, it is constant, determined primarily by the temperature and salinity of the water, and the ambient hydrostatic pressure, which in turn is a function of depth below the sea surface.

Pressure waves under water are measured with a hydrophone. At any location, the hydrophone senses not only the pressure wave from one specific source, but instead measures a superposition of sound from multiple sources distributed in space. Such a superposition can be considered a summation of pressure waves arriving from different directions. At any point in time, these pressure waves will most likely differ in amplitude, frequency, and phase. If the instantaneous pressure wave from one source (call it the ith source)

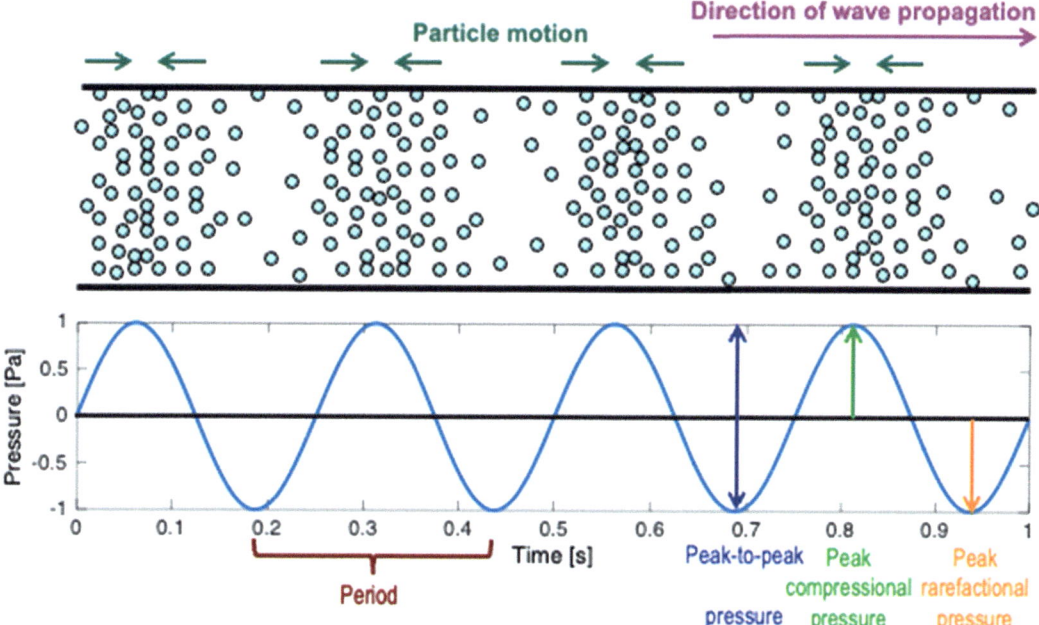

Fig. 1.1 Sketch of alternating compression and rarefaction of water particles and the resulting pressure wave

can be expressed as a waveform $p_i(t)$, then the superposition of pressure waves from N sources becomes $p(t) = \sum_{i=1}^{N} p_i(t)$. It can be difficult to tell apart the contributions from the different sources. A Fourier transform is commonly applied to decompose the recorded pressure time series into many oscillations at different frequencies. But to tell from which direction the sounds arrived, a beamforming sensor is needed. Beamforming is typically done with an array of hydrophones phased so that they are most sensitive in just one direction, the direction of the beam.

Sorting received sound to sources is important when studying the effects of noise on marine animals. Dedicated fieldwork to quantify the sound emitted by a specific source such as a dredger is needed yielding a set of characteristic quantities that can be used to predict received levels in other environments and scenarios. Such quantities could be a source level and a directivity index for directional sources. It is not trivial to determine a source level because the ocean is not an infinite, uniform, loss-less space. Instead, the ocean is stratified, leading to refraction, that is, bending sound propagation paths, and potential channeling of sound over very long ranges. The ocean absorbs sound, and the rate of absorption differs with frequency (i.e., greater absorption at higher frequencies). Many interactions of anthropogenic operations with marine animals take place in shallow water, where sound propagation includes many interactions with the sea surface and seafloor, leading to reflection, scattering, and absorption. To compute source levels from received levels that were measured at some range, the effect of the environment needs to be considered, adequately accounting for the site-specific transmission loss (termed propagation loss in ISO 18405). These terms will be defined more precisely later but loosely refer to the change in sound level as a receiver moves away from the source.

The computation of propagation loss requires knowledge of the local environment, including the bathymetry, upper seafloor geology (e.g., depth-profile of sea-bottom material, thickness of sedimentary layers, density, porosity, and sound speed), and ocean water conditions (e.g., depth-profile of temperature and salinity). While there are databases for some of this information, the exact conditions at the time of sound measurement are seldom known. There is a lot of variability and hence uncertainty. A variety of sound propagation models have been developed that are applicable to different scenarios (high versus low frequency, shallow versus deep water, changing conditions with range versus range-independence, etc.). The following sections explain how sound propagates from the source to a receiver and introduce common approaches to sound propagation modeling—to predict, for example, received levels at a marine mammal.

The quantities, abbreviations, symbols, and units used in this chapter are given at the end of the chapter. Definitions related to underwater sound are given in the textbox.

> ***Sound, underwater***—A disturbance of the ambient, hydrostatic pressure that travels through the water by oscillation of its particles. A sound wave in water travels as a longitudinal wave that consists of alternating compressions and rarefactions of ensembles of water particles.
>
> ***Frequency*** (symbol: f, unit: Hz)—The rate of oscillation of a sound wave measured as the number of cycles per second. The unit of frequency is hertz: $1\text{ Hz} = \frac{1}{s}$.
>
> ***Wavelength*** (symbol: λ, unit: m)—The spatial distance (measured in meters) between two successive peaks in a pressure wave.
>
> ***Speed of sound*** (symbol: c, unit: m/s)—The distance traveled per unit time. If τ is the period of the sound wave (i.e., the duration of one cycle), then the speed of sound can be expressed as $c = \frac{\lambda}{\tau} = \lambda f$. The speed of sound in fresh water at 25 °C is 1497 m/s. The speed of sound in salt water at 25 °C, with a salinity of 33 ppt, at 1 m depth is 1532 m/s.

(continued)

Particle velocity—The speed at and the direction in which the water particles move during their oscillations as a sound wave passes through the water. Velocity is a vector quantity having both magnitude and direction. Particle velocity is the first derivative with respect to time of the particle displacement. Particle acceleration is the second derivative with respect to time of the particle displacement. Water particles also move due to waves and currents; however, in this book, particle velocity relates only to sound.

Pressure (symbol: p, unit: Pa)—Hydrostatic pressure at a given depth under water is due to the weight of the liquid above plus the weight of the atmosphere above the sea surface. Acoustic (or sound) pressure is due to a deviation from the hydrostatic pressure as a result of a sound wave passing through. Sound pressure $p(t)$ at any location varies with time t and is measured with a hydrophone in water. The greatest magnitude of $p(t)$ is called the peak sound pressure p_{pk} (or zero-to-peak sound pressure). It is the greater of the greatest magnitude during compression and the greatest magnitude during rarefaction. The peak-to-peak sound pressure $p_{pk\text{-}pk}$ is the sum of the greatest magnitude during compression and the greatest magnitude during rarefaction. The root-mean-square sound pressure p_{rms} is literally the square root of the time-average of the squared pressure: $p_{rms} = \sqrt{\left(\int_{t_1}^{t_2} p^2(t) dt\right)/(t_2 - t_1)}$, where t_1 and t_2 are the start and end times, respectively. The unit of all pressures is pascal (Pa).

Sound pressure level (SPL, symbol: L_p)—The level of the root-mean-square sound pressure: $L_p = 20\log_{10}(p_{rms}/p_0)$ relative to a reference value $p_0 = 1$ µPa. SPL is expressed in dB re 1 µPa.

Peak sound pressure level (SPL$_{pk}$, symbol: $L_{p,pk}$)—The level of the peak sound pressure: $L_{p,\ pk} = 20\log_{10}(p_{pk}/p_0)$ relative to a reference value $p_0 = 1$ µPa. SPL$_{pk}$ is also expressed in dB re 1 µPa.

Sound exposure level (SEL, symbol: $L_{E,p}$)—The level of sound exposure: $L_{E,p} = 10\log_{10}\left(\frac{E_p}{E_0}\right)$ relative to a reference value $E_0 = 1$ µPa^{2}s. Sound exposure E_p is computed as the integral of the sound pressure p over some time interval [t_1, t_2]: $E_p = \int_{t_1}^{t_2} p^2(t) dt$.

Radiated noise level (RNL, symbol: L_{RN})—The level of the product of (1) the root-mean-square sound pressure measured at some range r and (2) the range r from the sound source: $L_{RN} = 20.\log_{10}\frac{p_{rms}(r)r}{p_0 r_0}$ The reference value is $p_0 r_0 = 1$ µPa m and so RNL is expressed in dB re 1 µPa m. As the logarithm of a product is equal to the sum of the logarithms of the factors, RNL can also be computed as $L_{RN} = L_p + 20\log_{10} r$, which is the sum of the received sound pressure level and a spherical propagation loss term. This equation does not yield a source level that is independent of the environment through which the sound propagated until it was measured at range r. Therefore, RNL is also referred to as the "affected source level."

Source level (SL, symbol: L_S)—Ideally, the SL would be characteristic of the sound source and independent of the environment into which it is placed. The determination of SL is difficult in practice because many sources occur near a boundary (i.e., mostly the sea surface, but also the seafloor), where reflection, scattering, absorption, and phase changes may occur, and because

(continued)

the ocean is not a uniform, loss-less medium, but instead is stratified (leading to refraction and potential channeling of sound) and it absorbs sound (differently at different frequencies). With knowledge of the site-specific hydro- and geoacoustic parameters (including absorption and sound speed profiles for the water column and the seafloor), sound propagation models can be applied to compute a monopole SL, which can then be used to predict received levels in other environments. SL is expressed in dB and referenced to a distance of 1 m. For continuous sources, the SL is typically expressed as SPL in dB re 1 µPa m (formerly, dB re 1 µPa @ 1 m). For a transient source, the SL is often expressed as SEL in dB re 1 µPa^{2}s m^2 (formerly, dB re 1 µPa^{2}s @ 1 m).

1.2.1 The Decibel (dB)

Quantities that have large dynamic ranges, that is, that vary by orders of magnitude, are typically expressed as logarithms relative to a reference value denoted by a subscript 0. These logarithmically defined quantities are referred to as levels. The levels of so-called field quantities F (such as pressure and particle velocity) are defined as $L_F = 20 \log_{10}\left(\frac{F}{F_0}\right)$ while the levels of so-called power quantities P are defined as $L_P = 10 \log_{10}\left(\frac{P}{P_0}\right)$ are unit-less and expressed in dB. It is critically important to always state the reference value.

The reference value can depend on the medium. For example, the sound pressure level under water is referenced to 1 µPa, while in air, it is referenced to 20 µPa. An SPL of 0 dB re 20 µPa is equal to a SPL of 26 dB re 1 µPa because $20\log_{10}20 \approx 26$. To convert from dB to linear field or power quantities, the inverse of the logarithm (i.e., the exponent) needs to be taken.

For example, an SPL of 140 dB re 1 µPa corresponds to a root-mean-square sound pressure of $10^{140/20}$ µPa $= 10^7$ µPa.

The doubling of a field quantity is equivalent to the addition of 6 dB to its level because $20 \log_{10}(2F/F_0) = 20\log_{10}(2) + 20\log_{10}(2F/F_0) \approx 6 + 20\log_{10}(F/F_0)$. For example, if a sound pressure of 10^7 µPa (SPL of 140 dB re 1 µPa) is doubled, the SPL becomes 146 dB re 1 µPa.

The doubling of a power quantity results in the addition of about 3 dB to its level because $10\log_{10}(2P/P_0) = 10\log_{10}(2) + 10\log_{10}(P/P_0) \approx 3 + 10\log_{10}(P/P_0)$. A tenfold increase of a power quantity results in the addition of 10 dB to its level: $10\log_{10}(10P/P_0) = 10\log_{10}(10) + 10\log_{10}(P/P_0) = 10 + 10\log_{10}(P/P_0)$.

When calibrating hydrophone systems, one often needs to handle levels relative to different reference values. For example, the sensitivity of a hydrophone might be -180 dB re 1 V/µPa and the digitization gain of the data acquisition board might be -20 dB re FS/V, where FS is the full-scale value and V stands for volts. The system response becomes -200 dB re FS/µPa by adding the decibels under careful consideration of the product of the reference values and units.

The inversion of the reference units changes the sign. For example, -180 dB re 1 V/µPa $= +180$ dB re 1 µPa/V.

The reference value and unit do not change if a gain (e.g., amplifier gain) is added: 120 dB re 1 µPa + 40 dB = 160 dB re 1 µPa.

Transmission loss (TL) or propagation loss (PL; ISO 18405) is roughly speaking a dB characterization of the reduction in the sound intensity at 1 m from the source as we move away from the source to greater ranges (Urick 1983). Thus, we can take an arbitrary source level and reduce it by the transmission loss to get the field at other locations. For most readers, that loose idea is sufficient, and it is embedded in the standard propagation models and predominant in the literature (see Jensen et al. 2012). Intensity in this context is a scalar conventionally taken to be the pressure squared divided by the local impedance ρc.

There are many approximations buried in this convention. In practice, the propagation models typically ignore the impedance variation as the models were predominantly developed for sources and receivers in the ocean volume. Impedance does not vary significantly in the ocean. However, for sources and receivers in the air or in the sub-bottom there can be large variations. Similarly, some of the models are derived based on a reduced pressure (e.g., pressure divided by the square root of the density). Again, the density is often ignored because the sound field in the ocean is typically the quantity of interest. It is also fairly common that TL is plotted with a sign error to orient a plot axis. Transmission loss should generally be a positive quantity.

Another subtlety is that the reference for transmission loss is taken as the spherical range of 1 m *for the equivalent source in free space with a homogeneous sound speed and density.* Thus, the computed TL will not in general be 0 dB at a range of 1 m in a particular environment because the field at that range includes a direct path as well as surface and bottom reflections, etc., that are not present in the free field. The waveguide itself changes the field, although often not very much. There is more that could be said on this subject, but it is typical that acoustic models will provide a pressure field scaled in inconsistent ways. An ISO standard has been put forth and attained some currency in the literature of environmental compliance, but not, as of this writing, in the broader literature. We will generally use the term "propagation loss" here to be consistent with the terminology in Chap. 2.

1.3 Basics of Sound Propagation

To introduce some basic concepts of sound propagation, we begin with one of the simplest models of the ocean consisting of a fluid overlying an infinitely thick seabed. The seabed is treated also as a fluid but with a higher sound speed and density. The sound speed is assumed constant in both media. This is called the Pekeris waveguide in honor of one of the earlier developers of normal mode methods in ocean acoustics. The specific case we consider here represents a shallow-water scenario with a sound speed of 1500 m/s in

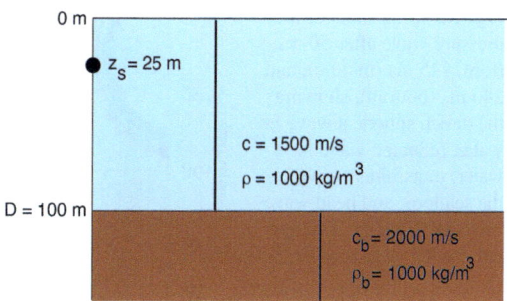

Fig. 1.2 Schematic of the Pekeris waveguide

the ocean and 2000 m/s in the seabed. The ocean depth is 100 m and the density has been set to 1000 kg/m^3 in both ocean and seabed as shown in Fig. 1.2. (A real sediment would have a much higher density.) This is a very idealized scenario but a useful starting point to introduce the basic physics.

In the following, we will gradually work our way up to the Pekeris waveguide starting first with free space, that is, with the top and bottom boundaries removed and then adding those features into the problem. Consider now a sound source in free space introduced at a depth of 25 m, which generates a simple impulse. One could imagine an explosive and so-called SUS (signal, underwater sound) charges are often used for that purpose. Oddly enough, common household lightbulbs (the old ones with filaments) are often used and they can be made to implode by just tying a weight to them and letting them sink to a depth where they suddenly crush (implode) under the weight of the water. Airguns are another impulsive source.

Regardless of the source, the result is a spherical wave that expands initially at the speed of sound in water, that is, about 1500 m/s. Captured in a snapshot 50 ms later, the acoustic pressure in the ocean would look like that in Fig. 1.3 having traveled about 30 m. The source here has been assumed to generate both a positive and negative pressure impulse. Positive pressures are shown in red and negative pressures in blue. The actual levels are not of great interest here; however, they obviously are proportional to the source strength. Satisfying conservation of energy, the

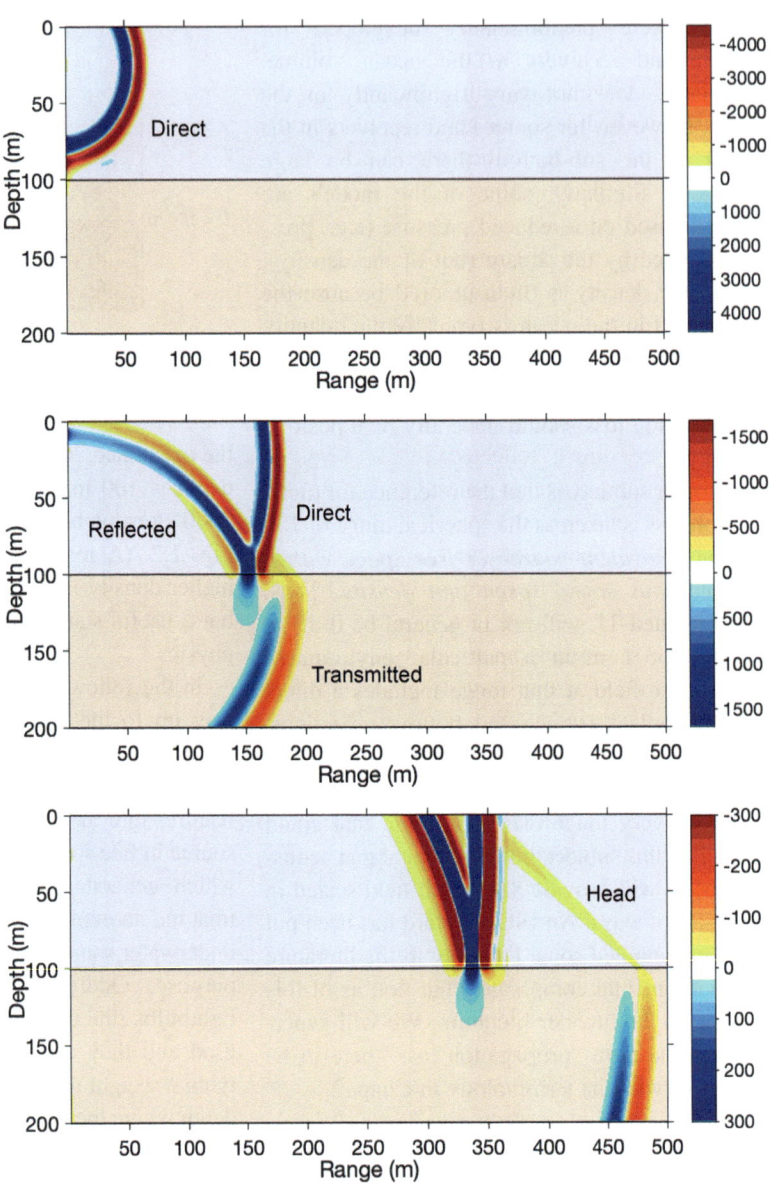

Fig. 1.3 Snapshots of the pressure wave after 50 ms (top), 125 ms (middle), and 240 ms (bottom), showing the direct, spherical wave in water, reflected wave in water, transmitted wave in the seafloor, and head wave in water. Note the different color scale in each panel

magnitude of the pressure (be it a positive overpressure or a negative under-pressure) decays in proportion to $1/R$, where R is the radius of the spherical wave.

At this stage, the wave is identified as a direct wave. That terminology is to suggest that receivers are hearing a wave that has not hit any boundaries nor had any other complicated modifications worthy to be called out.

As time progresses, the expanding spherical wave hits the seabed generating a reflected wave. Part of the energy also continues as a transmitted wave. Part of the original spherical wave that has not hit the bottom continues to propagate as a direct wave. These features are captured in a snapshot at 125 ms shown in Fig. 1.3b.

There are interesting features in each of these components. The direct wave has a lower amplitude because of the continued spherical spreading. The reflected wave has weak energy at steep grazing angles (i.e., short ranges). This is the so-called critical angle effect, which we will discuss later. In brief, it turns out that the strength of the reflected wave varies with angle. For a simple half-space without loss, one gets a perfect, mirror-like reflection up to the critical angle. For steeper angles, the reflection becomes progressively weaker.

Turning next to the transmitted wave that propagates in the sea bottom, we see that it is broadened; that is, it has a longer wavelength in the seabed because of the higher sound speed in the seabed. To summarize, the incident wave and its energy have been split into three parts consisting of direct, reflected, and transmitted waves. The partition of energy between the reflected and transmitted waves is determined by the bottom reflection coefficient, which in turn is a function of angle. In more general cases, the reflection coefficient can be a function of the source frequency and other parameters. Meanwhile, the direct and transmitted waves are racing ahead with the transmitted wave having the speed advantage of the seabed.

The pressure field after 240 ms is shown in Fig. 1.3c. Here, we can see the reflected and direct waves continuing ahead at the speed of sound in the water. Likewise, the transmitted wave moves ahead at the speed of sound in the seabed and has overtaken the waves in the water. A noteworthy new feature is the head wave in the ocean, which connects to the transmitted wave in the seabed. It can be understood as sound energy that is continuously radiated by the transmitted wave into the ocean. As such, its leading edge travels at the speed of sound in the seabed. It may be considered a ghost wave in the sense that it is typically weak because it is a leakage from the transmitted wave.

The head wave is also known as the conical wave or the lateral wave, depending on the literature. It is called the head wave because it often comes ahead of the direct wave in the water column. However, one can see in the snapshot that it is not always ahead; it depends on the depth of the receiver. It is called the conical wave because it forms a cone if you imagine this snapshot rotated to form the full three-dimensional picture. It is called the lateral wave because it involves a propagation path (a ray) that travels laterally along the ocean/seabed boundary. The head wave is typically very weak, and one should note the varying color scale in the plots, which was selected to make the head wave visible and compensate for the geometric spreading.

A general feature to note is that the sound field never cuts off instantaneously. For instance, the transmitted wave gradually tapers off toward the ocean/seabed interface. Likewise, the direct and reflected waves taper off into the seabed. Their influence into the seabed is sometimes called the direct-wave root. Finally, notice that the head wave produces a single band from the doubled-banded incident and transmitted waves. Mathematically it is the Hilbert transform of the incident wave, which is equivalent to a version that has been phase-shifted by 90°. The mathematics behind this are very complicated. It is interesting how the simple problem of reflection by a halfspace—that you might have thought was clear from an early physics class—presents such complexity.

These figures also introduce some features that will be important later when we talk about ray theory. The leading edge of the various types is referred to as a wavefront. Rays will be constructed as curves that are perpendicular to the wavefronts and indicate the direction of propagation (and energy transport).

1.4 Spherical Spreading (Point Source in Free Space in the Frequency Domain)

In the previous section, we have seen how a point source generates a spherical wave in the near field. By convention, this wave has become a key reference to describe the propagation modeling since it almost always represents the near field. In free space, it would continue indefinitely. Then deviations from it in the far-field represent the propagation effects of the waveguide. Some of the next few sections are very mathematical and will be accordingly difficult for readers without that background. However, it is difficult to use the sound propagation models reliably without understanding a bit about their formulation. If the mathematics is hard to follow, then it is hoped the reader will benefit from the examples, which explore typical ocean scenarios.

The pressure field due to a harmonic (single frequency) point source in free space (i.e., an idealized, unbounded ocean) is a simple, spherical wave whose pressure is proportional to

$$p(r,z) = \frac{e^{ikR}}{R},$$

where

$$R = \sqrt{r^2 + (z-z_s)^2},$$

r is the horizontal range, from the source, z_s is the source depth, R is the spherical range (also called the slant range) from the source, $k = \frac{\omega}{c_0}$ is the wavenumber (i.e., 2π times the number of wavelengths per meter, which motivates the terminology), ω is the angular frequency, and $f = \frac{\omega}{2\pi}$ is the frequency of the source. The medium in which the sound propagates is assumed to have a constant sound speed denoted by c_0. This spherical wave field is sometimes referred to as a monopole to suggest a single isotropic (i.e., radiating uniformly in angle) source. We use the term "proportional" above in defining the monopole because the important issue is often the general behavior, not the strength or amplitude.

Some readers may be confused trying to reconcile this point source solution with the earlier discussion about spherical waves. The difference between the two representations comes from considering an impulse in the time domain vs. a single tone, which can be viewed as a single-frequency component of the time-domain solution. Typical sound sources may be characterized by a sharp impulse (e.g., an explosion) or a persistent hum or tone (e.g., the noise of a ship's propeller). However, Fourier showed the remarkable fact that any arbitrary waveform can be expressed as a sum or integral over tones of different frequencies. As we discuss in Sect. 1.10, most numerical ocean acoustic models take advantage of this insight and address purely the problem of modeling a tonal source.

The intensity (neglecting the impedance) is proportional to the magnitude squared of the complex pressure, so here the intensity is $\frac{1}{R^2}$ (i.e., inversely proportional to the square of the spherical range from the source). Such spherical waves may be scaled in arbitrary ways; however, for the choice we have made, the intensity at 1 m is simply unity (or 0 dB on a logarithmic scale). Thus, with this scaling the propagation or transmission loss (TL) is simply $-20\log_{10}|p(r, z)|$.

1.5 Seawater Attenuation

The above discussion presents a basic description of the propagation physics mostly in terms of spherical waves whose intensity decreases as the spread in range. As we get into the far field and these spherical waves are repeatedly reflected by the boundaries, the summed intensity tends to follow a cylindrical spreading law instead. This is because the loss of individual spherical waves is countered by an increase in the number of such reflected waves. Then the rate of decrease in energy in range becomes much less rapid.

A separate loss factor that comes into play is the intrinsic medium attenuation and there are different effects involved in different media (the ocean, sediment, basement, etc.). Early work was done by Ewing and Worzel (1948), Thorp (1965), and Fisher and Simmons (1977) using long-range propagation measurements. This work led to a formula often called Thorp attenuation, which is given by

$$\alpha^{Sea} = 0.0033 + \frac{0.11f^2}{1+f^2} + \frac{44f^2}{4100+f^2} + 0.0003f^2$$

where α^{Sea} is the seawater attenuation in dB/km and f is the frequency in kHz. The terms in this formula incorporate two key loss mechanisms, where the sound causes ionic dissociation of (1) magnesium sulfate and (2) boric acid. This formula is widely used in ocean propagation models partly because of its simplicity. However, more accurate formulae by François and Garrison (1982) are also available and consider temperature, salinity, and depth. These variations can produce significant differences in the loss for different oceans and seas. A simplified version was later developed by Ainslie and McColm (1998). Seawater attenuation can often be neglected at lower frequencies and/or shorter ranges.

In summary, the TL in shallow water is often well represented by a near-field zone with spherical spreading, a far-field zone with cylindrical spreading, and an additional volume loss at a nearly constant dB/km. One seeks a formula that comprises these three effects. Urick (1983) suggests the following approximations:

$$TL = 20 \log_{10} r + \frac{\alpha^{Sea} r}{1000} - k_L \text{ for } r < D$$

$$TL = 15 \log_{10} r + \frac{\alpha^{Sea} r}{1000} + a_{Bdry}\left(\frac{r}{D} - 1\right)$$
$$+ 5 \log_{10} D - k_L \text{ for } D \leq r \leq 8D$$

$$TL = 10 \log_{10} r + \frac{\alpha^{Sea} r}{1000} + a_{Bdry}\left(\frac{r}{D} - 1\right)$$
$$+ 10 \log_{10} D + 4.5 - k_L \text{ for } r > 8D$$

where D is the bottom depth and a_{Bdry} represents boundary losses (surface and bottom). The latter depends on the bottom types (e.g., mud or sand). Urick also suggests values for k_L and a_{Bdry}; however, our purpose here is just to convey the physics. Note how these formulae transition smoothly from spherical to cylindrical spreading and include the ocean volume attenuation. Besides the ocean volume attenuation, the term involving a_{Bdry} incorporates losses from each subsequent bottom reflection, which are also roughly linear in range in dB.

These types of formulae were widely used up to the 1960s; however, in later years, and with increasing computer power, they were largely replaced with more accurate methods based on numerical solutions of the acoustic wave equation. It should be noted that they represent averaged values of the TL (e.g., over a one-third octave band), rather than the coherent field, which has a complicated interference pattern in range.

These formulae make it clear that one should not apply a simple spherical spreading formula to estimate the TL in the far-field. Similarly, if one makes a far-field measurement and simply removes cylindrical spreading, then one will get an erroneous estimate of the near-field levels.

1.6 Lloyd's Mirror Pattern

If the source is now placed below the reflecting ocean surface, the field is the sum of the direct propagated signal and its reflected image as illustrated in Fig. 1.4.

$$p(r,z) = \frac{e^{ikR_{direct}}}{R_{direct}} - \frac{e^{ikR_{image}}}{R_{image}},$$

where,

$$R_{direct} = \sqrt{r^2 + (z - z_s)^2},$$

is the spherical range from the source and

$$R_{image} = \sqrt{r^2 + (z + z_s)^2},$$

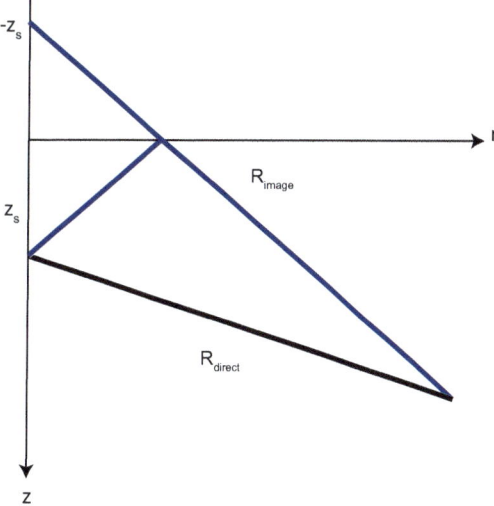

Fig. 1.4 Direct and reflected rays for a source below a perfectly reflecting surface (Lloyd's mirror)

is the spherical range from its image. The quantity z_s is the source depth. The sign of the image source is negative so that the sum of the two fields produces vanishing pressure at the ocean surface. We denote R as the range from the origin, that is,

$$R = \sqrt{r^2 + z^2}.$$

Then, for large R, the radii from the direct and image source are nearly equal to R, and the resulting sum can be approximated by

$$p(r,z) = -\frac{2i}{R}\sin(kz_s\sin\theta)e^{ikR}.$$

(This is not obvious.) Thus, the intensity (magnitude squared of the pressure) that is proportional to (magnitude squared of the pressure) is

$$I(r,z) = \frac{4}{R^2}\sin^2(kz_s\sin\theta).$$

This is the so-called Lloyd's mirror pattern (the term Lloyd mirror pattern is also used) where the intensity shows a directional radiation pattern resulting from the interference of the direct and image sources. It is also referred to as a dipole pattern, where dipole refers to a pair of monopoles. Note that the formula depends on the product of wavenumber and source depth so that increasing the frequency has the same effect as increasing the source depth. Figure 1.4 shows the resulting field pattern for a source depth of 5 m and three different frequencies of 50, 250, and 750 Hz. Note that the number of lobes increases as the frequency increases. For low frequencies (or equivalently as the source approaches the surface), one obtains a simple dipole pattern with a single lobe directed toward the bottom. We will encounter dipoles again in talking about the radiation pattern due to near-surface sources such as ships and winds. Once the source gets to less than about a quarter-wavelength, that dipole pattern changes little in shape; however, its intensity drops to zero as the source approaches the surface. This is sometimes called the surface decoupling effect, which implies that a source of the same strength tends to produce a quieter field when the source gets very close to the surface. Similarly, a receiver (or listener) will find a given source sounds quieter as the receiver approaches the surface. Said another way, the ocean surface is a pressure-release boundary, so the sound field vanishes as it is approached.

Will a marine mammal notice this beam pattern? For humans, it is not unusual to hear these odd variations in levels as we move through a room when there is a narrowband source (a tonal) for instance due to the whine of a motor. In a room, the physics is more complicated because of the various walls; however, whether a room or a half-space, the common feature is that of an interference pattern. For broadband sources, for instance, a voice in a room, these interference effects are present but less obvious since the interference pattern is different for each frequency. In short, such interference patterns are more noticeable for narrowband than broadband sources (Fig. 1.5).

To complete our study of the Pekeris waveguide, we return to the time domain and now include both the surface and bottom reflections. Figure 1.6 shows snapshots of the pressure field at the same times as before. With all these reflections, the picture rapidly becomes complicated; however, we can point out a few features of interest. As we know, the ocean surface acts like an acoustic mirror producing an inverted waveform. Thus, in the upper panel the direct wave is followed by an inverted waveform coming off the surface.

From our previous modeling of the head wave, we know to expect that there will be a simple bottom reflection as well as a head wave as seen in the middle panel. That same physics applies to the surface-reflected wave, and we see its echo producing all the same effects but with the phase inverted. As time progresses, we see these waves propagating down the waveguide. Eventually these many reflections will produce a cascade of surface, bottom, and head waves reflecting off the various boundaries in a sort of barber-shop mirror effect.

1 Fundamentals of Ocean Acoustics

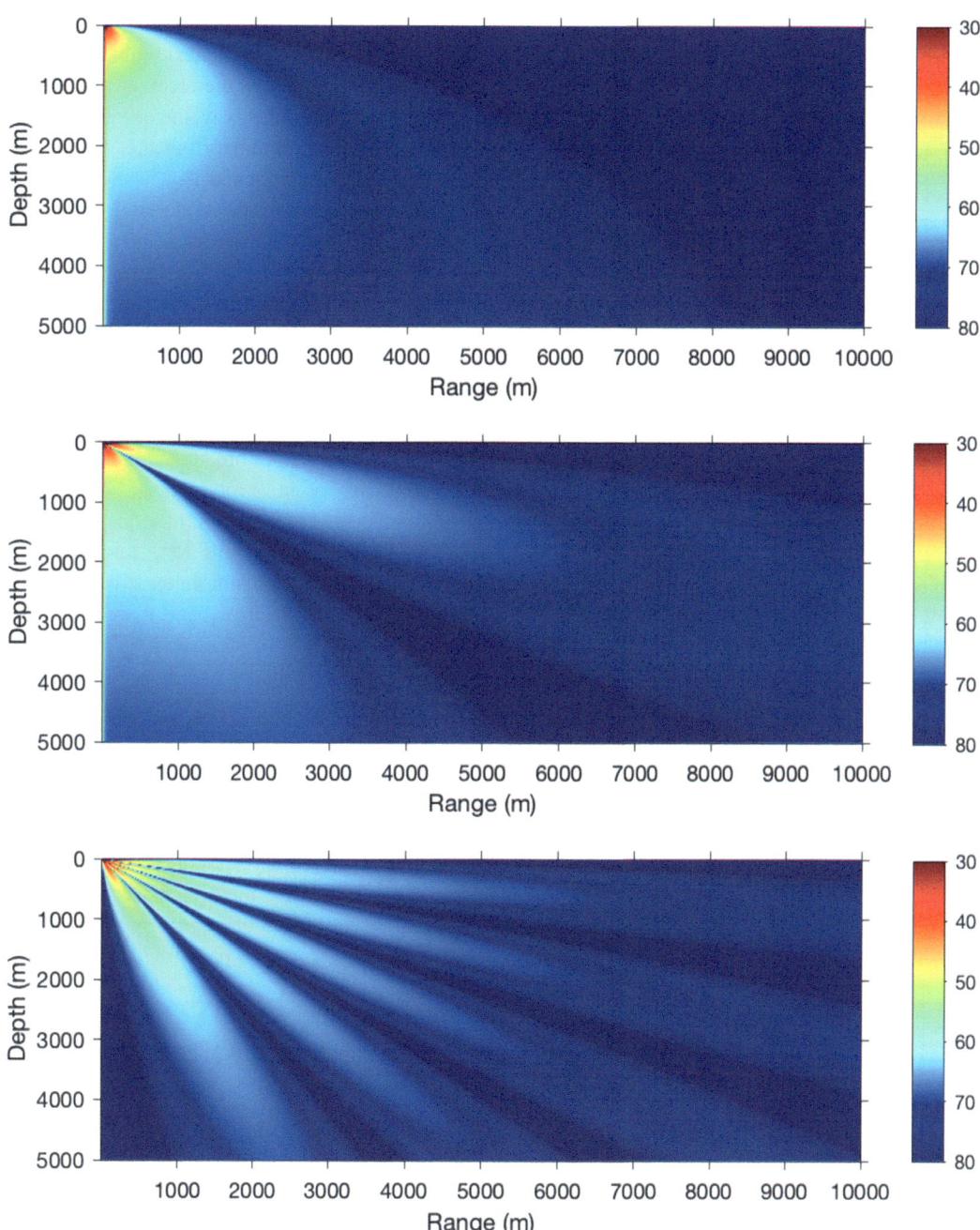

Fig. 1.5 Transmission loss showing the Lloyd's mirror pattern for a source depth of 5 m and three different frequencies: 50 Hz (**a**), 250 Hz (**b**), and 750 Hz (**c**)

1.7 Deep-Water Propagation

The terms "deep water" and "shallow water" are not precisely defined in any accepted convention; however, deep water is typically more than a few thousand meters and shallow water less than a few hundred meters. A representative deep-water sound speed profile is the Munk (1974) profile given by the formula

Fig. 1.6 Snapshots of the pressure field for the Pekeris waveguide after 50 ms (top), 125 ms (middle), and 240 ms (bottom). Note the different color scale in each panel

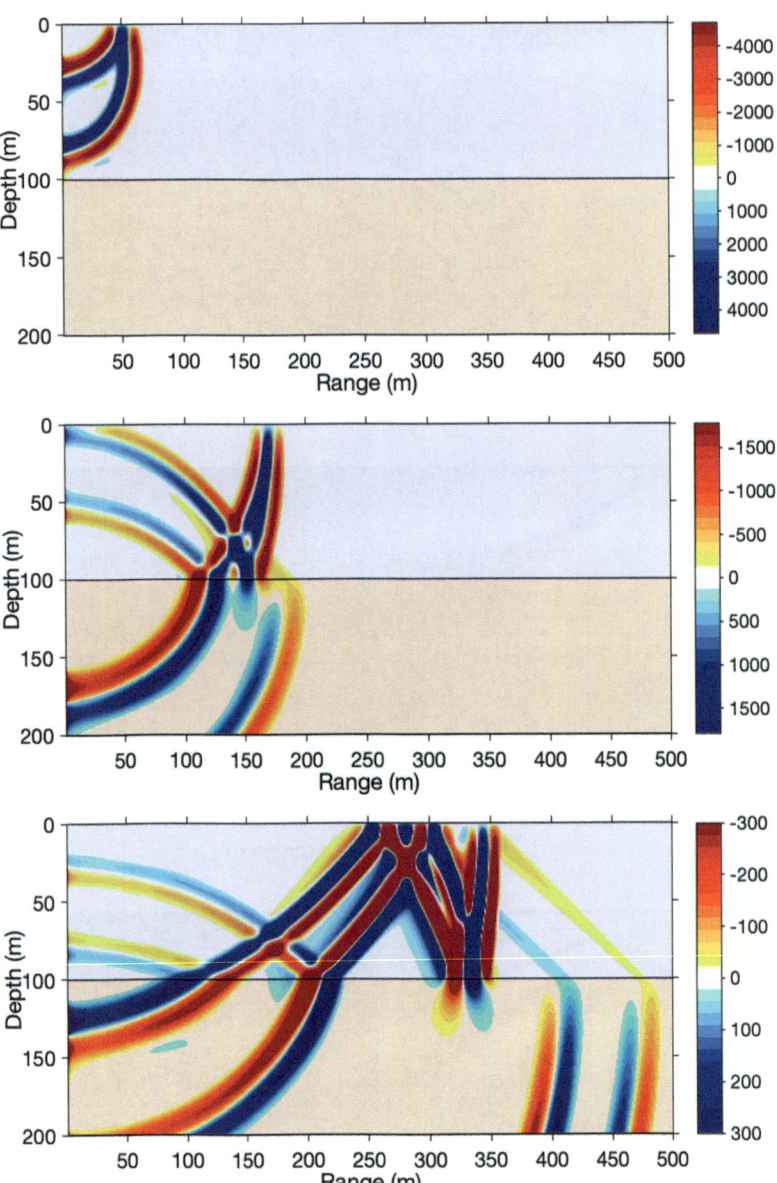

$$c = c_0 \left[1 + 0.00737 \left(\tilde{z} - 1 + e^{-\tilde{z}} \right) \right],$$

where $\tilde{z}$ is a scaled depth given by

$$\tilde{z} = \frac{2(z - z_c)}{z_c},$$

and c_0 is a reference sound speed chosen to be 1500 m/s, and z_c is the depth of the sound-channel axis where the sound speed minimum occurs. It is chosen here as 1300 m. (The Munk profile is a general class of deep-water profiles—specific parameter choices have been made here.) The resulting sound speed is shown in Fig. 1.7. We present this formula because it is often handy to have an analytic representation to describe the sound speed profile concisely. Real-world profiles will often have a similar shape; however, it is only an exemplar.

A key feature of this profile is its bowl shape with a minimum sound speed at the so-called

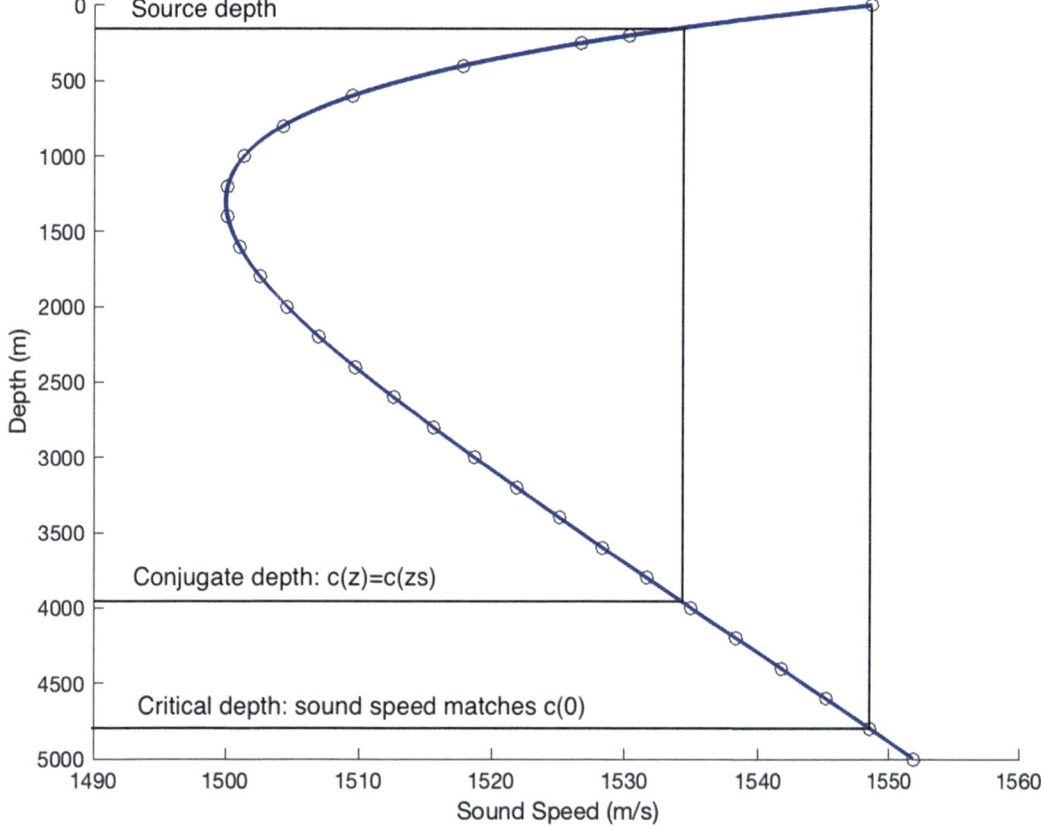

Fig. 1.7 A typical deep-water sound speed profile

sound channel axis. This shape arises from a competition of factors. As we move down from the sea surface, the water tends to get colder and that acts to reduce the sound speed. However, the weight of the water creates tremendous pressure that acts to increase the sound speed. If the water is deep enough, the pressure overwhelms the temperature effect producing a sound speed that is higher than that at the surface. This variation of the sound speed in m/s is conveniently approximated by the formula

$$c = 1449.2 + 4.6T - 0.055\,T^2 + 0.00029\,T^3 + (1.34 - 0.01T)(S - 35) + 0.016z,$$

where T is the temperature in degrees Celsius, S is the salinity in parts per thousand, and z is the depth in meters. However, through long-range propagation measurements (often for tomography), more accurate formulae have been developed.

Temperature and salinity are also functions of depth; however, if the ocean is well-mixed so that they are roughly constant, then the depth variation is simply the last term in this equation, which shows an increase in the sound speed of 1.6 m/s over each 100 m in depth. This amount of upward refraction is found in an idealized mixed layer (which forms a surface duct), which is often found in the upper ocean. It is also typical in shallow waters during stormy seasons of the year.

Inexpensive devices (CTDs) measure the conductivity (implying salinity) and temperature as a function of depth and are widely used for estimating the sound speed. However, the sound speed can also be measured directly.

The depth where the sound speed near the bottom exactly equals that at the sea surface is called the critical depth. The water depth beyond that is called excess depth. At the critical depth a horizontally launched ray at the surface will be refracted back to a horizontal angle before turning back (upward) toward the sea surface. (This is a consequence of Snell's law.) Thus, sources at the surface will produce purely refracted paths that can propagate effectively to large distances because they are not weakened by energy losses due to bottom reflection The presence of depth excess is then one coarse indicator of good propagation conditions for sources on the surface. Another useful term is the conjugate depth, which is the depth where the sound speed in the deeper part of the ocean matches that at the source. By the same arguments, a horizontally launched ray from the source will turn (become horizontal) at the conjugate depth and the presence of a conjugate depth is an indicator that there will be good propagation conditions from the source to long ranges.

The motivation for this terminology will become clearer as we look at the sound field. We consider a source frequency of 200 Hz and a source depth of 25 m. This could correspond to a navy sonar, noise from the propeller on a ship, or a single air gun on a seismic survey vessel. However, the depth and frequency in our example are not particularly representative of any of these sources. There are several features of interest in the propagation loss plot shown in Fig. 1.8. In the near field, we see a pattern of beams radiating down from the source. As discussed earlier in ocean acoustics, this is called the Lloyd's mirror pattern, and it results from the interference pattern due to the source and its reflection in the mirror of

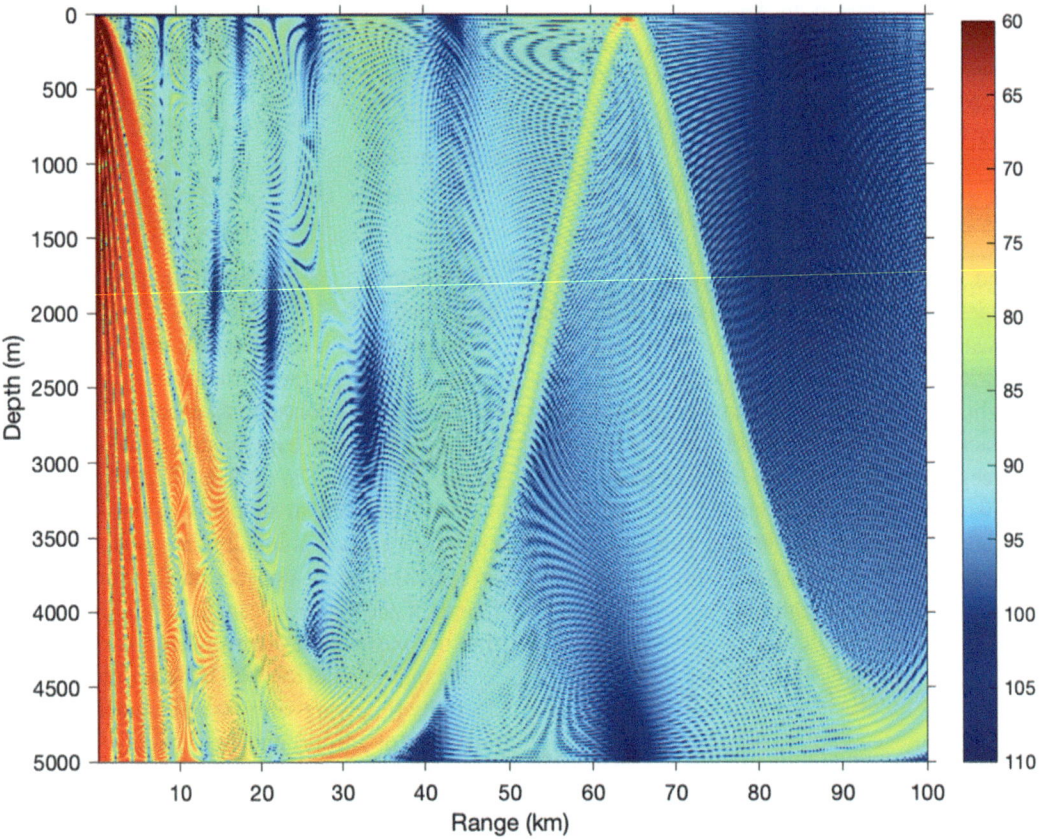

Fig. 1.8 An example of propagation loss in deep water (Munk profile). The source depth is 25 m, and the frequency is 200 Hz

the ocean surface. (In that sense, the origin of the beams is more properly assigned to the ocean surface.)

The Lloyd's mirror beams are seen to reflect off the ocean bottom and then off the ocean surface. The steepest beams hit the bottom above the critical angle and lose a lot of energy with every bounce. The shallower beams (smaller grazing angle) will have lower loss, depending on the bottom reflection coefficient. In deep water, the sediment is a pelagic (from the Greek: pelagios = "of the sea") ooze and typically has a sound speed *less* than that in the ocean, in which case there is no critical angle. However, depending on the sediment thickness, the energy may reflect off a basalt layer below, or refract within the sediment and get returned into the ocean.

Another prominent feature in this figure is the convergence zone (CZ) energy. This is the band of sound that appears in the near field to just be the first beam of the Lloyd's mirror pattern. However, it is launched at a shallow angle and refracted by the ocean sound speed so that it just misses the ocean bottom. Then it returns to the surface and the pattern is repeated cyclically in range. The sound energy returns near the surface roughly every 60 km in this example and is said to refocus there. That motivates calling the refocus region a convergence zone; however, saying it refocuses is a bit of an over-simplification. In any case, the CZ pattern is a typical deep-water phenomenon and of great significance, partly because the sound paths involved incur very low losses allowing (for better or for worse) sound to travel across ocean basins with little more than the cylindrical spreading losses. They are a common feature in deep water with cycle distances typically varying between 50 km and 60 km.

To understand better the propagation physics, it is useful to look at the corresponding ray trace, shown in Fig. 1.9. The ray launch angles have been spaced irregularly to produce a less cluttered plot. Different types of ray paths are shown in different colors. The ray paths that are purely refracted (not hitting either the surface or the bottom) are called RR (refracted–refracted) paths. These will not exist if there is no conjugate depth in the water column. The ray paths that hit the sea surface but are refracted before hitting the bottom are called RB (refracted–bottom reflected) paths. The ray paths that hit both the surface and the bottom are called SB (surface reflected–bottom reflected) paths. There are no SR (surface reflected–refracted) paths in this case. Rays that hit the boundaries suffer reflection loss. The ocean surface is usually (depending on sea state and frequency) very reflective; however, there may be significant loss on bottom reflection depending on the bottom type.

The region outside the band of RR and RB paths is an example of what may sometimes be called a shadow zone. Formally a shadow zone is a zone where there are no rays. The rays, of course, convey the sound energy so these are quiet zones. Of course, the black bottom bounce paths are populating the zone within the first 65 km. On the other hand, they typically will lose a lot of energy because of the bottom reflection. After the first CZ (beyond 65 km), the accumulated losses due to bottom reflection become still greater producing a more pronounced shadow. In summary, the term shadow zone may be used a little loosely depending on whether one neglects certain groups of rays and on how deep the shadow is. A further subtlety is that complex rays (not shown) may also leak sound into the shadows.

1.8 Environmental Information and Databases

The examples above have shown that the environment has a critical effect on the propagation of sound. The environmental information one needs may be divided into ocean and seabed information. Within the ocean, the key parameters are the sound speed and volume attenuation. However, one may also add bubble densities, and fish types and densities. Bubbles and fish are both scatterers. Wave height is also important in terms of surface scattering.

For the seabed, one needs a similar characterization of the sound speed, volume attenuation, and scattering parameters. There is also an

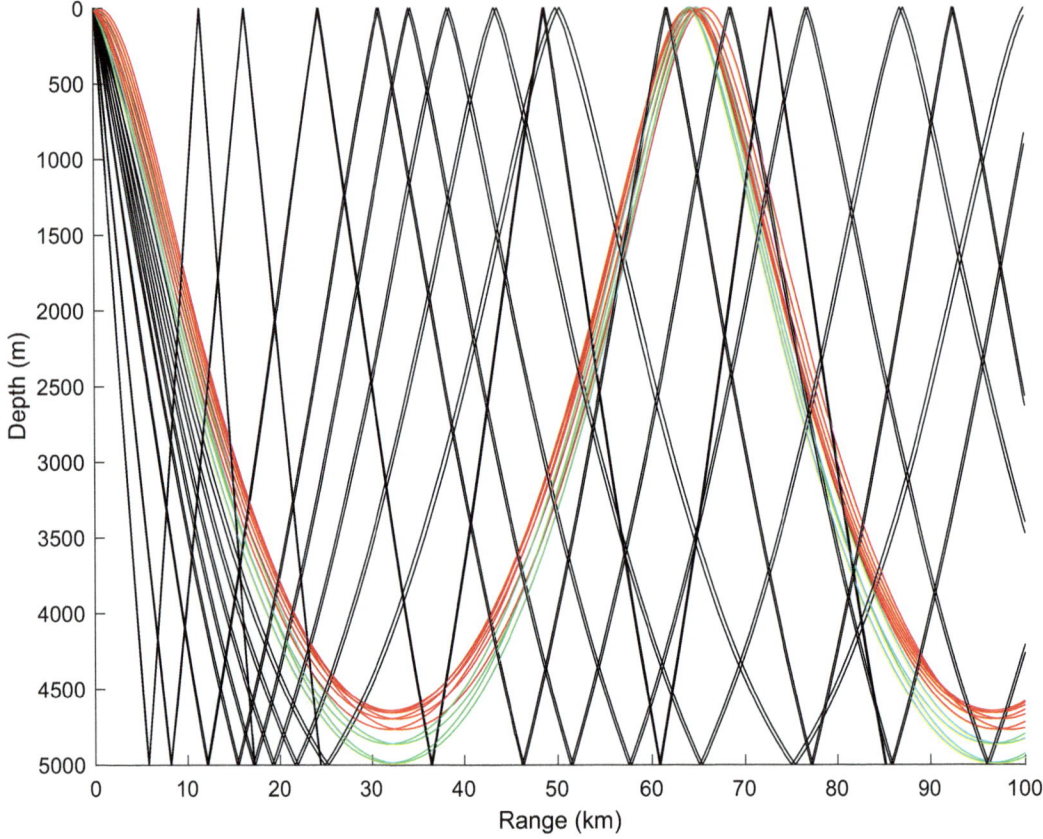

Fig. 1.9 Ray trace for the deep-water profile. RR paths in red; RB paths in green; SB paths in black

added complication in the form of the bathymetry and similar data for the varying sub-bottom layers.

Where then can one get all this information? In general, it is simply not available, and compromises are needed. One option is for the researcher to make their own direct measurements. For instance, expendable bathythermographs (XBTs) and/or self-recording thermistors are cheaply available to record the ocean temperature, which in turn provides a good estimate of the sound speed. Echo sounders can be used to profile the bathymetry and grab samplers can be used to collect a sample of the sea-bottom material. An alternative is to go to existing databases or environmental models as we discuss in the following.

1.8.1 Bathymetry

There are several widely used global bathymetry databases that are publicly available.; for example, Digital Bathymetric Data Base Variable Resolution (DBDB-V) developed by the U.S. Naval Oceanographic Office (NAVOCEANO) and distributed by both NASA https://cmr.earthdata.nasa.gov/search/concepts/C1214614815-SCIOPS.html and the USGS https://pubs.usgs.gov/of/2011/1127/dbdbv.html. Grid resolutions vary from 0.5, 1, 2, 5 minutes of latitude to about 1 km.

The Shuttle Radar Topography Mission elevation data was produced originally by NASA and is currently released by the USGS. The global bathymetry sub-sampled to a 1° grid is shown in Fig. 1.10.

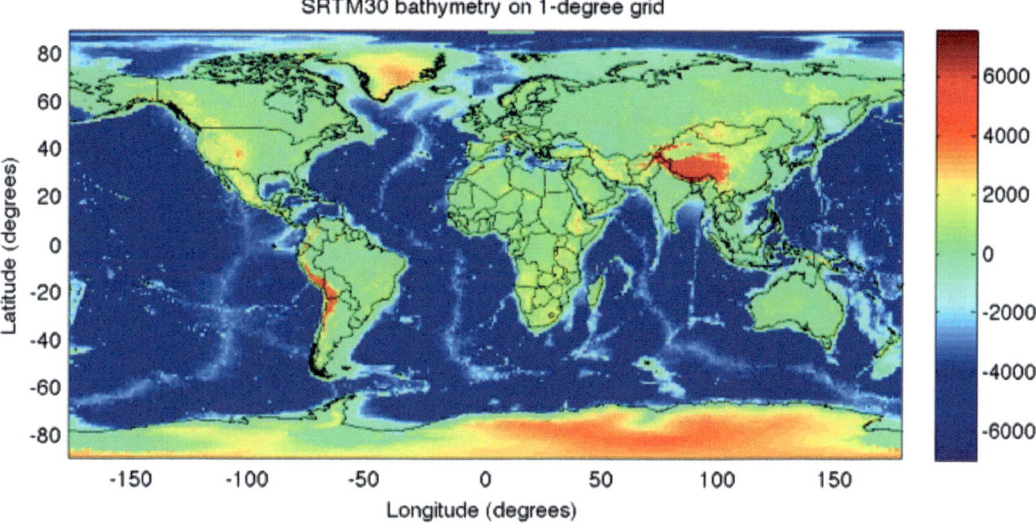

Fig. 1.10 Bathymetry from Shuttle Radar Topography Mission (SRTM) on a 1° grid

Another widely used database is GEBCO (Generalized Bathymetric Chart of the Oceans) available from https://www.gebco.net, with the current version providing data on a 15 arc-second interval. Lastly, we mention the ETOPO 2022 (Earth TOPOgraphy) available from the National Centers for Environmental Information at NOAA (https://www.ncei.noaa.gov/products/etopo-global-relief-model) currently providing data on a 15 arc-second grid.

1.8.2 Geo-acoustic Databases

Just as a room is brightened or darkened by a choice of paint color or presence of mirrors, ocean soundscapes can be dramatically affected by the acoustic reflectivity of the ocean surface and bottom boundaries. For the ocean surface, that reflectivity is usually about the roughness (or bubble clouds at higher frequencies). For the bottom, it clearly depends on the sub-bottom material. (Boundary and interfacial roughness are additional factors.) In U.S. Navy databases for frequencies below 1 kHz, the bottom is typically characterized as a multilayered medium with depth-dependent sound speed, density, and attenuation in each layer.

Estimating this sub-bottom information on a global scale is clearly an enormous challenge. Often it is done by geoacoustic inversion in which measurements of the sound propagation are used to estimate a complicated sub-bottom structure that allows a model to predict sound fields that match the measured data. In deeper water, as we will see in subsequent examples, the sound speed structure of the ocean will often refract the sound away from the bottom. Then, once we get away from the near field, the bottom losses will strip off the bottom interacting energy leaving a field dominated by the propagation in the ocean volume. In these cases, a precise knowledge of the bottom reflectivity is not very important, and the oceanography instead becomes critical. In contrast, in shallow water it is typical for the dominant energy to be bottom interacting. The obvious corollary is that the shallow-water soundscape is sensitive to the bottom loss. In summary, sometimes the sub-bottom properties are important and sometimes not. Collecting this information in particular areas can be a bit of a scavenger hunt with many different institutional sources having more detailed information for more localized areas.

For global soundscape modeling, we mention a few useful databases. First is a sediment

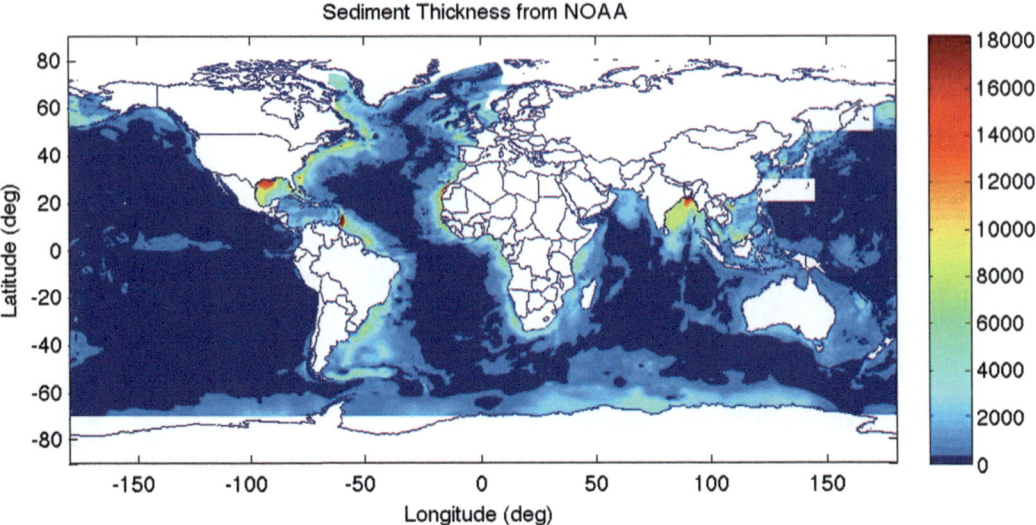

Fig. 1.11 Sediment thickness in meters from the NOAA National Geophysical Data Center database

thickness database available from NOAA (https://www.ngdc.noaa.gov/mgg/sedthick/) (Straume et al. 2019) and shown in Fig. 1.11. Deposition from major rivers such as the Mississippi and the Ganges is easily seen. However, one should realize that such maps present a simplified view of the ocean sub-bottom. In more detailed databases for localized areas, such sediments might be divided into multiple layers considering lithification of the various strata. Depending on the definition of the sediment, one may find wildly varying estimates of the sediment thickness.

The U.S. Navy provides a Bottom Sediment Type (BST) database that can be a first cut at the surficial sediment type in many areas of interest. Similarly, US SEABED and dbSEABED provide this information on a global scale but of necessity in a coarsely sampled way. For our global soundscape modeling, we combined these into a single database depicted in Fig. 1.12 with the colorbar presented in units of grain size (phi units; $\phi = \log_2 S$, where S is the grain size in mm).

1.8.3 Oceanographic Databases and Models

The World Ocean Atlas from NOAA National Center for Environmental Information provides global climatological fields of temperature and salinity, which can be converted to sound speed (https://www.ncei.noaa.gov/products/world-ocean-atlas). Data are provided on a 0.25° grid and at a variety of depths. Averages are provided for monthly, quarterly, and annual periods. Figure 1.13 shows an example for the month of January and for a slice of the sound speed taken at the ocean surface. Such information is hugely valuable; however, it should be noted that the averaging process inherent in such climatologies can distort important features of the ocean sound speed structure. For example, ocean surface ducts, which are often very important acoustic features that can evolve significantly over a week or month, may be averaged into a single profile that bears little resemblance to any daily realization.

Another very different approach to getting characterizations of the oceanography is to use ocean circulation models, which incorporate a variety of measurement systems such as global ocean observations, for example, the Argos network (https://argo.ucsd.edu), as well as sea-surface temperature, salinity, and altimetry from satellites. This is essentially a parallel of atmospheric weather forecasting but oriented to the ocean "weather." The HYCOM consortium (https://www.hycom.org) provides one widely

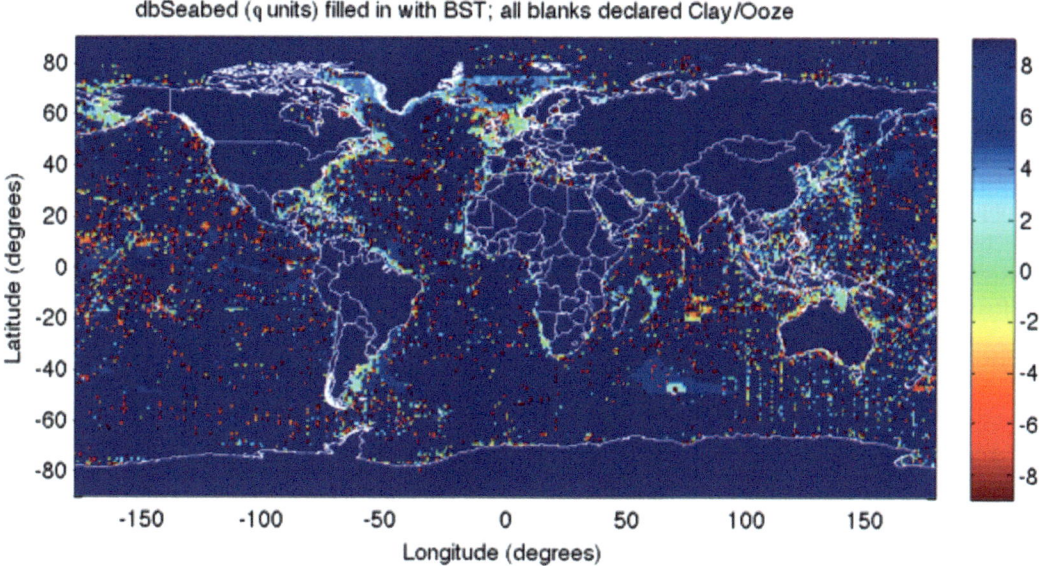

Fig. 1.12 Merge of Navy's Bottom Sediment Type database and dbSEABED

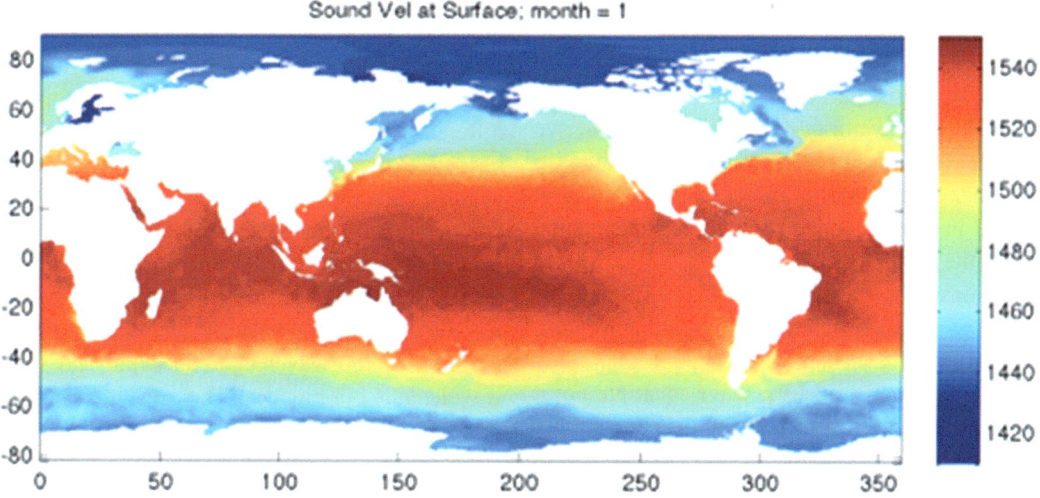

Fig. 1.13 Ocean sound speed at the surface derived from the World Ocean Atlas

used global source for such information along with the Copernicus Marine Environment Monitoring Service (CMEMS https://marine.copernicus.eu/access-data). These organizations incorporate the measured data in state-of-the-art numerical solutions representing the ocean dynamics. More recent work is also being done using artificial intelligence solutions to complement these analyses to produce the best estimates of the ocean state and associated uncertainties.

The wind and resulting ocean waves are a major source of underwater background noise (Urick 1986) and, along with ice coverage, can impact how sound is scattered and reflected near the surface. In higher latitudes, the ice roughness also promotes surface scattering and changes the reflection. When melting the ice can create layers of fresh water that also impact the near-surface sound speed. As such, one needs to use weather predictions such as those produced by meteorological services (e.g., https://mag.ncep.noaa.gov) to obtain reliable estimates of these environmental conditions when predicting how sound and noise are distributed in the ocean.

Recent developments include the coupling between all these fields (atmosphere–waves–ice–ocean), such that in the future operational systems all the relevant parameters will be described and linked in self-consistent ways.

1.9 Numerical Models

Ocean acoustic modeling is fundamentally about solving the acoustic wave equation:

$$\nabla^2 p - \frac{1}{c^2} p_{tt} = -s(t)\delta(\mathbf{x} - \mathbf{x_0}),$$

where $p(x, t)$ is the acoustic pressure and $s(t)$ is the source waveform. In the first section, we talked about the sound as a small perturbation to the background field. There is a lot of subtlety buried in that statement. At a more fundamental level, the fluid motion follows the Navier–Stokes equations, which are too complicated to solve for many problems of interest. However, if we consider a limit of an infinitesimal perturbation we get this far simpler acoustic wave equation. For it to be useful, the source term cannot be too strong. Explosives can, of course, be very strong and in those cases one may use a more complicated model (sometimes called Hydro codes) to capture the near field and then propagate that out in range using the acoustic wave equation.

For simplicity, we have assumed the density is constant; however, a slightly modified version is often used for problems with density variations. We have also assumed a point source located at $\mathbf{x_0}$. The sequential terms in this equation represent the spatial curvature of the pressure, the curvature in time of the pressure, and the time-space distribution of the source. Here, we present a brief overview of the mathematical and numerical techniques used to solve this equation. This is a complicated subject. For more details, we refer the reader to Jensen et al. (2012).

As mentioned earlier, the source waveform is typically assumed to be harmonic (meaning the waveform varies sinusoidally over time), where $s(t) = e^{i\omega t}$ and the resulting pressure field responds sinusoidally, $p(\mathbf{x}, t) = p(\mathbf{x})e^{i\omega t}$. (As we discuss later, any general waveform can then be built up as a sum of these harmonics.) Making this substitution in the wave equation yields the Helmholtz equation or reduced wave equation:

$$\nabla^2 p + \frac{\omega^2}{c^2} p = \delta(\mathbf{x} - \mathbf{x_0}).$$

This is the first key simplification, which reduces the complexity by eliminating time from the problem.

This equation can be solved directly by full finite-element methods (FEM) and finite-difference methods (FDM). These methods have been used in an exploratory fashion in underwater acoustics since the 1980s; however, they require computational grids with around 10 points per wavelength in range and depth. For 3D problems, one must take that further and consider a grid in all dimensions. Typical problems of interest involve 10–100 wavelengths in depth and 1000 wavelengths in range. The resulting linear systems of equations have historically been too large to be practical for most cases. However, with the ongoing improvements in computer performance (including Graphical Processing Units), the FDM and FEM are seeing wider use.

In general, though, it has been necessary to use a variety of approximations to simplify the acoustic models. If we make a high-frequency approximation, we are led to ray tracing methods, which in their current incarnation are often implemented by tracing beams. Hence, we refer to ray and beam tracing together.

A separate key approach is to assume that the ocean environment is stratified or range-independent (i.e., it does not vary in range). Sometimes the term "layered" is used here as well; however, the strata or layers can have a sound speed or density continuously varying in depth within each layer. This approximation leads to two main types of models called (1) normal modes and (2) wavenumber integration. (One can combine both the high-frequency approximation and range-independence to derive ray/beam methods for stratified problems.)

Finally, one can make an approximation based on outgoing propagation (i.e., there is no backscatter). This leads to the so-called parabolic equation that enables a solution by marching outward in range. In contrast, the original Helmholtz equation simultaneously couples the field at all ranges, leading to difficult linear systems of equations.

If we presented these models in a historical manner, we would start first with empirical formulae for propagation loss, which were widely used in the 1960s. Those models were followed by ray methods in the 1970s, which evolved into beam methods. Wavenumber integration and normal mode methods became popular in the 1970s. Subsequently, parabolic equation methods were introduced.

This history is not so much a transition from one model type to another as a flowering of new approaches. The empirical formulae have largely fallen out of use; however, the other model types have all found value for different applications and needs. To summarize, the four main types of models used for ocean acoustics today are (1) rays and beams, (2) wavenumber integration, (3) normal modes, and (4) parabolic equation.

In the following sections, we will discuss each of these alternatives. It should be emphasized that all the examples and the discussion of the algorithms in this chapter have been simplified or idealized to introduce the general concepts. In realistic environments, the ocean sub-bottom is typically modeled with many layers, sometimes incorporating elasticity or poro-elasticity. The numerical algorithms are correspondingly more complicated. Software for these various model types is available from the Ocean Acoustics Library (https://oalib-acoustics.org).

1.9.1 Rays and Beams

Rays are curves that are normal to the wavefronts of sound. One of the simplest cases to introduce the concept and define some terms is a point source in free space. The wavefronts are spherical in 3D or circular in 2D. Therefore, the rays, as indicated in Fig. 1.14 are straight lines radiating from the location of the point source. There is an infinite number of such rays; we can generate one for any arbitrary take-off angle.

If we think of every adjacent pair of rays as forming a ray tube, then the spreading of the ray tube tells us about the change in pressure. Thus, the pressure is highest near the origin where the ray tube vanishes. In fact, the pressure goes to infinity at the origin. The phase of the pressure along the ray is determined by the path length. Recall that the field due to a point source in free space is

$$p(r) = \frac{e^{ikR}}{R},$$

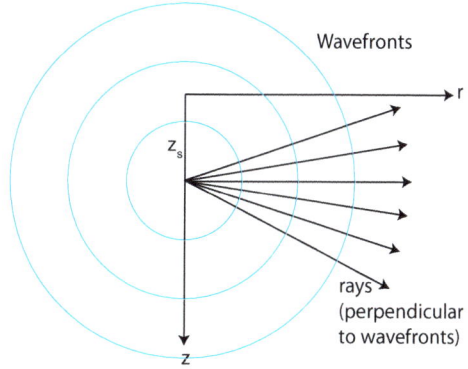

Fig. 1.14 Rays and wavefronts for a point source in a homogeneous medium

where R is the spherical range from the source and k is the wavenumber. In the framework of rays, we represent this as

$$p(s) = A(s)e^{i\phi(s)}$$

where $A(s)$ is an amplitude,

$$A(s) = \frac{1}{s},$$

$\phi(s)$ is a phase,

$$\phi(s) = ks$$

and s is the arclength along the ray. The arclength is the same as the spherical range.

We consider next the situation where the sound speed, c, is no longer constant but varies with depth, z. We have seen that the rays refract in an intuitive way based on the sound speed profile. One may then wonder what equations describe that refraction and how they can be derived. The derivation is very involved and relies on a high-frequency assumption. The rays are found to be governed by a simple set of ordinary differential equations. Defining $r(s)$ and $z(s)$ as the range and depth of the ray as a function of arclength, one finds

$$\frac{dr}{ds} = c\xi$$

$$\frac{d\xi}{ds} = \frac{\left(-\frac{1}{c^2}\right)\partial c}{\partial r}$$

$$\frac{dz}{ds} = c\zeta$$

$$\frac{d\zeta}{ds} = \frac{\left(-\frac{1}{c^2}\right)\partial c}{\partial z}$$

where the vector (ξ, ζ) is the tangent to the ray, $\frac{\partial c}{\partial r}$ is the partial derivative of c with respect to range, r, and $\frac{\partial c}{\partial z}$ is the partial derivative of c with respect to depth, z. The initial conditions required are that the ray starts off at the source position and with a specified take-off angle α. Often, we use a coordinate system centered around the source range. Then we have the initial conditions

$$r(0) = 0$$

$$z(0) = z_s$$

$$\xi(0) = c_s \cos \alpha$$

$$\zeta(0) = c_s \sin \alpha$$

where c_s denotes the sound speed at the source. This initial value problem is solved repeatedly with different take-off angles to produce a plot of ray paths in the ocean. Standard numerical methods such as Runge–Kutta are suitable. We refer the reader back to Fig. 1.9 to see an example of such rays in a deep-water environment. Note that the gradient of the sound speed acts a bit like a force bending the rays toward areas with a slower sound speed. Analogies may be drawn to a rank of soldiers marching forward with some on hard ground and some in the rough. Similarly, one may think of a cart with the wheels on one side in mud.

The rays may be viewed as a framework for constructing the pressure field. A formal derivation of the ray equations reveals that the sound intensity is inversely proportional to the spread (cross-sectional area) of the ray tube and the phase is proportional to the phase delay due to the travel time along the ray.

In general-purpose ray models, the sound speed and the bathymetry are generally provided as matrices and vectors, respectively. Thus, they are discretely sampled and often on irregular grids. Converting that discrete information to smoothly varying functions without introducing artifacts can be quite difficult. Conversely, solving the differential equations with functions that are not smoothly varying requires care in handling the places where the rays cross interfaces that are not smooth. For this reason, amongst others, many of the early ray models produced various artifacts. In ocean acoustics, these problems have been mitigated by using an approach where beams are constructed around the central rays of the beams.

In the U.S. Navy, the GRAB model has been accepted as a standard in the Ocean and

Atmospheric Master Library. An alternative is BELLHOP, which is available as open source from the Ocean Acoustics Library. All the ray/beam results presented in this chapter were generated using BELLHOP.

1.9.2 Wavenumber Integration

If the ocean waveguide is range-independent, then the acoustic pressure can be represented as a wavenumber integral:

$$p(r,z) = \int_0^\infty G(z, z_s; k_r) J_0(k_r r) \, dk_r$$

where $G(z, z_s; k_r)$ is a Green's function of the depth-separated wave equation:

$$\rho(z) \frac{d}{dz}\left[\frac{1}{\rho(z)} \frac{dG}{dz}\right] + \left[\frac{\omega^2}{c^2(z)} - k_r^2\right] G = -\frac{\delta(z - z_s)}{2\pi},$$

and $J_0(k_r r)$ is the Bessel function and ρ is the density. For readers not familiar with Bessel functions, they are roughly sinewaves with a factor for cylindrical spreading. The depth-separated wave equation is an ordinary differential equation in depth. To complete its specification, we require boundary conditions at the sea surface and at some bottom depth. For instance, if the surface is an air/water interface, then typically the surface is treated as a pressure-release boundary, that is, it is assumed that the pressure must vanish. Bottom boundaries are more complicated, but for the sake of introducing the method, we assume a rigid bottom boundary, which could be used to represent a very hard sea bottom such as a basalt. These two boundary conditions become

$$G(0; k_r) = 0$$

$$\left.\frac{G(z; k_r)}{dz}\right|_{z=D} = 0$$

where D is the depth where the rigid boundary occurs. The integral can be done with any standard quadrature method such as the trapezoidal rule. In that case, the continuous integral is replaced by a discrete sum for equally spaced wavenumbers in a finite interval $[0, k_{max}]$. In principle, the integral should go to infinity; however, the integrand becomes negligible for large wavenumbers. Typically, $k_{max} = 1.2 \frac{\omega}{c_{min}}$ is sufficient, where c_{min} is the minimum sound speed in the ocean and sub-bottom. It is also found that the spacing between wavenumbers should be about $\frac{1}{R_m}$ where R_{max} is the largest receiver range being considered. This implies that the execution time increases in proportion to receiver range. The integral generally should also be done on a contour in the complex plane slightly shifted from the real axis. This shift is essentially equivalent to including extra attenuation, which is later removed. It avoids the problem of integrating across poles (singularities) in the Green's function that occur at special resonant *wavenumbers*.

The real work is typically not in the final integral but rather in evaluating the integrand and, in particular, $G(z; k_r)$. That results from solving the above boundary value problem, which can be done using finite-difference or finite-element methods. This approach is used in the SCOOTER model, which is part of the open-source Acoustics Toolbox (http://oalib.hlsresearch.com/AcousticsToolbox/). All the results in this chapter using wavenumber integration were generated using SCOOTER. More commonly, the boundary value problem has been solved by coefficient approximation, in which the sound speed is approximated by piecewise linear layers. Then analytic solutions in each layer can be written in terms of Airy functions. This latter approach is used in the open-source OASES package (https://acoustics.mit.edu/faculty/henrik/oases.html).

To give an example, we consider an Arctic profile. These are commonly upward refracting and the case we consider is shown in Fig. 1.15. Taking a source frequency of 50 Hz and a source depth of 100 m, we obtain the Green's function shown in Fig. 1.16. (We are plotting magnitude.) One can see that for higher wavenumbers, the Green's function is very low at certain depths. This is because the term in square brackets changes from positive to negative below those depths. Then the Green's function changes from a sinusoidal shape to a decaying exponential.

After performing the final integral, one obtains the propagation loss plot shown in Fig. 1.17.

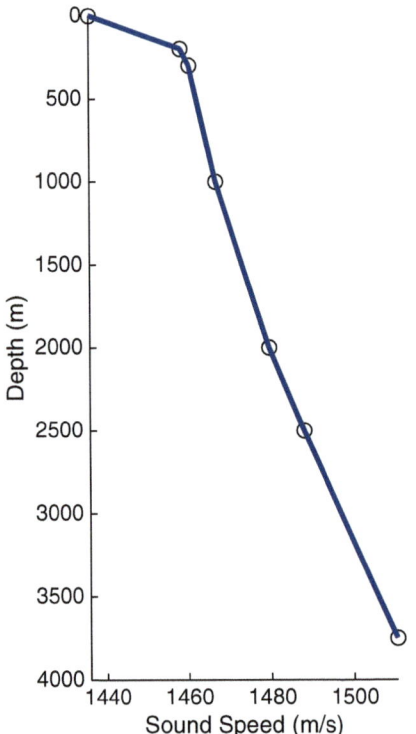

Fig. 1.15 Sound-speed profile for the Arctic scenario

Nothing in the Green's function plot provides an easy intuition into the result. However, the TL is consistent with an intuition based on ray theory showing upward refracting paths and bottom bounce energy interacting with each other.

This example is characteristic of Arctic propagation; however, one should be aware of the important role that the ice canopy can play. It has not been included here. When present, the ice canopy presents rough interfaces at both the ice/water interface and the air/ice interface. The upper boundary is then less like a perfect mirror and more like a broken mirror with significant scattering loss.

Wavenumber integration is perhaps the least used of the acoustic model types; however, it has much to commend it. First, the solution is essentially exact for range-independent problems. Second, it is extremely robust, and, in fact, we have not seen any failures. It is easy to implement and trivially parallelizable across wavenumber. It is also easily generalized to more complicated

problems with elasticity or poro-elasticity. With these comments, one may wonder why you would use anything else. However, for high-frequency or long-range problems, the execution time can be excessive. Furthermore, it is possible but difficult to generalize it to handle range dependence.

1.9.3 Normal Modes

The method of normal modes is closely related to wavenumber integration. Both methods rely on the assumption of range independence to construct a solution by separation of variables. That is, the solution is built up as a sum or integral of solutions that are of the form $\Psi(z)R(r)$ (i.e., a product of functions of one variable each). This is a very special class of functions that are separable in this way, and, in fact, the final pressure field is not separable. However, the pressure field is the *sum* of separable functions.

The Helmholtz equation, a partial differential equation involving range and depth, is then found to split into independent ordinary differential equations, one each for range and depth. The equation in range has an exact solution in terms of Bessel functions, which is exactly what was used in the wavenumber integral. The equation in depth is

$$\rho(z) \frac{d}{dz}\left[\frac{1}{\rho(z)} \frac{d\Psi}{dz}\right] + \left[\frac{\omega^2}{c^2(z)} - k_r^2\right]\Psi = 0$$

$$\Psi(0) = 0$$

$$\left.\frac{d\Psi}{dz}\right|_{z=D} = 0$$

which is the equation for the modes. The terminology is a bit sloppy here in that the modes can refer to the shapes in depth or the range-depth function that is the product of range and depth solutions. Dozens of models have been developed over the years to solve for the modes. The results in the chapter have been generated using the KRAKEN model, which is open source. The modes are arbitrary up to a scale factor. In this chapter, we assume they are scaled to have unit square norm, meaning the integral of the mode

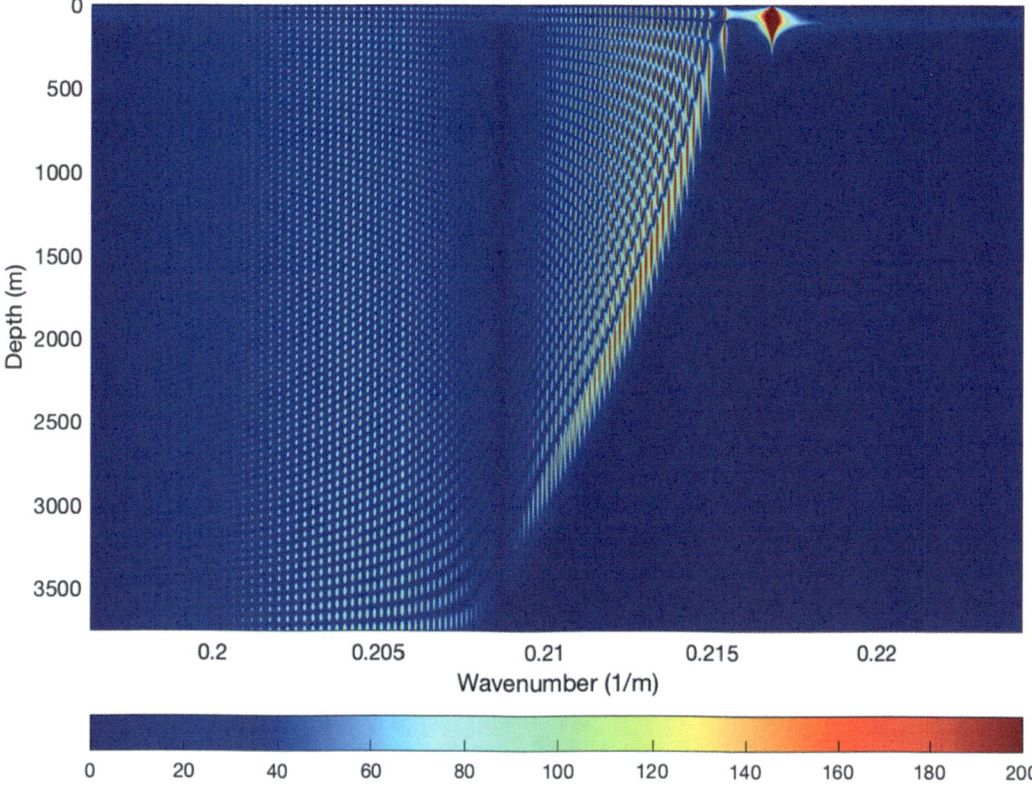

Fig. 1.16 Magnitude of the depth-separated Green's function for the Arctic scenario

squared is one. This simplifies all the equations, including the representation of the pressure field.

Perhaps the first difference one will note compared to the Green's function equation is the change in notation from G to Ψ. However, the important change that motivated that change in notation is the removal of the forcing term on the right-hand side. In addition, k_r in this equation is now an unknown. The analogy is a bit strained but roughly speaking, we have gone from the problem of a forced (plucked) guitar string to one where we are considering free vibrations. (The sound speed in that analogy is related to the local tension on the string, which would vary if the string had a variable diameter, for instance.)

This modal equation is solved for its free vibrations yielding a set of eigenfunctions (mode functions), $\Psi_m(z)$, and corresponding eigenvalues (mode wavenumbers), k_{rm}, where m is simply an index. In the analogy to a guitar string, these are the vibrating shapes and their frequencies. The Green's function, representing the response of the plucked string, can then be written as

$$G(z) = \frac{1}{2\pi\rho(z_s)} \sum_m \frac{\Psi_m(z_s)\Psi_m(z)}{k_r^2 - k_{rm}^2}$$

Note that there is a singularity or resonance, where the wavenumber k_r equals k_{rm}. However, the Green's function is not usually the main interest. The acoustic pressure in terms of that Green's function becomes

$$p(r,z) = \frac{i}{4\rho(z_s)} \sum_m \Psi_m(z_s)\Psi_m(z) H_0^{(1)}(k_{rm}r)$$

where $H_0^{(1)}$ is the zeroth-order Hankel function of the first kind. This result is obtained by substituting the modal expression for the Green's function into the wavenumber integral. Methods from complex variable analysis show that the

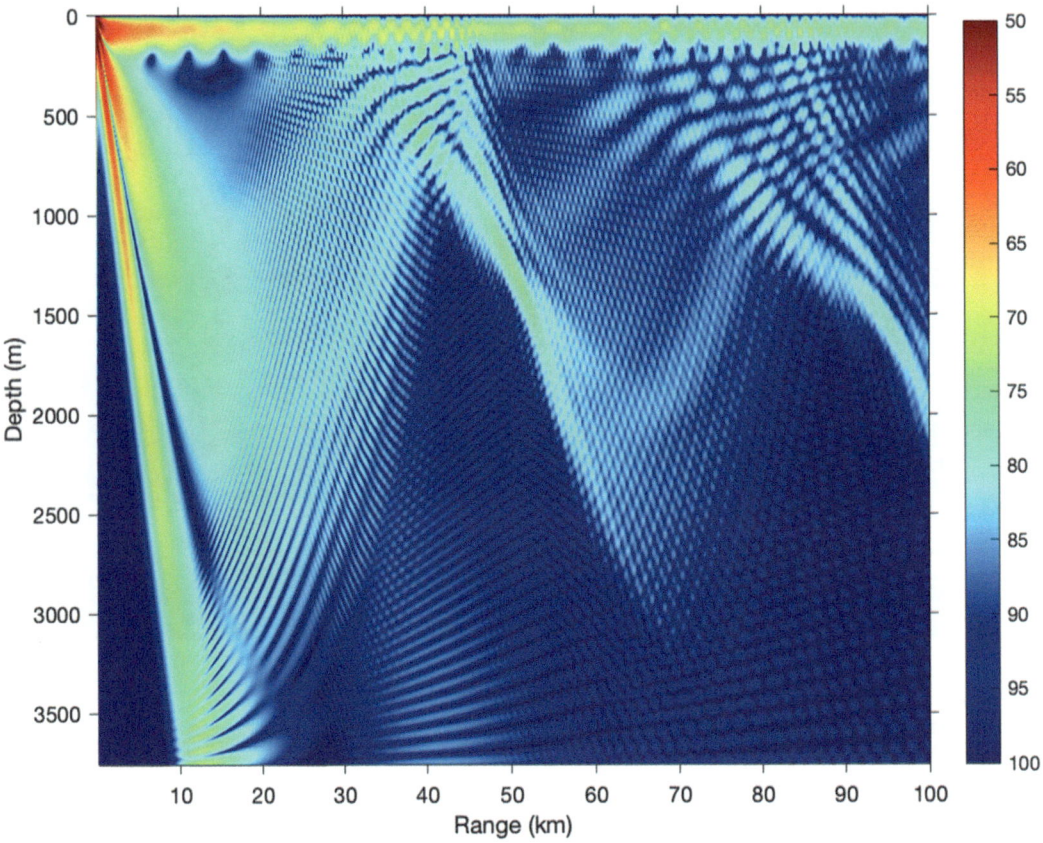

Fig. 1.17 Propagation loss for the Arctic profile calculated using wavenumber integration for a source depth of 100 m and a frequency of 50 Hz

integral can be represented by the contributions of each of the resonances. Thus, the wavenumber integral becomes a discrete sum of modes. Each mode is a separable function of range and depth; however, the resulting pressure field is a sum of such modes and is not itself separable. The number of such modes is typically infinite; however, an accurate answer can be obtained using just the lower order modes. The number of such modes needed; however, tends to increase with frequency.

Were we to plot the Green's function and resulting pressure for the Arctic case considered above, there would be no visible difference compared to the wavenumber integration results. Instead, we consider a surface-duct problem with the sound speed profile shown in Fig. 1.18. This is a deep-water problem, and we can see the typical bowl shape, where the sound speed drops in the thermocline, and then the temperature effects are overcome by the pressure, and the sound speed increases again. This is like the Munk profile considered previously. A new feature here is the zone in the upper 100 m, where the sound speed is increasing, thereby forming an upward refracting layer or surface duct. This type of feature is very common in the ocean and results from wind mixing. The resulting mixed layer has a roughly uniform temperature and so the sound speed is driven by the increase in pressure with water depth. In areas of the world and times of the year when there are strong storms, the mixed layer can be as deep as 500 m; however, a more typical value would be 25 m. In some cases, it is not present at all.

It should also be noted that this description of the mixed layer is really an idealization. More generally, the mixed layer is formed through a

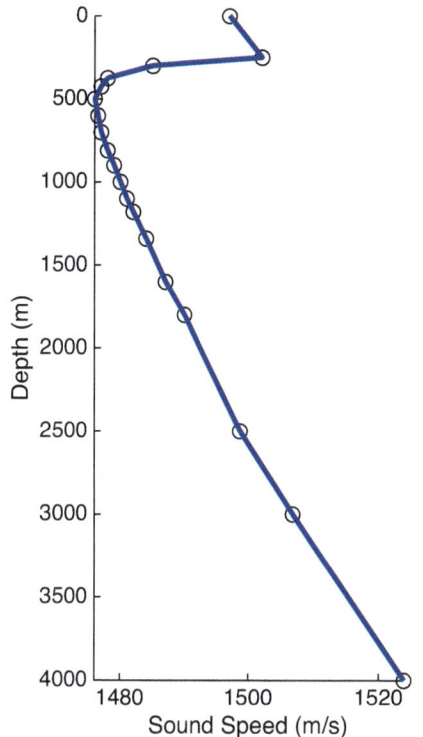

Fig. 1.18 Sound speed profile for the surface-duct problem

modes, each successive mode function having one extra zero crossing (at least for problems with negligible attenuation)

We can see many of these features in this mode plot. The first mode is trapped in the SOFAR channel and has no zero crossings. Mode 20 has 19 zero crossings, is trapped in the SOFAR channel, and oscillates in a domain of depths that goes both closer to the surface and to the bottom. Mode 41 is a surface duct mode in the sense that it is strongest in the surface duct. Note that its shape in the surface duct resembles the shape of Mode 1 in the SOFAR duct. However, it stays faithful to the rule about increasing zero crossings by having low-level oscillations in the SOFAR duct. In ray terms, this is a form of leakage of energy out of the surface duct into the SOFAR channel.

In going just from mode 41 to mode 42, there is a radical change. The number of zero crossings is incremented, but now mode 42 is a SOFAR mode with the largest amplitude of the mode in the SOFAR channel rather than the surface duct. Mode 80 shows the progression to a mode that is now oscillatory from the surface down to about 3800 m. In ray terms, this energy corresponds to paths that reflect off the surface but are refracted away from the bottom before hitting it. Finally, at mode 180, the mode is oscillatory throughout the entire water column. In ray terms, this energy corresponds to rays that reflect off both the surface and the bottom.

These line plots in individual modes show the shapes very clearly. However, to see the full set of modes, we need a different display. In Fig. 1.20, we have plotted all the modes in a color format. Here, you can see more clearly how each successive mode picks up an extra zero-crossing. You can also see where two or three of the modes are surface-duct modes and interspersed with the other modes.

The normal modes can be viewed as a pair; there is the mode shape vs. depth, $\Psi_m(z)$, also called (drawing on the mathematical terminology) an *eigenfunction* and the associated horizontal wavenumber, k_{rm}, also called an *eigenvalue*. The horizontal wavenumber appears in the Hankel function and governs the oscillatory shape of the mode in range. The eigenvalues are

series of mixing events due to multiple storms. The temperature structure also depends on the sun and the cloud cover. So-called mixed-layer or upper ocean models are used to predict the time-space evolution of this zone (Fig. 1.19).

We consider here a source frequency of 100 Hz and display a subset of the modes in Fig. 1.19 Note the oscillatory shape of the modes. The Wentzel–Kramers–Brillouin (WKB) approximation provides an approximate formula for these shapes that gives some intuition about their behavior. Roughly it says that in a duct, the modes will oscillate in the middle of the duct and then decay exponentially outside the duct. In our case, we have two ducts: (1) the main duct or sound-fixing-and-ranging (SOFAR) channel, and (2) the surface duct. Then the WKB theory becomes more complicated, and we find that the modes can be trapped in one or the other duct (or sometimes be strong in both ducts). Following Sturm–Liouville theory, there is a succession of

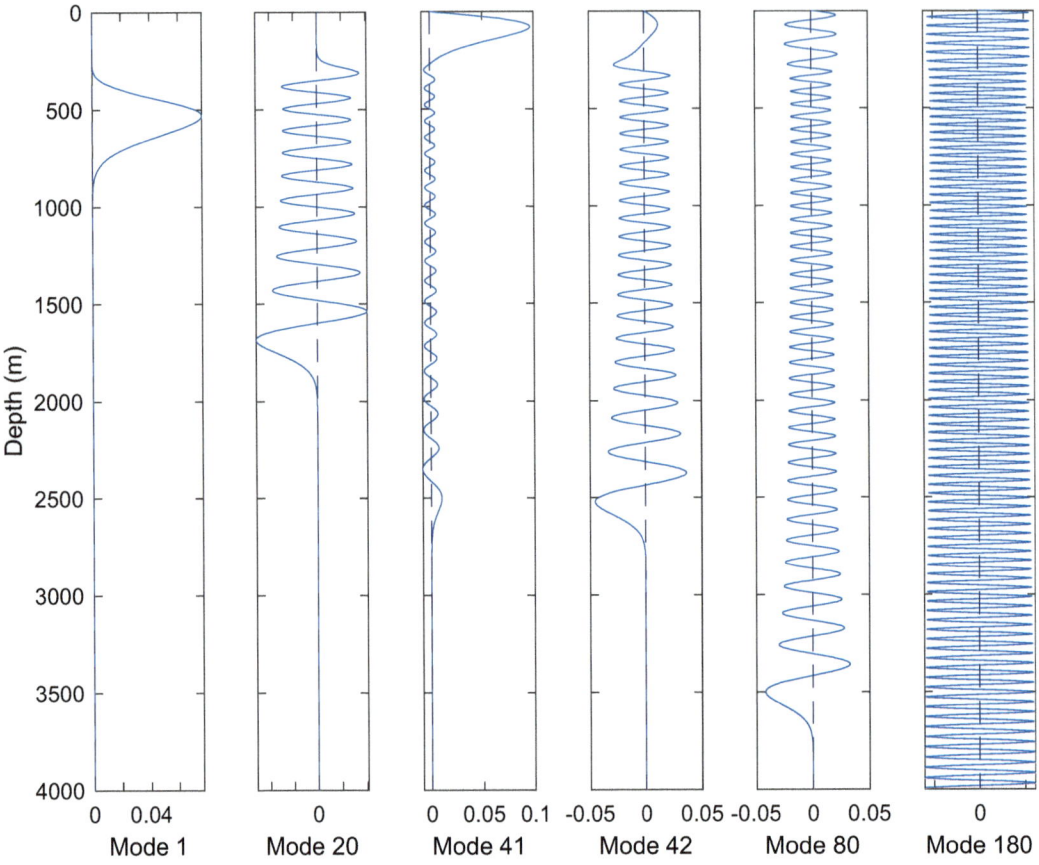

Fig. 1.19 Line plots of selected modes for the surface duct problem for a frequency of 100 Hz

normally complex numbers where the real part gives the rate of oscillation in range, while the imaginary part gives the rate of decay in range. This decay is caused by losses due to boundary reflections or volume attenuation in the ocean or the sub-bottom.

In Fig. 1.21, we show a plot of these wavenumbers in the complex plane. Again Sturm–Liouville theory tells us that (under some simplifying assumptions) there will be an infinite number of these wavenumbers with the largest wavenumber corresponding to the first mode at the right-hand side of the plot. The wavenumbers of each successive mode become smaller. This description is a bit imprecise because the nature of the spectrum depends in a subtle way on the boundary conditions; however, it is a useful starting point.

Note that there is a line of wavenumbers with a negligible imaginary component along the top of the plot. These correspond to modes that are not interacting much with the bottom. There is no surface or bottom loss, so the corresponding modes have little loss and the wavenumbers have almost no imaginary part. Then there is a sharp transition to the bottom-interacting modes and the wavenumbers echo that loss by assuming a more negative imaginary part. These highest modes correspond to steep ray paths and have a lot of bounces (reflections) off the bottom, so their loss is higher. However, this is not always the case. In some profiles, higher-order modes may have fewer bottom or surface bounces and so the wavenumbers can produce complicated plots in the complex plane.

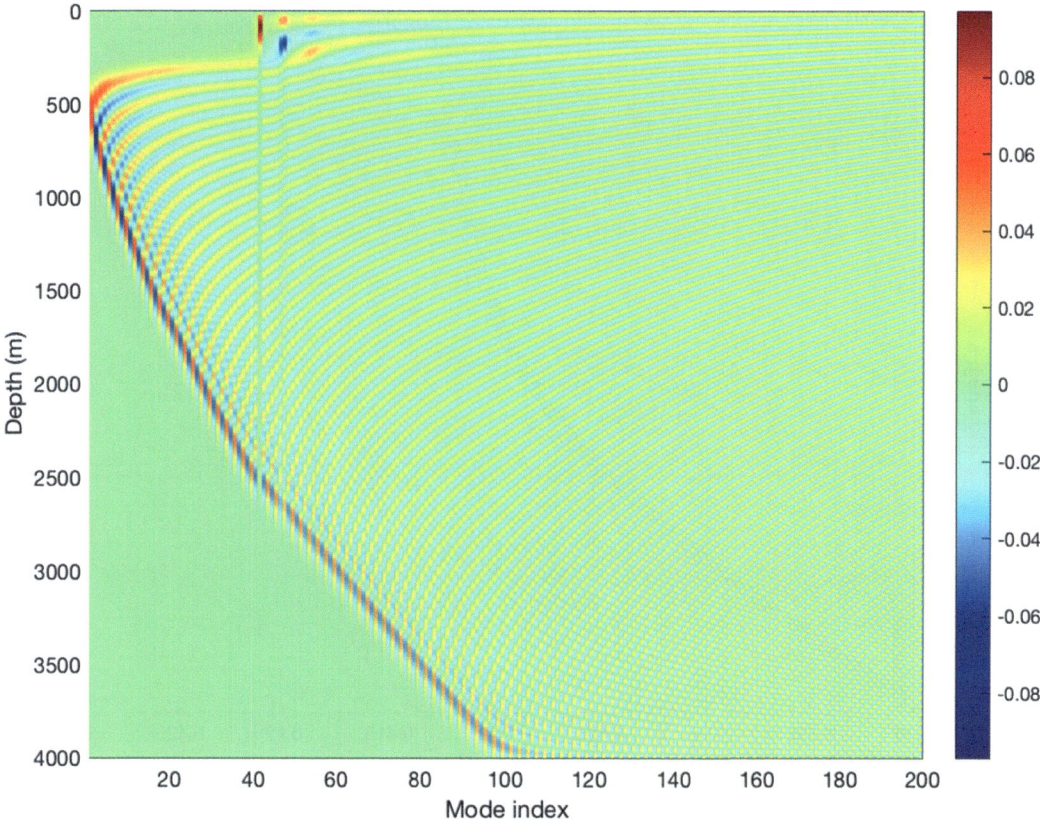

Fig. 1.20 Plots of the first 200 modes for the surface duct problem with a source frequency of 100 Hz

There is little in these mode plots that would prepare one to expect the propagation loss plot shown in Fig. 1.22 for a source depth of 25 m. In ray terms, we understand that there is a convergence zone pattern due to the SOFAR duct as well as surface-ducted energy. We can also see that there is a partial shadow in the area between the convergence zones. However, energy is continuously leaking out of the surface duct into those shadows. It is remarkable that the modal sum reproduces all these effects. It does so through constructive or destructive interference amongst families of modes. Note also that the source depth only enters at this final stage where we are summing the modes. The summing of the modes requires little computation so large numbers of source and receiver depths are easily handled.

In the context of marine mammals, it should be noted that the surface duct results in very good propagation conditions and ensonifies the upper part of the shadow zone that otherwise would be present. However, if there is a rough sea surface, then much of the energy can be scattered out of the surface duct. In the Bahamas stranding incident, it was hypothesized that the combination of calm seas and a surface duct might have led to increased received levels and ultimately, the stranding (Ocean Studies Board 2003).

As noted above, normal mode and wavenumber integration methods are closely linked as one may see also from the governing equations. However, they have different characteristics. The eigenvalue problems involved in the modal equation are harder to solve and the resulting models are often less robust. However, once the modes have been computed, the field can be computed for arbitrarily large ranges with little computational difficulty. Some people have suggested modal methods are appropriate beyond 10 water depths.

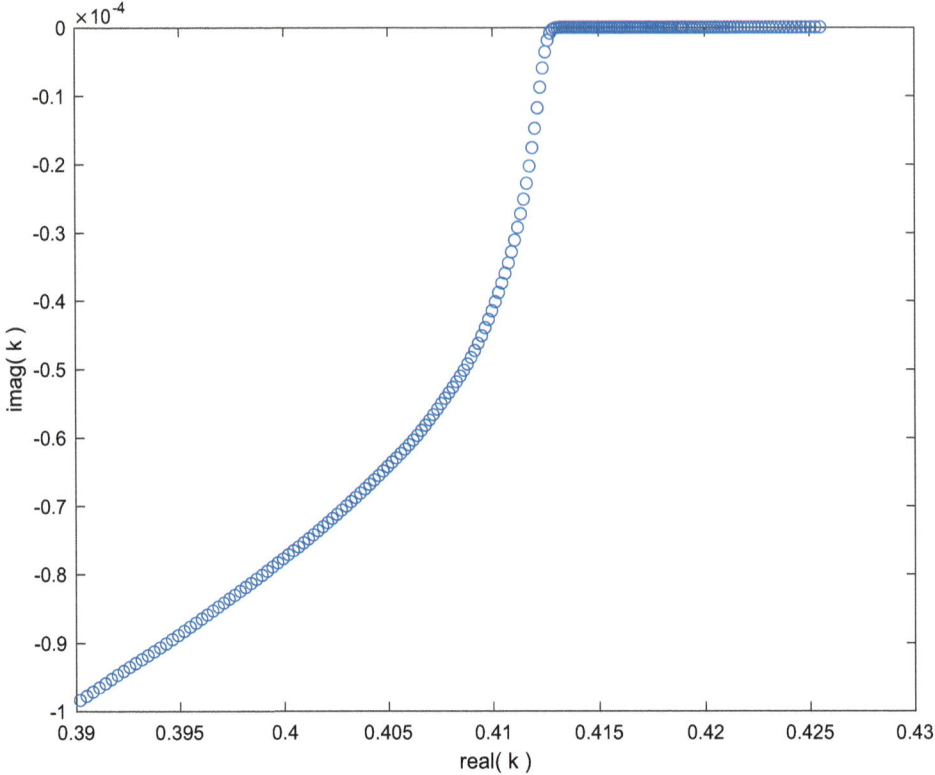

Fig. 1.21 Horizontal wavenumbers for the surface-duct problem

However, that notion is tied to subtle issues including an assumption that only "trapped" modes are included. In any case, it can be difficult to use normal modes to get the pressure in the near field. Another consideration is the number of source and receiver depths. The modal approach is insensitive to that. The execution time for wavenumber integration tends to increase with the number of source and receiver depths.

1.9.4 Parabolic Equation

As we have just seen, both wavenumber integration and normal mode methods rely on the problem being range independent. Then they use the idea of separation of variables together with the exact solution in range (a Bessel or Hankel) function to produce an essentially exact solution.

When the scenario is range dependent, either due to bathymetry or oceanographic variations, those methods are no longer applicable. There are range-dependent extensions of both those methods that involve breaking the problem into range segments, which are individually range-independent. However, those approaches have not found widespread use. The most popular full-wave solution is the so-called parabolic equation (PE). (In its modern implementations, it is no longer properly classified as a parabolic equation.)

The parabolic equation was applied to underwater acoustics problems by Hardin and Tappert in the 1970s following earlier work by Leontovich and Fock in electromagnetics. It is widely viewed as one of the most important developments in computational ocean acoustics. Tappert presented his work in a workshop on modeling in 1975, solving a surface duct problem that had been given as a test case for the participants. In a bit of a flourish, he then made the surface duct range dependent and showed that

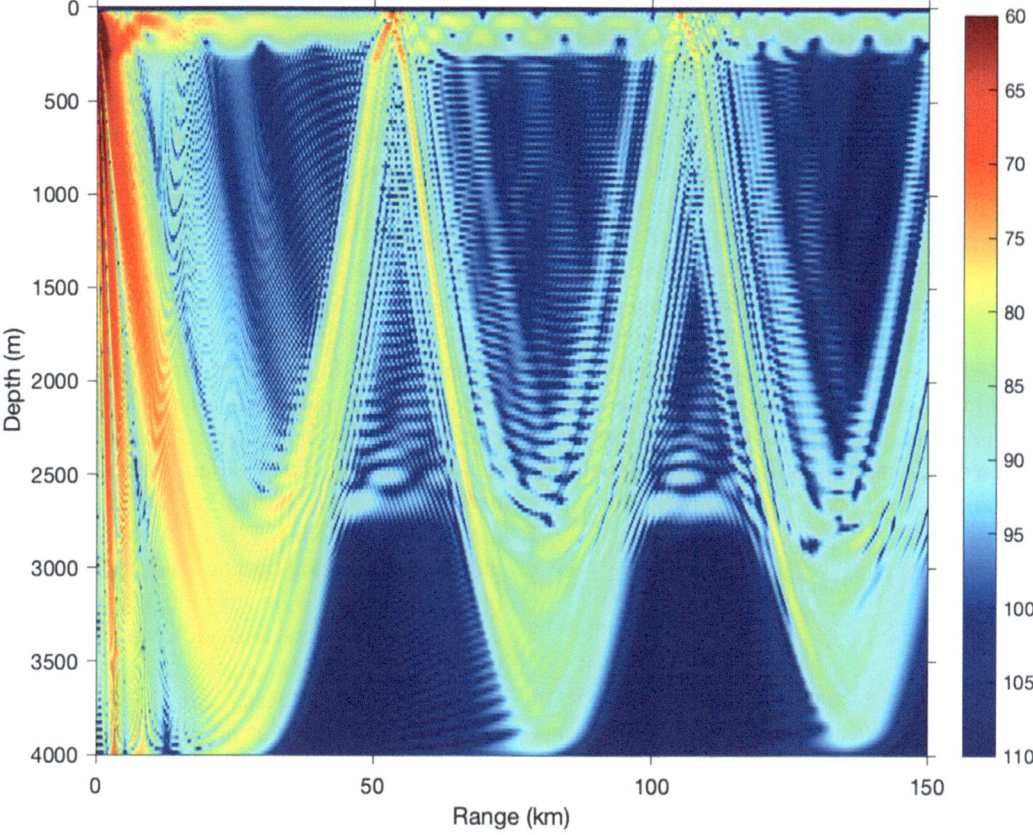

Fig. 1.22 Propagation loss for the surface-duct problem calculated using normal modes for a source depth of 25 m and frequency of 100 Hz

the PE could be applied to that with no changes, producing credible solutions. People were impressed. There was talk of limiting open distribution of the PE and questions about releasing it outside of Bell Labs, where the work had been done.

We will omit the full derivation, but in brief, one starts with the Helmholtz equation, which written out in cylindrical coordinates is

$$\frac{\partial^2 p}{\partial r^2} + \frac{1}{r}\left(\frac{\partial p}{\partial r}\right) + \frac{\partial^2 p}{\partial z^2} + k_0^2 \, n^2 \, (r,z) p = 0,$$

where $k_0 = \frac{\omega}{c_0}$ and $n(r,z) = \frac{c_0}{c(r,z)}$. Here, c_0 is a reference sound speed and $n(r, z)$ is the index of refraction relative to that reference sound speed. In this Helmholtz equation, the reference sound speed is irrelevant since the product $k_0 \, n(r, z)$ is insensitive to it. However, the PE is an expansion around that reference and so it can be sensitive to the choice. It can reasonably be set to either the minimum sound speed in the ocean or the sound speed at the source.

Even though this Helmholtz equation can easily be discretized directly, the pressure values on a discrete grid are coupled through a large system of linear equations, which can be very time-consuming to solve.

The PE approximation to this Helmholtz equation is

$$2ik_0 \, \frac{\partial \psi}{\partial r} + \partial^2 \frac{\psi}{\partial z^2} + k_0^2 \left(n^2 \, (r,z) - 1\right)\psi = 0.$$

After solving for ψ, the pressure p is given by

$$p(r,z) = \psi(r,z) H_0^{(1)}(k_0 r)$$

The term PE comes from the fact that second-order partial differential equations are classified as elliptic, hyperbolic, or parabolic based on a correspondence between the partial derivatives and the forms of those same curves. The original Helmholtz equation is elliptic. If one is new to this, then both equations (Helmholtz and PE) may appear equally complicated. However, there is a huge difference. As mentioned above, the elliptic problem leads to the large system of equations, where every value of the pressure on a discrete grid is connected to its neighbors and must be solved globally. In contrast, the PE is solved by a marching method. We set up an initial field at zero range and then march the solution out in range. To be specific, we can use a first-order difference formula for the range derivative, yielding

$$2ik_0 \left(\psi(r+\delta r) - \psi(r)\right)/\delta r + \frac{\partial^2 \psi(r)}{\partial z^2} + \left[k_0^2 n^2(r,z) - 1\right]\psi(r) = 0$$

which provides a simple formula for $\psi(r + \delta r)$ in terms of $\psi(r)$. We are omitting a lot of details and simplifying the presentation, but this gives a general feel for the method and would yield a working PE model. However, modern PEs use a more sophisticated factorization of the Helmholtz equation and the finite-difference discretization we used here is very basic.

We illustrate the method here with a range-dependent surface duct, just as Tappert did in the early workshop. The sound speed profile we consider is shown in Fig. 1.23. The only change is in the surface duct where the surface sound speed increases linearly in range. As a result, the surface duct essentially ceases to exist at larger ranges and becomes instead a downward refracting layer.

The resulting propagation loss is shown in Fig. 1.24, and we would identify it as a reference solution in the sense that it is expected to be highly accurate. As one would expect, the disappearance of the surface duct eliminates the energy in the upper ocean and instead it is dumped into the deep sound channel. In the near field, there is very little difference compared to the range-independent case in Fig. 1.22 that was solved by normal modes. Thus, we are comparing two completely different modeling approaches. That close agreement between the models is an indicator of their accuracy.

Earlier we mentioned that Tappert had considered a similar problem and produced "credible" results. It is interesting to note that this surface duct problem presents some very interesting physics and, in fact, revealed flaws in many PE models. The flaws resulted in discrepancies of some 15 dB in large zones in the surface duct. One can see that significant energy is leaking out of the surface duct and then cycling in the deep ocean channel. It turned out that energy reenters the surface duct and then interferes constructively or destructively with the surface duct energy. The standard PE and many variants were known to have phase errors, which caused these large discrepancies. Modern PEs have eliminated this problem by reducing the phase errors. The sensitivity to the phase is, in any case, not so much of computational interest as an indicator of an underlying sensitivity in the physics of the problem. It indicates that small changes in the sound speed profile can also have a dramatic effect on the pressure field.

It is interesting to consider the alternative modeling approaches for this sort of problem. As discussed above, range-dependent extensions of normal modes and wavenumber integration would work and provide an accurate solution. However, they tend to be slower. The main alternative is ray/beam tracing, which, like the PE, requires no special treatment to be applied to range-dependent problems (although, specializing ray/beam methods to range-independent problems can yield advantages in execution speed).

Using a ray/beam tracing method, we obtain the propagation loss plot shown in Fig. 1.25. We remind that the PE result presented above is the reference solution. The ray/beam result can be seen as both good and bad depending on one's threshold for accuracy. The ray/beam result does not precisely render the leakage out of the surface duct. This can be remedied in various ways and those options are a matter of current research. However, the ray/beam solution does replicate the broad features, showing how the surface duct ceases to function as a duct. The steeper

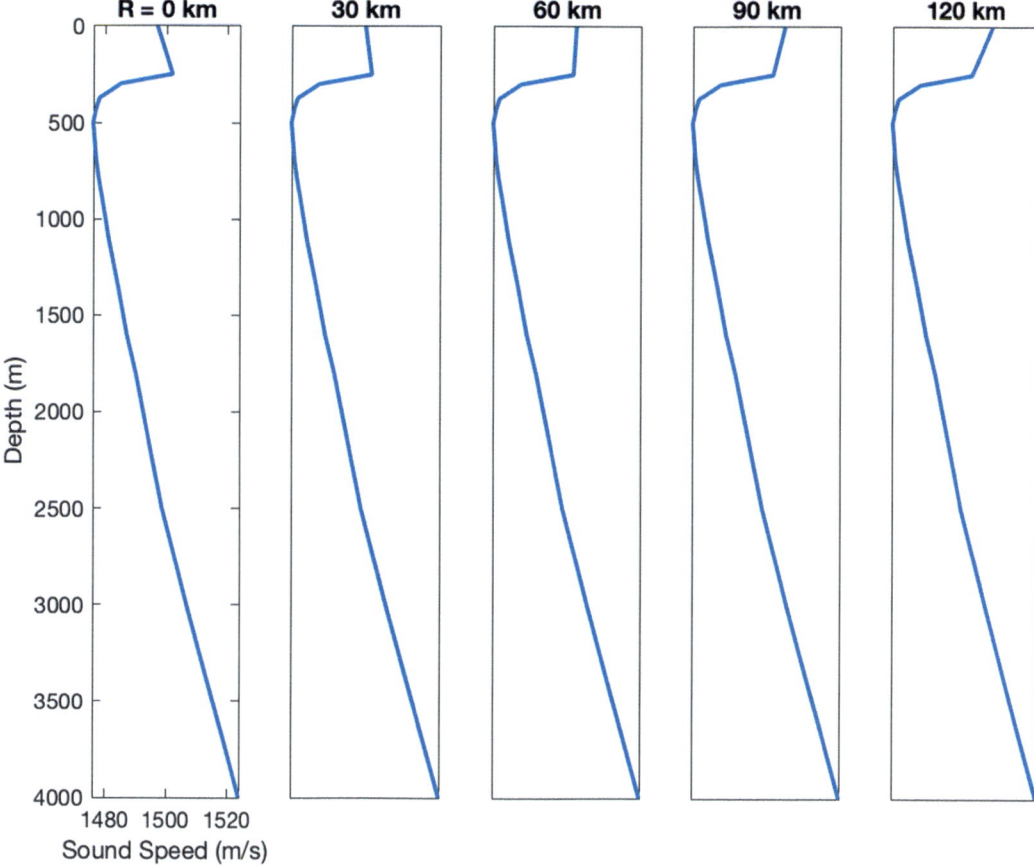

Fig. 1.23 Sound speed profile at successive ranges for the surface duct profile

angle paths are accurately rendered. It should also be noted that the ray/beam solution runs vastly quicker. In short, we trade off accuracy for execution speed in switching from a PE to a ray/beam solution.

1.10 Model Selection

The 2D models presented above are the core ones used to predict TL on range-depth slices. It can be difficult to learn how to use the four model types, so often a user will become familiar with one and try to use it for all applications. However, they all have a certain currency because of their different strengths and weaknesses. It is difficult to summarize those, partly because it is a function not just of the model type but of model implementations that are available. In any case, we present some considerations in the following.

1.10.1 Range Independent or Range Dependent

Normal mode and wavenumber integration models are often the best choice for range-independent problems, while parabolic equation and ray methods easily handle range-dependent problems. Of course, a range-dependent model includes range-independent cases as a special case—you can run range-dependent models on range-independent cases. However, the range-independent models often provide exact solutions and sometimes have a simpler input structure. Meanwhile, the environmental information may

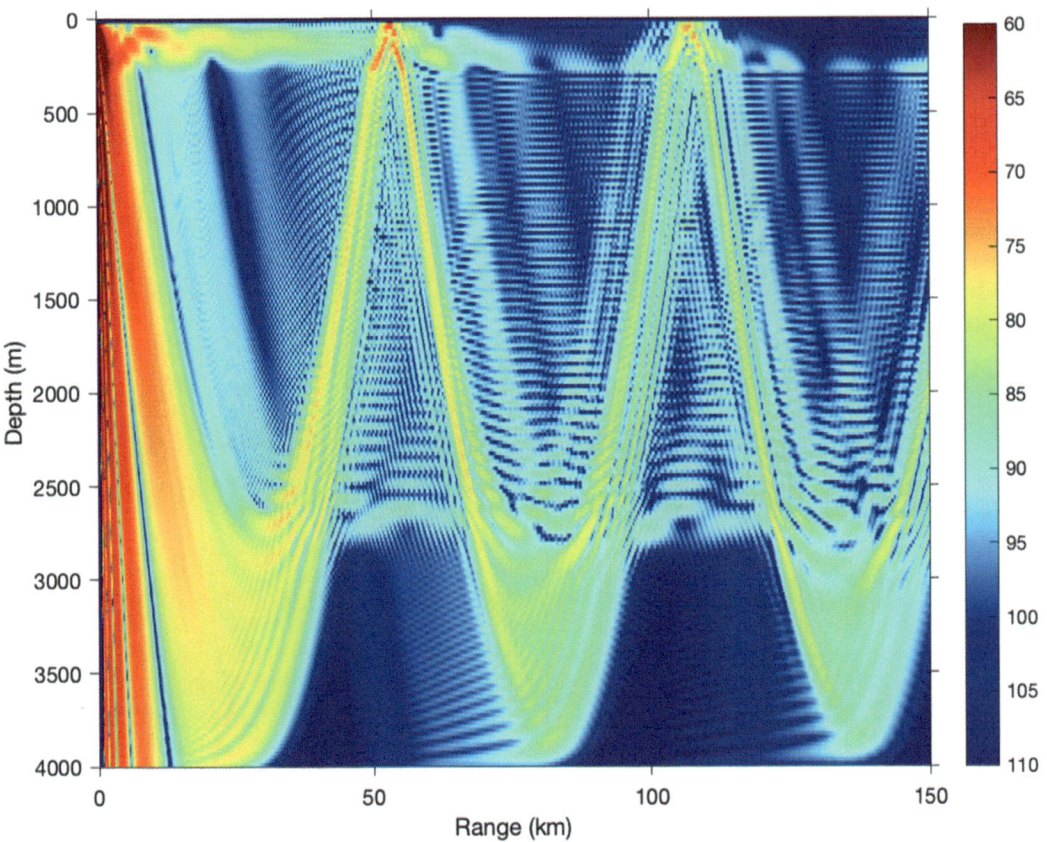

Fig. 1.24 Propagation loss for the range-dependent surface duct scenario computed by the parabolic equation for a source depth of 25 m and frequency of 100 Hz

not be available to provide the range-dependent inputs. There are range-dependent extensions of normal mode and wavenumber integration methods; however, they can be awkward and/or computationally intensive.

1.10.2 Low Frequency or High Frequency (Shallow Water or Deep Water)

All the full-wave models (normal mode, wavenumber integration, parabolic equation) have an execution time that increases strongly with frequency. In contrast, the ray/beam methods are not very sensitive to frequency and their accuracy often increases with higher frequencies. With modern computers, it is possible to run all the models at 10 kHz in deep water; however, when there are more than perhaps 1000 wavelengths in depth, one may favor a ray/beam method. That involves a trade-off with accuracy as the full-wave methods are generally more accurate.

In purely mathematical treatments, a dimensionless parameter kD is often used to characterize problems in terms of a wavenumber and the water depth. In this view, increasing the water depth is seen as equivalent to increasing the frequency and this does roughly characterize the complexity of the acoustic field. However, due to the oceanography shallow water and deep-water problems typically have quite different characteristics in terms of the propagation of sound. For instance, the SOFAR duct is characteristic of deep-water scenarios. Shallow-water problems are often downward refracting to that the propagation may be dominated by bottom-

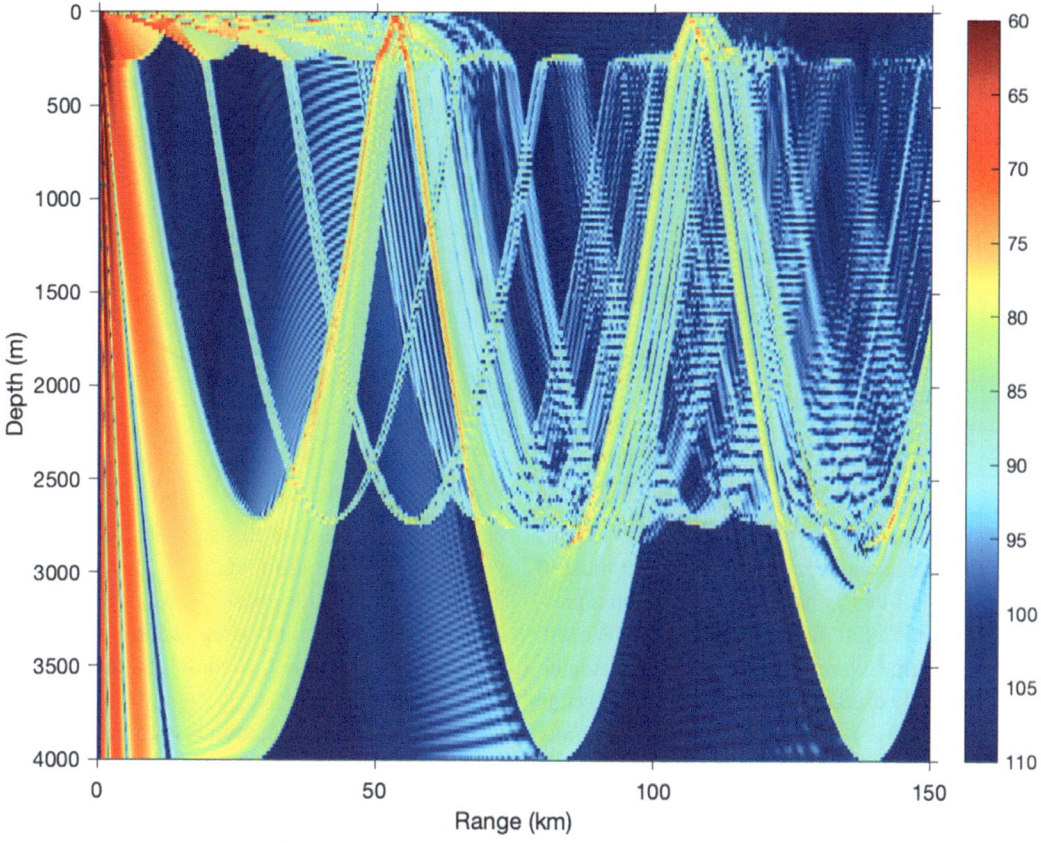

Fig. 1.25 Propagation loss for the range-dependent surface duct computed by ray/beam tracing for a source depth of 25 m and frequency of 100 Hz

reflected paths. As a result, shallow-water problems are often considered to be particularly difficult to model as they can be very sensitive to the bottom type.

1.10.3 Broadband or Narrowband

Typically, broadband calculations require that the acoustic models are run for a sequence of source frequencies. The execution time is then roughly proportional to the number of frequencies. In contrast, ray/beam methods can produce broadband results using a single, central frequency. Therefore, they may offer a significant speed advantage in such cases.

1.11 3D Effects

A full 3D spatial representation of the pressure can easily be constructed by running a 2D acoustic model along a series of bearings from the source. It is understood that along each bearing one uses the slice of the sound speed and bathymetry that is appropriate for that bearing. This is the uncoupled azimuth approach often called colloquially the Nx2D approach or sometimes the 2.5D approach. It is, of course, an approximation. When we do the calculation on a single slice, we have explicitly assumed cylindrical symmetry and then we violate that same assumption on the next bearing by replacing it with a different sound

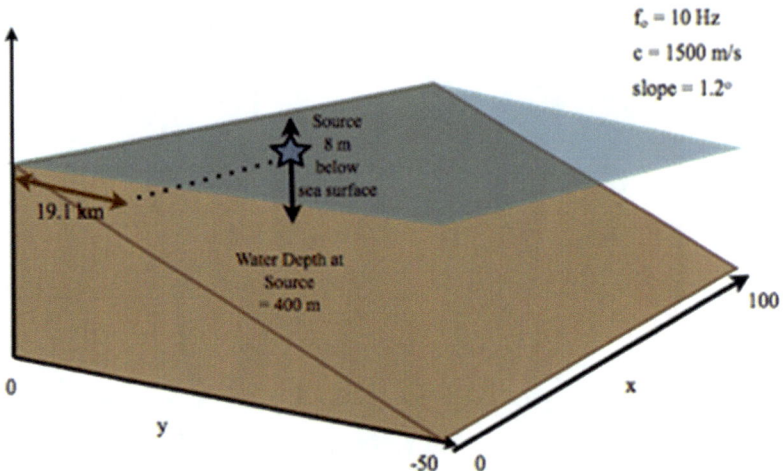

Fig. 1.26 Schematic of the wedge problem

speed profile and bathymetry. Nevertheless, this Nx2D approach is surprisingly good and most of the time is all that is needed. The obvious question is when is a full 3D solution needed? We will first show some examples and return to that later.

One widely used test case is the perfect wedge illustrated in Fig. 1.26. The wedge runs to infinity in the x-direction, that is, parallel to the apex of the wedge. The term "perfect" here refers to the use of perfectly reflecting top and bottom boundaries. (Real boundaries have a reflection loss that is a complicated function but generally increases with the angle of incidence.) That simplification permits an analytic solution for purposes of benchmarking. This scenario is a highly idealized version of a continental slope. The 10-Hz source has been placed 19.1 km offshore at a depth of 8 m.

A ray trace for this scenario is shown in Fig. 1.27. This plot was produced using a fan covering about +/− 80° in declination angle and +/−90° in bearing. In this perspective view, one can see how the glancing rays are deflected out of the incident plane on each reflection.

This same ray trace viewed from directly overhead is shown in, Fig. 1.28. From this vantage, the rays appear to refract smoothly in the x-y plane (latitude/longitude); however, we saw previously that horizontal refraction is the result of many bottom reflections. To keep the plot simple, the rays were traced every 10° in bearing. For a field calculation, we do calculations very finely in bearing, in which case this ray picture would be very dense. One should note that the rays launched toward the coast (apex of the wedge) are refracted back into deep water, while rays launched away from the coast continue in that direction.

The resulting propagation loss calculated from the exact analytical representation is shown in Fig. 1.29. This plot represents an interference between rays that have been launched toward the coast (but refracted back) with those that have been launched away from the coast. This pattern also has a simple modal interpretation.

The effects of the horizontal refraction are dramatic here, causing energy traveling upslope to the coast to turn completely away into the downslope direction. However, one should keep in mind that these effects are strongest at the longer ranges. But then, depending on the source level and the bottom reflectivity, these longer ranges may no longer have significant energy.

In general, features of the ocean environment that vary with latitude and longitude cause horizontal refraction. Bathymetry variations, as we have just seen, are one key mechanism. Another mechanism is variations in the oceanography. A particularly interesting example involves nonlinear internal waves (NLIW). Internal waves of the ocean may be compared to the ordinary waves we see at the beach. Those waves occur at the air/water interface and their physics is likely well known to you. They are excited by distant

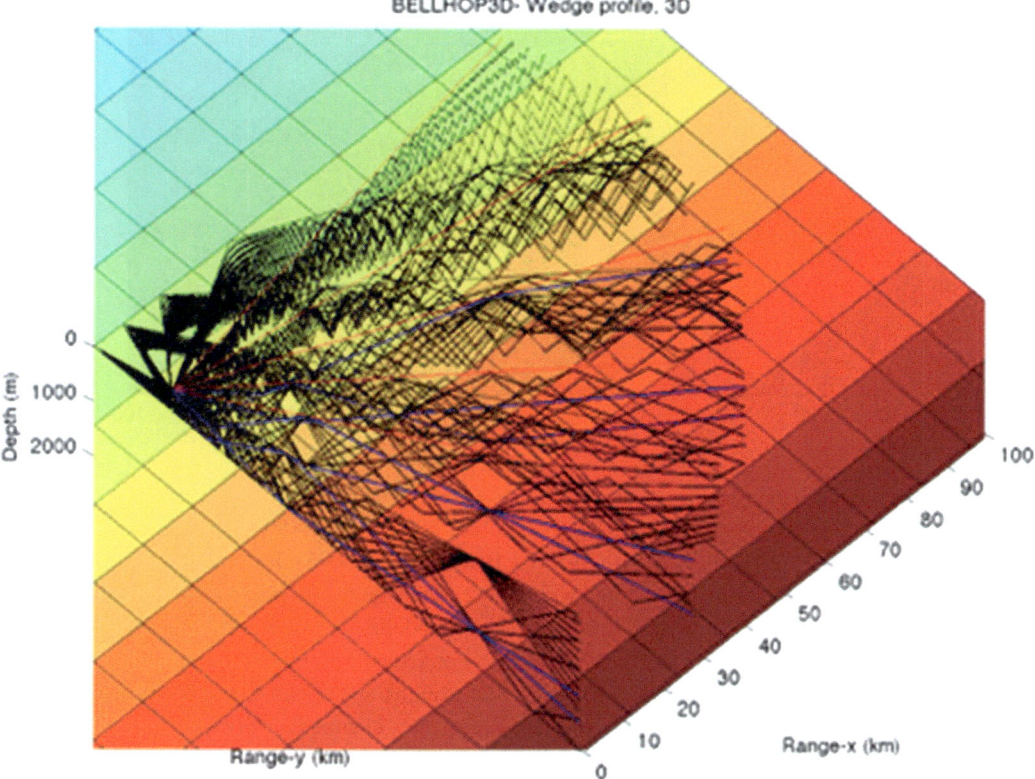

Fig. 1.27 Three-dimensional ray trace for the wedge problem

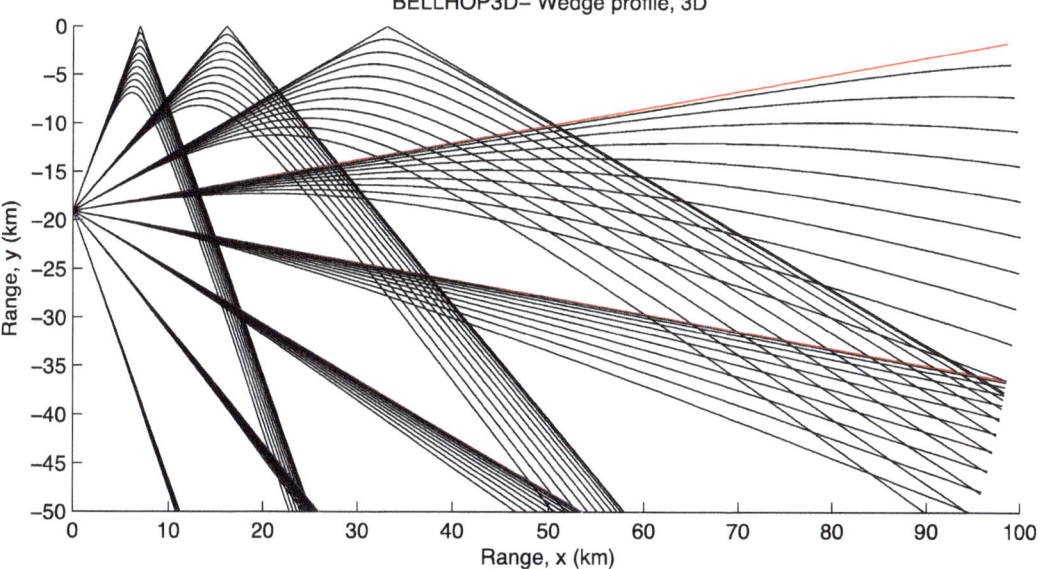

Fig. 1.28 Three-dimensional ray trace for the wedge problem seen from overhead

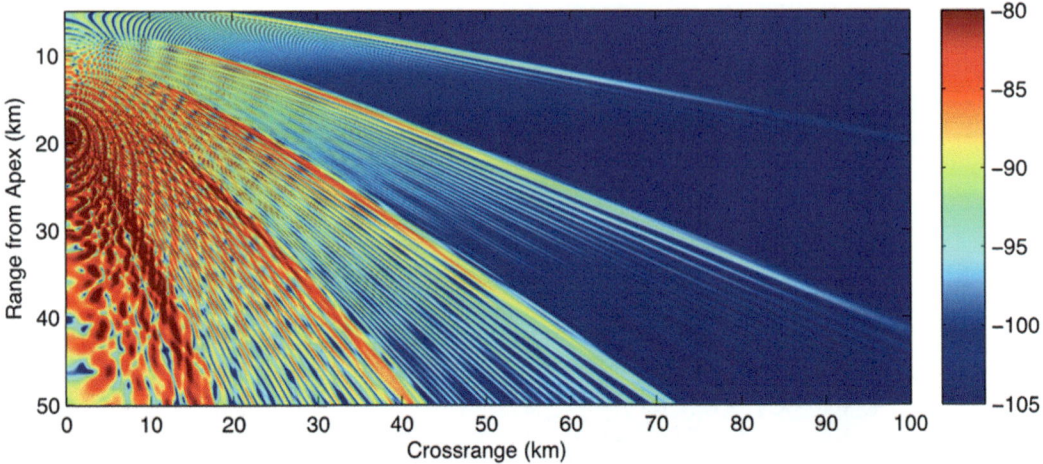

Fig. 1.29 Propagation loss for the wedge problem at a receiver depth of 80 m

storms and refract and reflect at the coast. If the waves are small, then they can be understood well in terms of the linear physics of a wave. However, at large amplitude, nonlinear effects distort their shapes and they may eventually break.

Internal waves are very similar. In a simple model, one considers a layer of fluid over another layer with a different sound speed and density. This two-layer model encompasses the air/water case we just discussed. However, in the ocean, we need to consider multiple layers, and, passing to a limit of an infinite number of layers, we obtain a continuously varying situation. Nevertheless, the same features we are familiar with for waves at the beach are found in these internal waves. At low amplitudes, they are linear sinusoids. At large amplitudes, nonlinear effects appear, and they may break. They are also refracted. The main difference is the excitation mechanism. Surface waves are driven by winds; internal waves are driven by ocean currents. In particular, currents passing over bathymetric features such as seamounts and ridges can excite the internal waves.

While the internal waves can be modeled purely within the volume of the ocean, they cause small disturbances on the ocean surface that reveal their presence. Those disturbances can be measured using radar from satellites or surface ship. An example is shown in Fig. 1.30 for an area off New Jersey, USA, as discussed by Badiey et al. (2005). It is evident that their structure can be quite complicated. We briefly describe their modeling and interpretation in the following.

For acoustic paths traveling roughly parallel to the troughs and crests of the internal waves, ray paths are channeled. For instance, Fig. 1.31 shows modeling results for the ray plots in the left panels and the resulting pressure levels in the right panels. The internal waves can cause either defocusing (upper panels) or focusing (lower panels) depending on how they change the curvature of the sound speed in the lateral direction.

An experiment was done at sea and the resulting transmission total time-integrated intensity is shown in Fig. 1.32 at a range of about 15 km from the source located at a depth of 12 m. The horizontal axis is time, and the vertical axis is depth, so these plots are showing the change in propagation loss over a 1-hour period as observed on a vertical array. The top plot is the measurement and shows quite remarkable variations in the 1-hour period. While this measurement was done for a narrowband source, it is not a simple interference pattern but a broadband feature. The middle plot shows the modeled TL using a 3D model. It confirms that the qualitative features can be reproduced; however, without a detailed reconstruction of the full 3D sound speed, it is not possible to precisely reproduce the focusing and de-focusing due to the internal waves. The bottom plot shows the modeled TL

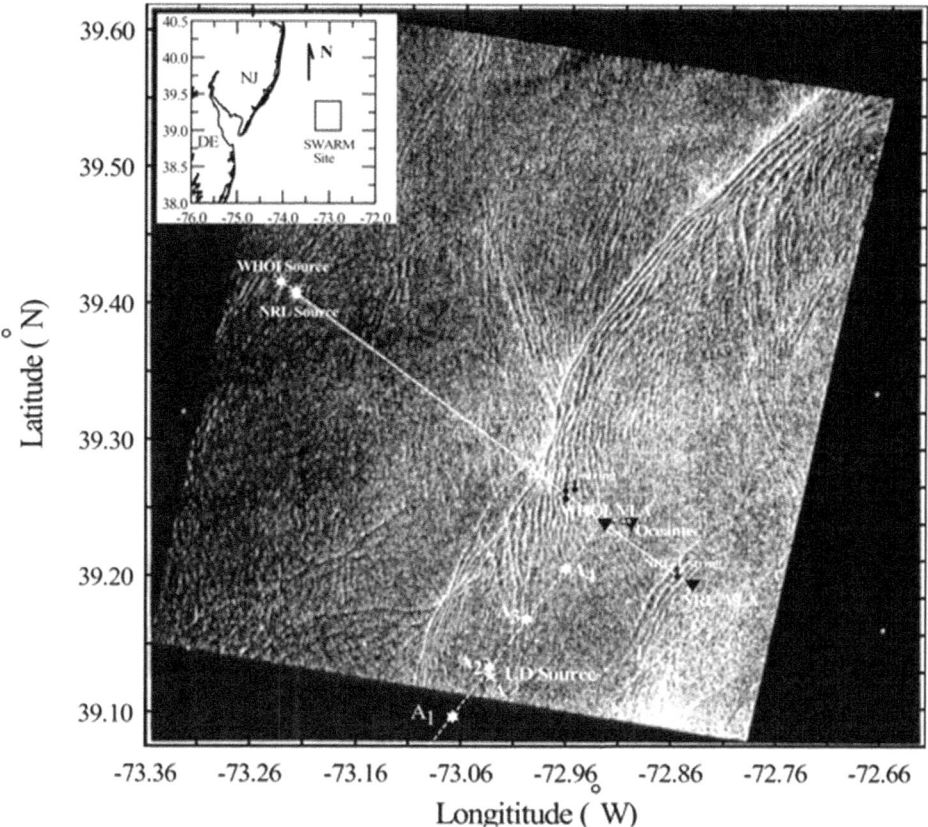

Fig. 1.30 Soliton packets off New Jersey. Reprinted from Badiey M, Katsnelson GG, Lynch JF, Pereselkov S, Siegmann WL (2005) Measurement and modeling of three-dimensional sound intensity variations due to shallow-water internal waves. J Acoust Soc Am 117(2): 613–625. https://doi.org/10.1121/1.1828571. With permission from the ASA. © Acoustical Society of America, 2005. All rights reserved

using a 2D model. It makes clear that the treatment of horizontal refraction effects is necessary.

A subsequent paper (Badiey et al. 2011) developed this work further. Interestingly they show experimentally and theoretically that the horizontal effects due to nonlinear internal waves can generate and interference pattern that is analogous to the Lloyd mirror pattern but turned on its side, that is, in the lat./long. plane.

The above examples have been selected to show the effects of horizontal refraction separately for bathymetry and oceanography and with different features of interest (continental slope and nonlinear internal waves). We end with a final example that shows how a 3D model may be used in a general complicated environment. The site considered is east of Taiwan. A map of the sound speed at a depth of 50 m is shown in Fig. 1.33. This information comes from HYCOM for a specific day as described previously.

Three-dimensional ray traces are shown in Fig. 1.34 for six different source locations in the area (the source depth is 100 m) and plotted in relation to the bathymetry derived from the ETOPPO1 database. For each source, a fan of rays has been traced covering four different bearing angles. The key feature to note in this plot is that some of the source locations lead to rays that stay in planes defined by the bearing at the source, that is, there is no horizontal refraction. For other locations, the horizontal refraction is quite strong.

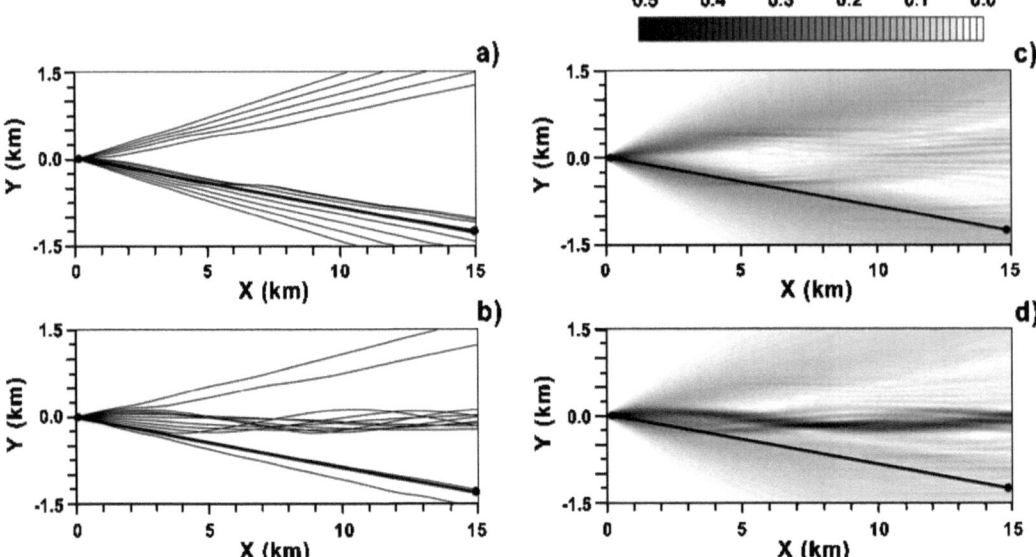

Fig. 1.31 Horizontal refraction due to soliton packets. The ray paths are shown in the left panels. The intensity calculated by a parabolic equation model is shown in the right panels. Results for a defocusing (upper panels) and a focusing (lower panels) are shown. Reprinted from Badiey M, Katsnelson GG, Lynch JF, Pereselkov S, Siegmann WL (2005) Measurement and modeling of three-dimensional sound intensity variations due to shallow-water internal waves. J Acoust Soc Am 117(2): 613–625. https://doi.org/10.1121/1.1828571. With permission from the ASA. © Acoustical Society of America, 2005. All rights reserved

Thus, the answer to the question of whether 3D effects matter is "it depends." Figure 1.35 shows the resulting propagation loss for each of these six locations and for a receiver depth of 500 m. To sample this properly, one needs to use a much finer fan than was shown in the ray trace. In these plots, we can see that the propagation loss is very dependent on the source location.

It is important to keep in mind that the illustrations shown above for 3D scenarios show how horizontal refraction plays a role, but do not provide a definitive answer about its importance. Typically, 3D effects accumulate in range. The wedge problem clearly showed the refraction but also plotted the field out to ranges that might be well beyond those where the signal is still strong. That same consideration is relevant to the scenario east of Taiwan. Another important consideration is that the clarity of 3D effects depends on the signal processing. If the acoustic field is observed through an array that can distinguish the direction of arrival, then the change in angle may be obvious. On the other hand, a point sensor aggregates the energy from all directions and will often be less sensitive to horizontal refraction.

1.12 Time-Domain Calculations

As discussed above, ocean acoustic models have generally been developed assuming a tonal source, that is, a single frequency. Some sources, such as motors and propellers on ships, naturally produce strong tonals. However, other sources, such as pile drivers, airguns, and explosives, produce broadband waveforms (i.e., they are characterized by a wide spectrum of frequencies). Such waveforms may be impulsive. However, they may also be continuous, such as the noise due to propeller cavitation.

There are two main approaches to modeling broadband sources. The first is to use Fourier

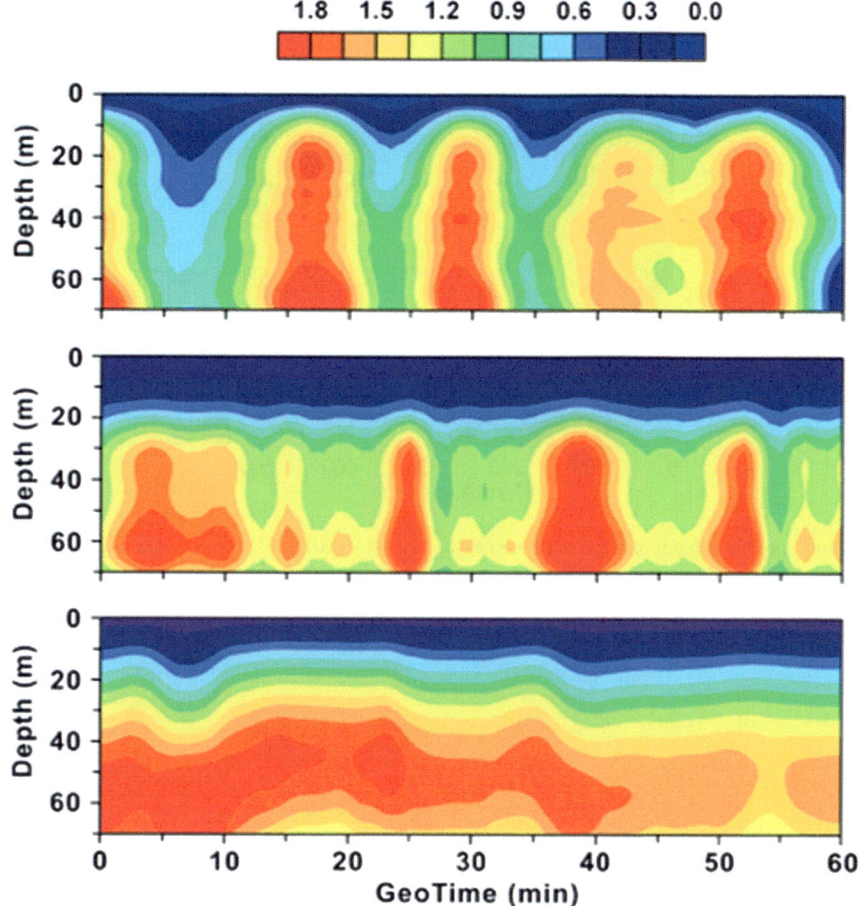

Fig. 1.32 Variations in total time-integrated intensity at a range of 15 km due to solitons. Experimental results (upper panel), acoustic model results including horizontal refraction (middle panel), and acoustic model results neglecting horizontal refraction (lower panel). Reprinted from Badiey M, Katsnelson GG, Lynch JF, Pereselkov S, Siegmann WL (2005) Measurement and modeling of three-dimensional sound intensity variations due to shallow-water internal waves. J Acoust Soc Am 117(2): 613–625. https://doi.org/10.1121/1.1828571. With permission from the ASA. © Acoustical Society of America, 2005. All rights reserved

synthesis. Thus, we do a Fourier transform to find the spectrum of the source waveform:

$$\widehat{s}(\omega) = \int s(t) e^{i\omega t}$$

where the hat in $\widehat{s}$ is used to distinguish the Fourier transformed function from its waveform in time. This integral is typically done using a discrete Fourier transform yielding a vector where each element represents the amplitude at a discrete set of frequencies $f_i = \frac{\omega_i}{2\pi}$. Then one simply runs an acoustic model for each individual frequency yielding the acoustic pressure field $\widehat{p}(r, z; f_i)$. Finally, one sums up the pressure fields over the frequency band, weighted by the spectral amplitude and phase of the source to get the received waveform:

$$r_{ts}(r, z, t) = \int \widehat{p}(r, z, \omega) e^{-i\omega t} d\omega.$$

The simple process just described is, in practice, not routinely done. The problem is that a

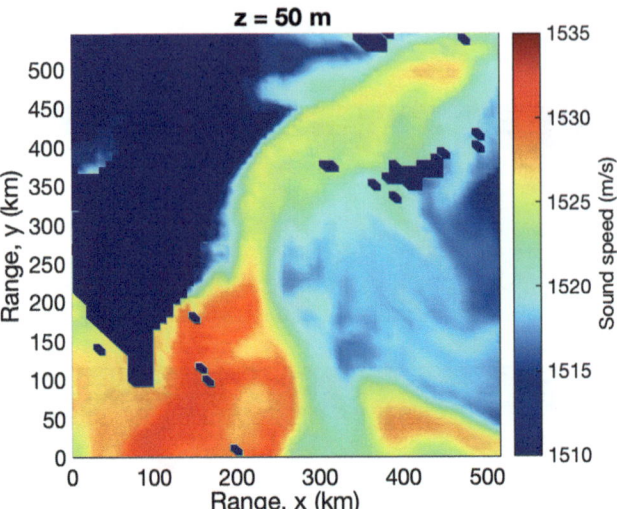

Fig. 1.33 Sound speed at a depth of 50 m in an area east of Taiwan (from the HYCOM database/model). Reprinted from Porter MB (2019) Beam tracing for two- and three-dimensional problems in ocean acoustics. J Acoust Soc Am 146:2016–2029. https://doi.org/10.1121/1.5125262. With permission from the ASA. © Acoustical Society of America, 2019. All rights reserved

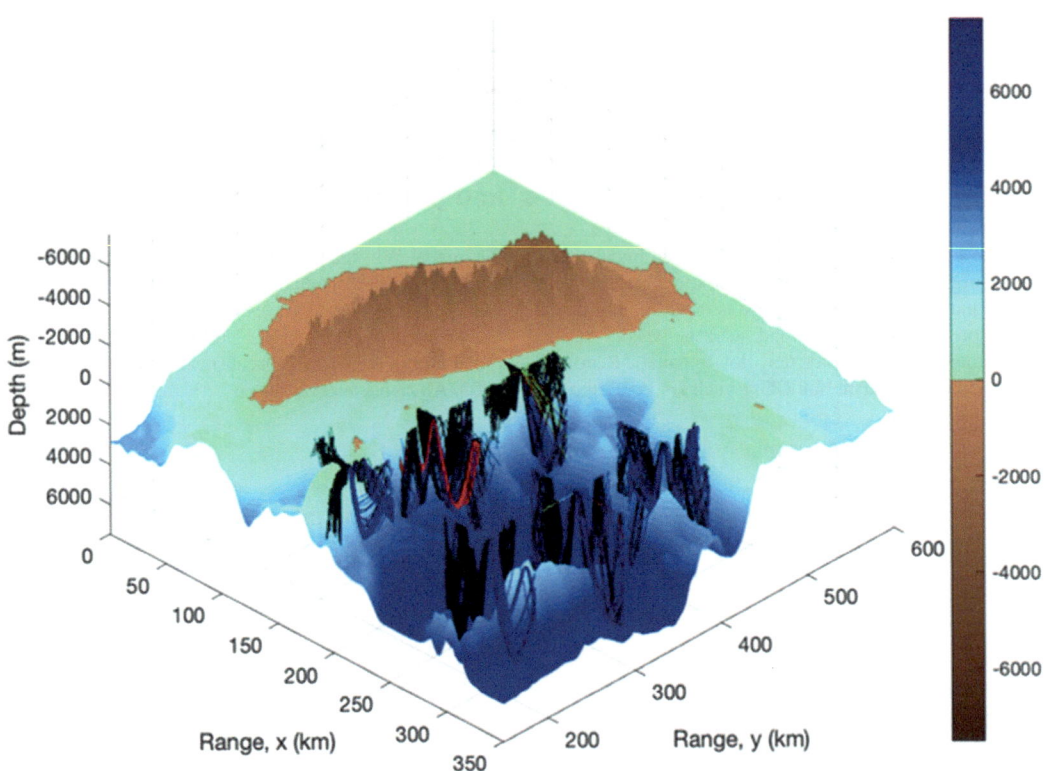

Fig. 1.34 Ray trace for six source locations east of Taiwan with a source depth of 100 m. The bathymetry is from the ETOPO1 Database). Reprinted from Porter MB (2019) Beam tracing for two- and three-dimensional problems in ocean acoustics. J Acoust Soc Am 146: 2016–2029. https://doi.org/10.1121/1.5125262. With permission from the ASA. © Acoustical Society of America, 2019. All rights reserved

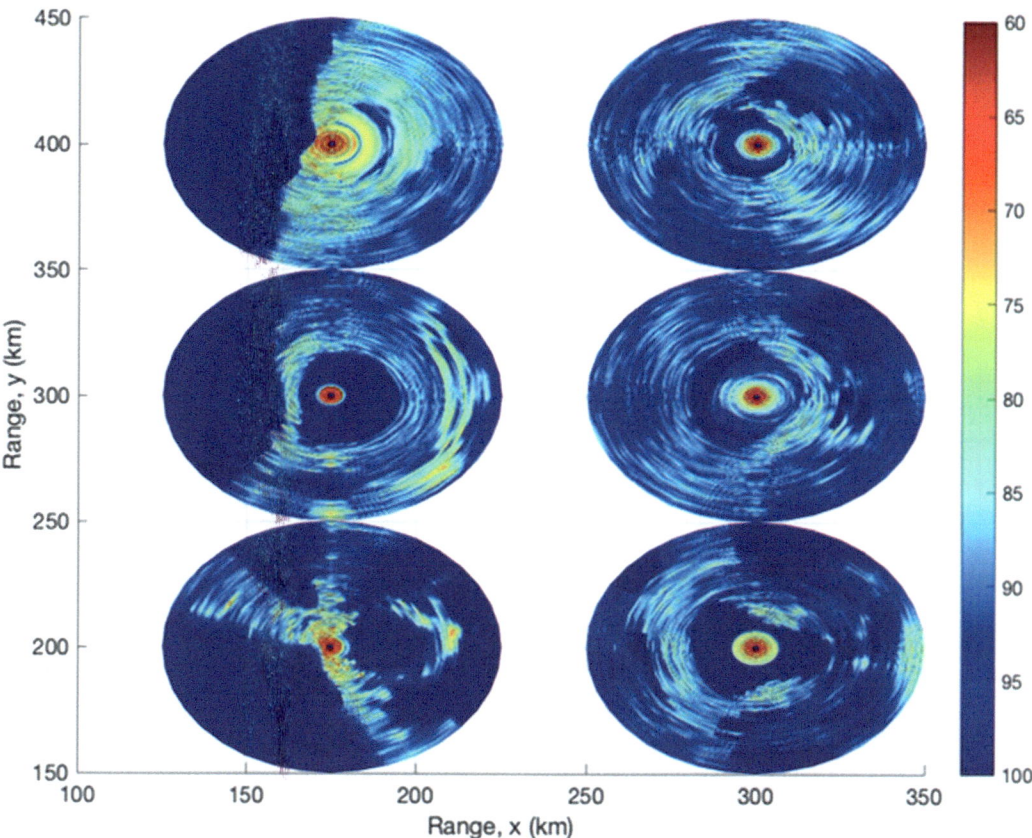

Fig. 1.35 Propagation loss on discs centered on six source locations. The source frequency if 250 Hz, the source depth is 100 m, and the receiver depth is 500 m. Reprinted from Porter MB (2019) Beam tracing for two- and three-dimensional problems in ocean acoustics, J Acoust Soc Am 146: 2016–2029. https://doi.org/10.1121/1.5125262. With permission from the ASA. © Acoustical Society of America, 2019. All rights reserved

broadband waveform may involve a large number, N, of time-samples. The number of frequencies required is then also N, which means the acoustic models must be run for a very large number of frequencies. That leads to very long computation times. In short, this means the Fourier synthesis approach can only be done for waveforms that can be characterized with a modest number of samples, say 1000 on today's computers. It should be noted here that the sampling requirements are based on both the source and the received waveforms. The multipath in the ocean typically produces multiple echoes of the source waveform and the received time series must be sampled to capture all the echoes as well as the detail in the echoes. (This is a description of the simplest process. For source waveforms that are longer than the multipath spread of the waveguide, the pressure fields can be interpolated in frequency so that the acoustic model can be run for fewer frequencies.)

Despite these computational issues, it is often useful and practical to be able to produce results in the time domain. As an example, we return to the Munk profile considered above and consider a simple Gaussian pulse with waveform and associated spectrum shown in Fig. 1.36. (We show only the positive part of the spectrum; it is symmetric.) This impulse lasts for 0.1 s. The spectrum of a Gaussian pulse is also Gaussian with a nominal upper frequency defined by the reciprocal of the pulse duration. In terms of the

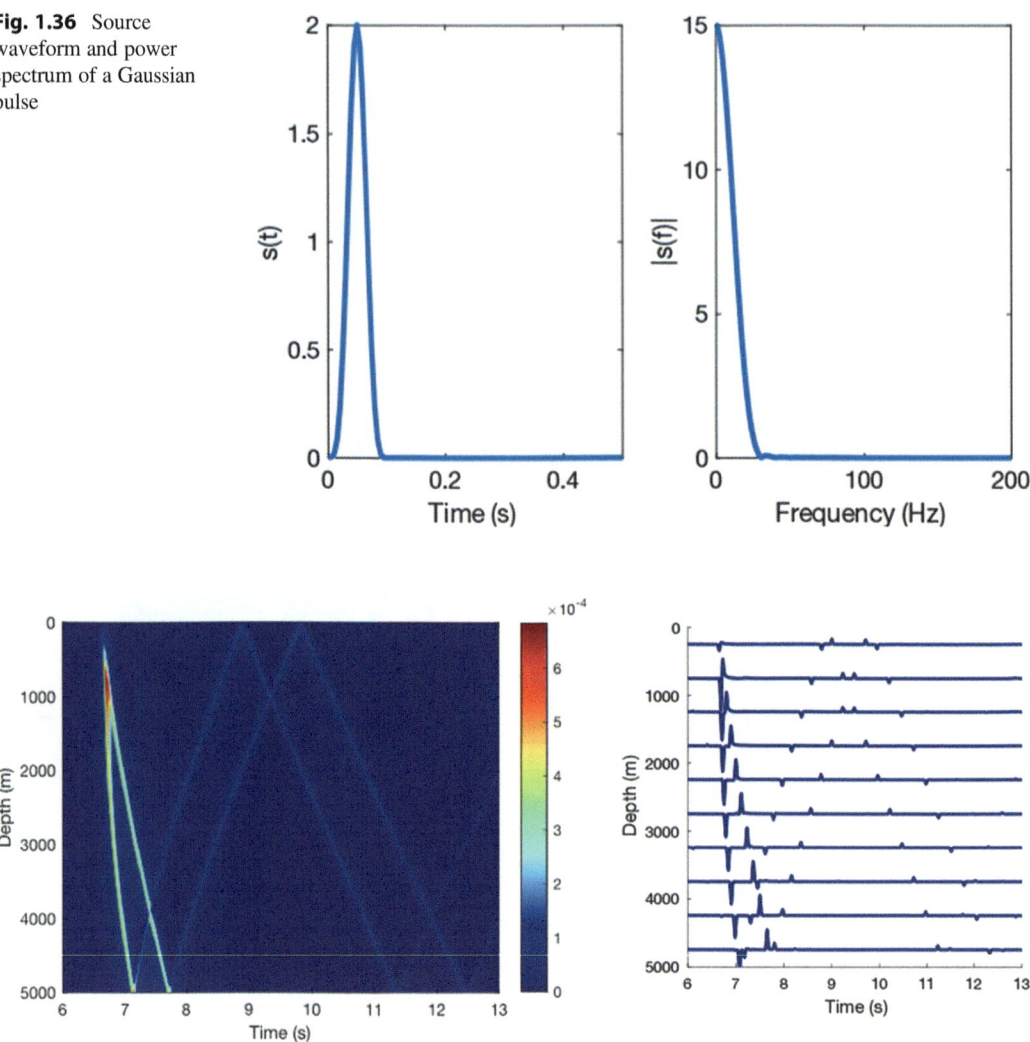

Fig. 1.36 Source waveform and power spectrum of a Gaussian pulse

Fig. 1.37 Pressure as a function of depth and time at a range of 10 km. Left panel is the envelope displayed in color; right panel is the waveform displayed as line plots

source then, we can say that we have a waveform with frequencies from 0 to 30 Hz roughly. For the model run, we need to think about the received waveform. Based on experience (or results from a ray model), we conclude that we will need to see a received waveform lasting for about 10 s. Thus, we will use 0.1-Hz frequency sampling over the 30-Hz band of the source. However, to get a well-sampled picture of the received waveform, we elect to produce 100 samples per second, implying an upper frequency of 100 Hz. In summary, the acoustic model is run 1000 times in the 0–100 Hz band. This could be reduced since the source spectrum vanishes at a lower frequency.

The resulting pressure field is a function of range, depth, and time; so, to render it on a 2D page requires slicing it in some fashion. In Fig. 1.37, we show the envelope of the received waveform over depth and at a range of 10 km. We also show the waveform as a line plot in Fig. 1.37. Note the changes in the polarity (and more generally the phase) of the waveform due to boundary reflections. Caustics also cause phase changes.

For more complicated waveforms, the Fourier synthesis approach becomes impractical. The method of choice often then becomes a delay and sum approach based on the echoes as computed by a ray/beam tracing model. If the strengths, A_i, of the echoes are real, then that formula for the received waveform is

$$r(t) = \sum_{i=1}^{NArrivals} A_i \, s(t - \tau_i)$$

where τ_I is the travel time delay to a particular receiver location. This is the often-quoted result. Unfortunately, it is incomplete since the source waveforms are generally distorted when they reflect from the bottom. The distortion is caused by the phase change in the reflection coefficient, and those phase changes accumulate with each bottom reflection. A further complication is that the source waveform also undergoes a 90° phase change as it passes through a caustic. These phase changes distort the echoes and require a slightly modified formula for the received waveform.

On the other hand, we can see flaws in the ray/beam result. The amplitudes are not precise and at some receiver depths, the shape of the waveform is in error. These are the limits of ray/beam models. However, the run time of the ray/beam approach is less than a second versus about an hour for the result with wavenumber integration. That speed difference becomes increasingly compelling as the upper frequency band increases or the number of frequencies in the band increases. For instance, in modeling acoustic modem waveforms, one may have waveforms lasting several seconds sampled at 48 kHz. These types of calculations are very practical with ray/beam models; there are no other alternatives.

In summary, the two main ways of producing time-domain results are (1) Fourier synthesis and (2) delay-and-sum processing of ray/beam arrivals. However, these are not the only approaches available. There are special time-domain versions of wavenumber integration models and parabolic equation models. However, they have few advantages and are not widely used. Their main advantage is that they produce the solution directly in the time domain avoiding doing a discrete Fourier transform, which is often confusing. Similarly, there is also a time-domain formulation of normal modes, which also has not found widespread use.

1.13 Aero-Acoustics and Air-to-Water Propagation

Aircraft are common sources of noise in the atmosphere. There is also currently interest in detonating unexploded ordinance on barges. In terms of the impact on marine mammals, it is fortunate that the air–water interface is highly reflective because of the contrast in both the sound speed and density. Nevertheless, the signals in water can be quite loud.

In terms of the propagation physics, the air is just another layer for multilayered propagation models. However, it has two distinct characteristics. First, the wind carries (convects) the sound. Ocean currents have a similar effect; however, they are usually below a few meters per second in comparison to sound speeds of around 1500 m/s. It turns out that these convective effects are like changes in the sound speed (except that they make the sound travel faster downstream and slower upstream). In any case, when they are small compared to the basic sound speed, they can often be neglected. In air, the wind speeds are typically higher, and the sound speed is around 340 m/s. Thus, the convective effects can be important.

The sensitivity to wind or ocean currents is also a function of the observation system. In ocean acoustic tomography, transmissions are often done bidirectionally between two source/receiver positions. The mean travel time between the two receivers is primarily a function of the ocean sound speed. However, when the difference in travel times is computed it depends strictly on the ocean currents. Thus, the ocean currents in this scenario are a critical feature.

A second consideration is that turbulence in the atmosphere leads to great spatial and temporal variability. Again, analogous features are present in the ocean but typically not so prominently. Having mentioned these considerations, we will neglect them in the following discussion and use a stationary medium to introduce the propagation physics.

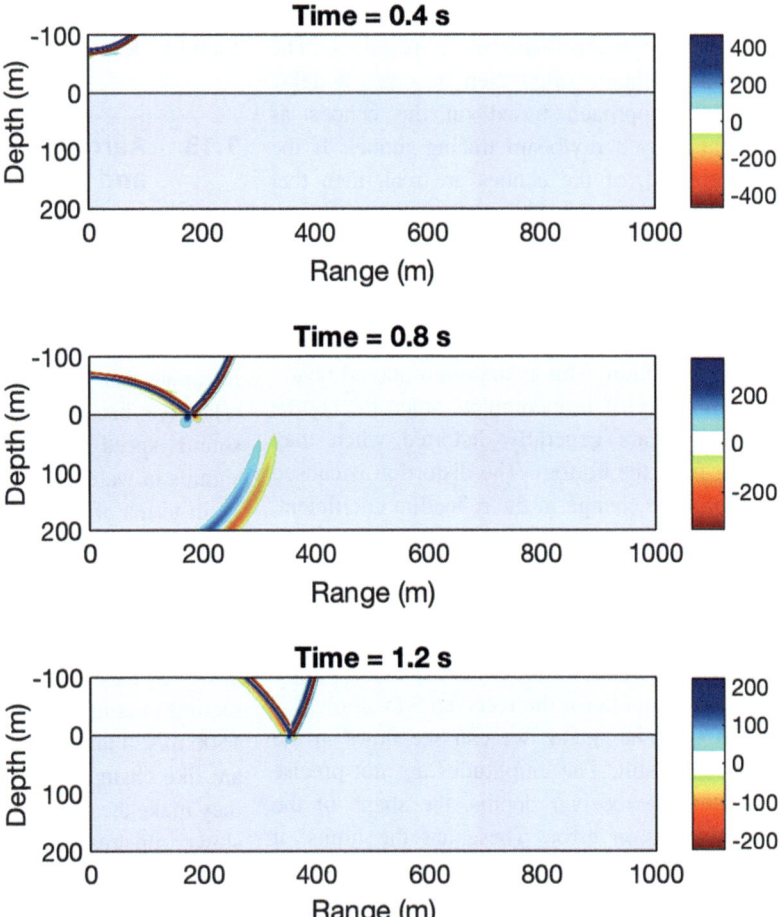

Fig. 1.38 Snapshot of the pressure field at three times: (**a**) 0.4 s, (**b**) 0.8 s, and (**c**) 1.2 s

For the source waveform, we consider an N-wave (i.e., one cycle of a sinewave); its central frequency was set at 20 Hz to allow us to easily see the footprint of the wave. The N-wave is often a good choice to understand the propagation effects. The environment consists of air (sound speed 344 m/s, density 1.225×10^{-3} g/cm^3 = 1.225 kg/m^3, and attenuation of 0.0009 dB/m = 0.9 dB/km. This attenuation was originally calculated for a higher carrier frequency but is not significant for the ranges we are considering.

Snapshots of the resulting pressure field are shown in Fig. 1.38. After 0.4 s, the direct blast is seen in the atmosphere, with the blue and red colors showing positive and negative pressure. White colors have been used where the absolute value of the pressure is near zero. After 0.8 s, the direct blast has hit the air–water interface, generating a transmitted wave in the ocean volume. In the atmosphere, you can also see a reflected wave, which meets the incident wave at the air–water interface. The waves in the atmosphere travel at about 344 m/s, while the waves in the ocean travel at about 1500 m/s and rapidly pull ahead of the atmospheric blast.

The physics suggested by these plots should look familiar as it is very similar to the results shown in Fig. 1.3 with a direct, reflected, head, and transmitted wave. However, on our scale the head wave is too weak to see. Also because of the large impedance contract at the air–water interface, the transmitted wave is much weaker. We also draw attention to a component known as the "direct wave root," which is an evanescent field leaking into the ocean and forming a root to the point where the reflected and incident waves occur. The root is just below the air–water interface starting at a depth of 0 m and on these plots can be seen for just a few meters into the water. (The term evanescent is used in wave propagation to refer to a quantity that decays exponentially in some direction.)

The transmitted wave "exists" in a 13° cone, where the cone width is calculated by Snell's law based on the ratio of the sound speed in air over water. Because of the difference in sound speed, the transmitted wave is both faster and has a longer wavelength. Of course, the energy does not immediately cut off outside the cone and we can see that in the plots. However, it is understood to be a weak arrival.

A useful approximate formula for the loss was derived by Eller and Cavanaugh (2000). The approximate formula uses spherical spreading to calculate the field at any point on the air–water interface. Then it replaces the field by a plane wave with an angle of incidence matching the spherical wave at that point. Finally, it uses the plane-wave transmission coefficient to find how the incident plane wave couples to the "transmitted" plane wave, which is evanescent in the water column. This is a generally a good approximation, but a more careful analysis would yield some correction terms.

1.14 Background on Noise (Soundscape) Modeling

We have so far presented fundamental approaches to modeling sound propagation in the ocean. As is typically done, an omni-directional point source was assumed, and models were developed for a single tone as well as a broadband time series. The development of these underwater propagation models was largely driven by passive sonar applications, where the target of interest (e.g., a submarine) is often adequately represented as a point source. However, many noise sources of interest in terms of their effects on marine mammals do not fit that mold. We show how the process can be modified to treat specific types of noise sources.

Some of these noise sources radiate sound in a directional way. For instance, seismic exploration generally uses arrays of airguns that radiate most of their energy vertically downward to ensonify the ocean bottom. Wind noise presents yet another source that is typically modeled as a sheet of point sources just below the ocean surface. Pile driving often involves a hammer strike that launches an elastic wave down the pile that in turn radiates a complicated wave into the ocean. Chapter 2 provides a much more complete discussion of different noise mechanisms. Here, we focus selectively on a few examples with particular attention on how they couple to the propagation models.

An additional complication or dimension of noise modeling is the issue of how to characterize the noise field as opposed to the noise source. There is no single right answer to that question, but several approaches have become common. For chronic noise sources such as shipping, one may wish to calculate an average noise level over a long period, for example, a year. However, clearly seasons, months, and days may be of interest as well. At the shortest time scales, one often considers a deterministic situation with specific ship trajectories; at the longest time scales, the sources are modeled as a distributed cloud producing a sheet with the noise distribution just as one does for winds. Ships produce a continuous hum for which the sound energy in one-third octave bands (or other frequency bands) is a typical metric. For impulsive sources such as the hammer strike in pile driving or the explosion of an airgun, one may also be interested in the peak energy level. The choice of measures of the sound field also affects the modeling process.

So different noise sources will be modeled in different ways. But which noise sources need to be modeled at all? This depends on the intersection of a noise footprint with the habitat of marine life and the range of frequencies that can be heard by species present in the area. However, if we focus just on the broadband noise levels, we are inevitably led to the Wenz curves shown in Fig. 1.39. These updated earlier work by Knudsen and elegantly present a huge amount of information about typical noise levels.

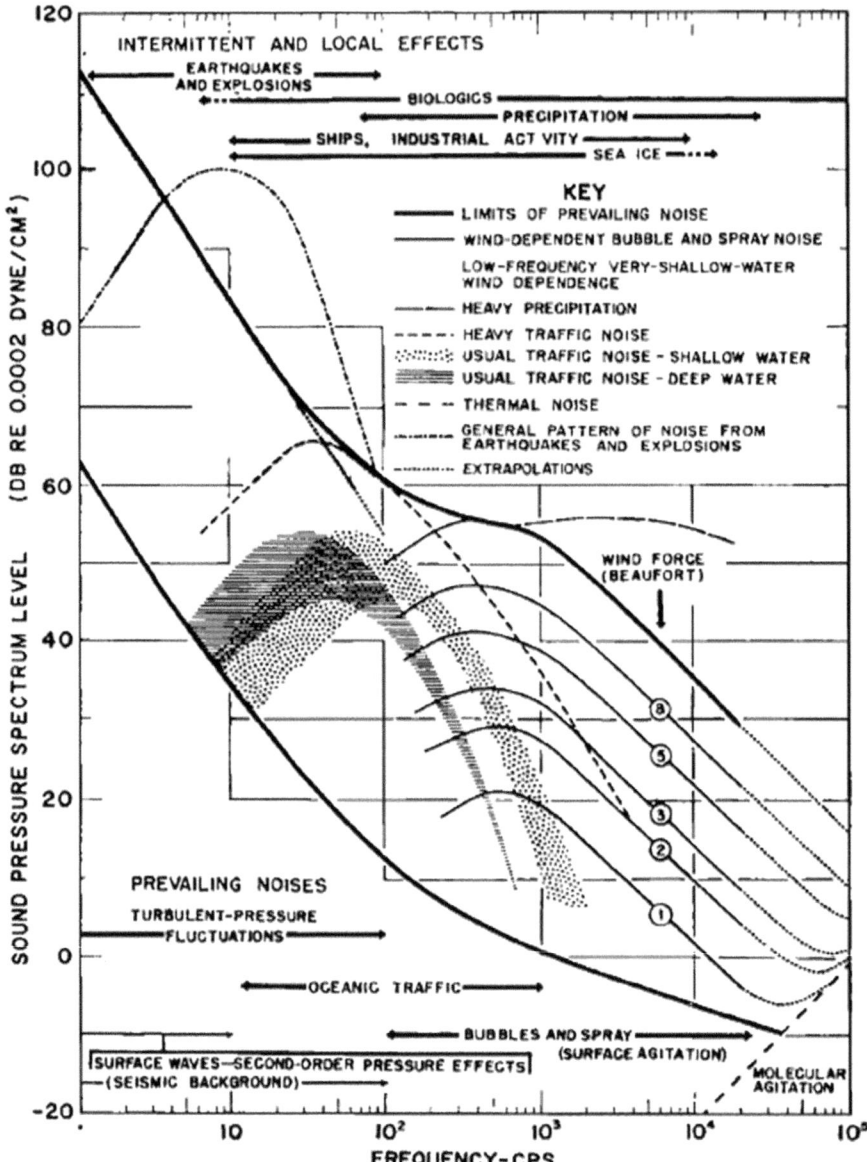

Fig. 1.39 Wenz curves showing received noise levels. Add 26 dB to yield dB re 1 μPa on the y-axis. Reprinted from Wenz GM (1962) Acoustic ambient noise in the ocean: spectra and sources. J Acoust Soc Am 34 (12): 1936–1956. https://doi.org/10.1121/1.1909155. © Acoustical Society of America, 1962; all rights reserved

For instance, we can see that ships and winds (which drive the sea state) are important sources of noise. Unsurprisingly their relative contribution depends on the nearby shipping and the local wind speed. These curves also illustrate that ship noise tends to be dominant in the band from 10 to 150 Hz, with wind noise taking over at higher frequencies. Note that these curves do not provide information about two other significant sources of sound: navy sonar and seismic exploration (e.g., for oil). The sounds from these sources vary widely with location and are not so easily characterized. In general, one should remember that these noise level curves are simply a statistical characterization of many measurements. A more complete approach to

Table 1.1 Comparison of anthropogenic underwater sound sources ordered by their total annual energy output

Sound source	Intensity (dB re 1 W/m²)	Directionality	Power (dB re 1 W)	Number of sources	Operations (days/year)	Repetitions (pings/day)	Total energy (J)
Underwater nuclear explosions	146	4π	157	1	0.05	1	2.6E+15
Airgun arrays	61	π	66	90	80	4320	3.9E+13
Military sonar (53C)	53	$\pi/2$	55	100	30	4320	8.5E+12
Super tankers	3.2	2π	11	11,000	300	86,400	3.7E+12
Ship shock trials	117	4π	128	1	0.5	1	3.3E+12
Military sonar (SURTASS/LFA)	53	π	58	1	30	175	1.7E+11
Merchant vessels	−17	2π	−8.8	40,000	300	86,400	1.4E+11
Navigation sonar	−1.8	π	3.2	100,000	100	86,400	3.6E+010
Research sonar	13	4π	24	10	4	86,400	9.1E+08
Fishing vessel 12 m long (7 knots)	−42	2π	−34	25,000	150	86,400	1.3E+08

noise modeling requires not simply these statistics of observations, but instead a careful characterization of the noise source. This in turn must be coupled to propagation models that consider the environmental conditions to predict the noise level. For instance, the observed wind noise is not simply a function of the wind speed or sea state, but also depends on the depth and reflectivity of the ocean bottom.

A very interesting question for the marine ecologist is how these various sources contribute to the global noise budget. Hildebrand (2004) attempted to systematically estimate the annual, global energy output of various sources. His result is shown in Table 1.1. The entry for underwater nuclear explosions serves just as an indicator of their potential influence; there has not been one in many years.

One should keep in mind that some sources, such as airgun arrays used in seismic surveys, radiate most of their energy into the ocean sub-bottom. Further, their impact depends on various biological parameters of the affected animals (see Chap. 8) and their location. In short, this table is about noise injection, not noise impact. Thus, one is driven to more careful modeling to understand regional effects.

As an example, we consider the Gulf of Mexico. Figure 1.40a shows a model of ship noise. Figure 1.40b considers the noise due to ships that are used to service rigs. Levels in these plots represent the L_{eq} in one-third octave bands, accumulated over a year's worth of activity. The quantity L_{eq} (equivalent continuous sound pressure level, often written as Leq) is one of many metrics used to characterize ambient noise. It is defined as a time average presented in dB:

$$L_{eq} = 10 \log_{10} \left(1/T \int_0^T |p(t)/p_0(t)|^2 dt \right)$$

The averaging time, T, is chosen based on the application of interest. Here, we consider an entire year. The waveform is also typically filtered to frequency bands of interest. Here, we select third-octave bands.

The acoustic models were run to calculate L_{eq} for a variety of central frequencies; these results are for a band around 200 Hz. Figure 1.40c shows the noise due to airguns used for seismic exploration. Finally, Figure 1.40d shows the sum of all these noise sources. It is evident that the overall noise in the Gulf of Mexico is dominated by seismic exploration. The modeling process behind these sound maps is described in later sections.

We should also emphasize that these are preliminary results that serve here more to illustrate the process than to be quantitative. They are

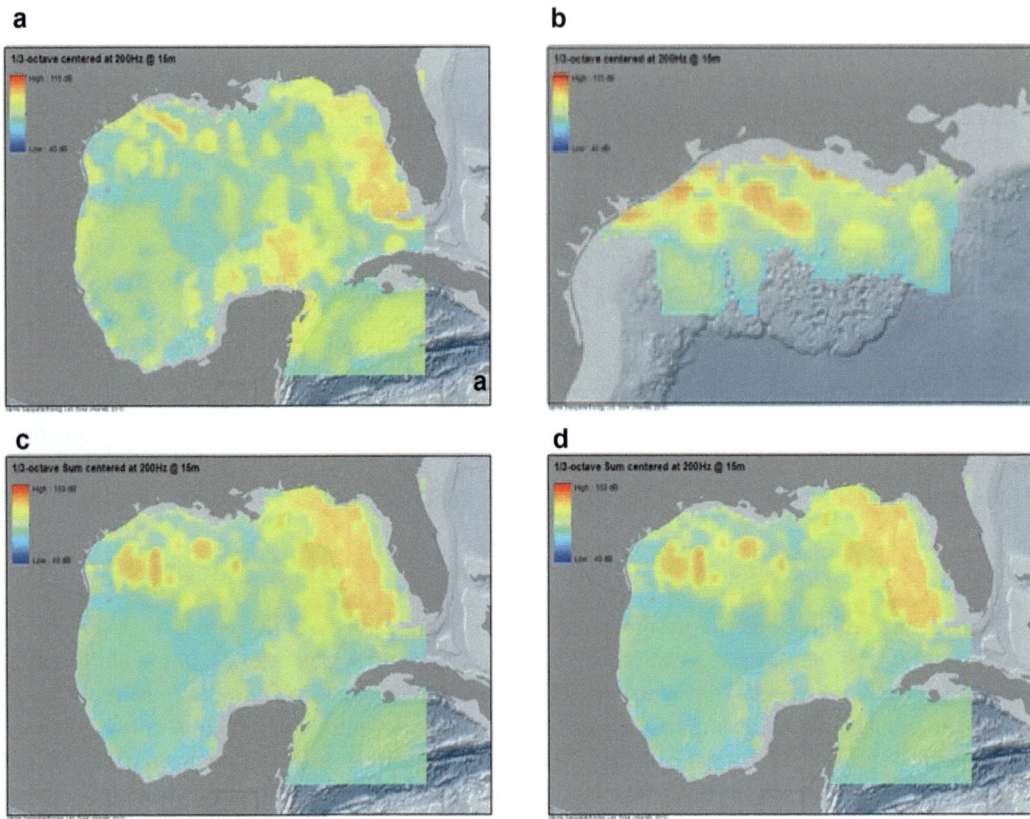

Fig. 1.40 Noise in a band around 200 Hz due to (**a**) shipping, (**b**) rig-servicing vessels, (**c**) air guns in seismic surveys, and (**d**) sum of all the noise sources

limited in various ways. For instance, the area near Florida is known to have a lot of limestone that supports shear waves. These waves require special modeling approaches and were neglected. The papers by Wiggins et al. (2016) and Thomsen et al. (2011) provide a great deal of additional information about the role of exploration and production in the Gulf of Mexico.

These results were part of the Undersea Soundfield Mapping study that considered a variety of noise sources around the U.S. Exclusive Economic Zone (EEZ). (See Gedamke et al. 2013) The noise level in L_{eq} was calculated for a one-third octave band centered at 100 Hz (as well as many other bands). The resulting sound map for the United States is shown in Fig. 1.41.

1.15 Noise in a Homogeneous Halfspace with a Sheet of Monopoles

A homogeneous halfspace (such as an infinitely deep ocean below a perfectly reflecting ocean surface) is almost always too idealized to be used to estimate the noise field in the ocean. However, it is an important starting point that provides key insights about noise modeling in more realistic situations.

Noise due to ships and wind are both often idealized as due to sheets of monopoles. For ships, the monopoles are placed at a nominal depth of the propeller (typically 6 or 8 m). For

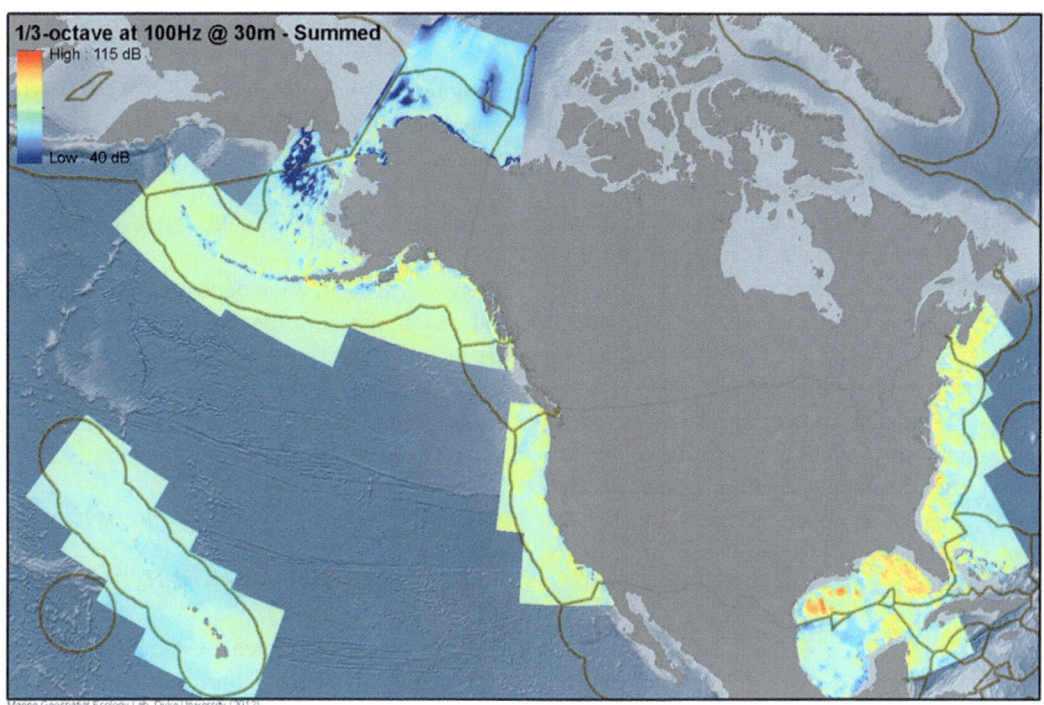

Fig. 1.41 Sound levels around the U.S. EEZ for a one-third octave band centered at 100 Hz. Different acoustic sources were considered in different areas

wind, the monopoles are normally placed at the depth $\frac{\lambda}{4}$, where λ is the acoustic wavelength. We have seen that the monopoles often produce dipole patterns. Some of the literature elects to use a dipole strength instead. If the source is close enough to the sea surface, then using dipoles is an equivalent process to monopoles once one gets the units sorted out.

We are interested in the noise due to an infinite distribution of uncorrelated monopoles at some depth z_s. Thus, we integrate the intensity, I, over that entire sheet. It can be shown (see Kewley et al. 1990 and references therein) that the integral can be written in terms of the angle as

$$I_{Total} = 2\pi \int_0^{\pi/2} 4 \sin^2(2\pi k z_s \sin\theta) \frac{\cos\theta}{\sin\theta} d\theta$$

Here, the integration is over the angle between the receiver and a location on the sheet of monopoles. The sweep over $[0, \pi/2]$ sweeps over all angles. The $1/R$ term has been essentially canceled by the $2\pi R\, dr$ area of an annulus (a ring) of noise. If one then makes the trigonometric substitution $t = \sin\theta$, this reduces to the Cosine Integral, Ci(x), and we get

$$I = 4\pi\, \text{Ci}(4\pi k z_s),$$

where,

$$\text{Ci}(x) = \int_x^\infty \frac{\cos t}{t} dt$$

If we then assume the monopoles are at a quarter-wavelength in depth, then $kz_s = \frac{1}{4}$ and $I = 4\pi\text{Ci}(\pi) \approx 6.6\pi$. (We have also assumed the wavenumber is real, meaning there is no volume attenuation.) Thus, in dB, the field for a sheet of monopoles at a quarter-wavelength below a free surface is about 13.2 dB higher than a monopole in free space. This is an important observation, but it must be qualified in many ways. First, the units for the noise source are in dB/m^2 referenced to 1 µPa. Thus, the monopoles have an areal density (i.e., they are a strength per unit area). In contrast, the received level is simply in dB referenced to 1 μPa. There is a tendency for people to jump casually between received levels and source levels and, in fact, Wenz curves (guesses at received levels based on field observations) have been used as source levels despite the difference in units.

One should also note that the received noise level in this simple model is independent of the proximity of the listener to the sheet of noise sources. This assumes there is no volume attenuation in the halfspace that would attenuate the sound as it propagates vertically down through the water volume. Loosely speaking, as the listener moves away from the sheet, he or she also hears better the more distant noise sources. They are weaker due to spherical spreading, but that loss is balanced by the increased area of the annulus of noise sources.

1.16 Stratified (Range-Independent) Media with a Sheet of Monopoles

In earlier sections, we discussed different approaches to propagation modeling, each with its own pros and cons in terms of accuracy and speed under different conditions. These propagation models are used to calculate the soundscape due to distributed noise fields by treating each point in the field of noise sources as an independent source. Then one can integrate (sum) over all the source positions to get the final received pressure field. Since all these different propagation models have different advantages, we present the results for each of them.

For most readers, the interpretation of the received field as a sum of all the contributing noise sources is natural. However, because of the principle of reciprocity, sources and receivers can be interchanged and we can propagate from hydrophone positions out to all the positions of noise sources and think in terms of how well one can *hear* the noise sources at a given hydrophone position. In short, you can position things in terms of the speaker (the source) or the listener (the receiver).

1.16.1 Normal Mode Representation

Previously we presented two analytic representations for the pressure field due to a point source. The normal mode representation is

$$p(r,z) = \frac{i}{4}\sum_m \Psi_m(z_s)\Psi_m(z)H_0^{(1)}(k_{rm}r),$$

where ψ and k_{rm} are the depth-dependent modes and wavenumbers, respectively. We remind that the mode shapes, ψ, are arbitrary up to a scale factor. We assume they are scaled such that the integral of the mode value squared is unity.

The intensity is then proportional to

$$|p(r,z)|^2 = \frac{1}{16}\sum_m\sum_n \Psi_m(z_s)\Psi_m(z)\Psi_n^*(z_s)$$
$$\times \Psi_n^*(z)H_0^{(1)}(k_{rm}r)\left(H_0^{(1)}(k_{rn}r)\right)^*.$$

To be precise, the intensity should include ρc factors defining the medium impedance. However, as is often done for convenience, we neglect those factors. (The conjugates in this equation and the following were neglected in some of the earlier literature that assumed real mode shapes. Many of the earlier normal mode models calculated purely real modes as an approximation.)

It will turn out to be useful to have this integrated over a disc of radius, R, so that we can understand the "noise footprint," that is, the radius to which one should integrate to include all noise sources that make a significant contribution to the field. The term "significant" here is based on the desired accuracy in the calculation. The integrals of these intensity terms are a bit tricky because of the Hankel functions. However, if one replaces them with their far-field approximations, they are integrals of exponentials. The result (see Carey and Evans 2011) is

$$|p(r,z)|^2 = 4\pi^2 \sum_m \sum_n \frac{\Psi_m(z_s)\Psi_m(z)\Psi_n^*(z_s)\Psi_n^*(z)}{\sqrt{k_{rm}k_{rn}^*}\,i(k_{rm}-k_{rn}^*)}$$
$$e^{i(k_{rm}-k_{rn}^*)R}.$$

This is the final result for the noise level due to a disc of noise sources of radius R, depth, z_s and unit strength. If we take the limit as R goes to infinity, we get the result for an infinite plane of noise sources:

$$|p(r,z)|^2 = 4\pi^2 \sum_m \sum_n \frac{\Psi_m(z_s)\Psi_m(z)\Psi_n^*(z_s)\Psi_n^*(z)}{\sqrt{k_{rm}k_{rn}^*}\,i(k_{rm}-k_{rn}^*)}$$

This double sum is sometimes time consuming to compute. Kuperman and Ingenito (1980) suggested keeping only the diagonal terms where $m = n$, arguing that the off-diagonals tended to cancel. That formula is then

$$|p(r,z)|^2 = 2\pi^2 \sum_m \frac{|\Psi_m(z_s)|^2 |\Psi_m(z)|^2}{|k_{rm}|\text{Im}(k_{rm})}$$

However, this approximation is sometimes inaccurate and modern computers can usually do the full double-sum in reasonable time. Nevertheless, this simpler form does help one to see a very important feature of the noise field. If the imaginary part of the wavenumber (corresponding to the attenuation of the mode in range) vanishes, then the noise level goes to infinity. Loosely speaking, the field due to a point source in a waveguide experiences cylindrical spreading in the far-field. That is counter-balanced by the increasing area of a noise annulus. So, if there is no additional loss mechanism, the noise level continues to build indefinitely as we increase the radius of the noise disc.

Of course, there is always some attenuation mechanism. However, in deep water there are typically modes that propagate in the SOFAR channel, and their loss is governed by the volume attenuation and the sea surface roughness. That loss can be extremely low, in which case the noise footprint can be hundreds of kilometers. In general, this balance implies that the noise footprint can vary wildly with the environmental conditions, including sea surface wind (and resulting waves), ocean sound speed profile, bottom topography, and bottom composition, etc.

The above discussion provides the intensity due to a uniform sheet of monopoles with a 0-dB source level. To get the true received level, one needs to add the source level. This statement applies also to the more sophisticated models described in the following sections.

1.16.2 Wavenumber Integral Representation

As discussed in Sec. 1.9.2, the wavenumber integration result for the pressure field is

$$p(r,z) = \int_0^\infty G(z,z_s;k_r) J_0(k_r r)\, k_r\, dk_r$$

where $G(z, z_s; k_r)$ is a Green's function of the depth-separated wave equation. In terms of this Green's function, the field due to a sheet of noise sources with unit strength is

$$I(z) = 2\pi \int_0^\infty |G(z,z_s;k_r)|^2\, k_r\, dk_r$$

The derivation of this equation may be found in Jensen et al. (2012), Ch. 9.

1.16.3 Field Representation

The above representations are convenient for mode and wavenumber integral models. However, sometimes either a ray/beam tracing or a parabolic equation model is preferred. One can derive special formulae for those cases as well; however, an easy solution is to use the acoustic pressure fields that all the models can produce and sum up the fields with a factor accounting for the increasing size with range of a noise annulus.

The noise is simply an integral of the pressure field over range. We may think of this as modeling how an array accumulates the energy from noise sources in a disc whose radius increases to the far field. There is one small detail in that such an array receives noise from annuli (rings); therefore, one needs to scale the pressure field up by a factor of $r\sqrt{2\pi\delta_r}$, where δ_r is the thickness of each ring. That factor represents the area of the ring. Passing to the limit of infinitesimal ring thickness, one obtains the final result:

$$I(z) = \int_0^\infty |p(r,z)|^2\, r\, 2\pi\, dr$$

The factor of r basically cancels out the cylindrical spreading, so the remaining attenuation effects are critical in producing a convergent integral. This integral can be done using standard numerical methods such as the trapezoidal rule. However, if the acoustic model permits an irregular mesh, we suggest it is more efficient to sample the pressure field using a logarithmic grid in range so that the density of samples decreases with range.

1.16.4 Examples

Following Kuperman and Ingenito (1980), we consider three environments. They are all 50 m deep with the same ocean sub-bottom. However, the sound speed profiles are (1) iso-velocity, (2) downward refracting, and (3) upward refracting (a surface duct scenario). The sound speed profiles for these cases are shown in the left panels of Fig. 1.42. The associated noise levels due to a sheet of monopoles are shown in the right panels. Note that the noise-level plots have been done for frequencies of 200 Hz, 400 Hz, and 800 Hz, resulting in three groups of curves. It is also important to note that the noise source level is set to unity to the results to not attempt to model realistic noise levels.

As discussed above, we can generate the noise field using either the analytic formulae for each model or by using the field representation and doing a direct numerical integral over the field due to a sheet of noise sources. Here, we have used wavenumber integration as the basic modeling approach.

To match the original paper, the source depth for the noise sheet was set at 0.5 m. This is roughly a quarter-wavelength at the highest frequency. Because of the surface decoupling effect, the noise level is reduced as the source gets closer to the surface in terms of wavelengths. (This results from the cancellation between a near-surface source and its out-of-phase reflection by the surface.) With the fixed source depth, this means that the noise level drops as a function of frequency since the wavelength is then lower. If we had used the more common quarter-wavelength depth for all frequencies, this effect would not be seen.

For the iso-velocity case, the noise level does not vary much with depth, and it is slightly higher than the 13.2 dB figure that one would get for a halfspace. This is because the seafloor is reflecting some of the noise into the water column.

The next case with downward refraction shows very similar results as seen in the middle panel of Fig. 1.42. However, there is a slight decrease in the noise level with depth. (Downward refraction refers to the sound speed decreasing with depth, which causes the sound rays to refract or bend toward the bottom.) This is probably since the rays from the near-surface noise sheet are bent more to the vertical (per Snell's law) where they encounter a higher loss per bounce.

The final case is upward refracting with a surface duct as shown in the bottom panel of Fig. 1.42. As discussed above, this sort of situation is difficult to model in that it has several modes trapped in the surface duct. Those modes are associated with ray paths that continually refract back to the surface without hitting the bottom. They therefore experience very little bottom loss. There is some loss because of the full-wave effects that the models include. Thus, if we do not include some additional loss mechanism, the noise level would go to infinity. The two commonly modeled loss mechanisms are surface scatter and volume attenuation. It is not clear precisely what was done in the original Kuperman and Ingenito paper. In this example, we used a surface roughness of 0.5 m root-mean-square.

Note that the excellent propagation conditions in the surface duct that channels the noise. This produces increased levels, with a peak in the surface duct. The strength of that peak is controlled by the sea surface roughness and can be quite high on a calm day. The channeling effect of the surface duct was believed to explain the Bahamas stranding incident. However, that circumstance is different in that it involved a single sound source (a Navy sonar), rather than a noise sheet as modeled here.

Many normal mode models take a shortcut to volume attenuation based on perturbation theory. In that case, they first solve the problem without

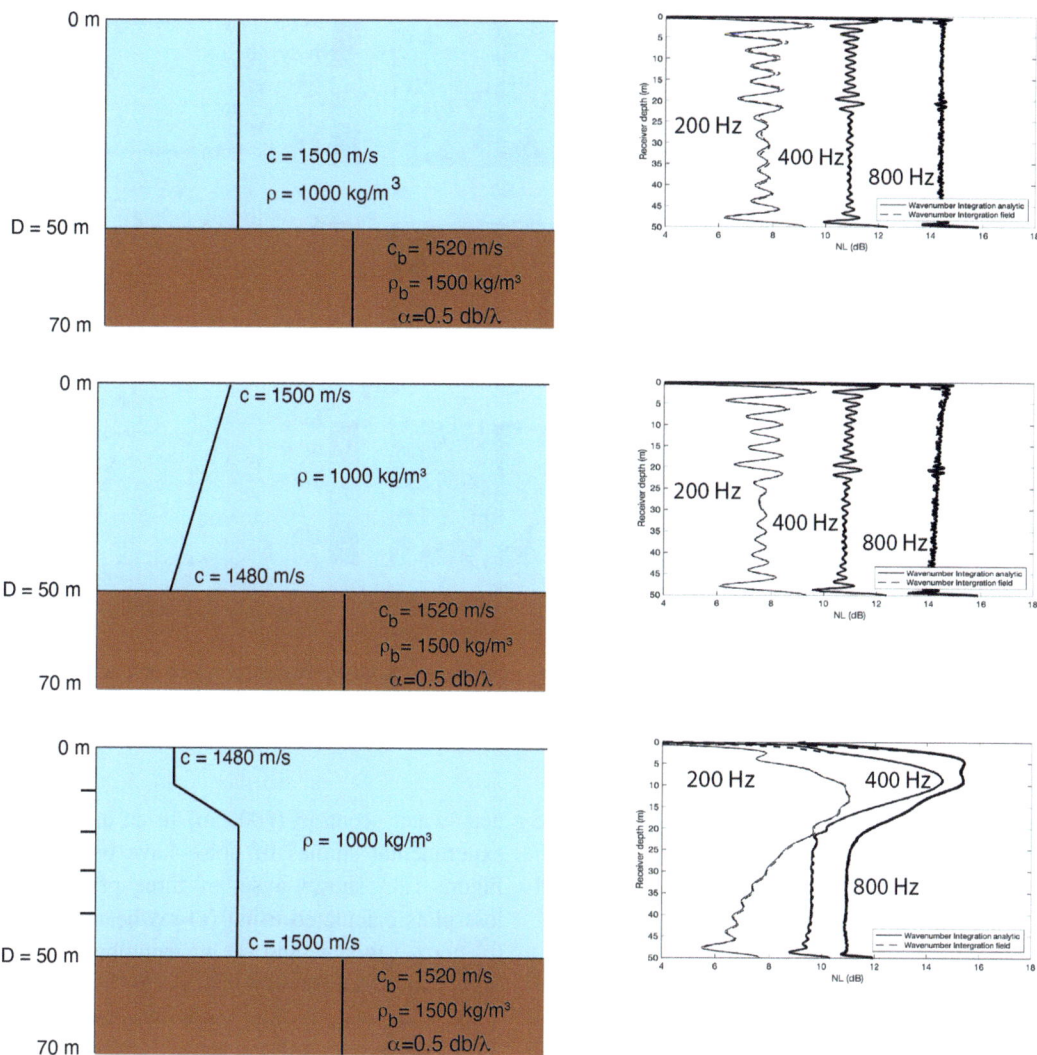

Fig. 1.42 Sound speed profiles and associated noise levels for three profiles and three frequencies. From top to bottom, we have (**a**) iso-velocity, (**b**) downward refracting, and (**c**) upward refracting (surface duct)

any loss mechanisms resulting in purely real modes. Then they make a correction to the wavenumbers to estimate their imaginary parts (the loss). However, it turns out that the imaginary part of the mode shapes is critically important for the noise modeling. Therefore, models using this approximation are not well suited to this application.

One may wonder why perturbation theory has been an established standard procedure based on this discrepancy. After all, the noise level at any depth is just an average of the received level over range after compensating for cylindrical spreading. Thus, the discrepancy says the average levels are incorrect. However, this type of noise calculation, motivated by wind-generated noise is unusual because the source is so close to the sea surface. This puts most of the energy into nearly vertical paths (or equivalently, higher-order modes) and the perturbation approach is much less accurate for such modes.

A last point of interest is that the noise field has been plotted using both the analytic formulae (presented previously in Sects. 1.16.1 and

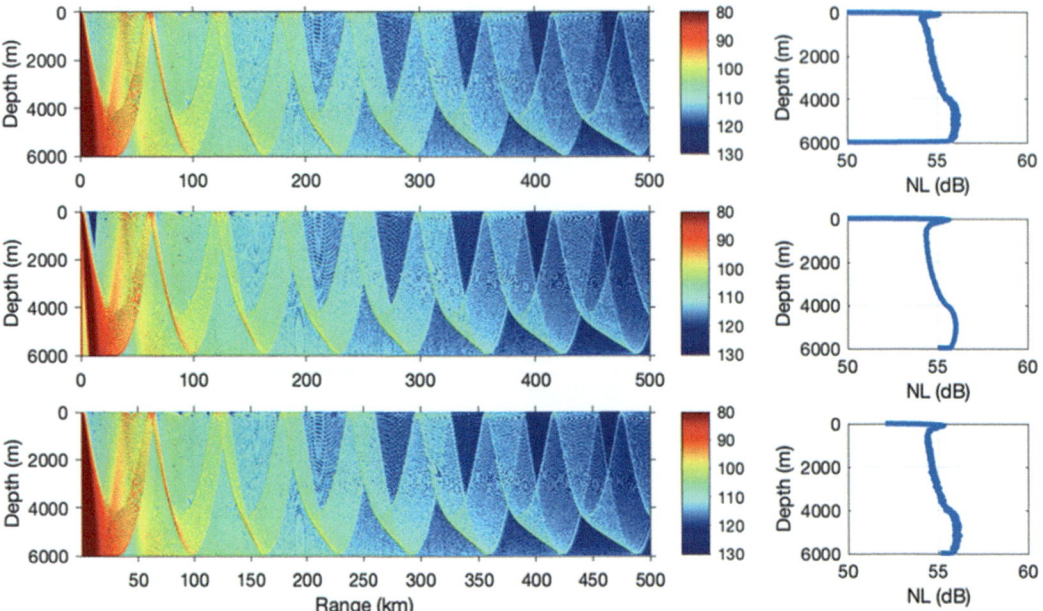

Fig. 1.43 Transmission loss computed using three different types of acoustic models (left panels). The source depth is 1.25 m and the source frequency is 300 Hz. From top to bottom, these were ray/beam tracing, normal modes, and wavenumber integration. Corresponding noise level vs. depth for those acoustic models (right panels)

1.16.2) and the method of integrating the pressure field over range (Sect. 1.16.3). As expected, both formulae yield nearly identical results. The small differences that do remain are due to the truncation of the field calculations at a finite range.

In summary, the noise level often does not vary much with depth, but with features in the sound speed profile, such as ducts, the noise level can vary a lot. Furthermore, when there is upward refraction that produces paths that do not interact with the bottom, the noise level can be extremely sensitive to the sea state and volume attenuation. Of course, the noise level does drop off in a shallow zone near the ocean surface due to the surface decoupling effect, that is, the cancellation of the sound field due to destructive interference.

The propagation conditions in shallow water are typically very different from those in deep water, so one may wonder how that affects the noise levels. We consider a deep-water profile constructed from the mean sound speed profile on a propagation path near Hawaii. The site selection was somewhat arbitrary here but reflects a deep-water scenario (6000 m) in an area where experimental studies of noise have been done. Figure 1.43 shows a set of three propagation loss plots calculated using (a) ray/beam tracing, (b) normal modes, and (c) wavenumber integration. The source frequency was 300 Hz.

A geoacoustic bottom model was included in this example with a sound speed of 1600 m/s and attenuation of 0.2 dB/wavelength. However, energy trapped in the main SOFAR channel does not interact significantly with the bottom. As a result, noise injected near the sea surface can contribute significantly for thousands of kilometers. To model this accurately, one must also consider the volume attenuation in the ocean and we have used standard formulae to account for that.

In the transmission loss (TL) plots in Fig. 1.43, one should note the presence of a typical convergence zone pattern with energy cycling every 55 km or so. However, we are also modeling steeper angle propagation to include the bottom bounce energy. One can see that contribution in what would otherwise be a shadow zone, between the convergence zones. These TL plots are the result of three completely different acoustic propagation models, so their agreement should not be taken for granted. The astute reader will notice some minor flaws in the ray/beam result. That is a characteristic of the high-frequency approximation in ray/beam models as discussed in Sect. 1.9.1.

The panels in Fig. 1.43 show the noise level as a function of depth due to a sheet of noise sources with a source-level density of 40 dB / m^2 re 1 µPa. This is a reasonable level for 10-knot wind noise at 300 Hz. We can see that there is not a lot of variation in the noise level over depth. Near the sea surface, the pressure always goes to zero since the ocean surface is a pressure-release boundary. The noise level also decreases below the critical depth since the shallow angle rays are refracted before getting there. (The critical depth is the depth below the main sound channel axis where the sound speed equals the sound speed at the surface. See Figs. 1.1–1.7)

Another very interesting feature of the noise field is its directionality. In Fig. 1.44, we have plotted the beamformer output vs. declination angle for a simulated vertical line array near the sound channel axis. The declination angle is defined as positive when pointing to the bottom. (A vertical array of hydrophones has no directionality in azimuth, so we are interested only in the vertical directionality. Note also that this sort of plot is affected by the beamforming process, which in turn is also dependent on the array configuration.) There is a pronounced dip of about 40 dB in the noise level near horizontal. This is called the noise notch and is a typical feature of noise in the deep ocean (see, for instance, Carey and Evans 2011). It results because noise near the sea surface is refracted (Snell's law again) to steeper angles, so there is no way (in the ray approximation) to get energy into horizontal angles. Thus, the directional characteristics of the hearing of a marine mammal can be very important in terms of its perception of the loudness of the noise field. The noise notch varies with the depth of the listener as well. Marine mammals may very well be sufficiently intelligent to exploit these features of the ocean channel.

Note that the noise is generally louder from the direction of the surface than from the bottom. See, for instance, the noise peaks at −13° (pointing to the surface) and + 13° (pointing to the bottom). It is not obvious, but a mathematical analysis (Harrison and Simons 2002) shows that the noise in a beam steered to the bottom has effectively one additional bottom bounce relative to a beam steered to the surface. Thus, the ratio of the noise between an uplooking to a downlooking

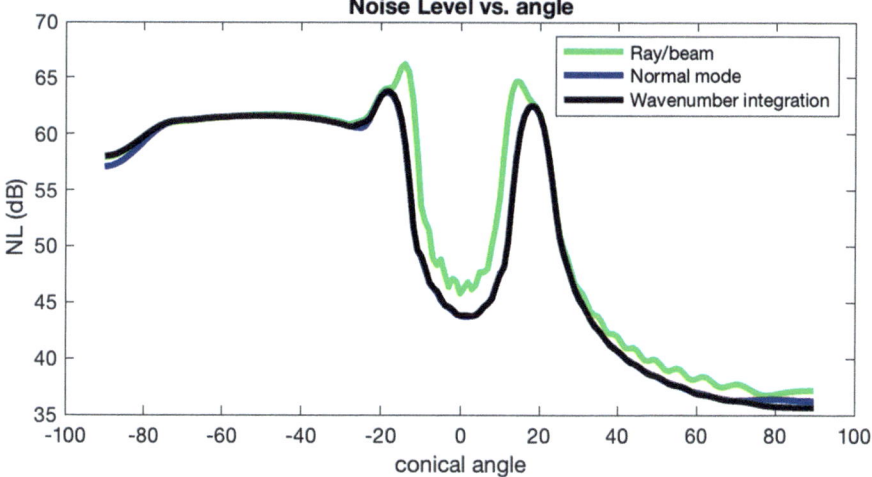

Fig. 1.44 Noise level vs. vertical (declination) angle calculated using three different models

beam is a measure of the bottom reflection coefficient. Since the bottom is typically less reflective than the surface, the down-pointing beams show lower levels. This feature has led to a lot of recent work using the noise field to learn about bottom reflectivity. See, for instance, Siderius and Gebbie (2019) and references therein.

1.17 Modeling Specific Noise Sources

1.17.1 Wind Noise

Wind noise is often the dominant noise for frequencies between a few hundred Hz and 20 kHz. One might think to ignore it in the context of its effect on marine mammals—it is clearly a natural phenomenon out of our control. On the other hand, it provides a reference level in comparing how much an anthropogenic source contributes beyond the natural noise background and therefore whether it is detectable and whether it can mask other signals (Sertlek et al. 2019; Farcas et al. 2020; Erbe et al. 2021). See also the discussion in Chap. 2.

Buckingham and Potter (1996), provide an interesting perspective on the understanding of wind noise in the preface to their workshop proceedings. Their workshop was a follow on to an earlier one organized by Kerman (1987), on Sea Surface Sound and the Natural Mechanisms of Surface Generate Noise in the Ocean. At that time, it was not very clear how wind generated noise. Today it is understood or at least widely accepted that wind forcing is the first stage of a complex cascade. When the wind becomes strong enough, the surface waves start to break, causing white-capping and bubble entrainment. These bubbles can exist from seconds to a few minutes; however, the bubbles are acoustically active only for about a second when they are freshly formed (Deane and Stokes 2002). The air in the bubble gradually dissipates into the ocean; however, in its early stages the bubble is an oscillator that radiates sound.

The bubble cloud is, of course, a complicated phenomenon with a distribution of bubbles at different depths and with different sizes cascading down from the breaking waves. Apart from irradiating the ocean, the bubble cloud is also an important part of the environment for sound propagation. Small concentrations of bubbles cause enormous changes in the effective sound speed and attenuation of the ocean. As a result, bubble clouds cause strong returns (echoes) for active sonar systems. In addition, the bubbles are powerful attenuators of sound at frequencies above 4 kHz. For this reason, bubble curtains are often used to mask unwanted sound sources. The attenuation effect of bubble clouds has been known for many years. However, recent work by Yang et al. (2023) has shown how this leads to big changes in the wind noise as a function of wind speed implying modifications to the Wenz curves. As seen in Fig. 1.45, there is a crossover at around 4 kHz where higher wind speeds can actually produce lower wind noise. This is believed to be due to the masking effect of the bubble clouds. For more on the acoustics of bubbles, see the book by Leighton (1994).

For lack of better alternatives, wind noise is often represented as a planar distribution of monopoles that ensonify the channel producing the received noise field. As a somewhat arbitrary convention, the monopoles are often placed at a quarter-wavelength below the surface, so they produce a dipole. If one wishes to use a shallower depth, one need only increase the source level proportionally to compensate for the surface-decoupling effect, that is, the cancellation between direct and surface-reflected energy. (Sorting out the units becomes a bit complicated.) Kuperman and Ingenito (1980) presented the first paper to show this for realistic ocean waveguides, building on earlier work for very simplified ocean models. One way to estimate the source level of these monopoles is to reverse the process. To do this, one measures the received noise level and determines what source level for the sheet of monopoles would have produced that field. At first blush, this seems like a pointlessly circular process; however, one learns about the source level in one environment, but is then able to apply it in other environments. Another method to estimate the source level uses a vertical array

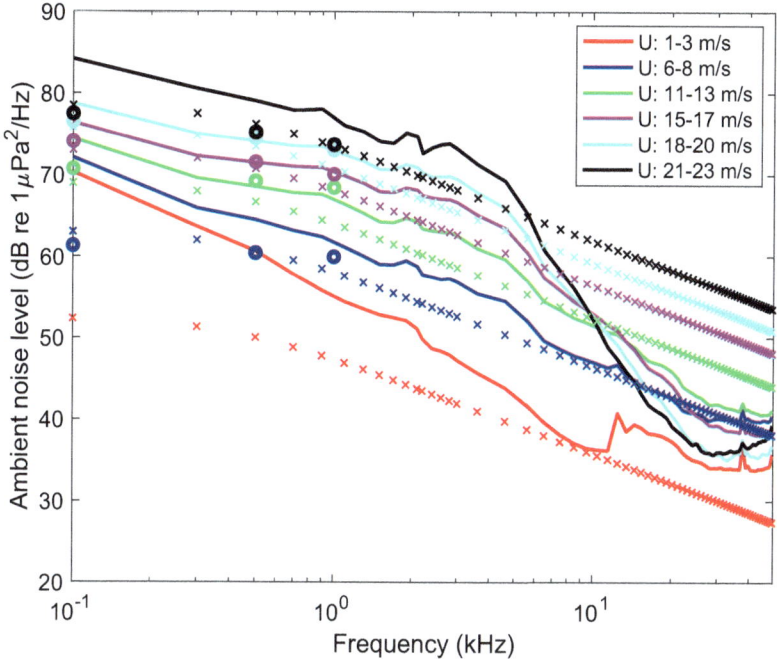

Fig. 1.45 Variation in wind noise as a function of wind speed, U, and frequency. Solid lines are measured data; circles and crosses are modeled using a sheet of monopoles but neglecting the shielding due to the bubble clouds; Note the cross-over at about 1 kHz where increasing wind speed can actually reduce the noise level.

Reprinted from Yang J, Nystuen JA, Riser SC, and Thorsos, EI (2023) Open-ocean ambient noise data in the frequency band of 100 Hz–50 kHz from the Pacific Ocean. J Acoust Soc Am EL. https://doi.org/10.1121/10.0017349. With permission from the ASA. © Acoustical Society of America, 2023. All rights reserved

and steers up- and down-looking beams to capture the energy overhead. (See the discussion in Kewley et al. 1990.)

These techniques have been applied by various investigators. Kewley et al. (1990; Fig. 6) present the following values for the source-level densities. They are intended to define a piecewise-linear fit between the three specified frequencies. Kewley et al. (1990) present these in terms of a "vertical source level" in dB. To get the monopole source level in our formulation, we have subtracted 6 dB to get the results in our table. The monopole is assumed placed at a depth of $\lambda/4$ (Table 1.2).

Wilson (1983) has another set of values given for a dipole source. A sheet of dipoles at the ocean surface is equivalent to a set of monopoles infinitesimally below the surface. However, there are different scale factors. The equivalent monopole level at a quarter-wavelength is obtained by subtracting $20\log_{10}\pi \approx 10$ dB. The resulting source-level density vs. wind speed and frequency is given in Table 1.3.

There are differences of a few dB (and more) between the Kewley, et al. and Wilson values, which gives an indication of their uncertainties. Chapman and Cornish (1993) provided additional measurements of wind-generated noise which generally agreed well with Kewley's results. Overall, these differences speak to the difficulty in estimating the source level of wind.

One may wonder what happens to the noise levels between the sea surface and the ¼ wavelength depth of the monopoles. For instance, at 50 Hz, the wavelength is about 30 m, so the monopoles would be placed at a depth of 7.5 m. Consider first that wind noise is rarely the dominant noise mechanism at low frequencies. Meanwhile, the noise generation mechanism for wind involves breaking waves generating bubbles at

Table 1.2 Source-level density (dB re 1 µPa m/m^2) vs. wind speed and frequency (Kewley et al. 1990). The monopole source depth is $\lambda/4$

Wind speed	5	10	20	30	40	knots
	2.6	5.1	10.3	15.4	20.6	m/s
Freq						
35 Hz	41	44	49	53	57	
600 Hz	32	39	49	53	57	
1500 Hz	30	35	41	45	48	

Table 1.3 Source-level density (dB re 1 µPa m / m^2) vs. wind speed and frequency (Wilson 1983). The monopole source depth is $\lambda/4$

Wind speed	10	15	20	25	30	35	40	45	50	60	70	80	90	knots
	5.14	7.72	10.3	12.9	15.4	18	20.6	23.1	25.7	30.9	36	41.2		m/s
Freq (Hz)														
5	45	50	53	57	58	61	63	65	66	68	71	73	75	
7.5	39	46	50	54	55	58	60	62	63	65	68	70	72	
10	37	43	49	52	54	57	60	61	63	65	67	70	72	
20	37	43	50	53	55	59	61	62	64	66	70	72	72	
50	39	45	52	54	58	61	62	63	65	66	67	68	70	
100	40	47	53	56	59	60	62	63	64	65	66	67	67	
500	39	46	52	54	57	59	60	60	61	62	63	64	65	
1000	38	46	52	54	55	57	58	59	59	61	62	63	63	

different depths. In other words, the monopole sheet is an idealization that endures a lack of more precise information about the noise generation and is incapable of providing details in the very near-surface layer.

1.17.2 Shipping

One general approach to modeling ship noise is to first calculate the sound field for a distribution of *virtual* sources around the globe and then sum up those fields with a weight depending on the injected ship noise at each virtual source point. In this fashion, the noise field can be dynamically updated with different shipping distributions.

In our calculations, the sources are positioned at a depth of 6 *m* below the ocean surface representing a nominal depth of the propeller of a large ship. The propeller is typically the dominant noise source. By splitting the process into these two stages, we can reuse the noise fields with different noise sources, for example, different classes of ships, and produce noise maps for a variety of sources with minimal additional effort.

Figure 1.46 shows how we make a grid of the ocean using nodes spaced roughly every 100 km apart. This plot is just a subset of the North Pacific. Details of the process may be found in Porter and Henderson (2011, 2013); normal modes of the ocean are calculated at each node and the sound field is calculated along a fan of radials using adiabatic mode theory. Adiabatic mode theory is an extension of the standard normal mode method that can handle range dependence in an approximate way. The modes are interpolated using bilinear interpolation within each triangular tile of the grid. The red fan in the lower-left corner of this figure is just one example of the roughly 65,000 source points that are used to cover the globe.

In Fig. 1.47, you can see how each node of the grid is used to calculate a noise field (approximately every tenth node in latitude and longitude from Fig. 1.46 is shown). We are displaying transmission loss plots here, so the noise field is obtained by subtracting the transmission loss from the source level in dB. The dark blue areas (very high TL) are typically regions where there is a bathymetric blockage. This plot is schematic

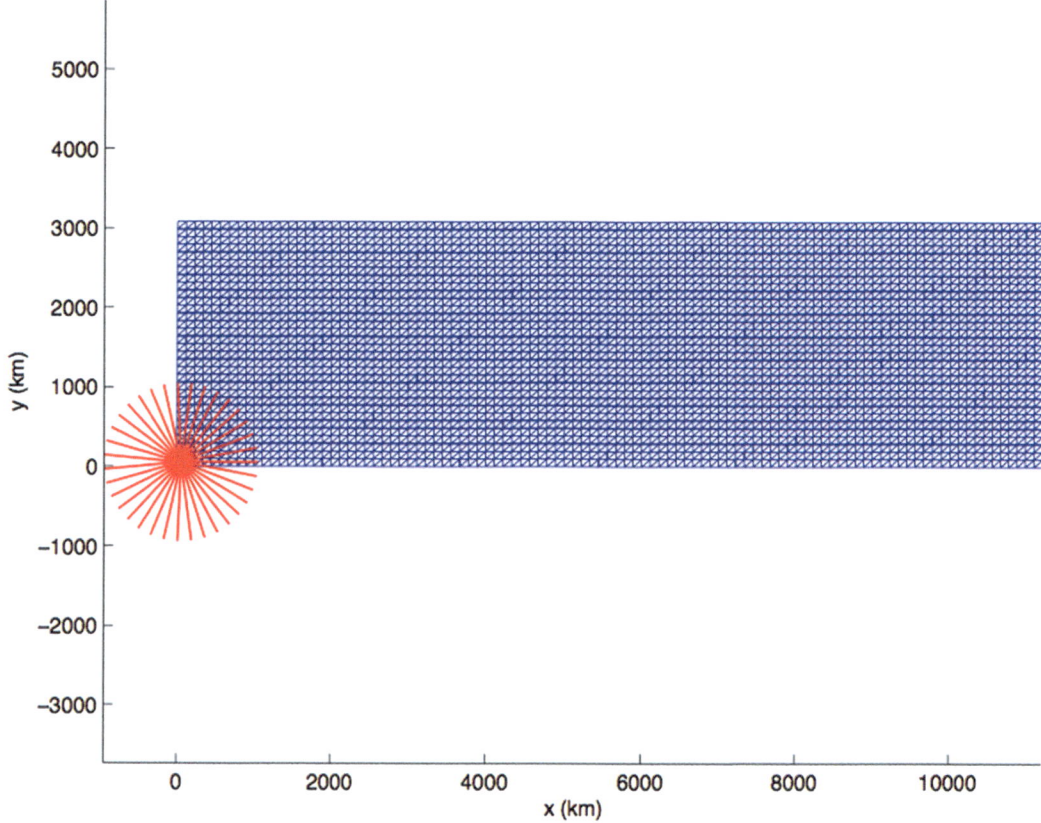

Fig. 1.46 Sound propagation calculations are done for a fan of radials and for a grid of noise sources spaced roughly every degree around the globe. There are about 65,000 (360 × 180) such virtual noise sources for the entire globe and a fan of radials is generated for each

in the sense that the real calculation is done with a finer grid of 1°, or very roughly 100 km around the world. However, it is presented here on a much coarser scale to make the calculations for individual nodes clear. Many of the terms in this process are frequency dependent such as the source level and the attenuation in the sediment. As a result, the transmission loss is also frequency dependent. A calculation is done for each frequency of interest.

This modeling approach was first described in Kuperman et al. (1991) and is referred to as KRAKEN3D, being an extension to the earlier KRAKEN normal mode model. The software that combines the propagation model for all the source locations to get the noise field is a simple post-processor called Soundscape (Henderson and Porter 2015). Earlier work used KRAKEN3D to generate a full noise covariance matrix on general hydrophone arrays (see Jensen et al. 2012). However, the emphasis here is simply on a noise level (Figs. 1.48 and 1.49).

1.17.3 Quick Approximate Solutions for Range-Dependent Problems

The process described above involves a very large number of transmission loss calculations to estimate how the noise from every point in the area of interest propagates to every other location. A very useful shortcut uses a "locally flat approximation" in which the noise at each point in latitude and longitude is treated independently.

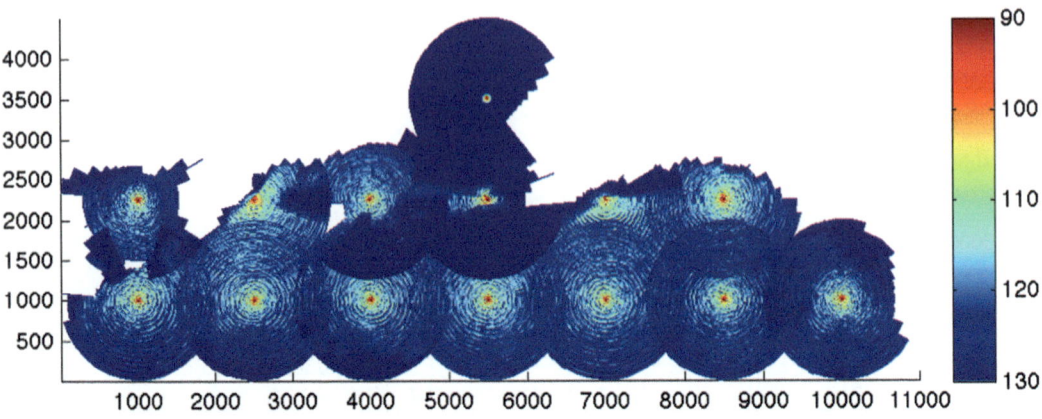

Fig. 1.47 Sample transmission loss plots in dB for each virtual noise source. The fine rings in each disc are convergence zone patterns caused by refraction in the water column. The example here shows virtual sources spaced every 1000 km. In the actual calculation, they were done every 100 km

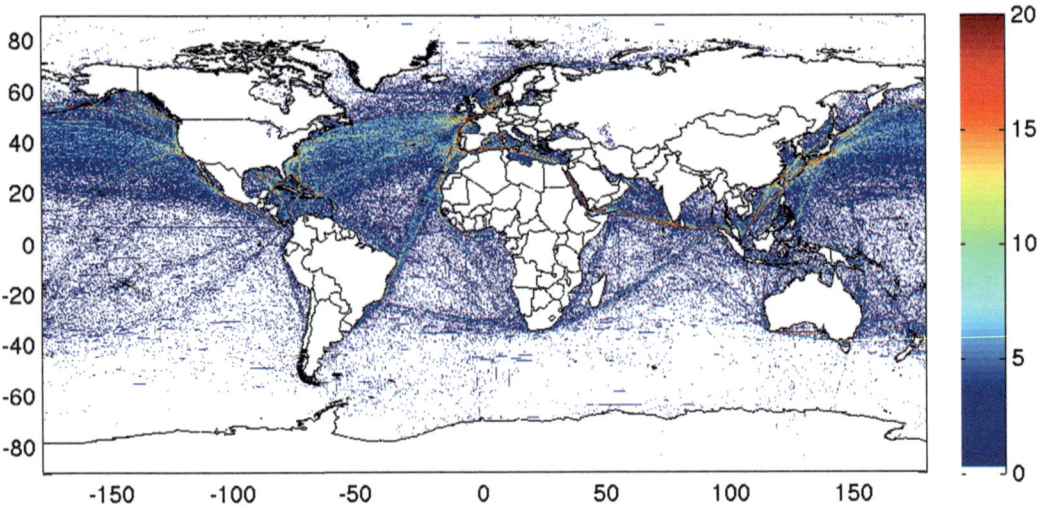

Fig. 1.48 Global shipping in units of km of track per unit is derived from the Voluntary Observing Ship Program

Thus, the local injected noise (e.g., wind noise) is assumed to be constant and a simple range-independent model is used to estimate the resulting received level. Any of the very fast models described in Sect. 1.16 may be applied. The local environmental model is created using the ocean sound speed profile and sub-bottom properties valid for that particular latitude and longitude.

As an example, we consider the Santa Barbara Channel off southern California, USA, where noise due to shipping is modeled (see Redfern et al. 2017). The bathymetry and ocean bottom type are shown in Fig. 1.50. The bottom type here has been classified simply as a hard or soft bottom as there was limited geoacoustic information available.

The upper panel in Fig. 1.51 shows a full calculation of the noise field using the 3D approach described in Sect. 1.17.2. The lower panel shows the result with the locally flat approximation. Note that this approximation shows much greater sensitivity to the bottom type resulting in patches with sharp boundaries

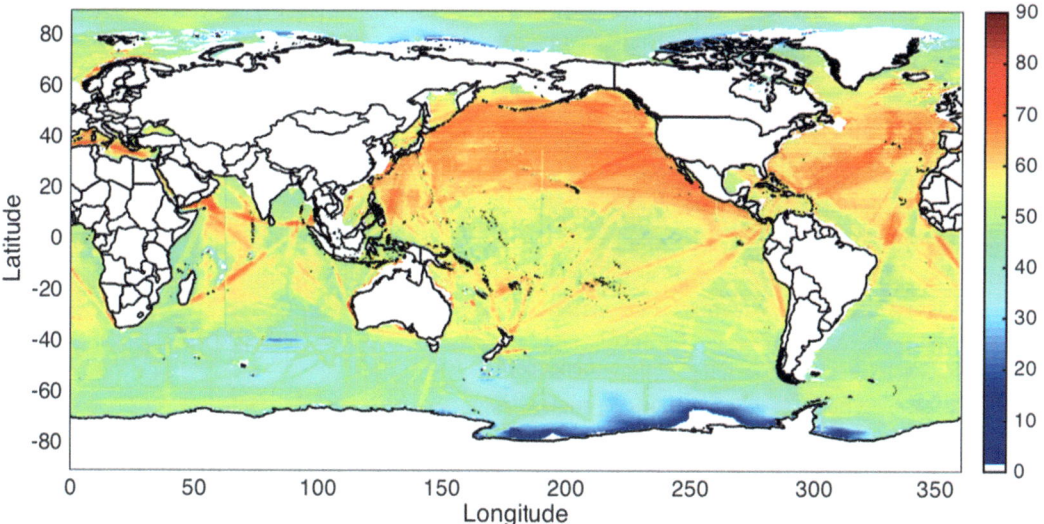

Fig. 1.49 A final soundscape or noise map for the entire globe for one frequency (200 Hz) and one receiver depth (200 m). The noise levels are presented in dB using the standard reference of 1 μPa and for a 1-Hz band.

where the noise is loud corresponding to places where the bottom type is hard and therefore acoustically very reflective. In contrast, the full 3D result is much smoother, reflecting the averaging process that results from the range-dependent modeling. In short, the locally flat approximation trades off accuracy for a big reduction in computational cost and a great simplification of the modeling process.

The locally flat approximation is also very useful in calculating the needed radius for a noise disc. As background, recall that the full 3D noise modeling process requires one to integrate over a noise disc out to a range where the noise field has converged to the correct level (within some tolerance). Because the locally flat approximation is so quick, it can be used to rapidly estimate the noise level as a function of the radius of the noise disc. Once the levels stop changing significantly as we increase the radius, we may feel satisfied with the results. In general, the required radius can vary widely depending on the ocean sound speed profile, water depth, etc., as discussed previously. Once this radius has been estimated, it can then be used in a more accurate fully 3D calculation of the noise field.

1.17.4 Estimating Shipping Noise Levels from Track Data

As discussed above, one way of predicting the soundscape due to ships (or other moving sources) is to sum up the sound fields at discrete steps in time. This can be useful or necessary if detailed information about the time evolution of the soundscape is needed. However, if we are only interested in long-term averages, then it is more efficient to convert the ship tracks to an average density of ships in discrete cells in latitude and longitude. Then one can incorporate the ship sound level with a single propagation run for each cell.

An example is illustrated in Fig. 1.52 for seismic surveys in the Gulf of Mexico that were active in 2009. This information was provided by the Bureau of Ocean Energy Management and shows permits based on survey areas as seen most clearly in the western part of the Gulf of Mexico. It is easy to convert these patches to a source-level density based on the area of the patches, the duration of the survey, the number of ships, and typical source levels for the airgun arrays.

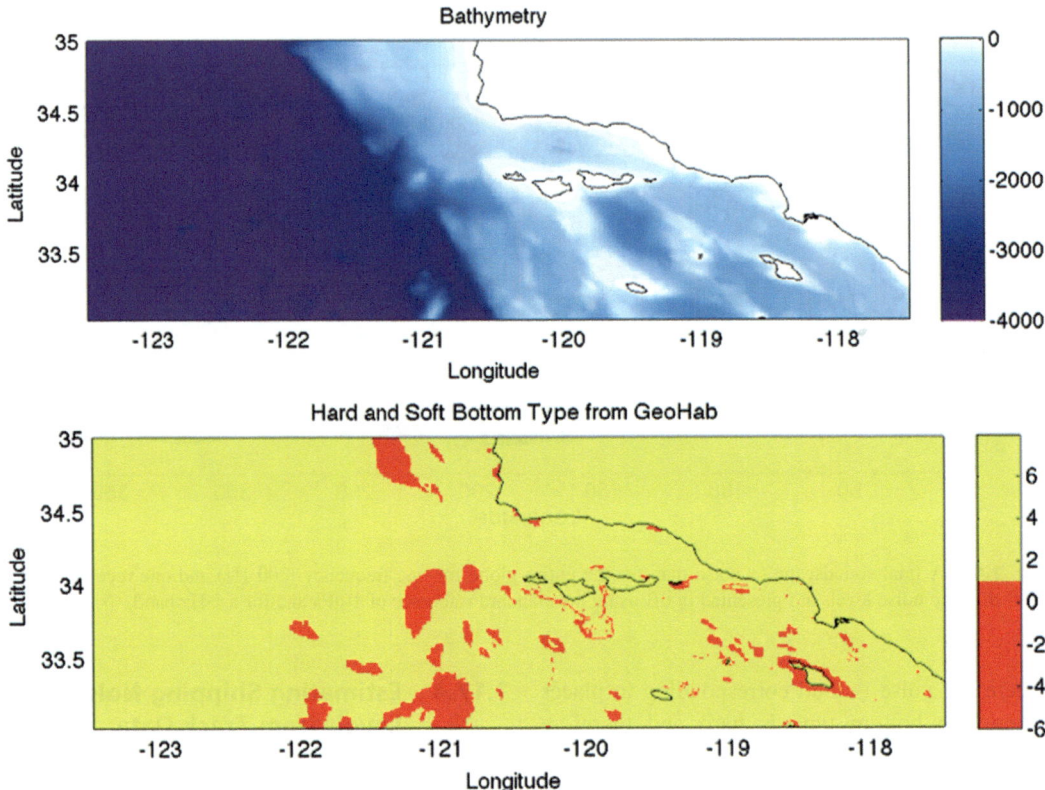

Fig. 1.50 Bathymetry (upper) and bottom type (lower) for the Santa Barbara Channel

A slightly more complicated process is required for the ship tracks shown most clearly in the eastern part of the figure. One needs a process for converting those tracks to an average ship density. The data shown in this figure were used in estimating the sound levels due to the seismic survey. However, as another example, we consider rig-support vessel traffic. These vessels travel from shore to one or more oil rigs in the Gulf of Mexico. The actual rig locations—there are several thousand—are shown in Fig. 1.53.

Our first step is to build a ship traffic model for these rig-support vessels. Our purpose here is not to promote a specific approach, but to demonstrate a process. Here, we assumed that the ships left from a hypothetical port located on a vertical line above each rig and that the vessel sequentially visited the three nearest rigs as well. Finally, it returns on a vertical path to shore. The vessel speed was assumed to be 12 knots. The resulting tracks are shown in Fig. 1.54. Clearly one can do a much more precise or realistic modeling of the vessel tracks.

At this point, we have defined the vessel tracks in terms of their waypoints. To get a ship density (number of ships per km^2), we create a simple simulation that steps along the polygon train defined by the waypoints. Looping first over the track segments, we calculate a unit tangent for each segment. Between waypoint i and waypoint $i + 1$, the tangent vector is $t = x_{i+1} - x_i$, where x_i and x_{i+1} are vectors containing the (x, y) coordinates of the waypoints. The length of that tangent vector is $\|t\|$ so a unit-tangent is $t_u = \frac{t}{\|t\|}$. Next, we step along the track, calculating a coordinate for the vessel as

$$x_t = x_t + \delta\, t_u$$

where δ is the length of the step along the track during the time step, which of course is

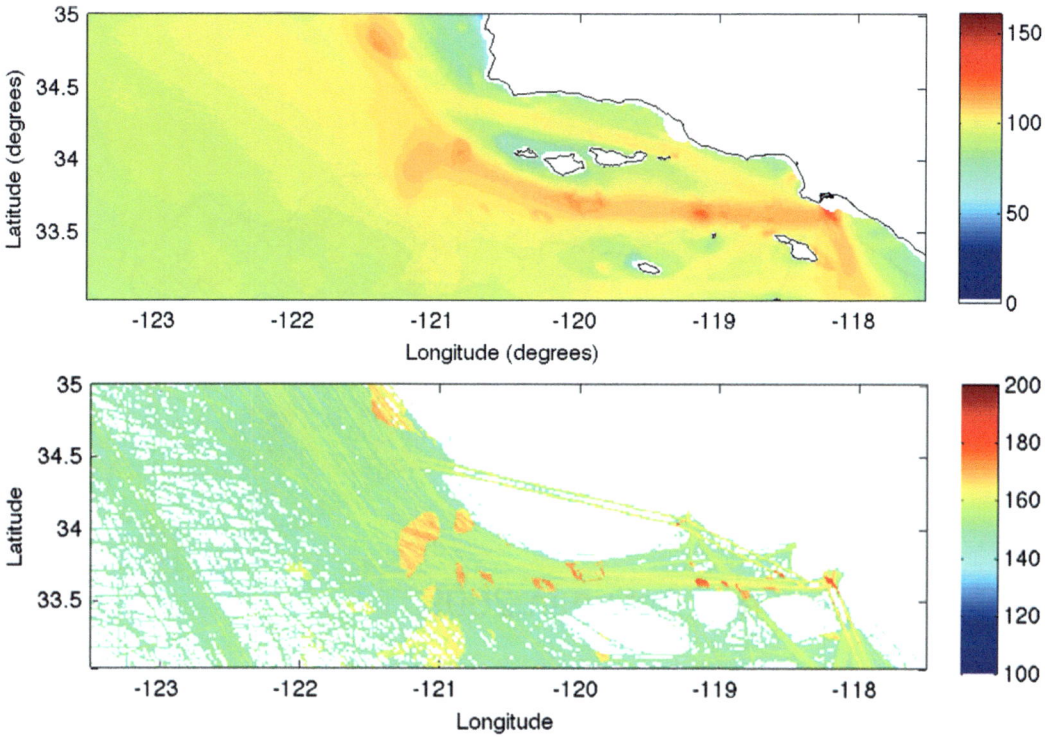

Fig. 1.51 Noise level for a third-octave band centered at 50 Hz, calculated using the 3D modeling approach (upper) and a range-dependent propagation model using the locally flat approximation with a noise disc of radius 320 km (lower)

proportional to the speed of the vessel. In other words, the spatial step along the track is defined based on a user-specified time step. Here, we have assumed a 12-knot vessel speed and a 10-s time step. Over that time, the vessel moves about 60 m. At each time step, we identify the spatial cell in latitude and longitude where the vessel is located. A matrix is set up to store the accumulated dwell time and incremented by the time step (here 10 s) during each step of the simulation. In this way, we build up an accumulation of the total time that a vessel spends in each cell. With that information, the ship density in each cell is simply its dwell time in the cell divided by the total time of interest for the simulation. This provides the ship density as shown in Fig. 1.55. The ship density combined with an estimate of the vessel noise level is the necessary input to calculate the final noise level.

1.17.5 Pile Driving

Piles provide a substructure for a variety of marine platforms. Piles have a variety of shapes, but are often cylindrical, and they are often driven into the seabed with a sequence of very large hammer impacts. However, they may be driven by alternate means such as vibratory pile driving, which is typically quieter. Sometimes mitigation effects, such as bubble screens, are used to reduce the noise propagated down the channel.

Pile driving clearly involves a complicated source mechanism first elucidated by Reinhall and Dahl (2011). As discussed by Dahl et al. (2015), it is useful to think of an idealized case with a cylindrical pile. The hammer strike generates a bulge in the pile, which then propagates downward. The propagating bulge

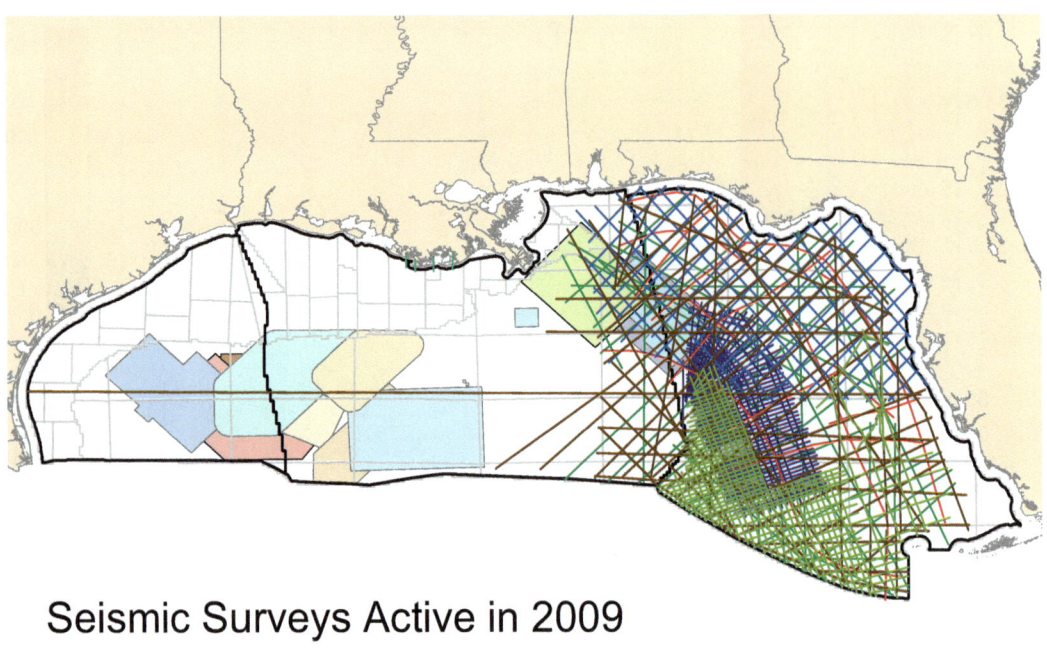

Fig. 1.52 Seismic surveys active in 2009. Data source: Bureau of Ocean Energy Management

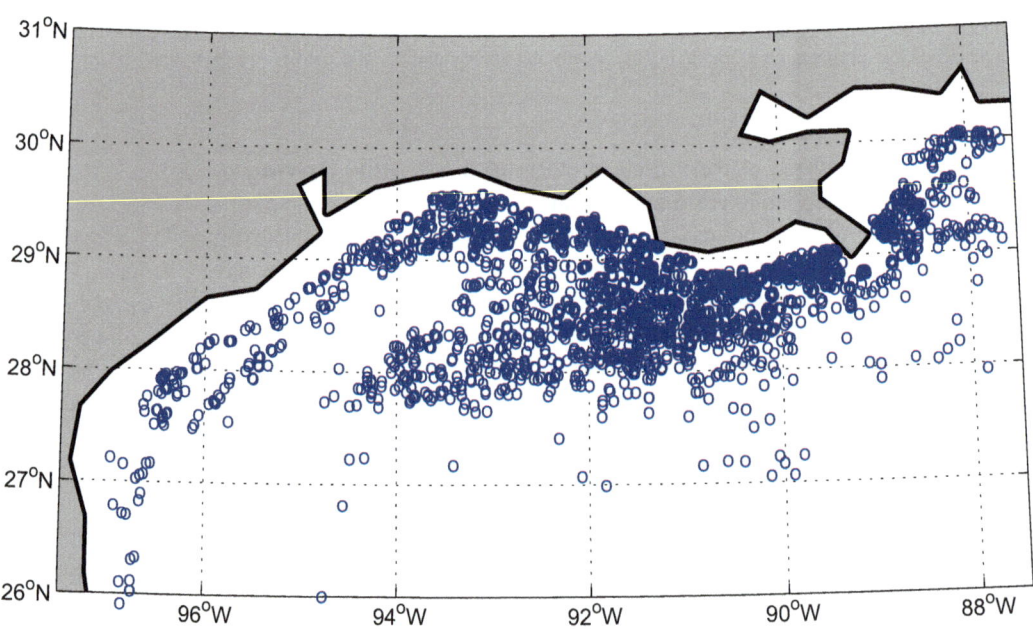

Fig. 1.53 Locations of oil rigs in the Gulf of Mexico

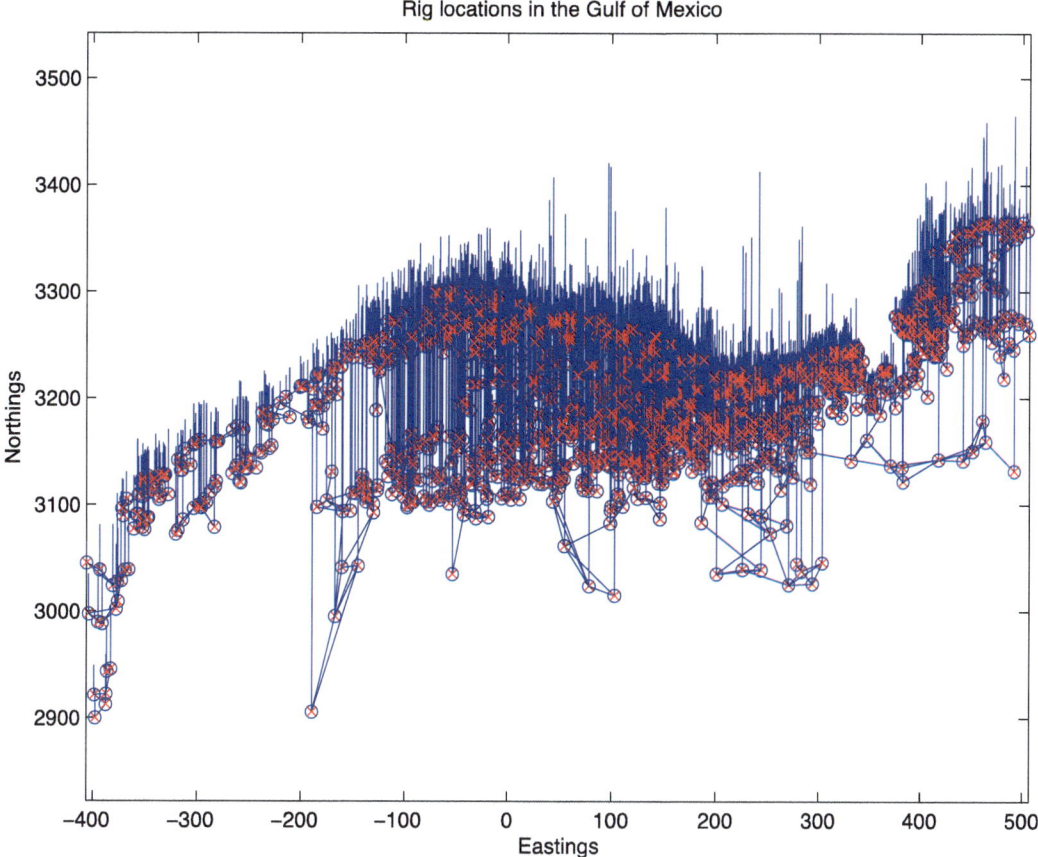

Fig. 1.54 Modeled tracks for rig-support vessels in the Gulf of Mexico

generates what is sometimes called a "Mach wave," which in turn radiates sound into the ocean as indicated in Fig. 1.56. The sound from the bulge is radiated in a conical wave whose angle is determined by the ratio of the propagation speed in water to that in the pile.

Finite element models have proven useful in modeling these effects; however, they generally are not practical to model the full problem out to the ranges of interest due to the computational burden. Therefore, as in the cases of the chemical explosives and airgun arrays, one needs to couple the results of the finite element model to an acoustic model for the channel, which can be done using the Helmholtz–Kirchhoff integral. This is an area of ongoing research.

1.17.6 Explosives

There are many explosive sources of sound in the ocean. Unexploded ordnance (abbreviated UXO or UO) is regularly found. In Europe, these are often weapons from World Wars I and II that simply failed to detonate. In other parts of the world, they are found in military training grounds, such as off the island of Vieques, Puerto Rico. In the United States alone, about 400 sites have been identified with potential UXOs. Efforts are ongoing to clear these areas, typically by open detonation. This then raises obvious questions about the impact on the marine environment.

Separately, particularly large detonations are done in ship-shock trials designed to test the

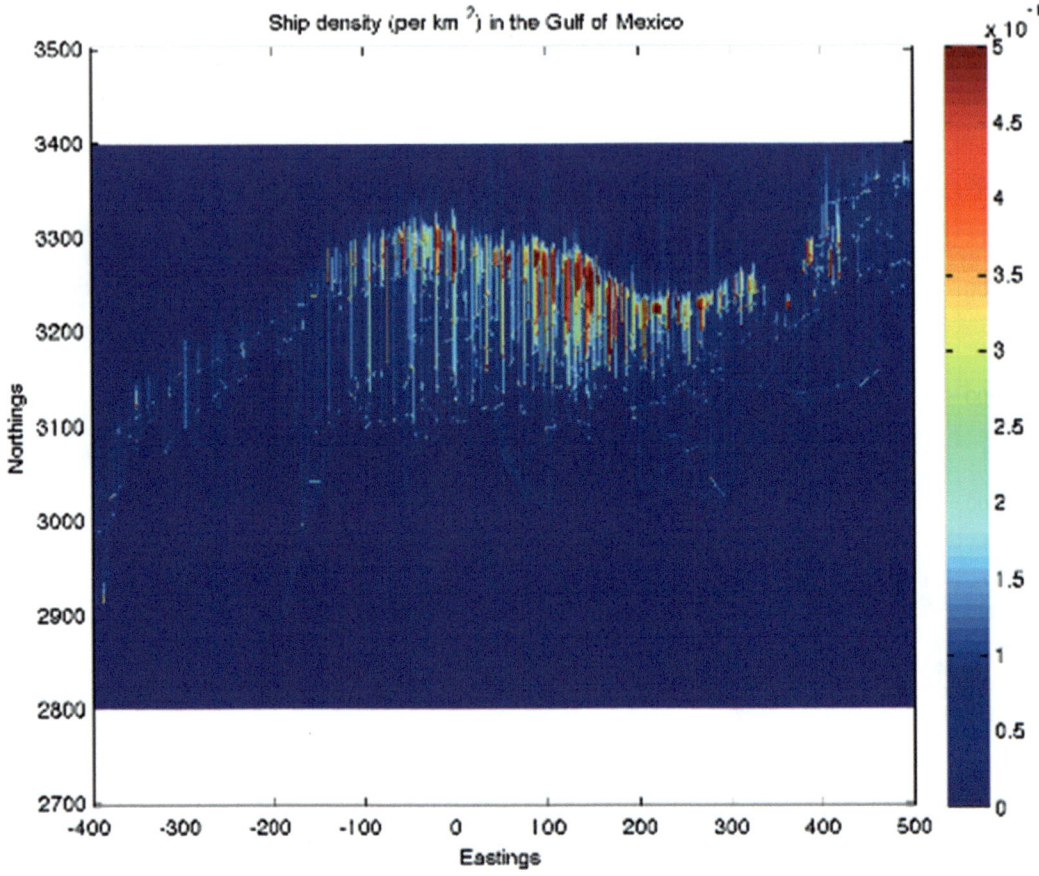

Fig. 1.55 Ship density in the Gulf of Mexico derived from the ship tracks

survivability of navy ships in combat. Explosives are also used in marine construction and in rig decommissioning. Finally, explosive charges called SUS (signal underwater sound) are routinely used in surveys to derive information about the reflectivity of the ocean bottom for subsequent use in propagation models.

These sources warrant particular attention because of several unique characteristics. First, even though they produce sound, they are not adequately characterized purely as acoustic sources. This is because they are so loud that the ocean no longer responds in a strictly linear way—acoustics is about small perturbations to the pressure field. One may compare this loosely to a small wave on the ocean that propagates as an undistorted sinewave vs. a large ocean wave that breaks. The physics is very different. However, as we get further from the origin of the explosion, the level weakens due to spherical spreading. At distances of perhaps hundreds of meters, the pulse is weak enough that it can be modeled with the standard acoustic codes. Thus, we have a nonlinear zone near the explosive where special models (hydrocodes) are required, which may be connected to an outer zone where acoustic models can be used. The term "finite amplitude" is also used to suggest that the sources are not just small perturbations.

A second interesting feature of explosive sources (shared with airguns) is that the explosion first forms an expanding gas (from the explosive and water vapor) bubble, which expands to a certain radius before collapsing. In fact, the gas bubble may go through several cycles of expansion and collapse. Because the bubble is buoyant,

Fig. 1.56 Bulge in the pile wall due to a hammer strike at the top of the pile. The red and blue circles depict the bulge at an early time and a later time when the bulge has propagated a distance L down the pile. Reprinted from Dahl PH, de Jong CAF, and Popper AN (2015) The underwater sound field from impact pile driving and its potential effects on marine life. Acoustics Today, 11(2). With permission from the ASA. © Acoustical Society of America, 2015. All rights reserved

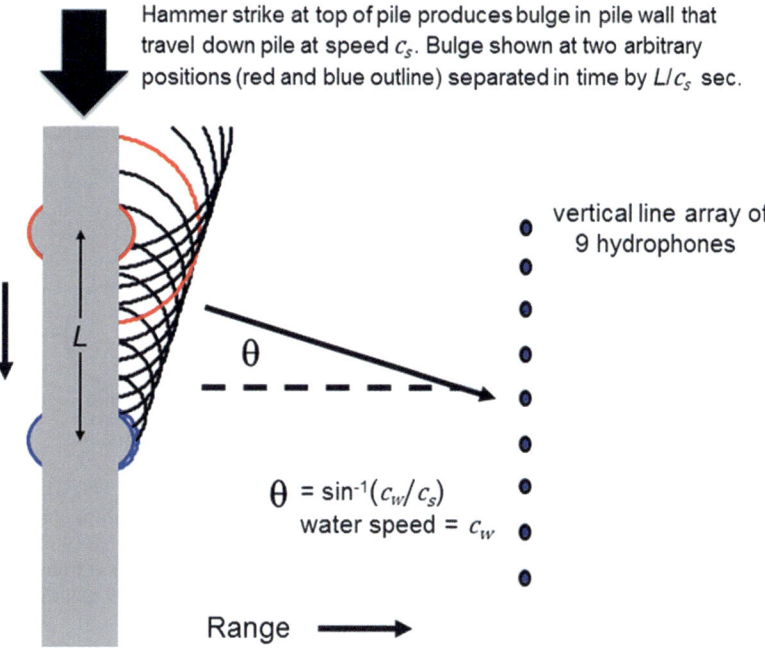

it rises to the surface and is therefore not characterized by a single source depth. This results in a series of impulses as shown in Fig. 1.57 where each impulse corresponds to a cycle of a collapsing and expanding gas bubble. Third, the wave that is reflected from the ocean surface often produces a huge drop in pressure. Water, of course, boils at a lower temperature at high altitudes where the pressure is low. In a similar way, with these explosive conditions where parts of the wave have extremely low pressures, gas boils out of the water causing many small bubbles. This phenomenon is called *cavitation*. The same effect often happens near ship propellers in zones of low pressure around the blades (Fig. 1.58). This cavitated region illustrated in Fig. 1.59 produces peculiar propagation effects since the wave is propagating through a bubbly liquid.

A precise modeling of all these complicated features is difficult. A widely used shortcut is based on the principle of *similarity*, which is basically a statement of simple scaling laws that describe the near-field evolution of the shock pulse in free space. These scaling laws are

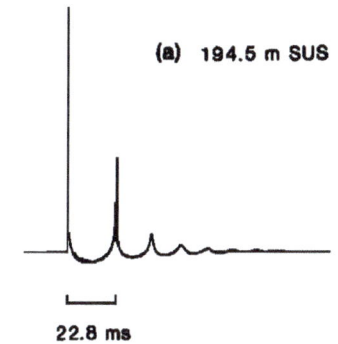

Fig. 1.57 Pressure history from a SUS charge showing the initial shockwave pulse and a subsequent series of bubble pulses. Reprinted from Chapman NR (1985) Measurement of the waveform parameters of shallow explosive charges. J Acoust Soc Am 78(2): 672–681. https://doi.org/10.1121/1.392436. With permission from the ASA. © Acoustical Society of America, 2019. All rights reserved

observed features and not derived directly from fundamental physics. We omit the full details here; however, Chapman (1985) suggests for the first shockwave, a peak pressure as

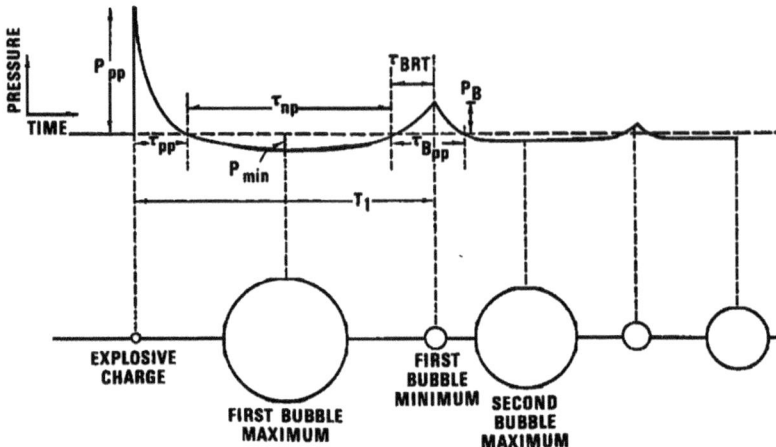

Fig. 1.58 Idealized pressure vs. time. The waveform due to a SUS charge is characterized by time constants characterizing the decay and rise time of various parts, the peak levels as well as the time delays between segments. Reprinted from Gaspin JB, Goertner JA, and Blatstein I (1979) The determination of acoustic source levels for shallow underwater explosions. J Acoust Soc Am 66(5):1453–1462. https://doi.org/10.1121/1.383539. With permission from the ASA. © Acoustical Society of America, 1979. All rights reserved

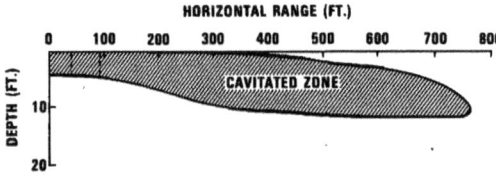

Fig. 1.59 Cavitated zone around a 3.5 kg charge detonated at 20 m. Reprinted from Gaspin JB, Goertner JA, and Blatstein I (1979) The determination of acoustic source levels for shallow underwater explosions. J Acoust Soc Am 66(5):1453–1462. https://doi.org/10.1121/1.383539. With permission from the ASA. © Acoustical Society of America, 1979. All rights reserved

$$P_s = 5.04 \times 10^{13} \left(w^{\frac{1}{3}} R \right)^{1.13}$$

and a decay time as

$$\tau_s = 8.12 \times 10^{-5} w^{\frac{1}{3}} \left(\frac{w^{1/3}}{R} \right)^{-0.14}$$

where w is the mass of the charge in kg, and R is the slant range (spherical range) from the charge in meters. Note that the peak level of the explosive depends on the cube root of the mass of the charge but varies in range in a way that deviates from the usual spherical spreading. These scaling formulas have been shown to be remarkably accurate over a wide range of conditions as shown in Fig. 1.60 reprinted from Soloway and Dahl (2014).

These formulae characterize an explosion in a homogeneous ocean without boundaries (free space). To couple them to a propagation model, we evaluate this free space field at a distance where the pressure should be low enough to treat the propagation using acoustic models. Then we can increase that level to remove spherical spreading. This provides a modified source level for the explosive, which, when propagated through the acoustic model, yields the levels suggested by the above formulae. However, the propagation model can then include all the boundary reflections and multipath effects. In summary, we use the nonlinear formulae above to understand what the field would be in free space but use the linear (acoustic) formulae to reverse those back to the source location, so that the acoustic formulae will give the correct far-field result. This process can be done with any of the time-domain modeling approaches; however, for computational reasons, it is most relevant to ray/beam methods.

Another approach to explosives that has some currency is based on an extension of the acoustic parabolic equation to include the nonlinear effects

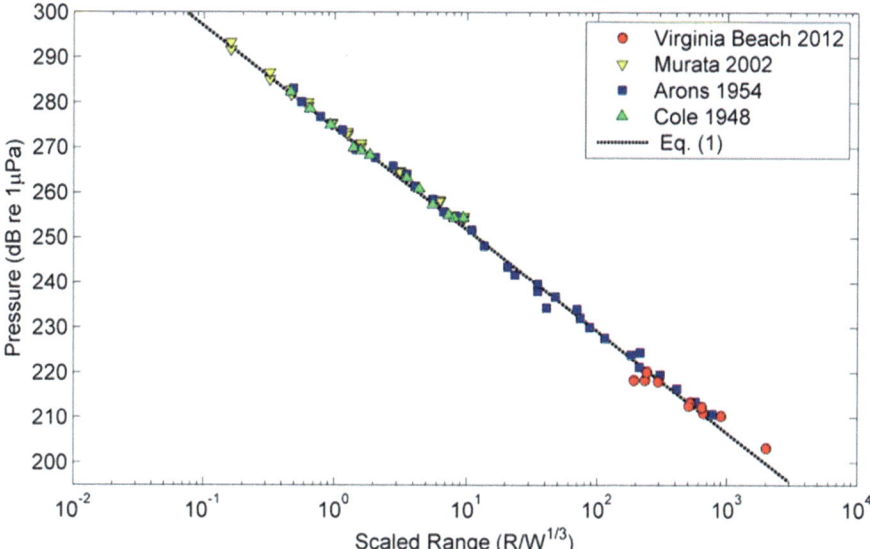

Fig. 1.60 Peak pressure from a variety of explosives plotted against a range scaled by the size of the charge. Reprinted from Soloway AG, Dahl PH (2014) Peak sound pressure and sound exposure level from underwater explosions in shallow water. J Acoust Soc Am 136(3): EL218–223. https://doi.org/10.1121/1.4892668. With permission from the ASA. © Acoustical Society of America, 2019. All rights reserved

of an explosive. This nonlinear parabolic equation (PE) is described by McDonald and Kuperman (1987). Finally, we mention the REFMS model (Britt et al. 1991) that has been used extensively to model large explosions such as those used in ship shock trials.

The above discussion describes how to model the waveform due to an explosion in a waveguide. This is relevant in cases where one is interested in metrics such as peak pressure. In other cases, the metric of interest is the energy across a certain frequency band, for example, one-third octaves, in which case it can be easier to work directly with the spectral levels and use frequency-domain acoustic models. Measurements of explosive source levels vs. frequency and depth are given by Chapman (1988) and Gaspin et al. (1979).

1.17.7 Seismic Exploration and Directional Sources

As discussed above, seismic exploration is widely used to characterize the sub-bottom for oil and gas exploration and is a significant source of sound in the marine environment. This is discussed more completely in Chap. 2. The receiving arrays typically involve tens of thousands of elements on multiple towed arrays, and as many as four ships. The towed arrays can be many kilometers in length and may be towed in complicated patterns performing a complex ballet. For instance, in coil surveys, the arrays are towed in loops with continuously shifting centers.

For our interest here, the source arrays are more important. Typically, these are airgun arrays where a chamber of pressurized air is suddenly opened. The resulting gas bubble behaves very much like the chemical explosive bubbles described previously, with much of the same complicated physics. New systems that use long-duration waveforms are under development. These waveforms distribute the sound energy in time, thereby reducing the peak sound level. Through signal processing techniques (pulse compression), the distributed energy can be mathematically recombined to be equivalent to a single loud pulse. It is assumed that there is less impact on the marine environment from a waveform that has the same energy distributed in time.

Relative to a chemical explosive, a key difference with airgun arrays is that they are not point sources, but rather distributions of point sources. Further, since the arrays are designed to explore the sub-bottom of the seafloor, they are specifically designed to radiate energy vertically downward, rather than horizontally along the ocean channel. This is accomplished by timing the airgun explosions so that they produce a beam pattern that mostly points downward.

The modeling of these airgun arrays is complicated. We refer the reader to the paper by Ainslie et al. (2016), which reports the results of a recent international workshop devoted to this topic. Two widely used models for the airgun arrays are *Gundalf* and *Nucleus*. These are commercially licensed. Other proprietary models have been developed. An open-source model called *Agora* was developed by Sertlek and Ainslie (2015). These models deal with the complicated nonlinear effects of the airgun; however, they generally consider an environment where the airgun is in a halfspace bounded by the ocean surface. Loosely speaking, we may say they consider the direct wave and the surface reflection.

To insert the information from these airgun models into the channel models, one must be careful not to effectively incorporate the ocean twice since the airgun model is already assuming a simple halfspace model of the ocean. The formally correct treatment of this problem requires that the near-field model supplies the pressure and its normal derivative on a surface. That surface should be at sufficient range that the propagation is accurately treated by an acoustic model. (In some cases, the near-field model may not be capable of propagating the field to a range where nonlinear effects are negligible.)

The Helmholtz–Kirchhoff integral expresses the field further down range in terms of standard propagation models. Loosely speaking, this Helmholtz–Kirchhoff integral is a statement of Huygens' principle; if we know the field on a surface, then we can propagate it onward by treating each point on the surface as an acoustic radiator. However, Huygens' principle is a heuristic that neglects the terms involving the normal derivative of the pressure. In either case, one may use the near-field model to calculate the pressure field on a vertical line (equivalent to a cylinder in three dimensions) and use that to initialize another acoustic model. Other shortcuts involve using a point source with a radiation pattern. If the ranges of interest are large, one may further simplify the computation by averaging the radiation pattern over the shallow angles that are relevant for long-range propagation.

1.18 Other Applications of Propagation Models

1.18.1 Animats

Quantitative predictions of a marine mammal's exposure to the operation of a noise source are sometimes required by government agencies (e.g., in environmental compliance) or are of interest to the scientific community. Multiple methods have been proposed for modeling marine mammal noise exposures over the last 25 years with the methods increasing in sophistication over time. Early models assumed animals were distributed in the environment according to some density function that was subsequently compared to a sound propagation model to estimate the number of animals exposed and the level of the noise exposure. The simplest models assumed that animals either received the highest level of sound that occurred at a particular range from the modeled sound source (i.e., the highest level as predicted across all depths at that range) or split the water column into a deep and shallow layer and applied the same maximum level of received sound within the layer (U.S. Department of the Navy 1998; U.S. Department of the Navy and NMFS 2001).

The use of *animats*, which are computer-simulated marine mammals, is an alternative and intuitive means by which noise exposure can be modeled for individual receivers as opposed to approximating via density distributions. An animat can be considered a spatially dynamic receiver within a range-dependent or independent

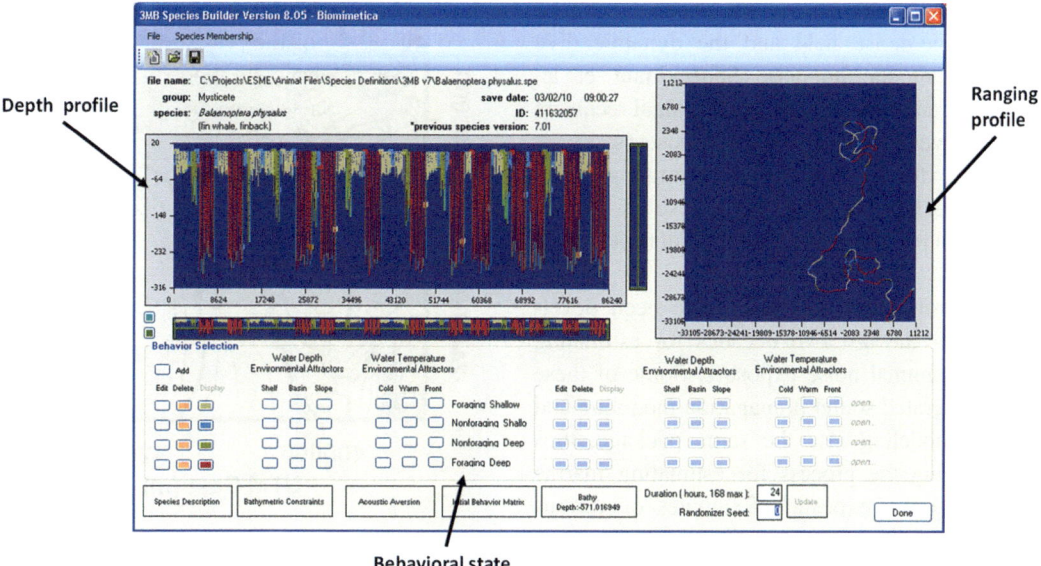

Fig. 1.61 Example of an animat model created with the Marine Mammal Movement and Behavior (3 MB) program (Houser and Cross 2011). This model is for a fin whale and has four defined behavioral states: foraging shallow, non-foraging shallow, foraging deep, and non-foraging deep. The depth (dive) profile and horizontal movement (ranging profile) vary as a function of the behavioral state, each of which contains a set of parameters that define the limits of the behavior. Each behavior state can be considered a bound stochastic process with user-defined transitional probabilities between behavioral states

environment that has movement governed by known or assumed diving and ranging behaviors of a particular marine mammal species (Houser 2006). The degree to which the behavior emulates that of a real marine mammal is based largely on the type of information that is available for the species. Some species have been studied extensively with deployable data loggers that measure dive depth, swim speed, orientation, and at-sea location. For these species, fairly detailed behavior models can be developed (see Fig. 5.27 as an example). Other species remain relatively unstudied and many assumptions about their behavior must be made, thus increasing the uncertainty in the model (Fig. 1.61).

The simulated environments in which sound propagation models are created can be seeded with animats. Distribution can be nonuniform based on dive behavior and known or estimated species' densities, distributions, and grouping behavior. The animats then range and dive over the time period during which the modeled sound sources are assumed to operate. Upon each noise event (source transmission), the location of the animat in depth and distance from the source can be compared to the transmission loss at the same location. Since records of exposure at the animat can be kept over time, estimates of exposure often include both instantaneous and cumulative metrics. The most common exposure metrics include the maximum instantaneous sound pressure level and the sound exposure level, as they are often used in regulatory estimates of impact.

The inclusion of time-varying receiver depth was a critical advancement in exposure estimation as the noise field typically varies more dramatically with depth than with range, and marine mammals can regularly and significantly change their depth over relatively short periods of time (e.g., ascent and descent rates of ~2 m/s are common in marine mammals). Comparison of animat models with static distribution models demonstrated that the latter underestimated exposure levels because it failed to account for the interaction between the depth-dependent variation in the noise field and the dynamic dive behavior of the animals (Schecklman et al. 2011). In addition, the use of individual receivers has allowed for worst-case exposures to be explored, for example, by determining when avoidance behavior and the course of mobile sound source might interact to increase noise exposure.

Numerous animat models have been developed over the last two decades for estimating marine mammal noise exposure. Some of these have integrated sound propagation modeling and animat modeling into the same package, thus streamlining the process for estimating marine mammal noise exposure. Examples of various animat-based models include the U.S. Navy Acoustic Effects Model (NAEMO), the Marine Mammal Movement and Behavior (3 MB), the Effects of Sound on the Marine Environment (ESME), the Acoustic Integration Model (AIM), and the Statistical Algorithms for Estimating the Sonar Influence on Marine Megafauna (SAFESIMM). Animat-based approaches are now widely implemented in regulatory compliance and scientific investigation and are commercially available.

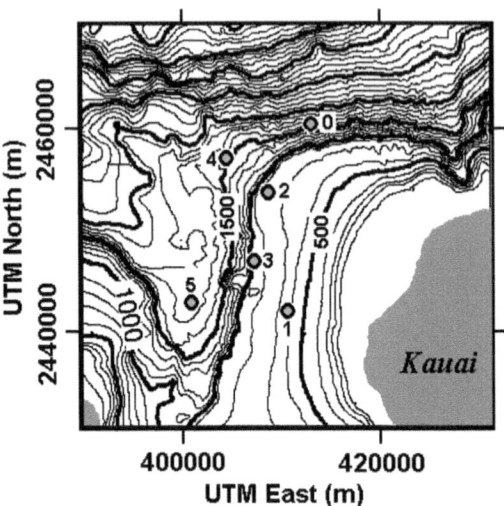

Fig. 1.62 Bathymetry contours (depths in meters) and hydrophone locations (0–5) at the Pacific Missile Range Facility. Axes are for Universal Transverse Mercator (UTM) Zone 4 coordinates. Reprinted from Tiemann CO, Porter MB, Frazer LN (2004) Localization of marine mammals near Hawaii using an acoustic propagation model. J Acoust Soc Am 115(6):2834–2843. https://doi.org/10.1121/1.1643368. With permission from the ASA. © Acoustical Society of America, 2004. All rights reserved

1.18.2 Tracking Marine Mammals

Marine mammals can be tracked acoustically by many methods depending largely on the sensor systems that are available. Here, we address one of the more common approaches, which is sometimes called *hyperbolic fixing*. Our example is based on measurements conducted at the Pacific Missile Range Facility off Kauai (Tiemann et al. 2004). Six bottom-mounted hydrophones were available as shown in Fig. 1.62.

The calls of humpback whales were audible in the recordings. However, because an individual whale is at different distances from particular hydrophones, there is a time lag between when the same call is heard on different phones. For instance, Fig. 1.63 shows the spectrograms of a sequence of calls on hydrophones 2 and 4. There is a clear correspondence in the calls, but with an offset in time. This offset can be estimated by a human visually or estimated mathematically by cross-correlating the two spectrograms and choosing the lag that maximizes the correlation. (The correlation can also be done directly with the time series.)

We digress a moment here to point out that these spectrograms do not show pristine copies of the whale calls. Instead, they show the spectrogram of the received waveform with all its echoes. Technically, that received waveform is a convolution of the ocean channel response with the source waveform (the whale call). So, some of the structure is due to the whale call and some of the structure due to propagation effects. The propagation effects are frequency-dependent, leading to what is sometimes called multi-path fading or frequency-selective filtering.

It is useful to consider briefly the propagation conditions for this site. Figure 1.64 shows the ray paths that connect the whale to a couple of

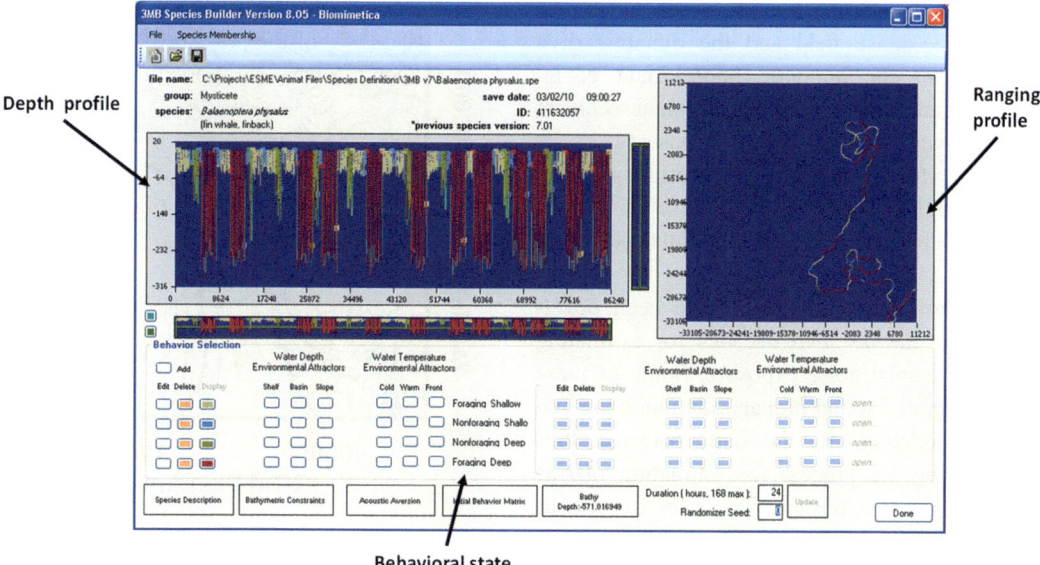

Fig. 1.61 Example of an animat model created with the Marine Mammal Movement and Behavior (3 MB) program (Houser and Cross 2011). This model is for a fin whale and has four defined behavioral states: foraging shallow, non-foraging shallow, foraging deep, and non-foraging deep. The depth (dive) profile and horizontal movement (ranging profile) vary as a function of the behavioral state, each of which contains a set of parameters that define the limits of the behavior. Each behavior state can be considered a bound stochastic process with user-defined transitional probabilities between behavioral states

environment that has movement governed by known or assumed diving and ranging behaviors of a particular marine mammal species (Houser 2006). The degree to which the behavior emulates that of a real marine mammal is based largely on the type of information that is available for the species. Some species have been studied extensively with deployable data loggers that measure dive depth, swim speed, orientation, and at-sea location. For these species, fairly detailed behavior models can be developed (see Fig. 5.27 as an example). Other species remain relatively unstudied and many assumptions about their behavior must be made, thus increasing the uncertainty in the model (Fig. 1.61).

The simulated environments in which sound propagation models are created can be seeded with animats. Distribution can be nonuniform based on dive behavior and known or estimated species' densities, distributions, and grouping behavior. The animats then range and dive over the time period during which the modeled sound sources are assumed to operate. Upon each noise event (source transmission), the location of the animat in depth and distance from the source can be compared to the transmission loss at the same location. Since records of exposure at the animat can be kept over time, estimates of exposure often include both instantaneous and cumulative metrics. The most common exposure metrics include the maximum instantaneous sound pressure level and the sound exposure level, as they are often used in regulatory estimates of impact.

The inclusion of time-varying receiver depth was a critical advancement in exposure estimation as the noise field typically varies more dramatically with depth than with range, and marine mammals can regularly and significantly change their depth over relatively short periods of time (e.g., ascent and descent rates of ~2 m/s are common in marine mammals). Comparison of animat models with static distribution models demonstrated that the latter underestimated exposure levels because it failed to account for the interaction between the depth-dependent variation in the noise field and the dynamic dive behavior of the animals (Schecklman et al. 2011). In addition, the use of individual receivers has allowed for worst-case exposures to be explored, for example, by determining when avoidance behavior and the course of mobile sound source might interact to increase noise exposure.

Numerous animat models have been developed over the last two decades for estimating marine mammal noise exposure. Some of these have integrated sound propagation modeling and animat modeling into the same package, thus streamlining the process for estimating marine mammal noise exposure. Examples of various animat-based models include the U.S. Navy Acoustic Effects Model (NAEMO), the Marine Mammal Movement and Behavior (3 MB), the Effects of Sound on the Marine Environment (ESME), the Acoustic Integration Model (AIM), and the Statistical Algorithms for Estimating the Sonar Influence on Marine Megafauna (SAFESIMM). Animat-based approaches are now widely implemented in regulatory compliance and scientific investigation and are commercially available.

1.18.2 Tracking Marine Mammals

Marine mammals can be tracked acoustically by many methods depending largely on the sensor systems that are available. Here, we address one of the more common approaches, which is sometimes called *hyperbolic fixing*. Our example is based on measurements conducted at the Pacific Missile Range Facility off Kauai (Tiemann et al. 2004). Six bottom-mounted hydrophones were available as shown in Fig. 1.62.

The calls of humpback whales were audible in the recordings. However, because an individual whale is at different distances from particular hydrophones, there is a time lag between when the same call is heard on different phones. For instance, Fig. 1.63 shows the spectrograms of a sequence of calls on hydrophones 2 and 4. There is a clear correspondence in the calls, but with an

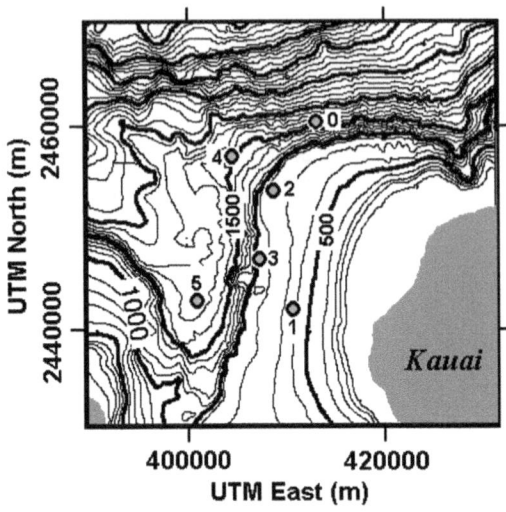

Fig. 1.62 Bathymetry contours (depths in meters) and hydrophone locations (0–5) at the Pacific Missile Range Facility. Axes are for Universal Transverse Mercator (UTM) Zone 4 coordinates. Reprinted from Tiemann CO, Porter MB, Frazer LN (2004) Localization of marine mammals near Hawaii using an acoustic propagation model. J Acoust Soc Am 115(6):2834–2843. https://doi.org/10.1121/1.1643368. With permission from the ASA. © Acoustical Society of America, 2004. All rights reserved

offset in time. This offset can be estimated by a human visually or estimated mathematically by cross-correlating the two spectrograms and choosing the lag that maximizes the correlation. (The correlation can also be done directly with the time series.)

We digress a moment here to point out that these spectrograms do not show pristine copies of the whale calls. Instead, they show the spectrogram of the received waveform with all its echoes. Technically, that received waveform is a convolution of the ocean channel response with the source waveform (the whale call). So, some of the structure is due to the whale call and some of the structure due to propagation effects. The propagation effects are frequency-dependent, leading to what is sometimes called multi-path fading or frequency-selective filtering.

It is useful to consider briefly the propagation conditions for this site. Figure 1.64 shows the ray paths that connect the whale to a couple of

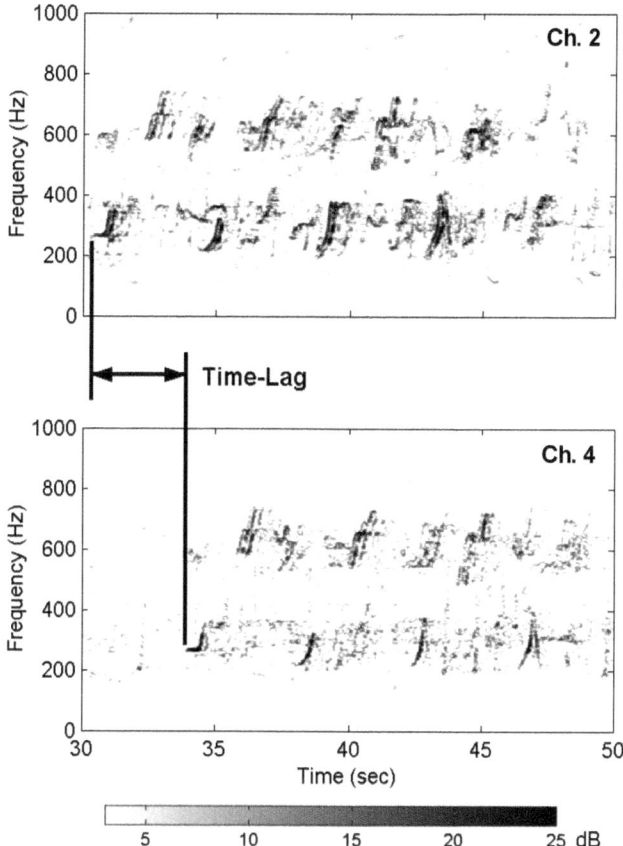

Fig. 1.63 Spectrograms of acoustic data from hydrophones 2 (top) and 4 (bottom) starting at time 20:16:30 on 3/22/01. A 3.5-s time-lag for spectral transients is apparent between the two spectrograms. Spectral patterns resemble those of humpback whale calls. Reprinted from Tiemann CO, Porter MB, Frazer LN (2004) Localization of marine mammals near Hawaii using an acoustic propagation model. J Acoust Soc Am 115(6):2834–2843. https://doi.org/10.1121/1.1643368. With permission from the ASA. © Acoustical Society of America, 2004. All rights reserved

example hydrophones. We can see that there are direct paths that involve no boundary reflections. However, there are also quite a few paths with reflections off the boundaries. This physics suggests several approaches to estimating the whale location that consider varying levels of sophistication in the propagation modeling. The simplest approach assumes some nominal propagation speed, for example, 1500 m/s and converts the time lag (the difference in time) to a difference in distance. The curves with a constant difference in distance are hyperbolas and there is one hyperbola for each pair of hydrophones. One then plots the hyperbolas as shown in Figure 1.65a, and their point of intersection identifies the whale's location. This is the standard form of hyperbolic fixing. In practice, the hyperbolas will not all intersect at a single point since the effective propagation speed will not be the same for all the receivers. However, if the differences are not great, then there will be a cluster of intercepts in a small area.

To improve further on this basic process, we need to bring in a more sophisticated propagation model. Our first improvement is to use a range-independent ray/beam model that predicts the arrival times for all the echoes of the whale call. (Range-independent here means that the bottom is

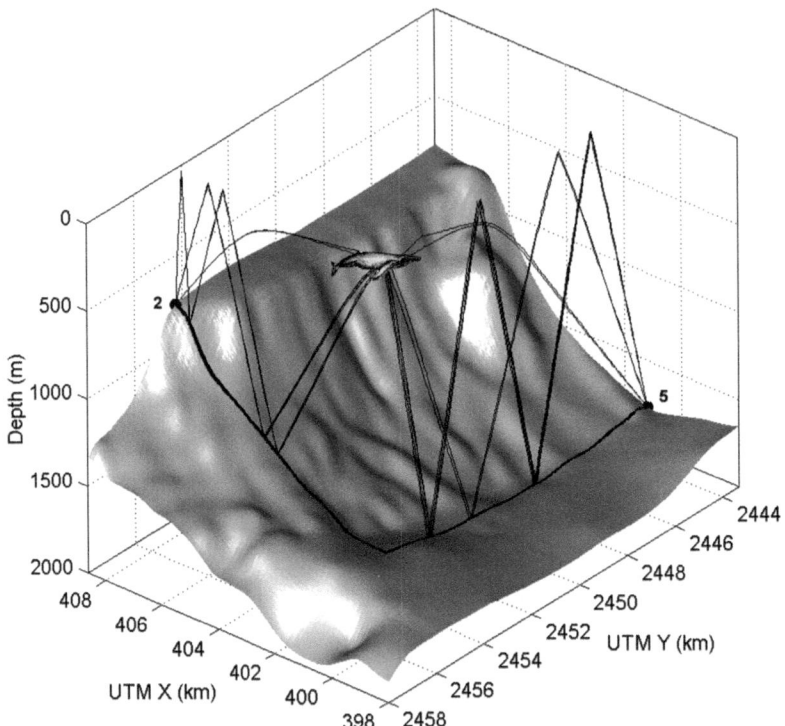

Fig. 1.64 Predicted direct and reflected acoustic ray paths between a shallow whale and hydrophones #2 and #5 along two perpendicular bathymetry slices. Whale not drawn to scale. Reprinted from Tiemann CO, Porter MB, Frazer LN (2004) Localization of marine mammals near Hawaii using an acoustic propagation model. J Acoust Soc Am 115(6):2834–2843. https://doi.org/10.1121/1.1643368. With permission from the ASA. © Acoustical Society of America, 2004. All rights reserved

flat, but we include the depth dependence of the sound speed.) These arrival times will, of course, depend on the whale's range (and depth). The acoustic model predicts many echoes; however, the process used on the recordings of the whale call generates a single time lag that is sort of an aggregate of all the echoes. To compare to the acoustic model then, we do a similar aggregation, averaging the various time delays. The averaging is actually a weighted average that takes into account the loudness of the different echoes. With these averaged travel times to each hydrophone, we can then calculate a predicted time lag between any pair of hydrophones. The final step is to plot an *ambiguity surface* that indicates the similarity between the modeled and observed time lags for each pair of hydrophones. The result is shown in Fig. 1.65b; we can see how this process produces a clear peak at the location of the whale.

An obvious next step up in the sophistication of the propagation model is to incorporate the bathymetry. This is clearly more complicated in that we must predict the travel time for every possible latitude, longitude, and depth of the whale. The result of this process is shown in Fig. 1.65c.

In this case, all three approaches are effective in localizing the whale. However, in general, the simplest model (hyperbolic fixing) will be degraded by its ignorance of the multipath structure. These are just a few of many approaches that can be used to localize a source. If vertical line arrays are available, then one can compare arrival angles between model and data to find a position with the best match. Similarly, if horizontal arrays are available, one can look for the intersection of beam angles in azimuth.

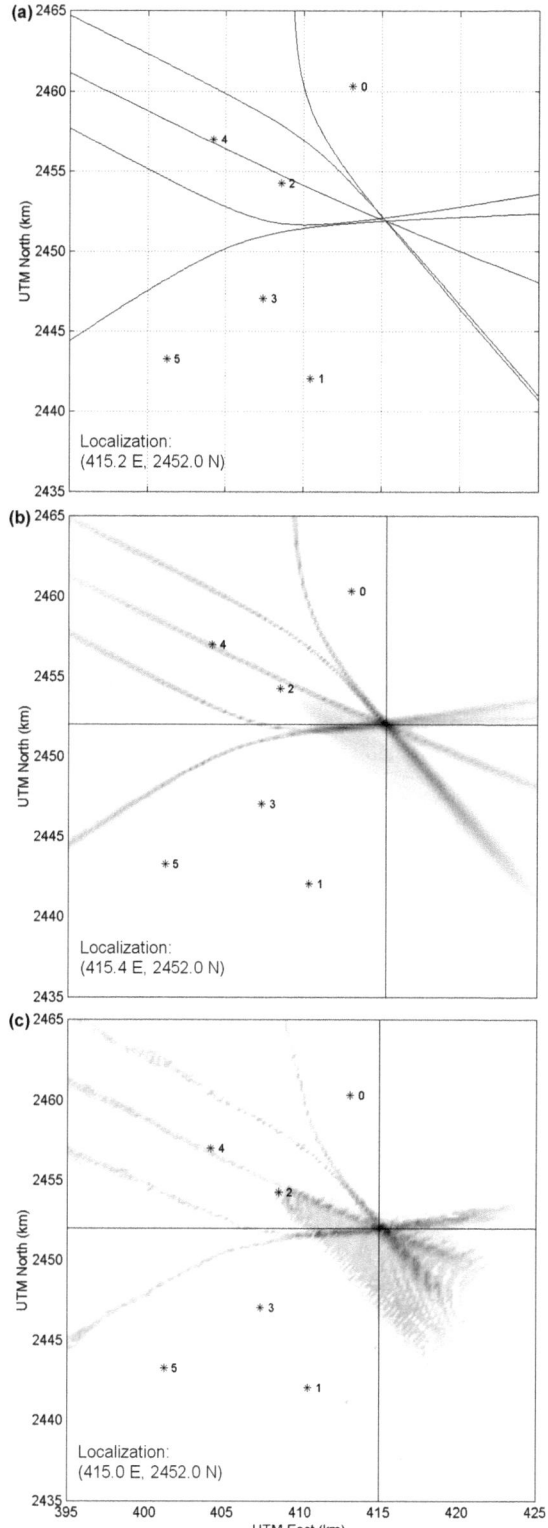

Fig. 1.65 Plan views of the waters around the PMRF array with hydrophone positions (0–5) indicated. Axes are for UTM Zone 4. Sub-plots show whale position estimated by (**a**) hyperbolic fixing, (**b**) range-independent modeling, (**c**) range-dependent modeling. Reprinted from Tiemann CO, Porter MB, Frazer LN (2004) Localization of marine mammals near Hawaii using an acoustic propagation model. J Acoust Soc Am 115(6):2834–2843. https://doi.org/10.1121/1.1643368. With permission from the ASA. © Acoustical Society of America, 2004. All rights reserved

1.19 Summary

The propagation of sound in the ocean involves many complicated effects. At a basic level, the sound reflects off the ocean surface and ocean bottom, producing a sort of barber-shop mirror effect that leads to multiple images of the sound source. The strength of these echoes depends on the quality of the surface and bottom reflectors. A rough sea is more like a broken mirror that scatters sound energy; a soft silty bottom is a weak reflector that may be compared to a dark mirror. However, it gets still more complicated. The ocean sound speed depends on temperature, salinity, and depth. Gradients in these environmental parameters cause the sound to bend leading, for instance, to "convergence zone" propagation in which the sound refocuses near the ocean surface in loops spaced some 55 km apart. Modeling these effects is the province of computational ocean acoustics. This is a field that has been extensively studied over decades for navy sonar applications; however, it continues to be an active area of research. The basic physics of sound propagation in the ocean and suitable computational approaches were explained in this chapter.

1.20 Abbreviations and Symbols

Table 1.4 lists the quantities defined in this chapter. For some of the more common quantities, abbreviations are given. The symbols that are used in equations and the measurement units are also listed. Several of the definitions are based on ISO 18405 (International Organization for Standardization, 2017) and ANSI/ASA S1.1 (American National Standards Institute, 2013).

Table 1.4 Quantities, abbreviations, symbols, and units used in this chapter

Quantity	Abbreviation	Symbol	Unit
Frequency		f	Hz
Angular frequency		ω	rad/s
Wavelength		λ	m
Wavenumber		k	
Amplitude		A	
Phase		ϕ	
Speed of sound		c	m/s
Density		ρ	kg/m^3
Temperature		T	°C
Salinity		S	psu
Water depth		D	m
Depth variable		z	m
Source depth		z_S	m
Range variable		r	m
Spherical range		R	m
Euler's number, exponential function		e	
Imaginary unit		i	
Hankel function, first kind, 0 order		$H_0^{(1)}$	
Bessel function, first kind, 0 order		J_0	
Eigenfunction		Ψ	
Nabla, vector differential operator		∇	
Partial differential		∂	
Dirac or Kronecker delta		δ	
Absorption coefficient		α	
Period of oscillation		τ	s

(continued)

Table 1.4 (continued)

Quantity	Abbreviation	Symbol	Unit
Time variable		t	s
Sound pressure		$p(t)$	Pa
Peak sound pressure		p_{pk}	Pa
Peak-to-peak sound pressure		$p_{pk\text{-}pk}$	Pa
Root-mean-square sound pressure		p_{rms}	Pa
Sound pressure level[a]	SPL	L_p	dB re 1 µPa
Peak sound pressure level[a]	SPLpk	$L_{p,pk}$	dB re 1 µPa
Sound exposure level[a]	SEL	$L_{E,p}$	dB re 1 µPa^{2}s
Radiated noise level[a]	RNL	L_{RN}	dB re 1 µPa m
Source level[a]	SL	L_S	dB re 1 µPa m
Propagation loss[a]	PL	N_{PL}	dB re 1 m
Intensity		I	W/m^2
Equivalent continuous SPL		L_{eq}	dB re 1 µPa

[a]Level quantities are logarithms of ratios and as such unit-less. They are expressed in decibel (dB) relative to a reference value

Acknowledgments This work was supported in part by the U.S. Office of Naval Research Grant N00014-22201-2052. We have benefitted from comments on this manuscript from various reviewers and would like to particularly thank Ross Chapman (Univ. of Victoria), Emanuel Coelho (Applied Ocean Sciences, Inc.), Peter Dahl (Univ. of Washington), Alec Duncan (Curtin Univ.), Richard Evans (formerly Science Applications International, Inc.), William Farrell (formerly Science Applications International), Sergio Jesus (Univ. of the Algarve), William Kuperman (Scripps Institution of Oceanography), and Jie Yang (Univ. of Washington).

References

Ainslie MA, McColm JG (1998) A simplified formula for viscous and chemical absorption in sea water. Acoust Soc Am 103:1671–1672. https://doi.org/10.1121/1.421258

Ainslie M et al (2016) Verification of airgun sound field models for environmental impact assessment. Proc Meet Acoust 27:1. https://doi.org/10.1121/2.0000339

Badiey M, Katsnelson GG, Lynch JF, Pereselkov S, Siegmann WL (2005) Measurement and modeling of three-dimensional sound intensity variations due to shallow-water internal waves. J Acoust Soc Am 117(2):613–625. https://doi.org/10.1121/1.1828571

Badiey M, Katsnelson BG, Lin YT, Lynch JF (2011) Acoustic multipath arrivals in the horizontal plane due to approaching nonlinear internal waves. J Acoust Soc Am 129: EL141–EL147. https://doi.org/10.1121/1.3553374

Britt JR, Eubanks RJ, Lumsden MG (1991) Underwater shock wave reflection and refraction in deep and shallow water: volume I – a user's manual for the REFMS code (version 4.0). Science Applications International Corp. Technical report, DNR-TR-91-15-VI

Buckingham MJ, Potter JR (eds) (1996) Sea surface Sound'94-proceedings of the III international meeting on natural physical processes related to sea surface sound. World Scientific

Carey WM, Evans RB (2011) Ocean ambient noise: measurement and theory. Springer

Chapman NR (1985) Measurement of the waveform parameters of shallow explosive charges. J Acoust Soc Am 78(2):672–681. https://doi.org/10.1121/1.392436

Chapman NR (1988) Source levels of shallow explosive charges. J Acoust Soc Am 84:697–702. https://doi.org/10.1121/1.396849

Chapman NR, Cornish JS (1993) Wind dependence of deep ocean ambient noise at low frequencies. J Acoust Soc Am 93:782–789. https://doi.org/10.1121/1.405440

Dahl PH, de Jong CAF, Popper AN (2015) The underwater sound field from impact pile driving and its potential effects on marine life. Acoustics Today 11(2):18–25

Deane GB, Stokes MD (2002) Scale dependence of bubble creation mechanisms in breaking waves. Nature 418:839–844. https://doi.org/10.1038/nature00967

Eller AI, Cavanaugh RC (2000) Subsonic aircraft noise at and beneath the ocean surface: estimation of risk for effects on marine mammals. SAIC report for the US-Air Force Research Laboratory, AFRL-H-WP_TR_2000-0156

Erbe C, Schoeman RP, Peel D, Smith JN (2021) It often howls more than it chugs: wind versus ship noise under water in Australia's maritime regions. J Mar Sci Eng 9(5):472. https://doi.org/10.3390/jmse9050472

Ewing M, Worzel JL (1948) Long-range sound transmission. Geol Soc Am, Mem, p 27. https://doi.org/10.1130/MEM27-3-p1

Farcas A, Powell CF, Brookes KL, Merchant ND (2020) Validated shipping noise maps of the Northeast Atlantic. Sci Total Environ 735:139509. https://doi.org/10.1016/j.scitotenv.2020.139509

Fisher FH, Simmons VP (1977) Sound absorption in sea water. J Acoust Soc Am 62:558–564. https://doi.org/10.1121/1.381574

François RE, Garrison GR (1982) Sound absorption based on ocean measurements. Part II: boric acid contribution and equation for total absorption. J Acoust Soc Am 72:1879–1990. https://doi.org/10.1121/1.388673

Gaspin JB, Goertner JA, Blatstein I (1979) The determination of acoustic source levels for shallow underwater explosions. J Acoust Soc Am 66(5):1453–1462. https://doi.org/10.1121/1.383539

Gedamke J, et al. (2013) Predicting anthropogenic noise contributions to U.S. waters. In: Popper A, Hawkins A (eds) The effects of noise on aquatic life (conference proceedings), 2013. https://doi.org/10.1007/978-1-4939-2981-8_40

Harrison CH, Simons DG (2002) Geoacoustic inversion of ambient noise: a simple method. J Acoust Soc Am 112(4):1377–1389. https://doi.org/10.1121/1.1506365

Henderson L, Porter M (2015) Soundscape user guide. Heat, light, and sound research technical report

Hildebrand JA (2004) Sources of anthropogenic sound in the marine environment. Report to the policy on sound and marine mammals: an international workshop. US Marine Mammal Commission and Joint Nature Conservation Committee, UK

Houser DS (2006) A method for modeling marine mammal movement and behavior for environmental impact assessment. IEEE J Ocean Eng 31:76–81. https://doi.org/10.1109/JOE.2006.872204

Houser DS, Cross MJ (2011) Marine mammal movement and behavior (3MB). User Manual. National Marine Mammal Foundation, San Diego

Jensen F, Kuperman W, Porter M, Schmidt H (2012) Computational ocean acoustics, American Institute of Physics, New York (1994), reprint Springer-Verlag, (2000), second edition Springer-Verlag

Kerman BR (ed) (1987) Sea surface sound: natural mechanisms of surface generated noise in the ocean, vol 238. Springer

Kewley DJ, Browning DG, Carey WM (1990) Low-frequency wind-generated ambient noise source levels. J Acoust Soc Am 88(4):1894–1902. https://doi.org/10.1121/1.400212

Kuperman WA, Ingenito F (1980) Spatial correlation of surface generated noise in a stratified ocean. J Acoust Soc Am 67(6):1988–1996. https://doi.org/10.1121/1.384439

Kuperman WA, Porter MB, Perkins JS, Evans RB (1991) Rapid computation of acoustic fields in three-dimensional ocean environments. J Acoust Soc Am 89(1):125–133. https://doi.org/10.1121/1.400518

Leighton TG (1994) The acoustic bubble, Academic, London

McDonald BE, Kuperman WA (1987) Time domain formulation for pulse propagation including nonlinear behavior at a caustic. J Acoust Soc Am 81(5):1406–1417. https://doi.org/10.1121/1.394546

Munk WH (1974) Sound channel in an exponentially stratified ocean with applications to SOFAR. J Acoust Soc Am 55:220–226. https://doi.org/10.1121/1.1914492

Ocean Studies Board (2003) Ocean noise and marine mammals. National Research Council, The National Academies Press

Porter MB (2019) Beam tracing for two- and three-dimensional problems in ocean acoustics. J Acoust Soc Am 146:2016–2029. https://doi.org/10.1121/1.5125262

Porter MB, Henderson L (2011) Modeling and prediction of soundscapes. Open Science Meeting for an International Quiet Ocean Experiment, 30 August–1 September 2011, UNESCO Headquarters, Paris, France

Porter MB and Henderson LJ (2013) Global ocean soundscapes. Proc. Mtgs Acoust 19:010050. https://doi.org/10.1121/1.4801016

Redfern J, Hatch L, Caldow C, DeAngelis M, Gedamke J, Hastings S, Henderson L, McKenna M, Moore TJ, Porter M (2017) Assessing the risk of chronic noise from commercial ships to large whale acoustic habitat. Endanger Species Res 32:153–167. https://doi.org/10.3354/esr00797

Reinhall PG, Dahl PH (2011) Underwater Mach wave radiation from impact pile driving: theory and observation. J Acoust Soc Am 130(3):1209–1216. https://doi.org/10.1121/1.3614540

Schecklman S, Houser DS, Cross M, Hernandez D, Siderius M (2011) Comparison of methods used for computing the impact of sound on the marine environment. Mar Environ Res 71:342–350. https://doi.org/10.1016/j.marenvres.2011.03.002

Sertlek HÖ, Ainslie MA (2015) AGORA: Airgun source signature model: its application of seismic survey maps in the North Sea, UAC 2015, Crete, Greece

Sertlek HÖ, Slabbekoorn H, ten Cate C, Ainslie MA (2019) Source specific sound mapping: spatial, temporal and spectral distribution of sound in the Dutch North Sea. Environ Pollut. https://doi.org/10.1016/j.envpol.2019.01.119

Siderius M, Gebbie J (2019) Environmental information content of ocean ambient noise. J Acoust Soc Am 146(3):1824–1833. https://doi.org/10.1121/1.5126520

Soloway AG, Dahl PH (2014) Peak sound pressure and sound exposure level from underwater explosions in shallow water. J Acoust Soc Am 136(3):EL218–223. https://doi.org/10.1121/1.4892668

Straume EO, Gaina C, Medvedev S, Hochmuth K, Gohl K, Whittaker JM et al (2019) GlobSed: updated total sediment thickness in the world's oceans. Geochem Geophys Geosyst 20. https://doi.org/10.1029/2018GC008115

Thomsen F, McCully S, Weiss L, Wood D, Warr K, Barry J, Law R (2011) Cetacean stock assessments in relation to exploration and production industry activity and other human pressures: review and data needs. Aquat Mamm 37:1–93. https://doi.org/10.1578/am.37.1.2011.1

Thorp WH (1965) Deep-ocean sound attenuation in the sub- and low-kilocycle-per-second region. J Acoust Soc Am 38:648–654. https://doi.org/10.1121/1.1909768

Tiemann CO, Porter MB, Frazer LN (2004) Localization of marine mammals near Hawaii using an acoustic propagation model. J Acoust Soc Am 115(6):2834–2843. https://doi.org/10.1121/1.1643368

U.S. Department of the Navy (1998) Final environmental impact statement, shock testing the SEAWOLF submarine. Department of the Navy, Washington, DC

U.S. Department of the Navy and National Marine Fisheries Service (2001) Final environmental impact statement for the shock trial of the Winston S. Churchill (DDG 81)

Urick RJ (1983) Principles of underwater sound. McGraw Hill, New York

Urick RJ (1986) Ambient noise in the sea. Naval Sea Systems Command technical report, Department of the Navy, 1984. Also available from Peninsula Publishing, Los Altos

Wiggins SM, Hall JM, Thayre BJ, Hildebrand JA (2016) Gulf of Mexico low-frequency ocean soundscape impacted by airguns. J Acoust Soc Am 140(1):176–183. https://doi.org/10.1121/1.4955300

Wilson JH (1983) Wind-generated noise modeling. J Acoust Soc Am 73(1):211–216. https://doi.org/10.1121/1.388841

Yang J, Nystuen JA, Riser SC, Thorsos EI (2023) Open ocean ambient noise data in the frequency band of 100 Hz–50 kHz from the Pacific Ocean. J Acoust Soc Am Express Lett 3:036001. https://doi.org/10.1121/10.0017349

Open Access This chapter is licensed under the terms of the Creative Commons Attribution 4.0 International License (http://creativecommons.org/licenses/by/4.0/), which permits use, sharing, adaptation, distribution and reproduction in any medium or format, as long as you give appropriate credit to the original author(s) and the source, provide a link to the Creative Commons license and indicate if changes were made.

The images or other third party material in this chapter are included in the chapter's Creative Commons license, unless indicated otherwise in a credit line to the material. If material is not included in the chapter's Creative Commons license and your intended use is not permitted by statutory regulation or exceeds the permitted use, you will need to obtain permission directly from the copyright holder.

Sources of Underwater Noise

2

Christine Erbe, Alec J. Duncan, Alexander Gavrilov, Montserrat Landero, Robert D. McCauley, Iain Parnum, Chandra Salgado-Kent, and Evgeny Sidenko

Contents

2.1	**Introduction**	86
2.2	**Aquatic Soundscapes**	86
2.3	**Geophysical Sound Sources**	88
2.3.1	Wind	88
2.3.2	Precipitation	90
2.3.3	Surf	90
2.3.4	Subsea Earthquakes and Volcanoes	92
2.3.5	Polar Ice	93
2.3.6	Thermal Agitation	95
2.3.7	Current-Induced Noise Artifacts	95
2.4	**Biotic Sound Sources**	95
2.4.1	Marine Invertebrates	95
2.4.2	Fishes	97
2.4.3	Marine Mammals	98
2.5	**Anthropogenic Sound Sources**	100
2.5.1	Ship Traffic	101
2.5.2	Marine Seismic Surveys	114
2.5.3	Drilling	124
2.5.4	Dredging	128
2.5.5	Pile Driving	133
2.5.6	Offshore Renewable Energy Plants in Operation	139

C. Erbe (✉) · A. J. Duncan · A. Gavrilov · M. Landero · R. D. McCauley · I. Parnum · E. Sidenko
Centre for Marine Science and Technology, Curtin University, Perth, WA, Australia
e-mail: c.erbe@curtin.edu.au; A.J.Duncan@curtin.edu.au; A.Gavrilov@curtin.edu.au; R.McCauley@cmst.curtin.edu.au; I.Parnum@curtin.edu.au; evgeny.sidenko@curtin.edu.au

C. Salgado-Kent
Centre for Marine Ecosystems Research, School of Science, Edith Cowan University, Joondalup, WA, Australia
e-mail: c.salgadokent@ecu.edu.au

© The Author(s) 2025
C. Erbe et al. (eds.), *Marine Mammal Acoustics in a Noisy Ocean*,
https://doi.org/10.1007/978-3-031-77022-7_2

2.5.7	Geotechnical Site Investigations	140
2.5.8	Sonar and Echosounder Systems	146
2.5.9	Explosions	155
2.5.10	Acoustic Mitigation Devices	157
2.6	**Summary**	159
2.7	**Glossary**	160
	References	162

2.1 Introduction

Marine mammals live in habitats that are naturally noisy. Wind blowing over the sea surface, waves breaking in strong winds and in the surf zone, rain falling onto the sea surface, sea ice cracking, icebergs calving, subsea earthquakes and volcanoes rumbling—all may result in noise under water. There are additional noises produced by marine fauna, such as snapping shrimp, munching urchins, chorusing fish, and other marine mammals. Since the industrial revolution, human activities have added anthropogenic noise to the oceans, notably through the mechanization of shipping and the commencement of industrial activities in and around the oceans. Marine mammals live in complex soundscapes (e.g., Schoeman et al. 2022a) and the same species of marine mammal may be found in greatly different soundscapes (e.g., Marley et al. 2017). Soundscapes are dynamic and vary in space and time. They provide context for potential responses by marine mammals to specific sources.

The concern is that noise might interfere with marine mammal communication, environmental sensing, acoustic and non-acoustic behaviors, and ultimately, life functions. In order to estimate the impacts of noise on marine fauna, the acoustic characteristics of the different sources of noise need to be quantified. Noise specifications may then be used in environmental impact assessments of underwater noise, in environmental management plans, and in future soundscape predictions. For example, a modeling study of cumulative sound exposure of humpback whales from cruise and tour vessels in Glacier Bay (AK), USA, showed that reducing vessel speed significantly reduced the cumulative sound exposure, even though vessels needed longer to transit (Frankel and Gabriele 2017). A probabilistic model of future expected increases in shipping in the Arctic (made possible by decreasing sea ice) showed that significant changes to the soundscape are to be expected with increasing risk to marine fauna (Aulanier et al. 2017). The sound emission and operational specifications of hydroacoustic instrumentation have been used in Antarctic marine soundscape planning to manage the potentially competing uses of acoustic space and to minimize the mutual interference between the equipment and marine mammals (van Opzeeland and Boebel 2018). In this chapter, we summarize information on underwater noise from abiotic, biotic, and anthropogenic sources, which might be used in predictive modeling for environmental management.

2.2 Aquatic Soundscapes

A soundscape constitutes sounds from all of the sources near and far that contribute to the total sound field in any particular location. While the ISO 12913-1 definition of soundscape (International Organization for Standardization 2014) requires a receiver (specifically a person or people) that perceives the sounds, in underwater acoustics, a soundscape is defined as the physical superposition of all sounds, whether there is a listener or not (ISO 18405; International Organization for Standardization 2017b).

The sources that contribute to a soundscape are commonly grouped into geophysical sources

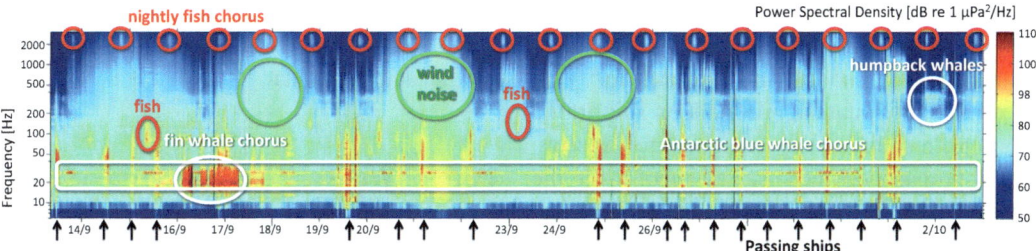

Fig. 2.1 Three-week spectrogram of the soundscape in the Perth Canyon during spring, showing a variety of sound sources of geophysical, biotic, and anthropogenic origin. Occurrences of ship noise are indicated by black arrows. Recording from the Australian Integrated Marine Observing System's (IMOS) passive acoustic observatories, described in Erbe et al. (2015)

making up the geophony, biotic sources making up the biophony, and anthropogenic sources making up the anthropophony (Krause 2008). In the underwater environment, the geophony includes sounds from bubbles produced by wind and breaking waves, precipitation acting upon the water surface, subsea earthquakes and volcanoes, and polar ice. The biophony includes sounds from crustaceans, fishes, and marine mammals. The anthropophony includes sounds from human activities, such as shipping, seismic surveying, sonar imaging, etc. Schoeman and coauthors (2022a) provide a high-level introduction to soundscapes. An example of the marine soundscape in the Perth Canyon is illustrated in Fig. 2.1. Even in the deepest parts of the ocean, at Challenger Deep in the Mariana Trench, at nearly 11 km depth, wind and waves, mysticetes and odontocetes, and ships, sonars, and seismic airguns have been recorded (Dziak et al. 2017).

As sound propagates through the ocean, its spectro-temporal features change due to frequency-dependent absorption, reverberation, or dispersion. An introduction to these sound propagation phenomena is given in Larsen et al. (2022). A very basic introduction to sound propagation under water is given in Erbe et al. (2022b), with a focus on sound propagation scenarios relevant to the effects of noise on marine mammals in Chap. 1. Moreover, many sources of sound move (e.g., marine mammals and ships) and most sources emit sound intermittently or at specific times rather than continuously (e.g., evening fish choruses). Therefore, the soundscape at any location can be highly variable, and soundscapes may vary greatly in space.

The field of soundscape research has grown substantially in recent years. It is a holistic way of studying underwater sound. It does not consider sounds in isolation, but rather the myriad of sounds that originate at different places and that propagate to overlap at one location. Soundscape research may involve identifying the sound sources, mapping their geospatial distribution, examining temporal patterns in sound production, and investigating the relationships between different sources, such as the effects of anthropogenic sources on biological sources. Soundscape research is a data-hungry field of research. Not only do off-the-shelf recorders easily collect tens or hundreds of thousands of samples every second, but in order to understand a soundscape, simultaneous data streams are often needed, such as environmental parameters that affect sources, their sound emission and movement, and the sound propagation conditions of the environment. For example, sea surface temperature, wind speed, precipitation rate, tidal height, light intensity, lunar phase, primary productivity (chlorophyll-a), and automatic identification system (AIS) logs of ship location and speed are common covariates that explain features and patterns in soundscapes (e.g., McWilliam et al. 2017; Menze et al. 2017). In addition, visual observations sometimes accompany soundscape recording to confirm specific sources.

Attempts have been made to model and predict underwater soundscapes using various input data and models. For example, the soundscape of the Perth Canyon (Fig. 2.1) was modeled using meteorological data, models and forecasts for wind speed and precipitation, AIS data on ship traffic along with a library of vessel source spectra, sound propagation models, and models for the major local biological sources (i.e., whales and fish) based on their seasonal presence, diel vocal activity, and vocalization spectra (Gavrilov et al. 2019). The soundscape model demonstrated reasonably accurate prediction results, especially for the hourly medians of the 1/3-octave bands corresponding to wind, ships, and fish choruses, and the daily and weekly medians for whale sounds. Understanding the acoustic characteristics of the sources of sound within a soundscape, their variability, and dependencies is important for the study of the effects of noise on marine mammals. The following sections give an overview of the most common sources of sound.

2.3 Geophysical Sound Sources

Geophysical sources of underwater sound are part of the natural, physical environment, including the seafloor, the water, and the atmosphere above it, provided that the sound can propagate into the aquatic habitat.

2.3.1 Wind

Wind blowing over the water surface and breaking wind-driven waves create air bubbles, which oscillate and collapse and thus generate broadband acoustic pressure waves under water (Medwin and Beaky 1989). Wind-driven sound increases with wind speed and is therefore also related to Sea State and Beaufort scale (Table 2.1). As wind speed increases, waves start to break, generating oscillating bubble plumes (Ding and Farmer 1994; Kerman 1984).

There is a series of famous curves that give the spectra of wind-driven underwater sound at different wind speeds. The "Knudsen curves" were published by Knudsen et al. (1948) based on recordings in coastal waters. They are reliable above 1 kHz. Wenz (1962) synthesized a larger data set, including deep-water recordings, and extended the spectra to 100 Hz, now known as the "Wenz curves." Cato (2008), working with

Table 2.1 Relationship between wind speed, Beaufort scale, sea state, and sea conditions. WMO: World Meteorological Organization

Wind speed [m/s]	Wind speed [kn]	Beaufort scale	Sea state (SS)	Wave height [m]	WMO meteorological terms	Description
<0.5	<1	0	0	0	Calm	Flat, glassy
0.5–1.5	1–3	1	0.5	0.1	Light air	Ripples without crests
1.6–3.3	4–6	2	1	0.2	Light breeze	Small wavelets, glassy crests
3.4–5.5	7–10	3	2	0.6	Gentle breeze	Large wavelets, crests begin to break
5.6–7.9	11–16	4	3	1	Moderate breeze	Small waves, whitecaps
8.0–10.7	17–21	5	4	2	Fresh breeze	Moderate waves, some foam + spray
10.8–13.8	22–27	6	5	3	Strong breeze	Large waves with foam crests
13.9–17.1	28–33	7	6	4	Near gale	Sea heaps up, foam begins to be blown
17.2–20.7	34–40	8	6	5.5	Gale	High waves, breaking crests form spindrift, streaks of foam
20.8–24.4	41–47	9	7	7	Strong gale	High waves with dense foam, crests roll over
24.5–28.4	48–55	10	7	9	Storm	Considerable tumbling of waves with heavy impact, large amounts of airborne spray reduce visibility
28.5–32.6	56–63	11	8	11.5	Violent storm	Very large patches of foam driven before the wind cover much of the sea surface
>32.7	>64	12	9	> 14	Hurricane	Sea completely white with foam, air filled with spray

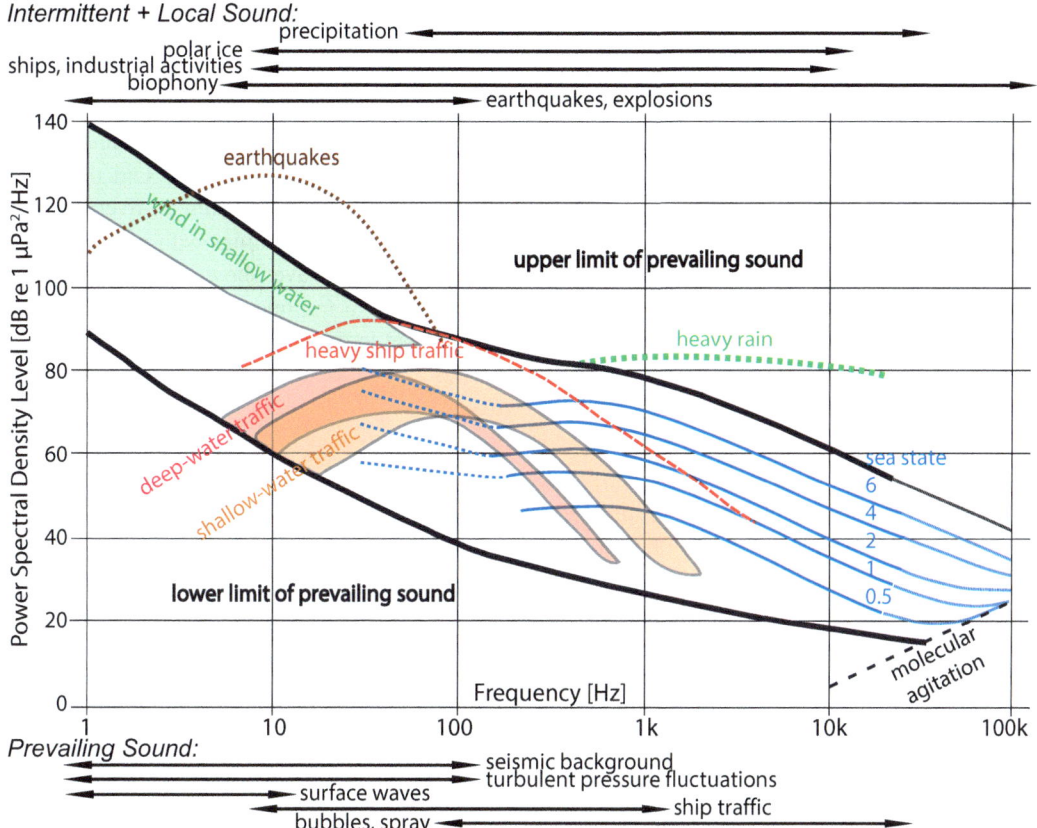

Fig. 2.2 Spectra of prevailing and local underwater sound sources between 1 Hz and 100 kHz, drawn after Wenz (1962) and Cato (2008): Wenz GM (1962) Acoustic ambient noise in the ocean: spectra and sources. J Acoust Soc Am 34 (12):1936-1956; https://doi.org/10.1121/1.1909155. Cato DH (2008) Ocean ambient noise: Its measurement and its significance to marine animals. In: Proceedings of the Institute of Acoustics - Underwater Noise Measurement, Impact and Mitigation, Southampton, UK, 14-15 October 2008. vol 5. Institute of Acoustics, pp 1–9

Southern Hemisphere underwater sound recordings that had much less contribution of sound from distant shipping below 100 Hz (Cato and Tavener 1997), was able to extend the wind spectra to 30 Hz. The wind spectra peak at around 300–500 Hz. For higher frequencies, the levels of mean-square sound pressure spectral density (commonly shortened to power spectral density in the literature) drop at somewhat less than 20 dB per decade of frequency. In other words, for every increase in frequency by a factor of 10, the power spectral density level (i.e., the level in 1 Hz bands) decreases by ~20 dB. At lower frequencies, there is a subtle notch around 200 Hz, with levels increasing for even lower frequencies (Fig. 2.2). The shape of the spectral curves hardly changes with wind speed, but the levels increase with increasing wind speed by ~5 dB per doubling in wind speed between 1.3 and 20.6 m/s (2.5–40 knots). The complete set of "Cato curves" also includes spectra for distant shipping sound and biological choruses. The "Wenz curves" also include spectra for shipping and rain sound. Whether it's Knudsen's, Wenz's, or Cato's empirical model of underwater sound, the spectral density levels are not universally applicable to the entire ocean. For example, in waters over the Northwest Shelf of Australia, the spectral density levels of wind-driven underwater sound are noticeably (by ~10 dB) lower than those predicted by these models at the same wind speeds. The main cause of such difference are the geoacoustic properties of the seabed, which are considerably more acoustically absorptive than those in other parts of the ocean (Gavrilov 2021).

2.3.2 Precipitation

Raindrops, hail grains, even individual snowflakes falling onto the water surface generate sound under water (Scrimger 1985). This is partly and initially due to the physical impact, but more so to the entrainment of gas and subsequent oscillation of bubbles (Crum et al. 1999; Medwin et al. 1992; Nystuen 1986). The sound level increases with the rate of precipitation and the drop, grain, or flake size (Ma et al. 2005; Schwock and Abadi 2021; Scrimger et al. 1989). The rain spectrum peaks at 13–25 kHz for light rain with a rate of <10 mm/h and a drop diameter of ~1 mm. More acoustic power is produced at lower frequencies (down to 500 Hz) in the case of heavy rain with higher rainfall rate and greater drop size (Ma et al. 2005). The hail spectrum peaks between 2 and 5 kHz (Scrimger et al. 1987), the snow spectrum above 50 kHz (Fig. 2.3; Crum et al. 1999).

Wind affects the rain (or other precipitation) spectrum in several ways, in particular when the rain is light. At increasing wind speeds, the angle of rain impact on the water surface is increasingly obtuse (i.e., less vertical and more horizontal), reducing the generation of bubbles and hence the noise level (Medwin et al. 1990). At strong winds >10 m/s, a bubble layer forms below the water surface that attenuates the sound of precipitation at the surface, affecting the spectrum above 10 kHz (Nystuen et al. 1993). Schwock and Abadi (2021) present many graphs of underwater sound at numerous combinations of wind speed and rain rate, up to 30 kHz, at both a shallow and a deep hydrophone.

The overall directivity of wind and precipitation sound is downward. This is because the sound originates from an area at the surface (and not a single point) and so radiates like a sheet of incoherent sources rather than a point source. Furthermore, the main source of sound is bubbles just below the water surface, which, together with their Lloyd's mirror image, form a dipole. As a result, this noise can be detected at considerable depth below the sea surface (Barclay and Buckingham 2013a, b). Having said this, there are interesting deviations from an overall downward directivity. Recordings with a vertical array of hydrophones have shown a noise notch at angles close to the horizontal (i.e., at a grazing angle equal to the arccosine of the ratio of sound speeds at the sea surface c_S and at the middle of the array c_M: $\Theta = \cos^{-1}(c_S/c_M)$); and there is a local maximum near the seabed in shallow water due to bottom reflection and at great depth in deep water due to refraction (e.g., Carey et al. 1990; Clark 2007).

2.3.3 Surf

The sound of surf is a feature of coastal regions. Strong winds produce long-period waves, which may travel hundreds or thousands of kilometers before breaking as they travel into shallow water. Such waves, which originate far away, are referred to as swell. Although the sound of surf is ultimately a result of waves produced by wind, it has received separate attention, as the breaking mechanism is different from the white-capping responsible for wind-driven sound, and the characteristics of swell, and hence the surf it creates when it breaks, are often unrelated to the local wind speed.

As a wave rolls toward shore, a breaking wave crest appears, elongating perpendicular to the direction of travel. Air is entrained in bubbles forming at the crest. As the crest breaks, bubble plumes develop, which reach well into the water column. A bubbly water residue is left behind the breaking crest partly absorbing sound from the newly created, resonating bubble plumes. As a result, a breaking wave has a distinct radiation pattern and, to a listener behind the wave, appears as two active hot spots traveling in opposite directions. The sound during breaking exceeds the sound before breaking in the 0.4–

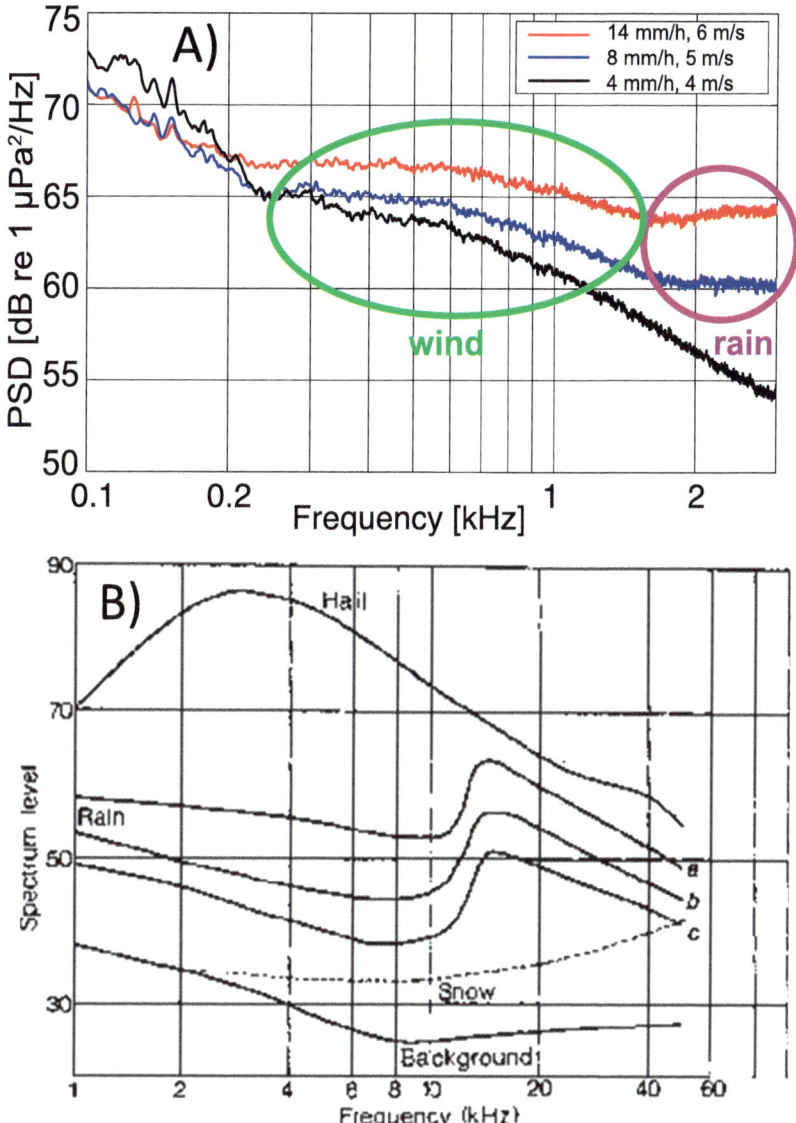

Fig. 2.3 (a) Wind and rain sound recorded in the Perth Canyon (WA), Australia, at 450 m depth, showing how, with increasing wind speed and rainfall rate, two distinct peaks form: one at about 500 Hz from wind and one above 3 kHz from rain. PSD: Power spectral density level. Sampling frequency was 6 kHz and hence the full rain spectrum could not be documented. Modified from Erbe et al. (2015). © Erbe et al., 2015; https://doi.org/10.1016/j.pocean.2015.05.015. Licensed CC BY 4.0; https://creativecommons.org/licenses/by/4.0/. (b) Underwater sound spectra of rain, hail, and snow at wind speeds >1.5 m/s. Rain spectra: *a* (1.1 mm/h rain, 2.7 m/s wind), *b* (0.29 mm/h rain, 2.7 m/s wind), and *c* (0.13 mm/h rain, 2.23 m/s wind). Reprinted by permission from Springer Nature. Scrimger JA, Underwater noise caused by precipitation. Nature 318 (6047):647; https://doi.org/10.1038/318647a0. © Springer, 1985. All rights reserved

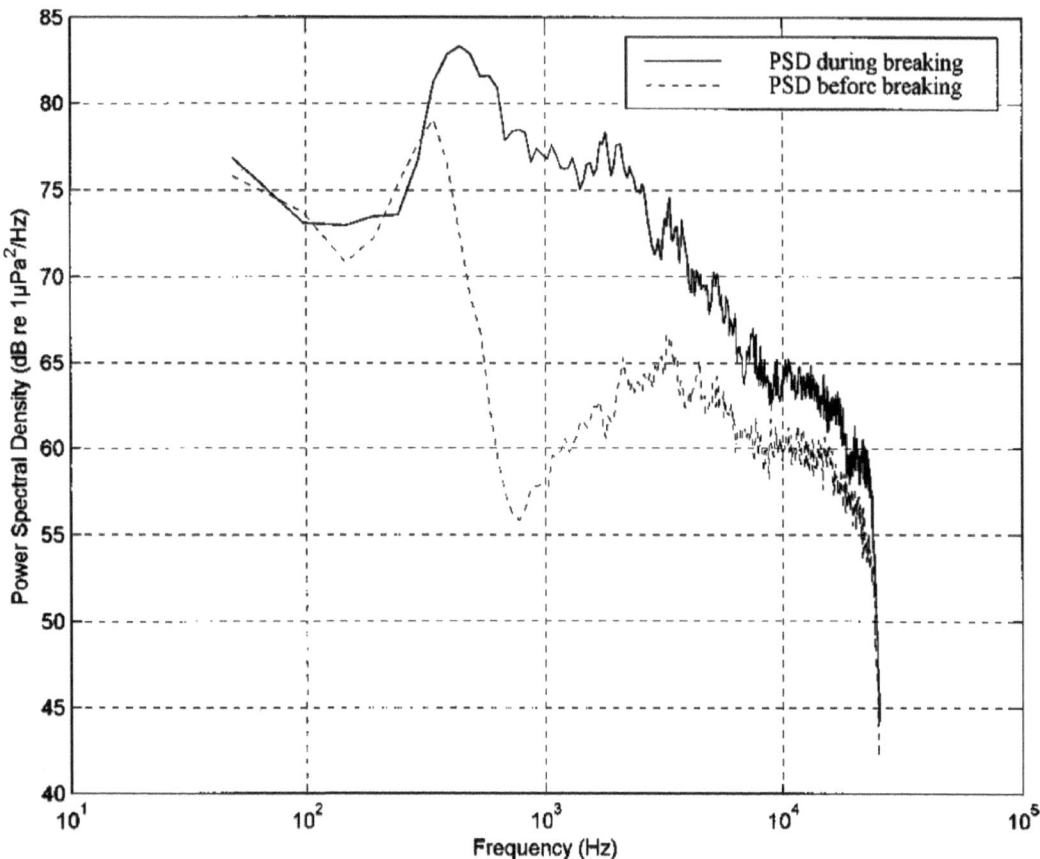

Fig. 2.4 Underwater sound spectra before and during the breaking of waves. Levels rapidly decreased above the anti-aliasing filter at 22.5 kHz. Reproduced from Deane GB (1999). Acoustic hot-spots and breaking wave noise in the surf zone. J Acoust Soc Am 105 (6):3151–3167; https://doi.org/10.1121/1.424646. With permission from the ASA. © Acoustical Society of America, 1999. All rights reserved

>22.5 kHz band (Fig. 2.4; Deane 1999). The noise level is primarily determined by wave height (Deane 2000).

2.3.4 Subsea Earthquakes and Volcanoes

Subsea earthquakes (also called seaquakes) and volcanoes are common near tectonic plate boundaries, such as the Mid-Atlantic Ridge, East Pacific Ridge, or Mid-Indian Ocean Ridge. The sound travels as seismic waves through the seafloor: as compressional waves, which travel at a high speed and are therefore called primary (P) waves, and as shear waves, which travel at a lower speed and are therefore called secondary (S) waves. Both seafloor compressional and shear waves can give rise to sound pressure waves in the water. Additional tertiary (T) waves may be recorded from distant subsea earthquakes. These T-waves travel through the SOFAR channel, where the speed of sound in the water column is at its minimum, and thus arrive last, after the P- and S-waves (Dziak et al. 2004; Park et al. 2001).

Underwater sound from subsea earthquakes can last from a few tens of seconds to an hour or more, as from the Great Sumatra-Andaman earthquake on December 26, 2004, when the rupture of subsea earth crust propagated over nearly 1500 km (Tolstoy and Bohnenstiehl 2005). Sounds from seaquakes may significantly exceed

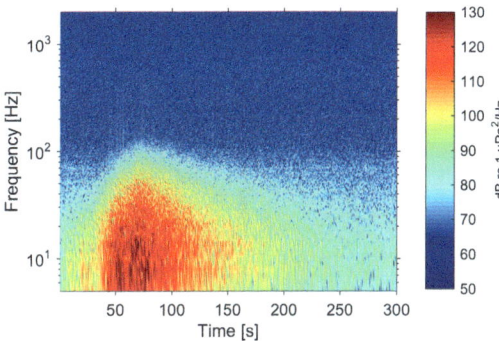

Fig. 2.5 Spectrogram (fs = 6 kHz, NFFT = 6000, Hann window, 0% overlap) of an earthquake recorded at the Australian Integrated Marine Observing System (IMOS) in the Perth Canyon (WA), Australia

ambient levels (by 20–60 dB) at low frequencies (1–40 Hz; e.g., McGrath 1976). In another study, the most common source level was 212 dB re 1 µPa m in the frequency band of 1–40 Hz (Fox et al. 2001). A range of source levels from 225 to 275 dB re 1 µPa m (4–8 Hz) was reported by Hanson and Bowman (2006). The spectrogram of an earthquake recorded in the Perth Canyon is shown in Fig. 2.5.

Sound energy from volcanoes is mostly contained at frequencies below 100 Hz, and sound may last tens of seconds or longer (Northrop 1974). A glider recorded sound from an erupting submarine volcano of the Lau Basin in the 2 Hz–1.5 kHz band. At 40 m range, levels in the band 10–60 Hz exceeded ambient levels of Sea State 0 by 50 dB. Source mean-square sound pressure spectral density levels were on average 166 ± 4 dB re 1 µPa^{2}m^2/Hz in the 100–200 Hz band (Matsumoto et al. 2011). Volcanic tremors exhibiting tones at 10 Hz of 225 dB re 1 µPa m source level, and harmonic overtones at 20, 30, and 40 Hz, lasting up to 5 minutes, have also been recorded (Dziak and Fox 2002).

2.3.5 Polar Ice

Ice is a significant contributor to underwater sound in the polar regions at frequencies ranging from <10 Hz to >10 kHz. Underwater sounds of ice-related events result from breakups of ice shelves and icebergs, ridging of sea ice, collision of large icebergs, interaction of icebergs with the seabed, and ice melting—due to wind, waves, currents, and changing temperatures. Underwater ambient noise in polar regions is highly variable, producing some of the lowest and highest noise levels in both Northern and Southern Hemispheres (Mikhalevsky 2001; Fig. 2.6a).

Under the sea ice sheet, in stable conditions, ambient noise power spectral density levels can be as low as 20 dB below the Knudsen open-ocean Sea State 0 levels (Milne and Ganton 1964). Under-ice sound increases with wind speed above the ice (Greene and Buck 1979). The marginal ice zone (i.e., the transition zone between the open sea and dense ice cover) is sometimes noisiest, because pack ice breaks into ice floes due to wind, waves, currents, and thermal stresses; and the floes collide, compress, shear, fracture, and break up. In the study by Yang et al. (1987), sound levels dropped either side of the ice edge, into the open ocean and under the ice (Fig. 2.6b). The interior of the sea ice cover is also a regular source of intense low-frequency underwater noise because of ice ridging under wind and current stress (e.g., Kinda et al. 2015).

The frequency spectrum of ice sound depends on the primary mechanism of sound generation. Pressure ridging in multi-year sea ice cover generates 5–10 s pulses with peak power at 1–30 Hz and source levels in excess of 180 dB re 1 µPa^{2}m^2/Hz at 10 Hz (Greening and Zakarauskas 1994a). Thermal cracking of multi-year ice generates <1 s pulses with peak power at 300–700 Hz and source levels of 110–180 dB re 1 µPa^{2}m^2/Hz at 100 Hz (Greening and Zakarauskas 1994b; Xie and Farmer 1991). Thermal ice sound is typically strongest during periods of rapid cooling (Milne and Ganton 1964). Ice floes rubbing against each other generate tones at hundreds of hertz depending on ice thickness and ice shear-wave speed (Xie and Farmer 1992). Melting icebergs generate lower-level sizzling sound due to the breaking of air bubble cavities (Urick 1971). Traveling icebergs colliding with ice shelves or seabed emit variable harmonic tremors or broadband sounds lasting up

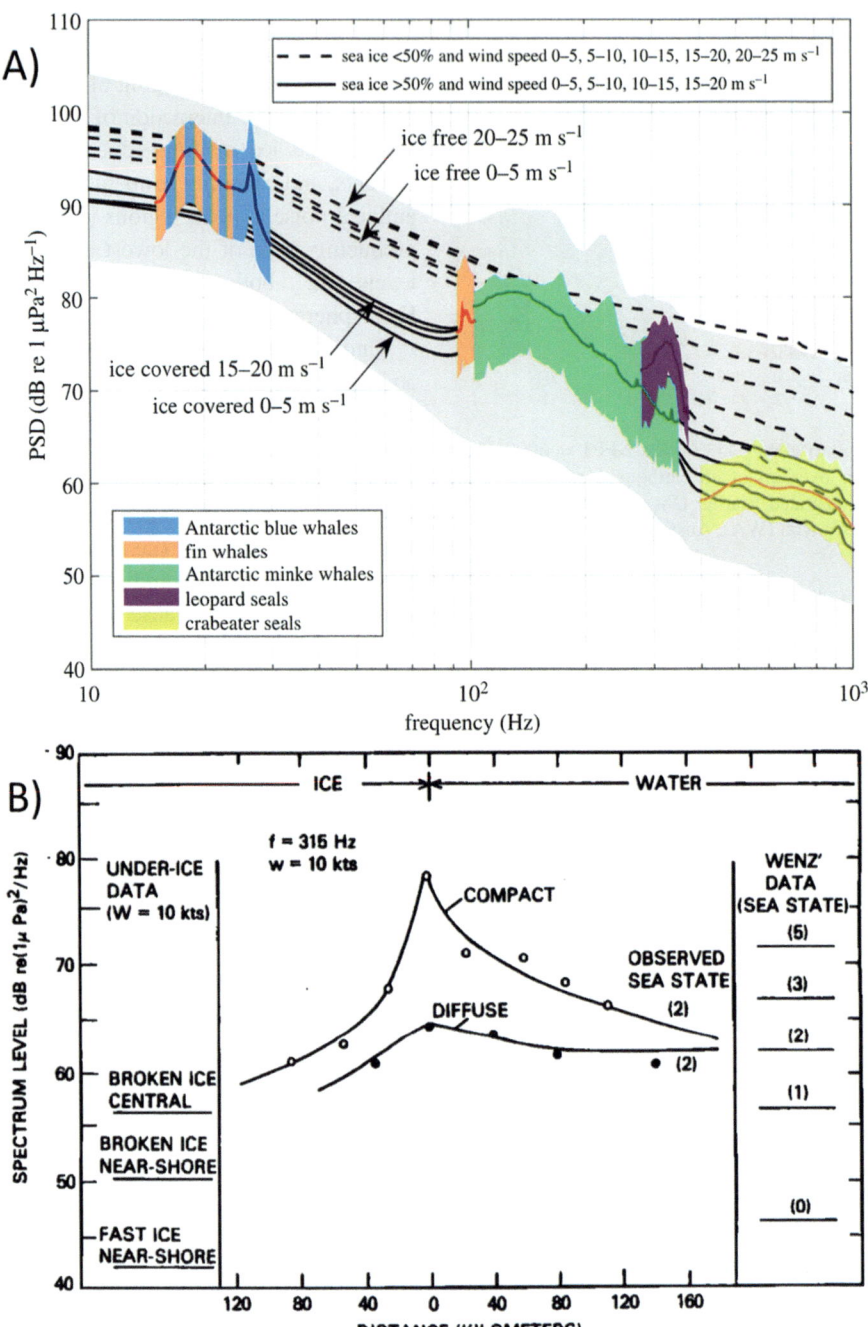

Fig. 2.6 (a) Spectra of underwater ambient noise in the Southern Ocean. Gray area: range of noise levels, 1st–99th percentiles; black lines: ambient noise at different wind speeds and ice coverage; colored areas: marine mammal contributions, 10th–90th percentiles; colored lines: mean marine mammal spectra. Reprinted from Menze S, Zitterbart DP, van Opzeeland I, Boebel O (2017) The influence of sea ice, wind speed and marine mammals on Southern Ocean ambient sound. R Soc Open Sci 4 (1): 160370; https://royalsocietypublishing.org/doi/10.1098/rsos.160370. © Menze et al., 2017. Licensed under CC BY 4.0; https://creativecommons.org/licenses/by/4.0/. (**b**) Arctic underwater ambient noise in the Greenland Sea, at 315 Hz, 10 kn wind, as a function of distance from the ice edge (which was compact versus diffuse). For comparison, typical levels under the ice are indicated as horizontal lines on the left, and those of the open ocean at different Sea States on the right. Reprinted from Yang TC, Giellis GR, Votaw CW, Diachok OI (1987) Acoustic properties of ice edge noise in the Greenland Sea. J Acoust Soc Am 82 (3): 1034–1038; https://doi.org/10.1121/1.395377. With permission from the ASA. © Acoustical Society of America, 2014. All rights reserved

to several hours with power concentrating in the 1–100 Hz band propagating through the SOFAR channel over thousands of miles (Gavrilov and Li 2007; Talandier et al. 2002; Tolstoy et al. 2004). The strongest ice-related sound was reported by Podolskiy et al. (2022): A "kilometer-scale iceberg calving" generated an "underwater-detonation-like signal" with a peak-to-peak source level of 225 ± 10 dB re 1 µPa m.

2.3.6 Thermal Agitation

At high frequencies >30 kHz, thermal agitation of the molecules in seawater limits how low ambient noise levels can become (Mellen 1952). The power spectral density of thermal noise increases with frequency at a rate of ~6 dB per octave (Fig. 2.2).

2.3.7 Current-Induced Noise Artifacts

Ocean currents by themselves do not emit sound. However, they can set ropes, stakes, chains, or any cavities into motion, generating sounds like clonking chains or vibrating strings. Poorly designed moorings of acoustic recorders or recorders deployed near anchored buoys will record such sound and this might interfere with the goal of the acoustic deployment. High currents near the seabed can move debris (e.g., sand, shells, and rocks) around, into, and over the recorder, producing noise.

Currents can cause turbulence by interacting with the seafloor or structures within the water column, such as moorings or the hydrophone itself. As this turbulence moves past the hydrophone or is generated right at the hydrophone, the transducer registers a pressure fluctuation. This is not acoustic pressure, and the vortices created in the turbulent flow do not generate a traveling acoustic wave. Rather, this is local hydrodynamic pressure, which may also result from rapidly varying hydrostatic pressure due to vertical motion of an acoustic sensor, producing pseudo noise (Strasberg 1979). This is not part of the soundscape, yet does appear in spectrograms.

Shielding the recorder or a protruding hydrophone with an acoustically transparent shield can help move the turbulent flow away from the hydrophone, hence reducing pseudo noise, as can keeping the hydrophone up-current.

Pseudo noise is most common at low frequencies up to a few tens of hertz, yet may reach into the kilohertz range in severe cases. If concurrent measurements of currents exist, then the acoustic recordings can be correlated with current speed in order to identify pseudo noise in spectrograms (e.g., Erbe et al. 2015; Willis and Dietz 1961). Alternatively, if two hydrophones are deployed in close proximity, the two acoustic time series can be correlated to identify periods of strong pseudo noise. Acoustic noise will be coherent across the two hydrophones, but pseudo noise will not. Therefore, a high cross-correlation indicates low or no pseudo noise, while a low cross-correlation indicates high pseudo noise. The acoustic intensity I_a free of pseudo noise can be computed from the intensities at the two hydrophones (I_1 and I_2) and the cross-correlation coefficient r_{12} (Buck and Greene 1980; Eq. 2.1).

$$I_a = r_{12}\sqrt{I_1 I_2} \quad (2.1)$$

Other noise artifacts are a result of poor deployment configurations. For example, hydrophones suspended over the side of a boat might record waves splashing against the boat or sounds occurring on the boat and radiating into the water through its hull. The hydrophone might move rapidly up and down as the vessel rolls in the waves, creating hydrostatic pressure fluctuations. Ideally, these artifacts would be minimized at the time of deployment by carefully designing moorings under consideration of the local environment (water depth, currents, weather, etc.) (Fig. 2.7).

2.4 Biotic Sound Sources

2.4.1 Marine Invertebrates

Marine invertebrates can be prolific contributors to ambient noise, in particular in coastal waters.

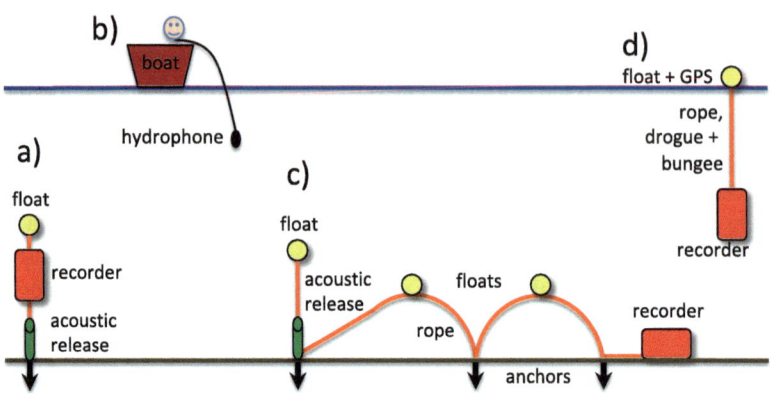

Fig. 2.7 Different recorder deployment configurations: (**a**) has the recorder in line with the mooring making this setup susceptible to vibrating strings and pseudo noise; (**b**) is susceptible to the recording of waves splashing against the boat or noise onboard transmitting into the water through the hull; (**c**) can work well if all joints are fastened and insulated as the recorder is physically removed from the main anchor; and (**d**) minimizes pseudo noise by letting the recorder freely drift with the currents (Erbe et al. 2016b)

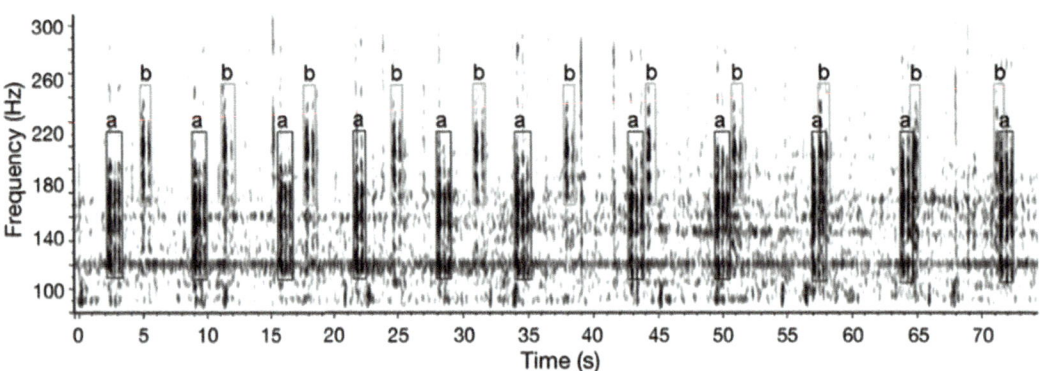

Fig. 2.8 Series of rumbles produced by (at least) two mantis shrimp (*a* and *b*). Reprinted from Staaterman E, Clark C, Gallagher A, deVries M, Claverie T, Patek S (2011) Rumbling in the benthos: acoustic ecology of the California mantis shrimp *Hemisquilla californiensis*. Aquat Biol 13 (2):97–105. https://doi.org/10.3354/ab00361. Reprinted with permission from Inter-Research. © Inter-Research, 2011. All rights reserved

Marine invertebrate noise is generally produced by stridulation of hard body parts, although it can also be created by gas bubble cavitation. Clawed lobsters (family Nephropidae) produce 0.07–1.7 s tonal sounds at 87–261 Hz with harmonics up to 600 Hz (Henninger and Watson 2005). Spiny lobsters (family Palinuridae) generate rapid pulse trains, called rasps, which last 15–303 ms and consist of 24–192 pulses/s, with peak frequency between 300 and 2100 Hz (Patek et al. 2009). Hermit crabs (family Diogenidae) produce pulse trains with peak frequency at 6–8 kHz, in which pulses last 2.6–8.1 ms and are repeated at a rate of 36–303 pulses/s (Field et al. 1987). Freshwater crayfish (family Parastacidae) hiss at 4–12 kHz, with each signal lasting 150–350 ms (Sandeman and Wilkens 1982). Mantis shrimp (family Hemisquillidae) produce a low-frequency rumble (53–257 Hz) consisting of 0.06–0.6 s pulses emitted in packages of up to 4 (Staaterman et al. 2011). A chorus forms during morning and evening twilight when so many mantis shrimp vocalize that ambient noise levels are raised by up to 20 dB in the 100–300 Hz band (Fig. 2.8). Snapping shrimp (family Alpheidae) produce a snapping sound by

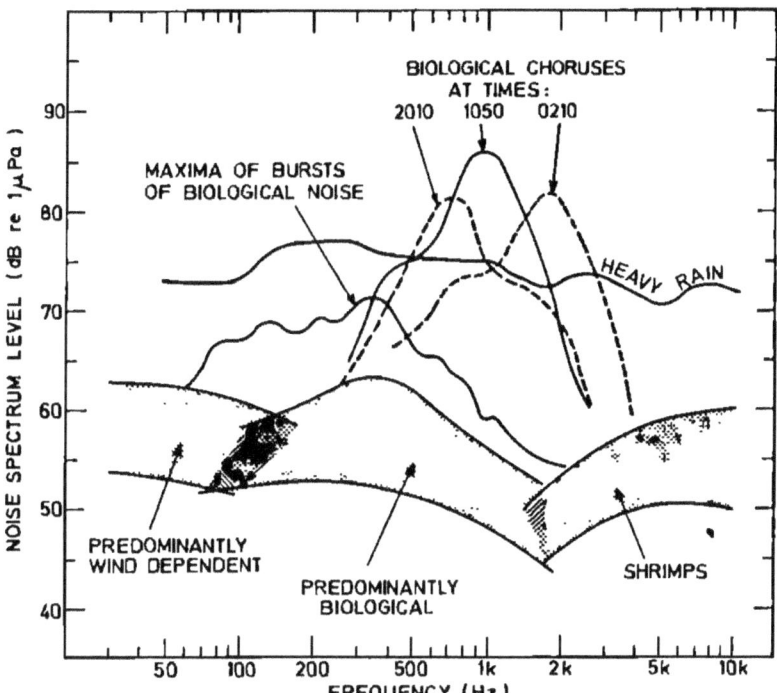

Fig. 2.9 Mean-square sound pressure spectral density levels [dB re 1 µPa^2/Hz] of underwater ambient noise in tropical water (Timor Sea) showing various animal choruses. Shaded areas: range of prevailing noise; lines: maximum chorus levels. Reprinted from Cato DH (1980) Some unusual sounds of apparent biological origin responsible for sustained background noise in the Timor Sea. J Acoust Soc Am 68 (4):1056–1060; https://doi.org/10.1121/1.384989. With permission from the ASA. © Acoustical Society of America, 1980. All rights reserved

suddenly creating and imploding a gas bubble using a modified claw arrangement (Shimu et al. 2019). The spectrum peaks between 2 and 5 kHz, but can extend up to 200 kHz, with a source level of 190 dB re 1 µPa m pk-pk (Au and Banks 1998). Aggregations of snapping shrimp produce a sizzling chorus, which exceeds ambient levels by about 20 dB between 2 and 20 kHz (Cato 1980; Fig. 2.9). Sea urchins make clicking sounds during feeding by rasping their calcareous feeding parts (Aristotle's lantern) on rock. This sound is amplified by their resonating ovoid skeleton, such that aggregations of urchins produce a characteristic night-time chorus, which lasts for a couple of hours and exceeds ambient levels by up to 30 dB at 400–4000 Hz (Radford et al. 2008).

2.4.2 Fishes

Over 800 species of fish from 109 families have been documented to emit sound (Slabbekoorn et al. 2010). The most powerful and complex fish signals are produced by oscillations of the fish swimbladder, an internal gas-filled organ used for buoyancy control and sound production. Not all fish species or all life cycle stages of fish may possess such a bubble. Other signal types are produced by rasping of hard body parts or by banging fins on the fish body wall, often setting the swimbladder in motion. Fish sounds have been associated with reproductive activity (mediating the advertisement of males, male–male competition, mate-choice by females, courtship behavior, assembly of spawning groups, and synchronized release of gametes), feeding, aggressive behavior, disturbance, defense of territory, etc. (Gannon 2008). Sounds are often emitted as a series of pulses, where each pulse can have a strong tonal character (including harmonics) or more broadband character. Fish sounds have been onomatopoeically labeled click, cluck, croak, grunt, hoot, hum, knock, pop, purr, whistle, etc. Most of spectral power commonly lies below 1 kHz (Fish and Mowbray 1970; Gannon 2008), but can extend to more than 4 kHz (Parsons et al. 2016a). Pulse duration and inter-pulse intervals are typically of the order of tens of milliseconds, with overall call durations ranging from less than 1 s to 20 s (Hawkins and

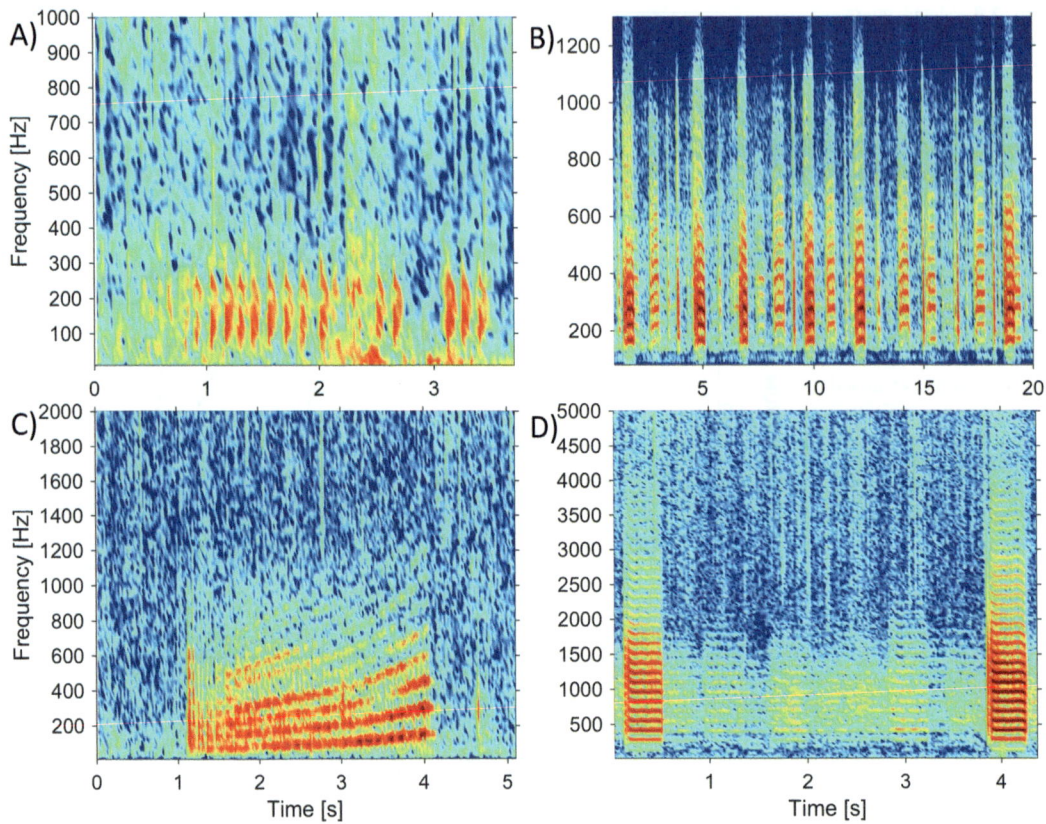

Fig. 2.10 Sounds of (**a**) a dhufish (*Glaucosoma hebraicum*; fs = 5 kHz, NFFT = 256, 50% overlap), (**b**) several mulloway calling in a chorus (*Argyrosomus japonicas*; fs = 5.2 kHz, NFFT = 512, 50% overlap), (**c**) a batfish (*Ogcocephalus darwini*; fs = 6 kHz, NFFT = 256, 50% overlap), and (**d**) a Terapontidae chorus with two nearby calls (fs = 12 kHz, NFFT = 512, 50% overlap). Note the different x- and y-axis ranges. Sounds courtesy of Miles Parsons and Rob McCauley (Parsons et al. 2013a, c, 2016b)

Rasmussen 1978; Parsons et al. 2016b; Fig. 2.10). Call source levels as high as 165 dB re 1 µPa m and 158 dB re 1 µPa^{2}m^{2}s have been reported (Locascio and Mann 2011; McCauley and Cato 2000; Parsons et al. 2016b).

When fish aggregate and emit sound together, choruses form, which raise ambient noise levels in specific frequency bands by as much as 50 dB (McCauley 2012; Parsons et al. 2016b; Fig. 2.9). Choruses follow seasonal, lunar, and diel cycles, typically occurring at night and lasting for an hour or more at a time (McWilliam et al. 2018; Parsons et al. 2016b; Ruppé et al. 2015), although some choruses occur during daylight hours. Multiple choruses can be heard at the same location, but they are generally separated in frequency and/or time (Fig. 2.9, Fig. 2.11), in line with the Acoustic Niche Hypothesis, which states that animals evolved their acoustic repertoires and calling behavior so as to minimize interference by others and by environmental noise, and to maximize communication success (Krause 2012).

2.4.3 Marine Mammals

The Acoustic Niche Hypothesis has also been applied to marine mammals (e.g., Kyhn et al. 2013; Mossbridge et al. 1999). Marine mammals produce a variety of sounds, which range from narrow- to broadband and from short to long duration, which exhibit different types and

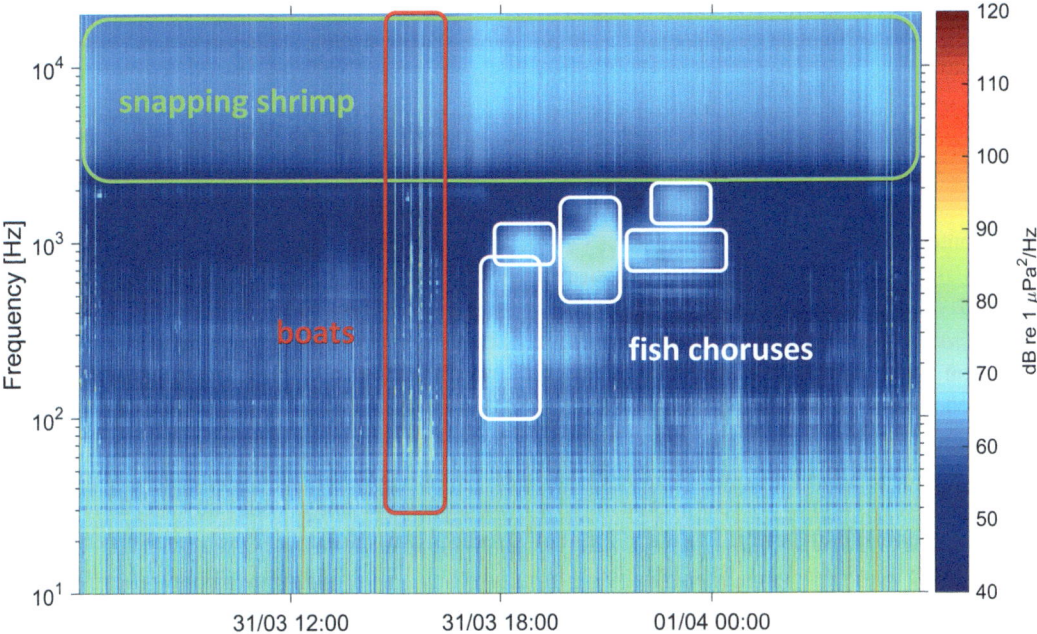

Fig. 2.11 Spectrogram of a 24 h recording in tropical water showing how biological choruses of shrimp and fish are separated in frequency and time, however boat sound can overlap with all of them. Modified from Williams R, Erbe C, Dewantama IMI, Hendrawan IG (2018) Effect on ocean noise: Nyepi, a Balinese day of silence. Oceanography 31 (2):16–18. © Williams et al. 2018; https://doi.org/10.5670/oceanog.2018.207. Published under CC BY 4.0; https://creativecommons.org/licenses/by/4.0/

degrees of modulation, and which might be emitted singly or in packages. Sounds may be described as constant-frequency, frequency-modulated, amplitude-modulated, or pulsed. Tonal sounds may be emitted with or without harmonics. Sounds of different types may be merged into one call; for example, starting as a constant-frequency sound, then introducing amplitude modulation, and eventually becoming pulsed. Pulses themselves may be of constant-frequency, frequency-modulated, or truly broadband type.

Mysticetes produce all of these sound types (see Chap. 3). True constant-frequency calls without any form of modulation are rare, but have been reported for right (*Eubalaena* spp.), sei (*Balaenoptera borealis*), Bryde's (*Balaenoptera edeni*), and blue whales (*Balaenoptera musculus*), ranging in frequency from 10 to 500 Hz and lasting 0.2–20 s (McDonald et al. 2005, 2006; Oleson et al. 2003; Webster et al. 2016). Frequency-modulated sounds are more common and are mostly simple downsweeps or upsweeps, with more complex contours being rarer. Across all species, the fundamental tones of these sounds range from 10 to 2000 Hz and last 0.1–10 s (e.g., Baumgartner et al. 2008; Parks et al. 2007; Rankin et al. 2005; Stafford et al. 2012; Thompson et al. 1992). Amplitude-modulated sounds are very common in gray whales (*Eschrichtius robustus*) and humpback whales (*Megaptera novaeangliae*) (Dunlop et al. 2007; Frouin-Mouy et al. 2020). Pulsive sounds may be emitted singly or in series. One example is the right whale gunshot, which covers 30–8400 Hz and lasts 0.1–0.4 s (e.g., Webster et al. 2016). Most of the mysticete species have been documented to arrange their sounds into songs, which can last many hours (e.g., Croll et al. 2002; Payne and McVay 1971).

Odontocete sounds are most commonly classified as whistles, burst-pulse sounds, and clicks in the literature (see Chap. 4). Whistles can be constant-frequency or frequency-modulated. The frequency of the fundamental contour can be quite low, below 100 Hz, as reported for killer whales (*Orcinus orca*) and bottlenose dolphins (*Tursiops truncatus*) (Gridley et al. 2015; Samarra et al. 2016). Upper limits of the fundamental have been reported at 20–30 kHz in various odontocete species such as humpback (*Sousa chinensis*), common (*Delphinus delphis*), striped (*Stenella coeruleoalba*), and Risso's dolphins (*Grampus griseus*) (Ansmann et al. 2007; Corkeron and Van Parijs 2001; Papale et al. 2013; Wang et al. 2013). Some species' whistles reach rather high frequencies: 40 kHz in bottlenose dolphins (Hiley et al. 2017) and 75 kHz in killer whales (Samarra et al. 2010). Whistles typically last from 0.1 to 1 or 2 s, but exceptionally long ones have been reported: 13 s in a Risso's dolphin (Corkeron and Van Parijs 2001) and 18 s in a killer whale (Thomsen et al. 2001). Burst-pulse sounds consist of a rapid series of broadband pulses. In spectrograms, depending on the length of the window used in the Fourier transform, these sounds often appear as tonal with a large number of overtones and sidebands. Burst-pulse sounds range in frequency from 200 Hz to over 100 kHz (Branstetter et al. 2012; Lammers et al. 2003) and last from 0.1 to 1 or 2 s, with exceptions of 8 s and 13 s for humpback and Risso's dolphins (Corkeron and Van Parijs 2001; Van Parijs and Corkeron 2001). Clicks are always emitted in series and the inter-click interval (i.e., time between two successive clicks) is greater than the inter-pulse interval (i.e., time between two successive pulses) in burst-pulse sounds. Clicks are broadband, ranging from 1 kHz to 200 kHz. The frequency of peak power and the bandwidth vary by species. Clicks are typically tens to hundreds of microseconds long. Click trains can last several seconds. The inter-click interval can be a few seconds as in the "slow clicks" of sperm whales (*Physeter macrocephalus*) (Madsen et al. 2002) or of the order of tens of milliseconds as in most odontocete echolocation (e.g., Baumann-Pickering et al. 2010; Neves 2013; Wahlberg et al. 2011).

Pinnipeds produce sounds on land and under water, with the early literature only addressing terrestrial sounds (see Chap. 5). Whether on land or under water, pinnipeds mostly make amplitude-modulated or pulsed sounds. Under water, these range from 50 Hz to >10 kHz and last from 0.1 to >1 s (e.g., Hanggi and Schusterman 1994; McCreery and Thomas 2009; Rogers et al. 1995). Underwater tonal sounds have been reported for some of the polar seals, ranging in frequency of the fundamental from 100 Hz to 14 kHz and lasting up to tens of seconds (Cleator et al. 1989; Thomas and Kuechle 1982).

Sirenians mostly produce tonal, frequency-modulated sounds, but amplitude-modulated and pulsive sounds have also been reported (see Chap. 7). Fundamental frequencies of tonal sounds range from 1 to 8 kHz, and these sounds last 0.1–3 s (de Sousa Lima et al. 1999; Parsons et al. 2013b; Sousa-Lima et al. 2002). Amplitude-modulated and pulsed sounds can be as broadband as <300 Hz to >22 kHz. Some pulses can be very brief (<0.1 s; Anderson and Barclay 1995; Evans and Herald 1970).

When many animals of one species aggregate in one region, distinct choruses may be formed, such as those of sperm whales (Cato 1978), blue whales (McCauley et al. 2018), or fin whales (Leroy et al. 2016). Choruses of singing humpback whales raised ambient levels off Hawaii by up to 20 dB in February–March with peak frequencies of 320 and 630 Hz; chorus levels were higher at night than during the day (Au et al. 2000). Chorus levels of humpback and pygmy blue whales (*Balaenoptera musculus brevicauda*) were used to document the seasonal presence of these migrating species in the Perth Canyon (Erbe et al. 2015), while Antarctic blue whale (*B. m. intermedia*) choruses have been used to infer population growth rates (McCauley et al. 2018).

2.5 Anthropogenic Sound Sources

The sources of anthropogenic underwater noise are diverse, and their acoustic characteristics vary greatly in frequency and time. Moreover, any offshore development typically proceeds through a number of distinct phases, each involving a multitude of sources: site selection might involve, inter alia, geophysical surveys, geotechnical investigations, test drilling, etc. Construction might involve foundation installation, mooring, cable laying, etc. Operation involves sources specific to the industry, such as drilling or pumping. Maintenance of offshore infrastructure involves vessels and site-specific activities. Finally, decommissioning involves vessels, underwater cutting, or perhaps explosives, etc. It can therefore be quite difficult to model the overall noise field and to assess its potential impacts. The following sections summarize some of the acoustic characteristics by source.

2.5.1 Ship Traffic

Vessels are omnipresent in the world's oceans. They range from small personal watercraft to supertankers (Fig. 2.12). There is a long history of studying ship noise, much more so than in the case of other, newer anthropogenic sources, such as pile driving. Research on ship noise was historically driven by defense, and much of the measurements from World War II and resulting empirical equations were summarized by Ross (1976) and Urick (1983). Ships have changed a lot in design, size, and propulsion since World War II. Ships have gotten bigger, carry more load, and can travel faster, and these characteristics typically correlate with an increase in noise generation. However, ship design and operations have improved significantly for various reasons, including fuel efficiency, greenhouse gas emissions, and noise.

Over the last decades, shipping has increased in terms of the number of ships, the size of ships (indicated by dead weight), and the total weight of goods carried (which is dependent on the number and size of ships, and the number of trips; Fig. 2.13). Ship traffic is a significant contributor to ambient noise. In fact, the 20–100 Hz din in deep-water ambient noise is attributed to distant shipping (Wenz 1962; Fig. 2.2). These ships might be thousands of kilometers away from the recorder, but their noise couples into the SOFAR channel (e.g., by bottom-bounce over the continental downslope) and travels long distances with little attenuation. The noise from many ships then combines to produce the continuous background din that dominates this frequency band. As a result of increasing ship traffic, ambient noise has been increasing in many regions over the second half of the twentieth century and the first two decades of the twenty-first (Andrew et al. 2002; Andrew et al. 2011; Chapman and Price 2011; Frisk 2012; Miksis-Olds et al. 2013). However, analysis of the more recent data collected at the Comprehensive Nuclear-Test-Ban Treaty (CTBT) hydroacoustic station off Cape Leeuwin (WA) has revealed that the levels of ocean ambient noise have been decreasing in the Indian and Southern Oceans over the last decade (Harris et al. 2019).

The noise emitted by a single vessel may have different sources (Urick 1983). First, propeller-driven ships will generate propeller noise, which may include propeller singing (by blade vibration), propeller-induced hull resonance, and

Research
Oceanographic Research Ships
Hydrographic Research Ships
Icebreakers

Oil & Gas
Seismic Survey Ships
Drillships
Tankers
Tenders
Workboats
Accommodation Ships
Tugs
FPSO
FLNG
Pipe Layers

Construction
Dredges
Barges
Supply Vessels
Jack-up Rigs

Non-Profit
Activist Ships
Medical Ships

Transport
Cruise Ships
Ferries
Cargo Vessels
Container Ships
Bulk Carriers
Livestock Transporters

Government
Coast Guard Vessels
Surveying Vessels
Patrol Boats

Private
Jetskis
Power Boats
Yachts

Fisheries
Stock Surveying Vessels
Seafood Processing Vessels
Trawlers
Crabbers
Longliners
Purse Seiners

Military
Battleships
Frigates
Patrol Ships
Submarines
Aircraft Carriers

Fig. 2.12 Examples of the types of ships used by different sectors

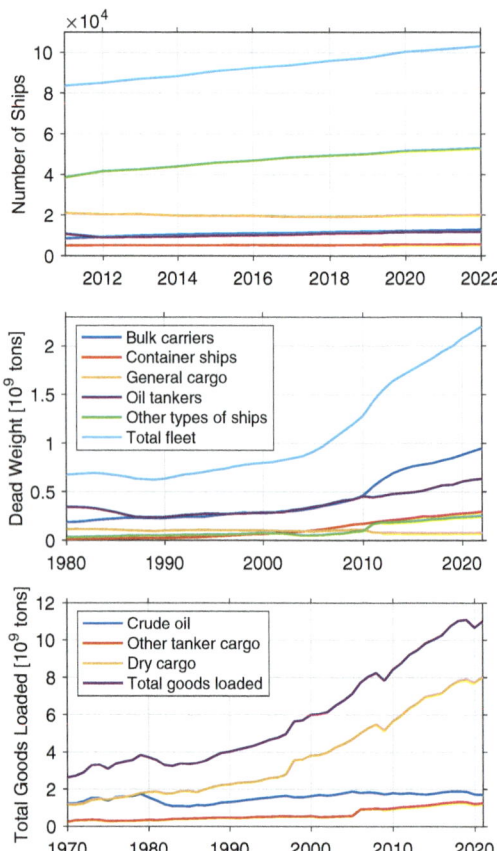

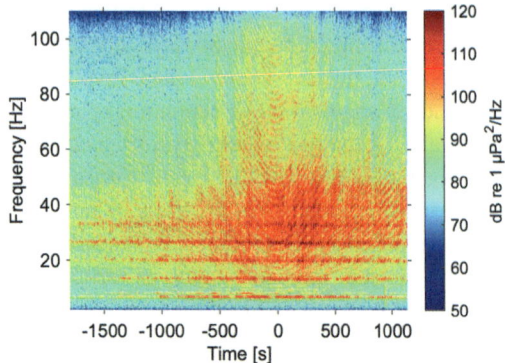

Fig. 2.14 Spectrogram of a large vessel passing by the Comprehensive Nuclear-Test-Ban Treaty Organization's hydrophones off Western Australia, showing the broadband cavitation spectrum (appearing U-shaped due to the Lloyd's mirror effect) superposed by engine or propeller tones (fs = 250 Hz, NFFT = 250, Hann window, 50% overlap)

Fig. 2.13 Graphs of world seaborne trade. The top and center plots share the legend. Data from the United Nations Conference on Trade and Development; https://unctadstat.unctad.org/wds/TableViewer/tableView.aspx; accessed 3 Jan. 2023

propeller cavitation. Propeller cavitation is the dominant source for many ships. Second, any noise generated onboard the ship (e.g., by machinery, engines, pumps, hydraulics, or any onboard activities) can transmit directly through the hull into the water. Third, hydrodynamic flow past the ship can cause vibration and resonance of appendages, plates, or cavities, and potentially cavitation at appendages. Hull vibrations may also be induced through coupling with the propeller.

An example spectrogram of a large vessel recorded by a seafloor-mounted hydrophone is shown in Fig. 2.14. There were a 7 Hz tone and harmonic overtones (at integer multiples: 14 Hz, 21 Hz, ...). The harmonic overtone at 28 Hz was received with the strongest level. These tones were overlain by the broadband propeller cavitation spectrum. As the source of the cavitation spectrum was close to the sea surface, sound propagating along the direct travel path to the recorder repeatedly interfered constructively and destructively with the surface-reflected path (Lloyd's mirror effect) yielding the characteristic U-shaped spectrogram as the ship traveled past the recorder. The closest point of approach was at the minimum of the U of the broadband cavitation spectrogram, occurring at 0 s in Fig. 2.14.

Propeller cavitation noise is broadband because it involves the creation, growth, vibration, and collapse of gas bubbles of diverse sizes in the water behind the propeller. There are regions of high and low pressure about a rotating propeller in accordance with hydrodynamic principles (specifically, Bernoulli's principle). The low-pressure regions occur at the tip and suction-side surface of the blades. If the pressure drops below the vapor pressure of water, bubbles are formed. The effect is similar to water boiling, with the difference that boiling is a result of an increase in temperature, whereas cavitation is a result of a decrease in pressure. Cavitation noise can range from as low as 5 Hz to over 100 kHz. At low frequencies, the power increases with

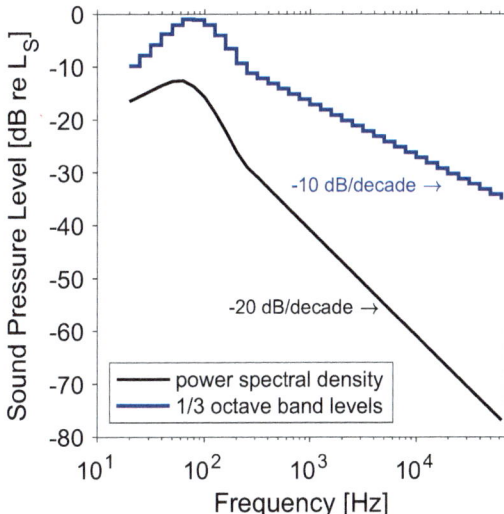

Fig. 2.15 Generalized propeller cavitation spectrum (after Ross 1976) in terms of both power spectral density and 1/3 octave band levels. Levels are referenced to the broadband source level (L_S)

frequency by ~9 dB/octave. There is a peak in power between 50 and 150 Hz. For higher frequencies, the power decreases by about −6 dB/octave (−20 dB/decade) (Ross 1976; Fig. 2.15).

Propeller cavitation noise can be amplitude-modulated by the propeller blade rate, which corresponds to a frequency f_b [Hz] equal to the number of blades b times the rotations per second (Eq. 2.2). The latter is typically expressed as the rotations per minute (rpm), which needs to be divided by 60 s/minute to convert to Hz. Large ship propellers may rotate at 60–300 rpm. With 3–6 blades, the blade-rate frequency lies between 3 and 30 Hz. Smaller boats typically have higher rpm and hence higher blade-rate frequencies.

$$f_b = b \times \frac{\text{rpm}}{60} \quad (2.2)$$

When a propeller cavitates, then this is generally the dominant source. At some speed, called the cavitation inception speed, every propeller will begin to cavitate. Traveling at low speed or deep depth (i.e., at greater hydrostatic pressure, as in the case of submarines) reduces the occurrence of cavitation. The spectral level of ship noise may further depend on the vessel's size, speed, draught, engine type, propulsion type, propeller design, propeller depth, etc. Levels are higher when the ship operates under load (e.g., in the case of tug boats) and during turns and acceleration (e.g., Trevorrow et al. 2008; Urick 1983). Damaged propellers are typically noisier than well-maintained propellers (Urick 1983).

As ship noise originates close to the sea surface, its radiation pattern has dipole character, with the dipole formed by the actual noise source and its image in air. Noise is therefore radiated primarily downward. In the horizontal plane, a frequency-dependent directivity pattern exists, often with stronger low-frequency noise recorded from the sides, shielded toward the front by the hull of the vessel and toward the back by the wake (Arveson and Vendittis 2000; Gassmann et al. 2017; Trevorrow et al. 2008).

Many studies were carried out on the sound signatures emitted by ships during World War II, yielding empirical equations for source levels. The lack of standardization, however, makes a synthesis and comparison of source signatures difficult. Ships were recorded in different environments (e.g., shallow versus deep water), at different hydrophone depths and ranges, at different azimuths, and under different operational modes. Data analysis methods differed (e.g., averaging time window and spectral resolutions). Some studies applied a geometric spreading loss (20 $\log_{10}(r)$ for spherical spreading, where r symbolizes range) while others applied a more sophisticated sound propagation model—both to yield sound levels at a nominal distance of 1 m. While the literature at the time referred to all of these as "source level," recent standardization attempts use different terminology depending on how propagation loss was treated. A sound propagation model (e.g., parabolic equation, see Chap. 1) can account for features of the local environment (e.g., hydroacoustic parameters of the water and geoacoustic parameters of the seafloor) and the specific experiment geometry (e.g., hydrophone depth and range) and thus yield levels that are independent of the specific environment in which the ship was recorded and that are a true characteristic of the ship. On the contrary, adding 20 $\log_{10}(r)$ to levels recorded in the far-field ignores both the specifics of the environment and the experiment geometry. The greatest difference between these two approaches results from the way the sea surface is treated, at which sound reflects and experiences a phase shift (see Chap. 1). At shallow inclinations (i.e., hydrophone at long range from the ship and at shallow depths below the sea surface), propagation loss is greater than 20 $\log_{10}(r)$ and so, application of a spherical spreading term yields lower levels (referenced to a nominal 1 m distance) than an environment-specific sound propagation model. To differentiate between these two methods, the geometric spreading correction yields "surface-affected source levels" by ANSI/ASA S12.64 (American National Standards Institute 2009) terminology and "radiated noise levels" by ISO 17208-1 (International Organization for Standardization 2016) terminology. By ISO standard, only the sound propagation modeling yields a true source level as a characteristic of the source, unaffected by the environment (International Organization for Standardization 2017b). Radiated noise levels and source levels are expressed in dB re 1 $\mu Pa^2 m^2$ (as the level of a power quantity; see International Organization for Standardization 2017b) or in dB re 1 μPa m (as the level of a field quantity; also see Erbe et al. 2022a for an introduction to these quantities and units). Many publications used dB re 1 μPa @ 1 m as notation to quantify source levels—now deprecated. Finally, as sound propagation modeling assumes the sound recorded in the far-field originated from a point source (i.e., a monopole), some literature stresses that a sound propagation model was applied by reporting "monopole source levels."

So, the broadband radiated noise level L_{RN} (i.e., surface-affected source level by ANSI terminology and merely "source level" in the original literature at the time) [dB re 1 μPa m] of large ships as a function of ship speed u [kn] and displacement tonnage T [t] was estimated as Eq. 2.3 from 157 recordings of 77 ships of the classes warships, freighters, and tankers (Ross 1976; Urick 1983). Using L_{RN} as the reference source level, the source spectrum could then be estimated from Fig. 2.15.

$$L_{RN} = 134 + 60 \log_{10} \frac{u}{10 \text{ kn}} + 9 \log_{10} \frac{T}{1 \text{ t}} \quad (2.3)$$

This equation may be interpreted as a power law. With L_{RN} computed as 10 $\log_{10}$ of the recorded mean-square pressure times the square of the measurement distance, the mean-square pressure is proportional to ship speed to the power of 6 ($\sim u^6$) and to ship tonnage to the power of 0.9 ($\sim T^{0.9}$).

As a function of the propeller tip speed u_{tip} [m/s] and the number of blades b, the broadband radiated noise level L_{RN} [dB re 1 μPa m] of World War II ships was estimated as Eq. 2.4 (Ross 1976). This equation was considered valid for vessels longer than 100 m and u_{tip} speeds of 15–50 m/s. Tonnage was indirectly included in Eq. 2.4, as heavier vessels had more blades and higher u_{tip} speed.

$$L_{RN} = 175 + 60 \log_{10} \frac{u_{tip}}{25 \text{ m/s}} + 10 \log_{10} \frac{b}{4} \quad (2.4)$$

Compared to the time of World War II, today's ships typically have more power, greater tonnage, increased length, and more propeller blades, and they operate at higher rpm and speed, theoretically resulting in more noise. However, naval architecture and propeller design have progressed so as to reduce cavitation and increase the cavitation inception speed. Deliberate efforts to quieten ships have been made to improve overall efficiency, to reduce environmental impact, and to undertake unbiased research on fauna, which might otherwise respond to the noise of the research vessel (Bahtiarian and Fischer 2006; de Robertis et al. 2013; Fischer and Brown 2005; Leaper 2019; Mitson 1995; Palomo et al. 2014).

2.5.1.1 Merchant Vessels

Merchant vessels are used for the transportation of goods, material, animals, cars, and people. This vessel class includes container ships, bulk carriers, tankers, livestock transporters, vehicle

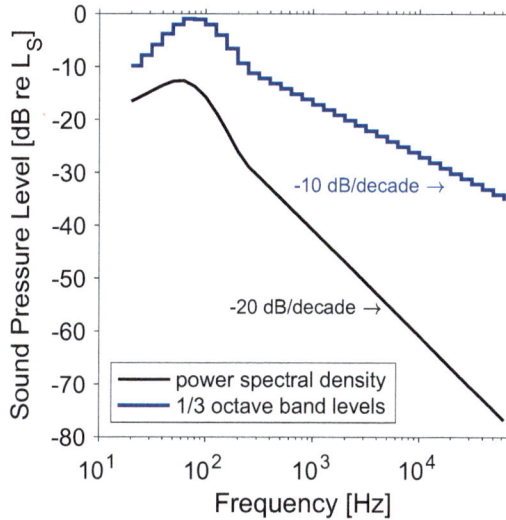

carriers, and passenger ferries. A study of merchant vessels recorded in 1985 confirmed that the generalized cavitation spectrum of Fig. 2.15 together with Eq. 2.4 from World War II ships matched the mean source spectra of 50 merchant vessels within <4 dB over the bandwidth 70–700 Hz (Scrimger and Heitmeyer 1991). The computed monopole source spectra (source depth 6 m) were independent of vessel type (passenger vessel, cargo vessel, tanker), but dependent on the vessel's speed and length. Using the mean source power spectral density level $\overline{L_{S,f}(f)}$ of the 50 merchant vessels, the dependence of the power spectral density level $L_{S,f}(f)$ on speed u [kn] and length l [ft] was derived as Eq. 2.5, which is used in the Research Ambient Noise Directionality model (RANDI-2; Breeding et al. 1996; Hamson 1997).

$$L_{S,f}(f) = \overline{L_{S,f}(f)} + 60 \log_{10} \frac{u}{12\,\text{kn}} + 20 \log_{10} \frac{l}{300\,\text{ft}} \quad (2.5)$$

While the generalized spectrum of Fig. 2.15 addresses the broadband propeller cavitation noise, separate models have been developed for the tonal components related to the propeller blade rate and engine. The empirical Eq. 2.6 gives the monopole source level $L_{S,b}$ [dB re 1 µPa m] of the blade rate harmonics f_b, where ρ is the density of water and d the propeller diameter [m]. The depth of this monopole source was taken as the ship draught minus 85% of the propeller diameter; that is, the depth was 15% of the propeller diameter below the top of the propeller blade arc (Gray and Greeley 1980). This model agreed well with the blade-rate overtones measured from 14 merchant vessels, but underestimated the source level of the blade-rate fundamental (Wright and Cybulski 1983).

$$L_{S,b} = 20 \log_{10} \left(\frac{3}{2\sqrt{2}} 10^{-5} \pi \rho d^3 f_b^2 \right) \quad (2.6)$$

A detailed investigation of the radiated noise of a coal carrier (built in 1977) was published from the *M/V Overseas Harriette* (173 m length, 25,515 t) (Arveson and Vendittis 2000). Measurements were taken in deep water, at keel aspect, and radiated noise levels (i.e., surface-affected source levels by ANSI terminology) were computed by correcting for spherical spreading loss. Recordings showed a typical, broadband cavitation spectrum with a peak around 50 Hz and superposed tones (Fig. 2.16). The cavitation spectrum increased with speed as 104 log (rpm). The diesel generator produced tones at 6 Hz and harmonics. The radiated noise levels of the harmonics at 24 and 30 Hz were

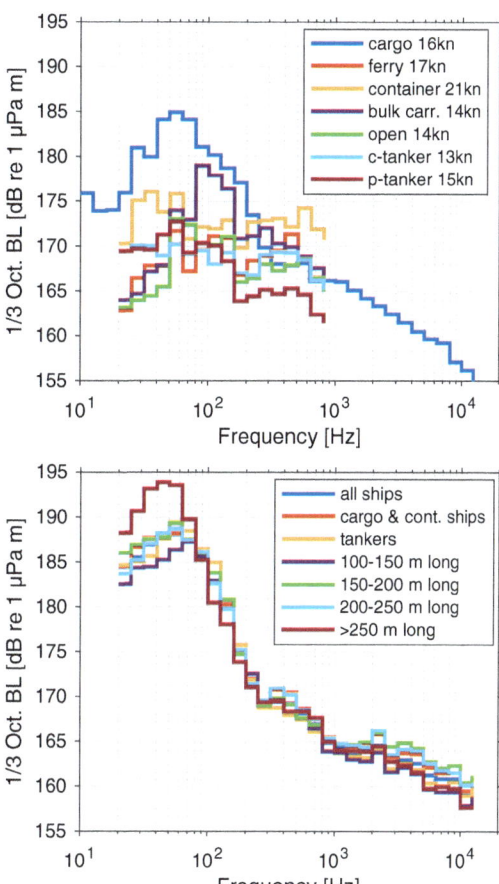

Fig. 2.16 (Top) Radiated noise levels (i.e., surface-affected source levels) in 1/3 octave band. Levels for merchant vessels: a cargo ship (Arveson and Vendittis 2000) and six types of commercial ships: car ferry, container ship, bulk carrier, open hatch, chemical tanker, and products tanker (McKenna et al. 2012). McKenna et al. recorded at low inclination angles, where propagation loss was likely greater than $20\log_{10}r$, yielding lower surface-affected source levels. (Bottom) Median (surface-corrected) source levels of different merchant ship classes (Simard et al. 2016)

179 and 168 dB re 1 μPa m, respectively. These frequencies and their levels were independent of speed. While the diesel generator was the primary source of noise at low speed (8 kn), the diesel engine firing rate became significant at higher speed. The 6-cylinder diesel engine yielded six diesel-cylinder firing cycles per propeller revolution, hence a firing rate of rpm/10 Hz. The firing-rate source level at 16 kn (140 rpm) was 174 dB re 1 μPa m. Overall, the firing-rate level increased with speed, though not monotonically. Resonance modes of the hull at low frequency influenced the level and directivity of the engine tones. Propeller blade-rate tones were dominant at low frequency above the cavitation inception speed of 10 kn (6 rpm). The radiated noise level of the propeller blade-rate tone at 9 Hz was 174 dB re 1 μPa m, when the ship traveled at a speed of 16 kn.

A different study found negligible dependence of the monopole source level on ship type, size, and routine traveling speed of a variety of merchant ships (perhaps due to the vessels in this location traveling at a narrow range of speeds), and a mean source spectrum was derived for frequencies f between 30 and 1200 Hz (Eq. 2.7) (Wales and Heitmeyer 2002). Source depth was computed per Gray and Greeley 1980.

$$L_{S,f}(f) = 230 - 35.9 \log_{10} \frac{f}{1\,\text{Hz}} + 9.17 \log_{10}\left(1 + \left(\frac{f}{340\,\text{Hz}}\right)^2\right) \quad (2.7)$$

McKenna et al. (2013) recorded 500 ship passes in the Santa Barbara Channel (CA) at 500 m depth and correlated radiated noise levels with 19 vessel- and environment-specific covariates in a generalized additive model, finding ship speed and size as the strongest predictors of radiated noise level. One individual container ship was recorded at four speeds and only showed a small increase of the broadband radiated noise level L_{RN} [dB re 1 μPa m] with speed u [kn] according to Eq. 2.8 ($r^2 = 0.9$). Others found a moderate (Allen et al. 2012) or much greater dependence on speed, to the power of 4–7, based on perhaps a greater variety of vessels in their database (Jiang et al. 2020).

$$L_S = 170 + 1.1 \log_{10} \frac{u}{1\,\text{kn}} \quad (2.8)$$

Simard et al. (2016) deployed a vertical hydrophone array in the St Lawrence Seaway (QC), Canada, to estimate the source spectra of 191 merchant vessels over 255 passes. ANSI/ASA S12.64 and ISO 17208-1 standards (American National Standards Institute 2009; International Organization for Standardization 2016) were followed as closely as possible. Propagation loss was estimated by a combination of sound propagation modeling (i.e., wavenumber integration to 300 m range) and an empirical function based on measurements at longer range. Ships were 15.4 ± 12.5 years old, 192 ± 51 m long, 26.5 ± 7.1 m wide, with a draught of 8.6 ± 2.2 m, traveling at 13.4 ± 2.9 kn. Source levels correlated with ship speed, length, breadth, then draught—in order of decreasing correlation. The broadband (20–500 Hz) source level L_S [dB re 1 μPa m] increased with speed u [kn] as Eq. 2.9 ($r^2 = 0.18$).

$$L_S = 186 + 0.827\,u \quad (2.9)$$

Large variability of >30 dB was observed in source level within a ship class, with 20–40 dB between the 1st and 99th percentile 1/3 octave band levels. The median broadband (20–500 Hz) source level was 197 dB re 1 μPa m (range 176–214 dB re 1 μPa m). By ship class, the median levels were 197 dB re 1 μPa m for tankers, cargo vessels, and container ships, and highest (median 201 dB re 1 μPa m) for vessels >250 m in length (Fig. 2.16). Equations 2.5 and 2.7 underestimated levels by 5–13 dB (Simard et al. 2016). Radiated noise levels (i.e., surface-affected source levels) reported by McKenna et al. (2012) were about 15 dB less and those reported by Veirs et al. (2016) were about 5–25 dB less, likely due to an underestimation of propagation loss at low inclination angles.

In 1999, several cruise ships built between 1958 and 1999 were recorded at the U.S. Navy Southeast Alaska Acoustic Measurement Facility (SEAFAC) with a 3-hydrophone vertical array in 370 m deep water. Measurement setup and procedure matched the more than a decade later published ANSI/ASA S12.64 and ISO 17208 standards. The deep-water setup, greater inclination angles, and averaging over the three hydrophones reduced the Lloyd's mirror (surface) and seafloor effects on the resulting source spectra (except at the lowest frequencies; Fig. 2.17; Kipple 2002, 2004).

Gassmann et al. (2017) deployed two vertical hydrophone arrays in the Santa Barbara Channel (CA) and computed source spectra of container ships according to ANSI/ASA S12.64 and ISO 17208. Surface-corrected source levels were up to 20 dB larger than surface-affected source levels (i.e., radiated noise levels), depending on frequency and angle. The *CSCL South China Sea* had a broadband source level of 195 and 209 dB re 1 µPa m, surface-affected and surface-corrected, respectively. Directionality patterns in the vertical and horizontal plane were presented. As expected, the surface-affected radiation pattern resembled a dipole and the surface-corrected radiation pattern had more monopole characteristics.

Chion et al. (2019) compiled 2275 radiated noise and source-level measurements of merchant vessels (https://www.frontiersin.org/articles/10.3389/fmars.2019.00714/full#supplementary-material; accessed 6 Jan. 2023) and correlated these with vessel-specific and methodological covariates (i.e., water depth, hydrophone depth, hydrophone inclination angle, and whether or not surface effects were corrected). They found that speed and size were the strongest predictors of noise level within any one vessel class. They further highlighted what is to be expected: that radiated noise levels (i.e., surface-affected source levels) are less than monopole source levels if the vessel is recorded at low inclination angles and long ranges.

MacGillivray and de Jong (2021) had access to 1862 vessel recordings at different speeds from the Canadian Enhancing Cetacean Habitat and Observation (ECHO) program within the framework of which vessels voluntarily slowed down near Vancouver, British Columbia, in 2017. The range of speeds at which vessels of the same type were recorded was great, with the faster speeds 3–4 times as high as the slower speeds. The authors kept the dependence of source level on speed and length (Eq. 2.5) from the RANDI model, but produced a new monopole reference spectrum $\overline{L_{S,f}(f)}$ with new mean speeds and lengths as reference for each vessel type (Fig. 2.18). The resulting JOMOPANS-ECHO source-level model yields a source spectrum in decidecade bands (between 10 Hz and 31.5 kHz) as a function of frequency, speed, and length for each vessel type. The model is publicly available and allows users to input their own vessel

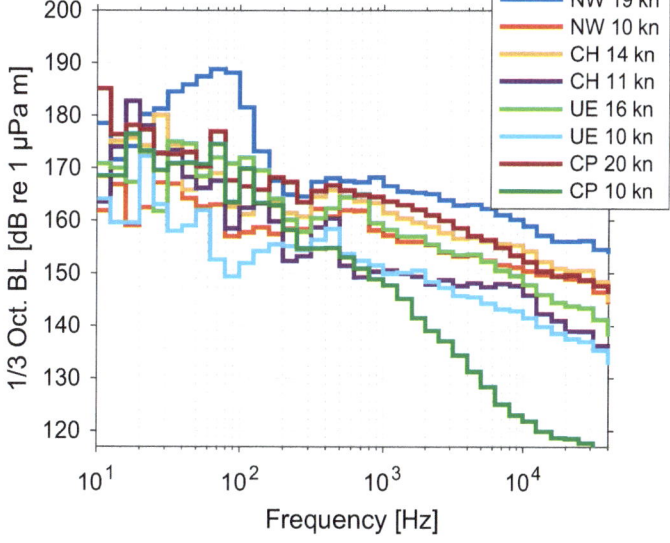

Fig. 2.17 Surface-affected (albeit minimized) source spectra in 1/3 octave bands for cruise ships *Norwegian Wind, Crystal Harmony, Universe Explorer,* and *Coral Princess*—each at two speeds (Kipple 2002, 2004)

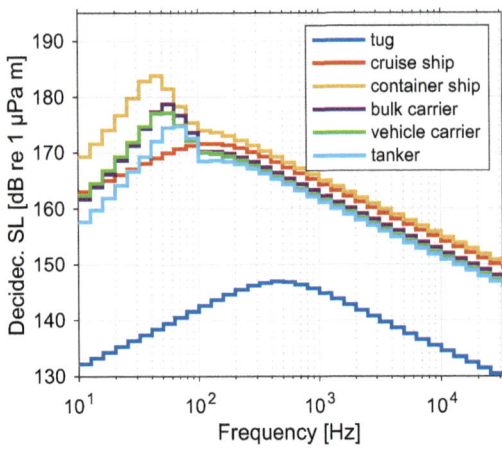

Fig. 2.18 Reference monopole source spectra of the JOMOPANS-ECHO model in decidecade bands. Reference speeds and lengths are bulk carrier: 13.9 kn, 211 m; container ship: 18.0 kn, 294 m; cruise ship: 17.1 kn, 268 m; tanker: 12.4 kn, 186 m; tug: 3.7 kn, 28 m; vehicle carrier: 15.8 kn, 194 m. Source depth = 6 m (MacGillivray and de Jong 2021)

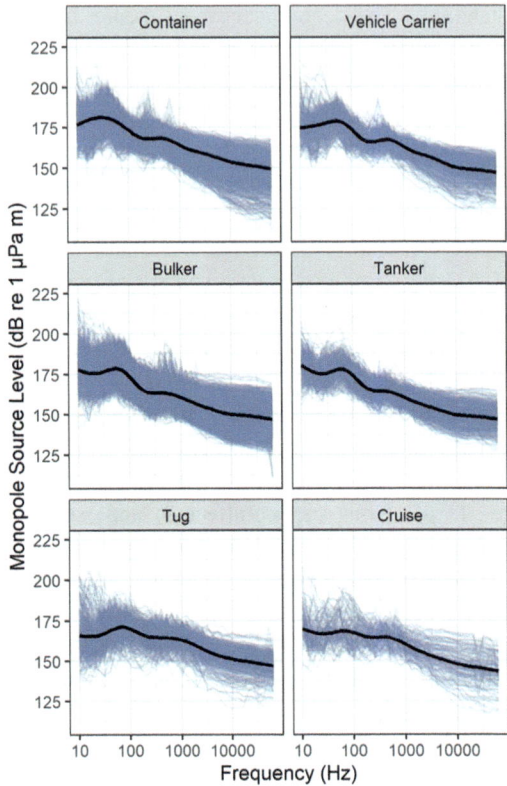

Fig. 2.19 Decidecade-band monopole source levels from >20,000 source level measurements in the ECHO database by vessel type (blue: measurements, black: median). Reprinted from MacGillivray AO, Ainsworth LM, Zhao J, Dolman JN, Hannay DE, Frouin-Mouy H, Trounce KB, White DA (2022) A functional regression analysis of vessel source level measurements from the Enhancing Cetacean Habitat and Observation (ECHO) database. J Acoust Soc Am 152 (3):1547–1563; https://doi.org/10.1121/10.0013747. Published CC BY 4.0; https://creativecommons.org/licenses/by/4.0/. © MacGillivray et al. 2022

specifications to yield the source spectrum (https://www.mdpi.com/article/10.3390/jmse9040369/s1; accessed 3 Jan. 2023). The uncertainty was estimated to be 6 dB (computed as the standard deviation of the differences between measurements and model). The source depth was fixed at 6 m. By 2020, the ECHO program had collected >20,000 source-level measurements of ships (Fig. 2.19). MacGillivray et al. (2022) developed a functional regression model to predict the source spectrum as a function of vessel type, age, draft, speed, length, main engine rpm, and main engine power. This model is also available for download, so users may estimate their own source spectrum based on their own specifications (https://asa.scitation.org/doi/suppl/10.1121/10.0013747; accessed 6 Jan. 2023).

ZoBell et al. (2023) were able to document the changes in radiated noise of 12 refitted container vessels that had their capacity increased by 20% and their draft by 1.5 m. Radiated noise levels increased by up to 2 dB at mid and high frequencies, but source levels decreased by over 5 dB at low frequencies, likely due to improvements in propeller and bow design.

There are not many merchant vessels without external propellers. The radiated noise level of the jet-propelled high-speed car and passenger ferry *M/V Alakai* increased with speed, peaked in the broadside direction, and had a maximum of 197 dB re 1 µPa m (broadband, 1–22,000 Hz) (Rudd et al. 2015).

2.5.1.2 Icebreakers

Icebreakers are powerful vessels that can travel through polar sea ice, often clearing a path for commercial or survey vessels following. Icebreakers have a strengthened hull, shaped to break ice and push it away from the sides of the

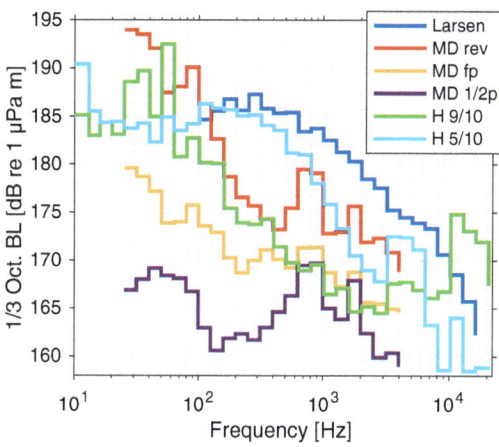
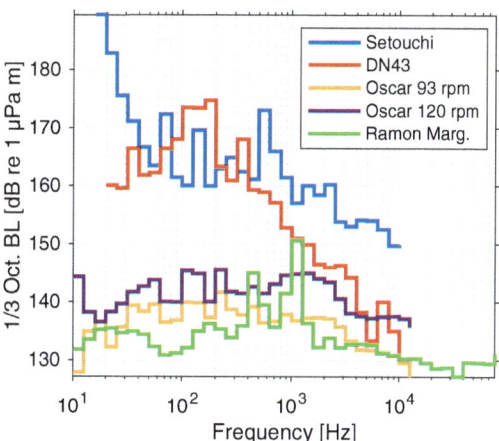

Fig. 2.20 Source levels and radiated noise levels in 1/3 octave bands for icebreakers: *Henry Larsen* while ramming an ice-ridge (surface-corrected; Erbe and Farmer 2000); *John A. MacDonald* reversing, forward at full power, and forward at half power (surface-affected; Thiele 1988); and *Healy* in 9/10 and 5/10 ice cover (surface-affected; Roth et al. 2013)

Fig. 2.21 Source levels and radiated noise levels in 1/3 octave bands of research and survey vessels: *Setouchi* (surface-corrected; Hannay et al. 2004 cited in Wyatt 2008), *DN43* (surface-corrected; Hannay et al. 2004), *Oscar Dyson* (surface-affected; de Robertis et al. 2013), and *Ramón Margalef* (surface-affected; Palomo et al. 2014). Note that the *Oscar Dyson* and *Ramón Margalef* were specifically designed to be quiet for fisheries research

ship. Some icebreakers like the *Henry Larsen* are equipped with a bubbler system that blows high-pressure air into the water to push floating ice away from the ship sides, producing continuous noise with a fairly uniform spectrum between 100 and 1000 Hz. Propeller cavitation noise from icebreakers is amongst the highest of all ships, in particular when backing and ramming ice (Erbe and Farmer 2000; Roth et al. 2013; Thiele 1988; Fig. 2.20).

2.5.1.3 Survey and Research Ships

Survey vessels are used for mapping. They might tow a seismic airgun array when surveying for subsea oil and gas, or operate a hydrographic echosounder, multibeam sonar, or sidescan sonar to map the bathymetry, or operate a multibeam fisheries sonar to survey fish stocks. Research vessels might employ similar equipment, or tow a hydrophone array for passive acoustic monitoring, or conduct visual observations of marine fauna. Efforts have been made to quieten fisheries research vessels in order to avoid data bias. Considerable reductions in noise emission have been achieved (Bahtiarian and Fischer 2006; de Robertis et al. 2013; Fischer and Brown 2005; Mitson 1995; Palomo et al.

2014). Example source spectra are given in Fig. 2.21. The broadband source level of the research vessel *CCGS Vector* increased with speed u [m/s] as Eq. 2.10 (Trevorrow et al. 2008):

$$L_S = 142 + 37.4 \log_{10} \frac{u}{1 \text{ m/s}} \quad (2.10)$$

The horizontal directivity index DI as a function of azimuth α from bow was derived for the same vessel (Eq. 2.11) (Trevorrow et al. 2008).

$$DI = 10 \log_{10} \left(\cos^{1.95}(\alpha - 97^\circ) + 0.08 \right) \quad (2.11)$$

2.5.1.4 Boats

Small boats come in many varieties, with propellers or impellers, with outboard or inboard engines, with steel, aluminum, fiberglass, or inflatable hulls, ranging from a few meters to tens of meters. They are mostly used in riverine, estuarine, and coastal areas. They might be used by industry, government, research, or private persons. Rigid-hull inflatable boats (RHIB or RIB) are common as sea rescue lifeboats, coast guard or military patrol boats, support boats in marine and offshore industry, whale-watching boats, and personal watercraft. They have rigid floorboards or hulls with inflatable gunwales and outboard motors. Studies collating 1400 previously published source-level measurements showed that source levels ranged widely between 135 and 189 dB re 1 µPa m (Lagrois et al. 2023; Parsons et al. 2021). From lowest to highest source levels, boat types were electric ferry, skiff, sailing boat, monohull, RHIB, catamaran, fishing boat, and landing craft. The same order occurred in the study by Wladichuk et al. (2019), although they did not have all of these types. Example source spectra are shown in Fig. 2.22.

Whale-watching vessels, in particular, have had much attention in the marine-mammal impacts literature, and their source spectra and levels have been determined in several studies over the years (Au and Green 2000; Wladichuk et al. 2019). Five different types of RHIB (differing in RHIB length and motor configuration) were recorded in British Columbia, Canada, in 1999, within the framework of a study that assessed the impact of whale-watching on killer whales (Erbe 2002). Pooled with Australian recordings of a smaller RHIB (Erbe et al. 2016a), the broadband (10–48,000 Hz) monopole source levels ranged from 134 to 171 dB re 1 µPa m, increasing with speed u [km/h] as Eq. 2.12, where the 95% confidence intervals for the first and second coefficient were 130–140 and 10–17, respectively ($F = 63$, $p = 8 \times 10^{-13}$). Propeller and engine tones as well as their harmonics were clearly identifiable in the spectra and increased with speed (Fig. 2.23).

$$L_S = 135 + 13 \log_{10} \frac{u}{1 \text{ km/h}} \quad (2.12)$$

Parsons and coauthors (2021) ran a generalized additive mixed model on their collated boat recordings with boat-specific and methodology-specific predictor variables. Speed and closest point of approach were the strongest predictors, with the latter hinting at issues with propagation loss estimates and seafloor and surface effects—perhaps not surprising given that nearly all of these recordings were collected in shallow water. Interestingly and plausibly, the relationship between source level and speed was not always monotonic (i.e., steadily increasing)—in particular in the case of planing hull and semi-displacement vessels.

Research boats have been measured and were not quieter than other boats (Brooker and Humphrey 2016; Parsons and Meekan 2020). A small, 10 m, electric passenger ferry was fairly quiet with a median monopole source level of 146 dB re 1 µPa m (Parsons et al. 2020). Below that, jetskis have been recorded with monopole source levels of 137 dB re 1 µPa m (Erbe 2013). One hot summer's day, we had fun recording ourselves during non-motorized water sports. Recordings were made in the near-field with each of us passing several times over a hydrophone 1 m below. Kayaking, scuba-diving, snorkeling, kicking a boogie board, freestyle, and breaststroke swimming each had distinct sound signatures, and received levels peaked between 110 and 131 dB re 1 µPa (Erbe et al. 2016c, 2017). Standup paddleboarding was not in fashion here yet.

2.5.1.5 Measuring Ship Noise

One difficulty when comparing ship noise levels across studies and when applying noise measurements from one study to model predictions in another study is the inconsistency in measurement, analysis, and reporting approaches. Critical information on how recordings were obtained and analyzed is often missing in reports. There are three commonly referred-to standards for measuring, analyzing, and reporting the noise of ships in deep water (ANSI/ASA S12.64, American National Standards Institute 2009; and ISO 17208-1 and 2, International Organization for Standardization 2016, 2019), in addition to others. The ANSI standard gives three options for measurement and analysis: a survey method, an engineering method, and a precision method (i.e., grades C, B, and A, respectively). The ISO standard only specifies a precision method, which contains aspects of the precision and engineering methods of the ANSI standard. Both standards estimate the

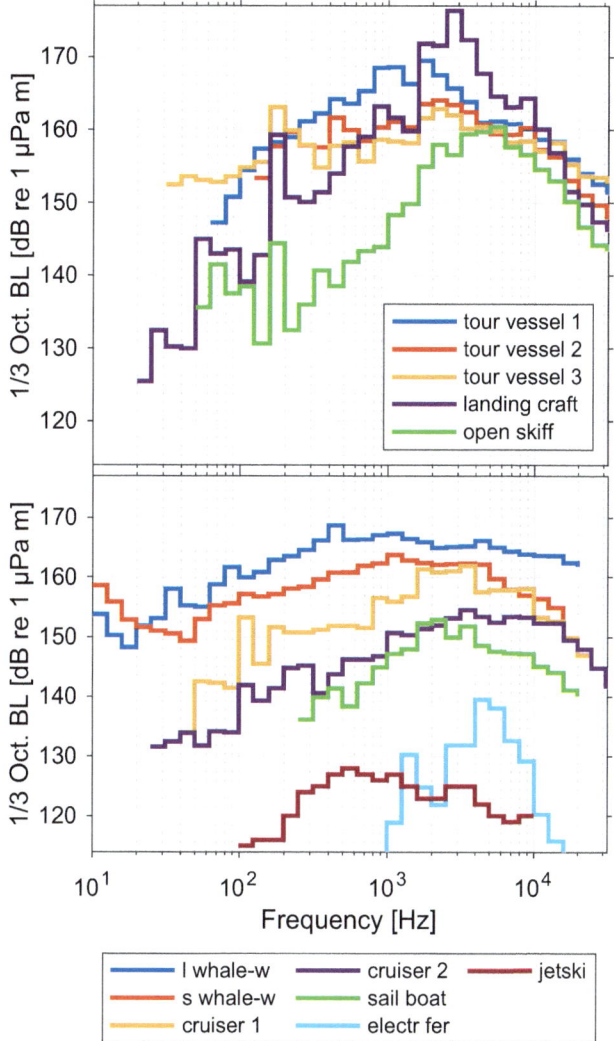

Fig. 2.22 Source levels and radiated noise levels in 1/3 octave bands for small to large boats: (top) tour vessel 1 (*Sea Lion*, 46 m, 800 hp., 10.9 kn, surface-affected; Kipple and Gabriele 2004); tour vessel 2 (*Sea Bird*, 46 m, 800 hp., 10.7 kn, surface-affected; Kipple and Gabriele 2004); tour vessel 3 (*Wilderness Adventurer*, 48 m, 475 hp., 10 kn, surface-affected; Kipple and Gabriele 2004); landing craft (*Capelin*, 8 m, 350 hp., 20 kn, surface-affected; Kipple and Gabriele 2003); open skiff (*Ursa*, 4 m, 25 hp., 16.3 kn, surface-affected; Kipple and Gabriele 2003). (Bottom) Large whale-watching boat median (23–40.2 m, 2×375–3×500 hp., 10–15 kn, surface-affected; Gervaise et al. 2012); small whale-watching boat median (6–12.2 m, 1×100–2×370 hp., 20–25 kn, surface-affected; Gervaise et al. 2012); cabin cruiser 1 (*Arete*, 8 m, 13 kn, surface-affected; Kipple and Gabriele 2004); cabin cruiser 2 (*Talus*, 9 m 420 hp., diesel jet, 13 kn, surface-affected; Kipple and Gabriele 2003); sail boat (*Quintessence*, 9 m, 25 hp., 5 kn, surface-affected; Kipple and Gabriele 2004); electric ferry (surface-corrected; Parsons et al. 2020); and jetski median (surface-corrected; Erbe 2013)

broadside radiated noise level (surface-affected source level by ANSI terminology). The ship has to sail a controlled path in deep water. Hydrophones are deployed in the geometric far-field at depths corresponding to specific inclination angles from the ship (Fig. 2.24). The radiated noise level (= surface-affected source level) is computed using free-field geometric

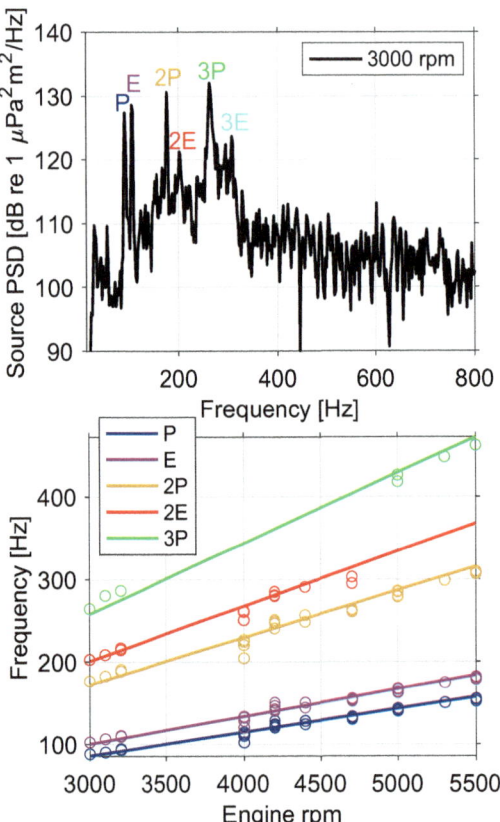

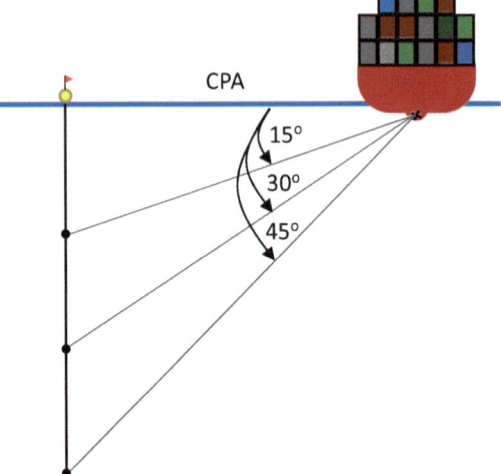

Fig. 2.23 (Top) Example monopole source spectrum of a RHIB operating at 3000 rpm, identifying propeller blade rate (P) and engine firing rate (E), and overtones. (Bottom) These tones increased with engine rpm (Erbe et al. 2016a)

Fig. 2.24 Setup to record ship noise to ANSI/ASA S12.64 and ISO 17208 standards. The depth of the three hydrophones is determined by the required inclination angles of 15°, 30°, and 45°. The distance at closest point of approach (CPA) is the greater of 100 m and the ship length. The vessel goes past the array from both port and starboard sides

(spherical) spreading. For the ANSI grades A and B and the ISO standard, the radiated noise level is computed as the mean of the product of mean-square sound pressure and distance squared over all of the hydrophone recordings at the different depths, so as to average over the interference pattern of the Lloyd's mirror effect. Levels are reported as 1/3 octave band levels. Part 2 of the ISO standard (International Organization for Standardization 2019) explains how to derive monopole source levels based on the measurements described in Part 1, at a standardized source depth of 70% of the ship draft. Alternatively, sound propagation models that are able to account for the specific environment (see Chap. 1) need to be run to calculate monopole source levels. Monopole source levels may then be applied to sound propagation models in new, different environments to predict the noise field. Sound propagation modeling requires knowledge of the monopole source depth, which can be estimated from the measurements taken under these standards using the received levels at the hydrophones in the vertical array (Gassmann et al. 2017).

In shallow water, the determination of vessel SL is complicated by repeated interactions of the propagating sound with the sea surface and seafloor, and multi-path arrivals at the hydrophone (s). Using both vertical and horizontal line arrays, MacGillivray et al. (2023) showed how reliable SL measurements may be performed; they validated their approach with 12079 SL measurements from 1880 individual vessels in BC, Canada. A standard (ISO 17208-3) for the measurement of underwater sound from ships is in development.

2.5.1.6 Mapping Ship Noise

Large ocean-going vessels are typically equipped with an Automatic Identification System (AIS), which sends vessel-identifying information on location at regular intervals (approximately every few minutes). While the purpose is safety (e.g., situational awareness and collision avoidance), the data can also be used to track vessels, compute their density by vessel type, and monitor changes over the years. Using the aforementioned source spectra and source models, underwater noise maps may be produced. This is not a small modeling exercise, unless the model only needs to cover a small area and short period of time. Basically, sound propagation needs to be modeled from every ship position to every receiver on a 3D grid. And a minute later, all ships are in new places. Several shortcuts have been implemented, such as integrating vessel positions over time, to yield vessel density maps, and then producing sound energy maps cumulatively over a certain period (e.g., a season of particular whale presence and activity, or an entire year; Erbe et al. 2012). While the reliability of such shortcuts varies, these maps may be useful for marine spatial planning. For example, they can highlight noise hotspots, and if overlain with animal habitat maps, may inform marine spatial planners on regions of more versus less concern (e.g., Williams et al. 2015). They can be used to compare anthropogenic stressors across regions (e.g., in two neighboring bays where one is being industrially developed and the other is not; Schoeman et al. 2022b). They can be used to compare the input of acoustic energy into the ocean across noise sources and thus put ship noise into context; for example, off South Australia (although not yet the Southern Ocean), wind noise energy dominates ship noise energy except right in the shipping lanes (Erbe et al. 2021).

Jalkanen et al. (2022) used global AIS maps and the source model from Wittekind (2014), but avoided sound propagation modeling by comparing the insertion of noise energy by vessel type, region, and year. They highlighted steady noise increases in some areas, but not others. They identified a COVID-19-related dip in ship noise. They showed that container ships, dry-bulk carriers, and liquid-cargo tankers together emitted 75% of global underwater ship noise in the 63 Hz 1/3 octave band during 2014–2020. The 63 Hz band is one of the two indicator bands in European noise guidelines (Dekeling et al. 2014).

Boats are often excluded from noise maps. They are difficult to model because their numbers, specifications, and locations are mostly unknown. They do not transmit AIS information as large vessels do. However, other systems have been employed to track them, such as marine radar and cameras (e.g., O'Hara et al. 2023). The Marine Monitor (M2) vessel tracking system combines AIS data with radar and photography (Cope et al. 2020). Cope et al. (2021) showed that boats may contribute significant noise under water, in particular near urban areas.

2.5.1.7 Reducing Ship Noise

Ship noise may be reduced through technical design measures or through operational measures. For newly built ships, architecture might include low-noise features in hull design (shape, structure), machinery isolation, or propeller design. Maintenance is another important technical aspect. Operational measures include rerouting from critical habitat during critical seasons or speed limitations. Technical design options were reviewed by Smith and Rigby (2022). More discussed in the marine bioacoustics literature have been operational measures.

All of these measures might not affect all ships equally. Williams et al. (2019) showed through measurement and modeling that removing or retrofitting the noisiest vessels may reduce ambient noise levels near shipping routes significantly while only affecting a small number of vessels. Rerouting or reducing speed would affect the majority of vessels. The authors further considered convoying; grouping ships into convoys would allow for quiet times in between trips, during which animals might recover from the potential effects of noise exposure. Veirs and collaborators, based on their database of 1582 ship source levels at the time, demonstrated that half of the total noise power emitted by ships in Haro Strait (BC, Canada, and WA, USA) came from 15% of the noisiest vessels (Veirs et al. 2016, 2018). MacGillivray et al. (2019) demonstrated, based on the ECHO database, that slowing down to 11 kn reduced vessel source

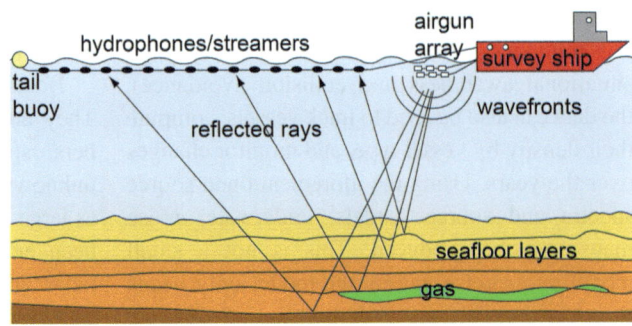

Fig. 2.25 Sketch of a marine seismic survey

levels of container ships, cruise ships, vehicle carriers, tankers, and bulkers by 11.5, 10.5, 9.3, 6.1, and 5.9 dB, respectively. Vessel length (design feature) and speed and draft (operational features) were the strongest correlators with noise emission of large commercial ships (MacGillivray et al. 2022). Joy and team (2019) demonstrated from the ECHO voluntary slowdown that noise levels could be reduced by up to 3.1 dB in the 10–100 Hz band (less at higher frequencies) in critical southern resident killer whale habitat, 2.3 km away from the nearest shipping lane. Even though slower vessels spent more time traversing the area, their model showed that the slowdown yielded a 22% reduction in lost foraging time of killer whales.

2.5.2 Marine Seismic Surveys

Marine seismic surveys are carried out for geophysical research, geotechnical investigations, and petroleum exploration (to detect oil and gas reservoirs within the seabed), and they can provide images of the Earth's crust down to 10 km or more below the seafloor. These surveys involve using a suitable sound source to regularly transmit pulses or bursts of low-frequency sound and then receiving the resulting reflected and/or refracted energy on an array of hydrophones. Analysis of the hydrophone signals is then carried out to provide information about the seabed. Seismic surveys may be conducted by various groups, such as a research institution wishing to understand crustal geology, an engineering consultant seeking information on shallow seabed geology for offshore structural purposes, a petroleum company aiming to locate potential petroleum fields in their lease area, or a geophysical contractor running a seismic survey across multiple petroleum leases (multi-client survey).

A number of different seismic survey configurations can be used, but in the most common, a survey vessel tows the seismic source and the acoustic receivers (usually one or more streamer cables, each of which can be several kilometers in length and contain hundreds of hydrophones). The source directs a series of acoustic pulses into the seafloor where they are reflected off layers of different acoustic impedance. The reflected pulses travel back into the water and are recorded by the streamers (Fig. 2.25). The depth and impedance of the seafloor layers can be estimated from the reflected pulses (i.e., their arrival time and spectral power).

Surveys of this basic configuration include 2D surveys, which use a single source and streamer to produce a vertical cross-section through the seabed along widely separated survey lines, and 3D surveys, which use two (flip-flop in seismic parlance) or three (flip-flop-flap) alternating sources, multiple (typically 14 or more) streamers (spaced perhaps ~100 m apart), and closely spaced survey lines to produce a 3D picture of the seabed geology. Such surveys may take weeks or, for large 3D surveys, months to complete. Multi-client surveys might take even longer.

Surveys in which the hydrophones are located on the seafloor and only the source is towed are often used for 4D surveys for purposes such as monitoring the extraction of oil or gas from a well as a function of time. Other common techniques include vertical seismic profiling (VSP), in which the hydrophones are located in a vertical borehole, and deep-towed seismic surveys, in which the source and hydrophones are towed close to the seafloor to gather information about the acoustic wave speeds in the near-surface geology.

The vast majority of these surveys are carried out using arrays of individual sources called airguns, which are described next.

2.5.2.1 Single-Airgun Acoustics

An airgun is a cylindrical container, about 10–20 cm in diameter. It contains an air chamber, the volume of which is typically measured in cubic inches and may range from 10 to 800 in^3 (0.00016–0.013 m^3). Highly compressed air (~2000 psi or 14 MPa) is pumped into the air chamber during charging. The air chamber is opened and the air rapidly released during firing. This generates a large bubble, which rapidly expands until the pressure inside is below the ambient water pressure, at which stage it collapses again. It overshoots the ambient pressure during each bubble pulse and hence oscillates around ambient pressure, generating an acoustic wave (Fig. 2.26a). Given that the bubble is created below the sea surface, its sound is reflected off the sea surface where it experiences a 180° phase shift (positive pressures become negative, and vice versa). The direct and the surface-reflected waves overlap and at a receiver below the bubble, the received pressure goes from a sharp positive peak to a sharp negative peak, followed by a series of decaying oscillations called bubble pulses (Fig. 2.26b). The acoustic signature of a single airgun is characterized by its peak pressure and bubble period (i.e., time between two consecutive bubble pulses). These depend on the airgun size, firing pressure, and depth below the surface where it is fired. The peak pressure is proportional to the firing pressure and to the airgun volume to the power of 1/3. The bubble period is proportional to the firing pressure to the power of 1/3 and to the airgun volume to the power of 1/3 (Caldwell and Dragoset 2000; Dragoset 2000). The sound from a single airgun in deep water would be omni-

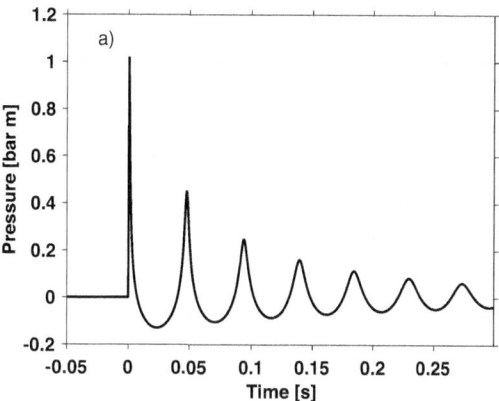

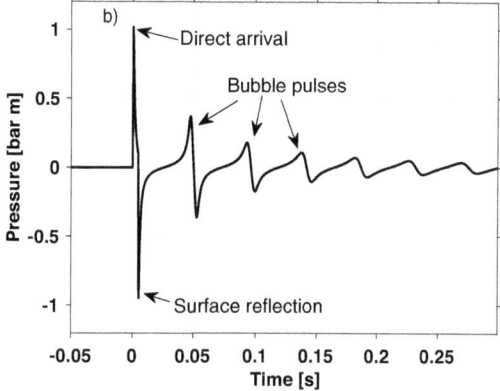

Fig. 2.26 Modeled acoustic signature of a single 20 in^3 airgun at 3.6 m depth without surface reflection (**a**) and with surface reflection (**b**) having a peak pressure of 1.0 bar m and a bubble period of just under 50 ms

directional. However, fired a few meters below the water surface, it is mostly directed downward due to surface reflection.

2.5.2.2 Airgun Array Acoustics

Airgun arrays may consist of two or more subarrays each containing about a dozen airguns. Common volumes of airgun arrays are 2000–8000 in^3. Airgun arrays are typically towed 5–8 m below the water surface, firing every 5–20 s with firing interval set by the depth of water to traverse and the maximum target depth required.

For seismic imaging, the primary pulse from its onset to its peak pressure, then negative pressure, and back to ambient is the useful part; the bubble pulse is undesired as it smears the image.

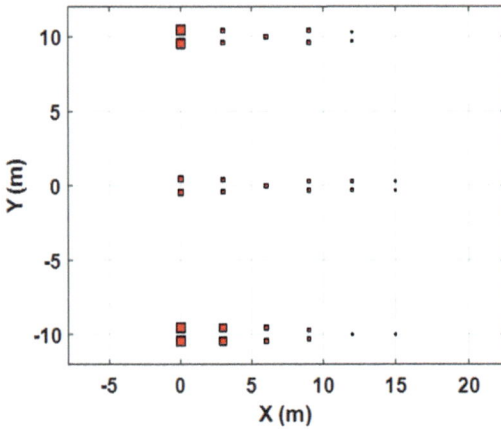

Fig. 2.27 Plan view showing the layout of a notional 3480 in^3 airgun array typical of those used for 3D marine seismic surveys. Airguns are shown much larger than, but proportional to, their actual size. The direction of tow would be from right to left (along track). This array consists of 30 individual airguns arranged in three sub-arrays, with most of the guns in clusters

A stronger (i.e., having higher peak pressure) and sharper (i.e., reducing the bubble pulse), downward-focused seismic signal is achieved by arranging a number of airguns with different volumes into a horizontal, rectangular array (Fig. 2.27). Such arrays can be tuned to maximize the primary-to-bubble ratio, which is the ratio of the primary peak-to-peak amplitude to the bubble-pulse peak-to-peak amplitude. The primary pulse is enhanced by simultaneous firing and constructive superposition of the primary pulses of the individual airguns. The bubble pulse is reduced because the periods of the differently sized guns differ and so the bubble pulses from the different guns cancel each other out. Another strategy used to reduce the bubble pulse is to arrange airguns in closely spaced pairs known as clusters, in which the acoustic and hydrodynamic interactions between the two bubbles increase the rate of decay of their oscillations. The peak of the primary pulse is directly proportional to the number of airguns in the array but only to the cube-root of the total volume of the array. So to increase the pressure pulse available for seismic imaging, it is much more efficient to use many small airguns rather than few large airguns (Caldwell and Dragoset 2000; Dragoset 2000). A limitation of how small the individual airguns in an array can be is the lower frequency required for imaging, with only the larger airguns (>250 in^3) producing peak energy at frequencies below ~10 Hz.

An airgun array is highly directional with the radiated sound being a function of azimuth (i.e., angle in the horizontal plane measured from the tow direction), inclination (i.e., angle in the vertical plane measured from the horizontal), and frequency. Figure 2.28a shows the modeled waveform and Fig. 2.28b the spectrum of the airgun array shown in Fig. 2.27 in several different directions. In the vertically downward direction, the maximum spectral levels occur below 200 Hz. While the highest levels occur vertically downward, airgun arrays also radiate broadband sound at near-horizontal inclination angles, and this sound can travel over long ranges in the water and potentially poses a risk to marine mammals. Figure 2.28 also shows the waveforms and spectra close to horizontal (15° inclination) in the along-track direction (i.e., toward the rear of the array) and in the across-track direction (i.e., perpendicular to the tow direction). Sound emission close to the horizontal can be considerably less than vertically down, particularly at low frequencies; however, it is quite common for levels in the across-track direction in the 100–400 Hz band to be similar to those vertically downward, as is the case in the example shown in Fig. 2.28. This example also illustrates that at near-horizontal inclination angles, airgun arrays typically produce higher sound levels in the across-track direction than along-track. The corresponding horizontal beam pattern as a function of frequency is shown in Fig. 2.28c. Example waveform, spectra, and directivity patterns of a rather different, smaller array, consisting of generator-injector (GI) guns that have reduced bubble-pulse amplitudes can be found in Erbe and King (2009).

The spectrum below the array in Fig. 2.28b has a deep notch at 125 Hz. This is called the ghost notch and it is due to the surface reflection. With a 180° phase shift at the sea surface, the direct wave and the surface-reflected wave cancel out for frequencies that are integer multiples of the

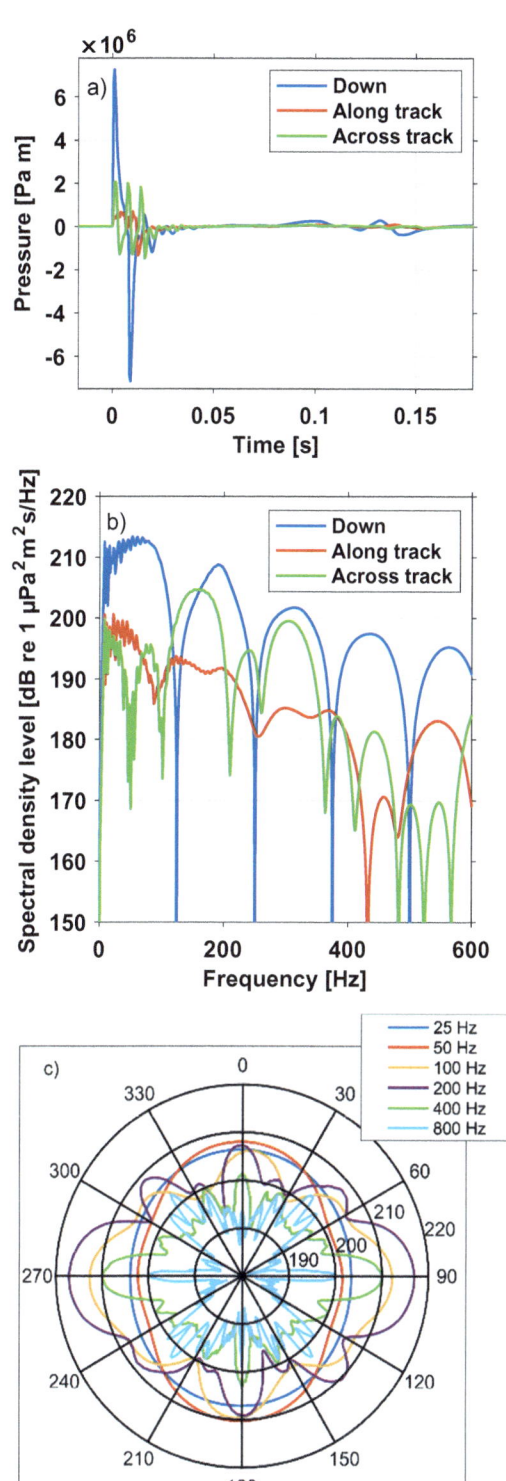

Fig. 2.28 Modeled airgun array waveforms (**a**) and sound exposure spectral density levels (**b**) for the array shown in Fig. 2.27, including the surface reflections. Plots are given

speed of sound divided by twice the tow depth (Caldwell and Dragoset 2000). The frequency at which the notch appears can be tuned by changing the tow depth of the airgun array. Towing at a greater depth yields a lower-frequency notch. Towing closer to the surface ideally moves the notch above the frequency band of interest to the geophysicist, but, due to the increased destructive interference at low frequencies, results in a reduction in the radiated sound power, reducing the maximum depth into the seabed that can be imaged.

Figure 2.28b and c show the spectra that relate to transmission into the far-field. An airgun array is not a point source. Quite the opposite, it is an extended source of tens of meters in both horizontal directions. By definition, the acoustic far-field commences at a range where the received pressure drops as $1/r$ with range, which in practice requires the distance from the center of the array to the receiver to be many times the dimensions of the array. This is far enough so that the contributions from individual airguns arrive with time differences that depend only on direction. In directions in which the receiver is effectively the same distance from multiple airguns (e.g., vertically downward or across-track), their signals add constructively, leading to higher sound levels. Backpropagating the sound received in the far-field to a nominal 1 m range from an infinitesimally small point source yields a far-field equivalent source level, meaning a level that can be used in sound propagation models to estimate received levels in the far-field. A different approach that generates a far-field solution is given in MacGillivray and Chapman (2012). The far-field equivalent source level cannot be used to estimate received levels in the near-field;

Fig. 2.28 (continued) for the vertically downward direction and for near-horizontal radiation in the along-track and across-track directions. The array depth was 6 m. Near-horizontal (15° inclination) beam patterns modeled for the same array (**c**). Tow direction is up (0°). Circumferential grid lines are in 10 dB steps and show one-third octave sound exposure band levels for the six selected center frequencies in dB re 1 $\mu Pa^2 m^2 s$. The beam patterns include the effect of the surface reflection. All levels are source levels (i.e., referenced to 1 m)

in fact, it would overpredict levels in the near-field. This is because in the near-field, the pulses from the different airguns arrive at different times, reducing the combined peak pressure below that predicted from the far-field equivalent source level. The maximum pressure that can be measured very close to an array is likely to be that due to whichever airgun is closest to the receiver.

2.5.2.3 Propagation of Airgun Signals

The spectral characteristics of sound received from an airgun array depend on the array design (i.e., number of airguns, airgun volumes and spacing), tow-depth, firing timing, and firing pressure, as well as sound-propagation conditions of the environment (i.e., bathymetry, hydro- and geoacoustic profiles). Far-field equivalent peak-to-peak source levels of up to 259 dB re 1 µPa m vertically down, 230 dB re 1 µPa m horizontally, and source sound exposure levels (SEL) of up to 224 dB re 1 $\mu Pa^2 m^2$s horizontally have been estimated (Caldwell and Dragoset 2000; MacGillivray 2006). Detailed sound source measurements were made up to 50 kHz of the most commonly used airguns covering a range of volumes and firing pressures (Mattsson 2010). At long ranges of 750 m from a 2120 in^3, 2D seismic survey off the British Isles, spectral density levels of 140 dB re 1 μPa^2/Hz at 200 Hz and 90 dB re 1 μPa^2/Hz at 20 kHz were measured (Goold and Fish 1998).

Long-range detections across ocean basins have been reported. Autonomous hydrophones were moored along the Mid-Atlantic Ridge for the monitoring of earthquakes and recorded repetitive (every 10–20 s) sounds from distant seismic surveys all year round, originating from eastern Canada, western Africa, and Brazil, often >3000 km away (Nieukirk et al. 2004, 2012). Similarly, a seismic survey off Australia was recorded >3000 km away by hydrophones in Antarctica (Gavrilov 2017).

Seismic pulses that propagate over long ranges may experience significant spectral change and can change from brief, broadband pulses to longer-lasting, narrow-band, frequency-modulated sounds (Greene and Richardson 1988; Guerra et al. 2011). Regional geoacoustic properties of the seabed influence the sound propagation loss and its frequency dependence to a great extent. Figure 2.29 compares spectrograms of series of impulsive airgun signals recorded at ranges of 40–50 km from offshore seismic surveys conducted over two different parts of the Australian continental shelf: on the Northwest Shelf (top) and in Bass Strait (bottom). The top panel illustrates acoustic waveguide and frequency dispersion characteristics of low-frequency impulsive sound signals propagating in the water column over soft, fluid-like seafloor sediments, which is considered in detail in a model of sound propagation in shallow water suggested by Pekeris (1948). The arrival of at least three normal modes propagating in the underwater sound channel can be distinguished in the spectrogram, with the modal group velocity decreasing with the frequency decrease, which is consistent with Pekeris' model. The bottom panel of Fig. 2.29 also shows distinct arrivals of four waveguide normal modes, although the intensity of modes 3 and 4 seen at about 25 and 32 Hz is noticeably lower than that of modes 1 and 2 at lower frequencies. However, the frequency dispersion characteristic of individual modes looks opposite to that seen on the top panel: The group velocity of individual normal modes decreases with increasing frequency. Moreover, the arrival of individual modes can be seen only in relatively narrow frequency bands. Even in those frequency bands of noticeably higher intensity of received airgun signals, the sound propagation loss is as high as that expected from the spherical spreading of acoustic energy. These effects are thoroughly modeled and explained in Duncan et al. (2013), where sound propagation was modeled over a layered, highly elastic solid model of seafloor sediments, which was based on geotechnical data from boreholes collected in the measurement area in Bass Strait.

2.5.2.4 Modeling Airgun Signals

The ability to numerically model sound production by airgun arrays has been of considerable interest to the offshore seismic survey industry since airguns began replacing explosives as the source of choice for marine seismic surveys in the

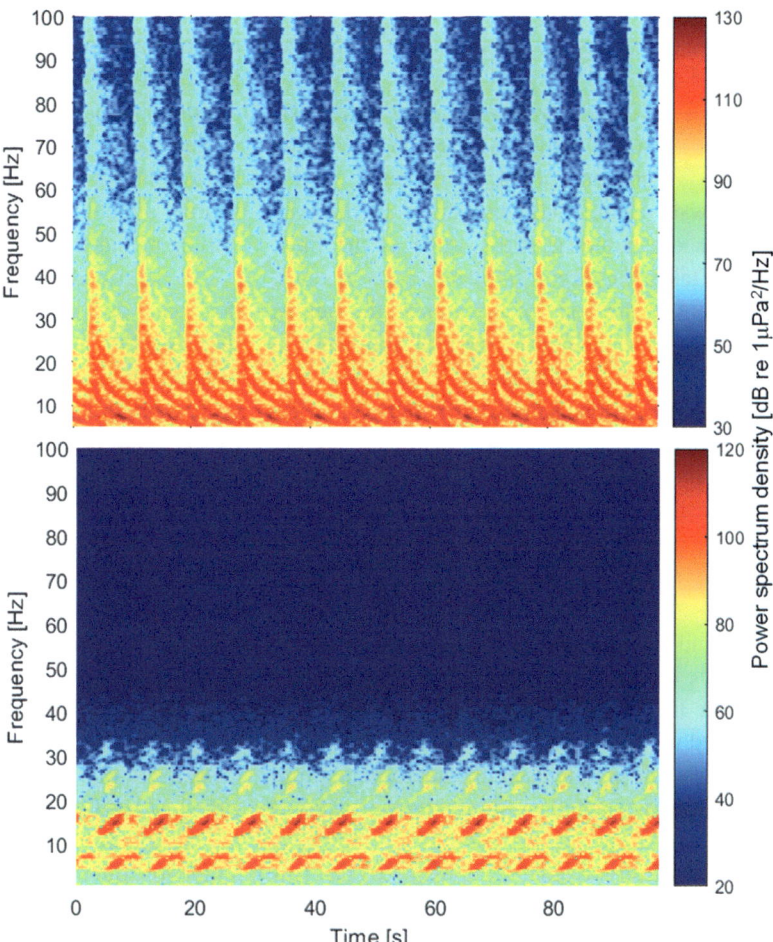

Fig. 2.29 Spectrogram of airgun signals received from offshore seismic exploration surveys at a range of ~50 km on the Northwest Shelf of Australia (top) and ~ 40 km in Bass Strait (bottom)

1970s. The motivation for numerical model development was twofold: to allow the design of tuned airgun arrays that produced the highest possible primary-to-bubble ratio in the vertically downward direction, and to provide the accurate acoustic waveform predictions required to produce high-resolution images of the subsurface geology. This led to the development of several commercial numerical models optimized for the needs of the seismic survey industry, including Gundalf (https://www.gundalf.com) and Nucleus (Long 2022). One characteristic of these early models is that they were calibrated over the frequency range of interest for seismic surveying and therefore to maximum frequencies of a few hundred Hz.

In the early 2000s, the growing concern about the potential impact of the sound produced by marine seismic surveys on marine life led to a need to predict airgun array output over frequency ranges extending to 1 kHz and beyond, which led to the development of several new numerical models optimized for this purpose, including AASM (MacGillivray 2019), Cagam (Duncan and Gavrilov 2019), and AGORA (Sertlek and Ainslie 2015; Sertlek and Blacquière 2019). In the meantime, both Gundalf and Nucleus were extended to allow modeling to the higher frequencies required by environmental impact assessments. Results produced by most of these models were compared at the International Airgun Array Modeling Workshop held in

Dublin, Ireland, 2016 (Ainslie et al. 2019), and at a second workshop held in Cambridge, England, 2022.

All of the models are based on the same basic physics of sound radiation from an oscillating bubble (Johnson 1994; Rayleigh 1917; Ziolkowski et al. 1982) but differ in the way they handle nonideal effects such as the finite time it takes the airgun's exhaust ports to open, the effect of the body of the gun on the bubble oscillation, and the development of turbulence in the bubble wall. Interaction effects between nearby guns are also handled differently in different models. The result is a set of coupled, nonlinear differential equations that is solved numerically to calculate the motion of each airgun's bubble wall, which is then used to calculate the sound field radiated by that particular airgun.

Sound levels within a few kilometers of the array are best calculated by using a suitable propagation model (see Chap. 1) to calculate the received waveform due to each individual airgun, offsetting this in time using the appropriate propagation delay, and then summing the received waveforms from all the airguns in the array (e.g., MacGillivray and Chapman 2012). This approach includes the array nearfield effects and, depending on the choice of propagation model, can be accurate even within the array itself (providing the receiver is outside the bubbles). However, this approach is computationally demanding, and not usually suitable for modeling propagation over ranges of tens to hundreds of kilometers.

For longer ranges, it is sufficiently accurate and much more computationally efficient to treat the array as a directional point source and then use this together with a suitable propagation model to calculate the acoustic field at the receivers. The directional source spectrum is obtained using an airgun array model by summing the received signals from the individual airguns at a position a large distance from the array in the desired direction. The distance is arbitrary but must be much larger than the dimensions of the array. The received signal is calculated assuming spherical spreading, with a propagation delay appropriate to the gun location and propagation direction. The combined signal is scaled to an equivalent source signal by multiplying by the distance, and then its spectrum is calculated. Note that the surface reflection (ghost) is not included when carrying out the source directionality calculation, as its effect will be included later by the propagation model.

As described previously, airgun arrays have horizontal and vertical directionality. Horizontal directionality is readily taken into account by running the propagation model separately for each azimuth and using the source spectrum for that azimuth in the corresponding received level calculation. The vertical directionality of airgun arrays is typically weak over the relatively small range of inclination angles around the horizontal that are responsible for most long-range propagation, but if that is not the case, then it can be included by using a propagation model that can model a source with vertical directionality.

An example of a modeling setup using this second approach is shown in Fig. 2.30. In this case, the notional 3480 in^3 airgun array shown in Fig. 2.27 has been arbitrarily placed about 50 km off the southwest coast of Australia and oriented so the tow direction is from south to north (parallel to the bathymetry contours as per normal 3D seismic survey practice). The array is at a depth of 6 m, so the near-horizontal beam pattern plot shown in Fig. 2.28c gives a good indication of its directionality, and the across-track maxima shown in that plot are oriented offshore and inshore. Propagation modeling was carried out using the parabolic equation model, *RAMGeo* (see Chap. 1).

A map of the maximum computed SEL at any depth is shown in Fig. 2.31, and a west-to-east vertical cross-section through the modeled acoustic field is shown in Fig. 2.32. These plots show a stark contrast between received levels offshore and inshore from the source, with the inshore levels reducing much more rapidly with increasing range than the offshore levels do. This is a propagation effect and a result of the seabed reflections that occur as sound travels upslope steepening the rays, leading to more frequent and more lossy bottom interactions, whereas

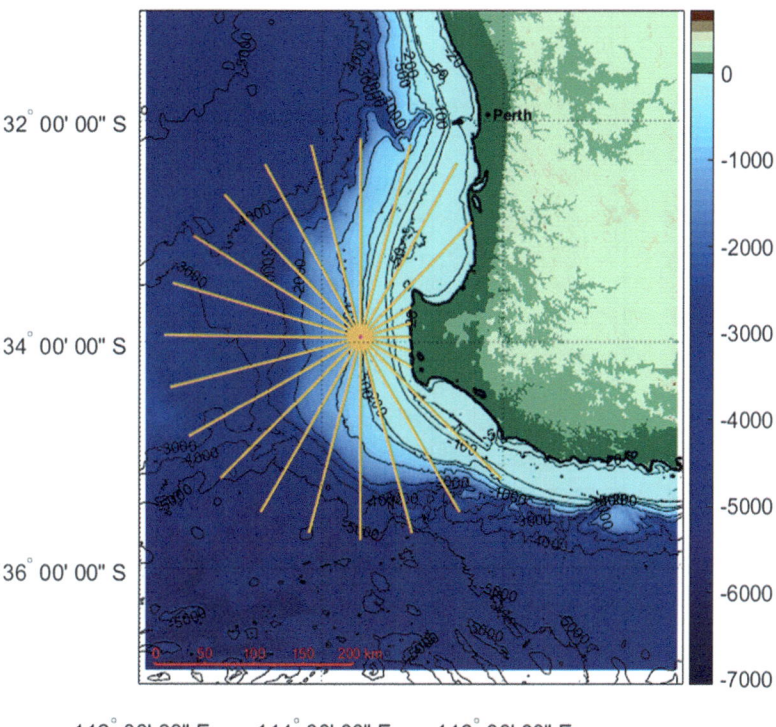

Fig. 2.30 Propagation modeling setup for calculating received levels from the notional 3480 in³ seismic survey source shown in Fig. 2.27. The source is assumed to be off the southwest coast of Australia and the tow direction is assumed to be from south to north. The propagation model *RAMGeo* was run for bathymetry profiles corresponding to the yellow lines that radiate out from the source location

downslope the bottom reflections flatten the rays, which can then couple efficiently into the deep sound channel (Gavrilov 2017). The directionality of the array itself is superimposed on the strong left–right asymmetry caused by the propagation, as can be seen in Fig. 2.31.

2.5.2.5 Reducing Far-Field Received Levels

The seismic survey industry has taken two main approaches to developing marine seismic sources with less environmental impact than traditional airgun arrays. In both cases, the aim is to maintain the same acoustic energy as a traditional airgun array in the frequency band used for seismic imaging (typically 5–100 Hz), while minimizing the source output at higher frequencies.

The first approach has been to modify the design of a traditional airgun to increase the rise time of the initial pressure pulse, which has the effect of reducing the high-frequency content of the signal. An example of this approach is the Teledyne Bolt eSource (Coste et al. 2014), which is marketed as being a drop-in replacement for a conventional airgun, requiring minimal modification to existing seismic survey equipment and operating procedures (Teledyne 2017). The eSource can be configured in three different ways to provide different trade-offs between the reduction of high-frequency energy and the total acoustic energy produced by the source. From source spectra provided by the manufacturer, the reduction in source spectral density level achieved by the eSource at 200 Hz for the A-, B-, and C-configurations is approximately 2.1 dB, 13 dB, and 16 dB respectively. These reductions do, however, come at the cost of some reduction in energy in the band of interest for seismic surveying, particularly in the C-configuration.

Li and Bayly (2017) carried out a modeling study of the likely reductions in marine mammal exposure levels that would be achieved if a proposed seismic survey using a 3147 in³ airgun array was switched from conventional airguns to eSource airguns in the A-configuration. Ignoring propagation effects, they predicted reductions in M-weighted SEL of 0.8 dB, 6.4 dB, and 7.6 dB for LF, MF, and HF cetaceans, respectively. They

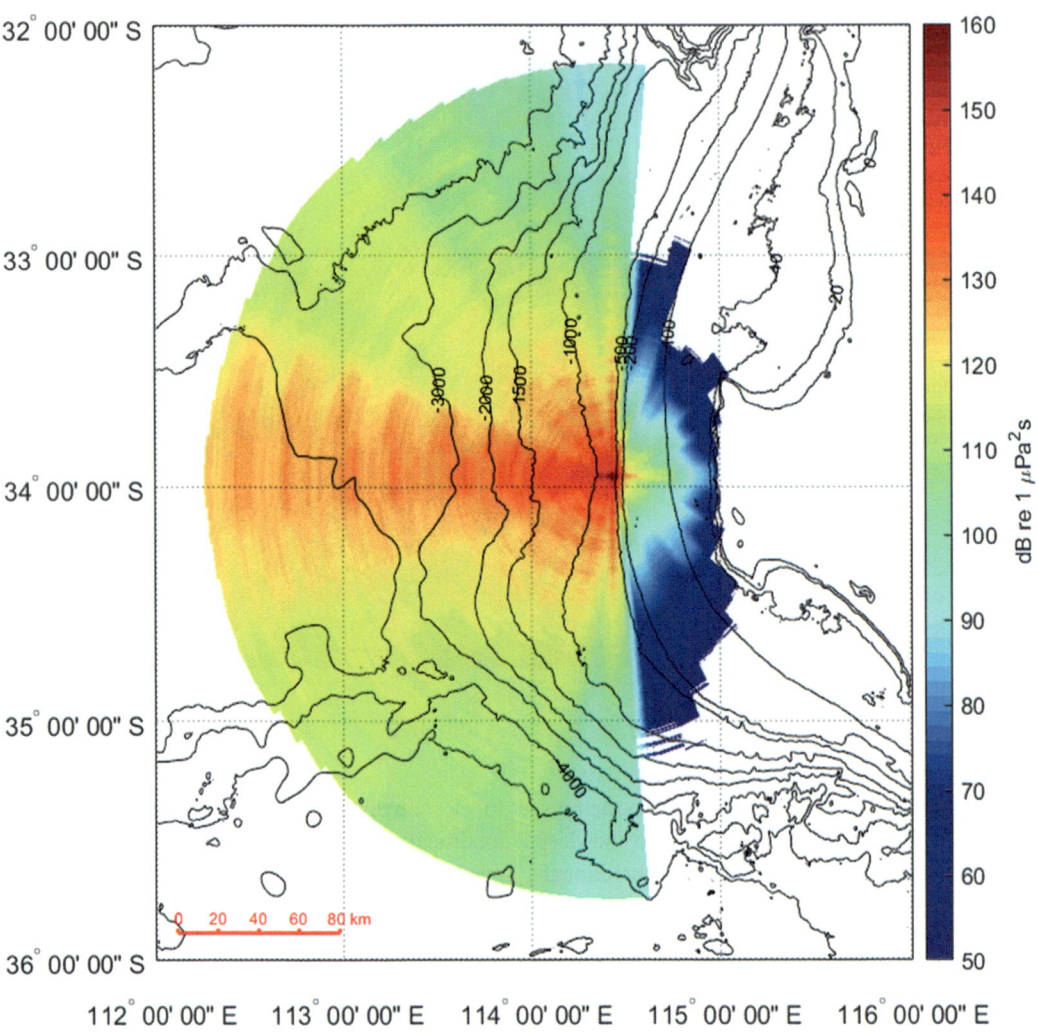

Fig. 2.31 A map showing the maximum predicted sound exposure level at any depth (unweighted) for the setup shown in Fig. 2.30 with water depth contours overlaid. Modeling extended to a maximum range of 200 km from the source, or shorter if the coastline was met. Levels below 50 dB re 1 µPa²s are not plotted

also combined their array model with propagation modeling for a shallow-water environment (31 m water depth), which indicated that the eSource would reduce maximum unweighted SELs by about 1.3 dB at 200 m range, 2.4 dB at 500 m range, 4.2 dB at 1 km range, and 5.3 dB at 2 km range. The greater reductions at longer ranges are a result of interference between the direct and surface-reflected acoustic paths, which shift the peak in the received spectrum to higher frequencies as range increases, particularly in shallow water. It should be noted that this study was carried out using the A-configuration of the eSource, which gives the smallest reduction in high-frequency energy of the three configurations, substantially larger reductions in long-range levels would be expected if either the B- or C-configurations were used.

The second approach to reducing high-frequency energy is to move away from airguns completely and instead use an electrically powered source that can transmit any desired waveform. This type of source is often referred to as a marine vibroseis as it is analogous to the

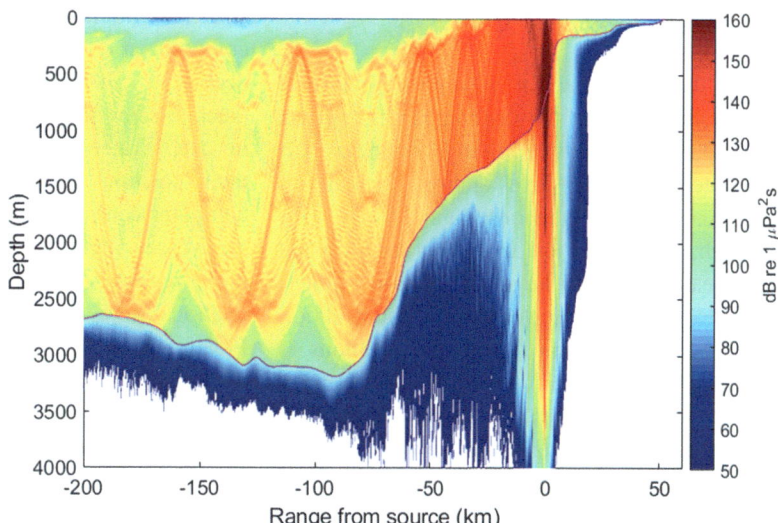

Fig. 2.32 West-to-east vertical cross-section through the predicted sound field. The source is at zero range and 6 m depth. Negative ranges are to the west of the source, positive ranges are to the east. The magenta line shows the seafloor. Colors represent predicted received sound exposure level in dB re 1 µPa²s

vibratory sources that have been used for land-based seismic surveys for many years. Instead of a short, high-amplitude pulse, these sources produce a much longer, lower-amplitude sinusoidal signal that sweeps through the desired range of frequencies. The received signals are then processed digitally to compress the energy contained in the sweep into a very short time interval, and the compressed signals are subsequently used for seismic imaging.

From an environmental perspective, the advantages of marine vibroseis relative to airguns are much reduced peak sound pressure levels (SPL_{pk}) leading to a greatly reduced risk of hearing damage to animals close to the source, and the practical elimination of transmitted energy at frequencies outside of the seismic band of interest. The disadvantage is that the increased signal duration could lead to a greater risk of masking for animals in some circumstances (LGL and MAI 2011).

Duncan et al. (2017) carried out a modeling study comparing a proposed marine vibroseis array based on sources developed by Petroleum Geo Services (PGS) to a conventional airgun array that produced similar acoustic energy in the downward direction at frequencies less than 100 Hz. This study considered three different propagation scenarios (shallow, deep, and slope) and found that the use of marine vibroseis would reduce peak-to-peak sound pressure levels (SPL_{pk-pk}) by about 20 dB at 100 m range and 12 dB at 5 km range. Similar trends in unweighted SELs were found to those reported by Li and Bayly (2017) for the eSource, with little if any improvement at short range but improvements of up to 8 dB at 100 km range, with the shallow-water scenario showing the greatest improvement of the three.

Despite their potential advantages, and several companies having put substantial resources into their development, marine vibroseis sources are yet to gain widespread acceptance by the seismic survey industry. Likely reasons for this include the fact that they are yet to prove themselves sufficiently reliable for long-duration surveys (a chicken and egg problem), and that the very different technology requires significant changes to equipment and procedures on the part of the seismic survey operators. The seismic survey industry is also conservative when it comes to adopting new technology due to the high cost of typical surveys and the large losses that can be incurred if something goes wrong. It is likely marine vibroseis will first find regular application in relatively small-scale niche applications, such as shallow-water transition zone surveys, which attempt to make a connection between a land-based survey and an offshore survey and involve operating in small boats in very shallow water.

2.5.3 Drilling

Ocean drilling is part of a range of industrial operations. There is the drilling of oil and gas wells, drilling for subsea mining, drilling as part of geotechnical site investigations, drilling of moorings and foundations, and drilling for research. Drilling for oil and gas production and for research involves large and deep boreholes, down to several kilometers below the seafloor. Drilling for geotechnical site investigations is shallow (tens of meters), but may involve many boreholes (several dozen) to map an area. Drilling for foundations may be both shallow and small in number, unless dozens to hundreds of foundations are needed for an entire wind farm. The following sections present examples of underwater sound recordings from drilling.

2.5.3.1 Drilling for Oil and Gas

Drilling for hydrocarbons can be undertaken from different platforms, with the longevity of the operation, water depth, and environmental conditions determining the choice of platform (Fig. 2.33). Artificial islands and caissons are used only in shallow water. Fixed legged platforms (such as jack-up rigs, gravity-based structures, metal jacket, or lattice structures are common on continental shelves that are not too deep (i.e., <300 m). Offshore, floating structures (such as semi-submersibles or drillships) are used in deep water. These structures can maintain their position to within a few meters using computer-controlled propellers placed around the vessel (dynamic positioning). In cyclone-prone regions, drillships as well as Floating Production Storage and Offloading (FPSO) and Floating Liquefied Natural Gas (FLNG) facilities are used as they can demobilize when severe weather is forecast. Drillships are also used at smaller fields where production is expected to last only a few years.

Drilling platforms are typically self-contained, generate their own power, and accommodate staff. Operations are accompanied by support vessels, crew boats, and sometimes helicopters. There will be regular tankers if the product is offloaded to ships rather than fed into pipelines. Dynamic positioning thrusters might be used by floating platforms (i.e., semi-submersibles and drillships) and by tankers during offloading.

The sound field about a rig is complex due to multiple sources on the platform and under water. Sound from the platform (e.g., from generators,

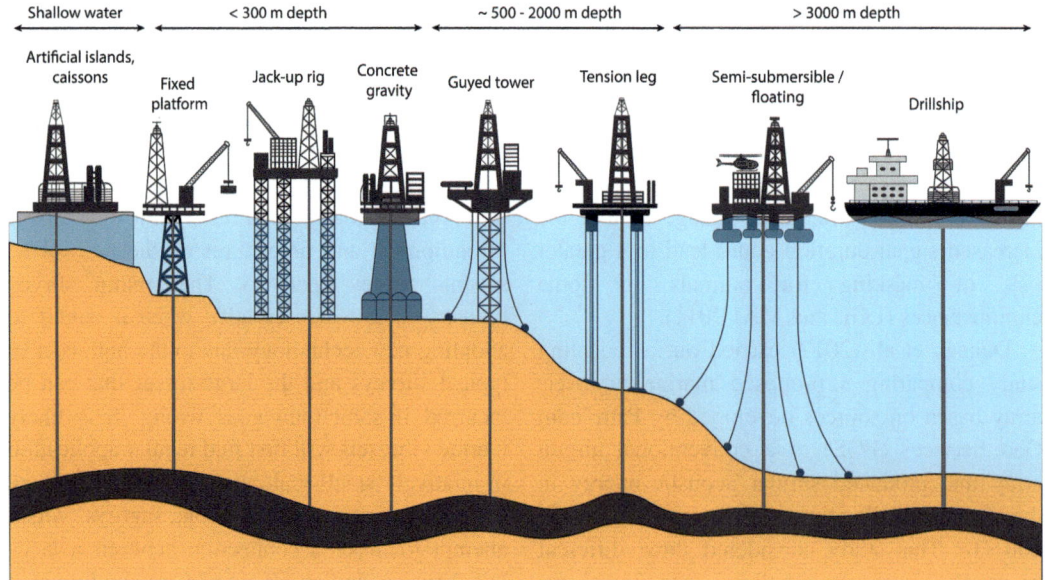

Fig. 2.33 Drilling platforms used in different water depths

compressors, and pumps) can transmit directly into the water through the supporting structures or a drillship's hull. Near the rig, sound may also travel from the air into the water at incidence angles <13° from the vertical. At greater angles (more toward the horizontal plane), airborne sound reflects off the sea surface and does not transmit into the water. The drill string and its casing vibrate during drilling, sending an acoustic wave through the water. The drill bit acting under ground creates sound that may travel back into the water. It can be difficult to isolate the actual sound from drilling in the superposition of sound from all the other sources.

Sounds from drillrigs in quiet ambient noise conditions have been detected up to 40 km away, with radiated noise levels determined to be 184 and 190 dB re 1 µPa m during drilling and maintenance, respectively (Kyhn et al. 2014). Drill islands tend to yield lower received levels in the water, partly because they are constructed in shallow water where sound is strongly attenuated. Legged platforms also occur in shallower water and with the platform above the sea, sound coupling into the water is poor. Drill ships tend to yield higher received levels as they are deployed in deeper water and because the sound from the platform couples well into the water through the hull. Figure 2.34 summarizes recordings from various drillrigs in different environments. Several rigs were recorded at 4–5 km range. Over the recording durations (which varied between 5 and 361 days), a diversity of activities would have occurred at and near the rigs, partly explaining the great variability in levels observed. Rig tenders are always present, tend to move around the rig, sometimes employ dynamic positioning, and other times drift out to 5 nautical mile range, then reposition.

The sound spectra recorded from operating drillrigs exhibit distinct tones at low frequencies of about 5–40 Hz, likely from drill-string vibrations, and tones from power generation, plus overtones up to 2000 Hz (Gales 1982; Kyhn et al. 2014; Nedwell and Edwards 2004; Todd et al. 2020). One-third octave radiated noise levels (computed by adding $20\log_{10} r$ to received levels measured at some range) of dominant tones

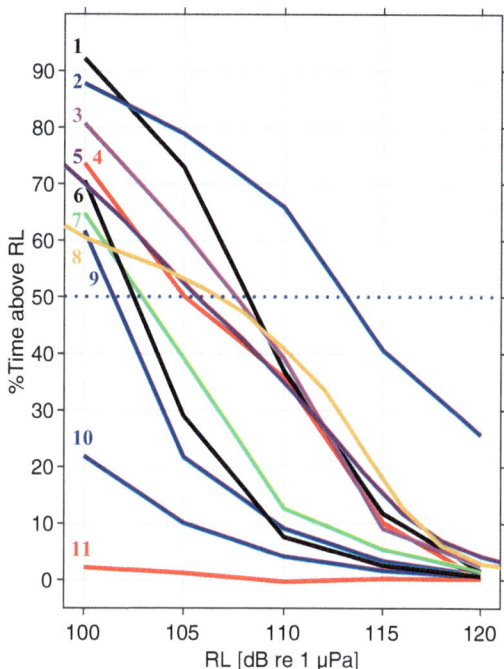

Fig. 2.34 Noise from drillrigs recorded at various distances for different durations (d: days). Exceedance curves are shown, plotting the percent of time that certain received levels (RL) were exceeded. 1: drilling @ 8.5 km + vessel (15 d). 2: drilling @ 4.6 km + vessel (94 d). 3: drilling @ 1.88 km (22 d). 4: drilling @ 4.6 km + vessel (5 d). 5: drilling @ 4.6 km + various vessels (361 d). 6: drilling @ 5.0 km (25 d). 7: drilling @ 4.1 km (22 d). 8: drilling @ 4.6 km (12 d). 9: drilling @ 5.0 km (169 d). 10: drilling @ 8.0 km (22 d). 11: drilling @ 28.0 km (22 d)

from three different drillrigs were 128–141 dB re 1 µPa m between 12 and 72 Hz, 123–131 dB re 1 µPa m between 125 and 630 Hz, and 109–117 dB re 1 µPa m between 1600 and 5000 Hz (Gales 1982). Noise levels referenced to a nominal distance of 1 m from the source at specific tone frequencies from a semi-submersible in 114 m of water were estimated as 149, 137, and 136 dB re 1 µPa m at 60, 181, and 301 Hz, respectively, through regression over received levels at various ranges (Greene 1986).

There are several problems with estimating source levels for drilling. While the sound from a ship or seismic airgun array might appear in the far-field to have originated from a point source just below the sea surface, for drilling, the geometric arrangement is more complicated. The vibrating drill string might be a line source, spanning the entire water column (from the platform at the sea surface, right into the seafloor), but there is also a point source underground where the drill bit grinds, and an extended source at the surface (i.e., the platform with everything operating on it). And drilling often happens in shallow water (though exploration drilling is moving into deeper waters). Many studies have therefore taken the approach to report received levels at a number of ranges from the drill string or platform and perform a linear regression on the logarithm of range. Some studies have also allowed for a linear (absorption-like) term in their regression equations. The approximated "source level" is then taken as the extrapolated received level at 1 m range (e.g., Greene 1986). Note that this procedure does not yield a source level as defined in ISO 18405. Received levels often do not decrease monotonically with range, but might be higher at some greater range, due to the complexity of the source geometry and sound propagation environment (Blackwell et al. 2004a; Greene 1986). The regression coefficients depend on frequency and hydrophone depth (Greene 1986). Figure 2.35 shows the regressions for drillrigs recorded in the Arctic with specifications in Table 2.2.

Noise from a dr`illship, a circular drill barge without propulsion, and a semi-submersible drilling platform in 33–46 m of water was measured with a hydrophone 30 cm above the seafloor (Austin et al. 2018). Exploration drilling with a 50 cm diameter drillbit and excavation of mudline cellars with a 6–7 m drillbit were recorded. With a monopole depth of half the draft of each rig, source levels were 169–175 dB re 1 µPa m for drilling and 192–193 dB re 1 µPa m for mudline cellar excavation (Fig. 2.36). The *Kulluk* had been recorded 30 years earlier (Greene 1987) and based on the regression equation in Table 2.2, its expected source level was 11 dB greater at that time, at that site.

2.5.3.2 Floating Production Storage and Offloading

Floating Production Storage and Offloading (FPSO) vessels are used for the production,

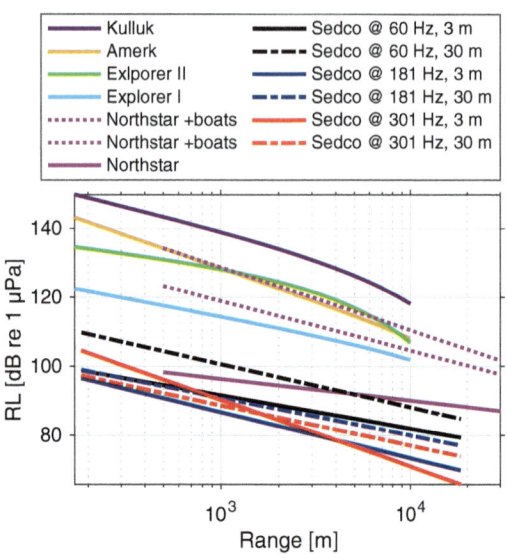

Fig. 2.35 Regressions of received level (RL) measurements against range. All rigs were drilling, except the *Explorer I* was well-logging. All levels are broadband, except the *Sedco* levels correspond to tones only. *Northstar* was an artificial gravel island, and so its levels were lower, except when support boats were present. Note how the *Sedco* regressions differ for different hydrophone depths (3 vs. 30 m) and different tone frequencies. Specifications in Table 2.2

processing, and storing of hydrocarbons. They do not drill boreholes but are connected to production wells via flowlines and risers, and possibly to drillrigs nearby. They are large ships, which sail or are towed to their offshore petroleum field, where they are moored ideally for years at a time. FPSOs rotate freely about the riser at their bow, hence adjusting to wind and sea conditions. They contain a processing plant on deck and storage tanks below. Processed petroleum is regularly (daily to every few days) offloaded to tankers. FPSOs are common in cyclone- and hurricane-prone areas, because of their ability to disconnect and relocate, as well as at smaller and more short-term fields. Underwater noise stems from the processing plant on deck, with sound being transmitted through the fluid-filled hull into the water; from dynamic positioning thrusters and propellers commonly used during offloading; from service vessels and tug boats; or from the FPSO's own propeller, which is often left slowly turning to avoid it seizing up. Six FPSOs were recorded off

2 Sources of Underwater Noise

Table 2.2 Regressions of received level with range r of drillrigs recorded in the Arctic

Platform	Size	Activity	Water depth [m]	Hydrophone depth [m]	Recording distance [m]	Regression	Frequency band [Hz]	References
Drillship *Canmar Explorer II*	115 m long	Drilling	27	9	200–8000	$149.9 - 6.74 \log(r) - 1.621\, r/1000$	20–1000	Greene (1987)
Drill barge *Kulluk*	80 m diameter	Drilling	31	18	980–8000	$179.6 - 13.3 \log(r) - 0.836\, r/1000$	20–1000	Greene (1987)
Drillship *Canmar Explorer I*	115 m long	Well logging	17	9	170–10,000	$145.4 - 10.3 \log(r) - 0.254\, r/1000$	20–1000	Greene (1987)
Drill island *Amerk*	Caisson, sand-filled	Drilling	26	18	220–8000	$185.4 - 18.9 \log(r) - 0.174\, r/1000$	20–1000	Greene (1987)
Drill island *Northstar*	Artificial gravel island in 12 m of water	Drilling with support boats present, day 1	10–37	10	500–10,000	$183.6 - 18.3 \log(r)$	10–10,000	Blackwell and Greene (2006)
Drill island *Northstar*	Artificial gravel island in 12 m of water	Drilling with support boats present, day 2	10–37	10	500–30,000	$162.1 - 14.4 \log(r)$	10–10,000	Blackwell and Greene (2006)
Drill island *Northstar*	Artificial gravel island in 12 m of water	Drilling w/o boats	10–37	10	500–4000	$115.2 - 6.3 \log(r)$	10–10,000	Blackwell and Greene (2006)
Semi-submersible *SEDCO 708*	90 m² platform	Drilling	114	2.5	185–18,500	$120.7 - 9.7 \log(r)$	Tone @ 60	Greene (1986)
Semi-submersible *SEDCO 708*	90 m² platform	Drilling	114	30	185–18,500	$138.1 - 12.5 \log(r)$	Tone @ 60	Greene (1986)
Semi-submersible *SEDCO 708*	90 m² platform	Drilling	114	2.5	185–18,500	$126.8 - 13.4 \log(r)$	Tone @ 181	Greene (1986)
Semi-submersible *SEDCO 708*	90 m² platform	Drilling	114	30	185–18,500	$123.8 - 11 \log(r)$	Tone @ 181	Greene (1986)
Semi-submersible *SEDCO 708*	90 m² platform	Drilling	114	2.5	185–18,500	$148.5 - 19.4 \log(r)$	Tone @ 301	Greene (1986)
Semi-submersible *SEDCO 708*	90 m² platform	Drilling	114	30	185–18,500	$123.8 - 11.7 \log(r)$	Tone @ 301	Greene (1986)

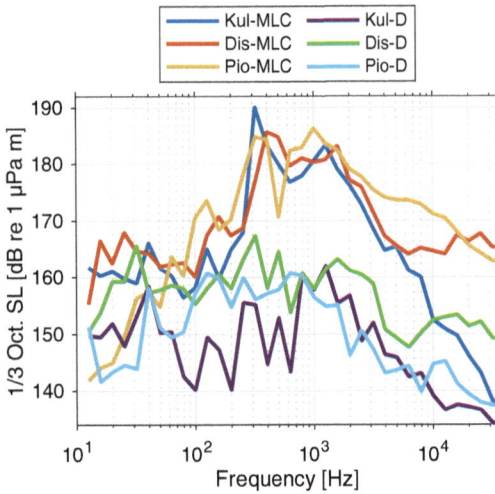

Fig. 2.36 Monopole source levels of drill barge *Kulluk*, drillship *Discoverer*, and semi-submersible *Polar Pioneer* during exploration drilling and drilling for mudline cellar excavation (Austin et al. 2018). One-third octave band levels are shown, but lines are not stepped over frequency bands for easier visibility

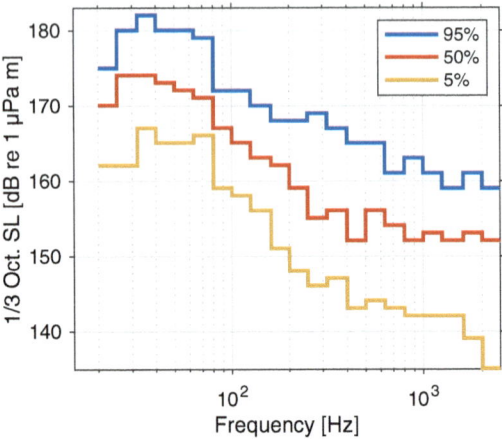

Fig. 2.37 Monopole median and 5th and 95th percentile source spectra of six FPSOs (Erbe et al. 2013)

Australia. Operations on and about an FPSO create a highly variable noise field, which is why a statistical approach was taken to quantify the radiated noise spectrum (Erbe et al. 2013). One-third octave monopole source levels are shown in Fig. 2.37. Noise levels are highest <100 Hz. Floating Liquefied Natural Gas (FLNG) vessels are a new type of platform and underwater noise assessments have not been published yet.

2.5.3.3 Diamond Coring

Drilling is also used for mineral exploration. Diamond coring has been recorded in an ice-covered lake from a rig resting on top of the ice. Drilling was done through the ice (1.63 m thickness), with a 6.35 cm drill bit, in 14.6 m of water. Recordings were made at 5 m range from the coring drill string and showed peak energy at 88 Hz with a broadband (1 Hz–22 kHz) level of 127.8 dB re 1 µPa (Mann et al. 2009).

2.5.4 Dredging

Dredging involves the excavation of lake-, river-, or seabed material and its transfer to a deposit site elsewhere. It is used for the creation or deepening of channels and harbors, land reclamation, construction of dikes, construction of submerged platforms, harvesting of construction material (e.g., sand and gravel), shellfish fishing, and subsea mining. Dredges come in different designs, depending on the specific purpose, environment (e.g., coastal or deep offshore), and material to be dredged (e.g., soft sand versus rock) (Fig. 2.38). In shallow waters with soft sediments (sand or mud), bed lever dredges tow a bar across the ground in order to level it (similar to a snow groomer on ski hills). Water injection dredges squirt water into soft sediment, making it waterborne and flushing it away. Clamshell dredges lower an opened clamshell bucket from a crane to the seafloor, where it closes, grabbing seafloor material before the crane hoists it. Bucket ladder dredges use a whole series of buckets attached to a rotating wheel or belt, which are scraped across the seafloor one after the other. A backhoe dredge operates like a digger on land. A trailing suction dredge tows a drag head over the seafloor, sucking up material. For the removal of hard seafloor material, cutter suction dredges cut the surface rock, which is then sucked through a pipe onto the dredge. The dredged material may be stored temporarily on the dredge or pumped straight into a barge for transfer to a deposit site. The particular dumping process when a dredge ejects the material in a high arc through the air to the dumping location is called rainbowing.

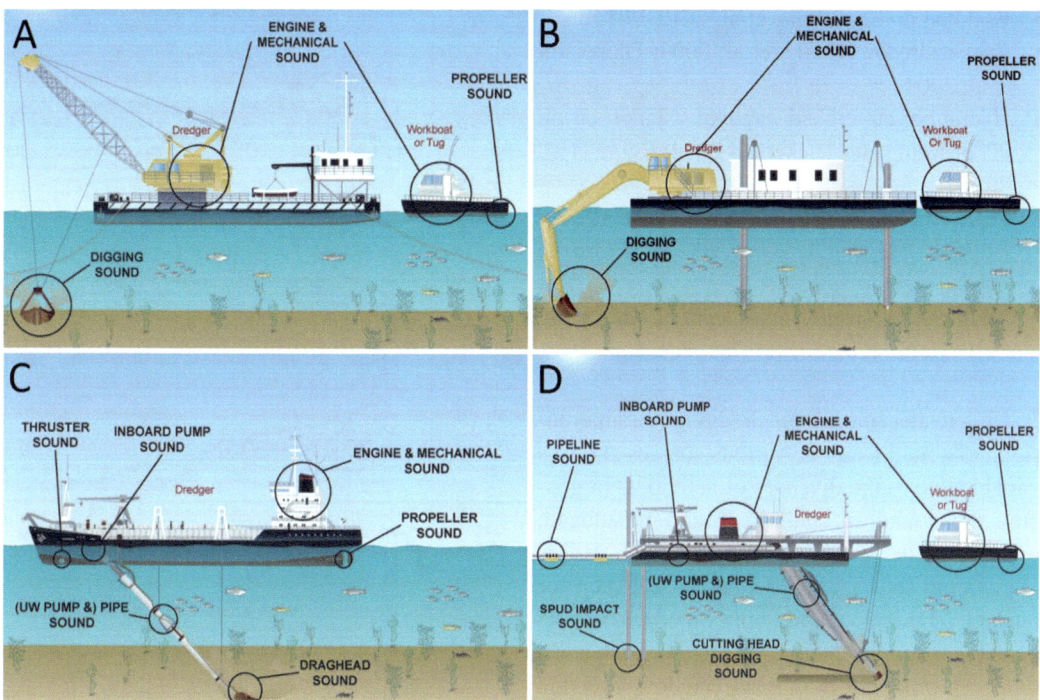

Fig. 2.38 Sketches of a (**A**) clamshell dredge, (**B**) backhoe dredge, (**C**) trailing suction hopper dredge, and (**D**) cutter suction dredge. Reprinted with permission from the Central Dredging Association

Underwater sound is produced by the excavation, transfer, and deposition of material. As the material may range from soft sand and mud to gravel and rock, the method of excavation and thus sound vary. Mechanical dredges, such as clamshell, bucket ladder, and backhoe dredges, remove material mechanically, which generates underwater noise (Clarke et al. 2003). Bottom contact was the noisiest activity of a bucket dredge (Dickerson et al. 2001). If the material needs to be cut or crushed before it can be uplifted, then this process adds additional noise. Similarly, the noisiest aspects of a backhoe dredge were bottom impact and spud walking (Reine et al. 2012, 2014b). Hydraulic dredges use water pumps to remove material; and some dredges combine mechanical and hydraulic aspects, such as cutter-suction transfer dredges and trailing suction hopper dredges. As the material is pumped from the seafloor to the vessel through a pipe, sound is generated throughout the water column. Larger and harder material, such as rock fragments and gravel, creates more sound in the suction pipe than does sand (Robinson et al. 2011). In deep water, there might be multiple pumps needed throughout the water column to uplift the material to the vessel. Each pump generates sound. On deck, power generators and engines create sound, which transfers through the hull into the water. Dredges may be self-propelled or be towed into location by tugs, which generates noise typical for ships (e.g., propeller cavitation). Dredges in shallow water may be anchored, but in deeper water, use thrusters, which generate cavitation noise (from the propellers). Dredges might transport the material to the dump site or else transfer it to a transportation barge, so there is noise related to reloading the material and then typical ship noise during transportation. At the deposit site, there is sound from expelling, pumping, or dumping of material.

Underwater sound from dredges has been measured in several studies, for example:

- Backhoe dredge: Reine et al. 2012, 2014b
- Bucket dredge: Clarke et al. 2003; Dickerson et al. 2001
- Hydraulic cutterhead dredge: Clarke et al. 2003; Reine and Dickerson 2014; Reine et al. 2014b
- Cutter suction transfer dredge: Greene 1987
- Hopper dredge: Clarke et al. 2003; Greene 1987
- Trailing suction hopper dredge: de Jong et al. 2010; Reine et al. 2014a; Robinson et al. 2011

Underwater sound from dredging is a superposition of sound from the various sources and it is variable due to the diversity of activities. In general, the sound is continuous and broadband, overlain with tones and occasionally pulses (e.g., from bottom impact of a backhoe or bucket). Trailing suction hopper dredges, in particular, are similar to large vessels, with the addition of a pipe and drag head scraping over the seafloor. And so, sound levels below 500 Hz are comparable to those of large vessels, even though dredges move more slowly, but then they incur greater load due to drag over the seafloor. At higher frequencies, dredge levels are typically higher than those of ships, likely related to the additional noise of handling the material (Robinson et al. 2011). Example source spectra of dredges are shown in Fig. 2.39 during different types of activities and in one instance, removing different material (i.e., sand versus gravel). A trailer suction hopper dredge source spectrum is further included in the updated JOMOPANS-ECHO source-level model (MacGillivray and de Jong 2021).

Computation of source levels for dredges during dredging is difficult, if not undefined (unless the dominant sound comes from the vessel propulsion or positioning). Sound sources are distributed from the vessel at the sea surface, through the water column (e.g., pipes for pumping water and sucking up material), to the seafloor (where sediment is crushed and collected). Dredging is not a point source. There might be multiple support boats at the site also generating noise. Dredging mostly occurs in

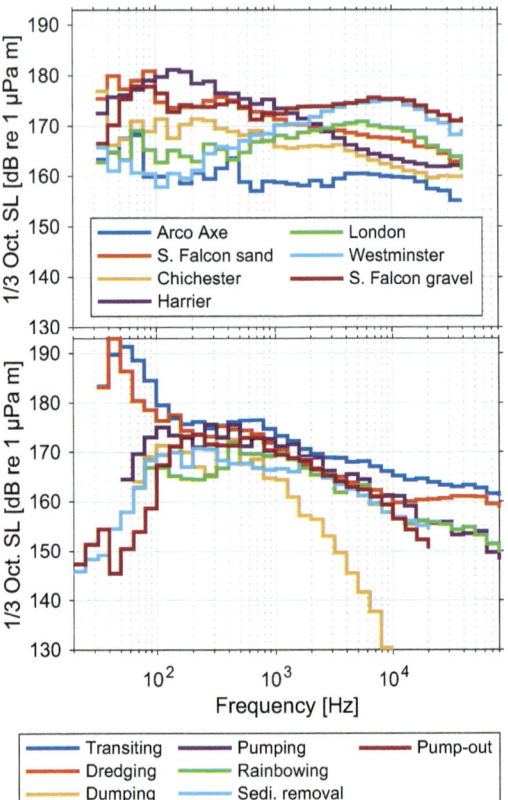

Fig. 2.39 Monopole source levels in 1/3 octave bands of trailing suction hopper dredges in shallow water (~20 m). Monopole depth 4 m. Top: *Arco Axe*, *Sand Falcon* removing sand, *City of Chichester*, *Sand Harrier*, *City of London*, *City of Westminster*, and *Sand Falcon* removing gravel (Robinson et al. 2011). Bottom: upper envelope of up to 7 dredges transiting, dredging, dumping, pumping, and rainbowing (de Jong et al. 2010), and of 3 dredges removing sediment and pumping out (Reine et al. 2014a). Note the Reine et al. levels were obtained from regression fitting and hence do not represent source levels by ISO 18405

shallow water, complicating accurate sound propagation estimation. Different studies have taken different approaches to quantify the source. Some authors have recorded received levels at numerous ranges and then fitted a linear regression of the received level against the logarithm of range (e.g., Greene 1987; Reine et al. 2014a). Others modeled propagation (de Jong et al. 2010; Robinson et al. 2011). Example source levels are tabulated in Table 2.3.

Table 2.3 Example source levels (SL) estimated for dredges. ASL: boundary-affected SL; MSL: monopole SL

Platform	Size	Power [MW]	Activity	Water depth [m]	Seafloor	Estimated SL [dB re 1 μPa m]	ASL vs MSL (source depth)	Frequency band [Hz]	Reference
Backhoe dredge *New York*	61 m long	2.6	Dredging	<15	Fractured limestone	164–179	ASL	20–20,000	Reine et al. (2014b)
Cutter suction dredge *Florida*	159 m long	19.0	Dredging	<15	Limestone rock	170–175	ASL	20–20,000	Reine et al. (2014b)
Cutter suction dredge *JFJ de Nul*	141 m long	27.2	Dredging	9		183	MSL	30–10,000	Wyatt (2008)
Cutter suction transfer dredge *Aquarius*	107 m long, capacity 2500 m³	12.9	Dredging	46		165	ASL	20–1000	Greene (1987)
Cutter suction transfer dredge *Aquarius*	107 m long, capacity 2500 m³	12.9	Dredging	46		186	MSL	20–1000	Richardson et al. (1995)
Cutter suction transfer dredge *Beaver Mackenzie*	86.5 m long	1.3	Dredging	13		157	ASL	20–1000	Greene (1987)
Cutter suction transfer dredge *Beaver Mackenzie*	86.5 m long	1.3	Dredging	13		173	MSL	20–1000	Richardson et al. (1995)
Hopper dredge *Geopotes X*	Capacity 8000 m³	15.4	Traveling, loaded, damaged propeller	25	Gravel	205	ASL	20–1000	Greene (1987)
Trailing suction hopper dredge *Arco Axe*	98 m long, capacity 2890 m³	2.94	Dredging	20–30	Gravelly sand	176	MSL (4 m)	30–40,000	Robinson et al. (2011)
Trailing suction hopper dredge *Atchafalaya*	Capacity 2300 m³	2.2	Dredging		Sand	173–180	ASL	20–20,000	Reine et al. (2014a)
Trailing suction hopper dredge *City of Chichester*	72 m long, capacity 1418 m³	2.7	Dredging	27	Gravelly sand	183	MSL (4 m)	30–40,000	Robinson et al. (2011)
Trailing suction hopper dredge *City of London*	100 m long, capacity 2652 m³	4.1	Dredging	37	Sandy gravel	182	MSL (4 m)	30–40,000	Robinson et al. (2011)
Trailing suction hopper dredge *City of Westminster*	100 m long, capacity 2999 m³	4.1	Dredging	35	Sandy gravel	186	MSL (4 m)	30–40,000	Robinson et al. (2011)
Trailing suction hopper dredge *Dodge Island*	Capacity 2754 m³	7.0	Traveling	7–11	Sand	174	ASL	20–20,000	Reine et al. (2014a)
Trailing suction hopper dredge *Dodge Island*	Capacity 2754 m³	7.0	Pumping out sand	7–11	Sand	167	ASL	20–20,000	Reine et al. (2014a)
Trailing suction hopper dredge *Dodge Island*	Capacity 2754 m³	7.0	Sediment removal	8–9	Sand	175	ASL	20–20,000	Reine et al. (2014a)

(continued)

Table 2.3 (continued)

Platform	Size	Power [MW]	Activity	Water depth [m]	Seafloor	Estimated SL [dB re 1 μPa m]	ASL vs MSL (source depth)	Frequency band [Hz]	Reference
Trailing suction hopper dredge *Gerardus Mercator*	153 m long, capacity 18,000 m³		Dredging	33		188	MSL	10–10,000	Wyatt (2008)
Trailing suction hopper dredge *Liberty Island*	Capacity 5003 m³	12.4	Traveling	7–11	Sand	179	ASL	20–20,000	Reine et al. (2014a)
Trailing suction hopper dredge *Liberty Island*	Capacity 5003 m³	12.4	Pumping out sand	7–11	Sand	176	ASL	20–20,000	Reine et al. (2014a)
Trailing suction hopper dredge *Liberty Island*	Capacity 5003 m³	12.4	Sediment removal	8–9	Sand	174	ASL	20–20,000	Reine et al. (2014a)
Trailing suction hopper dredge *Padre Island*	Capacity 2754 m³	7.0	Traveling	7–11	Sand	170	ASL	20–20,000	Reine et al. (2014a)
Trailing suction hopper dredge *Padre Island*	Capacity 2754 m³	7.0	Pumping out sand	7–11	Sand	172	ASL	20–20,000	Reine et al. (2014a)
Trailing suction hopper dredge *Padre Island*	Capacity 2754 m³	7.0	Sediment removal	8–9	Sand	173	ASL	20–20,000	Reine et al. (2014a)
Trailing suction hopper dredge *Sand Falcon*	120 m long, capacity 4832 m³	4.9	Dredging	28–32	Gravelly sand	189	MSL (4 m)	30–40,000	Robinson et al. (2011)
Trailing suction hopper dredge *Sand Harrier*	99 m long, capacity 2700 m³	3.8	Dredging	27	Gravelly sand	190	MSL (4 m)	30–40,000	Robinson et al. (2011)
Trailing suction hopper dredge *Taccola*	95 m long, capacity 4400 m³	6.3	Dredging			180	MSL	10–2000	Wyatt (2008)
Trailing suction hopper dredges, max of dredges 1–7	Capacity 3000–20,000 m³	8–30	Dredging	<20	Sand	195	MSL (4 m)	30–80,000	de Jong et al. (2010)
Trailing suction hopper dredges, max of dredges 1–7	Capacity 3000–20,000 m³	8–30	Transit	<20	Sand	196	MSL (4 m)	30–80,000	de Jong et al. (2010)
Trailing suction hopper dredges, max of dredges 1–7	Capacity 3000–20,000 m³	8–30	Dumping	<20	Sand	181	MSL (4 m)	30–80,000	de Jong et al. (2010)
Trailing suction hopper dredges, max of dredges 1–7	Capacity 3000–20,000 m³	8–30	Pumping ashore	<20	Sand	185	MSL (4 m)	30–80,000	de Jong et al. (2010)
Trailing suction hopper dredges, max of dredges 1–7	Capacity 3000–20,000 m³	8–30	Rainbowing	<20	Sand	180	MSL (4 m)	30–80,000	de Jong et al. (2010)

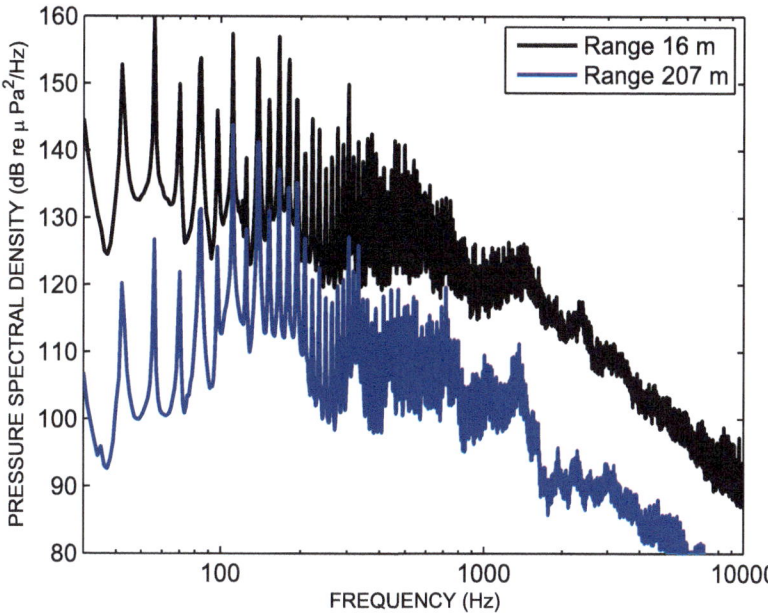

Fig. 2.40 Spectra recorded from vibratory piling showing distinct tones and harmonics in this 1 Hz resolution. Reprinted from Dahl PH, Dall'Osto DR, Farrell DM (2015) The underwater sound field from vibratory pile driving. J Acoust Soc Am 137 (6): 3544–3554, https://doi.org/10.1121/1.4921288. With permission from the ASA. © Acoustical Society of America, 2015. All rights reserved

2.5.5 Pile Driving

Pile driving in the marine environment is used in a diversity of projects: for example, to install towers supporting wind turbines in offshore windfarms, pillars for bridges, platforms for offshore petroleum production, and moorings for offshore industries. Piles come in different shapes (e.g., cylindrical piles, H-cross-sections, sheet piles; hollow versus solid piles) and materials (mostly wood, steel, or concrete), and can be driven vertically or slanted. There are also different methods for pile driving, with the most common ones being vibratory pile driving and percussive (also called impact) pile driving. Additional methods include pile drilling.

2.5.5.1 Vibratory Pile Driving

Vibratory pile drivers are used in soft substrates. Placed on top of the pile, counter-rotating eccentric weights create vibrations in the vertical direction that drive the pile into the ground. The sound recorded under water is a superposition of sound traveling along different propagation paths. In vibratory pile driving, sound is generated in air by the vibrator. This sound travels through the air and enters the water at inclination angles <13° from the vertical (i.e., close to the pile). Vibrations transmit through the pile into the ground, generating water- and ground-borne acoustic waves. The sound from vibratory pile driving is continuous with most energy below 1 kHz. The spectrum typically features tones related to the frequency of vibration (a few tens of Hz; Fig. 2.40). Table 2.4 gives example measurements from a number of published pile-driving recordings. The maximum reported root-mean-square sound pressure level (SPL$_{rms}$) for vibratory pile driving is 167 dB re 1 µPa at 16 m range (Dahl et al. 2015).

2.5.5.2 Percussive Pile Driving

Percussive pile drivers (Fig. 2.41) consist of a hammer with a supporting casing. Modern hummers include a ram (weight) of a large mass dropping onto an anvil, which is placed on the top of the pile to be driven. A cushion of certain stiffness is often placed above the anvil to reduce the peak impact on the pile and hence the peak sound pressure emitted into the air and water. The maximum mechanical force applied to the pile top depends on ram and anvil masses, cushion stiffness, velocity of the ram at the impact time, as well as area of the pile cross-section at its head

Table 2.4 Example levels recorded from vibratory pile driving. SPLpk: peak sound pressure level; SPLrms: root-mean-square sound pressure level; SEL: sound exposure level; r: range

Pile info	Pile size / diameter [m]	Hammer info	Mitigation	Seabed	Water depth [m]	Distance [m]	SPLpk [dB re 1 µPa]	SPLrms [dB re 1 µPa]	SEL [dB re 1 µPa²s]	Ref
Steel H-pile	0.25	Vibratory	None		2	10	161	147		Illinworth and Rodkin Inc. (2007)
Steel H-pile	0.25	Vibratory	None		2	20	152	137		Illinworth and Rodkin Inc. (2007)
Steel pipe	0.33	Vibratory	None		5	10	171	155	155	Illinworth and Rodkin Inc. (2007)
Steel pipe		PTC 60HD	None			16		144		Nedwell and Edwards (2002)
Steel pipe		PTC 60HD	None			24		152		Nedwell and Edwards (2002)
Steel pipe		PTC 60HD	None		0.5	80		151		Nedwell and Edwards (2002)
Steel pipe		PTC 60HD	None			82		132		Nedwell and Edwards (2002)
Steel pipe, wall 25 mm	0.91	APE400E + 400B	None			56		162–164		Blackwell (2005)
Steel pipe, wall 25 mm	0.76	Vibratory	None	Sand	7.5	16		167		Dahl et al. (2015)
Steel pipe, wall 25 mm	0.76	Vibratory	None	Sand	7.5	207		150		Dahl et al. (2015)
Steel pipe, wall 25 mm	0.76	Vibratory	None	Sand	7.5	417		142		Dahl et al. (2015)
Steel sheet		APE200	None	Driven into gravel island	0	100		143; 221–39 log r		Greene et al. (2008)
Steel sheet	0.61	Vibratory	None		15	10	175–177	162–163	162–163	Illinworth and Rodkin Inc. (2007)
Steel sheet	0.61	Vibratory	None		15	20	166			Illinworth and Rodkin Inc. (2007)

Fig. 2.41 Photo of pile driving for bridge construction in Queensland, Australia (Erbe 2009)

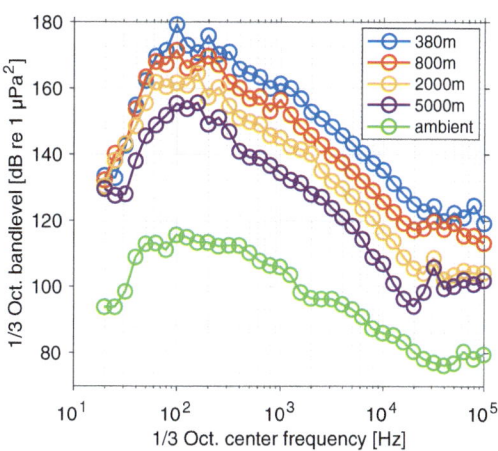

Fig. 2.42 One-third octave band levels from pile driving recorded at different ranges. Pile diameter 5.2 m, hammer energy 1 MJ, water depth 15–20 m, seafloor consisting of sand and gravel over chalk. Image created from data in Robinson et al. 2012, licensed under CC BY

and mechanical properties of the pile material, such as Young's modulus and Poisson's ratio. The impact energy E, which is commonly measured and recorded by hammer operators, is a function of ram mass m and velocity v at the impact: $E = v^2 m/2$. The means of raising the ram differ amongst percussive pile drivers. In a diesel-fueled pile driver, the falling ram compresses and heats the air above the pile to the ignition point of diesel fuel, which is injected into the cylinder. This detonation sends the ram back up. Other pile drivers use hydraulic or steam raisers.

In percussive marine pile driving, sound is generated in air by the impact of the hammer onto the top of the pile and the mechanism of raising the ram. The impact of the hammer causes a small deformation of the pile in the form of a bulge of outward radial expansion traveling down the pile at supersonic speed, compared to the sound speed in surrounding water, and then into the ground. It radiates a sound wave directly into the water in the form of a Mach cone (Reinhall and Dahl 2011). When the deformation wave hits the bottom of the pile, it is reflected and travels back up the pile with a lower amplitude due to partial energy losses through underwater sound generation and absorption in the ground, again radiating sound into the water as a Mach cone directed to the water surface. Top and bottom reflections of the traveling bulge continue until the energy of elastic waves traveling up and down along the pile has been fully attenuated. The next impact of the ram repeats the sound generation process.

The sound from percussive pile driving is pulsed and broadband with peak energy typically between 100 and 1000 Hz (Fig. 2.42). Over the course of driving one pile, received levels commonly increase as the resistance of the ground and the hammer energy increase (Fig. 2.43; also see Robinson et al. 2012). There is evidence of proportionality between the hammer energy and the acoustic energy under water (Bailey et al. 2010; Betke 2008; Robinson et al. 2012). Due to reverberation and sounds arriving via different propagation paths at slightly different times, the duration of the acoustic pulse from percussive pile driving increases with range from the pile (e.g., Erbe 2009).

2.5.5.2.1 Measuring Impulsive Noise from Percussive Pile Driving

For the purposes of monitoring noise levels for regulatory requirements, comparing levels across different scenarios, and validating pile driving noise models, the international standard ISO 18406 (International Organization for Standardization 2017a) sets pile driving

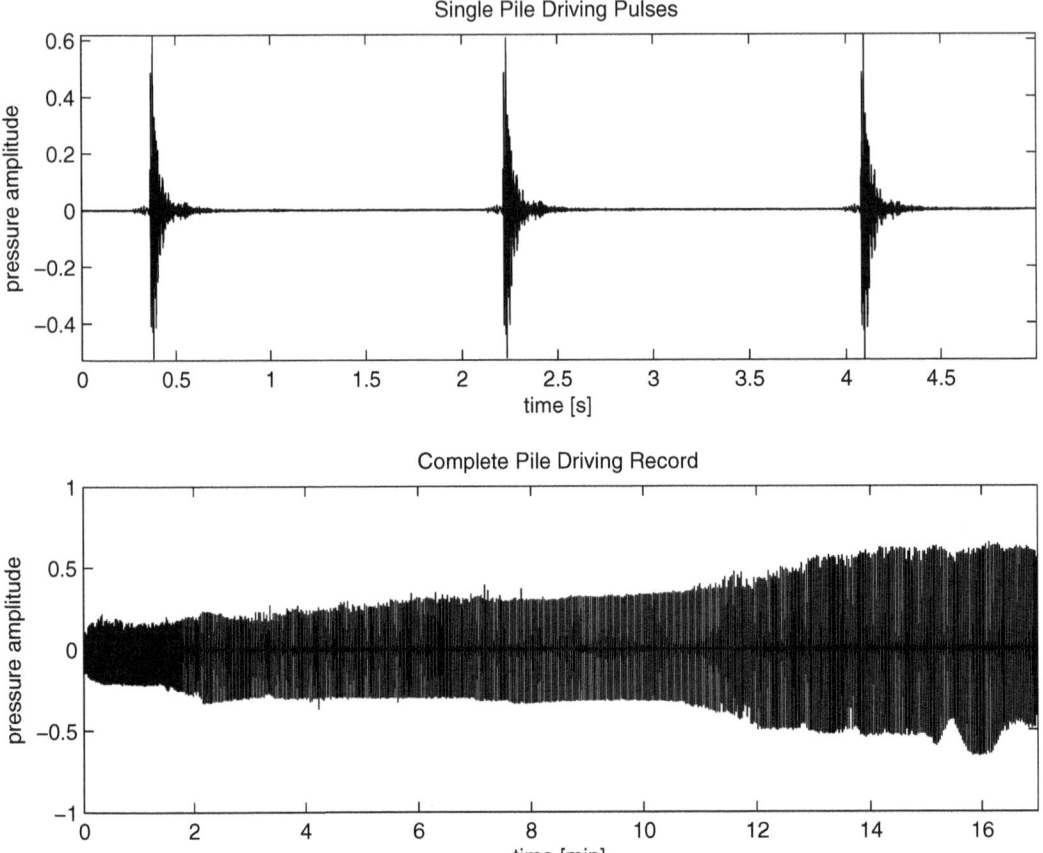

Fig. 2.43 Time series of (yet uncalibrated) pressure recorded under water near percussive pile driving. The hammer dropped every 1.8 s (top panel). Over the 17-minute duration of driving one pile, the pressure increased (bottom panel) (Erbe 2009)

recording, analysis, and reporting procedures. This standard does not define a source level. Acoustic quantities to be reported include SPL_{pk}, the single-strike sound exposure level (SEL_{SS}), and the signal-to-noise ratio (at 750 m range, or closer in very shallow water). Additional useful quantities are the cumulative sound exposure level (SEL_{cum}), the broadband sound pressure level (SPL), and the pulse duration.

There are plenty of example measurements from pile driving in the literature (e.g., Amaral et al. 2020; Bellmann et al. 2014; Betke 2008; Betke and Matuschek 2011; Blackwell 2005; Blackwell et al. 2004b; Caltrans 2001; Dahl et al. 2012; de Jong and Ainslie 2008; Duncan et al. 2010; Fricke and Rolfes 2015; Gündert et al. 2015; Hastings and Popper 2005; MacGillivray 2013, 2018; MacGillivray and Racca 2005; Nedwell et al. 2003, 2007; Nehls et al. 2007; Reinhall and Dahl 2011; Tougaard et al. 2009a; Vagle 2003; von Estorff et al. 2016; Wilkes and Gavrilov 2017; Yang et al. 2015) and some present quite comprehensive data sets (e.g., Bellmann et al. 2015; Illinworth and Rodkin Inc. 2007). Piles can be up to 10 m and more in diameter. For percussive pile driving, some of the higher reported levels are SPL_{pk} of 227 dB re 1 µPa and SEL_{SS} of 201 dB re 1 µPa^{2}s (Hastings and Popper 2005; Illinworth and Rodkin Inc. 2007) at 5 m range. The received levels vary with pile size,

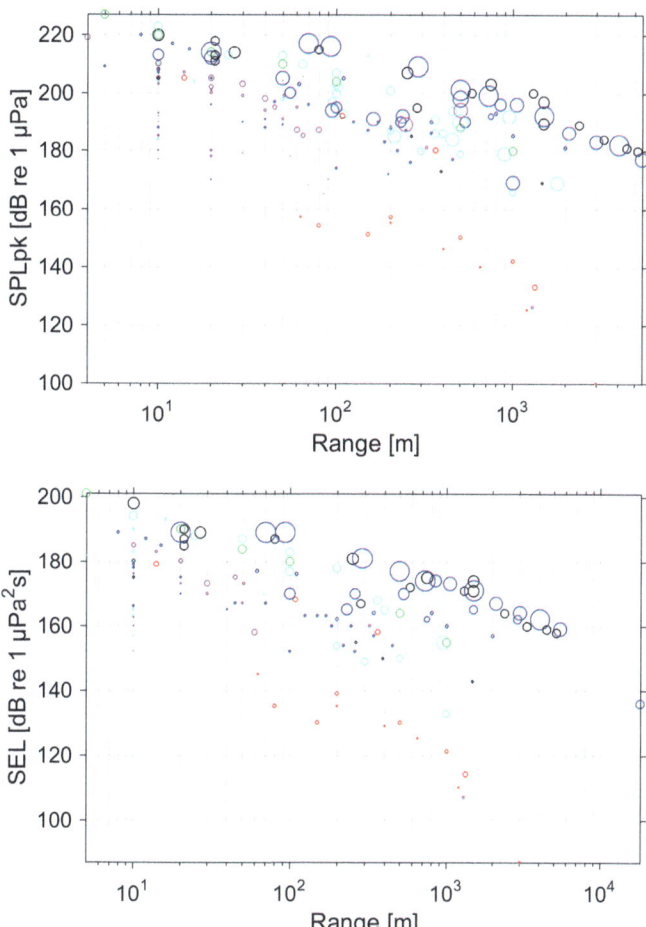

Fig. 2.44 Peak sound pressure levels (top) and average sound exposure levels (bottom) reported from percussive pile driving. The circles' diameters linearly scale with the diameters of the driven piles. The intention was for 1 mm of printed diameter to correspond to 2 m of pile diameter—which obviously did not work. Pile diameters in this figure range from 0.2 m to 5.9 m. The colors correspond to the water depth at the pile: ≤1 m (red), >1 m and ≤5 m (pink), >5 m and ≤10 m (light blue), >10 m and ≤30 m (dark blue), >30 m (black); water depth not listed (green). Based on data in the references cited in this section

shape, and material; hammer type and energy; water depth at the pile; bathymetry toward the recorder; seafloor geology; hydrophone depth; etc. Some patterns, however, can be seen in Fig. 2.44: The higher received levels at any range come from the larger piles in deeper water. Von Pein et al. (2022b) developed scaling laws to predict sound levels from pile driving as a function of strike energy, ram weight, pile diameter, and water depth.

2.5.5.2.2 Modeling Underwater Sound from Percussive Pile Driving

The challenge that marine pile driving presents to bioacoustic impact assessments, in particular to the modeling of the sound field in water for the prediction of impact zones, is that a driven pile is not a point source. Rather, the source extends from above the water to below the seafloor and the source is intrinsically linked to the environment (the water depth, the seafloor geology, etc.). Levels recorded under water cannot simply be converted into those from a monopole source, which is needed for common sound propagation models. Therefore, a lot of effort has recently been spent on modeling the sound near- and far-fields (e.g., Lippert et al. 2016; Peng et al. 2021; Tsouvalas 2020).

Underwater sound in the near-field from impact-driven piles is usually modeled using either a finite-element model (FEM) (e.g., Reinhall and Dahl 2011; Wilkes et al. 2016; Zampolli et al. 2013) or a finite-difference model (FDM) (e.g., MacGillivray 2013). An attempt to model the sound field from an impact-driven pile has also been made using an analytical approach (Hall 2015). Both FEM and FDM methods are computationally heavy, so an alternative approach was developed. Numerical predictions in the near-field of sound emission from driven marine piles are extrapolated into the far-field, at ranges much larger than the water depth, using one of the available underwater sound propagation models, such as the parabolic equation (Lippert et al. 2016) or the normal mode approach (Wilkes et al. 2016; see Chap. 1).

Various models of impulsive noise produced by impact-driven marine piles were assessed for accuracy and consistency at the COMPILE workshops held at the Hamburg University of Technology (TUHH). Seven (in 2014) and then 11 (in 2017) research groups from around the world presented their modeling results of critical acoustic quantities, such as SPL_{pk} and SEL_{SS}, for the same benchmark scenarios. Despite the differences in approach, all models demonstrated similar results with respect to variations of SPL_{pk} and SEL_{SS} with range (Lippert et al. 2016, 2018). Simple scaling laws are available for quick, back-of-the-envelope calculations of noise levels for impact piling as a function of strike energy, ram weight, pile diameter, and water depth, with and without bubble curtains (von Pein et al. 2022a, b).

Slanted piles (commonly referred to as raked piles) are also often used in near-shore marine constructions, such as piers and bridges. Measurements of underwater sound intensity from impact-driven raked piles show that sound characteristics vary noticeably with azimuth angle of the sound propagation direction from the pile. The difference between the noise levels radiated and propagated in water in the direction of pile inclination and in the opposite direction can be as large as 10 dB or more, with higher levels of underwater noise radiated by a slant pile in the opposite direction. The difference increases with the increase of the pile tilt angle relative to the vertical. This effect was measured, modeled, and explained in detail (Amaral et al. 2020; Wilkes and Gavrilov 2017).

Another important factor of potential environmental impact of marine pile driving is ground vibration. An interface wave (i.e., Scholte wave; Zhu et al. 2004) travels along the interface between water and an elastic substrate of the seabed. Such waves produce both radial and vertical vibrations in the seabed. The Scholte wave propagates along the interface with a speed significantly lower than that in water. Consequently, it is not reradiated back into the water column as a sound wave. However, it generates sound pressure oscillations near the bottom, which decay exponentially as a sound receiver moves away from it. Therefore, benthic animal species, such as crabs, scallops, and other potential marine-mammal prey species living on or near the seabed might be affected. Seabed vibration due to offshore impact pile driving has been analyzed in a few studies (Hazelwood and Macey 2012; Reimann and Grabe 2014; Wilkes et al. 2016).

2.5.5.2.3 Noise Reduction

A cushioning pile cap between the hammer and the pile has historically been used in order to evenly distribute the force but also to reduce in-air and more recently underwater noise. Common cushion materials include wood, metal, and plastics. The main effect is a decrease in peak level and an increase in the duration of the pulse. Under water, SPL_{pk} was reduced by 5–7 dB, SEL_{SS} by up to 14 dB (Elmer 2007; Laughlin 2006; Nehls et al. 2007). Cushioning reduces sound particularly at high frequencies (Deng et al. 2016).

Bubble curtains are the most common means of reducing received levels from percussive pile driving. A hose is laid on the seafloor. Holes are predrilled into the hose. High-pressure air is pumped through the hose leading to a curtain of rising bubbles. These bubbles act as sound absorbers near their resonance frequency and as sound scatterers over a broad range of frequencies. The resonance frequency of an air bubble in seawater near the sea surface depends on the bubble's size. With a diameter of 1 mm, the resonance frequency is about 6 kHz, at 10 mm it is about 600 Hz, decreasing by one decade in frequency for every one decade increase in diameter. A bubble curtain typically consists of an ensemble of bubbles of different sizes. Also, bubbles increase in size as they rise. Multiple hoses or more than one row of holes can be used to increase the thickness of the bubble curtain, effectively increasing the sound attenuation. Bubble curtains have proven effective at reducing noise above 100 Hz. Attenuations of up to 20 dB have been achieved in SPL_{pk}, SPL, and SEL_{SS} (Bellmann 2014; Caltrans 2001; Dähne et al. 2017; Lucke et al. 2011; Vagle 2003; Würsig et al. 2000). Currents in the environment will make the bubbles drift, tilting and potentially destroying the bubble curtain. In such situations, multiple hoses installed at different depths have been used. Also, current shields have been installed to avoid bubble curtain disruption.

Alternatives to bubble curtains are sleeves, such as foam sleeves. Foam sleeves have achieved an attenuation of 10–20 dB at frequencies between 1 kHz and 10 kHz (Schultz-von Glahn et al. 2006). Similar attenuations have been achieved with Hydro Sound Dampers that consist of sound-scattering elements of different sizes and shapes, fixed to a fishing net that surrounds the pile being driven (Elmer and Savery 2014). Cofferdams are steel sleeves that are installed around piles after which the water inside is removed, creating an air barrier. Due to the large difference in acoustic impedance between air and water, an air sleeve achieves the highest attenuation. Reviews of different pile installation, foundation, and noise mitigation methods have been published (e.g., Koschinski and Lüdemann 2011, 2020). The different sound-attenuation options come at different costs and are suitable in different environments.

In an attempt to reduce the likelihood of impact on marine fauna, soft starts are common practice, whereby the impact energy of the hammer is reduced by a half or more at the piling start, gradually increasing to its nominal value. The idea is to allow fish or marine mammals to move away from a source of intense underwater noise.

2.5.5.3 Drilling Moorings and Foundations

Depending on the substrate, offshore moorings and foundations (e.g., piles for wind turbine foundations) might be drilled. Levels are lower than from percussive pile driving, peaking between 100 and 500 Hz at 100 dB re 1 μPa^2/Hz at 500 m range (Broudic et al. 2014).

2.5.6 Offshore Renewable Energy Plants in Operation

Windfarms are the most common form of renewable energy collection offshore. They may consist of over a hundred individual wind turbines covering an area of hundreds of square kilometers (James 2013). Each turbine commonly has three blades, rotating in the wind about a horizontal axis. The rotor and electric generator are contained in the nacelle at the top of the tower. The types of foundation may differ and include (1) monopiles that are typically made of steel and that are driven about 10–20 m into the seafloor; (2) concrete gravity-based foundations that consist of a large base resting on the seafloor; (3) tripods that require three piles to be driven into the seafloor; (4) jacket or lattice foundations that consist of three or four corner piles that are driven into the seafloor and connected by a grid of bracings; and (5) floating and moored foundations. Gravity-based and floating foundations do not require pile driving (though mooring of floating platforms might). An overview of wind turbine installation methods is provided by Jiang (2021).

The sound from operational turbines is due to mechanical vibrations from the turbine (i.e., mostly the nacelle), which travel down the tower into the water and—unless it is a floating foundation—into the seabed. Vibrations of the tower might add to the noise. The sound under water is continuous and exhibits tones between tens of hertz to 1 kHz, related to the blade rotation, gearbox, and generator.

Nedwell et al. (2007) recorded underwater noise at four different operational windfarms finding that levels were low and only exceeded ambient levels by up to a few dB in narrow frequency bands. Mostly, operational windfarm noise was barely discernible from ambient noise in these high-wind seas that also have a lot of ship traffic. Similar findings resulted from other studies (see Madsen et al. 2006). Tone levels at about 180 Hz exceeded ambient levels by 20–30 dB elsewhere (Betke and Schultz-von Glahn 2004; Betke et al. 2004; Lindell 2003). SPL_{rms} from an operating wind turbine with jacket foundation were 137, 128, and 122 dB re 1 µPa at 40 m, 60 m, and 150 m; SPL_{rms} from an operating wind turbine with monopile foundation were 135 and 133 dB re 1 µPa at 40 m and 150 m; the transformer station generated 139 and 120 dB re 1 µPa at 60 m and 150 m (Thomsen et al. 2015). SPL_{rms} ranged from 109 to 127 dB re 1 µPa at 14–20 m distance from three different turbines ranging from 450 kW to 2 MW in power (Tougaard et al. 2009b). There are indications that the type of turbine foundation affects the water-borne noise level, with a 5 MW turbine with gravity-based foundation emitting less noise than a 3 MW turbine on a monopile foundation (Norro et al. 2015).

Pangerc et al. (2016) recorded two common Siemens SWT-3.6-107 turbines (3.6 MW, 3-blade rotor, 3-stage planetary/helical gearbox, with 1:119 ratio), which are commonly supported by 4.2–5.2 m diameter steel monopiles. Rotor speed varied with wind speed and ranged from 5 to 13 rpm. The cut-in wind speed at which the turbine started operating was 3–5 m/s and the maximum nominal wind energy of 3.6 MW was generated at wind speeds of 13–14 m/s. At wind speeds >25 m/s, the turbine cut out. These turbines were recorded over the range of operational wind speeds, in 20 m of water with a sand-clay seafloor, at a range of 50 m. The recorded sound exhibited distinct tones at 100–170 Hz related to turbine rotation, plus harmonics. The maximum mean-square sound pressure spectral density level was 126 dB re 1 μPa^2/Hz and occurred at 162 Hz. The average broadband SPL was 126–128 dB re 1 µPa at a wind speed of 10 m/s. The mean-square sound pressure spectral density level and the SPL increased with wind speed until the turbine reached its nominal maximum power.

Tougaard et al. (2020) summarized published data on underwater noise from operating offshore windfarms. Noise increased with wind speed and turbine size. Using a single-turbine source level of 156 dB re 1 µPa m and an 81-turbine windfarm source level of 175 dB re 1 µPa m, they estimated that cumulative noise levels from all the turbines within a farm might exceed ambient noise over a few kilometers range (Fig. 2.45).

Other offshore renewable energy installations have also been recorded, such as tidal stream turbines. Risch and team reported underwater noise levels from two 1.5 MW tidal turbines. Noise was strongest in the 50–1000 Hz band. Levels reached 140–150 dB re 1 µPa within 60 m from the turbine (Risch et al. 2020; Risch et al. 2023). Regression-fitted 1/3 octave band levels of 118–152 dB re 1 µPa at a nominal distance of 1 m between 40 and 8192 Hz were reported by Lossent et al. (2018); the strongest 1/3 octave band was centered at 128 Hz (152 dB re 1 µPa m).

2.5.7 Geotechnical Site Investigations

Offshore geotechnical site investigations are undertaken at the early stages of industrial development such as windfarms, drill rigs, ports, etc. High-resolution geophysical surveys are typically conducted as part of geotechnical site investigations to map geological hazards and to obtain information about the seafloor properties and geological structures below. These geophysical surveys are typically implemented in 2D configuration using a single source and a single hydrophone streamer (or a single receiver) towed behind the survey vessel.

Geophysical surveys use acoustic sources to transmit sound into the seafloor. Peak-to-peak source levels produced by geophysical equipment vary in the range of 170–240 dB re 1 µPa m (Crocker and Fratantonio 2016) and depend on

Fig. 2.45 Trends of sound pressure level (SPL) measurements from operating wind turbines based on published data collated by Tougaard et al. (2020). (**a**) All data were normalized to 10 m/s wind speed and 1 MW turbine size; (**b**) data were normalized to 100 m distance and 1 MW turbine size; (**c**) data were normalized to 100 m distance and 10 m/s wind speed. Linear fit: solid line; standard error: dotted lines. Reprinted from Tougaard J, Hermannsen L, Madsen PT (2020) How loud is the underwater noise from operating offshore wind turbines? J Acoust Soc Am 148 (5):2885–2893; https://doi.org/10.1121/10.0002453 © Tougaard et al. 2020. Published CC BY 4.0; https://creativecommons.org/licenses/by/4.0/

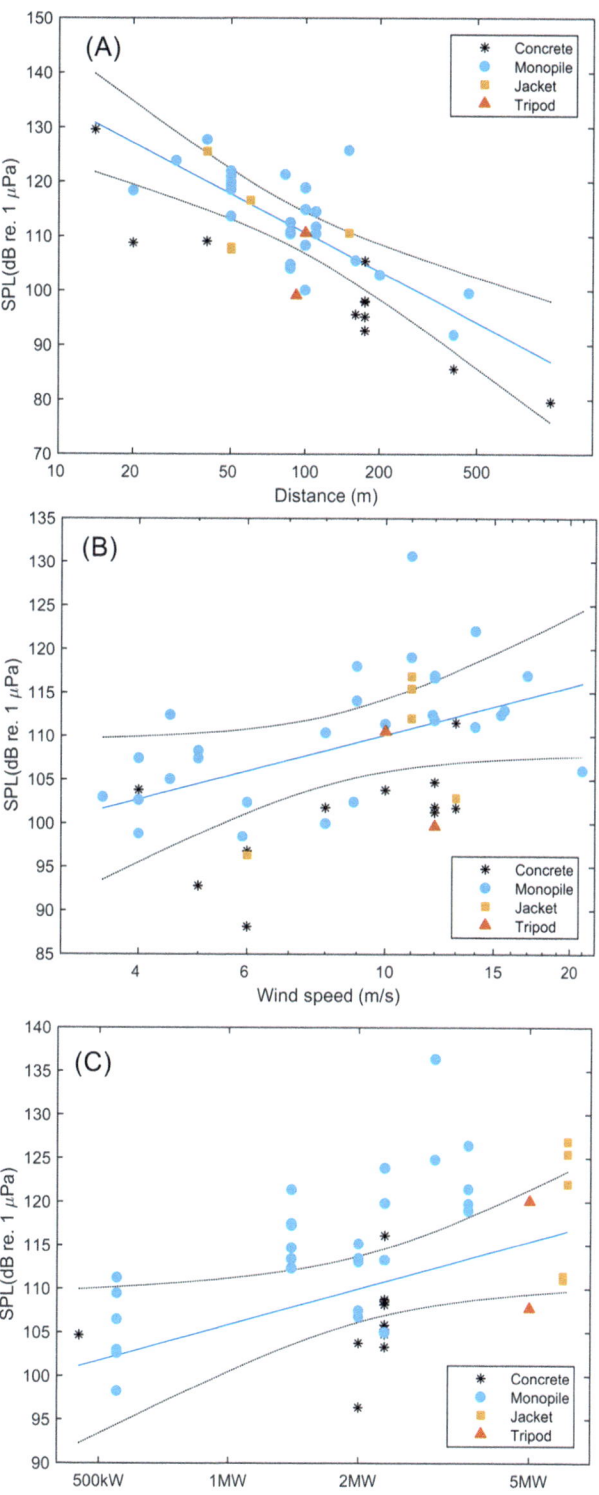

the source type and settings. The geophysical sources can be divided into two broad categories: sub-bottom profiling systems (i.e., airguns, sparkers, boomers, and sub-bottom profilers), which are used to obtain cross-sections of the seabed; and seafloor mapping systems (i.e., multibeam sonars and sidescan sonars), which are used to obtain images of the seafloor surface (Crocker et al. 2019).

2.5.7.1 Sub-Bottom Profiling Systems

Sub-bottom profiling systems consist of an acoustic source that sends sound signals into the seafloor. The signal reflects off the interfaces between different geological formations due to the difference in their acoustic impedance. The reflected signal is collected by the recorders (hydrophones). The collected data are then analyzed to provide information about the seabed.

Sub-bottom profiling systems can be characterized by the type of the source: pulse sources and frequency-modulated (FM) chirp sources (Crocker et al. 2019). Pulse sources use compressed air discharge (airgun), high-voltage discharge (sparker), or accelerated water mass (boomer) to generate an acoustic signal. The FM chirp signals are generated by piezoelectric transducers (chirp sub-bottom profilers or sub-bottom profilers). A schematic of a geophysical survey with different types of sources is shown in Fig. 2.46. Boomers and sub-bottom profilers radiate sound predominantly downward. A single airgun and a single sparker are omnidirectional sources. However, airguns and sparkers are typically towed close to the sea surface which significantly influences their directivity patterns. Sparkers, boomers, and sub-bottom profilers typically operate at higher frequencies and lower power providing higher resolution but smaller penetration depths than airguns.

2.5.7.1.1 Airguns

Airguns are pneumatic seismic sources that are typically deployed in arrays (groups) of two or more airguns. Arrays of airguns are mainly used for oil and gas exploration surveys deep into the seabed (see Sect. 2.5.2.2). Small airguns (<60 in^3) are mostly deployed singly or in doublets for geotechnical studies of the upper seafloor (Fig. 2.47). Airguns produce an acoustic pulse by the quick release of compressed air into the water. The main, strong, desired pulse is always followed by the oscillating bubble pulse, which is unwanted. To reduce bubble oscillation, some vendors utilize a 2-chamber airgun. The generator chamber is discharged first and produces the main

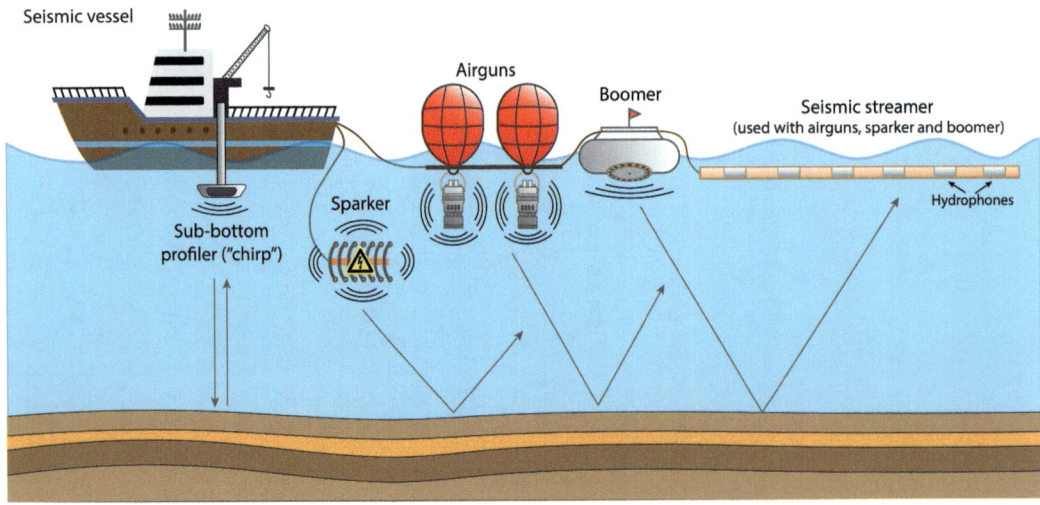

Fig. 2.46 A schematic of a geophysical survey with different sources

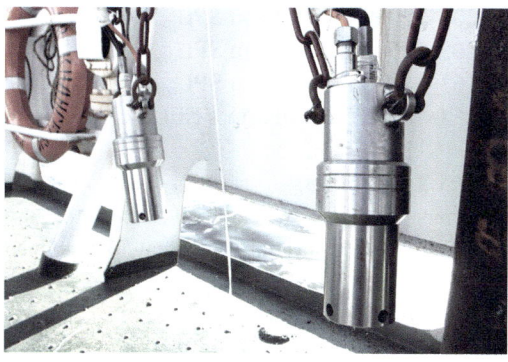

Fig. 2.47 Two airguns onboard a survey vessel, ready for deployment

pulse. The injector chamber is discharged halfway through the generator pulse period (i.e., out of phase), so as to dampen the oscillation (e.g., Crocker et al. 2019; Erbe and King 2009).

Figure 2.48 shows the modeled source signatures for a generator-injector (GI) gun with and without active injector. The generator and injector chambers each had 30 in^3 volume. The bubble pulse is considerably lower when the injector is used. The injector also smooths the frequency spectrum (Fig. 2.48c).

Example source levels determined from measurements of GI-guns (37 in^3 GI-gun with 13 in^3 generator plus 24 in^3 injector, 60 in^3 GI-gun with 30 in^3 generator plus 30 in^3 injector, and tandem arrangements; 1300–2500 psi; 1.5–4.5 m deployment depth) were (Crocker and Fratantonio 2016)

- 228–239 dB re 1 µPa m SPL$_{pk-pk}$
- 223–235 dB re 1 µPa m SPL$_{pk}$
- 218–228 dB re 1 µPa m SPL$_{rms}$
- 193–206 dB re 1 µPa^{2}m^{2}s SEL

Peak energy occurred between 40 and 440 Hz, with more low-frequency energy at deeper deployment depths.

2.5.7.1.2 Sparkers

A sparker is an electrostatic source that utilizes a high-voltage discharge to create an acoustic pulse. The acoustic pulse is broadband (50–4000 Hz); it can penetrate several hundred meters into the seabed and provide a vertical

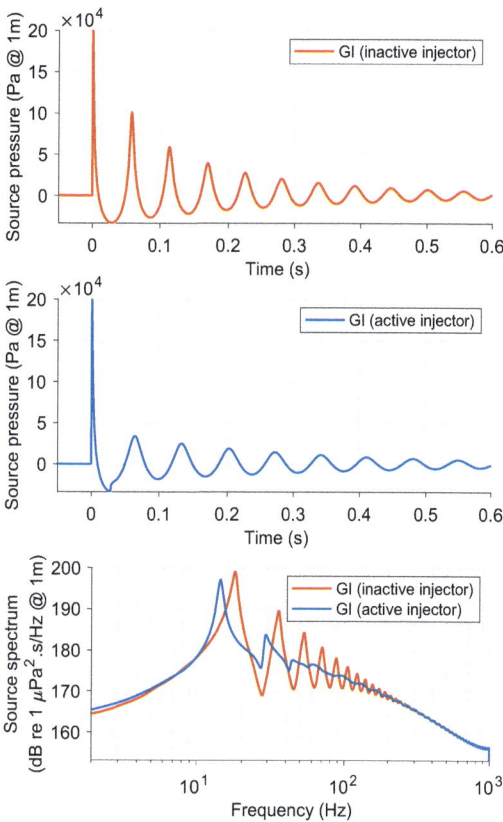

Fig. 2.48 Modeled acoustic waveforms and spectral characteristics of a 60 in^3 generator-injector airgun placed at 4 m depth: (top) source waveform of an airgun without injector; (middle) source waveform of an airgun with injector fired at half the generator bubble pulse period; (bottom) source spectra of both. The model (Duncan and Gavrilov 2019) excluded sea surface reflection

image resolution of ~0.3 m. The simplest sparker consists of two electrodes (anode and cathode) connected to a high-voltage bank of capacitors. An electric spark is created by the high-voltage discharge between the two electrodes. The spark immediately vaporizes the surrounding water and creates a quickly expanding water bubble, which generates an acoustic pulse in the water. The initial pulse is followed by the water bubble pulsation (similar to airguns). An example sparker waveform is shown in Fig. 2.49. The tow depth affects the notch in the frequency spectrum. Sparker systems vary in design to optimize penetration depth and image resolution.

- 206–229 dB re 1 µPa m SPL$_{pk-pk}$
- 203–225 dB re 1 µPa m SPL$_{pk}$
- 163–188 dB re 1 µPa m SPL$_{rms}$
- 163–188 dB re 1 µPa^{2}m^{2}s SEL

2.5.7.1.3 Boomers

Boomers are electro-mechanical sources that use water mass acceleration to create broadband acoustic signals (100–10,000 Hz). A single boomer plate consists of two circular pistons separated by a spiral coil. Similar to a sparker, a boomer discharges energy stored in capacitors to create a signal (Crocker et al. 2019). The energy discharge through the coil creates the resultant magnetic field, which causes the metal plate (circular piston) to vibrate and radiate an acoustic waveform into the surrounding water (Nedwell and Edwards 2004). Depending on the design, boomers consist of one or more plates, which can act simultaneously and thus increase the energy transmitted into the water (Fig. 2.50). Figure 2.51 shows the source characteristics of the Applied Acoustics S-BOOM for 300–700 J operating energies.

The sound levels of commercially available boomers (i.e., Applied Acoustics AA200, AA25, and S-Boom were tested with different discharge energy levels and reported to generate source signals in the ranges (Crocker and Fratantonio 2016):

- 199–219 dB re 1µPa m SPL$_{pk-pk}$
- 196–216 dB re 1µPa m SPL$_{pk}$
- 185–207 dB re 1µPa m SPL$_{rms}$
- 155–176 dB re 1µPa^{2}m^{2}s SEL
- 47–98° beamwidths at −3 dB (half-power)

2.5.7.1.4 Sub-Bottom Profilers (Chirp)

Chirp sub-bottom profilers use piezo-electric transducers to generate a frequency-modulated acoustic signal—a chirp. Compared to sparkers and boomers, sub-bottom profilers generate high-frequency (up to 25 kHz) signals, which provide greater resolution, but sacrifice penetration depth (Nedwell and Howell 2004). Unlike airguns, sparkers, and boomers, which require a hydrophone streamer (towed behind the vessel) to record reflected acoustic signals, sub-bottom

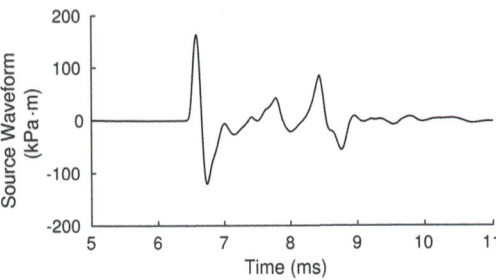

Fig. 2.49 Applied Acoustics Dura-Spark UHD 400 + 400 sparker with two decks of 400 tips (reprinted with permission from Applied Acoustics; https://www.aaetechnologiesgroup.com/applied-acoustics/) and the source waveform measured at 2400 J (reproduced from Crocker et al. 2019)

Multi-electrode sparker systems are used for surveys that require a higher energy output. Sparker electrodes wear off with repeated discharges because of ionization of the electrode material. Multi-tip electrode sparkers require less energy per electrode and hence last longer. Multi-electrode sparkers create a shorter signal as the bubble pulsation is reduced by the destructive interferences of bubbles from different electrodes (Duchesne et al. 2007).

The noise levels of some of the commercially available sparkers (i.e., SIG ELC 820 Sparker, Applied Acoustics Dura-Spark, and Applied Acoustics Delta-Spark) were measured at a range of discharge energies and operational depths (Crocker and Fratantonio 2016):

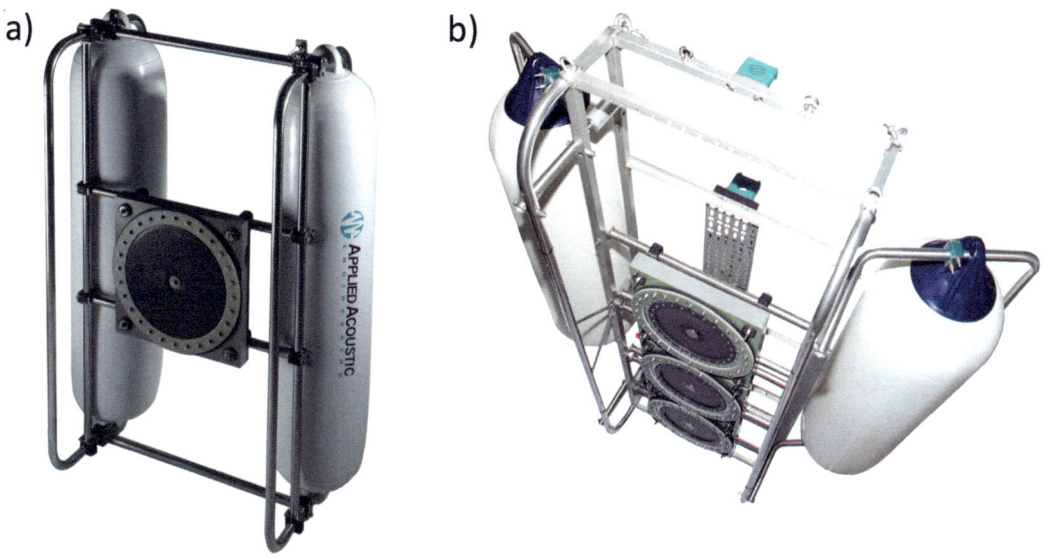

Fig. 2.50 (**a**) Applied Acoustics AA251 boomer; (**b**) Applied Acoustics S-Boom. Reprinted with permission from Applied Acoustics; https://www.aaetechnologiesgroup.com/applied-acoustics/products/sub-bottom-profiling/

Fig. 2.51 Source characteristics of Applied Acoustics S-Boom operating at 300, 400, 500, 600, and 700 J energies. (**a**) Pressure waveforms; (**b**) power spectral density (PSD); (**c**) directivity patterns. Reprinted with permission from Crocker and Fratantonio 2016

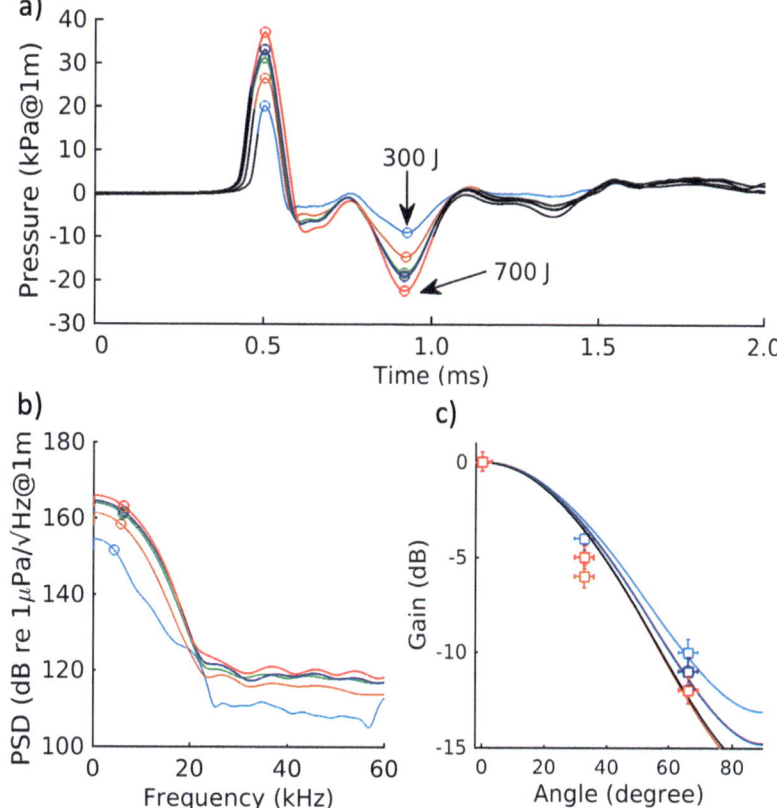

Fig. 2.52 Photo of a chirp sub-bottom profiler being deployed. Photo by Janet Watt, USGS Pacific Coastal and Marine Science Center. Public domain; https://www.usgs.gov/media/images/chirp-sub-bottom-profiler-ready-deployment

profilers combine the source and the receiver in a single unit (Fig. 2.52). An example of the source signature and spectrum of a sub-bottom profiler (EdgeTech SB-424) is shown in Fig. 2.53a, b. The beam pattern of the same profiler is shown in Fig. 2.53c.

Sound levels produced by some commercially available sub-bottom profilers (EdgeTech 424, EdgeTech 512i, and Knudsen 3202 Echosounder) were measured with different power setting and produced levels in the following ranges (Crocker and Fratantonio 2016):

- 173–220 dB re 1μPa m $SPL_{pk\text{-}pk}$
- 167–214 dB re 1μPa m SPL_{pk}
- 161–210 dB re 1μPa m SPL_{rms}
- 128–193 dB re 1μPa^{2}m^{2}s SEL
- 36–83° beamwidths at −3 dB (half-power)

2.5.7.2 Geotechnical Drilling

Geotech drilling for upper seafloor investigations uses small cores and less power, and thus produces less in-water noise. For example, an 83-mm-diameter drillbit operated at 120 kW power, in 16 m of sand and mudstone had source levels of 142–145 dB re 1 μPa m (30–2000 Hz; Erbe and McPherson 2017). A 91-mm-diameter drillbit operating 10 m into bedrock generated 155.9 ± 1.4 dB re 1 μPa m (Huang et al. 2023).

2.5.8 Sonar and Echosounder Systems

2.5.8.1 Active Sonar System Characteristics

Sonar, an acronym for "SOund NAvigation and Ranging," is a technique to detect and locate objects under water using sound waves. First termed in World War I, sonar systems have since been commonly used on ships, submarines, and underwater robots to map the seafloor, locate underwater mines and other hazards, and track other vessels (Lurton 2004). Sonar systems may be separated into "passive" versus "active." Passive sonar systems merely listen for sounds (e.g., from vessels), whereas active sonar systems emit sound (usually as a series of pulses) and then listen for echoes reflecting off targets (e.g., hazards, fish, or the seafloor). This section focuses on active sonar systems, in particular seafloor mapping sonar and military sonar. Seafloor mapping sonar systems include echosounders, which are acoustic devices designed to calculate the water depth by measuring the two-way travel time of sound between the water surface and the seafloor; and imaging and sidescan sonar systems, which are used to visualize the seafloor or water column. Military sonar systems considered in this section include those used in anti-submarine warfare. Before going into details of specific sonar systems, it is worth considering the features common to all active sonar systems, namely pulse type, frequency and bandwidth, pulse duration, ping rate, beam geometry, and source level.

Active sonar systems emit either sinusoidal pulses, also referred to as continuous waves (CW) or tones; or frequency-modulated (FM) pulses, also referred to as sweeps or chirps. Sonar systems that use CW pulses have a central frequency f_0, and those that transmit FM pulses have a lower and upper frequency: f_{min} and f_{max}. Although sound energy is transmitted outside of these main operating frequencies, these "side lobes" rapidly decrease in level either side of f_0 of a CW pulse, as can be seen in the spectra of a 200 kHz CW pulse from a multibeam

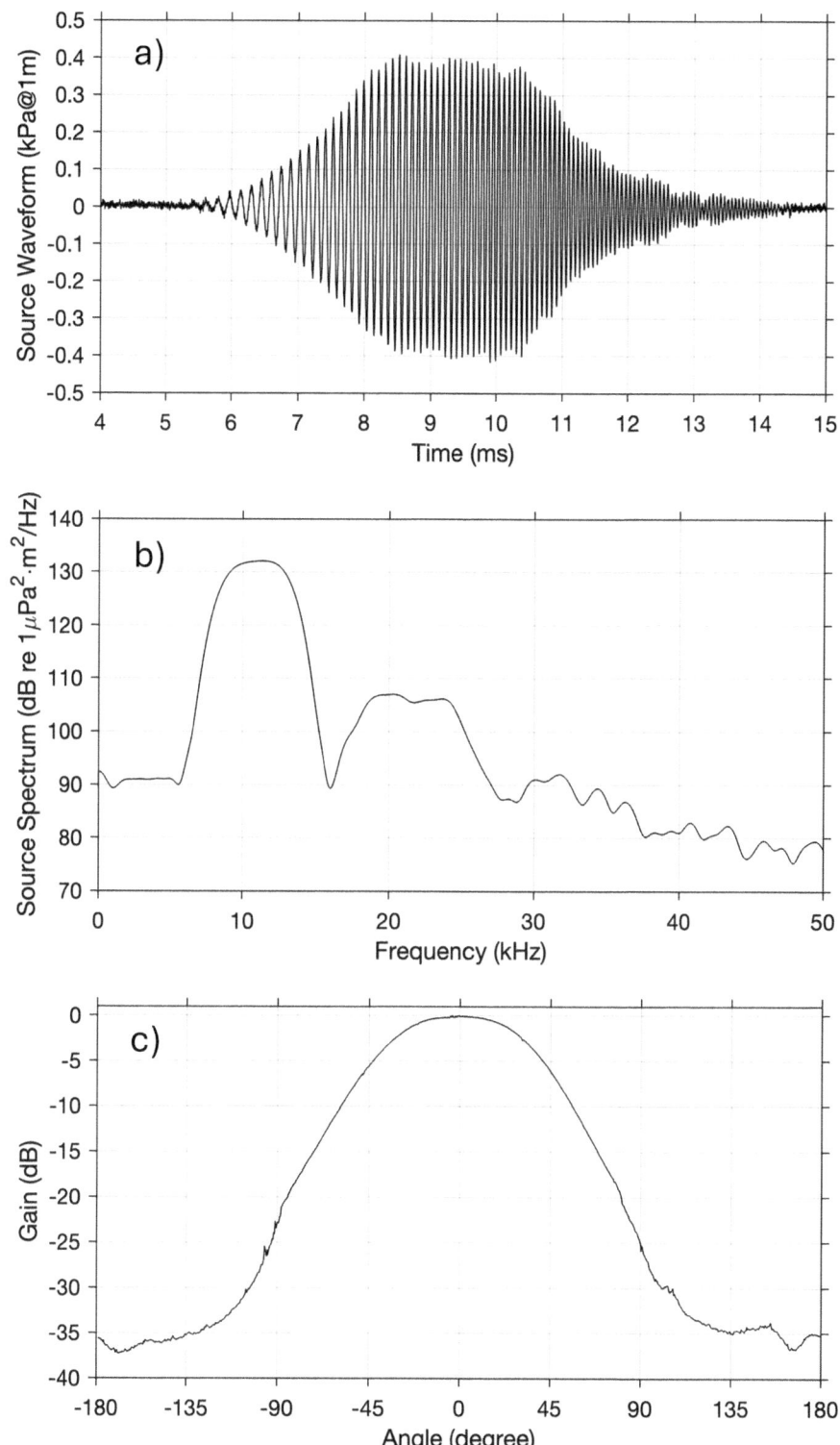

Fig. 2.53 Waveform (**a**), spectrum (**b**), and beam pattern (**c**) of an EdgeTech SB-424 sub-bottom profiler. Thank-you to Steven Crocker for kindly sharing data and code for us to create these plots (Crocker et al. 2019; Crocker and Fratantonio 2016)

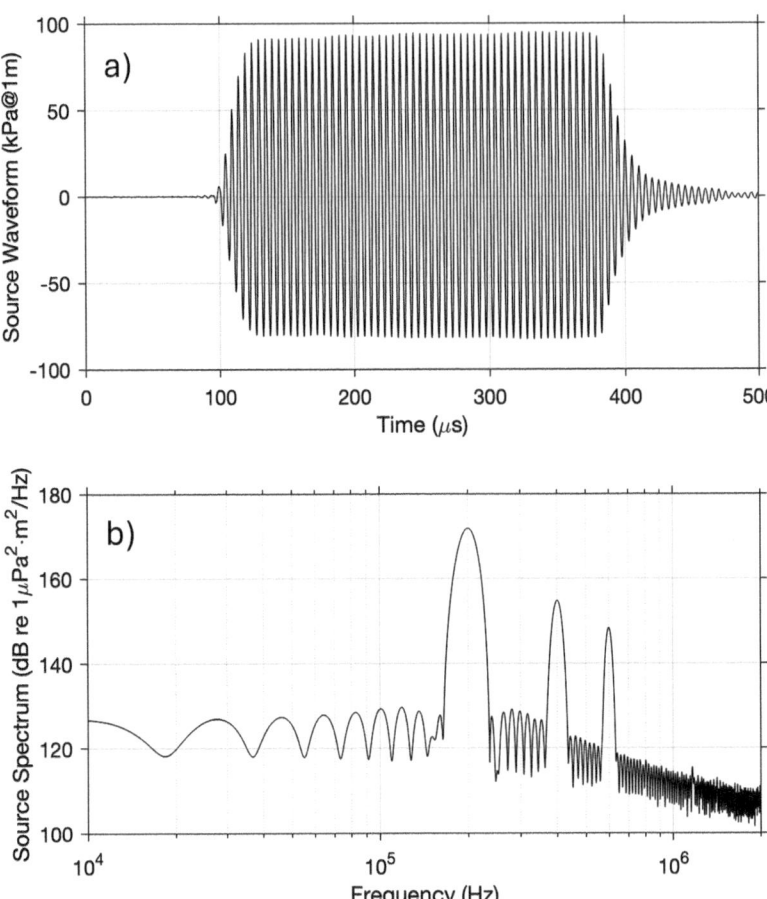

Fig. 2.54 Teledyne RESON SeaBat T20-P CW signal transmitted at 200 kHz with a 300 μs pulse duration and source level of 220 dB re 1 μPa²m². (**a**) Waveform; (**b**) spectrum. Plots created with data and code most kindly provided by Steven Crocker (Crocker et al. 2019)

echosounder in Fig. 2.54, or below f_{min} and above f_{max} of an FM pulse, but can increase at higher frequencies as harmonics (e.g., at $2f_0$, $3f_0$, etc. as seen in Fig. 2.54). The frequency bandwidth of CW pulses depends on the pulse duration; longer pulses have a narrow spectrum of frequencies, while shorter pulses have a wider bandwidth. The pulse duration used by a sonar system depends on the pulse type, frequency, and operational range. For instance, in general for CW pulses, the pulse duration is inversely proportional to the frequency, as decreasing f_0 of a CW pulse requires the pulse duration to be increased to accommodate the wavelength becoming longer. Increasing the pulse duration τ of a CW pulse is sometimes done to extend the operational range or increase the signal-to-noise ratio of returning echoes, but at the cost of decreasing the along-beam resolution $\Delta r = c\tau/2$, where c is the speed of sound. The pulse duration of an FM pulse depends on f_{min}, f_{max}, and the sweep rate of the chirp. The pulse repetition frequency, commonly referred to as the ping rate, describes how often pulses are transmitted by the sonar system. The ping rate is usually limited by the range set by the sonar operator (or automatically by the system), which is calculated from the two-way travel time using a measured or assumed speed of sound. The theoretical maximum ping rate (in Hz) for a particular range r is $c/2r$; however, computational limitations often mean the actual maximum ping rate is lower than the one set in the system (usually of the order of tens of hertz). The maximum range selected depends on the sonar's beam geometry and the ocean environment it ensonifies. For example, for seafloor mapping sonars, the maximum range is a function of the transducer's beamwidth and the depth below it. A

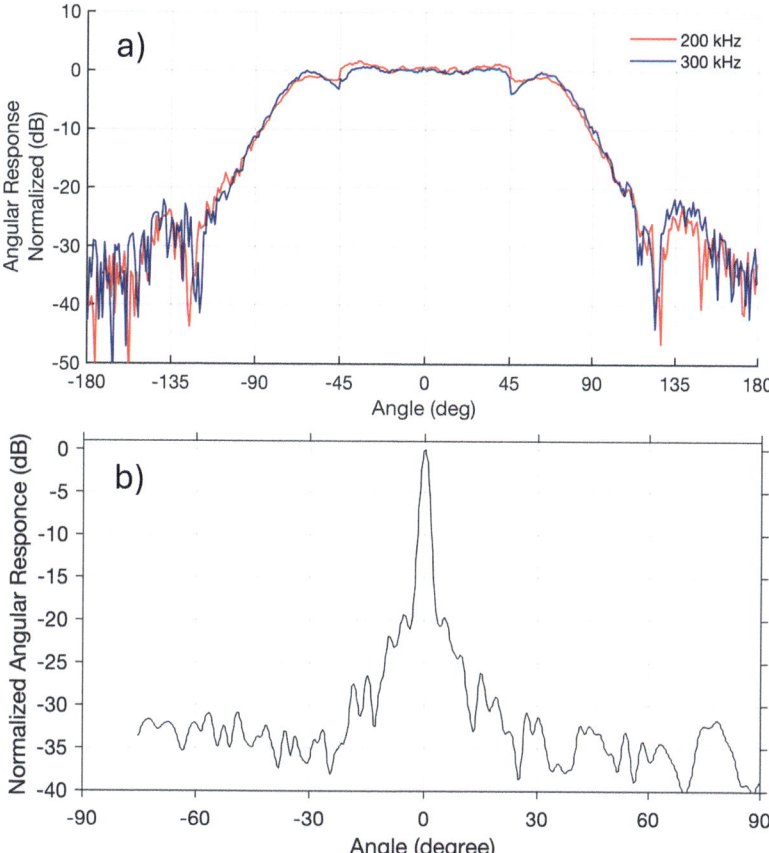

Fig. 2.55 Across- (**a**; 150° beamwidth) and along-track (**b**; 1.9° beamwidth) beam patterns for Teledyne RESON SeaBat T20-P multibeam echosounder. Plots created with data and code kindly provided by Steven Crocker (Crocker et al. 2019)

transducer's beamwidth is defined as the angle at half intensity (i.e., 3 dB down from the peak) on either side of the main axis of the main beam, which is illustrated on the across-track beam patterns shown in Fig. 2.55 for 200 and 300 kHz CW pulses transmitted from a multibeam echosounder. In specifications, the source level of sonar systems is usually quoted as L_S in dB re $1\mu Pa^2 m^2$ of the main beam. Although sound is transmitted at angles wider than the beamwidth, these are much lower in intensity (Fig. 2.55). The source level is sometimes required to be increased when the range increases, to maintain a high signal-to-noise ratio, or decreased to avoid saturation of any returning acoustic targets. Changes in source level are usually carried out by the operator, but some sonar systems automatically do this with changes in range setting. In addition, L_S can also change if the beam geometry is changed (e.g., if the beamwidth is increased, L_S might be reduced as the energy is spread over a larger area).

2.5.8.2 Seafloor Mapping Sonar Systems

The objective of seafloor mapping sonar systems is to depict the shape and/or composition of the seafloor, for a range of applications, including hydrographic surveying (e.g., charting and engineering works), scientific studies (e.g., marine geology and fisheries), and natural resource exploration (e.g., offshore oil and gas) (Lurton 2004). The main types of seafloor mapping sonars are single-beam echosounders (SBESs), sidescan sonars (SSSs), interferometric sidescan sonars, multibeam echosounders (MBES), and imaging sonars (Kenny et al. 2003; Penrose et al. 2005). Typical characteristics of the main seafloor mapping sonar systems are given in Table 2.5, and their beam geometry (i.e., coverage) is

Table 2.5 Typical beam geometries, frequencies, pulse durations, and source levels of the main seafloor mapping sonar systems using data collected from manufacturers and Bjørnø 2017; Clarke 2009; Crocker et al. 2019; Hammerstad 2005; Hastie 2012; Lurton 2016

Sonar system	Typical transmit beam geometry	Typical frequencies (and associated depth/ranges)	Typical CW pulse durations	Typical source levels
Single-beam echosounders	A narrow cone beam, pointing vertically down toward the seafloor, beamwidth: 5–30°	30–200 kHz (most applications)	10 µs– 5 ms	L_p 175–230 dB re 1 µPa m $L_{E,p}$ 138–180 dB re 1 µPa²m²s
		< 30 kHz (fisheries and deep-ocean mapping)	0.1–10 ms	L_p 230–240 dB re 1 µPa m
		>200 kHz (shallow hydrographic surveys and zooplankton studies)	Tens to hundreds µs	L_p 175–230 dB re 1 µPa m
Sidescan sonar	Two narrow swaths (one port, one starboard) across the seafloor, beamwidths: 0.2–3° along-track, 40–80° across-track	75–300 kHz (long range: 100 m–1 km)	20 µs to a few ms	L_p 195–228 dB re 1 µPa m $L_{E,p}$ 165–176 dB re µPa²m²s
		300–900 kHz (medium range: 50–200 m)	Tens of µs	L_p 198–220 dB re 1 µPa m $L_{E,p}$ 150–176 dB re 1 µPa²m²s
		900–1600 kHz (short range: < 50 m)	<20 µs	L_p 198–220 dB re 1 µPa m
Multibeam echosounders	One narrow swath directed across the seafloor, beamwidths: 0.25–3° along-track, 90–150° across-track	10–70 kHz (deep-ocean mapping)	1–100 ms	L_p 230–240 dB re 1 µPa m
		70–300 kHz (continental slope and shelf)	0.1–10 ms	L_p 218–230 dB re 1 µPa m $L_{E,p}$ 150–190 dB re 1 µPa²m²s
		300–710 kHz (shallow coastal waters)	Tens to hundreds µs	L_p 185–220 dB re 1 µPa m $L_{E,p}$ 150–190 dB re 1 µPa²m²s
Imaging sonar	Medium thick swath horizontally tilted down toward seafloor, beamwidths: 10–30° vertical, 30–130° horizontal	750–3000 kHz (max range: 100–5 m)	Tens to hundreds µs	L_p 195–198 dB re 1 µPa m

depicted in Fig. 2.56. Seafloor mapping sonar systems are generally deployed from vessels (Fig. 2.56), but they can also be deployed on underwater robots, such as tethered remotely operated vehicles (ROVs) and untethered autonomous underwater vehicles (AUVs). Apart from SBESs, which have a narrow cone beam, most seafloor mapping sonars transmit sound as narrow swaths. The frequency of sound transmitted depends on the depth or range for which the sonar is designed, but is somewhere between 10 kHz (for deep-ocean or long-range applications) and a few MHz (shallow-water or short-range applications). The sonar frequency also dictates the available resolution. Historically, seafloor mapping sonar systems mainly used CW pulses, but FM/chirp pulses are now common, as they can improve the signal-to-noise ratio. The length of transmitted pulses depends on sonar frequency, pulse type and depth or range, but typically spans from tens of microseconds to a few milliseconds, with longer pulses typically used for FM and deep-water operations (Table 2.5). The maximum ping rate is limited

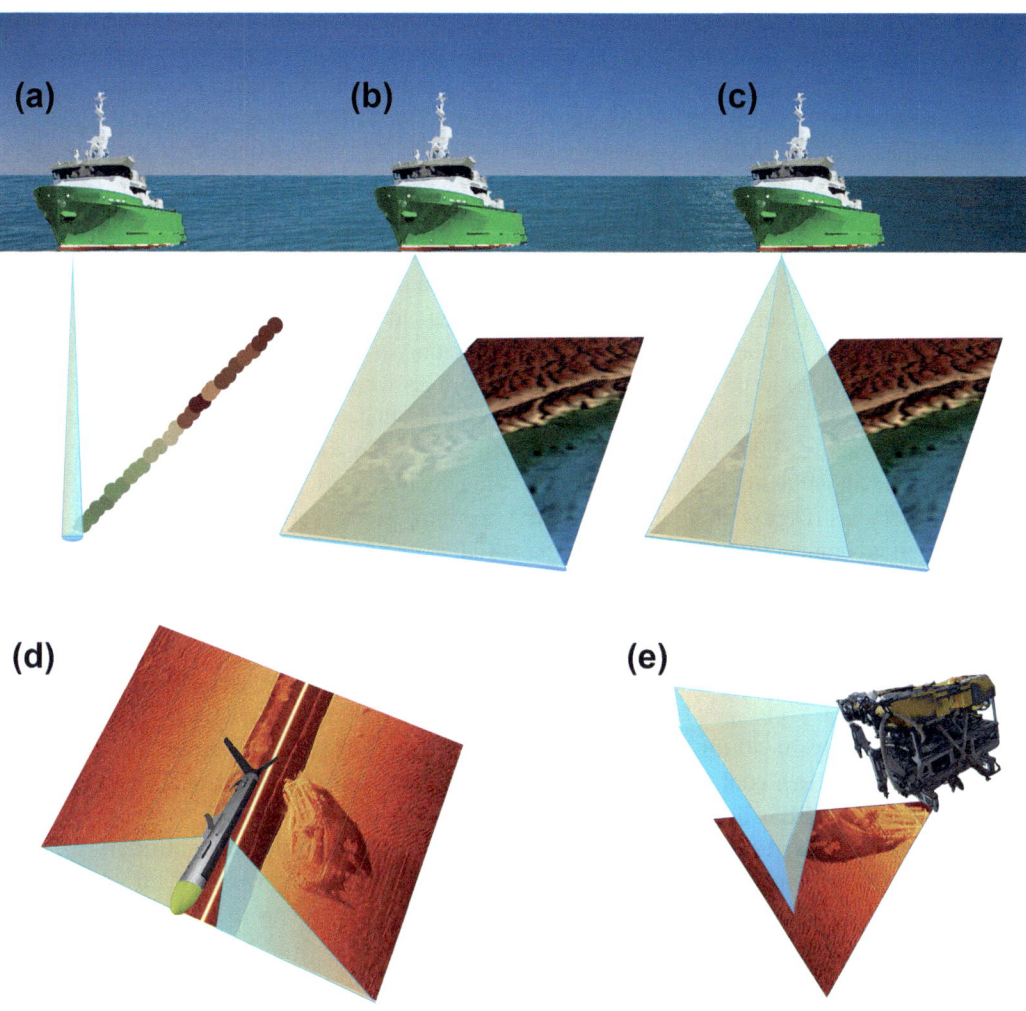

Fig. 2.56 Seafloor mapping sonar beam geometry and example data outputs of (**a**) single-beam echosounder, (**b**) multibeam echosounder, (**c**) multi-swath multibeam echosounder, (**d**) sidescan sonar, and (**e**) imaging sonar single-beam echosounders

to the two-way travel time of the set range, but computational limitations can further limit ping rates. The source level of seafloor mapping sonars depends on the frequency, coverage, and operational depth or range, but most specifications and studies have reported L_p values between 180 and 230 dB re 1 µPa m and $L_{E,\,p}$ between 140 and 190 re 1 µPa²m²s. Higher source levels have been reported, but usually just for deep-water mapping systems utilized on large ocean-going scientific vessels, with L_p between 230 and 245 dB re 1 µPa m (Table 2.5). Overall, seafloor mapping sonar systems transmit short pulses ($<< 1$ s) of ultrasonic sound, over narrow, highly directional areas toward the seafloor.

SBESs are one of the simplest seafloor mapping sonar systems and are widely used in fisheries, hydrographic surveys, and small recreational boats (Bjørnø 2017; Simmonds and MacLennan 2005). Although there is a range of SBES systems available, most have a cone beam with a width of 5–30° (Table 2.5). One advantage of SBES is, in addition to measuring the depth below a vessel, that they also can show the location of any strong acoustic targets in the water column, such as fish (Bjørnø 2017; Simmonds and MacLennan 2005). SBES can be found operating at frequencies between 10 and 1000 kHz, but most recreational and shipping echosounders operate between 30 and 200 kHz (Bjørnø 2017; Penrose et al. 2005). SBES systems that operate at frequencies below 30 kHz are designed for deep-water mapping or fisheries applications (Bjørnø 2017; Simmonds and MacLennan 2005), whereas SBES operating at frequencies above 200 kHz are usually employed for shallow hydrographic surveys and zooplankton studies (Depew et al. 2009; Warren et al. 2016). It is common for SBES systems to be able to operate at two or more different frequencies, to cover different water depths (e.g., 30 and 200 kHz is a common pairing). Source levels of SBESs are usually between 180 and 230 dB re 1 µPa m (Table 2.5), and side lobes are usually 20–40 dB below the main lobe (Hastie 2012; Risch et al. 2017). The pulse duration used for a SBESs depends on the pulse type, frequency, and water depth, but is usually somewhere between tens of microseconds to a few milliseconds (Table 2.5). Longer pulses can be used in broadband echosounders producing FM signals (Demer et al. 2016).

2.5.8.2.1 Sidescan and Interferometric Sidescan Sonar

SSSs can create an acoustic image of the seafloor as they move through the water by transmitting two wide narrow swaths (0.2–3° along-track × 40–90° across-track), one port and one starboard, usually slightly tilted out, and so do not measure directly under the vessel (Table 2.5, Fig. 2.56). Designed as an acoustic imaging device, a SSS does not measure depth. Nevertheless, SSSs have been used for a wide range of applications, including mine countermeasures, geological mapping, and marine archeology (Blondel 2009). SSSs are usually deployed on a towfish close to the seafloor, to increase resolution and decouple it from surface motion, but SSSs can also be found mounted on vessels, ROVs, and AUVs. There is a range of SSS systems, operating anywhere between 75 and 1600 kHz, but most operate at hundreds of kHz. The frequency of the SSS is normally selected based on the range or resolution desired. As frequency increases, resolution broadly increases (along with decreasing along-track beamwidth), but range decreases due to signal attenuation. Hence, SSS systems operating between 900 and 1600 kHz are usually limited to ranges <50 m; whereas a SSS operating at 75 kHz can potentially range up to 1 km. As with SBESs, it is common to find SSS systems that can operate at two different frequencies to cover changes in range or resolution requirements. Source levels of SSSs have been measured between 200 and 230 dB re 1 µPa m and 165 and 180 dB re 1 $\mu Pa^2 m^2$s (Clarke 2009; Crocker et al. 2019; Stanton et al. 2010). Interferometric SSSs transmit in a similar way to SSSs with comparable source levels (Crocker et al. 2019), but have an additional receiver array (on both sides) to use the phase of the returning signal to measure the along-track depth of the reflectors (Lurton 2004).

2.5.8.2.2 Multibeam Echosounders

MBESs have become the standard sonar system to map the seafloor as they can measure depth (and reflectivity or backscatter) at discrete intervals across a wide (90–150° across-track) swath (Kenny et al. 2003; Lurton 2004; Penrose et al. 2005). While MBES's transmit geometry is similar to that of SSS systems, MBES systems use a receiver array to carry out beam forming, to accurately determine the location of depth measurements across the transmit sector. Like other seafloor mapping sonar systems, MBESs operate at a wide range of frequencies depending on the application (between 10 and 710 kHz). MBESs operating at 400–710 kHz are used to map shallow water (<70 m); these systems offer high resolution and small transducer size. Typical CW pulse durations for high-frequency MBESs operating in shallow water can be between tens and hundreds of µs. Source levels for an MBES operating at 400 kHz were measured between 185 and 220 dB re 1 µPa m, and 150 and 190 dB re 1 $\mu Pa^2 m^2$s (depending on settings;

Crocker et al. 2019). In shallow water, sometimes two MBESs are mounted side-by-side to create a dual-head system to increase coverage. As water depth increases, so does coverage, but so also does signal attenuation, resulting in higher source levels, longer pulses, and lower frequencies being required (at the cost of lower resolution). For instance, MBES systems operating at frequencies between 90 and 300 kHz are mainly used on the continental slope and shelf, with a source level between 185 and 230 dB re 1 µPa m, depending on the system and settings (Crocker et al. 2019; Hammerstad 2005; Lurton 2016). The lowest frequencies, longest pulses, and highest source levels are used to map the deepest areas of the ocean, with typical frequencies between 10 and 70 kHz, pulse durations of 10–100 ms, and source levels between 230 and 245 dB re 1 µPa m (Bjørnø 2017; Hammerstad 2005; Lurton 2016). Also in deep water, the ping rate can be slow, which reduces the along-track resolution. One way to improve the ping rate is to use a dual or multi-swath MBES system, where multiple pings can be in the water simultaneously. This is done by dividing their transmit fan into multiple swaths, whereby each sector transmits at different frequency, duration, bandwidth, source level, and beam pattern (Teng 2011).

Lurton (2016) modeled the received levels from three different MBESs operating at 12, 30, and 100 kHz, with one of the outputs shown in Fig. 2.57. The highest received levels were directly below the transducer and in narrow strips port and starboard of it, and the received levels outside of the defined beam dropped off quickly.

2.5.8.2.3 Imaging Sonars

Imaging sonar systems are usually deployed on ROVs to visualize the seafloor and to aid navigation and avoid hazards. They transmit sound horizontally and down toward the seafloor. There are two main ways in which an imaging sonar operates: mechanically scanning and beamforming. Mechanically scanning imaging sonar systems transmit a narrow swath (similar to a single SSS beam), which is slowly rotated to build up a 2D image similar to one produced by a SSS. Beamforming imaging sonar systems operate similarly to MBESs, but unlike MBESs, transmit horizontally (15–130° across), have a wider vertical angle (~10–30°) which is tilted down to the seafloor, and do not measure depth. Imaging sonar systems typically operate at short ranges (<100 m) and at several hundreds to thousands of kHz (Table 2.5). Pulse durations are usually short to optimize resolution and are of the order of tens to hundreds of µs. Source levels of a 375 kHz and a 720 kHz sonar were measured at 195 and 198 re 1 µPa m, respectively (Hastie 2012).

2.5.8.3 Military Sonars

Anti-submarine warfare active sonars are used in defense exercises to detect submarines, underwater vehicles, moored mines, or underwater obstacles (Lurton 2004). They typically operate at frequencies between 100 Hz and 20 kHz, transmitting CW and FM pulses seconds long in duration. They can cover up to 360° of the field of view, and source levels have been reported between 160 and 235 dB re 1 µPa m (Table 2.6). Military sonars can be classified as low-frequency active sonar (LFAS) or mid-frequency active sonar (MFAS). The long-range LFASs usually operate at frequencies between 200 Hz and 2 kHz (Antunes et al. 2014; Popper et al. 2007). Tactical mid-range MFASs operate at between 2 and 20 kHz (Halvorsen et al. 2012; Isojunno et al. 2016). Both LFAS and MFAS can produce a range of pulse types, depending on the operational needs (Popper et al. 2007; Tyack et al. 2011). Narrowband CW pulses are used to detect movement, while FM signals are used in reverberant environments to improve SNR (Evans and England 2001). Typically, a sequence of CW and/or FM signals is emitted followed by a waiting period to listen to the returning echoes (Fig. 2.58; Acevedo and Smultea 1995; D'Spain et al. 2006; Halvorsen et al. 2012). The number, order, frequency, and beamwidth of pulses may vary with the purpose of the exercise.

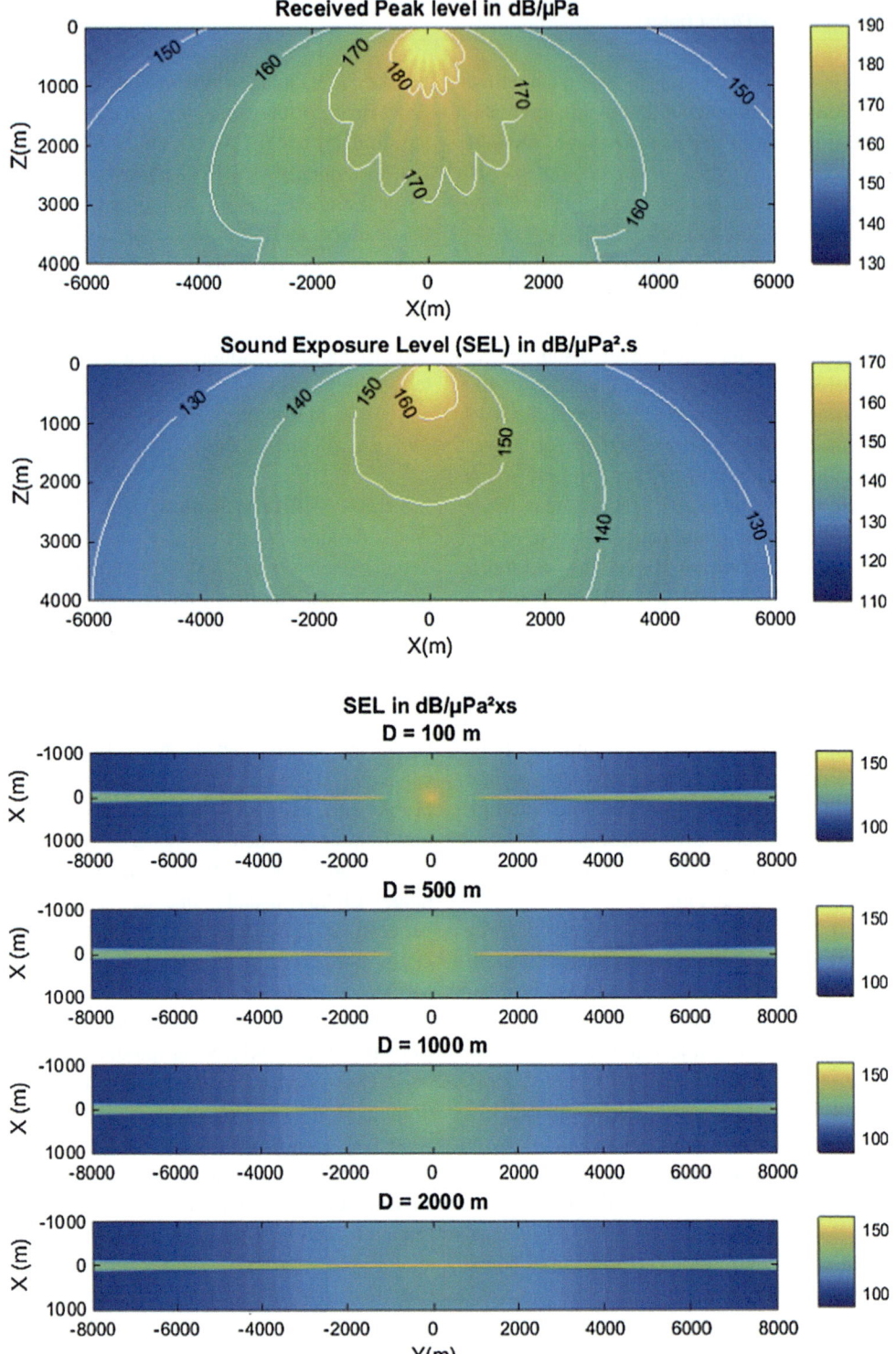

Fig. 2.57 Radiation pattern modeled for a MBES transmitting a 10 ms CW pulse at 12 kHz over a 100° × 1° swath with a source level of 240 dB re 1 μPa m. The upper part depicts the field in the vertical plane in terms of peak sound pressure level and sound exposure level (SEL). The lower part gives horizontal cross-sections of the SEL at different depths. Reprinted from Lurton X (2016) Modelling of the sound field radiated by multibeam echosounders for acoustical impact assessment. Appl Acoust 101:201–221; https://doi.org/10.1016/j.apacoust.2015.07.012 © Lurton 2016; https://doi.org/10.1016/j.apacoust.2015.07.012. Published under CC BY-NC-ND 4.0; https://creativecommons.org/licenses/by-nc-nd/4.0/. Reprinted with permission from Elsevier and Lurton

Table 2.6 Typical frequencies, pulse durations, inter-pulse intervals, and source levels for military sonar systems. Data from Acevedo and Smultea (1995), Cox et al. (2006), D'Spain et al. (2006), Doksaeter et al. (2009), Doksæter et al. (2012), Evans and England (2001), Halvorsen et al. (2012), and Sivle et al. (2012)

Military sonar type	Frequency	Pulse duration	Inter-pulse interval	Source level (L_p)
Low-frequency active sonar (LFAS)	100–3300 Hz	0.3–12 s	20–60 s	182–235 dB re 1 µPa m
Mid-frequency active sonar (MFAS)	3000–8000 Hz	0.5–4 s	20–60 s	160–226 dB re 1 µPa m

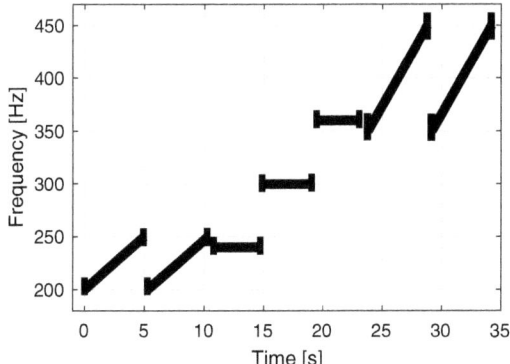

Fig. 2.58 Sketched spectrogram of an active low-frequency sonar transmission sequence consisting of CW and FM tones, similar to those reported elsewhere for LFAS and MFAS (Doksæter et al. 2012; Halvorsen et al. 2012; Popper et al. 2007; Southall et al. 2012; Tyack et al. 2011)

2.5.9 Explosions

Underwater explosions from construction, demolition, fishing, and military activities are encountered in the underwater soundscape as sound impulses of high amplitude. Explosive charges were used extensively as sources for underwater acoustics research up until the 1990s and are still used occasionally in situations where the required signal characteristics cannot be obtained by other means (Dall'Osto et al. 2023). As a result, their acoustic characteristics are well understood.

When an underwater explosive is detonated, a spherically symmetrical shock wave is generated due to the release of large quantities of energy in a short time. The increase in pressure in the shock wave is practically instantaneous, followed by a decay that is initially exponential with a time constant typically of a fraction of a millisecond. The gas bubble created by the explosion expands rapidly and overshoots its equilibrium radius because of the water's inertia. It reaches a maximum radius and then begins to contract, rebounding again when the pressure in the bubble becomes sufficient to counteract the momentum of the incoming water. The result is a series of damped radial oscillations, with the number of cycles depending on the speed at which the bubble breaks up into smaller parts. At each bubble minimum, a pressure pulse is radiated, of smaller amplitude but longer duration than the shock wave. The first bubble pulse has an impulse comparable with that of a shock wave. The bubble pulse pressure and impulse are comparable to those of the shock pulse, but it is not itself a shock wave.

Figure 2.59 shows examples of signals received at close range from explosions from charges of different sizes. These were both signal underwater sound (SUS) charges, which were designed as sound sources for anti-submarine warfare, with the Mk64 and Mk82 charges containing 0.031 kg and 0.82 kg of high explosives, respectively. Note the different measurement ranges (12 m and 70 m, respectively, for the two different SUS charges).

Chapman (1985) gave semi-empirical relationships that allow prediction of many of the properties of the acoustic signals generated by uncontained underwater explosions based on the charge weight w (kg), range r (m), and depth d (m). For example, the relationship for the peak pressure, in Pa, of the initial shock wave is

$$p_{pk} = 5.04 \times 10^7 \left(\frac{w^{\frac{1}{3}}}{r}\right)^{1.13} \quad (2.13)$$

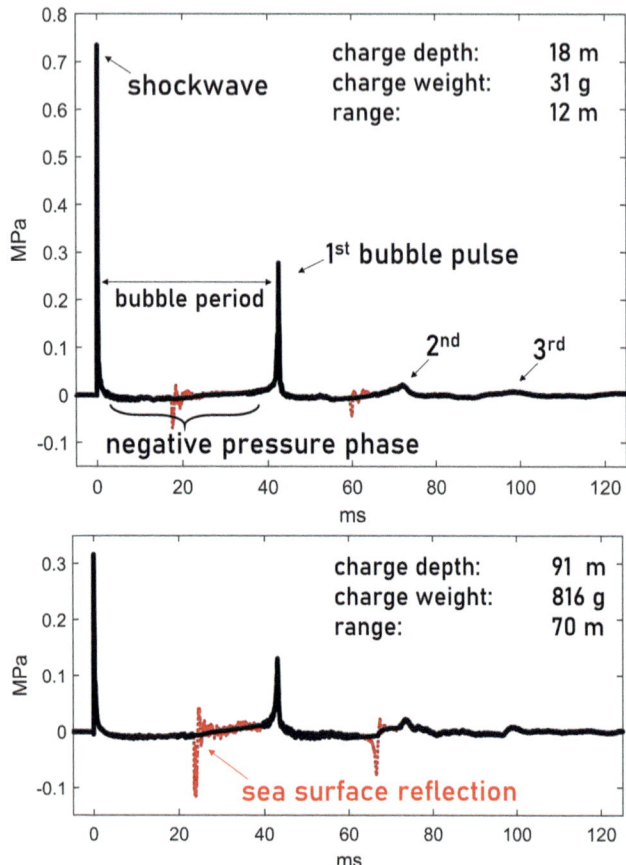

Fig. 2.59 Pressure signal from a Mk64 (top) and Mk82 (bottom) SUS charge measured by a hydrophone suspended 20 m below the surface. Red lines: reflections from the sea surface. Reproduced with permission from the ASA; Dall'Osto DR, Dahl PH, Chapman NR (2023) The sound from underwater explosions. Acoustics Today 19 (1):12–19; https://doi.org/10.1121/AT.2023.19.1.12. © Acoustical Society of America, 2023. All rights reserved

which shows that, as the shock wave propagates, the peak pressure decays slightly faster than the $1/r$ dependence expected from linear acoustics. Conversely, other relationships show that the decay time constant of the shock wave, and hence its duration, increases as it propagates. The initial shock wave is very short, so the probability of interference between arrivals at a receiver that have traveled by different paths is often small, and as a result, Eq. 2.13 for the peak pressure is quite robust in a wide variety of propagation conditions (Dall'Osto et al. 2023). It can be used for uncontained in-water explosions for many types and sizes of charges, providing w is interpreted as an equivalent charge weight that takes account of the type of explosive used. An exception is fully or partially contained explosions, in which the charge is detonated inside a structure or within the seafloor, which produce lower peak pressures than uncontained explosions of the same charge weight. The amount of reduction depends on the details of the individual scenario and is difficult to predict; however, this effect results in US environmental protection guidelines for explosive removal of offshore structures being less stringent if the charges are detonated below the seafloor than if they are detonated above the seafloor (Viada et al. 2008). These guidelines also encourage the use of shaped charges, which direct the explosive energy at the object to be cut and therefore can use a smaller charge weight to achieve the same effect.

The sound from detonations of unexploded ordnance has been well studied in the European North Sea. Salomons et al. (2021) compared model with measurements from 140 and 325 kg TNT-equivalent detonations in shallow water.

2 Sources of Underwater Noise

Table 2.6 Typical frequencies, pulse durations, inter-pulse intervals, and source levels for military sonar systems. Data from Acevedo and Smultea (1995), Cox et al. (2006), D'Spain et al. (2006), Doksaeter et al. (2009), Doksæter et al. (2012), Evans and England (2001), Halvorsen et al. (2012), and Sivle et al. (2012)

Military sonar type	Frequency	Pulse duration	Inter-pulse interval	Source level (L_p)
Low-frequency active sonar (LFAS)	100–3300 Hz	0.3–12 s	20–60 s	182–235 dB re 1 µPa m
Mid-frequency active sonar (MFAS)	3000–8000 Hz	0.5–4 s	20–60 s	160–226 dB re 1 µPa m

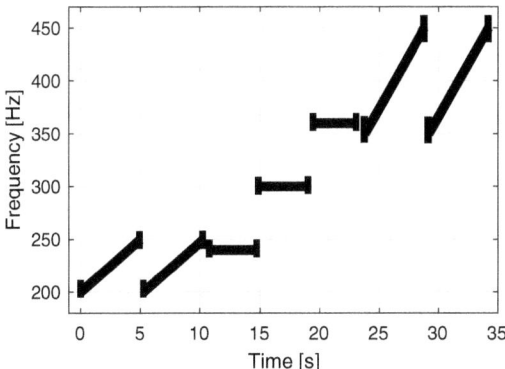

Fig. 2.58 Sketched spectrogram of an active low-frequency sonar transmission sequence consisting of CW and FM tones, similar to those reported elsewhere for LFAS and MFAS (Doksæter et al. 2012; Halvorsen et al. 2012; Popper et al. 2007; Southall et al. 2012; Tyack et al. 2011)

2.5.9 Explosions

Underwater explosions from construction, demolition, fishing, and military activities are encountered in the underwater soundscape as sound impulses of high amplitude. Explosive charges were used extensively as sources for underwater acoustics research up until the 1990s and are still used occasionally in situations where the required signal characteristics cannot be obtained by other means (Dall'Osto et al. 2023). As a result, their acoustic characteristics are well understood.

When an underwater explosive is detonated, a spherically symmetrical shock wave is generated due to the release of large quantities of energy in a short time. The increase in pressure in the shock wave is practically instantaneous, followed by a decay that is initially exponential with a time constant typically of a fraction of a millisecond. The gas bubble created by the explosion expands rapidly and overshoots its equilibrium radius because of the water's inertia. It reaches a maximum radius and then begins to contract, rebounding again when the pressure in the bubble becomes sufficient to counteract the momentum of the incoming water. The result is a series of damped radial oscillations, with the number of cycles depending on the speed at which the bubble breaks up into smaller parts. At each bubble minimum, a pressure pulse is radiated, of smaller amplitude but longer duration than the shock wave. The first bubble pulse has an impulse comparable with that of a shock wave. The bubble pulse pressure and impulse are comparable to those of the shock pulse, but it is not itself a shock wave.

Figure 2.59 shows examples of signals received at close range from explosions from charges of different sizes. These were both signal underwater sound (SUS) charges, which were designed as sound sources for anti-submarine warfare, with the Mk64 and Mk82 charges containing 0.031 kg and 0.82 kg of high explosives, respectively. Note the different measurement ranges (12 m and 70 m, respectively, for the two different SUS charges).

Chapman (1985) gave semi-empirical relationships that allow prediction of many of the properties of the acoustic signals generated by uncontained underwater explosions based on the charge weight w (kg), range r (m), and depth d (m). For example, the relationship for the peak pressure, in Pa, of the initial shock wave is

$$p_{pk} = 5.04 \times 10^7 \left(\frac{w^{\frac{1}{3}}}{r}\right)^{1.13} \quad (2.13)$$

Fig. 2.59 Pressure signal from a Mk64 (top) and Mk82 (bottom) SUS charge measured by a hydrophone suspended 20 m below the surface. Red lines: reflections from the sea surface. Reproduced with permission from the ASA; Dall'Osto DR, Dahl PH, Chapman NR (2023) The sound from underwater explosions. Acoustics Today 19 (1):12–19; https://doi.org/10.1121/AT.2023.19.1.12.
© Acoustical Society of America, 2023. All rights reserved

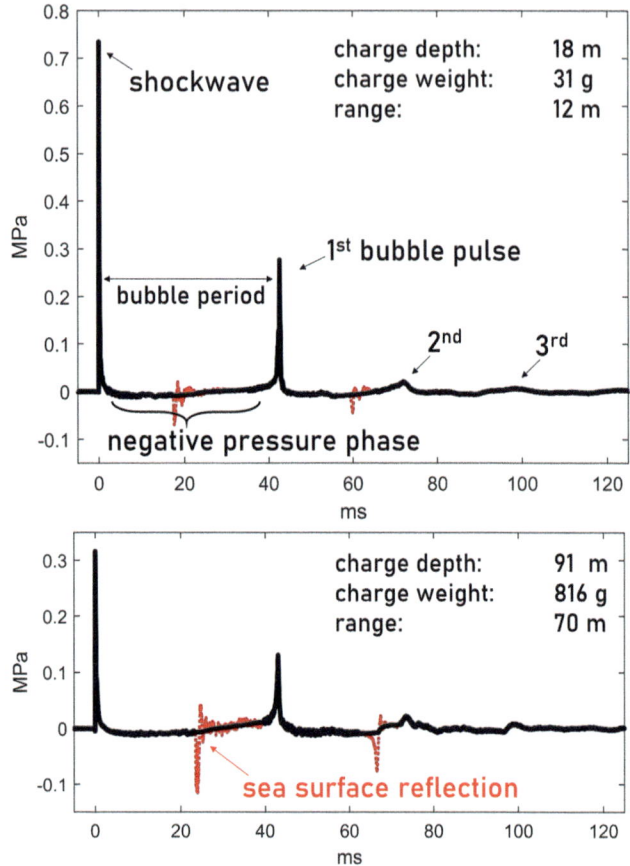

which shows that, as the shock wave propagates, the peak pressure decays slightly faster than the $1/r$ dependence expected from linear acoustics. Conversely, other relationships show that the decay time constant of the shock wave, and hence its duration, increases as it propagates. The initial shock wave is very short, so the probability of interference between arrivals at a receiver that have traveled by different paths is often small, and as a result, Eq. 2.13 for the peak pressure is quite robust in a wide variety of propagation conditions (Dall'Osto et al. 2023). It can be used for uncontained in-water explosions for many types and sizes of charges, providing w is interpreted as an equivalent charge weight that takes account of the type of explosive used. An exception is fully or partially contained explosions, in which the charge is detonated inside a structure or within the seafloor, which produce lower peak pressures than uncontained explosions of the same charge weight. The amount of reduction depends on the details of the individual scenario and is difficult to predict; however, this effect results in US environmental protection guidelines for explosive removal of offshore structures being less stringent if the charges are detonated below the seafloor than if they are detonated above the seafloor (Viada et al. 2008). These guidelines also encourage the use of shaped charges, which direct the explosive energy at the object to be cut and therefore can use a smaller charge weight to achieve the same effect.

The sound from detonations of unexploded ordnance has been well studied in the European North Sea. Salomons et al. (2021) compared model with measurements from 140 and 325 kg TNT-equivalent detonations in shallow water.

Received levels at 1.5, 6, and 12 km range were 195, 178, and 170 dB re 1 µPa²s, respectively, for the smaller charge and 212, 198, and 187 dB re 1 µPa (peak pressure), respectively, for the larger charge in these environments. The likelihood of hearing loss in harbor porpoises (*Phocoena phocoena*) was modeled.

The use of explosives in fishing, also known as destructive fishing or fish blasting, is common practice in shallow coral reefs around Southeast Asia. They are usually weak shock blasts produced by a mix of ammonium nitrate fertilizer and kerosine (fuel oil) of ~0.5 kg equivalent charge weight (Woodman et al. 2003). The authors showed a received signal with a peak pressure of 0.8 kPa, which corresponded to a compressional SPL_{pk} of 218 dB re 1 µPa at a range of 250 m. This agreed well with the predicted level of 217.6 dB re 1 µPa for this charge weight and range. The signal at 3000 m range was complicated by many seabed and surface reflections. The measured compressional SPL_{pk} of 194.0 dB re 1 µPa was close to the predicted level of 193.2 dB re 1 µPa; however, the measured rarefactional SPL_{pk} was somewhat larger at 198.6 dB re 1 µPa, which indicated that in this case the signal was long enough to allow some constructive interference between different acoustic paths (Woodman et al. 2003).

2.5.10 Acoustic Mitigation Devices

Acoustic mitigation devices (AMDs) use sound to promote behavioral change in target species (primarily marine mammals) to reduce their interaction with human activities (Dawson et al. 2013; Mackay and Knuckey 2013). Historically, application was directed primarily at fisheries and mariculture activities; however, AMDs have increasingly also been used to reduce the risks of collisions with vessels, and hearing impairment and injury from pile driving and underwater explosives (Gerstein 2002; McGarry et al. 2022). Interactions between marine mammals and fisheries have a long history of being considered problematic, with damage to gear and fish, and taking of fish (i.e., depredation) causing conflict between fishers and marine mammals (Mate and Harvey 1986). Historically, approaches to manage these conflicts have been at the discretion of fishers, including lethal means to decrease marine mammal populations or remove problematic individuals by placing bounties or hiring hunters (Pearson and Verts 1970; also see references cited in Mate and Harvey 1986). Over more recent decades, increasing awareness of the extent of impacts on marine mammals has led to their protection and regulation of human activities in many jurisdictions. Over time, as bycatch has been recognized as a serious conservation problem (e.g., in trawl and gill nets Barlow et al. 1995; Blaylock et al. 1995; Perrin et al. 1994), acoustic mitigation devices have been developed and used increasingly.

Early tests of acoustic mitigation devices in the 1970s involved playback of sounds of predatory killer whales to deter belugas and gray whales, and while they were effective initially, without the presence of killer whales, the sounds' effectiveness decreased (Cummings and Thompson 1971; Fish and Vania 1971; Mate and Harvey 1986). While inconclusive tests were met with skepticism (Jefferson and Curry 1994), more controlled experiments indicated significant reductions in bycatch in certain fisheries (Gearin et al. 1996; Kraus et al. 1995). In the 1980s, for example, tests of 4 kHz low-powered acoustic devices in Newfoundland were reportedly successful at reducing whale entanglements in fishing gear (Lien et al. 1992). Research over the decades, however, has confirmed variability in effectiveness, which largely depends upon species, human activity (e.g., fishery and gear involved), location, and acoustic device (Bordino et al. 2002; Reeves et al. 2001). Furthermore, some AMDs may cause impacts on species by deterring them from critical habitats or causing hearing impairment from high-power devices (Dolman et al. 2022; Schaffeld et al. 2019).

AMDs, thus, have commonly been classed into two types depending upon their targeted application and power: acoustic deterrent devices (ADDs) and acoustic harassment devices (AHDs) (Reeves et al. 2001). While there is some variability in the use of the terminology, ADDs (commonly known as pingers) often refer to low-amplitude acoustic devices designed to prevent marine mammal (commonly cetacean) bycatch and entanglement, or reduce depredation (Dawson et al. 2013; Reeves et al. 1996, 2001). In contrast, AHDs largely refer to high-amplitude devices designed to scare, annoy, or cause discomfort or pain (Dawson et al. 2013; Götz and Janik 2013; Lepper et al. 2014). AHDs, also known as seal scarers or scrammers, were created primarily to reduce pinniped (seal and sea lion) depredation around aquaculture pens. Even small explosives (so-called seal bombs) with source levels of 233 dB re 1 μPa m and 203 dB re 1 μPa^{2}m^{2}s have been used to chase pinnipeds away (Krumpel et al. 2021; Wiggins et al. 2021).

While ADDs have successfully reduced bycatch in some fisheries, they have been associated with increased prey removal from nets in others (Bordino et al. 2002), hypothesized to be a learned response (Bordino et al. 2002). The design of AHDs, with the intention of causing discomfort or pain, has been intended to counter this dinner-bell effect (Wilson and Carter 2013). The use of ADDs and AHDs also varies in that ADDs have tended to be more variable in their deployments (time, duration, and location) as they are often associated with fishing gear (Johnston and Woodley 1998), while AHDs have tended to be more permanently installed at aquaculture facilities (Reeves et al. 2001). ADDs and AHDs have been tested and are used in many regions of the world (Dawson et al. 2013; McGarry et al. 2022). For a recent comprehensive list of devices, including the names of the AMDs, manufacturers, and their technical specifications, see McGarry et al. (2022).

2.5.10.1 Acoustic Deterrent Devices

Commercially available ADDs can produce a variety of sounds in a range of frequencies, pulse types, durations, pulse intervals, and intensities (Kastelein et al. 2007; McGarry et al. 2022; Shapiro et al. 2009). Most pingers produce frequencies between 3 and 160 kHz, and can include long tones, intermittent sounds at single or multiple frequencies, FM sweeps, and more complex sounds (Kastelein et al. 2007; Shapiro et al. 2009). Many intermittent ADDs have variable ping durations and intervals, with some having different pulse rates that can be selected and many having random inter-pulse intervals. The duration of intermittent pulses is usually on the order of a fraction of a second, and the pulse interval may vary between 2 and 40 s (Kraus et al. 1997; McGarry et al. 2022; McPherson et al. 2004). To counter habituation by marine mammals, the dinner-bell effect, and noise pollution from ongoing emissions by ADDs, interactive devices have been developed that only transmit the deterring signal when dolphin echolocation clicks are detected (Ceciarini et al. 2023). ADDs produce relatively low-intensity sounds with source levels generally ≤165 dB re 1 μPa m (Kastelein et al. 2007, 2008; Kraus et al. 1997; McGarry et al. 2022; Shapiro et al. 2009). The highest power spectral density level might not occur at the fundamental, but rather at overtones (e.g., Erbe and McPherson 2012) (Fig. 2.60).

2.5.10.2 Acoustic Harassment Devices

AHDs can produce a variety of sounds including sinusoidal tonal bursts, trains of pulses with harmonics, or FM sweeps, in regular or randomized sequences (Johnston 2002; Lepper et al. 2004; McGarry et al. 2022). AHDs operate at a range of frequencies, mostly between ~1 kHz and 70 kHz (Lepper et al. 2004, 2014; McGarry et al. 2022). AHDs have higher source levels than ADDs, reaching 190 dB re 1 μPa m or more (Götz and Janik 2013; Olesiuk et al. 2002). As with ADDs, the highest source power spectral density level might not occur at the fundamental but at overtones. One of the programs of the DSMS-4, for example, produces random sequences of tones with fundamental frequencies between 1.8 kHz and 3.8 kHz, while the maximum source power spectral density level was observed at 6.6 kHz (Lepper et al. 2004).

2 Sources of Underwater Noise 159

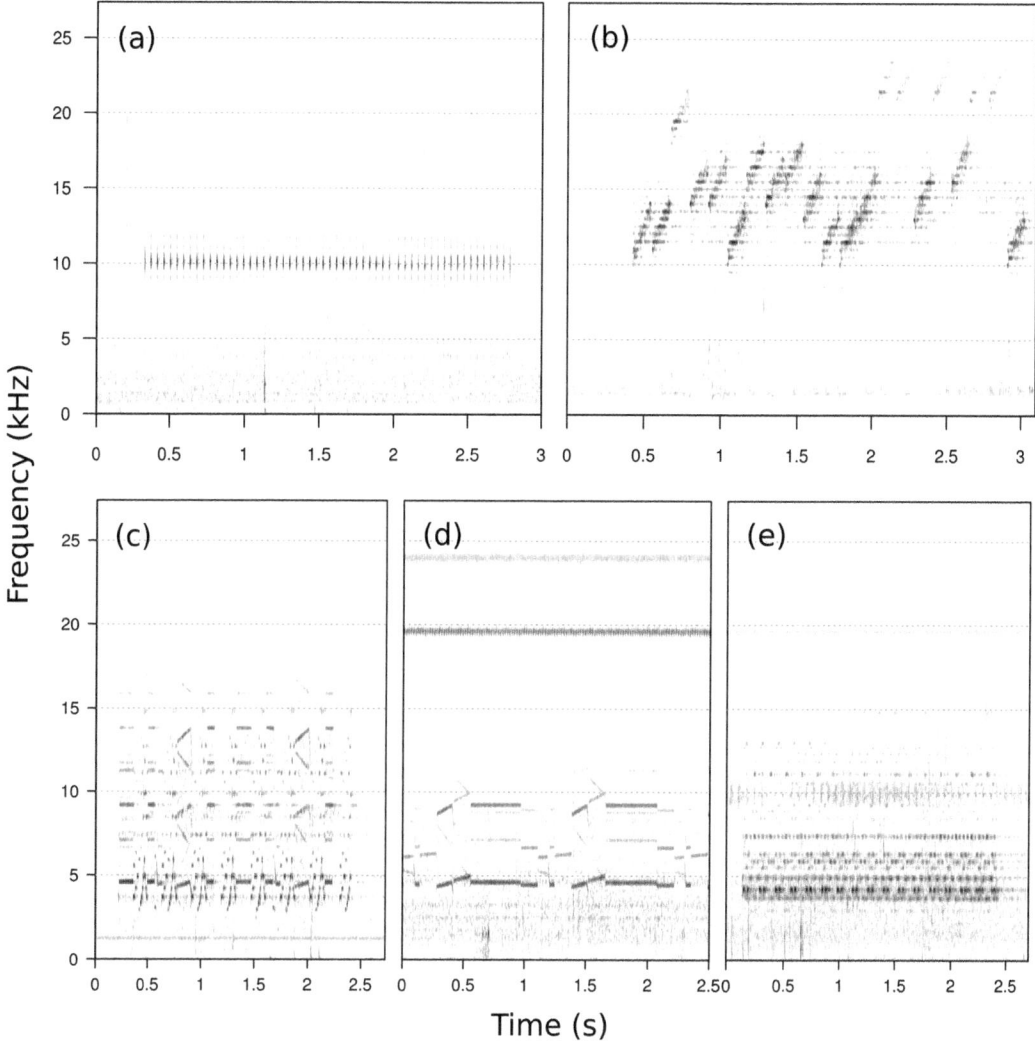

Fig. 2.60 Example spectrograms of acoustic deterrent device (ADD) types recorded on Hebridean Whale and Dolphin Trust cetacean surveys (2011–2015; Findlay et al. 2018). FFT size 1024 points, overlap 50%, sampling frequency 96 kHz; resulting in frequency and time resolution of 93.8 Hz and 10.67 ms, respectively. (**a**) Airmar™ (dB Plus II); (**b**) Ace Aquatec™ (US3); (**c**) Terecos™ (Type DSMS-4) Programme 4; (**d**) Terecos™ (Type DSMS-4) Programme 2; (**e**) Terecos™ (Type DSMS-4) Programme 3. Reprinted with permission from Elsevier. Findlay CR, et al. Mapping widespread and increasing underwater noise pollution from acoustic deterrent devices. Mar Pollut Bull 135:1042–1050. https://doi.org/10.1016/j.marpolbul.2018.08.042 © Elsevier, 2018. All rights reserved

2.6 Summary

The marine soundscape, whether in coastal areas or in the deepest ocean, whether in the tropics or at the poles, contains a myriad of sounds. Sounds may be grouped by their origin into geophony (i.e., wind, precipitation, waves, earthquakes, volcanoes, and ice), biophony (e.g., invertebrates, fishes, and marine mammals), and anthropophony (e.g., port construction, mineral and hydrocarbon exploration and production, renewal energy installation, and shipping). This chapter gave a brief overview of the geophony and biophony, and then focused on the anthropophony. The sounds of boats and ships of various types, marine seismic surveys, drilling, dredging, pile driving, windfarms, geotechnical site investigations, sonars, echosounders, explosions, and acoustic mitigation devices were presented, with their characteristic source levels and spectra. Approaches to modeling in particular the sounds emitted by ships, seismic airguns, and pile driving were discussed.

2.7 Glossary

Soundscape: In terrestrial acoustics, a soundscape is an acoustic environment as perceived by a listener (i.e., a person or people) (ISO 12913-1; International Organization for Standardization 2014). Soundscape is a perceptual construct, which exists through human perception of the physical (acoustic) environment. In underwater acoustics, the definition does not require a listener. Rather, a soundscape is the total sound field that results from contributions of a number of sources, synonymously to acoustic environment (ISO 18405; International Organization for Standardization 2017b).

Geophony: The sounds of the geophysical, natural environment. Under water, these are due to wind, breaking waves, rain impacting on the water surface, polar ice break-up, subsea earthquakes and volcanoes, etc.

Biophony: The sounds of biological origin in an acoustic environment. Under water, these include marine mammals, fishes, and crustaceans.

Anthropophony: The sounds made by humans. Under water, these include ship traffic, marine seismic surveys, pile driving, sonars, etc.

Acoustic environment: The superposition of sounds from any sources, which are distributed in space and of which sound propagates through the environment (and is modified by the environment due to absorption, reverberation, etc.), eventually overlapping at the location of the receiver (ISO 12913-1; International Organization for Standardization 2014).

Ambient noise: All of the sounds from sources that may be distributed in space, both near and far. Often a specific signal is studied relative to the ambient noise at a particular location. In this case, the signal is not considered part of the ambient noise (ANSI/ASA S1.1; American National Standards Institute 2013).

Background noise: All sources of acoustic interference. In studies aiming to record specific signals, the ambient noise is part of the background noise. In addition, background noise includes electric interference by the recording system and its power supply (such as a 50 Hz hum; ANSI/ASA S1.1; American National Standards Institute 2013).

Frequency (symbol: f, unit: Hz): The rate of oscillation of a sound wave measured as the number of cycles per second. The unit of frequency is hertz: $1\,\text{Hz} = \frac{1}{s}$.

Wavelength (symbol: λ, unit: m): The spatial distance (measured in meters) between two successive peaks in a pressure wave.

Speed of sound (symbol: c, unit: m/s): The distance traveled per unit time. If τ is the period of the sound wave (i.e., the duration of one cycle), then the speed of sound can be expressed as $c = \frac{\lambda}{\tau} = \lambda f$.

Sound pressure (symbol: p, unit: Pa): Sound pressure $p(t)$ at any location varies with time t and is measured with a hydrophone in water. The greatest magnitude of $p(t)$ is called the peak sound pressure p_{pk} (or zero-to-peak sound pressure). It is the greater of the greatest magnitude during compression and the greatest magnitude during rarefaction. The peak-to-peak sound pressure p_{pk-pk} is the sum of the greatest magnitude during compression and the greatest magnitude during rarefaction. The root-mean-square sound pressure p_{rms} is literally the root of the time-average of the squared pressure: $p_{rms} = \sqrt{\frac{\int_{t_1}^{t_2} p^2(t)dt}{t_2 - t_1}}$, where t_1 and t_2 are the start and end times, respectively. The unit of all pressures is pascal (Pa).

Sound pressure level (SPL, symbol: L_p): The level of the root-mean-square sound pressure: $L_p = 20 \log_{10}\left(\frac{p_{rms}}{p_0}\right)$ relative to a reference value $p_0 = 1$ μPa. SPL is expressed in dB re 1 μPa.

Peak sound pressure level (SPL$_{pk}$, symbol: $L_{p,\,pk}$): The level of the peak sound pressure: $L_{p,pk} = 20 \log_{10}\left(\frac{p_{pk}}{p_0}\right)$ relative to a reference value $p_0 = 1$ μPa. SPL$_{pk}$ is also expressed in dB re 1 μPa.

Power spectral density level (PSD, symbol: $L_{p,\,f}$): The level of the mean-square sound pressure spectral density relative to 1 μPa^2/Hz. The term power spectral density is sloppy, but it is the most commonly used term in the literature, and it rolls off the tongue better than mean-square sound pressure spectral density, even though the latter represents exactly how it is computed.

Sound exposure level (SEL, symbol: $L_{E,\,p}$): The level of sound exposure: $L_{E,p} = 10 \log_{10}\left(\frac{E_p}{E_0}\right)$ relative to a reference value $E_0 = 1$ μPa^{2}s. Sound exposure E_p is computed as the integral of the sound pressure p over time t: $E_p = \int p^2(t)dt$.

Particle velocity: The speed at and the direction in which the water particles move during their oscillations as a sound wave passes through the water.

Pulse duration: Defined as the percentage energy signal duration; that is, the duration of time during which a certain percentage of sound exposure is delivered. For example, T90% is the time between the 5% and the 95% points on the cumulative energy curve (ISO 18405; International Organization for Standardization 2017b).

Sonar: Method or equipment relying on sound to study objects under water. Abbreviated from SOund Navigation And Ranging. Sonar can be either passive (listening to the object's sounds) or active (impinging on the object with sound generated by the active sonar equipment; see ANSI/ASA S1.1; American National Standards Institute 2013).

Far-field: The far-field of a sound source is the region that is far enough so that the particle velocity and pressure are effectively in phase.

Near-field: The near-field is the region closer to the source where particle velocity and pressure become out of phase, either because sound from different parts of the source arrives at different times (this is the case of an extended source) or because the curvature of the spherical wavefront from the source is too great to be ignored (this is the case of a source small enough to be considered a point source). See Erbe et al. (2022a).

Lloyd's mirror: For a sound source below the water surface, the sound reflects off the surface, where it experiences a 180° phase shift, and then the surface-reflected rays overlap with the direct rays creating a pattern of constructive and destructive interference near the source, the Lloyd's mirror pattern. At the receiver, the surface-reflected ray appears to have originated at an image source in air that is 180° out of phase with the underwater source. For receivers near the surface at some range, the direct ray and the surface-reflected ray become similarly long and cancel out. Marine mammals near the surface thus might not hear an approaching ship. See also ANSI/ASA S12.64 (American National Standards Institute 2009).

Monopole: A monopole source is a point source. If it is placed in a homogeneous medium and far away from any boundary, then sound radiates equally in all directions.

Dipole: A sound source consisting of a pair of monopoles radiating 180° out of phase. A sound source near the sea surface (e.g., an outboard motor) radiates like a dipole, where one the monopoles is the propeller and the other is the image in air.

Source level (SL, symbol: L_S): The source level is a far-field quantity. It is a feature of the sound source and as such ideally independent of the environment in which the source was recorded. In praxis, the sound source is recorded in the far-field and a sound propagation model applied (e.g., a parabolic equation) to compute the sound level at a nominal distance of 1 m from a hypothetical point source (i.e., radiating as a monopole). The sound propagation model should account for (or correct for) the hydroacoustic and geoacoustic parameters of the water and seafloor, respectively. It should also take the experiment geometry into account (source depth, hydrophone depth and range). The resulting monopole source level may then be used to model the sound field of the same source in a different environment. The SL may be expressed as the level of a field quantity (root-mean-square sound pressure) in dB re 1 µPa m or as the level of a power quantity (mean-square sound pressure) in dB re 1 µPa^{2}m^2 (International Organization for Standardization 2017b). It was previously common in the literature to find the notation dB re 1 µPa @ 1 m. The numerical value in all three notations should be the same. For transient sources (e.g., airguns, pile driving), it is more common to report the sound exposure SL in dB re 1 µPa^{2}s m^2 (formerly, dB re 1 µPa^{2}s @ 1 m).

- **Radiated noise level** (RNL, symbol: L_{RN}): This is also a far-field quantity. It is determined by recording the SPL of a sound source in the far-field and then adding a spherical spreading correction ($20\log_{10}r$) to yield a level at a nominal distance of 1 m. The environment is not properly accounted for, and so, this quantity is called surface-affected source level in ANSI/ASA S12.64. It may also be affected by the seafloor, yet the surface will have a stronger effect in most settings where the source is near the surface (e.g., ships, marine seismic surveys). The RNL is also expressed in dB re 1 µPa m.
- **Received level** (RL): Can be either SPL or SEL.
- **Propagation loss** (PL, symbol: N_{PL}): Difference between the SL in a specified direction and the RL at a specified location.
- **SOFAR channel**: Also called the deep sound channel, corresponds to a layer of water in the ocean, at which the speed of sound has a minimum. It acts as a waveguide. Sound may travel over very long ranges through the channel by refraction, and so does not interact with the sea surface or seafloor, where great losses would otherwise occur.

References

Acevedo A, Smultea MA (1995) First records of humpback whales including calves at Golfo Dulce and Isla del Coco, Costa Rica, suggesting geographical overlap of Northern and Southern hemisphere populations. Mar Mamm Sci 11(4):554–560. https://doi.org/10.1111/j.1748-7692.1995.tb00677.x

Ainslie MA, Laws RM, Sertlek HÖ (2019) International airgun modeling workshop: validation of source signature and sound propagation models—Dublin (Ireland), July 16, 2016—problem description. IEEE J Ocean Eng 44(3):565–574. https://doi.org/10.1109/JOE.2019.2916956

Allen K, Peterson M, Sharrard G, Wright D, Todd S (2012) Radiated noise from commercial ships in the Gulf of Maine: implications for whale/vessel collisions. J Acoust Soc Am 132 (3):EL229–EL235. https://doi.org/10.1121/1.4739251

Amaral JL, Miller JH, Potty GR, Vigness-Raposa KJ, Frankel AS, Lin Y-T, Newhall AE, Wilkes DR, Gavrilov AN (2020) Characterization of impact pile driving signals during installation of offshore wind turbine foundations. J Acoust Soc Am 147(4):2323–2333. https://doi.org/10.1121/10.0001035

American National Standards Institute (2009) Quantities and procedures for description and measurement of underwater sound from ships—Part 1: general requirements (ANSI/ASA S12.64-2009/Part 1). Acoustical Society of America, Melville, New York, USA

American National Standards Institute (2013) Acoustical Terminology (ANSI/ASA S1.1-2013). Acoustical Society of America, Melville

Anderson PK, Barclay RMR (1995) Acoustic signals of solitary dugongs: physical characteristics and behavioural correlates. J Mammal 76(4):1226–1237. https://doi.org/10.2307/1382616

Andrew R, Bruce MH, James AM (2002) Ocean ambient sound: comparing the 1960s with the 1990s for a receiver off the California coast. Acoust Res Lett Online 3(2):65–70. https://doi.org/10.1121/1.1461915

Andrew RK, Howe BM, Mercer JA (2011) Long-time trends in ship traffic noise for four sites off the North American West Coast. J Acoust Soc Am 129(2):642–651. https://doi.org/10.1121/1.3518770

Ansmann IC, Goold JC, Evans PGH, Simmonds M, Keith SG (2007) Variation in the whistle characteristics of short-beaked common dolphins, *Delphinus delphis*, at two locations around the British Isles. J Mar Biolog

Assoc 87(1):19–26. https://doi.org/10.1017/S0025315407054963

Antunes R, Kvadsheim PH, Lam FPA, Tyack PL, Thomas L, Wensveen PJ, Miller PJO (2014) High thresholds for avoidance of sonar by free-ranging long-finned pilot whales (*Globicephala melas*). Mar Pollut Bull 83(1):165–180. https://doi.org/10.1016/j.marpolbul.2014.03.056

Arveson PT, Vendittis DJ (2000) Radiated noise characteristics of a modern cargo ship. J Acoust Soc Am 107(1):118–129. https://doi.org/10.1121/1.428344

Au WWL, Banks K (1998) The acoustics of the snapping shrimp *Synalpheus parneomeris* in Kaneohe Bay. J Acoust Soc Am 103(1):41–47. https://doi.org/10.1121/1.423234

Au WWL, Green M (2000) Acoustic interaction of humpback whales and whale-watching boats. Mar Env Res 49(5):469–481. https://doi.org/10.1016/s0141-1136(99)00086-0

Au WL, Mobley J, Burgess WC, Lammers MO, Nachtigall PE (2000) Seasonal and diurnal trends of chorusing humpback whales wintering in waters off Western Maui. Mar Mamm Sci 16(3):530–544. https://doi.org/10.1111/j.1748-7692.2000.tb00949.x

Aulanier F, Simard Y, Roy N, Gervaise C, Bandet M (2017) Effects of shipping on marine acoustic habitats in Canadian Arctic estimated via probabilistic modeling and mapping. Mar Pollut Bull 125(1–2):115–131. https://doi.org/10.1016/j.marpolbul.2017.08.002

Austin ME, Hannay DE, Bröker KC (2018) Acoustic characterization of exploration drilling in the Chukchi and Beaufort seas. J Acoust Soc Am 144(1):115–123. https://doi.org/10.1121/1.5044417

Bahtiarian M, Fischer R (2006) Underwater radiated noise of the NOAA ship Oscar Dyson. Noise Contr Eng J 54(4):224–235. https://doi.org/10.3397/1.2219891

Bailey H, Senior B, Simmons D, Rusin J, Picken G, Thompson PM (2010) Assessing underwater noise levels during pile-driving at an offshore windfarm and its potential effects on marine mammals. Mar Pollut Bull 60(6):888–897. https://doi.org/10.1016/j.marpolbul.2010.01.003

Barclay D, Buckingham M (2013a) Depth dependence of wind-driven, broadband ambient noise in the Philippine Sea. J Acoust Soc Am 133(1):62–71. https://doi.org/10.1121/1.4768885

Barclay DR, Buckingham MJ (2013b) The depth-dependence of rain noise in the Philippine Sea. J Acoust Soc Am 133(5):2567–2585. https://doi.org/10.1121/1.4799341

Barlow J, Swartz SL, Eagle TC, Wade PR (1995) U.-S. marine mammal stock assessments: guidelines for preparation, background, and a summary of the 1995 assessments. U.S. Dep. Commer., NOAA Tech. Memo. NMFS-OPR-6, 73 pp. https://www.arlis.org/docs/vol1/D/33420422.pdf

Baumann-Pickering S, Wiggins SM, Hildebrand JA, Roch MA, Schnitzler HU (2010) Discriminating features of echolocation clicks of melon-headed whales (*Peponocephala electra*), bottlenose dolphins (*Tursiops truncatus*), and Gray's spinner dolphins (*Stenella longirostris longirostris*). J Acoust Soc Am 128(4):2212–2224. https://doi.org/10.1121/1.3479549

Baumgartner MF, Van Parijs SM, Wenzel FW, Tremblay CJ, Esch HC, Warde AM (2008) Low frequency vocalizations attributed to sei whales (*Balaenoptera borealis*). J Acoust Soc Am 124(2):1339–1349. https://doi.org/10.1121/1.2945155

Bellmann MA (2014) Overview of existing noise mitigation systems for reducing pile-driving noise. In: Internoise 2014, Melbourne, Australia, 16–19 November 2014

Bellmann M, Gündert S, Remmers P (2014) Offshore Messkampagne 1 (OMK 1) für das Projekt BORA im Windpark BARD Offshore 1, Oldenburg. Institut für technische und angewandte Physik (itap) GmbH

Bellmann M, Gündert S, Remmers P (2015) Offshore Messkampagne 2 (OMK 2) für das Projekt BORA im Offshore-Windpark Global Tech I, Oldenburg, Germany. Institut für technische und angewandte Physik (itap) GmbH

Betke K (2008) Measurement of wind turbine construction noise at Horns Rev II (Report for BioConsultSH), Oldenburg. Institut für technische und angewandte Physik GmbH (itap)

Betke K, Matuschek R (2011) Messungen von Unterwasserschall beim Bau der Windenergieanlagen im Offshore-Testfeld "alpha ventus" (Report for Stiftung Offshore-Windenergie). Oldenburg, Germany: Institut für technische und angewandte Physik GmbH (itap)

Betke K, Schultz-von Glahn M (2004) Underwater noise measurements on an offshore wind turbine--Utgrunden I. Oldenburg: Institut für technische und Angewandte Physik GmbH (ITAP)

Betke K, Schultz-von Glahn M, Matuschek R (2004) Underwater noise emissions from offshore wind turbines. In: CFA/DAGA, Strasbourg, France

Bjørnø L (2017) Chapter 10 – sonar systems. In: Neighbors TH, Bradley D (eds) Applied Underwater Acoustics. Elsevier, Amsterdam, pp 587–742. https://doi.org/10.1016/B978-0-12-811240-3.00010-2

Blackwell SB (2005) Underwater measurements of pile driving sounds during the Port MacKenzie Dock Modifications, 13–16 August 2004 (Report from Greeneridge Sciences, Inc., Goleta, CA, and LGL Alaska Research Associates, Inc., Anchorage, AK, in association with HDR Alaska, Inc., Anchorage, AK, for Knik Arm Bridge and Toll Authority, Anchorage, AK, Department of Transportation and Public Facilities, Anchorage, AK, and Federal Highway Administration, Juneau, AK)

Blackwell SB, Greene CR, Richardson WJ (2004a) Drilling and operational sounds from an oil production Island in the ice-covered Beaufort Sea. J Acoust Soc

Am 116(5):3199–3211. https://doi.org/10.1121/1.1806147

Blackwell SB, Lawson JW, Williams MT (2004b) Tolerance by ringed seals (*Phoca hispida*) to impact pipe-driving and construction sounds at an oil production Island. J Acoust Soc Am 115(5):2346–2357. https://doi.org/10.1121/1.1701899

Blackwell SB, Greene CR (2006) Sounds from an oil production Island in the Beaufort Sea in summer: characteristics and contribution of vessels. J Acoust Soc Am 119(1):182–196. https://doi.org/10.1121/1.2140907

Blaylock RA, Hain LJ, Hansen JW, Palka DL, Waring GT (1995) U.S. Atlantic and Gulf of Mexico marine mammal stock assessments:211

Blondel P (2009) The handbook of sidescan sonar. Springer, Berlin https://doi.org/10.1007/978-3-540-49886-5

Bordino P, Kraus S, Albareda D, Fazio A, Palmerio A, Mendez M, Botta S (2002) Reducing incidental mortality of Franciscana dolphin *Pontoporia blainvillei* with acoustic warning devices attached to fishing nets. Mar Mamm Sci 18(4):833–842. https://doi.org/10.1111/j.1748-7692.2002.tb01076.x

Branstetter BK, Moore PW, Finneran JJ, Tormey MN, Aihara H (2012) Directional properties of bottlenose dolphin (*Tursiops truncatus*) clicks, burst-pulse, and whistle sounds. J Acoust Soc Am 131(2):1613–1621. https://doi.org/10.1121/1.3676694

Breeding JE, Pflug LA, Bradley M, Walrod MH, McBride W (1996) Research Ambient Noise Directionality (RANDI) 3.1 physics description (Report NRL/FR/7176–95-9628). Naval Research Laboratory, Stennis Space Center

Brooker A, Humphrey V (2016) Measurement of radiated underwater noise from a small research vessel in shallow water. Ocean Eng 120:182–189. https://doi.org/10.1016/j.oceaneng.2015.09.048

Broudic M, Berggren P, Laing S, Blake L, Pace F, Neves S, Voellmy I, Dobbins P, Radford A, Simpson S, Robinson S, Lepper PA, Bruintjes R (2014) Underwater noise emission from the NOAH's drilling operation at the narec site, Blyth, UK. In: 2nd international conference on Environmental Interations of Marine Renewable Energy Technologies (EIMR2014), Isle of Lewis. Scotland, 28 April–2 May 2014

Buck BM, Greene CR (1980) A two-hydrophone method of eliminating the effects of nonacoustic noise interference in measurements of infrasonic ambient noise levels. J Acoust Soc Am 68(5):1306–1308. https://doi.org/10.1121/1.385097

Caldwell J, Dragoset W (2000) A brief overview of seismic air-gun arrays. Lead Edge 19(8):898–902. https://doi.org/10.1190/1.1438744

Caltrans (2001) San Francisco–Oakland Bay Bridge East Span Seismic Safety Project: Pile Installation Demonstration Project. Marine Mammal Impact Assessment (PIDP EA 012081, PIDP 04-ALA-80-0.0/0.5, Caltrans Contract 04A0148, Task Order 205.10.90)

Carey WM, Evans RB, Davis JA, Botseas G (1990) Deep-ocean vertical noise directionality. IEEE J Ocean Eng 15(4):324–334. https://doi.org/10.1109/48.103528

Cato DH (1978) Marine biological choruses observed in tropical waters near Australia. J Acoust Soc Am 64(3):736–743. https://doi.org/10.1121/1.382038

Cato DH (1980) Some unusual sounds of apparent biological origin responsible for sustained background noise in the Timor Sea. J Acoust Soc Am 68(4):1056–1060. https://doi.org/10.1121/1.384989

Cato DH (2008) Ocean ambient noise: its measurement and its significance to marine animals. In: Proceedings of the Institute of Acoustics – underwater noise measurement, impact and mitigation, Southampton, UK, 14–15 October 2008. vol 5. Institute of Acoustics, pp 1–9

Cato DH, Tavener S (1997) Ambient sea noise dependence on local, regional and geostrophic wind speeds: implications for forecasting noise. Appl Acoust 51(3):317–338. https://doi.org/10.1016/S0003-682X(97)00001-7

Ceciarini I, Franchi E, Capanni F, Consales G, Minoia L, Ancora S, D'Agostino A, Lucchetti A, Li Veli D, Marsili L (2023) Assessment of interactive acoustic deterrent devices set on trammel nets to reduce dolphin–fishery interactions in the Northern Tyrrhenian Sea. Sci Rep 13(1):20680. https://doi.org/10.1038/s41598-023-46836-z

Chapman NR (1985) Measurement of the waveform parameters of shallow explosive charges. J Acoust Soc Am 78(2):672–681. https://doi.org/10.1121/1.392436

Chapman NR, Price A (2011) Low frequency deep ocean ambient noise trend in the Northeast Pacific Ocean. J Acoust Soc Am 129 (5):EL161–EL165:EL161. https://doi.org/10.1121/1.3567084

Chion C, Lagrois D, Dupras J (2019) A meta-analysis to understand the variability in reported source levels of noise radiated by ships from opportunistic studies. Front Mar Sci 6:714. https://doi.org/10.3389/fmars.2019.00714

Clark CA (2007) Vertical directionality of midfrequency surface noise in downward-refracting environments. IEEE J Ocean Eng 32(3):609–619. https://doi.org/10.1109/JOE.2007.903450

Clarke PA (2009) Mitigation modelling of the Leeuwin Class hydrographic sonars in Shoalwater Bay. Technical report DSTO-TR-2121. Defence Science and Technology Organisation, Edinburgh

Clarke D, Dickerson C, Reine K (2003) Characterization of underwater sounds produced by dredges. In: Dredging'02: key technologies for global prosperity. American Society of Civil Engineers, Orlando

Cleator HJ, Stirling I, Smith TG (1989) Underwater vocalizations of the bearded seal (*Erignathus barbatus*). Can J Zool 67(8):1900–1910. https://doi.org/10.1139/z89-272

Cope S, Hines E, Bland R, Davis JD, Tougher B, Zetterlind V (2020) Application of a new shore-based vessel traffic monitoring system within San Francisco Bay. Front Mar Sci 7:86. https://doi.org/10.3389/fmars.2020.00086

Cope S, Hines E, Bland R, Davis JD, Tougher B, Zetterlind V (2021) Multi-sensor integration for an assessment of underwater radiated noise from common vessels in San Francisco Bay. J Acoust Soc Am 149(4):2451–2464. https://doi.org/10.1121/10.0003963

Corkeron PJ, Van Parijs SM (2001) Vocalizations of eastern Australian Risso's dolphins, *Grampus griseus*. Can J Zool 79(1):160–164. https://doi.org/10.1139/z00-180

Coste E, Gerez D, Groenaas H, Hopperstad J-F, Larsen OP, Laws R, Norton J, Padula M, Wolfstirn M (2014) Attenuated high-frequency emission from a new design of air-gun. Paper presented at the 2014 SEG annual meeting, Denver, Colorado, USA, October 2014. https://doi.org/10.1190/segam2014-0445.1

Cox TM, Ragen TJ, Read AJ, Vos E, Baird RW, Balcomb K, Barlow J, Caldwell J, Cranford T, Crum L, Amico AD, Spain GD, Fernández A, Finneran J, Gentry R, Gerth W, Gulland F, Hidebrand J, Houser D, Hullar T, Jepson PD, Ketten D, MacLeod CD, Miller P, Moore S, Mountain DC, Palka D, Ponganis P, Rommel S, Rowles T, Taylor B, Tyack P, Wartzok D, Gisiner R, Mead J, Benner L (2006) Understanding the impacts of anthropogenic sound on beaked whales. J Cetacean Res Manag 7(3):177–187. https://doi.org/10.47536/jcrm.v7i3.729

Crocker SE, Fratantonio FD (2016) Characteristics of sounds emitted during high-resolution marine geophysical surveys (Report number TR 12,203). Naval Undersea Warfare Center Division, Newport

Crocker SE, Fratantonio FD, Hart PE, Foster DS, O'Brien TF, Labak S (2019) Measurement of sounds emitted by certain high-resolution geophysical survey systems. IEEE J Ocean Eng 44(3):796–813. https://doi.org/10.1109/JOE.2018.2829958

Croll DA, Clark CW, Acevedo A, Tershy B, Flores S, Gedamke J, Urban J (2002) Only male fin whales sing loud songs. Nature 417 (20 June 2002):809. https://doi.org/10.1038/417809a

Crum LA, Pumphrey HC, Roy RA, Prosperetti A (1999) The underwater sounds produced by impacting snowflakes. J Acoust Soc Am 106(4):1765–1770. https://doi.org/10.1121/1.427925

Cummings WC, Thompson PO (1971) Gray whales, *Eschrichtius robustus*, avoid the underwater sounds of killer whales, *Orcinus orca*. Fish Bull 69(3):525–530. https://doi.org/10.1121/1.407772

D'Spain GL, D'Amico A, Fromm DM (2006) Properties of the underwater sound fields during some well documented beaked whale mass stranding events. J Cetacean Res Manag 7(3):223–238. https://doi.org/10.47536/jcrm.v7i3.733

Dahl P, Reinhall P, Farrell D (2012) Transmission loss and range, depth scales associated with impact pile driving. In: 11th European Conference on Underwater Acoustics (ECUA 2012), pp 1860–1867

Dahl PH, Dall'Osto DR, Farrell DM (2015) The underwater sound field from vibratory pile driving. J Acoust Soc Am 137(6):3544–3554. https://doi.org/10.1121/1.4921288

Dähne M, Tougaard J, Carstensen J, Rose A, Nabe-Nielsen J (2017) Bubble curtains attenuate noise from offshore wind farm construction and reduce temporary habitat loss for harbour porpoises. Mar Ecol Prog Ser 580:221–237. https://doi.org/10.3354/meps12257

Dall'Osto DR, Dahl PH, Chapman NR (2023) The sound from underwater explosions. Acoustics Today 19(1):12–19. https://doi.org/10.1121/AT.2023.19.1.12

Dawson S, Northridge SP, Waples D, Read A (2013) To ping or not to ping: the use of active acoustic devices in mitigating interactions between small cetaceans and gillnet fisheries. Endang Species Res 19:201–221. https://doi.org/10.3354/esr00464

de Jong CAF, Ainslie MA (2008) Underwater radiated noise due to the piling for the Q7 Offshore Wind Park. In: Acoustics 08, Paris, France, 29.6.-4.7.2008, pp 117–122

de Jong C, Ainslie M, Dreschler J, Jansen E, Heemskerk E, Groen W (2010) Underwater noise of trailing suction hopper dredgers at Maasvlakte 2: analysis of source levels and background noise. TNO, The Hague

de Robertis A, Wilson CD, Furnish SR, Dahl PH (2013) Underwater radiated noise measurements of a noise-reduced fisheries research vessel. ICES J Mar Sci 70(2):480–484. https://doi.org/10.1121/1.3660550

de Sousa Lima RS, Paglia AP, Fonseca GAB (1999) Vocal discrimination of two species of manatees (*Trichechus inunguis*) and (*T. manatus manatus*) in Brazil. J Acoust Soc Am 106 (4):2164–2164:2164. https://doi.org/10.1121/1.427204

Deane GB (1999) Acoustic hot-spots and breaking wave noise in the surf zone. J Acoust Soc Am 105(6):3151–3167. https://doi.org/10.1121/1.424646

Deane GB (2000) Long time-base observations of surf noise. J Acoust Soc Am 107(2):758–770. https://doi.org/10.1121/1.428259

Dekeling R, Tasker M, van der Graaf S, Ainslie M, Andersson M, André M, Borsani JF, Brensing K, Castellote M, Cronin D, Dalen J, Folgegot T, Leaper R, Pajala J, Redman P, Robinson S, Sigray P, Sutton G, Thomsen F, Werner S, Wittekind DK, Young J (2014) Monitoring guidance for underwater noise in European Seas, Part II: monitoring guidance specifications; report EUR 26557 EN (RC scientific and policy report EUR 26557 EN). Publications Office of the European Union, Luxembourg

Demer D, Andersen L, Bassett C, Berger L, Chu D, Condiotty J, Cutter G (2016) USA–Norway EK80 workshop report: evaluation of a wideband echosounder for fisheries and marine ecosystem science. ICES Cooperative Research Report 336:69

Deng Q, Jiang W, Zhang W (2016) Theoretical investigation of the effects of the cushion on reducing underwater noise from offshore pile driving. J Acoust Soc Am 140(4):2780–2793. https://doi.org/10.1121/1.4963901

Depew DC, Stevens AW, Smith REH, Hecky RE (2009) Detection and characterization of benthic filamentous algal stands (*Cladophora* sp.) on rocky substrata using a high-frequency echosounder. Limnol Oceanogr 7 (10):693–705:693. https://doi.org/10.4319/lom.2009.7.693

Dickerson C, Reine KJ, Clarke DG (2001) Characterization of underwater sounds produced by bucket dredging operations (DOER Technical Notes Collection ERDC TN-DOER-E14). U.S. Army Engineer Research and Development Center, Vicksburg

Ding L, Farmer DM (1994) On the dipole acoustic source level of breaking waves. J Acoust Soc Am 96(5):3036–3044. https://doi.org/10.1121/1.411375

Doksaeter L, Godo OR, Handegard NO (2009) Behavioural responses of herring (*Clupea harengus*) to 1-2 and 6-7 kHz sonar signals and killer whale feedings sounds. J Acoust Soc Am 125(1):554–564. https://doi.org/10.1121/1.3021301

Doksæter L, Handegard N, Godø O, Kvadsheim P, Nordlund N (2012) Behaviour of captive herring exposed to naval sonar transmissions (1.0–1.6 kHz) throughout a yearly cycle. J Acoust Soc Am 131(2):1632–1642. https://doi.org/10.1121/1.3675944

Dolman SJ, Breen CN, Brakes P, Butterworth A, Allen SJ (2022) The individual welfare concerns for small cetaceans from two bycatch mitigation techniques. Mar Policy 143:105126. https://doi.org/10.1016/j.marpol.2022.105126

Dragoset B (2000) Introduction to air guns and air-gun arrays. Lead Edge 19(8):892–897. https://doi.org/10.1190/1.1438741

Duchesne MJ, Bellefleur G, Galbraith M, Kolesar R, Kuzmiski R (2007) Strategies for waveform processing in sparker data. Mar Geophys Res 28(2):153–164. https://doi.org/10.1007/s11001-007-9023-8

Duncan AJ, Gavrilov AN (2019) The CMST airgun array model—a simple approach to modeling the underwater sound output from seismic airgun arrays. IEEE J Ocean Eng 44(3):589–597. https://doi.org/10.1109/JOE.2019.2899134

Duncan AJ, McCauley RD, Parnum I, Salgado Kent C (2010) Measurement and modelling of underwater noise from pile driving. In: 20th International Congress on Acoustics (ICA 2010), Sydney, NSW, 23–27 August 2010

Duncan AJ, Gavrilov AN, McCauley RD, Parnum IM, Collis JM (2013) Characteristics of sound propagation in shallow water over an elastic seabed with a thin cap-rock layer. J Acoust Soc Am 134:207–215. https://doi.org/10.1121/1.4809723

Duncan AJ, Weilgart LS, Leaper R, Jasny M, Livermore S (2017) A modelling comparison between received sound levels produced by a marine Vibroseis array and those from an airgun array for some typical seismic survey scenarios. Mar Pollut Bull 119(1):277–288. https://doi.org/10.1016/j.marpolbul.2017.04.001

Dunlop RA, Noad MJ, Cato DH, Stokes D (2007) The social vocalization repertoire of east Australian migrating humpback whales (*Megaptera novaeangliae*). J Acoust Soc Am 122(5):2893–2905. https://doi.org/10.1121/1.2783115

Dziak RP, Fox CG (2002) Evidence of harmonic tremor from a submarine volcano detected across the Pacific Ocean basin. J Geophys Res 107 (B5, 2085):1–11. https://doi.org/10.1029/2001JB000177

Dziak RP, Bohnenstiehl DR, Matsumoto H, Fox CG, Smith DK, Tolstoy M, Lau T-K, Haxel JH, Fowler MJ (2004) P- and T-wave detection thresholds, Pn velocity estimate, and detection of lower mantle and core P-waves on ocean sound-channel hydrophones at the Mid-Atlantic Ridge. Bull Seismol Soc Am 94(2):665–677. https://doi.org/10.1785/0120030156

Dziak RP, Haxel JH, Matsumoto H, Lau T-K, Heimlich S, Nieukirk S, Mellinger DK, Osse J, Meinig C, Delich N, Stalin S (2017) Ambient sound at challenger deep, Mariana Trench. Oceanography 30(2):186–197. https://doi.org/10.5670/oceanog.2017.240

Elmer K (2007) Schallemissionen beim Rammen von Offshore-Fundamenten. In: 2nd scientific conference on the use of offshore wind energy. Federal Environment Ministry, Berlin, pp 20–21

Elmer K-H, Savery J (2014) New hydro sound dampers to reduce piling underwater noise. In: Internoise 2014, Melbourne, Australia, 16–19 November 2014, vol 2. Institute of Noise Control Engineering, pp 5551–5560

Erbe C (2002) Underwater noise of whale-watching boats and its effects on killer whales (*Orcinus orca*). Mar Mamm Sci 18(2):394–418. https://doi.org/10.1111/j.1748-7692.2002.tb01045.x

Erbe C (2009) Underwater noise from pile driving in Moreton Bay. Qld Acoust Aust 37(3):87–92

Erbe C (2013) Underwater noise of small personal watercraft (jet skis). J Acoust Soc Am 133 (4):EL326–EL330. https://doi.org/10.1121/1.4795220

Erbe C, Farmer DM (2000) Zones of impact around icebreakers affecting beluga whales in the Beaufort Sea. J Acoust Soc Am 108(3):1332–1340. https://doi.org/10.1121/1.1288938

Erbe C, King AR (2009) Modelling cumulative sound exposure around marine seismic surveys. J Acoust Soc Am 125(4):2443–2451. https://doi.org/10.1121/1.3089588

Erbe C, McPherson C (2012) Acoustic characterisation of bycatch mitigation pingers on Queensland Shark Control nets. Endang Species Res 19(2):109–121. https://doi.org/10.3354/esr00467

Erbe C, McPherson C (2017) Underwater noise from geotechnical drilling and standard penetration testing. J Acoust Soc Am 142 (3):EL281–EL285. https://doi.org/10.1121/1.5003328

Erbe C, MacGillivray AO, Williams R (2012) Mapping cumulative noise from shipping to inform marine spatial planning. J Acoust Soc Am 132 (5): EL:423–428. https://doi.org/10.1121/1.4758779

Erbe C, McCauley RD, McPherson C, Gavrilov A (2013) Underwater noise from offshore oil production vessels. J Acoust Soc Am 133 (6):EL465–EL470. https://doi.org/10.1121/1.4802183

Erbe C, Verma A, McCauley R, Gavrilov A, Parnum I (2015) The marine soundscape of the Perth canyon. Prog Oceanogr 137:38–51. https://doi.org/10.1016/j.pocean.2015.05.015

Erbe C, Liong S, Koessler MW, Duncan AJ, Gourlay T (2016a) Underwater sound of rigid-hulled inflatable boats. J Acoust Soc Am 139 (6):EL223–EL227. https://doi.org/10.1121/1.4954411

Erbe C, McCauley R, Gavrilov A, Madhusudhana S, Verma A (2016b) The underwater soundscape around Australia. Proceedings of Acoustics 2016, 9–11 November 2016, Brisbane, Australia

Erbe C, Parsons M, Duncan AJ, Allen K (2016c) Underwater acoustic signatures of recreational swimmers, divers, surfers and kayakers. Acoust Aust 44(2): 333–341. https://doi.org/10.1007/s40857-016-0062-7

Erbe C, Parsons M, Duncan AJ, Lucke K, Gavrilov A, Allen K (2017) Underwater particle motion (acceleration, velocity and displacement) from recreational swimmers, divers, surfers and kayakers. Acoust Aust 45:293–299. https://doi.org/10.1007/s40857-017-0107-6

Erbe C, Schoeman RP, Peel D, Smith JN (2021) It often howls more than it chugs: wind versus ship noise under water in Australia's maritime regions. J Mar Sci Eng 9(5):472. https://doi.org/10.3390/jmse9050472

Erbe C, Duncan A, Hawkins L, Terhune JM, Thomas JA (2022a) Introduction to acoustic terminology and signal processing. In: Erbe C, Thomas JA (eds) Exploring animal behavior through sound: volume 1: methods. Springer, Cham, pp 111–152. https://doi.org/10.1007/978-3-030-97540-1_4

Erbe C, Duncan A, Vigness-Raposa KJ (2022b) Introduction to sound propagation under water. In: Erbe C, Thomas JA (eds) Exploring animal behavior through sound: volume 1: methods. Springer, Cham, pp 185–216. https://doi.org/10.1007/978-3-030-97540-1_6

Evans DL, England GR (2001) Joint interim report of The Bahamas marine mammal stranding event of 14–16 March 2000. National Oceanic and Atmospheric Administration, Washington, DC, p 61

Evans WE, Herald ES (1970) Underwater calls of a captive Amazon manatee, *Trichechus inunguis*. J Mammal 51(4):820–823. https://doi.org/10.2307/1378319

Field LH, Evans A, Macmillan DL (1987) Sound production and stridulatory structures in hermit crabs of the genus *Trizopagurus*. J Mar Biolog Assoc 67(1): 89–110. https://doi.org/10.1017/S0025315400026382

Findlay CR, Ripple HD, Coomber F, Froud K, Harries O, van Geel NCF, Calderan SV, Benjamins S, Risch D, Wilson B (2018) Mapping widespread and increasing underwater noise pollution from acoustic deterrent devices. Mar Pollut Bull 135:1042–1050. https://doi.org/10.1016/j.marpolbul.2018.08.042

Fischer RW, Brown NA (2005) Factors affecting the underwater noise of commercial vessels operating in environmentally sensitive areas. In: IEEE Oceans conference, 2005. IEEE, pp 1982–1988

Fish MP, Mowbray WH (1970) Sounds of Western North Atlantic fishes: a reference file of biological underwater sounds. The Johns Hopkins University Press, Baltimore

Fish JF, Vania JS (1971) Killer whale, *Orcinus orca*, sounds repel white whales, *Delphinapterus leucas*. Fish Bull 69(3):531–535

Fox CG, Matsumoto H, Lau TKA (2001) Monitoring Pacific Ocean seismicity from an autonomous hydrophone array. J Geophys Res 106(B3):4183–4206. https://doi.org/10.1029/2000JB900404

Frankel AS, Gabriele CM (2017) Predicting the acoustic exposure of humpback whales from cruise and tour vessel noise in Glacier Bay, Alaska, under different management strategies. Endang Species Res 34:397–415. https://doi.org/10.3354/esr00857

Fricke MB, Rolfes R (2015) Towards a complete physically based forecast model for underwater noise related to impact pile driving. J Acoust Soc Am 137(3): 1564–1575. https://doi.org/10.1121/1.4908241

Frisk G (2012) Noiseonomics: the relationship between ambient noise levels in the sea and global economic trends. Sci Rep 2:437. https://doi.org/10.1038/srep00437

Frouin-Mouy H, Tenorio-Hallé L, Thode A, Swartz S, Urbán J (2020) Using two drones to simultaneously monitor visual and acoustic behaviour of gray whales (*Eschrichtius robustus*) in Baja California, Mexico. J Exp Mar Biol Ecol 525:151321. https://doi.org/10.1016/j.jembe.2020.151321

Gales RS (1982) Effects of noise of offshore oil and gas operations on marine mammals – an introductory assessment (NTIS AD-A123699 + AD-A123700). U.S. Naval Ocean Systems Center, San Diego

Gannon DP (2008) Passive acoustic techniques in fisheries science: a review and prospectus. Trans Am Fish Soc 137(2):638–656. https://doi.org/10.1577/T04-142.1

Gassmann M, Wiggins SM, Hildebrand JA (2017) Deepwater measurements of container ship radiated noise signatures and directionality. J Acoust Soc Am 142(3): 1563–1574. https://doi.org/10.1121/1.5001063

Gavrilov A (2017) Propagation of underwater noise from an offshore seismic survey in Australia to Antarctica: measurements and modelling. In: Acoustics 2017 conference, Perth, WA, Australia

Gavrilov A (2021) Feasibility and accuracy of multi-component models of ocean ambient noise to predict noise spectra on a short-time scale and assess long-term trends in noise components. Report for Office of Naval Research Global (Report for Office of Naval Research Global). Perth, Western Australia. http://cmst.curtin.edu.au/wp-content/uploads/sites/4/2023/03/ONRG_Curtin_Progress_report_2_online_ submission.pdf. Centre for Marine Science and Technology, Curtin University

Gavrilov A, Li B (2007) Antarctica as one of the major sources of noise in the ocean. Paper presented at the Underwater Acoustic Measurements: Technologies & Results, 2nd International Conference and Exhibition, Heraklion, Crete, 25–29 June 2007

Gavrilov A, McCauley R, Zhang ZY (2019) Feasibility and accuracy of prediction of ocean noise on a short-time scale. In: Proceedings of the underwater acoustics conference and exhibition, Crete, Greece, 1–5 July, pp 375–382. https://www.uaconferences.org/docs/Past_Proceedings/UACE2019_Proceedings.pdf

Gearin PJ, Gosho ME, Cooke L, Delong R, Laake J, Greene D (1996) Acoustic alarm experiment in the 1995 northern Washington marine setnet fishery. National Marine Mammal Laboratory Report

Gerstein ER (2002) Manatees, bioacoustics and boats: hearing tests, environmental measurements and acoustic phenomena may together explain why boats and animals collide. Am Sci 90(2):154–163

Gervaise C, Simard Y, Roy N, Kinda B, Menard N (2012) Shipping noise in whale habitat: characteristics, sources, budget, and impact on belugas in Saguenay–St. Lawrence Marine Park hub. J Acoust Soc Am 132 (1):76–89:76. https://doi.org/10.1121/1.4728190

Goold JC, Fish PJ (1998) Broadband spectra of seismic survey air-gun emissions, with reference to dolphin auditory thresholds. J Acoust Soc Am 103(4): 2177–2184. https://doi.org/10.1121/1.421363

Götz T, Janik VM (2013) Acoustic deterrent devices to prevent pinniped depredation: efficiency, conservation concerns and possible solutions. Mar Ecol Prog Ser 492:285–302. https://doi.org/10.3354/meps10482

Gray LM, Greeley DS (1980) Source level model for propeller blade rate radiation for the world's merchant fleet. J Acoust Soc Am 67(2):516–522. https://doi.org/10.1121/1.383916

Greene CR (1986) Underwater sounds from the semi-submersible drill rig SEDCO 708 drilling in the Aleutian Islands (report for the American Petroleum Institute). Santa Barbara: Polar Research Laboratory, Inc.

Greene CR (1987) Characteristics of oil industry dredge and drilling sounds in the Beaufort Sea. J Acoust Soc Am 82(4):1315–1324. https://doi.org/10.1121/1.395265

Greene CR, Buck BM (1979) Influence of atmospheric pressure gradient on under-ice ambient noise. J Acoust Soc Am 66(S1):S25–S25. https://doi.org/10.1121/1.2017678

Greene CR, Richardson WJ (1988) Characteristics of marine seismic survey sounds in the Beaufort Sea. J Acoust Soc Am 83(6):2246–2254. https://doi.org/10.1121/1.396354

Greene CR, Blackwell SB, McLennan MW (2008) Sounds and vibrations in the frozen Beaufort Sea during gravel Island construction. J Acoust Soc Am 123(2):687–695. https://doi.org/10.1121/1.2821970

Greening MV, Zakarauskas P (1994a) Pressure ridging spectrum level and a proposed origin of the infrasonic peak in arctic ambient noise spectra. J Acoust Soc Am 95(2):791–797. https://doi.org/10.1121/1.408389

Greening MV, Zakarauskas P (1994b) Spatial and source level distributions of ice cracking in the Arctic Ocean. J Acoust Soc Am 95(2):783–790. https://doi.org/10.1121/1.404948

Gridley T, Nastasi A, Kriesell HJ, Elwen SH (2015) The acoustic repertoire of wild common bottlenose dolphins (*Tursiops truncatus*) in Walvis Bay, Namibia. Bioacoustics 24(2):153–174. https://doi.org/10.1080/09524622.2015.1014851

Guerra M, Thode A, Blackwell S, Macrander M (2011) Quantifying seismic survey reverberation off the Alaskan North Slope. J Acoust Soc Am 130(5): 3046–3058. https://doi.org/10.1121/1.3628326

Gündert S, Bellmann M, Remmers P (2015) Offshore Messkampagne 3 (OMK 3) für das Projekt BORA im Offshore-Windpark Borkum Riffgrund 01, Oldenburg. Institut für technische und angewandte Physik (itap) GmbH

Hall MV (2015) An analytical model for the underwater sound pressure waveforms radiated when an offshore pile is driven. J Acoust Soc Am 138(2):795–806. https://doi.org/10.1121/1.4927034

Halvorsen M, Zeddies D, Ellison W, Chicoine D, Popper A (2012) Effects of mid-frequency active sonar on hearing in fish. J Acoust Soc Am 131(1):599–607. https://doi.org/10.1121/1.3664082

Hammerstad E (2005) Sound levels from Kongsberg Multibeams (Kongsberg technical note). Kongsberg Technical Note, Norway

Hamson RM (1997) The modelling of ambient noise due to shipping and wind sources in complex environments. Appl Acoust 51(3):251–287. https://doi.org/10.1016/S0003-682X(97)00003-0

Hanggi EB, Schusterman RJ (1994) Underwater acoustic displays and individual variation in male harbour seals, *Phoca vitulina*. Anim Behav 48(6):1275–1283. https://doi.org/10.1006/anbe.1994.1363

Hannay D, MacGillivray A, Laurinolli M, Racca R (2004) Source level measurements from 2004 acoustics program (Report for Sakhalin Energy). JASCO Research Ltd, Victoria

Hanson JA, Bowman JR (2006) Methods for monitoring hydroacoustic events using direct and reflected T waves in the Indian Ocean. J Geophys Res 111: B02305. https://doi.org/10.1029/2004JB003609

Harris P, Sotirakopoulos K, Robinson S, Wang L, Livina V (2019) A statistical method for the evaluation of long term trends in underwater noise measurements. J Acoust Soc Am 145(1):228–242. https://doi.org/10.1121/1.5084040

Hastie G (2012) Tracking marine mammals around marine renewable energy devices using active sonar (SMRUL-DEC-2012-002.v2 SMRUL-DEC-2012-002.v2). SMRUL-DEC-2012-002.v2: report by SMRU consulting for UK Department of Energy and Climate Change (DECC), 99 p

Hastings M, Popper AN (2005) Effects of sound on fish (Report for California Department of Transportation). Jones & Stokes, Sacramento

Hawkins AD, Rasmussen KJ (1978) The calls of gadoid fish. J Mar Biolog Assoc 587(4):891–911

Hazelwood RA, Macey PC (2012) Seabed vibration induced by impacts – theory and practice. In: 11th European Conference on Underwater Acoustics, Edinburgh, 2–6 July 2012, pp 1563–1570

Henninger HP, Watson WH (2005) Mechanisms underlying the production of carapace vibrations and associated waterborne sounds in the American lobster, *Homarus americanus*. J Exp Biol 208(17):3421. https://doi.org/10.1242/jeb.01771

Hiley HM, Perry S, Hartley S, King SL (2017) What's occurring? Ultrasonic signature whistle use in Welsh bottlenose dolphins (*Tursiops truncatu*s). Bioacoust Intl J Anim Sound Record 26 (1):25–35. https://doi.org/10.1080/09524622.2016.1174885

Huang L-F, Xu X-M, Yang L-L, Huang S-Q, Zhang X-H, Zhou Y-L (2023) Underwater noise characteristics of offshore exploratory drilling and its impact on marine mammals. Front Mar Sci 10:1097701. https://doi.org/10.3389/fmars.2023.1097701

Illinworth and Rodkin Inc. (2007) Compendium of pile driving sound data (Report for the California Department of Transportation). Petaluma, CA

International Organization for Standardization (2014) Acoustics – soundscape – part 1: definition and conceptual framework (ISO 12913-1:2014(E)). ISO, Geneva, Switzerland

International Organization for Standardization (2016) Underwater acoustics—quantities and procedures for description and measurement of underwater sound from ships—part 1: requirements for precision measurements in deep water used for comparison purposes (ISO 17208-1). Geneva, Switzerland

International Organization for Standardization (2017a) Underwater acoustics—measurement of radiated underwater sound from percussive pile driving (ISO 18406). Geneva, Switzerland

International Organization for Standardization (2017b) Underwater acoustics—terminology (ISO 18405). Geneva, Switzerland

International Organization for Standardization (2019) Underwater acoustics—quantities and procedures for description and measurement of underwater sound from ships—Part 2: determination of source levels from deep water measurements (ISO 17208-2). Switzerland, Geneva

Isojunno S, Curé C, Kvadsheim PH, Lam F-PA, Tyack PL, Wensveen PJ, Miller PJOM (2016) Sperm whales reduce foraging effort during exposure to 1–2 kHz sonar and killer whale sounds. Ecol Appl 26(1): 77–93. https://doi.org/10.1890/15-0040

Jalkanen J-P, Johansson L, Andersson MH, Majamäki E, Sigray P (2022) Underwater noise emissions from ships during 2014–2020. Environ Pollut 311:119766. https://doi.org/10.1016/j.envpol.2022.119766

James V (2013) Marine renewable energy: a global review of the extent of marine renewable energy developments, the developing technologies and possible conservation implications for cetaceans. Whale and Dolphin Conservation, Wiltshire

Jefferson TA, Curry BE (1994) Review and evaluation of potential acoustic methods of reducing or eliminating marine mammal-fishery interactions

Jiang Z (2021) Installation of offshore wind turbines: a technical review. Renew Sust Energ Rev 139:110576. https://doi.org/10.1016/j.rser.2020.110576

Jiang P, Lin J, Sun J, Yi X, Shan Y (2020) Source spectrum model for merchant ship radiated noise in the Yellow Sea of China. Ocean Eng 216:107607. https://doi.org/10.1016/j.oceaneng.2020.107607

Johnson DT (1994) Understanding air-gun bubble behavior. Geophysics 59(11):1729–1734. https://doi.org/10.1190/1.1443559

Johnston DW (2002) The effect of acoustic harassment devices on harbour porpoises (*Phocoena phocoena*) in the Bay of Fundy. Canada Biol Conserv 108(1): 113–118. https://doi.org/10.1016/s0006-3207(02)00099-x

Johnston DW, Woodley TH (1998) A survey of acoustic harassment device (AHD) use in the Bay of Fundy, NB, Canada. Aquat Mamm 24:51–61

Joy R, Tollit D, Wood J, MacGillivray A, Li Z, Trounce K, Robinson O (2019) Potential benefits of vessel slowdowns on endangered southern resident killer whales. Front Mar Sci 6:344. https://doi.org/10.3389/fmars.2019.00344

Kastelein RA, van der Heul S, van der Veen J, Verboom WC, Jennings N, Haan D, Reijnders PJH (2007) Effects of acoustic alarms, designed to reduce small cetacean bycatch in gillnet fisheries, on the behaviour of North Sea fish species in a large tank. Mar Env Res 64:160–180. https://doi.org/10.1016/j.marenvres.2006.12.012

Kastelein RA, Verboom WC, Jennings N, de Haan D (2008) Behavioural avoidance threshold level of a harbour porpoise (*Phocoena phocoena*) for a continuous 50 kHz pure tone (L). J Acoust Soc Am 123(4): 1858–1861. https://doi.org/10.1121/1.2874557

Kenny AJ, Cato I, Desprez M, Fader G, Schüttenhelm RTE, Side J (2003) An overview of seabed-mapping technologies in the context of marine habitat classification. ICES J Mar Sci 60(2):411–418. https://doi.org/10.1016/S1054-3139(03)00006-7

Kerman BR (1984) Underwater sound generation by breaking wind waves. J Acoust Soc Am 75(1): 149–165. https://doi.org/10.1121/1.390409

Kinda GB, Simard Y, Gervaise C, Mars JI, Fortier L (2015) Arctic underwater noise transients from sea ice deformation: characteristics, annual time series, and forcing in Beaufort Sea. J Acoust Soc Am 138(4):2034. https://doi.org/10.1121/1.4929491

Kipple B (2002) Southeast Alaska cruise ship underwater acoustic noise (Technical report NSWCCD-71-TR-2002/574). Naval Surface Warfare Center, Bremerton/Washington, DC

Kipple B (2004) Coral Princess underwater acoustic levels (Technical report). Naval Surface Warfare Center, Bremerton/Washington, DC

Kipple B, Gabriele C (2003) Glacier Bay Watercraft Noise (Technical report NSWCCD-71-TR-2003/522). Naval Surface Warfare Center, Bremerton

Kipple B, Gabriele C (2004) Glacier Bay Watercraft Noise–Noise characterization for tour, charter, private, and government vessels (Technical Report NSWCCD-71-TR-2004/545). Naval Surface Warfare Center, Bremerton, pp 1–47

Knudsen VO, Alford RS, Emling JW (1948) Underwater ambient noise. J Mar Res 7(3):410–429

Koschinski S, Lüdemann K (2011) Stand der Entwicklung schallminimierender Maßnahmen beim Bau von Offshore-Windenergieanlagen. Bundesamt für Naturschutz, Nehmten, Germany

Koschinski S, Lüdemann K (2020) Noise mitigation for the construction of increasingly large offshore wind turbines. https://tethys.pnnl.gov/sites/default/files/publications/Koschinskietal2020.pdf: Report for Bundesamt für Naturschutz, BfN, Germany, 40 pp

Kraus S, Read A, Anderson E, Baldwin K, Solow A, Spradlin T, Williamson J (1995) A field test of the use of acoustic alarms to reduce incidental mortality of harbour porpoises and gill nets. Draft Final Report to NMFS:1–29

Kraus SD, Read AJ, Solow A, Baldwin K, Spradlin T, Anderson E, Williamson J (1997) Acoustic alarms reduce porpoise mortality. Nature 388(6642):525. https://doi.org/10.1038/41451

Krause BL (2008) Anatomy of the soundscape: evolving perspectives. J Audio Eng Soc 56(1/2):73–80

Krause BL (2012) The great animal orchestra: finding the origins of music in the world's wild places. Profile Books, London

Krumpel A, Rice A, Frasier KE, Reese F, Trickey JS, Simonis AE, Ryan JP, Wiggins SM, Denzinger A, Schnitzler H-U, Baumann-Pickering S (2021) Long-term patterns of noise from underwater explosions and their relation to fisheries in Southern California. Front Mar Sci 8:796849. https://doi.org/10.3389/fmars.2021.796849

Kyhn LA, Tougaard J, Beedholm K, Jensen FH, Ashe E, Williams R, Madsen PT (2013) Clicking in a killer whale habitat: Narrow-band, high-frequency biosonar clicks of harbour porpoise (*Phocoena phocoena*) and Dall's porpoise (*Phocoenoides dalli*). PLoS One 8(5):e63763. https://doi.org/10.1371/journal.pone.0063763

Kyhn LA, Sveegaard S, Tougaard J (2014) Underwater noise emissions from a drillship in the Arctic. Mar Pollut Bull 86(1):424–433. https://doi.org/10.1016/j.marpolbul.2014.06.037

Lagrois D, Kowalski C, Sénécal J-F, Martins CCA, Chion C (2023) Low-to-mid-frequency monopole source levels of underwater noise from small recreational vessels in the St. Lawrence Estuary beluga critical habitat. Sensors 23(3):1674. https://doi.org/10.3390/s23031674

Lammers MO, Au WWL, Herzing DL (2003) The broadband social acoustic signaling behavior of spinner and spotted dolphins. J Acoust Soc Am 114(3):1629–1639. https://doi.org/10.1121/1.1596173

Larsen ON, Gannon WL, Erbe C, Pavan G, Thomas JA (2022) Source-path-receiver model for airborne sounds. In: Erbe C, Thomas JA (eds) Exploring animal behavior through sound: volume 1: methods. Springer, Cham, pp 153–183. https://doi.org/10.1007/978-3-030-97540-1_5

Laughlin J (2006) Underwater sound levels associated with pile driving at the Cape Disappointment boat launch facility, wave barrier project. Washington State Department of Transportation, Seattle

Leaper R (2019) The role of slower vessel speeds in reducing greenhouse gas emissions, underwater noise and collision risk to whales. Front Mar Sci 6:505. https://doi.org/10.3389/fmars.2019.00505

Lepper PA, Turner VLG, Goodson AD, Black KD (2004) Source levels and spectra emitted by three commercial aquaculture anti-predation devices. In: Proceedings of the 7th European Conference on Underwater Acoustics, ECUA 2004, Delft, Netherlands, 5–8 July 2004

Lepper PA, Gordon J, Booth C, Theobald P, Robinson SP, Northridge S, Wang L (2014) Establishing the sensitivity of cetaceans and seals to acoustic deterrent devices in Scotland (Report for Scottish Natural Heritage No. 517). Loughborough University

Leroy EC, Samaran F, Bonnel J, Royer J-Y (2016) Seasonal and diel vocalization patterns of Antarctic blue whale (*Balaenoptera musculus intermedia*) in the southern Indian Ocean: a multi-year and multi-site study. PLoS One 11(11):e0163587. https://doi.org/10.1371/journal.pone.0163587

LGL and MAI (2011) Environmental Assessment of Marine Vibroseis (Report TA4604-1). https://gisserver.intertek.com/JIP/DMS/ProjectReports/Cat1/JIP-Proj1.6_EAofMarVibr_LGL&MAI_2011.pdf. International Association of Oil & Gas Producers, Joint Industry Programme, E&P Sound and Marine Life. Report TA4604-1

Li B, Bayly M (2017) Quantitative analysis on the environmental impact benefits from the bandwidth-controlled marine seismic source technology. Paper presented at the Acoustics 2017 conference, Perth, WA, Australia, 19–22 November, 2017

Lien J, Barney W, Todd S, Seton R, Guzzwell J (1992) Effects of adding sounds to cod traps on the probability of collisions by humpback whales. In: Thomas JA, Kastelein RA, Supin AY (eds) Marine mammal sensory systems. Plenum Press, New York, pp 701–708

Lindell H (2003) Utgrunden off-shore wind farm – measurements of underwater noise (NEI-SE--598). Ingemansson Technology AB, Goeteborg

Lippert S, Nijhof M, Lippert T, Wilkes D, Gavrilov A, Heitmann K, Ruhnau M, von Estorff O, Schäfke A, Schäfer I (2016) COMPILE—a generic benchmark case for predictions of marine pile-driving noise. IEEE J Ocean Eng 41(4):1061–1071. https://doi.org/10.1109/JOE.2016.2524738

Lippert S, Nijhof M, Lippert T, von Estorff O (2018) COMPILE II – a benchmark of pile driving noise models against offshore measurements. In: Proceedings of inter-noise 2018, Chicago, IL, pp 2984–3995

Locascio JV, Mann DA (2011) Localization and source level estimates of black drum (*Pogonias cromis*) calls. J Acoust Soc Am 130(4):1868–1879. https://doi.org/10.1121/1.3621514

Long A (2022) Environmental modeling of acoustic marine seismic sources. PGS Industry Insights 2022–06. https://www.pgs.com/globalassets/technical-library/tech-lib-pdfs/industry_insights2022_06_source_design_final.pdf

Lossent J, Lejart M, Folegot T, Clorennec D, Di Iorio L, Gervaise C (2018) Underwater operational noise level emitted by a tidal current turbine and its potential impact on marine fauna. Mar Pollut Bull 131, Part A:323–334. https://doi.org/10.1016/j.marpolbul.2018.03.024, 323

Lucke K, Lepper PA, Blanchet MA, Siebert U (2011) The use of an air bubble curtain to reduce the received sound levels for harbour porpoises (*Phocoena phocoena*). J Acoust Soc Am 130(5):3406–3412. https://doi.org/10.1121/1.3626123

Lurton X (2004) An introduction to underwater acoustics – principles and applications. Springer, Berlin

Lurton X (2016) Modelling of the sound field radiated by multibeam echosounders for acoustical impact assessment. Appl Acoust 101:201–221. https://doi.org/10.1016/j.apacoust.2015.07.012

Ma BB, Nystuen JA, Lien R-C (2005) Prediction of underwater sound levels from rain and wind. J Acoust Soc Am 117(6):3555–3565. https://doi.org/10.1121/1.1910283

MacGillivray AO (2006) An acoustic modelling study of Seismic Airgun noise in Queen Charlotte Basin. MSc thesis, University of Victoria, BC

MacGillivray A (2013) A model for underwater sound levels generated by marine impact pile driving. Proc Meet Acoust 20(1):045008. https://doi.org/10.1121/2.0000030

MacGillivray A (2018) Underwater noise from pile driving of conductor casing at a deep-water oil platform. J Acoust Soc Am 143(1):450–459. https://doi.org/10.1121/1.5021554

MacGillivray AO (2019) An airgun array source model accounting for high-frequency sound emissions during firing—solutions to the IAMW source test cases. IEEE J Ocean Eng 44(3):582–588. https://doi.org/10.1109/JOE.2018.2853199

MacGillivray AO, Chapman NR (2012) Modeling underwater sound propagation from an airgun array using the parabolic equation method. Can Acoust 40(1):19–25

MacGillivray A, de Jong C (2021) A reference spectrum model for estimating source levels of marine shipping based on Automated Identification System data. J Mar Sci Eng 9(4):369. https://doi.org/10.3390/jmse9040369

MacGillivray A, Racca R (2005) Sound pressure and particle velocity measurements from marine pile driving at Eagle Harbor maintenance facility, Bainbridge Island, WA (report for Washington State Department of Transportation). JASCO Research Ltd, Victoria

MacGillivray AO, Li Z, Hannay DE, Trounce KB, Robinson OM (2019) Slowing deep-sea commercial vessels reduces underwater radiated noise. J Acoust Soc Am 146(1):340–351. https://doi.org/10.1121/1.5116140

MacGillivray AO, Ainsworth LM, Zhao J, Dolman JN, Hannay DE, Frouin-Mouy H, Trounce KB, White DA (2022) A functional regression analysis of vessel source level measurements from the Enhancing Cetacean Habitat and Observation (ECHO) database. J Acoust Soc Am 152(3):1547–1563. https://doi.org/10.1121/10.0013747

MacGillivray AO, Martin SB, Ainslie MA, Dolman JN, Li Z, Warner GA (2023) Measuring vessel underwater radiated noise in shallow water. J Acoust Soc Am 153(3):1506–1524. https://doi.org/10.1121/10.0017433

Mackay A, Knuckey I (2013) Mitigation of marine mammal bycatch in gillnet fisheries using acoustic devices, literature review. Final Report to the Australian Fisheries Management Authority 25pp

Madsen PT, Wahlberg M, Møhl B (2002) Male sperm whale (*Physeter macrocephalus*) acoustics in a high-latitude habitat: implications for echolocation and communication. Behav Ecol Sociobiol 53(1):31–41. https://doi.org/10.1007/s00265-002-0548-1

Madsen PT, Wahlberg M, Tougaard J, Lucke K, Tyack P (2006) Wind turbine underwater noise and marine mammals: implications of current knowledge and data needs. Mar Ecol Prog Ser 309:279–295. https://doi.org/10.3354/meps309279

Mann D, Cott P, Horne B (2009) Under-ice noise generated from diamond exploration in a Canadian sub-arctic lake and potential impacts on fishes. J Acoust Soc Am 126(5):2215–2222. https://doi.org/10.1121/1.3203865

Marley SA, Salgado Kent CP, Erbe C, Thiele D (2017) A tale of two soundscapes: comparing the acoustic characteristics of urban versus pristine coastal dolphin habitats in Western Australia. Acoust Aust 45(2):159–178. https://doi.org/10.1007/s40857-017-0106-7

Mate BR, Harvey JT (1986) Acoustical deterrents in marine mammal conflicts with fisheries. Oregon State University, Newport

Matsumoto H, Haxel JH, Dziak RP, Bohnenstiehl DR, Embley RW (2011) Mapping the sound field of an erupting submarine volcano using an acoustic glider. J Acoust Soc Am 129 (3):EL94–EL99. https://doi.org/10.1121/1.3547720

Mattsson A (2010) Svein Vaage broadband Airgun study (Report for OGP, E&P Sound and Marine Life, JIP). PGS Geophysical, London

McCauley RD (2012) Fish choruses from the Kimberley, seasonal and lunar links as determined by long term sea noise monitoring. In: McMinn T (ed) Annual conference of the Australian Acoustical Society, Fremantle, Western Australia, 21–23 November 2012

McCauley RD, Cato DH (2000) Patterns of fish calling in a nearshore environment in the Great Barrier Reef. Philos Trans R Soc Lond B 355:1289–1293. https://doi.org/10.1098/rstb.2000.0686

McCauley RD, Gavrilov AN, Jolliffe CD, Ward R, Gill PC (2018) Pygmy blue and Antarctic blue whale presence, distribution and population parameters in southern Australia based on passive acoustics. Deep-Sea Res II Top Stud Oceanogr:154–168. https://doi.org/10.1016/j.dsr2.2018.09.006

McCreery L, Thomas JA (2009) Acoustic analysis of underwater vocalizations from crabeater seals (*Lobodon carcinophagus*): not so monotonous. Aquat Mamm 35(4):490–501. https://doi.org/10.1578/AM.35.4.2009.490

McDonald MA, Hildebrand JA, Wiggins SM, Thiele D, Glasgow D, Moore SE (2005) Sei whale sounds recorded in the Antarctic. J Acoust Soc Am 118(6): 3941–3945. https://doi.org/10.1121/1.2130944

McDonald MA, Mesnick SL, Hildebrand JA (2006) Biogeographic characterisation of blue whale song worldwide: using song to identify populations. J Cetacean Res Manag 8(1):55–65. https://doi.org/10.47536/jcrm.v8i1.702

McGarry T, De Silva R, Canning S, Mendes S, Prior A, Stephenson S, Wilson J (2022) Evidence base for application of Acoustic Deterrent Devices (ADDs) as marine mammal mitigation (Version 4) (JNCC Report No. 615). Joint Nature Conservation Committee, Peterborough

McGrath JR (1976) Infrasonic sea noise at the mid-Atlantic ridge near 37°N. J Acoust Soc Am 60(6):1290–1299. https://doi.org/10.1121/1.381243

McKenna M, Ross D, Wiggins S, Hildebrand J (2012) Underwater radiated noise from modern commercial ships. J Acoust Soc Am 131(1):92–103. https://doi.org/10.1121/1.3364100

McKenna MF, Wiggins SM, Hildebrand JA (2013) Relationship between container ship underwater noise levels and ship design, operational and oceanographic conditions. Sci Rep 3:1760. https://doi.org/10.1038/srep01760

McPherson GR, Ballam D, Stapley J, Peverell S, Cato DH, Gribble N, Clague C, Lien J (2004) Acoustic alarms to reduce marine mammal bycatch from gillnets in Queensland waters:optimising the alarm type and spacing. In: Acoustics 2004 conference, Gold Coast, Australia, 3–5 November 2004. pp 363–368

McWilliam JN, McCauley RD, Erbe C, Parsons MJG (2017) Patterns of biophonic periodicity on coral reefs in the Great Barrier Reef. Sci Rep 7(1):17459. https://doi.org/10.1038/s41598-017-15838-z

McWilliam JN, McCauley RD, Erbe C, Parsons MJG (2018) Soundscape diversity in the Great Barrier Reef: Lizard Island, a case study. Bioacoustics 27(3):295–311. https://doi.org/10.1080/09524622.2017.1344930

Medwin H, Beaky MM (1989) Bubble sources of the Knudsen sea noise spectra. J Acoust Soc Am 86(3): 1124–1130. https://doi.org/10.1121/1.398104

Medwin H, Kurgan A, Nystuen JA (1990) Impact and bubble sound from raindrops at normal and oblique incidence. J Acoust Soc Am 88(1):413–418. https://doi.org/10.1121/1.399918

Medwin H, Nystuen JA, Jacobus PW, Ostwald LH, Snyder DE (1992) The anatomy of underwater rain noise. J Acoust Soc Am 92(3):1613–1623. https://doi.org/10.1121/1.403902

Mellen RH (1952) The thermal-noise limit in the detection of underwater acoustic signals. J Acoust Soc Am 24(5):478–480. https://doi.org/10.1121/1.1906924

Menze S, Zitterbart DP, van Opzeeland I, Boebel O (2017) The influence of sea ice, wind speed and marine mammals on Southern Ocean ambient sound. R Soc Open Sci 4(1):160370. https://doi.org/10.1098/rsos.160370

Mikhalevsky PN (2001) Acoustics, Arctic. In: Steele JH, Thorpe SA, Turekian KK (eds) Encyclopedia of Ocean Sciences, vol. 1 (A-C), pp 53–61

Miksis-Olds JL, Bradley DL, Niu XM (2013) Decadal trends in Indian Ocean ambient sound. J Acoust Soc Am 134(5):3464–3475. https://doi.org/10.1121/1.4821537

Milne AR, Ganton JH (1964) Ambient noise under Arctic-sea ice. J Acoust Soc Am 36(5):855–863. https://doi.org/10.1121/1.1919103

Mitson RB (1995) Underwater noise of research vessels (Cooperative Res Report 209). International Council for the Exploration of the sea, Copenhagen

Mossbridge JA, Shedd JG, Thomas JA (1999) An "acoustic niche" for antarctic killer whale and leopard seal sounds. Mar Mamm Sci 15(4):1351–1357. https://doi.org/10.1111/j.1748-7692.1999.tb00897.x

Nedwell J, Edwards B (2002) Measurements of underwater noise in the Arun River during piling at county wharf, Littlehamptopn (report for David Wilson Homes Ltd.). Subacoustech Ltd, Soberton Heath

Nedwell JR, Edwards B (2004) A review of measurements of underwater man-made noise carried out by Subacoustech Ltd, 1993–2003 (report for Chevron Texaco Ltd., TotalFinaElf exploration UK PLC, DSTL, Department of Trade and Industry, and Shell UK Exploration and Production Ltd. 534R0109). Subacoustech Ltd, Hampshire

Nedwell J, Howell D (2004) A review of offshore windfarm related underwater noise sources. Cowrie Rep 544:1–57

Nedwell J, Langworthy J, Howell D (2003) Assessment of sub-sea acoustic noise and vibration from offshore wind turbines and its impact on marine wildlife; initial measurements of underwater noise during construction of offshore windfarms, and comparison with background noise (Report for the Crown Estates Office 544R0424). Subacoustech Ltd, Southampton

Nedwell J, Parvin S, Edwards B, Workman R, Brooker A, Kynoch J (2007) Measurement and interpretation of underwater noise during construction and operation of offshore windfarms in UK waters (Report to COWRIE Ltd 544R0738). Subacoustech, Hampshire

Nehls G, Betke K, Eckelmann S, Ros M (2007) Assessment and costs of potential engineering solutions for the mitigation of the impacts of underwater noise arising from the construction of offshore windfarms. BioConsult report, Husum, Germany. For COWRIE Ltd., COWRIE ENG-01-2007

Neves S (2013) Acoustic behaviour of Risso's dolphins, *Grampus griseus*, in the Canary Islands, Spain. PhD thesis, University of St Andrews, St Andrews

Nieukirk SL, Stafford KM, Mellinger DK, Dziak RP, Fox CG (2004) Low-frequency whale and seismic airgun sounds recorded in the mid-Atlantic Ocean. J Acoust Soc Am 115(4):1832–1843. https://doi.org/10.1121/1.1675816

Nieukirk S, Mellinger D, Moore S, Klinck K, Dziak R, Goslin J (2012) Sounds from airguns and fin whales recorded in the mid-Atlantic Ocean, 1999–2009. J Acoust Soc Am 131(2):1102–1112. https://doi.org/10.1121/1.3672648

Norro A, Rumes B, Degraer S (2015) Characterisation of the operational noise, generated by offshore wind farms in the Belgian part of the North Sea. In: 3rd Underwater Acoustics Conference and Exhibition (UACE 2015), Crete, pp 507–514

Northrop J (1974) Detection of low-frequency underwater sounds from a submarine volcano in the Western Pacific. J Acoust Soc Am 56(3):837–841. https://doi.org/10.1121/1.1903334

Nystuen JA (1986) Rainfall measurements using underwater ambient noise. J Acoust Soc Am 79(4):972–982. https://doi.org/10.1121/1.393695

Nystuen JA, McGlothin CC, Cook MS (1993) The underwater sound generated by heavy rainfall. J Acoust Soc Am 93(6):3169–3177. https://doi.org/10.1121/1.405701

O'Hara PD, Serra-Sogas N, McWhinnie L, Pearce K, Le Baron N, O'Hagan G, Nesdoly A, Marques T, Canessa R (2023) Automated identification system for ships data as a proxy for marine vessel related stressors. Sci Total Environ 865:160987. https://doi.org/10.1016/j.scitotenv.2022.160987

Olesiuk P, Nichol LM, Snowden MJ, Ford JKB (2002) Effect of the sound generated by an acoustic harassment device on the relative abundance and distribution of harbor porpoises (*Phocoena phocoena*) in Retreat Passage, British Columbia. Mar Mamm Sci 18(4): 843–862. https://doi.org/10.1111/j.1748-7692.2002.tb01077.x

Oleson EM, Barlow J, Gordon J, Rankin S, Hildebrand JA (2003) Low frequency calls of Bryde's whales. Mar Mamm Sci 19(2):160–172. https://doi.org/10.1111/j.1748-7692.2003.tb01119.x

Palomo PB, Mullor RS, Rodriguez AM (2014) Reduction of the underwater radiated noise by ships: new shipbuilding challenge. The vessels "Ramón Margalef" and "Ángeles Alvariño" as technological references of how to build silent vessels. In: Transport Research Arena 2014 Conference, Paris, France

Pangerc T, Theobald PD, Wang LS, Robinson SP, Lepper PA (2016) Measurement and characterisation of radiated underwater sound from a 3.6 MW monopile wind turbine. J Acoust Soc Am 140 (4):2913–2922. https://doi.org/10.1121/1.4964824

Papale E, Azzolin M, Cascão I, Gannier A, Lammers MO, Martin VM, Oswald J, Perez-Gil M, Prieto R, Silva MA, Giacoma C (2013) Geographic variability in the acoustic parameters of striped dolphin's (*Stenella coeruleoalba*) whistles. J Acoust Soc Am 133(2): 1126–1134. https://doi.org/10.1121/1.4774274

Park M, Odom RI, Soukup DJ (2001) Modal scattering: a key to understanding oceanic T-waves. Geophys Res Lett 28(17):3401–3404. https://doi.org/10.1029/2001GL013472

Parks SE, Clark CW, Tyack PL (2007) Short- and long-term changes in right whale calling behavior: the potential effects of noise on acoustic communication. J Acoust Soc Am 122(6):3725–3731. https://doi.org/10.1121/1.2799904

Parsons M, Meekan M (2020) Acoustic characteristics of small research vessels. J Mar Sci Eng 8(12):970. https://doi.org/10.3390/jmse8120970

Parsons MJ, McCauley RD, Mackie MC (2013a) Characterisation of mulloway *Argyrosomus japonicus* advertisement sounds. Acoust Aust 41(3):196–201

Parsons MJG, Holley D, McCauley RD (2013b) Source levels of dugong (*Dugong dugon*) vocalizations recorded in Shark Bay. J Acoust Soc Am 134(3): 2582–2588. https://doi.org/10.1121/1.4816583

Parsons MJG, Longbottom S, Lewis P, McCauley RD, Fairclough DV (2013c) Sound production by the West Australian dhufish (*Glaucosoma hebraicum*). J Acoust Soc Am 134(4):2701–2709. https://doi.org/10.1121/1.4818775

Parsons MJG, Recalde-Salas A, Salgado-Kent CP, McCauley RD (2016a) Fish choruses off Port Hedland, Western Australia. Bioacoustics 26(2):135–152. https://doi.org/10.1080/09524622.2016.1227940

Parsons MJG, Salgado-Kent CP, Marley SA, Gavrilov AN, McCauley RD (2016b) Characterizing diversity and variation in fish choruses in Darwin Harbour. ICES J Mar Sci 73(8):2058–2074. https://doi.org/10.1093/icesjms/fsw037

Parsons MJG, Duncan AJ, Parsons SK, Erbe C (2020) Reducing vessel noise: an example of a solar-electric passenger ferry. J Acoust Soc Am 147(5):3575–3583. https://doi.org/10.1121/10.0001264

Parsons MJG, Erbe C, Meekan MG, Parsons SK (2021) A review and meta-analysis of underwater noise radiated by small (<25 m length) vessels. J Mar Sci Eng 9(8): 827. https://doi.org/10.3390/jmse9080827

Patek SN, Shipp LE, Staaterman ER (2009) The acoustics and acoustic behavior of the California spiny lobster (*Panulirus interruptus*). J Acoust Soc Am 125(5): 3434–3443. https://doi.org/10.1121/1.3097760

Payne RS, McVay S (1971) Songs of humpback whales. Science 173(3997):585–597. https://doi.org/10.1126/science.173.3997.585

Pearson JP, Verts BJ (1970) Abundance and distribution of harbor seals and northern sea lions in Oregon. Murrelet 51:1–5

Pekeris CL (1948) Theory of propagation of explosive sound in shallow water. Mem-Geol Soc Am 27:1–117

Peng Y, Tsouvalas A, Stampoultzoglou T, Metrikine A (2021) A fast computational model for near- and far-field noise prediction due to offshore pile driving. J Acoust Soc Am 149(3):1772–1790. https://doi.org/10.1121/10.0003752

Penrose J, Siwabessy P, Gavrilov A, Parnum I, Hamilton L, Bickers A, Brooke B, Ryan D, Kennedy P (2005) Acoustic techniques for seabed classification. Cooperative Research Centre for Coastal Zone Estuary and Waterway Management, Technical Report 32:11

Perrin WF, Donovan GP, Barlow J (1994) Gillnets and cetaceans: incorporating the proceedings of the symposium and workshop on the mortality of cetaceans in passive fishing nets and traps. Cambridge, UK, p 629

Podolskiy EA, Murai Y, Kanna N, Sugiyama S (2022) Glacial earthquake-generating iceberg calving in a narwhal summering ground: the loudest underwater sound in the Arctic? J Acoust Soc Am 151(1):6–16. https://doi.org/10.1121/10.0009166

Popper AN, Halvorsen MB, Kane A, Miller DL, Smith ME, Song J, Stein P, Wysocki LE (2007) The effects of high-intensity, low-frequency active sonar on rainbow trout. J Acoust Soc Am 122(1):623–635. https://doi.org/10.1121/1.2735115

Radford C, Jeffs A, Tindle C, Montgomery JC (2008) Resonating sea urchin skeletons create coastal choruses. Mar Ecol Prog Ser 362:37–43. https://doi.org/10.3354/meps07444

Rankin S, Ljungblad D, Clark CW, Kato H (2005) Vocalisations of Antarctic blue whales, *Balaenoptera musculus intermedia*, recorded during the 2001/2002 and 2002/2003 IWC/SOWER circumpolar cruises, Area V Antarctica. J Cetacean Res Manag 7(1): 13–20. https://doi.org/10.47536/jcrm.v7i1.752

Rayleigh L (1917) On the pressure developed in a liquid during the collapse of a spherical cavity. Philos Mag 34(200):94–98. https://doi.org/10.1080/14786440808635681

Reeves RR, Hofman RJ, Silber GK, Wilkinson D (1996) Acoustic deterrence of harmful marine mammal-fishery interactions (proceedings of a workshop held 20–22 March 1996 NMFS-OPR-10 NOAA technical memorandum). U.S. Department of Commerce, Seattle/Washington, DC

Reeves RR, Read AJ, Notarbartolo di Sciara G (2001) Report of the Workshop on Interactions between Dolphins and Fisheries in the Mediterranean: Evaluation of Mitigation Alternatives; SC/53/SM3, London

Reimann K, Grabe J (2014) Soil vibration due to offshore pile driving and induced underwater noise. In: 2nd Underwater Acoustics Conference and Exhibition, Rhodes, Greece, 22–27 June 2014, pp 279–294

Reine K, Dickerson C (2014) Characterization of underwater sounds produced by a hydraulic cutterhead dredge during maintenance dredging in the Stockton deepwater shipping channel, California (report ERDC TN-DOER-E38). U.S. Army Engineer Research and Development Center, Vicksburg

Reine KJ, Clarke DG, Dickerson C (2012) Characterization of underwater sounds produced by a backhoe dredge excavating rock and gravel (DOER technical notes collection ERDC TN-DOER-E36). U.S. Army Engineer Research and Development Center, Vicksburg

Reine KJ, Clarke D, Dickerson C, Wikel G (2014a) Characterization of underwater sounds produced by trailing suction hopper dredges during sand mining and pump-out operations (Final Report ERDC/EL TR-14-3). Engineer Research and Development Center (ERDC), US Army, Washington, DC

Reine KJ, Clarke DG, Dickerson C (2014b) Characterization of underwater sounds produced by hydraulic and mechanical dredging operations. J Acoust Soc Am 135(6):3280–3294. https://doi.org/10.1121/1.4875712

Reinhall PG, Dahl PH (2011) Underwater Mach wave radiation from impact pile driving: theory and observation. J Acoust Soc Am 130(3):1209–1216. https://doi.org/10.1121/1.3614540

Richardson WJ, Greene CR, Malme CI, Thomson DH (1995) Marine mammals and noise. Academic, San Diego. https://doi.org/10.1016/C2009-0-02253-3

Risch D, Wilson B, Lepper P (2017) Acoustic assessment of SIMRAD EK60 high frequency echo sounder signals (120 & 200 kHz) in the context of marine mammal monitoring. Scottish Marine Freshw Sci 8(13):1–24. https://doi.org/10.7489/1978-1

Risch D, van Geel N, Gillespie D, Wilson B (2020) Characterisation of underwater operational sound of a tidal stream turbine. J Acoust Soc Am 147(4): 2547–2555. https://doi.org/10.1121/10.0001124

Risch D, Marmo B, van Geel N, Gillespie D, Hastie G, Sparling C, Onoufriou J, Wilson B (2023) Underwater noise of two operational tidal stream turbines: a comparison. In: Popper AN, Sisneros J, Hawkins AD, Thomsen F (eds) The effects of noise on aquatic life: principles and practical considerations. Springer, Cham, pp 1–22. https://doi.org/10.1007/978-3-031-10417-6_135-1

Robinson SP, Theobald PD, Hayman G, Wang LS, Lepper PA, Humphrey V, Mumford S (2011) Measurement of underwater noise arising from marine aggregate dredging operations (Final report MEPF 09/P108). Centre for Environment, Fisheries & Aquaculture Science (CEFAS): Marine Aggregate Levy Sustainability Fund (MALSF), Suffolk

Robinson SP, Theobald PD, Lepper PA (2012) Underwater noise generated from marine piling. Proc Meet Acoust 17(1):070080. https://doi.org/10.1121/1.4790330

Rogers TL, Cato DH, Bryden MM (1995) Underwater vocal repertoire of the leopard seal, *Hydrurga leptonyx*, in Prydz Bay, Antarctica. In: Kastelein RA, Thomas JA, Nachtigal PE (eds) Sensory abilities of aquatic animals. DeSpil Publishers, Amsterdam

Ross D (1976) Mechanics of underwater noise. Pergamon Press, New York

Roth EH, Schmidt V, Hildebrand JA, Wiggins SM (2013) Underwater radiated noise levels of a research icebreaker in the Central Arctic Ocean. J Acoust Soc Am 133(4):1971–1980. https://doi.org/10.1121/1.4790356

Rudd AB, Richlen MF, Stimpert AK, Au WWL (2015) Underwater sound measurements of a high-speed jet-propelled marine craft: implications for large whales. Pac Sci 69(2):155–164. https://doi.org/10.2984/69.2.2

Ruppé L, Clément G, Herrel A, Ballesta L, Décamps T, Kéver L, Parmentier E (2015) Environmental constraints drive the partitioning of the soundscape in fishes. Proc Natl Acad Sci 112(19):6092–6097. https://doi.org/10.1073/pnas.1424667112

Salomons EM, Binnerts B, Betke K, von Benda-Beckmann AM (2021) Noise of underwater explosions in the North Sea. A comparison of experimental data and model predictions. J Acoust Soc Am 149(3):1878–1888. https://doi.org/10.1121/10.0003754

Samarra FIP, Deecke VB, Vinding K, Rasmussen MH, Swift RJ, Miller PJO (2010) Killer whales (*Orcinus orca*) produce ultrasonic whistles. J Acoust Soc Am 128:EL205–EL210. https://doi.org/10.1121/1.3462235

Samarra FIP, Deecke VB, Miller PJO (2016) Low-frequency signals produced by Northeast Atlantic killer whales (*Orcinus orca*). J Acoust Soc Am 139(3):1149–1157. https://doi.org/10.1121/1.4943555

Sandeman DC, Wilkens LA (1982) Sound production by abdominal stridulation in the Australian Murray River crayfish, *Euastacus armatus*. J Exp Biol 99(1):469–472. https://doi.org/10.1242/jeb.99.1.469

Schaffeld T, Ruser A, Woelfing B, Baltzer J, Kristensen JH, Larsson J, Schnitzler JG, Siebert U (2019) The use of seal scarers as a protective mitigation measure can induce hearing impairment in harbour porpoises. J Acoust Soc Am 146(6):4288–4298. https://doi.org/10.1121/1.5135303

Schoeman RP, Erbe C, Pavan G, Righini R, Thomas JA (2022a) Analysis of soundscapes as an ecological tool. In: Erbe C, Thomas JA (eds) Exploring animal behavior through sound: volume 1: methods. Springer, Cham, pp 217–267. https://doi.org/10.1007/978-3-030-97540-1_7

Schoeman RP, Erbe C, Plön S (2022b) Underwater chatter for the win: a first assessment of underwater soundscapes in two bays along the Eastern Cape Coast of South Africa. J Mar Sci Eng 10(6):746. https://doi.org/10.3390/jmse10060746

Schultz-von Glahn M, Betke K, Nehls G (2006) Underwater noise reduction of pile driving for offshore wind turbines–evaluation of several techniques under offshore conditions (UFOPLAN 205 53 113). Umweltbundesamt, Berlin, p 53

Schwock F, Abadi S (2021) Characterizing underwater noise during rain at the Northeast Pacific continental margin. J Acoust Soc Am 149(6):4579–4595. https://doi.org/10.1121/10.0005440

Scrimger JA (1985) Underwater noise caused by precipitation. Nature 318(6047):647–649. https://doi.org/10.1038/318647a0

Scrimger P, Heitmeyer RM (1991) Acoustic source-level measurements for a variety of merchant ships. J Acoust Soc Am 89(2):691–699. https://doi.org/10.1121/1.1894628

Scrimger P, Evans DJ, McBean GA, Farmer DM, Kerman BR (1987) Underwater noise due to rain, hail, and snow. J Acoust Soc Am 81(1):79–86. https://doi.org/10.1121/1.394936

Scrimger JA, Evans DJ, Yee W (1989) Underwater noise due to rain—open ocean measurements. J Acoust Soc Am 85(2):726–731. https://doi.org/10.1121/1.397598

Sertlek HÖ, Ainslie MA (2015) Airgun source model (AGORA): its application for seismic surveys sound maps in the dutch north sea. Proc UACE, Crete, Greece, 21–26 June 2015

Sertlek HÖ, Blacquière G (2019) Effects of the rough sea surface on the signature of a single air gun. IEEE J Ocean Eng 44(3):575–581. https://doi.org/10.1109/JOE.2018.2890464

Shapiro AD, Tougaard J, Jorgensen PB, Kyhn LA, Balle JD, Bernardez C, Fjalling A, Karlsen J, Wahlberg M (2009) Transmission loss patterns from acoustic harassment and deterrent devices do not always follow geometrical spreading predictions. Mar Mamm Sci 25(1):53–67. https://doi.org/10.1111/j.1748-7692.2008.00243.x

Shimu QIN, Yang Y, Junqi QIN, Changchun DI (2019) Research on the cavitation in the snapping shrimp: a review. IOP Conf Ser Earth Environ Sci 310 (5): 052057. https://doi.org/10.1088/1755-1315/310/5/052057

Simard Y, Roy N, Gervaise C, Giard S (2016) Analysis and modeling of 255 source levels of merchant ships from an acoustic observatory along St. Lawrence Seaway. J Acoust Soc Am 140(3):2002–2018. https://doi.org/10.1121/1.4962557

Simmonds J, MacLennan DN (2005) Fisheries acoustics: theory and practice, 2nd edn. Wiley, Hoboken

Sivle LD, Kvadsheim PH, Fahlman A, Lam FPA, Tyack PL, Miller PJO (2012) Changes in dive behavior during naval sonar exposure in killer whales, long-finned pilot whales, and sperm whales. Front Physiol 3(400):1–11. https://doi.org/10.3389/fphys.2012.00400

Slabbekoorn H, Bouton N, van Opzeeland I, Coers A, ten Cate C, Popper AN (2010) A noisy spring: the impact of globally rising underwater sound levels on fish. Trends Ecol Evol 25(7):419–427. https://doi.org/10.1016/j.tree.2010.04.005

Smith TA, Rigby J (2022) Underwater radiated noise from marine vessels: a review of noise reduction methods and technology. Ocean Eng 266:112863. https://doi.org/10.1016/j.oceaneng.2022.112863

Sousa-Lima RS, Paglia AP, Da Fonseca GAB (2002) Signature information and individual recognition in the isolation calls of Amazonian manatees, *Trichechus inunguis* (Mammalia: Sirenia). Anim Behav 63:301–310. https://doi.org/10.1006/anbe.2001.1873

Southall BL, Moretti D, Abraham B, Calambokidis J, DeRuiter SL, Tyack PL (2012) Marine mammal behavioral response studies in Southern California: advances in technology and experimental methods. Mar Technol Soc J 46(4):48–59. https://doi.org/10.4031/MTSJ.46.4.1

Staaterman E, Clark C, Gallagher A, deVries M, Claverie T, Patek S (2011) Rumbling in the benthos: acoustic ecology of the California mantis shrimp *Hemisquilla californiensis*. Aquat Biol 13(2):97–105. https://doi.org/10.3354/ab00361

Stafford K, Moore S, Berchok C, Wiig O, Lydersen C, Hansen E, Kalmbach D, Kovacs K (2012) Spitsbergen's endangered bowhead whales sing through the polar night. Endang Species Res 18(2):95–103. https://doi.org/10.3354/esr00444

Stanton TK, Chu D, Jech JM, Irish JD (2010) New broad-band methods for resonance classification and high-resolution imagery of fish with swimbladders using a modified commercial broadband echosounder. ICES J Mar Sci 67(2):365–378. https://doi.org/10.1093/icesjms/fsp262

Strasberg M (1979) Nonacoustic noise interference in measurements of infrasonic ambient noise. J Acoust Soc Am 66(5):1487–1493. https://doi.org/10.1121/1.383543

Talandier J, Hyvernaud O, Okal EA, Piserchia P-F (2002) Long-range detection of hydroacoustic signals from large icebergs in the Ross Sea, Antarctica. Earth Planet Sci Lett 203(1):519–534. https://doi.org/10.1016/S0012-821X(02)00867-1

Teledyne (2017) Calmer waters: reducing sound exposure to marine mammals during seismic surveys. Teledyne marine product case study; www.teledynemarine.com

Teng Y-T (2011) Sector-specific beam pattern compensation for multi-sector and multi-swath multibeam sonars. MSc thesis, University of New Brunswick, Department of Geodesy and Geomatics Engineering, New Brunswick, Canada

Thiele L (1988) Underwater noise study from the Ice-breaker "John A. MacDonald" (Report 85.133). Odegaard & Danneskiold-Samsoe ApS, Copenhagen

Thomas JA, Kuechle VB (1982) Quantitative analysis of Weddell seal (*Leptonychotes weddelli*) underwater vocalizations at McMurdo Sound, Antarctica. J Acoust Soc Am 72(6):1730–1738. https://doi.org/10.1121/1.388667

Thompson PO, Findley LT, Vidal O (1992) 20-Hz pulses and other vocalizations of fin whales, *Balaenoptera physalus*, in the Gulf of California. Mexico J Acoust Soc Am 92(6):3051–3057. https://doi.org/10.1121/1.404201

Thomsen F, Franck D, Ford JKB (2001) Characteristics of whistles from the acoustic repertoire of resident killer whales (*Orcinus orca*) off Vancouver Island, British Columbia. J Acoust Soc Am 109(3):1240–1246. https://doi.org/10.1121/1.1349537

Thomsen F, Gill A, Kosecka M, Andersson M, Andre M, Degraer S, Folegot T, Gabriel J, Judd A, Neumann T, Norro A, Risch D, Sigray P, Wood DT, Wilson B (2015) MaRVEN–environmental impacts of noise, vibrations and electromagnetic emissions from marine renewable energy (Final study report RTD-KI-NA-27-738-EN-N). DHI, Brussels

Todd VLG, Williamson LD, Jiang J, Cox SE, Todd IB, Ruffert M (2020) Proximate underwater soundscape of a North Sea offshore petroleum exploration jack-up drilling rig in the Dogger Bank. J Acoust Soc Am 148(6):3971–3979. https://doi.org/10.1121/10.0002958

Tolstoy M, Bohnenstiehl DR (2005) Hydroacoustic constraints on the rupture duration, length, and speed of the great Sumatra-Andaman earthquake. Seismol Res Lett 76(4):419–425. https://doi.org/10.1785/gssrl.76.4.419

Tolstoy M, Bohnenstiehl DR, Chapp E (2004) Long range acoustic propagation of high frequency energy in the Indian Ocean from icebergs and earthquakes. Paper presented at the 26th Annual Seismic Research Review Conference, Orlando, Florida

Tougaard J, Carstensen J, Teilmann J (2009a) Pile driving zone of responsiveness extends beyond 20 km for harbor porpoises (*Phocoena phocoena* (L.)). J Acoust Soc Am 126 (1):11-14:11. https://doi.org/10.1121/1.3132523

Tougaard J, Henriksen OD, Miller LA (2009b) Underwater noise from three types of offshore wind turbines: estimation of impact zones for harbor porpoises and harbor seals. J Acoust Soc Am 125(6):3766–3773. https://doi.org/10.1121/1.3117444

Tougaard J, Hermannsen L, Madsen PT (2020) How loud is the underwater noise from operating offshore wind turbines? J Acoust Soc Am 148(5):2885–2893. https://doi.org/10.1121/10.0002453

Trevorrow MV, Vasiliev B, Vagle S (2008) Directionality and maneuvering effects on a surface ship underwater acoustic signature. J Acoust Soc Am 124(2):767–778. https://doi.org/10.1121/1.2939128

Tsouvalas A (2020) Underwater noise emission due to offshore pile installation: a review. Energies 13(12): 3037. https://doi.org/10.3390/en13123037

Tyack PL, Zimmer WMX, Moretti D, Southall BL, Claridge DE, Durban JW, Clark CW, D'Amico A, DiMarzio N, Jarvis S, McCarthy E, Morrissey R, Ward J, Boyd IL (2011) Beaked whales respond to simulated and actual navy sonar. PLoS One 6(3): e17009. https://doi.org/10.1371/journal.pone.0017009

Urick RJ (1971) The noise of melting icebergs. J Acoust Soc Am 50:337–341. https://doi.org/10.1121/1.1912637

Urick RJ (1983) Principles of underwater sound, 3rd edn. McGraw Hill, New York

Vagle S (2003) On the impact of underwater pile-driving noise on marine life (Internal report). Institute of Ocean Sciences, Fisheries & Oceans Canada, Sidney

van Opzeeland I, Boebel O (2018) Marine soundscape planning: seeking acoustic niches for anthropogenic sound. J Ecoacoust 2:5GSNT8. https://doi.org/10.22261/JEA.5GSNT8

Van Parijs SM, Corkeron PJ (2001) Vocalizations and behaviour of Pacific humpback dolphins *Sousa chinensis*. Ethology 107(8):701–716. https://doi.org/10.1046/j.1439-0310.2001.00714.x

von Pein J, Lippert T, Lippert S, von Estorff O (2022a) Scaling offshore pile driving noise: examples for scenarios with and without a big bubble curtain. Proc Meet Acoust 47(1):070015. https://doi.org/10.1121/2.0001622

Veirs S, Veirs V, Wood JD (2016) Ship noise extends to frequencies used for echolocation by endangered killer whales. PeerJ 4:e1657. https://doi.org/10.7717/peerj.1657

Veirs S, Veirs V, Williams R, Jasny M, Wood JD (2018) A key to quieter seas: half of ship noise comes from 15% of the fleet. PeerJ Preprints 6:e26525v26521. https://doi.org/10.7287/peerj.preprints.26525v1

Viada ST, Hammer RM, Racca R, Hannay D, Thompson MJ, Balcom BJ, Phillips NW (2008) Review of potential impacts to sea turtles from underwater explosive removal of offshore structures. Environ Impact Assess Rev 28:267–285

von Estorff O, Chmelnizkij A, Grabe Jr, Heins E, Heitmann K, Stephan Lippert TL, Ruhnau M, Sieg K, Bohne T, Grießmann T, Rolfes R, Rustemeie JR, Podolski C, Rabbel W, Wilken D (2016) Schlussbericht des Verbundprojektes BORA: Entwicklung eines Berechnungsmodells zur Vorhersage des Unterwasserschalls bei Rammarbeiten zur Gründung von OWEA (FKZ 0325421A/B/C). Technische Universität Hamburg-Harburg, Leibniz Universität Hannover, Christian-Albrechts-Universität zu Kiel

von Pein J, Lippert T, Lippert S, von Estorff O (2022b) Scaling laws for unmitigated pile driving: dependence of underwater noise on strike energy, pile diameter, ram weight, and water depth. Appl Acoust 198: 108986. https://doi.org/10.1016/j.apacoust.2022.108986

Wahlberg M, Jensen F, Soto N, Beedholm K, Bejder L, Oliveira C, Rasmussen M, Simon M, Villadsgaard A, Madsen P (2011) Source parameters of echolocation clicks from wild bottlenose dolphins (*Tursiops aduncus* and *Tursiops truncatus*). J Acoust Soc Am 130(4):2263–2274. https://doi.org/10.1121/1.3624822

Wales SC, Heitmeyer RM (2002) An ensemble source spectra model for merchant ship-radiated noise. J Acoust Soc Am 111(3):1211–1231. https://doi.org/10.1121/1.1427355

Wang Z, Fang L, Shi W, Wang K, Wang D (2013) Whistle characteristics of free-ranging Indo-Pacific humpback dolphins (*Sousa chinensis*) in Sanniang Bay, China. J Acoust Soc Am 133(4):2479–2489. https://doi.org/10.1121/1.4794390

Warren JD, Leach TH, Williamson CE (2016) Measuring the distribution, abundance, and biovolume of zooplankton in an oligotrophic freshwater lake with a 710 kHz scientific echosounder. Limnol Oceanogr 14(4):231–244. https://doi.org/10.1002/lom3.10084

Webster TA, Dawson SM, Rayment WJ, Parks SE, Van Parijs SM (2016) Quantitative analysis of the acoustic repertoire of southern right whales in New Zealand. J Acoust Soc Am 140(1):322–333. https://doi.org/10.1121/1.4955066

Wenz GM (1962) Acoustic ambient noise in the ocean: spectra and sources. J Acoust Soc Am 34(12): 1936–1956. https://doi.org/10.1121/1.1909155

Wiggins SM, Krumpel A, Dorman LM, Hildebrand JA, Baumann-Pickering S (2021) Seal bomb explosion sound source characterization. J Acoust Soc Am 150(3):1821–1829. https://doi.org/10.1121/10.0006101

Wilkes DR, Gavrilov AN (2017) Sound radiation from impact-driven raked piles. J Acoust Soc Am 142(1): 1–11. https://doi.org/10.1121/1.4990021

Wilkes DR, Gourlay TP, Gavrilov AN (2016) Numerical modeling of radiated sound for impact pile driving in offshore environments. IEEE J Ocean Eng 41(4): 1072–1078. https://doi.org/10.1109/JOE.2015.2510860

Williams R, Erbe C, Ashe E, Clark CW (2015) Quiet(er) marine protected areas. Mar Pollut Bull 100(1): 154–161. https://doi.org/10.1016/j.marpolbul.2015.09.012

Williams R, Erbe C, Dewantama IMI, Hendrawan IG (2018) Effect on ocean noise: Nyepi, a Balinese day of silence. Oceanography 31(2):16–18. https://doi.org/10.5670/oceanog.2018.207

Williams R, Veirs S, Veirs V, Ashe E, Mastick N (2019) Approaches to reduce noise from ships operating in important killer whale habitats. Mar Pollut Bull 139: 459–469. https://doi.org/10.1016/j.marpolbul.2018.05.015

Willis J, Dietz FT (1961) Effect of tidal currents on 25 cps shallow water ambient noise measurements. J Acoust

Soc Am 33(11):1659. https://doi.org/10.1121/1.1936636

Wilson B, Carter C (2013) The use of acoustic devices to warn marine mammals of tidal-stream energy devices. Scottish Association for Marine Science, Scotland

Wittekind DK (2014) A simple model for the underwater noise source level of ships. J Ship Product Des 30:1–8

Wladichuk JL, Hannay DE, MacGillivray AO, Li Z, Thornton S (2019) Systematic source level measurements of whale watching vessels and other small boats. J Ocean Technol 14(3):108–126

Woodman GH, Wilson SC, Li VYF, Renneberg R (2003) Acoustic characteristics of fish bombing: potential to develop an automated blast detector. Mar Pollut Bull 46(1):99–106. https://doi.org/10.1016/S0025-326X(02)00322-3

Wright EB, Cybulski J (1983) Low-frequency acoustic source levels of large merchant ships (NRL report 8677). Naval Research Laboratory, Washington, DC

Würsig B, Greene CR Jr, Jefferson TA (2000) Development of an air bubble curtain to reduce underwater noise of percussive piling. Mar Env Res 49:79–93. https://doi.org/10.1016/S0141-1136(99)00050-1

Wyatt R (2008) Review of existing data on underwater sounds produced by the oil and gas industry (report by Seiche measurements ltd.). Joint Industry Programme on Sound and Marine Life, London

Xie Y, Farmer DM (1991) Acoustical radiation from thermally stressed sea ice. J Acoust Soc Am 89(5):2215–2231. https://doi.org/10.1121/1.400914

Xie Y, Farmer DM (1992) The sound of ice break-up and floe interaction. J Acoust Soc Am 91(3):1423–1428. https://doi.org/10.1121/1.402473

Yang TC, Giellis GR, Votaw CW, Diachok OI (1987) Acoustic properties of ice edge noise in the Greenland Sea. J Acoust Soc Am 82(3):1034–1038. https://doi.org/10.1121/1.395377

Yang L, Xu X, Huang Z, Tu X (2015) Recording and analyzing underwater noise during pile driving for bridge construction. Acoust Aust 43(2):159–167. https://doi.org/10.1007/s40857-015-0015-6

Zampolli M, Nijhof M, De Jong C, Ainslie M, Jansen E, Quesson B (2013) Validation of finite element computations for the quantitative prediction of underwater noise from impact pile driving. J Acoust Soc Am 133(1):72–81. https://doi.org/10.1121/1.4768886

Zhu J, Popovics JS, Schubert F (2004) Leaky Rayleigh and Scholte waves at the fluid–solid interface subjected to transient point loading. J Acoust Soc Am 116(4):2101–2110. https://doi.org/10.1121/1.1791718

Ziolkowski A, Parkes G, Hatton L, Haugland T (1982) The signature of an air gun array: computation from near-field measurements including interactions. Geophysics 47(10):1413–1421. https://doi.org/10.1190/1.1441289

ZoBell VM, Gassmann M, Kindberg LB, Wiggins SM, Hildebrand JA, Frasier KE (2023) Retrofit-induced changes in the radiated noise and monopole source levels of container ships. PLoS One 18(3):e0282677. https://doi.org/10.1371/journal.pone.0282677

Open Access This chapter is licensed under the terms of the Creative Commons Attribution 4.0 International License (http://creativecommons.org/licenses/by/4.0/), which permits use, sharing, adaptation, distribution and reproduction in any medium or format, as long as you give appropriate credit to the original author(s) and the source, provide a link to the Creative Commons license and indicate if changes were made.

The images or other third party material in this chapter are included in the chapter's Creative Commons license, unless indicated otherwise in a credit line to the material. If material is not included in the chapter's Creative Commons license and your intended use is not permitted by statutory regulation or exceeds the permitted use, you will need to obtain permission directly from the copyright holder.

Mysticete Sounds

3

Christine Erbe ⓘ, Anita Murray, Meghan Aulich, Ann Bowles, Ciara Browne, Brodie Elsdon, Emily K. Evans, Adam Frankel, Alexander Gavrilov, Corinna Gosby, Lauren Hawkins, Capri Jolliffe, Paul Nguyen Hong Duc, and Chong Wei

Contents

3.1	**Introduction**	180
3.2	**Sound Production**	181
3.3	**Song**	183
3.3.1	Song Description	183
3.3.2	Whale Choruses	185
3.3.3	Song Function	186
3.4	**Non-Song Sounds**	189
3.4.1	Description of Non-Song Sounds	189
3.4.2	Function of Non-Song Sounds	190
3.5	**Vocal Ontogeny**	192
3.6	**Non-Vocal Sounds**	193
3.7	**Echolocation**	194

C. Erbe (✉) · M. Aulich · C. Browne · B. Elsdon ·
E. K. Evans · A. Gavrilov · C. Gosby · L. Hawkins ·
C. Jolliffe · P. Nguyen Hong Duc · C. Wei
Centre for Marine Science and Technology, Curtin
University, Perth, WA, Australia
e-mail: c.erbe@curtin.edu.au; A.Gavrilov@curtin.edu.au;
paul.nguyenhongduc@curtin.edu.au; chong.wei@curtin.edu.au

A. Murray
Division of Marine Mammal Research, Bureau of Marine
Science, Maine Department of Marine Resources, West
Boothbay Harbor, ME, USA
e-mail: anita.murray@maine.gov

A. Bowles
Hubbs-Sea World Research Institute, San Diego, CA,
USA
e-mail: ABowles@hswri.org

A. Frankel
Hawai'i Marine Mammal Consortium, Kamuela, HI, USA

© The Author(s) 2025
C. Erbe et al. (eds.), *Marine Mammal Acoustics in a Noisy Ocean*,
https://doi.org/10.1007/978-3-031-77022-7_3

3.8	Geographic Variability	195
3.9	Diel Patterns	196
3.10	**Decadal Changes in Sound Features**	198
3.10.1	Blue Whales	199
3.10.2	Fin Whales	199
3.10.3	Bowhead Whales	201
3.10.4	Hypotheses for Long-Term Changes	201
3.11	**The Effects of Noise on Acoustic Behavior**	202
3.12	**Acoustic Niches**	204
3.13	**Taxonomic Overview of Sounds**	206
3.13.1	Balaenidae	220
3.13.2	Neobalaenidae	224
3.13.3	Balaenopteridae	224
3.13.4	Eschrichtiidae	244
3.14	**Unidentified Sounds**	246
3.15	**Summary**	247
3.16	**Abbreviations**	248
3.17	**Data Availability**	248
	References	248

3.1 Introduction

All the species of marine mammals studied-to-date produce sound. All marine mammals produce sounds for communication purposes (to coordinate group behavior, attract sexual mates, facilitate mother-offspring recognition, etc.). Some species use sound for navigational purposes; specifically, odontocetes emit echolocation signals for navigation and ranging. Other sounds are made by many species without the use of structures and airways adapted for sound production. Such non-vocal sounds include jaw clapping, baleen rattling, fluke slapping, and breaching, which also are likely to serve in communication.

The sounds produced by marine mammals cover nearly four orders of magnitude in frequency, from tens of hertz (in the case of some mysticete songs) to over 200 kHz (in the case of some odontocete echolocation clicks). The range in signal duration is even greater: six orders of magnitude, from tens of microseconds (in the case of echolocation clicks) to tens of seconds (in the case of some mysticete and polar seal vocalizations). Sounds range from brief and broadband to long and narrowband. Sounds can be tonal and constant-frequency or frequency-modulated. Individual sounds can be amplitude-modulated or pulsed and can occur in bursts or trains (i.e., a repetitive series of similar sounds). Sounds can be arranged into distinctive rhythmic patterns that can be repeated for minutes to hours (e.g., mysticete songs or sperm whale codas).

The variety of marine mammal sounds is great and has evolved over millions of years. Ecological drivers (relating to both physical and biological environments) have shaped both the diversity and the specialization of marine mammal sounds. For example, some small odontocetes that live in shallow, coastal habitats use narrowband high-frequency signals at relatively low source levels for echolocation (biosonar). Such signals optimize biosonar performance in acoustically cluttered environments (i.e., those having lots of sound scatterers like suspended sediments and entrained air bubbles). At the opposite end of the acoustic spectrum, Antarctic blue whales, which cross the deep ocean on their annual migration to and from Antarctica, mostly alone or in small groups, use very-low-frequency sounds at high source levels, maximizing communication range in their habitat. Moreover, marine habitats may exhibit characteristic ambient noise; for example, wind and thus breaking waves in shallow water result in low-frequency noise from a few hertz to a few hundred hertz. In the open ocean, the wind-driven noise produces a peak at 300–500 Hz. Animal sound production likely evolved to avoid such predominant peaks in the ambient spectrum and animals likely partitioned the available communication space amongst themselves. This is the idea behind the acoustic niche hypothesis. Communication space, in this hypothesis, is multi-dimensional and includes aspects of frequency, modulation, timing, and location, which may have been selected to avoid overlap with other sound-producing species and with the dominant ambient noise (e.g., Schoeman et al. 2022). With the relatively recent advent of anthropogenic underwater noise, changes to marine mammal sounds have been documented.

While there is a great variety among marine mammal sounds, there are acoustic features by which sounds can be attributed to species. Indeed, the repertoires of many marine mammal species have been well described. In several cases, even the sounds produced by populations belonging to the same species can be differentiated (e.g., in the case of blue whales). For a few species, we can even tell different family groups apart by their distinct sound features (e.g., in the case of killer whales and sperm whales), and among some dolphins and pinnipeds, individually distinctive "signature" vocalizations have been identified. The production and reception of acoustic signals are at the core of marine mammal communication systems in support of critical life functions such as reproduction and foraging. The following chapters describe the sounds of the various marine mammal species. This chapter focuses on the sounds made by mysticetes.

3.2 Sound Production

Both mysticetes and odontocetes generate underwater sounds using a pneumatic (i.e., air-driven) mechanism; however, the sound generation processes, internal sound propagation pathways, and resulting sound features differ. While odontocetes produce sounds using a pair of phonic lips located near the posterodorsal terminus of the melon, mysticetes produce sounds laryngeally, similar to terrestrial mammals. Given that mysticetes are not accessible in aquaria for controlled experiments, and the large heads of adults cannot fit into medical imaging scanners to obtain detailed geometrical information for acoustic numerical modeling, our knowledge of sound production mechanisms in mysticetes mostly relies on morphological and anatomical information obtained from stranded individuals (e.g., Cranford and Krysl 2015; Haldiman and Tarpley 1993; Hosokawa 1950; Quayle 1991; Sukhovskaya and Yablokov 1979).

Early anatomical work suggested that whales lacked vocal folds (also called vocal cords) that are the structures responsible for sound production in terrestrial mammals (Beauregard and Boulart 1882; Benham 1901). And so, mysticetes were considered incapable of vocal sound production until the first humpback whales were recorded (Schreiber 1952). Researchers then looked for the potential sound source in mysticetes but initially focused on the upper respiratory structures (as in odontocetes) rather than the larynx (Green 1972; Matthews 1978). Some studies suggested the larynx as the sound generator and the lungs as the resonator (Barham 1973), but later studies confirmed that vocal folds are consistently present in mysticete species (Reidenberg and Laitman 2007).

Mysticetes have two laryngeal components related to sound production: vocal folds and a laryngeal sac. Recent anatomical studies showed that mysticetes have a U-shaped pair of vocal folds (U-fold) located in the lumen of the larynx (Cazau et al. 2013; Reidenberg and Laitman 2007). After interspecific comparisons of the position, attachments, and composition, it was found that the U-fold was structurally homologous to the vocal folds of terrestrial mammals (Reidenberg and Laitman 2007). The laryngeal sac is an air sac located at the ventral aspect of the larynx (orange region in Fig. 3.1a). Two symmetric lungs are connected to the trachea, a short and broad cone-shaped canal. The vocal folds are oriented parallel to the long axis of the trachea

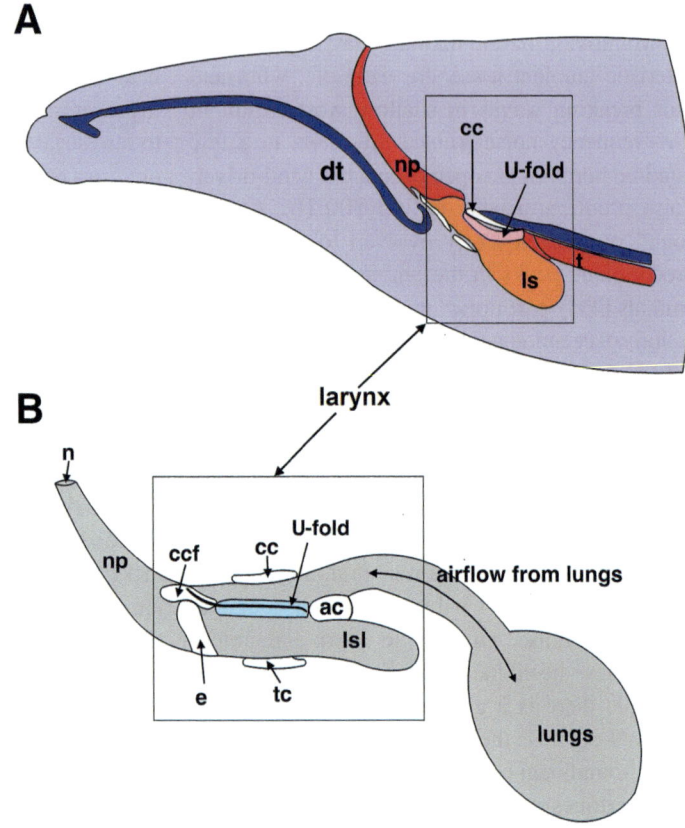

Fig. 3.1 Sound production in mysticetes. (**a**) Schematic diagram of the larynx and vocal tract in a typical mysticete whale. (**b**) Diagram of midsagittal anatomy of the respiratory system. *dt* digestive tract, *np* nasal passages (paired), *n* nasal plugs (paired), *ccf* corniculate cartilage flaps, *e* epiglottis, *cc* cricoid cartilage, *tc* thyroid cartilage, *ac* arytenoid cartilage, *ls* laryngeal sac, *lsl* laryngeal sac lumen, *t* trachea, *U-fold* vocal folds

(airflow direction), different from those of terrestrial mammals in which the vocal folds are perpendicular to airflow. Repeated rapid adduction/abduction and elevation/depression of the vocal folds may play an important role in controlling airflow. The vocal folds are controlled by tall cartilages (epiglottic and corniculate, Fig. 3.1b) to seal the entrance of the larynx (to protect the larynx from water incursions), leading to the diversion of airflow from the trachea into the flexible laryngeal sac. Airflow through the gap surrounded by the U-fold creates low-frequency vibrations on the edge of the vocal folds. The nasal cavities (main resonator) and the laryngeal sac act as acoustic resonators to resonate the sounds (Adam et al. 2013). The vibrations of the laryngeal sac walls are transferred through the overlying soft tissues (e.g., throat muscles, blubber, and skin) into the surrounding water (Adam et al. 2013; Cazau et al. 2013; Reidenberg and Laitman 2007). Sound production can continue without surfacing because the mysticetes recycle the air back into the lungs to produce sounds repeatedly.

Of course, not all mysticetes produce the same sounds. A large-sized larynx comes with larger resonating organs to generate louder and lower-frequency sounds, and hence, the body size imposes a limit to the frequency of sounds produced. There are additional differences between balaenopterids and balaenids with regards to laryngeal anatomy and physiology, contributing to intraspecies differences in mysticete sounds (Reidenberg 2022). For example, balaenopterids have longer and thinner vocal folds than balaenids. Balaenopterids appear to have specialized tissues around the vocal folds that might be able to direct air flow along different paths across the folds, resulting in different sound types.

To better relate the variations in laryngeal anatomy to the stereotypical sounds of mysticete species, researchers have used biomechanical and acoustic models based on anatomical data. Adam et al. (2013) constructed a theoretical acoustic model for the humpback whale sound production system based on anatomical studies to explain the vibrator's mechanics and the airflow directions. Their model suggested that the lungs and laryngeal sac were the two sound generators (air sources) for humpback whales to produce song units. In the model, the air was moved bidirectionally between them through the trachea to vibrate the vocal folds and the sound then resonated in the nasal cavities (the main resonator). Humpback whales might change the air volumes of the lungs and laryngeal sac to manipulate the airflow (e.g., volume, pressure, and velocity) in both directions. Therefore, the characteristics of humpback whale song units might be related to the physiological and biomechanical properties of the anatomical components. The latest research by Elemans et al. (2024) combined in vitro experiments and a fluid-structure interaction computational model of the larynges of subadult common minke, humpback, and sei whales to show that mysticetes evolved unique laryngeal structures for sound production. They found that the U-fold pushed against the cricoid cushion (a large wedge-shaped fat body that occupies the dorsal inner surface of the cricoid) on the inside of the larynx, creating vibration when the air from the lungs passed the cushion, allowing mysticetes to efficiently produce frequency-modulated, low-frequency calls.

For blue whales, Aroyan et al. (2000) modeled a monopole sound source to generate strong, low-frequency B-calls. They proposed that a Helmholtz resonator was more applicable than simple bubble resonance, since Helmholtz resonance provides efficient constant frequency over changing depths in dive and ascent by facilitating air movement back and forth through the nasal passages. Similar to humpback whales, blue whales might also apply muscular contractions to control the oscillating valve to modify the characteristics of the calls. Dziak et al. (2017) designed a pulse production model for blue whales and compared it to the previous humpback whale sound production models (Cazau et al. 2013). They proposed that the harmonic sounds of blue whales were mainly produced by pneumatic airflow through the respiratory valves: epiglottis, corniculate flaps, and U-fold. The laryngeal sac and nasal passages are likely contributors to the resonance during airflow. The airflow might be facilitated by changes in hydrostatic pressure during dive or ascent, allowing the whales to reduce energy to make B-calls while maintaining the sound intensity over a relatively long duration (~10–20 s).

In comparison, while the odontocetes generate sound with a highly specialized nasal region, the ancestors of modern mysticetes developed a highly modified larynx that produces low-frequency sound and infrasound, which is beneficial for communication over long distances.

3.3 Song

3.3.1 Song Description

The first technical description of mysticete song was published by Payne and McVay (1971) for the humpback whale. Song consists of sound units that are arranged into rhythmical patterns. A unit refers to a single sound that is continuous to the human ear; although units may be composed of subunits. A phrase refers to a distinct sequence of units that are repeated at fixed intervals. Multiple phrases make up a theme, and multiple themes make up a song (Payne and McVay 1971; Fig. 3.2).

The classification of units is often subjective, done either aurally or visually (by comparing spectrograms). While frequency and duration measurements can be used in clustering algorithms to make the process more objective and reproducible, there are several factors that are difficult to control. The same unit recorded in different environments might sound and look quite different. The ocean acts like a filter, and different environments attenuate energy at different frequencies (see Chap. 1 and Erbe et al. 2022b). The range of the animal to the recorder affects the spectrum, and therefore, far-away

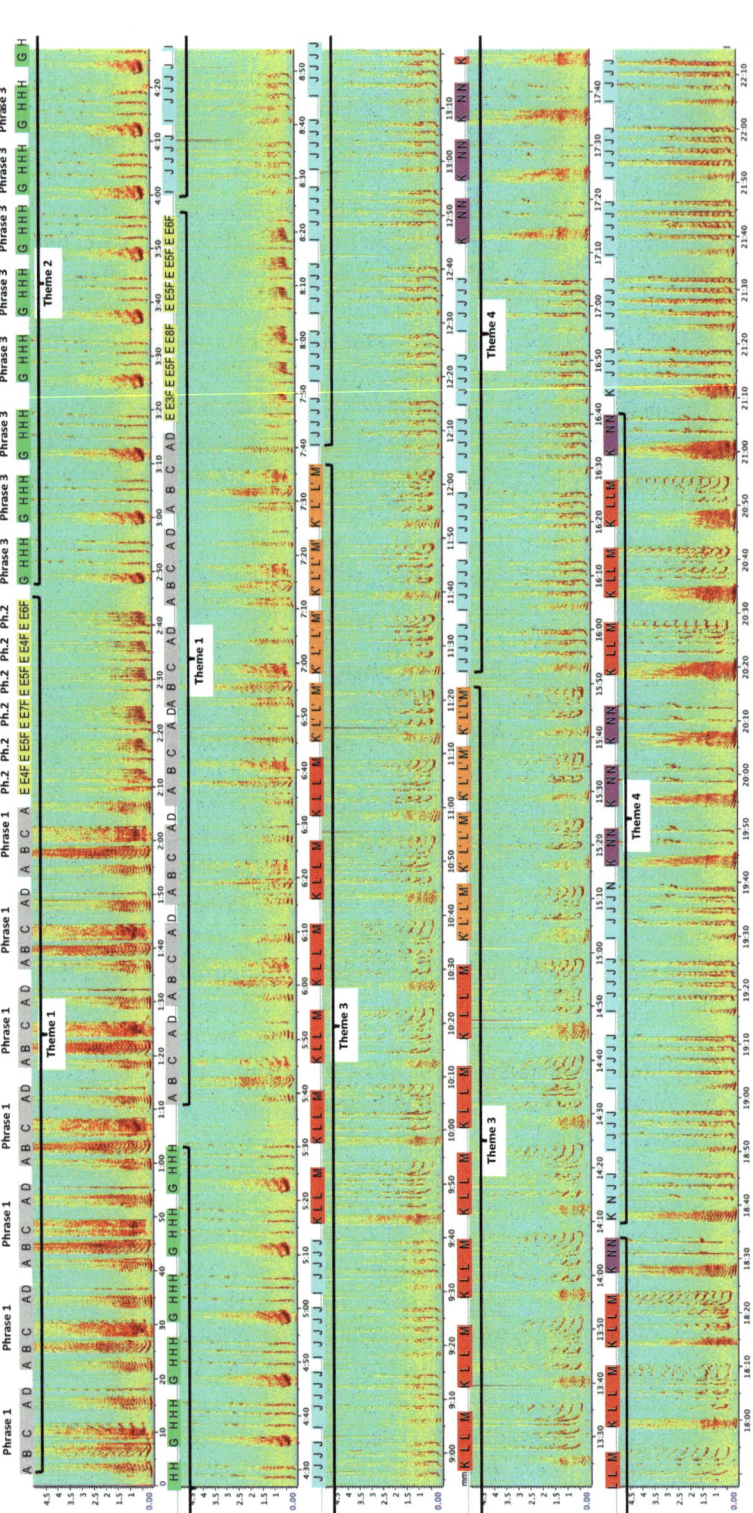

Fig. 3.2 Example of Western Australian humpback whale song from the winter of 2012. Units are labelled with capital letters. Phrase 1 (gray underlay) contained five units in the pattern ABCAD. On the occasion shown in the image, the final D was missing before the animal commenced Phrase 2. Phrase 2 (yellow underlay) always started with two E-units, followed by a packet of 3–8 F-units. Phrase 3 (green underlay) was a consistent GHHH. Phrase 4 (light blue underlay) was IJJJJ, though during the later recordings (bottom row), the initial I was replaced by a K, the first J by an N, and one of the Js was dropped. Phrase 5 (red underlay) was a consistent KLLM. Phrase 6 (orange underlay; K'L'L'M') was similar to Phrase 5 but slightly shorter in duration and with more emphasis on the lower frequencies. Phrase 6 was always preceded by Phrase 5, and sometimes Phrase 5 seemed to gradually change into Phrase 6. Phrase 7 (purple underlay) was KNN. Theme 1 consisted of at least seven repetitions of Phrase 1, followed by typically six repetitions of Phrase 2. Theme 2 only consisted of repetitions of Phrase 3. Theme 3 started with at least six repetitions of Phrase 4, followed by about six Phrases 5 and four Phrases 6. Theme 4 also started with repetitions of Phrase 4, but then had a new Phrase 7 before Phrases 5. Phrases 6 were dropped from Theme 4. Note that recordings were duty cycled (5 min every 15 min); ship noise often masked the song and the corresponding sections were removed. Recordings from three separate days were concatenated to produce this hypothetical song example for illustrative purposes only. While the sounds are displayed in the order in which they were recorded, the image does not reflect a complete song. Concatenated song snippets were likely produced by different individuals

sounds might contain much less energy at high frequencies. The location of the animal would need to be known and the propagation of its sound would need to be modeled to account for these effects—impractical in most settings. Grouping units into phrases and phrases into themes is equally subjective, and we do not know if animals perceive similar groupings.

Unfortunately, the terminology is not consistent within the mysticete song literature. For example, when describing blue whale song from the North Atlantic, Mellinger and Clark (2003) followed the Payne and McVay (1971) definition of unit; therefore, sounds that were separated by an interval of silence were categorized as two different units, although they used the term "parts" in place of units. On the other hand, McDonald et al. (2006, p. 56) split blue whale sounds into two units at a significant change in call character (e.g., frequency, sweep rate, or modulation rate), even if there was no interval of silence in the middle. Their definition of "unit" is, in fact, equivalent to the definition of "subunit" in the humpback whale literature. Subunits are components of a unit that may not be distinct to the human ear since the subunits are not separated by a significant interval of silence, but may be seen as discrete components of a unit within a spectrogram (e.g., delineated by a discontinuity in frequency; Pace et al. 2010; Payne and McVay 1971). What is fairly consistent though is that song units of tonal character are referred to as "notes" (e.g., Clark and Johnson 1984).

There has also been inconsistent terminology used to describe the arrangement of units into sequences. Some authors in accordance with the Payne and McVay (1971) definition use the term "phrase" to refer to unit sequences that are repeated within song (e.g., McDonald et al. 2006; Mellinger and Clark 2003). However, in the literature on blue and fin whale sound production, for example, units and phrases have historically been referred to as "calls", even if repeated in song (e.g., the blue whale Z-call and AB-call). This word choice might be due to these species having apparently more simplistic song (than, e.g., humpback whales) consisting of less than a handful of different sounds in a simple pattern, or because these sounds might also occur outside of song, or because early studies on these species were focused on stereotypical sound types, rather than song. An example of the latter would be the "20 Hz pulse" of fin whales, as the term was intended to signify a stereotypical single sound type (a pulse) rather than its role (e.g., a unit in song). Moreover, in the early literature, the term "song" was restricted to vocalizations recorded during the breeding season on breeding grounds.

Differences in terminology aside, song has been described for bowhead, right, minke, sei, blue, Omura's, fin, and humpback whales. While most species (as described so far) seem to use a very limited number of stereotypical units in their songs, one balaenid species (i.e., the bowhead whale) and one balaenopterid species (i.e., the humpback whale) have a great repertoire of units. Moreover, only bowhead and humpback songs seem to change over time on a scale of weeks, seasons, and years. A major distinction between the songs of humpback whales and those of bowhead whales is that from season to season, humpback song comprises similar units and phrases, whereas bowhead song units and phrases change, showing less inter-annual repetition. Also, all humpback whale singers within a particular population end up singing the same song within any one season, whereas in bowheads, individuals likely sing different songs and more than one song type in any one season (Stafford 2022).

3.3.2 Whale Choruses

A chorus of whale sounds can be heard and seen in spectrograms of ocean sound when many individuals of the same species sing within the same time window in the same region. While individual sounds are produced at different ranges, azimuths, and slightly different times, they are combined into a chorus at the sound receiver. In spectrographic displays, the longer the averaging time, the more regular the whale chorus looks. Whether chorusing in whales adds an additional functional advantage to the function of song alone (e.g., if singing together attracts more females than singing alone) is uncertain. A chorus raises ambient noise levels in the frequency band characteristic of the species-specific sounds making up the song. Distinctive individual whale sounds produced by nearby whales can sometimes be seen above the background chorus of sounds from distant whales (Fig. 3.3).

The band level of the chorus has often been assumed to relate to the abundance (or spatial density) of the singers. Examples in the literature include studies on humpback whales (e.g., Au et al. 2000; Seger et al. 2016), Antarctic blue whales, and fin whales (e.g., Širović et al. 2004). In fact, Kügler and team (2021) demonstrated the correlation between chorus level and humpback whale abundance and density on the Hawaiian breeding grounds. Elsewhere, changes in the chorus band level over the course of a year have been used to identify the seasonal presence of whales and the timing of migrations (i.e., start, peak, and end of migration; e.g., Erbe et al. 2015).

3.3.3 Song Function

Song function in mysticete whales remains an area of active research. It has been associated with sexual selection, social communication, navigation, and foraging.

3.3.3.1 Breeding Behavior

The most common hypothesis is that song is a mating strategy of males that acoustically displays and advertises the singer's fitness, so that females may select a desirable mate. In support of this hypothesis, all singers for whom the sex has been determined have been male (e.g., in blue whales, McDonald et al. 2001; fin whales, Croll et al. 2002; and humpback whales, Darling and Bérubé 2001). While females may produce similar sound types, they have not produced the patterns attributed to an advertisement song in field recordings. This sex-specific sound production is analogous to some songbird species in which males produce elaborate songs and females either do not sing or produce simple songs (Nowicki et al. 1998). Singing in birds is an example of sexual selection: Female birds repeatedly selected males with more complex vocalizations over many generations (Catchpole 1987). In humpback whales, singers were significantly more likely to join other humpback whale groups that contained a mother-calf pair than those that did not. Moreover, males started to sing after joining mother-calf pairs without escorts, and singers spent a greater percentage of time singing when with mother-calf pairs than in other group settings (Smith et al. 2008).

Singing to entice females might happen competitively or cooperatively. In bowhead and right whales, different individuals seem to sing different songs, and so, complexity or repertoire size could correlate with fitness (Crance et al. 2019; Tervo et al. 2011b). In humpback whales, individuals sing the same song in one season and at one location, perhaps as a form of lekking (Darling et al. 2006). According to the lekking hypothesis, a larger chorusing group of male whales will have more success at attracting females, compared to an individual singing male (Herman 2017). Moreover, despite song conformity, male humpback whales can produce individually distinctive patterns that could convey individual qualities to females (Lamoni et al. 2023; Murray et al. 2018). Bowhead and humpback whales have the complexity of their songs in common, in terms of the number of different units and their arrangements. Perhaps this is an example of convergent evolution for species mating in small-scale breeding aggregations (Tervo et al. 2012).

In further support of the breeding function hypothesis, song occurs more frequently during the breeding season and on breeding grounds (e.g., in bowhead whales, Tervo et al. 2009; fin whales, Watkins 1981; and humpback whales, Au et al. 2000). Here, as documented in some birds, male song might stimulate ovulation in females (Herman 2017).

Blue whales might not have localized, restricted, or exclusive breeding grounds, and reproductive activities might happen over large areas and, therefore, more opportunistically (e.g., Schall et al. 2020). In this species, communication may have to be effective over great distances,

3 Mysticete Sounds

Fig. 3.3 Spectrograms of whale choruses. (**a**) Four single spot calls (red spots at 22 Hz) standing out in a chorus of spot-calling whales (thin turquoise band at 22 Hz) (fs = 6 kHz, NFFT = 12,000, Hamming window, 50% overlap), South Australia. (**b**) Three Antarctic blue whale Z-calls standing out in an 18–28 Hz band of Z-call chorus (fs = 250 Hz, NFFT = 500, Hamming window, 50% overlap). (**c**) A long-term average spectrogram showing an Antarctic blue whale Z-call chorus identifiable by the strong bands of at 18 Hz and 28 Hz, and a 3 h fin whale chorus (20-30 Hz) in the late afternoon of 26 August (fs = 250 Hz, NFFT = 500, Hamming window, 50% overlap). Recordings B and C are from the Comprehensive Test Ban Treaty Organization's (CTBTO) International Monitoring System's hydroacoustic station at Cape Leeuwin, Western Australia

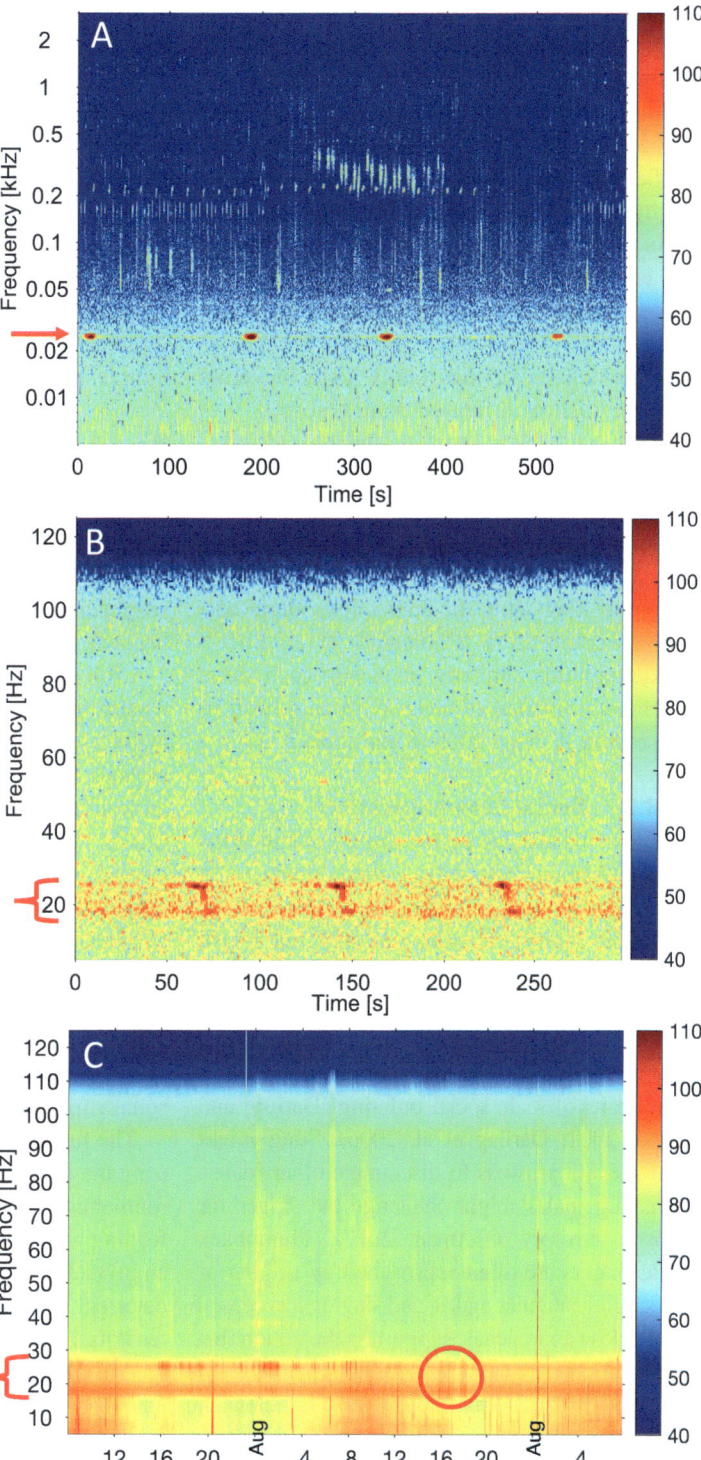

and repetitive sound production (song) may create reproductive opportunities, aiding discoverability by communicating location and readiness to breed. Blue whale songs have been recorded over hundreds of kilometers in the North Pacific, consistent with the long-distance attraction hypothesis (Stafford et al. 1998).

As part of breeding behavior, male song might also be directed at other males. Observations of male-male interactions, where at least one was a singer, have ranged from cooperative (i.e., "non-agonistic whales acting jointly to achieve an objective such as increased mating success") to agonistic (i.e., "clear agonistic threats and fights between whales including exaggerated lunges, throat expansion, body blocking, physical blows with flukes and often minor injuries"; Darling et al. 2006). The cooperative alliance might be a form of lekking; agonistic interactions might be a form of competition (Cholewiak et al. 2018). Some songbirds similarly challenge each other by temporal overlap of their songs as a form of competition (Hyman 2003; Kunc et al. 2006).

3.3.3.2 Social Communication

Singing could also have a broader communication function than direct sexual selection. Males might use song to generally moderate their interactions with other males. For example, male humpback whales altered their song presentation in the presence of other singers (changing song evenness and phrase-type switching rate; Cholewiak et al. 2018). Song might be used to display dominance and thus be a means of male social ordering (Darling and Bérubé 2001; Darling et al. 2006). Song might communicate prowess to discourage other males, or visiting males might challenge the singer for acoustic territory (Herman 2017). Humpback whale singers are often approached by singing or non-singing singular males, and singing has ceased during close approaches, supporting the notion that singing males communicate over greater-than-visual ranges in male-male interactions (Cholewiak et al. 2018). Alternatively, males that join singers may be prospecting for females; thus, a potential cost of singing is the attraction of competing males (Smith et al. 2008).

3.3.3.3 Navigational Tool

Whales that sing outside of the breeding season might utilize songs to coordinate their long-distance travel among individuals that are out of sight, so they may move together between their grounds. Humpback whales sing throughout their migration, including in their migratory corridors (e.g., Gosby et al. 2022). Different humpback whale songs propagate over different ranges and some units might serve as migratory beacons signaling routes of migration (Clapham and Mattila 1990).

3.3.3.4 Foraging Behavior

Some mysticete species are suspected of utilizing food-associated songs on their feeding grounds. The bowhead whale has distinctly different songs between the breeding and feeding season, suggesting that different songs are for different behaviors (Tervo et al. 2009).

Croll et al. (2002) genetically sexed singing fin whales in a foraging area and hypothesized that males use song to attract females to prey patches, perhaps reporting their ability to find prey as an indication of their fitness to a female. Clark et al. (2002) recorded both fin whale 20 Hz song and non-song 20 Hz calls in feeding areas; therefore, reproductive and foraging activity may not be temporally or spatially segregated in this species, and males may use song to convey information about food resources. Morano et al. (2012) suggested that males may use seasonal variability in song inter-note interval to communicate seasonal variability in food resources.

The functional significance of food-associated song may vary depending on the social and environmental conditions of the species. Songs in feeding areas may communicate to conspecifics an invitation to join, or they may be produced as a deterrent to avoid a food source, similar to a territorial song in birds. Songs that function to invite more conspecifics toward a food source may reduce the food-capture effort or may improve resource defense from competing predators (as seen in birds; Sridhar et al. 2009).

3.3.3.5 Multiple Functionality

Songs might serve different functions—in different areas, in different seasons, or at the same time (Parsons et al. 2008). Furthermore, different parts of the same song might serve different functions. Murray et al. (2018) hypothesized that complex phrases in humpback whale song that were structurally variable, individually unique, and composed primarily of mid-frequency units were used for short-range communication and could convey individual quality. Simple phrases that were stereotypical in structure, shared amongst multiple males, and composed of low-frequency units were for long-range communication and could convey group membership. Similarly, in sei whales, the more complex mid-frequency units might function over short ranges (where females evaluate singing males, or males assess each other), while low-frequency units might merely communicate the presence of the singer (Cerchio and Weir 2022). Blue whales use different song types inshore versus offshore in waters off southern California; offshore, more B-units occur and travel farther, and more solitary whales are observed—all hinting at differential functions of different units within their song (Lewis and Širović 2018).

3.3.3.6 How to Determine Function

To determine the functions of song or non-song sounds, multiple data streams are needed in addition to passive acoustics: simultaneous visual observations, passive acoustic tracking, tagging, and/or collection of tissue (biopsy) for sex determination. The context in which certain vocalizations are emitted requires characterization. A large sample size is needed, because the same sound types have been recorded in quite varied contexts. Moreover, given the relatively small number of sound categories that bioacousticians have described, we may be missing nuances that could distinguish more functional categories.

For example, through acoustic tagging, visual observations, and biopsying, Oleson et al. (2007a) showed that blue whale AB song was sung by lone, traveling males; non-song AB calls were recorded in the presence of pairs; and D-calls were recorded from both sexes during foraging. Through acoustic tracking and tagging, we know that blue (and other baleen) whales only produce sound—whether song or non-song—at relatively shallow depths (~10–40 m below the sea surface; Bouffaut et al. 2021; Lewis et al. 2018; Thode et al. 2000). Blue whales sang during highly stereotypical shallow dives. Non-song sounds, such as single A- or B-calls, were produced in between deep (likely feeding) lunges (Fig. 3.4; Lewis et al. 2018). In Australia, too, highly stereotypical patterns in dive profile (including dive depth, fluking, and pitch) have been described and correlated with song pattern (Davenport et al. 2022). Such links between physical and acoustical behavior may ultimately be used to infer functional behavior from passive or remote acoustic recordings.

3.4 Non-Song Sounds

3.4.1 Description of Non-Song Sounds

Sounds, when not arranged into song, may—in the most general sense—be referred to as non-song sounds. In the broader animal communication literature, in particular based on decades of research on acoustic communication in birds, these sounds are more commonly referred to as calls (Bradbury and Vehrencamp 2011). However, some mysticete researchers reject this term, as it implies a function (i.e., calling others), which remains unproven for the majority of mysticete sounds. In some articles, non-song sounds are called "social sounds" as they were recorded in social encounters, without demonstrating a calling function, and often without demonstrating a social function either (e.g., Dunlop et al. 2013).

Some sounds may be recorded both as standalone calls and as units arranged into song (e.g., blue whale A and B calls). But not every song unit also exists as a call, and not every call is incorporated into song. For example, the fin whale 40 Hz downsweep call has not be recorded as part of song. However, non-song sounds fall into the same broad sound types as song units;

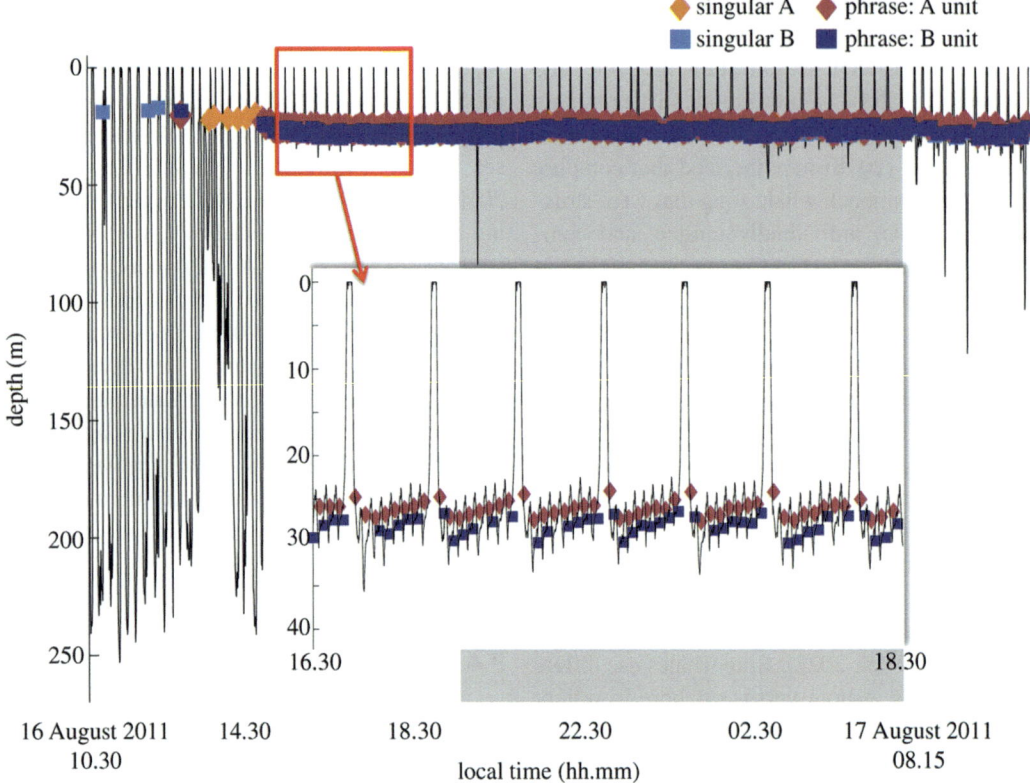

Fig. 3.4 Dive profile of a blue whale tagged in the Northeast Pacific, overlain with song and non-song production, which was limited to the top 40 m of water. Notice the stereotypical dive pattern matched to A- and B-units in song. Reprinted from Lewis LA, Calambokidis J, Stimpert AK, Fahlbusch J, Friedlaender AS, McKenna MF, Mesnick SL, Oleson EM, Southall BL, Szesciorka AR, Širović A (2018) Context-dependent variability in blue whale acoustic behaviour. R Soc Open Sci 5 (8):180241. https://doi.org/10.1098/rsos.180241 © Lewis et al. 2018. Published CC BY 4.0; https://creativecommons.org/licenses/by/4.0/

they can be tonal, frequency-modulated, amplitude-modulated, pulsive, or noisy (broadband, non-harmonic, longer-duration, without any temporal or spectral structure).

3.4.2 Function of Non-Song Sounds

Non-song sounds have been recorded in all behavioral contexts, including reproduction. Moreover, the same calls have often been recorded during different behaviors, with no unique match between call and behavior that is discernible. However, some calls seem to be more frequently produced in certain settings than in others.

3.4.2.1 Feeding Calls

Humpback whale behavior on their northern feeding grounds has been studied by many authors. Humpback whales engage in cooperative feeding events, during which prey (e.g., herring) is herded into a ball surrounded by a ring of exhaled bubbles and then engulfed by individuals that lunge through the ball of prey. During this coordinated activity, humpback whales emit cries in rhythmic patterns (Cerchio and Dahlheim 2001; D'Vincent et al. 1985; Qing et al. 2019).

In blue whales, the downsweeping D-call is commonly associated with feeding behavior. The D-call seems to be produced by all blue whale subspecies, by both sexes, and at irregular intervals. The rate of D-calling increased on feeding grounds, during foraging, in particular during shallow dives in between deep, lunge-feeding dives (McDonald et al. 2001; Oleson et al. 2007a, b). However, in a different study, D-calls were recorded in both diving and non-diving contexts and so might serve multiple functions (Lewis et al. 2018).

Another suspected feeding call is the fin whale 40 Hz call. Observations by Širović et al. (2013) indicated that 40 Hz calls were primarily produced by Pacific fin whales preoccupied with feeding. Furthermore, Romagosa et al. (2021) demonstrated a correlation between 40 Hz calling rates and prey biomass. However, Edds (1988) recorded 40 Hz calls from pairs of fin whales in the St. Lawrence, implying a social context.

3.4.2.2 Contact Calls

The right whale upcall is a candidate contact call, which keeps whales in acoustic contact with each other. It was used in call–counter-call sequences, after which lone whales were joined by others whales, upon which upcall production ceased (Clark 1982). Upcall production was high in surface-active groups, which suggests a social function (Parks and Tyack 2005). Fin whale 20 Hz calls have been used in call–counter-call sequences, presumably to maintain contact between fin whales over large distances (McDonald et al. 1995). Blue whale D-calls have also been recorded in call–counter-call exchanges (McDonald et al. 2001).

Fournet and team designed a playback experiment to specifically test whether the humpback whale "whup" was a contact call. Broadcasted whups elicited whup replies 100% of the time, but did not result in approaches of whupping whales to the speaker. Those observations indicate that while the whup likely is a contact call for acoustic contact between separated individuals, it is not an aggregation signal (Fournet et al. 2021).

3.4.2.3 Mother-Calf Calls

Mother-calf pairs generally exhibit lower calling rates and call at lower levels (Parks et al. 2019b). Acoustic tags deployed on humpback whale mothers and calves by Saloma and team (2022) recorded numerous call types. However, tagged whales were not followed during the study to avoid interference, and so group composition might have changed (e.g., by the addition of escorts), and the signaler could not be confidently identified. In a different study, only low-arousal, single calls were recorded from lone female-calf pairs, while in the presence of one or more escorts, call rate increased, calls occurred in bouts, and a greater diversity of calls was recorded (Cusano et al. 2022). Acoustic tags have also been deployed on females with calves, when no other adults were around, and group composition was monitored from shore (Indeck et al. 2020). Calls clustered into lower frequency and longer duration (presumably produced by the mothers) versus higher frequency and shorter duration (presumably produced by the calves). These observations indicate that spectral features might reflect age and body size. Contact calls between mothers and calves were also recorded from Bryde's whales in the Gulf of California (Edds et al. 1993). In right whales, the upcall might be used to maintain contact between mothers and calves, as it was the predominant call type on calving grounds (Clark 1982; Dombroski et al. 2016). Given the many different contexts in which upcalls have been recorded, we might be unaware of features that differentiate its functions.

3.4.2.4 Calls Signifying Motivational State and Arousal

Based on his research on southern right whales, Clark (1982) suggested a call's spectral features (i.e., peak frequency, frequency-modulation rate, amplitude-modulation rate, and pulsiveness) may indicate levels of excitement and aggression.

Through simultaneous visual and acoustic observations of humpback whales, non-song sounds were associated with differing behavioral contexts as defined by group size, composition, and behavior (Dunlop 2017; Silber 1986). Dunlop clustered sounds according to frequency and duration measurements, and matched clusters to motivational state. Observed groups included single adults, female-calf pairs, female-calf pairs with escorts, and various combinations. Low-arousal signals were of low frequency and unmodulated, or simple upsweeps. These sounds were recorded from single adults, female-calf pairs, or female-calf plus escort groups. Affiliating groups used higher-arousal signals, which were higher in frequency and bandwidth. Adult groups produced a higher proportion of aggressive signals (growls, screeches, and underwater blows; Dunlop 2017).

In both South Pacific and North Atlantic blue whales, D-calls were produced at high rates during rare observations of reproductive encounters involving one female and two males, with male-male aggression observed (Schall et al. 2020). Given the various contexts in which D-calls have been recorded, there might be differences between D-calls that have gone unnoticed by the human ear and eye.

3.4.2.5 Calls Signifying Individual Identity

It has been shown that mysticete calls differ across individuals of the same population and hence might communicate individual information. In fin and minke whales in the St. Lawrence, sweep rates of downsweeps varied among recordings with different individuals as did initial frequencies when multiple individuals were in the same area (Edds 1988; Edds-Walton 2000). Acoustic tags deployed on right whales showed great variability in upcall features across age classes and individuals (McCordic et al. 2016). Humpback whale feeding cries differ between individuals within a cooperative feeding group and so might carry individual information that helps coordinate group behavior (Cerchio and Dahlheim 2001). Cries within song also seem to vary between individuals (Hafner et al. 1979).

3.5 Vocal Ontogeny

Mysticetes are believed to develop their vocalizations (whether in song or not) through vocal learning. Vocal learning occurs when a calf learns the stereotypical vocalizations of its species, when a juvenile learns its species' song, and when adults change their songs from year to year. Vocal learning comprises three key interrelated elements (Carouso-Peck et al. 2021; Janik 2014; Janik and Knörnschild 2021; Verpooten 2021). First, vocal production learning involves the modification of vocalizations based on an experience with signals produced by conspecifics. Second, learning how to use these vocalizations in context is referred to as contextual vocal learning. Third, comprehension learning requires a level of perceptual and cognitive capacity to recognize, interpret, and respond to vocalizations produced by others. Vocal learning is considered to be the driver of vocal culture and song displays in mysticetes (Verpooten 2021).

Better documented in songbirds, vocal learning involves the imitation and learning of a song model during a critical development period. The individual forms an auditory template memory upon exposure to a sound, and the vocalization is practiced until it matches the template. Vocalizations are practiced from an early age until the stereotyped parameters and pattern of song have been achieved (Bottjer et al. 1985; Isaac and Marler 1963; Nottebohm 1970; Verpooten 2021).

Our knowledge of vocal learning in mysticetes remains relatively scarce. Evidence primarily stems from the elaborate song displays of male humpback whales and how they change over time (Carouso-Peck et al. 2021; Janik and Knörnschild 2021; Tyack 2020; Verpooten 2021). Within a humpback population, changes in song displays are synchronized. Changes to the song sequence are tracked and copied by each individual male, as all males in a breeding population eventually conform to the same song. The introduction of variants and modification of song displays is known to occur via social learning and cultural transmission (i.e., the spread of acoustic information between populations; Allen et al. 2018; Garland and McGregor 2020). For example, the eastern Australian humpback whale population is known to rapidly replace elements (or all) of its song display every few years with novel material learned from other humpback whale populations (e.g., the Western Australian population), which is then transmitted eastward to populations across the South Pacific (Allen et al. 2018; Garland et al. 2011; Noad et al. 2000).

A more recent example of vocal learning in humpback whales is from the western North Atlantic, where seasonal song ontogeny was documented within a single breeding season (Kowarski et al. 2022). Parallels between humpback whale song and bird song development were identified (e.g., canary, an open-ended learner that continues to learn novel songs as an adult), with both species producing fragments of song (or subsongs) at the onset of the breeding season, before progressing to non-stereotyped songs (plastic song or inconsistent theme order), and finally achieving stereotyped song during the peak breeding season (Kowarski et al. 2022; Nottebohm et al. 1986). This progression in male humpback whale song during the breeding season is comparable to the process of song-model learning suspected in juvenile whales and could be driven by testosterone levels (i.e., song may evolve progressively with increasing testosterone levels) or the need for communicating in different contexts, and therefore, could be important for song evolution and social learning (Cholewiak and Cerchio 2022; Kowarski et al. 2022).

Vocal learning is not restricted to complex song displays. Vocal ontogeny in the stereotyped contact calls of right whales has been observed across all age classes from calves to adults. Root-Gutteridge et al. (2018) observed calves producing chaotic vocalizations characterized by short duration and high frequency-modulation and non-linear phenomena. From the second year and with increasing age, increased control and durations of vocalizations were apparent, with biphonation and subharmonics being introduced, before reaching maturation and forming adult vocalization types after around 8 years of age. Variability in the acoustic behavior of calves and juveniles has been the topic of several studies. The distinct differences in calf and adult vocalizations in right whales (Parks et al. 2011a; Parks and Tyack 2005) and humpback whales (Indeck et al. 2020; Zoidis et al. 2008) suggest that vocalization production is refined with increasing age (and perhaps vocal learning) and that duration and frequency of vocalizations are reliable signals of age and physical condition (McCordic et al. 2016; Root-Gutteridge et al. 2018).

3.6 Non-Vocal Sounds

Mysticetes produce a variety of non-vocal sounds. These might be byproducts of physical behavior, such as baleen rattle during filter feeding, blow sounds during respiration, or flipper slap sounds and body slams during aerial displays at the sea surface. Some of these sounds might also serve in communication.

When mysticetes feed, the baleen may rattle, most likely as a result of water gushing through the baleen when the mouth is partly above and partly below water. As recorded from right, humpback, and gray whales on their feeding grounds, the sound is pulsive and noisy, with energy between 200 Hz and 4000 Hz, durations of 0.1–2 s, and estimated 100 dB re 1 μPa m source level (Charles 2011; Ollervides 2001; Thompson et al. 1986; Watkins and Schevill 1976).

Blow sounds are produced at the blowhole during exhalation. They may be produced above or below water. Similar sounds have also been reported during inhalation. Blow sounds are noisy and broadband, with energy between 10 Hz and 6000 Hz, duration up to several seconds, and source levels up to 140–150 dB re 1 μPa m (Clark 1982; Cummings et al. 1968; Dunlop et al. 2007, 2013; Frouin-Mouy et al. 2020; Parks and Tyack 2005; Webster et al. 2016). Watkins (1967) likened some humpback whale blow sounds to wheezes, which were recorded both above and below water. Thompson et al. (1986) recorded humpback whale blow sounds they described as shrieks, trumpets, and horn blasts. Source levels were 179–185 dB re 1 μPa m. Clark (1982) grouped southern right whale blow sounds into three types: normal, tonal, and growl. These sounds were produced when individuals were resting at the surface or swimming and by members of sexually active groups. Webster et al. (2016) sorted southern right whale blow sounds into three types called simple, tonal, and rumble. The rumble blows were produced mostly during times of elevated vocal activity. Parks and Tyack (2005) recorded North Atlantic right whale blow sounds in surface active groups. Various authors have speculated that in these settings, blow sounds may serve a communication function.

Some humpback whale populations perform bubble-net feeding, whereby a submerged whale creates a ring of bubbles while swimming in a spiral during a feeding maneuver. Krill and fishes are trapped inside and the whale rises through the center, engulfing its prey (Wiley et al. 2011). Multiple individuals may join in coordinated bubble-net feeding. The sound of bubble generation is broadband and pulsed (source level 162–181 dB re 1 µPa m); the sound of bubbles bursting at the sea surface can be quite impulsive (Thompson et al. 1986).

Emission of single bubbles, bubble blasts, and bubble trails has most commonly been described for humpback and gray whales. The associated sounds are broadband (~50–1000 Hz). Single bubbles are of short duration (<1 s), bubble blasts of longer duration (<5 s), and bubble trails of the longest duration (<10 s), with source levels of ~122–138 dB re 1 µPa m (Charles 2011; Crane and Lashkari 1996; Cummings et al. 1968; Dahlheim 1987; Frouin-Mouy et al. 2020; Ollervides 2001). Bubble blasts have been observed in gray whales during interactions with boats, possibly indicating disturbance or stress (Charles 2011; Ollervides 2001). They have also been recorded during tactile interactions between gray whale mother and calf (Frouin-Mouy et al. 2020) and during sexual interactions between humpback whale males and a female (Jones et al. 2022). Bubble trails have been reported in agonistic situations (Charles 2011; Dahlheim 1987; Moreno and Macgregor 2019; Ollervides 2001). Bubbles recorded under water might also be related to flatulence (Charles 2011; Ollervides 2001).

Sounds from impact of the whale's body at the sea surface are generated by breaching, tail fluke slapping, and pectoral fin slapping. The sounds are brief and broadband (<0.2 s, <12,000 Hz) with a great range in source level of 133–192 dB re 1 µPa m (Clark 1982; Dunlop et al. 2013; Thompson et al. 1986). Breaches are often singular events, whereas slaps occur in bouts (Dunlop et al. 2013). These energetic physical behaviors may have a communicative function.

3.7 Echolocation

Do mysticetes echolocate? Odontocetes are well known for their high-performance biosonar system. They actively emit ultrasonic clicks and listen for echoes. They use echolocation for orientation, navigation, and foraging (e.g., Au 1993). By contrast, the sounds produced by mysticetes are mainly known for intraspecific communication, mate attraction, mother-calf contact, feeding coordination, etc. (e.g., Edds-Walton 1997; Tyack and Clark 2000). Soon after the demonstration of echolocation in dolphins (Norris et al. 1961), researchers started to investigate whether mysticetes may also echolocate.

Initially, researchers looked for mysticete sounds that might be suitable for echolocation on small targets: brief, high-frequency clicks, possibly in trains. Candidate signals were reported from many species (Poulter 1968a), including minke whales (Beamish and Mitchell 1973; Winn and Perkins 1976), blue whales (Beamish and Mitchell 1971), fin whales (Perkins 1966), humpback whales (Stimpert et al. 2007), and gray whales (Asa-Dorian and Perkins 1967; Fish et al. 1974; Norris et al. 1977; Poulter 1968b; Wenz 1964). There was substantial speculation about whether the reported sounds might enable echolocation on prey (i.e., zooplankton and small fish). In addition, many bioacousticians in the marine mammal field argued that those studies failed to show convincingly that the source of the clicks were the mysticetes.

A controlled experiment to answer this question was undertaken by Beamish (1978). He performed a navigational maze experiment with a temporarily captive, 10 m female humpback whale using artificial targets (sonar-reflective aluminum poles) with diameters of 1.3–2.5 cm. The experiments were conducted under different conditions: with light (daytime and in the presence of an underwater flashlight at nighttime), without light (nighttime), and with or without blindfolds. Hydrophones were deployed at the time but no sounds were recorded. The humpback whale successfully navigated during light conditions, likely using sight. During darkness or when blindfolded, the animal collided with the targets. The results suggested that the humpback whale was unable to produce sounds for echolocation to detect and avoid objects.

Doubt remains whether any mysticete species can produce high-frequency (ultrasonic) clicks for small-target detection; however, focus has shifted to the potential for echo-ranging and navigation using their well-documented, low-frequency sounds, including tones, narrowband signals, pulses, and pulse trains (Norris 1969; Poulter 1968a; Thompson et al. 1979). Those low-frequency sounds are sufficiently intense to generate detectable echoes from large-sized (e.g., whale-sized and greater) objects, detectable above ambient noise over several kilometers. One of the stronger mysticete sounds is the right whale gunshot (~190 dB re 1 µPa m; Parks et al. 2005). In addition to playing an important role in communication, this pulsive sound might be useful for obtaining information on bottom topography or prey patches (Tyack and Clark 2000). Even the weaker humpback whale song units (150–170 dB re 1 µPa m; Dunlop et al. 2013) are commonly recorded with seafloor echoes that are well above ambient noise, and those echoes could be used for depth sounding (e.g., Tyack 1997; Winn and Winn 1978). George et al. (1989) observed that bowhead whales were able to avoid thick multi-year ice floes during swimming and find thinner ice layers to break through to breathe, suggesting that bowhead whales might use the echoes of their vocalizations to assess ice thickness and for under-ice navigation. Such hypotheses have been supported by acoustic models (e.g., Clark and Ellison 2004). Yi and Makris' (2016) model demonstrated that minke, blue, fin, and humpback whales could acoustically track the seafloor 1–2 km below. Yi and Makris further modeled that minke and humpback whales could echolocate herring shoals on the Gulf of Maine spawning grounds over 1 and 10 km, respectively. Frazer and Mercado (Frazer and Mercado III 2000) modeled the sound propagation of humpback whale song units showing that at least some units could generate potentially detectable echoes reflected from other whales.

Mercado and colleagues argue that the primary purpose of some mysticete vocalizations (including song) could be echolocation. Mercado explained how different units might serve different roles in a biosonar model of song. Narrowband (constant-frequency) units may generate persistent reverberation at distinct frequencies, useful for target detection. Whenever a target (e.g., another whale) enters the search space of the echolocating whale, the reverberant echo stream temporarily changes. Broadband units then may be used to locate and classify the target. These complementary sonar functions are simultaneously possible due to spectral interleaving in this duplex model of humpback whale song (Mercado III 2021). Over a series of papers, Mercado provides arguments for the sonar function of humpback whale song: for example, song consists of inter-unit silences during which echoes may be processed, singers on breeding grounds are more likely solitary than in groups, females tend to ignore or avoid singers, individuals approaching a singer are typically male, escorts (which may be males) often do not sing, singers stop singing to approach distant non-singing whales, the location of distant whales cannot be determined by vision, singing also occurs on feeding grounds and during feeding lunges, and vocalization bandlevels have correlated with prey density in places (Mercado III 2018, 2020, 2021). While counter-arguments have, of course, been published (Au et al. 2001a), we cannot simply ignore the possibility of some mysticete sounds having different or multiple functions, and more work is needed to understand these functions.

3.8 Geographic Variability

Mysticetes produce a diversity of vocalizations, the features of which are often stereotypical to species. Within species, vocalizations may further vary between populations or geographic regions—much like accents in humans. Such variations can identify different "acoustic populations" and subpopulations or stocks of cetaceans (e.g., Cerchio et al. 2001; Delarue et al. 2009b; Stafford et al. 2001). In mysticetes, geographic variation is most commonly observed in song, rather than non-song sounds (McDonald et al. 2006).

Geographic variation is greater across different oceans than within any one ocean, suggesting that dissimilarity results from geographically isolated populations—perhaps because of environmental differences that affect sound transmission and genetic isolation of a closed breeding population (Helweg et al. 1998; Murray et al. 2012). Whereas songs from different regions within the same ocean have both differences and similarities that are likely due to (acoustic) contact occurring at some point of the migration cycle (Helweg et al. 1990).

Payne and Guinee (1983) first used acoustic characteristics to distinguish two acoustic populations of humpback whales in the North Pacific and North Atlantic, as well as two different stocks within the eastern North Pacific. Geographic variation of song and non-song sounds to identify stocks or acoustic populations has since been observed for blue (Leroy et al. 2021; McDonald et al. 2006; Fig. 3.5), Bryde's (Figueiredo and Simao 2014; Heimlich et al. 2005; Viloria-Gómora et al. 2021), fin (Archer et al. 2020; Delarue et al. 2009b; Morano et al. 2012; Širović et al. 2017), minke (Risch et al. 2014a), and Omura's whales (Cerchio et al. 2019). Comparing a species' song from different regions over time may be used to study stock and population structure and evaluate the potential for whale movements within or between populations (Delarue et al. 2009b; Helweg et al. 1998; Noad et al. 2000).

3.9 Diel Patterns

Diel patterns of mysticete vocalizations (both song and non-song sounds) are frequently reported from underwater acoustic recordings, implying that animals change their acoustic behavior as a function of time of day. At times of no acoustic detections, it is often unknown whether animals were absent, or whether they remained present yet silent. In the case of continuous acoustic presence yet sound detection rates that change with time of day, it is often unknown whether all individuals changed their sound production rate, or whether some individuals joined or left the area while others kept vocalizing. Furthermore, data on whether the observed patterns are stable over the longer term (i.e., over multiple seasons or years) are often not available. Nonetheless, these patterns are not uncommon, as described for bowhead whales (Blackwell et al. 2007), right whales (Matthews et al. 2014; Munger et al. 2008), minke whales (Casey et al. 2022), sei whales (Baumgartner and Fratantoni 2008; Romagosa et al. 2020), Bryde's whales (Putland et al. 2018), blue whales (Gavrilov and McCauley 2013; Leroy et al. 2016; Stafford et al. 2005; Wiggins et al. 2005; Wingfield et al. 2022), fin whales (Fig. 3.6; Aulich et al. 2023; Romagosa et al. 2020), humpback whales (Au et al. 2000; Kowarski et al. 2018), and gray whales (Rannankari et al. 2018).

Based on the current data, diel vocalization patterns are species-dependent and may vary between and within ocean regions, as well as with season. For example, in the Northeast Pacific, diel patterns of fin whale vocalizations varied along the species' migratory route. In high-latitude waters, in the Bering Sea, a greater acoustic presence of both the fin whale 20 Hz and 40 Hz call types occurred during the day than during the night (Širović et al. 2013). In lower-latitude waters off southern California, this observed diel pattern persisted for the 40 Hz call but became insignificant for the 20 Hz call. However, in the northern Gulf of California, a greater acoustic presence of both fin whale call types occurred during the night than during the day. A difference in the seasonal peaks was also found between the two call types (Širović et al. 2013). Off South Africa, a greater acoustic presence of fin whales occurred during the night in winter and during the day in spring. At the same site, blue

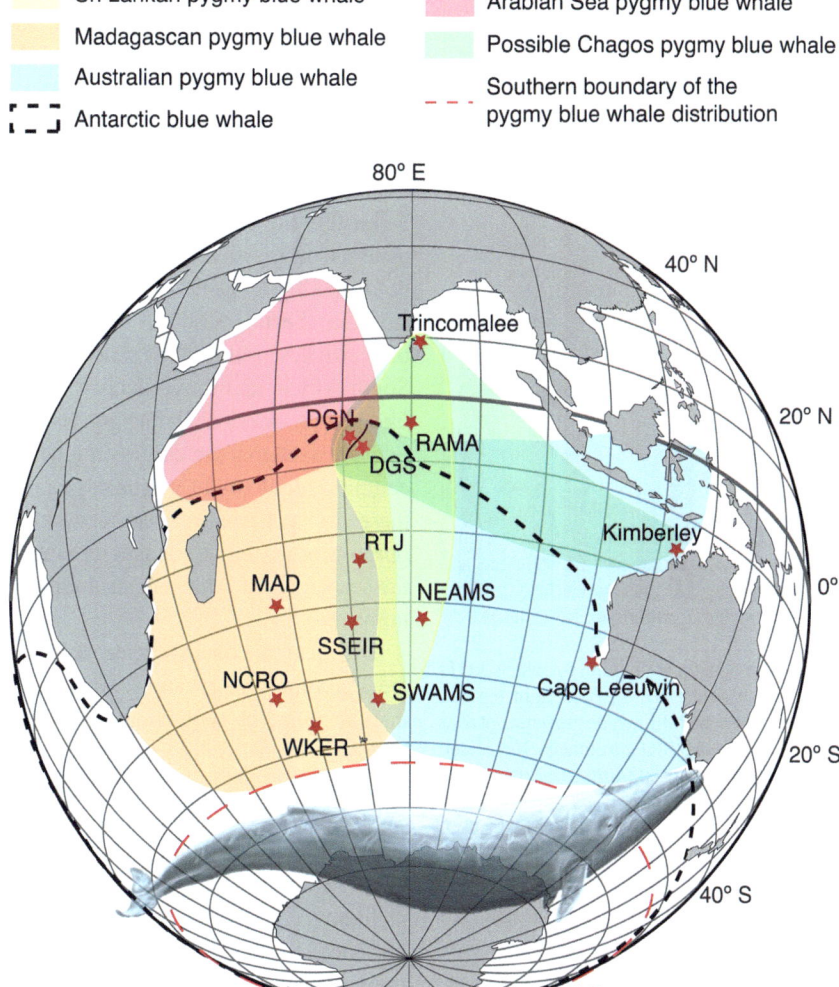

Fig. 3.5 Range of possible, distinct acoustic populations of pygmy blue whales in the Indian Ocean, based on their songs. Reprinted from Leroy EC, Royer J-Y, Alling A, Maslen B, Rogers TL (2021) Multiple pygmy blue whale acoustic populations in the Indian Ocean: whale song identifies a possible new population. Sci Rep 11 (1): 8762. https://doi.org/10.1038/s41598-021-88062-5. © Leroy et al. 2021. Published CC BY 4.0; https://creativecommons.org/licenses/by/4.0/

whale acoustic activity peaked during the day in both fall and winter, before switching to a greater acoustic presence during the night in spring (Shabangu et al. 2019).

Several hypotheses have been put forward to explain such diel patterns in mysticete acoustic detections. Different types of sound are thought to be related to different types of functional behavior, and so these observed diel patterns may be associated with diel behavioral activity of the animals, such as foraging or reproduction. In particular, the trend of increased call rates of certain call types during twilight or at nighttime suggests that these sounds play a role in foraging because the preferred prey for mysticetes tends to follow a diel vertical migration pattern, aggregating at depth during the day and rising to shallow surface waters at night (Baumgartner et al. 2003; Fiedler

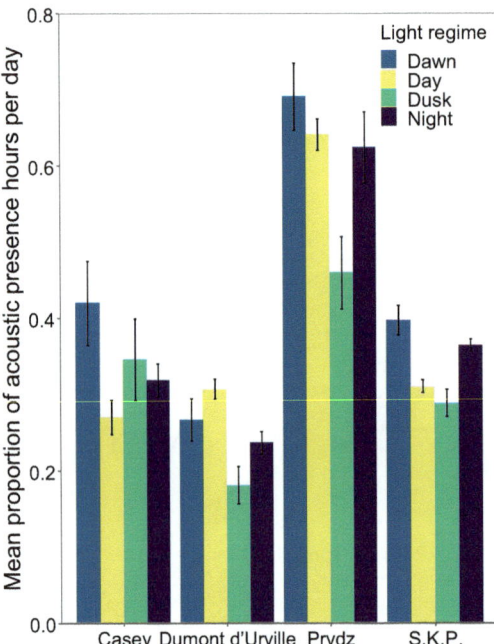

Fig. 3.6 Proportion of hours per day with fin whale 20 Hz pulses, normalized by the duration of each light regime, and averaged over all days with fin whale presence, at four sites in the Southern Ocean. SKP: Southern Kerguelen Plateau. Data from Aulich et al. 2023

et al. 1998; Zhou and Dorland 2004). Therefore, mysticetes might preferentially feed at nighttime, when prey occurs at shallower depths. Alternatively, feeding calls might not be necessary during daytime feeding when light conditions suffice. Conversely, suspected feeding calls, such as the fin whale 40 Hz call and the blue whale D-call, peaked during the day on the feeding grounds and might call conspecifics to prey patches (Oleson et al. 2007b; Širović et al. 2013).

Diel patterns in mysticete vocalizations might also be related to reproductive activity. Hawaiian humpback whale song peaked at night on the breeding grounds, where animals are fasting (Au et al. 2000). With perhaps less ambient noise at night, singers might be heard better at night. During the day, male humpback whale mating strategies might be based on visual interactions such as male-to-male physical aggression. However, at night, males might switch to acoustic strategies. Similarly, gray whale calling peaked in the early morning and evening on their breeding grounds (Ponce et al. 2012) and at night in a migratory corridor far from typical prey fields (Rannankari et al. 2018). Nighttime calling may help with group cohesion during migration.

Lastly, the observed seasonal shifts in diel patterns of mysticete acoustic detections might be driven by acoustic competition between species (see Sect. 3.12). Fin whales and Antarctic blue whales seasonally share the habitat off South Africa, and they vocalize within the same frequency band. Shabangu et al. (2019) hypothesized that the contrasting peaks in animal calling periods (night versus day) and the shift in these diel patterns between seasons are for the purpose of avoiding acoustic competition during the season when both species are present in large numbers.

Diel patterns in mysticete acoustic detection are a common observation, potentially providing insight into the behavioral ecology of these animals, such as region-specific foraging or breeding activity. However, some limitations exist due to the varying analysis methodologies implemented. For example, the unit of measurement of acoustic activity varies across studies, including call counts, vocalization rates, hourly presence/absence, and chorus band level. Implementing unified methodologies for future diel analysis would improve comparability and synthesis of findings.

3.10 Decadal Changes in Sound Features

Long-term (decadal) changes in the features of mysticete sounds, such as peak frequency and inter-pulse interval, have been reported from all oceans, based on passive acoustic recordings spanning several decades. The reasons for this phenomenon remain unknown. Several examples are discussed in the following sections.

3.10.1 Blue Whales

The peak frequency of blue whale song units is declining. The phenomenon has been reported for all blue whale subspecies. For example, in Bass Strait, Australia, from 2004 to 2016, the Antarctic blue whale Z-call 27 Hz subunit declined by 0.12 Hz/year and the 18 Hz subunit by 0.09 Hz/year (McCauley et al. 2018). At the same location over the same time, the Eastern Indian Ocean pygmy blue whale 22 Hz tone (Unit II) declined by 0.12 Hz/year and the 45 Hz overtone by 0.23 Hz/year, while the New Zealand pygmy blue whale 27 Hz tone (Unit I) declined by 0.12 Hz/year and the 18 Hz tone (Unit II, Subunit II) by 0.15 Hz/year (McCauley et al. 2018). Similar blue whale frequency decreases have been reported elsewhere in the Indian Ocean (Leroy et al. 2018; Miksis-Olds et al. 2018), in the Northeast Pacific (Carbaugh-Rutland et al. 2021; Rice et al. 2022), in the Southwest Pacific (Miller et al. 2014a), and in the Southeast Pacific (Malige et al. 2020). McDonald et al. (2009) collated blue whale call measurements from the Northeast, Northwest, and Southwest Pacific, North Atlantic, Northern Indian, Southeast Indian, and Southern Oceans, over 8–46 years, finding similar shifts. In the Northeast Pacific, from which the longest time series was available, blue whales decreased their B-call frequency by 31% over 46 years. Rice et al. (2022) added 11 years to the Northeast Pacific observations by McDonald et al. (2009) indicating that the percent change per year of the B-call is levelling off (being less in recent than earlier years).

The frequency decline differs between geographic regions, blue whale subspecies and populations, and song units. Overtones decline proportionally to fundamentals, of course, meaning that the drop in Hz/year of the second harmonic is twice that of the first harmonic (McCauley et al. 2018). But interestingly, different units in song may decline at different rates. In the New Zealand pygmy blue whale, the 18 Hz component of Unit II declined more than the 27 Hz component of Unit I (even though the former is lower in frequency than the latter; McCauley et al. 2018). Such differential shifts have also been reported for the Sri Lankan pygmy blue whale; however, the unit with the higher peak frequency (Unit III) decreased at a greater rate (expressed as the percent change per year) than Unit II, which has a lower peak frequency (Miksis-Olds et al. 2018). In the Northeast Pacific, on the contrary, both A- and B-units declined at the same rate (Rice et al. 2022). Similarly, in the Southeast Pacific, different song units changed at the same rate (Malige et al. 2020).

Intra-seasonal change, consistent over many years, has been observed in addition to the inter-annual decline (Gavrilov et al. 2012; Leroy et al. 2018; Miller et al. 2014b). Off Cape Leeuwin, WA, Australia, the first subunit of the Antarctic blue whale Z-call declined in frequency within a season by ~0.5 Hz. At the beginning of the next season, it recommenced at about the mean frequency of the previous season, before declining at the previous season's slope again. Over the now 20-year period, the linear regression shows a long-term decline by 0.14 Hz/year (Fig. 3.7).

In addition to a decline in peak frequency, blue whales also decrease their pulse rate. The Northeast Pacific A-unit is pulsed. The pulse rate decreased from 1.24 pulses/s in 2006 to 1.158 pulses/s in 2019 (Rice et al. 2022). Finally, the inter-song interval of Eastern Indian Ocean pygmy blue whales increased from 2003 to 2017 in the Perth Canyon, WA, Australia (Jolliffe et al. 2019). The inter-song interval of the one-unit phrase (P1 song) increased by 0.3 s/y, that of the two-unit phrase (P2 song) by 0.8 s/y, and that of the three-unit phrase (P3 song) by 1.73 s/y.

3.10.2 Fin Whales

The peak frequency of the 20 Hz pulse in fin whales declined at a rate of 0.17 Hz/year in the period 2003–2013 in the Northeast Pacific (Weirathmueller et al. 2017). In the Indian Ocean, the higher-frequency component at 99 Hz was measured and declined at 0.2 Hz/year (Fig. 3.7; Leroy et al. 2018). In the Southern

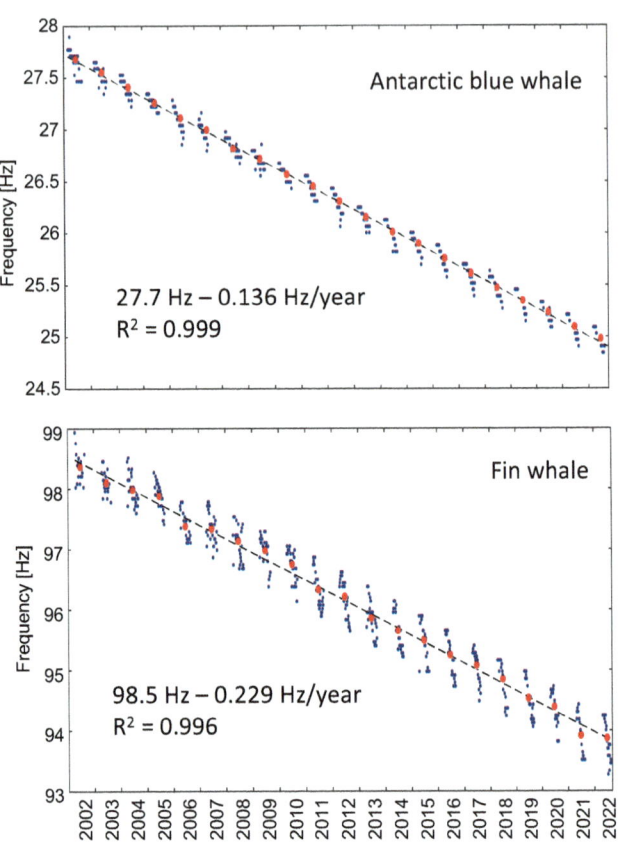

Fig. 3.7 Intra-seasonal and inter-annual frequency shifts in the first subunit of the Antarctic blue whale Z-call (top) and higher-frequency component of the fin whale 20 Hz pulse (bottom). Blue dots are 5-day averages of peak frequency; red dots show the frequency at the time of maximum chorus intensity each season. Linear regressions were drawn over the 21 years. Recordings from the Comprehensive Test Ban Treaty Organization's (CTBTO) International Monitoring System's hydroacoustic station at Cape Leeuwin, Western Australia

Ocean, the higher-frequency component at 89 Hz was measured in 2001 and declined by 0.19 Hz/year to 86 Hz in 2016 (Wood and Širović 2022).

In addition to the decline in frequency, the vocalization (pulse) rate has also declined. In the Southern Ocean, the inter-unit interval increased by 1.5 s between 2001 and 2016 (i.e., the pulse rate decreased; Wood and Širović 2022). In the Northeast Pacific, the inter-unit interval between two successive 20 Hz pulses in singlet song was 25 s in 2003, increasing to 30 s in 2013. The inter-unit interval of doublet song increased as well, at a greater rate (Weirathmueller et al. 2017). Doublet and triplet songs were analyzed from 2000 to 2013 off California, showing an inter-annual increase in inter-unit interval and also an intra-seasonal shift to greater inter-unit interval. At the beginning of the following season, the inter-unit interval reset to a shorter interval, but not as short as at the beginning of the previous season. Hence, over the years, the inter-unit interval kept increasing (Širović et al. 2017). Interestingly, in 2006/7, the long-doublet song ceased and a short-doublet song commenced. One hypothesis is that there is a limit to these shifts, and once the interval has become too great, animals reset to a value from decades before, only to restart annual hikes. Further, increasing fin whale inter-unit intervals have been described in the Central North Pacific (Helble et al. 2020a), North Atlantic (Morano et al. 2012), Mediterranean (Best et al. 2022), Arctic (Furumaki et al. 2021), and Antarctic (Wood and Širović 2022). Most of these studies also reported intra-seasonal shifts in frequency and inter-unit interval.

3.10.3 Bowhead Whales

A decadal frequency decline has been observed in bowhead whale calls in the Beaufort Sea. The proportion of calls that were recorded below 75 Hz increased from 27% to 41% between 2008 and 2014, representing a 10 Hz decrease in the mean value of the minimum frequency distribution from 94 Hz to 84 Hz (Thode et al. 2017). No specific call type was responsible for the shift in frequency, however the more complex calls accounted for the most significant decrease. The authors performed multivariate regression analyses looking for biological and environmental parameters that might explain this shift. No significant correlations were found with water depth, distance to the calling animal, calling depth below water, ambient noise level, signal-to-noise ratio, or airgun activity.

3.10.4 Hypotheses for Long-Term Changes

Several hypotheses for the observed shifts in mysticete sound features have been proposed, but each has short-comings. Hypotheses have focused on the frequency shifts.

3.10.4.1 Changes in Whale Size

The peak frequency of sound is related to the resonating sound-producing structures (see Sect. 3.2). A decrease in peak frequency could hence imply an increase in body size. While whaling targeted the larger individuals, leaving a greater proportion of smaller individuals in the population, it would only have taken a decade post-whaling for the proportion of large individuals to recover, given whales reach physical maturity within 10 years of age. There is no evidence of whales getting bigger from year to year in recent decades.

With regards to the intra-seasonal shifts, whale blubber thickness increases over the course of the feeding season, and a correlation of blubber thickness with peak frequency has been observed (Miller et al. 2014b). This correlation might be merely temporal but not causal. In migratory corridors, staggered migration (whereby different demographic cohorts of a population migrate at slightly different times) could result in steady frequency shifts, if juveniles arrived before adults, but this would not explain the inter-annual decline. Changes in whale size further cannot explain different shifts of different song units.

3.10.4.2 Doppler Effect

The Doppler effect explains the frequency shifts observed if a sound source and a recorder move relative to each other. With moored recorders, the frequency of a moving sound source would be higher during the approach, drop rapidly at the point of closest approach, and then continue at a lower frequency as the source departs from the recorder. This effect would only be possible in migratory corridors where whales travel along more-or-less straight paths, but not on feeding or breeding grounds where travel directions are not consistent (Miller et al. 2014b). Also, the slow swimming speed of whales would render the Doppler effect small (0.01 Hz for a whale swimming at 4 km/h and singing at 20 Hz). The same downsweep (Doppler shift) would be seen for every passing individual and frequency would reset after every individual and would not result in either the intra-seasonal or inter-annual decline.

3.10.4.3 Change in Migration Route

If whales were to shift their migration routes into deeper water, where energy at lower frequencies travels at less loss than energy at higher frequencies, more low-frequency energy might be recorded—over a season or across years (Thode et al. 2017). However, no seasonal inshore-offshore shifts, nor steady shifts away from the coasts have been documented.

3.10.4.4 Shift in Calling Depth

Changes in calling depth might affect the peak frequency of the emitted sound if increased pressure at depth reduces the volume of the resonating structures and increases the density of the gas within, leading to a rise in resonant frequency. Conversely, a decrease in calling depth (i.e., calling from shallower depths) would lead to a decrease in resonant frequency. However, while the lungs will ultimately collapse at great depth, the lungs are the air source, but the resonating structures are the nasal cavities, which are much firmer and more resistant to pressure changes (see Sect. 3.2; Adam et al. 2013; Gavrilov et al. 2011).

Calling at deeper depth even within an unchanged source bandwidth could still reduce the peak of the recorded energy at some range, because a deeper source is less affected by the Lloyd's mirror effect at lower frequencies (Thode et al. 2017; also see Chap. 1). However, a 30 Hz whale call has a wavelength of ~50 m, and so calling depths would need to change significantly to have an effect, yet whales sing only in the top ~40 m (Fig. 3.4). Moreover, there is no indication of repetitive changes in peak frequency coincident with calling depth.

3.10.4.5 Climate Change

Ocean warming and ocean acidification change the speed of sound and the frequency-dependent attenuation of sound. The effects would be too small to explain the observed frequency shifts (McDonald et al. 2009).

3.10.4.6 Ambient Noise

Could the frequency shift be a Lombard effect? The frequency shifts are too small to make a significant difference with respect to the level of ambient noise in the same frequency band. Analyses showed no evidence of an increase in ambient noise levels in the vocalization frequency bands of pygmy blue whales, nor a compensating increase in call amplitude, removing the likelihood of a Lombard effect (Miksis-Olds et al. 2018). Moreover, a previously observed response to increasing low-frequency ocean noise was an increase in tonal frequency in right whales (Parks et al. 2007).

3.10.4.7 Increasing Population Density

As whale populations increase post-whaling, sexual selection might favor males singing at lower frequency (which might indicate greater fitness), leading to a gradual change across the population (McDonald et al. 2009). This, however, does not explain the intra-seasonal pattern nor the different shifts of different units in song.

There is no single hypothesis that explains the observed changes. As no environmental drivers change at the time scales and rates as the sound features do, an environmental driver (such as water temperature) seems unlikely. More likely is a population-scale behavioral change for unknown underlying reasons. Different drivers might be at play for intra-seasonal versus inter-annual shifts, and for different song units. As pygmy blue whale Units II and III are shifting at different rates, it is suggested that each song unit may serve a different function, and thus could be responding to different selective pressures (Miksis-Olds et al. 2018).

It seems plausible that there will be a limit to these shifts, based on physics, anatomy, and physiology. Peak frequencies might level out, or gradually reverse, or jump to those from decades before. If these shifts are related to population increases, then a levelling-out might be seen as the population reaches environmental carrying capacity post-whaling. There are examples of levelling out, as in the peak frequency of blue whale B-calls (Rice et al. 2022). However, there are also examples of sudden reversals (jumps) in long-term observations. The steadily increasing inter-unit interval of Northeast Pacific fin whales suddenly dropped again (Širović et al. 2017) and the steadily decreasing peak frequency of tonal calls (spot call) from an unidentified whale suddenly increased again in the western Indian Ocean and off South Australia (Leroy et al. 2017; Ward et al. 2017; see Sect. 3.14). These resets occurred in at least two different species in three different oceans in both hemispheres in exactly the same season: 2006/7.

3.11 The Effects of Noise on Acoustic Behavior

As some regions of the ocean become increasingly noisy, population fitness depends on the ability of marine mammal species to adapt their acoustic signaling such that they can still be heard by conspecifics (Kunc et al. 2022). Acoustic behavior has been shown to be rather dynamic across taxa at timescales that indicate phenotypic plasticity as opposed to long-term genetic shifts. Phenotypic plasticity allows species to adapt their acoustic signaling to the continually changing soundscape to enable effective communication (Kunc et al. 2022). The Lombard effect is one such mechanism that is well described across a number of taxa, whereby a signaler increases the amplitude or—in a broader definition of the Lombard effect—the repetition of its signals, or shifts the frequency, in response to increases in ambient noise (Lombard 1911). Limitations in these kinds of acoustic adjustments exist with regard to a species' physical ability and species-typical sounds that must have certain temporal or frequency characteristics to be recognized by conspecifics.

Increases in mysticete source levels have been reported, for example, for right, blue, and gray whales in response to ship noise (Dahlheim and Castellote 2016; McKenna 2011; Parks et al. 2011b), humpback whales with increasing wind noise (Dunlop et al. 2014), and bowhead whales in the presence of both wind and seismic airgun noise (Thode et al. 2020). However, these studies hinted at a physiological constraint to this response, in that these animals initially increased their source level linearly with increases in noise level, but then plateaued, perhaps due to physiological constraints.

Increases in vocalization rate have been reported, for example, for bowhead and blue whales in seismic airgun noise (Blackwell et al. 2015; di Iorio and Clark 2010), bowhead whales in drilling noise (Blackwell et al. 2017), gray whales in ship and boat noise (Dahlheim and Castellote 2016), and bowhead whales in wind noise (Thode et al. 2020). The bowhead responses plateaued and then reversed, with call rates decreasing at the highest noise levels, potentially indicating another physiological or energetic constraint.

Changes in song structure might be related to changes in vocalization rate. Jolliffe et al. (2021) found significant variability in the structure and length of pygmy blue whale song phrases and suggested that variability may be linked to ambient noise levels with shorter songs being produced in stronger (more intense) noise. As the second unit of this species' song has the highest source level, production of shorter song phrases that included this unit would increase the likelihood of being heard.

Shifting the peak frequency of vocalizations away from the peak of the noise has been reported for right whales in ship noise (Parks et al. 2007). However, not all observations of vocal changes in the presence of noise are in line with the Lombard effect. In high levels of ship and seismic survey noise, fin whales shortened the duration of their 20 Hz notes, lowered center and peak frequencies, and decreased the bandwidth (Castellote et al. 2012a).

Changes in vocalization type with increases in ambient noise have also been reported from gray whales, which used more frequency-modulated sounds that might be more robust to masking (Dahlheim and Castellote 2016). This indicates that in higher levels of ambient noise, whales might shift to a communication strategy that optimizes the transmission of specific information. The confounding issue is that whales may have had a change in motivational state as a result of ship passes and noise.

The kinds of information whales may prioritize communicating and the nature of any response may vary not only with the type of noise but also with the general context, such as the specific habitat, behavior at the time of exposure, and demographic parameters. For example, while blue whales were observed to increase their call rate during seismic surveys and nearby ship passes (di Iorio and Clark 2010; Melcón et al. 2012), they reduced vocalization rates in the presence of mid-frequency active sonar (Melcón et al. 2012). Studies of gray whales also have shown highly variable acoustic responses that were dependent on the source of the noise (which included ship noise, boat noise, drilling noise, and predator sounds; Dahlheim and Castellote 2016). Some differences in responses were also found between exposures to real sources versus playback. While it is very difficult to reproduce the noise field of a complex source with a small, monopole speaker, the differential responses to real versus simulated sources of anthropogenic noise indicate that it is not purely the noise level but also the proximity and perhaps behavior (e.g., movement) of the source that influence the response (Dahlheim and Castellote 2016).

Finally, novelty of a noise type and perceived threat might explain differential responses. A cessation of vocal behavior (e.g., during playback of killer whale/predator sounds) has been associated with the presence of threats and is often accompanied by startle, fright, or stress responses. Consequently, it is likely that ceasing to vocalize in the presence of perceived or known threats is a survival strategy to avoid detection by predators (Dahlheim and Castellote 2016). Conversely, when exposed to increasing levels of noise from familiar sources, whales might be more likely to increase the detectability of their signals to conspecifics and effectively maintain the range over which they can communicate.

3.12 Acoustic Niches

The acoustic space refers to the range (or area, or volume) over which acoustic signals can be detected and interpreted by the desired receivers. The acoustic space is limited by the environment (i.e., the local sound propagation conditions and ambient noise). For the purposes of this discussion, communication space is the volume of ocean within which two animals can communicate. In their seminal article on mysticete communication space, Clark et al. (2009) showed through measurement and modeling how the communication spaces of right, fin, and humpback whales changed with location, frequency, and time due to changes in ambient noise.

On evolutionary time scales, anthropogenic noise is a recent source of ambient noise. Mysticetes evolved to communicate in an ocean naturally noisy with the sounds of the geophysical environment (e.g., wind, waves, rain, and ice) and of other animals (e.g., invertebrates, fishes, odontocetes, and other mysticete species). The sum of all sounds within a soundscape (e.g., Schoeman et al. 2022) limit the acoustic space available to any one species.

In species-rich acoustic communities, communication space may be considered a limited ecological resource, similar to food, water, or shelter. When resources are limited, species compete for these resources; some species may become competitively excluded; or species may partition the resource to coexist (Pocheville 2015). Acoustic space is partitioned through the formation of acoustic niches. A "niche", in the most general definition, refers to the position or functional role of a species within its respective ecosystem (Vandermeer 1972). The Acoustic Niche Hypothesis proposes that acoustic niches are formed through the partitioning of acoustic signals by frequency, modulation, time, or space (Krause 1987). The partitioning of acoustic signals within an acoustic community increases the Darwinian success of signalers and reduces the possibility of acoustic interference or masking from other biotic or abiotic sound sources.

Acoustic niches have been reported in terrestrial ecosystems for decades. The separation of acoustic signals in the domains of frequency, time, and space has been examined most commonly in acoustic communities of birds, insects, and frogs (e.g., Boncoraglio and Saino 2007; Brenowitz 1982; Chek et al. 2003; Hart et al. 2021; Stone 2000). Frequency partitioning occurs when signalers constrain their vocalizations to distinct frequency bands that are different from those of other signalers within their community. Luther and Gentry (2013) displayed the frequency partitioning that occurred during the dawn chorus of an acoustic community in El Verde, Puerto Rico. The frequency bands of the calls of three out of four bird species and a single frog species were distinct, with little overlap. Due to the richness of some acoustic communities, frequency partitioning of acoustic signals is not always possible and hence does not suffice. Temporal partitioning occurs when signalers stagger the timing of their vocalizations to avoid overlap with signals of a similar frequency band. Temporal separation of acoustic signals may occur over a variety of time scales, from season to hour, minute, second, and even millisecond (Garcia-Rutledge and Narins 2001). Temporal separation is common in insect and frog choruses, as well as in fish choruses (e.g., McWilliam et al. 2018; Ruppé et al. 2015; Williams et al. 2018). Spatial partitioning may also be employed to avoid acoustic interference. This type of partitioning occurs when signalers spatially stratify their vocalizations within their respective habitats. For example, insects called at different heights above the ground (Sueur 2002), and frogs aggregated and vocalized in different regions or microhabitats (Garcia-Rutledge and Narins 2001).

Literature that specifically examines the acoustic niches of mysticetes is scarcer. However, acoustic niche partitioning is apparent in many studies on marine soundscapes that include mysticetes (e.g., Oswald et al. 2016; Putland et al. 2017; Warren et al. 2021). Mysticete species recorded in the Artic and Antarctic exhibited acoustic partitioning of signals in frequency and time (i.e., season; Menze et al. 2017; van Opzeeland and Boebel 2018). Weiss and team (2021) illustrated how cetaceans (including mysticetes) partitioned their acoustic space in frequency and time (season) in three canyons in the Northwestern Atlantic Ocean (Weiss et al. 2021). Seasonal partitioning was particularly evident in the calling presence of right and humpback whales, which occupied similar frequency bands.

Several mysticete species have further exhibited temporal partitioning on a finer time scale. Diel temporal partitioning was evident in the calling behaviors of Antarctic blue and fin whales off the west coast of South Africa (Shabangu et al. 2019). The authors noted that while both species produced vocalizations within the same frequency band, coexistence within the same acoustic space was possible as both species vocalized at different times of the day during the period of the highest acoustic presence of both species (i.e., during the winter months): Antarctic blue whale calling activity peaked during the day, whereas fin whale calling presence was at its highest in the evening.

In Australia, the Perth Canyon is a migratory corridor and opportunistic feeding stop-over for pygmy blue whales (Feb.–Jun. on their northern migration and Nov.–Jan. on their southern migration), humpback whales (May–Jul. on their

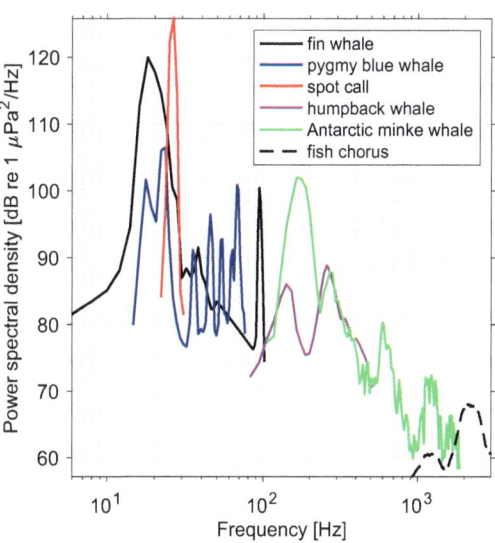

Fig. 3.8 Frequency spectra of the different biological sources in the Perth Canyon, WA, Australia. Three mysticete species shared the 10–100 Hz band, two shared the 100–1000 Hz band, and fishes called above 1 kHz. Recordings from Erbe et al. (2015)

northern migration and Aug.–Nov. on their southern migration), the spot-call whale (Jun.–Oct.), fin whales (Jun.–Aug.), and Antarctic minke whales (Jul.–Aug.) (Erbe et al. 2015). While there is some partitioning in time (season) and space (with some species on one of their migratory lags traveling farther offshore through the canyon than others), additional partitioning in frequency and modulation is obvious (Fig. 3.8). Within the 10–100 Hz band, fin whales produce 1 s downswept pulses at 30–20 Hz and 100–90 Hz. The spot call is a 12–15 s, 26 Hz tone. And pygmy blue whales produce a 3-unit phrase of tonal-to-pulsed modulations occupying the entire band. These stereotypical sounds, even when occurring simultaneously, can easily be told apart by their spectral and temporal features, where the temporal features refer to frequency and amplitude modulations over the duration of the sound. Similarly, in the 100–1000 Hz band shared between Antarctic minke whales and humpback whales, the former produce brief broadband pulses in packets, while the latter produce longer tonal sounds with frequency and amplitude modulation, overtones and sidebands (easily visible behind the pulses in the spectrogram in Erbe et al. 2015).

Finally, mysticetes have demonstrated short-term (compared to evolutionary time scales) acoustic adaptations, likely to increase their acoustic space. For example, right whales varied the frequency and frequency modulation rates of their signals in association with calling group size (Clark 1982). Right, minke, and humpback whales changed the frequency and level of their vocalizations in response to ambient noise (Dunlop et al. 2014; Helble et al. 2020b; Parks et al. 2007). Despite the benefits of utilizing acoustic adaptions, this type of partitioning likely comes with costs to the signaler (e.g., in terms of energy; Barber et al. 2010). Given the presence and intensity of anthropogenic noise in marine soundscapes, a greater understanding of how mysticetes utilize acoustic spaces is needed. Knowledge of the acoustic niches these animals occupy and the processes that drive the formation of these niches can inform marine soundscape planning strategies to reduce anthropogenic acoustic interference (van Opzeeland and Boebel 2018).

3.13 Taxonomic Overview of Sounds

We undertook a literature search for the sounds made by the various species of mysticete and for measurements of their acoustic features. We grouped sounds into song versus non-song. We collated measurements for song sequences, themes, phrases, and units. Recognizing that mysticete sounds lie along a continuum from tonal to pulsed, we classified sounds based on their predominant features (based on the published spectrograms, reported features, and authors' own classifications). We arrived at the following sound types:

1. Tone (i.e., constant-frequency or frequency-modulated, with or without overtones)
2. Pulse (including single, amplitude-modulated sounds that exhibit many tight sidebands in spectrograms)
3. Noisy narrowband sound (i.e., when tonal, harmonic, or sideband structure was not apparent but the sound appeared as a bandlimited blob in spectrograms)
4. Pulse train (i.e., series of distinct pulses with determinable inter-pulse interval)
5. Impulse (i.e., brief, strong pulses like right whale gunshots)
6. Click (i.e., very brief and high-frequency clicks likened to those of odontocetes; e.g., humpback whale megapclicks)

For each sound type, we listed the following measurements:

1. Peak frequency f_p (i.e., frequency of maximum power)
2. Bandwidth (meant to capture the dominant band; hence, 3 dB bandwidth, 10 dB bandwidth, or inter-quartile range—in this order of preference, depending on what was reported in the literature; see Erbe et al. 2022a for definitions and equations)
3. Minimum frequency (meant to reflect the lower edge of the dominant frequency band; hence, in the following order of preference, depending on what was reported in the literature: (a) taken as the lower reported range of f_p; (b) computed as mean(f_p) − std(f_p); (c) taken as the lower quartile or lowest percentile (sometimes 10th percentile); (d) taken as the mean minimum frequency; or (e) taken as the lower edge of the reported frequency range)
4. Maximum frequency (meant to reflect the upper edge of the dominant frequency band; hence, in the following order of preference: (a) taken as the upper reported range of f_p; (b) computed as mean(f_p) + std(f_p); (c) taken as the upper quartile or highest percentile (sometimes 90th percentile); (d) taken as the mean maximum frequency; or (e) taken as the upper edge of the reported frequency range)
5. Mean duration
6. Minimum duration (meant to reflect the lower edge of the dominant duration and hence computed as mean − std; however, this often ended up with negative durations as a result of durations not being Gaussian distributed; in

this case, the minimum duration was taken as the lower range of durations reported)
7. Maximum duration (meant to reflect the upper edge of the dominant duration and hence computed as mean + std)
8. Source level measurements (means not maximum levels reported) in terms of peak-to-peak sound pressure level, root-mean-square sound pressure level, and sound exposure level

We further tracked the nomenclature used in the literature for each sound type (often onomatopoeic names), season when the sound was recorded, ground (i.e., breeding ground, feeding ground, or migratory corridor), IUCN region, site, and reference. Our working table may be requested from the first author (see Sect. 3.17). A summary of mean features by species and call type is given in Table 3.1.

We reviewed 179 publications. Studies had occurred in 16 of the 18 IUCN marine regions (Hoyt 2011); no publications on mysticete vocalizations were found for the Baltic Sea and the East Asian Seas. The Baltic Sea is not a ground for mysticetes, whereas the East Asian Seas are (Cerchio et al. 2019; McGowen et al. 2021), but no data were retrieved from this region. The most represented regions are the Northeast Pacific (39 articles), the Northwest Atlantic (35), and the Arctic (32). These three regions surround North America. Ten references were retrieved from the wider Caribbean region, which makes it the least studied region of North America. Mysticete acoustics in Australia and New Zealand has been well published with 22 references, followed by the Antarctic (13) and South Pacific region (13), then Southeast Pacific (12), and South Atlantic (10). East Africa is represented with 9 publications from La Reunion and Madagascar. Three studies present data from animals in captivity (Fig. 3.9; Table 3.2).

Comparing sound features across species, the largest whales (blue and fin whales) produce the lowest-frequency and longest-duration sounds, with the strongest source levels. While such relationships have previously been identified in the literature (e.g., Martin et al. 2017), there is, of course, great variability in the literature. This variability is partly due to differences in study design (e.g., how recordings were obtained; how data were analyzed, including how sound types were defined, how the Fourier window was chosen for spectral analysis, how spectrographic features were defined, which measurements were chosen such as 3 dB or 10 dB bandwidth, or frequency range, and what statistics were reported such as means, medians, variance, or percentiles; and how results were presented), environment in which the observations and measurements were made (e.g., in deep versus shallow water, in noisy or pristine environments, and whether the sound propagation environment was considered), and species' ecologies (e.g., migratory or resident; solitary or social; seasonal or opportunistic feeders; recorded on breeding or feeding grounds, or in migratory corridors). Ignoring all these factors, we simply averaged over all studies for each species and compared acoustic features against body size (compiled from the Society for Marine Mammalogy species fact sheets, https://marinemammalscience.org/science-and-publications/species-information/facts/ accessed 11 Feb. 2024, as well as Agbayani et al. 2020; Kok et al. 2023; Pastene et al. 2020; Würsig et al. 2018). Table 3.3 lists the acoustic features and body sizes by species.

Figure 3.10 shows three acoustic features of mysticete calls (i.e., peak frequency, duration, and source level) against body size. Because mysticete songs are believed to be produced by males only, song unit features were plotted against male body sizes (~5–10% smaller than female body sizes). However, for non-song calls, female and male body sizes were averaged. For both song and non-song vocalizations, body sizes were averaged at a species level except for the blue whale species, where we treated the largest mysticete species, the Antarctic blue whale (*Balaenoptera musculus intermedia*), separately, and averaged over the other subspecies (grouped as *Balaenoptera musculus* in Fig. 3.10). We note that the number of vocalization types (either song or non-song) reported and the number of measurements available for each sound feature vary significantly among the different

Table 3.1 Overview of sound types by species. Types: T (tone), P (pulse), NN (noisy narrowband), PT (pulse train), I (impulse), C (click). Fp: mean of all reported peak frequencies. Fmin: mean of all reported lower frequencies of the dominant band. Fmax: mean of all reported upper frequencies of the dominant band. Dur: mean duration. Durmin: minimum of the reported lower edges of the dominant duration. Durmax: maximum of the reported upper edges of the dominant duration. SL: source level in terms of SPLpk-pk, SPLrms, and SEL

Species	Common name	Sound type	Fp [Hz]	Fmin [Hz]	Fmax [Hz]	Dur [s]	Durmin [s]	Durmax [s]	SL_SPLpk-pk [dB re 1 µPa m]	SL_SPLrms [dB re 1 µPa m]	SL_SEL [dB re 1 µPa² s m²]	References
Balaena mysticetus	Bowhead whale	T	604	329	1089	2.5	0.1		205	166	174	Clark and Johnson (1984), Cummings and Holliday (1987), Delarue et al. (2009a), Johnson et al. (2015), Ljungblad et al. (1982), Richardson et al. (1982), Stafford et al. (2008), Tervo et al. (2009, 2011a, 2012), and Thode et al. (2016)
Balaena mysticetus	Bowhead whale	P	697	288	2031	4.1	0.3	29				Clark and Johnson (1984), Delarue et al. (2009a), Ljungblad et al. (1982), Richardson et al. (1982), Stafford et al. (2008), and Tervo et al. (2009)
Balaenoptera acutorostrata	Common minke whale	T	150	50	250		0.2	0.3		154		Gedamke et al. (2001)
Balaenoptera acutorostrata	Common minke whale	P	235	90	4383	1.1		2				Erbe et al. (2017) and Gedamke et al. (2001)
Balaenoptera acutorostrata	Common minke whale	PT		50	9400		1	2.5		158		Gedamke et al. (2001)
Balaenoptera acutorostrata	North Atlantic minke whale	T	99	80	118	0.4	0.3	0.5				Edds-Walton (2000)
Balaenoptera acutorostrata acutorostrata	North Atlantic minke whale	PT	139	519	915	29	9.5	67	180	165	160	Beamish and Mitchell (1973), Mellinger et al. (2000), Risch et al. (2013, 2014c), and Wang et al. (2016)

3 Mysticete Sounds

Species	Common name	T/P									References
Balaenoptera acutorostrata scammoni	North Pacific minke whale	T	105	38	142	0.65	0.5	0.8			Nikolich and Towers (2020)
Balaenoptera acutorostrata scammoni	North Pacific minke whale	P	1400	1000	1800	0.28	0.15	1.6			Oswald et al. (2011)
Balaenoptera acutorostrata scammoni	North Pacific minke whale	PT	1015	777	2470	2.1	0.4	4.4	166		Delarue et al. (2013a), Helble et al. (2020b), Nikolich and Towers (2020), Oswald et al. (2011), and Rankin and Barlow (2005)
Balaenoptera bonaerensis	Antarctic minke whale	T	95	79	144	0.33	0.1	1.3	156		Casey et al. (2022), Dominello and Širović (2016), Risch et al. (2014b), and Schevill and Watkins (1972)
Balaenoptera bonaerensis	Antarctic minke whale	P	174	133	229	0.4	0.1	0.7			Casey et al. (2022)
Balaenoptera bonaerensis	Antarctic minke whale	PT	155	113	201	1.8	0.3	6.5	140		Dominello and Širović (2016), Risch et al. (2014b), Rossi-Santos et al. (2022), and Shabangu et al. (2020)
Balaenoptera borealis borealis	Northern sei whale	T	45	30	67	0.84	0.5	2	194	179	Baumgartner et al. (2008), Cusano et al. (2023), Macklin (2022), Newhall et al. (2012), and Rankin and Barlow (2007), Romagosa et al. (2015), Tremblay et al. (2019), and Wang et al. (2016)
Balaenoptera borealis borealis	Northern sei whale	P	135	72	273	0.16	0.1	0.6			Cusano et al. (2023)
Balaenoptera borealis schlegelii	Southern sei whale	T	123	73	179	1.2	0.15	2.3	156		Calderan et al. (2014), Cerchio and Weir (2022), Español-Jiménez et al. (2019), and McDonald et al. (2005)

(continued)

Table 3.1 (continued)

Species	Common name	Sound type	Fp [Hz]	Fmin [Hz]	Fmax [Hz]	Dur [s]	Durmin [s]	Durmax [s]	SL_SPLpk-pk [dB re 1 µPa m]	SL_SPLrms [dB re 1 µPa m]	SL_SEL [dB re 1 µPa² s m²]	References
Balaenoptera edeni	Bryde's whale	T	86	67	125	1.2	0.09	6.7		163		Cummings et al. (1986), Figueiredo and Simao (2014), Heimlich et al. (2005), Oleson et al. (2003), Viloria-Gómora et al. (2015), and Wang et al. (2022)
Balaenoptera edeni	Bryde's whale	P	402	311	701	0.01	0.025	51				Edds et al. (1993)
Balaenoptera musculus	Blue whale	T	31	26	39	9.9	1.4	24		180		Buchan et al. (2014), Cummings and Thompson (1971), Frank and Ferris (2011), Leroy et al. (2021), Miller et al. (2014a), Stafford et al. (1999b), and Thode et al. (2000)
Balaenoptera musculus	Blue whale	P	121	116	127	8.3	0.6	22		181		Buchan et al. (2010, 2014), Frank and Ferris (2011), Miller et al. (2014a), and Stafford et al. (1999b)
Balaenoptera musculus brevicauda	Pygmy blue whale	T	37	17	93	16	1.3	49		174		Bouffaut et al. (2021), Gavrilov et al. (2011), Leroy et al. (2021), McCauley et al. (2000), Recalde-Salas et al. (2014), and Samaran et al. (2010)
Balaenoptera musculus brevicauda	Pygmy blue whale	P	86	10	750	4.4						Recalde-Salas et al. (2014)
Balaenoptera musculus indica	Northern Indian Ocean pygmy blue whale	T	46	44	49	16	7.4	30				Cerchio et al. (2020), Leroy et al. (2021), and Stafford et al. (2011)

Balaenoptera musculus indica	Northern Indian Ocean pygmy blue whale	P	32	28	36	12	2.1	23		Leroy et al. (2021) and Stafford et al. (2011)
Balaenoptera musculus intermedia	Antarctic blue whale	T	29	19	39	9.5	1.4	18	188	Bouffaut et al. (2021), Miller et al. (2021), Rankin et al. (2005), Samaran et al. (2010), Širović et al. (2004, 2007), and Stafford et al. (1999b, 2004)
Balaenoptera musculus intermedia	Antarctic blue whale	P	57	23	85	2.5	1.8	3.2		Rankin et al. (2005)
Balaenoptera musculus musculus	Northern blue whale	T	30	24	36	12	0.5	26	168	Akamatsu et al. (2014), Berchok et al. (2006), Edds (1982), McDonald et al. (2001), Mellinger and Clark (2003), Oleson et al. (2007a), Rivers (1997), Stafford et al. (1998, 1999b, 2001), and Thompson et al. (1996)
Balaenoptera musculus musculus	Northern blue whale	P	60	49	102	10	0.6	21	178	Berchok et al. (2006), McDonald et al. (2001), Oleson et al. (2007a), Rivers (1997), Stafford et al. (2001), and Thompson et al. (1996)
Balaenoptera musculus musculus	Northern blue whale	NN	43	25	158	3.2	1.3	5.1		Berchok et al. (2006)
Balaenoptera omurai	Omura's whale	P	34	16	51	13	2.1	16		Cerchio (2022), Cerchio et al. (2015), Erbe et al. (2017), Leroy et al. (2021), Madhusudhana et al. (2020), Moreira et al. (2020), and Sousa and Harris (2015)
		T	47	42	64	0.76	0.27	1.3	177	

(continued)

Table 3.1 (continued)

Species	Common name	Sound type	Fp [Hz]	Fmin [Hz]	Fmax [Hz]	Dur [s]	Durmin [s]	Durmax [s]	SL_SPLpk-pk [dB re 1 µPa m]	SL_SPLrms [dB re 1 µPa m]	SL_SEL [dB re 1 µPa² s m²]	References
Balaenoptera physalus physalus	North Atlantic fin whale											Castellote et al. (2012b), Clark et al. (2002), Delarue (2008), Edds (1988), Garcia et al. (2019), Miksis-Olds et al. (2019), Sciacca et al. (2023), Simon et al. (2010), Wang et al. (2016), Watkins (1981), and Watkins et al. (1987)
Balaenoptera physalus quoyi	Southern fin whale	T	54	47	62	0.74	0.4	1.1		190		Constaratas et al. (2021), Miller et al. (2021), Širović et al. (2004, 2006, 2007), and Vega et al. (2023)
Balaenoptera physalus velifera	North Pacific fin whale	T	30	18	48	0.96	0.1	4.7		180		Archer et al. (2020), Charif et al. (2002), Cummings et al. (1986), Delarue et al. (2013b), McDonald et al. (1995), Miksis-Olds et al. (2019); Northrop et al. (1968), Širović et al. (2013), Thompson et al. (1992), Varga et al. (2018), Weirathmueller et al. (2013), and Wiggins and Hildebrand (2020)
Balaenoptera ricei	Rice's whale	T	107	78	134	11	0.2	25		155		Rice et al. (2014), Širović et al. (2014), and Soldevilla et al. (2022a, b)
Caperea marginata	Pygmy right whale	T	70	60	135	0.18	0.14	0.23		160		Dawbin and Cato (1992)
Eschrichtius robustus	Gray whale	T	190	134	379	1.4	0.03	6.6	175	155	159	Burnham et al. (2018), Charles (2011), Crane and Lashkari (1996), Cummings et al. (1968), Dahlheim (1987), Fish et al. (1974), Guazzo et al. (2017), López-Urbán et al. (2018), Moore and Ljungblad (1984),

3 Mysticete Sounds

Species	Common name	Type									References
Balaenoptera musculus indica	Northern Indian Ocean pygmy blue whale	P	32	28	36	12	2.1	23			Leroy et al. (2021) and Stafford et al. (2011)
Balaenoptera musculus intermedia	Antarctic blue whale	T	29	19	39	9.5	1.4	18		188	Bouffaut et al. (2021), Miller et al. (2021), Rankin et al. (2005), Samaran et al. (2010), Širović et al. (2004, 2007), and Stafford et al. (1999b, 2004)
Balaenoptera musculus intermedia	Antarctic blue whale	P	57	23	85	2.5	1.8	3.2			Rankin et al. (2005)
Balaenoptera musculus musculus	Northern blue whale	T	30	24	36	12	0.5	26		168	Akamatsu et al. (2014), Berchok et al. (2006), Edds (1982), McDonald et al. (2001), Mellinger and Clark (2003), Oleson et al. (2007a), Rivers (1997), Stafford et al. (1998, 1999b, 2001), and Thompson et al. (1996)
Balaenoptera musculus musculus	Northern blue whale	P	60	49	102	10	0.6	21		178	Berchok et al. (2006), McDonald et al. (2001), Oleson et al. (2007a), Rivers (1997), Stafford et al. (2001), and Thompson et al. (1996)
Balaenoptera musculus musculus	Northern blue whale	NN	43	25	158	3.2	1.3	5.1			Berchok et al. (2006)
Balaenoptera omurai	Omura's whale	P	34	16	51	13	2.1	16			Cerchio (2022), Cerchio et al. (2015), Erbe et al. (2017), Leroy et al. (2021), Madhusudhana et al. (2020), Moreira et al. (2020), and Sousa and Harris (2015)
		T	47	42	64	0.76	0.27	1.3		177	

(continued)

Table 3.1 (continued)

Species	Common name	Sound type	Fp [Hz]	Fmin [Hz]	Fmax [Hz]	Dur [s]	Durmin [s]	Durmax [s]	SL_SPLpk-pk [dB re 1 μPa m]	SL_SPLrms [dB re 1 μPa m]	SL_SEL [dB re 1 μPa² s m²]	References
Balaenoptera physalus physalus	North Atlantic fin whale											Castellote et al. (2012b), Clark et al. (2002), Delarue (2008), Edds (1988), Garcia et al. (2019), Miksis-Olds et al. (2019), Sciacca et al. (2023), Simon et al. (2010), Wang et al. (2016), Watkins (1981), and Watkins et al. (1987)
Balaenoptera physalus quoyi	Southern fin whale	T	54	47	62	0.74	0.4	1.1		190		Constaratas et al. (2021), Miller et al. (2021), Širović et al. (2004, 2006, 2007), and Vega et al. (2023)
Balaenoptera physalus velifera	North Pacific fin whale	T	30	18	48	0.96	0.1	4.7		180		Archer et al. (2020), Charif et al. (2002), Cummings et al. (1986), Delarue et al. (2013b), McDonald et al. (1995), Miksis-Olds et al. (2019); Northrop et al. (1968), Širović et al. (2013), Thompson et al. (1992), Varga et al. (2018), Weirathmueller et al. (2013), and Wiggins and Hildebrand (2020)
Balaenoptera ricei	Rice's whale	T	107	78	134	11	0.2	25		155		Rice et al. (2014), Širović et al. (2014), and Soldevilla et al. (2022a, b)
Caperea marginata	Pygmy right whale	T	70	60	135	0.18	0.14	0.23		160		Dawbin and Cato (1992)
Eschrichtius robustus	Gray whale	T	190	134	379	1.4	0.03	6.6	175	155	159	Burnham et al. (2018), Charles (2011), Crane and Lashkari (1996), Cummings et al. (1968), Dahlheim (1987), Fish et al. (1974), Guazzo et al. (2017), López-Urbán et al. (2018), Moore and Ljungblad (1984),

3 Mysticete Sounds

Eschrichtius robustus	Gray whale	P	342	141	1210	0.83	0.019	5.2	134	Ollervides (2001), and Wisdom et al. (2001)
										Burnham et al. (2018), Charles (2011), Crane and Lashkari (1996), Cummings et al. (1968), Dahlheim (1987), Frouin-Mouy et al. (2020), Guazzo et al. (2017), López-Urbán et al. (2018), Moore and Ljungblad (1984), Norris et al. (1977), Ollervides (2001), Stafford et al. (2007), Wisdom et al. (2001), and Youngson and Darling (2016)
Eschrichtius robustus	Gray whale	PT	1182	114	4202	1.7	0.25	5.6		Dahlheim (1987), Fish et al. (1974), Moore and Ljungblad (1984), and Wisdom et al. (2001)
Eschrichtius robustus	Gray whale	C		2000	6000	0.002	0.001	0		Dahlheim (1987) and Fish et al. (1974)
Eschrichtius robustus	Gray whale	NN	203	133	615	1.7	0.23	7.3	125	Burnham et al. (2018), Charles (2011), Crane and Lashkari (1996), Cummings et al. (1968), Dahlheim (1987), Fish et al. (1974), Frouin-Mouy et al. (2020), Moore and Ljungblad (1984), and Ollervides (2001)
Eubalaena australis	Southern right whale	T	204	98	448	1.3	0.2	6		Clark (1982), Cummings et al. (1972), Dombroski et al. (2016), Jacobs et al. (2019), Parks et al. 2007; Tellechea and Norbis (2012), and Webster et al. (2016)
Eubalaena australis	Southern right whale	P	205	61	916	1.2	0.2	3.7	180	Clark (1982), Cummings et al. (1972), Dombroski et al. (2016), Tellechea and Norbis (2012), and Webster et al. (2016)
Eubalaena australis	Southern right whale	I	795	69	3042	0.13	0.06	0.22		Cummings et al. (1972) and Webster et al. (2016)

(continued)

Table 3.1 (continued)

Species	Common name	Sound type	Fp [Hz]	Fmin [Hz]	Fmax [Hz]	Dur [s]	Durmin [s]	Durmax [s]	SL_SPLpk-pk [dB re 1 μPa m]	SL_SPLrms [dB re 1 μPa m]	SL_SEL [dB re 1 μPa^2 s m^2]	References
Eubalaena australis	Southern right whale	NN	312	74	1478	1.1	0.5	26				Clark (1982) and Webster et al. (2016)
Eubalaena glacialis	North Atlantic right whale	T	283	133	592	1	0.44	2.3	171	159		Parks et al. (2007, 2009, 2019a), Parks and Tyack (2005) and Trygonis et al. (2013)
Eubalaena glacialis	North Atlantic right whale	P	576	326	1626	0.71	0.005	2.3	169	156		Parks et al. (2007, 2019a), Parks and Tyack (2005), Trygonis et al. (2013), and Watkins and Schevill (1976)
Eubalaena glacialis	North Atlantic right whale	I	1019	52	8132	0.067	0.02	0.16	194	183		Parks et al. (2005), Parks and Tyack (2005), and Trygonis et al. (2013)
Eubalaena glacialis	North Atlantic right whale	NN	1640	160	6130	0.76	0.51	1	162	144		Parks and Tyack (2005)
Eubalaena japonica	North Pacific right whale	T	139	94	178	1	0.43	2.4	196	177	178	Crance et al. (2019), McDonald and Moore (2002), Mellinger et al. (2004), Munger et al. (2011), and Širović et al. (2015)
Eubalaena japonica	North Pacific right whale	P	133	65	200	1.3	0.5	2.3				Crance et al. (2019)
Eubalaena japonica	North Pacific right whale	I	630	50	5500	0.27	0.16	0.61				Crance et al. (2017)
Megaptera novaeangliae australis	Southern humpback whale	T	684	465	860	1.1	0.02	7.1	181	154	163	Cusano et al. (2022); D'Souza et al. (2023), Dunlop et al. (2007, 2013), Gascón (2021), Girola et al. (2019), Indeck et al. (2020), Pénitot et al. (2021), Recalde-Salas et al. (2020), Rekdahl et al. (2017), Saloma et al. (2022), and Tellechea et al. (2017)

Megaptera novaeangliae australis	Southern humpback whale	P	485	285	743	0.99	0.05	6.5	181	156	167	Cusano et al. (2022), D'Souza et al. (2023), Dunlop et al. (2007, 2013), Gascón (2021), Girola et al. (2019), Indeck et al. (2020), Pénitot et al. (2021), Recalde-Salas et al. (2020), Rekdahl et al. (2017), and Saloma et al. (2022)
Megaptera novaeangliae australis	Southern humpback whale	PT	643	520	856	0.58	0.04	3.5	183	162	159	Cusano et al. (2022), Dunlop et al. (2007, 2013), Gascón (2021), Girola et al. (2019), Pénitot et al. (2021), and Saloma et al. (2022)
Megaptera novaeangliae australis	Southern humpback whale	NN	645	448	820	0.98	0.01	3.6		150		Cusano et al. (2022), Dunlop et al. (2007, 2013), Gascón (2021), and Saloma et al. (2022)
Megaptera novaeangliae kuzira	North Pacific humpback whale	T	765	456	1261	1.7	0.024	41		169		Au et al. (2001b); Barlow et al. (2019), Cerchio and Dahlheim (2001), Chereskin et al. (2019), Darling et al. (2019), Fournet et al. (2015, 2018a, c), Perazio et al. (2018), Thompson et al. (1986), Vierling (2022), and Wild and Gabriele (2014)
Megaptera novaeangliae kuzira	North Pacific humpback whale	P	638	326	2156	1.2	0.1	9.1		171		Au et al. (2001b), Chereskin et al. (2019), Fournet et al. (2015, 2018a), Perazio et al. (2018), Thompson et al. (1986), and Vierling (2022)
Megaptera novaeangliae kuzira	North Pacific humpback whale	PT	383	217	717	0.67	0.2	2.4		172		Fournet et al. (2018a, c), Perazio et al. (2018), Thompson et al. (1986), and Vierling (2022)
Megaptera novaeangliae kuzira	North Pacific humpback whale	NN	507	347	783	2.2	0.5	5.4		166		Au et al. (2001b), Fournet et al. (2018a), Perazio et al. (2018) and Vierling (2022)

(continued)

Table 3.1 (continued)

Species	Common name	Sound type	Fp [Hz]	Fmin [Hz]	Fmax [Hz]	Dur [s]	Durmin [s]	Durmax [s]	SL_SPLpk-pk [dB re 1 µPa m]	SL_SPLrms [dB re 1 µPa m]	SL_SEL [dB re 1 µPa² s m²]	References
Megaptera novaeangliae novaeangliae	North Atlantic humpback whale	T	302	237	586	0.74	0.03	8.1				Epp (2019), Fournet et al. (2018c), Stimpert et al. (2007, 2011) and Watkins (1967)
Megaptera novaeangliae novaeangliae	North Atlantic humpback whale	P	450	329	415	0.77	0.2	2.8				Epp (2019)
Megaptera novaeangliae novaeangliae	North Atlantic humpback whale	PT	188	174	477	0.3	0.2	0.4				Epp (2019)
Megaptera novaeangliae novaeangliae	North Atlantic humpback whale	NN	114	80	1183	1.2	0.3	0.5				Epp (2019) and Watkins (1967)

3 Mysticete Sounds

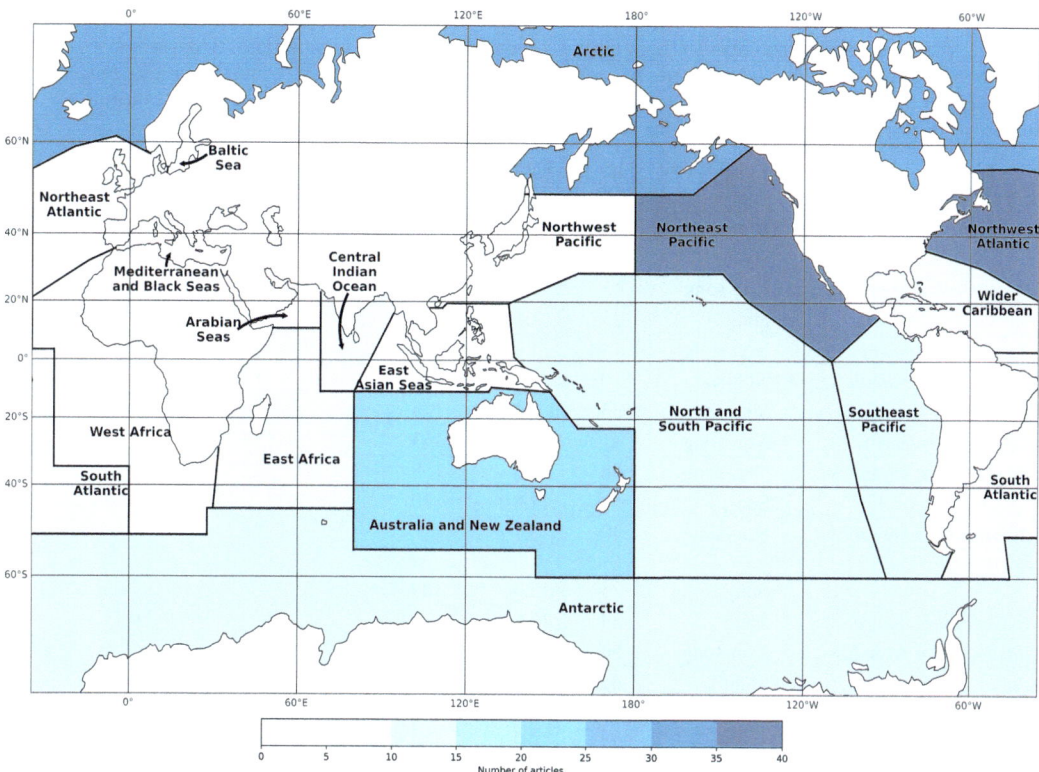

Fig. 3.9 Number of articles per IUCN marine region (Hoyt 2011)

Table 3.2 Number of articles on the features of mysticete vocalizations by IUCN marine region (Hoyt 2011)

IUCN region	ID	No. articles
Antarctic	1	13
Arctic	2	32
Mediterranean and Black Seas	3	3
Northwest Atlantic	4	35
Northeast Atlantic	5	3
Baltic	6	0
Wider Caribbean	7	10
West Africa	8	3
South Atlantic	9	10
Central Indian Ocean	10	7
Arabian Seas	11	1
East Africa	12	9
East Asian Seas	13	0
South Pacific	14	13
Northeast Pacific	15	39
Northwest Pacific	16	4
Southeast Pacific	17	12
Australia and New Zealand	18	22
Captivity		3

Table 3.3 Non-song and song vocalization features by species. Both female and male body sizes were averaged for non-song vocalizations, whereas only male body sizes were considered for song vocalizations. Body sizes were averaged at a species level except for the *Balaenoptera musculus* species. This species was divided into two groups: *Balaenoptera musculus intermedia*, the largest mysticete, the Antarctic blue whale, and *Balaenoptera musculus*, gathering all other subspecies. When no value was available, the cell was left empty. f_p: peak frequency; SL_{rms}: source level

Species	Vocalization type	f_p [Hz]	Duration [s]	SL_{rms} [dB re 1 µPa m]	Body size [m]	Male body size [m]
Balaena mysticetus	Non-song	778	1.4	159	16.2	
	Song	600	3.1	181		15.5
Balaenoptera acutorostrata	Non-song	433	14.1	170	8.3	
	Song	235	1.1			8
Balaenoptera bonaerensis	Non-song	140	1.4	160	8.8	
Balaenoptera borealis	Non-song	101	1.1	176	15.1	
	Song	33	0.4	178		14.1
Balaenoptera edeni	Non-song	112	1.2	163	13	
Balaenoptera musculus	Non-song	58	4.3	168	22.2	
	Song	52	12.8	180		21.7
Balaenoptera musculus intermedia	Non-song	40	2.7		25	
	Song	25	13.4	189		24.3
Balaenoptera omurai	Song	34	12.9			9.7
Balaenoptera physalus	Non-song	54	0.8	186	20.9	
	Song	35	0.8	186		20.1
Balaenoptera ricei	Non-song	107	10.9	155	9.2	
Caperea marginata	Non-song	70	0.2	160	6.1	
Eschrichtius robustus	Non-song	297	1.3	150	12.8	
Eubalaena australis	Non-song	228	1.2	180	14.3	
Eubalaena glacialis	Non-song	491	0.8	179	14.5	
Eubalaena japonica	Non-song	214	0.9	177	15.5	
	Song	150	1.2			13.2
Megaptera novaeangliae	Non-song	589	1.1	177	13.4	
	Song	890	1.8	165		13

species. To examine the possible relationship between each sound feature and body size, we attempted to fit a linear regression. For each sound feature, the slope, the F and p values, and the coefficient of determination R^2 are reported in the corresponding plot. Regression lines were drawn when the statistical significance was $p < 0.1$.

Peak frequency (Fig. 3.10 top row) was highest for bowhead and humpback whales (>500 Hz for both song and non-song vocalizations), but no significant relationship between peak frequency and body size was found. These two species, especially humpback whales, were often recorded close to shore, in the shallow waters of their calving and breeding grounds, or in their shallow, coastal migratory corridors, where the sound propagation environment might have favored mid-frequency propagation and attenuated energy at low frequencies. Alternatively, these species might have evolved to communicate at higher frequencies than other mysticetes, because of their shallow habitats (in which low-frequency sound propagates poorly), or because their smaller anatomy (body size) produces higher-frequency sound and is more suited for the reception of higher-frequency sound (Erbe 2002; Houser et al. 2001). For other mysticete species, a mild trend between peak frequency and body size is observed: the larger the mammal, the lower the peak frequency. The lowest mean peak frequency (25 Hz) corresponds to the largest mysticete: the Antarctic blue whale (25 m). The Omura's whale forms an interesting

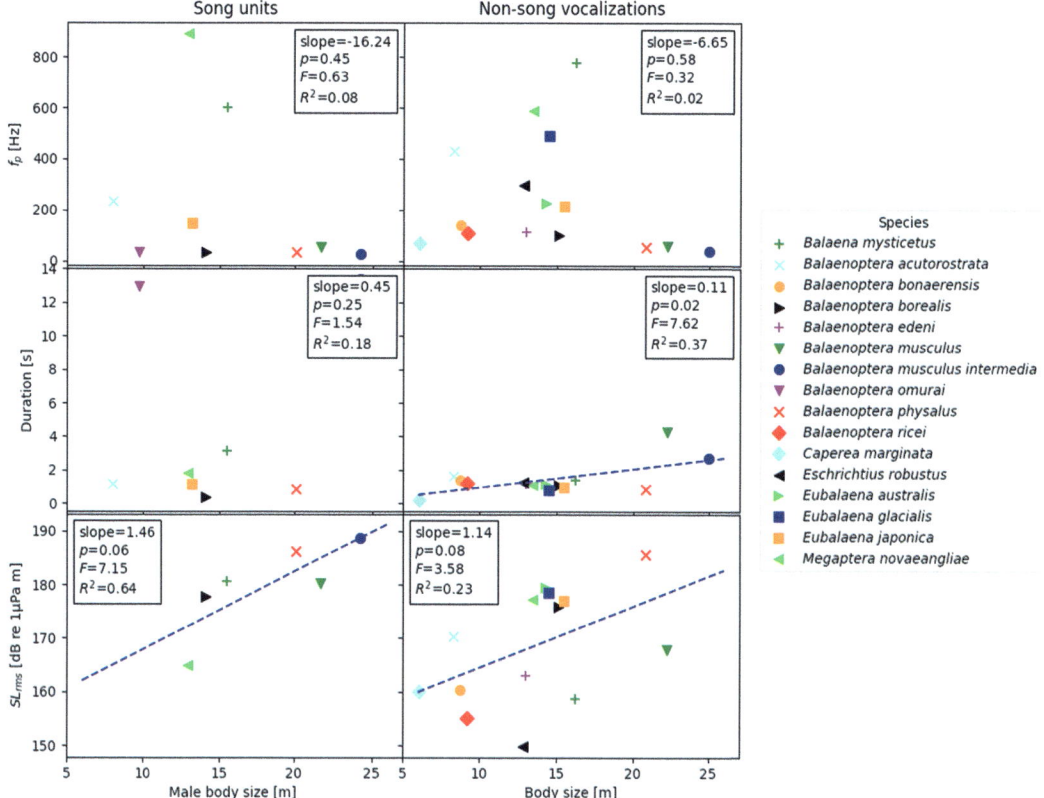

Fig. 3.10 Peak frequency (f_p, top row), duration (middle row), and source level (SL_{rms}, bottom row) for song units (left column) and non-song vocalizations (right column) compared to male and both female and male body sizes, respectively. Linear regressions were attempted for each sound feature and were plotted for statistically significant results ($p < 0.1$, blue dotted lines)

exception: It is one of the smallest species (10 m), yet produces some of the lowest-frequency sounds (peak frequency 34 Hz). Similarly, the smallest mysticete, the pygmy right whale (6 m), produces a peak frequency of 70 Hz.

The longest-duration sounds are produced by the largest mysticetes: blue whales (Fig. 3.10 middle row; both song and non-song vocalizations). Only the duration of non-song vocalizations had a statistically significant linear relationship with body size. The slow and long pulse trains of North Atlantic minke and Rice's whales (1–2 pulses/s, train duration typically >10 s, which some of the literature considers a single call) were excluded from this analysis. The small Omura's whale again presents an exception: Its song units last, on average, 13 s. Having said that, its song units may consist of singlets (~5–9 s) or doublets (in which each pulse is ~5–6 s, with a 1–2 s gap in between). The fin whale, despite its large body size (20 m), has only been reported to produce relatively short sounds (<1 s).

The dependence of source level on body size presents the strongest and statistically significant correlation (Fig. 3.10 bottom row; both song and non-song vocalizations): the larger the whale, the stronger its sounds. Blue and fin whales (i.e., the largest species) produce the strongest sounds. We note that most of the studies did not compute source levels according to the recent ISO standard (International Organization for Standardization 2017, also see Chap. 1). Instead, a variety of approaches has been used, including corrections for spherical and cylindrical spreading, or regressions of received level with range. With whales vocalizing near the sea surface, in the top ~10–40 m of water, and calls often being recorded at shallow inclinations and long ranges, propagation loss could often have been much greater than assumed (including >20 $\log_{10}$(range)), leading to great and variable uncertainty in the reported source levels, and likely often an underestimation.

3.13.1 Balaenidae

3.13.1.1 *Balaena mysticetus*: Bowhead Whale

Bowhead whales are an Arctic species, spending the summers in the High Arctic, and traveling south ahead of the ice edge in fall in order to pass the winter in more subarctic waters before heading north in spring again. Their vocal behavior has been studied extensively since the early 1980s, when concerns over the effects of oil and gas industry development, management requirements, and regulation of aboriginal subsistence hunts focused research efforts on their populations.

Bowhead whale sounds include frequency-modulated tones, amplitude-modulated tones, and pulsed sounds. Tonal sounds have often been grouped into downsweeps, upsweeps, constant-frequency, and combinations thereof. They often have harmonically related overtones. They may further be amplitude-modulated, exhibiting many sidebands. With this great variability of sounds, the bowhead whale acoustic repertoire presents a continuum from tonal to pulsed. Most sounds are of low frequency (~20–500 Hz) and last ~0.5–2 s. High-frequency sounds up to 6 kHz and longer-duration sounds up to ~20 s have also been documented (e.g., Clark and Johnson 1984; Cummings and Holliday 1987; Delarue et al. 2009a; Ljungblad et al. 1982; Richardson et al. 1982; Stafford et al. 2008; Tervo et al. 2011a). Biphonations (i.e., the simultaneous emission of two sounds of different frequency and different modulation) have been reported, particularly in song (e.g., Stafford and Clark 2021; Tervo et al. 2011a). Estimated source levels can be as high as 189 dB re 1 µPa m (Clark and Johnson 1984; Cummings and Holliday 1987; Tervo et al. 2012).

Bowhead sounds may occur singly, in short sequences of 3–25 sounds, or in songs, which may last hours at a time (Johnson et al. 2015; Stafford et al. 2012). Song structure may be broken down into different themes, each consisting of phrases, which in turn are made up of repetitions of units (Tervo et al. 2011b). Song production peaks during the breeding season (e.g., Stafford and Clark 2021). Non-song calls typically are present at the same time as song and into the feeding season (Delarue et al. 2009a; Stafford et al. 2012). The proportion of complex calls and complex songs seems to decrease towards the end of the breeding season, and more simple calls are heard on the feeding grounds (Blackwell et al. 2007; Ljungblad et al. 1982; Stafford and Clark 2021).

In Disko Bay, western Greenland, Stafford et al. (2008) identified three distinct song types during the breeding season. Songs were made up of phrases of up to six units each. Song structure changed over the season from winter to spring at this site (Tervo et al. 2009), with less complexity at the end of the breeding season. Tervo et al. (2011b) studied the changing song structure over four consecutive winters at this site. New song was not merely a rearrangement of units from a limited repertoire. Rather, new song mostly involved new units. Tervo et al. (2011b) determined that multiple individuals sang the same shared song type. Moreover, multiple simultaneous singers and multiple song types were recorded at the peak of the breeding season in winter (Tervo et al. 2009), in contrast to recordings in spring from the same site, when only one animal sang at a time. In spring, bowhead behavior shifts to feeding at this site (Tervo et al. 2009).

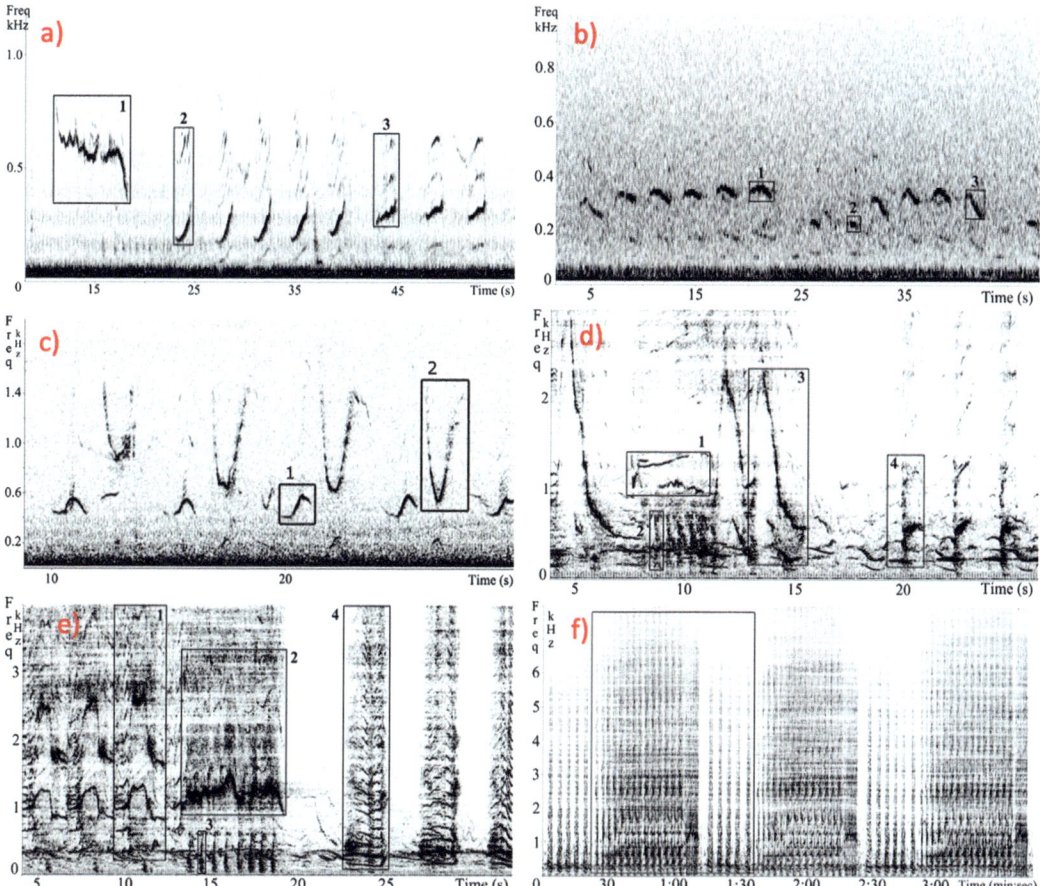

Fig. 3.11 Examples of simple and more complex bowhead whale songs: (**a**) upsweep song, (**b**) downsweep song, (**c**) U song, (**d**) whistle song, (**e**) shriek song, (**f**) shriek song bout. For each song, each unit type is identified by a numbered box. Reprinted from Delarue J, Laurinolli M, Martin B (2009) Bowhead whale (*Balaena mysticetus*) songs in the Chukchi Sea between October 2007 and May 2008. J Acoust Soc Am 126 (6): 3319–3328. https://doi.org/10.1121/1.3257201. With permission from the ASA. © Acoustical Society of America, 2009. All rights reserved

In the migratory corridor off Point Barrow, AK, Johnson et al. (2015) identified 12 different song types; however, there was typically only one whale signing at a time. Similarly, Cummings and Holliday (1987) noted that while >400 bowheads passed their acoustic array over a 3-day period, only one bowhead whale sang at a time, even if groups of up to five animals passed together.

In the southeast Chukchi Sea, Delarue et al. (2009a) described six song types, up to four units each, between fall and spring (Fig. 3.11). Multiple simultaneous song types were recorded. Song type shifted over the course of the year either because different groups of animals passed through or as a result of synchrony and cultural transmission of song in bowhead whales (Fig. 3.12; Delarue et al. 2009a; Johnson et al. 2015).

In Fram Strait, near Spitsbergen, Stafford and colleagues identified 66 different song types in one season (Stafford et al. 2012) and 184 song types over 3 years (Stafford et al. 2018). Song types were typically short-lived, with some only lasting hours to days, while other types lasted up to a month, and a few an entire season. New song types presented sequentially. No song type was

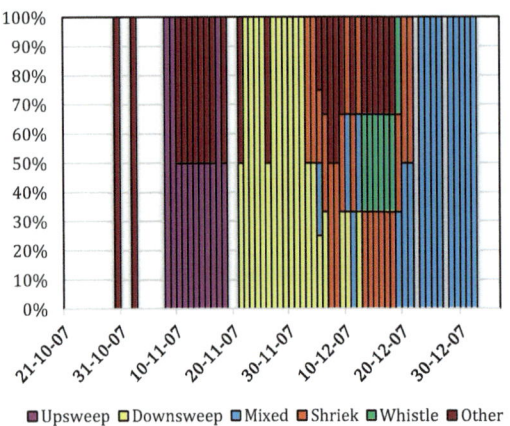

Fig. 3.12 Bowhead whale song change off Point Lay (AK) in the Chukchi Sea over the course of one season. Data kindly provided by Delarue et al. 2009a. Some of the song types are illustrated in Fig. 3.11

heard again in the following season. The authors discussed how this great diversity and rapid change might be driven by population expansion (with new individuals immigrating from different areas and populations), lack of need for interspecific identification in the Arctic, or selection pressure for novelty in a small bowhead population (Stafford et al. 2018). It remains unknown whether individual bowheads sing multiple song types within any one season, whether individuals sing the same song type in multiple years, whether individuals have individually distinctive song types, or whether individuals switch song types from season to season (Stafford et al. 2018; Tervo et al. 2011b).

3.13.1.2 *Eubalaena* spp.: Right Whales

Within the genus *Eubalaena*, there are three species of right whales: the southern right whale (SRW; *Eubalaena australis*), the North Atlantic right whale (*Eubalaena glacialis*), and the North Pacific right whale (*Eubalaena japonica*). They all migrate between winter breeding grounds and summer feeding grounds. The acoustic repertoire of right whales is qualitatively similar across the three species (e.g., Clark 1982; McDonald and Moore 2002; Parks and Tyack 2005) and contains a continuum of tonal to pulsive vocalizations, an impulsive vocalization, and noisy broadband sounds (Fig. 3.13; Crance et al. 2017; Parks et al. 2012; Payne and Payne 1972; Webster et al. 2016). The most extensively researched vocalizations are the upcall and gunshot, which are produced by all three species (e.g., Crance et al. 2017; McDonald and Moore 2002; Trygonis et al. 2013; Webster et al. 2016). At least one species (North Pacific right whale) produces song that consists of patterned repetitions of mostly gunshot vocalizations with a few pulsive and tonal vocalizations (Crance et al. 2019).

The upcall was first described by Clark (1982) and is a low-frequency, harmonic upsweep produced by adults, juveniles, and calves, all of both sexes (Dombroski et al. 2016; Parks et al. 2011a; Parks and Tyack 2005). A review of the right whale literature indicates that the upcall starts at ~50 Hz and sweeps up to ~250 Hz over an average duration of ~0.9 s, and has an average peak frequency of ~120 Hz. Source levels reported in the literature were 150, 155, and 177 dB re 1 µPa m (Munger et al. 2011; Parks and Tyack 2005; Trygonis et al. 2013).

Variability in the acoustic features of the upcall appears to be driven by multiple factors, including differences between species, age classes, individuals, and ambient noise. Generally, southern right whales produce upcalls at lower frequencies and slightly shorter durations than Northern Hemisphere right whales (Dombroski et al. 2016; Jacobs et al. 2019; Parks et al. 2007; Webster et al. 2016). However, this difference may partially be driven by a difference in ambient noise conditions between the Northern and Southern Hemispheres, in that right whales produce upcalls at higher frequencies and greater amplitude in environments with increased ambient noise (Parks et al. 2007, 2011b). Raising both the frequency and amplitude of an upcall increases its likelihood to be detected by conspecifics despite higher noise levels (Tennessen and Parks 2016). Age classes may be distinguished based on variability in the upcall: Adults produce calls with a longer duration and less spectral entropy (i.e., more tonal character) than juveniles, the calls of which also

Fig. 3.13 Spectrograms of right whale vocalizations from New Zealand (fs = 44–48 kHz, NFFT = 4096 top row and High, NFFT = 512 Gunshot, NFFT = 2048 elsewhere, Hann window, 50% overlap). Note the different x- and y-scales. Recordings courtesy of Webster et al. 2016

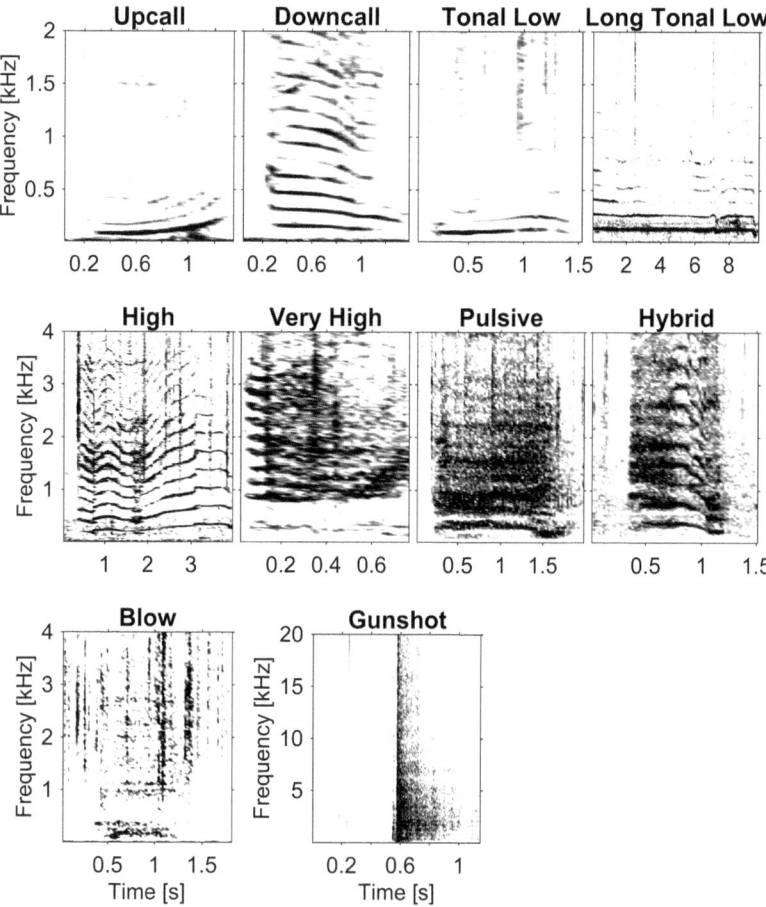

include nonlinear phenomena such as deterministic chaos, biphonation, and subharmonics (McCordic et al. 2016; Parks et al. 2011a; Root-Gutteridge et al. 2018). Finally, the upcalls of individual whales may have different fundamental frequencies and formant structure, potentially allowing individual identification (McCordic et al. 2016).

Clark (1982) suggested that the upcall functions as a contact call. This supposition was based on the robust and consistent structural features of the upcall and on the behavioral observation that lone whales appeared to counter-call with other whales via upcalls. These individuals were then joined by other whales, after which all individuals in the newly formed group stopped producing upcalls. The use of upcalls within the context of surface-active groups also supports the hypothesis that the upcall is a contact call (Parks and Tyack 2005). Male right whales produced upcalls when they first approached a surface-active group, they also produced upcalls while searching for a focal female after she left the surface-active group, and they produced upcalls while alone at the surface as the focal female dove below the surface. There is also evidence that the upcall is used to maintain contact in mother-calf pairs (Clark 1982; Dombroski et al. 2016): New-born calves already produce upcalls (Clark 1982), and upcalls were the prevailing call type on a southern right whale winter calving ground, where the predominant group type was mother-calf pairs (Dombroski et al. 2016).

The gunshot vocalization is a broadband (20 Hz–>24 kHz) impulsive sound ($\lesssim$0.1 s; 191–196 dB re 1 $\mu Pa_{pk\text{-}pk}$ m). The sound is presumably produced internally since it is not accompanied by the slapping of flippers or flukes against the water surface, or by jaw movements (Clark 1982; Parks et al. 2005). In contrast, slaps are broadband impulsive sounds made when a whale leaps out of the water, slaps its pectoral fin against the water, or strikes the water with its fluke (Clark 1982). Gunshot vocalizations and slap sounds are aurally similar; however, gunshots have a greater amplitude and frequency range than slaps when both sounds are recorded at the same distance (Parks et al. 2005). The proposed functions of gunshot vocalizations are a male mating display, male or female agonistic signal, mother-calf separation signal, and foraging signal (Crance et al. 2017; Gerstein et al. 2014; Parks et al. 2005; Parks and Tyack 2005; Trygonis et al. 2013).

Songs described for the North Pacific right whale mostly consist of gunshot units with frequency- and amplitude-modulated units interspersed (Fig. 3.14; Crance et al. 2019). Units were repeated and arranged into phrases, with phrases repeated to form song. Four different song types were reported over an 8-year time span. All sexed singers were males (Crance et al. 2019). Simultaneously recorded songs were always of different types, implying that multiple singers in the same location sing different songs. Conversely, multiple singers sang the same song at the same time in different locations (Crance et al. 2019).

Beyond the upcalls and gunshots, right whales produce a variety of vocalizations with tonal character, which have been grouped into numerous categories (including downcall, down-upcall, constant, tonal constant, tonal low, long tonal low, simple moan, high, very high, complex moan, tonal variable, and warble; Clark 1982; Dombroski et al. 2016; McDonald and Moore 2002; Parks and Tyack 2005; Širović et al. 2015; Trygonis et al. 2013; Webster et al. 2016). These sounds have peak or fundamental frequencies between ~80 Hz and ~400 Hz, durations of ~0.5–1.5 s (with some exceptions of longer tonal calls), and source levels of 150–170 dB re 1 µPa m.

In addition to tonal vocalizations, right whales produce amplitude-modulated and pulsive vocalizations, and hybrid vocalizations that are a combination of tonal and pulsive (Clark 1982; Dombroski et al. 2016; Trygonis et al. 2013). The frequency spectrum of these sounds is more broadband (<1 kHz). Durations are ~1 s and source levels ~150–170 dB re 1 µPa m. The scream vocalization was produced by the focal female within a surface-active group (Parks and Tyack 2005). Grunts, single and double pulses were recorded on calving grounds in the absence of males (Dombroski et al. 2020). Mothers and calves on calving grounds called at lower source levels and lower call rates, likely as a form of acoustic crypsis, avoiding detection by eavesdropping predators (Nielsen et al. 2019; Parks et al. 2019a, b).

3.13.2 Neobalaenidae

3.13.2.1 *Caperea marginata*: Pygmy Right Whale

Very little is known about the vocal repertoire of pygmy right whales, which occur in temperate waters of the Southern Hemisphere. To date, there is only one publication describing pygmy right whale vocalizations, which were produced by a juvenile that spent some months in Portland harbor, southeastern Australia (Dawbin and Cato 1992). The vocalizations consisted of a doublet or triplet of amplitude- and frequency-modulated downsweeps, which started between 80 and 135 Hz and decreased to 60 Hz (peak frequency 70 Hz) over a mean duration of 0.18 s (Fig. 3.15). The estimated source levels varied from 153 to 167 dB re 1 µPa m.

3.13.3 Balaenopteridae

3.13.3.1 *Balaenoptera acutorostrata*: Common Minke Whale

In the Northern Hemisphere, there are two subspecies of common minke whales: the North Atlantic minke whale (*B.a. acutorostrata*) and the North Pacific minke whale (*B.a. scammoni*).

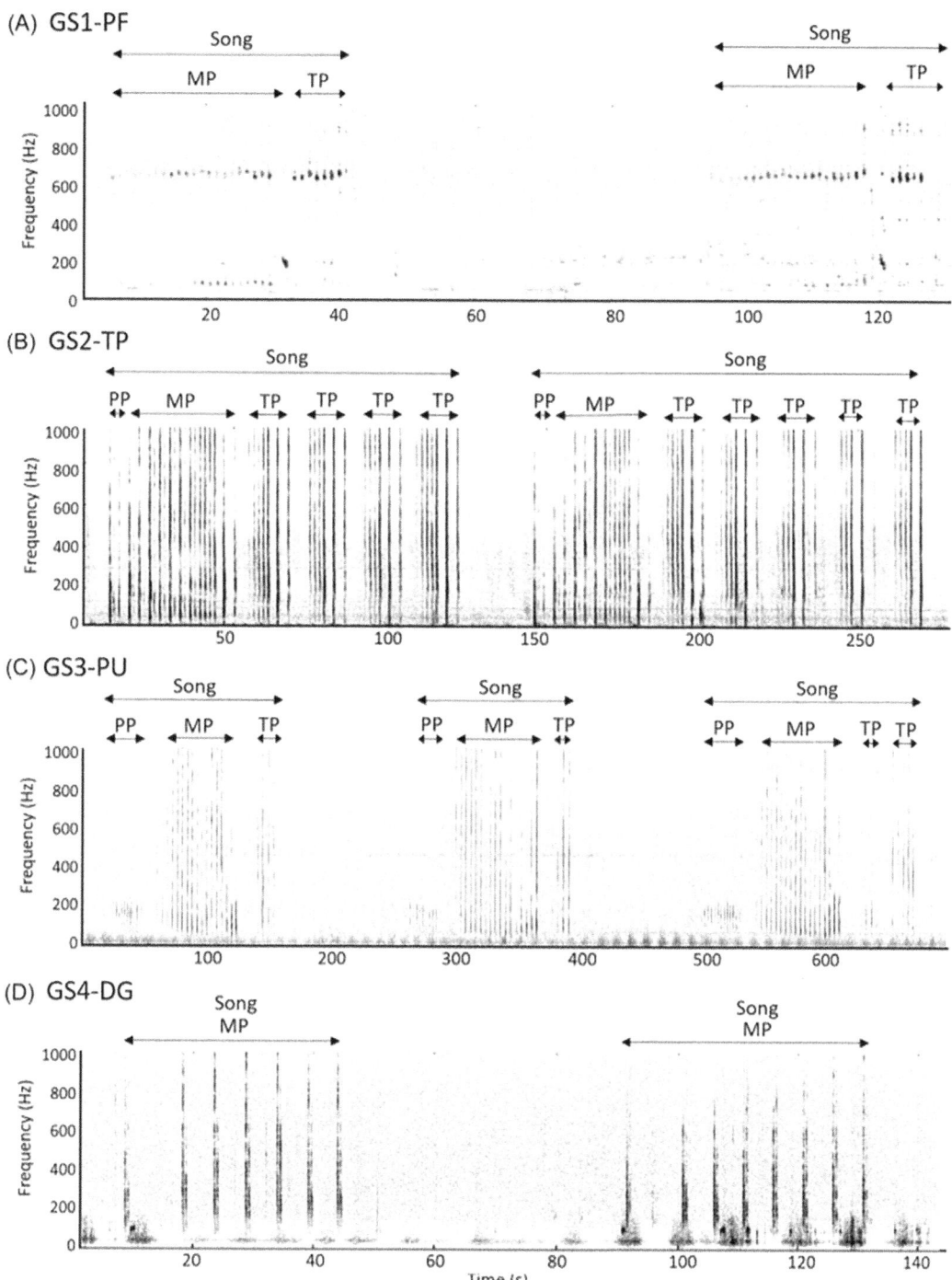

Fig. 3.14 North Pacific right whale song. Reprinted from Crance JL, Berchok CL, Wright DL, Brewer AM, Woodrich DF (2019) Song production by the North Pacific right whale, *Eubalaena japonica*. J Acoust Soc Am 145 (6):3467–3479. https://doi.org/10.1121/1.5111338. With permission from the ASA. © Acoustical Society of America, 2019. All rights reserved

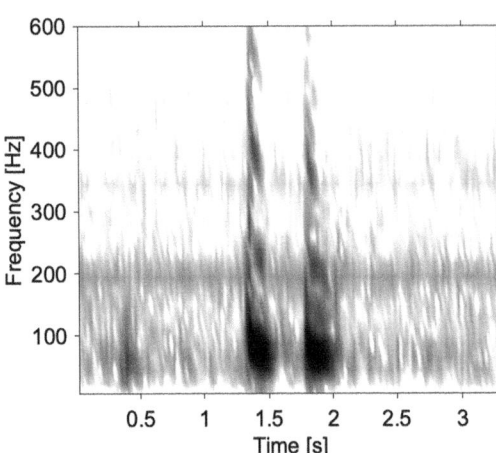

Fig. 3.15 Downsweep doublet recorded from a pygmy right whale (fs = 44 kHz, NFFT = 2048, Hann window, 50% overlap). Recording courtesy of Dawbin and Cato (1992)

In the Southern Hemisphere, a dwarf form of *B. acutorostrata* exists. There are possibly multiple Southern Hemisphere subspecies: one in the South Atlantic, one in the South Pacific, and one in the Indian Ocean. All of the subspecies are believed to undertake annual migrations between winter breeding grounds in low latitudes and summer feeding grounds in high latitudes (e.g., Risch et al. 2014a), however not as far as the Arctic or Antarctic.

3.13.3.2 *B. a. acutorostrata*: North Atlantic Minke Whale

In the North Atlantic, the most commonly described sounds of minke whales are low-frequency pulse trains recorded seasonally. Each pulse typically covers a frequency range from ~100 Hz to ~400 Hz and lasts <0.1 s (Mellinger et al. 2000; Risch et al. 2013). Individual pulses may be frequency-modulated (Mellinger et al. 2000). There may be 100 or more pulses in a train with pulse trains lasting 10–70 s. The amplitude of the pulses in a train typically changes, with trains fading in and out (~15 dB change), and peak amplitude in the center (Mellinger et al. 2000). Source levels of 170–180 dB re 1 µPa m have been reported (Risch et al. 2014c; Wang et al. 2016).

The pulse trains of North Atlantic minke whales fall into three classes, depending on the inter-pulse interval: a slow-down class, a constant class, and a speed-up class (Fig. 3.16; Mellinger et al. 2000; Risch et al. 2013). Within each class, up to three types have been identified by cluster analysis of inter-pulse interval, peak frequency, and duration (Risch et al. 2013). In the slow-down class, inter-pulse interval increased from 0.3 to 0.5 s (Types 1 and 2) and from 0.4 to 0.7 s (Type 3). In the constant class, inter-pulse interval was 0.4, 0.6, and 0.8 s for the three types, respectively. The speed-up type showed a decrease in inter-pulse interval from 0.5 to 0.4 s (Risch et al. 2013). Geographic variation within the North Atlantic has also been demonstrated (Risch et al. 2014a).

In the St. Lawrence Estuary feeding area, single, 0.4 s downsweeps have been reported from minkes in transit with frequency characteristics resembling those of individual pulses in pulse trains reported elsewhere, except for the longer duration (0.4 s versus 0.1 s; Edds-Walton 2000). Vocal activity was rare, sound production occurred in only about half of the sightings with underwater recording, calling rates did not correlate with number of animals present, and no sounds were recorded during feeding on capelin (Edds and Macfarlane 1987; Edds-Walton 2000). In contrast, higher-frequency pulse trains (4–7.5 kHz; 7 pulses/s) were reported from Artimon Bank with a mild directional beam pattern, being 3 dB stronger to the front than to the sides (Beamish and Mitchell 1973).

3.13.3.2.1 *B. a. scammoni*: North Pacific Minke Whale

In the North Pacific, the most commonly described sounds of minke whales are series of pulses that have been called "boings". A boing consists of a ~0.1–0.3 s broadband pulse (precursor), followed by a longer ringing sound, which is a rapid pulse train of ~1.5–4.5 s that appears in spectrograms as tonal with overtones, amplitude-, and frequency-modulation. The boing typically covers a frequency range of 1–5 kHz (Delarue et al. 2013a; Oswald et al. 2011; Rankin and Barlow 2005). Some boings ended in a

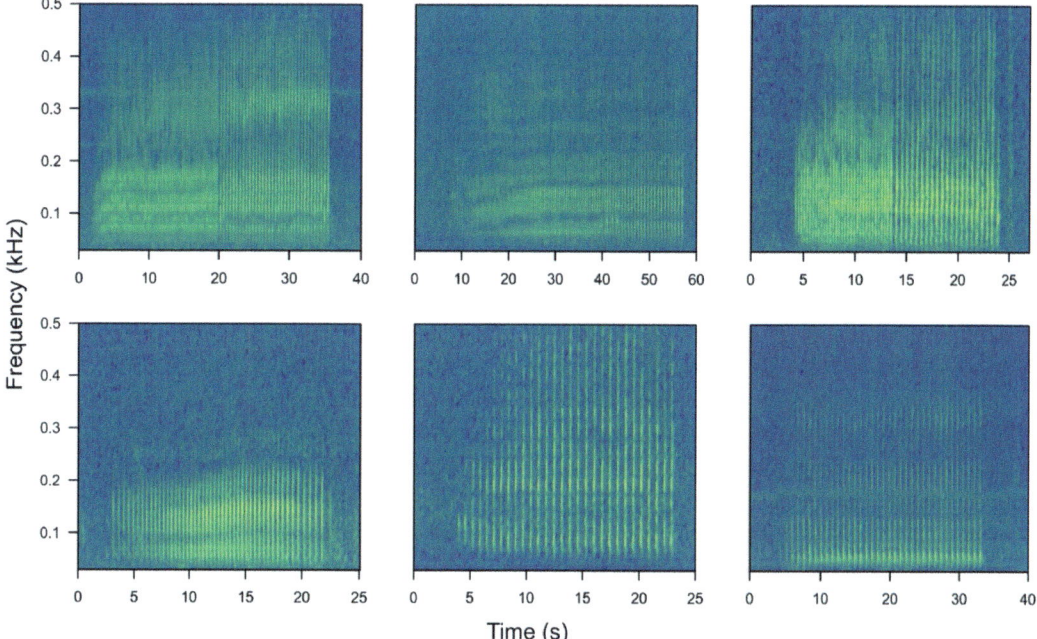

Fig. 3.16 Spectrograms of North Atlantic minke whale pulse trains. Top row: slow-down pulse trains. Bottom row: constant pulse trains (Risch 2022). Reprinted with permission from Springer Nature. Risch D, Mysterious minke whales: Acoustic diversity and variability. In: Clark CW, Garland EC (eds) Ethology and Behavioral Ecology of Mysticetes. Springer International Publishing, Cham, pp 329–348, https://doi.org/10.1007/978-3-030-98449-6_14. © Springer Nature, 2022. All rights reserved

low-frequency growl after the initial precursor and the main ringing part (Delarue et al. 2013a). The source level has been measured as ~170 dB re 1 μPa m; however, minke whales called louder in stronger, natural ambient noise (Helble et al. 2020b).

On the Hawaiian breeding grounds, minke whales produced boings at different repetition rates. Acoustically tracked minke whales exhibited a bimodal call rate centered around one boing per either 6.85 or 0.63 min; call rate increased with decreasing distance to other calling minke whales (Martin et al. 2022).

Geographic variability with different pulse repetition rates in the Eastern Pacific (92 pulses/s in a 3.6 s boing) versus Central Pacific (115 pulses/s in a 2.6 s boing) has been described (Rankin and Barlow 2005). Moreover, individual differences in boing peak frequency have been suggested (Martin et al. 2012).

Very brief pulse trains (<1 s) with energy between 300 Hz and 1400 Hz, with decreasing, constant, or increasing pulse rate were tentatively assigned to minke whales in British Columbia, Canada (Nikolich and Towers 2020).

3.13.3.2.2 *B. a.*: Southern Hemisphere Minke Whale

In the Southern Hemisphere, minke whales around Australia boing, too. In the Great Barrier Reef (QLD), the boing has a stuttered start, commencing with three broadband pulses (50–9400 Hz, 0.1 s per pulse, 20 ms between pulses; Gedamke et al. 2001). Assigning the label "Unit A" to each pulse, the immediately following Unit B consisted of two sounds: an 80 Hz tone with overtones and a narrowband (1.65–1.95 kHz), pulsed sound with peak energy at 1.8 kHz. Unit B lasted 0.24 s. Immediately following was Unit C, which continued the

1.8 kHz narrowband sound at a lower pulse repetition rate, while increasing the frequency of the tonal sound to 140 Hz (plus overtones) and pulsing the tonal sound. Altogether, the boing (named "star-wars vocalization" by Gedamke et al. 2001) lasted 1–2.5 s and mostly occurred in this AAABC arrangement, although shorter boings with a truncated or missing C-unit were also recorded. Boings were emitted in bouts (and likened to other mysticete species' song) with constant intervals up to 3–4 min. Rapid vocal activity with boings every 1–2 s and brief grunts in the middle of boing doublets were also reported (Gedamke et al. 2001). In addition to boings, other non-song sounds that occur singly have been documented, including 0.2–0.3 s downsweeps from 250 Hz to 50 Hz (Gedamke et al. 2001). Source levels of boings were 150–165 dB re 1 µPa m and those of downsweeps were 148–160 dB re 1 µPa m (Gedamke et al. 2001).

On the Western Australian coast, minke whales do not stutter, but—after a brief and broadband precursor—go straight into the ringing part of the boing (Unit C from the Great Barrier Reef). Constant boing repetition intervals of 2–3 min are frequently observed on Australia's north-western shelf, as are boing doublets with brief grunts in the middle (Erbe et al. 2017, and CE pers. obs.) (Fig. 3.17).

3.13.3.3 *Balaenoptera bonaerensis*: Antarctic Minke Whale

Antarctic minke whales are thought to migrate between their feeding grounds in the Antarctic and their breeding grounds at lower latitudes; however, groups overwintering in Antarctica have repeatedly been detected, calling activity in year-long Antarctic recordings peaked in the winter, and a positive correlation between the number of minke whale calls and sea ice cover has been shown (Dominello and Širović 2016; Filun et al. 2020). Moreover, acoustic detections of Antarctic minke whales in lower-latitude breeding grounds peaked during summer (Rossi-Santos et al. 2022; Shabangu et al. 2020).

The most stereotypical sound of Antarctic minke whales is the "bioduck" sound. It is a packet of 2–12 downswept pulses. The typical frequency range of the downswept pulses is ~300–60 Hz, yet overtones might reach >2 kHz. Each pulse lasts ~0.1–0.3 s. Inter-pulse intervals are ~0.1–0.4 s. One packet lasts ~1–5 s, and a packet may be emitted every ~3 s for several minutes at a time (Dominello and Širović 2016; Risch et al. 2014b; Rossi-Santos et al. 2022; Shabangu et al. 2020). The source level is ~140 dB re 1 µPa m (Risch et al. 2014b). Dominello and Širović (2016) identified four variants of the bioduck off the western Antarctic Peninsula during winter (with one variant having two subtypes), which differed in the number of pulses per package, frequency range, pulse duration, and inter-pulse interval (Fig. 3.18). Filún and van Opzeeland (2023) made a case for Antarctic minke whale song, presenting different unit

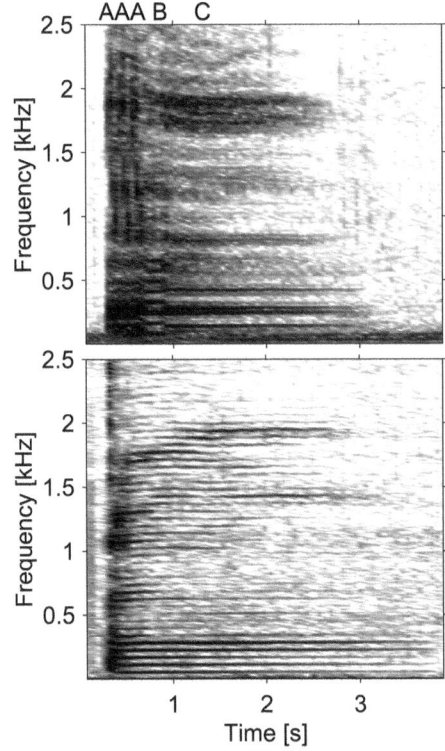

Fig. 3.17 Dwarf minke whale boings recorded off the Australian northeast coast (top; "star-wars" vocalization; Units A, B, and C; fs = 44,100 Hz, NFFT = 4096, Hann window, 50% overlap) and northwest coast (bottom; fs = 16,000 Hz, NFFT = 2048, Hann window, 50% overlap)

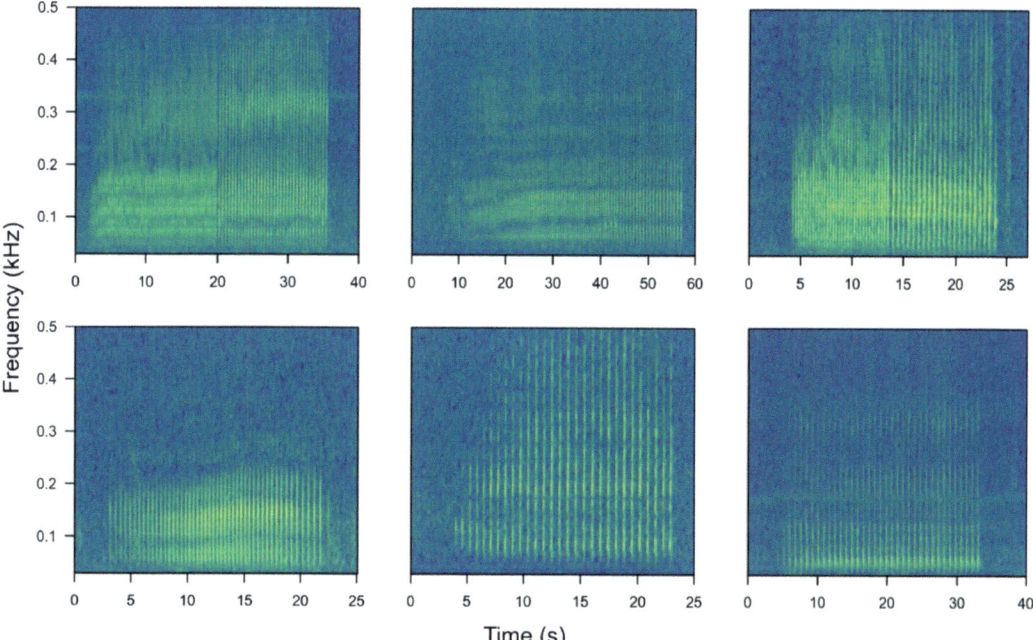

Fig. 3.16 Spectrograms of North Atlantic minke whale pulse trains. Top row: slow-down pulse trains. Bottom row: constant pulse trains (Risch 2022). Reprinted with permission from Springer Nature. Risch D, Mysterious minke whales: Acoustic diversity and variability. In: Clark CW, Garland EC (eds) Ethology and Behavioral Ecology of Mysticetes. Springer International Publishing, Cham, pp 329–348, https://doi.org/10.1007/978-3-030-98449-6_14. © Springer Nature, 2022. All rights reserved

low-frequency growl after the initial precursor and the main ringing part (Delarue et al. 2013a). The source level has been measured as ~170 dB re 1 µPa m; however, minke whales called louder in stronger, natural ambient noise (Helble et al. 2020b).

On the Hawaiian breeding grounds, minke whales produced boings at different repetition rates. Acoustically tracked minke whales exhibited a bimodal call rate centered around one boing per either 6.85 or 0.63 min; call rate increased with decreasing distance to other calling minke whales (Martin et al. 2022).

Geographic variability with different pulse repetition rates in the Eastern Pacific (92 pulses/s in a 3.6 s boing) versus Central Pacific (115 pulses/s in a 2.6 s boing) has been described (Rankin and Barlow 2005). Moreover, individual differences in boing peak frequency have been suggested (Martin et al. 2012).

Very brief pulse trains (<1 s) with energy between 300 Hz and 1400 Hz, with decreasing, constant, or increasing pulse rate were tentatively assigned to minke whales in British Columbia, Canada (Nikolich and Towers 2020).

3.13.3.2.2 *B. a.*: Southern Hemisphere Minke Whale

In the Southern Hemisphere, minke whales around Australia boing, too. In the Great Barrier Reef (QLD), the boing has a stuttered start, commencing with three broadband pulses (50–9400 Hz, 0.1 s per pulse, 20 ms between pulses; Gedamke et al. 2001). Assigning the label "Unit A" to each pulse, the immediately following Unit B consisted of two sounds: an 80 Hz tone with overtones and a narrowband (1.65–1.95 kHz), pulsed sound with peak energy at 1.8 kHz. Unit B lasted 0.24 s. Immediately following was Unit C, which continued the

1.8 kHz narrowband sound at a lower pulse repetition rate, while increasing the frequency of the tonal sound to 140 Hz (plus overtones) and pulsing the tonal sound. Altogether, the boing (named "star-wars vocalization" by Gedamke et al. 2001) lasted 1–2.5 s and mostly occurred in this AAABC arrangement, although shorter boings with a truncated or missing C-unit were also recorded. Boings were emitted in bouts (and likened to other mysticete species' song) with constant intervals up to 3–4 min. Rapid vocal activity with boings every 1–2 s and brief grunts in the middle of boing doublets were also reported (Gedamke et al. 2001). In addition to boings, other non-song sounds that occur singly have been documented, including 0.2–0.3 s downsweeps from 250 Hz to 50 Hz (Gedamke et al. 2001). Source levels of boings were 150–165 dB re 1 μPa m and those of downsweeps were 148–160 dB re 1 μPa m (Gedamke et al. 2001).

On the Western Australian coast, minke whales do not stutter, but—after a brief and broadband precursor—go straight into the ringing part of the boing (Unit C from the Great Barrier Reef). Constant boing repetition intervals of 2–3 min are frequently observed on Australia's north-western shelf, as are boing doublets with brief grunts in the middle (Erbe et al. 2017, and CE pers. obs.) (Fig. 3.17).

3.13.3.3 *Balaenoptera bonaerensis*: Antarctic Minke Whale

Antarctic minke whales are thought to migrate between their feeding grounds in the Antarctic and their breeding grounds at lower latitudes; however, groups overwintering in Antarctica have repeatedly been detected, calling activity in year-long Antarctic recordings peaked in the winter, and a positive correlation between the number of minke whale calls and sea ice cover has been shown (Dominello and Širović 2016; Filun et al. 2020). Moreover, acoustic detections of Antarctic minke whales in lower-latitude breeding grounds peaked during summer (Rossi-Santos et al. 2022; Shabangu et al. 2020).

The most stereotypical sound of Antarctic minke whales is the "bioduck" sound. It is a packet of 2–12 downswept pulses. The typical frequency range of the downswept pulses is ~300–60 Hz, yet overtones might reach >2 kHz. Each pulse lasts ~0.1–0.3 s. Inter-pulse intervals are ~0.1–0.4 s. One packet lasts ~1–5 s, and a packet may be emitted every ~3 s for several minutes at a time (Dominello and Širović 2016; Risch et al. 2014b; Rossi-Santos et al. 2022; Shabangu et al. 2020). The source level is ~140 dB re 1 μPa m (Risch et al. 2014b). Dominello and Širović (2016) identified four variants of the bioduck off the western Antarctic Peninsula during winter (with one variant having two subtypes), which differed in the number of pulses per package, frequency range, pulse duration, and inter-pulse interval (Fig. 3.18). Filún and van Opzeeland (2023) made a case for Antarctic minke whale song, presenting different unit

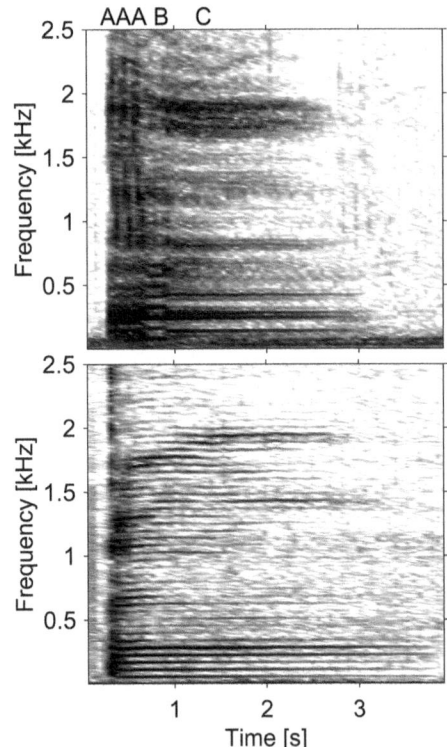

Fig. 3.17 Dwarf minke whale boings recorded off the Australian northeast coast (top; "star-wars" vocalization; Units A, B, and C; fs = 44,100 Hz, NFFT = 4096, Hann window, 50% overlap) and northwest coast (bottom; fs = 16,000 Hz, NFFT = 2048, Hann window, 50% overlap)

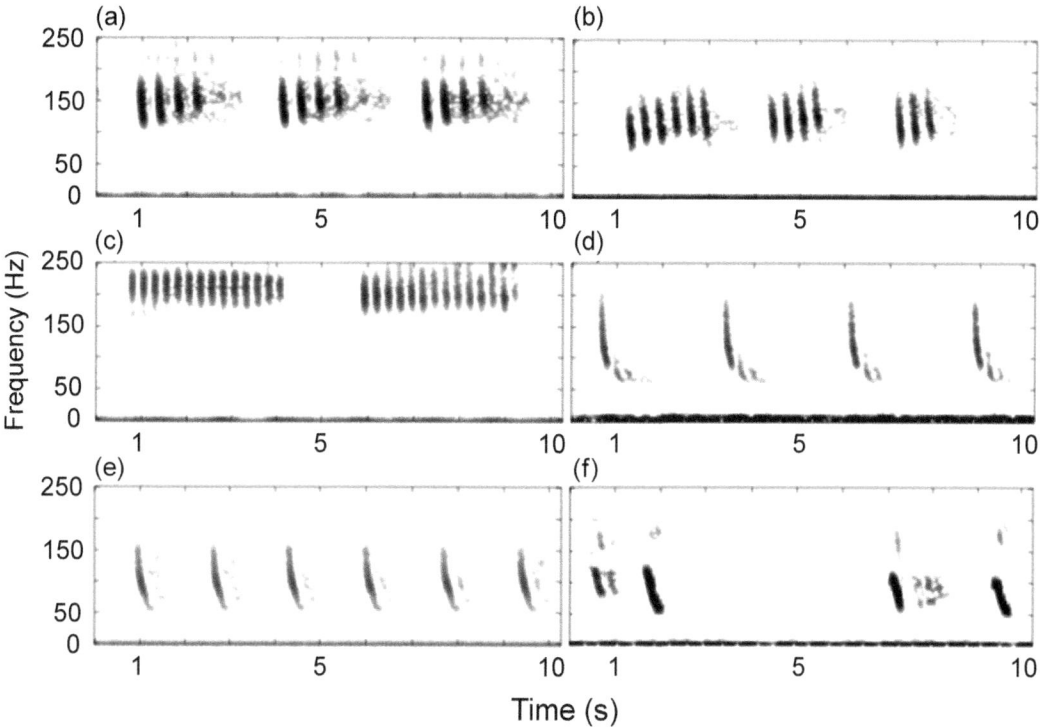

Fig. 3.18 Bioduck variants recorded off the western Antarctic Peninsula. Variants A1 (**a**), A2 (**b**), B (**c**), C (**d**), and D (**e**), as well as downsweeps (**f**). Reprinted with permission from John Wiley & Sons. Dominello T, Širović A, Seasonality of Antarctic minke whale (*Balaenoptera bonaerensis*) calls off the western Antarctic Peninsula. Mar Mamm Sci 32 (3):826–838; https://doi.org/10.1111/mms.12302. © Society for Marine Mammalogy, 2016. All rights reserved

types in the Southern Ocean. Rossi-Santos et al. (2022) identified nine variants grouped into four clusters on Brazilian breeding grounds during summer. They stated the following relationship between bioduck features: The more pulses there were in a packet, the shorter the inter-pulse interval; however, the packet duration still increased with the number of pulses.

The next most commonly described sound made by Antarctic minke whales is a 0.1–0.4 s downsweep from ~130 Hz to 60 Hz with a source level of up to 165 dB re 1 μPa m (Casey et al. 2022; Dominello and Širović 2016; Risch et al. 2014b; Schevill and Watkins 1972; Fig. 3.18f). Additional low-frequency sounds (<300 Hz, <1 s; rumble, boom, and growl) were described by Casey et al. (2022). With animal-borne audio-video tags, Casey and colleagues were able to associate call types with foraging dives (rumble, growl), increased presence of conspecifics, and day versus night (Fig. 3.19).

3.13.3.4 *Balaenoptera borealis*: Sei Whale

The sei whale is a global species comprising two subspecies: the northern sei whale (*B.b. borealis*) and the southern sei whale (*B.b. schlegelii*). Both subspecies undertake annual migrations from temperate, subtropical waters in winter to cooler, subpolar waters in summer. The most frequently reported sei whale sounds are downsweeps from ~100 Hz to ~30 Hz, lasting 1 s, as reported from the North and South Atlantic, as well as North and South Pacific (Baumgartner et al. 2008; Cerchio and Weir 2022; Español-Jiménez et al. 2019; Macklin 2022; Newhall et al. 2012; Rankin and Barlow 2007; Romagosa et al. 2015). Briefer, lower-frequency downsweeps (~50–30 Hz, <0.6 s) have been reported from the eastern

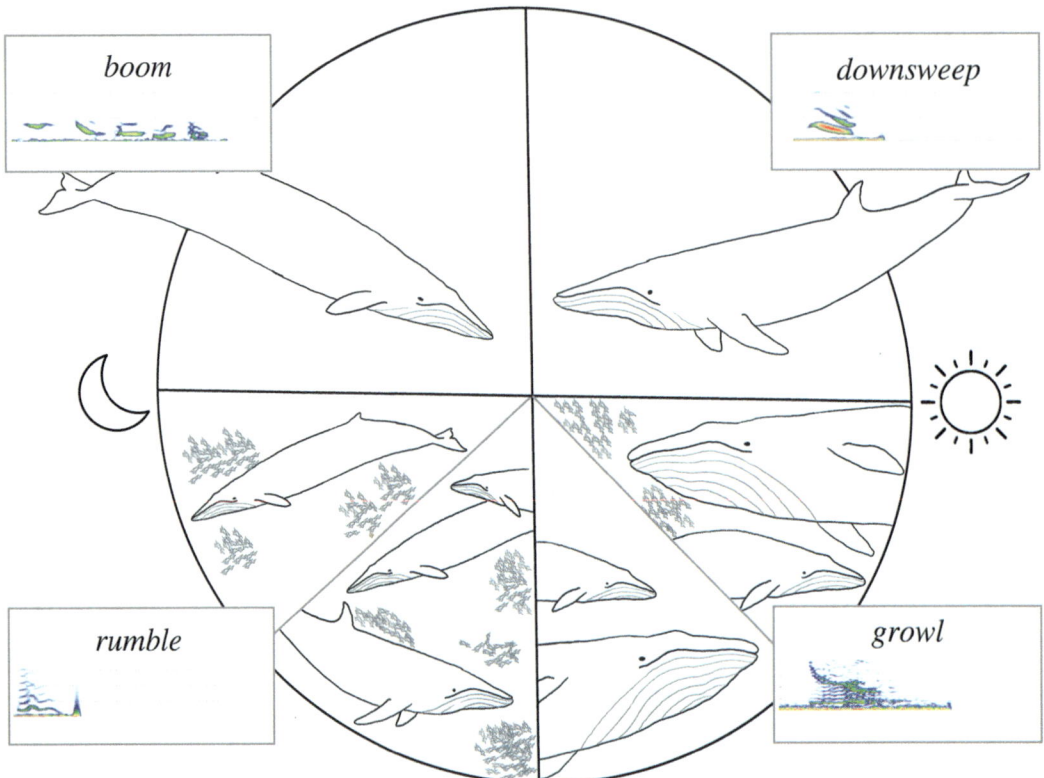

Fig. 3.19 "Schematic illustration depicting the association between each call type and specific behavioural and diel conditions. This figure represents the dominant diel condition and foraging state during which each call type was produced. The analysis of whether the call was produced in the presence of close conspecifics is limited to those calls produced during daylight hours when video data was available. Boom calls occurred most frequently at night while animals were in a non-foraging state; when they were observed during the day and concurrent video data was available, booms were produced when conspecifics were not present. Rumbles were similarly emitted most frequently at night and were predominantly made during foraging bouts. When concurrent video data was available, rumbles were produced both when conspecifics were present and absent. Downsweeps were produced most often during daylight hours and typically occurred in non-foraging contexts when no other whales were within the camera's field of view. Growls were also most common during the day and were associated with both foraging and non-foraging behavioural states while animals were in the presence of conspecifics. Line drawing by R. Jones." Reprinted from Casey CB, Weindorf S, Levy E, Linsky JMJ, Cade DE, Goldbogen JA, Nowacek DP, Friedlaender AS (2022) Acoustic signalling and behaviour of Antarctic minke whales (*Balaenoptera bonaerensis*). R Soc Open Sci 9 (7):211557. https://doi.org/10.1098/rsos.211557. © Casey et al. 2022. Published CC BY 4.0; https://creativecommons.org/licenses/by/4.0/

USA (Tremblay et al. 2019). Higher-frequency downsweeps (~600–300 Hz, ~1 s) have been reported from the Antarctic (McDonald et al. 2005). Upsweeps, up-downsweeps, and more variable frequency-modulated tones in this low- to mid-frequency range have been reported from the Antarctic (Calderan et al. 2014; McDonald et al. 2005) and Southwest Pacific (Cerchio and Weir 2022). Sei whales in the Antarctic further produce 0.5–2 s constant-frequency sounds at 200–500 Hz, with or without higher-frequency harmonics, including multi-part frequency stepping tonals (McDonald et al. 2005). They also produce broadband sounds, which the authors described as growls or whooshes (McDonald et al. 2005). Source levels of the low-frequency

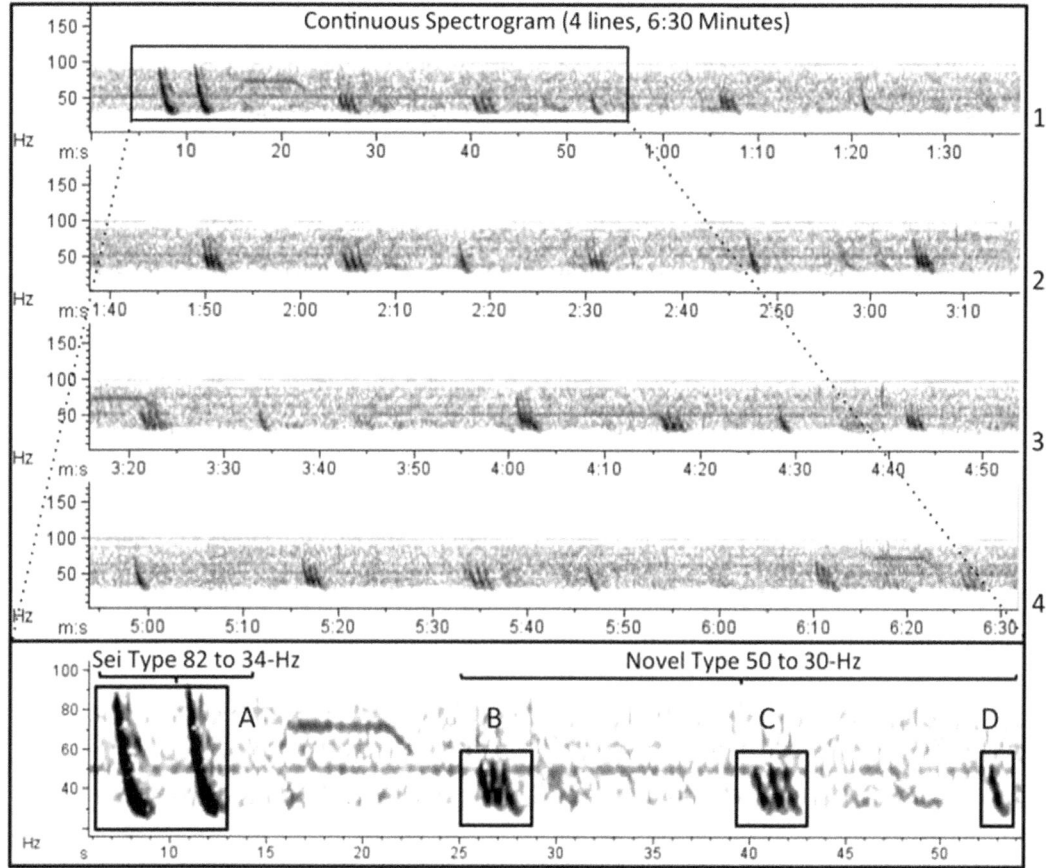

Fig. 3.20 Sei whale song of units A, B, C, and D. Reprinted from Tremblay CJ, Parijs SMV, Cholewiak D (2019) 50 to 30-Hz triplet and singlet down sweep vocalizations produced by sei whales (*Balaenoptera borealis*) in the western North Atlantic Ocean. J Acoust Soc Am 145 (6):3351–3358. https://doi.org/10.1121/1.5110713. With permission from the ASA. © Acoustical Society of America, 2019. All rights reserved

constant-frequency and frequency-modulated sounds are 156–179 dB re 1 μPa m (McDonald et al. 2005; Newhall et al. 2012; Romagosa et al. 2015; Tremblay et al. 2019; Wang et al. 2016). Sei whale low-frequency downsweeps may be emitted as singlets, doublets, or triplets (Baumgartner et al. 2008; Español-Jiménez et al. 2019; Newhall et al. 2012; Tremblay et al. 2019) and be arranged into song (e.g., of AABCDBD pattern, Fig. 3.20; Tremblay et al. 2019).

Mid-frequency (1.0–5.5 kHz) broadband sounds described as whooshes that lasted a few seconds and brief (<0.1 s) noisy squeaks and creaks that occurred in sequences have been reported from the North Atlantic (Knowlton et al. 1991; Thompson et al. 1979) and Southwestern Pacific (Cerchio and Weir 2022). Cerchio and Weir (2022) showed how these sounds were arranged, together with low-frequency downsweeps, into subphrases, phrases, and song.

Newhall et al. (2012) suggested, based on differences between two animals, that there might be individual stereotypical downsweep characteristics. There also appear to be differences based on activity state. Sei whale calls were heard more during the day than at night, attributed to social behavior (Baumgartner and Fratantoni 2008) because whales fed on copepods at night, at the same time that acoustic behavior was reduced. Downsweep calls might function as contact calls between sei whales, as they were recorded at daytime in the presence of sei whales that were not feeding (Baumgartner and Fratantoni 2008; McDonald et al. 2005; Romagosa et al. 2015). Near the Falkland Islands, low-frequency downsweeps were present from early December through late June. However, song only started in March and might thus indicate the onset of breeding behavior as the timing correlates with when males leave high-latitude feeding grounds for lower-latitude breeding grounds (Cerchio and Weir 2022).

3.13.3.5 *Balaenoptera edeni*: Bryde's Whale

Bryde's whales inhabit tropical to subtropical waters in both Northern and Southern Hemispheres. Two subspecies are currently recognized: the larger Bryde's whale (*B.e. brydei*), which occurs worldwide in offshore waters, and the smaller Eden's whale (*B.e. edeni*), which might be limited to the Indo-Pacific region. The literature describing the Bryde's whale vocal repertoire is limited and contains inconsistent terminology for sound types, making it difficult to summarize across different studies. To date, there is no evidence that Bryde's whales sing.

Bryde's whale populations from the North Pacific Ocean have been recorded at various sites: the Eastern Tropical Pacific, Gulf of California, Hawaii, Japan, and China (Cummings et al. 1986; Edds et al. 1993; Heimlich et al. 2005; Helble et al. 2016; Oleson et al. 2003; Viloria-Gómora et al. 2015; Wang et al. 2022). Bryde's whales in the Eastern Tropical Pacific produce call types that have tonal, harmonic, constant-frequency, frequency-modulated, and/or amplitude-modulated characteristics (Heimlich et al. 2005; Oleson et al. 2003). The constant-frequency calls had an average peak frequency of ~35 Hz and ranged in frequency between ~15 and 80 Hz, with an average duration of ~2 s (Fig. 3.21). The constant-frequency calls were split into classes Be1–Be5 by Oleson et al. (2003). Heimlich et al. (2005) found two subtypes for Be1 (swept alternating tonal and non-swept alternating tonal) and two subtypes for Be2 (high burst-tonal and low burst-tonal); they aptly described Be3 as the harmonic tone. Calls Be3 and Be4 were also recorded in Hawaii and the Gulf of California, respectively (Helble et al. 2016; Viloria-Gómora et al. 2015). In addition, Bryde's whales in the Eastern Tropical Pacific produced a downswept call (Be6), which, on average, started at ~200 Hz and decreased to 80 Hz over an average duration of ~3 s (Oleson et al. 2003; Fig. 3.21). Two additional call types (Be8a, b) were recorded off Japan, and had a peak frequency of ~45 Hz, ranged in frequency from ~40 to 190 Hz, and had an average duration of ~0.35 s (Oleson et al. 2003).

In the Gulf of California, Bryde's whale vocalizations were first described by Cummings et al. (1986) and Edds et al. (1993) who identified three call types (unpulsed moans, individual pulses, and pulsed moans), which ranged in frequency from 70 to 900 Hz and were ≤1.5 s in duration. Edds et al. (1993) observed the whales while recording in a feeding area and individual pulses were produced repeatedly in a descending series (IPI of 0.08–0.10 s) only when a calf remained at the surface as the mother dove, presumably to feed. The relatively high-frequency pulse series (900–700 Hz) ceased when the mother surfaced; therefore, these calls may be contact calls specific to mother-calf pairs. More recently, three more Gulf of California call types were identified (Be10–12; Viloria-Gómora et al. 2015). The Be10 and Be11 calls were downsweeping vocalizations that had short average durations (0.1–0.2 s) and occurred over two different frequency ranges: ~130–90 Hz and 240–110 Hz, respectively (Viloria-Gómora et al. 2015). The Be12 call was only recorded once. This call had a peak frequency of ~110 Hz and ranged in frequency from 90–150 Hz over 1.3 s (Viloria-Gómora et al. 2015). Downsweeps from ~150 Hz to 30 Hz, peaking at ~90 Hz, and lasting ~0.5 s were recorded off China (Wang et al. 2022). In the South Pacific, Bryde's whale calls similar to those described by Oleson et al. (2003) from the Eastern Tropical Pacific have been reported from New Zealand (Constantine et al. 2015; McDonald 2006; Putland et al. 2018).

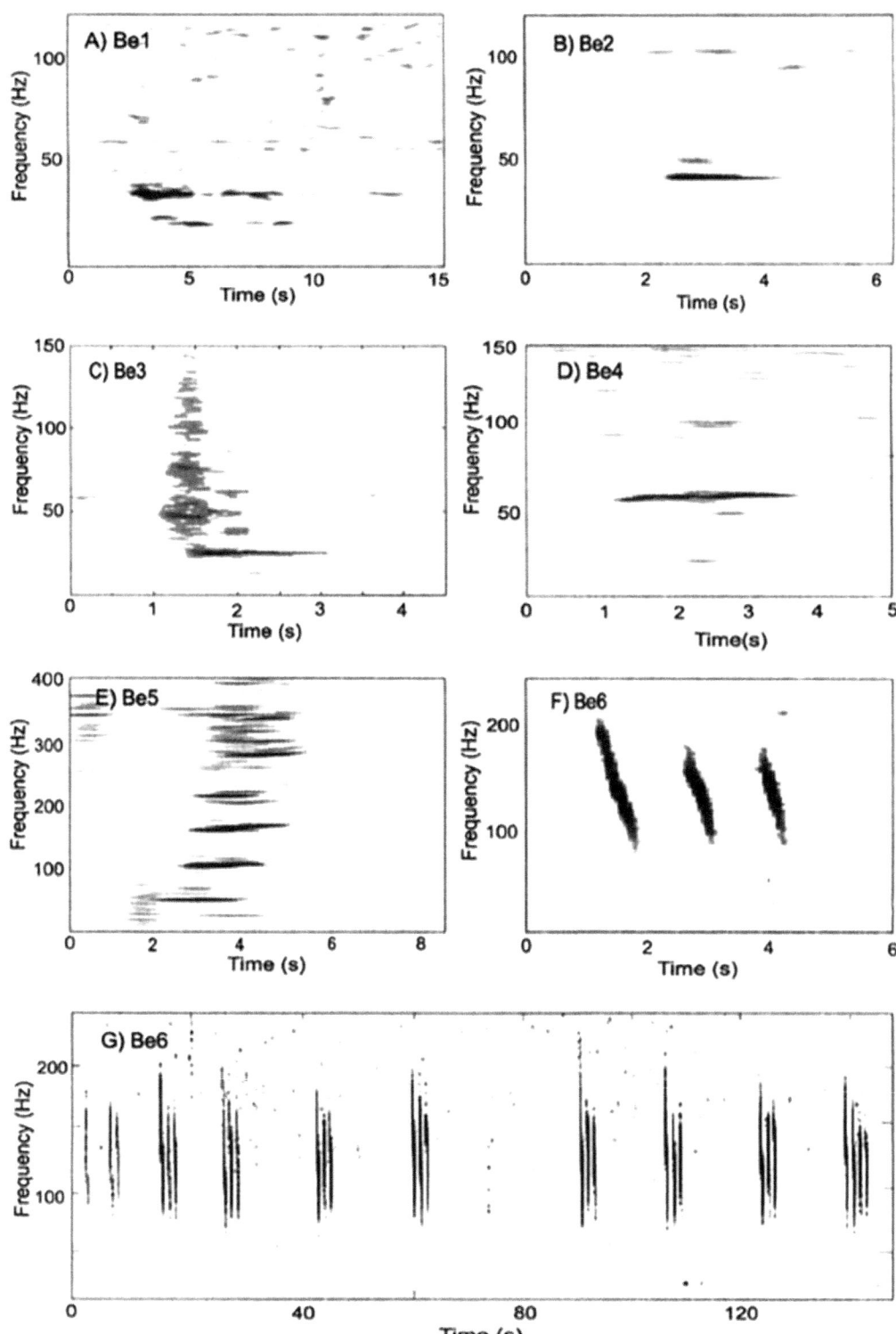

Fig. 3.21 Spectrograms of Bryde's whale calls (Oleson et al. 2003). Reprinted with permission from John Wiley & Sons. Oleson EM, Barlow J, Gordon J, Rankin S, Hildebrand JA, Low frequency calls of Bryde's whales, Mar Mamm Sci 19:160–172, https://doi.org/10.1111/j.1748-7692.2003.tb01119.x. © Society for Marine Mammalogy, 2003. All rights reserved

In the North Atlantic Ocean (more specifically, in the Caribbean Sea), call type Be7 has been reported as constant-frequency with an average peak frequency of ~43 Hz and duration of 1.6 s (Oleson et al. 2003). Bryde's-whale-like calls were reported from the Gulf of Mexico (Rice et al. 2014; Širović et al. 2014; Soldevilla et al. 2022a), but this species has been determined to be different and is now called the Rice's whale. In the South Atlantic, five Bryde's whale call types were recorded off the coast of Rio de Janeiro, Brazil: series of discrete pulses (PS1), low-frequency tonal calls (LFT), downsweep calls (FMT), and two biphonic calls (TM1 and TM2; Figueiredo and Simao 2014). The PS1 call, on average, ranged from ~230 Hz to 560 Hz over 0.8 s. Based on spectrographic comparison, Figueiredo and Simao concluded the LFT call was similar to the Be3 call reported by Oleson et al. (2003) and the harmonic tone call reported by Heimlich et al. (2005), despite having a lower average fundamental frequency than these calls. The LFT call, on average, ranged from ~10 Hz to 20 Hz over 1 s. The FMT call was a downsweep from ~670 Hz to ~420 Hz over 2 s, and thus higher in frequency than the Be6 call described by Oleson et al. (2003). The TM1 call ranged in frequency from ~90–120 Hz over 0.9 s, and the TM2 call had a frequency range of ~50–110 Hz with a duration of 1.2 s (Figueiredo and Simao 2014).

The function of Bryde's whale calls remains unclear, but some authors have proposed several possible functions. Oleson et al. (2003) was the first to suggest that the Be3 call may function as a contact call to maintain acoustic contact between Bryde's whales. Helble et al. (2016) used the Be3 call to acoustically track Bryde's whales off the coast of Hawaii. Based on their observations of whale movements and corresponding vocal activity, they concluded the Be3 call is a contact call that may play a role in group cohesion. Viloria-Gómora et al. (2015) found that the Be4 call was only produced by groups of Bryde's whales with two or more individuals, suggesting it may have a social function. In addition, the Be10, Be11, and Be12 calls were recorded when a Bryde's whale was feeding, and so, these calls may play a role in foraging.

3.13.3.6 *Balaenoptera musculus*: Blue Whale

The blue whale occurs in all of Earth's oceans. There are currently four subspecies of blue whale (*Balaenoptera musculus*): the pygmy blue whale (*B.m. musculus brevicauda*), the northern Indian Ocean blue whale (*B.m. indica*), the Antarctic blue whale (*B.m. intermedia*), and the northern blue whale (*B.m. musculus*). These subspecies migrate annually between summer feeding at high latitude and winter breeding grounds at low latitude. They produce song and non-song sounds. Blue whales have been further divided into regionally distinct "acoustic types" corresponding to distinct song types, which may be indicative of discrete populations (McDonald et al. 2006).

Blue whale song consists of structured sequences of song units. For the North Atlantic song, five units have been described (i.e., A, B, hybrid, arch sound, and 9 Hz sound), and phrases may contain 1–8 units (Berchok et al. 2006; Mellinger and Clark 2003; Nieukirk et al. 2004). Unit A is an ~18 Hz, ~14 s, constant-frequency tone. Unit B is a ~13 s, ~18–15 Hz downsweep. The hybrid song unit consist of units A and B sung as a single sound without an interval of silence between the units (Berchok et al. 2006; Mellinger and Clark 2003). The arch sound and the 9 Hz sound are song units that have only been reported by Mellinger and Clark (2003). The arch sound was named after its spectrographic contour, starting at ~56 Hz, rising to ~69 Hz, and then falling to ~35 Hz, over a duration of ~6 s. The 9 Hz sound lasts ~2–5 s. Song phrases may consist of a single unit (A, B, or the hybrid unit), exhibit a sequence of A and B units (e.g., A-A, A-B, A-A-B, or A-B-A), or incorporate the arch or 9 Hz sounds (Berchok et al. 2006; Mellinger and Clark 2003; Nieukirk et al. 2004).

For the North Pacific, two song types have been described: the Northwest Pacific song, referred to as the North Pacific song in McDonald et al. (2006), and the Northeast Pacific song (Stafford 2003; Stafford et al. 2001, 2005). The Northwest Pacific song consists of four song units labelled I, II, III, and IV (Stafford et al. 2001).

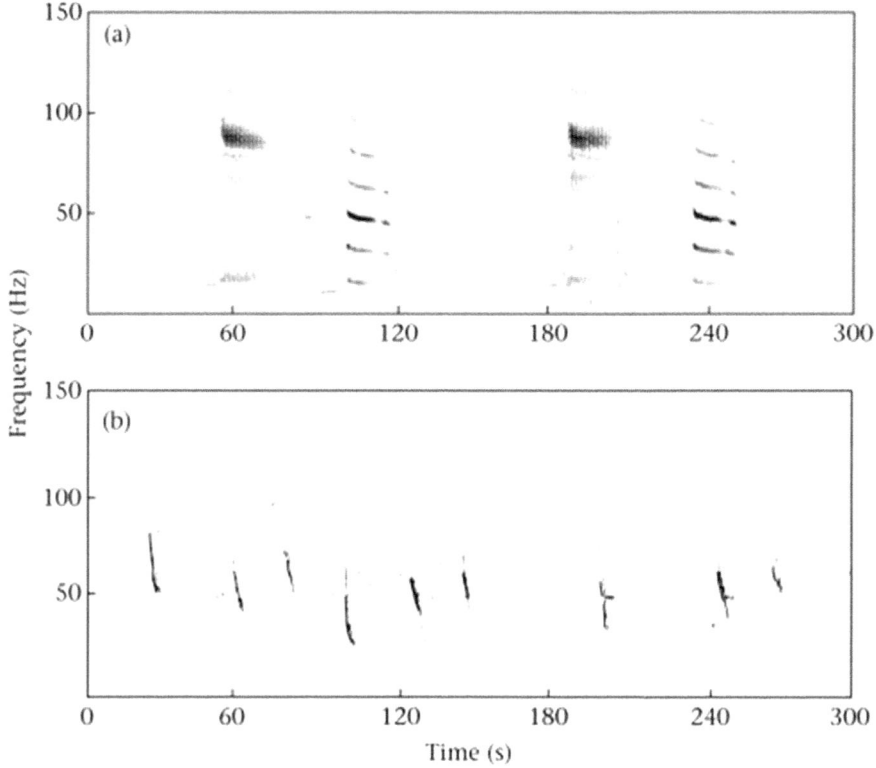

Fig. 3.22 Northeast Pacific blue whale (**a**) A and B units in song and (**b**) non-song D-calls (Oleson et al. 2007b). Reprinted by permission from Elsevier. Oleson EM, Wiggins SM, Hildebrand JA, Temporal separation of blue whale call types on a southern California feeding ground. Anim Behav 74(4), 881–894, https://doi.org/10.1016/j.anbehav.2007.01.022. © The Association for the Study of Animal Behaviour, 2007. All rights reserved

Units I–III are frequency modulated between ~18 and ~ 20 Hz, lasting ~23, 17, and 11 s, respectively. Unit IV sweeps down from ~56 to ~32 Hz over ~7 s. The Northeast Pacific song contains three song units: A, B, and C (Fig. 3.22; McDonald et al. 2001; Oleson et al. 2007b; Rivers 1997; Stafford et al. 1998, 1999b, 2001; Thompson et al. 1996). Unit A is amplitude modulated with a minimum frequency of ~16 Hz and a duration of ~17 s. Unit B is frequency modulated decreasing from ~18 Hz to ~17 Hz over ~16 s. Source levels of 178 dB re 1 µPa m and 186 dB re 1 µPa m (10–110 Hz) have been reported for the A and B units, respectively (McDonald et al. 2001). The C-unit is of a lower frequency than A and B, rising from ~11 to 12 Hz over ~8 s. Typically, song phrases consist of 2–3 units (A-B, A-C-B), although A might be followed by multiple Bs (Lewis and Širović 2018; McDonald et al. 2001; Stafford et al. 1998, 1999a, b, 2001, 2005.

For the South Pacific, four song types have been described: the Southwest Pacific song, the Solomon Sea song, and two types of Southeast Pacific song, with units A, B, C, D, and E (Buchan et al. 2014; Cummings and Thompson 1971; Frank and Ferris 2011; Kibblewhite et al. 1967; McDonald et al. 2006; Miller et al. 2014a; Patris et al. 2020; Stafford et al. 1999b). These units can be frequency and/or amplitude modulated, roughly cover a frequency range of 17–26 Hz, and last 3–14 s. Shorter (<1 s) and higher-frequency (<400 Hz) precursors have also been measured (Buchan et al. 2014). Source levels as high as 188 dB re 1 μPa m have been reported (Cummings and Thompson 1971).

Songs in the Indian Ocean have been described in several studies (Bouffaut et al. 2021; Cerchio et al. 2020; Jolliffe et al. 2019; McCauley et al. 2000; Panicker and Stafford 2021; Stafford et al. 2011). The songs most commonly consist of three units of tonal and/or pulsed nature with source levels of 167–179 dB re 1 μPa m (Bouffaut et al. 2021; Gavrilov et al. 2011; Samaran et al. 2010). Songs have been compared across regions (Fig. 3.23), and so, there might be at least five different acoustic populations of pygmy blue whales (Cerchio et al. 2020; Leroy et al. 2021; Stafford et al. 2011), in addition to Antarctic blue whales. Intra-regional variation was investigated by Jolliffe et al. (2019) for the Eastern Indian Ocean pygmy blue whale population. Song structure changed over the years. The inter-song interval increased by 0.3 s/y over 2003–2017 in the Perth Canyon. Jolliffe et al. (2019) postulated that song variability might be related to the number of background singers driving individuals to alter their song to stand out in the crowd.

The Southern Ocean song type, sung by Antarctic blue whales, generally consists of a vocalization referred to as the Z-call because of its Z-shape in spectrograms (e.g., Gavrilov et al. 2012; Leroy et al. 2016; Miller et al. 2014a; Shabangu et al. 2017; Fig. 3.3b). The Z call has been described as consisting of three parts (e.g., Stafford et al. 1999b, 2004), three components (e.g., Širović et al. 2004), or three units (e.g., McDonald et al. 2006; Rankin et al. 2005), which are repeated as a phrase to make a song. Using the terminology of Payne and McVay (1971), in which a song unit contains no silent gaps, the Z-call is a song unit that consists of three subunits. The first subunit is a ~9 s tonal sound at ~28 Hz. The second subunit is a short, ~1 s downsweep from ~28 Hz to 18 Hz. The third subunit is an ~8 s tonal sound at ~18 Hz. Some authors have also noted two additional song units, the 28 Hz downsweep and 28 Hz tone, also referred to as the 27 Hz tone (Rankin et al. 2005; Stafford et al. 2004, 1999b). The 28 Hz downsweep is a sound that includes the first and second subunits of the Z call, and starts at ~28 Hz and decreases to ~19 Hz (Rankin et al. 2005). The 28 Hz tone (or 27 Hz tone) is a sound with only the first subunit of the Z call that occurs at ~28 Hz. However, it is unclear if these sounds are distinct song units or degraded forms of the Z-call. Source levels as high as 191 dB re 1 μPa m have been reported (Bouffaut et al. 2021; Miller et al. 2021; Samaran et al. 2010; Širović et al. 2007). The peak frequency of song units of all blue whale subspecies is declining over time (see Sect. 3.10.1).

All subspecies of blue whale also produce non-song sounds, which occur singly or in bouts. Seven consecutive long (15–18 s), low-frequency moans (20–18 Hz) were produced by a single North Atlantic blue whale observed in transit through a known feeding area (Edds 1982). Those sounds were consistent with previous recordings from unidentified sources suspected to be Atlantic blue whales (e.g., Thompson et al. 1979), and very different from sounds reported from the Pacific.

In the Northeast Pacific, song units A and B may also occur outside of song (Oleson et al. 2007a). The most common non-song sound reported in several geographic regions is the downsweeping D-call (although it has not always been called that). It sweeps from ~100 to 30 Hz over ~1–2 s with a source level of <190 dB re 1 μPa m and it has commonly been associated with foraging (Fig. 3.22; Akamatsu et al. 2014; Barlow et al. 2023; Berchok et al. 2006; Gavrilov et al. 2011; McDonald et al. 2001; Miller et al. 2021; Oleson et al. 2007a; Rankin et al. 2005). Additional non-song sounds in the range ~10–120 Hz, lasting 1–4 s have been described for the Eastern Indian Ocean pygmy blue whale (Recalde-Salas et al. 2014).

3.13.3.7 *Balaenoptera omurai*: Omura's Whale

Omura's whales were only described as separate from Bryde's whales, which have similar physical appearances, in 2003. They occur in tropical waters in the Indian, Atlantic, and Western Pacific Oceans (Cerchio et al. 2019). The first description of Omura's whale song did not occur until

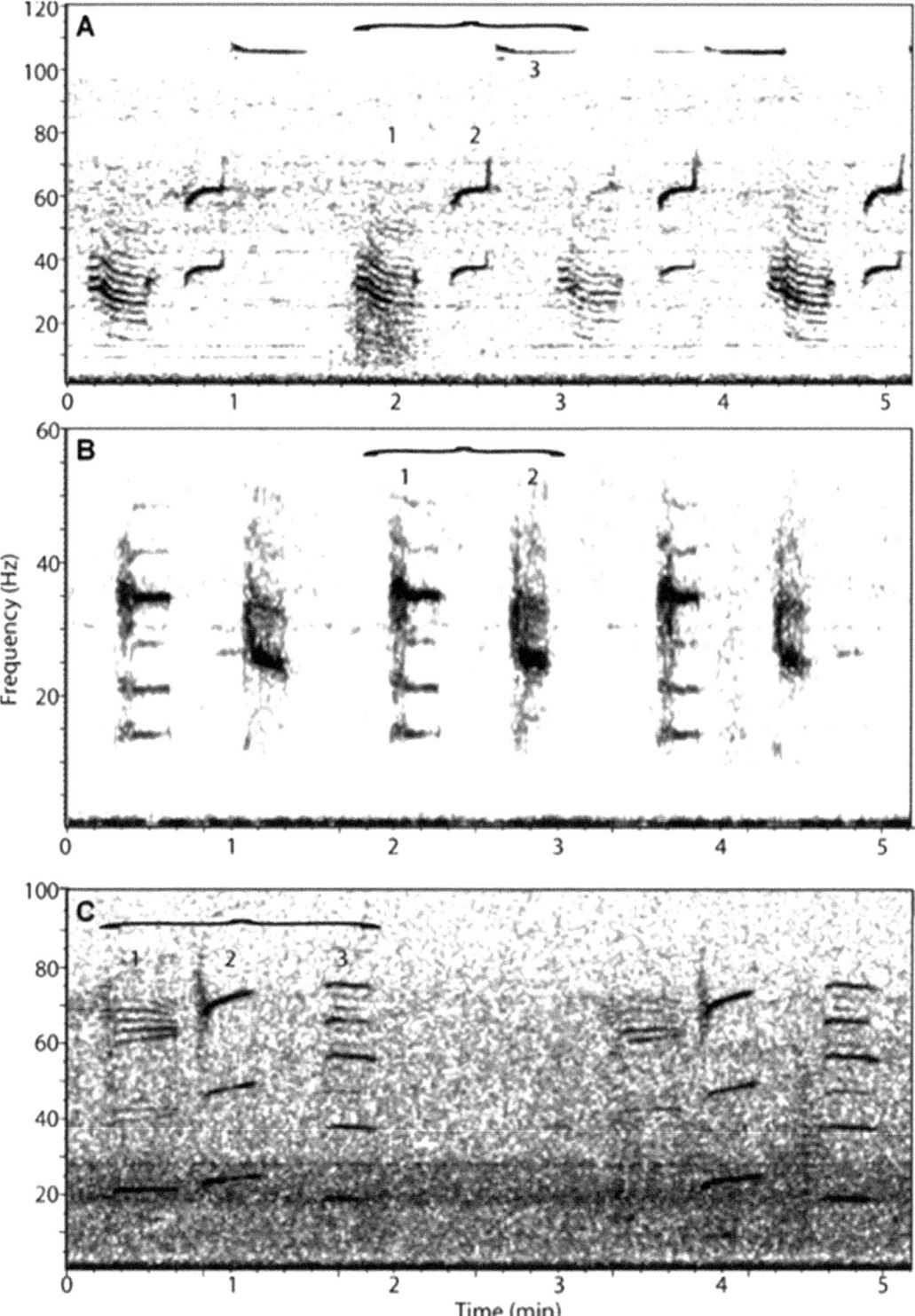

Fig. 3.23 Blue whale song types in the Indian Ocean. (**a**) Sri Lanka (Northern Indian Ocean) type recorded at Diego Garcia. (**b**) Madagascar (Southwestern Indian Ocean) type recorded at Diego Garcia. (**c**) Eastern Indian Ocean pygmy blue whales recorded off Cape Leeuwin, WA, Australia. Units are numbered; phrases are bracketed Stafford. Reprinted by permission from John Wiley and Sons. Stafford KM, Chapp E, Bohnenstiel DR, Tolstoy M, Seasonal detection of three types of "pygmy" blue whale calls in the Indian Ocean. Mar Mamm Sci 27 (4):828–840; https://doi.org/10.1111/j.1748-7692.2010.00437.x.
© Society for Marine Mammalogy, 2010. All rights reserved

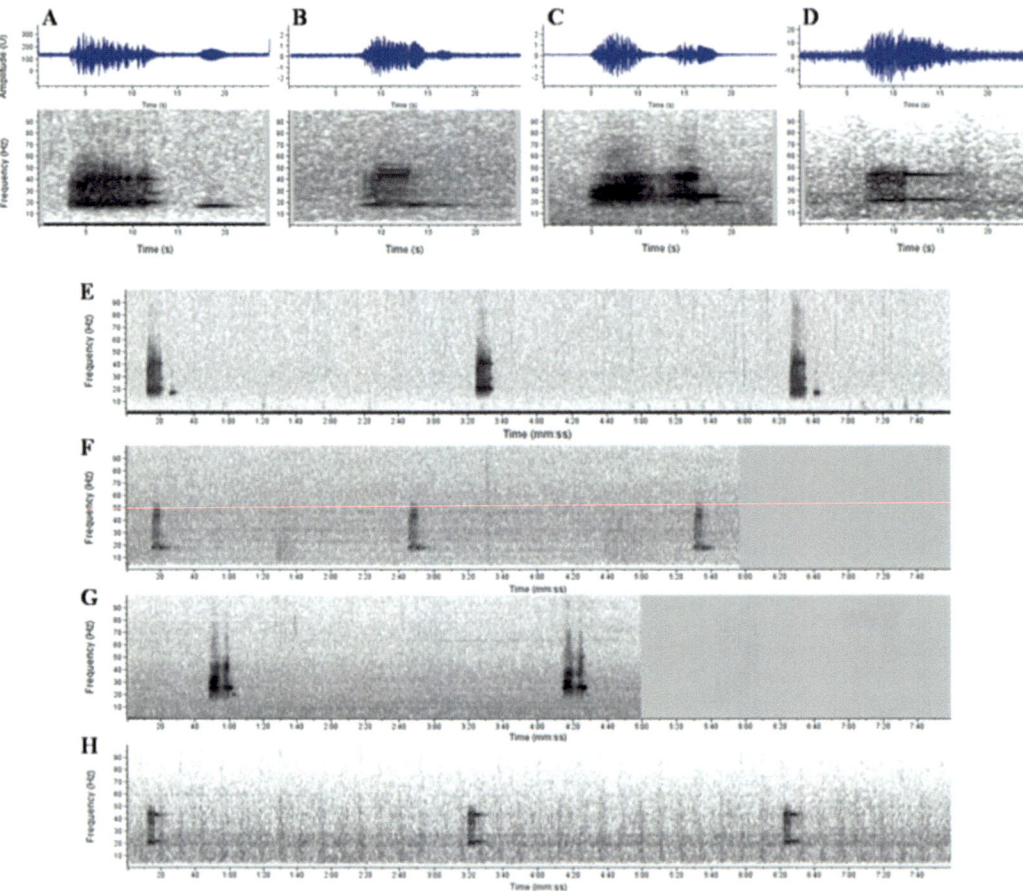

Fig. 3.24 Omura's whale song units (**a–d**) and song sections (**e–h**) from Madagascar (**a, e**), Diego Garcia (**b, f**), northwest Australia (**c, g**), and the mid-Atlantic Ridge (**d, h**) (Cerchio 2022). Reprinted with permission from Springer Nature. Cerchio S, The Omura's whale: Exploring the enigma. In: Clark CW, Garland EC (eds) Ethology and Behavioral Ecology of Mysticetes, pp 349–374; https://doi.org/10.1007/978-3-030-98449-6_15. © Springer Nature, 2022. All rights reserved

Cerchio et al. (2015), from Madagascar. Omura's whale song consists of one or two units: a narrowband, noisy vocalization, which, in some regions, is immediately followed by a more tonal, constant-frequency vocalization within the same frequency band. The frequency of the phrase ranges from ~15 Hz to ~50 Hz, and the duration ranges from ~8 s to 16 s, depending on whether it is the singlet or doublet phrase (Fig. 3.24; Cerchio 2022; Cerchio et al. 2015, 2019; Erbe et al. 2017; Leroy et al. 2021; Madhusudhana et al. 2020; Moreira et al. 2020; Sousa and Harris 2015). The phrase is repeated at relatively fixed intervals of 2–3 min, over hours at a time (Cerchio 2022). Based on passive acoustic monitoring, the Western Australian population does not seem to undertake long migrations between grounds (as other mysticete species in this region do), but appears resident all year round (Browne et al. 2024).

3.13.3.8 *Balaenoptera physalus*: Fin Whale

The fin whale occurs in all the world's oceans and migrates annually between its low-latitude breeding grounds and high-latitude feeding grounds. There are three recognized subspecies: the North Atlantic fin whale (*B.p. physalus*), the southern fin whale (*B.p. quoyi*), and the North Pacific fin whale (*B.p. velifera*).

The fin whale's acoustic repertoire consists of song and calls. Fin whale song may include three sound types: the 20 Hz note, the backbeat, and a higher-frequency component that occurs at the same time as either the 20 Hz note or the backbeat (Clark et al. 2002; Garcia et al. 2019; Širović et al. 2004; Watkins 1981). These sounds are often referred to as pulses, because of their short duration (~1 s). The most commonly studied non-song sound is the 40 Hz downsweep; however, the repertoire includes a variety of frequency-modulated downsweeps and upsweeps, or some combination of the two, as well as constant-frequency calls (Cummings et al. 1986; Delarue 2008; Edds 1988; Širović et al. 2006; Thompson et al. 1992; Watkins 1981). The majority of recent fin whale acoustic studies have focused on song rather than calls.

Fin whales produce the 20 Hz pulse in repeated stereotyped patterns that have fixed intervals between pulses (referred to as the 20 Hz note in song), as single pulses in a sporadic manner (referred to as the 20 Hz call), or in brief sequences of 2–5 pulses with irregular intervals between pulses (Watkins 1981). The 20 Hz pulse starts at ~35 Hz and sweeps downward to ~15 Hz (Fig. 3.25). The duration is ~1 s. Source levels as high as 192 dB re 1 µPa m have been reported (Charif et al. 2002; Garcia et al. 2019; Miksis-Olds et al. 2019; Miller et al. 2021; Širović et al. 2007; Varga et al. 2018; Wang et al. 2016; Watkins et al. 1987; Weirathmueller et al. 2013).

In addition to the 20 Hz notes, fin whale song may include a backbeat note (Castellote et al. 2012b; Clark et al. 2002; Constaratas et al. 2021; Delarue et al. 2013b; Garcia et al. 2019). The term backbeat was coined by Clark et al. (2002) based on the aural characteristics of the sound when recordings were played at a higher speed. Backbeats are shorter (<1 s) and have a narrower bandwidth than 20 Hz notes, sweeping down from ~22 Hz to ~14 Hz (Fig. 3.25b). Both the 20 Hz note and the backbeat may have a higher-frequency component somewhere between ~80 Hz and 130 Hz, with its frequency varying geographically (Constaratas et al. 2021;

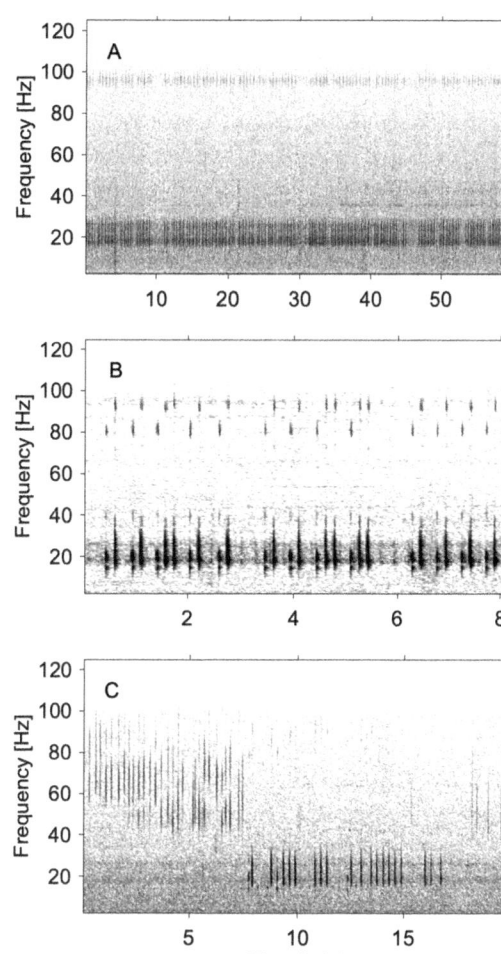

Fig. 3.25 Fin whale sounds: (**a**) 20 Hz song; note the higher-frequency component at 95–100 Hz. (**b**) Song consisting of backbeats and 20 Hz notes; note that both backbeats and 20 Hz notes have a higher-frequency component at 80 Hz and 95 Hz, respectively. (**c**) Many 40 Hz downsweeps (from 0 to 7 min into this recording), followed by a backbeat and 20 Hz notes. Fs = 250 Hz, NFFT = 1024, Hann window, 50% overlap. Recordings from the Comprehensive Test Ban Treaty Organization (CTBTO) hydroacoustic station at Cape Leeuwin, Western Australia (Aulich et al. 2019, 2022)

Erbe et al. 2015; Garcia et al. 2019; Simon et al. 2010; Širović et al. 2004, 2009). This higher-frequency component may also occur separately from 20 Hz notes or backbeats (Garcia et al. 2019).

Fin whales produce different song types that vary by the number of notes in the repeated sequence and by the inter-note interval (i.e., the duration of silence between song notes, also referred to as the inter-pulse interval in the fin whale literature; Clark et al. 2002; Delarue et al. 2013b; Helble et al. 2020a; Oleson et al. 2014; Širović et al. 2017; Thompson et al. 1992; Watkins 1981; Watkins et al. 1987; Weirathmueller et al. 2017). A singlet song consists of repetitions of a single note at a consistent inter-note interval. A doublet song consists of repetitions of a pair of notes, whereby the inter-note interval between the two notes of the pair is short and the inter-note interval to the next pair is long. The triplet song consists of repetitions of a three-note sequence. Within a singing bout, an individual may switch between different song types (Helble et al. 2020a; Thompson et al. 1992). For a given region and time period, there may be one predominant song type, but transitions to a different dominant song type may occur over time (Širović et al. 2017; Weirathmueller et al. 2017).

Fin whales exhibit geographic, seasonal, and decadal variability in song features; however, the functions of this variability remain unclear. The inter-note interval, note bandwidth, and production of 20 Hz notes with higher-frequency components are song features that distinguish different ocean basins (e.g., Furumaki et al. 2021; Širović et al. 2009; Thompson et al. 1992). Within ocean basins, the spectral structure of the higher-frequency component, the proportion of backbeat notes, and/or the inter-note interval vary geographically (Archer et al. 2020; Castellote et al. 2012b; Constaratas et al. 2021; Delarue et al. 2009b, 2013b; Širović et al. 2017). Within the North Atlantic, short song inter-note intervals occurred during the reproductive season and long inter-note intervals occurred when the predominant behavior was foraging (Morano et al. 2012). Both intra- and inter-annual decreases in frequency, increases in inter-note interval, and gradual shifts of the predominant song type have been reported from all oceans (Best et al. 2022; Furumaki et al. 2021; Helble et al. 2020a; Leroy et al. 2018; Morano et al. 2012; Oleson et al. 2014; Širović et al. 2017; Weirathmueller et al. 2017; Wood and Širović 2022).

In terms of non-song sounds, fin whales produce the so-called 40 Hz downsweep, which starts at ~80 Hz and ends at ~35 Hz, with 1 s duration and source levels up to 189 dB re 1 µPa m (Delarue 2008; Garcia et al. 2019; Miller et al. 2021; Širović et al. 2013; Watkins 1981; Wiggins and Hildebrand 2020; Fig. 3.25c). Additional higher-frequency (i.e., higher than the 20 Hz call) calls have been reported in the literature, such as brief (<1 s) downsweeps, upsweeps, or a combination of the two, between ~18 Hz and 310 Hz (Cummings et al. 1986; Edds 1988; Garcia et al. 2019; Širović et al. 2006; Thompson et al. 1992). Delarue (2008) categorized higher-frequency calls recorded in the Gulf of Maine and Gulf of Saint Lawrence into five call types: B, F, G, H, and I. Call type B was a 1 s downsweep between 25 Hz and 50 Hz. Call type F was of constant frequency at ~50–70 Hz and lasted <1 s. Call type G was a 1 s upsweep between ~120 Hz and 140 Hz. Call type H was an upsweep at lower frequencies (~60–80 Hz) than type G and had a shorter average duration of ~0.6 s. Call type I had an upsweeping component followed by a downsweeping component covering ~40–90 Hz over an average duration of ~1 s. Longer-duration calls also exist. For example, Cummings et al. (1986) described a ~3 s long moan with energy at 34 and 68 Hz, recorded in the Gulf of California. Watkins (1981) and Edds (1988) further described long (up to 30 s), 10–30 Hz growl or "rumble" from the North Atlantic, which was associated with agonistic behaviors between two adults.

In general, call detection does not peak at the same time as song detection, reflecting seasonal population movements and/or different functions of singing and calling. For example, off British Columbia, Canada, song dominated in winter and the 40 Hz call in spring (likely related to seasonal breeding versus feeding behaviors; Burnham 2019).

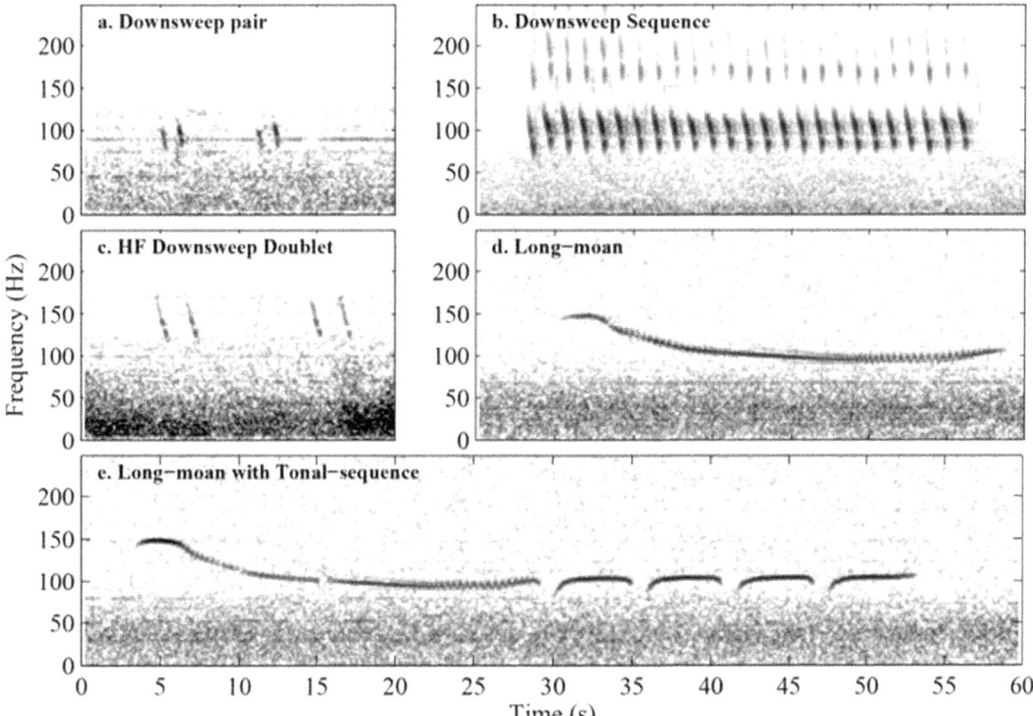

Fig. 3.26 Spectrograms of Rice's whale calls. Reprinted from Soldevilla MS, Ternus K, Cook A, Hildebrand JA, Frasier KE, Martinez A, Garrison LP (2022b) Acoustic localization, validation, and characterization of Rice's whale calls. J Acoust Soc Am 151 (6):4264–4278. https://doi.org/10.1121/10.0011677. © Soldevilla et al. 2022b. Published CC BY 4.0; https://creativecommons.org/licenses/by/4.0/

3.13.3.9 *Balaenoptera ricei*: Rice's Whale

Rice's whales occur in the Gulf of Mexico. They were previously classified as Bryde's whales but were recently recognized as a different species (Rosel et al. 2021). As a result, earlier acoustic recordings of Gulf of Mexico Bryde's whales have been reassigned to Rice's whales. Rice's whales produce brief downsweeps in doublets or longer sequences, in which each call sweeps from 110 Hz down to 70 Hz over 0.4 s (Rice et al. 2014; Širović et al. 2014; Fig. 3.26). A source level of 155 ± 14 dB re 1 μPa m was measured (Širović et al. 2014). Sequences of downsweeps extending down to 30 Hz were reported by Soldevilla et al. (2022b). Rice's whales further produce long-duration tonal sounds (long moans), which sweep from ~150 Hz to 70 Hz, albeit over ~15–20 s (Rice et al. 2014; Soldevilla et al. 2022a). Seven variations of this long moan have been identified, which exhibit brief frequency modulations (Soldevilla et al. 2022a).

3.13.3.10 *Megaptera novaeangliae*: Humpback Whale

Humpback whales are found in all the world's oceans. They can be split into three subspecies: the Southern humpback whale (*Megaptera n. australis*), the North Pacific humpback whale (*Megaptera n. kuzira*), and the North Atlantic humpback whale (*Megaptera n. novaeangliae*). Humpback whales have further been grouped into 15 different distinct population segments (Bettridge et al. 2015). Most of the populations are migratory, spending the winter reproductive season at low latitudes and migrating to mid-to-high latitudes to feed. One of these, the Pacific Central American population makes the longest known mammalian migration of 8300 km one way, from Antarctica to as far as 11°N (Rasmussen et al. 2007). However, within a migratory population, some individuals are known to overwinter in high latitudes, forgoing the migration (Straley et al. 2018). This may be the result of insufficient foraging during the summer and, therefore, not enough energy reserves to undertake migration. Only one of these, the Arabian Sea population, does not migrate. The seasonal upwelling in the area provides sufficient food resources obviating the need for migration (Mikhalev 1997; Minton et al. 2011).

Humpback whales are famous for being the first whale species in which song was described and the first species for which photographic identification of individuals was demonstrated. This ability revolutionized whale science as it introduced an entire new wave of non-invasive research methods that could be deployed to understand whale biology.

Payne and McVay (1971) first described humpback whale song—the first description of song in any whale species. They found the song to have a hierarchical structure (described in Sect. 3.3.1 and illustrated in Fig. 3.2). In this first paper, several songs were analyzed that had been recorded in different years. Payne and McVay (1971) reported variability between songs from the same animal in the number of repetitions of themes within the song. There were far greater differences between songs of different whales from different years. Further study revealed that the song structure of humpback whales gradually evolved over the years: The song at the beginning of a mating season was very similar to the song produced at the end of the previous year's mating season (Payne et al. 1983). Moreover, most of the individuals in a singing population made the same changes nearly simultaneously. The songs of three individually identified singers were compared to songs of other animals recorded at the same time (Guinee et al. 1983). These comparisons showed that all animals made the same basic evolutions at the same time and that the changes in the singing population were due to the evolution of the song within individuals, rather than the replacement of individuals singing song type "A" with individuals singing song type "B".

Not only was this gradual evolution of song structure occurring within a specific wintering ground such as Hawaii, the same changes in structure were being observed in Mexico, 4700 km away (Payne and Guinee 1983). Later work showed that the Western Pacific populations were also making changes largely in parallel with the central Hawaiian and Eastern Mexican groups (Darling et al. 2014).

This general model of the gradual evolution of humpback whale song was challenged by the observation of the first song "revolution" that occurred when Western Australian humpbacks apparently migrated to the eastern population grounds and the entire eastern population adopted the western humpback whale song (Noad et al. 2000). The second revolution was observed in 2003, when apparently the same transition of Western Australian whales to the eastern grounds occurred again (Garland et al. 2011). A song revolution has also been reported from Brazil (Gonçalves et al. 2023).

Furthermore, song evolution in the South Pacific did not appear to follow the parallel model observed in the North Pacific. Rather, songs were created along the east coast of Australia and propagated eastward, being adopted in subsequent years by singing populations farther and farther east of Australia and continuing to the west coast of south America (Schulze et al. 2022).

While no one knows what causes the difference in song learning between these ocean basins, the geography offers a viable hypothesis. The North Pacific is effectively bounded by the continents of Asia and North America and the migration of animals northward in the summer months brings these subpopulations closer together (Fig. 3.27). Furthermore, we know through individual identification that individuals move between Japanese, Hawaiian, and Mexican breeding grounds. This offers an additional potential mechanism for these subpopulations to remain largely in song synchrony. Conversely, the South Pacific has relatively few land masses that could lead to isolation of different groups.

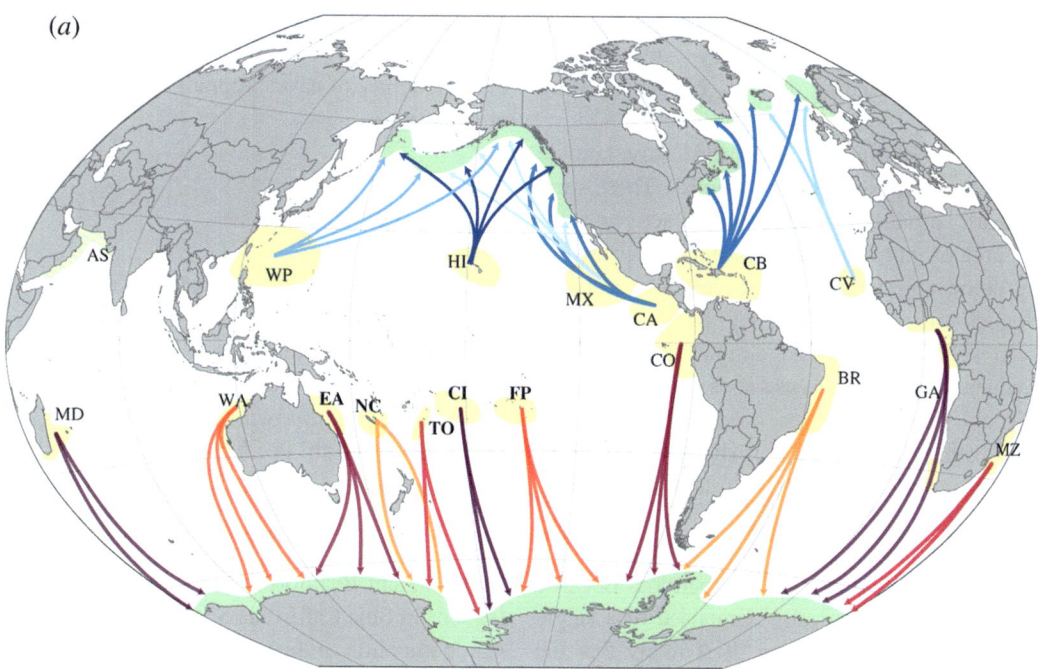

Fig. 3.27 Overview of the distribution and likely general migration patterns of humpback whales. Reprinted from Zandberg L, Lachlan RF, Lamoni L, Garland EC (2021) Global cultural evolutionary model of humpback whale song. Philos Trans R Soc B 376 (1836):20200242. https://doi.org/10.1098/rstb.2020.0242. © Zandberg et al. 2021. Published CC BY 4.0; https://creativecommons.org/licenses/by/4.0/

It is more likely that some individuals from an area migrate east or west, distributing the newer version of song.

While humpback whales are famously known for their songs, they also produce a wide repertoire of non-song sounds on feeding and reproductive grounds, and during migrations between the two areas. The first non-song sounds were reported in the Northeast Pacific summering waters from foraging whales that produced grunts, moans, and pulse trains, as well as blowhole sounds and mechanical sounds (Thompson et al. 1986).

Fournet et al. (2015) recorded and classified 299 Alaskan non-song calls into four classes and 16 call types. Most common were low-frequency harmonic calls, followed by pulsed calls, noisy/complex calls, and then tonal calls. Further analyses of six of these call types found that they were structurally stable over time, and five had similar structure in the geographically separated North Pacific and North Atlantic populations (Fournet et al. 2018c).

A 42–58 s, 500 Hz tonal signal has been associated with group lunge feeding on herring in Alaska (D'Vincent et al. 1985). This signal was initiated approximately 90 s prior to the group lunging to the surface. Hypotheses for this call include group coordination and prey manipulation (Baker and Herman 1985; Sharpe 2001). More recent recordings have found variation in the frequency of the call among individuals (Cerchio and Dahlheim 2001). Solitary whales, both males and females, were observed making this call, which indicates the function is more likely to manipulate prey (Fournet et al. 2018b).

A remarkable observation was made in the Northeast Atlantic. Two humpback whales tagged with acoustic recorders produced "megapclicks" while feeding: broadband (2 kHz bandwidth) pulses in trains with decreasing inter-pulse interval turning into buzzes (Stimpert et al. 2007). While the function of these sounds has not been ascertained, they (superficially) resemble the feeding buzzes of echolocating odontocetes.

Silber (1986) first described low-latitude non-song sounds as "social sounds". These were primarily produced by large surface-active groups of three or more adults, when males appeared to be competing for physical access to a female. Sounds ranged in frequency from 50 Hz to >10 kHz, typically had simple structure, and frequency-modulated upsweeps were common. Signals were most often produced as a single call with at least 10 s of silence before and after the call, or as a series of calls with a short inter-call interval. The numbers of calls increased with group size, suggesting that multiple animals were vocalizing during the competition. Because these groups are typically composed of a single female and multiple males, it is likely that the males were making the calls as agonistic signals.

Non-song sounds are also heard during migration. The first description of them included 34 call types, of which 21 were also found in that season's song; the other 13 types were not used as song units and remained structurally stable during a 3-year study (Dunlop et al. 2007). Later studies found a greater number of calls that resembled song units, but confirmed the existence of 12 stable non-song sounds (Rekdahl et al. 2013). Similar non-song sounds have been reported from Southern Africa (Rekdahl et al. 2017), Western Australia (Recalde-Salas et al. 2020), and the South Atlantic (Ross-Marsh et al. 2022).

Acoustic recordings of mother-calf groups showed that both male and female calves produced amplitude- and frequency-modulated tonal calls, as well as pulsed calls, from 140 to 4000 Hz with durations <1 s (Zoidis et al. 2008). Mother-calf calls recorded off Madagascar were acoustically similar to non-song sounds from other populations, and some resembled units in the local song (Saloma et al. 2022). Off Australia, calling rate increased when mother and calf were separated, but source levels did not increase (Indeck et al. 2022). The authors surmised that the calls were being used to maintain acoustic contact while physically separated, yet the calls remained quiet to reduce the probability of eavesdropping by other whales or predators.

3.13.4 Eschrichtiidae

3.13.4.1 *Eschrichtius robustus*: Gray Whale

The gray whale only occurs in the Northern Hemisphere—in fact, only in the North Pacific—and is separated into two stocks: the eastern and the western North Pacific stocks. The eastern North Pacific stock (EGW) makes one of the longest round-trip migrations of any mammal from its summer feeding grounds in the Bering, Chukchi, and Beaufort Seas to its winter calving lagoons in Baja California Sur, Mexico. The remnant population of the western Pacific gray whale (WGW) migrates along the western Pacific from feeding grounds around Sakhalin Island, Russia, to unknown breeding grounds somewhere near the Korean Peninsula. Although adults of both sexes can migrate to the eastern Pacific and back, no immigrants to the WGW have been reported (Mate et al. 2015). All the available data on gray whale vocal behavior come from the EGW population.

The gray whale repertoire consists of (1) low-frequency, modulated, relatively tonal moans, (2) burst-pulse sounds variously described as belches, groans, and grunts, and (3) variable trains of pulses, often with a metallic timbre, variously described as knocks, pops, and bongs (Fig. 3.28). None of these sounds are known to be stereotyped, nor have any repetitive patterns or songs been reported from gray whales.

Most authors have assigned a code to each call type, based on location along the migration route where the call was recorded (S = southern ground, M = migratory corridor, N = northern grounds). The same call type is heard in multiple areas, and so there are different codes for the same or similar calls.

Call type S1 is a pulse train of 2–60 pulses, lasting 0.3–5 s, covering 20–5000 Hz, and peaking at ~250 Hz (Charles 2011; Dahlheim

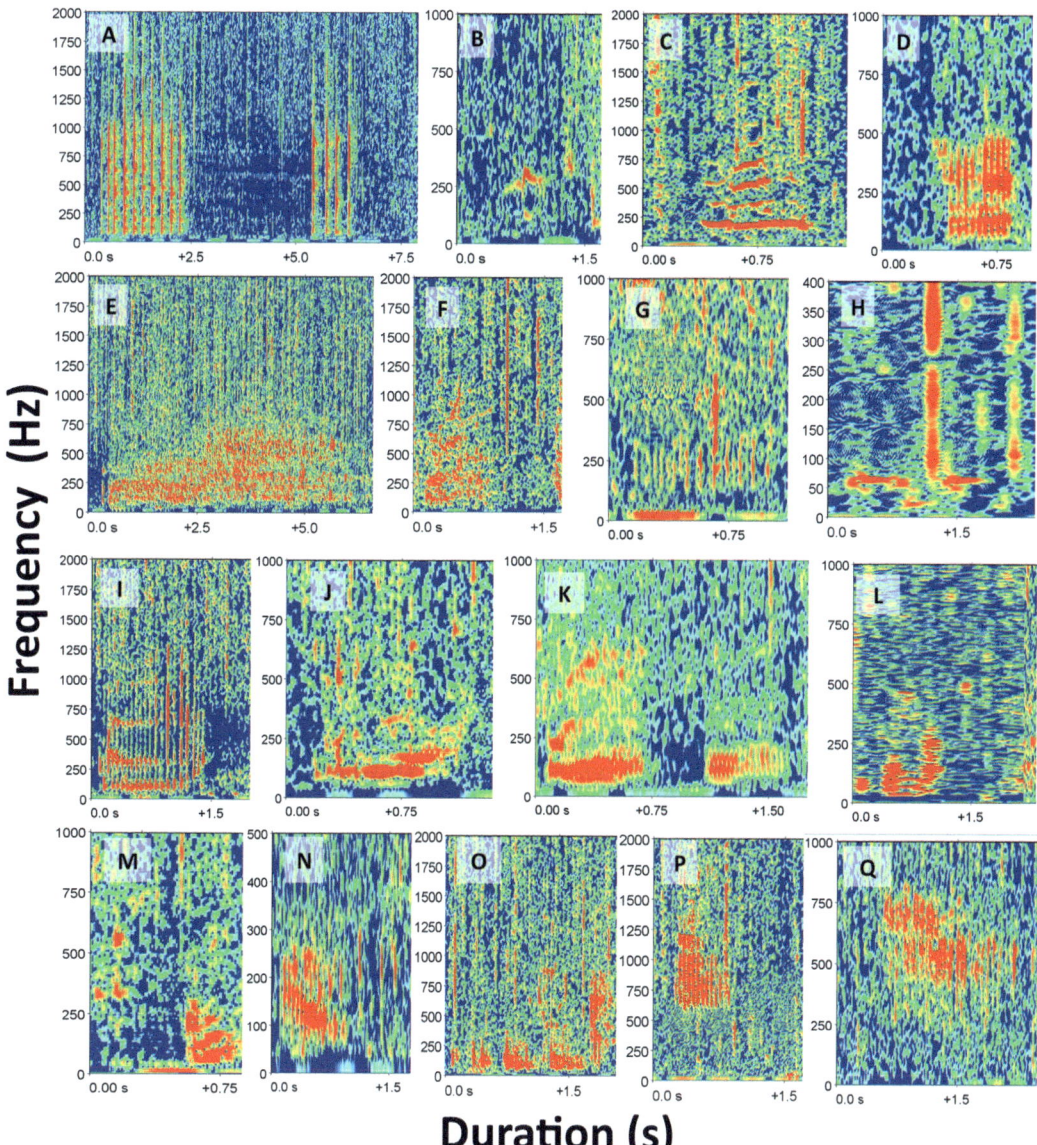

Fig. 3.28 Spectrograms of gray whale calls. **a**: S1. **b**: S2. **c**: S3. **d**: S4. **e**: S5, bubble blast. See Dahlheim 1987 for types S1–S5. **f**: Blow, surface exhalation recorded under water. **g**: S8. **h**: S9. See López-Urbán et al. 2018 for types S8 and S9. **i**: Croak. **j**: BMC2. **k**: BMC7. See Ollervides 2001 for BMC call types. **l**: Belch. **m**: Grunt. **n–q**: Unclassified gray whale sounds. Collage from Frouin-Mouy et al. 2020. Reprinted with permission from Elsevier. Frouin-Mouy H, Tenorio-Hallé L, Thode A, Swartz S, Urbán J, Using two drones to simultaneously monitor visual and acoustic behaviour of gray whales (*Eschrichtius robustus*) in Baja California, Mexico. J Exp Mar Biol Ecol 525:151321; https://doi.org/10.1016/j.jembe.2020.151321. © Elsevier, 2020. All rights reserved

1987; López-Urbán et al. 2018; Ollervides 2001; Wisdom et al. 2001). A single pulse lasts ~0.05 s (Dahlheim 1987). Only one study estimated the source level: 142 dB re 1 µPa m (Cummings et al. 1968). These calls have a metallic sound and are sometimes referred to as bongo or conga. They have also been reported from migratory corridors (M1; Burnham et al. 2018; Crane and Lashkari

1996; Cummings et al. 1968; Guazzo et al. 2017; Youngson and Darling 2016) and northern grounds (N1; Moore and Ljungblad 1984; Stafford et al. 2007).

Call type S2/M2 comprises frequency-modulated downsweeps and upsweeps, recorded on wintering grounds and in migratory corridors (0.2–7 s, 30–400 Hz, peak ~75 Hz; Burnham et al. 2018; Charles 2011; Dahlheim 1987; Ollervides 2001).

Call type S3/M3/N3 are tonal, amplitude-modulated moans, recorded on all grounds (0.1–4 s, 20–1300 Hz, peak ~100 Hz, 155 dB re 1 μPa m; Burnham et al. 2018; Charles 2011; Crane and Lashkari 1996; Cummings et al. 1968; Dahlheim 1987; Fish et al. 1974; Guazzo et al. 2017; López-Urbán et al. 2018; Moore and Ljungblad 1984; Ollervides 2001; Wisdom et al. 2001).

Call type S4/M4/N4 has mostly been referred to as grunts, appearing pulsed or noisy in spectrograms (0.2–3 s, 10–2000 Hz; Burnham et al. 2018; Charles 2011; Crane and Lashkari 1996; Dahlheim 1987; Fish et al. 1974; Moore and Ljungblad 1984; Ollervides 2001; Wisdom et al. 2001). It has been recorded on all grounds.

Call types S5 and S6 are non-vocal sounds, resulting from creation of bubbles. They have been reported from the southern and migratory areas. Type S5 corresponds to bubble blasts (0.2–5 s, 70–1000 Hz, peak ~200 Hz, 122–138 dB re 1 μPa m; Charles 2011; Crane and Lashkari 1996; Cummings et al. 1968; Dahlheim 1987; Frouin-Mouy et al. 2020; Ollervides 2001). Type S6 corresponds to longer-duration (up to 8 s) bubble trails, which also appear narrower in bandwidth (Charles 2011; Dahlheim 1987; Ollervides 2001).

Surface exhalations (blows) recorded under water are brief (~1 s) and narrowband (15–175 Hz), with a source level of 114 dB re 1 μPa m (Cummings et al. 1968; Frouin-Mouy et al. 2020).

Additional sound types have been reported, but often only in one study (e.g., Charles 2011; Dahlheim 1987; Frouin-Mouy et al. 2020; López-Urbán et al. 2018; Ollervides 2001). Some studies reported very broadband (<20 kHz) and brief pulses in trains (Fish et al. 1974; Norris et al. 1977).

Several studies have collected visual observations at the time of acoustic recordings, sometimes in unique settings (e.g., when only mother-calf groups were present). However, the association of acoustic with physical or functional behavior remains inconsistent (e.g., Charles 2011; Dahlheim 1987; Fish et al. 1974; Frouin-Mouy et al. 2020; López-Urbán et al. 2018; Norris et al. 1977; Youngson and Darling 2016). Off San Francisco, the call rate changed from 2 to 24 calls/whale/day over the course of the 2014/15 southern migration (Guazzo et al. 2019). In Puget Sound, where some northerly migrating gray whales (on their way from California breeding grounds to Alaskan feeding grounds) stop over to feed, more vocalizations occurred at the end of a feeding event than at the beginning, and at shallow depths and slow swimming speeds (Clayton et al. 2023).

3.14 Unidentified Sounds

While Sect. 3.13 seems to imply that we have a catalogue of species-specific sounds for every mysticete species, the identification of sounds to species has been a long process. Whales have been acoustically recorded since the early 1900s, in the first instance by some nations' navies. Several strong, repetitive, and widely distributed underwater sounds puzzled mariners and ocean scientists for decades before the source was identified.

One such sound was the 20-cycle pulse, 1 s in duration, high amplitude, repetitive, acoustically observed in all the world's oceans, albeit seasonally. It was described in the naval and geophysical literature; it was tracked on hydrophone arrays, showing single and, at times, multiple sources that travelled along erroneous lines at slow speeds (Jensvold and Wright 1959; Patterson and Hamilton 1964; Walker 1961). While a whale as the source was hypothesized early on (even a whale's heartbeat was suggested because of the doublet pattern of this sound; Walker 1963), it was not until 1964 that sufficient evidence (including from simultaneous visual and acoustic surveys) had been compiled for the world to begin to concur on the source: the fin whale (Schevill et al. 1964).

At the same time, the world was trying to identify the source of equally strong, 20–30 Hz moans, which also often occurred in doublets, yet lasted 10–20 s at a time (Kibblewhite et al. 1967; Northrop et al. 1971; Weston and Black 1965). Some discussion remained for a few years whether these strong 20 Hz signals that occurred in doublets and were repeated for hours, but only differed in their duration would be from the same source or not (Northrop et al. 1968). Cummings and Thompson (1971) were confident the moans off Chile were blue whales, based on simultaneous visual and acoustic observations. Evidence from other oceans including the North Atlantic followed a few years later (Edds 1982).

Right within the human hearing range, unidentified sounds received more flamboyant names, such as the boing, which challenged oceanographers for over 50 years. It was regularly recorded on naval arrays; it occurred seasonally; it was localized and tracked; yet the source remained elusive (Thompson 1982; Wenz 1964). Gedamke (2001) described a similar sound for common minke whales off Australia, and finally, in 2005, Rankin and Barlow (2005), during a combined visual and acoustic survey, were able to confirm the minke whale as the source of the North Pacific boing.

Another unidentified sound well within the human hearing range was the bioduck. It was only recorded in the southern hemisphere (i.e., in the Southern Ocean and off southern Western Australia; Klinck and Boebel 2007; Matthews et al. 2004; van Opzeeland 2010). Again, it took decades before the bioduck was attributed to the Antarctic minke whale (Risch et al. 2014b).

There are still unidentified sounds puzzling us today. One of them is the "spot call", a 26 Hz tone, 10 s in duration, repeated every 2–3 min (Fig. 3.3a). It has been documented from southern Australia and the Indian Ocean for more than two decades (Ward et al. 2017). The call occurs seasonally, during the austral winter. It is often accompanied by brief (<1 s) 200–50 Hz downsweeps. The peak frequency of the spot call has been declining within each season and from season to season, at different rates in different locations, at steeper rates than the Antarctic blue whale Z-call has, and after it had reached a minimum of 22.5 Hz in 2006/7, it jumped up to ~28 Hz and slowly decreased again thereafter (Ward et al. 2017). Interestingly, in 2006/7, both 22.5 Hz and 28 Hz versions occurred, possibly indicating that not all individuals switched at the same time. The same call was recorded in the Western Indian Ocean between 2007 and 2015 (Leroy et al. 2017). Interestingly, only in 2007, both a 22 Hz version and a 27 Hz version coexisted, with only the 27 Hz version recorded thereafter and its frequency decreasing over time. Assuming these tones were made by the same source, the switch in frequency occurred in the same season over a wide geographic region: from the Western Indian Ocean to southern Australia. The source level of the spot call was determined after localization (and propagation modeling) at the Australian Integrated Marine Observing System's (IMOS) passive acoustic monitoring array near the Perth Canyon: mean 179 dB re 1 μPa m, range 166–189 dB re 1 μPa m. The sound is likely produced by a very large mysticete, because of its song-like pattern, and because it moves and exhibits seasonality (likely related to long-range migration). It has very low frequency for biological sources, long duration, and very high intensity, and thus cannot be produced by a fish.

3.15 Summary

In this chapter, we have demonstrated that mysticetes produce a great variety of sounds ranging in frequency from as low as 10 Hz to as high as 20 kHz. These sounds have been classed based on their spectrographic features as tonal, frequency-modulated, amplitude-modulated, and pulsive. They have often been categorized based on what they sound like to the human ear (e.g., moan, grunt, screech, and boing). They might occur in rhythmic patterns (song) or as single, non-song sounds. We provided an overview of the sounds made by the various species of mysticete, with measurements of sound features. We discussed the likely functions of these sounds (e.g., as related to reproduction or foraging). We showed long-term changes of some of these sounds over multiple decades. We also summarized the effects of noise on the acoustic behavior of mysticetes, which are further detailed in the later chapters.

3.16 Abbreviations

Quantity	Abbreviation	Symbol	Unit
Frequency		f	Hz
Peak frequency (frequency of maximum power)		f_p	Hz
Sampling frequency		f_s	Hz
Number of Fourier components	NFFT		
Standard deviation	std		
Root-mean square sound pressure level	SPL_{rms}	L_p	dB re 1 µPa
Peak-to-peak sound pressure level	SPL_{pk-pk}	$L_{p,pk-pk}$	dB re 1 µPa
Sound exposure level	SEL	$L_{E,p}$	dB re 1 µPa²s

3.17 Data Availability

A detailed spreadsheet in which we collated mysticete sound features from the literature and based on which we compiled Tables 1–3 may be requested from the first author.

References

Adam O, Cazau D, Gandilhon N, Fabre B, Laitman JT, Reidenberg JS (2013) New acoustic model for humpback whale sound production. Appl Acoust 74(10): 1182–1190. https://doi.org/10.1016/j.apacoust.2013.04.007

Agbayani S, Fortune SME, Trites AW (2020) Growth and development of North Pacific gray whales (*Eschrichtius robustus*). J Mammal 101(3):742–754. https://doi.org/10.1093/jmammal/gyaa028

Akamatsu T, Rasmussen MH, Iversen M (2014) Acoustically invisible feeding blue whales in northern Icelandic waters. J Acoust Soc Am 136(2):939–944. https://doi.org/10.1121/1.4887439

Allen JA, Garland EC, Dunlop RA, Noad MJ (2018) Cultural revolutions reduce complexity in the songs of humpback whales. Proc Royal Soc B 285 (1891):20182088. https://doi.org/10.1098/rspb.2018.2088

Archer FI, Rankin S, Stafford KM, Castellote M, Delarue J (2020) Quantifying spatial and temporal variation of North Pacific fin whale (*Balaenoptera physalus*) acoustic behavior. Mar Mamm Sci 36(1):224–245. https://doi.org/10.1111/mms.12640

Aroyan JL, McDonald MA, Webb SC, Hildebrand JA, Clark D, Laitman JT, Reidenberg JS (2000) Acoustic models of sound production and propagation. In: Au WWL, Fay RR, Popper AN (eds) Hearing by whales and dolphins. Springer, New York, New York, NY, pp 409–469. https://doi.org/10.1007/978-1-4612-1150-1_10

Asa-Dorian PV, Perkins PJ (1967) The controversial production of sound by the California gray whale, *Eschrichtius gibbosus*. Norsk Hvalfangst-Tidende 56: 74–77

Au WWL (1993) The sonar of Dolphins. Springer, New York. https://doi.org/10.1007/978-1-4612-4356-4

Au WL, Mobley J, Burgess WC, Lammers MO, Nachtigall PE (2000) Seasonal and diurnal trends of chorusing humpback whales wintering in waters off Western Maui. Mar Mamm Sci 16(3):530–544. https://doi.org/10.1111/j.1748-7692.2000.tb00949.x

Au WWL, Frankel A, Helweg DA, Cato DH (2001a) Against the humpback whale sonar hypothesis. IEEE J Ocean Eng 26(2):295–300. https://doi.org/10.1109/48.922795

Au WWL, James D, Andrews K (2001b) High-frequency harmonics and source level of humpback whale songs. J Acoust Soc Am 110(5):2770. https://doi.org/10.1121/1.4777702

Aulich MG, McCauley RD, Saunders BJ, Parsons MJG (2019) Fin whale (*Balaenoptera physalus*) migration in Australian waters using passive acoustic monitoring. Sci Rep 9(1):8840. https://doi.org/10.1038/s41598-019-45321-w

Aulich MG, McCauley RD, Miller BS, Samaran F, Giorli G, Saunders BJ, Erbe C (2022) Seasonal distribution of the fin whale (*Balaenoptera physalus*) in Antarctic and Australian waters based on passive acoustics. Front Mar Sci 9:864153. https://doi.org/10.3389/fmars.2022.864153

Aulich MG, Miller BS, Samaran F, McCauley RD, Saunders BJ, Erbe C (2023) Diel patterns of fin whale 20 Hz acoustic presence in Eastern Antarctic waters. R Soc Open Sci 10(4):220499. https://doi.org/10.1098/rsos.220499

Baker CS, Herman L (1985) Whales that go to extremes. Nat Hist 10(85):52–61

Barber JR, Crooks KR, Fristrup KM (2010) The costs of chronic noise exposure for terrestrial organisms. Trends Ecol Evol 25(3):180–189. https://doi.org/10.1016/j.tree.2009.08.002

Barham EG (1973) Whales' respiratory volume as a possible resonant receiver for 20 Hz signals. Nature 245(5422):220–221. https://doi.org/10.1038/245220a0

Barlow DR, Fournet M, Sharpe F (2019) Incorporating tides into the acoustic ecology of humpback whales. Mar Mamm Sci 35(1):234–251. https://doi.org/10.1111/mms.12534

Barlow DR, Klinck H, Ponirakis D, Branch TA, Torres LG (2023) Environmental conditions and marine heatwaves influence blue whale foraging and reproductive effort. Ecol Evol 13(2):e9770. https://doi.org/10.1002/ece3.9770

Baumgartner MF, Fratantoni DM (2008) Diel periodicity in both sei whale vocalization rates and the vertical migration of their copepod prey observed from ocean gliders. Limnol Oceanogr 53(5):2197–2209. https://doi.org/10.4319/lo.2008.53.5_part_2.2197

Baumgartner MF, Cole TVN, Campbell RG, Teegarden GJ, Durbin EG (2003) Associations between North Atlantic right whales and their prey, *Calanus finmarchicus*, over diel and tidal time scales. Mar Ecol Prog Ser 264:155–166. https://doi.org/10.3354/meps264155

Baumgartner MF, Van Parijs SM, Wenzel FW, Tremblay CJ, Esch HC, Warde AM (2008) Low frequency vocalizations attributed to sei whales (*Balaenoptera borealis*). J Acoust Soc Am 124(2):1339–1349. https://doi.org/10.1121/1.2945155

Beamish P (1978) Evidence that a captive humpback whale (*Megaptera novaeangliae*) does not use sonar. Deep-Sea Res 25(5):469–472. https://doi.org/10.1016/0146-6291(78)90554-4

Beamish P, Mitchell E (1971) Ultrasonic sounds recorded in the presence of a blue whale *Balaenoptera musculus*. Deep-Sea Res 18:803–809

Beamish P, Mitchell E (1973) Short pulse length audio frequency sounds recorded in the presence of a minke whale (*Balanoptera acutorostrata*). Deep-Sea Res 20(4):375–386. https://doi.org/10.1016/0011-7471(73)90060-0

Beauregard H, Boulart R (1882) Recherches sur le larynx et la trachée des balaenides. Journal de l'Anatomie et de la Physiologie normales et pathologiques de l'Homme et des Animaux 18:611–637

Benham WB (1901) On the larynx of certain whales (*Cogia, Balaenoptera*, and *Ziphius*). Proc Zool Soc London 1:278–300

Berchok CL, Bradley DL, Gabrielson TB (2006) St. Lawrence blue whale vocalizations revisited: characterization of calls detected from 1998 to 2001. J Acoust Soc Am 120(4):2340–2354. https://doi.org/10.1121/1.2335676

Best P, Marxer R, Paris S, Glotin H (2022) Temporal evolution of the Mediterranean fin whale song. Sci Rep 12(1):13565. https://doi.org/10.1038/s41598-022-15379-0

Bettridge S, Baker CS, Barlow J, Clapham PJ, Ford M, Gouveia D, Mattila DK, Pace RM III, Rosel PE, Silber GK, Wade PR (2015) Status review of the humpback whale (*Megaptera novaeangliae*) under the Endangered Species Act. Southwest Fisheries Science Center, National Marine Fisheries Service, La Jolla

Blackwell SB, Richardson WJ, Greene CR, Streever B (2007) Bowhead whale (*Balaena mysticetus*) migration and calling behaviour in the Alaskan Beaufort Sea, autumn 2001-04: an acoustic localization study. Arctic 60(3):255–270

Blackwell SB, Nations CS, TL MD, Thode AM, Mathias D, Kim KH, Green CR, Macrander AM (2015) Effects of airgun sounds on bowhead whale calling rates: evidence for two behavioral thresholds. PLoS One 10(6):e0125720. https://doi.org/10.1371/journal.pone.0125720

Blackwell SB, Nations CS, Thode AM, Kauffman ME, Conrad AS, Norman RG, Kim KH (2017) Effects of tones associated with drilling activities on bowhead whale calling rates. PLoS One 12(11):e0188459. https://doi.org/10.1371/journal.pone.0188459

Boncoraglio G, Saino N (2007) Habitat structure and the evolution of bird song: a meta-analysis of the evidence for the acoustic adaptation hypothesis. Funct Ecol 21(1):134–142. https://doi.org/10.1111/j.1365-2435.2006.01207.x

Bottjer SW, Glaessner SL, Arnold AP (1985) Ontogeny of brain nuclei controlling song learning and behavior in zebra finches. J Neurosci 5(6):1556–1562. https://doi.org/10.1523/JNEUROSCI.05-06-01556.1985

Bouffaut L, Landrø M, Potter JR (2021) Source level and vocalizing depth estimation of two blue whale subspecies in the western Indian Ocean from single sensor observations. J Acoust Soc Am 149(6):4422–4436. https://doi.org/10.1121/10.0005281

Bradbury JW, Vehrencamp SL (2011) Principles of animal communication, 2nd edn. Sinauer Associates, Sunderland

Brenowitz EA (1982) The active space of red-winged blackbird song. J Comp Physiol 147(4):511–522. https://doi.org/10.1007/BF00612017

Browne CE, Erbe C, McCauley RD (2024) Distribution and seasonality of the Omura's whale (*Balaenoptera omurai*) in Australia based on passive acoustic recordings. Animals 14 (20):2944. https://doi.org/10.3390/ani14202944

Buchan SJ, Rendell LE, Hucke-Gaete R (2010) Preliminary recordings of blue whale (*Balaenoptera musculus*) vocalizations in the Gulf of Corcovado, northern Patagonia. Chile Mar Mamm Sci 26(2):451–459. https://doi.org/10.1111/j.1748-7692.2009.00338.x

Buchan SJ, Hucke-Gaete R, Rendell L, Stafford KM (2014) A new song recorded for blue whales in the Corcovado Gulf, Southern Chile, and an acoustic link to the Eastern Tropical Pacific. Endang Species Res 23: 241–252. https://doi.org/10.3354/esr00566

Burnham RE (2019) Fin whale call presence and type used to describe temporal distribution and possible area use of Clayoquot sound. Northw Sci 93(1):66–74. https://doi.org/10.3955/046.093.0106

Burnham R, Duffus D, Mouy X (2018) Gray whale (*Eschrichtius robustus*) call types recorded during migration off the west coast of Vancouver Island. Front Mar Sci 5(329). https://doi.org/10.3389/fmars.2018.00329

Calderan S, Miller B, Collins K, Ensor P, Double M, Leaper R, Barlow J (2014) Low-frequency vocalizations of sei whales (*Balaenoptera borealis*) in the Southern Ocean. J Acoust Soc Am 136(6): 418–423. https://doi.org/10.1121/1.4902422

Carbaugh-Rutland A, Have Rasmussen J, Sterba-Boatwright B, Širović A (2021) Geographically distinct blue whale song variants in the Northeast Pacific. Endang Species Res 46:19–33. https://doi.org/10.3354/esr01145

Carouso-Peck S, Goldstein MH, Fitch WT (2021) The many functions of vocal learning. Philos Trans R Soc B 376(1836):20200235. https://doi.org/10.1098/rstb.2020.0235

Casey CB, Weindorf S, Levy E, Linsky JMJ, Cade DE, Goldbogen JA, Nowacek DP, Friedlaender AS (2022) Acoustic signalling and behaviour of Antarctic minke whales (*Balaenoptera bonaerensis*). R Soc Open Sci 9(7):211557. https://doi.org/10.1098/rsos.211557

Castellote M, Clark C, Lammers M (2012a) Acoustic and behavioural changes by fin whales (*Balaenoptera physalus*) in response to shipping and airgun noise. Biol Conserv 147(1):115–122. https://doi.org/10.1016/j.biocon.2011.12.021

Castellote M, Clark CW, Lammers MO (2012b) Fin whale (*Balaenoptera physalus*) population identity in the western Mediterranean Sea. Mar Mamm Sci 28(2): 325–344. https://doi.org/10.1111/j.1748-7692.2011.00491.x

Catchpole CK (1987) Bird song, sexual selection and female choice. Trends Ecol Evol 2(4):94–97. https://doi.org/10.1016/0169-5347(87)90165-0

Cazau D, Adam O, Laitman JT, Reidenberg JS (2013) Understanding the intentional acoustic behaviour of humpback whales: a production-based approach. J Acoust Soc Am 134(3):2268–2273. https://doi.org/10.1121/1.4816403

Cerchio S (2022) The Omura's whale: exploring the enigma. In: Clark CW, Garland EC (eds) Ethology and behavioral ecology of mysticetes. Springer, Cham, pp 349–374. https://doi.org/10.1007/978-3-030-98449-6_15

Cerchio S, Dahlheim M (2001) Variation in feeding vocalizations of humpback whales (*Megaptera novaeangliae*) from Southeast Alaska. Bioacoustics 11(4):277–295. https://doi.org/10.1080/09524622.2001.9753468

Cerchio S, Weir CR (2022) Mid-frequency song and low-frequency calls of sei whales in The Falkland Islands. R Soc Open Sci 9(11):220738. https://doi.org/10.1098/rsos.220738

Cerchio S, Jacobsen JK, Norris TF (2001) Temporal and geographical variation in songs of humpback whales, *Megaptera novaeangliae*: synchronous change in Hawaiian and Mexican breeding assemblages. Anim Behav 62(2):313–329. https://doi.org/10.1006/anbe.2001.1747

Cerchio S, Andrianantenaina B, Lindsay A, Rekdahl M, Andrianarivelo N, Rasoloarijao T (2015) Omura's whales (*Balaenoptera omurai*) off Northwest Madagascar: ecology, behaviour and conservation needs. R Soc Open Sci 2(10):150301. https://doi.org/10.1098/rsos.150301

Cerchio S, Yamada TK, Brownell RL (2019) Global distribution of Omura's whales (*Balaenoptera omurai*) and assessment of range-wide threats. Front Mar Sci 6(67). https://doi.org/10.3389/fmars.2019.00067

Cerchio S, Willson A, Leroy EC, Muirhead C, Al Harthi S, Baldwin R, Cholewiak D, Collins T, Minton G, Rasoloarijao T, Rogers TL, Sarrouf Willson M (2020) A new blue whale song-type described for the Arabian Sea and Western Indian Ocean. Endang Species Res 43:495–515. https://doi.org/10.3354/esr01096

Charif RA, Mellinger DK, Dunsmore KJ, Fristrup KM, Clark CW (2002) Estimated source levels of fin whale (*Balaenoptera physalus*) vocalizations: adjustments for surface interference. Mar Mamm Sci 18(1):81–98. https://doi.org/10.1111/j.1748-7692.2002.tb01020.x

Charles SM (2011) Social context of gray whale *Eschrichtius robustus* sound activity. MSc. thesis, Texas A&M University, Galveston, TX, USA

Chek AA, Bogart JP, Lougheed SC (2003) Mating signal partitioning in multi-species assemblages: a null model test using frogs. Ecol Lett 6(3):235–247. https://doi.org/10.1046/j.1461-0248.2003.00420.x

Chereskin E, Beck L, Gamboa-Poveda MP, Palacios-Alfaro JD, Monge-Arias R, Chase AR, Coven BM, Gloria Guzmán A, McManus NW, Neuhaus AP, O'Halloran RA, Rosen SG, May-Collado LJ (2019) Song structure and singing activity of two separate humpback whales populations wintering off the coast of Caño Island in Costa Rica. J Acoust Soc Am 146 (6):EL509-EL515. https://doi.org/10.1121/1.5139205

Cholewiak D, Cerchio S (2022) Humpback whales: exploring global diversity and behavioral plasticity in an undersea virtuoso. In: Clark CW, Garland EC (eds) Ethology and behavioral ecology of mysticetes. Springer, Cham, pp 247–276. https://doi.org/10.1007/978-3-030-98449-6_11

Cholewiak DM, Cerchio S, Jacobsen JK, Urbán-R J, Clark CW (2018) Songbird dynamics under the sea: acoustic interactions between humpback whales suggest song mediates male interactions. R Soc Open Sci 5(2): 171298. https://doi.org/10.1098/rsos.171298

Clapham PJ, Mattila DK (1990) Humpback whale songs as indicators of migration routes (note). Mar Mamm Sci 6(2):155–160

Clark CW (1982) The acoustic repertoire of the southern right whale, a quantitative analysis. Anim Behav 30(4): 1060–1071. https://doi.org/10.1016/S0003-3472(82)80196-6

Clark CW, Ellison WT (2004) Potential use of low-frequency sounds by baleen whales for probing the environment: evidence from models and empirical measurements. In: Thomas JA, Moss C, Vater M (eds) Echolocation in bats and dolphins. The University of Chicago Press, Chicago, pp 564–582

Clark CW, Johnson JH (1984) The sounds of the bowhead whale, *Balaena mysticetus*, during the spring migrations of 1979 and 1980. Can J Zool 62(7): 1436–1441. https://doi.org/10.1139/z84-206

Clark CW, Borsani JF, Notarbartolo-di-Sciara G (2002) Vocal activity of fin whales, *Balaenoptera physalus*, in the Ligurian Sea. Mar Mamm Sci 18(1):286–295. https://doi.org/10.1111/j.1748-7692.2002.tb01035.x

Clark CW, Ellison WT, Southall BL, Hatch L, Van Parijs SM, Frankel A, Ponirakis D (2009) Acoustic masking in marine ecosystems: intuitions, analysis, and implication. Mar Ecol Prog Ser 395:201–222. https://doi.org/10.3354/Meps08402

Clayton H, Cade DE, Burnham R, Calambokidis J, Goldbogen J (2023) Acoustic behavior of gray whales tagged with biologging devices on foraging grounds. Front Mar Sci 10:1111666. https://doi.org/10.3389/fmars.2023.1111666

Constantine R, Johnson M, Riekkola L, Jervis S, Kozmian-Ledward L, Dennis T, Torres LG, Aguilar de Soto N (2015) Mitigation of vessel-strike mortality of endangered Bryde's whales in the Hauraki Gulf, New Zealand. Biol Conserv 186:149–157. https://doi.org/10.1016/j.biocon.2015.03.008

Constaratas AN, McDonald MA, Goetz KT, Giorli G (2021) Fin whale acoustic populations present in New Zealand waters: description of song types, occurrence and seasonality using passive acoustic monitoring. PLoS One 16(7):e0253737. https://doi.org/10.1371/journal.pone.0253737

Crance JL, Berchok CL, Keating JL (2017) Gunshot call production by the North Pacific right whale *Eubalaena japonica* in the southeastern Bering Sea. Endang Species Res 34:251–267. https://doi.org/10.3354/esr00848

Crance JL, Berchok CL, Wright DL, Brewer AM, Woodrich DF (2019) Song production by the North Pacific right whale, *Eubalaena japonica*. J Acoust Soc Am 145(6):3467–3479. https://doi.org/10.1121/1.5111338

Crane NL, Lashkari K (1996) Sound production of gray whales, *Eschrichtius robustus*, along their migration route: a new approach to signal analysis. J Acoust Soc Am 100(3):1878–1886. https://doi.org/10.1121/1.416006

Cranford TW, Krysl P (2015) Fin whale sound reception mechanisms: skull vibration enables low-frequency hearing. PLoS One 10(1):e0116222. https://doi.org/10.1371/journal.pone.0116222

Croll DA, Clark CW, Acevedo A, Tershy B, Flores S, Gedamke J, Urban J (2002) Only male fin whales sing loud songs. Nature 417 (20 June 2002):809. https://doi.org/10.1038/417809a

Cummings WC, Holliday DV (1987) Sounds and source levels from bowhead whales off Pt. Barrow, Alaska. J Acoust Soc Am 82(3):814–821. https://doi.org/10.1121/1.395279

Cummings WC, Thompson PO (1971) Underwater sounds from the blue whale, *Balaenoptera musculus*. J Acoust Soc Am 50(4):1193–1198. https://doi.org/10.1121/1.1912752

Cummings WC, Thompson PO, Cook R (1968) Underwater sounds of migrating gray whales, *Eschrichtius glaucus* (Cope). J Acoust Soc Am 44(5):1278–1281. https://doi.org/10.1121/1.1911259

Cummings WC, Fish JF, Thompson PO (1972) Sound production and other behavior of southern right whales, *Eubalaena glacialis*. Transactions of the San Diego Society of Natural History 17(1):1–13

Cummings WC, Thompson PO, Ha SJ (1986) Sounds from Bryde, *Balaenoptera edeni*, and finback, *B. physalus*, whales in the Gulf of California. Fish Bull 84(2):359–370

Cusano DA, Indeck KL, Noad MJ, Dunlop RA (2022) Humpback whale (*Megaptera novaeangliae*) social call production reflects both motivational state and arousal. Bioacoustics 31(1):17–40. https://doi.org/10.1080/09524622.2020.1858450

Cusano DA, Wiley D, Zeh JM, Kerr I, Pensarosa A, Zadra C, Shorter KA, Parks SE (2023) Acoustic recording tags provide insight into the springtime acoustic behavior of sei whales in Massachusetts Bay. J Acoust Soc Am 154(6):3543–3555. https://doi.org/10.1121/10.0022570

D'Souza ML, Bopardikar I, Sutaria D, Klinck H (2023) Arabian Sea humpback whale (*Megaptera novaeangliae*) singing activity off Netrani Island, India. Aquat Mamm 49(3):223–235. https://doi.org/10.1578/AM.49.3.2023.223

D'Vincent CG, Nilson RM, Hanna RE (1985) Vocalization and coordinated feeding behavior of the humpback whale in southeastern Alaska. Sci Rep Whales Res Inst 36:41–47

Dahlheim ME (1987) Bio-acoustics of the gray whale (*Eschrichtius robustus*). PhD thesis, University of British Columbia, Vancouver, BC, Canada

Dahlheim M, Castellote M (2016) Changes in the acoustic behavior of gray whales *Eschrichtius robustus* in response to noise. Endang Species Res 31:227–242. https://doi.org/10.3354/esr00759

Darling JD, Bérubé M (2001) Interactions of singing humpback whales with other males. Mar Mamm Sci 17(3):570–584. https://doi.org/10.1111/j.1748-7692.2001.tb01005.x

Darling JD, Jones ME, Nicklin CP (2006) Humpback whale songs: do they organize males during the breeding season? Behaviour 143(9):1051–1101

Darling JD, Acebes JMV, Yamaguchi M (2014) Similarity yet a range of differences between humpback whale songs recorded in the Philippines, Japan and Hawaii in 2006. Aquat Biol 21(2):93–107. https://doi.org/10.3354/ab00570

Darling JD, Goodwin B, Goodoni MK, Taufmann AJ, Taylor MG (2019) Humpback whale calls detected in tropical ocean basin between known Mexico and Hawaii breeding assemblies. J Acoust Soc Am 145 (6):EL534–EL540. https://doi.org/10.1121/1.5111970

Davenport AM, Erbe C, Jenner M-NM, Jenner KCS, Saunders BJ, McCauley RD (2022) Pygmy blue whale diving behaviour reflects song structure. J Mar Sci Eng 10(9):1227. https://doi.org/10.3390/jmse10091227

Dawbin WH, Cato DH (1992) Sounds of a pygmy right whale (*Caperea marginata*). Mar Mamm Sci 8(3): 213–219. https://doi.org/10.1111/j.1748-7692.1992.tb00405.x

Delarue J (2008) Northwest Atlantic fin whale vocalizations: geographic variations and implications for stock assessments. M. Phil. thesis, College of the Atlantic, Bar Harbor, ME, USA

Delarue J, Laurinolli M, Martin B (2009a) Bowhead whale (*Balaena mysticetus*) songs in the Chukchi Sea between October 2007 and May 2008. J Acoust Soc Am 126(6):3319–3328. https://doi.org/10.1121/1.3257201

Delarue J, Todd SK, Van Parijs SM, Di Iorio L (2009b) Geographic variation in Northwest Atlantic fin whale (*Balaenoptera physalus*) song: implications for stock structure assessment. J Acoust Soc Am 125(3): 1774–1782. https://doi.org/10.1121/1.3068454

Delarue J, Martin B, Hannay D (2013a) Minke whale boing sound detections in the northeastern Chukchi Sea. Mar Mamm Sci 29(3):E333–E341. https://doi.org/10.1111/j.1748-7692.2012.00611.x

Delarue J, Martin B, Hannay D, Berchok CL (2013b) Acoustic occurrence and affiliation of fin whales detected in the Northeastern Chukchi Sea, July to October 2007–10. Arctic 66(2):159–172. https://doi.org/10.14430/arctic4287

di Iorio L, Clark CW (2010) Exposure to seismic survey alters blue whale acoustic communication. Biol Lett 6: 334–335. https://doi.org/10.1098/rsbl.2009.0967

Dombroski JRG, Parks SE, Groch KR, Flores PAC, Sousa-Lima RS (2016) Vocalizations produced by southern right whale (*Eubalaena australis*) mother-calf pairs in a calving ground off Brazil. J Acoust Soc Am 140(3):1850–1857. https://doi.org/10.1121/1.4962231

Dombroski JRG, Parks SE, Flores PAC, López LMM, Shorter KA, Groch KR (2020) Animal-borne tags provide insights into the acoustic communication of southern right whales (*Eubalaena australis*) on the calving grounds. J Acoust Soc Am 147 (6):EL498-EL503. https://doi.org/10.1121/10.0001391

Dominello T, Širović A (2016) Seasonality of Antarctic minke whale (*Balaenoptera bonaerensis*) calls off the western Antarctic Peninsula. Mar Mamm Sci 32(3): 826–838. https://doi.org/10.1111/mms.12302

Dunlop RA (2017) Potential motivational information encoded within humpback whale non-song vocal sounds. J Acoust Soc Am 141(3):2204–2213. https://doi.org/10.1121/1.4978615

Dunlop RA, Noad MJ, Cato DH, Stokes D (2007) The social vocalization repertoire of east Australian migrating humpback whales (*Megaptera novaeangliae*). J Acoust Soc Am 122(5):2893–2905. https://doi.org/10.1121/1.2783115

Dunlop RA, Cato DH, Noad MJ, Stokes DM (2013) Source levels of social sounds in migrating humpback whales (*Megaptera novaeangliae*). J Acoust Soc Am 134(1):706–714. https://doi.org/10.1121/1.4807828

Dunlop RA, Cato DH, Noad MJ (2014) Evidence of a Lombard response in migrating humpback whales (*Megaptera novaeangliae*). J Acoust Soc Am 136(1): 430–437. https://doi.org/10.1121/1.4883598

Dziak RP, Haxel JH, Lau TK, Heimlich S, Caplan-Auerbach J, Mellinger DK, Matsumoto H, Mate B (2017) A pulsed-air model of blue whale B call vocalizations. Sci Rep 7(1):9122. https://doi.org/10.1038/s41598-017-09423-7

Edds PL (1982) Vocalizations of the blue whale, *Balaenoptera musculus*, in the St. Lawrence River J Mammal 63(2):345–347. https://doi.org/10.2307/1380656

Edds PL (1988) Characteristics of finback *Balaenoptera physalus* vocalizations in the St. Lawrence Estuary Bioacoustics 1(2–3):131–149. https://doi.org/10.1080/09524622.1988.9753087

Edds PL, Macfarlane JA (1987) Occurrence and general behaviour of balaenopterid cetaceans summering in the St. Lawrence Estuary, Canada. Can J Zool 65 (6): 1363–1376

Edds PL, Odell DK, Tershy BR (1993) Vocalizations of a captive juvenile and free-ranging adult-calf pairs of Bryde's whales, *Balaenoptera edeni*. Mar Mamm Sci 9(3):269–284. https://doi.org/10.1111/j.1748-7692.1993.tb00455.x

Edds-Walton PL (1997) Acoustic communication signals of mysticete whales. Bioacoustics 8(1–2):47–60. https://doi.org/10.1080/09524622.1997.9753353

Edds-Walton PL (2000) Vocalizations of minke whales *Balaenoptera acutorostrata* in the St. Lawrence estuary. Bioacoustics 11(1):31–50. https://doi.org/10.1080/09524622.2000.9753448

Elemans CPH, Jiang W, Jensen MH, Pichler H, Mussman BR, Nattestad J, Wahlberg M, Zheng X, Xue Q, Fitch WT (2024) Evolutionary novelties underlie sound production in baleen whales. Nature. https://doi.org/10.1038/s41586-024-07080-1

Epp M (2019) The call repertoire of humpback whales (*Megaptera novaeangliae*) on a Newfoundland foraging ground (2015, 2016) with comparison to a Hawaiian breeding ground (1981, 1982). M.Sc. thesis, University of Manitoba, Canada

Erbe C (2002) Hearing abilities of baleen whales (DRDC Atlantic Report CR2002-065). Defence R&D Canada—Atlantic, Dartmouth

Erbe C, Verma A, McCauley R, Gavrilov A, Parnum I (2015) The marine soundscape of the Perth Canyon. Prog Oceanogr 137:38–51. https://doi.org/10.1016/j.pocean.2015.05.015

Erbe C, Dunlop R, Jenner KCS, Jenner M-NM, McCauley RD, Parnum I, Parsons M, Rogers T, Salgado-Kent C (2017) Review of underwater and in-air sounds emitted by Australian and Antarctic marine mammals. Acoust Aust 45:179–241. https://doi.org/10.1007/s40857-017-0101-z

Erbe C, Duncan A, Hawkins L, Terhune JM, Thomas JA (2022a) Introduction to acoustic terminology and signal processing. In: Erbe C, Thomas JA (eds) Exploring animal behavior through sound: volume 1: methods. Springer, Cham, pp 111–152. https://doi.org/10.1007/978-3-030-97540-1_4

Erbe C, Duncan A, Vigness-Raposa KJ (2022b) Introduction to sound propagation under water. In: Erbe C, Thomas JA (eds) Exploring animal behavior through sound: volume 1: methods. Springer, Cham, pp 185–216. https://doi.org/10.1007/978-3-030-97540-1_6

Español-Jiménez S, Bahamonde PA, Chiang G, Häussermann V (2019) Discovering sounds in Patagonia: characterizing sei whale (*Balaenoptera borealis*) downsweeps in the south-eastern Pacific Ocean. Ocean Sci 15(1):75–82. https://doi.org/10.5194/os-15-75-2019

Fiedler PC, Reilly SB, Hewitt RP, Demer D, Philbrick VA, Smith S, Armstrong W, Croll DA, Tershy BR, Mate BR (1998) Blue whale habitat and prey in the California Channel Islands. Deep-Sea Res II Top Stud Oceanogr 45(8):1781–1801. https://doi.org/10.1016/S0967-0645(98)80017-9

Figueiredo LD, Simao SM (2014) Bryde's Whale (*Balaenoptera edeni*) vocalizations from Southeast Brazil. Aquat Mamm 40(3):225–231. https://doi.org/10.1578/AM.40.3.2014.225

Filún D, van Opzeeland I (2023) Spatial and temporal variability of the acoustic repertoire of Antarctic minke whales (*Balaenoptera bonaerensis*) in the Weddell Sea. Sci Rep 13(1):11861. https://doi.org/10.1038/s41598-023-38793-4

Filun D, Thomisch K, Boebel O, Brey T, Širović A, Spiesecke S, Opzeeland IV (2020) Frozen verses: Antarctic minke whales (*Balaenoptera bonaerensis*) call predominantly during austral winter. R Soc Open Sci 7(10):192112. https://doi.org/10.1098/rsos.192112

Fish JF, Sumich JL, Lingle GL (1974) Sounds produced by the gray whale, *Eschrichtius robustus*. Mar Fish Rev 36(4):38–45

Fournet ME, Szabo A, Mellinger DK (2015) Repertoire and classification of non-song calls in southeast Alaskan humpback whales (*Megaptera novaeangliae*). J Acoust Soc Am 137(1):1–10. https://doi.org/10.1121/1.4904504

Fournet MEH, Gabriele CM, Culp DC, Sharpe F, Mellinger DK, Klinck H (2018a) Some things never change: multi-decadal stability in humpback whale calling repertoire on Southeast Alaskan foraging grounds. Sci Rep 8(1):13186. https://doi.org/10.1038/s41598-018-31527-x

Fournet MEH, Gabriele CM, Sharpe F, Straley JM, Szabo A (2018b) Feeding calls produced by solitary humpback whales. Mar Mamm Sci 34(3):851–865. https://doi.org/10.1111/mms.12485

Fournet MEH, Jacobsen L, Gabriele CM, Mellinger DK, Klinck H (2018c) More of the same: allopatric humpback whale populations share acoustic repertoire. PeerJ 6:e5365. https://doi.org/10.7717/peerj.5365

Fournet MEH, Matthews LP, Bartlett A, Mastick N, Sharpe F, Doyle L, McCowan B, Crutchfield JP, Mellinger DK (2021) Using conspecific playbacks to investigate contact calling in southeast Alaskan humpback whales. bioRxiv:2021.2008.2025.457675. https://doi.org/10.1101/2021.08.25.457675

Frank SD, Ferris AN (2011) Analysis and localization of blue whale vocalizations in the Solomon Sea using waveform amplitude data. J Acoust Soc Am 130(2):731–736. https://doi.org/10.1121/1.3605550

Frazer LN, Mercado E III (2000) A sonar model for humpback whale song. IEEE J Ocean Eng 25(1):160–182. https://doi.org/10.1109/48.820748

Frouin-Mouy H, Tenorio-Hallé L, Thode A, Swartz S, Urbán J (2020) Using two drones to simultaneously monitor visual and acoustic behaviour of gray whales (*Eschrichtius robustus*) in Baja California, Mexico. J Exp Mar Biol Ecol 525:151321. https://doi.org/10.1016/j.jembe.2020.151321

Furumaki S, Tsujii K, Mitani Y (2021) Fin whale (*Balaenoptera physalus*) song pattern in the southern Chukchi Sea. Polar Biol 44:1021–1027. https://doi.org/10.1007/s00300-021-02855-y

Garcia HA, Zhu C, Schinault ME, Kaplan AI, Handegard NO, Godø OR, Ahonen H, Makris NC, Wang D, Huang W, Ratilal P (2019) Temporal–spatial, spectral, and source level distributions of fin whale vocalizations in the Norwegian Sea observed with a coherent hydrophone array. ICES J Mar Sci 76(1):268–283. https://doi.org/10.1093/icesjms/fsy127

Garcia-Rutledge EJ, Narins PM (2001) Shared acoustic resources in an old world frog community. Herpetologica 57(1):104–116

Garland Ellen C, Goldizen Anne W, Rekdahl Melinda L, Constantine R, Garrigue C, Hauser Nan D, Poole MM, Robbins J, Noad Michael J (2011) Dynamic horizontal cultural transmission of humpback whale song at the ocean basin scale. Curr Biol 21(8):687–691. https://doi.org/10.1016/j.cub.2011.03.019

Garland EC, McGregor PK (2020) Cultural transmission, evolution, and revolution in vocal displays: insights from bird and whale song. Front Psychol 11:544929. https://doi.org/10.3389/fpsyg.2020.544929

Gascón MP (2021) Humpback whale (*Megaptera novaeangliae*) social calls in the Southeast Pacific population: context, diversity, and call bouts analysis. M.Sc. thesis, Rheinische Friedrich-Wilhelms-Universität, Bonn, Germany

Gavrilov AN, McCauley RD (2013) Acoustic detection and long-term monitoring of pygmy blue whales over the continental slope in Southwest Australia. J Acoust Soc Am 134(3):2505–2513. https://doi.org/10.1121/1.4816576

Gavrilov A, McCauley R, Salgado-Kent C, Tripovich J, Burton C (2011) Vocal characteristics of pygmy blue whales and their change over time. J Acoust Soc Am 130(6):3651–3660. https://doi.org/10.1121/1.3651817

Gavrilov A, McCauley R, Gedamke J (2012) Steady inter and intra-annual decrease in the vocalization frequency of Antarctic blue whales. J Acoust Soc Am 131(6): 4476–4480. https://doi.org/10.1121/1.4707425

Gedamke J, Costa DP, Dunstan A (2001) Localization and visual verification of a complex minke whale vocalization. J Acoust Soc Am 109(6):3038–3047. https://doi.org/10.1121/1.1371763

George JC, Clark C, Carroll GM, Ellison WT (1989) Observations on the ice-breaking and ice navigation behavior of migrating bowhead whales (*Balaena mysticetus*) near Point Barrow, Alaska, spring 1985. Arctic 42(1):24–30. https://doi.org/10.14430/arctic1636

Gerstein ER, Trygonis V, McCulloch S, Moir J, Kraus S (2014) Female North Atlantic right whales produce gunshot sounds (Abstract). J Acoust Soc Am 135(4): 2369–2369. https://doi.org/10.1121/1.4877814

Girola E, Noad MJ, Dunlop RA, Cato DH (2019) Source levels of humpback whales decrease with frequency suggesting an air-filled resonator is used in sound production. J Acoust Soc Am 145(2):869–880. https://doi.org/10.1121/1.5090492

Gonçalves MIC, Djokic D, Baumgarten JE, Marcondes MCC, Padovese LR, Eugenio LDS, Sousa-Lima RS (2023) Abrupt change in humpback whale song from Brazil suggests cultural revolutions may occur in the South Atlantic. Mar Mamm Sci 40:e13093. https://doi.org/10.1111/mms.13093

Gosby C, Erbe C, Harvey ES, Figueroa Landero MM, McCauley RD (2022) Vocalizing humpback whales (*Megaptera novaeangliae*) migrating from Antarctic feeding grounds arrive earlier and earlier in the Perth Canyon, Western Australia. Front Mar Sci 9:1086763. https://doi.org/10.3389/fmars.2022.1086763

Green RF (1972) Observations on the anatomy of some cetaceans and pinnipeds. In: Ridgway SH (ed) Mammals of the sea: biology and medicine. Charles C. Thomas Publishers, Springfield, pp 247–297

Guazzo RA, Helble TA, D'Spain GL, Weller DW, Wiggins SM, Hildebrand JA (2017) Migratory behavior of eastern North Pacific gray whales tracked using a hydrophone array. PLoS One 12(10):e0185585. https://doi.org/10.1371/journal.pone.0185585

Guazzo RA, Weller DW, Europe HM, Durban JW, D'Spain GL, Hildebrand JA (2019) Migrating eastern North Pacific gray whale call and blow rates estimated from acoustic recordings, infrared camera video, and visual sightings. Sci Rep 9(1):12617. https://doi.org/10.1038/s41598-019-49115-y

Guinee LN, Chu K, Dorsey EM (1983) Changes over time in the songs of known individual humpback whales (*Megaptera novaeangliae*). In: Payne R (ed) Communication and behavior of Whales. Westview Press, Boulder, pp 59–80

Hafner GW, Hamilton CL, Steiner WW, Thompson TJ, Winn HE (1979) Signature information in the song of the humpback whale. J Acoust Soc Am 66(1):1–6. https://doi.org/10.1121/1.383072

Haldiman JT, Tarpley RJ (1993) Anatomy & physiology. In: Burns JJ, Montague JJ, Cowles CJ (eds) The Bowhead Whale. The Society for Marine Mammalogy, Lawrence, pp 90–97

Hart PJ, Ibanez T, Paxton K, Tredinnick G, Sebastián-González E, Tanimoto-Johnson A (2021) Timing is everything: acoustic niche partitioning in two tropical wet forest bird communities. Front Ecol Evol 9. https://doi.org/10.3389/fevo.2021.753363

Heimlich SL, Mellinger DK, Nieukirk SL, Fox CG (2005) Types, distribution, and seasonal occurrence of sounds attributed to Bryde's whales (*Balaenoptera edeni*) recorded in the eastern tropical Pacific, 1999-2001. J Acoust Soc Am 118(3):1830–1837. https://doi.org/10.1121/1.1992674

Helble TA, Henderson EE, Ierley GR, Martin SW (2016) Swim track kinematics and calling behavior attributed to Bryde's whales on the Navy's Pacific missile range facility. J Acoust Soc Am 140(6):4170–4177. https://doi.org/10.1121/1.4967754

Helble TA, Guazzo RA, Alongi GC, Martin CR, Martin SW, Henderson EE (2020a) Fin whale song patterns shift over time in the Central North Pacific. Front Mar Sci 7:587110. https://doi.org/10.3389/fmars.2020.587110

Helble TA, Guazzo RA, Martin CR, Durbach IN, Alongi GC, Martin SW, Boyle JK, Henderson EE (2020b) Lombard effect: Minke whale boing call source levels vary with natural variations in ocean noise. J Acoust Soc Am 147(2):698–712. https://doi.org/10.1121/10.0000596

Helweg DA, Herman LM, Yamamoto S, Forestell PH (1990) Comparison of songs of humpback whales (*Megaptera novaeangliae*) recorded in Japan, Hawaii, and Mexico during the winter of 1989. Scientific Reports of Cetacean Research 1(September):1–20

Helweg DA, Cato DH, Jenkins PF, Claire G, McCauley RD (1998) Geographic variation in South Pacific humpback whale songs. Behaviour 135(1):1–27. https://doi.org/10.1163/156853998793066438

Herman LM (2017) The multiple functions of male song within the humpback whale (*Megaptera novaeangliae*)

mating system: review, evaluation, and synthesis. Biol Rev 92(3):1795–1818. https://doi.org/10.1111/brv.12309

Hosokawa H (1950) On the cetacean larynx with special remarks on the laryngeal sack of the sei whale and the aryteno-epiglottideal tube of the sperm whale. Sci Rep Whales Res Inst 3:23–62

Houser DS, Helweg DA, Moore PWB (2001) A bandpass filter-bank model of auditory sensitivity in the humpback whale. Aquat Mamm 27:82–91

Hoyt E (2011) Habitat protection for cetaceans around the world: status and prospects in the 18 marine regions. Chapter 5. In: Marine protected areas for whales, dolphins and porpoises: a world handbook for cetacean habitat conservation and planning. Taylor & Francis, Oxford, pp 107–121

Hyman J (2003) Countersinging as a signal of aggression in a territorial songbird. Anim Behav 65(6):1179–1185. https://doi.org/10.1006/anbe.2003.2175

Indeck KL, Girola E, Torterotot M, Noad MJ, Dunlop RA (2020) Adult female-calf acoustic communication signals in migrating east Australian humpback whales. Bioacoustics 30(3):341–365. https://doi.org/10.1080/09524622.2020.1742204

Indeck KL, Noad MJ, Dunlop RA (2022) Humpback whale adult females and calves balance acoustic contact with vocal crypsis during periods of increased separation. Ecol Evol 12(2):e8604. https://doi.org/10.1002/ece3.8604

International Organization for Standardization (2017) Underwater acoustics—terminology (ISO 18405). International Organization for Standardization, Geneva

Isaac D, Marler P (1963) Ordering of sequences of singing behaviour of mistle thrushes in relationship to timing. Anim Behav 11(1):179–188. https://doi.org/10.1016/0003-3472(63)90027-7

Jacobs E, Duffy M, Magolan J, Galletti Vernazzani B, Cabrera E, Landea R, Buchan S, Sayigh L (2019) First acoustic recordings of critically endangered eastern South Pacific southern right whales (*Eubalaena australis*). Mar Mamm Sci 35(1):284–289. https://doi.org/10.1111/mms.12519

Janik VM (2014) Cetacean vocal learning and communication. Curr Opin Neurobiol 28:60–65. https://doi.org/10.1016/j.conb.2014.06.010

Janik VM, Knörnschild M (2021) Vocal production learning in mammals revisited. Philos Trans R Soc B 376(1836):20200244. https://doi.org/10.1098/rstb.2020.0244

Jensvold RD, Wright KA (1959) 20-Hz pulses in the North Atlantic. US Navy Journal of Underwater Acoustics 9:685–704

Johnson HD, Stafford KM, George JC, Ambrose WGJ, Clark CW (2015) Song sharing and diversity in the Bering-Chukchi-Beaufort population of bowhead whales (*Balaena mysticetus*), spring 2011. Mar Mamm Sci 31(3):902–922. https://doi.org/10.1111/mms.12196

Jolliffe CD, McCauley RD, Gavrilov AN, Jenner KCS, Jenner M-NM, Duncan AJ (2019) Song variation of the South Eastern Indian Ocean pygmy blue whale population in the Perth Canyon. Western Australia PLOS ONE 14(1):e0208619. https://doi.org/10.1371/journal.pone.0208619

Jolliffe CD, McCauley RD, Gavrilov AN, Jenner C, Jenner MN (2021) Comparing the acoustic behaviour of the Eastern Indian Ocean pygmy blue whale on two Australian feeding grounds. Acoust Aust 49:331–344. https://doi.org/10.1007/s40857-021-00229-2

Jones ME, Nicklin CP, Darling JD (2022) Female humpback whale (*Megaptera novaeangliae*) positions genital-mammary area to intercept bubbles emitted by males on the Hawaiian breeding grounds. Aquat Mamm 48(6):617–620. https://doi.org/10.1578/AM.48.6.2022.617

Kibblewhite AC, Denham RN, Barnes DJ (1967) Unusual low-frequency signals observed in New Zealand waters. J Acoust Soc Am 41(3):644–655. https://doi.org/10.1121/1.1910392

Klinck H, Boebel O (2007) Marine Mammal Automated Perimeter Surveillance (MAPS). Rep Polar Mar Res 568:95–98

Knowlton AR, Clark CW, Kraus SD (1991) Sounds recorded in the presence of sei whales, *Balaenoptera borealis*. J Acoust Soc Am 89 (4, Pt. 2):1968–1968. https://doi.org/10.1121/1.2029710

Kok ACM, Hildebrand MJ, MacArdle M, Martinez A, Garrison LP, Soldevilla MS, Hildebrand JA (2023) Kinematics and energetics of foraging behavior in Rice's whales of the Gulf of Mexico. Sci Rep 13(1):8996. https://doi.org/10.1038/s41598-023-35049-z

Kowarski K, Evers C, Moors-Murphy H, Martin B, Denes SL (2018) Singing through winter nights: seasonal and diel occurrence of humpback whale (*Megaptera novaeangliae*) calls in and around the gully MPA, offshore eastern Canada. Mar Mamm Sci 34(1):169–189. https://doi.org/10.1111/mms.12447

Kowarski K, Cerchio S, Whitehead H, Cholewiak D, Moors-Murphy H (2022) Seasonal song ontogeny in western North Atlantic humpback whales: drawing parallels with songbirds. Bioacoustics 32(3):325–347. https://doi.org/10.1080/09524622.2022.2122561

Krause B (1987) The niche hypothesis: how animals taught us to dance and sing. Whole Earth Review 57:14–16

Kügler A, Lammers MO, Zang EJ, Pack AA (2021) Male humpback whale chorusing in Hawai'i and its relationship with whale abundance and density. Front Mar Sci 8. https://doi.org/10.3389/fmars.2021.735664

Kunc HP, Amrhein V, Naguib M (2006) Vocal interactions in nightingales, *Luscinia megarhynchos*: more aggressive males have higher pairing success. Anim Behav 72(1):25–30. https://doi.org/10.1016/j.anbehav.2005.08.014

Kunc HP, Morrison K, Schmidt R (2022) A meta-analysis on the evolution of the Lombard effect reveals that amplitude adjustments are a widespread vertebrate mechanism. Proc Natl Acad Sci 119(30): e2117809119. https://doi.org/10.1073/pnas.2117809119

Lamoni L, Garland EC, Allen JA, Coxon J, Noad MJ, Rendell L (2023) Variability in humpback whale songs reveals how individuals can be distinctive when sharing a complex vocal display. J Acoust Soc Am 153(4):2238–2250. https://doi.org/10.1121/10.0017602

Leroy EC, Samaran F, Bonnel J, Royer J-Y (2016) Seasonal and diel vocalization patterns of Antarctic blue whale (Balaenoptera musculus intermedia) in the southern Indian Ocean: a multi-year and multi-site study. PLoS One 11(11):e0163587. https://doi.org/10.1371/journal.pone.0163587

Leroy EC, Samaran F, Bonnel J, Royer J-Y (2017) Identification of two potential whale calls in the southern Indian Ocean, and their geographic and seasonal occurrence. J Acoust Soc Am 142(3):1413–1427. https://doi.org/10.1121/1.5001056

Leroy EC, Royer J-Y, Bonnel J, Samaran F (2018) Long-term and seasonal changes of large whale call frequency in the Southern Indian Ocean. J Geophys Res Oceans 123(11):8568–8580. https://doi.org/10.1029/2018jc014352

Leroy EC, Royer J-Y, Alling A, Maslen B, Rogers TL (2021) Multiple pygmy blue whale acoustic populations in the Indian Ocean: whale song identifies a possible new population. Sci Rep 11(1):8762. https://doi.org/10.1038/s41598-021-88062-5

Lewis LA, Širović A (2018) Variability in blue whale acoustic behavior off southern California. Mar Mamm Sci 34(2):311–329. https://doi.org/10.1111/mms.12458

Lewis LA, Calambokidis J, Stimpert AK, Fahlbusch J, Friedlaender AS, McKenna MF, Mesnick SL, Oleson EM, Southall BL, Szesciorka AR, Širović A (2018) Context-dependent variability in blue whale acoustic behaviour. R Soc Open Sci 5(8):180241. https://doi.org/10.1098/rsos.180241

Ljungblad DK, Thompson PO, Moore SE (1982) Underwater sounds recorded from migrating bowhead whales, Balaena mysticetus, in 1979. J Acoust Soc Am 71(2):477–482. https://doi.org/10.1121/1.387419

Lombard É (1911) Le signe de l'élévation de la voix. Annales des Maladies de L'Oreille et du Larynx 37(2):101–109

López-Urbán A, Thode A, Durán CB, UrbáN-R J, Swartz S (2018) Two new grey whale call types detected on bioacoustic tags. J Mar Biolog Assoc 98(5):1169–1175. https://doi.org/10.1017/S0025315416001697

Luther D, Gentry K (2013) Sources of background noise and their influence on vertebrate acoustic communication. Behaviour 150(9–10):1045–1068. https://doi.org/10.1163/1568539X-00003054

Macklin G (2022) Spatiotemporal patterns in acoustic presence of sei whales (Balaenoptera borealis) in Atlantic Canada. Thesis, Dalhousie University, Halifax, NS, Canada, M. Sc

Madhusudhana S, Murray A, Erbe C (2020) Automatic detectors for low-frequency vocalizations of Omura's whales, Balaenoptera omurai: a performance comparison. J Acoust Soc Am 147(5):3078–3090. https://doi.org/10.1121/10.0001108

Malige F, Patris J, Buchan SJ, Stafford KM, Shabangu F, Findlay K, Hucke-Gaete R, Neira S, Clark CW, Glotin H (2020) Inter-annual decrease in pulse rate and peak frequency of Southeast Pacific blue whale song types. Sci Rep 10(1):8121. https://doi.org/10.1038/s41598-020-64613-0

Martin S, Marques T, Thomas L, Morrissey R, Jarvis S, Di Marzio N, Moretti D, Mellinger D (2012) Estimating minke whale (Balaenoptera acutorostrata) boing sound density using passive acoustic sensors. Mar Mamm Sci 29(1):142–158. https://doi.org/10.1111/j.1748-7692.2011.00561.x

Martin K, Tucker MA, Rogers TL (2017) Does size matter? Examining the drivers of mammalian vocalizations. Evolution 71(2):249–260. https://doi.org/10.1111/evo.13128

Martin CR, Guazzo RA, Helble TA, Alongi GC, Durbach IN, Martin SW, Matsuyama BM, Henderson EE (2022) North Pacific minke whales call rapidly when calling conspecifics are nearby. Front Mar Sci 9:897298. https://doi.org/10.3389/fmars.2022.897298

Mate BR, Ilyashenko VY, Bradford AL, Vertyankin VV, Tsidulko GA, Rozhnov VV, Irvine LM (2015) Critically endangered western gray whales migrate to the eastern North Pacific. Biol Lett 11(4):20150071. https://doi.org/10.1098/rsbl.2015.0071

Matthews LH (1978) The natural history of the Whale. Columbia University Press, New York Chichester, West Sussex. https://doi.org/10.7312/matt92090

Matthews D, Macleod R, McCauley RD (2004) Bio-duck activity in the Perth Canyon. An automatic detection algorithm. In: Proceedings of Acoustics 2004, 3–5 November, Gold Coast, Australia, 2004. Australian Acoustical Society, pp 63–66

Matthews L, McCordic JA, Parks SE (2014) Remote acoustic monitoring of North Atlantic right whales (Eubalaena glacialis) reveals seasonal and diel variations in acoustic behavior. PLoS One 9(3):e93167. https://doi.org/10.1371/journal.pone.0091367

McCauley RD, Jenner C, Bannister JL, Cato DH, Duncan AJ (2000) Blue whale calling in the Rottnest trench, Western Australia, and low frequency sea noise. In: Annual Conference of the Australian Acoustical Society, Joondalup, Western Australia, 15–17 Nov 2000. Australian Acoustical Society, pp 245–250

McCauley RD, Gavrilov AN, Jolliffe CD, Ward R, Gill PC (2018) Pygmy blue and Antarctic blue whale presence, distribution and population parameters in southern Australia based on passive acoustics. Deep-Sea Res II: Top Stud Oceanogr:154–168. https://doi.org/10.1016/j.dsr2.2018.09.006

McCordic JA, Root-Gutteridge H, Cusano DA, Denes SL, Parks SE (2016) Calls of North Atlantic right whales *Eubalaena glacialis* contain information on individual identity and age class. Endang Species Res 30:157–169. https://doi.org/10.3354/esr00735

McDonald MA (2006) An acoustic survey of baleen whales off Great Barrier Island, New Zealand. N Z J Mar Freshw Res 40(4):519–529. https://doi.org/10.1080/00288330.2006.9517442

McDonald MA, Moore SE (2002) Calls recorded from North Pacific right whales (*Eubalaena japonica*). J Cetacean Res Manag 4(3):261–266. https://doi.org/10.47536/jcrm.v4i3.838

McDonald MA, Hildebrand JA, Webb SC (1995) Blue and fin whales observed on a seafloor array in the Northeast Pacific. J Acoust Soc Am 98(2):712–721. https://doi.org/10.1121/1.413565

McDonald MA, Calambokidis J, Teranishi AM, Hildebrand JA (2001) The acoustic calls of blue whales off California with gender data. J Acoust Soc Am 109(4):1728–1735. https://doi.org/10.1121/1.1353593

McDonald MA, Hildebrand JA, Wiggins SM, Thiele D, Glasgow D, Moore SE (2005) Sei whale sounds recorded in the Antarctic. J Acoust Soc Am 118(6):3941–3945. https://doi.org/10.1121/1.2130944

McDonald MA, Mesnick SL, Hildebrand JA (2006) Biogeographic characterisation of blue whale song worldwide: using song to identify populations. J Cetacean Res Manag 8(1):55–65. https://doi.org/10.47536/jcrm.v8i1.702

McDonald M, Hildebrand J, Mesnick S (2009) Worldwide decline in tonal frequencies of blue whale songs. Endang Species Res 9:13–21. https://doi.org/10.3354/esr00217

McGowen MR, Vu L, Potter CW, Tho TA, Jefferson TA, Kuit SH, Abdel-Raheem ST, Hines E (2021) Whale temples are unique repositories for understanding marine mammal diversity in Central Vietnam. Raffles Bull Zool 69:481–496

McKenna MF (2011) Blue whale response to underwater noise from commercial ships. PhD thesis, University of California, San Diego

McWilliam JN, McCauley RD, Erbe C, Parsons MJG (2018) Soundscape diversity in the Great Barrier Reef: Lizard Island, a case study. Bioacoustics 27(3):295–311. https://doi.org/10.1080/09524622.2017.1344930

Melcón M, Cummins A, Kerosky S, Roche L, Wiggins S, Hildebrand J (2012) Blue whales respond to anthropogenic noise. PLoS One 7(2):e32681. https://doi.org/10.1371/journal.pone.0032681

Mellinger DK, Clark CW (2003) Blue whale (*Balaenoptera musculus*) sounds from the North Atlantic. J Acoust Soc Am 114(2):1108–1119. https://doi.org/10.1121/1.1593066

Mellinger DK, Carson CD, Clark CW (2000) Characteristics of minke whale (*Balaenoptera acutorostrata*) pulse trains recorded near Puerto Rico. Mar Mamm Sci 16(4):739–756. https://doi.org/10.1111/j.1748-7692.2000.tb00969.x

Mellinger DK, Stafford KM, Moore SE, Munger LM, Christopher G (2004) Detection of North Pacific right whale (*Eubalaena japonica*) calls in the Gulf of Alaska. Mar Mamm Sci 20(4):872–879. https://doi.org/10.1111/j.1748-7692.2004.tb01198.x

Menze S, Zitterbart DP, van Opzeeland I, Boebel O (2017) The influence of sea ice, wind speed and marine mammals on Southern Ocean ambient sound. R Soc Open Sci 4(1):160370. https://doi.org/10.1098/rsos.160370

Mercado E III (2018) The sonar model for humpback whale song revised. Front Psychol 9:1156. https://doi.org/10.3389/fpsyg.2018.01156

Mercado E III (2020) Humpback whale (*Megaptera novaeangliae*) sonar: ten predictions. J Comp Psychol 134(1):123–131. https://doi.org/10.1037/com0000199

Mercado E III (2021) Spectral interleaving by singing humpback whales: signs of sonar. J Acoust Soc Am 149(2):800–806. https://doi.org/10.1121/10.0003443

Mikhalev YA (1997) Humpback whales *Megaptera novaeangliae* in the Arabian Sea. Mar Ecol Prog Ser 149:13–21

Miksis-Olds JL, Nieukirk SL, Harris DV (2018) Two unit analysis of Sri Lankan pygmy blue whale song over a decade. J Acoust Soc Am 144(6):3618–3626. https://doi.org/10.1121/1.5084269

Miksis-Olds JL, Harris DV, Heaney KD (2019) Comparison of estimated 20-Hz pulse fin whale source levels from the tropical Pacific and Eastern North Atlantic Oceans to other recorded populations. J Acoust Soc Am 146(4):2373–2384. https://doi.org/10.1121/1.5126692

Miller BS, Collins K, Barlow J, Calderan S, Leaper R, Mark M, Ensor P, Olson PA, Olavarria C, Double MC (2014a) Blue whale vocalizations recorded around New Zealand: 1964–2013. J Acoust Soc Am 135(3):1616–1623. https://doi.org/10.1121/1.4863647

Miller BS, Leaper R, Calderan S, Gedamke J (2014b) Red shift, blue shift: investigating Doppler shifts, blubber thickness, and migration as explanations of seasonal variation in the tonality of Antarctic blue whale song. PLoS One 9(9):e107740. https://doi.org/10.1371/journal.pone.0107740

Miller BS, Calderan S, Leaper R, Miller EJ, Širović A, Stafford KM, Bell E, Double MC (2021) Source level of Antarctic blue and fin whale sounds recorded on sonobuoys deployed in the deep-ocean off Antarctica. Front Mar Sci 8:792651. https://doi.org/10.3389/fmars.2021.792651

Minton G, Collins T, Findlay K, Ersts P, Rosenbaum H, Berggren P, Baldwin R (2011) Seasonal distribution, abundance, habitat use and population identity of humpback whales in Oman. J Cetacean Res Manage:185–198. https://doi.org/10.47536/jcrm.vi3.329

Moore SE, Ljungblad DK (1984) Gray whales in the Beaufort, Chukchi, and Bering seas: distribution and sound production. In: Jones ML, Swartz SL, Leatherwood S (eds) The gray whale *Eschrichtius robustus*, vol 29. Academic, San Diego, pp 543–559

Morano J, Salisbury D, Rice A, Conklin K, Falk K, Clark C (2012) Seasonal and geographical patterns of fin whale song in the western North Atlantic Ocean. J Acoust Soc Am 132(2):1207–1212. https://doi.org/10.1121/1.4730890

Moreira SC, Weksler M, Sousa-Lima RS, Maia M, Sukhovich A, Royer J-Y, Marcondes MCC, Cerchio S (2020) Occurrence of Omura's whale, *Balaenoptera omurai* (Cetacea: Balaenopteridae), in the Equatorial Atlantic Ocean based on passive acoustic monitoring. J Mammal 101(6):1727–1735. https://doi.org/10.1093/jmammal/gyaa130

Moreno KR, Macgregor RP (2019) Bubble trails, bursts, rings, and more: a review of multiple bubble types produced by cetaceans. Anim Behav Cogn 6(2):105–126. https://doi.org/10.26451/abc.06.02.03.2019

Munger LM, Wiggins SM, Moore SE, Hildebrand JA (2008) North Pacific right whale (*Eubalaena japonica*) seasonal and diel calling patterns from long-term acoustic recordings in the southeastern Bering Sea, 2000–2006. Mar Mamm Sci 24(4):795–814. https://doi.org/10.1111/j.1748-7692.2008.00219.x

Munger LM, Wiggins SM, Hildebrand JA (2011) North Pacific right whale up-call source levels and propagation distance on the southeastern Bering Sea shelf. J Acoust Soc Am 129(6):4047–4054. https://doi.org/10.1121/1.3557060

Murray A, Cerchio S, McCauley R, Jenner CS, Razafindrakoto Y, Coughran D, McKay S, Rosenbaum H (2012) Minimal similarity in songs suggests limited exchange between humpback whales (*Megaptera novaeangliae*) in the southern Indian Ocean. Mar Mamm Sci 28(1):E41–E57. https://doi.org/10.1111/j.1748-7692.2011.00484.x

Murray A, Dunlop RA, Noad MJ, Goldizen AW (2018) Stereotypic and complex phrase types provide structural evidence for a multi-message display in humpback whales (*Megaptera novaeangliae*). J Acoust Soc Am 143(2):980–994. https://doi.org/10.1121/1.5023680

Newhall A, Lin Y-T, Lynch J, Baumgartner M, Gawarkiewicz G (2012) Long distance passive localization of vocalizing sei whales using an acoustic normal mode approach. J Acoust Soc Am 131(2):1814–1825. https://doi.org/10.1121/1.3666015

Nielsen MLK, Bejder L, Videsen SKA, Christiansen F, Madsen PT (2019) Acoustic crypsis in southern right whale mother–calf pairs: infrequent, low-output calls to avoid predation? J Exp Biol 222 (13):jeb190728. https://doi.org/10.1242/jeb.190728

Nieukirk SL, Stafford KM, Mellinger DK, Dziak RP, Fox CG (2004) Low-frequency whale and seismic airgun sounds recorded in the mid-Atlantic Ocean. J Acoust Soc Am 115(4):1832–1843. https://doi.org/10.1121/1.1675816

Nikolich K, Towers JR (2020) Vocalizations of common minke whales (*Balaenoptera acutorostrata*) in an eastern North Pacific feeding ground. Bioacoustics 29(1):97–108. https://doi.org/10.1080/09524622.2018.1555716

Noad MJ, Cato DH, Bryden MM, Jenner MN, Jenner KCS (2000) Cultural revolution in whale songs. Nature 408:537. https://doi.org/10.1038/35046199

Norris KS (1969) The echolocation of marine mammals. In: Andersen HT (ed) The biology of marine mammals. Academic, New York, pp 391–423

Norris KS, Prescott JH, Asa-Dorian PV, Perkins P (1961) An experimental demonstration of echo-location behavior in the porpoise, *Tursiops truncatus* (Montagu). Biol Bull 120(2):163–176

Norris KS, Goodman RM, Villa-Ramírez B, Hobbs L (1977) Behavior of California gray whale, *Eschrichtius robustus*, in southern Baja California, Mexico. Fish Bull 75(1):159–172

Northrop J, Cummings WC, Thompson PO (1968) 20-Hz signals observed in the Central Pacific. J Acoust Soc Am 43(2):383–384. https://doi.org/10.1121/1.1910799

Northrop J, Cummings WC, Morrison MF (1971) Underwater 20-Hz signals recorded near Midway Island. J Acoust Soc Am 49 (6, pt. 2):1909–1910. https://doi.org/10.1121/1.1912606

Nottebohm F (1970) Ontogeny of bird song. Science 167(3920):950–956. https://doi.org/10.1126/science.167.3920.950

Nottebohm F, Nottebohm ME, Crane L (1986) Developmental and seasonal changes in canary song and their relation to changes in the anatomy of song-control nuclei. Behav Neural Biol 46(3):445–471. https://doi.org/10.1016/S0163-1047(86)90485-1

Nowicki S, Peters S, Podos J (1998) Song learning, early nutrition and sexual selection in songbirds. Am Zool 38(1):179–190. https://doi.org/10.1093/icb/38.1.179

Oleson EM, Barlow J, Gordon J, Rankin S, Hildebrand JA (2003) Low frequency calls of Bryde's whales. Mar Mamm Sci 19(2):160–172. https://doi.org/10.1111/j.1748-7692.2003.tb01119.x

Oleson EM, Calambokidis J, Burgess WC, McDonald MA, LeDuc CA, Hildebrand JA (2007a) Behavioral context of call production by eastern North Pacific blue whales. Mar Ecol Prog Ser 330:269–284. https://doi.org/10.3354/meps330269

Oleson EM, Wiggins SM, Hildebrand JA (2007b) Temporal separation of blue whale call types on a southern California feeding ground. Anim Behav 74(4):881–894. https://doi.org/10.1016/j.anbehav.2007.01.022

Oleson EM, Širović A, Bayless AR, Hildebrand JA (2014) Synchronous seasonal change in fin whale song in the North Pacific. PLoS One 9(12):e115678. https://doi.org/10.1371/journal.pone.0115678

Ollervides FJ (2001) Gray whales and boat traffic: movement, vocal, and behavioral responses in Bahia

Magdalena, Mexico. PhD thesis, Texas A&M University, Galveston, TX, USA

Oswald JN, Au WWL, Duennebier F (2011) Minke whale (*Balaenoptera acutorostrata*) boings detected at the station ALOHA cabled observatory. J Acoust Soc Am 129(5):3353–3360. https://doi.org/10.1121/1.3575555

Oswald JN, Ou H, Au WWL, Howe BM, Duennebier F (2016) Listening for whales at the station ALOHA cabled observatory. In: Au WWL, Lammers OM (eds) Listening in the ocean. Springer, New York, pp 221–237. https://doi.org/10.1007/978-1-4939-3176-7_9

Pace F, Benard F, Glotin H, Adam O, White P (2010) Subunit definition and analysis for humpback whale call classification. Appl Acoust 71(11):1107–1112. https://doi.org/10.1016/j.apacoust.2010.05.016

Panicker D, Stafford KM (2021) Northern Indian Ocean blue whale songs recorded off the coast of India. Mar Mamm Sci 37(4):1564–1571. https://doi.org/10.1111/mms.12827

Parks SE, Tyack PL (2005) Sound production by North Atlantic right whales (*Eubalaena glacialis*) in surface active groups. J Acoust Soc Am 117(5):3297–3306. https://doi.org/10.1121/1.1882946

Parks SE, Hamilton PK, Kraus SD, Tyack PL (2005) The gunshot sound produced by male North Atlantic right whales (*Eubalaena glacialis*) and its potential function in reproductive advertisement. Mar Mamm Sci 21(3): 458–475. https://doi.org/10.1111/j.1748-7692.2005.tb01244.x

Parks SE, Clark CW, Tyack PL (2007) Short- and long-term changes in right whale calling behavior: the potential effects of noise on acoustic communication. J Acoust Soc Am 122(6):3725–3731. https://doi.org/10.1121/1.2799904

Parks SE, Urazghildiiev I, Clark CW (2009) Variability in ambient noise levels and call parameters of North Atlantic right whales in three habitat areas. J Acoust Soc Am 125(2):1230–1239. https://doi.org/10.1121/1.3050282

Parks S, Searby A, Célérier A, Johnson M, Nowacek D, Tyack P (2011a) Sound production behavior of individual North Atlantic right whales: implications for passive acoustic monitoring. Endang Species Res 15(1):63–76. https://doi.org/10.3354/esr00368

Parks SE, Johnson M, Nowacek D, Tyack PL (2011b) Individual right whales call louder in increased environmental noise. Biol Lett 7(1):33–35. https://doi.org/10.1098/rsbl.2010.0451

Parks S, Hotchkin C, Cortopassi K, Clark C (2012) Characteristics of gunshot sound displays by North Atlantic right whales in the Bay of Fundy. J Acoust Soc Am 131(4):3173–3179. https://doi.org/10.1121/1.3688507

Parks SE, Cusano DA, Parijs SMV, Nowacek DP (2019a) North Atlantic right whale (*Eubalaena glacialis*) acoustic behavior on the calving grounds. J Acoust Soc Am 146 (1):EL15-EL21. https://doi.org/10.1121/1.5115332

Parks SE, Cusano DA, Van Parijs SM, Nowacek DP (2019b) Acoustic crypsis in communication by North Atlantic right whale mother–calf pairs on the calving grounds. Biol Lett 15(10):20190485. https://doi.org/10.1098/rsbl.2019.0485

Parsons ECM, Wright AJ, Gore MA (2008) The nature of humpback whale (*Megaptera novaeangliae*) song. J Mar Anim Ecol 1(1):12–30

Pastene LA, Acevedo J, Branch TA (2020) Morphometric analysis of Chilean blue whales and implications for their taxonomy. Mar Mamm Sci 36(1):116–135. https://doi.org/10.1111/mms.12625

Patris J, Buchan SJ, Alosilla G, Balcazar-Cabrera N, Malige F, Glotin H (2020) Southeast Pacific blue whale song recorded off Isla Chañaral, northern Chile. Mar Mamm Sci 36(4):1339–1346. https://doi.org/10.1111/mms.12738

Patterson B, Hamilton GR (1964) Repetitive 20 cycle per second biological hydroacoustic signals at Bermuda. In: Tavolga WN (ed) Marine bio-acoustics. Pergamon, Oxford, UK, pp 125–145

Payne R, Guinee LN (1983) Humpback whale (*Megaptera novaeangliae*) songs as an indicator of "stocks". In: Payne R (ed) Communication and behavior of whales. Westview Press, Boulder, pp 333–368

Payne RS, McVay S (1971) Songs of humpback whales. Science 173(3997):585–597. https://doi.org/10.1126/science.173.3997.585

Payne R, Payne K (1972) Underwater sounds of southern right whales. Zoologica 56(4):159–165

Payne K, Tyack P, Payne R (1983) Progressive changes in the songs of humpback whales (*Megaptera novaeangliae*): a detailed analysis of two seasons in Hawaii. In: Payne R (ed) Communication and behavior of Whales. Westview Press, Boulder, pp 9–57

Pénitot A, Schwarz D, Nguyen Hong Duc P, Cazau D, Adam O (2021) Bidirectional interactions with humpback whale singer using concrete sound elements. Front Psychol 12:654314. https://doi.org/10.3389/fpsyg.2021.654314

Perazio CE, Zapetis ME, Roberson D, Botero N, Kuczaj S (2018) Humpback whale, *Megaptera novaeangliae*, song during the breeding season in the Gulf of Tribugá, Colombian Pacific. Madagascar Conserv Develop 13(1):83–90

Perkins PJ (1966) Communication sounds of finback whales. Norsk Hvalfangst-Tidende 55(10):199–200

Pocheville A (2015) The ecological niche: history and recent controversies. In: Heams T, Huneman P, Lecointre G, Silberstein M (eds) Handbook of evolutionary thinking in the sciences. Springer, Dordrecht, pp 547–586. https://doi.org/10.1007/978-94-017-9014-7_26

Ponce D, Thode A, Guerra M, Urbán J, Swartz S (2012) Relationship between visual counts and call detection rates of gray whales (*Eschrichtius robustus*) in Laguna

San Ignacio, Mexico. J Acoust Soc Am 131(4): 2700–2713. https://doi.org/10.1121/1.3689851

Poulter TC (1968a) Chapter 17: marine mammals. In: Sebeok TA (ed) Animal communication: techniques of study and results of research. Indiana University Press, Bloomington, pp 405–465. https://doi.org/10.2979/AnimalCommunicationT

Poulter TC (1968b) Vocalizations of the gray whales in Laguna Ojo de Liebre (Scammon's Lagoon) Baja California, Mexico. Norsk Hvalfangst-Tidende 57(3): 53–62

Putland RL, Constantine R, Radford CA (2017) Exploring spatial and temporal trends in the soundscape of an ecologically significant embayment. Sci Rep 7(1): 5713. https://doi.org/10.1038/s41598-017-06347-0

Putland RL, Ranjard L, Constantine R, Radford CA (2018) A hidden Markov model approach to indicate Bryde's whale acoustics. Ecol Indic 84:479–487. https://doi.org/10.1016/j.ecolind.2017.09.025

Qing X, White PR, Leighton TG, Liu S, Qiao G, Zhang Y (2019) Three-dimensional finite element simulation of acoustic propagation in spiral bubble net of humpback whale. J Acoust Soc Am 146(3):1982–1995. https://doi.org/10.1121/1.5126003

Quayle CJ (1991) A dissection of the larynx of a humpback whale calf with a review of its functional morphology. Memoirs Queensl Museum 30(2):352–354

Rankin S, Barlow J (2005) Source of the North Pacific 'boing' sound attributed to minke whales. J Acoust Soc Am 118(5):3346–3351. https://doi.org/10.1121/1.2046747

Rankin S, Barlow J (2007) Vocalizations of the sei whale Balaenoptera borealis off the Hawaiian islands. Bioacoustics 16(2):137–145. https://doi.org/10.1080/09524622.2007.9753572

Rankin S, Ljungblad D, Clark CW, Kato H (2005) Vocalisations of Antarctic blue whales, Balaenoptera musculus intermedia, recorded during the 2001/2002 and 2002/2003 IWC/SOWER circumpolar cruises, Area V, Antarctica. J Cetacean Res Manag 7(1): 13–20. https://doi.org/10.47536/jcrm.v7i1.752

Rannankari L, Burnham RE, Duffus DA (2018) Diurnal and seasonal acoustic trends in northward migrating Eastern Pacific gray whales (Eschrichtius robustus). Aquat Mamm 44(1):1–6. https://doi.org/10.1578/AM.44.1.2018.1

Rasmussen K, Palacios DM, Calambokidis J, Saborío MT, Rosa LD, Secchi ER, Steiger GH, Allen JM, Stone GS (2007) Southern hemisphere humpback whales wintering off Central America: insights from water temperature into the longest mammalian migration. Biol Lett 3(3):302–305. https://doi.org/10.1098/rsbl.2007.0067

Recalde-Salas A, Salgado Kent CP, Parsons MJG, Marley SA, McCauley RD (2014) Non-song vocalizations of pygmy blue whales in Geographe Bay, Western Australia. J Acoust Soc Am 135 (5):EL213-EL218. https://doi.org/10.1121/1.4871581

Recalde-Salas A, Erbe C, Salgado Kent C, Parsons M (2020) Non-song vocalizations of humpback whales in Western Australia. Front Mar Sci 7(141). https://doi.org/10.3389/fmars.2020.00141

Reidenberg JS (2022) Anatomy of sound production and reception. In: Clark CW, Garland EC (eds) Ethology and behavioral ecology of mysticetes. Springer, Cham, pp 45–69. https://doi.org/10.1007/978-3-030-98449-6_3

Reidenberg JS, Laitman JT (2007) Discovery of a low frequency sound source in Mysticeti (baleen whales): anatomical establishment of a vocal fold homolog. Anat Rec 290(6):745–759. https://doi.org/10.1002/ar.20544

Rekdahl ML, Dunlop RA, Noad MJ, Goldizen AW (2013) Temporal stability and change in the social call repertoire of migrating humpback whales. J Acoust Soc Am 133(3):1785–1795. https://doi.org/10.1121/1.4789941

Rekdahl M, Tisch C, Cerchio S, Rosenbaum H (2017) Common nonsong social calls of humpback whales (Megaptera novaeangliae) recorded off northern Angola, southern Africa. Mar Mamm Sci 33(1): 365–375. https://doi.org/10.1111/mms.12355

Rice AN, Palmer KJ, Tielens JT, Muirhead CA, Clark CW (2014) Potential Bryde's whale (Balaenoptera edeni) calls recorded in the northern Gulf of Mexico. J Acoust Soc Am 135(5):3066–3076. https://doi.org/10.1121/1.4870057

Rice A, Širović A, Hildebrand JA, Wood M, Carbaugh-Rutland A, Baumann-Pickering S (2022) Update on frequency decline of Northeast Pacific blue whale (Balaenoptera musculus) calls. PLoS One 17(4): e0266469. https://doi.org/10.1371/journal.pone.0266469

Richardson WJ, Buchanan RA, Clark CW, Dorsey EM, Fraker MA (1982) Behavior, disturbance responses, and feeding of bowhead whales Balaena mysticetus in the Beaufort Sea, 1980–1981 (PB-86-152170/XAB United States NTIS, PC A20/MF A01. GRA English). LGL Ecological Research Associates, Inc., Bryan, pages 461

Risch D (2022) Mysterious minke whales: acoustic diversity and variability. In: Clark CW, Garland EC (eds) Ethology and behavioral ecology of mysticetes. Springer, Cham, pp 329–348. https://doi.org/10.1007/978-3-030-98449-6_14

Risch D, Clark CW, Dugan PJ, Popescu M, Siebert U, Van Parijs SM (2013) Minke whale acoustic behavior and multi-year seasonal and diel vocalization patterns in Massachusetts Bay, USA. Mar Ecol Prog Ser 489:279–295. https://doi.org/10.3354/meps10426

Risch D, Castellote M, Clark CW, Davis GE, Dugan PJ, Hodge LEW, Kumar A, Lucke K, Mellinger DK, Nieukirk SL, Popescu CM, Ramp C, Read AJ, Rice AN, Silva MA, Siebert U, Stafford KM, Verdaat H, Van Parijs SM (2014a) Seasonal migrations of North Atlantic minke whales: novel insights from large-scale passive acoustic monitoring networks. Mov Ecol 2(1): 24. https://doi.org/10.1186/s40462-014-0024-3

Risch D, Gales NJ, Gedamke J, Kindermann L, Nowacek DP, Read AJ, Siebert U, Van Opzeeland IC, Van Parijs

SM, Friedlaender AS (2014b) Mysterious bio-duck sound attributed to the Antarctic minke whale (*Balaenoptera bonaerensis*). Biol Lett 10(4): 20140175. https://doi.org/10.1098/rsbl.2014.0175

Risch D, Siebert U, Van Parijs SM (2014c) Individual calling behaviour and movements of North Atlantic minke whales (*Balaenoptera acutorostrata*). Behaviour 151(9):1335–1360. https://doi.org/10.1163/1568539X-00003187

Rivers JA (1997) Blue Whale, *Balaenoptera musculus*, vocalizations from the waters off Central California. Mar Mamm Sci 13(2):186–195. https://doi.org/10.1111/j.1748-7692.1997.tb00626.x

Romagosa M, Boisseau O, Cucknell A-C, Moscrop A, McLanaghan R (2015) Source level estimates for sei whale (*Balaenoptera borealis*) vocalizations off the Azores. J Acoust Soc Am 138. https://doi.org/10.1121/1.4930900

Romagosa M, Baumgartner M, Cascão I, Lammers MO, Marques TA, Santos RS, Silva MA (2020) Baleen whale acoustic presence and behaviour at a Mid-Atlantic migratory habitat, the Azores Archipelago. Sci Rep 10(1):4766. https://doi.org/10.1038/s41598-020-61849-8

Romagosa M, Pérez-Jorge S, Cascão I, Mouriño H, Lehodey P, Pereira A, Marques TA, Matias L, Silva MA (2021) Food talk: 40-Hz fin whale calls are associated with prey biomass. Proc Royal Soc B 288(1954):20211156. https://doi.org/10.1098/rspb.2021.1156

Root-Gutteridge H, Cusano DA, Shiu Y, Nowacek DP, Van Parijs SM, Parks SE (2018) A lifetime of changing calls: North Atlantic right whales, *Eubalaena glacialis*, refine call production as they age. Anim Behav 137: 21–34. https://doi.org/10.1016/j.anbehav.2017.12.016

Rosel PE, Wilcox LA, Yamada TK, Mullin KD (2021) A new species of baleen whale (*Balaenoptera*) from the Gulf of Mexico, with a review of its geographic distribution. Mar Mamm Sci 37(2):577–610. https://doi.org/10.1111/mms.12776

Rossi-Santos MR, Filun D, Soares-Filho W, Paro AD, Wedekin LL (2022) "Playing the beat": occurrence of bio-duck calls in Santos Basin (Brazil) reveals a complex acoustic behaviour for the Antarctic minke whale (*Balaenoptera bonaerensis*). PLoS One 17(9): e0255868. https://doi.org/10.1371/journal.pone.0255868

Ross-Marsh EC, Elwen SH, Fearey J, Thompson KF, Maack T, Gridley T (2022) Detection of humpback whale (*Megaptera novaeangliae*) non-song vocalizations around the Vema Seamount, Southeast Atlantic Ocean. JASA Express Lett 2(4):041201. https://doi.org/10.1121/10.0010072

Ruppé L, Clément G, Herrel A, Ballesta L, Décamps T, Kéver L, Parmentier E (2015) Environmental constraints drive the partitioning of the soundscape in fishes. Proc Natl Acad Sci 112(19):6092–6097. https://doi.org/10.1073/pnas.1424667112

Saloma A, Ratsimbazafindranahaka MN, Martin M, Andrianarimisa A, Huetz C, Olivier A, Charrier I (2022) Social calls in humpback whale mother-calf groups off Sainte Marie breeding ground (Madagascar). PeerJ, Indian Ocean. https://doi.org/10.7717/peerj.13785

Samaran F, Guinet C, Adam O, Motsch JF, Cansi Y (2010) Source level estimation of two blue whale subspecies in southwestern Indian Ocean. J Acoust Soc Am 127(6):3800–3808. https://doi.org/10.1121/1.3409479

Schall E, Di Iorio L, Berchok C, Filún D, Bedriñana-Romano L, Buchan SJ, Van Opzeeland I, Sears R, Hucke-Gaete R (2020) Visual and passive acoustic observations of blue whale trios from two distinct populations. Mar Mamm Sci 36(1):365–374. https://doi.org/10.1111/mms.12643

Schevill WE, Watkins WA (1972) Intense low-frequency sounds from an Antarctic minke whale, *Balaenoptera acutorostrata*. Breviora 388:1–8

Schevill WE, Watkins WA, Backus RH (1964) The 20-cycle signals and *Balaenoptera* (fin whales). In: Tavolga WN (ed) Marine bio-acoustics. Pergamon, Oxford, pp 147–152

Schoeman RP, Erbe C, Pavan G, Righini R, Thomas JA (2022) Analysis of soundscapes as an ecological tool. In: Erbe C, Thomas JA (eds) Exploring animal behavior through sound: volume 1: methods. Springer, Cham, pp 217–267. https://doi.org/10.1007/978-3-030-97540-1_7

Schreiber OW (1952) Some sounds from marine life in the Hawaiian area. J Acoust Soc Am 24(1):116–116. https://doi.org/10.1121/1.1917427

Schulze JN, Denkinger J, Oña J, Poole MM, Garland EC (2022) Humpback whale song revolutions continue to spread from the central into the eastern South Pacific. R Soc Open Sci 9(8):220158. https://doi.org/10.1098/rsos.220158

Sciacca V, Morello G, Beranzoli L, Embriaco D, Filiciotto F, Marinaro G, Riccobene GM, Simeone F, Viola S (2023) Song notes and patterns of the Mediterranean fin whale (*Balaenoptera physalus*) in the Ionian Sea. J Mar Sci Eng 11(11):2057. https://doi.org/10.3390/jmse11112057

Seger KD, Thode AM, Urbán-R J, Martínez-Loustalot P, Jiménez-López ME, López-Arzate D (2016) Humpback whale-generated ambient noise levels provide insight into singers' spatial densities. J Acoust Soc Am 140(3):1581–1597. https://doi.org/10.1121/1.4962217

Shabangu FW, Yemane D, Stafford KM, Ensor P, Findlay KP (2017) Modelling the effects of environmental conditions on the acoustic occurrence and behaviour of Antarctic blue whales. PLoS One 12(2):e0172705. https://doi.org/10.1371/journal.pone.0172705

Shabangu FW, Findlay KP, Yemane D, Stafford KM, van den Berg M, Blows B, Andrew RK (2019) Seasonal occurrence and diel calling behaviour of Antarctic blue whales and fin whales in relation to environmental conditions off the west coast of South Africa. J Mar Syst. https://doi.org/10.1016/j.jmarsys.2018.11.002

Shabangu FW, Findlay K, Stafford KM (2020) Seasonal acoustic occurrence, diel-vocalizing patterns and bioduck call-type composition of Antarctic minke whales off the west coast of South Africa and the Maud Rise, Antarctica. Mar Mamm Sci 36(2):658–675. https://doi.org/10.1111/mms.12669

Sharpe FA (2001) Social foraging of the Southeast Alaskan humpback whale. PhD thesis. Simon Fraser University, Vancouver, BC, Canada

Silber GK (1986) The relationship of social vocalizations to surface behavior and aggression in the Hawaiian humpback whale (*Megaptera novaeangliae*). Can J Zool 64(10):2075–2080. https://doi.org/10.1139/z86-316

Simon M, Stafford KM, Beedholm K, Lee CM, Madsen PT (2010) Singing behavior of fin whales in the Davis Strait with implications for mating, migration and foraging. J Acoust Soc Am 128(5):3200–3210. https://doi.org/10.1121/1.3495946

Širović A, Hildebrand JA, Wiggins SM, McDonald MA, Moore SE, Thiele D (2004) Seasonality of blue and fin whale calls and the influence of sea ice in the Western Antarctic Peninsula. Deep-Sea Res II Top Stud Oceanogr 51(17–19):2327–2344. https://doi.org/10.1016/j.dsr2.2004.08.005

Širović A, Hildebrand JA, Theile D (2006) Baleen whales in the Scotia Sea during January and February 2003. J Cetacean Res Manag 8(2):161–171. https://doi.org/10.47536/jcrm.v8i2.712

Širović A, Hildebrand JA, Wiggins SM (2007) Blue and fin whale call source levels and propagation range in the Southern Ocean. J Acoust Soc Am 122(2):1208–1215. https://doi.org/10.1121/1.2749452

Širović A, Hildebrand JA, Wiggins SM, Thiele D (2009) Blue and fin whale acoustic presence around Antarctica during 2003 and 2004. Mar Mamm Sci 25(1):125–136. https://doi.org/10.1111/j.1748-7692.2008.00239.x

Širović A, Williams L, Kerosky S, Wiggins S, Hildebrand J (2013) Temporal separation of two fin whale call types across the eastern North Pacific. Mar Biol 160:47–57. https://doi.org/10.1007/s00227-012-2061-z

Širović A, Bassett HR, Johnson SC, Wiggins SM, Hildebrand JA (2014) Bryde's whale calls recorded in the Gulf of Mexico. Mar Mamm Sci 30(1):399–409. https://doi.org/10.1111/mms.12036

Širović A, Johnson SC, Roche LK, Varga LM, Wiggins SM, Hildebrand JA (2015) North Pacific right whales (*Eubalaena japonica*) recorded in the northeastern Pacific Ocean in 2013. Mar Mamm Sci 31(2):800–807. https://doi.org/10.1111/mms.12189

Širović A, Oleson EM, Buccowich J, Rice A, Bayless AR (2017) Fin whale song variability in southern California and the Gulf of California. Sci Rep 7(1):10126. https://doi.org/10.1038/s41598-017-09979-4

Smith JN, Goldizen AW, Dunlop RA, Noad MJ (2008) Songs of male humpback whales, *Megaptera novaeangliae*, are involved in intersexual interactions. Anim Behav 76:467–477. https://doi.org/10.1016/j.anbehav.2008.02.013

Soldevilla MS, Debich AJ, Garrison LP, Hildebrand JA, Wiggins SM (2022a) Rice's whales in the northwestern Gulf of Mexico: call variation and occurrence beyond the known core habitat. Endang Species Res 48:155–174. https://doi.org/10.3354/esr01196

Soldevilla MS, Ternus K, Cook A, Hildebrand JA, Frasier KE, Martinez A, Garrison LP (2022b) Acoustic localization, validation, and characterization of Rice's whale calls. J Acoust Soc Am 151(6):4264–4278. https://doi.org/10.1121/10.0011677

Sousa AG, Harris D (2015) Description and seasonal detection of two potential whale calls recorded in the Indian Ocean. J Acoust Soc Am 138:1379–1388. https://doi.org/10.1121/1.4928719

Sridhar H, Beauchamp G, Shanker K (2009) Why do birds participate in mixed-species foraging flocks? A large-scale synthesis. Anim Behav 78(2):337–347. https://doi.org/10.1016/j.anbehav.2009.05.008

Stafford KM (2003) Two types of blue whale calls recorded in the Gulf of Alaska. Mar Mamm Sci 19(4):682–693. https://doi.org/10.1111/j.1748-7692.2003.tb01124.x

Stafford KM (2022) Singing behavior in the bowhead whale. In: Clark CW, Garland EC (eds) Ethology and behavioral ecology of mysticetes. Springer, Cham, pp 277–295. https://doi.org/10.1007/978-3-030-98449-6_12

Stafford KM, Clark CW (2021) Chapter 22 – acoustic behavior. In: George JC, Thewissen JGM (eds) The bowhead whale. Academic, pp 323–338. https://doi.org/10.1016/B978-0-12-818969-6.00022-4

Stafford KM, Fox CG, Clark DS (1998) Long-range acoustic detection and localization of blue whale calls in the Northeast Pacific Ocean. J Acoust Soc Am 104(6):3616–3625. https://doi.org/10.1121/1.423944

Stafford KM, Nieukirk SL, Fox CG (1999a) An acoustic link between blue whales in the eastern tropical Pacific and the Northeast Pacific. Mar Mamm Sci 15(4):1258–1268. https://doi.org/10.1111/j.1748-7692.1999.tb00889.x

Stafford KM, Nieukirk SL, Fox CG (1999b) Low-frequency whale sounds recorded on hydrophones moored in the eastern tropical Pacific. J Acoust Soc Am 106(6):3687–3698. https://doi.org/10.1121/1.428220

Stafford KM, Nieukirk SL, Fox CG (2001) Geographic and seasonal variation of blue whale calls in the North Pacific. J Cetacean Res Manag 3(1):65–76. https://doi.org/10.47536/jcrm.v3i1.902

Stafford KM, Bohnenstiehl DR, Tolstoy M, Chapp E, Mellinger DK, Moore SE (2004) Antarctic-type blue whale calls recorded at low latitudes in the Indian and eastern Pacific Oceans. Deep-Sea Res I Oceanogr Res Pap 51(10):1337–1346. https://doi.org/10.1016/j.dsr.2004.05.007

Stafford KM, Moore SE, Fox CG (2005) Diel variation in blue whale calls recorded in the eastern tropical Pacific. Anim Behav 69(4):951–958. https://doi.org/10.1016/j.anbehav.2004.06.025

Stafford KM, Moore SE, Spillane M, Wiggins S (2007) Gray whale calls recorded near Barrow, Alaska, throughout the winter of 2003-04. Arctic 60(2): 167–172. https://doi.org/10.14430/arctic241

Stafford KM, Moore SE, Laidre KL, Heide-Jorgensen MP (2008) Bowhead whale springtime song off West Greenland. J Acoust Soc Am 124(5):3315–3323. https://doi.org/10.1121/1.2980443

Stafford KM, Chapp E, Bohnenstiel DR, Tolstoy M (2011) Seasonal detection of three types of "pygmy" blue whale calls in the Indian Ocean. Mar Mamm Sci 27(4):828–840. https://doi.org/10.1111/j.1748-7692.2010.00437.x

Stafford K, Moore S, Berchok C, Wiig O, Lydersen C, Hansen E, Kalmbach D, Kovacs K (2012) Spitsbergen's endangered bowhead whales sing through the polar night. Endang Species Res 18(2): 95–103. https://doi.org/10.3354/esr00444

Stafford KM, Lydersen C, Wiig Ø, Kovacs KM (2018) Extreme diversity in the songs of Spitsbergen's bowhead whales. Biol Lett 14(4):20180056. https://doi.org/10.1098/rsbl.2018.0056

Stimpert AK, Wiley DN, Au WWL, Johnson MP, Arsenault R (2007) 'Megapclicks': acoustic click trains and buzzes produced during night-time foraging of humpback whales (*Megaptera novaeangliae*). Biol Lett 3(5):467–470. https://doi.org/10.1098/rsbl.2007.0281

Stimpert AK, Au WWL, Parks SE, Hurst T, Wiley DN (2011) Common humpback whale (*Megaptera novaeangliae*) sound types for passive acoustic monitoring. J Acoust Soc Am 129(1):476–482. https://doi.org/10.1121/1.3504708

Stone E (2000) Separating the noise from the noise: a finding in support of the "niche hypothesis," that birds are influenced by human-induced noise in natural habitats. Anthrozoös 13 (4):225–231. https://doi.org/10.2752/089279300786999680

Straley JM, Moran JR, Boswell KM, Vollenweider JJ, Heintz RA, Quinn Ii TJ, Witteveen BH, Rice SD (2018) Seasonal presence and potential influence of humpback whales on wintering Pacific herring populations in the Gulf of Alaska. Deep-Sea Res II Top Stud Oceanogr 147:173–186. https://doi.org/10.1016/j.dsr2.2017.08.008

Sueur J (2002) Cicada acoustic communication: potential sound partitioning in a multispecies community from Mexico (Hemiptera: Cicadomorpha: Cicadidae). Biol J Linn Soc 75(3):379–394. https://doi.org/10.1046/j.1095-8312.2002.00030.x

Sukhovskaya LI, Yablokov AV (1979) Morpho-functional characteristics of the larynx in Balaenopteridae. Investigations Cetacea 10:205–214

Tellechea JS, Norbis W (2012) A note on recordings of Southern right whales (*Eubalaena australis*) off the coast of Uruguay. J Cetacean Res Manag 12:361–364

Tellechea JS, Lima M, Olsson D, Mendez V, Perez W (2017) Possible distress sounds from a stranded humpback whale (*Megaptera novaeangliae*). Aquat Mamm 43(3):299–301. https://doi.org/10.1578/AM.43.3.2017.299

Tennessen JB, Parks SE (2016) Acoustic propagation modeling indicates vocal compensation in noise improves communication range for North Atlantic right whales. Endang Species Res 30:225–237. https://doi.org/10.3354/esr00738

Tervo OM, Parks SE, Miller LA (2009) Seasonal changes in the vocal behavior of bowhead whales (*Balaena mysticetus*) in Disko Bay, Western-Greenland. J Acoust Soc Am 126(3):1570–1580. https://doi.org/10.1121/1.3158941

Tervo O, Christoffersen M, Parks S, Kristensen R, Madsen P (2011a) Evidence for simultaneous sound production in the bowhead whale (*Balaena mysticetus*). J Acoust Soc Am 130(4):2257–2262. https://doi.org/10.1121/1.3628327

Tervo OM, Parks SE, Christoffersen MF, Miller LA, Kristensen RM (2011b) Annual changes in the winter song of bowhead whales (*Balaena mysticetus*) in Disko Bay, Western Greenland. Mar Mamm Sci 27(3):E241–E252. https://doi.org/10.1111/j.1748-7692.2010.00451.x

Tervo O, Christoffersen M, Simon M, Miller L, Jensen F, Parks S, Madsen P (2012) High source levels and small active space of high-pitched song in bowhead whales (*Balaena mysticetus*). PLoS One. https://doi.org/10.1371/journal.pone.0052072

Thode AM, D'Spain GL, Kuperman WA (2000) Matched-field processing, geoacoustic inversion, and source signature recovery of blue whale vocalizations. J Acoust Soc Am 107(3):1286–1300. https://doi.org/10.1121/1.428417

Thode AM, Blackwell SB, Seger KD, Conrad AS, Kim KH, Macrander AM (2016) Source level and calling depth distributions of migrating bowhead whale calls in the shallow Beaufort Sea. J Acoust Soc Am 140(6): 4288–4297. https://doi.org/10.1121/1.4968853

Thode AM, Blackwell SB, Conrad AS, Kim KH, Macrander AM (2017) Decadal-scale frequency shift of migrating bowhead whale calls in the shallow Beaufort Sea. J Acoust Soc Am 142(3):1482–1502. https://doi.org/10.1121/1.5001064

Thode AM, Blackwell SB, Conrad AS, Kim KH, Marques T, Thomas L, Oedekoven CS, Harris D, Bröker K (2020) Roaring and repetition: how bowhead whales adjust their call density and source level (Lombard effect) in the presence of natural and seismic airgun survey noise. J Acoust Soc Am 147(3): 2061–2080. https://doi.org/10.1121/10.0000935

Thompson PWF (1982) A long term study of low frequency sounds from several species of whales off Oahu, Hawaii. Cetology 45:1–19

Thompson TJ, Winn HE, Perkins PJ (1979) Mysticete sounds. In: Winn HE, Olla BL (eds) Behavior of marine animals, Cetaceans, vol 3. Plenum Press, New York, pp 403–431. https://doi.org/10.1007/978-1-4684-2985-5_12

Thompson PO, Cummings WC, Ha SJ (1986) Sounds, source levels, and associated behavior of humpback whales, Southeast Alaska. J Acoust Soc Am 80(3): 735–740. https://doi.org/10.1121/1.393947

Thompson PO, Findley LT, Vidal O (1992) 20-Hz pulses and other vocalizations of fin whales, *Balaenoptera physalus*, in the Gulf of California, Mexico. J Acoust Soc Am 92(6):3051–3057. https://doi.org/10.1121/1.404201

Thompson PO, Findley LT, Vidal O, Cummings WC (1996) Underwater sounds of blue whales, *Balaenoptera musculus*, in the Gulf of California, Mexico. Mar Mamm Sci 12(2):288–293. https://doi.org/10.1111/j.1748-7692.1996.tb00578.x

Tremblay CJ, Parijs SMV, Cholewiak D (2019) 50 to 30-Hz triplet and singlet down sweep vocalizations produced by sei whales (*Balaenoptera borealis*) in the western North Atlantic Ocean. J Acoust Soc Am 145(6):3351–3358. https://doi.org/10.1121/1.5110713

Trygonis C, Gerstein E, Moir J, McCulloch S (2013) Vocalization characteristics of North Atlantic right whale surface active groups in the calving habitat, Southeastern United States. J Acoust Soc Am 134(6): 4518–4531. https://doi.org/10.1121/1.4824682

Tyack PL (1997) Studying how cetaceans use sound to explore their environment. In: Owings DH, Beecher MD, Thompson NS (eds) Communication. Perspectives in ethology, vol 12. Springer, Boston, pp 251–297. https://doi.org/10.1007/978-1-4899-1745-4_9

Tyack PL (2020) A taxonomy for vocal learning. Philos Trans R Soc B 375(1789):20180406. https://doi.org/10.1098/rstb.2018.0406

Tyack PL, Clark CW (2000) Communication and acoustic behavior of dolphins and whales. In: Au WWL, Popper AN, Fay RR (eds) Hearing by whales and dolphins. Springer-Verlag, New York, pp 156–224

van Opzeeland I (2010) Acoustic ecology of marine mammals in polar oceans. Rep Polar Mar Res 619:332

van Opzeeland I, Boebel O (2018) Marine soundscape planning: seeking acoustic niches for anthropogenic sound. J Ecoacoust 2:5GSNT8. https://doi.org/10.22261/JEA.5GSNT8

Vandermeer JH (1972) Niche theory. Annu Rev Ecol Syst 3(1):107–132. https://doi.org/10.1146/annurev.es.03.110172.000543

Varga LM, Wiggins SM, Hildebrand JA (2018) Behavior of singing fin whales *Balaenoptera physalus* tracked acoustically offshore of Southern California. Endang Species Res 35:113–124. https://doi.org/10.3354/esr00881

Vega M, Buchan S, Olavarria C, Ramos M, Valladares M (2023) Preliminary characterization and diel variation of fin whale (*Balaenoptera physalus*) downsweep calls off Isla Chañaral, northern Chile. Mar Mamm Sci 39(4):1313–1323. https://doi.org/10.1111/mms.13038

Verpooten J (2021) Complex vocal learning and three-dimensional mating environments. Biol Philos 36(2): 12. https://doi.org/10.1007/s10539-021-09786-2

Vierling E (2022) Non-song humpback vocalizations from the inland waters of Washington and British Columbia. Orcasound. Accessed 28 Dec 2023. https://www.orcasound.net/portfolio/humpback-catalogue/

Viloria-Gómora L, Romero-Vivas E, Urbán RJ (2015) Calls of Bryde's whale (*Balaenoptera edeni*) recorded in the Gulf of California. J Acoust Soc Am 138(5): 2722. https://doi.org/10.1121/1.4932032

Viloria-Gómora L, Urbán RJ, Leon-Lopez B, Romero-Vivas E (2021) Geographic variation in Bryde's whale Be4 calls in the Gulf of California: an insight to population dynamics. Front Mar Sci 8(877). https://doi.org/10.3389/fmars.2021.651469

Walker RA (1961) Twenty-cycle pulses at Nantucket. US Navy J Underw Acoust 11(3):489–501

Walker RA (1963) Some intense, low-frequency, underwater sounds of wide geographic distribution, apparently of biological origin. J Acoust Soc Am 35(11): 1816–1824. https://doi.org/10.1121/1.1918828

Wang D, Huang W, Garcia H, Ratilal P (2016) Vocalization source level distributions and pulse compression gains of diverse baleen whale species in the Gulf of Maine. Remote Sens 8(11):881. https://doi.org/10.3390/rs8110881

Wang Z-T, Duan P-X, Chen M, Mei Z-G, Sun X-D, Nong Z-W, Liu M-H, Akamatsu T, Wang K-X, Wang D (2022) Vocalization of Bryde's whales (*Balaenoptera edeni*) in the Beibu Gulf, China. Mar Mamm Sci 38(3): 1118–1139. https://doi.org/10.1111/mms.12917

Ward R, Gavrilov AN, McCauley RD (2017) "Spot" call: a common sound from an unidentified great whale in Australian temperate waters. J Acoust Soc Am 142(2): EL231–EL236. https://doi.org/10.1121/1.4998608

Warren VE, McPherson C, Giorli G, Goetz KT, Radford CA (2021) Marine soundscape variation reveals insights into baleen whales and their environment: a case study in Central New Zealand. R Soc Open Sci 8(3):201503. https://doi.org/10.1098/rsos.201503

Watkins WA (1967) Air-borne sounds of the humpback whale, *Megaptera novaeangliae*. J Mammal 48(4): 573–578. https://doi.org/10.2307/1377580

Watkins WA (1981) Activities and underwater sounds of fin whales. Sci Rep Whales Res Inst 33:83–117

Watkins WA, Schevill WE (1976) Right whale feeding and baleen rattle. J Mammal 57(1):58–66. https://doi.org/10.2307/1379512

Watkins WA, Tyack P, Moore KE, Bird JE (1987) The 20-Hz signals of finback whales (*Balaenoptera physalus*). J Acoust Soc Am 82(6):1901–1912. https://doi.org/10.1121/1.395685

Webster TA, Dawson SM, Rayment WJ, Parks SE, Van Parijs SM (2016) Quantitative analysis of the acoustic repertoire of southern right whales in New Zealand. J

Acoust Soc Am 140(1):322–333. https://doi.org/10.1121/1.4955066

Weirathmueller MJ, Wilcock WSD, Soule DC (2013) Source levels of fin whale 20 Hz pulses measured in the Northeast Pacific Ocean. J Acoust Soc Am 133(2): 741–749. https://doi.org/10.1121/1.4773277

Weirathmueller MJ, Stafford KM, Wilcock WSD, Hilmo RS, Dziak RP, Tréhu AM (2017) Spatial and temporal trends in fin whale vocalizations recorded in the NE Pacific Ocean between 2003–2013. PLoS One 12(10): e0186127. https://doi.org/10.1371/journal.pone.0186127

Weiss SG, Cholewiak D, Frasier KE, Trickey JS, Baumann-Pickering S, Hildebrand JA, Van Parijs SM (2021) Monitoring the acoustic ecology of the shelf break of Georges Bank, Northwestern Atlantic Ocean: new approaches to visualizing complex acoustic data. Mar Policy 130:104570. https://doi.org/10.1016/j.marpol.2021.104570

Wenz GM (1964) Curious noises and the sonic environment in the ocean. In: Tavolga WN (ed) Marine bio-acoustics. Pergamon, Oxford, pp 101–119

Weston DE, Black RI (1965) Some unusual low-frequency biological noises under water. Deep-Sea Res 12:295–298

Wiggins SM, Hildebrand JA (2020) Fin whale 40-Hz calling behavior studied with an acoustic tracking array. Mar Mamm Sci 36(3):964–971. https://doi.org/10.1111/mms.12680

Wiggins SM, Oleson EM, McDonald MA, Hildebrand JA (2005) Blue whale (*Balaenoptera musculus*) diel call patterns offshore of Southern California. Aquat Mamm 31(2):161–168. https://doi.org/10.1578/AM.31.2.2005.161

Wild LA, Gabriele C (2014) Putative contact calls made by humpback whales (*Megaptera novaeangliae*) in southeastern Alaska. Can Acoust 42(4):23–31

Wiley D, Ware C, Bocconcelli A, Cholewiak D, Friedlaender A, Thompson M, Weinrich M (2011) Underwater components of humpback whale bubble-net feeding behaviour. Behaviour 148(5–6):575–602. https://doi.org/10.1163/000579511X570893

Williams R, Erbe C, Dewantama IMI, Hendrawan IG (2018) Effect on ocean noise: Nyepi, a Balinese day of silence. Oceanography 31(2):16–18. https://doi.org/10.5670/oceanog.2018.207

Wingfield JE, Rubin B, Xu J, Stanistreet JE, Moors-Murphy HB (2022) Annual, seasonal, and diel patterns in blue whale call occurrence off eastern Canada. Endang Species Res 49:71–86. https://doi.org/10.3354/esr01204

Winn HE, Perkins PJ (1976) Distribution and sounds of the minke whale, with a review of mysticete sound. Cetology 19:1–12

Winn HE, Winn LK (1978) The song of the humpback whale *Megaptera novaeangliae* in the West Indies. Mar Biol 47:97–114

Wisdom S, Bowles A, Anderson K (2001) Development of behavior and sound repertoire of a rehabilitating gray whale calf. Aquat Mamm 27(3):239–255

Wood M, Širović A (2022) Characterization of fin whale song off the Western Antarctic Peninsula. PLoS One 17(3):e0264214. https://doi.org/10.1371/journal.pone.0264214

Würsig B, Thewissen JGM, Kovacs KM (eds) (2018) Encyclopedia of marine mammals, 3rd edn. Academic, Cambridge, MA

Yi DH, Makris NC (2016) Feasibility of acoustic remote sensing of large herring shoals and seafloor by baleen whales. Remote Sens 8(9):693. https://doi.org/10.3390/rs8090693

Youngson BT, Darling JD (2016) The occurrence of pulse, "knock" sounds amidst social/sexual behavior of gray whales (*Eschrichtius robustus*) off Vancouver Island. Mar Mamm Sci 32(4):1482–1490. https://doi.org/10.1111/mms.12325

Zandberg L, Lachlan RF, Lamoni L, Garland EC (2021) Global cultural evolutionary model of humpback whale song. Philos Trans R Soc B 376(1836): 20200242. https://doi.org/10.1098/rstb.2020.0242

Zhou M, Dorland RD (2004) Aggregation and vertical migration behavior of *Euphausia superba*. Deep-Sea Res II Top Stud Oceanogr 51(17):2119–2137. https://doi.org/10.1016/j.dsr2.2004.07.009

Zoidis AM, Smultea MA, Frankel AS, Hopkins JL, Day A, McFarland AS, Whitt AD, Fertl D (2008) Vocalizations produced by humpback whale (*Megaptera novaeangliae*) calves recorded in Hawaii. J Acoust Soc Am 123(3):1737–1746. https://doi.org/10.1121/1.2836750

Open Access This chapter is licensed under the terms of the Creative Commons Attribution 4.0 International License (http://creativecommons.org/licenses/by/4.0/), which permits use, sharing, adaptation, distribution and reproduction in any medium or format, as long as you give appropriate credit to the original author(s) and the source, provide a link to the Creative Commons license and indicate if changes were made.

The images or other third party material in this chapter are included in the chapter's Creative Commons license, unless indicated otherwise in a credit line to the material. If material is not included in the chapter's Creative Commons license and your intended use is not permitted by statutory regulation or exceeds the permitted use, you will need to obtain permission directly from the copyright holder.

Odontocete Sounds

Christine Erbe and Chong Wei

Contents

4.1	**Introduction**	267
4.1.1	Odontocete Taxonomy	269
4.2	**Sound Production**	269
4.3	**Echolocation**	271
4.3.1	What Is It and How Is It Used?	271
4.3.2	The SONAR Equation	272
4.3.3	Spectral, Temporal, and Directional Features of Echolocation Signals	272
4.3.4	Taxonomic Comparisons	281
4.3.5	Wild Versus Captive Studies	288
4.3.6	Dolphin Biosonar Performance	290
4.3.7	Biosonar Ontogeny	296
4.4	**Acoustic Communication**	297
4.4.1	Sound Types	298
4.4.2	Repeated Vocalization Sequences	309
4.4.3	Duetting, Chorusing, and Synchronous Rhythmic Calling	311
4.4.4	Correlation of Acoustic and Physical Behavior	311
4.4.5	Vocal Learning and Mimicry	316
4.4.6	The Effects of Noise on Acoustic Behavior	317
4.4.7	Taxonomic Comparisons	317
4.4.8	Recording Communication Sounds	323
4.5	**Taxonomic Overview of Sounds**	331
4.5.1	Clicks and Burst-Pulse Sounds	331
4.5.2	Whistles	332
4.6	**Summary**	332
4.7	**Abbreviations**	332
References		333

C. Erbe (✉) · C. Wei
Centre for Marine Science and Technology, Curtin University, Perth, WA, Australia
e-mail: c.erbe@curtin.edu.au; chong.wei@curtin.edu.au

© The Author(s) 2025
C. Erbe et al. (eds.), *Marine Mammal Acoustics in a Noisy Ocean*,
https://doi.org/10.1007/978-3-031-77022-7_4

4.1 Introduction

Odontocete sounds have commonly been grouped into three classes based on their spectrographic features: whistles, burst-pulse sounds, and clicks. Whistles are tonal signals. Burst-pulse sounds are bursts of rapid, broadband pulses. Clicks are very brief (tens of microseconds) and broadband. Clicks are emitted in trains with an inter-click interval (ICI) greater than the inter-pulse interval (IPI) in burst-pulse sounds. Separation into these three classes is not hard-limited; rather, odontocete sounds lie along a continuum from whistles to burst-pulse sounds, to clicks. Graded sound types are common. For example, whistles that are amplitude-modulated and pulsed lie along the continuum from whistles to burst-pulse sounds. Click trains, in which the ICI decreases and which end in a feeding buzz, are an example of the continuum from discrete clicks to burst-pulse sounds. Indeed, the waveform of a burst-pulse sound recorded at a sufficiently high digitization rate shows a string of very closely spaced click-like signals.

Odontocete sounds can also be grouped based on their function into two main classes: communication and echolocation. Echolocation is presumably exclusively done with clicks. However, communication sounds include whistles, burst-pulse sounds, and clicks. Some odontocetes do not whistle, or whistle extremely rarely, and these species communicate with click series of different spectro-temporal structure for the different purposes (Oliveira et al. 2013).

Echolocation and communication have historically been studied and hence described in a rather compartmentalized way, focusing on either the one or the other. Echolocation research has included studies on echolocation behavior, signal characteristics, and target detection and classification performance. Communication research has quantified and categorized sound repertoires and correlated acoustic behavior with physical behavior. Behavioral contexts for which specific communication vocalizations have been identified include: broadcasting of individual identity (e.g., signature whistles), mother-calf reunions and alloparental care, courtship, group cohesion, foraging, agonistic and aggressive behavior, distress, and excitement (Dawson 1991; Herzing 1996; Herzing 2000). However, there are indications that typical echolocation clicks also play a communicative role, specifically in the coordination of group foraging (e.g., Benoit-Bird and Au 2009).

Odontocete sounds differ not only in their function and spectrographic features, but also in their directionality. Whistles are typically of lower frequency and lower directionality, while clicks are high in frequency and directionality. High directionality is useful for echolocation, whereas more omnidirectional sounds are useful for communication because they broadcast over a wider angle. Whistles, however, can contain overtones and it has been shown that the higher overtones are more directional than the fundamental tone. As such, whistling odontocetes may provide more information about their location and orientation by producing whistles with higher (harmonic) overtones (Lammers and Au 2003).

Odontocetes further produce other sounds with their bodies, such as breaches (Finneran et al. 2000), tail slaps (Finneran et al. 2000; Marten et al. 1988), pectoral slaps (Deecke et al. 2005), jaw claps (Belikov and Bel'kovich 2008; Finneran et al. 2000; Lilly 1962), and bubble trails and rings (Marten et al. 1996). Dolphins also produce in-air vocalizations, such as constricted or explosive exhalations, commonly called chuffs (Wells et al. 1998).

Even though odontocete sounds can be grouped into a few broad classes based on their form and function, there is a lot of variability. Sounds differ by species and by geographic region. Sounds evolved to suit each species' environment and ecology. Even in sympatric species with very similar ecologies, communication sounds differ, perhaps for species identification. In many species, variations in sound features have been classified to populations, family groups, and individuals.

The following sections provide an overview of odontocete sound production, function, and features. Throughout this text, we use the term vocalization for sounds produced by a pneumatic apparatus, even though the apparatus does not contain vocal cords.

4.1.1 Odontocete Taxonomy

Throughout this chapter, we use the taxonomy as recognized by the Society for Marine Mammalogy (https://marinemammalscience.org/science-and-publications/list-marine-mammal-species-subspecies/) in December 2022. A complete table of marine mammal taxonomy is supplied in the Front Matter of this book. Note that taxonomy has changed in several cases since early publications on a species' sound repertoire, requiring careful checking of the geographic region in which recordings were obtained. Examples are:

- *Cephalorhynchus commersonii* (Lacépède, 1804)—Commerson's dolphin, is now split into two subspecies: *C. c. commersonii* (Lacépède, 1804), Commerson's dolphin, and *C. c. kerguelenensis* Robineau, Goodall, Pichler, and C. S. Baker, 2007, Kerguelen dolphin.
- *Cephalorhynchus hectori* (Van Beneden, 1881)—Hector's dolphin, is now split into two subspecies: *C. h. hectori* (Van Beneden, 1881), South Island Hector's dolphin, and *C. h. maui* A. Baker, Smith, and Pichler, 2002, Māui dolphin, North Island Hector's dolphin.
- *Delphinus delphis* Linnaeus, 1758—Common dolphin, now has four subspecies: *D. d. delphis* Linnaeus, 1758, Common dolphin, *D. d. bairdii* Dall, 1873, Eastern North Pacific long-beaked common dolphin, *D. d. ponticus* Barabash, 1935, Black Sea common dolphin, and *D. d. tropicalis* van Bree, 1971, Indo-Pacific common dolphin.
- *Globicephala melas* (Traill, 1809)—Long-finned pilot whale, called *G. melaena* until 1986 and prior to that, *Delphinus melas*. Furthermore, there are now two subspecies: *G. m. edwardii* (A. Smith, 1834), Southern long-finned pilot whale, and *G. m. melas* (Traill, 1809), North Atlantic long-finned pilot whale.
- *Lagenorhynchus obscurus* (Gray, 1828)—Dusky dolphin, is now split into three subspecies: *L. o. fitzroyi* (Waterhouse, 1838), Fitzroy's dolphin, *L. o. obscurus* (Gray, 1828), African dusky dolphin, and *L. o. posidonia* (Philippi, 1893), Peruvian/Chilean dusky dolphin.
- *Orcaella heinsohni* Beasley, Robertson and Arnold, 2005—Australian snubfin dolphin, previously same species as *Orcaella brevirostris* (Owen in Gray, 1866), Irrawaddy dolphin, pesut (until 2005).
- *Sotalia guianensis* (P.J. Van Beneden, 1864)—Guiana dolphin, costero, previously same species as *S. fluviatilis* (Gervais and Deville in Gervais, 1853), tucuxi.
- *Sousa chinensis* (Osbeck, 1765)—Indo-Pacific humpback dolphin, is now split into two subspecies: *S. c. chinensis* (Osbeck, 1765), Chinese humpback dolphin and *S. c. taiwanensis* Wang, Yang and Hung, 2015, Taiwanese humpback dolphin.
- *Sousa sahulensis* Jefferson and Rosenbaum, 2014—Australian humpback dolphin, previously same species as *S. chinensis* (Osbeck, 1765), Indo-Pacific humpback dolphin (until 2014).
- *Stenella attenuata* (Gray, 1846)—Pantropical spotted dolphin, is now split into two subspecies: *S. a. attenuata* (Gray, 1846), offshore pantropical spotted dolphin, and *S. a. graffmani* (Lönnberg, 1934), coastal pantropical spotted dolphin.
- *Stenella longirostris* (Gray, 1828)—Spinner dolphin, is now split into four subspecies: *S. l. centroamericana* Perrin, 1990, Central American spinner dolphin, *S. l. longirostris* (Gray, 1828), Gray's spinner dolphin, *S. l. orientalis* Perrin, 1990, Eastern spinner dolphin, and *S. l. roseiventris* (Wagner, 1846), dwarf spinner dolphin.
- *Inia geoffrensis* (Blainville, 1817)—Amazon river dolphin, now comprises two subspecies: *I. g. boliviensis* (d'Orbigny, 1834)—Bolivian bufeo, and *I. g. geoffrensis* (Blainville, 1817), common boto.

4.2 Sound Production

Whistles, burst-pulse sounds, and echolocation clicks of odontocetes are pneumatically

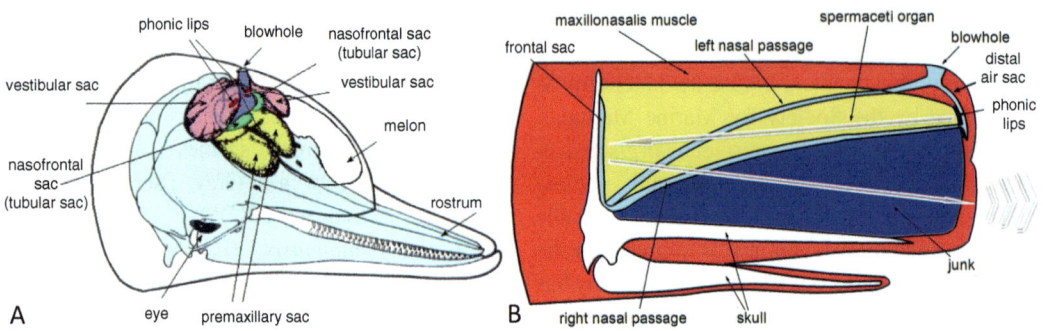

Fig. 4.1 Sound production anatomy in the dolphin (**a**) and sperm whale (**b**)

generated. Air within the nasal passages of odontocetes is pressurized just prior to sound production (Amundin and Andersen 1983; Ridgway et al. 1980); the air is forced past specialized labia, known as the phonic lips (Cranford 2000), which exist just below the nasal plug. Pressurized air rapidly opens the phonic lips, and their rapid closing induces self-sustained vibration and generates a short pulse (Madsen et al. 2023). A series of air sacs (pre-maxillary, accessory, and vestibular) are associated with the phonic lips and appear to support sound production, although the exact mechanism by which this occurs has not been empirically determined (Fig. 4.1).

Except for species within the family Physeteroidea (i.e., sperm whales, dwarf sperm whales, and pygmy sperm whales), all odontocetes have two pairs of phonic lips, one on each side of the terminal portions of the nares. The phonic lips have been observed to be used in both click and whistle production (Cranford et al. 2011). Dolphins may produce both sound types simultaneously, which suggests they use different sound generating apparati (Lilly and Miller 1961). The exact role of the phonic lips in click production has been hotly debated (Cranford 2011). It has been hypothesized that both pairs of phonic lips can produce echolocation clicks and can be timed such that a single, summed click is produced with a larger source level than either of the individual clicks, or that they can be acutely staggered in their actuation in order to achieve beam steering. In recent years, mounting evidence has indicated that echolocation clicks are produced primarily, if not exclusively, with the right pair of phonic lips. Acoustic recordings collected with contact hydrophones and electrophysiological potentials collected just prior to click production have pointed to this phenomenon in the species tested to date (Ames et al. 2020; Finneran et al. 2018; Madsen et al. 2010, 2013b, 2023).

Echolocation clicks are forward-projected through the various forehead structures of the dolphin, including a fatty body known as the melon, complex-shaped nasal air sacs, and the skull. The click, which is produced by the phonic lips, is reflected off the skull and associated air sacs, resulting in forward propagation (Aroyan et al. 1992; Cranford et al. 1996; Wei et al. 2014). The dense connective tissue, which envelops the posterior portion of the melon and forms the supporting elements of the phonic lips (Cranford et al. 1996), effectively transmits the signal into the melon after the reflection (Wei et al. 2018a). The melon is composed of specialized lipids, dubbed acoustic fats, which are structured such that the melon has a low sound-velocity core with a gradient of increasing sound velocity surrounding it (Koblitz et al. 2012; Norris and Harvey 1974). The sound-velocity gradient collimates the projected echolocation signal such that it is narrow and forward-projected in all species studied (see Table 4.1 in Koblitz et al. 2012; Wei et al. 2017, 2018a). The anatomy of the sound-generating apparatus, structures of the forehead soft tissues (such as the melon or, in the case of sperm whales, the junk), and arrangement of air sacs can vary considerably between species (e.g., Cranford 1999; Cranford et al. 2008; Harper et al. 2008; Houser et al. 2004), but all have been hypothesized to contribute to the forward-focusing of sound into the water.

4.3 Echolocation

4.3.1 What Is It and How Is It Used?

Echolocation is the detection and location of objects using reflected sound. Synonymously to biosonar, echolocation involves the emission of sound and the processing of echoes to detect, recognize, classify, locate, and track targets. The term "sonar" in biosonar is an abbreviation for sound navigation and ranging. Biosonar is thus similar to sonar use by humans (e.g., in naval or fisheries applications). All odontocetes studied to-date produce clicks suitable for echolocation. Actual echolocation function has been demonstrated in fewer species (e.g., bottlenose dolphins, false killer whales, beluga whales, harbor porpoises, some beaked whales, and sperm whales; Au 1993; Madsen et al. 2005b; Tønnesen et al. 2020; Wisniewska et al. 2012). Nevertheless, it is almost certain that echolocation serves a critical role in navigation and foraging in odontocetes. It is therefore a critical consideration when determining how anthropogenic sound might interfere with its function.

During echolocation, the animal emits brief clicks, which are forward-projected into the water through the fatty melon. As the sound propagates through the water, it interacts with physical objects (e.g., subsea structures or prey) causing some of the acoustic energy to reflect back to the echolocating animal. The animal then analyzes the echoes for information on target shape, composition, and range. The more echoes the animal receives, the easier it is to detect and identify targets of interest (e.g., Altes et al. 2003; Finneran et al. 2014b). Information on target movement (speed and direction) can be gained by integrating information across multiple echoes. Target detection ranges range from very near the animal (< 1 m) to in excess of 650 m, depending upon the specific biosonar system and the size and composition of the target (Akamatsu et al. 1998; Au et al. 2004, 2007; Finneran 2013; Ivanov 2004; Ladegaard et al. 2019).

The limits of target detection and discrimination are imposed by the biosonar frequencies, level, and beamwidth; the hearing sensitivity of the animal; the distance to the target (echoes lose energy, the farther they propagate); the ambient noise (at the biosonar frequencies); the water depth (as multipath reflections and reverberation in shallow water interfere with the echo); and the clutter in the environment (which also returns echoes). The echolocation signals of many species have been shown optimal for the specific habitat in which each species forages. For example, a narrow biosonar beam is preferable in reverberation and clutter-prone environments as it reduces the number of scatterers in the sonar beam. Additionally, a biosonar beam consisting of high frequencies (e.g., porpoises) is useful for hunting small prey. However, there are animals with wide and broadband biosonar beams (e.g., bottlenose dolphins) that also live in such environments, and so other selective pressures, such as avoidance of acoustic detection by predators (e.g., killer whales) and avoiding overlap with other echolocating species within the same habitat, may also play a role (e.g., Kyhn et al. 2010, 2013).

Odontocete biosonar behavior typically moves through three phases during foraging. To begin, the animal searches a wider area, emitting regular clicks at slow ICI (< 0.1 s) and high source level (Phase 1: scan and search). When a potential prey is found, the animal changes its acoustic gaze; it approaches the target and the ICI decreases as a function of range (Phase 2: approach). In the final phase (Phase 3: prey capture), odontocetes typically emit very rapid click trains with ICIs of 1–10 ms (click repetition rate ~100–1000 Hz) and lower level (10–20 dB below the level of the preceding regular clicks; Arranz et al. 2016; Verfuß et al. 2009), sounding like—and hence called—buzzes. The duration of such terminal buzzes is typically of the order of one to a few seconds (Wisniewska et al. 2014). Most odontocetes start the terminal buzz when they are one to a few body lengths away from the target (e.g., Arranz et al. 2016; Fais et al. 2016; Johnson et al. 2008; Tønnesen et al. 2020; Wisniewska et al. 2012). Sperm whales start to buzz earlier, at longer ranges (Tønnesen et al. 2020). After prey-capture, even faster bursts of clicks, termed victory squeals, have been recorded from bottlenose dolphins, false killer whales, and beluga whales (Ridgway et al. 2014, 2015; Wisniewska et al. 2014).

4.3.2 The SONAR Equation

The SONAR equation describes the echolocation process as a simple sum of intensity ratios, expressed in decibel (see Erbe et al. 2022b for a gentle introduction). The source level (SL) is computed as $10 \log_{10}$(source intensity/reference intensity); the received level (RL) is $10 \log_{10}$(received intensity/reference intensity). The propagation loss (PL) is $10 \log_{10}$(source intensity/received intensity). During echolocation, this loss is incurred twice, on the way from the source (i.e., the echolocating odontocete) to the target (e.g., a fish), and back from the fish to the listening odontocete. The target strength (TS) is defined as $10 \log_{10}$(echo intensity/incident intensity). The animal has a detection threshold (DT), which is expressed as the level of a signal-to-noise ratio: $10 \log_{10}$(intensity of a signal that is just detectable in noise/noise intensity). The ambient noise can mask the echolocation signal or raise the detection threshold. The noise level (NL) is computed as $10 \log_{10}$(noise intensity/reference intensity). With these terms, various forms of the SONAR equation may be derived. For example, the level of the returning echo at the location of the odontocete can be computed as the SL, less the PL to the target, plus the TS, less the PL from the target to the odontocete:

$$RL = SL - 2\,PL + TS$$

Whether the echo is detectable depends on the echo intensity compared to the level of ambient noise; the level difference has to exceed the detection threshold:

$$RL - NL > DT$$

In practice, the terms in the SONAR equation are more commonly computed as ratios of mean-square pressure rather than intensity.

4.3.3 Spectral, Temporal, and Directional Features of Echolocation Signals

4.3.3.1 Click Waveforms and Spectra

Individual clicks are ~10–100 μs in duration. The spectra may be broadband (~tens of kilohertz in bandwidth), narrowband (like tone pips), or frequency-modulated. The amplitude envelope is sometimes Gaussian. Sinusoidal signals that are amplitude-modulated by a Gaussian are called Gabor signals (where A is the amplitude, f_p the peak frequency, φ the phase, t_0 the time centroid,

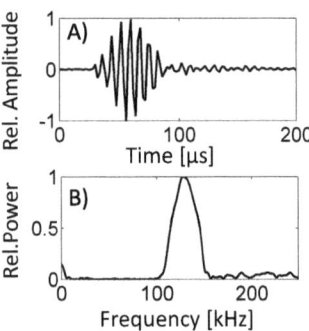

Fig. 4.2 (a) Waveform and (b) spectrum of a Gabor click from a Yangtze finless porpoise. Modified from Fang et al. (2015b). The source parameters of echolocation clicks from captive and free-ranging Yangtze finless porpoises (*Neophocaena asiaeorientalis asiaeorientalis*). PLOS ONE 10 (6):e0129143; https://doi.org/10.1371/journal.pone.0129143). © Fang et al. 2015b. Licensed CC BY 4.0; https://creativecommons.org/licenses/by/4.0/

and τ the root-mean-square (rms) duration of the signal $p(t)$):

$$p(t) = A \sin(2\pi f_p t + \varphi) e^{-\pi^2 \frac{(t-t_0)^2}{\tau^2}}$$

The echolocation signals of porpoises are good examples of Gabor signals (Fig. 4.2). Gabor signals have a small time-bandwidth product (Beedholm 2008; Beedholm and Møhl 2006), providing certain biosonar advantages. Most species of beaked whales use frequency-modulated biosonar signals, which have a larger bandwidth and hence time-bandwidth product. Such signals are advantageous in dealing with signal detection conflict (i.e., masking).

When describing the spectro-temporal features of biosonar clicks, various frequency and time measurements are useful. The 3 dB duration is determined by (1) computing the upper envelope of the waveform (using a Hilbert transform), (2) finding the maximum amplitude of the envelope, (3) finding the points on the envelope 3 dB down from the maximum (i.e., at half power), and (4) computing the time difference between these two points (Fig. 4.3a). The 10 dB duration is the time difference between the two points 10 dB down from the maximum of the envelope (i.e., at one-tenth of the maximum power). The percentage energy signal duration is the time during which a certain percentage of time-integrated squared sound pressure passes, commonly 90%. It is determined by (1) computing the cumulative (i.e., time-integrated) squared pressure, (2) marking the 5% and 95% energy on the y-axis, (3) finding the corresponding times on the x-axis, when the 5% and 95% energies were reached, and (4) computing the time difference between these points (Fig. 4.3b). Once the power spectrum has been computed, the peak frequency is the frequency of maximum power. The 3 dB bandwidth is the difference between the two frequencies (on either side of the peak frequency), at which the spectrum has dropped 3 dB below its peak (Fig. 4.3c). Similarly, the 10 dB bandwidth is measured as the difference in frequencies at which the spectrum has fallen 10 dB from its peak. The center frequency splits the power

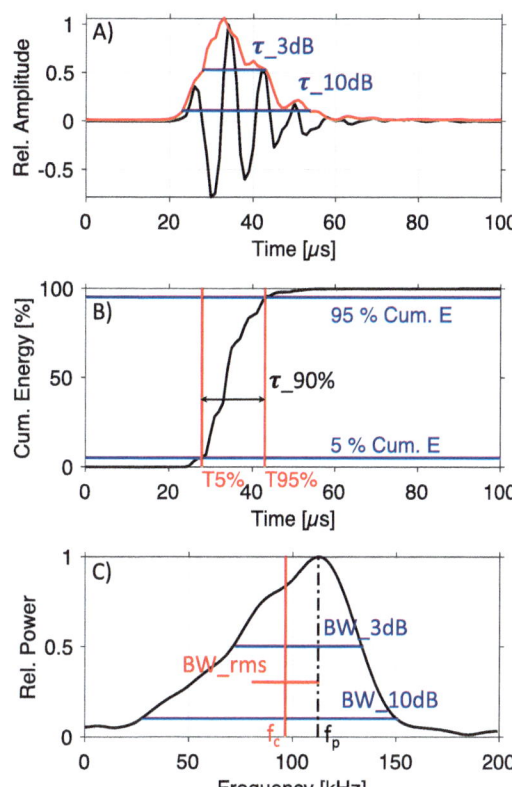

Fig. 4.3 Echolocation click from a bottlenose dolphin; (**a**) waveform and amplitude envelope, (**b**) cumulative energy, and (**c**) spectrum. Commonly measured parameters refer to duration (τ), such as the 3 dB and 10 dB durations and the duration over which 90% of energy passes, and to bandwidth (BW), such as the 3 dB, 10 dB, and rms bandwidths. Center frequency: f_c, peak frequency: f_p. Click recording courtesy of Whitlow Au

spectrum into two halves of equal power. The rms bandwidth is computed as the standard deviation about the center frequency (where f_c is the center frequency, $H(f)$ the Fourier transform, and BW_{rms} the rms bandwidth; see Erbe et al. 2022a for a more detailed introduction to acoustic quantities and units):

$$f_c = \frac{\int_{-\infty}^{\infty} f|H(f)|^2 df}{\int_{-\infty}^{\infty} |H(f)|^2 df}$$

$$\mathrm{BW_{rms}} = \sqrt{\frac{\int_{-\infty}^{\infty}(f-f_c)^2|H(f)|^2 df}{\int_{-\infty}^{\infty}|H(f)|^2 df}}$$

While many species have been documented to make broadband clicks, the entire spectrum is not necessarily utilized during echolocation, as has been shown with trained bottlenose dolphins and false killer whales listening to phantom targets and echoes that were artificially filtered and modified (Ibsen et al. 2010, 2011). In fact, in these studies, the lower half of the spectrum was more important in the set echolocation tasks than the higher half.

Bottlenose dolphins have been observed to produce clicks with single, narrow spectral peaks, clicks with very broad spectral bandwidths (3 dB bandwidths >85 kHz), and clicks with two spectral peaks, called bimodal clicks (Houser et al. 1999; Moore and Pawloski 1990). Dusky dolphins (Au et al. 2010), false killer whales (Au et al. 1995), Indo-Pacific humpback dolphins (Goold and Jefferson 2004), pygmy killer whales (Madsen et al. 2004), and Risso's dolphins (Smith et al. 2016) are examples of other species in which bimodal clicks have been observed. Reported click spectra from wild animals must be carefully considered due to the lack of control over animal location and orientation; for example, multi-modal spectra might result from off-axis clicks or multi-path arrivals at the recording hydrophone.

4.3.3.2 Source Levels

The SL of odontocete clicks varies by species. Levels as high as 229 dB re 1 µPa m pk and 236 dB re 1 µPa m rms have been reported for the sperm whale (Møhl et al. 2003; Zimmer et al. 2005b), 228 dB re 1 µPa m pk-pk in bottlenose dolphins (Au et al. 1974), and 225 dB re 1 µPa m pk-pk in the false killer whale (Thomas and Turl 1990). Note that different authors have used different measures. The SL of clicks generally increases asymptotically as a function of the range investigated or at which a target is expected (e.g., Atem et al. 2009; Beedholm and Miller 2007; Li et al. 2006; Malinka et al. 2021a; Rasmussen et al. 2002). Over successive clicks, as an animal approaches a target (i.e., decreasing range to target), both the ICI (see Sect. 3.3.5) and the SL decrease (Fig. 4.4). The reduction in SL partly compensates for a decrease in PL and a decrease in TS (if the target is a school of fish; Au and Würsig 2004), keeping the echo level somewhat constant. However, this is not always the case, as there are instances when odontocetes have been observed to either increase the SL or keep the SL constant while approaching a target (Atem et al. 2009; Jensen et al. 2013), which would presumably increase the level of the returned echo. Unfortunately, in most instances, no parallel measurement of the echolocation beam pattern (see below) exists to inform whether there is also a concomitant change in the beam.

Studies have shown a correlation between SL and center frequency in several species (e.g., Au et al. 1995; Finneran et al. 2014a; Gong et al. 2019; Ladegaard et al. 2015; Moore and Pawloski 1990; Smith et al. 2016; Wisniewska et al. 2012), which might be due to physical limitations on click production. However, Houser et al. (2005) showed that free-swimming bottlenose dolphins had a preferred center frequency (depending on the individual and their task), and these frequencies were fairly consistent, no matter the SL, which varied over 30 dB. Similarly, a consistency in the peak frequency of echolocation clicks has been reported for individual bottlenose dolphins, as well as certain odontocete populations and species (Baumann-Pickering et al. 2013a; Houser et al. 2005; Soldevilla et al. 2008, 2010b). Consistent patterns in biosonar features (such as SL and frequency) imply allometric relationships between overall body size and the size of the click generator in the various individuals and species recorded, as well as other selective pressures placed on the animals. Thus, species-specific biosonar features have been successfully used to identify species in passive acoustic datasets (e.g., Baumann-Pickering et al. 2010; Roch et al. 2011; Soldevilla et al. 2008).

But are biosonar features really so rigid? How much control do animals have over their biosonar

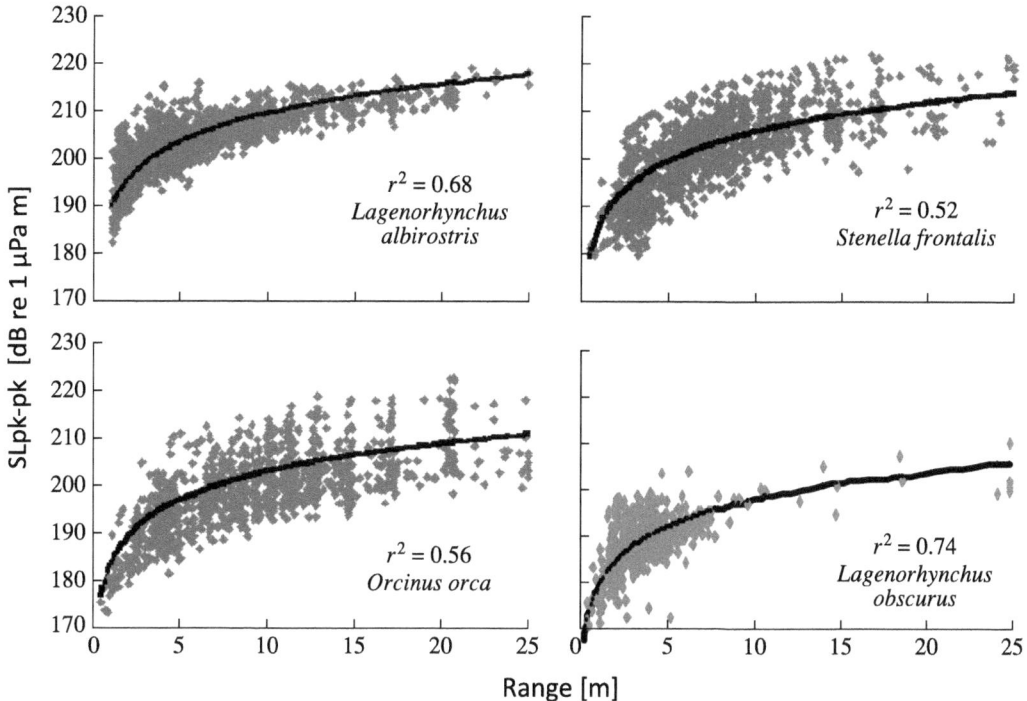

Fig. 4.4 Source level change as a function of range to target (Au 2004). Reprinted by permission from Springer Nature. Au WWL, Echolocation signals of wild dolphins, Acoust Phys 50(4), 454–462; https://doi.org/10.1134/1.1776224 © Au 2004. All rights reserved. Similar results are shown in Ladegaard et al. (2015) and others

features? Odontocetes can vary much of the spectral, temporal, and amplitude features of their clicks and adapt them to the task. For example, a beluga whale that was moved from San Diego Bay (CA) to Kaneohe Bay (HI) increased the peak frequency by an octave and the amplitude of its echolocation signals by 10 dB, likely to compensate for increased noise from snapping shrimp (at 2–20 kHz) in Hawaii (Au et al. 1985). False killer whales in reverberant tanks used lower frequencies and lower SL than those in an open pen in Kaneohe Bay (Thomas et al. 1988; Thomas and Turl 1990). Dolphins have been reported to change the frequency content of their clicks in response to changes in their frequency range of hearing, as a result of losing high-frequency hearing at old age (Houser et al. 1999; Ibsen et al. 2010; Moore et al. 2004).

4.3.3.3 Beam Patterns

Echolocation clicks are highly directional and forward-projected in a beam. The main lobe of the beam has been measured to be slightly elevated relative to the axis defined by the animal's jaw in the bottlenose dolphin and beluga whale (~5°; Au et al. 1986, 1987), at a similar elevation in the harbor porpoise (Au et al. 1999, 2006), and slightly lower in the false killer whale (from 0 to −5°; Au et al. 1995). Average 3 dB horizontal beamwidths are ~7° in the beluga whale (Au et al. 1987), 6–10° in the bottlenose dolphin (Au 1993; Finneran et al. 2014a), ~6° in the false killer whale (Au et al. 1995), 13–17° in the harbor porpoise (Au et al. 1999; Koblitz et al. 2012), and ~9° in the Risso's dolphin (Smith et al. 2016). Vertical beamwidths differ from horizontal ones, but are also narrow. These beam patterns were recorded in laboratory conditions with good control over the orientation of the animal relative to the recording hydrophone arrays. Beam patterns have also been estimated from other species in the wild, although there is greater uncertainty in these measurements because of uncertainty in animal orientation relative to the recording hydrophones, sparseness of arrays used, assumptions made, etc. Nevertheless, reported beam patterns appear consistent with those reported from laboratory settings. Estimates of the 3 dB horizontal beamwidth have been reported as ~10° in the Atlantic spotted dolphin (Jensen et al. 2015), 8–9° in the bottlenose dolphin (Wahlberg et al. 2011), ~13° in the Cuvier's beaked whale (Shaffer et al. 2013), and ~8° in the white-beaked dolphin (Rasmussen et al. 2004). Directivity indices, reported in dB and defined as $10\log_{10}$ of the ratio of the intensity of a directional source in the target direction and the intensity of an omnidirectional source of equal power, have also been used to characterize the directivity of the beam in these and other species (see Table 4.1 of Koblitz et al. 2012; Fig. 4.5).

The directivity pattern is, of course, frequency-dependent, with higher-frequency components having narrower beamwidths (e.g., Starkhammar et al. 2011). However, the beams are not static; it has been demonstrated that odontocetes can vary the width of their echolocation beams and steer the beams, at least to some degree (Finneran et al. 2014a; Jensen et al. 2015; Koblitz et al. 2012; Moore et al. 2008; Wisniewska et al. 2015). Bottlenose dolphins have been observed to more than double the beamwidth under the experimental challenge of detecting off-axis targets (Moore et al. 2008), and Atlantic spotted dolphins have been estimated to increase beamwidth by as much as 50% when coming into close range of a target (Jensen et al. 2015). The latter behavior has also been observed in other experimental situations with harbor porpoises (Wisniewska et al. 2012) and could potentially be related to increasing the acoustic field of view as an animal approaches mobile and elusive prey. During the prey capture phase, beamwidth widens and this can be controlled independently of an often simultaneous drop in center frequency, SL, and ICI (Jensen et al. 2015; Ladegaard et al. 2017; Wisniewska et al. 2015).

As previously noted, the waveform and thus spectrum and level of individual clicks depend on the angle about the animal at which the clicks are recorded. Clicks that are recorded off-axis in either the horizontal or vertical plane appear distorted relative to on-axis clicks (Fig. 4.5). Much of the early literature reporting on echolocation clicks of wild animals did not measure the angle to the animal and the reported measures are therefore an average over a potentially large range of angles. Off-axis clicks still have sufficient energy for nearby echoes to identify off-axis targets (Moore et al. 2008). The functional role of off-axis clicks in the wild, however, is unknown.

The measurement of off-axis clicks has informed hypotheses about the source of echolocation signals. Several studies of odontocete echolocation have noted that a double pulse is measurable with increasing time separation between the two pulses as recordings are made increasingly off the main axis of the echolocation beam (Au et al. 2012; Branstetter et al. 2012; Finneran et al. 2014a; Lammers and Castellote 2009). Different hypotheses have been presented to explain this observation: that the two signals represent the action of two sets of phonic lips operating simultaneously or near-simultaneously (Cranford et al. 2011; Lammers and Castellote 2009); or that the second of the two signals is due to reflections off the dolphin skull (or other anatomy) but represents a single signal generated by a single sound source (Au et al. 2012; Finneran et al. 2014a). Mounting evidence suggests that echolocation clicks are preferentially, if not almost exclusively, produced by the right set of phonic lips in odontocetes that have two sets (Finneran et al. 2018; Madsen et al. 2010, 2013b), but the mechanism by which the single click separates into two signals at off-axis angles is still unresolved. Recently, it has been proposed that the forward-projected (on-axis) click waveform can be decomposed into two overlaid transients (Starkhammar et al. 2019). The two components are hypothesized to be from one signal that has been slightly delayed at the more directional high frequencies due to the sound field gradient within the melon. If true, this finding

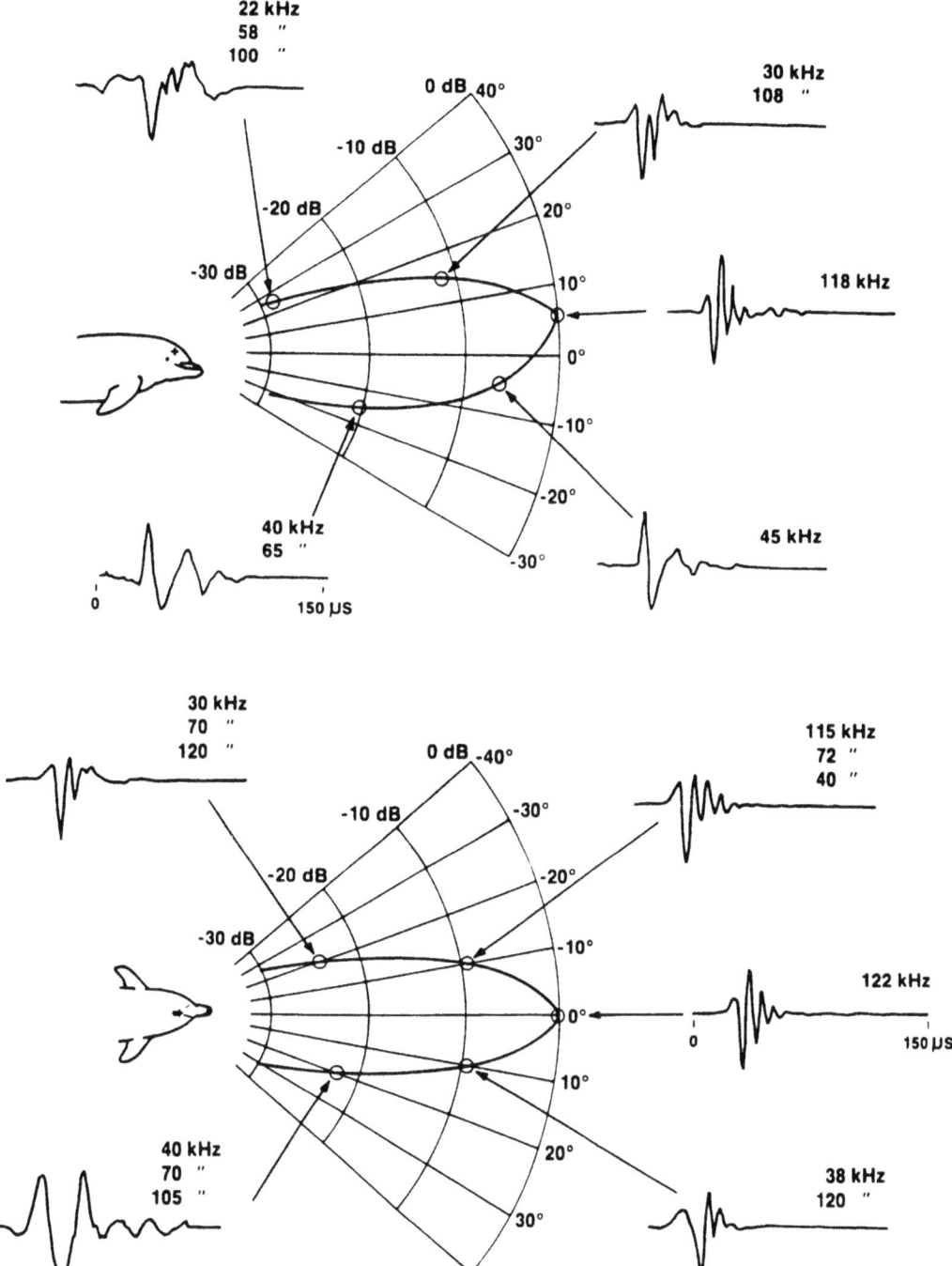

Fig. 4.5 Emission beam patterns in the vertical and horizontal planes about a bottlenose dolphin. Click waveforms on- and off-axis are also shown and the frequencies of peak power are listed with multiple frequencies given for multi-modal clicks, in the order of decreasing amplitude (Au 2000). Reprinted by permission from Springer Nature. Au WWL, Echolocation in dolphins. In WWL Au, AN Popper & RR Fay (Eds.), Hearing by Whales and Dolphins (pp. 364–408); https://doi.org/10.1007/978-1-4612-1150-1_9. © Springer Nature, 2000. All rights reserved

might help explain the observation of double pulses measured off the main axis of the echolocation beam.

4.3.3.4 Multi-Pulse Clicks

Clicks consisting of multiple pulses have been reported in much of the biosonar literature. Multiple pulses can be separated in time or overlap with no temporal separation (e.g., Dawson 1991; Kamminga and Wiersma 1981). There are signal-processing advantages of multi-pulses, such as Doppler processing to determine target speed, or sonar enhancement by coherent averaging to detect targets within bubble clouds (Finfer et al. 2012). However, in many cases, the observation is artefactual: Off-axis recordings can result in echolocation clicks having the appearance of multi-pulse clicks.

Sperm whales, in contrast, do produce a click with a multi-pulse structure that arises from internal reflections of the primary pulse within the whale's head. In sperm whales, these pulses, if recorded on-axis, are separated in time and do not overlap. The bent horn hypothesis states that the multi-pulsed nature of the sperm whale click is due to reflection of the primary click (produced by the phonic lips) back into the spermaceti organ, followed by reflection off the frontal air sac located at the posterior end of the spermaceti organ, and then collimation within the junk complex before projection into the water (Fig. 4.1; Madsen et al. 2003; Møhl et al. 2000, 2003; Schulz et al. 2009; Zimmer et al. 2005a). Energy from the initial pulse (p0) is weak but leads that of the second pulse (p1), which is of much greater amplitude, and which is followed by additional pulses.

Finally, the apparent multi-pulse structure of recorded clicks might be due to environmental factors, such as reflections off the sea surface, seafloor, or tank wall. With high-enough signal-to-noise ratio and high-enough sampling frequency, reflections off the sea surface can be identified as being 180° out of phase; no phase shift occurs at the water-solid interface at the seafloor or tank walls (e.g., Li et al. 2005). In addition, external reflections are expected to arrive later and appear at lower amplitude than direct arrivals. Recording with an array of hydrophones allows better identification of multi-path arrivals and identification of the primary click waveform (e.g., Kyhn et al. 2009). For example, double- and multi-pulse structure of finless porpoise clicks were found to be due to multipath propagation of the signal in shallow water (and near the water surface) rather than produced by the animal itself (Li et al. 2005).

4.3.3.5 Inter-Click Intervals

Echolocation clicks are emitted in series and each series is commonly called a click train. When the animal is in scan and search mode, click trains typically have a high ICI (~0.1 s). As an animal gets closer to a target such as potential prey, the ICI decreases. When focusing on a target (i.e., during both the approach and the prey-capture phases), the next click is typically only emitted after the echo of the former click has been received, plus a lag time (Au 1993). The two-way travel time is the time it takes for a click to travel from the animal to the target and back and is critical for the odontocete to determine the range to the target. In other words, the two-way travel time is directly related to distance by the speed of sound through water. The lag time is thought to be related to the neural processing of the echo. Lag times of 10–45 ms have been reported during the approach phase, decreasing to 1.5–4 ms during the prey capture phase (feeding buzzes) (Au 1980, 1993; Au et al. 1974; Evans and Powell 1967; Wisniewska et al. 2014).

During long-range echolocation tasks (generally exceeding >75 m), some animals have been recorded to emit packets of clicks and wait for the packet of echoes to return before emitting the next packet (Finneran 2013; Ladegaard et al. 2019). In other words, the two-way travel time corresponds to the beginning and end of click packets, not to individual clicks. The ICIs for clicks within a packet are less than the two-way travel time to the inspected target. This strategy is believed to relate to temporal limits of the dolphins' ability to process echoes (i.e., echoes need to be received within a certain period of time for the integration across echoes to occur). Dolphin biosonar performance improved with more clicks in a packet suggesting that dolphins were integrating information about the target across multiple echoes (Finneran et al. 2014b). Some dolphins also use click packets at short ranges (Ladegaard et al. 2019) or in certain experimental contexts (Finneran et al. 2019), although the reason for doing so is unclear. Measuring AEPs from a trained harbor porpoise, Beedholm et al. (2023) found additional AEP peaks with latencies greater than long-range echolocation click intervals. Echo information relating to the range to the target is processed with latencies shorter than the two-way travel time, while information relating to target characterization and auditory scene analysis is processed more slowly, with greater latencies, corresponding to integration over multiple clicks.

During foraging, porpoises have been shown to alter their acoustic gaze from long range to short range as they approach prey, and possibly back to long range if prey escaped. An example of such acoustic gaze changes is shown in Fig. 4.6. These gaze changes involve changes in ICI, SL, and beamwidth. Towards prey capture, the ICI and SL decrease, while the beamwidth has been reported to widen, possibly in order to widen the field of view to avoid prey escape just before capture (Jensen et al. 2015; Wisniewska et al. 2012, 2014, 2015).

During prey-capture events, it is common for the ICIs to be reduced to the point that the click train becomes a very rapid burst of clicks (ICI:

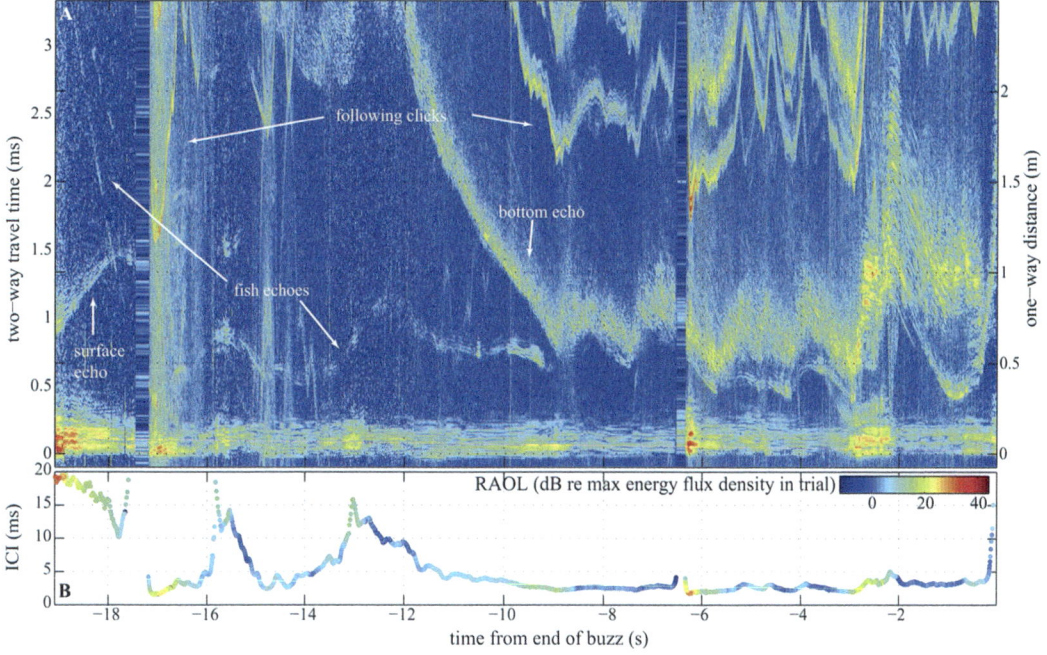

Fig. 4.6 Acoustic gaze changes in a foraging harbor porpoise (Wisniewska et al. 2015). X-axis represents time from the end of prey-capture. Prey escaped at −16 s and −13 s from final capture. (**a**) Echogram showing outgoing clicks (near 0 ms travel time and 0 m distance) and reflected clicks (off the water surface, bottom, and fish), recorded by an acoustic tag on the animal's head. Color represents click energy (blue: weak; red: strong). (**b**) Inter-click interval. Color represents relative apparent output level (RAOL). Reprinted from Wisniewska DM, Ratcliffe JM, Beedholm K, Christensen CB, Johnson M, Koblitz JC, Wahlberg M, Madsen PT (2015) Range-dependent flexibility in the acoustic field of view of echolocating porpoises (*Phocoena phocoena*). eLife: e05651. https://doi.org/10.7554/eLife.05651. © Wisniewska et al., 2015. Licensed CC BY 4.0; https://creativecommons.org/licenses/by/4.0/

1–10 ms), called a feeding or terminal buzz (e.g., DeRuiter et al. 2009; Verfuß et al. 2009). Feeding buzzes have been identified in numerous odontocete species, including bottlenose dolphins (Wisniewska et al. 2014), Chilean dolphins (Götz et al. 2010), Commerson's dolphins (Reyes Reyes et al. 2015), false killer whales (Wisniewska et al. 2014), Heaviside's dolphins (Martin et al. 2018), killer whales (Wright et al. 2021), pilot whales (Aguilar de Soto et al. 2008), Risso's dolphin (Arranz et al. 2016), boto (Ladegaard et al. 2017), beluga whales (Castellote et al. 2021; Ridgway et al. 2014), narwhals (Miller et al. 1995; Rasmussen et al. 2015), harbor porpoises (DeRuiter et al. 2009; Verfuß et al. 2009), sperm whales (Miller et al. 2004a), dwarf sperm whales (Malinka et al. 2021b), and Blainville's and Cuvier's beaked whales (Aguilar de Soto et al. 2012; Johnson et al. 2004, 2006; Madsen et al. 2013a). In some odontocetes, the transition from search clicks to approach clicks and finally feeding buzzes is smooth, with the ICI steadily decreasing (e.g., DeRuiter et al. 2009; Miller et al. 2004a). In other odontocetes, such as deep-diving Risso's dolphins and beaked whales, the transition from slow search clicks to feeding buzzes is very rapid with a sudden drop in ICI (Arranz et al. 2016; Madsen et al. 2005b). Such differences in biosonar usage may be related to habitat and prey type, or could simply be due to chance encounter of prey at close range. Nevertheless, the ICI during buzzing is variable, perhaps due to the size and maneuverability of the hunting odontocete and the agility of the prey (Wisniewska et al. 2014).

4.3.3.6 What Drives the Spectral, Temporal, and Directional Characteristics?

A good biosonar system generally will have a high SL, so it is effective at long range, and a narrow beam to facilitate clutter rejection (e.g., from the sea surface, seafloor, and objects not on the main axis of the sonar beam). But there are trade-offs and limitations. The higher the frequency of the biosonar signal, the easier it is to focus the beam, but the shorter the operational range due to scattering and absorption, both of which increase with frequency. There are anatomical design limitations on SL (i.e., how much energy an animal can put into its biosonar signals), signal duration, emitted frequency, and beamwidth. For any resonator (like gas cavities or strings), the larger the resonator is, the lower are the frequencies that can be produced. Also, a resonator of fixed size is more omnidirectional at its lower frequencies. For animal sound production, this means that larger sound production apparati can produce lower frequencies. And smaller sound production apparati need to use higher frequencies to achieve the same directionality as a larger sound production apparatus at lower frequencies. Body size is therefore expected to affect biosonar characteristics. Large animals like sperm whales can produce directional low-frequency clicks. But small porpoises need to use very high frequencies to make a directional beam with their small head.

Ecological drivers also have shaped biosonar characteristics. Odontocetes range from riverine, estuarine, and coastal habitats to pelagic and deep, offshore waters. They further range from warm equatorial to cold polar waters. Habitat and prey type exert pressures toward optimum echolocation parameters. In the open ocean, where prey is patchy and widely distributed, a biosonar system with a wider beam can rapidly search a larger area. High SL enables a long search range. Also, the lower the frequency, the farther the echolocation signal and its echoes travel due to decreased absorption of low-frequency energy. In coastal water, a wide search beam would be detrimental because of scattering from the sea surface and seafloor, as well as from suspended sediment and bubbles in the water column. As the outgoing echolocation beam hits either surface, the signal reflects, and returning echoes interfere with echoes of potential prey. A narrow beam reduces the amount of unwanted scattering (i.e., clutter). A higher SL does not help in environments with a lot of scattering, because a higher source level leads to higher echo levels and increased clutter. To improve sonar performance in cluttered environments, a narrower beam and narrower frequency band can help focus on targets.

The interplay of ecological drivers and environmental and biophysical constraints sets an optimization problem that should lead to similar biosonar design solutions in species with similar ecologies. Indeed, sperm whales are the largest odontocete species. They live in pelagic waters, covering a great range of depths. They emit echolocation signals at the lowest frequencies, lowest click rates (i.e., high ICI), and some of the highest SL, optimal for long-range searching (Madsen et al. 2002b; Møhl et al. 2003). Smaller odontocetes echolocate at higher frequencies, higher rates, and lower SL to find small prey over short ranges in coastal, acoustically cluttered environments (Au 1993; Jensen et al. 2013; Kyhn et al. 2009).

4.3.4 Taxonomic Comparisons

4.3.4.1 NBHF Clicks

Some species produce narrowband (3 dB bandwidth <20 kHz), high-frequency (peak frequency ~130 kHz) (NBHF) clicks. These species are the porpoises of the family Phocoenidae (i.e., Burmeister's, Chinese finless, Dall's, harbor, Indo-Pacific finless, and spectacled porpoise, and vaquita), the small dolphins of the genus *Cephalorhynchus* (i.e., Chilean, Commerson's, Heaviside's, and Hector's dolphins), at least two species of the genus *Lagenorhynchus* (i.e., hourglass and Peale's dolphin), the franciscana dolphin, the dwarf sperm whale, and the pygmy sperm whale. These species are from four different families and differ in morphology and habitat. Convergent evolution might have led to these species producing NBHF clicks.

Several hypotheses for the evolution of NBHF clicks have been suggested. The small dolphins and porpoises tend to occur in shallow, coastal habitats, where ambient noise is high and where reverberation and hence acoustic clutter is great (e.g., Miller and Wahlberg 2013). They feed on small prey, which they hunt over short ranges. Small prey requires short wavelengths (i.e., high biosonar frequencies) for detection. Absorption is greater at higher frequencies, and so echolocation ranges are short. High click rates (i.e., short ICI) are indicative of short-range echolocation behavior. A higher SL does not help increase echolocation ranges in shallow, coastal waters, because interfering echoes would also have a higher level. But a narrow beamwidth (in addition to a narrow bandwidth) helps reduce clutter. And so, NBHF clicks fit the foraging ecologies of coastal dolphins and porpoises.

Furthermore, NBHF species tend to be smaller and are preyed upon by larger odontocetes, which are typically more sensitive at lower frequencies. Avoiding the emission of signals that predators can hear has been termed acoustic crypsis (Morisaka and Connor 2007). The frequencies of NBHF clicks are above the frequencies of best sensitivity of their typical predators. Interestingly, NBHF species mostly do not whistle. But they do produce burst-pulse sounds down to 1 kHz (Dawson 1991; Dziedzic and Debuffrenil 1989; Watkins and Schevill 1980; Watkins et al. 1977; Yoshida et al. 2014). Reyes Reyes et al. (2016) hypothesized that these types of signals would be safe to produce in sheltered bays absent of larger odontocetes (primarily killer whales) that prey on NBHF species in deeper water.

NBHF species that occur outside coastal areas are the hourglass dolphin and the dwarf and pygmy sperm whales. Hourglass dolphins use higher SL than do coastal NBHF species, which increases target detection in the open ocean where clutter is not an issue (Kyhn et al. 2009). The dwarf sperm whale produces NBHF clicks at higher SL, slower click rates (i.e., greater ICI), and higher directionality than porpoise and dolphin NBHF clicks (Malinka et al. 2021b). These adaptations allow them to forage on prey in reliable, mesopelagic prey layers that are known to be acoustically cluttered. Also, NBHF biosonar may have evolved independently in these species to exploit a notch in ocean ambient noise at 100 kHz (Madsen et al. 2005a). Exploitation of such a notch in ambient noise also requires narrow auditory filters, which these species likely have.

While NBHF species share common features of their clicks, there are also inter-species differences. Kyhn et al. (2010) measured clicks from two sympatric NBHF species. The centroid frequency was higher and the SL lower for Commerson's than Peale's dolphins, while bandwidth and directivity index were the same. The significant albeit slight differences in centroid frequency and SL may be characteristic of species separation allowing sympatric species identification.

4.3.4.2 River Dolphin Clicks

There are currently three families of river dolphins, all of which are endangered: Iniidae, Platanistidae, and Pontoporiidae. The Lipotidae (baiji) are now extinct. These species occur in limited and isolated home ranges, and have evolved convergently in morphology, ecology, and biosonar. Botos, Ganges river dolphins, and the now extinct baijis were shown to produce short broadband clicks at high repetition rates, low ICI, and low SL (Jensen et al. 2013; Ladegaard et al. 2015), compared to bottlenose dolphins in the wild. Rivers have shallow water and thus a high level of reverberation and clutter. Echolocation ranges are thus short, and high repetition rates provide fast updates for successful navigation and prey tracking. The franciscana dolphin is the exception among river dolphins, because it produces NBHF clicks, which might be due to it inhabiting estuarine and coastal habitats as well, where predation by killer whales is a factor shaping franciscana echolocation clicks. In fact, the franciscana in Babitonga Bay (Brazil), which is free of killer whales, exhibit slightly wider echolocation frequency bands, than the open-sea population off Itapirubá Beach (Paitach et al. 2021) (Fig. 4.7).

4.3.4.3 Sperm Whale Clicks

Sperm whale clicks have been categorized into five sound types based on their emission pattern (Goold 1999; Madsen et al. 2002b; Mullins et al. 1988): (1) slow or single clicks with 3–8 s ICI; (2) usual clicks with 0.5–1 s ICI; (3) creaks of up to 220 clicks/s; (4) squeals of up to 1600 clicks/s; and (5) codas, which are stereotyped click sequences that can last for hours. Slow clicks have lower frequency content, lower directionality, and longer duration than usual or creak clicks and possibly play a communication role (Madsen et al. 2002b). Usual clicks were only recorded during deep foraging dives and are therefore likely used to echolocate prey. Sperm whales adjusted ICI during the descent such that they may have tracked the seafloor or deep prey field (Madsen et al. 2002a; Thode et al. 2002). Creaks occurred at the end of a foraging sequence of usual clicks and have been likened to feeding buzzes in smaller odontocetes. Click trains with high numbers of clicks/s (squeals) appear like typical odontocete burst-pulse sounds in spectrograms (Weir et al. 2007).

Due to internal reflections inside the sperm whale head, a recorded click may consist of a handful of separate pulses and the IPI is indicative of the size of the sperm whale (head) (Fig. 4.8; Møhl et al. 2003; Rhinelander and Dawson 2004). Pulses are numbered (p1, p2, ...); counting starts with the dominant on-axis pulse (p1). Precursors (p0) also exist. Off-axis, there can be p1/2 pulses between p1 and p2. Multi-pulse structure is only evident in slow clicks, usual clicks, and coda clicks, but not in creak clicks (Madsen et al. 2002a, b). The waveform of the outgoing click and hence the spectrum depend on the angle about the head at which the click is recorded. In front of the head, the main pulse p1 can be 40 dB stronger than the internal reflections (p2,...), making on-axis clicks appear mono-pulsed, yet off-axis clicks multi-pulsed (Møhl et al. 2003; Schulz et al. 2009; Zimmer et al. 2005a). A directivity index of 27 dB has been calculated (Møhl et al. 2000, 2003). As the orientation of the whale to the recorder is often unknown, differences in published source levels and spectral characteristics of clicks occur.

Sperm whales produce the most powerful biosonar signals, which has led to the hypothesis that they debilitate prey with sound (Norris and Møhl 1983). Tønnesen et al. (2020) deployed an acoustic and accelerometer tag on the tip of the nose of a sperm whale. The whale used strong clicks at a steady, slow click rate, indicative of long-range sensing. Echoes from the

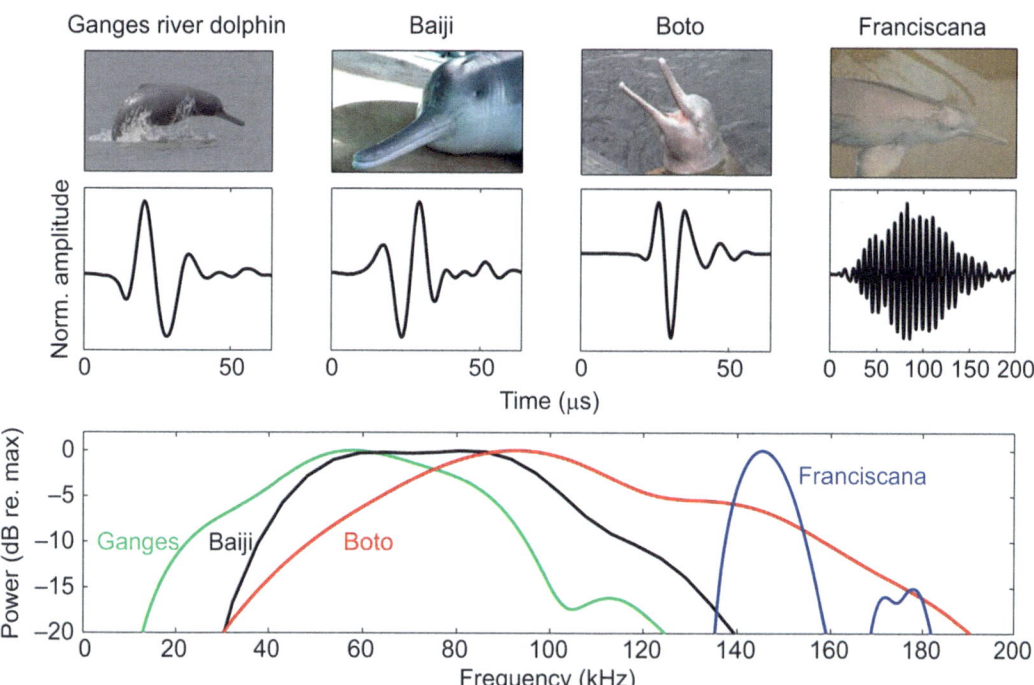

Fig. 4.7 Comparison of echolocation signals by river dolphins: waveforms and spectra $f_s = 500$ kHz, NFFT = 1024; except for the baji with $f_s = 5512.5$ kHz). Note that the baji click is from a dataset, where this click had one of the lowest peak frequencies, suggesting a potential under-representation of high-frequency energy in the power spectrum shown compared with the mean for this species. Ganges river dolphin data: Jensen et al. 2013; baji data: Akamatsu et al. 1998; boto data: Ladegaard et al. 2015; franciscana data: Melcón et al. 2012. Photo credits: Ganges river dolphin: E. and R. Mansur, WCS; baiji: Institute of Hydrobiology, Chinese Academy of Sciences; boto: Jorge Andrade; franciscana: Miguel Iniguez, WDC. Reprinted with permission from The Company of Biologists. Ladegaard M, Jensen FH, de Freitas M, da Silva VMF, Madsen PT, Amazon river dolphins (*Inia geoffrensis*) use a high-frequency short-range biosonar. J Exp Biol 218 (19): 3091–3101; https://doi.org/10.1242/jeb.120501. © The Company of Biologists, 2015. All rights reserved

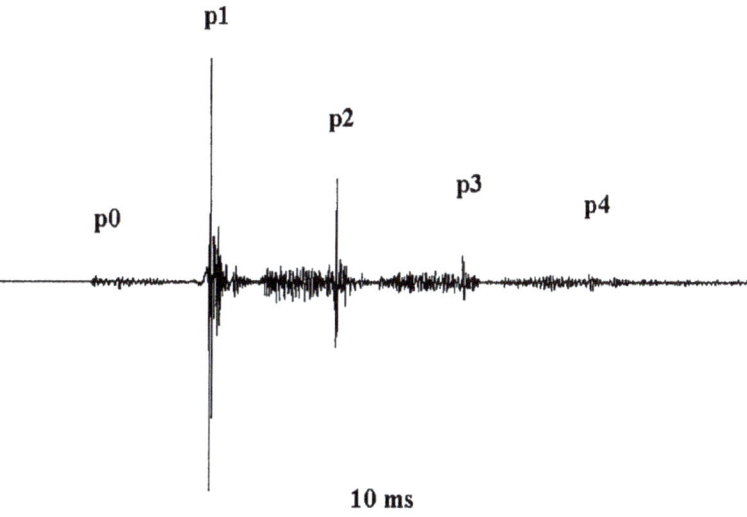

Fig. 4.8 Sperm whale click waveform showing five pulses at constant time interval, which relates to the size of the animal. Reprinted from Møhl B, Wahlberg M, Madsen PT, Heerfordt A, Lund A (2003) The monopulsed nature of sperm whale clicks. J Acoust Soc Am 114 (2):1143–1154. https://doi.org/10.1121/1.1586258. With permission from the ASA. © Acoustical Society of America, 2003. All rights reserved

seafloor were identified in all recordings, showing that this powerful biosonar provides acoustic beacons of the seafloor bathymetry and allows sperm whales to navigate deep-sea prey fields. The whale started buzzing much sooner than smaller odontocetes do (i.e., several body lengths away), possibly because of its massive size and hence reduced maneuverability compared to smaller predators. The buzz provides more rapid updates on prey movement at a longer range in sperm whales so that the whale can respond to changes in prey movement. Given the drop in SL from usual clicks to buzz and the propagation loss over the longer range to the prey, sperm whale biosonar hits the prey at levels well below potential debilitation (Benoit-Bird et al. 2006; Tønnesen et al. 2020; Wilson et al. 2007).

The dwarf and pygmy sperm whales also produce clicks at different ICI, comparable to usual clicks and creaks (Merkens et al. 2018). Dwarf and pygmy sperm whale clicks have NBHF character (Madsen et al. 2005a; Malinka et al. 2021b).

4.3.4.4 Beaked Whale Clicks

Beaked whales forage in deep offshore water (at depths of 500–1000 m) on pelagic and bentho-pelagic fish and invertebrates. Cuvier's beaked whales hold the marine mammal deep-dive record (down to 3 km over 140 minutes; Schorr et al. 2014). They only echolocate during foraging at depths in excess of 475 m (Johnson et al. 2004). Blainville's beaked whales rarely echolocate on shallow dives, nor during deep-dive descents and ascents; they mostly echolocate only at deep foraging depths. In near-surface water (i.e., within 170 m below the sea surface), they were silent 80% of the time (Aguilar de Soto et al. 2012). This behavior can be explained by the acoustic crypsis hypothesis, given beaked whales are hunted by killer whales (Wellard et al. 2016), which would be able to hear their clicks. Beaked whale echolocation clicks are upward frequency-modulated (Fig. 4.9). Peak frequencies range between 40 and 50 kHz; though some variability is obvious in the stacked spectrograms of Fig. 4.9 and was further quantified by Keating et al. (2016).

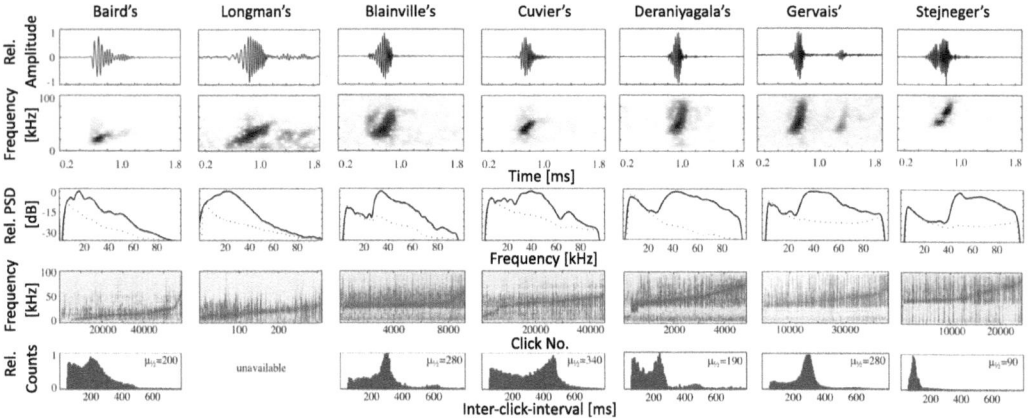

Fig. 4.9 Beaked whale click comparison by species. The top two rows show the waveforms and spectrograms ($f_s = 200$ kHz, NFFT $= 60$, Hann window, 98% overlap) of individual clicks. The third row shows mean spectra (solid line) and ambient noise at the time (dashed). The fourth row presents a stacked spectrogram (NFFT $= 512$, Hann window, no overlap) of all clicks recorded from each species, sorted by increasing peak frequency. The final row presents histograms of ICI, with medians noted in the top right corner. Collated from Baumann-Pickering S, McDonald MA, Simonis AE, Solsona Berga A, Merkens KPB, Oleson EM, Roch MA, Wiggins SM, Rankin S, Yack TM, Hildebrand JA (2013) Species-specific beaked whale echolocation signals. J Acoust Soc Am 134 (3): 2293–2301. https://doi.org/10.1121/1.4817832. With permission from the ASA. © Acoustical Society of America, 2013. All rights reserved

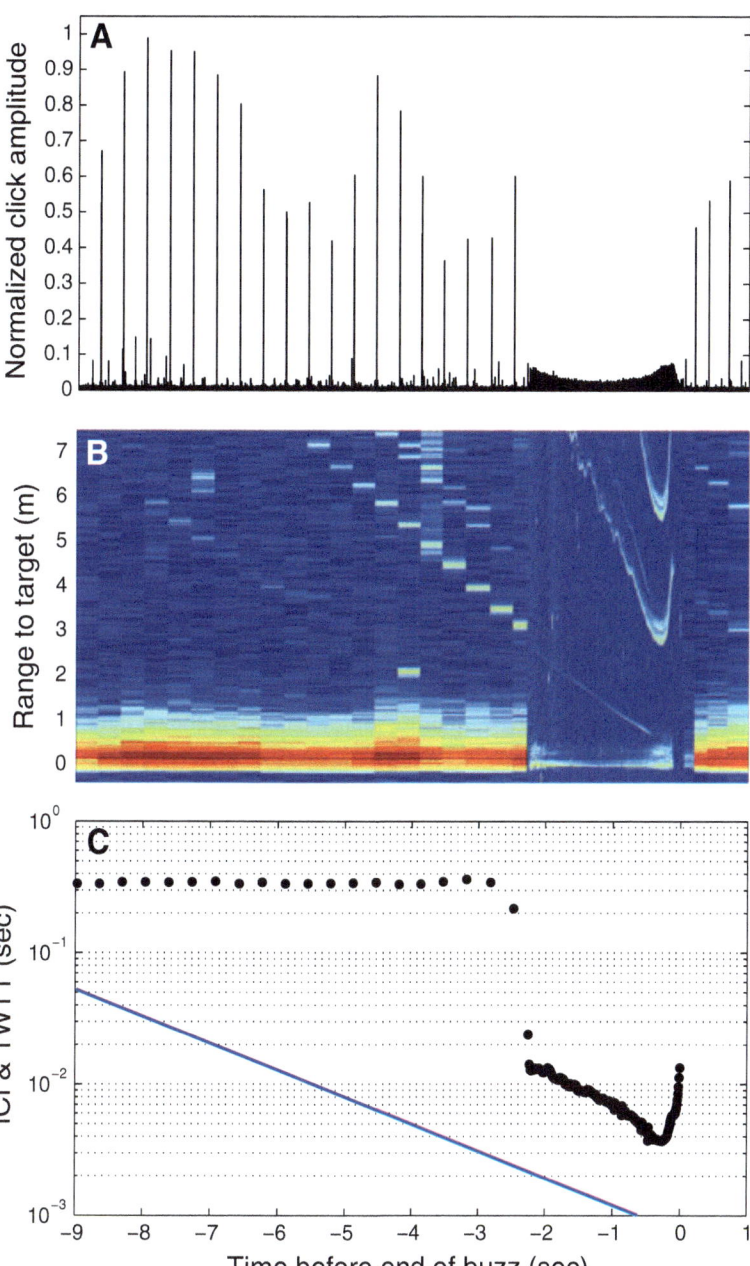

Fig. 4.10 (**a**) Amplitude of search and buzz clicks. Time relative to the end of the buzz. The amplitude drops by two orders of magnitude from search clicks to buzz. (**b**) Echogram of the clicks from A showing the range to all reflectors over time. Multiple reflectors (at different ranges) exist, but the beaked whale buzz only focusses on one (visible as the first stepped and then near continuous downwards sloping line). (**c**) Inter-click interval and two-way travel time showing consistently slow clicking during the search phase and a rapid drop (by two orders of magnitude) during the buzz phase. Reprinted by permission from Springer Nature. Madsen PT, Aguilar de Soto N, Arranz P, Johnson M. Echolocation in Blainville's beaked whales (*Mesoplodon densirostris*). J Comp Physiol 199(6), 451–469; https://doi.org/10.1007/s00359-013-0824-8. © Springer Nature, 2013. All rights reserved

Beaked whales use low SL in general, hence do not attempt to echolocate over long ranges. They do not adjust SL and ICI as a function of range to target as do other odontocetes. Instead, tagged Blainville's beaked whales abruptly switched from search clicks (upsweep clicks) to buzz (consisting of constant-wave clicks) when prey was within ~1 body length of the whale (Madsen et al. 2005b). Both SL and ICI dropped by two orders of magnitude from search clicks to buzz (Fig. 4.10; Madsen et al. 2013a). This implies that beaked whales keep a wide acoustic gaze until just before prey capture, different from other odontocetes. Rather than targeting one prey animal at a time, beaked whales might use a wide acoustic gaze to image several targets for sequential capture or to select the preferred animal out of a patch of prey.

Hunting at great depth limits the time available for prey capture. Beaked whales evolved unique foraging tactics under tight physiological constraints. They target patches of prey. A high ICI (~0.2–0.4 s; i.e., low click rate) is useful when searching for a patch of prey. Once within a patch, tags with motion sensors have shown that beaked whales turn rapidly, likely exploring the patch and catching multiple prey animals sequentially (Johnson et al. 2008). A low ICI (0.003–0.05 s; i.e., high click rate; buzz) allows rapid updates on prey distribution within the patch, and together with frequent head movement, widens the acoustic gaze (Madsen et al. 2013a). In beaked whales, it seems ICI is not a function of the range to target but rather related to their behavioral foraging tactics. The upswept nature of beaked whale search clicks adds to the wide acoustic gaze and is advantageous when foraging in prey patches. Exploring a prey patch with a narrowband pulse (in the extreme case, a single-frequency signal) will lead to multiple overlapping echoes, all at the same frequency but different phase. It is extremely difficult to resolve an image from such a returned signal. However, if the outgoing signal consists of a frequency sweep (in the beaked whale case, an upsweep), then the returned echoes all begin at the lower frequency, but overlap with the higher-frequency parts of echoes from closer-range reflectors. By Fourier transform, it is then possible to simultaneously solve for different reflectors (targets, prey) at different ranges (and locations). The frequency-modulated nature of beaked whale clicks thus supports the hypothesis that they do not pursue individual prey animals, but instead take multiple animals sequentially from the same prey patch.

4.3.4.5 Inter-Species Comparison of Click Characteristics

Comparisons of the waveforms of odontocete clicks across different species have been previously published (e.g., Akamatsu et al. 1998; Baumann-Pickering et al. 2013a; Ladegaard et al. 2015) and examples were shown in the previous sections. There have been several attempts to group and classify click features. For example, based on click bandwidth alone, Fang et al. (2015a) grouped echolocation clicks into four broad categories: broadband low-frequency clicks (sperm whales), broadband high-frequency clicks (most delphinids and river dolphins), NBHF clicks (porpoises and small, coastal dolphins), and frequency-modulated clicks (beaked whales).

When plotting different click features against each other, some species (or groups of species) separate out (Fig. 4.11; Suppl. Table 1; note that measurements were not filtered for on- versus off-axis clicks). The *Cephalorhynchus* and *Lagenorhynchus* species cluster with the porpoises and *Kogia* to form the NBHF group. These species have higher peak frequency, lower bandwidth, shorter duration, and lower SL than most of the other species. There is one exception (orange triangle at 44 kHz f_p in Fig. 4.11a) amongst the *Cephalorhynchus* studies. However, within the NBHF cluster, the river dolphin is not an outlier but represents the franciscana. Coastal dolphins emit very brief clicks with relatively high peak frequencies, but much greater bandwidth than NBHF species do, albeit at similarly low SL. At the opposite end of the spectrum, sperm whales emit the strongest clicks (highest SL) at the lowest peak frequencies in narrow bands. Beaked whales follow with higher peak frequencies than sperm whales. Beaked whale clicks also have the longest durations. River dolphins and Monodontidae fall in the middle on the peak frequency and bandwidth scales. Offshore dolphins seem to cover the greatest range in peak frequency, bandwidth, duration, and SL measurements. Across the species, there appears an inverse relationship between click bandwidth and duration, which has been identified by others (Beedholm 2008; Wiersma 1988).

Certain features of echolocation clicks (such as frequency, bandwidth, source level, and beamwidth) have been shown to correlate with body size (e.g., Au 1993; Jensen et al. 2018; Madsen and Surlykke 2014; Fig. 4.12). Smaller odontocetes tend to emit clicks at higher frequency, greater bandwidth, lower SL, and lower

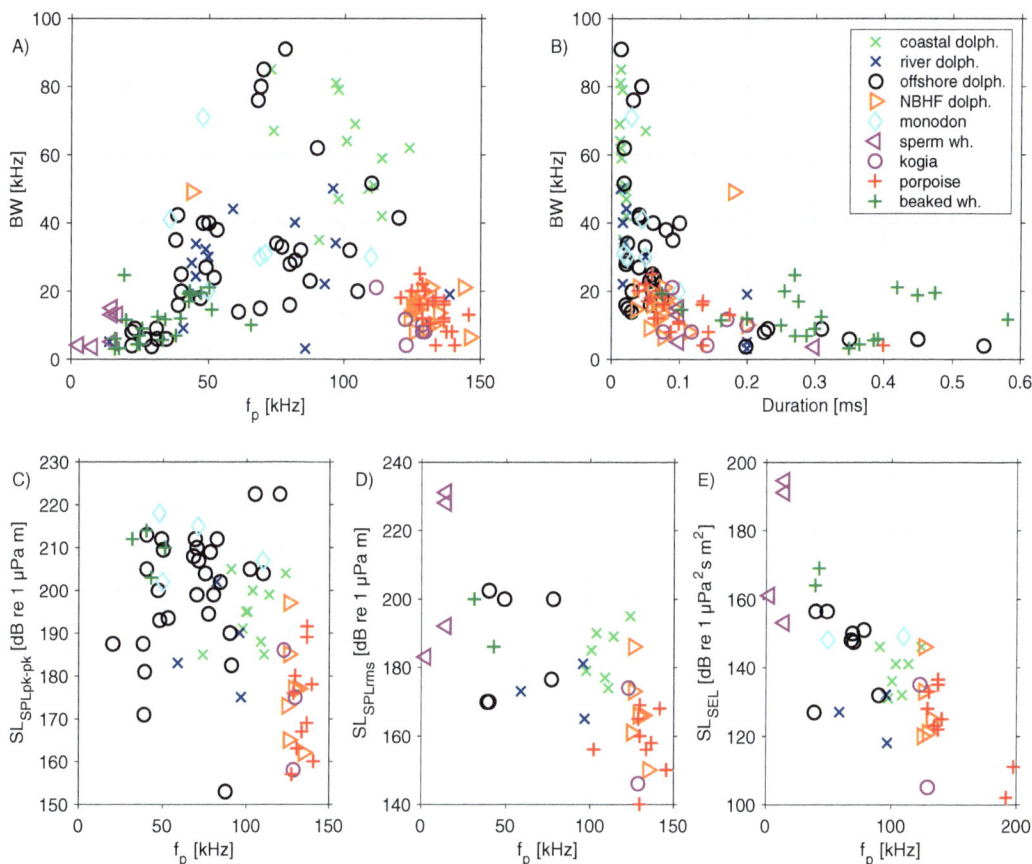

Fig. 4.11 Scatter plots of peak frequency (f_p), bandwidth (BW), duration, and source level (SL in terms of SPL_{pk-pk}, SPL_{rms}, and SEL) for coastal dolphins, river dolphins, offshore dolphins, *Cephalorhynchus* and *Lagenorhynchus* dolphins (NBHF dolphins), beluga and narwhal (monodon), sperm whales, pygmy and dwarf sperm whales (kogia), porpoises, and beaked whales, based on Suppl. Table 1

ICI, with some exceptions such as NBHF species' narrow bandwidth. Beamwidth and directivity are related to the ratio of head circumference and wavelength of the peak frequency (Fig. 4.13). The Risso's dolphin has a greater beam directivity and narrower beamwidth, which may be related to its unique forehead cleft. However, our recent study showed that the cleft had minimal effect on the outgoing beam features (Wei et al. 2022, 2024). Moreover, there is a great spread in the measurements across studies, which might have a number of reasons, such as not all studies using on-axis clicks exclusively, studies reporting click features for different behavioral states or tasks, studies not accounting for environmental effects on click features and not deriving SL in consistent ways, etc. Therefore, Fig. 4.12 does not fully support any linear relationship (e.g., an inverse relationship between body size and peak frequency or bandwidth, or a linear relationship between body size and SL), but rather indicates that there is an upper limit to these click features as a function of body size. Jensen et al. (2018) more carefully selected the studies to compare and found that SL increases steeply with body mass, "indicating an evolutionary hyperallometric investment into sound production structures that may be driven by a strong selective pressure for long-range biosonar", and that frequency is inversely proportional to body

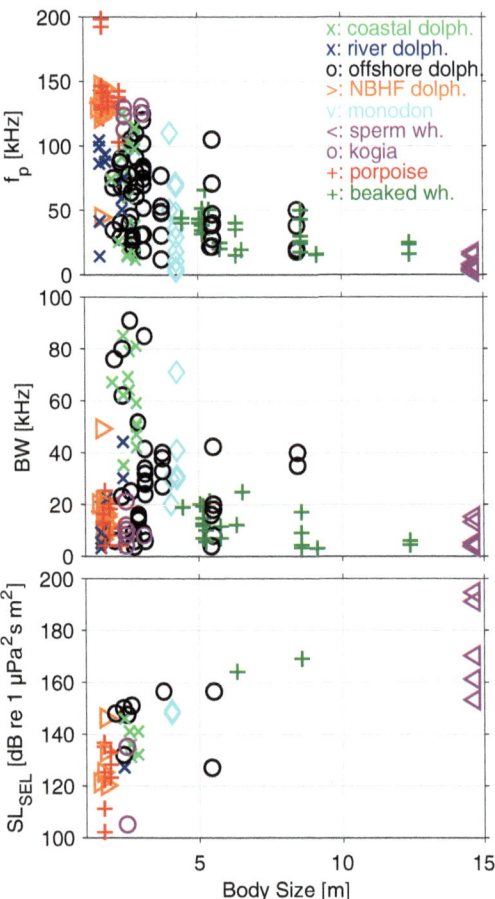

Fig. 4.12 Scatter plots of click peak frequency (f_p), bandwidth (BW), and SL versus odontocete body size, ranging from the smallest dolphins and porpoises to large sperm whales, based on Suppl. Table 1

mass, "resulting in remarkably stable biosonar beamwidth that is independent of body size". The authors proposed "that a narrow acoustic field of view, analogous to the fovea of many visual predators, is the primary evolutionary driver of biosonar frequency in toothed whales, serving as a spatial filter to reduce clutter levels and facilitate long-range prey detection" (Jensen et al. 2018).

4.3.4.6 Intra-Species Variability of Click Characteristics

Geographic variability in biosonar signals of the same species has been reported for a few species, such as Irrawaddy dolphins (Niu et al. 2019), Pacific white-sided dolphins (Soldevilla et al. 2010b), pantropical spotted dolphins (Gong et al. 2019), Risso's dolphins (Soldevilla et al. 2010a, 2017), Amazon river dolphins (Melo et al. 2021), and dwarf sperm whales (Malinka et al. 2021b). For example, dwarf sperm whales use clicks with greater ICI, greater SL, and longer duration in deep water than in shallow, coastal water, albeit with the same peak frequency and bandwidth. Such differences likely relate to different foraging ecologies in the different environments. Careful experimental design is needed to tease out such differences, and factors of the sound propagation environment may need to be accounted for as they may affect the biosonar measurements. Intra-species differences in biosonar features have been used to argue for splitting populations into separate species (e.g., in the case of Amazon river dolphins; Melo et al. 2021), as differences in acoustic features could be indicative of differences in underlying morphology.

4.3.5 Wild Versus Captive Studies

Historically, echolocation has been studied with captive dolphins performing carefully designed echolocation tasks. Some of this research was driven by naval interests and for the development of biomimetic sonar (e.g., Houser et al. 2003). Echolocation tasks therefore generally involved artificial targets rather than prey, and focused on the great capabilities as well as limits of the biosonar system.

Echolocation has been measured from wild animals perhaps offering a more realistic picture of the natural use of echolocation, but such measurements often lack the control available in laboratory settings. For example, in order to characterize and quantify echolocation signals, the location and orientation of the animal need to be known as the signal characteristics change with angle of measurement about the animal. Studies with simultaneous visual and passive acoustic observation and studies with acoustic and movement tags have also provided valuable information on biosonar during foraging in the wild. For example, prey-capture events can be identified from movement tags as sudden jerk

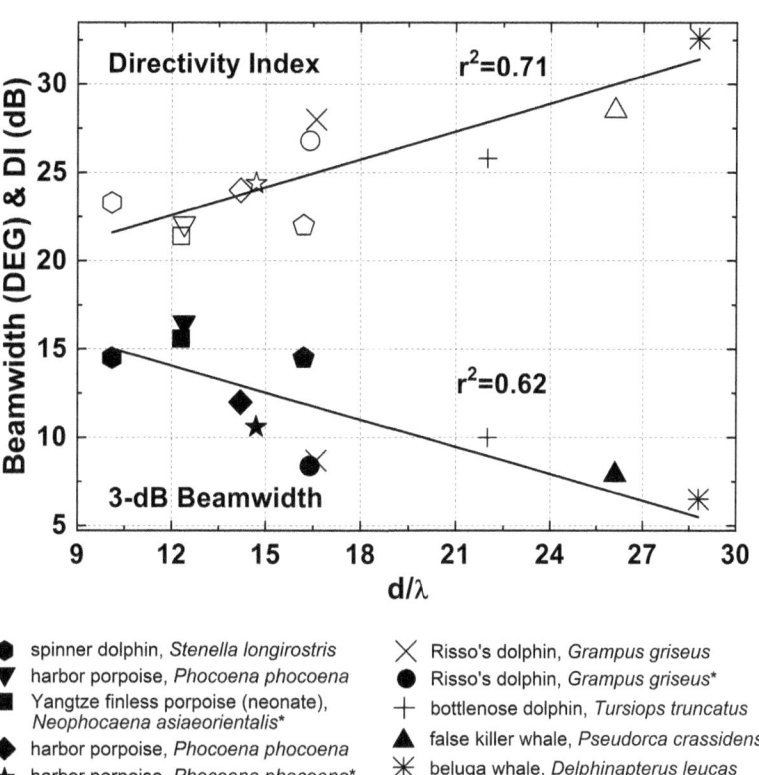

Fig. 4.13 Comparison of directivity index (DI) and 3 dB beamwidth as a function of head diameter (d) to wavelength λ of the peak frequency. Data from Au 1993; Au et al. 1986, 1987, 1995, 1999; Koblitz et al. 2012; Smith et al. 2016; Smith et al. 2019; Wei et al. 2014, 2016, 2017, 2018a, b, 2022. Reprinted from Wei C, Gill LG, Erbe C, Smith AB, Yang W-C (2022) The distinctive forehead cleft of the Risso's dolphin (*Grampus griseus*) hardly affects biosonar beam formation. Animals 12 (24): 3472. https://doi.org/10.3390/ani12243472. © Wei et al. 2022. Licensed CC BY 4.0; https://creativecommons.org/licenses/by/4.0/

movements (e.g., Johnson et al. 2004) and these can be correlated with changes in echolocation behavior. These studies have verified the association of buzzes with prey-capture in the wild.

However, there may be differences in echolocation behavior, performance, and features between captive and wild studies. Bottlenose dolphins housed in a tank used lower peak frequency and lower SL than bottlenose dolphins kept in an open pen (Kaneohe Bay, HI, USA) with strong snapping shrimp noise (Au 2000; Fig. 4.14). Niu and team (2021) reported 16 dB lower source levels from captive (in a 30 m diameter pool) versus wild Indo-Pacific humpback dolphins.

4.3.5.1 Recording Clicks

In captivity, if animals are trained to station, it is straightforward to position hydrophones and record on- and off-axis clicks as desired. In the wild, animal position relative to hydrophone position cannot be controlled. Researchers have used hydrophone arrays (with or without concurrent visual observations) and developed a set of criteria for identifying on-axis clicks in recordings (Madsen and Wahlberg 2007; Rasmussen et al. 2004; Villadsgaard et al. 2007):

- The animal was approaching the array (based on its acoustic track or from simultaneous video from a camera on the array).
- The animal was more than 1 m from the array to avoid near-field effects.
- The vertical and horizontal angles between the animal and the center hydrophone were less than a specified angle (e.g., 35°).
- The potential on-axis click was recorded on all hydrophones in the array.

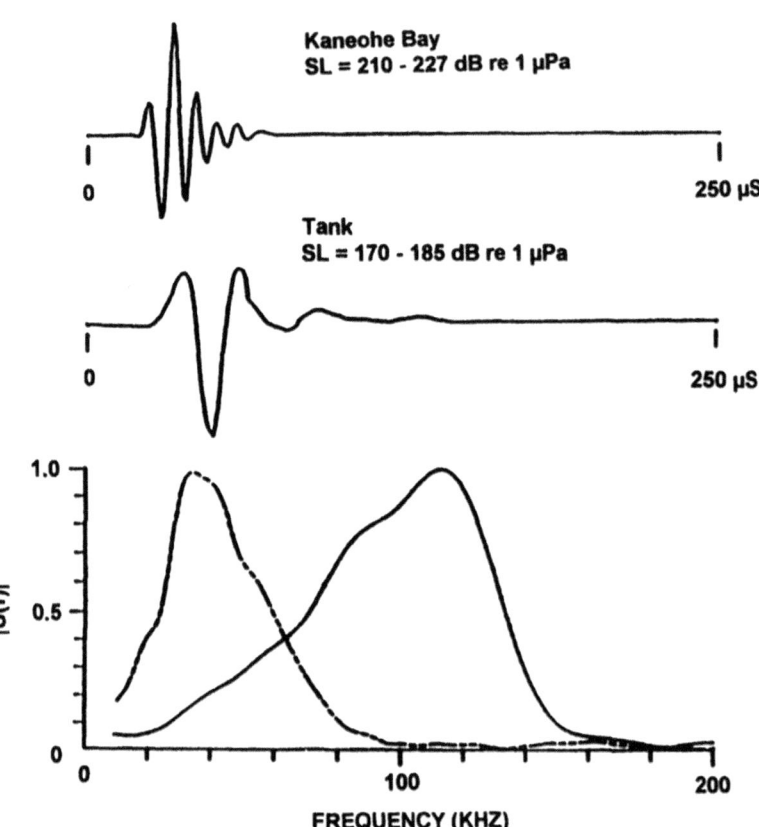

Fig. 4.14 Waveforms and spectra of echolocation clicks of bottlenose dolphins in an open ocean pen versus a tank. The spectrum of the click from the tank had a lower frequency peak at 40 kHz and a lower source level of 170–185 dB re 1 µPa m. Note that relative, not absolute, amplitude is plotted. Reprinted by permission from Springer Nature. Au WWL, Echolocation in dolphins. In Au WWL, Popper AN, Fay RR (eds) Hearing by Whales and Dolphins, pp 364–408; https://doi.org/10.1007/978-1-4612-1150-1_9. © Springer, 2000. All rights reserved

- The click was part of a scan indicated by rising then falling amplitude of clicks in the train; the click was selected from the middle part of the scan where amplitude was highest. Or, the click was part of a train aimed at the hydrophone; simultaneous video from a camera on the array confirmed the animal was facing the array.
- The click had maximum amplitude on the center hydrophone.
- The direct arrival of the click was stronger than the surface- or bottom-reflected arrivals.

The on- and off-axis clicks recorded on a symmetric cross-shaped 9-element array are beautifully illustrated by Fang et al. (2017; Fig. 4.15).

The SL can be estimated from the RL by localizing the click using time-of-arrival differences among the elements of the array, and then correcting for PL. The aspect of the animal, however, remains unknown, which is why this SL estimate is sometimes called the Apparent SL (ASL) of candidate on-axis clicks (Madsen and Wahlberg 2007; Møhl et al. 2000): $ASL = RL + PL$ with $PL = 20\log_{10} r + \alpha r$, where r is the range from the animal to the hydrophone and α is the absorption coefficient.

4.3.6 Dolphin Biosonar Performance

The capabilities of the odontocete biosonar system are usually divided into two categories: target detection and target discrimination. Various types of acoustic experiments have been performed with trained individuals to study these capabilities. Bottlenose dolphins have been the most studied species in almost every area of biosonar performance research. Dolphin biosonar systems have been quoted as outperforming man-made sonar systems, in particular in

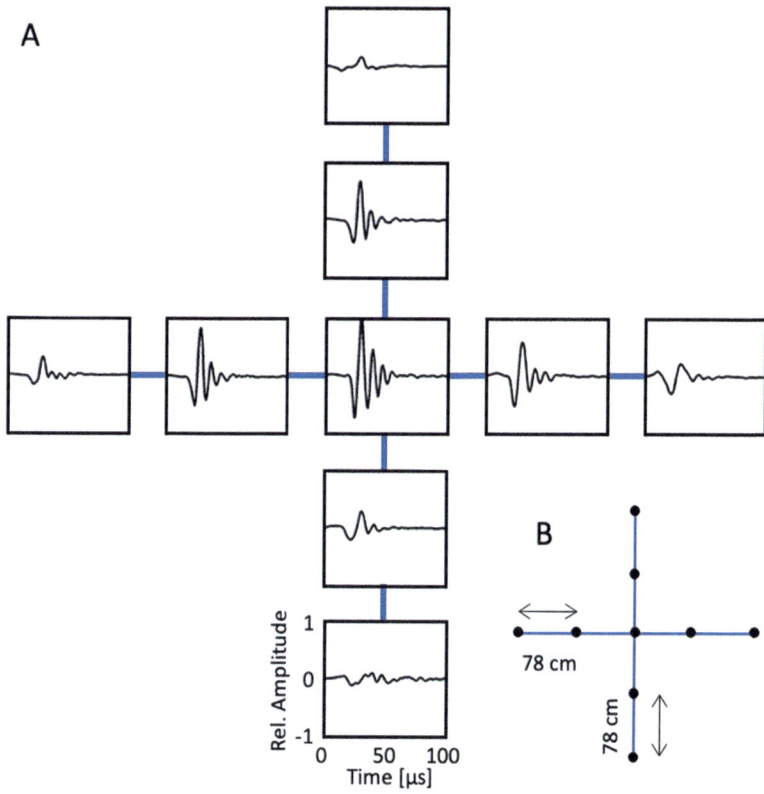

Fig. 4.15 Waveforms (**a**) of on- and off-axis clicks recorded from free-ranging Indo-Pacific humpback dolphins on a 9-element symmetric cross array (**b**). Recordings courtesy of Fang et al. 2017

shallow-water target detection and discrimination tasks (Martin et al. 2005; Sigurdson 1997).

4.3.6.1 Detection Range

Au and colleagues used a 7.62 cm diameter, water-filled, stainless-steel sphere as the target to determine the maximum target detection range of an Atlantic bottlenose dolphin (Au 1993; Au and Snyder 1980). In the experiment, the target was moved progressively away from the animal until it could no longer be detected. Using the same target, similar experiments were conducted with a false killer whale and a harbor porpoise (Kastelein et al. 1999; Thomas and Turl 1990). Figure 4.16 shows the experiment setup in the study of Kastelein et al. (1999). The 50% correct-detection thresholds for the bottlenose dolphin, false killer whale, and harbor porpoise were at a target range of 113 m, 119 m, and 26 m, respectively. The detection ranges of the bottlenose dolphin and false killer whale were considerably longer than that of the harbor porpoise, even though the background noise level was higher in the bottlenose dolphin and false killer whale experiments. The main reason was difference in SL, with the bottlenose dolphin clicks 50–60 dB stronger than those of the harbor porpoise.

4.3.6.2 Target Detection in Noise

Au and Penner (1981) studied the capability of bottlenose dolphins to detect targets in masking noise. The target detection threshold was determined as a function of the ratio of the echo energy flux density and the estimated received noise power spectral density. The percentage of the animal's correct responses correlated with the echo-to-noise ratio. The target detection thresholds of two bottlenose dolphins occurred at 7.4 and 12.8 dB echo-to-noise ratio. Using an integration time of 264 μs and an auditory bandwidth of 16.7 kHz, Au and colleagues recalculated the ratio of echo energy to noise

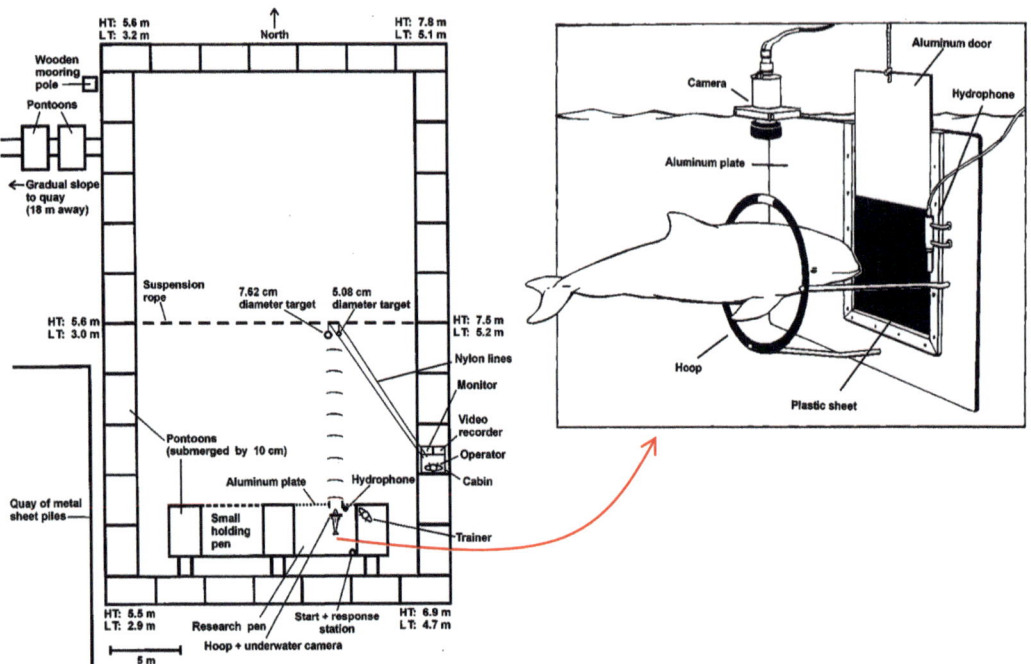

Fig. 4.16 Experiment setup of target-detection range measurement for a harbor porpoise in the study of Kastelein RA, Au WWL, Rippe HT, Schooneman NM (1999) Target detection by an echolocating harbor porpoise (*Phocoena phocoena*). J Acoust Soc Am 105 (4): 2493–2498. https://doi.org/10.1121/1.426951. Reprinted with permission. Right-panel drawing by Rijkent Vleeshouwer. © Acoustical Society of America, 1999. All rights reserved

energy at the target detection threshold, which was ~1 dB. The results suggested that the dolphin biosonar system might act as a typical energy detector by integrating sound energy over a short time window and characteristic frequency band (Au 2014; Au et al. 1988).

4.3.6.3 Buried Target Discrimination

The seafloor is highly reverberant due to its rough surface and typically mixed material composition. The amount of energy in returning biosonar echoes is affected by the size of the scatterers, and the angle and frequency of incident signals, making echolocation in these environments an extremely complex acoustic process. Therefore, it is very challenging for any technological sonar system to detect the presence of targets on or just under the seafloor and effectively distinguish the targets from the clutter. Dolphins, however, seem to have evolved mechanisms to deal with these problems.

Herzing and colleagues (Herzing 1996; Rossbach and Herzing 1997) observed wild Atlantic bottlenose and spotted dolphins foraging for prey buried under the sand by performing echolocation scans (clicks were heard while the dolphins foraged off the bottom). To emulate a dolphin detecting and recognizing targets buried under sediment, Roitblat et al. (1995) designed a biomimetic system, which transmitted dolphin-like clicks, recorded the echoes, and processed digitized echoes with a neural network. Four cylinders of the same size but different material composition were buried in mud at a depth of <10 cm. The system was able to successfully discriminate a small stainless-steel cylinder from the others. Zaitseva et al. (2022) trained two dolphins to find targets (brass cylinder) in the water, on the seafloor, and buried 10 cm into the sediment. The dolphins used brief and broadband clicks (10–12 μs, peak ~120 kHz) while searching in the water, longer clicks at lower frequencies (14–15 μs, peak ~50 kHz) while searching near the seafloor, and the strongest, longest, and lowest-frequency clicks (15–20 μs, peak ~20 kHz) while searching in the sediment—maximizing penetration depth and echo-to-noise ratio.

4.3.6.4 Cylinder Wall Thickness Discrimination

Au and Pawloski (1992) studied a dolphin's ability to tell cylinders of different wall thickness apart. The cylinders were made of aluminum; the wall thickness of the standard cylinder was 6.35 mm, and comparison cylinders had thinner or thicker walls, which incrementally differed from the standard one by 0.2–0.8 mm. The dolphin's 75% correct response threshold appeared at a wall thickness 0.23 mm thinner and 0.27 mm thicker than that of the standard target, implying asymmetric performance in cylinder wall thickness discrimination. Feng et al. (2018) applied finite-element analysis to create a bottlenose-dolphin-like echolocation signal and simulated the process of discriminating cylinders of different thicknesses. They compared the waveforms and frequency spectra of the returning echoes from a standard cylinder and two comparison cylinders with the wall thickness differing by ±0.3 mm. According to wave acoustics, the circumferential surface waves in the cylinder wall are frequency-dispersive and travel more slowly in a thinner cylinder wall, causing the asymmetric discrimination performance. The cues by which the dolphin discriminates wall thickness could include arrival time differences between echoes reflected from the front and back walls of the cylinder (Au 1993). If the dolphin were to rely on this cue, it would need to be able to resolve time intervals of 0.5–0.6 μs. Another cue could be the spectral shift in echoes. Au and Pawloski (1992) showed the average frequency differences were 3.2 kHz and 2.2 kHz for a wall thickness difference of −0.3 mm and − 0.2 mm, respectively. If this was the cue, then the bottlenose dolphin can perceive a frequency shift of 2–3 kHz in the broadband echoes.

4.3.6.5 Target Feature Discrimination

Dolphins are able to discriminate features of the target, such as shape, size, and material composition. Nachtigall et al. (1980) reported on a blindfolded bottlenose dolphin discriminating between a cylinder and a cube located 2 m from the animal. Au et al. (1980) reported that a dolphin could easily (with 94% accuracy) discriminate between foam spheres and cylinders, which were presented 6 m away from the hoop station. In order to study the possible acoustic cues, the spheres and cylinders had different sizes but overlapping target strength, so that target strength was eliminated as a possible cue. There was a secondary reflection in the echoes only from the sphere, suggesting that the presence of circumferential waves following the specular reflection off the target front could be the distinguishing cue for the dolphin. Roitblat et al. (1990) undertook a matching-to-sample experiment, in which a dolphin was required to match a sample target from three alternative objects with simple geometry (sphere, tube, and cone). The dolphin emitted on average 37 clicks to identify the sample and the average matching accuracy exceeded 94%. Delong et al. (2006) performed a matching-to-sample experiment to study which echo features a dolphin used to discriminate objects differing in shape, size, material, and texture. After analyzing the measured echoes, they demonstrated that dolphins used multiple features and integrated information across the reflected echoes from a range of object orientations to identify the targets. Finless porpoises have similar discrimination capabilities, distinguishing time differences between echo features of ~1 μs, frequency shifts of ~7 kHz in a broadband echo (with a peak frequency near 140 kHz), and target strength differences of ~1 dB (Nakahara et al. 1997).

4.3.6.6 Cross-Modal Matching-to-Sample

Some researchers hypothesized that dolphins have the ability to integrate information across the senses of vision and echolocation. To investigate this ability, Pack and Herman's research group trained a couple of Atlantic bottlenose dolphins for cross-modal matching-to-sample experiments, specifically echoic-to-visual (E-V) and visual-to-echoic (V-E) cross-modal matching tasks (Herman and Pack 1992; Herman et al. 1998; Pack and Herman 1995; Pack et al. 2002). In the experiments, polyvinyl chloride (PVC) pipes were used to build objects of various shapes. These PVC pipes were all filled with dry sand to achieve negative buoyancy for immersion in water and to reduce internal reflections. In the E-V trials, the dolphins were able to use their echolocation sense alone to inspect a complex-shaped sample object and subsequently find the match between several alternative objects by using their visual sense alone without any prior exposure to these objects (vice versa for V-E trials). For example, in one experiment, the dolphin was required to match the sample with one of three alternatives. The matching accuracies for V-E and E-V trials were both above 90%. Then they added a no-match option as the fourth alternative, yet matching accuracies remained at 74% and 76% for V-E and E-V trials, respectively (Pack et al. 2002). These studies suggested that dolphins could directly perceive shape through echolocation, and that the representation of the object echolocated on was then also accessible to their visual sense. The reverse was also true. The hypothesis that dolphins can "see through sound" was formulated. Harley et al. (1996) performed three-alternative matching-to-sample experiments in different modality conditions showing that the dolphin's performance accuracy rose to 95% when it was allowed to use both echolocation and visual senses. Unfortunately, Pack and Herman's group did not publish any recorded acoustic data from the experiment (i.e., dolphin echolocation signals, returning echoes from the sample object and alternatives), based on which one could have investigated what information was carried by the echoic cues.

Hoffmann-Kuhnt and colleagues trained Indo-Pacific bottlenose dolphins for cross-modal matching-to-sample experiments focusing on problem-solving skills and cognitive processes. Figure 4.17 shows their E-V experiment setup. A sample object of complex shape was concealed in a box under water and available for investigation by echolocation. A visually opaque black plexiglass sheet 3 mm thick, which was acoustically transparent in water, was placed in front of the sample box to block the dolphin's vision of the sample objects. The dolphin was then presented with 2–4 alternative objects simultaneously in display boxes in air. Only one object matched the sample. The dolphin had to make a decision using its visual sense only (Dolphin echolocation does not work in air due to the impedance mismatch between water and air.) and respond by touching one of the response paddles underneath the display boxes. An array with 16 hydrophones directly behind the plexiglass recorded the acoustic signals in the underwater inspection process showing how the dolphin scanned the sample objects in each trial. With the recorded acoustic data, Hoffmann-Kuhnt et al. (2011) documented both beam steering and beam shaping while the dolphin was acoustically interrogating the sample objects and found that the dolphin scanned more widely in the horizontal than vertical plane. This was consistent with the findings of a previous target detection study (Moore et al. 2008).

Wei et al. (2021) examined whether the dolphin's E-V cross-modal matching ability was affected by the material composition of the objects while the shape, size, and position of the objects remained the same. Four compositions were used: (1) air-filled PVC pipes (AF), (2) water-filled PVC pipes (WF), (3) foam ball arrangements (FB) yielding the same overall shape with foam ball diameter equal to that of the PVC pipes, and (4) soft, closed-cell foam-wrapped PVC objects (SF), which were air-filled PVC pipes of slightly thinner diameter than AF, wrapped by a thin layer of soft, closed-cell foam. The same shapes as in Fig. 4.17c were used, and the same compositions were used in the sample and display boxes (i.e., if a FB object was the sample, then the display choices were also all made of FB). The dolphin's matching accuracy was highest in AF and SF trials (92.7% and 88.5% across four objects, respectively) and poorer in WF and FB trials (62.5% and 58.3

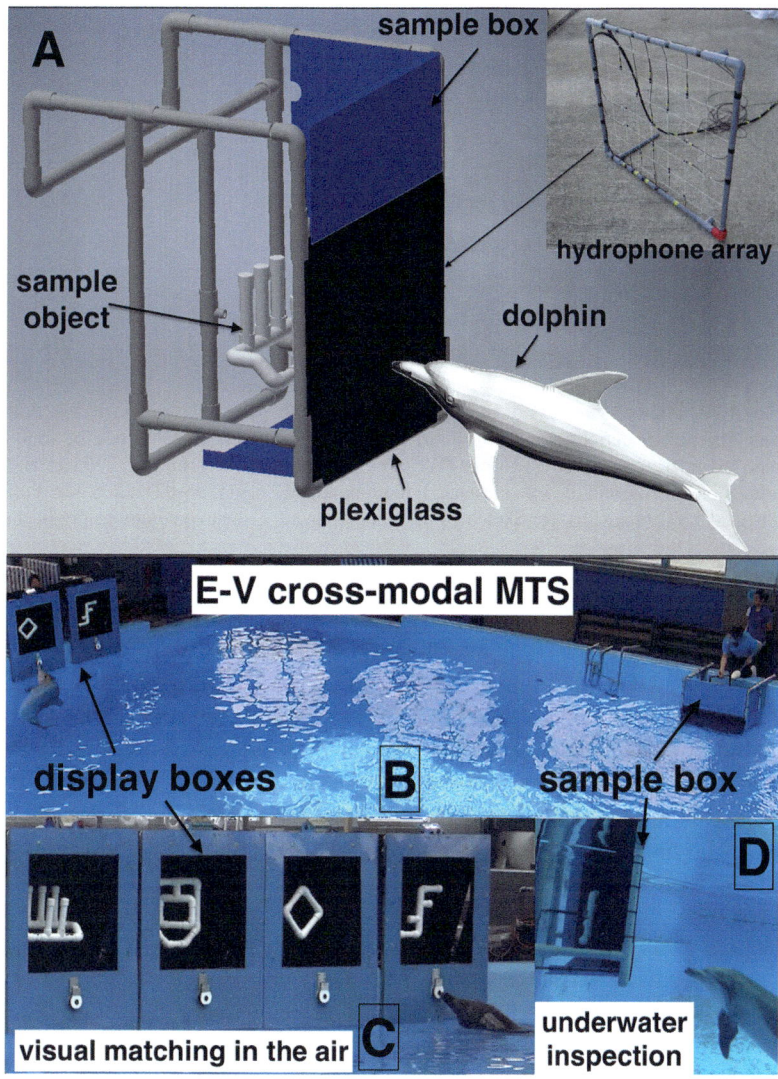

Fig. 4.17 Echoic-to-visual cross-modal matching-to-sample experiment setup. (**a**) 3D illustration of the experiment setup. The hydrophone array (top right insert) was mounted on the back of the plexiglass (black). (**b**) Photo of the experiment tank with sample box on the right and multiple-choice display boxes on the left side. (**c**) Photo of the dolphin choosing one of the display boxes. (**d**) Photo of the dolphin echolocating on the sample. Reprinted from Wei C, Hoffmann-Kuhnt M, Au WWL, Ho AZH, Matrai E, Feng W, Ketten DR, Zhang Y (2021) Possible limitations of dolphin echolocation: a simulation study based on a cross-modal matching experiment. Sci Rep 11 (1): 6689. https://doi.org/10.1038/s41598-021-85063-2. © Wei et al. 2021. Published CC BY 4.0; https://creativecommons.org/licenses/by/4.0/

across two objects, respectively). The authors constructed a finite-element model to explain the observations. The AF and SF objects created strong and highly directional echoes, while the WF object produced a very weak echo, and the FB object scattered echo energy over a broad range of angles (Figs. 4.18 and 4.19).

4.3.6.7 Notes on Finite-Element Modeling

Finite-element models have been constructed for the entire echolocation process, that is, not just the target reflection process but also the click generation, emission, and reception processes at the dolphin (e.g., Aroyan 2001; Aroyan et al. 1992; Cranford et al. 2014; Wei et al. 2018b). Models of the acoustic processes at the dolphin have been based on computed tomography (CT) images of dolphin heads, accompanied by measurements of the acoustic properties of the different tissues inside a dolphin's head. The CT scans and tissue measurements are combined into digital, 3D anatomical models. The models are dynamic and so can be excited at specific locations (e.g., at the phonic lips) to simulate the creation and then propagation of an echolocation click through the head, and similarly, its reception. Such models have been validated with

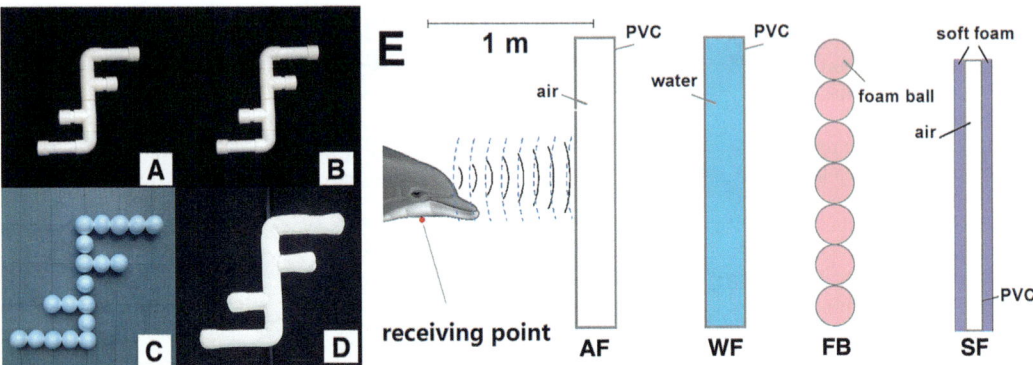

Fig. 4.18 Photos of objects of the same shape but with four different material compositions: (**a**) air-filled PVC pipe (AF); (**b**) water-filled PVC pipe (WF); (**c**) foam ball arrangement (FB); and (**d**) soft, closed-cell foam-wrapped, air-filled PVC pipe (SF). (**e**) Sketch of the finite-element model, in which the dolphin emits a click, which is reflected off one of four materials and received at the lower jaw. © Wei et al. 2021; https://doi.org/10.1038/s41598-021-85063-2. Published CC BY 4.0; https://creativecommons.org/licenses/by/4.0/

recordings from echolocating dolphins and porpoises (Wei et al. 2017, 2018a, 2020, 2022). They allow us to investigate biosonar processes and capabilities.

For example, dolphins stationing at a bite plate inside a hoop during live biosonar experiments, while allowing control over the measurement setup, limits physical biosonar behaviors seen in free-swimming (wild or captive) dolphins: Dolphins commonly rotate along their longitudinal axis while echolocating. The effects of rotational behavior on target ensonification and returned echo properties have been investigated with finite-element models (Wei et al. 2023), demonstrating a wider ensonification area and a wider receiving area. Thus, simultaneous rotation may compensate for the dolphin's narrow biosonar beam, asymmetries in sound reception, and limitations on the pointing direction imposed by little head movement ability.

While finite-element models are typically developed based on CT scans of deceased animals, some scans have been collected from live animals. Comparative finite-element analysis demonstrated the relevance and applicability of models from deceased specimens to live animals (Cranford et al. 2014; Wei et al. 2023).

4.3.7 Biosonar Ontogeny

The ontogeny of sound production in newborn-to-juvenile dolphins has been relatively well studied with regards to social sounds: whistles and burst-pulse sounds (e.g., Eskelinen and Jones 2021; Killebrew et al. 2001; McCowan and Reiss 1995b; Morisaka et al. 2005b). Fewer studies have been able to determine biosonar ontogeny—perhaps because of difficulties isolating the calf from its mother when both might be echolocating, and controlling the position of the calf that has not been trained to station at a bite plate for on-axis measurements.

Nonetheless, several studies have reported on neonatal biosonar production. Reiss (1988) investigated the click development in two captive male infant bottlenose dolphins from birth through 40 days old. Early pulsed signals were lower in frequency. Adult-like clicks developed between 24 and 40 days. Hendry (2004) recorded the first click train from a bottlenose dolphin calf at 22 days postpartum and found that the number of echolocation attempts increased steadily with growth. Harder et al. (2016) reported that click duration and the number of clicks per train were lower in calves than in adults but increased with age, whereas click train density (clicks/s) and inter-click intervals were rather consistent. Most

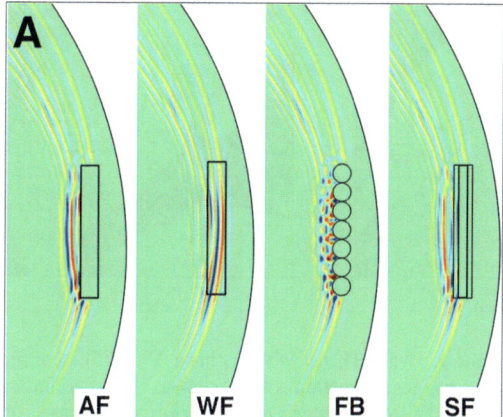

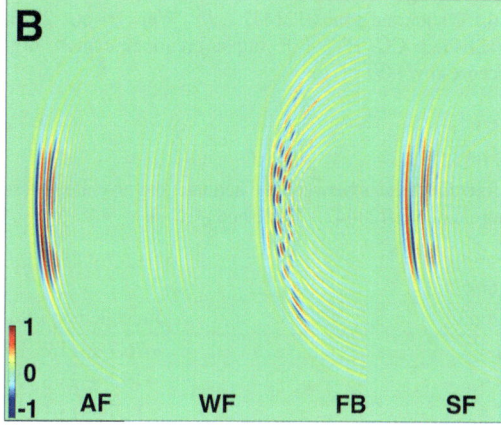

Fig. 4.19 Snapshots of the simulated signal wavefronts in the finite-element model. (**a**) Wavefronts of an echolocation click just after it impinged on each of the four objects showing how the echoes are formed. (**b**) Wavefronts of the echoes at a location half-way between the target and the listening dolphin (i.e., 0.5 m from the dolphin). © Wei et al. 2021; https://doi.org/10.1038/s41598-021-85063-2. Published CC BY 4.0; https://creativecommons.org/licenses/by/4.0/

physiological maturation and learning processes. Wei et al. (2015) had access to a stranded <1-month-old Yangtze finless porpoise, which died the next day; as well as archived measurements of both a 1-year-old juvenile and an adult finless porpoise. The acoustic properties (sound speed, density, and acoustic impedance) of head tissues were measured after dissection. Acoustic property reconstruction (Fig. 4.20) then highlighted that the neonate porpoise's melon lacked a significant sound speed gradient when compared to adults (Norris and Harvey 1974). Wei et al. (2015) statistically compared the Hounsfield unit values (HU, which is linearly related to sound speed and density) of the main forehead tissues between the neonate and juvenile finless porpoises (Fig. 4.20; Table 4.1). The data suggested that the acoustic properties of the forehead tissues change with growth and the lipid components of the melon may not be completely laid down until later stages of maturity. Similar results were reported by Gardner and Varanasi (2003), whose biochemical analysis of the melons of bottlenose dolphins and harbor porpoises demonstrated an increased proportion of isovalerate fatty acid with age. Fetal and neonatal animals have a lower concentration of acoustic fats in the melon (related to biosonar sound propagation) and mandible (related to biosonar sound reception), indicating the biosonar system is not fully developed in the younger animals.

click trains were recorded from 5 to 6 month-old calves, especially when separated and swimming independently from their mothers or in a social group with other calves. Faveoro et al. (2013) suggested that the first 2 months postpartum were likely to be critical for the calves to develop their echolocation abilities and behaviors.

Based on such evidence, it is likely that dolphin calves innately produce biosonar signals soon after birth, although the biosonar system will not be fully developed yet. Biosonar ability develops over time, through a combination of

4.4 Acoustic Communication

Odontocetes communicate acoustically in a variety of behavioral contexts. Their sounds may convey certain behavioral states, such as aggression, excitement, distress, or readiness to mate (e.g., Herzing 2000; Rehn et al. 2011). Their sounds may convey group or individual identity (e.g., Caldwell and Caldwell 1971; Ford 1991; Weilgart and Whitehead 1997). The communication sounds of odontocetes have commonly been classed as whistles, burst-pulse sounds, or clicks. Clicks can serve both communication and echolocation purposes, but their spectro-temporal

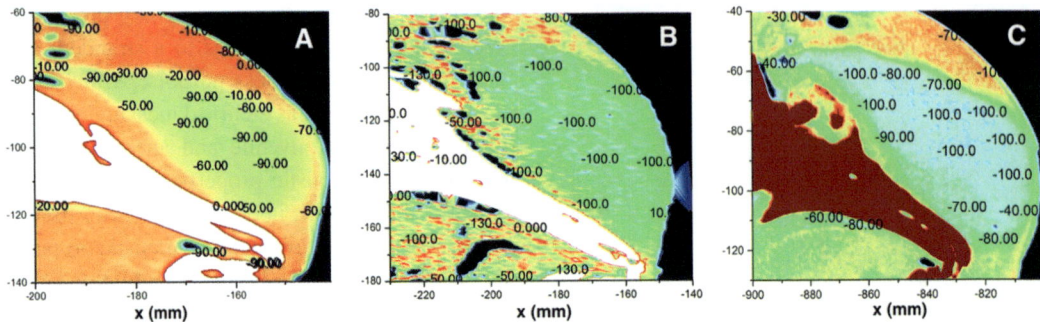

Fig. 4.20 Comparison of the Hounsfield unit (HU) distributions of the melon in the heads of a neonate (<1-month-old) finless porpoise (**a**), a juvenile (1-year-old) finless porpoise (**b**), and an adult finless porpoise (**c**). The numbers in the images represent the HU values at each position. Data for panels B and C courtesy of Zhitao Wang. Panels A and B from Wei C, Wang Z, Song Z, Wang K, Wang D, Au WWL, Zhang Y (2015) Acoustic property reconstruction of a neonate Yangtze finless porpoise's (*Neophocaena asiaeorientalis*) head based on CT imaging. PLOS ONE 10 (4):e0121442. https://doi.org/10.1371/journal.pone.0121442. © Wei et al. 2015. Published CC BY 4.0; https://creativecommons.org/licenses/by/4.0/

Table 4.1 Comparison of Hounsfield unit values of the main forehead tissues between the neonate and juvenile finless porpoises. © Wei et al. 2015; https://doi.org/10.1371/journal.pone.0121442. Published CC BY 4.0; https://creativecommons.org/licenses/by/4.0/

	Hounsfield unit	
Forehead tissues	Neonate porpoise	Juvenile porpoise
Melon	−63.3 ± 24.4	−76.7 ± 43.7
Blubber	−68 ± 19.9	−76.7 ± 35.6
Muscle	9.6 ± 7.4	11.7 ± 13.1
Mandibular fat	−59.7 ± 30.1	−70.3 ± 39.8
Connective tissue	55.8 ± 20.1	66.3 ± 18.1

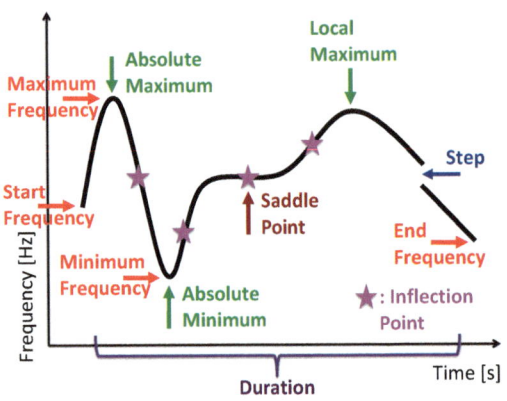

Fig. 4.21 Commonly measured features of a fundamental whistle contour

characteristics may vary with function (e.g., Martin et al. 2021).

4.4.1 Sound Types

4.4.1.1 Whistles

Whistles are tonal sounds that can be constant in frequency (called constant-wave) or frequency-modulated. Harmonically related overtones may or may not be present. Several features of the whistle's fundamental contour have been measured in order to describe recorded whistles (Fig. 4.21). These include the frequency at the start of the fundamental contour (start frequency), the frequency at the end of fundamental contour (end frequency), the minimum frequency, the maximum frequency, and the duration. Additional measures include the numbers of local extrema, inflection points, and saddle points, but there is some confusion in the literature about how these are defined, with extrema commonly called inflections. Referring to differential calculus (Bronshtein et al. 2015), a local extremum occurs at a so-called stationary point, where the first derivative with regards to time is 0 and changes sign. The first derivative gives the slope of the curve (in this case, the curve is the fundamental whistle contour). At a local extremum, the slope is 0 (i.e., the tangent to the curve is horizontal). If the second derivative with regards to time is negative, then the local extremum is a maximum and the slope changes from positive to negative. If the second derivative is positive, then the local extremum is a minimum and the slope changes from negative to positive. At an inflection point, the second derivative with regards to time is 0 and changes sign. The second derivative gives the curvature of the curve. At an inflection point, the curvature changes from clock-wise to counter-clock-wise (i.e., the curvature changes from negative to positive) or vice versa (i.e., the curvature changes from positive to negative). The tangent to the curve crosses the curve at an inflection point. Whistle contours might not always pass through local extrema and inflection points quickly, but rather plateau (i.e., the frequency does not change over some time). Such plateaus are sometimes called saddle points. In differential calculus, multiple derivatives with regards to time are 0 at such saddle points. If the first non-zero derivative is the 2nd or 4th (or higher even order), then the whistle has a local extremum at this point. If the first non-zero derivative is the 3rd or 5th (or higher odd order), then the whistle has an inflection point at this point. Breaks in a whistle contour refer to a discontinuity where the frequency jumps without a gap in time. Such breaks are sometimes called steps in the literature, yet other authors mean saddle points when they talk about steps. Rough-toothed dolphins possibly hold the record in the number of steps in a whistle contour (Fig. 4.22).

It is often assumed that whistles are emitted omnidirectionally, but the directivity pattern depends on the frequency, with whistle energy at higher frequencies dropping off to the sides of the dolphin head and the most broadband spectrum being recorded in front of the animal (Fig. 4.23; Branstetter et al. 2012). In other words, higher-frequency harmonics have greater directivity than lower-frequency harmonics, which are rather omnidirectional. Such mixed-directional whistles provide cues about the

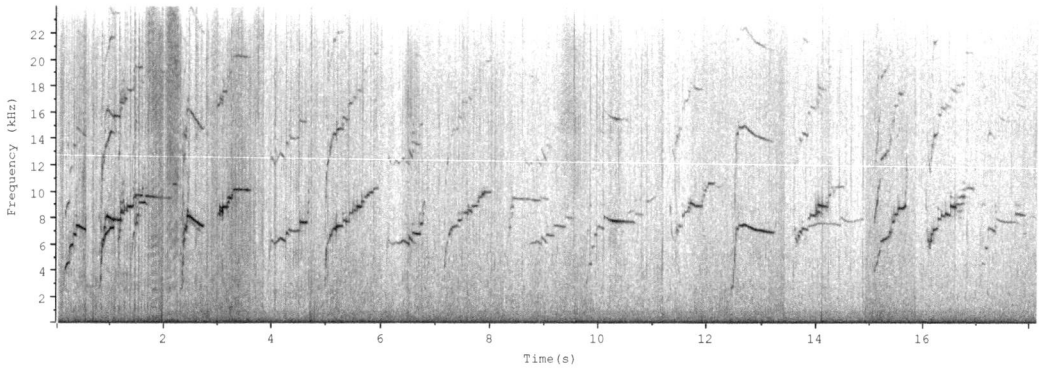

Fig. 4.22 Spectrograms of stepped whistles from rough-toothed dolphins ($f_s = 48$ kHz, NFFT $= 1024$, Hann window, 50% overlap) recorded by Boisseau et al. (2010). Quiet periods and signals at low signal-to-noise ratio were removed, and so, displayed sounds did not occur at the displayed timing

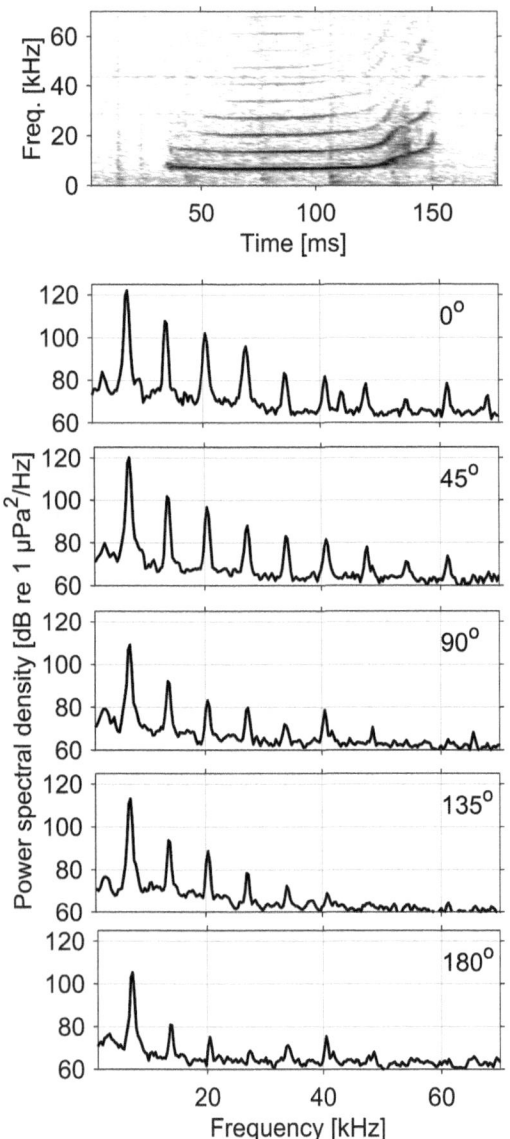

Fig. 4.23 Spectrogram (top) of a whistle recorded on-axis (f_s = 300 kHz, NFFT = 1000, Hann window, 50% overlap). Spectra recorded at different horizontal angles. Higher-frequency harmonics are attenuated more than lower-frequency harmonics at greater off-axis angles. Recording courtesy of Branstetter et al. 2012

Kogiidae, and the genus *Cephalorhynchus* (see Au et al. 2010). However, narrowband frequency-modulated sounds have been documented for several of these—albeit rarely (Goold 1999; Reyes Reyes et al. 2016; Thomas et al. 1990; Verboom and Kastelein 1995). One hypothesis for the rare occurrence of whistles in these species is predation pressure by killer whales, which might hear the whistles (Morisaka and Connor 2007). Other hypotheses include anatomical differences and behavioral-ecological factors (Oswald et al. 2008). Many studies agree that frequently whistling species tend to have a high degree of sociality (Herman and Tavolga 1980; May-Collado et al. 2007).

4.4.1.2 Burst-Pulse Sounds

Burst-pulse sounds are rapid series of pulses with a higher pulse-repetition rate (PRR) than click trains. In click trains, clicks are typically spaced >10 ms apart, whereas in burst-pulse sounds, pulses are <10 ms apart. For example, Lammers et al. (2004) found a bimodal distribution of ICIs in spinner dolphins, separating series of pulses into burst-pulse sounds and click trains. The number of pulses/s of burst-pulse sounds is too large for the human ear to hear the individual pulses. Instead, these are perceived as buzzes or as having tonal structure, where the tones are related to the PRR. Burst-pulse sounds have therefore often been given onomatopoeic names such as moans, screams, squawks, trumpets, or squeals (Fig. 4.24).

In his seminal article, Watkins (1967) explained the tonal structure seen in spectrograms of burst-pulse sounds. He presented a series of spectrograms of simple, artificially created burst-pulse sounds and real, recorded odontocete burst-pulse sounds. If each pulse is a very brief, broadband click, then the spectrum shows lines at the PRR and higher harmonics. As Watkins stated, the PRR may be read off the spectrogram as the harmonic interval between the lines. Tightly spaced spectral lines correspond to a low PRR (i.e., slow series of pulses), while widely spaced spectral lines indicate a high PRR (i.e., fast series of pulses). If the clicks occur in packages (e.g., 2 clicks, 2 pauses, 2 clicks, ...; or, 3 clicks, 1 pause, 3 clicks, ...), then the relative amplitudes

sender's location and direction of movement. They might thus play an important role in coordinating group behavior such as cooperative foraging (Lammers and Au 2003).

Not all odontocetes whistle, or some whistle very rarely. Rarely-whistling odontocetes are from the Families Phocoenidae, Physeteridae,

4 Odontocete Sounds

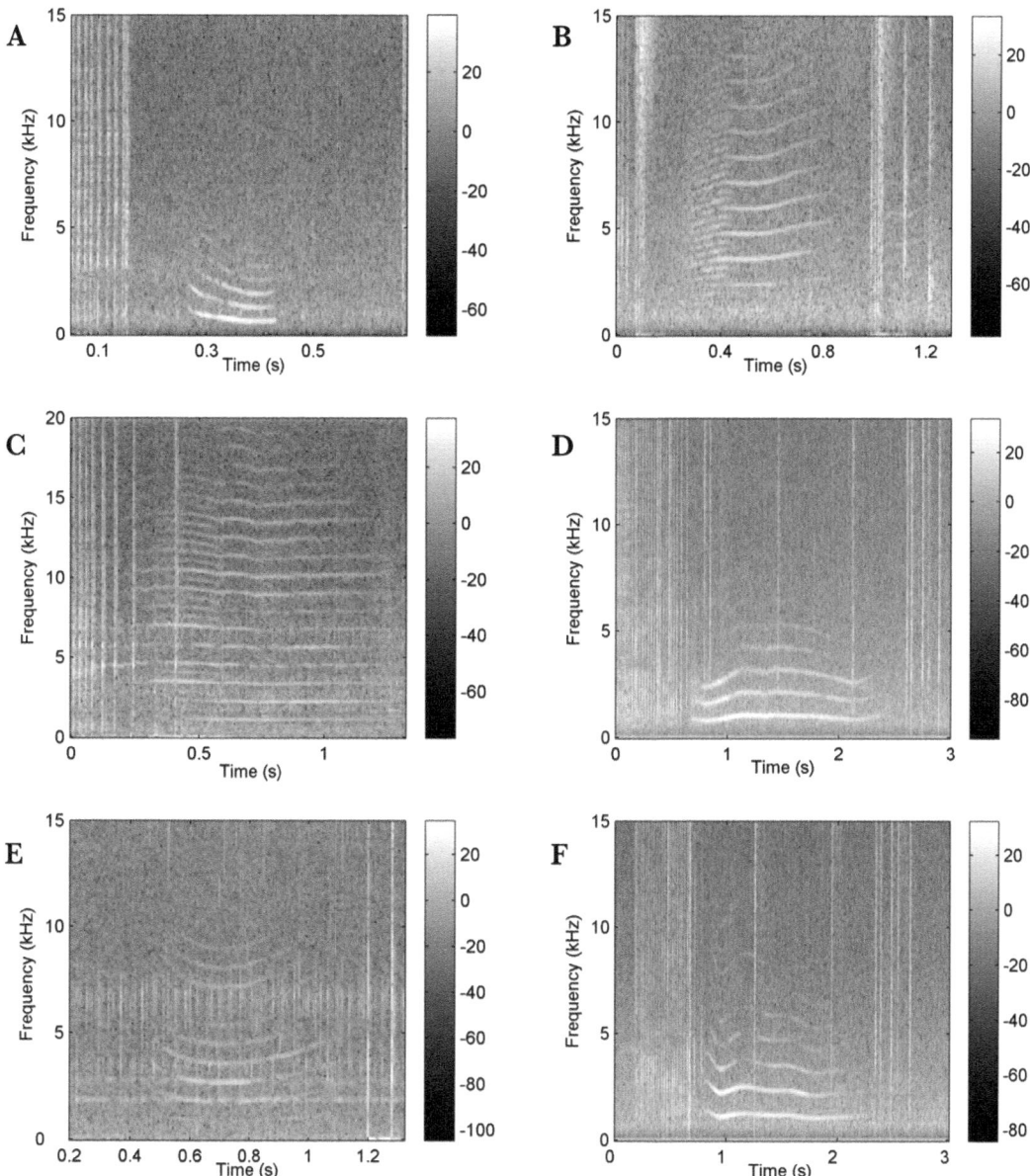

Fig. 4.24 Spectrograms of sperm whale burst-pulse sounds (squeals; $f_s = 48$ kHz, NFFT $= 512$, Hann window; Weir et al. 2007). Reprinted with permission from Cambridge University Press. Weir CR, Frantzis A, Alexiadou P, Goold JC, The burst-pulse nature of 'squeal' sounds emitted by sperm whales (*Physeter macrocephalus*). J Mar Biolog Assoc 87 (1):39–46; https://doi.org/10.1017/S0025315407054549. © Marine Biological Association of the United Kingdom, 2007. All rights reserved

of the spectral lines will differ (Watkins 1967). Relative amplitudes also differ when pulses are extended from infinitesimally sharp clicks to extended square-wave pulses. Depending on the symmetry within pulse trains (i.e., number of pulses on and off, and duration of pulses versus duration of gaps in between pulses), certain spectral lines disappear. Brown (2008) provided the equations for the spectra of such burst-pulse sounds. If the pulses in a burst-pulse sound are

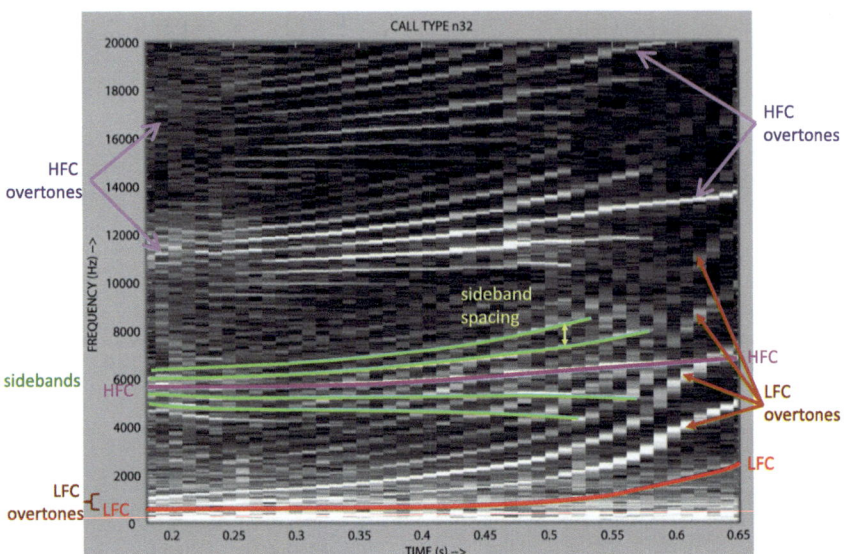

Fig. 4.25 Spectrogram of a killer whale vocalization. The LFC (PRR) starts at about 400 Hz and increases over the duration of the vocalization. At least 5 harmonics of the LFC are visible (fundamental and 4 overtones). The HFC starts at 5800 Hz and sweeps up. Harmonics are visible at 11,600 Hz and 17,400 Hz. The LFC creates sidebands about each of the 3 harmonics of the HFC. These sidebands occur above and below the HFC harmonics and the sideband spacing equals the frequency of the LFC. The PRR (and thus sideband spacing) increases over time. Modified from Brown JC (2008) Mathematics of pulsed vocalizations with application to killer whale biphonation. J Acoust Soc Am 123 (5):2875–2883. https://doi.org/10.1121/1.2890745. With permission from the ASA. © Acoustical Society of America, 2008. All rights reserved

brief tones rather than broadband clicks, additional lines corresponding to the tone frequency and harmonics will appear in the spectrum (Watkins 1967). Such burst-pulse sounds can be thought of as having been created by two sound-producing mechanisms: The first produces the sine wave at the tone frequency and the second produces the waveform that amplitude-modulates the sine wave. Brown (2008) succinctly derived the equations and spectrograms for these burst-pulse sounds and presented examples from killer whales.

In some of the killer whale literature, the PRR has been referred to as the low-frequency component (LFC) and the frequency of the tone has been referred to as the high-frequency component (HFC) (Miller and Bain 2000). Interesting spectrograms occur when both have multiple harmonics and are frequency-modulated (i.e., the PRR and the tone frequency change over the duration of the vocalization; Fig. 4.25; Brown 2008).

Burst-pulse sounds are omnidirectional at low frequencies and increasingly directional at higher frequencies. This mixed directionality might provide cues about the sender's location and travel direction. The full, broadband spectrum at maximum level is emitted in the forward direction (Fig. 4.26; Branstetter et al. 2012). There are several scenarios in which this projection of the message in a targeted direction might be useful; for example, if the message relates to aggression or threat.

4.4.1.3 Biphonations

Biphonation refers to the emission of two different sounds simultaneously. These may be two whistles with different frequency modulation emitted simultaneously, or a whistle emitted together with a burst-pulse sound. Killer whale vocalizations with low- and high-frequency

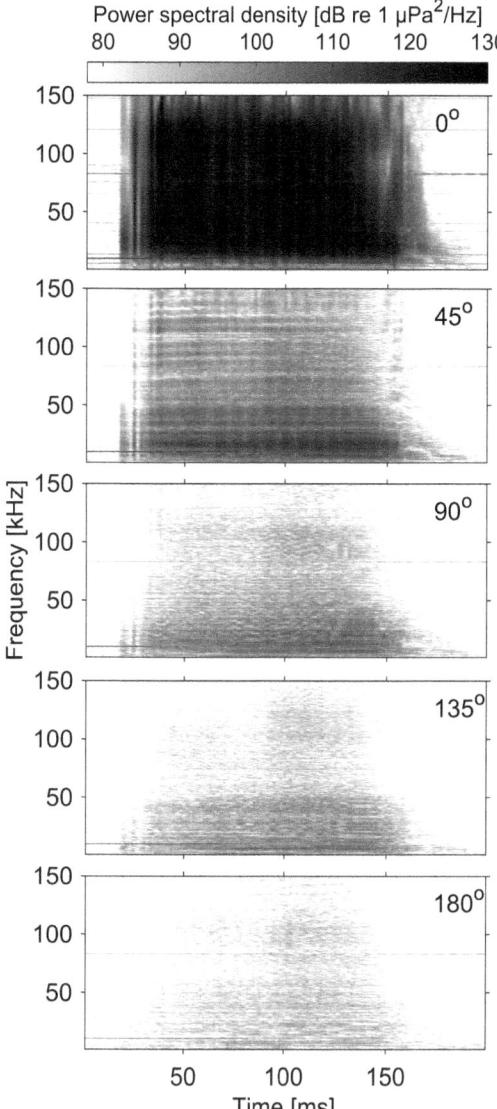

Fig. 4.26 Spectrograms of a bottlenose dolphin burst-pulse sound recorded at different horizontal angles. Amplitude is greatest in the forward direction. High-frequency energy decreases more than low-frequency energy at greater off-axis angles ($f_s = 300$ kHz, NFFT = 1000, Hann window, 50% overlap). Recordings courtesy of Branstetter et al. 2012

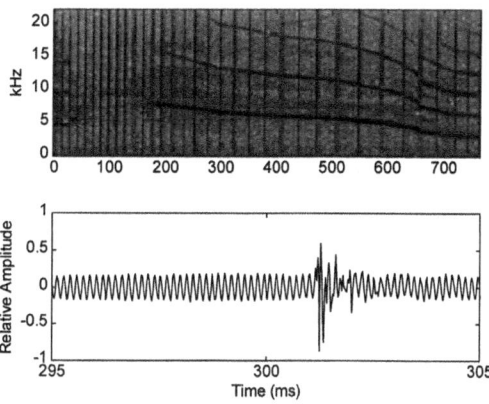

Fig. 4.27 Spectrogram of simultaneous whistle and clicks from a false killer whale (top) and waveform at 300 ms showing the whistle and one click ($f_s = 44$ kHz, NFFT = 256). Reprinted from Murray SO, Mercado E, Roitblat HL (1998) Characterizing the graded structure of false killer whale (*Pseudorca crassidens*) vocalizations. J Acoust Soc Am 104 (3):1679–1688. https://doi.org/10.1121/1.424380. With permission from the ASA. © Acoustical Society of America, 1998. All rights reserved

components are biphonations. While the literature mostly considers only whistles and burst-pulse sounds in biphonations, the simultaneous emissions of clicks and whistles or clicks and burst-pulse sounds have also been documented (Fig. 4.27; Lilly and Miller 1961; Murray et al. 1998). Biphonations have been reported from Atlantic spotted dolphins (Kaplan et al. 2018), bottlenose dolphins (Kriesell et al. 2014; Papale et al. 2015a), killer whales (e.g., Filatova et al. 2007; Sharpe et al. 2019; Tyson et al. 2007; Wellard et al. 2020), long-finned pilot whales (Busnel and Dziedzic 1966; Courts et al. 2020; Nemiroff and Whitehead 2009), short-finned pilot whales (Caldwell and Caldwell 1969; Quick et al. 2018; Sayigh et al. 2013), Risso's dolphins (Corkeron and Van Parijs 2001), beluga whales (e.g., Chmelnitsky and Ferguson 2012; Panova et al. 2017), and narwhals (Ames et al. 2021; Shapiro 2006). Biphonations are not an artifact of sound production (or recording), but likely serve distinct communicative functions; for example, as a broadcast of group or individual identity (Kriesell et al. 2014; Miller and Bain 2000; Papale et al. 2015a) or to coordinate group behavior (Samarra and Miller 2015). In killer whales, where the HFC is much higher in frequency than the LFC, the two components might also have different directionality and hence may convey information on the direction and orientation of the vocalizing animal (Miller 2002).

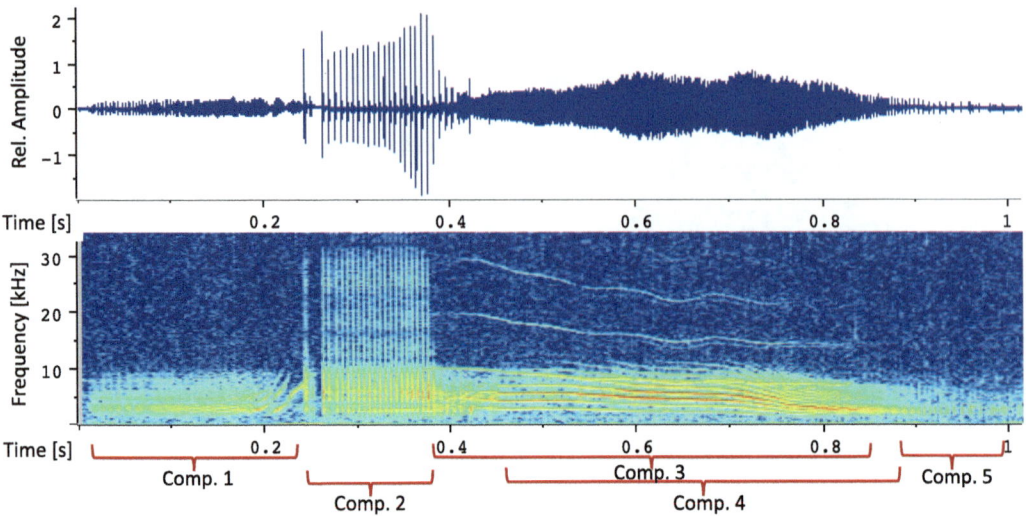

Fig. 4.28 Waveform (top) and spectrogram (bottom) of the multi-component vocalization McM1 recorded 101 times from Antarctic Type C killer whales (f_s = 96 kHz, NFFT = 512, Hann window, 50% overlap; recording from Wellard et al. 2020). The first component is a burst-pulse sound at increasing PRR. Individual pulses are visible at the beginning of the waveform before the IPI quickly decreases. Towards the end of this component (at 0.23 s), the PRR is >3 kHz (i.e., at least 3000 pulses/s). Component 1 ends abruptly when Component 2 commences, which is a series of slow pulses (~20 pulses per 0.1 s), clearly seen in the waveform plot at the top. Component 3 is a down-sweeping whistle, starting at 10 kHz at 0.38 s. Two overtones are visible. Component 4 is a burst-pulse sound, starting at 0.42 s. While the pulses of Component 2 had a low PRR, those of Component 4 start at a high PRR of ~1500 pulses/s (sideband spacing 1.5 kHz). The PRR decreases gradually until individual, slow pulses are seen in the waveform and spectrogram at 0.87–1 s (Component 5). Component 4 amplitude-modulates Component 3, which is seen as sidebands around the whistle harmonics. Components 3 and 4 make up the biphonation

4.4.1.4 Multi-component Vocalizations

In particular, killer whales (Ford 1989; Wellard et al. 2020) and pilot whales (long-finned; Courts et al. 2020; Nemiroff and Whitehead 2009; and short-finned; Pérez et al. 2017) produce vocalizations that consist of multiple components (also called elements) in immediate succession (i.e., without a gap in time). Up to six components in one vocalization have been measured (Nemiroff and Whitehead 2009). The order of components is always the same, although the first and last components are sometimes missing (Wellard et al. 2020). Multi-component calls often include a biphonation as well (Fig. 4.28; Nemiroff and Whitehead 2009; Wellard et al. 2020). The demarcation between components is not always clearly defined, in particular if sounds smoothly transition from one type to another (e.g., the burst-pulse Component 4 transitions to a series of slow pulses in Component 5 in Fig. 4.28).

4.4.1.5 Graded Sound Types

Graded sounds are those that fall in between two categories or gradually change from one category to another. For example, as a whistle becomes amplitude-modulated, its spectrogram shows sidebands. As its amplitude-modulation becomes stronger, the whistle becomes pulsed, thus turning into a burst-pulse sound. As the PRR decreases (i.e., the IPI increases), the sound turns into a series of discrete pulses or click trains. Coming the other way, click trains at decreasing ICI and increasing PRR become buzzes or burst-pulse sounds. As the IPI goes to zero, the sound becomes continuous but still consists of joined damped oscillations. Further, the damping decreases until a pure sinusoidal signal is reached.

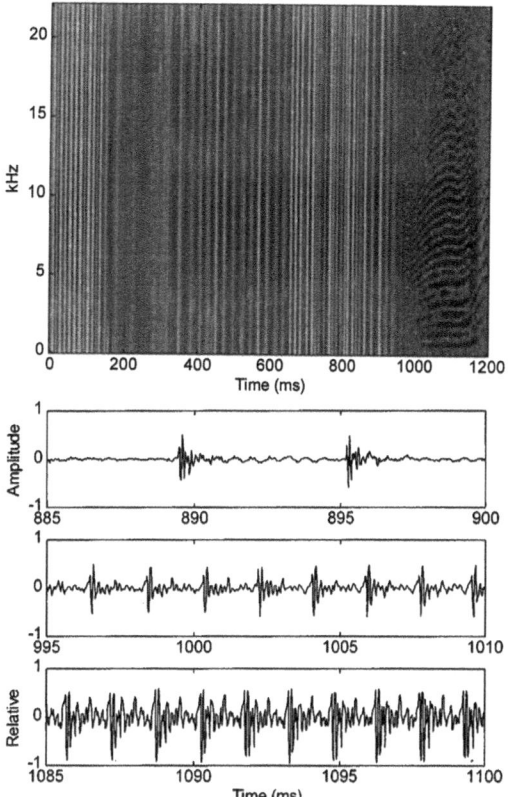

Fig. 4.29 Spectrogram (top) and waveform (bottom) of a false killer whale vocalization that transitions from a series of stand-alone pulses to a burst-pulse sound as the IPI decreases ($f_s = 44$ kHz, NFFT = 256; Murray et al. 1998; https://doi.org/10.1121/1.424380). With permission from the ASA. © Acoustical Society of America, 1998. All rights reserved

There is a continuum from whistles via burst-pulse sounds to clicks. Marine mammal sounds lie somewhere along this continuum (see false killer whale examples in Figs. 4.29 and 4.30). From clicks to whistles, the duty cycle (DC; i.e., the time occupied by the signal as a percentage of the total recorded time) changes from low to 100%. Murray et al. (1998) normalized the DC of false killer whale sounds by the DC of a pure sine wave, so that sounds that are continuous but still have damped oscillations end up with a DC <100%. Murray et al. proposed a DC-based criterion for odontocete sound classification. In their false killer whale examples, typical click trains had a DC of 0.05, burst-pulse sounds had a DC of

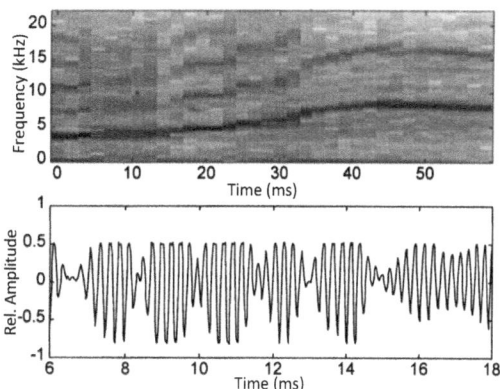

Fig. 4.30 Spectrogram (top) and waveform (bottom) of a false killer whale vocalization that transitions from a burst-pulse sound to a whistle as the IPI decreases until the pulses appear joint, after which the amplitude modulation reduces and finally ceases ($f_s = 44$ kHz, NFFT = 128; Murray et al. 1998; https://doi.org/10.1121/1.424380). With permission from the ASA. © Acoustical Society of America, 1998. All rights reserved

0.3–0.6 and whistles had a DC >0.8, although there was no clear demarcation. Rather, scatter plots of DC showed a continuous distribution hence graded structure of sound types. In terms of sound duration, sounds with low DC (i.e., pulse trains) had longer duration. Murray et al. also found a correlation with peak frequency: High-DC signals (i.e., whistles) had lower peak frequency than did low-DC signals (i.e., clicks).

Beluga whales also have been shown to own a graded vocal repertoire (Garland et al. 2015; Karlsen et al. 2002). Whistles with a high degree of amplitude modulation have been recorded from Atlantic spotted and spinner dolphins (Lammers et al. 2003), suggesting that this graded continuum model is applicable to several species.

4.4.1.6 To Whistle or Not

The NBHF species (mostly) do not whistle. The exceptions in the literature are a report on Commerson's dolphins that described whistles, although some of the spectrograms had the fundamental missing and had non-harmonic overtones, which is indicative of rapid pulse series rather than pure tones (Reyes Reyes et al. 2016); a report of whistles recorded from a stranded, captive harbor porpoise (Verboom and

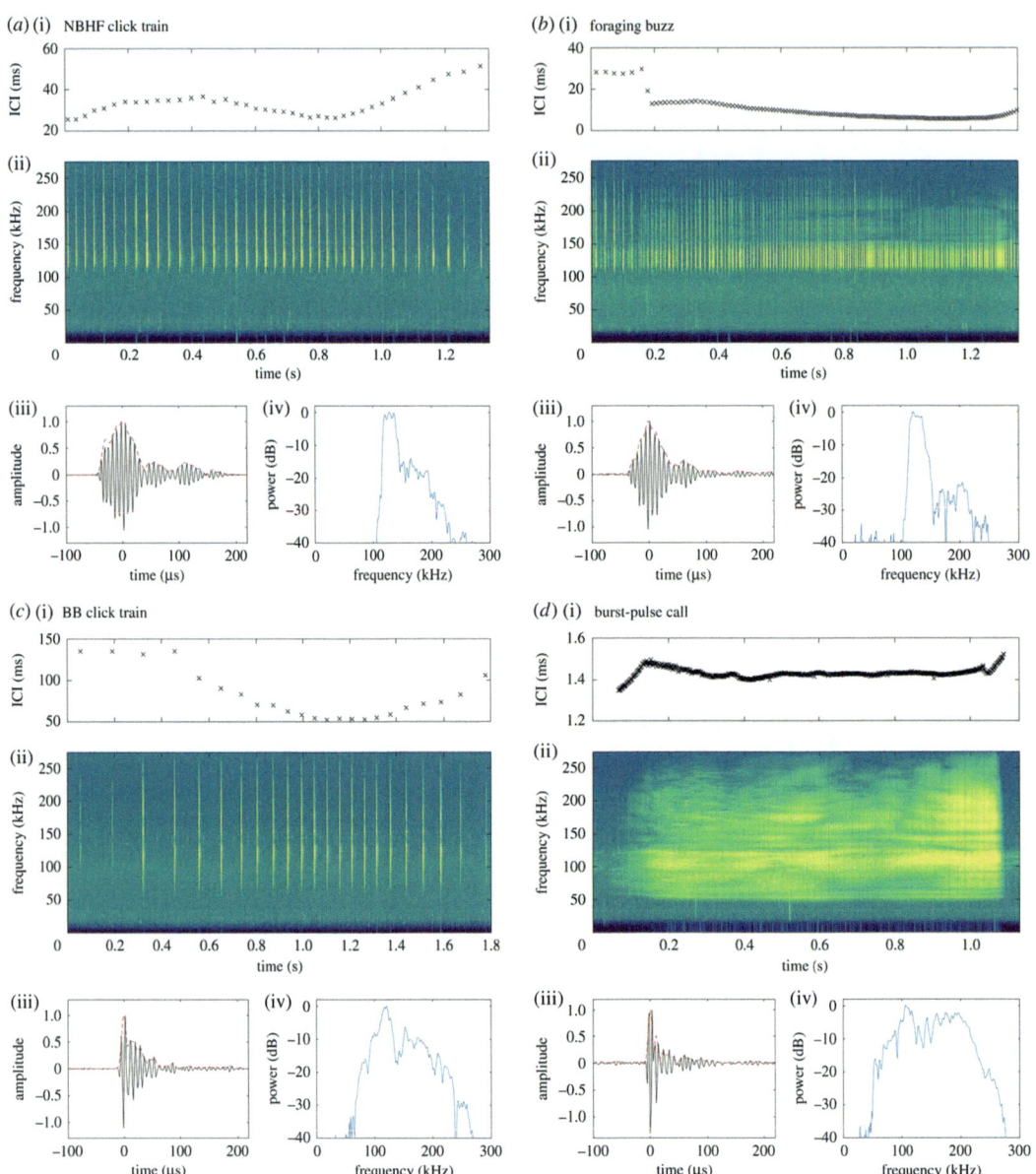

Fig. 4.31 Examples of Heaviside's dolphin sounds. (**a**) NBHF click train, (**b**) foraging buzz, (**c**) broadband click train, and (**d**) burst-pulse sound. For each sound, the panels give (i) ICI, (ii) spectrogram (f_s = 576 kHz, NFFT = 512, Hamming window, 50% overlap), (iii) waveform and envelope, and (iv) spectrum. Reprinted by permission from the Royal Society; Martin MJ, Gridley T, Elwen SH, Jensen FH (2018) Heaviside's dolphins (*Cephalorhynchus heavisidii*) relax acoustic crypsis to increase communication range. Proc Royal Soc B 285 (1883). https://doi.org/10.1098/rspb.2018.1178. © Martin et al. 2018. All rights reserved

Kastelein 1995); and a report of one tonal sound from a stranded, captive pygmy sperm whale (Thomas et al. 1990). In the absence of whistles, NBHF species use clicks for both echolocation and communication. NBHF species further make broadband clicks that extend to lower frequencies, and they emit burst-pulse sounds, both of which likely serve a communication

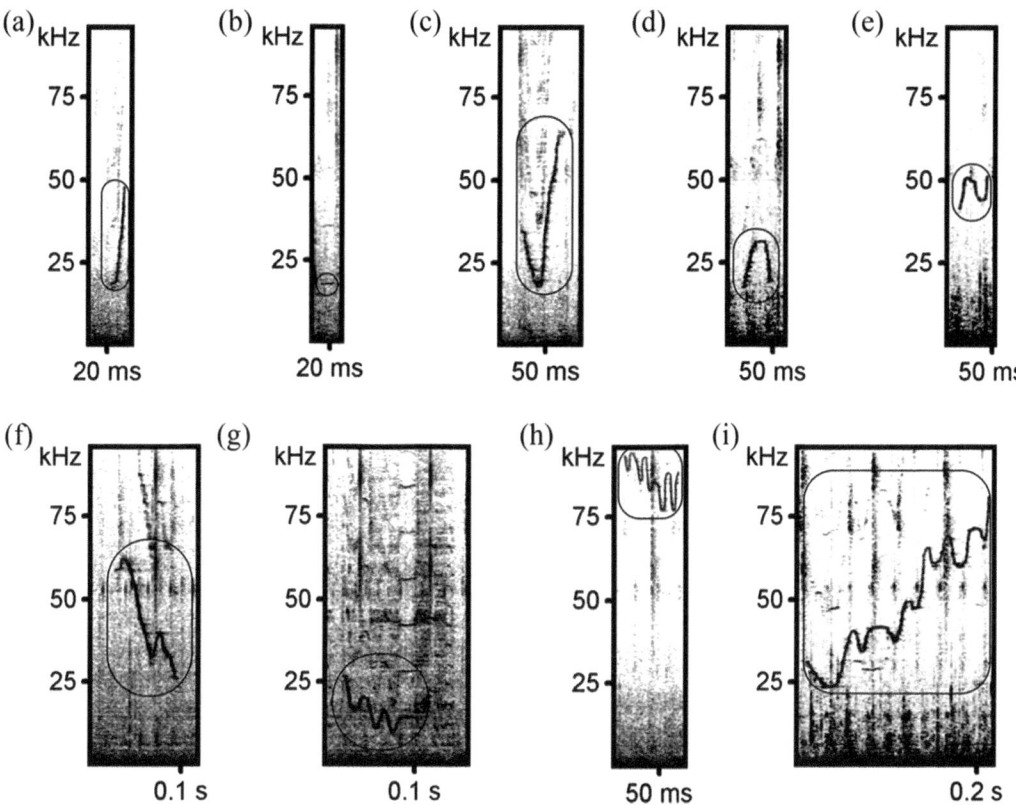

Fig. 4.32 Spectrograms of franciscana dolphin whistles recorded from a tagged, wild animal ($f_s = 44$ kHz, NFFT = 512, Hamming window, 50% overlap). Reprinted from Cremer MJ, Holz AC, Bordino P, Wells RS, Simões-Lopes PC (2017) Social sounds produced by franciscana dolphins, *Pontoporia blainvillei* (Cetartiodactyla, Pontoporiidae). J Acoust Soc Am 141 (3):2047–2054. https://doi.org/10.1121/1.4978437. With permission from the ASA. © Acoustical Society of America, 2017. All rights reserved

function (Fig. 4.31; Martin et al. 2018; Martin et al. 2021).

Within the *Cephalorhynchus* species, Commerson's dolphins used click trains with increasing ICI and bursts of clicks with small ICI to initiate social behaviors, while click trains with decreasing ICI were used during echolocation (Yoshida et al. 2014). All four *Cephalorhynchus* species have been shown to produce burst-pulse sounds at high PRR (>200 pulses/s), with one such sound having been termed "cry" (Dawson 1991; Dawson and Thorpe 1990; Dziedzic and Debuffrenil 1989; Watkins and Schevill 1980; Watkins et al. 1977; Yoshida et al. 2014). Cries were emitted mostly during social and aggressive behavior (Dawson 1991). Broadband clicks and burst-pulse sounds increase communication space over clicks with NBHF characteristics (Martin et al. 2018; Reyes Reyes et al. 2016).

In the case of porpoises, which are all NBHF species, stereotyped NBHF clicks of different click repetition rates are used in social interactions and thus likely serve a communication function, whereby information is encoded in the PRR. Context-specific click repetition patterns have been documented in several studies (Amundin 1991; Clausen et al. 2010; Nakamura et al. 1998; Sørensen et al. 2018). Furthermore, low-frequency (2–3 kHz) burst-pulse sounds have been recorded from neonatal finless porpoises with and without simultaneous clicks

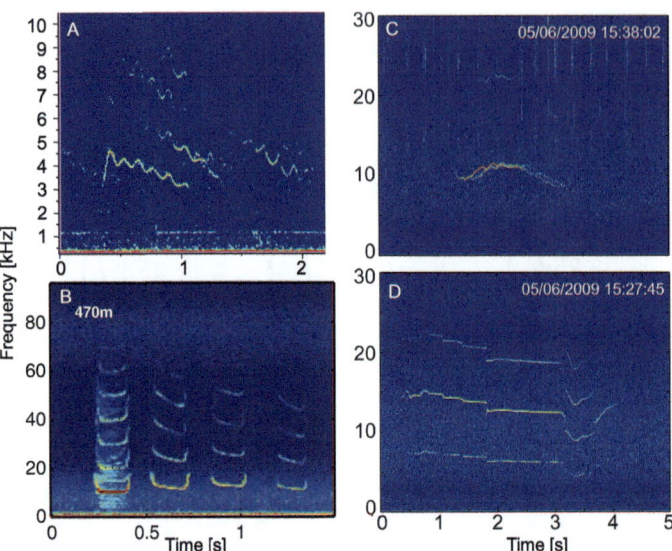

Fig. 4.33 Whistle spectrograms of beaked whale whistles. (**a**) Arnoux's beaked whale (f_s = 192 kHz, NFFT = 768, Hann window, 50% overlap; recording from Rogers and Brown 1999). (**b**) Blainville's beaked whale (f_s = 96 kHz, NFFT = 1024, Hamming window, 50% overlap). Reprinted with permission from John Wiley & Sons; Aguilar de Soto N, Madsen P, Tyack P, Arranz P, Marrero J, Fais A, Revelli E, Johnson M (2011), No shallow talk: Cryptic strategy in the vocal communication of Blainville's beaked whales. Mar Mamm Sci 28 (2):E75–E92; https://doi.org/10.1111/j.1748-7692.2011.00495.x. © Society for Marine Mammalogy, 2011. All rights reserved. (**c**, **d**) Baird's beaked whale (f_s = 480 kHz, Hann window). Reprinted from Baumann-Pickering S, Yack TM, Barlow J, Wiggins SM, Hildebrand JA (2013) Baird's beaked whale echolocation signals. J Acoust Soc Am 133 (6):4321–4331. https://doi.org/10.1121/1.4804316. With permission from the ASA. © Acoustical Society of America, 2013. All rights reserved

(Li et al. 2008). The acoustic features, however, might have been due to an underdeveloped clicking apparatus. Porpoise burst-pulse sounds might serve as contact calls, with most of these sounds recorded when the neonates were separated from their mother. Other low-frequency burst-pulse sounds have been reported from adult captive finless porpoises (Mizue et al. 1968).

In the case of river dolphins, the franciscana is the only species that produces NBHF clicks. In contrast to the other river dolphin species, it does not exclusively live in rivers, but also in estuaries and coastal waters, as do other NBHF species. The franciscana dolphin whistles, as do the other river dolphins (Fig. 4.32).

Sperm whales do not whistle but communicate with long, low-frequency, omnidirectional slow clicks, burst-pulse sounds, and codas (Madsen et al. 2002b; Weilgart and Whitehead 1993; Weir et al. 2007). The pygmy and dwarf sperm whales are NBHF species.

In the case of beaked whales, much of the research literature has focused on their unique, frequency-modulated clicks, which might leave the impression that they do not whistle. However, whistles have been reported for some species (Fig. 4.33), such as Arnoux's beaked whale (Rogers and Brown 1999), Baird's beaked whale (Baumann-Pickering et al. 2013b; Dawson et al. 1998), Blainville's beaked whale (Aguilar de Soto et al. 2012; Rankin and Barlow 2007), Hubbs' beaked whale (Lynn and Reiss 1992), and Northern bottlenose whale (Winn et al. 1970). In addition, burst-pulse sounds have been reported, which might serve a communication role (e.g., Baumann-Pickering et al. 2013b; Cholewiak et al. 2013; Dawson et al. 1998; DeAngelis et al. 2023; Leunissen et al. 2018; Rankin et al. 2011).

The absence or rarity of whistles in NBHF species and beaked whales might be a form of

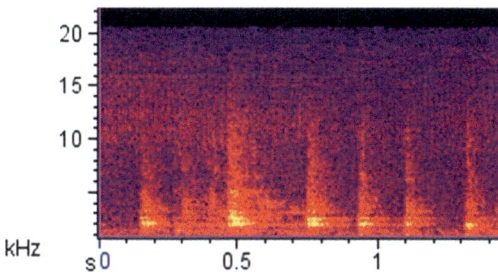

Fig. 4.34 Spectrogram of thunks recorded from bottlenose dolphins ($f_s = 44$ kHz, NFFT = 512, 80% overlap; Jones et al. 2020). Reprinted with permission from Taylor & Francis. Jones B, Zapetis M, Samuelson MM, Ridgway S, Sounds produced by bottlenose dolphins (*Tursiops*): a review of the defining characteristics and acoustic criteria of the dolphin vocal repertoire. Bioacoustics 29 (4):399–440; https://doi.org/10.1080/09524622.2019.1613265. © Taylor & Francis, 2020. All rights reserved

acoustic crypsis, intended to hide their presence from their predators—killer whales (e.g., Wellard et al. 2016). Similarly, killer whales that prey on marine mammals (delphinids, beaked whales, and pinnipeds) go into acoustic stealth mode during hunting, so as to avoid being detected by their prey (Deecke et al. 2005). While the wealth of information we have compiled in this chapter on odontocete sounds might appear as if these animals were making sounds all the time, that is not the case. There are good reasons to be quiet at times.

4.4.1.7 Additional Sound Types

Apart from typical whistles, burst-pulse sounds, and clicks, dolphins of various species and beluga whales also make low-frequency (<3 kHz) narrowband sounds (sometimes called noisy sounds; Chmelnitsky and Ferguson 2012; de Melo et al. 2021; Jayathilaka and Arulananthan 2018; Kreb 2004; Schultz et al. 1995; Van Parijs and Corkeron 2001c), low-frequency thunks (Fig. 4.34; McCowan and Reiss 1995a), and moans as low in frequency as those of baleen whales (van der Woude 2009). A single pulse or a single click may be emitted, not as part of an echolocation click train. Single broadband, metallic-sounding pulses have been termed knock, crack, or blast (dos Santos et al. 1990; Schevill and Lawrence 1956). Low-frequency single pulses have been called bangs and pops (Caldwell et al. 1966; Connor and Smolker 1996; dos Santos et al. 1990; Marten et al. 1988). Entirely non-vocal sounds include jaw claps (Belikov and Bel'kovich 2008; Caldwell et al. 1966; Herzing 1996; McCowan and Reiss 1995a; Overstrom 1983), flipper or fluke slaps (Corkeron and Van Parijs 2001; Henderson et al. 2011; Herzing 1996; Simon et al. 2006), and bubbles from the blowhole (Ames et al. 2017; Marten et al. 1996; Pryor and Kang 1980). Jones et al. (2020) provide an illustrated overview of the variety of sounds recorded from bottlenose dolphins.

4.4.2 Repeated Vocalization Sequences

The vast majority of studies on odontocete vocalizations have focused on the acoustic features and behavioral context of individual sounds. Some studies noted, however, that these sounds sometimes appear in structured sequences. Sequences may include whistles, burst-pulse sounds, and other sounds, and comprise up to a dozen or more vocalizations (Zwamborn and Whitehead 2017b). Patterned sequences of vocalizations have been reported from Araguaian river dolphins (Melo-Santos et al. 2020), Atlantic spotted dolphins (Herzing 2015), bottlenose dolphins (e.g., dos Santos et al. 1995; Herzing 2015), Guiana dolphins (Figueiredo and Simão 2009), killer whales (Ford 1989; Selbmann et al. 2023; Wellard et al. 2020), melon-headed whales (Kaplan et al. 2014), northern right whale dolphins (Rankin et al. 2007), Pacific white-sided dolphins (Mishima et al. 2019), long-finned pilot whales (Courts et al. 2020; Zwamborn and Whitehead 2017b), short-finned pilot whales (Sayigh et al. 2013), and narwhals (Walmsley et al. 2020; see Fig. 4.35). The repeated vocalizations are not necessarily identical, but can be embellished or morphed (Zwamborn and Whitehead 2017a).

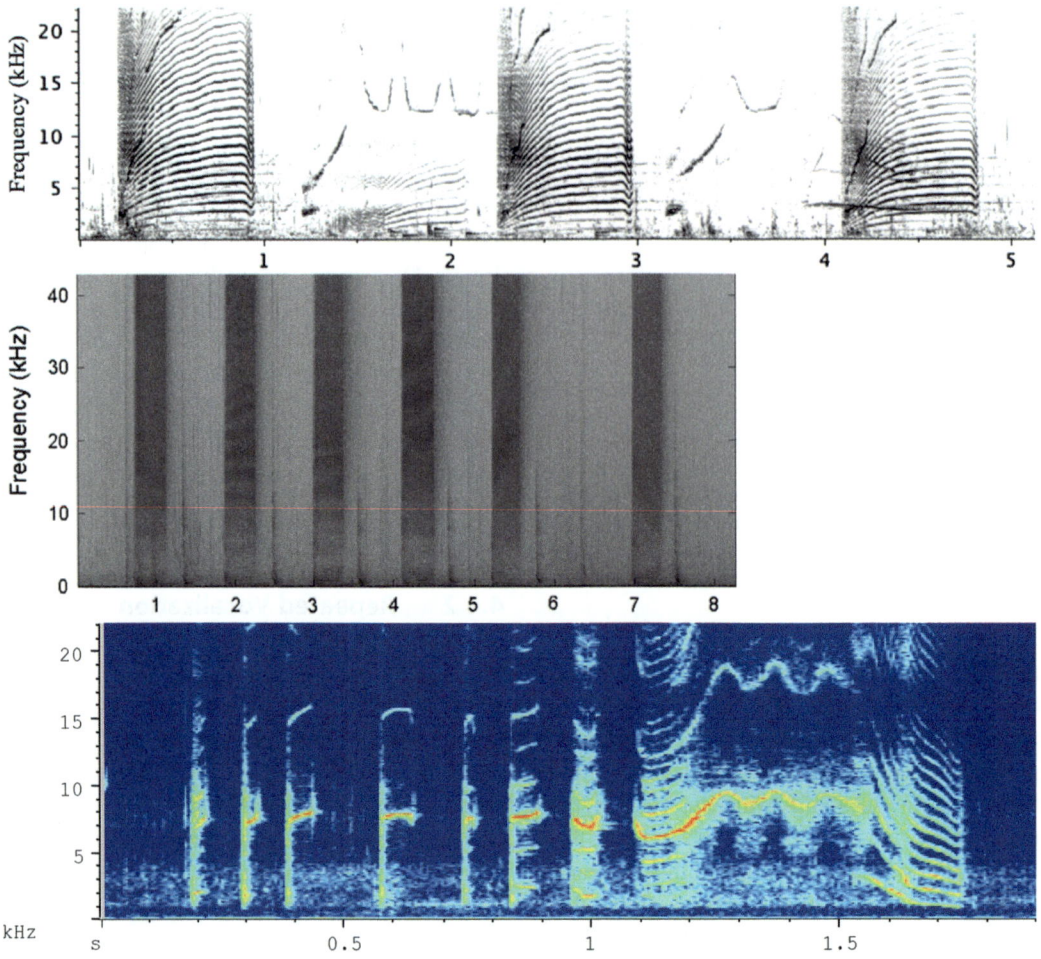

Fig. 4.35 Spectrograms of vocalization sequences; x-axes are time [s]; y-axes are frequency [kHz]. **Top**: Patterned sequence of long-finned pilot whale vocalizations (f_s = 44 kHz, NFFT = 1024, Hann window, 50% overlap). Reprinted by permission from Taylor & Francis. Zwamborn EMJ, Whitehead H, Repeated call sequences and behavioural context in long-finned pilot whales off Cape Breton, Nova Scotia, Canada. Bioacoustics 26 (2):169–183; https://doi.org/10.1080/09524622.2016.1233457. © Taylor & Francis, 2017. All rights reserved. **Middle**: Bottlenose dolphin bray sequence (f_s = 96 kHz, NFFT = 1024, Hann window, 87.5% overlap). Reprinted by permission from Springer Nature. King SL, Janik VM, Come dine with me: Food-associated social signalling in wild bottlenose dolphins (*Tursiops truncatus*). Anim Cogn 18 (4):969–974; https://doi.org/10.1007/s10071-015-0851-7. © Springer Nature, 2015. All rights reserved. **Bottom**: Vocalization type McM10 of Antarctic Type C killer whales, recorded 95 times (f_s = 96 kHz, NFFT = 512, Hann window, 50% overlap). The sequence of whistles and burst-pulse sounds leading up to the final component varied in duration and number of components. Data from Wellard et al. 2020

Two Pacific white-sided dolphins housed in separate pools emitted pulsed call sequences of three or more pulsed calls, which function to make contact (Mishima et al. 2023). Swimming freely within their pool, dolphins increased the call amplitude as they swam farther away from the other's pool, implying they control the source level so as to reach dispersed conspecifics.

Patterned sequences have been noted commonly in socializing odontocetes in large group sizes, and so their role might relate to the coordination of contacts and group cohesion, and to

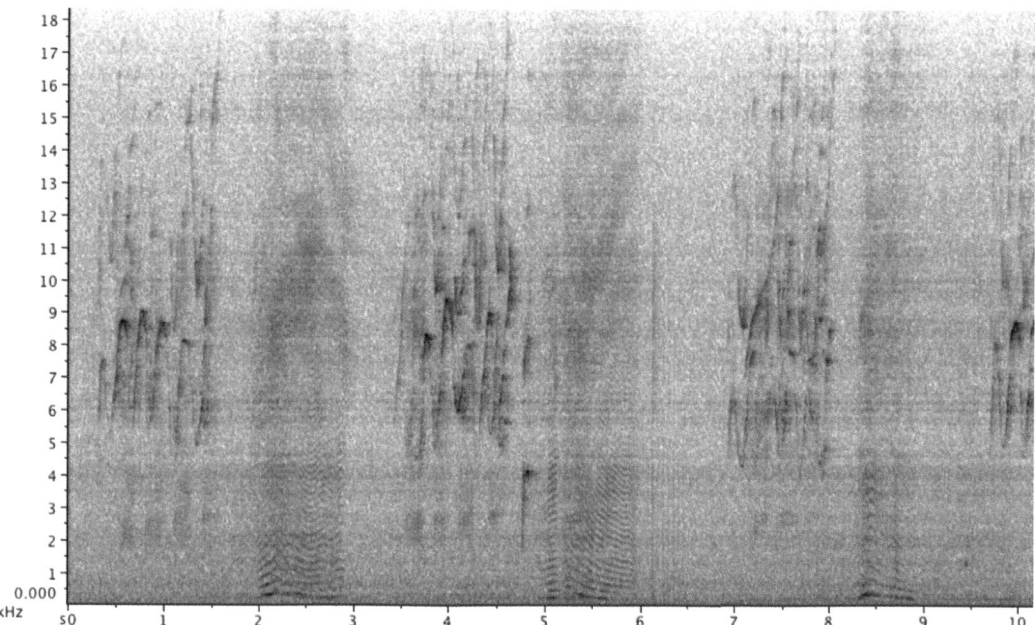

Fig. 4.36 Spectrogram of synchronized whistles and burst-pulse sounds recorded from bottlenose dolphins. Reprinted from Herzing DL (2015) Synchronous and rhythmic vocalizations and correlated underwater behavior of free-ranging Atlantic spotted dolphins (*Stenella frontalis*) and bottlenose dolphins (*Tursiops truncatus*) in the Bahamas. Anim Behav Cogn 2 (1):14–29. https://doi.org/10.12966/abc.02.02.2015. © Herzing 2015. Licensed CC BY 3.0 Unported; https://creativecommons.org/licenses/by/3.0/

group or individual identification (Zwamborn and Whitehead 2017b).

The low-frequency bray series of bottlenose dolphins is a single-unit or multi-unit series including broadband burst-pulse sounds, that have been labeled grunts, gulps, and squeaks (dos Santos et al. 1995), and that have been associated with feeding (dos Santos et al. 1990; Janik 2000a; King and Janik 2015). Thirteen different bray sequences were described from Italy (Pace et al. 2022).

4.4.3 Duetting, Chorusing, and Synchronous Rhythmic Calling

Synchronous calling and rhythmic patterning by two or more individuals has been described for Atlantic spotted and bottlenose dolphins (dos Santos et al. 1995; Herzing 2015). A sequence of two alternating vocalizations was also recorded from long-finned pilot whales, where the received level changed differently for the two sounds in the sequence, indicating two separate individuals (Courts et al. 2020). Duets involve two individuals. These may vocalize simultaneously or in an alternating, interfacing pattern (e.g., bottlenose dolphins; Janik et al. 2011). A chorus involves more than two individuals (e.g., bottlenose dolphins; Janik et al. 2011). The hypothesized functions of such acoustic behavior include the coordination of synchronous physical behavior, support of social cohesion, attraction of females, etc. (Fig. 4.36; Herzing 2015).

4.4.4 Correlation of Acoustic and Physical Behavior

Many studies have attempted to correlate acoustic with visual observations to derive the function of certain vocalization types (e.g., in killer whales, Ford 1989; beluga whales, Panova et al. 2012;

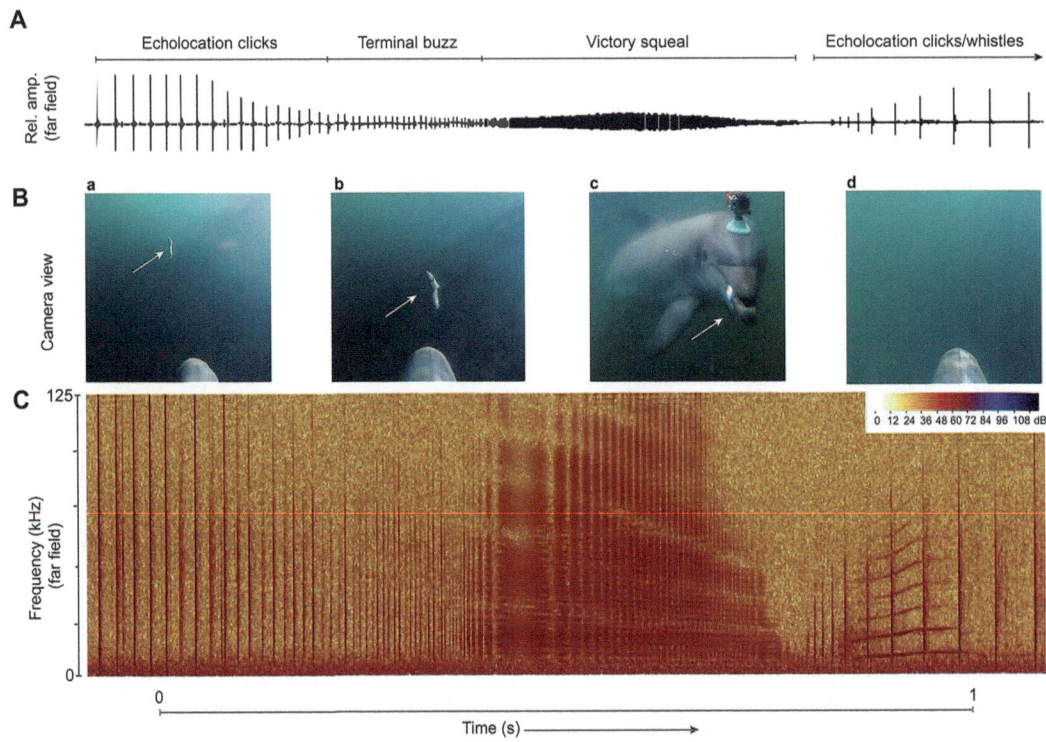

Fig. 4.37 Audio-video presentation of a dolphin's foraging process. (**a**) acoustic waveform, (**b**) video images, (**c**) acoustic spectrogram, showing how the dolphin progresses from echolocation search clicks to feeding buzz and victory squeal, before slowly clicking for navigation or search again. Note the click-train-and-whistle biphonation from this individual during the final 0.2 s. Reprinted with permission from The Company of Biologists. Ridgway S, Dibble DS, Alstyne KV, Price D, On doing two things at once: dolphin brain and nose coordinate sonar clicks, buzzes and emotional squeals with social sounds during fish capture. J Exp Biol 218: 3987–3995; https://doi.org/10.1242/jeb.130559. © The Company of Biologists, 2015. All rights reserved

Sjare and Smith 1986; and harbor porpoises, Clausen et al. 2010). Gallo et al. (2023) studied the association of acoustic with physical behavior in a group of seven captive bottlenose dolphins. Whistles occurred more during solitary swimming than social encounters and thus might have functioned as distant contact calls; burst-pulse sounds were recorded during high arousal (agonistic and social) contexts. Click production peaked during affiliative interactions, possibly playing a communicative role (rather than mere biosonar). Mixed-type vocalizations occurred as well: A combination of burst-pulse sound, whistle, and click occurred during sexual behavior; a combination of burst-pulse sound and whistle during solitary play, and a whistle-click combination during affiliative behavior (Gallo et al. 2023).

Herzing (2000) summarized contextual information on wild dolphin vocalizations. Based on her own work with wild Atlantic spotted dolphins, Herzing identified vocalizations (whistles and burst-pulse sounds) associated with excitement, distress, and alarm. Low-frequency buzzes were associated with pursuit, herding, and discipline. The most common vocalization type during agonistic or aggressive encounters was the burst-pulse type. One specific burst-pulse sound, called a squawk, possibly mediated synchronized group swimming behavior (Herzing 1996, 2000). Díaz López (2022) showed how the vocalizations of pairs of adult bottlenose dolphins engaged in feeding differed with season (meteorological season and mating season) and sex (male-male versus male-female pairs).

Ridgway et al. (2015) instrumented seven bottlenose dolphins with an audio-video recorder, while having additional audio and video recorders in the surrounding area. Data were collected both in a floating net enclosure as well as in the wild in San Diego Bay. Matching audio with video, the authors showed how dolphins searched for prey with click trains at high ICI, switched to buzzes just before prey capture (with clear images of the fish finally in the dolphin's mouth), sounded a victory squeal after prey capture, and then went on to a new series of slow echolocation clicks, with a simultaneous whistle (biphonation) on at least one occasion (Fig. 4.37).

Playback experiments have also been used to derive information on the function of certain (played-back) sounds. Through playback, Barluet de Beauchesne et al. (2022) were able to demonstrate eavesdropping in Risso's dolphins. They recorded, then played back foraging sounds, male social sounds, and female-calf social sounds, and documented differential responses of wild Risso's dolphins. The authors concluded that "by acoustically eavesdropping on conspecifics, dolphins can discriminate between social and behavioural contexts and anticipate potential threatening or beneficial situations".

4.4.4.1 Signature Whistles and Burst-Pulse Sounds

It has been shown in some delphinid species that each animal emits an individually distinctive whistle, called a signature whistle. Identity is believed to be encoded in the spectro-temporal features as signature whistles exhibit categorically different contours across individuals. Signature whistles are emitted when an individual becomes isolated from its group, during reunion, or during mother-calf contact. Sayigh et al. (2017) showed that while dolphins responded strongly to signature whistles of conspecifics, they did not seem to transmit individual voice cues in non-signature whistles. Dolphins may also copy the signature whistle of another individual, possibly as a form of addressing each other (King and Janik 2013). The fact that dolphins use signature whistles to coordinate group behavior was shown in a cooperation experiment where two dolphins had to press a paddle at opposite sides of the enclosure within 1 s of each other (and they were sent off on their task with variable delays between them). The dolphins used a mixture of signature whistles and other whistles to coordinate pressing their paddles simultaneously (King et al. 2021).

Signature whistle research has focused on the unique frequency modulations within an individual's whistle. Additional information (e.g., on identity or context) could be encoded in amplitude modulation, as has been shown in other taxa (see examples discussed in Jones et al. 2022). While amplitude along a signature whistle was mostly not constant, but increasing or decreasing, Jones and team (2022) failed to find amplitude stereotypy in their bottlenose dolphins' signature whistles. It seems plausible that recipients perceive and respond to changes in amplitude along a sound—something that future research might investigate.

How do signature whistles develop? In their first month after birth, dolphin calves commonly produce bubble streams from their blowhole, coincident with whistles, which might be a precursor to individual identification by signature whistles (Eskelinen and Jones 2021). A juvenile individual later forms its own signature whistle (within the first few months after birth) by vocal learning from other whistles to which it has been exposed, such as those of conspecifics, but not necessarily from its mother (Bebus and Herzing 2015; Janik and Sayigh 2013). There is evidence that environmental sounds, such as trainers' whistles in the case of captive juveniles, are incorporated in signature whistle formation (Miksis et al. 2002). The signature whistles of males may change over time, for example when strong male-only associations are formed and individual signature whistles converge (Smolker and Pepper 1999). Female signature whistles are more stable over long durations such as decades (Sayigh et al. 1990). However, it is also more common in females than in males for whistles to be differentiated from those of their mothers, possibly because matrilineally related females

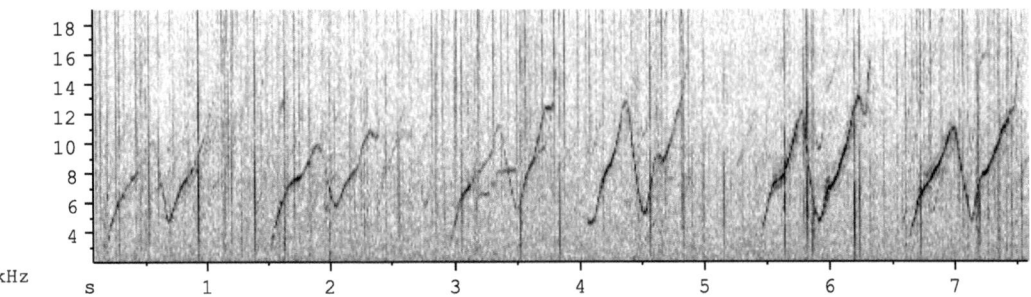

Fig. 4.38 Spectrogram of a repeated signature whistle of an Indian-Ocean bottlenose dolphin in the Swan River, Perth, WA, Australia ($f_s = 96$ kHz, NFFT = 2048, Hann window, 50% overlap). Note the increase in maximum frequency from whistle to whistle and the embellishments in the forms of inserted inflections near the end of some whistles and sidebands in the final two whistles. Recording from Ward et al. 2016

may associate for extended periods of time and must distinguish themselves in order to maintain contact with their own offspring (Sayigh et al. 1995).

Dolphins can intentionally change the features of their signature whistles. Mothers increased the maximum frequency and the overall frequency range of their signature whistles in the presence of their own calves versus others' calves (Sayigh et al. 2023). In analogy to child-directed communication ("motherese") in humans, such temporary modifications might stimulate attention, support bonding, and enhance vocal learning in dolphin calves.

Signature whistles are thus different from shared vocalization types with individual variability of acoustic parameters. While this distinction seems plausible conceptually, there is no consensus as to when two signals are categorically different. How different do the acoustic parameters of a whistle have to be in order to be categorically different? Some studies sort whistles based on the shape of the fundamental contour. Shifts in frequency or changes in duration are accepted variability within a class. Contours are commonly classed by the degree of frequency modulation, captured by the number of inflections, but these can vary even within a rapid repetition of the same whistle, as animals add embellishments, often to the end of their whistles (Fig. 4.38; also see Zwamborn and Whitehead 2017a for other types of embellishments). Furthermore, dolphins may at times add non-linear phenomena (i.e., biphonation, deterministic chaos, sidebands, and subharmonics) to their stereotyped, individual signature whistle contours (see the spectrogram examples of contours with and without such additions in Sportelli et al. 2023).

Determining signature whistles in the wild is not an easy task as animals are not commonly encountered in isolation and they emit plenty of non-signature whistles. Research with wild dolphins has shown that signature whistles are typically emitted in bouts, where the same whistle is repeated several times with 1–10 s in between. Looking for such temporal patterns can identify signature whistle candidates. This signature identification (SIGID) method has been applied in several studies to determine signature whistles in wild dolphin populations (Janik et al. 2013).

Associating signature whistles with individual animals is the next challenge. For several coastal dolphin communities, photo-ID catalogues exist, identifying individuals by unique markings on their dorsal fin and bodies. Obtaining underwater acoustic recordings simultaneously to visual observations is needed to make the link between visual markings and acoustic signatures. As dolphins are rarely encountered in isolation, the problem becomes one of computing and comparing association statistics (how often are specific vocalizations recorded together and how often are specific individuals sighted together). Over a sufficiently large sample of individual dolphins in different associations, the probability of an

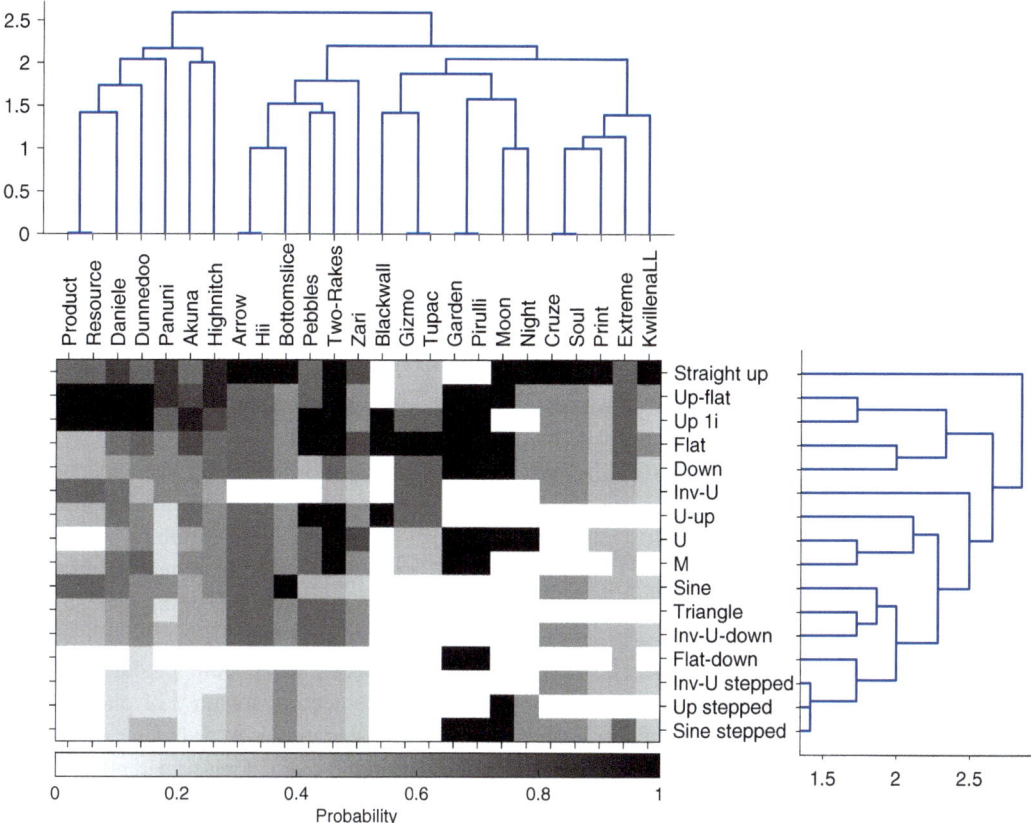

Fig. 4.39 Matrix of probabilities that an individual dolphin is present when a specific whistle is recorded. The dendrogram at the top shows the probability of sighting animals together. The dendrogram to the right gives the probability of recording different whistles together. Reprinted by permission from Springer Nature. Erbe C, Salgado-Kent C, de Winter S, Marley S, Ward R, Matching signature whistles with photo-identification of Indo-Pacific bottlenose dolphins (*Tursiops aduncus*) in the Fremantle Inner Harbour, Western Australia. Acoust Aust 48:23–38; https://doi.org/10.1007/s40857-020-00178-2. © Australian Acoustical Society, 2020. All rights reserved

individual producing a specific signature vocalization may be derived (Fig. 4.39; Erbe et al. 2020).

Signature whistles have been identified in Atlantic spotted dolphins (Bebus and Herzing 2015; Caldwell et al. 1973), Atlantic white-sided dolphins (Cones et al. 2023), bottlenose dolphins (Caldwell and Caldwell 1965), common dolphins (Caldwell and Caldwell 1968; Cones et al. 2023), Guiana dolphins (Lima and Le Pendu 2014), Indo-Pacific bottlenose dolphins (Gridley et al. 2014; Probert et al. 2023), Indo-Pacific humpback dolphins (Van Parijs and Corkeron 2001b), Pacific white-sided dolphins (Caldwell and Caldwell 1971), Pantropical spotted dolphins (Silva et al. 2017), rough-toothed dolphins (Ramos et al. 2023), spinner dolphins (Rio 2023), and narwhals (Shapiro 2006).

Burst-pulse sounds have also been suggested as signature vocalizations, in which the PRR varies significantly with individual. Captive beluga whales at times of isolation, reunion, birth or death of a calf, and husbandry behaviors, as well as wild belugas at times of calf death or temporary restraint produced individually distinctive sounds of pulsed nature, likely as a means of maintaining contact and providing individual identification (Mishima et al. 2015; Morisaka et al. 2013; Panova et al. 2017; Vergara et al. 2010). Biphonations having a whistle and a burst-pulse component or two simultaneous whistle components have also been shown to carry individually distinctive acoustic signatures and are used in contexts similar to that of signature whistles in bottlenose dolphins (Papale et al. 2015a), beluga whales (Panova et al. 2017), and narwhals (Ames et al. 2021). In sperm whales, individuality might be encoded in click patterns (Antunes et al. 2011; Gero et al. 2016; Oliveira et al. 2016; Schulz et al. 2011).

4.4.5 Vocal Learning and Mimicry

Vocal learning refers to the development or modification of sounds based on sounds from the environment including those of other animals (Janik 2014). In killer whales, for example, vocal repertoires are passed on vertically, whereby children learn from their parent (i.e., mother in this case; e.g., Deecke et al. 2000; Yurk et al. 2002). In captivity, two juvenile male killer whales learned new vocalization types from an adult male killer whale while temporarily housed together (Crance et al. 2014). A Risso's dolphin housed with bottlenose dolphins developed a potential signature whistle that was spectrographically similar to whistles from bottlenose dolphins, rather than whistles from wild Risso's dolphins (Favaro et al. 2016). A beluga whale housed with bottlenose dolphins started to produce dolphin-type whistles after 2 months (Panova and Agafonov 2017). A comparative study of killer whales kept with and without bottlenose dolphins showed that killer whales copied dolphin sounds, produced sounds with intermediate characteristics, and developed an acoustic repertoire intermediate between those of bottlenose dolphins and killer whales (Musser et al. 2014). In the wild, a juvenile killer whale, separated from its pod, started to produce barks similar to those of surrounding Californica sea lions (Foote et al. 2006). A female common dolphin that frequently socialized with harbor porpoises rather than its own species produced NBHF clicks (Cosentino et al. 2022). These species therefore exhibit a degree of vocal plasticity together with a drive towards vocal conformity with social associates. These are examples of vocal learning, which is different from vocal mimicry.

Odontocetes also imitate and copy sounds; for example, dolphins copy each other's signature whistles in social interactions (Janik 2000b). Killer whales match the vocalization type of others within the same group demonstrated by stereotyped vocalizations being recorded in bouts yet originating from different individuals. In a series of vocalizations, subsequent vocalizations matched preceding vocalization types (Miller et al. 2004b).

Killer whales can be both predator of pilot whales and competitor for prey of pilot whales. In Norway, fish-eating killer whales live sympatrically with squid-eating long-finned pilot whales. Playbacks of killer whale feeding sounds to pilot whales attracted the latter to the sound source (Curé et al. 2012). Elsewhere, in Spain, long-finned pilot whales exhibited mobbing behavior towards killer whales (de Stephanis et al. 2015). Interestingly, southern Australian long-finned pilot whales produce some vocalization types that are to the human ear and eye (looking at spectrograms) identical to southern Australian killer whale vocalizations (Courts et al. 2020; Wellard et al. 2015). Courts et al. hypothesized an anti-predator mechanism, whereby the pilot whales acoustically mask themselves while scavenging food remnants from killer whales, avoiding to be eaten themselves.

Finally, false killer whales mimicked mid-frequency active sonar sounds, increasing their whistle rate and producing more sonar-like sound immediately after sonar exposure (DeRuiter et al. 2013). Long-finned pilot whales have also been recorded mimicking sonar signals (Alves et al. 2014).

4.4.6 The Effects of Noise on Acoustic Behavior

Ambient noise can also affect the features of odontocete vocalizations. For example, ship traffic has been shown to affect the vocal behavior of beluga whales, in that they reduced their overall sound production rate, while increasing the emission of particular downsweeps and pulsed sounds, and shifting sound frequency from a mean of 3.6 kHz to 5.2–8.8 kHz (Lesage et al. 1999). A concomitant increase in the level of beluga vocalizations with an increase in ambient noise has also been documented (Scheifele et al. 2005). Similarly, raised source levels in noise have been published for bottlenose dolphins, beluga whales, and killer whales (Au et al. 1974, 1985; Holt et al. 2009, 2011). Shifts in frequency have further been reported for Atlantic spotted dolphins, bottlenose dolphins, common dolphins, false killer whales, and striped dolphins (Ansmann et al. 2007; Au et al. 1985; Fouda et al. 2018; Heiler et al. 2016; La Manna et al. 2013; May-Collado and Quinones-Lebron 2014; Moore and Pawloski 1990; Papale et al. 2015b; Rako Gospić and Picciulin 2016; Romanenko and Kitain 1992; Thomas and Turl 1990; van Ginkel et al. 2018). Increases in vocalization rate have been reported for Guiana dolphins and humpback dolphins in vessel noise (Bittencourt et al. 2017; Van Parijs and Corkeron 2001a), as well as killer whales and long-finned pilot whales in naval sonar (Miller et al. 2012; Rendell and Gordon 1999). Lengthening of vocalizations has been reported for bottlenose dolphins and killer whales in vessel noise (Foote et al. 2004; May-Collado and Quinones-Lebron 2014). Noise-induced changes in vocalization level, frequency, duration, and repetition constitute the Lombard effect (see Chap. 8, sections on masking).

One expects a physiological and anatomical limit to such adjustments, as well as a limit to the effectiveness of such adjustments. Sørensen et al. (2023) demonstrated how ambient noise can impair cooperative behavior in bottlenose dolphins. In their earlier study, two dolphins communicated with whistles in order to press a paddle at the same time at opposite sides of a pool. The authors then increased the ambient noise (1–20 kHz) in the pool. As a result, the dolphins increased the whistle level by ~0.1 dB per 1 dB noise increase and the duration of their whistles by 3–7 ms per 1 dB noise increase, nearly doubling whistle duration in the highest noise conditions (Lombard response of the signaler). One animal reduced the whistle rate, both animals reduced their echolocation rate. Both dolphins changed their swimming behavior by orientating more towards their partner, possibly as an anti-masking strategy of the listener. Despite these adaptations, the success rate (of simultaneously hitting their paddles) decreased with increasing noise, indicating limits to compensatory mechanisms (Sørensen et al. 2023).

4.4.7 Taxonomic Comparisons

4.4.7.1 Inter-Species Comparison of Communication Sounds

Hundreds of studies describe the vocal repertoires of odontocete species. The majority have addressed species-specific features of whistles. The similarity of features has been shown to correlate with the taxonomic relatedness of species (Steiner 1981), where congeneric species have more similar whistle features than species of other genera. Some studies further imply that allopatric species have more similar whistle features than sympatric species (Steiner 1981). Sympatric dolphin species have successfully been told apart by automatic classification algorithms based on a number of whistle features (Oswald et al. 2007). Similarly, canonical variables based on linear combinations of whistle features have pulled different odontocete species apart (Ding et al. 1995a). Figure 4.40 presents scatter plots of peak frequency, duration, and SL for both whistles and burst-pulse sounds of the various odontocete groups. In the whistle peak frequency versus duration plot (panel T1), the Monodontidae, coastal dolphins, and offshore dolphins clearly cluster, but the clusters overlap

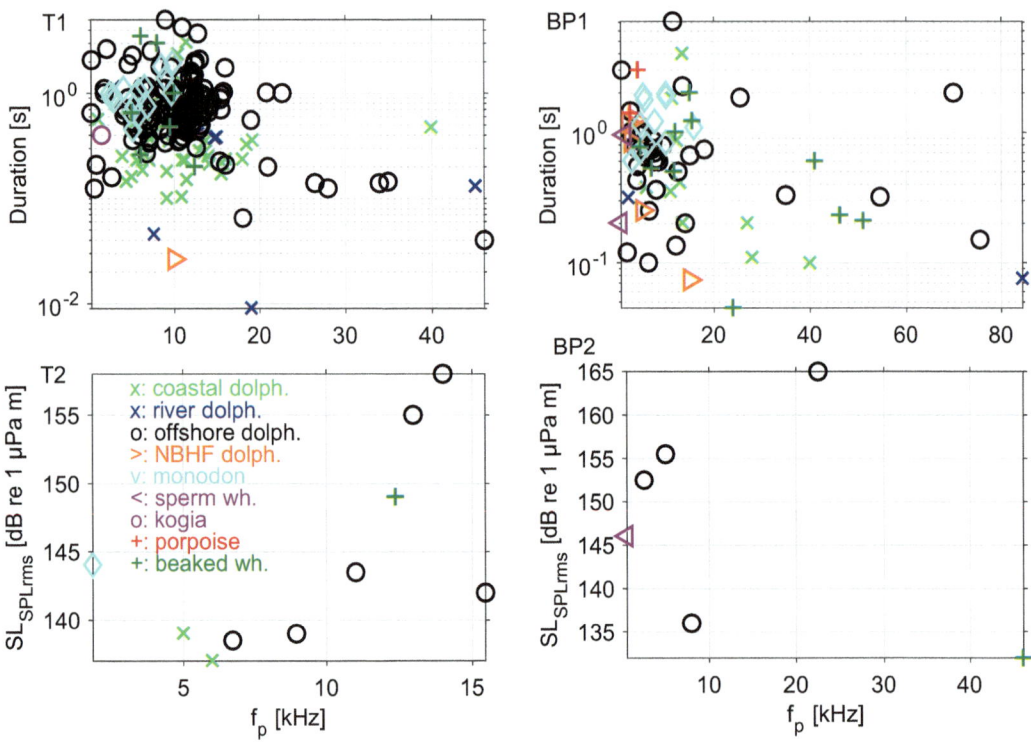

Fig. 4.40 Scatter plots of peak frequency (i.e., frequency of maximum energy), duration, and SL (in terms of SPLrms) for odontocete whistles (panels T1 and T2) and burst-pulse sounds (panels BP1 and BP2), based on the data in Suppl. Table 1

greatly. Matthews et al. (1999) created the same plot, albeit for individual species rather than species groups. While their sample size at the time was limited, odontocete species did cluster based on whistle frequency and duration measurements. In the burst-pulse sound plot (panel BP1, Fig. 4.40), only the Monodontidae appear clustered, with the other dolphin groups covering a greater range. Most of the other odontocetes are underrepresented in sample size to see clear clusters. The peak frequency versus SL plots imply a possible correlation whereby SL increases with peak frequency—at least for offshore dolphins.

Body size has also been correlated with whistle frequency, and some studies have found a good correlation with smaller animals whistling at higher frequencies (Ding et al. 1995a; Matthews et al. 1999; Podos et al. 2002). This relationship, however, is violated by the ultrasonic whistle frequencies of some killer whale populations (Filatova et al. 2012b). The scatter plots in Fig. 4.41 imply that the larger odontocetes produce whistles and burst-pulse sounds at lower frequencies than some of the smaller odontocetes do; however, there is great variability. No pattern emerged in the body size versus duration and SL plots for either whistles or burst-pulse sounds.

4.4.7.2 Intra-Species Variability of Communication Sounds

Populations of the same species may exhibit distinct vocalization features. For example, beluga whale sounds differ in PRR across populations and the degree of variation was related to the geographic distance between populations, while over time, the PRR of each population remained

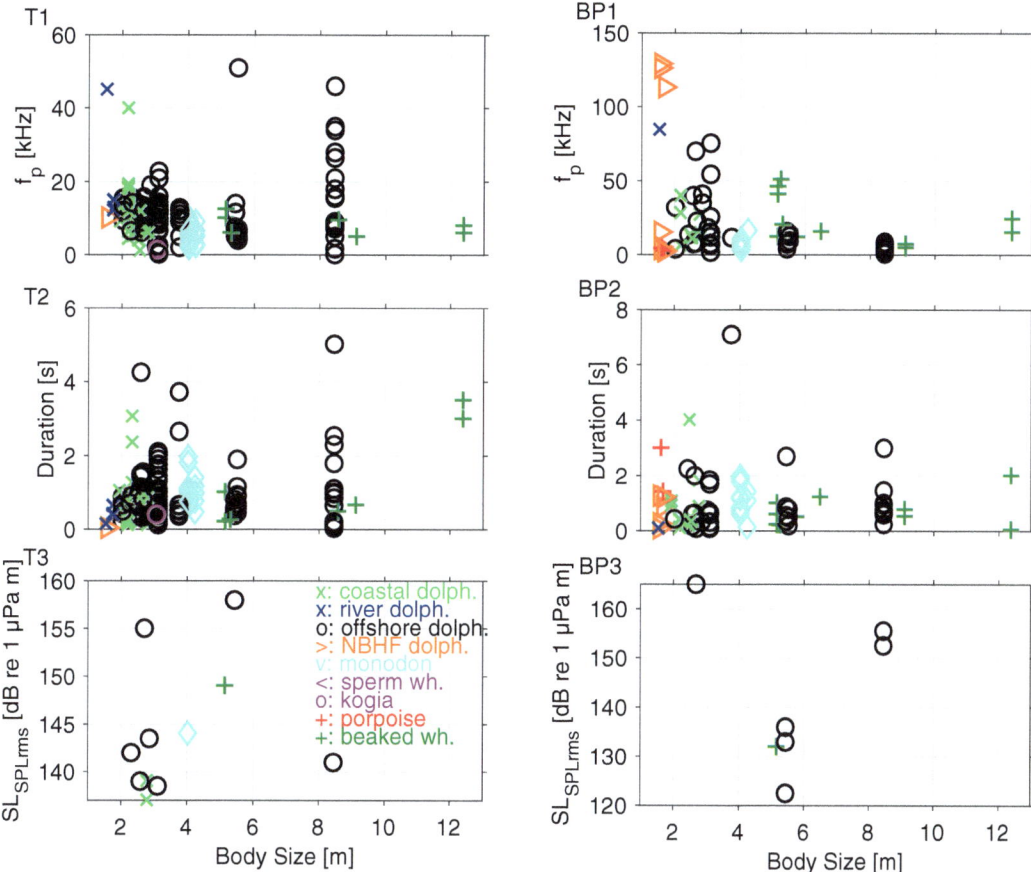

Fig. 4.41 Scatter plots of frequency, duration, and SL versus body size for odontocete whistles (panels T) and burst-pulse sounds (panels BP). In panel T1, the mean frequency was computed as the mean of measured minimum and maximum frequencies of the whistle fundamental contours. In panel BP1, peak frequency refers to the frequency of maximum energy. Plots are based on the data in Suppl. Table 1

constant (Panova et al. 2016). Variability in spectro-temporal features within a species and across populations living in different geographic regions has also been shown for bottlenose dolphins (Ding et al. 1995b; Hawkins 2010; Morisaka et al. 2005a), Indo-Pacific humpback dolphins (Fig. 4.42; Dong et al. 2021; Dong et al. 2019; Yuan et al. 2021), killer whales (Samarra et al. 2015), short-beaked common dolphins (Azzolin et al. 2021; Papale et al. 2014), short-finned pilot whales (van Cise et al. 2017), spinner dolphins (Bazua-Duran and Au 2004), striped dolphins (Azzolin et al. 2013; Papale et al. 2013), beluga whales (Panova et al. 2019), and sperm whales (Masao et al. 2014; Rendell and Whitehead 2003; Weilgart and Whitehead 1997).

Difference in acoustic features might be related to geographic and ecological differences (e.g., Filatova et al. 2015a; Oswald et al. 2008; Rendell et al. 1999; Yuan et al. 2021), which can ultimately lead to genetic divergence if the populations stop interbreeding, and so the populations split into separate species. If populations are geographically separated, then they cannot interbreed, and so they evolve independently, and their acoustic behavior evolves independently, potentially diverging. Even if geography is no barrier for population mixing (e.g., males of population 1 roam to mate with

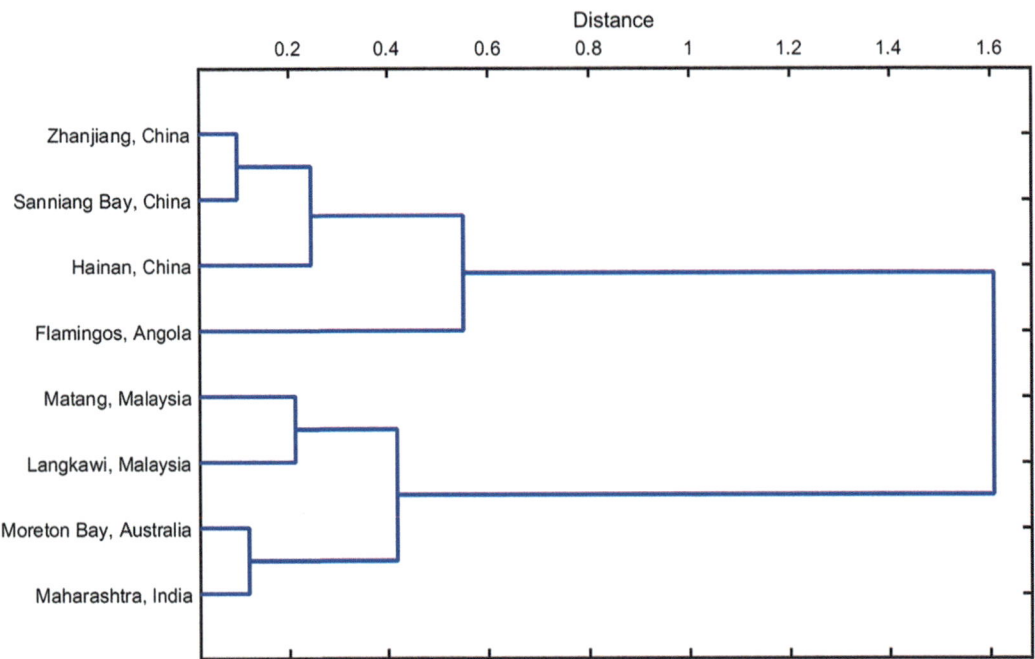

Fig. 4.42 Dendrogram of whistle feature similarity across *Sousa* populations in Zhanjiang, Sanniang, Hainan, Matang, and Langkwai (*S. chinensis*), Moreton Bay (*S. sahulensis*), Flamingos (*S. teuszii*), and Maharashtra (*S. plumbea*). Reprinted with permission from John Wiley & Sons. Dong L, Caruso F, Dong J, Liu M, Lin M, Li S, Whistle characteristics of a newly recorded Indo-Pacific humpback dolphin (*Sousa chinensis*) population in waters southwest of Hainan Island, China, differ from other humpback dolphin populations. Mar Mamm Sci 37 (4): 1341–1362; https://doi.org/10.1111/mms.12816. © Society for Marine Mammalogy, 2021. All rights reserved

females of population 2), then acoustic divergence might still lead to genetic divergence if mates no longer recognize the other's call. Divergence in acoustic behavior (including vocalization features) may thus indicate population divergence.

For example, *Sotalia fluviatilis* used to comprise two ecotypes: a riverine and a coastal. Morphology and genetics differ significantly and the two ecotypes are now classed as separate species, with the coastal form being *S. guianensis*. Interestingly, whistle parameters between the two forms differ as well (Azevedo and Sluys 2005). Across South America, a tendency towards longer and higher-frequency whistles in northern populations, and in fact, nearly linear correlations with latitude have been reported (Azevedo and Sluys 2005; Rossi-Santos and Podos 2006). Some of the early studies, however, used lower sampling rates than the more recent studies, which would have affected maximum frequencies reported. Some of the more recent studies did not find latitudinal correlations with whistle parameters but instead pointed at the influence of environmental and ecological parameters (Deconto and Monteiro-Filho 2013; Leão et al. 2016; Moron et al. 2019). Similarly, discussions about the speciation of the genus *Sousa* have been based on not only morphology and genetics, but also acoustics. The whistles of six populations of Indo-Pacific and Australian humpback dolphins were significantly different and matched "population-, regional-, and species-level differences [...], based on morphological and genetic data, as well as geographic distance and barriers to movement" (Hoffman et al. 2017).

It is not just geographic separation that has led to differences in vocalization features of populations of the same species. Ecological drivers also play a role, for example, in killer

Fig. 4.43 Sperm whale codas in Indian Ocean. (1) 3 + 1 coda from the Maldives; (2) 1 + 1 + 2 + 1 coda from the Cocos Islands; (**a**) waveforms; (**b**) spectrograms ($f_s = 48$ kHz, NFFT = 1024, Hann window, 50% overlap). Recordings from the *Voyage of the Odyssey* expedition 2002–2004. Courtesy of Chris Johnson, Ocean Alliance

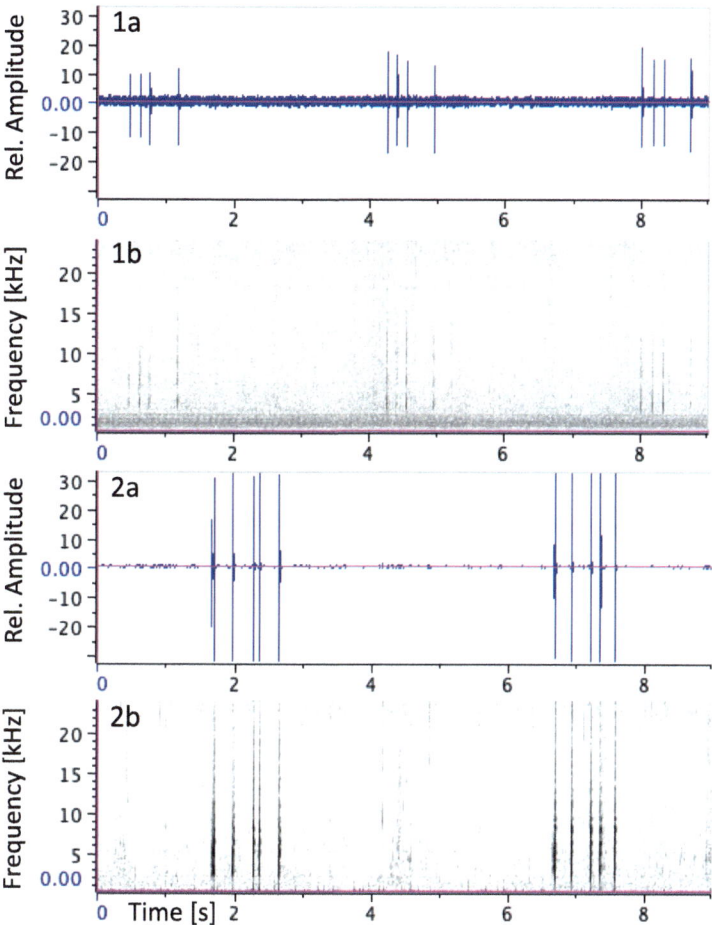

whales in the north-eastern Pacific, where different ecotypes live sympatrically. The transient killer whale ecotype feeds on marine mammals; the resident killer whale ecotype feeds on fish. Not only the acoustic features of the sounds differ, but also the vocal behavior, including vocalization rates. Transients vocalize almost exclusively after a successful hunt and during associated surface-active behavior (Deecke et al. 2005; Riesch and Deecke 2011). Transients vocalize significantly less than residents, likely because the prey of transients would be able to hear its predator's sounds and try to avoid being predated. In fact, harbor seals can tell the different killer whale ecotypes apart from their sounds (Deecke et al. 2002).

We talk about dialects, when different populations of the same species live in the same geographic area but exhibit distinct vocalization features. For example, sperm whale dialects manifest in differences in the rhythmic pattern of codas (Fig. 4.43; Rendell and Whitehead 2003; Weilgart and Whitehead 1997). Short-finned pilot whales in Hawaii show evidence of dialects in that social clusters exhibit distinct compositions of stereotyped vocalizations and distinct acoustic parameters (van Cise et al. 2017). In his pioneering work on killer whale dialects, John Ford demonstrated how different pods within the same clan of killer whales use variations of the same vocalization type (Ford 1991). For example, Fig. 4.44 shows spectrograms of vocalization number N1, which is used by seven of the eight pods that form Clan A. N1 consists of three components: Part 1 is a burst-pulse sound with a PRR of 25–100 Hz. Part 2 is a tone that starts at

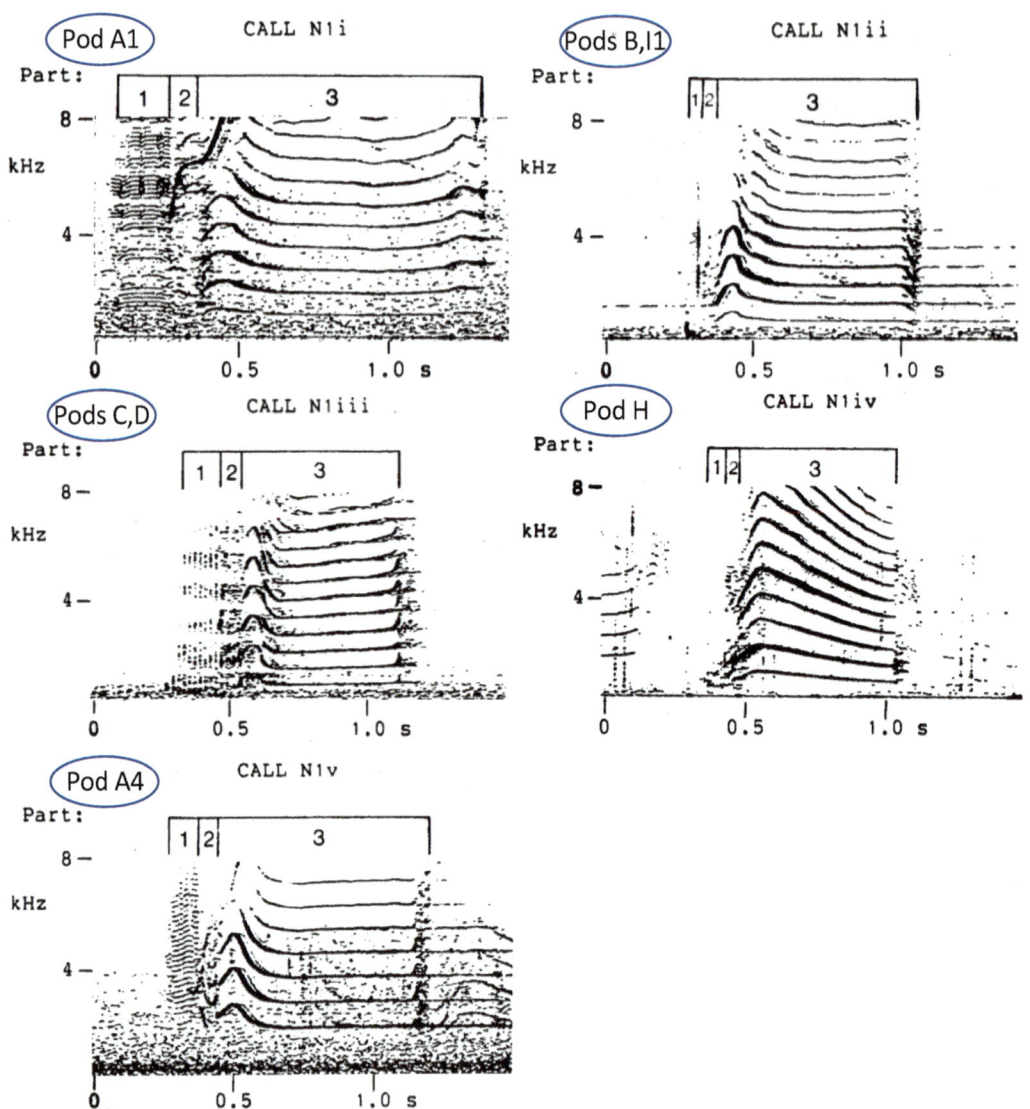

Fig. 4.44 Spectrograms of the same vocalization type (N1) used by sympatric matrilineal groups of killer whales (Pods A1, A4, B, C, D, H, and I1), showing dialects (i.e., variations i, ii, iii, iv, and v) by which pods can be identified. Figure created from spectrograms in Ford 1984, with kind permission from John Ford

2.5–4.5 kHz and sweeps up to >8 kHz. Part 3 is a burst-pulse sound with first increasing, then decreasing PRR, so that the contour has a peak in spectrograms. The spectrograms of N1 differ, depending on which pod makes it. The dialects manifest in the relative durations, frequencies, and levels (emphasis) of the three parts in the vocalization. Variation N1i made by Pod A1 has the longest and strongest Part 1 and ends Part 3 on another contour peak. Variation N1iv made by Pod H has the slowest decrease in PRR in Part 3 (appearing like a stretched-out decrease in contour). These vocalization variations are transmitted vertically; that is, calves learn the vocal repertoire from the members in their matriline. The repertoires of sister matrilines diverge over time due to random mistakes and innovations, yet horizontal transmission across different matrilines might also occur (Filatova et al. 2012a, 2013, 2015b).

4.4.8 Recording Communication Sounds

4.4.8.1 Historic Recordings

Older recordings were often done at low sampling frequency and hence missed higher-frequency signals. In addition, high-frequency signals might have been aliased, with their energy appearing at lower frequencies if no anti-aliasing filter was used (see spectrogram examples in Erbe et al. 2022a). There is a bit of discussion in the literature over low-frequency sounds from typically high-frequency species. For example, prominent low-frequency sounds were reported for porpoises in the early literature (Busnel and Dziedzic 1966; Møhl and Andersen 1973; Schevill et al. 1969; Verboom and Kastelein 1995) that were absent from the newer literature (Au et al. 1999; Teilmann et al. 2002; Villadsgaard et al. 2007). There could be many reasons for this and it would be a mistake to disregard the older literature. Equipment would have changed over the years, but also, animals were recorded in different settings (i.e., in different behavioral states, or in captivity versus the wild), which could have affected the recorded sound features (e.g., animals might have produced different sounds in the different studies or the sound propagation environment might have filtered some of the spectral features). Even the focus of the literature might have changed, with newer literature focusing more on high-frequency clicks (in particular, porpoise NBHF clicks). As Hansen et al. (2008) wrote, the low-frequency components in some porpoise sounds are genuine—albeit comparatively weak. Whether they are a by-product of high-frequency sound production and whether they play a functional role is unclear. Some lower-frequency components described in the early literature were not recorded by Hansen et al. (2008), "but we can reproduce them as artifacts by clipping analog tape recorders with the HF component of porpoise-like clicks."

4.4.8.2 Wild Versus Captive Recordings

Studies done in captivity offer a high level of control. Animals trained can be asked to perform tasks designed specifically to investigate sound production, reception, and utility. However, due to the reverberant nature of tanks and the typically short ranges involved, animals might use lower source levels and modify other spectro-temporal features of their sounds (e.g., Kastelein et al. 1999; Villadsgaard et al. 2007). Due to the artificial ecological settings in a captive environment, animals might display different behaviors, including different acoustic behaviors. The sound repertoire and features recorded in captivity may also be different from the natural sound repertoire in the wild (e.g., Miksis et al. 2002).

4.4.8.3 Effects of the Environment on Spectro-temporal Features

As sound propagates through water, its acoustic features may change. Energy at high frequencies diminishes faster with range than energy at low frequencies does, due to scattering and absorption. Energy at low frequencies does not travel well through shallow water. If animals are recorded close to the sea surface, reflections off the sea surface may overlap with direct-path recordings and this might explain some of the multi-pulse nature of small-cetacean clicks (e.g., porpoises and *Cephalorhynchus*) as reported in the literature. Reverberation can extend the signal duration and might explain the often long durations reported in the literature. As energy at different frequencies propagates differently, recordings of broadband signals such as pulses and clicks may exhibit patterns of frequency peaks and notches. This might explain some of the variability in the peak and notch frequencies reported for the same species in different locations. It is paramount to understand the acoustic environment in which recordings are made.

4.4.8.4 Labeling Sounds

Sound propagation effects on the spectro-temporal features of sounds are mostly ignored in the literature and hence grouping sounds based

Table 4.2 Features measured off clicks (C), burst-pulse sounds (BP), single pulses (P), and whistles (T), summarized from all published data

Species	Sound type	fp [kHz]	3 dB BW [kHz]	Fmin [Hz]	Fmax [Hz]	Dur [ms]	Durmin [s]	Durmax [s]	SL_SPLpk-pk [dB re 1 µPa m]	SL_SPLrms [dB re 1 µPa m]	SL_SEL [dB re 1 µPa² s m²]
Berardius arnuxii	C	15.8	3	12,000	20,300						
Berardius arnuxii	BP	5.8		1000	11,000	650	0.63	0.91			
Berardius arnuxii	T	4.9		2800	7400	650	0.37	0.93			
Berardius bairdii	C	21.5	5	9000	43,000	0.464	0.00012	0.00095			
Berardius bairdii	BP	19.5		5000	35,000	1023	0.094	0.62			
Berardius bairdii	T	7		4000	12,000	3250	2	4			
Cephalorhynchus commersonii	C	122	18	29,000	181,000	0.096	0.00002	0.00035	170	158	125
Cephalorhynchus commersonii	BP	55.7	15	100	139,500	524	0.1	2			
Cephalorhynchus commersonii	T	10		4000	16,000	26	0.013	0.0397			
Cephalorhynchus eutropia	C	126	18	109,000	143,000	0.083	0.000053	0.000113	165		120
Cephalorhynchus heavisidii	C	130	16	12,000	141,000	0.06	0.000017	0.000129	173	161	
Cephalorhynchus heavisidii	BP	57.5	16	100	133,000	1200	0.4	2			
Cephalorhynchus hectori	C	129	20	117,000	135,000	0.06	0.000041	0.000065	177	166	121
Delphinapterus leucas	C	80	25	40,000	120,000	0.075			205		149
Delphinapterus leucas	BP	6.7	3	100	20,000	1274	0.01	3.9		144	
Delphinapterus leucas	T	5.6		200	20,700	1067	0.01	3.9			
Delphinus capensis	T	13.4		5800	21,100	675	0.3	1.1			
Delphinus delphis	C	75		20,000	190,000	0.125	0.00005	0.00025		147	
Delphinus delphis	BP	39		2000	180,000	1313	0.5	0.75			
Delphinus delphis	T	12.5		3600	27,900	827	0.01	3			
Delphinus delphis	P	1.6		100	3000	200	0.1	0.3			
Feresa attenuata	C	70		20,000	120,000	0.03	0.00002	0.00004	210		148
Feresa attenuata	BP	13.5		3000	24,000	2250	1	3.5			

4 Odontocete Sounds

Species										
Feresa attenuata	T	11.5		5000	18,000	550	0.3	0.8		
Globicephala macrorhynchus	C	30.5	10	7800	55,000	0.301	0.000034	0.001225	181	170	127
Globicephala macrorhynchus	BP	8.7		700	30,000	969	0.1	6.8		131	
Globicephala macrorhynchus	T	8.4		440	22,400	532	0.1	1.2		158	
Globicephala melas	BP	11.7		700	22,000	495	0.1	1.1			
Globicephala melas	T	9.6		1000	100,000	939	0.02	3			
Grampus griseus	C	44.9	35	2000	110,000	0.058	0.000026	0.000118	198	188	157
Grampus griseus	BP	11.5		1000	22,000	7100	2	13			
Grampus griseus	T	9.5	5	100	22,000	1049	0.1	12.5			
Hyperoodon ampullatus	C	31.7	8	4400	90,000	0.321	0.0002	0.0007	203	186	169
Hyperoodon ampullatus	T	9.5		3000	16,000	483	0.115	0.85			
Indopacetus pacificus	C	22.4	9	12,700	37,700	0.23	0.000086	0.000511			
Indopacetus pacificus	BP	11.7		10,100	18,600	500					
Inia geoffrensis	C	71.8	33	23,100	170,000	0.037	0.000011	0.0001	189	173	125
Inia geoffrensis	BP	2.6		100	10,100	608	0.01	1.6			
Inia geoffrensis	T	9.8		1800	48,830	398	0.002	2.2			
Inia geoffrensis	P	1.3		90	2110	260	0.03	0.6			
Kogia breviceps	C	128	8	60,000	200,000	0.28	0.0001	0.00014	175		
Kogia breviceps	T	1.4				400					
Kogia sima	C	123	11	91,000	140,000	0.136	0.00001	0.00025	172	160	120
Lagenodelphis hosei	T	12.2		7600	16,900	527	0.06	2			
Lagenorhynchus acutus	T	12.9		5900	20,000	700	0.23	0.9			
Lagenorhynchus albirostris	C	82.2	24	41,000	122,000	0.023	0.000012	0.000045	190		
Lagenorhynchus albirostris	BP	38	8	1500	48,000	330	0.16	0.5			

(continued)

Table 4.2 (continued)

Species	Sound type	fp [kHz]	3 dB BW [kHz]	Fmin [Hz]	Fmax [Hz]	Dur [ms]	Durmin [s]	Durmax [s]	SL_SPLpk-pk [dB re 1 µPa m]	SL_SPLrms [dB re 1 µPa m]	SL_SEL [dB re 1 µPa² s m²]
Lagenorhynchus albirostris	T	13.7		3000	35,000	605	0.06	1.05		144	
Lagenorhynchus australis	C	126	15	120,000	133,000	0.092	0.000065	0.000153	185	173	133
Lagenorhynchus australis	BP	2.7		300	5000	850	0.6	1.1			
Lagenorhynchus cruciger	C	126	8	122,000	131,000	0.115	0.000079	0.000176	197	186	146
Lagenorhynchus obliquidens	C	52.3	23	22,000	80,000	0.059	0.000025	0.0001	153	170	
Lagenorhynchus obliquidens	T	6.5		1000	12,000	700	0.2	1.2			
Lagenorhynchus obscurus	C	74	67	30,000	130,000	0.05	0.00004	0.00007	185		
Lagenorhynchus obscurus	BP					980	0.01	2.1			
Lagenorhynchus obscurus	T	10.9		2000	19,200	778	0.29	1.4			
Lipotes vexillifer	C	75		50,000	120,000	0.018	0.00001	0.000028			
Lipotes vexillifer	T	5.6		4400	6700	1000	0.7	1.3			
Lissodelphis borealis	C	31	9	23,000	41,000	0.31	0.0002	0.0004		153	
Lissodelphis borealis	BP	18	10	6300	37,000	740	0.18	1.3			
Lissodelphis borealis	T	2.4		1800	3000						
Mesoplodon bidens	C	49.6	8	27,400	76,500	0.318	0.0002	0.000667			
Mesoplodon bidens	BP	51		35,100	66,800	209	0.132	0.285			
Mesoplodon carlhubbsi	BP	20.2		300	40,000						
Mesoplodon carlhubbsi	T	6		4200	7700	261	0.156	0.45			
Mesoplodon densirostris	C	39.7	13	26,000	63,000	0.268	0.00005	0.00095	211	200	

Species										
Mesoplodon densirostris	BP	33.1		2000	80,000	610	0.09	0.36		132
Mesoplodon densirostris	T	11.2		11,400	13,400	600	0.17	0.23		149
Mesoplodon europaeus	C	41.9	19	30,000	55,900	0.325	0.00026	0.000765		
Mesoplodon hotaula	C	47.3	20	28,900	69,100	0.475	0.0003	0.00072		
Mesoplodon mirus	C	41.7	20	35,000	55,000	0.255	0.000183	0.00039		
Mesoplodon stejnegeri	C	50.4	21	45,700	73,800	0.42	0.000245	0.000746		
Monodon monoceros	C	32.1	43	500	112,000	0.029	0.000014	0.000049	217	
Monodon monoceros	BP	15.9		1650	36,550	866	0.05	2.3		
Monodon monoceros	T	5.1		300	12,500	946	0.1	2.2		
Neophocaena asiaeorientalis	C	130	18	87,000	149,000	0.064	0.00003	0.00012	174	163
Neophocaena phocaenoides	C	132	16	113,000	142,000	0.084	0.00003	0.000143	168	
Neophocaena phocaenoides	BP	2.5		2000	3000	1400	1.2	1.6		
Orcaella brevirostris	C	79.8	72	30,000	130,000	0.019	0.00001	0.00003	194	182
Orcaella brevirostris	BP	11.1		100	22,000	308	0.05	0.9		
Orcaella brevirostris	T	7.2		900	23,000	438	0.02	1.6		
Orcaella brevirostris	P	2.9		500	5000	290	0.05	0.75		
Orcaella heinsohni	C	37	69	6000	123,000	0.012	0.000011	0.000013	200	190
Orcaella heinsohni	BP	13.3		4000	22,000	1537	0.07	6		
Orcaella heinsohni	T	4.7		2000	9000	240	0.1	0.3		
Orcinus orca	C	28.8	38	10,000	108,000	0.098	0.00004	0.00014	195	
Orcinus orca	BP	4.9	0	250	16,000	971	0.05	4.1	166	154

(continued)

Table 4.2 (continued)

Species	Sound type	fp [kHz]	3 dB BW [kHz]	Fmin [Hz]	Fmax [Hz]	Dur [ms]	Durmin [s]	Durmax [s]	SL_SPLpk-pk [dB re 1 µPa m]	SL_SPLrms [dB re 1 µPa m]	SL_SEL [dB re 1 µPa² s m²]
Orcinus orca	T	16.1	7	90	72,000	1022	0.01	11.3	189	141	
Peponocephala electra	C	26.2	5	13,000	41,000	0.324	0.000019	0.0007			
Peponocephala electra	BP	22.5		5000	40,000	100		0.2		165	
Peponocephala electra	T	9.8	3	1800	24,500	533	0.1	0.9		155	
Phocoena phocoena	C	145	14	110,000	215,000	0.096	0.000029	0.000206	173	152	122
Phocoena phocoena	BP	4.1		200	8000	3000	1	18			
Phocoena sinus	C	133	17	128,000	139,000	0.136	0.000095	0.000177			
Phocoena spinipinnis	C	138	8	129,000	186,000	0.144	0.00008	0.00021			
Phocoenoides dalli	C	129	8	90,000	149,000	0.204	0.000015	0.000804		162	
Physeter macrocephalus	C	6.9	9	200	26,000	14.086	0.0005	0.1	191	199	174
Physeter macrocephalus	BP	0.9		500	3600	575	0.05	1.4		146	
Platanista gangetica gangetica	C	54.5	44	44,700	73,300	0.067	0.00002	0.00015	183	157	127
Platanista gangetica minor	C	50	30			0.05					
Pontoporia blainvillei	C	80.5	9	11,000	149,000	2.737	0.000018	0.017			
Pontoporia blainvillei	BP	84.5		73,000	96,000	75	0.02	0.14			
Pontoporia blainvillei	T	45		1000	89,000	130	0.01	0.25			
Pseudorca crassidens	C	50.2	22	7800	111,000	0.089	0.000018	0.00053	203	203	157
Pseudorca crassidens	BP	12.5		3000	22,000	500					
Pseudorca crassidens	T	6.7		4000	8700	505	0.2	0.78			

Species	Type										
Sotalia fluviatilis	C	90.3	22	47,000	137,000	0.04	0.000011	0.0001			
Sotalia fluviatilis	T	13.6		6100	22,400	449	0.12	0.95			
Sotalia guianensis	C	25.2		9300	41,000						
Sotalia guianensis	BP	24.7	15	900	100,000	247	0.006	0.925			
Sotalia guianensis	T	14.8	24	400	51,900	292	0.009	2.3			
Sousa chinensis	C	109	48	30,000	200,000	0.021	0.000015	0.000028	184	176	132
Sousa chinensis	BP	14.6		4100	25,000					138	
Sousa chinensis	T	6.9		1600	17,200	466	0.02	1			
Sousa plumbea	C	52.7	81	1000	119,000	0.069	0.000011	0.00015			
Sousa plumbea	BP	19.3		3000	42,000	200	0.1	1			
Sousa plumbea	T	9.9		3000	20,000	283	0.01	1			
Sousa plumbea	P	1		400	2500	100					
Sousa sahulensis	C	47.2	59	1000	135,000	0.015	0.000013	0.000017	199	189	141
Sousa sahulensis	BP	11.9		800	22,000	1330	0.07	3.5			
Sousa sahulensis	T	8.7		900	22,000	343	0.05	1.2			
Sousa sahulensis	P	1.2		600	1800	90	0.06	0.4			
Sousa teuszii	T	6.9		2500	11,300	760	0.29	1.23			
Stenella attenuata	C	79.5	71	38,000	111,000	0.031	0.000014	0.000058	201		141
Stenella attenuata	T	14.1		6000	24,000	733	0.2	1.3		142	
Stenella coeruleoalba	T	12.9		5800	20,800	869	0.3	2.2			
Stenella frontalis	C	59	58	30,000	109,000	0.037	0.00001	0.00007	207	200	151
Stenella frontalis	BP	23.8	18	100	15,000	390	0.08	1			
Stenella frontalis	T	11.2		2400	21,300	971	0.07	8		139	
Stenella longirostris	C	59.2	41	20,000	130,000	0.191	0.000019	0.0006	206		148
Stenella longirostris	BP	17.9	21	200	7500	430	0.05	0.9			
Stenella longirostris	T	13		6100	23,000	597	0.036	1.22			
Steno bredanensis	C	43.5	9	15,000	112,000	0.23	0.000141	0.000345			
Steno bredanensis	T	7.6		4100	12,600	553	0.1	1.1			
Tasmacetus shepherdi	C	19.3	25	12,200	25,900	0.27	0.000074	0.000436			
Tasmacetus shepherdi	BP	15.5	26	8900	23,600	1220	0.27	2.67			
Tursiops aduncus	C	96	61	45,000	141,000	0.015	0.000011	0.000024	205	195	146

(continued)

Table 4.2 (continued)

Species	Sound type	fp [kHz]	3 dB BW [kHz]	Fmin [Hz]	Fmax [Hz]	Dur [ms]	Durmin [s]	Durmax [s]	SL_SPLpk-pk [dB re 1 µPa m]	SL_SPLrms [dB re 1 µPa m]	SL_SEL [dB re 1 µPa² s m²]
Tursiops aduncus	T	9.7		1000	22,000	1039	0.05	5.9			
Tursiops aduncus	P	1.7		300	3000	70	0.04	0.1			
Tursiops truncatus	C	76.4	35	7000	205,000	0.091	0.00001	0.0007	207		
Tursiops truncatus	BP	24.8		100	150,000	626	0.04	3.61	195		
Tursiops truncatus	T	10.6	10	30	43,000	991	0.01	4		139	
Tursiops truncatus	P	2.7		50	6000	57.5	0.03	0.09			
Ziphius cavirostris	C	32.6	12	13,000	49,200	0.51	0.000306	0.0016	214		164

fp: mean of all reported peak frequencies. 3 dB BW: mean of all reported 3 dB bandwidths. Fmin: minimum of all reported lower frequencies of the dominant band for C, BP, and P sounds, and of the fundamental for T sounds. Fmax: maximum of all reported upper frequencies of the dominant band for C, BP, and P sounds, and of the fundamental for T sounds. Dur: mean duration. Durmin: minimum of the reported lower edges of the dominant duration (where the dominant duration was computed as mean(Dur)-std(Dur) for each sound type and study). Durmax: maximum of the reported upper edges of the dominant duration, which was computed as mean(Dur) + std(Dur). SL: source level in terms of SPL$_{pk-pk}$, SPL$_{rms}$, and SEL. Published data by species and sound type, as well as all references, are collated in Suppl. Table 1 on the features in affected recordings can be problematic, in particular if recordings from different sites are compared. Accounting for sound propagation effects is often not possible, because the location of the animal, its orientation, and the hydro- and geoacoustic parameters of the environment are not known.

The use of onomatopoeic names (e.g., grunt, moan, scream, squawk, squeal, or thunk) for sound types is common in the literature. Usage, however, is not consistent, and sound names imply different sound types to different readers (in particular with different language backgrounds).

Labeling sounds by their generation mechanism or "right at the source" would be independent of sound propagation and recording artifacts and independent of analyst perception and native language, but we often do not understand how differently the different sounds are produced.

4.5 Taxonomic Overview of Sounds

Measurements of the most common frequency, duration, and SL features of odontocete sounds (clicks, whistles, and burst-pulse sounds) have been compiled from the literature and are available in Suppl. Table 1. A summary of mean features by species and call type is given in Table 4.2.

4.5.1 Clicks and Burst-Pulse Sounds

The full table (Suppl. Table 1) gives:

- f_p: peak frequency (i.e., frequency of maximum power)
- f_c: center frequency, which divides the power spectrum into two halves of equal power
- $bw3$: 3 dB bandwidth (i.e., 3 dB below maximum power at f_p)
- $bw10$: 10 dB bandwidth (i.e., 10 dB below maximum power at f_p)
- $bwrms$: root-mean-square bandwidth (i.e., standard deviation about f_c)
- *Fmin* is meant to reflect the lower edge of the dominant frequency band and was, in the following order of preference:
  1. Taken as the lower reported range of f_p
  2. Computed as the average f_p minus half of the average bw (in the order of $bw10$, then $bw3$—depending on availability)
  3. Computed as mean(f_p) – std(f_p)
  4. Taken as the lower quartile or lowest percentile (sometimes 10th percentile)
  5. Taken as the lower edge of the reported frequency range
- *Fmax* is meant to reflect the upper edge of the dominant frequency band and was, in the following order of preference:
  1. Taken as the upper reported range of f_p
  2. Computed as the average f_p plus half of the average bw (in the order of $bw10$, then $bw3$—depending on availability)
  3. Computed as mean(f_p) + std(f_p)
  4. Taken as the upper quartile or highest percentile (sometimes 90th percentile)
  5. Taken as the upper edge of the reported frequency range
- Q-value: f_c / $bwrms$. This is sometimes given and was occasionally used to compute f_c.
- *Durmin* is meant to reflect the lower edge of the dominant duration and was computed as the mean – std. Extremely often, this ended up with negative durations as a result of durations not being Gaussian distributed. In these cases, *Durmin* was taken as the lower range of durations reported.
- *Durmax* is meant to reflect the upper edge of the dominant duration and was computed as the mean + std.
- *Dur10*: 10 dB duration (i.e., time between the −10 dB points of the signal envelope)
- *ICI*: inter-click interval
- *PRR*: pulse-repetition rate
- Source level measurements are means (not maximum levels reported):
  1. *SLpp*: SL in terms of SPLpk-pk
  2. *SLrms*: SL in terms of SPLrms
  3. *SEL*: SL in terms of SEL

4.5.2 Whistles

Reports of whistle features commonly list minimum, maximum, start, and end frequencies of the fundamental, giving averages, ranges, and statistics for each (e.g., mean ± std, median, quartiles, and other percentiles). We computed the dominant frequency range as the band from (1) the mean($Fmin$) − std($Fmin$) to mean($Fmax$) + std ($Fmax$), or (2) the lowest reported percentile to the highest reported percentile—in this order, depending on the measurements that were reported.

Often, reports give all measures as mean ± std, even though the underlying distributions are not Gaussian. This is always the case with counts of extrema, inflection points, and steps, as these are Poisson distributed. This is also too often the case with durations, when duration is log-norm distributed. This can also occur with $Fmin$. We tabulated the actual range of frequency and time measures in cases where mean − std resulted in negative values.

4.6 Summary

This chapter has provided numerous examples for odontocetes producing and listening to sound in support of their major life functions, including socializing, mating, rearing of young, navigating, and foraging. The majority of odontocete species appear to be highly social animals and acoustic communication plays an important role in all social encounters and in the coordination of group behaviors. Many species, such as bottlenose dolphins, beluga whales, and sperm whales seem to produce sound prolifically. Out of the ~80 species of odontocete, only ~10 of the most cryptic have not had their vocalizations described. In addition to emitting sounds for communication, odontocetes also emit biosonar for echolocation. In this chapter, we have explained how echolocation works and how well the dolphin biosonar system performs. We have discussed biosonar design features and how they differ between species. We then covered odontocete communication sounds, their acoustic features and functions. We have provided a comparative taxonomic overview of odontocete sounds in tables and figures.

4.7 Abbreviations

Quantity	Abbreviation	Symbol	Unit
Frequency		f	Hz
Sampling frequency		f_s	Hz
Peak frequency (frequency of maximum power)		f_p	Hz
Center frequency (splits a spectrum into two halves of equal power)		f_c	Hz
Bandwidth	BW	bw	Hz
Duration	Dur	τ	s
Inter-click interval	ICI		s
Inter-pulse interval	IPI		s
Pulse-repetition rate	PRR		1/s
Duty cycle	DC		%
Time variable		t	s
Sound pressure		$p(t)$	Pa
Root-mean square sound pressure level	SPL$_{rms}$	L_p	dB re 1 μPa
Peak-to-peak sound pressure level	SPL$_{pk-pk}$	$L_{p,pk-pk}$	dB re 1 μPa
Sound exposure level	SEL	$L_{E,p}$	dB re 1 μPa^{2}s
Source level (sound pressure)	SL	L_S	dB re 1 μPa m
Propagation loss	PL	N_{PL}	dB re 1 m
Received level	RL		
Narrowband high-frequency	NBHF		

Acknowledgments We thank Ann Bowles and Anita Murray for helping with the literature searches. Anita Murray further kindly compiled the odontocete body sizes from the literature.

Data Availability We undertook a literature review and compiled measurements of features of odontocete sounds from 387 articles in a spreadsheet: Supplementary Table 1.

This spreadsheet may be requested from the first author or accessed here: https://cmst.curtin.edu.au/research/marine-mammal-bioacoustics/

References

Aguilar de Soto N, Johnson MP, Madsen PT, Diaz F, Dominguez I, Brito AE, Tyack P (2008) Cheetahs of the deep sea: deep foraging sprints in short-finned pilot whales off Tenerife (Canary Islands). J Anim Ecol 77: 936–947. https://doi.org/10.1111/j.1365-2656.2008.01393.x

Aguilar de Soto N, Madsen P, Tyack P, Arranz P, Marrero J, Fais A, Revelli E, Johnson M (2012) No shallow talk: cryptic strategy in the vocal communication of Blainville's beaked whales. Mar Mamm Sci 28(2):E75–E92. https://doi.org/10.1111/j.1748-7692.2011.00495.x

Akamatsu T, Wang D, Nakamura K, Wang K (1998) Echolocation range of captive and free-ranging baiji (*Lipotes vexillifer*), finless porpoise (*Neophocaena phocaenoides*), and bottlenose dolphin (*Tursiops truncatus*). J Acoust Soc Am 104(4):2511–2516. https://doi.org/10.1121/1.423757

Altes RA, Dankiewicz LA, Moore PW, Helweg DA (2003) Multiecho processing by an echolocating dolphin. J Acoust Soc Am 114(2):1155–1166. https://doi.org/10.1121/1.1590969

Alves A, Antunes R, Tyack PL, Miller PJO, Lam F-PA, Kvadsheim PH (2014) Vocal matching of naval sonar signals by long-finned pilot whales (*Globicephala melas*). Mar Mamm Sci 30(3):1248–1257. https://doi.org/10.1111/mms.12099

Ames AE, Zapetis ME, Witlicki KL, Wielandt SJ, Cameron DM, Walker RT, Kuczaj SA (2017) Thunks: evidence for varied harmonic structure in an Atlantic bottlenose dolphin (*Tursiops truncatus*) sound. Int J Comp Psychol 30:4489132c. https://doi.org/10.46867/ijcp.2017.30.00.08

Ames AE, Beedholm K, Madsen PT (2020) Lateralized sound production in the beluga whale (*Delphinapterus leucas*). J Exp Biol 223 (17):jeb226316. https://doi.org/10.1242/jeb.226316

Ames AE, Blackwell SB, Tervo OM, Heide-Jørgensen MP (2021) Evidence of stereotyped contact call use in narwhal (*Monodon monoceros*) mother-calf communication. PLoS One 16(8):e0254393. https://doi.org/10.1371/journal.pone.0254393

Amundin M (1991) Sound production in odontocetes with emphasis on the harbour porpoise *Phocoena phocoena*. PhD thesis, University of Stockholm, Stockholm

Amundin M, Andersen SH (1983) Bony nares air pressure and nasal plug muscle activity during click production in the harbour porpoise, *Phocoena phocoena*, and the bottlenosed dolphin, *Tursiops truncatus*. J Exp Biol 105(1):275–282. https://doi.org/10.1242/jeb.105.1.275

Ansmann IC, Goold JC, Evans PGH, Simmonds M, Keith SG (2007) Variation in the whistle characteristics of short-beaked common dolphins, *Delphinus delphis*, at two locations around the British Isles. J Mar Biolog Assoc 87(1):19–26. https://doi.org/10.1017/S0025315407054963

Antunes R, Schulz T, Gero S, Whitehead H, Gordon J, Rendell L (2011) Individually distinctive acoustic features in sperm whale codas. Anim Behav 81(4): 723–730. https://doi.org/10.1016/j.anbehav.2010.12.019

Aroyan JL (2001) Three-dimensional modelling of hearing in *Delphinus delphis*. J Acoust Soc Am 110(6):3305–3318. https://doi.org/10.1121/1.1401757

Aroyan JL, Cranford TW, Kent J, Norris KS (1992) Computer modelling of acoustic beam formation in *Delphinus delphis*. J Acoust Soc Am 92(5):2539–2545. https://doi.org/10.1121/1.404424

Arranz P, DeRuiter SL, Stimpert AK, Neves S, Friedlaender AS, Goldbogen JA, Visser F, Calambokidis J, Southall BL, Tyack PL (2016) Discrimination of fast click series produced by tagged Risso's dolphins (*Grampus griseus*) for echolocation or communication. J Exp Biol 219:2898–2907. https://doi.org/10.1242/jeb.144295

Atem ACG, Rasmussen MH, Wahlberg M, Petersen HC, Miller LA (2009) Changes in click source levels with distance to targets: studies of free-ranging white-beaked dolphins *Lagenorhynchus albirostris* and captive harbour porpoises *Phocoena phocoena*. Bioacoustics 19(1–2):49–65. https://doi.org/10.1080/09524622.2009.9753614

Au WWL (1980) Echolocation signals of the Atlantic bottlenose dolphin (*Tursiops truncatus*) in open waters. In: Busnel RG, Fish JF (eds) Animal sonar systems, vol 17. vol 4. Plenum Press, pp 251–282. https://doi.org/10.1007/978-1-4684-7254-7_10

Au WWL (1993) The sonar of dolphins. Springer, New York https://doi.org/10.1007/978-1-4612-4356-4

Au WWL (2000) Echolocation in dolphins. In: Au WWL, Popper AN, Fay RR (eds) Hearing by whales and dolphins. Springer, New York, pp 364–408. https://doi.org/10.1007/978-1-4612-1150-1_9

Au WWL (2004) Echolocation signals of wild dolphins. Acoust Phys 50(4):454–462. https://doi.org/10.1134/1.1776224

Au WWL (2014) Dolphin biosonar target detection in noise: wrap up of a past experiment. J Acoust Soc Am 136(1):9–12. https://doi.org/10.1121/1.4883476

Au WWL, Pawloski DA (1992) Cylinder wall thickness difference discrimination by an echolocating Atlantic bottlenose dolphin. J Comp Physiol 170(1):41–47. https://doi.org/10.1007/BF00190399

Au WW, Penner RH (1981) Target detection in noise by echolocating Atlantic bottlenose dolphins. J Acoust

Soc Am 70(3):687–693. https://doi.org/10.1121/1.386931

Au WWL, Snyder KJ (1980) Long-range target detection in open waters by an echolocating Atlantic bottlenose dolphin (*Tursiops truncatus*). J Acoust Soc Am 68(4):1077–1084. https://doi.org/10.1121/1.384993

Au WWL, Würsig B (2004) Echolocation signals of dusky dolphins (*Lagenorhynchus obscurus*) in Kaikoura, New Zealand. J Acoust Soc Am 115(5):2307–2313. https://doi.org/10.1121/1.1690082

Au WWL, Floyd RW, Penner RH, Murchison AE (1974) Measurement of echolocation signals of the Atlantic bottlenose dolphin, *Tursiops truncatus* Montagu, in open waters. J Acoust Soc Am 56(4):1280–1290. https://doi.org/10.1121/1.1903419

Au WWL, Schusterman RJ, Kersting DA (1980) Sphere-cylinder discrimination via echolocation by *Tursiops truncatus*. In: Busnel R-G, Fish JF (eds) Animal sonar systems. Vol series A: life sciences. Plenum Press, New York, pp 859–862. https://doi.org/10.1007/978-1-4684-7254-7_41

Au WWL, Carder DA, Penner RH, Scronce BL (1985) Demonstration of adaptation in beluga whale echolocation signals. J Acoust Soc Am 77(2):726–730. https://doi.org/10.1121/1.392341

Au WWL, Moore PWB, Pawloski D (1986) Echolocation transmitting beam of the Atlantic bottlenose dolphin. J Acoust Soc Am 80(2):688–691. https://doi.org/10.1121/1.394012

Au WWL, Penner RH, Turl CW (1987) Propagation of beluga echolocation signals. J Acoust Soc Am 82(3):807–813. https://doi.org/10.1121/1.395278

Au WWL, Moore PB, Pawloski DA (1988) Detection of complex echoes in noise by an echolocating dolphin. J Acoust Soc Am 83(2):662–668. https://doi.org/10.1121/1.396161

Au WWL, Pawloski JL, Nachtigall PE, Blonz M, Gisiner RC (1995) Echolocation signals and transmission beam pattern of a false killer whale (*Pseudorca crassidens*). J Acoust Soc Am 98(1):51–59. https://doi.org/10.1121/1.413643

Au WWL, Kastelein RA, Rippe T, Schooneman NM (1999) Transmission beam pattern and echolocation signals of a harbor porpoise (*Phocoena phocoena*). J Acoust Soc Am 106(6):3699–3705. https://doi.org/10.1121/1.428221

Au WWL, Ford JKB, Horne JK, Allman KAN (2004) Echolocation signals of free-ranging killer whales (*Orcinus orca*) and modelling of foraging for Chinook salmon (*Oncorhynchus tshawytscha*). J Acoust Soc Am 115(2):901–909. https://doi.org/10.1121/1.1642628

Au WWL, Kastelein RA, Benoit-Bird KJ, Cranford TW, McKenna MF (2006) Acoustic radiation from the head of echolocating harbour porpoises (*Phocoena phocoena*). J Exp Biol 209(14):2726–2733. https://doi.org/10.1242/Jeb.02306

Au WWL, Benoit-Bird KJ, Kastelein RA (2007) Modelling the detection range of fish by echolocating bottlenose dolphins and harbour porpoises. J Acoust Soc Am 121(6):3954–3962. https://doi.org/10.1121/1.2734487

Au WWL, Lammers MO, Yin S (2010) Acoustics of dusky dolphins (*Lagenorhynchus obscurus*). In: Würsig B, Würsig M (eds) The dusky dolphin: master acrobats off different shores. Elsevier, Amsterdam, pp 75–97. https://doi.org/10.1016/B978-0-12-373723-6.00004-7

Au W, Branstetter B, Moore P, Finneran J (2012) Dolphin biosonar signals measured at extreme off-axis angles: insights to sound propagation in the head. J Acoust Soc Am 132(2):1199–1206. https://doi.org/10.1121/1.4730901

Azevedo AF, Sluys MV (2005) Whistles of tucuxi dolphins (*Sotalia fluviatilis*) in Brazil: comparisons among populations. J Acoust Soc Am 117(3):1456–1464. https://doi.org/10.1121/1.1859232

Azzolin M, Papale E, Lammers MO, Gannier A, Giacoma C (2013) Geographic variation of whistles of the striped dolphin (*Stenella coeruleoalba*) within the Mediterranean Sea. J Acoust Soc Am 134(1):694–705. https://doi.org/10.1121/1.4808329

Azzolin M, Gannier A, Papale E, Buscaino G, Mussi B, Ardizzone G, Giacoma C, Pace DS (2021) Whistle variability of the Mediterranean short beak common dolphin. Aquat Conserv 31(S1):36–50. https://doi.org/10.1002/aqc.3168

Barluet de Beauchesne L, Massenet M, Oudejans MG, Kok ACM, Visser F, Curé C (2022) Friend or foe: Risso's dolphins eavesdrop on conspecific sounds to induce or avoid intra-specific interaction. Anim Cogn 25:287–296. https://doi.org/10.1007/s10071-021-01535-y

Baumann-Pickering S, Wiggins SM, Hildebrand JA, Roch MA, Schnitzler HU (2010) Discriminating features of echolocation clicks of melon-headed whales (*Peponocephala electra*), bottlenose dolphins (*Tursiops truncatus*), and Gray's spinner dolphins (*Stenella longirostris longirostris*). J Acoust Soc Am 128(4):2212–2224. https://doi.org/10.1121/1.3479549

Baumann-Pickering S, McDonald MA, Simonis AE, Solsona Berga A, Merkens KPB, Oleson EM, Roch MA, Wiggins SM, Rankin S, Yack TM, Hildebrand JA (2013a) Species-specific beaked whale echolocation signals. J Acoust Soc Am 134(3):2293–2301. https://doi.org/10.1121/1.4817832

Baumann-Pickering S, Yack TM, Barlow J, Wiggins SM, Hildebrand JA (2013b) Baird's beaked whale echolocation signals. J Acoust Soc Am 133(6):4321–4331. https://doi.org/10.1121/1.4804316

Bazua-Duran C, Au WWL (2004) Geographic variations in the whistles of spinner dolphins (*Stenella longirostris*). J Acoust Soc Am 116(6):3757–3769. https://doi.org/10.1121/1.4779792

Bebus SE, Herzing DL (2015) Mother-offspring signature whistle similarity and patterns of association in Atlantic spotted dolphins (*Stenella frontalis*). Anim Behav

Cogn 2(1):71–87. https://doi.org/10.12966/abc.02.06.2015

Beedholm K (2008) Harbor porpoise clicks do not have conditionally minimum time bandwidth product. J Acoust Soc Am 124(2):EL15–EL20. https://doi.org/10.1121/1.2947623

Beedholm K, Miller LA (2007) Automatic gain control in harbour porpoises (*Phocoena phocoena*)? Central versus peripheral mechanisms Aquat Mamm 33(1):69–75. https://doi.org/10.1578/AM.33.1.2007.69

Beedholm K, Møhl B (2006) Directionality of sperm whale sonar clicks and its relation to piston radiation theory. J Acoust Soc Am 119 (2):El14–El19. https://doi.org/10.1121/1.2161799

Beedholm K, Ladegaard M, Madsen PT, Tyack PL (2023) Latencies of click-evoked auditory responses in a harbor porpoise exceed the time interval between subsequent echolocation clicks. J Acoust Soc Am 153(2):952–960. https://doi.org/10.1121/10.0017163

Belikov RA, Bel'kovich VM (2008) Communicative pulsed signals of beluga whales in the reproductive gathering off Solovetskii Island in the White Sea. Acoust Phys 54(1):115–123. https://doi.org/10.1134/S1063771008010168

Benoit-Bird KJ, Au WWL (2009) Phonation behavior of cooperatively foraging spinner dolphins. J Acoust Soc Am 125(1):539–546. https://doi.org/10.1121/1.2967477

Benoit-Bird KJ, Au WWL, Kastelein R (2006) Testing the odontocete acoustic prey debilitation hypothesis: no stunning results. J Acoust Soc Am 120(2):1118–1123. https://doi.org/10.1121/1.2211508

Bittencourt L, Lima IMS, Andrade LG, Carvalho RR, Bisi TL, Lailson-Brito J Jr, Azevedo AF (2017) Underwater noise in an impacted environment can affect Guiana dolphin communication. Mar Pollut Bull 114(2):1130–1134. https://doi.org/10.1016/j.marpolbul.2016.10.037

Boisseau O, Lacey C, Lewis T, Moscrop A, Danbolt M, McLanaghan R (2010) Encounter rates of cetaceans in the Mediterranean Sea and contiguous Atlantic area. J Mar Biolog Assoc 90(8):1589–1599. https://doi.org/10.1017/S0025315410000342

Branstetter BK, Moore PW, Finneran JJ, Tormey MN, Aihara H (2012) Directional properties of bottlenose dolphin (*Tursiops truncatus*) clicks, burst-pulse, and whistle sounds. J Acoust Soc Am 131(2):1613–1621. https://doi.org/10.1121/1.3676694

Bronshtein IN, Semendyayev KA, Musiol G, Mühlig H (2015) Handbook of mathematics, 6th edn. Springer, Berlin. https://doi.org/10.1007/978-3-662-46221-8

Brown JC (2008) Mathematics of pulsed vocalizations with application to killer whale biphonation. J Acoust Soc Am 123(5):2875–2883. https://doi.org/10.1121/1.2890745

Busnel R-G, Dziedzic A (1966) Acoustic signals of the pilot whale *Globicephala melaena* and of the porpoises *Delphinus delphis* and *Phocoena phocoena*. In: Norris KS (ed) Whales, dolphins, and porpoises. University of California Press, Berkeley, pp 607–646. https://doi.org/10.1525/9780520321373-036

Caldwell MC, Caldwell DK (1965) Individualized whistle contours in bottle-nosed dolphins (*Tursiops truncatus*). Nature 207:434–435. https://doi.org/10.1038/207434a0

Caldwell MC, Caldwell DK (1968) Vocalization of naive captive dolphins in small groups. Science 159(3819):1121–1123. https://doi.org/10.1126/science.159.3819.1121

Caldwell MC, Caldwell DK (1969) Simultaneous but different narrow-band sound emissions by a captive eastern Pacific pilot whale, *Globicephala scammoni*. Mammalia 33(3):505–508. https://doi.org/10.1515/mamm.1969.33.3.505

Caldwell MC, Caldwell DK (1971) Statistical evidence for individual signature whistles in the Pacific whitesided dolphin, *Lagenorhynchus obliquidens*. Cetology 3:1–9.

Caldwell MC, Caldwell DK, Evans WE (1966) Sounds and behaviour of captive Amazon freshwater dolphins, *Inia geoffrensis*. Los Angeles County Museum Contributions in Science 108:1–24. https://doi.org/10.5962/p.241097

Caldwell MC, Caldwell DK, Miller JF (1973) Statistical evidence for individual signature whistles in the spotted dolphin, *Stenella plagiodon*. Cetology 16:1–21

Castellote M, Mooney A, Andrews R, Deruiter S, Lee W-J, Ferguson M, Wade P (2021) Beluga whale (*Delphinapterus leucas*) acoustic foraging behavior and applications for long term monitoring. PLoS One 16(11):e0260485. https://doi.org/10.1371/journal.pone.0260485

Chmelnitsky E, Ferguson S (2012) Beluga whale, *Delphinapterus leucas*, vocalizations from the Churchill River, Manitoba. Canada J Acoust Soc Am 131(6):4821–4835. https://doi.org/10.1121/1.4707501

Cholewiak D, Baumann-Pickering S, Van Parijs S (2013) Description of sounds associated with Sowerby's beaked whales (*Mesoplodon bidens*) in the western North Atlantic Ocean. J Acoust Soc Am 134(5):3905–3912. https://doi.org/10.1121/1.4823843

Clausen KT, Wahlberg M, Beedholm K, Deruiter S, Madsen PT (2010) Click communication in harbour porpoises *Phocoena phocoena*. Bioacoustics 20(1):1–28. https://doi.org/10.1080/09524622.2011.9753630

Cones S, Dent M, Walkes S, Bocconcelli A, DeWind C, Arjasbi K, Rose K, Silva T, Sayigh L (2023) Probable signature whistle production in Atlantic white-sided (*Lagenorhynchus acutus*) and short-beaked common (*Delphinus delphis*) dolphins near Cape Cod, Massachusetts. Mar Mamm Sci 39(1):338–344. https://doi.org/10.1111/mms.12976

Connor RC, Smolker RA (1996) 'Pop' goes the dolphin: a vocalization male bottlenose dolphins produce during consortships. Behaviour 133:643–662. https://doi.org/10.1163/156853996X00404

Corkeron PJ, Van Parijs SM (2001) Vocalizations of eastern Australian Risso's dolphins, *Grampus griseus*. Can J Zool 79(1):160–164. https://doi.org/10.1139/z00-180

Cosentino M, Nairn D, Coscarella M, Jackson JC, Windmill JFC (2022) I beg your pardon? Acoustic behaviour of a wild solitary common dolphin who interacts with harbour porpoises. Bioacoustics 31(5): 517–534. https://doi.org/10.1080/09524622.2021.1982005

Courts R, Erbe C, Wellard R, Boisseau O, Jenner KC, Jenner M-N (2020) Australian long-finned pilot whales (*Globicephala melas*) emit stereotypical, variable, biphonic, multi-component, and sequenced vocalisations, similar to those recorded in the northern hemisphere. Sci Rep 10(1):20609. https://doi.org/10.1038/s41598-020-74111-y

Crance JL, Bowles AE, Garver A (2014) Evidence for vocal learning in juvenile male killer whales, *Orcinus orca*, from an adventitious cross-socializing experiment. J Exp Biol 217(8):1229–1237. https://doi.org/10.1242/jeb.094300

Cranford TW (1999) The sperm whale's nose: sexual selection on a grand scale? Mar Mamm Sci 15(4): 1133–1157. https://doi.org/10.1111/j.1748-7692.1999.tb00882.x

Cranford TW (2000) In search of impulse sound sources in Odontocetes. Hearing by whales and dolphins:109–155. https://doi.org/10.1007/978-1-4612-1150-1_3

Cranford TW (2011) Biosonar sources in odontocetes: considering structure and function. J Exp Biol 214(8): 1403. https://doi.org/10.1242/jeb.053660

Cranford TW, Amundin M, Norris KS (1996) Functional morphology and homology in the odontocete nasal complex: implications for sound generation. J Morphol 228(3):223–285. https://doi.org/10.1002/(SICI)1097-4687(199606)228:3<223::AID-JMOR1>3.0.CO;2-3

Cranford TW, Mckenna MF, Soldevilla MS, Wiggins SM, Goldbogen JA, Shadwick RE, Krysl P, Leger JAS, Hildebrand JA (2008) Anatomic geometry of sound transmission and reception in Cuvier's beaked whale (*Ziphius cavirostris*). Anat Rec 291(4):353–378. https://doi.org/10.1002/Ar.20652

Cranford TW, Elsberry WR, Van Bonn WG, Jeffress JA, Chaplin MS, Blackwood DJ, Carder DA, Kamolnick T, Todd MA, Ridgway SH (2011) Observation and analysis of sonar signal generation in the bottlenose dolphin (*Tursiops truncatus*): evidence for two sonar sources. J Exp Mar Biol Ecol 407(1):81–96. https://doi.org/10.1016/j.jembe.2011.07.010

Cranford TW, Trijoulet V, Smith CR, Krysl P (2014) Validation of a vibroacoustic finite element model using bottlenose dolphin simulations: the dolphin biosonar beam is focused in stages. Bioacoustics 23(2):161–194. https://doi.org/10.1080/09524622.2013.843061

Cremer MJ, Holz AC, Bordino P, Wells RS, Simões-Lopes PC (2017) Social sounds produced by franciscana dolphins, *Pontoporia blainvillei* (Cetartiodactyla, Pontoporiidae). J Acoust Soc Am 141(3):2047–2054. https://doi.org/10.1121/1.4978437

Curé C, Antunes R, Samarra F, Alves AC, Visser F, Kvadsheim P, Miller PJO (2012) Pilot whales attracted to killer whale sounds: acoustically-mediated interspecific interactions in cetaceans. PLoS One 7(12): e52201. https://doi.org/10.1371/journal.pone.0052201

Dawson SM (1991) Clicks and communication: the behavioural and social contexts of Hector's dolphin vocalizations. Ethology 88(4):265–276. https://doi.org/10.1111/j.1439-0310.1991.tb00281.x

Dawson SM, Thorpe CW (1990) A quantitative analysis of the sounds of Hector's dolphin. Ethology 86(2): 131–145. https://doi.org/10.1111/j.1439-0310.1990.tb00424.x

Dawson S, Barlow J, Ljungblad D (1998) Sounds recorded from Baird's beaked whale, *Berardius bairdii* (note). Mar Mamm Sci 14(2):335–344. https://doi.org/10.1111/j.1748-7692.1998.tb00724.x

de Melo JF, Amorim TOS, Andriolo A (2021) Delving deep into unheard waters: new types of low frequency pulsed sounds described for the boto (*Inia geoffrensis*). Mamm Biol 101(4):429–437. https://doi.org/10.1007/s42991-021-00134-1

de Stephanis R, Giménez J, Esteban R, Gauffier P, García-Tiscar S, Sinding M-HS, Verborgh P (2015) Mobbing-like behavior by pilot whales towards killer whales: a response to resource competition or perceived predation risk? Acta Ethol 18(1):69–78. https://doi.org/10.1007/s10211-014-0189-1

DeAngelis AI, Barlow J, Gillies D, Ballance LT (2023) Echolocation depths and acoustic foraging behavior of Baird's beaked whales (*Berardius bairdii*) based on towed hydrophone recordings. Mar Mamm Sci 39(1): 289–298. https://doi.org/10.1111/mms.12958

Deconto LS, Monteiro-Filho ELA (2013) High initial and minimum frequencies of *Sotalia guianensis* whistles in the southeast and south of Brazil. J Acoust Soc Am 134(5):3899–3904. https://doi.org/10.1121/1.4823845

Deecke VB, Ford JKB, Spong P (2000) Dialect change in resident killer whales: implications for vocal learning and cultural transmission. Anim Behav 60(5):629–638. https://doi.org/10.1006/anbe.2000.1454

Deecke VB, Slater PJB, Ford JKB (2002) Selective habituation shapes acoustic predator recognition in harbour seals. Nature 420(6912):171–173. https://doi.org/10.1038/nature01030

Deecke VB, Ford JKB, Slater PJB (2005) The vocal behaviour of mammal-eating killer whales: communicating with costly calls. Anim Behav 69(2):395–405. https://doi.org/10.1016/j.anbehav.2004.04.014

DeLong CM, Au WWL, Lemonds DW, Harley HE, Roitblat HL (2006) Acoustic features of objects matched by an echolocating bottlenose dolphin. J Acoust Soc Am 119(3):1867–1879. https://doi.org/10.1121/1.2161434

DeRuiter SL, Bahr A, Blanchet MA, Hansen SF, Kristensen JH, Madsen PT, Tyack PL, Wahlberg M (2009) Acoustic behaviour of echolocating porpoises during prey capture. J Exp Biol 212(19):3100–3107. https://doi.org/10.1242/Jeb.030825

DeRuiter SL, Boyd IL, Claridge DE, Clark CW, Gagnon C, Southall BL, Tyack PL (2013) Delphinid whistle production and call matching during playback of simulated military sonar. Mar Mamm Sci 29(2): E46–E59. https://doi.org/10.1111/j.1748-7692.2012.00587.x

Díaz López B (2022) Context-dependent and seasonal fluctuation in bottlenose dolphin (*Tursiops truncatus*) vocalizations. Anim Cogn 25:1381–1392. https://doi.org/10.1007/s10071-022-01620-w

Ding W, Würsig B, Evans W (1995a) Comparisons of whistles among seven odontocete species. In: Kastelein RA, Thomas JA, Nachtigall PE (eds) Sensory systems of aquatic mammals. De Spil Publishers, Woerden, pp 299–323

Ding W, Würsig B, Evans WE (1995b) Whistles of bottlenose dolphins: comparisons among populations. Aquat Mamm 21(1):65–77

Dong L, Caruso F, Lin M, Liu M, Gong Z, Dong J, Cang S, Li S (2019) Whistles emitted by Indo-Pacific humpback dolphins (*Sousa chinensis*) in Zhanjiang waters. China J Acoust Soc Am 145(6):3289–3298. https://doi.org/10.1121/1.5110304

Dong L, Caruso F, Dong J, Liu M, Lin M, Li S (2021) Whistle characteristics of a newly recorded Indo-Pacific humpback dolphin (*Sousa chinensis*) population in waters southwest of Hainan Island, China, differ from other humpback dolphin populations. Mar Mamm Sci 37(4):1341–1362. https://doi.org/10.1111/mms.12816

dos Santos ME, Caporin G, Moreira HO, Ferreira AJ, Coelho JLB (1990) Acoustic behavior in a local population of bottlenose dolphins. In: Thomas JA, Kastelein RA (eds) Sensory abilities of cetaceans: laboratory and field evidence. Springer US, Boston, pp 585–598. https://doi.org/10.1007/978-1-4899-0858-2_41

dos Santos ME, Ferreira AJ, Harzen S (1995) Rhythmic sound sequences emitted by aroused bottlenose dolphins in the Sado estuary, Portugal. In: Sensory systems of aquatic mammals. de Spil, Woerden, pp 325–334

Dziedzic A, Debuffrenil V (1989) Acoustic signals of the Commerson's dolphin, *Cephalorhynchus commersonii*, in the Kerguelen Islands. J Mammal 70(2):449–452. https://doi.org/10.2307/1381541

Erbe C, Salgado-Kent C, de Winter S, Marley S, Ward R (2020) Matching signature whistles with photo-identification of Indo-Pacific bottlenose dolphins (*Tursiops aduncus*) in the Fremantle inner harbour, Western Australia. Acoust Aust 48:23–38. https://doi.org/10.1007/s40857-020-00178-2

Erbe C, Duncan A, Hawkins L, Terhune JM, Thomas JA (2022a) Introduction to acoustic terminology and signal processing. In: Erbe C, Thomas JA (eds) Exploring animal behavior through sound: volume 1: methods. Springer, Cham, pp 111–152. https://doi.org/10.1007/978-3-030-97540-1_4

Erbe C, Duncan A, Vigness-Raposa KJ (2022b) Introduction to sound propagation under water. In: Erbe C, Thomas JA (eds) Exploring animal behavior through sound: volume 1: methods. Springer, Cham, pp 185–216. https://doi.org/10.1007/978-3-030-97540-1_6

Eskelinen HC, Jones BL (2021) Acoustic characteristics of bubblestream-associated whistles produced by Atlantic bottlenose dolphins (*Tursiops truncatus*) during the first thirty days of life. Aquat Mamm 47(4):337–348. https://doi.org/10.1578/AM.47.4.2021.337

Evans WE, Powell BA (1967) Discrimination of different metallic plates by an echolocating delphinid. Les Systèmes Sonars Animaux:363–382

Fais A, Johnson M, Wilson M, Soto NA, Madsen PT (2016) Sperm whale predator-prey interactions involve chasing and buzzing, but no acoustic stunning. Sci Rep 6:28562. https://doi.org/10.1038/srep28562

Fang L, Li S, Wang K, Wang Z, Shi W, Wang D (2015a) Echolocation signals of free-ranging Indo-Pacific humpback dolphins (*Sousa chinensis*) in Sanniang Bay, China. J Acoust Soc Am 138:1346–1352. https://doi.org/10.1121/1.4929492

Fang L, Wang D, Li Y, Cheng Z, Pine MK, Wang K, Li S (2015b) The source parameters of echolocation clicks from captive and free-ranging Yangtze finless porpoises (*Neophocaena asiaeorientalis asiaeorientalis*). PLoS One 10(6):e0129143. https://doi.org/10.1371/journal.pone.0129143

Fang L, Wu Y, Wang K, Pine MK, Wang D, Li S (2017) The echolocation transmission beam of free-ranging Indo-Pacific humpback dolphins (*Sousa chinensis*). J Acoust Soc Am 142(2):771–779. https://doi.org/10.1121/1.4996499

Favaro L, Neves S, Furlati S, Pessani D, Martin V, Janik VM (2016) Evidence suggests vocal production learning in a cross-fostered Risso's dolphin (*Grampus griseus*). Anim Cogn 19:847–853. https://doi.org/10.1007/s10071-016-0961-x

Feng W, Zhang Y, Wei C (2018) A study on the asymmetric cylinder wall thickness difference discrimination by dolphins. J Acoust Soc Am 144(2):1018–1027. https://doi.org/10.1121/1.5051330

Figueiredo LD, Simão SM (2009) Possible occurrence of signature whistles in a population of *Sotalia guianensis* (Cetacea, Delphinidae) living in Sepetiba Bay. Brazil J Acoust Soc Am 126(3):1563–1569. https://doi.org/10.1121/1.3158822

Filatova OA, Fedutin ID, Burdin AM, Hoyt E (2007) The structure of the discrete call repertoire of killer whales *Orcinus orca* from Southeast Kamchatka. Bioacoustics 16(3):261–280. https://doi.org/10.1080/09524622.2007.9753581

Filatova O, Deecke V, Ford J, Matkin C, Barrett-Lennard L, Guzeev M, Burdin A, Hoyt E (2012a) Call diversity in the North Pacific killer whale populations: implications for dialect evolution and population history. Anim Behav 83(3):595–603. https://doi.org/10.1016/j.anbehav.2011.12.013

Filatova O, Ford J, Matkin C, Barrett-Lennard L, Burdin A, Hoyt E (2012b) Ultrasonic whistles of killer whales (*Orcinus orca*) recorded in the North Pacific (L). J Acoust Soc Am 132(6):3618–3621. https://doi.org/10.1121/1.4764874

Filatova OA, Burdin AM, Hoyt E (2013) Is killer whale dialect evolution random? Behav Process 99:34–41. https://doi.org/10.1016/j.beproc.2013.06.008

Filatova OA, Miller PJO, Yurk H, Samarra FIP, Hoyt E, Ford JKB, Matkin CO, Barrett-Lennard LG (2015a) Killer whale call frequency is similar across the oceans, but varies across sympatric ecotypes. J Acoust Soc Am 138:251–257. https://doi.org/10.1121/1.4922704

Filatova OA, Samarra FIP, Deecke VB, Ford JKB, Miller PJO, Yurk H (2015b) Cultural evolution of killer whale calls: background, mechanisms and consequences. Behaviour 152(15):2001–2038. https://doi.org/10.1163/1568539X-00003317

Finfer DC, White PR, Chua GH, Leighton TG (2012) Review of the occurrence of multiple pulse echolocation clicks in recordings from small odontocetes. IET Radar Sonar Navig 6(6):545–555. https://doi.org/10.1049/iet-rsn.2011.0348

Finneran JJ (2013) Dolphin "packet" use during long-range echolocation tasks. J Acoust Soc Am 133(3): 1796–1810. https://doi.org/10.1121/1.4788997

Finneran JJ, Oliver CW, Schaefer KM, Ridgway SH (2000) Source levels and estimated yellowfin tuna (*Thunnus albacares*) detection ranges for dolphin jaw pops, breaches, and tail slaps. J Acoust Soc Am 107(1): 649–656. https://doi.org/10.1121/1.428330

Finneran JJ, Branstetter BK, Houser DS, Moore AW, Mulsow J, Martin C, Perisho S (2014a) High-resolution measurement of a bottlenose dolphin's (*Tursiops truncatu*s) biosonar transmission beam pattern in the horizontal plane. J Acoust Soc Am 136(4): 2025–2038. https://doi.org/10.1121/1.4895682

Finneran JJ, Schroth-Miller M, Borror N, Tormey M, Brewer A, Black A, Bakhtiari K, Goya G (2014b) Multi-echo processing by a bottlenose dolphin operating in "packet" transmission mode at long range. J Acoust Soc Am 136(5):2876–2886. https://doi.org/10.1121/1.4898043

Finneran JJ, Mulsow J, Jones R, Houser DS, Accomando AW, Ridgway SH (2018) Non-auditory, electrophysiological potentials preceding dolphin biosonar click production. J Comp Physiol 204(3):271–283. https://doi.org/10.1007/s00359-017-1234-0

Finneran JJ, Jones R, Mulsow J, Houser DS, Moore PW (2019) Jittered echo-delay resolution in bottlenose dolphins (*Tursiops truncatus*). J Comp Physiol 205(1):125–137. https://doi.org/10.1007/s00359-018-1309-6

Foote AD, Osborne RW, Hoelzel AR (2004) Whale-call response to masking boat noise. Nature 428:910. https://doi.org/10.1038/428910a

Foote AD, Griffin RM, Howitt D, Larsson L, Miller PJO, Hoelzel AR (2006) Killer whales are capable of vocal learning. Biol Lett 2(4):509–512. https://doi.org/10.1098/rsbl.2006.0525

Ford JKB (1984) Call traditions and dialects of killer whales (*Orcinus orca*) in British Columbia. PhD thesis, University of British Columbia, Vancouver

Ford JKB (1989) Acoustic behaviour of resident killer whales (*Orcinus orca*) off Vancouver Island. British-Columbia Can J Zool 67(3):727–745. https://doi.org/10.1139/z89-105

Ford JKB (1991) Vocal traditions among resident killer whales (*Orcinus orca*) in coastal waters of British Columbia. Can J Zool 69(6):1454–1483. https://doi.org/10.1139/z91-206

Fouda L, Wingfield JE, Fandel AD, Garrod A, Hodge KB, Rice AN, Bailey H (2018) Dolphins simplify their vocal calls in response to increased ambient noise. Biol Lett 14(10). https://doi.org/10.1098/rsbl.2018.0484

Gallo A, De Moura LA, Böye M, Hausberger M, Lemasson A (2023) Study of repertoire use reveals unexpected context-dependent vocalizations in bottlenose dolphins (*Tursiops truncatus*). The Science of Nature 110(6):56. https://doi.org/10.1007/s00114-023-01884-3

Gardner SC, Varanasi U (2003) Isovaleric acid accumulation in odontocete melon during development. Naturwissenschaften 90:528–531. https://doi.org/10.1007/s00114-003-0472-x

Garland EC, Castellote M, Berchok CL (2015) Beluga whale (*Delphinapterus leucas*) vocalizations and call classification from the eastern Beaufort Sea population. J Acoust Soc Am 137(4):3054–3066. https://doi.org/10.1121/1.4919338

Gero S, Whitehead H, Rendell L (2016) Individual, unit and vocal clan level identity cues in sperm whale codas. R Soc Open Sci 3(1):150372. https://doi.org/10.1098/rsos.150372

Gong Z, Dong L, Caruso F, Lin M, Liu M, Dong J, Li S (2019) Echolocation signals of free-ranging pantropical spotted dolphins (*Stenella attenuata*) in the South China Sea. J Acoust Soc Am 145(6):3480–3487. https://doi.org/10.1121/1.5111742

Goold JC (1999) Behavioural and acoustic observations of sperm whales in Scapa Flow, Orkney Islands. J Mar Biol Assoc 79(3):541–550. https://doi.org/10.1017/S0025315498000666

Goold JC, Jefferson TA (2004) A note on clicks recorded from free-ranging Indo-Pacific humpback dolphins, *Sousa chinensis*. Aquat Mamm 30(1):175–178. https://doi.org/10.1578/AM.30.1.2004.175

Götz T, Antunes R, Heinrich S (2010) Echolocation clicks of free-ranging Chilean dolphins (*Cephalorhynchus eutropia*). J Acoust Soc Am 128(2):563–566. https://doi.org/10.1121/1.3353078

Gridley T, Cockcroft VG, Hawkings ER, Blewitt ML, Morisaka T, Janik VM (2014) Signature whistles in free-ranging populations of Indo-Pacific bottlenose dolphins. Mar Mamm Sci 30(2):512–527. https://doi.org/10.1111/mms.12054

Hansen M, Wahlberg M, Madsen PT (2008) Low-frequency components in harbor porpoise (*Phocoena phocoena*) clicks: communication signal, by-products, or artifacts? J Acoust Soc Am 124(6): 4059–4068. https://doi.org/10.1121/1.2945154

Harder JH, Hill HM, Dudzinski KM, Sanabria KT, Guarino S, Kuczaj SA II (2016) The development of echolocation in bottlenose dolphins. Int J Comp Psychol 29. https://doi.org/10.46867/ijcp.2016.29.00.17

Harley HE, Roitblat HL, Nachtigall PE (1996) Object representation in the bottlenose dolphin (*Tursiops truncatus*): integration of visual and echoic information. J Exp Psychol Anim Behav Process 22(2):164–174. https://doi.org/10.1037/0097-7403.22.2.164

Harper CJ, McLellan WA, Rommel SA, Gay DM, Dillaman RM, Pabst DA (2008) Morphology of the melon and its tendinous connections to the facial muscles in bottlenose dolphins (*Tursiops truncatus*). J Morphol 269(7):820–839. https://doi.org/10.1002/jmor.10628

Hawkins ER (2010) Geographic variations in the whistles of bottlenose dolphins (*Tursiops aduncus*) along the east and west coasts of Australia. J Acoust Soc Am 128(2):924–935. https://doi.org/10.1121/1.3459837

Heiler J, Elwen SH, Kriesell HJ, Gridley T (2016) Changes in bottlenose dolphin whistle parameters related to vessel presence, surface behaviour and group composition. Anim Behav 117:167–177. https://doi.org/10.1016/j.anbehav.2016.04.014

Henderson EE, Hildebrand JA, Smith MH (2011) Classification of behaviour using vocalizations of Pacific white-sided dolphins (*Lagenorhynchus obliquidens*). J Acoust Soc Am 130(1):557–567. https://doi.org/10.1121/1.3592213

Hendry JL (2004) The ontogeny of echolocation in the Atlantic bottlenose dolphin (*Tursiops truncatus*). PhD thesis, The University of Southern Mississippi, Mississippi

Herman LM, Pack AA (1992) Echoic-visual cross-modal recognition by a dolphin. In: Thomas JA, Kastelein RA, Supin AY (eds) Marine mammal sensory systems. Springer US, Boston, pp 709–726. https://doi.org/10.1007/978-1-4615-3406-8_44

Herman LM, Tavolga WN (1980) The communication systems of cetaceans. In: Herman LM (ed) Cetacean behaviour: mechanisms and functions. Wiley, New York, pp 149–209

Herman LM, Pack AA, Hoffmann-Kuhnt M (1998) Seeing through sound: dolphins (*Tursiops truncatus*) perceive the spatial structure of objects through echolocation. J Comp Psychol 112(3):292–305. https://doi.org/10.1037/0735-7036.112.3.292

Herzing DL (1996) Vocalizations and associated underwater behaviour of free-ranging Atlantic spotted dolphins, *Stenella frontalis* and bottlenose dolphins, *Tursiops truncatus*. Aquat Mamm 22(2):61–79

Herzing DL (2000) Acoustics and social behavior of wild dolphins: implications for a sound society. In: Au WWL, Popper AN, Fay RR (eds) Hearing by whales and dolphins, Springer handbook of auditory research. Springer, New York, pp 225–272. https://doi.org/10.1007/978-1-4612-1150-1_5

Herzing DL (2015) Synchronous and rhythmic vocalizations and correlated underwater behavior of free-ranging Atlantic spotted dolphins (*Stenella frontalis*) and bottlenose dolphins (*Tursiops truncatus*) in The Bahamas. Anim Behav Cogn 2(1):14–29. https://doi.org/10.12966/abc.02.02.2015

Hoffman JM, Hung SK, Wang JY, White BN (2017) Regional differences in the whistles of Australasian humpback dolphins (genus *Sousa*). Can J Zool 95(7):515–526. https://doi.org/10.1139/cjz-2016-0204

Hoffmann-Kuhnt M, Chitre M, Wellard R, Lee J, Abel G, Yeo K, Jee-Loong C (2011) Is synthetic aperture an essential tool for echoic shape recognition in dolphins? In: OCEANS'11 MTS/IEEE KONA, 19–22 September 2011. pp 1–7. https://doi.org/10.23919/OCEANS.2011.6107034

Holt MM, Noren DP, Veirs V, Emmons CK, Veirs S (2009) Speaking up: killer whales (*Orcinus orca*) increase their call amplitude in response to vessel noise. J Acoust Soc Am 125 (1):El27-El32. https://doi.org/10.1121/1.3040028

Holt M, Noren D, Emmons C (2011) Effects of noise levels and call types on the source levels of killer whale calls. J Acoust Soc Am 130(5):3100–3106. https://doi.org/10.1121/1.3641446

Houser DS, Helweg DA, Moore PW (1999) Classification of dolphin echolocation clicks by energy and frequency distributions. J Acoust Soc Am 106(3):1579–1585. https://doi.org/10.1121/1.427153

Houser D, Martin S, Phillips M, Bauer E, Herrin T, Moore P (2003) Signal processing applied to the dolphin-based sonar system. In: Oceans 2003. Celebrating the past ... teaming toward the future (IEEE Cat. No.03CH37492), 22–26 September 2003, vol 291, pp 297–303. https://doi.org/10.1109/OCEANS.2003.178572

Houser DS, Finneran J, Carder D, Van Bonn W, Smith C, Hoh C, Mattrey R, Ridgway S (2004) Structural and functional imaging of bottlenose dolphin (*Tursiops truncatus*) cranial anatomy. J Exp Biol 207(21):3657–3665. https://doi.org/10.1242/Jeb.01207

Houser D, Martin SW, Bauer EJ, Phillips M, Herrin T, Cross M, Vidal A, Moore PW (2005) Echolocation characteristics of free-swimming bottlenose dolphins during object detection and identification. J Acoust Soc Am 117(4):2308–2317. https://doi.org/10.1121/1.1867912

Ibsen SD, Au WWL, Nachtigall PE, DeLong CM, Breese M (2010) Changes in consistency patterns of click frequency content over time of an echolocating Atlantic bottlenose dolphin. J Acoust Soc Am 127(6):3821–3829. https://doi.org/10.1121/1.3419905

Ibsen S, Krause-Nehring J, Nachtigall P, Au W, Breese M (2011) Similarities in echolocation strategy and click characteristics between a *Pseudorca crassidens* and a *Tursiops truncatus*. J Acoust Soc Am 130(5):3085–3089. https://doi.org/10.1121/1.3621716

Ivanov MP (2004) Dolphin's echolocation signals in a complicated acoustic environment. Acoust Phys 50(4):469–479. https://doi.org/10.1134/1.1776226

Janik VM (2000a) Food-related bray calls In wild bottlenose dolphins (*Tursiops truncatus*). Proc Royal Soc B 267(1446):923–927. https://doi.org/10.1098/rspb.2000.1091

Janik VM (2000b) Whistle matching in wild bottlenose dolphins (*Tursiops truncatus*). Science 2000(289):1355–1357. https://doi.org/10.1126/science.289.5483.1355

Janik VM (2014) Cetacean vocal learning and communication. Curr Opin Neurobiol 28:60–65. https://doi.org/10.1016/j.conb.2014.06.010

Janik VM, Sayigh LS (2013) Communication in bottlenose dolphins: 50 years of signature whistle research. J Comp Physiol 199(6):479–489. https://doi.org/10.1007/s00359-013-0817-7

Janik VM, Simard P, Sayigh LS, Mann D, Frankel A (2011) Chorussing in delphinids. J Acoust Soc Am 130(4):2322. https://doi.org/10.1121/1.3654281

Janik V, King S, Sayigh L, Wells R (2013) Identifying signature whistles from recordings of groups of unrestrained bottlenose dolphins (*Tursiops truncatus*). Mar Mamm Sci 29 (1):109-122. https://doi.org/10.1111/j.1748-7692.2011.00549.x

Jayathilaka RMRM, Arulananthan K (2018) Vocalization patterns of Indo-Pacific humpback dolphins (*Sousa plumbea*) in Puttalam Lagoon, Sri Lanka. J Nat Aquat Resour Res Develop Agency 45–47:38–51

Jensen FH, Rocco A, Mansur RM, Smith BD, Janik VM, Madsen PT (2013) Clicking in shallow rivers: short-range echolocation of Irrawaddy and Ganges river dolphins in a shallow, acoustically complex habitat. PLoS One 8(4):e59284. https://doi.org/10.1371/journal.pone.0059284

Jensen FH, Wahlberg M, Beedholm K, Johnson M, Aguilar De Soto N, Madsen PT (2015) Single-click beam patterns suggest dynamic changes to the field of view of echolocating Atlantic spotted dolphins (*Stenella frontalis*) in the wild. J Exp Biol 218(9):1314–1324. https://doi.org/10.1242/jeb.116285

Jensen FH, Johnson M, Ladegaard M, Wisniewska DM, Madsen PT (2018) Narrow acoustic field of view drives frequency scaling in toothed whale biosonar. Curr Biol 28(23):3878–3885. https://doi.org/10.1016/j.cub.2018.10.037

Johnson M, Madsen PT, Zimmer WMX, de Soto NA, Tyack PL (2004) Beaked whales echolocate on prey. Proc Royal Soc B 271:S383–S386. https://doi.org/10.1098/rsbl.2004.0208

Johnson M, Madsen PT, Zimmer WMX, de Soto NA, Tyack PL (2006) Foraging Blainville's beaked whales (*Mesoplodon densirostris*) produce distinct click types matched to different phases of echolocation. J Exp Biol 209(24):5038–5050. https://doi.org/10.1242/Jeb.02596

Johnson M, Hickmott LS, Aguilar Soto N, Madsen PT (2008) Echolocation behaviour adapted to prey in foraging Blainville's beaked whale (*Mesoplodon densirostris*). Proc Royal Soc B 275:133–139. https://doi.org/10.1098/rspb.2007.1190

Jones B, Zapetis M, Samuelson MM, Ridgway S (2020) Sounds produced by bottlenose dolphins (*Tursiops*): a review of the defining characteristics and acoustic criteria of the dolphin vocal repertoire. Bioacoustics 29(4):399–440. https://doi.org/10.1080/09524622.2019.1613265

Jones B, Tufano S, Daniels R, Mulsow J, Ridgway S (2022) Non-stereotyped amplitude modulation across signature whistle contours. Behav Process 194:104561. https://doi.org/10.1016/j.beproc.2021.104561

Kamminga C, Wiersma H (1981) Investigations on Cetacean Sonar II. Acoustical similarities and differences in odontocete sonar signals. Aquat Mamm 8(2):41–61

Kaplan MB, Mooney TA, Sayigh LS, Baird RW (2014) Repeated call types in Hawaiian melon-headed whales (*Peponocephala electra*). J Acoust Soc Am 136(3):1394–1401. https://doi.org/10.1121/1.4892759

Kaplan JD, Melillo-Sweeting K, Reiss D (2018) Biphonal calls in Atlantic spotted dolphins (*Stenella frontalis*): bitonal and burst-pulse whistles. Bioacoustics 27(2):145–164. https://doi.org/10.1080/09524622.2017.1300105

Karlsen JD, Bisther A, Lydersen C, Haug T, Kovacs KM (2002) Summer vocalisations of adult male white whales (*Delphinapterus leucas*) in Svalbard. Norway Polar Biol 25(11):808–817. https://doi.org/10.1007/s00300-002-0415-6

Kastelein RA, Au WWL, Rippe HT, Schooneman NM (1999) Target detection by an echolocating harbor porpoise (*Phocoena phocoena*). J Acoust Soc Am 105(4):2493–2498. https://doi.org/10.1121/1.426951

Keating JL, Barlow J, Rankin S (2016) Shifts in frequency-modulated pulses recorded during an encounter with Blainville's beaked whales (*Mesoplodon densirostris*). J Acoust Soc Am 140 (2):EL166–EL171. https://doi.org/10.1121/1.4959598

Killebrew DA, Mercado E III, Herman LM, Pack AA (2001) Sound production of a neonate bottlenose dolphin. Aquat Mamm 27(1):34–44

King SL, Janik VM (2013) Bottlenose dolphins can use learned vocal labels to address each other. Proc Natl Acad Sci 110(32):13216–13221. https://doi.org/10.1073/pnas.1304459110

King SL, Janik VM (2015) Come dine with me: food-associated social signalling in wild bottlenose dolphins (*Tursiops truncatus*). Anim Cogn 18(4):969–974. https://doi.org/10.1007/s10071-015-0851-7

King SL, Guarino E, Donegan K, McMullen C, Jaakkola K (2021) Evidence that bottlenose dolphins can communicate with vocal signals to solve a cooperative task. R Soc Open Sci 8(3):202073. https://doi.org/10.1098/rsos.202073

Koblitz J, Wahlberg M, Stilz P, Madsen P, Beedholm K, Schnitzler H-U (2012) Asymmetry and dynamics of a narrow sonar beam in an echolocating harbor porpoise. J Acoust Soc Am 131(3):2315–2324. https://doi.org/10.1121/1.3683254

Kreb D (2004) Facultative river dolphins: conservation and social ecology of freshwater and coastal Irrawaddy dolphins in Indonesia. Ph.D. Thesis,. University of Amsterdam

Kriesell HJ, Elwen SH, Nastasi A, Gridley T (2014) Identification and characteristics of signature whistles in wild bottlenose dolphins (*Tursiops truncatus*) from Namibia. PLoS One 9(9):e106317. https://doi.org/10.1371/journal.pone.0106317

Kyhn LA, Tougaard J, Jensen F, Wahlberg M, Stone G, Yoshinaga A, Beedholm K, Madsen PT (2009) Feeding at a high pitch: source parameters of narrow band, high-frequency clicks from echolocating off-shore hourglass dolphins and coastal Hector's dolphins. J Acoust Soc Am 125(3):1783–1791. https://doi.org/10.1121/1.3075600

Kyhn LA, Jensen FH, Beedholm K, Tougaard J, Hansen M, Madsen PT (2010) Echolocation in sympatric Peale's dolphins (*Lagenorhynchus australis*) and Commerson's dolphins (*Cephalorhynchus commersonii*) producing narrow-band high-frequency clicks. J Exp Biol 213(11):1940–1949. https://doi.org/10.1242/Jeb.042440

Kyhn LA, Tougaard J, Beedholm K, Jensen FH, Ashe E, Williams R, Madsen PT (2013) Clicking in a killer whale habitat: narrow-band, high-frequency biosonar clicks of harbour porpoise (*Phocoena phocoena*) and Dall's porpoise (*Phocoenoides dalli*). PLoS One 8(5): e63763. https://doi.org/10.1371/journal.pone.0063763

La Manna G, Manghi M, Pavan G, Lo Mascolo F, Sara G (2013) Behavioural strategy of common bottlenose dolphins (*Tursiops truncatus*) in response to different kinds of boats in the waters of Lampedusa Island (Italy). Aquat Conserv 23(5):745–757. https://doi.org/10.1002/aqc.2355

Ladegaard M, Jensen FH, de Freitas M, da Silva VMF, Madsen PT (2015) Amazon river dolphins (*Inia geoffrensis*) use a high-frequency short-range biosonar. J Exp Biol 218(19):3091–3101. https://doi.org/10.1242/jeb.120501

Ladegaard M, Jensen FH, Beedholm K, da Silva VMF, Madsen PT (2017) Amazon river dolphins (*Inia geoffrensis*) modify biosonar output level and directivity during prey interception in the wild. J Exp Biol 220(14):2654–2665. https://doi.org/10.1242/jeb.159913

Ladegaard M, Mulsow J, Houser DS, Jensen FH, Johnson M, Madsen PT, Finneran JJ (2019) Dolphin echolocation behaviour during active long-range target approaches. J Exp Biol 222 (2):jeb189217. https://doi.org/10.1242/jeb.189217

Lammers MO, Au WWL (2003) Directionality in the whistles of Hawaiian spinner dolphins (*Stenella longirostris*): A signal feature to cue direction of movement? Mar Mamm Sci 19(2):249–264. https://doi.org/10.1111/j.1748-7692.2003.tb01107.x

Lammers MO, Castellote M (2009) The beluga whale produces two pulses to form its sonar signal. Biol Lett 5(3):297–301. https://doi.org/10.1098/rsbl.2008.0782

Lammers MO, Au WWL, Herzing DL (2003) The broadband social acoustic signaling behavior of spinner and spotted dolphins. J Acoust Soc Am 114(3):1629–1639. https://doi.org/10.1121/1.1596173

Lammers MO, Au WWL, Aubauer R, Nachtigall PE (2004) A comparative analysis of the pulsed emissions of free-ranging Hawaiian spinner dolphins (*Stenella longirostris*). In: Thomas J, Moss C, Vater M (eds) Echolocation in bats and dolphins. University of Chicago Press, Chicago, IL, USA, pp 414–419

Leão DT, Monteiro-Filho ELA, Silva FJL (2016) Acoustic parameters of sounds emitted by *Sotalia guianensis*: dialects or acoustic plasticity. J Mammal 97 (2): 611–618. ://doi.org/https://doi.org/10.1093/jmammal/gyv208

Lesage V, Barrette C, Kingsley MCS, Sjare B (1999) The effect of vessel noise on the vocal behaviour of belugas in the St. Lawrence River Estuary, Canada. Mar Mamm Sci 15 (1):65–84. https://doi.org/10.1111/j.1748-7692.1999.tb00782.x

Leunissen EM, Webster T, Rayment W (2018) Characteristics of vocalisations recorded from free-ranging Shepherd's beaked whales, *Tasmacetus shepherdi*. J Acoust Soc Am 144(5):2701–2708. https://doi.org/10.1121/1.5067380

Li S, Wang K, Wang D, Akamatsu T (2005) Origin of the double- and multi-pulse structure of echolocation signals in Yangtze finless porpoise (*Neophocaena phocaenoides asiaeorientialis*). J Acoust Soc Am 118(6):3934–3940. https://doi.org/10.1121/1.2126919

Li S, Wang D, Wang K, Akamatsu T (2006) Sonar gain control in echolocating finless porpoises (*Neophocaena phocaenoides*) in an open water. J Acoust Soc Am 120(4):1803–1806. https://doi.org/10.1121/1.2335674

Li S, Wang K, Wang D, Dong S, Akamatsu T (2008) Simultaneous production of low- and high-frequency sounds by neonatal finless porpoises. J Acoust Soc Am 124(2):716–718. https://doi.org/10.1121/1.2945152

Lilly JC (1962) Vocal behaviour of the bottlenose dolphin. Proc Am Philos Soc 106(6):520–529

Lilly JC, Miller AM (1961) Sounds emitted by the bottlenose dolphin. Science 133(3465):1689–1693. https://doi.org/10.1126/science.133.3465.1689

Lima A, Le Pendu Y (2014) Evidence for signature whistles in Guiana dolphins (*Sotalia guianensis*) in Ilheus, northeastern Brazil. J Acoust Soc Am 136(6): 3178–3185. https://doi.org/10.1121/1.4900829

Lynn SK, Reiss DL (1992) Pulse sequence and whistle production by two captive beaked whales, *Mesoplodon* species. Mar Mamm Sci 8(3):299–305. https://doi.org/10.1111/j.1748-7692.1992.tb00413.x

Madsen PT, Surlykke A (2014) Echolocation in air and water. In: Surlykke A, Nachtigall PE, Fay RR, Popper AN (eds) Biosonar. Springer, New York, pp 257–304. https://doi.org/10.1007/978-1-4614-9146-0_9

Madsen PT, Wahlberg M (2007) Recording and quantification of ultrasonic echolocation clicks from free-ranging toothed whales. Deep-Sea Res I Oceanogr Res Pap 54(8):1421–1444. https://doi.org/10.1016/j.dsr.2007.04.020

Madsen PT, Payne R, Kristiansen NU, Wahlberg M, Kerr I, Møhl B (2002a) Sperm whale sound production studied with ultrasound time/depth-recording tags. J Exp Biol 205(13):1899–1906. https://doi.org/10.1242/jeb.205.13.1899

Madsen PT, Wahlberg M, Møhl B (2002b) Male sperm whale (*Physeter macrocephalus*) acoustics in a high-latitude habitat: implications for echolocation and communication. Behav Ecol Sociobiol 53(1):31–41. https://doi.org/10.1007/s00265-002-0548-1

Madsen PT, Carder DA, Au WWL, Nachtigall PE, Møhl B, Ridgway SH (2003) Sound production in neonate sperm whales (L). J Acoust Soc Am 113(6):2988–2991. https://doi.org/10.1121/1.1572137

Madsen PT, Kerr I, Payne R (2004) Source parameter estimates of echolocation clicks from wild pygmy killer whales (*Feresa attenuata*) (L). J Acoust Soc Am 116(4):1909–1912. https://doi.org/10.1121/1.1788726

Madsen PT, Carder DA, Bedholm K, Ridgway SH (2005a) Porpoise clicks from a sperm whale nose – convergent evolution of 130 kHz pulses in toothed whale sonars? Bioacoustics 15(2):195–206. https://doi.org/10.1080/09524622.2005.9753547

Madsen PT, Johnson M, de Soto NA, Zimmer WMX, Tyack P (2005b) Biosonar performance of foraging beaked whales (*Mesoplodon densirostris*). J Exp Biol 208(2):181–194. https://doi.org/10.1242/Jeb.01327

Madsen PT, Wisniewska D, Beedholm K (2010) Single source sound production and dynamic beam formation in echolocating harbour porpoises (*Phocoena phocoena*). J Exp Biol 213(18):3105–3110. https://doi.org/10.1242/jeb.044420

Madsen PT, Aguilar de Soto N, Arranz P, Johnson M (2013a) Echolocation in Blainville's beaked whales (*Mesoplodon densirostris*). J Comp Physiol 199(6):451–469. https://doi.org/10.1007/s00359-013-0824-8

Madsen PT, Lammers M, Wisniewska D, Beedholm K (2013b) Nasal sound production in echolocating delphinids (*Tursiops truncatus* and *Pseudorca crassidens*) is dynamic, but unilateral: clicking on the right side and whistling on the left side. J Exp Biol 216(21):4091–4102. https://doi.org/10.1242/jeb.091306

Madsen PT, Siebert U, Elemans CPH (2023) Toothed whales use distinct vocal registers for echolocation and communication. Science 379(6635):928–933. https://doi.org/10.1126/science.adc9570

Malinka CE, Rojano-Doñate L, Madsen PT (2021a) Directional biosonar beams allow echolocating harbour porpoises to actively discriminate and intercept closely spaced targets. J Exp Biol 224(16):jeb242779. https://doi.org/10.1242/jeb.242779

Malinka CE, Tønnesen P, Dunn CA, Claridge DE, Gridley T, Elwen SH, Teglberg Madsen P (2021b) Echolocation click parameters and biosonar behaviour of the dwarf sperm whale (*Kogia sima*). J Exp Biol 224(6):jeb240689. https://doi.org/10.1242/jeb.240689

Marten K, Norris KS, Moore PWB, Englund KA (1988) Loud impulse sounds in odontocete predation and social behavior. In: Nachtigall PE, Moore PWB (eds) Animal sonar: processes and performance. Springer US, Boston, pp 567–579. https://doi.org/10.1007/978-1-4684-7493-0_57

Marten K, Shariff K, Psarakos S, White DJ (1996) Ring bubbles of dolphins. Sci Am 275(2):82–87

Martin SW, Phillips M, Bauer EJ, Moore PW, Houser DS (2005) Instrumenting free-swimming dolphins echolocating in open water. J Acoust Soc Am 117(4):2301–2307. https://doi.org/10.1121/1.1867913

Martin MJ, Gridley T, Elwen SH, Jensen FH (2018) Heaviside's dolphins (*Cephalorhynchus heavisidii*) relax acoustic crypsis to increase communication range. Proc Royal Soc B 285(1883). https://doi.org/10.1098/rspb.2018.1178

Martin MJ, Torres Ortiz S, Reyes Reyes MV, Marino A, Iñíguez Bessega M, Wahlberg M (2021) Commerson's dolphins (*Cephalorhynchus commersonii*) can relax acoustic crypsis. Behav Ecol Sociobiol 75(6):100. https://doi.org/10.1007/s00265-021-03035-y

Masao A, Kourogi A, Aoki K, Yoshioka M, Kori K (2014) Differences in sperm whale codas between two waters off Japan: possible geographic separation of vocal clans. J Mammal 95(1):169–175. https://doi.org/10.1644/13-MAMM-A-172

Matthews JN, Rendell LE, Gordon JCD, Macdonald DW (1999) A review of frequency and time parameters of cetacean tonal calls. Bioacoustics 10(1):47–71. https://doi.org/10.1080/09524622.1999.9753418

May-Collado LJ, Quinones-Lebron SG (2014) Dolphin changes in whistle structure with watercraft activity depends on their behavioral state. J Acoust Soc Am 135 (4):EL193–EL198. https://doi.org/10.1121/1.4869255

May-Collado LJ, Agnarsson I, Wartzok D (2007) Phylogenetic review of tonal sound production in whales in relation to sociality. BMC Evol Biol 7:136. https://doi.org/10.1186/1471-2148-7-136

McCowan B, Reiss D (1995a) Maternal aggressive contact vocalizations in captive bottlenose dolphins (*Tursiops truncatus*): wide-band, low-frequency signals during mother/aunt-infant interactions. Zoo Biol 14(4):293–309. https://doi.org/10.1002/zoo.1430140402

McCowan B, Reiss D (1995b) Whistle contour development in captive-born infant bottlenose dolphins (*Tursiops truncatus*): role of learning. J Comp Psychol 109(3):242–260. https://doi.org/10.1037/0735-7036.109.3.242

Melcón M, Failla M, Iñíguez M (2012) Echolocation behavior of franciscana dolphins (*Pontoporia blainvillei*) in the wild. J Acoust Soc Am 131(6):E448–E453. https://doi.org/10.1121/1.4710837

Melo JF, Amorim TOS, Paschoalini M, Andriolo A (2021) The biosonar of the boto: evidence of differences among species of river dolphins (*Inia* spp.) from the

Amazon. PeerJ 9:e11105. https://doi.org/10.7717/peerj.11105

Melo-Santos G, Walmsley SF, Marmontel M, Oliveira-da-Costa M, Janik VM (2020) Repeated downsweep vocalizations of the Araguaian river dolphin, *Inia araguaiaensis*. J Acoust Soc Am 147(2):748–756. https://doi.org/10.1121/10.0000624

Merkens K, Mann D, Janik VM, Claridge D, Hill M, Oleson E (2018) Clicks of dwarf sperm whales (*Kogia sima*). Mar Mamm Sci 34(4):963–978. https://doi.org/10.1111/mms.12488

Miksis JL, Tyack PL, Buck JR (2002) Captive dolphins, *Tursiops truncatus*, develop signature whistles that match acoustic features of human-made model sounds. J Acoust Soc Am 112(2):728–739. https://doi.org/10.1121/1.1496079

Miller PJO (2002) Mixed-directionality of killer whale stereotyped calls: a direction of movement cue? Behav Ecol Sociobiol 52(3):262–270. https://doi.org/10.1007/s00265-002-0508-9

Miller PJO, Bain DE (2000) Within-pod variation in the sound production of a pod of killer whales, *Orcinus orca*. Anim Behav 60(5):617–628. https://doi.org/10.1006/anbe.2000.1503

Miller L, Wahlberg M (2013) Echolocation by the harbour porpoise: life in coastal waters. Front Physiol 4(52). https://doi.org/10.3389/fphys.2013.00052

Miller L, Pristed J, Møhl B, Surlykke A (1995) The click-sounds of narwhals (*Monodon monoceros*) in Inglefield Bay. Northwest Greenland Mar Mamm Sci 11(4):491–502. https://doi.org/10.1111/j.1748-7692.1995.tb00672.x

Miller PJO, Johnson MP, Tyack PL (2004a) Sperm whale behaviour indicates the use of echolocation click buzzes "creaks" in prey capture. Proc Royal Soc B 271:2239–2247. https://doi.org/10.1098/rspb.2004.2863

Miller PJO, Shapiro AD, Tyack PL, Solow AR (2004b) Call-type matching in vocal exchanges of free-ranging resident killer whales, *Orcinus orca*. Anim Behav 67:1099–1107. https://doi.org/10.1016/j.anbehav.2003.06.017

Miller PJ, Kvadsheim PH, Lam F-PA, Wensveen PJ, Antunes R, Alves AC, Visser F, Kleivane L, Tyack PL, Sivle LD (2012) The severity of behavioral changes observed during experimental exposures of killer (*Orcinus orca*), long-finned pilot (*Globicephala melas*), and sperm (*Physeter macrocephalus*) whales to naval sonar. Aquat Mamm 38(4):362–401. https://doi.org/10.1578/AM.38.4.2012.362

Mishima Y, Morisaka T, Itoh M, Matsuo I, Sakaguchi A, Miyamoto Y (2015) Individuality embedded in the isolation calls of captive beluga whales (*Delphinapterus leucas*). Zool Lett 1:27–40. https://doi.org/10.1186/s40851-015-0028-x

Mishima Y, Morisaka T, Ishikawa M, Karasawa Y, Yoshida Y (2019) Pulsed call sequences as contact calls in Pacific white-sided dolphins (*Lagenorhynchus obliquidens*). J Acoust Soc Am 146(1):409–424. https://doi.org/10.1121/1.5116692

Mishima Y, Matsuo I, Karasawa Y, Ishii M, Morisaka T (2023) Directional and amplitude characteristics of pulsed call sequences in captive free-swimming Pacific white-sided dolphins (*Lagenorhynchus obliquidens*). J Acoust Soc Am 154(5):2974–2987. https://doi.org/10.1121/10.0022377

Mizue K, Takemura A, Nakasai K (1968) Studies on the little toothed whales in the west sea area of Kyushu -- XV: underwater sound of the Chinese finless porpoise caught in the Japanese coastal sea. Bulletin of the Faculty of Fisheries, Nagasaki University 25 (August):25–32

Møhl B, Andersen S (1973) Echolocation: high frequency component in the click of the harbour porpoise (*Phocoena ph.* L.). J Acoust Soc Am 54(5):1368–1372. https://doi.org/10.1121/1.1914435

Møhl B, Wahlberg M, Madsen PT, Miller LA, Surlykke A (2000) Sperm whale clicks: directionality and source level revisited. J Acoust Soc Am 107(1):638–648. https://doi.org/10.1121/1.428329

Møhl B, Wahlberg M, Madsen PT, Heerfordt A, Lund A (2003) The monopulsed nature of sperm whale clicks. J Acoust Soc Am 114(2):1143–1154. https://doi.org/10.1121/1.1586258

Moore PWB, Pawloski DA (1990) Investigations on the control of echolocation pulses in the dolphin (*Tursiops truncatus*). In: Thomas JA, Kastelein RA (eds) Sensory abilities of cetaceans. Plenum Press, New York, pp 305–316. https://doi.org/10.1007/978-1-4899-0858-2_19

Moore PW, Finneran J, Houser DS (2004) Hearing loss and echolocation signal change in dolphins. J Acoust Soc Am 116(4):2503–2503. https://doi.org/10.1121/1.4784989

Moore PW, Dankiewicz LA, Houser DS (2008) Beamwidth control and angular target detection in an echolocating bottlenose dolphin (*Tursiops truncatus*). J Acoust Soc Am 124(5):3324–3332. https://doi.org/10.1121/1.2980453

Morisaka T, Connor RC (2007) Predation by killer whales (*Orcinus orca*) and the evolution of whistle loss and narrow-band high frequency clicks in odontocetes. J Evol Biol 20(4):1439–1458. https://doi.org/10.1111/j.1420-9101.2007.01336.x

Morisaka T, Shinohara M, Nakahara F, Akamatsu T (2005a) Geographic variations in the whistles among three Indo-Pacific bottlenose dolphin *Tursiops aduncus* populations in Japan. Fish Sci 71:568–576. https://doi.org/10.1111/j.1444-2906.2005.01001.x

Morisaka T, Shinohara M, Taki M (2005b) Underwater sounds produced by neonatal bottlenose dolphins (*Tursiops truncatus*): i. acoustic characteristics. Aquat Mamm 3(2):248–257. https://doi.org/10.1578/AM.31.2.2005.248

Morisaka T, Yoshida Y, Akune Y, Mishima H, Nishimoto S (2013) Exchange of "signature" calls in captive belugas (*Delphinapterus leucas*). J Ethol 31(2):141–149. https://doi.org/10.1007/s10164-013-0358-0

Moron JR, Lopes NP, Reis SS, Mamede N, Reis SS, Toledo G, Corso G, Sousa-lima RS, Andriolo A (2019) Whistle variability of Guiana dolphins in

South America: latitudinal variation or acoustic adaptation? Mar Mamm Sci 35(3):843–874. https://doi.org/10.1111/mms.12572

Mullins J, Whitehead H, Weilgart LS (1988) Behavior and vocalizations of two single sperm whales, *Physeter macrocephalus*, off Nova Scotia. Can J Fish Aquat Sci 45(10):1736–1743. https://doi.org/10.1139/f88-205

Murray SO, Mercado E, Roitblat HL (1998) Characterizing the graded structure of false killer whale (*Pseudorca crassidens*) vocalizations. J Acoust Soc Am 104(3):1679–1688. https://doi.org/10.1121/1.424380

Musser WB, Bowles AE, Grebner DM, Crance JL (2014) Differences in acoustic features of vocalizations produced by killer whales cross-socialized with bottlenose dolphins. J Acoust Soc Am 137(4):1990–2002. https://doi.org/10.1121/1.4893906

Nachtigall PE, Murchison AE, Au WWL (1980) Cylinder and cube discrimination by an echolocating blindfolded bottlenose dolphin. In: Busnel R-G, Fish JF (eds) Animal sonar systems. Plenum, New York, pp 945–947. https://doi.org/10.1007/978-1-4684-7254-7_64

Nakahara F, Takemura A, Koido T, Hiruda H (1997) Target discrimination by an echolocating finless porpoise, *Neophocaena phocaenoides*. Mar Mamm Sci 13(4):639–649. https://doi.org/10.1111/j.1748-7692.1997.tb00088.x

Nakamura K, Akamatsu T, Shimazaki K (1998) Threat clicks of captive harbor porpoises, *Phocoena phocoena*. Bull Facult Fisher Hokkaido Univ 49:91–105

Nemiroff L, Whitehead H (2009) Structural characteristics of pulsed calls of long-finned pilot whales *Globicephala melas*. Bioacoustics 19:67–92. https://doi.org/10.1080/09524622.2009.9753615

Niu FQ, Yang YM, Zhou ZM, Wang XY, Monanunsap S, Junchompoo C (2019) Echolocation clicks of free-ranging Irrawaddy dolphins (*Orcaella brevirostris*) in Trat Bay, the eastern Gulf of Thailand. J Acoust Soc Am 145(5):3031–3037. https://doi.org/10.1121/1.5100619

Niu F, Kittiwattanawong K, Wang X, Xue R, Sakornwimon W, Wu F, Yang Y (2021) A comparative study of echolocation parameters of wild and captive Indo-Pacific humpback dolphins (*Sousa chinensis*). Aquat Mamm 47(6):574–584. https://doi.org/10.1578/AM.47.6.2021.574

Norris K, Harvey G (1974) Sound transmission in the porpoise head. J Acoust Soc Am 56(2):659–664. https://doi.org/10.1121/1.1903305

Norris KS, Møhl B (1983) Can odontocetes debilitate prey with sound? Am Nat 122(1):85–104. https://doi.org/10.1086/284120

Oliveira C, Wahlberg M, Johnson M, Miller PJO, Madsen PT (2013) The function of male sperm whale slow clicks in a high latitude habitat: communication, echolocation, or prey debilitation? J Acoust Soc Am 133(5):3135–3144. https://doi.org/10.1121/1.4795798

Oliveira C, Wahlberg M, Silva MA, Johnson M, Antunes R, Wisniewska DM, Fais A, Gonçalves J, Madsen PT (2016) Sperm whale codas may encode individuality as well as clan identity. J Acoust Soc Am 139(5):2860–2869. https://doi.org/10.1121/1.4949478

Oswald JN, Rankin S, Barlow J, Lammers MO (2007) A tool for real-time acoustic species identification of delphinid whistles. J Acoust Soc Am 122(1):587–595. https://doi.org/10.1121/1.2743157

Oswald JN, Rankin S, Barlow J (2008) To whistle or not to whistle? Geographic variation in the whistling behavior of small odontocetes. Aquat Mamm 34(3):288–302. https://doi.org/10.1578/AM.34.3.2008.288

Overstrom NA (1983) Association between burst-pulse sounds and aggressive behavior in captive Atlantic bottlenosed dolphins (*Tursiops truncatus*). Zoo Biol 2(2):93–103. https://doi.org/10.1002/zoo.1430020203

Pace DS, Tumino C, Silvestri M, Giacomini G, Pedrazzi G, Pavan G, Papale E, Ceraulo M, Buscaino G, Ardizzone G (2022) Bray-call sequences in the Mediterranean common bottlenose dolphin (*Tursiops truncatus*) acoustic repertoire. Biology 11(3):367. https://doi.org/10.3390/biology11030367

Pack AA, Herman LM (1995) Sensory integration in the bottlenosed dolphin: immediate recognition of complex shapes across the senses of echolocation and vision. J Acoust Soc Am 98(2):722–733. https://doi.org/10.1121/1.413566

Pack AA, Herman LM, Hoffmann-Kuhnt M, Branstetter BK (2002) The object behind the echo: dolphins (*Tursiops truncatus*) perceive object shape globally through echolocation. Behav Process 58(1):1–26. https://doi.org/10.1016/S0376-6357(01)00200-5

Paitach RL, Amundin M, Teixeira G, Cremer MJ (2021) Echolocation variability of franciscana dolphins (*Pontoporia blainvillei*) between estuarine and open-sea habitats, with insights into foraging patterns. J Acoust Soc Am 150(5):3987–3998. https://doi.org/10.1121/10.0007277

Panova EM, Agafonov AV (2017) A beluga whale socialized with bottlenose dolphins imitates their whistles. Anim Cogn 20(6):1153–1160. https://doi.org/10.1007/s10071-017-1132-4

Panova E, Belikov R, Agafonov A, Belkovich V (2012) The relationship between the behavioral activity and the underwater vocalization of the beluga whale (*Delphinapterus leucas*). Oceanology 52(1):79–87. https://doi.org/10.1134/S000143701201016X

Panova EM, Belikov RA, Agafonov AV, Kirillova OI, Chernetsky AD, Bel'kovich VM (2016) Intraspecific variability in the "vowel"-like sounds of beluga whales (*Delphinapterus leucas*): intra- and interpopulation comparisons. Mar Mamm Sci 32(2):452–465. https://doi.org/10.1111/mms.12266

Panova E, Agafonov A, Belikov R, Melnikova F (2017) Vocalizations of captive beluga whales, *Delphinapterus leucas*: additional evidence for contact signature "mixed" calls in belugas. Mar Mamm Sci 33(3):889–903. https://doi.org/10.1111/mms.12393

Panova E, Agafonov A, Belikov R, Melnikova F (2019) Characteristics and microgeographic variation of whistles from the vocal repertoire of beluga whales (*Delphinapterus leucas*) from the White Sea. J Acoust Soc Am 146(1):681–692. https://doi.org/10.1121/1.5119249

Papale E, Azzolin M, Cascão I, Gannier A, Lammers MO, Martin VM, Oswald J, Perez-Gil M, Prieto R, Silva MA, Giacoma C (2013) Geographic variability in the acoustic parameters of striped dolphin's (*Stenella coeruleoalba*) whistles. J Acoust Soc Am 133(2):1126–1134. https://doi.org/10.1121/1.4774274

Papale E, Azzolin M, Cascao I, Gannier A, Lammers MO, Martin VM, Oswald J, Perez-Gil M, Prieto R, Silva MA, Giacoma C (2014) Macro- and micro-geographic variation of short-beaked common dolphin's whistles in the Mediterranean Sea and Atlantic Ocean. Ethol Ecol Evol 26:392–404. https://doi.org/10.1080/03949370.2013.851122

Papale E, Buffa G, Filiciotto F, Maccarrone V, Mazzola S, Ceraulo M, Giacoma C, Buscaino G (2015a) Biphonic calls as signature whistles in a free-ranging bottlenose dolphin. Bioacoustics 24(3):223–231. https://doi.org/10.1080/09524622.2015.1041158

Papale E, Gamba M, Perez-Gil M, Martin VM, Giacoma C (2015b) Dolphins adjust species-specific frequency parameters to compensate for increasing background noise. PLoS One 10(4):e0121711. https://doi.org/10.1371/journal.pone.0121711

Pérez JM, Jensen FH, Rojano-Doñate L, Aguilar de Soto N (2017) Different modes of acoustic communication in deep-diving short-finned pilot whales (*Globicephala macrorhynchus*). Mar Mamm Sci 33(1):59–79. https://doi.org/10.1111/mms.12344

Podos J, Da Silva VMF, Rossi-Santos MR (2002) Vocalizations of Amazon river dolphins, *Inia geoffrensis*: insights into the evolutionary origins of delphinid whistles. Ethology 108(7):601–612. https://doi.org/10.1046/j.1439-0310.2002.00800.x

Probert R, Gullan A, Rocha D, Dines S, Gridley T (2023) Evidence of signature whistles produced by Indian Ocean bottlenose dolphins (*Tursiops aduncus*) in Mozambique. Bioacoustics 32(5):580–600. https://doi.org/10.1080/09524622.2023.2229290

Pryor K, Kang I (1980) Social behavior and school structure in pelagic porpoises (*Stenella attenuata* and *S. longirostris*) during purse seining for tuna. NOAA/NMFS, Southwest Fisheries Center Administrative Report LJ-80-11C, Contract #01-78-027-1043:119

Quick N, Callahan H, Read AJ (2018) Two-component calls in short-finned pilot whales (*Globicephala macrorhynchus*). Mar Mamm Sci 34(1):155–168. https://doi.org/10.1111/mms.12452

Rako Gospić N, Picciulin M (2016) Changes in whistle structure of resident bottlenose dolphins in relation to underwater noise and boat traffic. Mar Pollut Bull 105(1):193–198. https://doi.org/10.1016/j.marpolbul.2016.02.030

Ramos EA, Jones BL, Austin M, Eierman L, Collom KA, Melo-Santos G, Castelblanco-Martínez N, Arreola MR, Sánchez-Okrucky R, Rieucau G (2023) Signature whistle use and changes in whistle emission rate in a rehabilitated rough-toothed dolphin. Front Mar Sci 10. https://doi.org/10.3389/fmars.2023.1278299

Rankin S, Barlow J (2007) Sounds recorded in the presence of Blainville's beaked whales, *Mesoplodon densirostris*, near Hawai'i (L). J Acoust Soc Am 122(1):42–45. https://doi.org/10.1121/1.2743159

Rankin S, Oswald J, Barlow J, Lammers M (2007) Patterned burst-pulse vocalizations of the northern right whale dolphin, *Lissodelphis borealis*. J Acoust Soc Am 121(2):1213–1218. https://doi.org/10.1121/1.2404919

Rankin S, Baumann-Pickering S, Yack T, Barlow J (2011) Description of sounds recorded from Longman's beaked whale, *Indopacetus pacificus*. J Acoust Soc Am 130 (5):EL339–EL344. https://doi.org/10.1121/1.3646026

Rasmussen MH, Miller LA, Au WWL (2002) Source levels of clicks from free-ranging white-beaked dolphins (*Lagenorhynchus albirostris* Gray 1846) recorded in Icelandic waters. J Acoust Soc Am 111(2):1122–1125. https://doi.org/10.1121/1.1433814

Rasmussen MH, Wahlberg M, Miller LA (2004) Estimated transmission beam pattern of clicks recorded from free-ranging white-beaked dolphins (*Lagenorhynchus albirostris*). J Acoust Soc Am 116(3):1826–1831. https://doi.org/10.1121/1.1775274

Rasmussen MH, Koblitz JC, Laidre KL (2015) Buzzes and high-frequency clicks recorded from narwhals (*Monodon monoceros*) at their wintering ground. Aquat Mamm 41(3):256–264. https://doi.org/10.1578/AM.41.3.2015.256

Rehn N, Filatova OA, Durban JW, Foote AD (2011) Cross-cultural and cross-ecotype production of a killer whale 'excitement' call suggests universality. Naturwissenschaften 98(1):1–6. https://doi.org/10.1007/s00114-010-0732-5

Reiss D (1988) Observations on the development of echolocation in young bottlenose dolphins. In: Nachtigall PE, Moore PWB (eds) Animal sonar: processes and performance. Springer US, Boston, pp 121–127. https://doi.org/10.1007/978-1-4684-7493-0_14

Rendell LE, Gordon JCD (1999) Vocal response of long-finned pilot whales (*Globicephala melas*) to military sonar in the Ligurian Sea. Mar Mamm Sci 15(1):198–204. https://doi.org/10.1111/j.1748-7692.1999.tb00790.x

Rendell LE, Whitehead H (2003) Vocal clans in sperm whales (*Physeter macrocephalus*). Proc Royal Soc B 270(1512):225–231. https://doi.org/10.1098/rspb.2002.2239

Rendell LE, Matthews JN, Gill A, Gordon JCD, Macdonald DW (1999) Quantitative analysis of tonal calls from five odontocete species, examining interspecific and intraspecific variation. J Zool 249(4):403–410. https://doi.org/10.1111/j.1469-7998.1999.tb01209.x

Reyes Reyes MV, Iñíguez MA, Hevia M, Hildebrand JA, Melcón ML (2015) Description and clustering of echolocation signals of Commerson's dolphins (*Cephalorhynchus commersonii*) in Bahía San Julián. Argentina J Acoust Soc Am 138(4):2046–2053. https://doi.org/10.1121/1.4929899

Reyes Reyes MV, Tossenberger VP, Iñiguez MA, Hildebrand JA, Melcón ML (2016) Communication sounds of Commerson's dolphins (*Cephalorhynchus commersonii*) and contextual use of vocalizations. Mar Mamm Sci 32(4):1219–1233. https://doi.org/10.1111/mms.12321

Rhinelander MQ, Dawson SM (2004) Measuring sperm whales from their clicks: stability of interpulse intervals and validation that they indicate whale length. J Acoust Soc Am 115(4):1826–1831. https://doi.org/10.1121/1.1689346

Ridgway SH, Carder DA, Green RF, Gaunt AS, Gaunt SLL, Evans WE (1980) Electromyographic and pressure events in the nasolaryngeal system of dolphins during sound production. In: Busnel RG, Fish JF (eds) Animal sonar systems. Plenum Press, New York, pp 239–250. https://doi.org/10.1007/978-1-4684-7254-7_9

Ridgway SH, Moore PW, Carder DA, Romano TA (2014) Forward shift of feeding buzz components of dolphins and belugas during associative learning reveals a likely connection to reward expectation, pleasure and brain dopamine activation. J Exp Biol 217(16):2910–2919. https://doi.org/10.1242/jeb.100511

Ridgway S, Dibble DS, Alstyne KV, Price D (2015) On doing two things at once: dolphin brain and nose coordinate sonar clicks, buzzes and emotional squeals with social sounds during fish capture. J Exp Biol 218:3987–3995. https://doi.org/10.1242/jeb.130559

Riesch R, Deecke VB (2011) Whistle communication in mammal-eating killer whales (*Orcinus orca*): further evidence for acoustic divergence between ecotypes. Behav Ecol Sociobiol 65:1377–1387. https://doi.org/10.1007/s00265-011-1148-8

Rio R (2023) First acoustic evidence of signature whistle production by spinner dolphins (*Stenella longirostris*). Anim Cogn. https://doi.org/10.1007/s10071-023-01824-8

Roch MA, Klinck H, Baumann-Pickering S, Mellinger DK, Qui S, Soldevilla MS, Hildebrand JA (2011) Classification of echolocation clicks from odontocetes in the Southern California Bight. J Acoust Soc Am 129(1):467–475. https://doi.org/10.1121/1.3514383

Rogers TL, Brown SM (1999) Acoustic observations of Arnoux's beaked whale (*Berardius arnuxii*) off Kemp Land, Antarctica (Note). Mar Mamm Sci 15(1):192–198. https://doi.org/10.1111/j.1748-7692.1999.tb00789.x

Roitblat HL, Penner RH, Nachtigall PE (1990) Matching-to-sample by an echolocating dolphin (*Tursiops truncatus*). J Exp Psychol Anim Behav Process 16(1):85–95. https://doi.org/10.1037/0097-7403.16.1.85

Roitblat HL, Au WWL, Nachtigall PE, Shizumura R, Moons G (1995) Sonar recognition of targets embedded in sediment. Neural Netw 8(7):1263–1273. https://doi.org/10.1016/0893-6080(95)00052-6

Romanenko EV, Kitain VY (1992) The functioning of the echolocation system of *Tursiops truncatus* during noise masking. In: Thomas JA, Kastelein RA, Supin AY (eds) Marine mammal sensory systems. Plenum, New York, pp 415–419

Rossbach KA, Herzing DL (1997) Underwater observations of benthic-feeding bottlenose dolphins (*Tursiops truncatus*) near Grand Bahama Island. Bahamas Mar Mamm Sci 13(3):498–504. https://doi.org/10.1111/j.1748-7692.1997.tb00658.x

Rossi-Santos MR, Podos J (2006) Latitudinal variation in whistle structure of the estuarine dolphin *Sotalia guianensis*. Behaviour 143:347–346. https://doi.org/10.1163/156853906775897905

Samarra FIP, Miller PJO (2015) Prey-induced behavioural plasticity of herring-eating killer whales. Mar Biol 162(4):809–821. https://doi.org/10.1007/s00227-015-2626-8

Samarra FIP, Deecke VB, Simonis AE, Miller PJO (2015) Geographic variation in the time-frequency characteristics of high-frequency whistles produced by killer whales (*Orcinus orca*). Mar Mamm Sci 31(2):688–706. https://doi.org/10.1111/mms.12195

Sayigh LS, Tyack PL, Wells RS, Scott MD (1990) Signature whistles of free-ranging bottlenose dolphins *Tursiops truncatus*: stability and mother-offspring comparisons. Behav Ecol Sociobiol 26(4):247–260. https://doi.org/10.1007/BF00178318

Sayigh LS, Tyack PL, Wells RS, Scott MD, Irvine AB (1995) Sex differences in signature whistle production of free-ranging bottlenose dolphins, *Tursiops truncatus*. Behav Ecol Sociobiol 36:171–177. https://doi.org/10.1007/BF00177793

Sayigh L, Quick N, Hastie G, Tyack P (2013) Repeated call types in short-finned pilot whales, *Globicephala macrorhynchus*. Mar Mamm Sci 29(2):312–324. https://doi.org/10.1111/j.1748-7692.2012.00577.x

Sayigh LS, Wells RS, Janik VM (2017) What's in a voice? Dolphins do not use voice cues for individual recognition. Anim Cogn 20(6):1067–1079. https://doi.org/10.1007/s10071-017-1123-5

Sayigh LS, El Haddad N, Tyack PL, Janik VM, Wells RS, Jensen FH (2023) Bottlenose dolphin mothers modify signature whistles in the presence of their own calves. Proc Natl Acad Sci 120(27):e2300262120. https://doi.org/10.1073/pnas.2300262120

Scheifele PM, Andrew S, Cooper RA, Darre M, Musiek FE, Max L (2005) Indication of a Lombard vocal response in the St. Lawrence River beluga. J Acoust Soc Am 117(3):1486–1492. https://doi.org/10.1121/1.1835508

Schevill WE, Lawrence B (1956) Food-finding by a captive porpoise (*Tursiops truncatus*). Breviora 53:1–13

Schevill WE, Watkins WA, Ray C (1969) Click structure in the porpoise, *Phocoena phocoena*. J Mammal 50(4):721–728. https://doi.org/10.2307/1378247

Schorr GS, Falcone EA, Moretti DJ, Andrews RD (2014) First long-term behavioral records from Cuvier's beaked whales (*Ziphius cavirostris*) reveal record-breaking dives. PLoS One 9(3):e92633. https://doi.org/10.1371/journal.pone.0092633

Schultz KW, Cato DH, Corkeron PJ, Bryden MM (1995) Low-frequency narrow-band sounds produced by bottle-nosed dolphins. Mar Mamm Sci 11(4): 503–509. https://doi.org/10.1111/j.1748-7692.1995.tb00673.x

Schulz TM, Whitehead H, Rendell L (2009) Off-axis effects on the multi-pulse structure of sperm whale coda clicks. J Acoust Soc Am 125(3):1768–1773. https://doi.org/10.1121/1.3075598

Schulz TM, Whitehead H, Gero S, Rendell L (2011) Individual vocal production in a sperm whale (*Physeter macrocephalus*) social unit. Mar Mamm Sci 27(1):149–166. https://doi.org/10.1111/j.1748-7692.2010.00399.x

Selbmann A, Miller PJO, Wensveen PJ, Svavarsson J, Samarra FIP (2023) Call combination patterns in Icelandic killer whales (*Orcinus orca*). Sci Rep 13(1): 21771. https://doi.org/10.1038/s41598-023-48349-1

Shaffer JW, Moretti D, Jarvis S, Tyack P, Johnson M (2013) Effective beam pattern of the Blainville's beaked whale (*Mesoplodon densirostris*) and implications for passive acoustic monitoring. J Acoust Soc Am 133(3):1770–1784. https://doi.org/10.1121/1.4776177

Shapiro AD (2006) Preliminary evidence for signature vocalizations among free-ranging narwhals (*Monodon monoceros*). J Acoust Soc Am 120(3):1695–1705. https://doi.org/10.1121/1.2226586

Sharpe DL, Castellote M, Wade PR, Cornick LA (2019) Call types of Bigg's killer whales (*Orcinus orca*) in western Alaska: using vocal dialects to assess population structure. Bioacoustics:1–26 https://doi.org/10.1080/09524622.2017.1396562

Sigurdson JE (1997) Analyzing the dynamics of dolphin biosonar behavior during search and detection tasks. Proceedings of the Institute of Acoustics 19:123–132

Silva TL, Aran Mooney T, Sayigh LS, Baird RW, Tyack PL (2017) Successful suction-cup tagging of a small delphinid species, *Stenella attenuata*: insights into whistle characteristics. Mar Mamm Sci 33(2): 653–668. https://doi.org/10.1111/mms.12376

Simon M, Ugarte F, Wahlberg M, Miller LA (2006) Icelandic killer whales *Orcinus orca* use a pulsed call suitable for manipulating the schooling behaviour of herring *Clupea harengus*. Bioacoustics 16(1):57–74. https://doi.org/10.1080/09524622.2006.9753564

Sjare BL, Smith TG (1986) The relationship between behavioral activity and underwater vocalizations of the white whale, *Delphinapterus leucas*. Can J Zool 64(12):2824–2831. https://doi.org/10.1134/S000143701201016X

Smith AB, Kloepper LN, Yang W-C, Huang W-H, Jen I-F, Rideout BP, Nachtigall PE (2016) Transmission beam characteristics of a Risso's dolphin (*Grampus griseus*). J Acoust Soc Am 139(1):53–62. https://doi.org/10.1121/1.4937752

Smith AB, Pacini AF, Nachtigall PE, Laule GE, Aragones LV, Magno C, Suarez LJA (2019) Transmission beam pattern and dynamics of a spinner dolphin (*Stenella longirostris*). J Acoust Soc Am 145(6):3595–3605. https://doi.org/10.1121/1.5111347

Smolker R, Pepper JW (1999) Whistle convergence among allied male bottlenose dolphins (Delphinidae, *Tursiops* sp.). Ethology 105 (7):595–617. https://doi.org/10.1046/j.1439-0310.1999.00441.x

Soldevilla MS, Henderson EE, Campbell GS, Wiggins SM, Hildebrand JA, Roch MA (2008) Classification of Risso's and Pacific white-sided dolphins using spectral properties of echolocation clicks. J Acoust Soc Am 124(1):609–624. https://doi.org/10.1121/1.2932059

Soldevilla MS, Wiggins SM, Hildebrand JA (2010a) Spatial and temporal patterns of Risso's dolphin echolocation in the Southern California Bight. J Acoust Soc Am 127(1):124–132. https://doi.org/10.1121/1.3257586

Soldevilla MS, Wiggins SM, Hildebrand JA (2010b) Spatio-temporal comparison of Pacific white-sided dolphin echolocation click types. Aquat Biol 9(1): 49–62. https://doi.org/10.3354/Ab00224

Soldevilla MS, Baumann-Pickering S, Cholewiak D, Hodge LEW, Oleson EM, Rankin S (2017) Geographic variation in Risso's dolphin echolocation click spectra. J Acoust Soc Am 142(2):599–617. https://doi.org/10.1121/1.4996002

Sørensen PM, Wisniewska DM, Jensen FH, Johnson M, Teilmann J, Madsen PT (2018) Click communication in wild harbour porpoises (*Phocoena phocoena*). Sci Rep 8(1):9702. https://doi.org/10.1038/s41598-018-28022-8

Sørensen PM, Haddock A, Guarino E, Jaakkola K, McMullen C, Jensen FH, Tyack PL, King SL (2023) Anthropogenic noise impairs cooperation in bottlenose dolphins. Curr Biol 33 (4):P749-754.E744. https://doi.org/10.1016/j.cub.2022.12.063

Sportelli JJ, Jones BL, Ridgway SH (2023) Non-linear phenomena: a common acoustic feature of bottlenose dolphin (*Tursiops truncatus*) signature whistles. Bioacoustics 32(3):241–260. https://doi.org/10.1080/09524622.2022.2106306

Starkhammar J, Moore P, Talmadge L, Houser D (2011) Frequency-dependent variation in the two-dimensional beam pattern of an echolocating dolphin. Biol Lett 7(6):836–839. https://doi.org/10.1098/rsbl.2011.0396

Starkhammar J, Reinhold I, Moore PW, Houser DS, Sandsten M (2019) Detailed analysis of two detected overlaying transient components within the echolocation beam of a bottlenose dolphin (*Tursiops truncatus*). J Acoust Soc Am 145(4):2138–2148. https://doi.org/10.1121/1.5096640

Steiner WW (1981) Species-specific differences in pure tonal whistle vocalizations of five western North Atlantic dolphin species. Behav Ecol Sociobiol 9:241–246. https://doi.org/10.1007/BF00299878

Teilmann J, Miller LA, Kirketerp T, Kastelein RA, Madsen PT, Nielsen BK, Au WW (2002) Characteristics of echolocation signals used by a harbour porpoise (*Phocoena phocoena*) in a target detection experiment. Aquat Mamm 28(3):275–284

Thode A, Mellinger DK, Stienessen S, Martinez A, Mullin K (2002) Depth-dependent acoustic features of diving sperm whales (*Physeter macrocephalus*) in the Gulf of Mexico. J Acoust Soc Am 112(1):308–321. https://doi.org/10.1121/1.1482077

Thomas JA, Turl CW (1990) Echolocation characteristics and range detection threshold of a false killer whale (*Pseudorca crassidens*). In: Thomas JA, Kastelein RA (eds) Sensory abilities of cetaceans. De Spil Publishers, New York, pp 321–333

Thomas J, Stoermer M, Bowers C, Anderson L, Garver A (1988) Detection abilities and signal characteristics of echolocating false killer whales (*Pseudorca crassidens*). In: Nachtigall PE, Moore PWB (eds) Animal sonar: processes and performance. Springer US, Boston, pp 323–328. https://doi.org/10.1007/978-1-4684-7493-0_35

Thomas JA, Moore PWB, Nachtigall PE, Gilmartin WG (1990) A new sound from a stranded pygmy sperm whale. Aquat Mamm 16(1):28–30

Tønnesen P, Oliveira C, Johnson M, Madsen PT (2020) The long-range echo scene of the sperm whale biosonar. Biol Lett 16(8):20200134. https://doi.org/10.1098/rsbl.2020.0134

Tyson RB, Nowacek DP, Miller PJO (2007) Nonlinear phenomena in the vocalizations of North Atlantic right whales (*Eubalaena glacialis*) and killer whales (*Orcinus orca*). J Acoust Soc Am 122(3):1365–1373. https://doi.org/10.1121/1.2756263

van Cise AM, Roch MA, Baird RW, Mooney TA, Barlow J (2017) Acoustic differentiation of Shiho- and Naisa-type short-finned pilot whales in the Pacific Ocean. J Acoust Soc Am 141(2):737–748. https://doi.org/10.1121/1.4974858

van der Woude SE (2009) Bottlenose dolphins (*Tursiops truncatus*) moan as low in frequency as baleen whales. J Acoust Soc Am 126(3):1552–1562. https://doi.org/10.1121/1.3177272

van Ginkel C, Becker DM, Gowans S, Simard P (2018) Whistling in a noisy ocean: bottlenose dolphins adjust whistle frequencies in response to real-time ambient noise levels. Bioacoustics 27(4):391–405. https://doi.org/10.1080/09524622.2017.1359670

Van Parijs SM, Corkeron P (2001a) Boat traffic affects the acoustic behaviour of Pacific humpback dolphins, *Sousa chinensis*. J Mar Biolog Assoc 81(3):533–538. https://doi.org/10.1017/S0025315401004180

Van Parijs SM, Corkeron PJ (2001b) Evidence for signature whistle production by a Pacific humpback dolphin, *Sousa chinensis*. Mar Mamm Sci 17(4):944–949. https://doi.org/10.1111/j.1748-7692.2001.tb01308.x

Van Parijs SM, Corkeron PJ (2001c) Vocalizations and behaviour of Pacific humpback dolphins *Sousa chinensis*. Ethology 107(8):701–716. https://doi.org/10.1046/j.1439-0310.2001.00714.x

Verboom W, Kastelein R (1995) Acoustic signals by harbour porpoises (*Phocoena phocoena*). In: Nachtigall PE, Lien J, Au WWL, Read AJ (eds) Harbour porpoises: laboratory studies to reduce bycatch. DeSpil Publishers, Woerden, pp 1–39

Verfuß UK, Miller LA, Pilz PKD, Schnitzler H-U (2009) Echolocation by two foraging harbour porpoises (*Phocoena phocoena*). J Exp Biol 212(6):823–834. https://doi.org/10.1242/jeb.022137

Vergara V, Michaud R, Barrett-Lennard L (2010) What can captive whales tell us about their wild counterparts? Identification, usage, and ontogeny of contact calls in belugas (*Delphinapterus leucas*). Int J Comp Psychol 23(3):278–309. https://doi.org/10.46867/ijcp.2010.23.03.08

Villadsgaard A, Wahlberg M, Tougaard J (2007) Echolocation signals of wild harbour porpoises, *Phocoena phocoena*. J Exp Biol 210(1):56–64. https://doi.org/10.1242/Jeb.02618

Wahlberg M, Jensen F, Soto N, Beedholm K, Bejder L, Oliveira C, Rasmussen M, Simon M, Villadsgaard A, Madsen P (2011) Source parameters of echolocation clicks from wild bottlenose dolphins (*Tursiops aduncus* and *Tursiops truncatus*). J Acoust Soc Am 130(4):2263–2274. https://doi.org/10.1121/1.3624822

Walmsley SF, Rendell L, Hussey NE, Marcoux M (2020) Vocal sequences in narwhals (*Monodon monoceros*). J Acoust Soc Am 147(2):1078–1091. https://doi.org/10.1121/10.0000671

Ward R, Parnum I, Erbe C, Salgado-Kent CP (2016) Whistle characteristics of Indo-Pacific bottlenose dolphins (*Tursiops aduncus*) in the Fremantle Inner Harbour, Western Australia. Acoust Aust 44(1):159–169. https://doi.org/10.1007/s40857-015-0041-4

Watkins WA (1967) The harmonic interval: fact or artifact in spectral analysis of pulse trains. In: Tavolga WN (ed) Marine bio-acoustics, vol 2, pp 15–43. https://doi.org/10.1575/1912/2726

Watkins WA, Schevill WE (1980) Characteristic features of the underwater sounds of *Cephalorhynchus commersonii*. J Mammal 61(4):738–739. https://doi.org/10.2307/1380327

Watkins WA, Schevill WE, Best PB (1977) Underwater sounds of *Cephalorhynchus heavisidii* (Mammalia: Cetacea). J Mammal 58(3):316–320. https://doi.org/10.2307/1379330

Wei C, Zhang Y, Au WWL (2014) Simulation of ultrasound beam formation of baiji (*Lipotes vexillifer*) with a finite element model. J Acoust Soc Am 136(1):423–429. https://doi.org/10.1121/1.4883597

Wei C, Wang Z, Song Z, Wang K, Wang D, Au WWL, Zhang Y (2015) Acoustic property reconstruction of a neonate Yangtze finless porpoise's (*Neophocaena asiaeorientalis*) head based on CT imaging. PLoS One 10(4):e0121442. https://doi.org/10.1371/journal.pone.0121442

Wei C, Au WWL, Song Z, Zhang Y (2016) The role of various structures in the head on the formation of the biosonar beam of the Baiji (*Lipotes vexillifer*). J Acoust Soc Am 139(2):875–880. https://doi.org/10.1121/1.4941780

Wei C, Au WWL, Ketten DR, Song Z, Zhang Y (2017) Biosonar signal propagation in the harbor porpoise's (*Phocoena phocoena*) head: the role of various structures in the formation of the vertical beam. J Acoust Soc Am 141(6):4179–4187. https://doi.org/10.1121/1.4983663

Wei C, Au WWL, Ketten DR, Zhang Y (2018a) Finite element simulation of broadband biosonar signal propagation in the near- and far-field of an echolocating Atlantic bottlenose dolphin (*Tursiops truncatus*). J Acoust Soc Am 143(5):2611–2620. https://doi.org/10.1121/1.5034464

Wei C, Song Z, Au WWL, Zhang Y, Wang D (2018b) A numerical evidence of biosonar beam formation of a neonate Yangtze finless porpoise (*Neophocaena asiaeorientalis*). J Theoret Comput Acoustics 26(02):1850009. https://doi.org/10.1142/S2591728518500093

Wei C, Au WWL, Ketten DR (2020) Modeling of the near to far acoustic fields of an echolocating bottlenose dolphin and harbor porpoise. J Acoust Soc Am 147(3):1790–1801. https://doi.org/10.1121/10.0000918

Wei C, Hoffmann-Kuhnt M, Au WWL, Ho AZH, Matrai E, Feng W, Ketten DR, Zhang Y (2021) Possible limitations of dolphin echolocation: a simulation study based on a cross-modal matching experiment. Sci Rep 11(1):6689. https://doi.org/10.1038/s41598-021-85063-2

Wei C, Gill LG, Erbe C, Smith AB, Yang W-C (2022) The distinctive forehead cleft of the Risso's dolphin (*Grampus griseus*) hardly affects biosonar beam formation. Animals 12(24):3472. https://doi.org/10.3390/ani12243472

Wei C, Houser D, Erbe C, Matrai E, Ketten D, Finneran JJ (2023) Does rotation increase the acoustic field of view? Comparative models based on CT data of a live versus a dead dolphin. Bioinspir Biomim 18(3):035006. https://doi.org/10.1088/1748-3190/acc43d

Wei C, Erbe C, Smith AB, Yang W-C (2024) Validated 3D finite-element model of the Risso's dolphin (*Grampus griseus*) head anatomy demonstrates gular sound reception and channelling through the mandibular fats. Bioinspiration & Biomimetics 19(5):056025. https://doi.org/10.1088/1748-3190/ad7344

Weilgart L, Whitehead H (1993) Coda communication by sperm whales (*Physeter macrocephalus*) off the Galapagos Islands. Can J Zool 71:744–752. https://doi.org/10.1139/z93-098

Weilgart L, Whitehead H (1997) Group-specific dialects and geographical variation in coda repertoire in South Pacific sperm whales. Behav Ecol Sociobiol 40(5):277–285. https://doi.org/10.1007/s002650050343

Weir CR, Frantzis A, Alexiadou P, Goold JC (2007) The burst-pulse nature of 'squeal' sounds emitted by sperm whales (*Physeter macrocephalus*). J Mar Biolog Assoc 87(1):39–46. https://doi.org/10.1017/S0025315407054549

Wellard R, Erbe C, Fouda L, Blewitt M (2015) Vocalisations of killer whales (*Orcinus orca*) in the Bremer Canyon. Western Australia PLOS ONE 10(9):e0136535. https://doi.org/10.1371/journal.pone.0136535

Wellard R, Lightbody K, Fouda L, Blewitt M, Riggs D, Erbe C (2016) Killer whale (*Orcinus orca*) predation on beaked whales (*Mesoplodon* spp.) in the Bremer Sub-Basin, Western Australia. PLoS One 11(12):e0166670. https://doi.org/10.1371/journal.pone.0166670

Wellard R, Pitman RL, Durban J, Erbe C (2020) Cold call: the acoustic repertoire of Ross Sea killer whales (*Orcinus orca*, Type C) in McMurdo Sound. Antarctica R Soc Open Sci 7(2):191228. https://doi.org/10.1098/rsos.191228

Wells RS, Bassos-Hull K, Norris KS (1998) Experimental return to the wild of two bottlenose dolphins. Mar Mamm Sci 14(1):51–71. https://doi.org/10.1111/j.1748-7692.1998.tb00690.x

Wiersma H (1988) The short-time-duration narrow-bandwidth character of odontocete echolocation signals. Animal sonar processes and permormance NATO ASI series, Series A: life sciences, vol 156 (223):115–120

Wilson M, Hanlon RT, Tyack PL, Madsen PT (2007) Intense ultrasonic clicks from echolocating toothed whales do not elicit anti-predator responses or debilitate the squid *Loligo pealeii*. Biol Lett 3(3):225–227. https://doi.org/10.1098/rsbl.2007.0005

Winn HE, Perkins PJ, Winn L (1970) Sounds and behavior of the northern bottle-nosed whale. In: Proceedings of the 7th annual conference on biological sonar and diving animals. Stanford Research Institute, Menlo Park, CA, pp 53–59

Wisniewska DM, Johnson M, Beedholm K, Wahlberg M, Madson P (2012) Acoustic gaze adjustments during active target selection in echolocating porpoises. J Exp Biol 215(24):4358–4373. https://doi.org/10.1242/jeb.074013

Wisniewska DM, Johnson M, Nachtigall PE, Madsen PT (2014) Buzzing during biosonar-based interception of prey in the delphinids *Tursiops truncatus* and *Pseudorca crassidens*. J Exp Biol 217(24):4279–4282. https://doi.org/10.1242/jeb.113415

Wisniewska DM, Ratcliffe JM, Beedholm K, Christensen CB, Johnson M, Koblitz JC, Wahlberg M, Madsen PT (2015) Range-dependent flexibility in the acoustic field of view of echolocating porpoises (*Phocoena phocoena*). eLife:e05651. https://doi.org/10.7554/eLife.05651

Wright BM, Deecke VB, Ellis GM, Trites AW, Ford JKB (2021) Behavioral context of echolocation and prey-handling sounds produced by killer whales (*Orcinus orca*) during pursuit and capture of Pacific salmon (*Oncorhynchus* spp.). Mar Mamm Sci 37(4): 1428–1453. https://doi.org/10.1111/mms.12836

Yoshida YM, Morisaka T, Sakai M, Iwasaki M, Wakabayashi I, Seko A, Katsamatsu M, Akamatsu T, Kohshima S (2014) Sound variation and function in captive Commerson's dolphins (*Cephalorhynchus commersonii*). Behav Process 108:11–19. https://doi.org/10.1016/j.beproc.2014.08.017

Yuan J, Wang Z, Duan P, Xiao Y, Zhang H, Huang Z, Zhou R, Wen H, Wang K, Wang D (2021) Whistle signal variations among three Indo-Pacific humpback dolphin populations in the South China Sea: a combined effect of the Qiongzhou Strait's geographical barrier function and local ambient noise? Integr Zool 16(4):499–511. https://doi.org/10.1111/1749-4877.12531

Yurk H, Barrett-Lennard L, Ford JKB, Matkin CO (2002) Cultural transmission within maternal lineages: vocal clans in resident killer whales in southern Alaska. Anim Behav 63(6):1103–1119. https://doi.org/10.1006/anbe.2002.3012

Zaitseva KA, Korolev VI, Akhi AV, Akhi AA (2022) Adaptation of dolphins' (*Tursiops truncatus*) location signals when searching for and identifying objects hidden by sea sediments. Bioacoustics 31(5): 535–544. https://doi.org/10.1080/09524622.2021.1994467

Zimmer WMX, Madsen PT, Teloni V, Johnson MP, Tyack PL (2005a) Off-axis effects on the multipulse structure of sperm whale usual clicks with implications for sound production. J Acoust Soc Am 118(5): 3337–3345. https://doi.org/10.1121/1.2082707

Zimmer WMX, Tyack PL, Johnson MP, Madsen PT (2005b) Three-dimensional beam pattern of regular sperm whale clicks confirms bent-horn hypothesis. J Acoust Soc Am 117(3):1473–1485. https://doi.org/10.1121/1.1828501

Zwamborn EMJ, Whitehead H (2017a) The baroque potheads: modification and embellishment in repeated call sequences of long-finned pilot whales. Behaviour 154(9–10):963–979. https://doi.org/10.1163/1568539X-00003451

Zwamborn EMJ, Whitehead H (2017b) Repeated call sequences and behavioural context in long-finned pilot whales off Cape Breton, Nova Scotia. Canada Bioacoustics 26(2):169–183. https://doi.org/10.1080/09524622.2016.1233457

Open Access This chapter is licensed under the terms of the Creative Commons Attribution 4.0 International License (http://creativecommons.org/licenses/by/4.0/), which permits use, sharing, adaptation, distribution and reproduction in any medium or format, as long as you give appropriate credit to the original author(s) and the source, provide a link to the Creative Commons license and indicate if changes were made.

The images or other third party material in this chapter are included in the chapter's Creative Commons license, unless indicated otherwise in a credit line to the material. If material is not included in the chapter's Creative Commons license and your intended use is not permitted by statutory regulation or exceeds the permitted use, you will need to obtain permission directly from the copyright holder.

Pinniped Sounds

5

Sylvia K. Parsons, Christine Erbe, Sarah A. Marley, and Miles J. Parsons

Contents

5.1	**Introduction**	352
5.2	**Sound Production**	353
5.2.1	Vocal Sounds	353
5.2.2	Nonvocal Sounds	355
5.2.3	Sound Types	356
5.2.4	Pattern Sequencing and Song	357
5.3	**Associated Behavioral Functions**	358
5.3.1	Affiliation	359
5.3.2	Courtship and Mating	361
5.3.3	Competition and Agonistic Behavior	362
5.3.4	Feeding	362
5.3.5	Echolocation	363
5.3.6	Navigation	363
5.4	**Call Variability**	363
5.4.1	Phylogeny (Odobenids, Otariids, and Phocids)	363
5.4.2	Polar vs. Nonpolar Species	364
5.4.3	Geographic Variation	365
5.4.4	Environmental Drivers	367
5.4.5	Ontogeny and Allometry	368
5.4.6	Sex	369
5.4.7	Individual Identity	369
5.4.8	Vocal Development, Plasticity, and Learning	371
5.4.9	Anti-masking Strategies	372

S. K. Parsons (✉) · C. Erbe
Centre for Marine Science and Technology, Curtin University, Perth, WA, Australia
e-mail: c.erbe@curtin.edu.au

S. A. Marley
Scotland's Rural College (SRUC), Aberdeen, UK
e-mail: sarah.marley@sruc.ac.uk

M. J. Parsons
Australian Institute of Marine Science, Crawley, WA, Australia
e-mail: m.parsons@aims.gov.au

5.5	**Taxonomic Overview of Pinniped Sounds**	372
5.5.1	Family Odobenidae: Walrus	373
5.5.2	Family Otariidae: Eared Seals	387
5.5.3	Family Phocidae: True, Earless, Hair Seals	401
5.6	**Summary**	423
5.7	**Data Availability**	423
	References	424

5.1 Introduction

Pinnipeds are a diverse and widely distributed clade of carnivorous, amphibious mammals, comprising the Odobenidae (walruses), Otariidae (eared seals), and Phocidae (earless seals) families (Campagna and Harcourt 2021; Costa and McHuron 2022). For clarity and simplicity, we refer to the odobenids as walruses but eared seals as "otariids" and earless seals as "phocids" throughout this chapter. Walruses and otariids share enough traits that pinnipeds have been considered diphyletic, with phocids in one group and walruses and otariids in the other (Berta et al. 2015); presently, they are treated as three separate families (Committee on Taxonomy 2023). They are highly vocal, using sound in most of their communication (Charrier 2021; Charrier and Casey 2022; Insley et al. 2003). In contrast to fully aquatic fauna, such as cetaceans, which conduct all life functions at sea, these semiaquatic animals spend time under water and in air (i.e., on land or ice). All pinnipeds use the water largely for feeding, while rearing of pups mostly occurs in air (Churchill and Clementz 2016). However, reproductive behavior differs between the phocids and the walruses and otariids, which conduct courtship primarily at sea and on land, respectively (Campagna and Harcourt 2021; Costa and McHuron 2022). As the use of acoustic cues is often function- and life stage-specific, this spatial and temporal separation is critical when assessing the sounds pinnipeds produce. Further, sound propagates differently in air and under water, and producing sound in each medium holds different physiological requirements. As a result, the sound source, signal characteristics, and the received acoustic information or effects need to be considered differently. Therefore, we characterize in-air and underwater sounds produced by pinnipeds separately.

Studies on vocalizations produced in air typically predate those on underwater vocalizations, for reasons of access. For example, a handful of pinniped species are common in zoos and aquaria, and their aerial vocalizations have been well studied in captivity, long before those of their counterparts in the wild. In contrast, polar pinnipeds have historically been studied the least, due to their inaccessibility, yet these species might be at greatest risk because of environmental and anthropogenic stressors (e.g., climate change, receding sea ice, increased ship traffic—in particular in the Arctic). Miksis-Olds et al. (2016) reviewed habitats of polar pinniped species (including mating substrate and whelping habitat) and the number of call types, the existence of dialects, and sex-specific and individually distinctive vocalizations, linking ecological aspects to key bioacoustic observations.

Here, we present an overview of the present literature available on in-air and underwater sounds produced by all pinniped species; summarize their call types, variability, and behavioral functions; provide some insights into ecological drivers of acoustic behavior by way of notable example studies; and offer a tabulated summary of reported temporal and spectral characteristics for each species (Table S1, which may be requested from the first author). For selected cases, we present some comparisons of the call characteristics among species groups, sex, and age. In our taxonomic description of known sounds for each species, we provide a note on whether in-air or underwater vocalizations have been reported, a summary of the typical frequency band that encompasses all reported call spectra, and anecdotal insights into their acoustic ecology. Scientific names are given in the Table of Contents, and common names are used in the main body text.

To complete this review, we conducted literature searches on each extant pinniped species. Using Google Scholar and Scopus, we searched for articles that included "species scientific name" *or* "species common name" *and* "sound" *or* "call" *or* "acoustic" *or* "vocal" *or* "noise" in the title or keywords. We then assessed citations found within these publications to see if they included articles missed in the original search and added those that provided additional information on the target or other pinniped species. This search provided a dataset of 249 papers with quantitative information on acoustic characteristics and an additional 61 papers with descriptions of sounds. We compiled the quantitative data, including spectral peak, minimum, and maximum frequencies; bandwidth; minimum, maximum, and mean duration; and source level, if reported (Table S1).

There are no standards for categorizing pinniped calls, and, similar to other taxa, their sounds lie along a continuum, intergrading multiple types of calls with different acoustic properties (Parsons et al. 2024). Further, the characteristics gleaned from spectrograms, either by the original authors or the authors of this chapter, are affected by the respective processing settings (e.g., recording environment and equipment) which can alter the appearance of any spectrograms produced. Therefore, although we have attempted to standardize the extraction of acoustic characteristics from spectrograms and minimize variation, the line between tonal and pulsed sounds has occasionally been drawn in different places among articles. Data on acoustic characteristics for each species were collated and figures produced using purpose-written code in MATLAB (version 2020b; The MathWorks Inc., Natick, MA, USA). Given the variety of methods for call categorization by previous studies, we have not compared the complexity or diversity of species' acoustic repertoires, as this is likely affected by observer bias.

Finally, similar to cetaceans, species names have changed over time. For example, the South American sea lion (*Otaria flavescens*) was formerly known as *O. byronia*. Hence, species names in this chapter might not match those of the cited literature.

5.2 Sound Production

Pinnipeds generate sound by various mechanisms. They produce vocal calls, such as an agonistic roar (e.g., McCann 1981; Sandegren 1976), sounds of inhalation and exhalation (e.g., McCann 1981; Sandegren 1976), and mechanical sounds through the stridulation or knocking of body parts together or on the ground (e.g., Wahlberg et al. 2002).

5.2.1 Vocal Sounds

Here, we provide a brief overview of example mechanisms of vocal sound production by pinnipeds. For further details on these physiological systems and how differences in their application can vary the acoustic characteristics of pinniped sounds, see more focused articles (e.g., de Reus et al. 2022; Reidenberg and Laitman 2010; Tyack and Miller 2002).

In air, terrestrial mammals typically produce sound by channeling air across their vocal folds or cords, through an open larynx, along the vocal tract, and exhale through oral or nasal passages, as reflected in the recordings of pinnipeds vocalizing with open or closed mouths. Open vocal cords release pressure from the lungs and allow respiration, while closed vocal cords and larynx increase internal pressure and protect from incursion of water or foreign objects. When the vocal cords are partially open, air exhaled from the lungs passes across them and causes their edges to oscillate. The vibrating folds generate pressure waves producing sound that moves along the vocal tract. In terrestrial animals, vocal cords are oriented perpendicular to the airflow, and in some fully aquatic mammals, such as odontocetes, they are parallel to the airflow (Reidenberg 2017). In pinnipeds, however, their orientation varies, positioned perpendicular to the airflow in some species (e.g., gray and harbor seals, walrus), yet parallel in others (e.g., elephant seal and California sea lion). Orientation of the vocal cords does not follow phylogenetic affinity, suggesting an adaptive component to the trait.

The combination of the vocal cords and tract determines the spectral content of the sound. The frequency at which the vocal cords vibrate produces the fundamental frequency, and the resonances of the vocal tract filter the sound, creating regions of frequency prominence, known as formants. These formants differ from harmonics in that the vocal tract is not considered the source of the sound, as would be the case for harmonics. In allometrically (size-dependent) "honest" calls, the structure of formants provides significant information on the age, size, and resource-holding potential of individuals, so that vocal displays may be used to resolve interactions without physical contact (Sanvito et al. 2007c). However, as discussed below, many pinnipeds can control their vocal cords, tracts, and associated tissues, thus varying characteristics of their sounds and producing "dishonest" sounds that are unexpectedly lower or higher in frequency than anticipated for the size and physical attributes of the individual (de Reus et al. 2022). For example, similar to humans, sound characteristics can be varied by the position of the tongue, whether at the back of the mouth or at the teeth (Frankel 2009). South American sea lions push their lower jaw down and extend their tongue to varying degrees during exhalation while keeping their nostrils closed, and Australian sea lions vibrate the posterior of their tongue against the soft palate to produce the sound of their barks (Marlow 1975; Tyack and Miller 2002). Changes to the shape and size of the throat or the tongue can be utilized. Elephant seals can double the diameter of their vocal tract by opening their mouth, significantly changing the spectra of the calls and, therefore, the formant frequencies (Sanvito et al. 2007c), and walruses can effect subpharyngeal filtering by varying the shape of a dexterous tongue, such that even brief calls of this species are highly diverse (Kastelein et al. 1997; Miller 1985).

Under water, the larynx is closed and the vocal cords usually are sealed against each other to avoid drowning, preventing the release of air. Sound production by passing air over the vocal cords is only possible until the internal air pockets (oral and nasal) are filled with air and can no longer be compressed by further transfer from the lungs. This only allows short, quiet sounds to be produced before the air must be transferred back to the lungs. As a result, an alternative method of air transfer, storage, or vibration is required for effective sound production. To achieve this, many marine mammals pass air from their lungs into other elastic structures, such as tracheal, pharyngeal, laryngeal, and nasal air sacs (Reidenberg and Laitman 2010). These reservoirs may act as secondary vibrators, enabling open-mouthed sound production under water, despite the closed larynx and vocal cords. This was first observed in California sea lions as they barked under water without emitting bubbles, simultaneously producing two different types of sound (Brauer et al. 1966).

There are multiple variants of these mechanisms. Weddell and Ross seals vibrate an anterior tracheal membrane (Piérard 1969; Tyack and Miller 2002), while bearded seals possess an extensive dorsal tracheal membrane that extends down toward the lungs and is used to produce long (>1 min) sounds (Burns 1981; Ray et al. 1969). The recycling of air to the lungs facilitates the production of long-duration sounds without needing to surface to breathe, as would be required if the air was exhaled (Reidenberg and Laitman 2010). Reidenberg (2017) hypothesized that sound transmission from vibrating pharyngeal or tracheal tissues may occur through the adjoining neck tissues, meaning ontogenic (age-/life-stage-related) and sex-driven differences in the thickness and density of neck tissues may affect sound production and characteristics. Other species possess more visible mechanisms. Male ribbon seals, for example, have a large sac on the right-hand side of the trachea that is either small or not present in females and may be used for sound production (Tyack and Miller 2002).

The clearest of all adaptations might be in the hooded seal, which has an inflatable nasal septum. Part of this septum is highly elastic and can be extruded through one nostril as a large, brightly colored, air-filled bladder (Berland 1958; Porov 1961; Tyack and Miller 2002) that vibrates during inflation and deflation, producing distinctive "bloop," "ping," and "whoosh" sounds.

Using these mechanisms, pinnipeds have been found to actively control their sounds in both media. For example, harp and Weddell seals produce vocalizations both in air and under water; the latter are produced between 10 and 35 m below the water's surface, at a similar frequency to those in air, and without changing call frequency with depth (Moors and Terhune 2005). These observations support the use of adapted tissues and active control of the vocal cords across the two media, to maintain the fundamental frequency against varying external pressure.

Pinniped heads and throats are not omnidirectional sound sources, and, as might be expected, the directivity index of their calls has a positive relationship with frequency. Holt et al. (2010) reported significantly different one-third octave levels above 1 kHz at more than 45° from directly in front of the head of northern elephant seals and that the directivity index of one-third octave bands ranged from 0.75 to 7.3 dB. Thus, sound spectra may vary with the orientation of the listener to the calling pinniped, and this directionality is dependent on the medium in which the sound is produced. This array of potential mechanisms and subtle differences in sound production contribute somewhat to the significant inter- and intraspecies variation, yet it is interesting to note that some call types occur both in air and under water, where the requirements for producing sound are significantly different (e.g., Terhune et al. 1994; Thomas and Kuechle 1982).

5.2.2 Nonvocal Sounds

Not all pinniped acoustic communication stems from sounds produced by vocalization, nor are the sounds detected by purely auditory mechanisms. Flipper slaps are commonly used to communicate information. For example, a female Steller sea lion may slap her belly, while in a standing position, to pose a threat to other cows, and young bulls often slap wet rocks with their front flippers to scare off surrounding animals (Bishop et al. 2014; Sandegren 1970). Spotted seals in captivity slap their flippers against the water surface to communicate with keepers (Beier and Wartzok 1979), and harbor seals, gray seals, and walruses (among several species) have been observed conveying information by slapping their flippers together in their natural environments (e.g., Hocking et al. 2020; Larsen and Reichmuth 2021; Sabinsky et al. 2017; Wahlberg et al. 2002).

Another common mechanism of sound production is via the snapping or chattering of jaws. Weddell seals, for example, have been observed producing "rapid clattering of teeth" and "teeth chatter" (as a high-intensity threat), jaw snaps, and jaw claps (Russell et al. 2016; Terhune et al. 1994; Thomas 1979). These sounds may also be produced with vocal sounds, such as the trills and knocks of a male Weddell seal in the presence of females within his territory, which may be reciprocated by female jaw claps (Russell et al. 2016). Some signals have been recorded without a known source mechanism, though several are often hypothesized, such as the teeth clacking, tongue movement, or suction for sounds produced by walruses (Larsen and Reichmuth 2021; Reichmuth et al. 2009; Sjare and Stirling 1996; Sjare et al. 2003). Less commonly reported communication techniques include bubble blowing by harbor seals (Sabinsky et al. 2017) and the loud, splashing "fall into the water" sound Baikal seals may use to alert conspecifics to danger (Petrov et al. 2022). Some vocal signals may have similar characteristics to nonvocal sounds (Asselin et al. 1993; Ballard and Kovacs 1995; Hocking et al. 2020; Kunnasranta et al. 1996; Larsen and Reichmuth 2021; McCulloch 2000; Mizuguchi et al. 2016a; Møhl et al. 1975; Rautio et al. 2009; Ray and Watkins 1975; Sills and Reichmuth 2022; Stirling et al. 1987).

Finally, ground-borne signals have been reported from the movement and vocalizations of northern elephant seals on the beach, producing seismic vibrations that can propagate up to 20 m at levels above background noise, providing important information about the presence and behavior of potential competing males on a breeding beach (Bishop et al. 2014; Shipley et al. 1992). Such multimodal approaches to communication can reach the receiver visually, acoustically, or haptically, allowing multisensory integration, and, thus, have advantages but also energetic considerations (Verga and Ravignani 2021).

5.2.3 Sound Types

There are no guidelines or standards for sorting pinniped calls into classes or types, and historically, similar to other marine fauna (e.g., Chaps. 6 and 7), researchers have applied their own methods of grouping sounds using techniques that have changed with time (Parsons et al. 2024). Pinniped call types are sometimes named onomatopoeically by using the closest human phonetic description of the sounds (e.g., if it sounds like a bark, call it a bark); by a musical term, such as a trill; by an associated behavior, such as a pup-attraction call (PAC) or a female-attraction call (FAC) that is also often described as sounding like a bleat; and less often by a physics-based assessment of the acoustic characteristics (e.g., tonal, constant frequency, frequency-modulated, amplitude-modulated, pulsive, or impulsive; Fig. 5.1). As a result, numerous overlapping call types have been created, often within individual studies. Van Opzeeland et al. (2010), for example, studied four species (crabeater, leopard, Ross, and Weddell seals) and reported a total of 26 different underwater call types across three species, while Terhune et al. (2008) categorized 30 different types of Weddell seal trill alone.

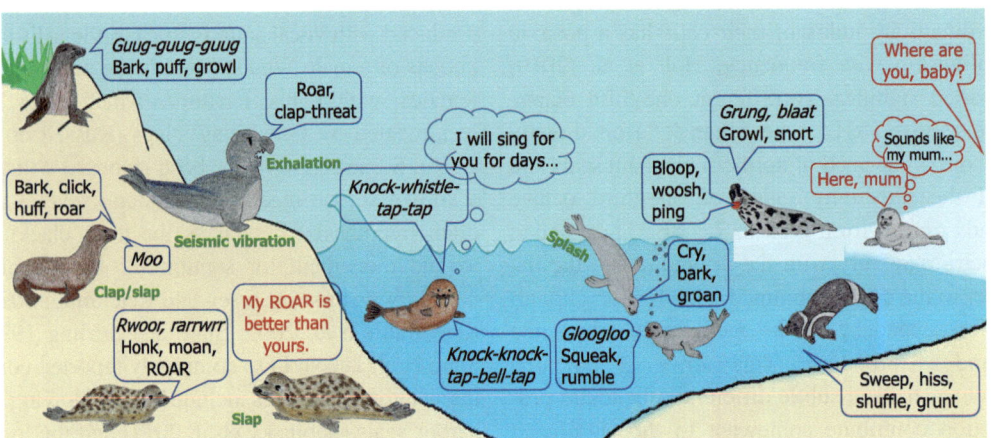

Fig. 5.1 Example sounds produced by a selection of pinniped species in their environment (on land, on ice, or under water). Sounds, as categorised in the literature, are shown in speech bubbles, with some qualitative onomatopoeic descriptions in italics, inferred meanings denoted by red text, and the example of interaction between a mother and pup (Weddell seal, *Leptonychotes weddellii*) shown by red outlines. Behaviors associated with the sounds (e.g., recognition by *L. weddellii* pup) are denoted by thought bubbles. Some examples of non-vocal sounds are shown in green. Note, these examples are not an exhaustive list of sound types reported for each species, nor the number of species reported to produce sound. From left to right: South American fur seal (*Arctocephalus australis*) bark, generic sea lion bark and clap, harbor seal (*Phoca vitulina*) roars, elephant seal (*Mirounga*) seismic vibration, walrus (*Odobenus rosmarus*) song, Mediterranean monk seal (*Monachus monachus*) splash and cry, hooded seal (*Cystophora cristata*) bloop, ribbon seal (*Histriophoca fasciata*) sweep, and Weddell seal mother-pup calls

Intergradation between call types occurs when there is graded variation in structure within and among different call types of the same species (i.e., a call may fit the criteria for categorization into multiple call types; Brady et al. 2020). This phenomenon has been observed in several aquatic and terrestrial species and can be problematic for the categorization of calls (Barklow 1997; Charlton 2015; Fischer et al. 2017; Fitch et al. 2002). If the calls are being categorized via visual scrutiny of spectrograms, intergradation is exacerbated by the influence of the settings applied when generating the spectrogram. Further, like many cetaceans, pinnipeds can vary the method and type of sound they produce within a timeframe, which may be considered a single call (Schevill and Watkins 1971). This can cause confusion about the category a sound may fit (potentially more than one) and is typically reported according to the preference of the authors. For example, in Fig. 5.2, it appears that the start of the trill is tonal (very narrowband), before amplitude modulation is introduced after a few seconds (the tone appears more broadband and smudged, with sidebands), and it finally becomes pulsed (rapid pulses from 15 s onward that slow and then remain consistent toward the end). What the animal may perceive as a unitary call type, the observer may perceive differently. The change in the appearance of the signal is a function of the pulse rate of the seal's call, as well as the parameters used to produce the spectrogram.

We categorized in-air and underwater pinniped sounds into five main categories: (1) tonal (a sinusoidal signal appearing in spectrograms as a narrowband contour with or without harmonic overtones), (2) pulsed (short-duration, broadband pulses), (3) burst-pulsed (rapid series of pulses which are typically not resolved in spectrograms but instead appear as a tonal call, often with many overtones and sidebands), (4) click-like (very sharp, <1 s, broadband, high frequency into tens of kilohertz), and (5) noisy (broadband and "blurred" without clear features in spectrograms). We also retained the original classifications and labels from the literature (e.g., bark, clack, clap, growl, grunt, moan, roar, or whistle). We summarized our literature search in Table S1.

5.2.4 Pattern Sequencing and Song

Vocalizations can be emitted as individual calls or sequences of sounds, often in particular repeated rhythms. Patterned call sequences have been described for many pinniped species, including bearded seals (Cleator et al. 1989), California sea lions (Schusterman 1977), and South American sea lions (Fernández-Juric et al. 2003). These sequences can encode information related to the individual in the rhythm of the calls rather than call spectra, and, as they are often less affected by signal attenuation, rhythms can provide greater information than the acoustic characteristics of individual calls alone (Rogers and Cato 2002), especially at distance.

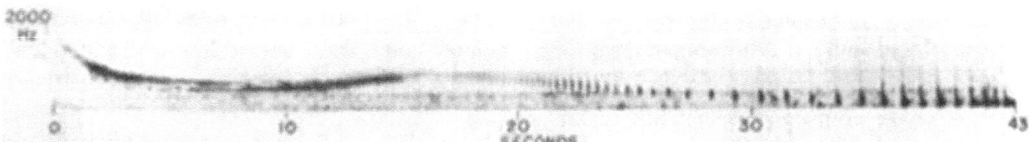

Fig. 5.2 Spectrogram of a Weddell seal trill showing a graded structure from tonal to pulsed over its 43-s duration. Reproduced from Schevill WE, Watkins WA (1971) Directionality of the sound beam in *Leptonychotes weddelli* (Mammalia: Pinnipedia). Antarctic Research Series 18:163–168. https://doi.org/10.1029/AR018p0163. With permission from Wiley. © American Geophysical Union. All rights reserved

Complex sequences of learned vocalizations have been defined as song (Fitch 2006). Several studies of pinnipeds have found sequences that are sufficiently complex to be deemed songs, as in walruses (Reichmuth and Casey 2014), leopard seals (Rogers and Cato 2002), and Weddell seals (e.g., Thomas and Kuechle 1982; Thomas et al. 1988), as well as ringed, Ross, bearded, Hawaiian monk, and ribbon seals (Jones et al. 2014; Ray et al. 1969; Terhune 2019; Van Parijs et al. 2003b). The term "song" is often used to describe these call sequences in pinnipeds.

Songs are predominantly reported for male pinnipeds, in relation to courtship (Van Parijs et al. 1997). These call sequences typically develop in species that display in large groups with complex social structures and vary significantly in complexity (Terhune 2019). As defined, song is believed to show evidence of vocal plasticity, but in many cases, evidence for learning is limited or unavailable. Male walruses, for example, have been suggested to learn song as adult males during the breeding season, and their sound production via nontypical mammalian mechanisms has been suggested to rival that of odontocetes (Reichmuth and Casey 2014), potentially changing and learning new songs between years (Sjare et al. 2003). In contrast, harbor seals are semi-solitary animals and yet are still speculated by some to learn songs with minimal exposure (Reichmuth and Casey 2014).

Duengen et al. (2023) hypothesized that harbor seal mothers give birth around the time male calling starts and that pups are sensitive learners during weaning periods, which coincides with foraging trips and the peak singing of adult males. This combination may give solitary males the opportunity to learn songs, acquiring sequence elements as pups. This may explain evidence of versatility and potentially regional variation in calls of pinniped species (discussed in sections by species below). However, whether the species vocalizations meet the criteria for "song" is left to the reporting scientist, and the evidence for species-specific vocal learning requires validation. Given the vocal flexibility of pinnipeds, and that they are often presented as examples of singers that demonstrate vocal plasticity, exploring and confirming this as a species trait is an important topic of future research.

5.3 Associated Behavioral Functions

Pinnipeds use sound to convey cues associated with a variety of life functions, notably recognition (e.g., mother–pup communication), feeding, courtship, and competition (e.g., Insley et al. 2003; Nel 1966; Rogers 2003; Sanvito and Galimberti 2000a; Sanvito et al. 2007c; Shabangu et al. 2022; Fig. 5.3). Some call types are used in the context of multiple behaviors, and different species may use the same sound type for different behaviors or different sounds for the same behaviors. In addition, some broad behavioral categories (e.g., mating) may include multiple behaviors and call types, and the validation of these types is nontrivial. For example, in species that mate in the water, males vocalize under water during the mating season (e.g., Parnell 2018). There are different hypotheses for the function of these vocalizations: (1) to ward off other males, (2) to defend territory, (3) as male–male competition, and (4) to attract females. It is possible that they have more than one function. Male vocalizations may be directed at males and/or females (Van Parijs 2003). In species that mate on land, male vocalizations are more likely directed at other males to ward off competitors, convey boundaries, defend territory, and defend harems. Here, we briefly describe some of these categories. For further reading, see Campagna and Harcourt (2021), Charrier (2021), Charrier and Casey (2022), and Costa and McHuron (2022).

Fig. 5.3 Photos of hauled-out pinnipeds during vocal communication in the Perth metropolitan area, Western Australia. A female southern elephant seal (*Mirounga leonina*) out of her usual range, visiting Point Peron, Western Australia (top left). Male Australian sea lions (*Neophoca cinerea*) on Seal and Carnac Island, Western Australia

5.3.1 Affiliation

Call recognition, or contact calling, is the ability of an individual to discriminate the calls of another individual from a group of conspecific sounds and use the information to maintain contact. It requires not only distinctiveness of the call but also the ability of the receiver to discriminate individual differences. Individual information is encoded within the call and is discussed in the section below on individual variability. For pinnipeds, two main associated functions of recognition are typically reported: communication between mother and pup and communication between neighbors or competitors. Studies have typically either validated these functions through playback experiments, such as for Antarctic fur seals (Aubin et al. 2015), Australian sea lions (Charrier et al. 2009; Pitcher et al. 2012), harbor seals (Sauve et al. 2015b), northern elephant seals (Casey et al. 2015; Linossier et al. 2021), and Atlantic walruses (Charrier et al. 2010), or inferred them from long-term observation such as for Galapagos fur seals (Trillmich 1981), northern elephant seals (Linossier et al. 2021; Petrinovich 1974), subantarctic fur seals (Charrier et al. 2003a; Insley 1992; Roux and Jouventin 1987), and Atlantic walruses (Charrier et al. 2010; Stirling et al. 1987), though these examples do not form an exhaustive list.

Call recognition is particularly beneficial to mothers returning from foraging (or in breeding colonies) that emit PACs and to their pups that emit FACs (Roux and Jouventin 1987). Mothers further respond acoustically to pup calls (e.g., Le Boeuf et al. 1972). These interactions indicate that there are acoustically distinctive features in the calls, which have been observed in several species, for example, Australian sea lions (Charrier et al. 2001, 2009; Marlow 1975; Pitcher et al. 2010b), California sea lions (Hanggi 1992; Peterson and Bartholomew 1969; Schusterman et al. 1992), New Zealand fur seals (Stirling 1971b), northern elephant seals (Klopfer and

Gilbert 1966; Linossier et al. 2021; Petrinovich 1974), northern fur seals (Insley 1992, 2000, 2001), Steller sea lions (Campbell et al. 2000; Higgins 1984), and subantarctic fur seals (Charrier et al. 2002, 2003a, c).

Aubin et al. (2015) observed this recognition in Antarctic fur seals and hypothesized that pups use modulation of the fundamental frequency at long range and then additional features (such as the energy spectrum or amplitude modulation) at close range to finalize the identification of their returning and approaching mothers. For South American fur seals, characteristics of the fundamental frequency were most important for distinguishing among mothers; however, pup calls, which typically contained less harmonic structure, could be differentiated by formants (Phillips and Stirling 2000). In addition, although calls of both mothers and pups of South American fur seals showed low variability within and high variability across individuals, mothers' calls were more individually distinct, suggesting that calls were more diverse among adult females than among pups and that even the youngest pups exhibit vocal discrimination (Phillips 2003; Phillips and Stirling 2000).

Subantarctic fur seal pups are thought to use spectral differences and ascending frequency modulation at the beginning of each call to recognize their mothers, while mothers rely on temporal patterns to find their pups (Charrier et al. 2002, 2003c). Pup FACs are complex tonal sounds comprising a fundamental frequency and up to 15 harmonics, and these calls are modulated in frequency and vary in duration. Therefore, information about individual identity is likely to be encoded by both spectral and temporal patterns (Charrier et al. 2002). In Australian fur seals, pup FACs exhibit variation between individuals in fundamental frequency, parts per call, duration, and "quavering" modulation (Tripovich et al. 2006, 2009a). Cape fur seal mothers and pups emit bark sequences to communicate with each other (Martin et al. 2021a, b). Osiecka et al. (2024) investigated the isochrony (temporal uniformity) of these sequences in both age classes. They concluded that while both age classes "bark in an isochronous rhythm," mothers are more precise in maintaining their rhythm than pups and that, for both classes, the faster the barks are produced, the less precise the rhythm. That study did not assess the stability of these calls for each individual over time, but the implication was that these calls become more consistent as the individual ages and that if the mothers use the inter-bark timing to recognize pups, then they must be able to account for the lack of precision in their pup's calls (Osiecka et al. 2024).

The time and number of repetitions pups need to recognize their mother's calls varies among species. Subantarctic fur seals, for example, can learn to respond specifically to their mother's call within 2–5 days of birth; in observations by Charrier et al. (2001), recognition was always established before the mother departed 2–10 days later. In contrast, while Pitcher et al. (2010b) found Australian sea lion mothers could recognize their pups within 48 hours, a previous study showed that the pups required longer to respond to their mother alone, frequently following other females (Marlow 1975). However, by the time they were 2 months old, Australian sea lion pups reacted more strongly to their mother's calls than to others (Pitcher et al. 2009), and did so by paying attention to both amplitude and frequency modulations and to the exact frequency values of the call to identify their mother (Charrier et al. 2009). Mothers may only take hours to recognize the call of their pup (e.g., Galápagos sea lions; Trillmich 1981). Cape fur seals are the champions in this regard, as both mothers and pups can recognize each other within 2–6 h after birth (Martin et al. 2022a). This recognition between adult and pup can be enduring. Female northern fur seals and their pups, for example, learn to recognize each other's calls during a breeding season and still remember them over 4 years later (Insley 1992, 2000, 2001). Further, despite several acoustic characteristics of subantarctic fur seal pup calls changing with age, females can still recognize immature and mature versions of their calls years after hearing them, suggesting that they are likely to be capable of permanently learning their acoustic characteristics (Charrier et al. 2003a). A similar pattern is observed in Australian sea lions (Pitcher et al. 2010a).

5.3.2 Courtship and Mating

The use of acoustic cues in courtship and mating has been reported for many pinniped species. Female pinnipeds have been shown to vocalize when in estrus and males when in an elevated reproductive state—traits that are more easily validated in captive studies than in the wild. For example, during a study on two captive individual leopard seals (one male, one female), six types of "broadcast" calls were recorded when the seals were alone. The female only produced these when she was sexually receptive, and the mature male only produced them during months believed to be the breeding season for wild leopard seals (Rogers et al. 1995; Rogers et al. 1996). Sills and Reichmuth (2022) observed that captive male spotted seals' vocal activity peaked in springtime, just before the annual molt, and coincided with the presence of a "notable musky odor, and urogenital swelling indicative of heightened reproductive status." Indeed, a positive relationship between reproductive hormones and vocal behavior has been observed in multiple species, including female Weddell seals (Bartsh et al. 1992), a male Pacific walrus (Hughes et al. 2011), and male Australian fur seals (Tripovich et al. 2009b), collectively suggesting a strong hormonal component to sound production (Parnell 2018).

The seasonal timing of vocal activity by both male and female, captive and free-roaming individuals has supported the relationship between acoustic cues and reproduction in multiple other species. For several years, Casey et al. (2021) recorded a male harbor seal, housed without conspecifics, that produced vocalizations similar to those of wild harbor seals during the breeding season along the Californian coast. The same study found a female northern elephant seal that was raised primarily without access to conspecifics producing an unusual (not recorded in the wild) call type at an age and time of year when breeding commences in the wild (Casey et al. 2021). Among other examples, Kunc and Wolf (2008) observed territorial male Galápagos sea lions increasing their rate of calling when females were present in their territory. During the breeding season, Van Parijs et al. (1999) observed male harbor seals in the Moray Firth and Orkney Islands (Scotland), noting the adaptation of spatiotemporal calling to match female availability, to the point of targeting tidal patterns and female travel routes.

Pinniped call characteristics not only convey information about the physical and social status of the caller (see Sect. 5.4.7) but also the motivational state (Martin et al. 2022b). In relation to courtship and mating, encoding physical information has been observed in multiple species' calls. For example, in bearded seals, territorial males have significantly longer trills than roaming males and show marked individual variation, suggesting these calls are indicative of condition and potential reproductive success (Van Parijs et al. 2003b). Russell et al. (2016) created an artificial under shore-ice courtship arena to monitor Weddell seal interactions and found that male vocalizations changed quantitatively and qualitatively with context. They speculated that these calls encode physical information about the male caller.

As a further step, a few studies have tested whether pinnipeds encode motivational state within these calls. Tripovich et al. (2008a) showed that male Australian fur seal bark sequences are slower when males are stationary and advertising their territorial status than when herding females or interacting with other males. Similarly, this encoding of arousal state was observed in male Cape fur seals, where increased calling rate and fundamental frequency of bark sequences elicited higher rates of vigilance by other males, indicating the information was decoded by potentially competing subadults (Martin et al. 2022b). Further studies may elucidate whether this information is also decoded by females and used in mate selection or courtship behaviors (Martin et al. 2022b) and whether barks of other otariids encode similar motivational information.

In some circumstances, individuals may alter their vocalizations in exchange for better supervision of their surroundings. For example, male Atlantic walruses attending a herd of females or singing close to females reduced their song duration and spent more time at the surface compared to when they sang alone (Sjare et al. 2003). This allows a greater number of full songs to be produced, per hour, while presumably allowing greater monitoring at the surface and minimizing potential interactions between females in the herd and younger males (Sjare et al. 2003).

5.3.3 Competition and Agonistic Behavior

Aggressive, agonistic, and male threat calls emitted during male standoffs have been reported for multiple species, and vocalizations used in territorial disputes can be situation-specific. For example, threat calls of male northern elephant seals are individually distinctive and thus might avoid combat and injury (Shipley et al. 1981). Even in captivity, spotted seals use these calls in mate attraction and territory maintenance (Beier and Wartzok 1979). Male Australian fur seals respond significantly more to territorial calls of unknown males than to those of neighbors and do not need to hear the whole bark to recognize them (Tripovich et al. 2008b). Indeed, they may recognize a neighbor from just 25–75% of their call, although an increase in strength and intensity of responses with an increase in number of barks suggests the importance of repetition in discrimination (Tripovich et al. 2008b). A male may even make itself sound more threatening to potential rivals, by producing bark series with accelerated rhythmicity and higher formants to elicit a stronger response (Charrier et al. 2011a). Male subantarctic fur seals also respond less to territorial calls of a neighbor than to those of a non-familiar caller, notably with agonistic reactions, such as threat calls, open mouth displays, and movements (Roux and Jouventin 1987), suggesting both species become habituated to neighbors over time. Martin et al. (2023a) used a playback experiment to show that male Cape fur seals recognize the calls of their neighbors, together with position relative to their own harem. The strongest response occurred when the sound of a familiar neighbor came from an unexpected location, such as outside of its territory (Martin et al. 2023a).

Discrimination of familiar competitors by male northern elephant seals based on calls has been well studied (e.g., Casey et al. 2015; Insley and Holt 2012; Shipley et al. 1986). These seals remember individual sound signatures from previous contests and may vary their responses based on relative dominance status, familiarity of the caller, and conditions at the time of the interaction. Indeed, these males can not only acoustically discriminate between individuals but also infer the position and behavior of their competitors. Holt et al. (2010), for example, found that subordinate males retreated more strongly when a dominant male called while facing the receiver, compared to when facing away or at right angles from the receiver. This implies discrimination of the frequency-dependent directivity discussed above.

5.3.4 Feeding

Until recently, there has been a lack of reports directly linking sound production with feeding and foraging behaviors. Schusterman (2008) hoped that future studies would investigate the association of sound production with foraging behavior, and van Opzeeland et al. (2010) could only suggest that feeding was not the reason for reduced vocalizations by Weddell seals during foraging periods. More recently, Stansbury and Janik (2021) used playback experiments to show that captive gray seal pups can associate gray seal sounds with feeding. Chevallay et al. (2024) fitted passive acoustic monitoring and tracking tags to Antarctic fur seals and recorded them while they produced series of low-frequency pulses that occurred only during foraging dives. They hypothesized that these calls might be used as an acoustic lure to confuse or attract fish prey. At the other end of the frequency scale, Weddell and leopard seals reportedly produce ultrasonic sounds, which could potentially be useful in the location of underwater targets when foraging (Awbrey et al. 2003; Cziko et al. 2020).

5.3.5 Echolocation

Nocturnal and deep-diving behaviors exhibited by many pinniped species (e.g., feeding, foraging, and navigation) and the benefits of acoustic communication over other sensory modalities would suggest that underwater echolocation may be advantageous in certain circumstances, though it also entails multiple energetic and predator-related costs. Indeed, multiple reports between the 1960s and 1980s suggested that pinnipeds do echolocate (e.g., Poulter 1963; Renouf and Davis 1982; Schevill et al. 1963; Thomas et al. 1983). However, this contentious assertion was refuted in subsequent studies, including experimental work, highlighting that pinnipeds lack elaborate echolocation abilities (e.g., Au and Hastings 2008; Schusterman et al. 2000; Scronce and Ridgway 1980). Nevertheless, a report of Weddell seals producing ultrasonic sounds under water (Cziko et al. 2020) may reinvigorate this discussion.

5.3.6 Navigation

Although pinnipeds may not echolocate, they have been hypothesized to collaboratively use sound to assist in navigation. Both gray and ringed seals have been reported calling beneath sea ice, potentially as an aid for under-ice orientation during the winter, presumably by navigating toward the calls of conspecifics (Asselin et al. 1993; Jones et al. 2014; Prawirasasra et al. 2021).

5.4 Call Variability

Significant variation in pinniped calls has been reported at the individual, life-stage, population, species, family, and clade levels. This suggests that the use of sound by pinnipeds is not only a derived trait but also acquired, adapted to ecological constraints, and potentially learned. Variation may be present in call spectral or temporal characteristics, occurrence and time of use, or associated behavioral function (Campagna and Harcourt 2021; Costa and McHuron 2022). The potential drivers of call variation can arise from allometric (size-related), ontogenic, or sex-driven physical differences; differential evolution of species sound production; vocal plasticity; responses to environmental drivers or constraints; or other factors acting individually or in combination (Campagna and Harcourt 2021; Costa and McHuron 2022). Further, as a clade, pinnipeds display a broad diversity of social structures, breeding systems, and selection pressures (e.g., colonial species vs. solitary species), contributing to variation in complexity of vocal repertoire and call recognition (Charrier 2020).

When comparing sound production among pinnipeds, choosing the species and behaviors to include in the comparison is nontrivial. For example, a handful of studies have investigated behaviors at a regional level (e.g., those of polar and nonpolar species; Miksis-Olds et al. 2016) or among the three families of phocids, otariids, and odobenids (e.g., Charrier 2020; Martin et al. 2017). We further these group comparisons using a broader dataset of sound characteristics; however, for the most part, we have restricted our discussion to drivers of variability at the species level. Qualitative descriptions of species characteristics can be found in the taxon-by-taxon overview in Sect. 5.5.

5.4.1 Phylogeny (Odobenids, Otariids, and Phocids)

Closely related mammal species often display similar call types; however, this is not typically the case with phocids (Terhune 2019). Harp and Weddell seals produce large repertoires of calls and are distantly related, yet their nearest relatives (gray and crabeater seals, respectively) exhibit low diversity (Terhune 2019). Further, highly variable characteristics within calls can be similar among different species, such as the large frequency changes (more than five octaves in 5 s) produced by bearded, ribbon, and Weddell seals (Cleator et al. 1989; Mizuguchi et al. 2016a; Thomas and Kuechle 1982). Collectively, these factors suggest that most seal vocalizations are derived traits (Terhune 2019) and that environmental drivers have significant impact on the number and variety of species call types.

In general, the social structures, breeding systems, and selective pressures of phocids are considered less complex than those of otariids and odobenids (Charrier et al. 2001; Shabangu et al. 2022). Thus, their in-air communication has greater focus on other sensory modalities, such as olfactory and spatial cues (Insley et al. 2003; Kovacs 1995; Shabangu and Rogers 2021; Terhune et al. 1979). The repertoires of in-air vocalizations are smaller and more simplistic than those of the underwater calls—for most species (Terhune 2019). There have been several attempts to group pinniped vocal behaviors by species, most commonly using repertoire size as the driving factor (e.g., Rogers 2003; Stirling and Thomas 2003), and to relate these groups to social structure and selection pressures. However, these efforts do not account for call characteristics and diversity within the repertoire, which may provide more information on species vocal complexity (Terhune 2019).

Terhune (2019) assessed the repertoires of 13 phocid species and computed vocal complexity scores based on waveform type, repertoire size, repetition and rhythm patterns, frequency, and call duration, drawing information from 39 previous studies. Terhune found no relationship between phylogeny and vocal complexity but did find that species groups of low vocal complexity had small repertoires and used burst-pulsed or noisy waveforms, while the intermediate group had a range of "clean" (presumably less broadband) and noisy waveforms and produced songs and repeated calls in a "single rhythm pattern." The group of greatest vocal complexity had the largest repertoire, clean and noisy waveforms, songs, and two or more rhythm patterns in repeated-element calls. Species falling into the low-complexity group were monogamous, did not establish breeding groups, bred on beaches or pack ice, and were at greater predation risk. Species with greater vocal complexity were promiscuous or polygamous, had established breeding groups on pack or land-fast ice, and were at lower predation risk. Further, Charrier (2020) examined the complexity of mother–pup calls, inferred from an index of vocal stereotypy (IVS), developed by standardizing classification rates of calls for individuals in previous reports by the number of individuals contributing sounds within the respective study. Although this provided only one IVS estimate for each species, when compared to the selective pressures the species face, Charrier (2020) observed a positive relationship between IVS and pressure in that those species with more complex social and breeding systems developed more individualistic sounds for mothers and pups to recognize each other.

In terms of basic spectral characteristics, we compare peak, minimum, and maximum frequencies (Fpeak, Fmin, Fmax); bandwidth (BW); and mean, minimum, and maximum duration (Dur, Durmin, Durmax) across all call types in air and under water between phocids and otariids (Fig. 5.4). Phocids have a lower median peak frequency and minimum frequency than otariids and a higher maximum frequency, leading to greater bandwidth. However, the spread in measurements (with the box edges corresponding to the 25th and 75th percentiles) reported in the literature is much greater for phocids. Phocids also typically produce calls of longer duration than otariids (Fig. 5.4). Martin et al. (2017) reported similar minimum frequencies for phocid and otariid calls but noted that, in general, phocids had a higher maximum frequency for both in-air and underwater calls. These results were thought to be a result of the general difference in body size between the families. In addition, different ecologies and environments (e.g., polar vs. temperate) and selective pressures drive call features (Charrier 2020; Terhune 2019).

5.4.2 Polar vs. Nonpolar Species

Polar species face the annual challenge of surviving periods (in which they engage in social activities, such as establishing and defending territories) in prolonged darkness and in obstructed environments (ice cover), where

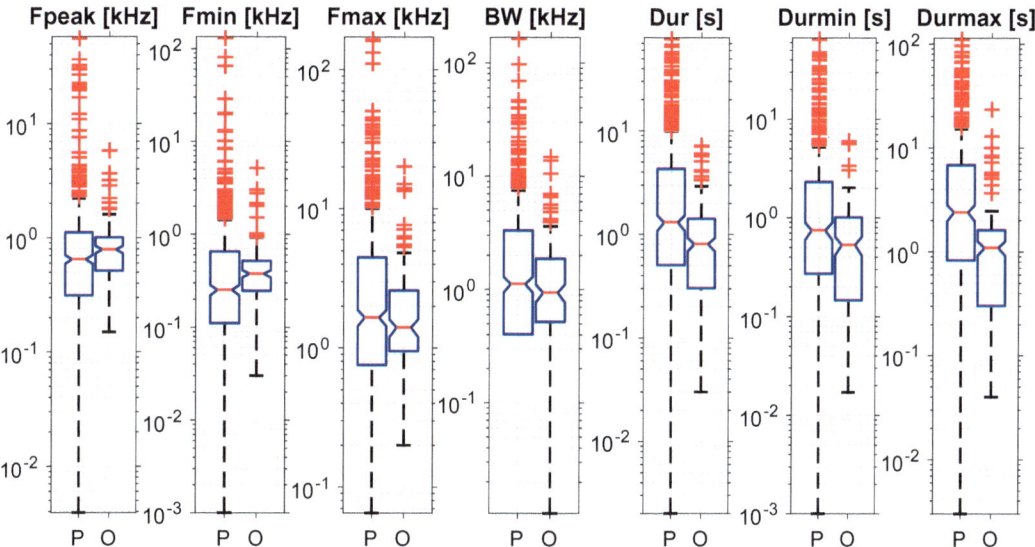

Fig. 5.4 Comparison of call features between phocids (P) and otariids (O) based on data in Table S1. Boxes show the medians (red line) and 25th and 75th percentiles (box edges). Outliers are represented by red +. The notches illustrate confidence intervals; in comparing phocids to otariids, the medians are significantly different (95% confidence) if the notches do not overlap

acoustic cues are likely the most feasible method of communication. Miksis-Olds et al. (2016) summarized the call characteristics of polar pinniped species. Given polar species' more remote and solitary nature, it might be hypothesized that they have developed less complex calls over a smaller bandwidth. However, they also live in an environment where both visual and acoustic propagation tends to be impaired at some times of year, so that a more complex repertoire or one suited to propagation under ice might be favored. We conducted a simple comparison of call features based on the literature summarized in Table S1. Underwater vocalizations of polar pinnipeds have higher peak, minimum, and maximum frequencies; greater bandwidth; and longer mean, minimum, and maximum durations than nonpolar species (Fig. 5.5 top row). All of these differences (with the exception of maximum duration) are statistically significant and might be an anti-masking adaptation related to increased ambient noise levels in the marginal ice zone (i.e., the transition zone between the open sea and dense ice cover; Yang et al. 1987). In-air vocalizations differed significantly only in peak frequency and bandwidth, both of which were lower in polar pinnipeds (Fig. 5.5 bottom row).

5.4.3 Geographic Variation

At a global scale, the evolution of pinniped sound production has partly stemmed from key biogeographic events as the clade has spread around the world (e.g., see Fig. 1.4 in Berta et al. 2022 for the global biogeographic spread of phocids). However, acoustic variation at broad- and microgeographic scales has been observed in both cosmopolitan and local species, for several species of otariids and phocids, with some reports noting significantly different dialects. Here, a dialect may be considered to be "similarities in vocal behavior that are specific to geographical regions or social groups that typically do not intersect" (Casey et al. 2018), whether due to learned behaviors, genetic drift, or habitat adaptation. Further, call characteristics at the population level have been observed as they evolve through time, becoming more complex and displaying more variation among individuals than

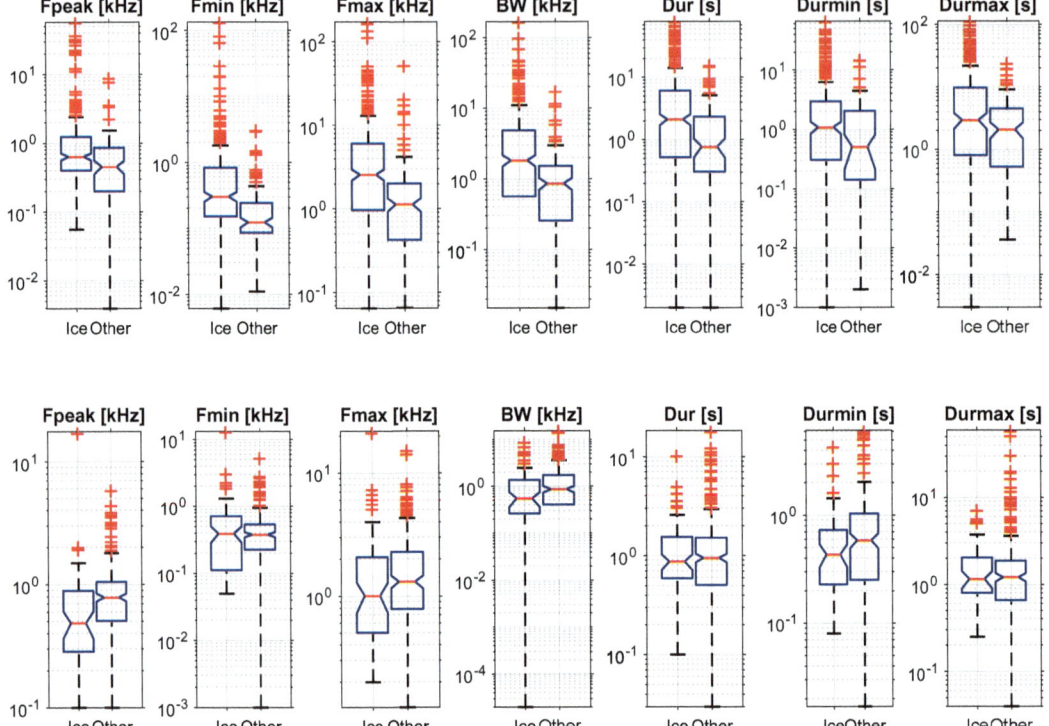

Fig. 5.5 Comparison of call features between polar (ice) seals and other pinnipeds based on data in Table S1. Top row: underwater call features. Bottom row: in-air call features. Boxes show the medians (red line) and 25th and 75th percentiles (box edges). Outliers are represented by red +. The notches illustrate confidence intervals; in comparing polar to other pinnipeds, the medians are significantly different (95% confidence) if the notches do not overlap

populations. For example, 50 years ago, different rookeries of northern male elephant seals displayed dialects in their in-air calls (Le Boeuf and Peterson 1969). Le Boeuf and Petrinovich (1974) later observed drift in call characteristics and hypothesized that it resulted from movements among rookeries. Four generations later, the differences were no longer detectable. Indeed, there was greater variation among individuals, and contemporary males displayed more call complexity than their earlier counterparts, including features not present in earlier recordings (Casey et al. 2018).

Geographic variation has not only been seen in adults but juveniles and pups as well. For example, calls of harp seal pups from Newfoundland, Canada and the Greenland Sea differed notably in the length of their call (van Opzeeland et al. 2009). These variations have been observed in recordings taken of some species in air (e.g., Australian sea lions, Ahonen et al. 2014, 2018; Cape fur seals, Martin et al. 2023b; harp seals, van Opzeeland et al. 2009; South American sea lions, Trimble and Charrier 2011; southern elephant seals, Sanvito and Galimberti 2000b) or under water (e.g., harbor seals, Bjorgesaeter et al. 2004, Sabinsky et al. 2017; harp seals, Turnbull and Terhune 1994; ribbon seals, Mizuguchi et al. 2016a) and in some cases reported in both media. Weddell seals, for example, display geographic differences in aerial calls (e.g., PACs; Collins and Terhune 2007; Terhune et al. 2008) and underwater sound types (e.g., trills) and their associated functions (Abgrall et al. 2003; Collins and Terhune 2007; Morrice et al. 1994; Pahl et al. 1997; Terhune et al. 2008).

Leopard seals have also had geographic variation reported for in-air (Rogers et al. 1995; Stirling and Siniff 1979; Thomas and Golladay 1995) and underwater calls (Thomas and Golladay 1995).

This spatial variation in sound production has not only been observed in acoustic analysis but also in the responses of the receiving animals. Nunavut and Greenland male bearded seals, for example, responded more strongly to trills from local males than those from a distant area (Charrier et al. 2013)—an expected response given the trill has been associated with territorial defense (Charrier et al. 2013; Cleator et al. 1989; Risch et al. 2007). Similarly, male Australian sea lions responded more strongly to barks from a local male than to barks from a male of a distant colony (Ahonen et al. 2018; Attard et al. 2010). However, these differences need not always imply sensitivity to dialect differences—they may also be explained by familiarity with local rivals versus all others.

5.4.4 Environmental Drivers

Call types and rates may also be linked to temporal environmental patterns (e.g., diel, lunar, and seasonal cycles), various environmental conditions (e.g., geophysical factors, such as temperature), or specific behavioral or physiological states (e.g., breeding). Diel patterns in call rates, for example, are often noted and differ among species. Elephant seal bulls produce their highest vocalization rates at night, reduce rates around crepuscular periods, and exhibit their lowest call rate in the early afternoon (Shipley and Strecker 1986). In contrast, one Hawaiian monk seal produced few vocalizations after sunset (Parnell 2018), though this may not be stereotypical for the species. In fact, many seals have displayed temporal patterns in vocalizations that have been related to diurnal or tidal cycles, such as leopard and crabeater seals (Thomas and Demaster 1982), Ross seals (e.g., Seibert et al. 2007), and ribbon seals (Frouin-Mouy et al. 2019). Bearded seals produce significantly more calls during the early mornings of the breeding season, decreasing to minimum levels during the afternoon (Boye et al. 2020).

At a seasonal scale, and likely linked to the breeding cycle, bearded seal call detections for the Baffin Bay and Davis Strait had a limited period of calling that started in winter, peaked in spring, and ceased during the second half of June (Boye et al. 2020). Indeed, the influence of time of year on sound production has been reported for bearded, gray, and ribbon seals and walruses, among others (Chou et al. 2020; Prawirasasra et al. 2021). Seasonal variation in sea surface temperature and ice cover has been shown to drive crabeater seal and walrus calling behavior (Chou et al. 2020; Nagaraj et al. 2021), and seasonal changes in ice cover drive the geographical variation in rates of trills produced by bearded seals (Llobet et al. 2021). Indeed, longitude and latitude were the most important predictors of Ross seal acoustic occurrence, and month of the year highly predicted acoustic detection of leopard seals (Shabangu and Rogers 2021). Moors and Terhune (2005) reported that, in ice-covered waters, harp and Weddell seals called predominately from positions in the water column where light would likely penetrate but still avoided sea-ice interference to some extent.

Other factors can affect sound production. For example, underwater calls by Weddell seals can be heard by humans on the ice (Thomas and Kuechle 1982) and, therefore, potentially by predators as well. The risk of predation has been reported as a likely reason for reduced vocalization in pinnipeds (Terhune 2019; van Opzeeland et al. 2010). Captivity has been shown to impact the behavior of many fauna, and while numerous studies of captive pinnipeds have been conducted and provided a wealth of information on vocal behavior, they do not necessarily reflect natural behavior in the wild. Leopard seals, for example, are solitary and hence emit broadcast calls (Rogers et al. 1995, 1996); yet in captivity, they reportedly made "local calls" all throughout the year but "broadcast calls" only during the mating season. Captive Steller sea lions produced calls that are shorter in duration and of higher frequency than those of their wild counterparts

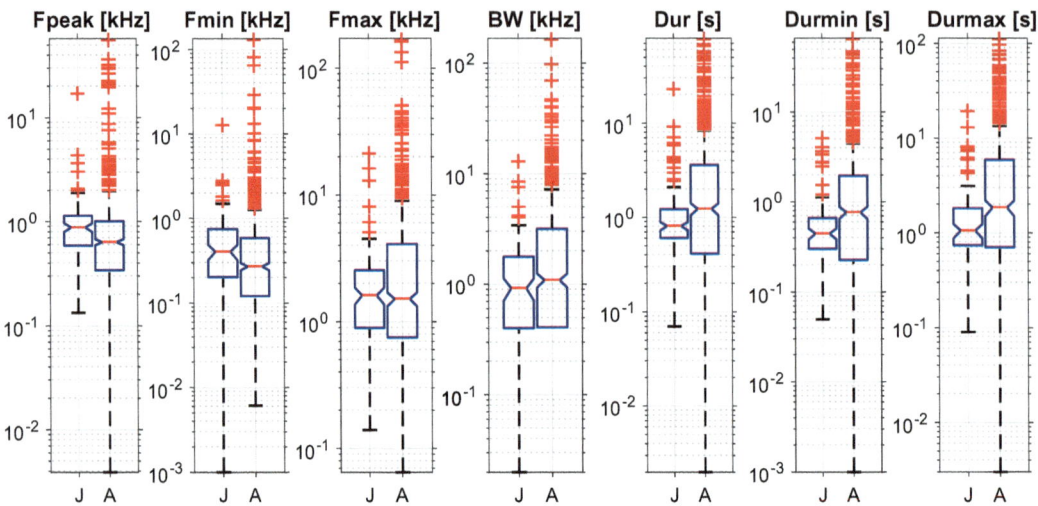

Fig. 5.6 Comparison of call features between juvenile (J) and adult (A) pinnipeds based on data in Table S1. Boxes show the medians (red line) and 25th and 75th percentiles (box edges). Outliers are represented by red +. The notches illustrate confidence intervals; in comparing juveniles to adults, the medians are significantly different (95% confidence) if the notches do not overlap

(Park et al. 2006). So, there is potential for the combination of environmental, resource, and predation risk factors to affect the number and type of calls detected by observers, and while these factors may be controlled in captivity, removing the individual from the wild comes with behavioral implications.

5.4.5 Ontogeny and Allometry

The acoustic characteristics of sounds relate to the structure of the sound production mechanism and the vessels (i.e., the combination of tissues and airspaces) within which the sound travels before leaving the animal's body. From a modeling perspective, larger vessels have lower resonant frequencies as do longer, looser strings. In pinnipeds, one would expect sounds produced with larger throats, thicker throat linings, longer vocal cords, and bigger lungs of larger animals to have different characteristics to those of smaller counterparts. In comparing the sounds of pinnipeds across different life stages, we found that juveniles tend to produce calls at higher peak and minimum frequencies and shorter durations than adults do (Fig. 5.6).

The fundamental frequency, structure of formants, and harmonics of calls can provide significant information on the ontogeny, allometry, and resource-holding potential of individuals, typically showing a negative relationship with the dominant frequencies (de Reus et al. 2022). For instance, Martin et al. (2017) observed a negative relationship between call frequency and body mass for pinnipeds and inferred the same trend for minimum and maximum reported call frequencies for phocids, though this relationship was derived from few data points. Casey et al. (2015) observed that call frequency was lower with increasing size of northern elephant seals, and in Steller sea lions calls of the males are not only of lower frequency but also longer duration than those of females, potentially as a result of size (Park et al. 2006). Calls of Australian fur seal pups increase in duration, decrease in number of parts per call, and decrease in peak frequency as the pups age, suggesting that these call features are related to growth and maturational changes (Tripovich et al. 2009a). Thus, recipients can infer physical traits from call spectra. As a consequence, vocal displays may allow individuals to attract mates or resolve interactions without physical contact (Sanvito et al. 2007a). In southern elephant seal vocalizations, for example, the fifth formant and the minor formant convey the most information about phenotype, as they decrease with age (Sanvito et al. 2007c, 2008). However, this relationship relies on the calls produced being an "honest" reflection of the caller, which is not always the case, particularly if there are selective advantages to being perceived as larger, healthier, etc. Although Casey et al. (2021) found age and body mass to be highly correlated, they found no relationship between body size and spectral features of captive harbor seal (ages 1–6 years) and northern elephant seal (ages 7–18 years) calls in their longitudinal studies. However, in both species, calls changed in duration, becoming longer as they aged. Further, call complexity may develop or reduce with age. In Weddell seals, for example, calls of pups are often also less complex than those of their mothers (Collins et al. 2005), implying the development of sound production as a skill, through time. In contrast, Phillips and Stirling (2000) reported that calls of South American fur seal pups were more complex than those of the elder females. The reasons behind such differences in pinniped call complexity, and indeed individuality, are thought to be context-specific. An increasing number of studies have reported a positive correlation between social and vocal complexity, including in pinnipeds (e.g., Stansbury and Janik 2019; Stirling and Thomas 2003; also see Chap. 6), leading to the concept that despite evolved biological differences and differences in habitat, underwater vocal complexity in phocids is related to breeding behaviors and risk of predation (Terhune 2019; see Sect. 5.4.1).

5.4.6 Sex

Acoustic characteristics of pup calls may differ between males and females. In harbor seal pups, males have been observed to emit lower-frequency calls (e.g., Khan et al. 2006), possibly as a result of sex-driven body size or thickness of throat walls, structuring the formants as the sound is emitted. In terms of call individuality, van Opzeeland et al. (2009) observed that female harp seal pups had a significantly higher proportion of correctly classified vocalizations than male pups, suggesting that vocalizations of female pups are considerably more individually specific than those of males.

In comparing the sounds of pinnipeds between sexes, the only statistically significant difference we found was that males produce lower minimum frequencies (Fig. 5.7), likely related to body size.

5.4.7 Individual Identity

In air, many pinniped species reportedly produce individually specific calls in common call types, such as individual barks, chirps, and screams, in both the wild (e.g., Antarctic fur seals, St Clair Hill et al. 2001; Stirling and Warneke 1971; Australian sea lions, Charrier et al. 2009; Galápagos fur seals, Trillmich 1981; leopard seals, Rogers and Cato 2002; South American fur seals, Phillips 2003; Phillips and Stirling 2000; South American sea lions, Fernández-Juricic et al. 1999; Trimble and Charrier 2011; Steller sea lions, Campbell et al. 2002) and captivity (e.g., northern fur seal pups, Takemura et al. 1983). Such variation has been observed to persist for long periods, on the scale of decades. For example, when observing male threat calls of New Zealand fur seals, "sound-spectrographic analysis revealed that individual territorial males, when motivated to respond several times to the calls of other males, gave vocalizations that were precisely reproducible and unique to each individual" (Stirling 1971a). This variability can occur across all age groups. Roux and Jouventin (1987) found that subantarctic fur seal inter-individual variability of PACs and FACs allowed for individual recognition by the pup and the mother, respectively. Within the same age group, calls of closely related species, including hybrids, have been shown to vary; Page et al. (2002a) found PACs to be more variable among individual New Zealand and subantarctic fur seals

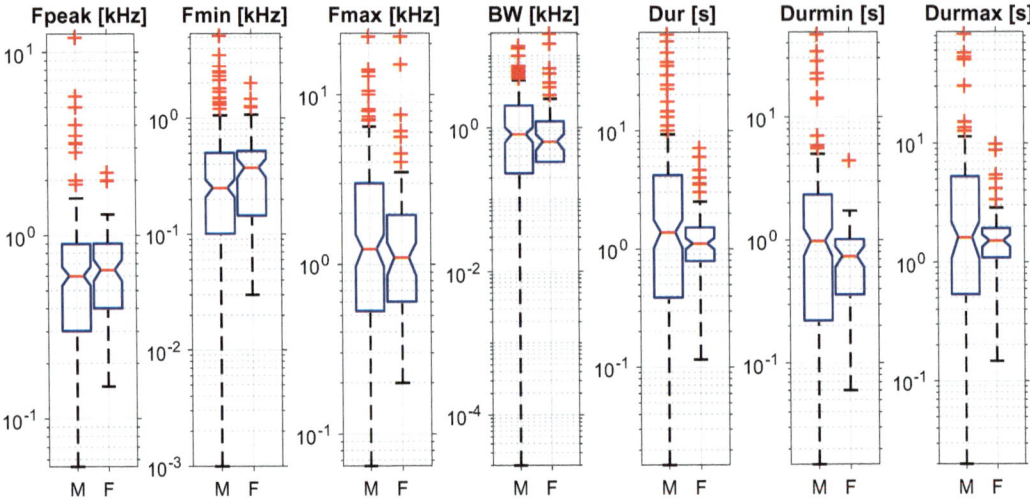

Fig. 5.7 Comparison of call features between male (M) and female (F) pinnipeds based on data in Table S1. Boxes show the medians (red line) and 25th and 75th percentiles (box edges). Outliers are represented by red +. The notches illustrate confidence intervals; in comparing males to females, the medians are significantly different (95% confidence) if the notches do not overlap

compared to the same calls of Antarctic fur seals or in the fur seal hybrids of these three species.

The number and type of features required to discriminate between calls of individuals varies among species (and analysis technique). Calls of South American fur seals, for example, can be "discriminated on the basis of a combination of frequency, temporal, and amplitude-related characteristics" (Phillips and Stirling 2000). Charrier and Harcourt (2006) measured 11 acoustic parameters of Australian sea lion calls and determined that fundamental frequency, energy spectrum, amplitude, and frequency modulation were individually specific. This is echoed by Gwilliam et al. (2008) and Ahonen et al. (2018; 2014), who found calls of the same species were individually distinctive, based on five acoustic parameters. Charrier and Harcourt (2006) determined that the spectral features and frequency modulation were the most important of these, and enough features were contained in both mother and pup calls to discriminate between individuals. Fundamental frequency, spectral differences, amplitude, and frequency modulation all represented key components of mother–pup recognition (Charrier and Harcourt 2006). In contrast, individual variation in trills of territorial bearded seals (Cleator et al. 1989) was driven by duration, with the length thought to be an indicator of male quality in courtship display (Van Parijs et al. 2003b), while for northern elephant seals, the combination of call pulse rate and centroid of the call spectra was used to "characterize the unique acoustic space of each individual" (Casey et al. 2015).

Mediterranean monk seal barks and screams are individually specific (Charrier et al. 2017), and in Hawaiian monk seals, although pup vocalizations tended to be "simple in structure," they contained energy spectra that varied within and among pups (Eliason et al. 1990; Job et al. 1995). Indeed, Hawaiian monk seal pup calls of all age groups were individually specific, with fundamental frequency and, to a lesser extent, duration and spectral peak contributing the most to individuality (Job et al. 1995). However, while barks of Australian fur seals are stereotyped and individually different, they are easier to identify when produced as a series, lasting between 2 and 9.8 s, than a single bark (Tripovich et al. 2005). Thus, in short sounds of limited spectral variation, sequences of calls may have evolved specifically to allow better discrimination between calls of individuals.

Given the level of variation and number of categorization methods applied to reports of pinniped sound production, perhaps future research could aim to fully understand variation among pinniped species by encouraging standardized techniques across recordings from as many species, age classes, and geographic regions as possible.

5.4.8 Vocal Development, Plasticity, and Learning

Here, we consider vocal development to be the progression of an individual's sound production through time, which may be innate. Vocal plasticity and vocal learning may be considered related concepts and additional steps along a developmental pathway. In this case, plasticity is the ability to change characteristics of vocalizations in response to external stimuli, such as environmental conditions (e.g., background noise) or social context. Vocal learning is then the ability to modify output through vocal coordination, production variability, and vocal versatility (Wirthlin et al. 2019), often based on the experience of being exposed to the sounds of others (Janik and Slater 1997), typically to produce new and complex sounds, used in context, with purpose. Few mammal species have been found to exhibit clear vocal production learning capabilities (Janik and Knörnschild 2021), and the diverse social structures, call variation, and recognition capacity of pinnipeds mean that they provide an excellent model for understanding mammalian acoustic plasticity (Charrier 2020). However, the extent to which tests and categorizations of species' vocalizations are classed as plasticity and learning appears to be a gray area.

Vocal development of specific sounds is commonplace and has been reported in many pinniped species, including, but not limited to, Australian sea lions (Pitcher et al. 2009, 2010b), Galápagos sea lions (Trillmich 1981), harbor seals (de Reus et al. 2022; Torres Borda et al. 2021), and Steller sea lions (Sandegren 1970). Shipley et al. (1986) found that male juvenile northern elephant seals exhibited developmental changes in their calls, beginning as "highly variable versions of the adult repertoire" and stabilizing with increasing age. Miller (1985) observed the changes in walrus pup barks over months; in captivity, the species has developed novel sounds and sound combinations (Schusterman and Reichmuth 2008). In bearded seals, "individual capacity for vocal production appears to develop gradually, showing plasticity in form development over time," with frequency range changing most in the first years of vocalizing and lengthening of the call duration becoming more predominant in later years (Davies et al. 2006). However, some of these sounds may not necessarily be learned or intentionally modified, but rather the result of developmental changes. In some cases, such sounds have developed in individuals that have rarely been exposed to conspecifics or not at all (e.g., Casey et al. 2021).

In few cases, plasticity has been experimentally proven. For example, Torres Borda et al. (2021) exposed captive 1–3-week-old harbor seal pups to frequency-tailored noise (i.e., bandpass filtered to match the fundamental frequency of their call) to see if they would alter their call frequency range to avoid being masked by the noise. In response, the pups lowered their fundamental frequencies, with one pup increasing amplitude and "flattening spectral tilt," which are unusual responses, indicative of vocal plasticity. Schusterman and Reichmuth (2008) conditioned one male and one female 12-year-old captive Pacific walrus to produce specific sounds in air, using food reinforcement; they also conditioned them to produce novel variants. When the experiment shifted to underwater sounds, both individuals spontaneously produced novel "knocks" and "soft bells" (components of male "song" in the wild) without additional conditioning, suggesting that they had learned the task and were capable of vocal creativity.

In a demonstration of vocal learning, Stansbury and Janik (2019) used a playback experiment to test whether juvenile gray seals could copy melodies and formants, finding that they could copy vowels and the peak frequencies of simple melodies. They improved their frequency matching over time, becoming most adept at 2–3 weeks of age, at the conclusion of the playback test (Stansbury and Janik 2021).

Perhaps the most prominent example of pinniped vocal plasticity and potentially vocal learning was Hoover, a captive male harbor seal that developed the ability to produce multiple human words (e.g., "Hoova! Hoova! Hey!") in a New England accent (Reichmuth and Casey 2014). In this case, the learned signals were remembered over considerable time periods before being used. Hoover was exposed to auditory stimuli from his carer in the first few months of his life, with occasional responses (Swallow 2001). After years in an aquarium, it was upon reaching sexual maturity that Hoover started to reproduce the sounds he had heard as a pup (Ralls et al. 1985). However, although call development has been reported for decades, this level of ability has not been replicated elsewhere.

5.4.9 Anti-masking Strategies

Several pinniped species reportedly exhibit the Lombard effect at times of increased background noise, including bearded seals (Fournet et al. 2021) and harbor seals (Matthews 2017), but others do not (e.g., northern elephant seals; Southall et al. 2019). Harp seals increase the number of elements per call to avoid masking by conspecifics (Serrano and Terhune 2001), and calls that overlap are often not of the same type, implying that individuals actively listen to each other and change their own call type (and therefore call spectra) to be heard and avoid masking each other (Serrano and Terhune 2002). These two factors, however, operate in opposition; while longer calls help detection within gaps, shorter calls provide more gaps for individuals to insert their calls.

Torres Borda et al. (2021) observed unusual mammalian behavior by harbor seal pups. When exposed to band-pass filtered sounds that spanned and masked the typical fundamental frequency range of their calls, the pups lowered their call frequencies in response to increasing noise, in a precise and frequency-adaptive manner, rather than increasing their call source level. Such a downward frequency shift is not normally displayed by mammals, as the additional effort to increase amplitude results in tighter vocal cords and increased frequency content (Lombard 1911). The ability of pinnipeds to produce size-dishonest calls due to their vocal plasticity likely accounts for this; however, it has otherwise not been observed in response to noise, nor for such a young age group.

5.5 Taxonomic Overview of Pinniped Sounds

Our literature review revealed that of the 34 IUCN-listed extant pinniped species, 32 have recordings of sound production in air and 22 under water. Those (sub)species not yet recorded are deemed likely to produce sound in both media based on biomorphophysiology and lineage, under the same assessment conditions used in Looby et al. (2023). The reports originate from 13 of the 18 International Union for Conservation of Nature (IUCN) global regions. The exceptions are the Caribbean, Central Indian, Arabian, East African, and East Asian regions (regions 7 and 10–13, respectively), which are beyond the presently known extents of pinniped distribution (Berta et al. 2022). Most studies reporting sound features were undertaken in the Arctic (IUCN region 1), then Antarctic (IUCN region 2), and captivity (Fig. 5.8).

Table 5.1 gives an overview of the acoustic features of pinniped sounds measured in air and under water—by species. We compiled this summary table based on the features reported in the literature (Table S1). The peak frequency in Table 2 is the mean of all reported peak frequencies in Table S1. Similarly, the bandwidth, duration, and source levels represent the means of those reported across individual studies. The minimum frequency is the minimum of the mean minimum frequencies reported by individual studies; the maximum frequency is the

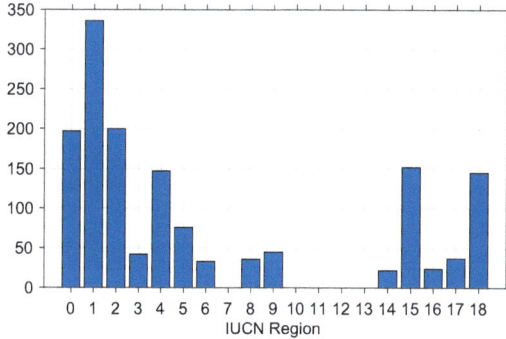

Fig. 5.8 Number of publications on acoustic features of pinnipeds by IUCN region (https://www.cetaceanhabitat.org/directory.php; accessed Feb 4, 2024). 1, Antarctic; 2, Arctic; 3, Mediterranean; 4, Northwest Atlantic; 5, Northeast Atlantic; 6, Baltic; 7, wider Caribbean; 8, West Africa; 9, South Atlantic; 10, central Indian Ocean; 11, Arabian Seas; 12, East Africa; 13, East Asian Seas; 14, South Pacific; 15, Northeast Pacific; 16, Northwest Pacific; 17, Southeast Pacific; 18, Australia and New Zealand; and 0, captivity. Data in Table S1

maximum of all mean maximum frequencies published; and the same with minimum and maximum durations.

5.5.1 Family Odobenidae: Walrus

There is one species of walrus (*Odobenus rosmarus*) with two subspecies: *O. r. divergens*, the Pacific walrus, and *O. r. rosmarus*, the Atlantic walrus. Despite their geographic separation, their vocal repertoire appears to be similar, and their acoustic features are, therefore, summarized together.

Walruses emit sounds in air and under water. Males produce stereotyped, individually distinctive, repetitive songs consisting of a surface and an underwater part lasting for, on average, 3.9–9.3 min combined (Fig. 5.9). There are mainly four types of songs: diving vocalization, intermediate, coda, and aberrant songs, plus variations of these. They differ in compositions and sequences of knocks, bell-like (church bell) sounds, taps, and strums (as "strummed over the strings of a guitar or zither") emitted under water and knocks, whistles, and taps emitted between breaths at the surface. Males may sing for hours at a time, up to 2 or 3 days (Sjare et al. 2003; Stirling et al. 1983, 1987).

Under water, walruses also produce soft bell sounds, burps, buzzes, soft chuffs, chugging sounds, clicks, glubs, grunts (chimp sounds), knocks, moans, rasps, and roars between 100 and 4800 Hz with a peak energy of 200–2000 Hz (bell sounds up to 5000 Hz) for 0.015–5.8 s (Fig. 5.10). Source levels of knocks were measured as 168–197 dB re 1 µPa pk-pk and 147–194 dB re 1 µPa rms, while maximum source levels for bell sounds and moans were 147 and 166 dB re 1 µPa, respectively (Denes 2014; Denes et al. 2015; Fay et al. 1984; Hughes et al. 2011; Jézéquel et al. 2022; Madan et al. 2020; Mouy et al. 2012; Ray and Watkins 1975; Rideout et al. 2013; Schevill et al. 1966; Schusterman and Reichmuth 2008; Sjare et al. 2003; Stirling et al. 1983, 1987).

In air, barks, clacks, flatulent sounds, grunts, motorboat calls, sneezes, snorts, short harsh whistles, and rutting whistles are emitted between 50 and 1350 Hz with peak energy from 100 to 1900 Hz, lasting from 0.1 to 10 s (Asselin et al. 1993; Charrier et al. 2010, 2011b; Fay 1982; Fay et al. 1984; Miller 1985; Schevill et al. 1966; Sjare et al. 2003; Stirling et al. 1983, 1987; Verboom and Kastelein 1995). Source levels for rutting whistle sounds were measured in captivity as 111 dB re 20 µPa m (Verboom and Kastelein 1995).

Barks are highly variable in all age and sex groups. Apart from females barking to their pups, adult walruses nearly always emit barking sounds in submissive circumstances, for which it is characteristic, while pups bark in various situations (Miller 1985). As pups grow, their barks become longer and deeper (fundamental frequency decreases), harmonic structure diminishes, barks become noisier, barks merge, and pulsing becomes more pronounced—they become more similar to adult barks (Miller 1985). Females' and pups' barking sounds in the Atlantic walrus are individually different, and females react more to their own pup's bark than to others' (Charrier et al. 2010).

Table 5.1 Overview of sound characteristics by species. Sound types were tonal (T), burst-pulsed (BP), pulse (P), click (C), or noisy (N). Reported are peak frequency (Fpeak), bandwidth (BW), minimum frequency (Fmin), maximum frequency (Fmax), duration (Dur), minimum duration (Durmin), maximum duration (Durmax), and source level as root-mean-square pressure (SL_SPLrms) or peak-to-peak (SL_SPLpk-pk). Based on data in Table S1

Species	Common	Medium	Sound type	Fpeak [Hz]	BW [Hz]	Fmin [Hz]	Fmax [Hz]	Dur [s]	Durmin [s]	Durmax [s]	SL_SPLrms [dB re 1 (water) or 20 (air) µPa m]	SL_SPLpk-pk [dB re 1 (water) or 20 (air) µPa m]	References
Arctocephalus australis	South American fur seal	Air	T	1100		1000		0.7					Trillmich and Majluf (1981)
Arctocephalus australis	South American fur seal	Air	BP	1061	1012	710	1640	1.1	0.57	1.4			Phillips and Stirling (2000, 2001), Trillmich and Majluf (1981)
Arctocephalus australis	South American fur seal	Air	P	841	799	573	1372	0.3	0.04	1.3			Phillips and Stirling (2001), Trillmich and Majluf (1981)
Arctocephalus australis	South American fur seal	Air	N	721	453	598	1051	1.9	1.4	2.5			Phillips and Stirling (2001)
Arctocephalus forsteri	New Zealand fur seal	Air	T	1678	1535	1258	2793	1.8	1.4	1.6			Page et al. (2001), Stirling (1970, 1971a)
Arctocephalus forsteri	New Zealand fur seal	Air	BP	1170	3658	274	3931	1.3	0.1	2			Page et al. (2001, 2002a), Stirling (1970, 1971a), Stirling and Warneke (1971)
Arctocephalus forsteri	New Zealand fur seal	Air	P	900	2838	466	3304	0.72	0.1	1.6			Miller (1971), Page et al. (2002b), Stirling and Warneke (1971)
Arctocephalus forsteri	New Zealand fur seal	Air	N	496	1127	141	1268	1.6	1.5	1.7			Page et al. (2002a)
Arctocephalus galapagoensis	Galápagos fur seal	Air	BP	1800	850	650	1500						Trillmich (1981)
Arctocephalus galapagoensis	Galápagos fur seal	Air	P	1000									Trillmich (1981)
Arctocephalus galapagoensis	Galápagos fur seal	Water	P		850	100	950	0.05					Merlen (2000)
Arctocephalus gazella	Antarctic fur seal	Air	T	1102	1375	460	1835	1.2	0.4	1.5			Page et al. (2001, 2002b), St Clair Hill et al. (2001)
Arctocephalus gazella	Antarctic fur seal	Air	BP	1864	2090	1262	3352	1.4	0.3	5.7			Aubin et al. (2015), Page et al. (2001, 2002a), St Clair Hill et al. (2001), Stirling and Warneke (1971)

Species	Common name	Medium	Code							References
Arctocephalus gazella	Antarctic fur seal	Air	P	802	3419	536	3955	0.54	0.06	1.5	Page et al. (2002b), St Clair Hill et al. (2001), Stirling and Warneke (1971)
Arctocephalus gazella	Antarctic fur seal	Water	P	435	1075	35	1110	0.03	0.03	0.04	Chevallay et al. (2024)
Arctocephalus philippii	Juan Fernández fur seal	Air	P								Norris and Watkins (1971)
Arctocephalus philippii	Juan Fernández fur seal	Water	P	150	80	120	200	0.07	0.05	0.11	Norris and Watkins (1971)
Arctocephalus pusillus	Cape fur seal	Air	T	318	272	182	454	0.77	0.29	1.2	Tripovich et al. (2008a)
Arctocephalus pusillus	Cape fur seal	Air	BP	767	1743	288	2031	1.1	0.03	9.9	Martin et al. (2021b), Stirling and Warneke (1971), Tripovich et al. (2005, 2006, 2008a, b, 2009a)
Arctocephalus pusillus	Cape fur seal	Air	P	300	1385	154	1539	0.19	0.1	0.29	Stirling and Warneke (1971), Tripovich et al. (2008a)
Arctocephalus townsendi	Guadalupe fur seal	Air	BP	550				1.4			Peterson et al. (1968), Pierson (1978, 1987)
Arctocephalus townsendi	Guadalupe fur seal	Air	P					2			Peterson et al. (1968), Pierson (1978, 1987), Reynoso (1994)
Arctocephalus townsendi	Guadalupe fur seal	Air	N	800				1.7			Peterson et al. (1968)
Arctocephalus tropicalis	Subantarctic fur seal	Air	T	512	517	257	774	1.2	1.2	1.2	Page et al. (2001)
Arctocephalus tropicalis	Subantarctic fur seal	Air	BP	1159	886	742	1628	1.3	0.1	8.3	Charrier et al. (2002, 2003a, b, c), Page et al. (2001, 2002a, 2002b), Roux and Jouventin (1987), St Clair Hill et al. (2001)
Arctocephalus tropicalis	Subantarctic fur seal	Air	P	800	863	368	1231	0.22	0.05	0.63	Gwilliam et al. (2008), St Clair Hill et al. (2001)
Callorhinus ursinus	Northern fur seal	Air	T		200	400	600	0.65	0.4	0.9	Lisitsyna (1973)
Callorhinus ursinus	Northern fur seal	Air	BP	1280	1543	500	2043	1.7	0.55	2.2	Antonelis and York (1983), Insley (1992, 2001), Lisitsyna (1973), Peterson (1968), Poulter (1968), Takemura et al. (1983)
Callorhinus ursinus	Northern fur seal	Air	P	800	410	240	680	0.84	0.02	1.6	Insley (2001), Lisitsyna (1973)

(continued)

Table 5.1 (continued)

Species	Common	Medium	Sound type	Fpeak [Hz]	BW [Hz]	Fmin [Hz]	Fmax [Hz]	Dur [s]	Durmin [s]	Durmax [s]	SL_SPLrms [dB re 1 (water) or 20 (air) µPa m]	SL_SPLpk-pk [dB re 1 (water) or 20 (air) µPa m]	References
Callorhinus ursinus	Northern fur seal	Air	N	300				0.65	0.1	1.2			Lisitsyna (1973), Peterson (1968)
Cystophora cristata	Hooded seal	Air	T	525	629	379	1007	0.44	0.5	1			Frouin-Mouy and Hammill (2021), Terhune and Ronald (1973)
Cystophora cristata	Hooded seal	Air	BP		4500	500	4000	3.3	0.1	6.9			Ballard and Kovacs (1995), Terhune and Ronald (1973)
Cystophora cristata	Hooded seal	Air	P	450	1996	669	2761	2	0.08	5.3			Ballard and Kovacs (1995), Frouin-Mouy and Hammill (2021), Terhune and Ronald (1973)
Cystophora cristata	Hooded seal	Water	T	381	2225	164	3448	0.88	0.49	1.5			Frouin-Mouy and Hammill (2021)
Cystophora cristata	Hooded seal	Water	BP	1200	550	150	2467	0.75	0.5	1			Ballard and Kovacs (1995), Terhune and Ronald (1973)
Cystophora cristata	Hooded seal	Water	P	678	1934	178	3305	0.74	0.44	1.3			Ballard and Kovacs (1995), Frouin-Mouy and Hammill (2021), Schevill et al. (1963)
Erignathus barbatus	Bearded seal	Water	T	1193	3250	373	3623	18	0.09	83			Charrier et al. (2013), Cleator et al. (1989), Davies et al. (2006), De Vreese et al. (2018), Frouin-Mouy et al. (2016), Halliday et al. (2020b), Heimrich et al. (2021), Jenkins et al. (2022), Jones et al. (2014), Lomac-MacNair et al. (2019), MacIntyre et al. (2013, 2015), Madan et al. (2020), Parisi et al. (2017), Ray et al. (1969), Risch et al. (2007), Southall et al. (2020), Stirling et al. (1983), Terhune (1999), Van Parijs et al. (2001)
Erignathus barbatus	Bearded seal	Water	BP	3200	7350	400	7750	6.5	1.3	14			Davies et al. (2006), Van Parijs et al. (2001)
Erignathus barbatus	Bearded seal	Water	P	217	107	117	223	1.5	0.9	3.8			Cleator et al. (1989), Heimrich et al. (2021), Parisi et al. (2017)
Eumetopias jubatus	Steller sea lion	Air	T	400	71	366	437	1.2	0.25	1.9			Park et al. (2006)
Eumetopias jubatus	Steller sea lion	Air	BP		850	150	1000	1.3	1	1.5			Campbell et al. (2002), Orr and Poulter (1967)

Latin name	Common name	Medium	Code									References
Eumetopias jubatus	Steller sea lion	Air	P	318	27	304	330	0.89	0.41	1.7		Park et al. (2006)
Eumetopias jubatus	Steller sea lion	Air	N									Orr and Poulter (1967)
Eumetopias jubatus	Steller sea lion	Water	P	400			1600					Orr and Poulter (1967), Schusterman et al. (1970)
Eumetopias jubatus	Steller sea lion	Water	N									Orr and Poulter (1967)
Halichoerus grypus	Gray seal	Air	T	380	475	161	636	3	0.27	19		McCulloch (2000), Nowak (2021)
Halichoerus grypus	Gray seal	Air	BP	624	338	309	647	1.7	0.41	5.1		Caudron et al. (1998), McCulloch (2000), McCulloch et al. (1999), Stansbury et al. (2015)
Halichoerus grypus	Gray seal	Air	P	522	716	164	880	1.3	0.73	2		McCulloch (2000), Stansbury et al. (2015)
Halichoerus grypus	Gray seal	Air	N		160	103	263	2.7	1.5	4		McCulloch (2000)
Halichoerus grypus	Gray seal	Water	T	659	962	169	1130	0.75	0.08	19		Asselin et al. (1993), Nowak (2021), Pérez Tadeo et al. (2023), Pozo Galván et al. (2024)
Halichoerus grypus	Gray seal	Water	BP	2082	3134	353	3487	0.96	0.04	4.3		Asselin et al. (1993), Hocking et al. (2020), McCulloch (2000), Nowak (2021), Pérez Tadeo et al. (2023), Prawirasasra et al. (2021)
Halichoerus grypus	Gray seal	Water	P	1075	2527	172	5181	0.71	0.04	3		Asselin et al. (1993), Hocking et al. (2020), McCulloch (2000), Nowak (2021), Pérez Tadeo et al. (2023), Pozo Galván et al. (2024), Prawirasasra et al. (2021)
Halichoerus grypus	Gray seal	Water	C									Schevill et al. (1963)
Halichoerus grypus	Gray seal	Water	N		1861	48	1908	1.4	0.69	2.5		McCulloch (2000)
Histriophoca fasciata	Ribbon seal	Water	T	675	1762	624	2386	1.2	0.19	4.7		Frouin-Mouy et al. (2019), Jones et al. (2014), Mizuguchi et al. (2016a), Otsuki et al. (2018), Watkins and Ray (1977)
Histriophoca fasciata	Ribbon seal	Water	BP	929	3152	186	3338	1.5	0.31	3.1	174	Frouin-Mouy et al. (2019), Jones et al. (2014), Miksis-Olds and Parks (2011), Mizuguchi et al. (2016a), Otsuki et al. (2018), Watkins and Ray (1977)
Histriophoca fasciata	Ribbon seal	Water	P	750	1351	110	1461	0.64	0.19	2.2		Frouin-Mouy et al. (2019), Jones et al. (2014), Otsuki et al. (2018)

(continued)

Table 5.1 (continued)

Species	Common	Medium	Sound type	Fpeak [Hz]	BW [Hz]	Fmin [Hz]	Fmax [Hz]	Dur [s]	Durmin [s]	Durmax [s]	SL_SPLrms [dB re 1 (water) or 20 (air) µPa m]	SL_SPLpk-pk [dB re 1 (water) or 20 (air) µPa m]	References
Histriophoca fasciata	Ribbon seal	Water	N	1573	2032	463	2495	2.6	0.2	15			Frouin-Mouy et al. (2019), Jones et al. (2014), Miksis-Olds and Parks (2011), Mizuguchi et al. (2016a), Watkins and Ray (1977)
Hydrurga leptonyx	Leopard seal	Air	T										Stirling and Siniff (1979)
Hydrurga leptonyx	Leopard seal	Water	T	5190	22,354	11,447	33,801	3.3	0.01	8.2	165		Awbrey et al. (2003), Gedamke and Robinson (2010), Klinck (2008), Kreiss et al. (2013), Meister (2017), Rogers (2007, 2014), Rogers et al. (1996), Schwenke (2013), Stirling and Siniff (1979), Thomas et al. (1982), Thomas and Golladay (1995), van Opzeeland et al. (2010)
Hydrurga leptonyx	Leopard seal	Water	BP	989	6354	5542	11,897	3.2	0.04	9.5			Awbrey et al. (2003), Klinck (2008), Meister (2017), Rogers (2017), Rogers et al. (1995), Schwenke (2013), Stirling and Siniff (1979)
Hydrurga leptonyx	Leopard seal	Water	P	1200	939	6347	7286	5.3	0.27	9.5			Awbrey et al. (2003), Klinck (2008), Mossbridge and Thomas (1999), Rogers (2007), Rogers et al. (1995), Schwenke (2013), van Opzeeland et al. (2010)
Hydrurga leptonyx	Leopard seal	Water	N	1910	4251	1379	5629	4.2	1	3.5			Dziak et al. (2017), Rogers (2017), Rogers et al. (1995), van Opzeeland et al. (2010)
Leptonychotes weddellii	Weddell seal	Air	T		4360	640	5000	1.4	0.72	3.3			Terhune et al. (1994)
Leptonychotes weddellii	Weddell seal	Air	BP	323	333	131	464	0.91	0.22	2.9			Collins et al. (2005, 2006), Collins and Terhune (2007), Schevill and Watkins (1965), Thomas (1979), van Opzeeland et al. (2012)
Leptonychotes weddellii	Weddell seal	Air	P		778	798	1575	0.35	0.08	1.7			Terhune et al. (1994), Thomas (1979)
Leptonychotes weddellii	Weddell seal	Air	N										Thomas (1979)

Leptonychotes weddellii										
Leptonychotes weddellii	Weddell seal	Water	T	8867	7470	2773	10,458	8	0.03	43	160	Cziko et al. (2020), Doiron et al. (2012), McCarthy (2023), Morrice et al. (1994), Pahl et al. (1997), Ray (1967), Russell et al. (2016), Schevill and Watkins (1971), Terhune (2016, 2017), Terhune and Dell'Apa (2006), Terhune et al. (2001), Thomas (1979), Thomas and Kuechle (1982), van Opzeeland et al. (2010)
Leptonychotes weddellii	Weddell seal	Water	BP	6680	8069	1830	13,729	6.6	0.01	76		Cziko et al. (2020), Doiron et al. (2012), Evans et al. (2004), Morrice et al. (1994), Pahl et al. (1997), Russell et al. (2016), Terhune and Dell'Apa (2006), Thomas (1979), Thomas and Kuechle (1982)
Leptonychotes weddellii	Weddell seal	Water	P	509	7815	1351	9671	3.4	0.6	42	172	Doiron et al. (2012), Evans et al. (2004), McCarthy (2023), Morrice et al. (1994), Pahl et al. (1997), Russell et al. (2016), Schevill and Watkins (1965), Terhune (2016, 2017), Terhune and Dell'Apa (2006), Terhune et al. (2001), Thomas (1979), Thomas and Kuechle (1982), van Opzeeland et al. (2010)
Leptonychotes weddellii	Weddell seal	Water	C									Evans et al. (2004)
Leptonychotes weddellii	Weddell seal	Water	N		6440	640	7080	8.8			170	Doiron et al. (2012), Terhune and Dell'Apa (2006), Thomas (1979), Thomas and Kuechle (1982)
Lobodon carcinophagus	Crabeater seal	Air	BP	1036	883	613	1496	0.92	0.09	2.3		Shabangu et al. (2022)
Lobodon carcinophagus	Crabeater seal	Air	N	16.733	8354	12,556	20,910	1.8	1	2.5		Shabangu et al. (2022)
Lobodon carcinophagus	Crabeater seal	Water	T	935	1384	696	1405	0.86	0.54	0.62		McCreery and Thomas (2009)
Lobodon carcinophagus	Crabeater seal	Water	BP	1046	1845	459	2290	2.2	0.13	6		Klinck et al. (2010), McCreery and Thomas (2009), Meister (2017), Nagaraj et al. (2021), Shabangu and

(continued)

Table 5.1 (continued)

Species	Common	Medium	Sound type	Fpeak [Hz]	BW [Hz]	Fmin [Hz]	Fmax [Hz]	Dur [s]	Durmin [s]	Durmax [s]	SL_SPLrms [dB re 1 (water) or 20 (air) µPa m]	SL_SPLpk-pk [dB re 1 (water) or 20 (air) µPa m]	References
Mirounga angustirostris	Northern elephant seal	Air	BP	926	697	448	1145	2.2	4.4	9.8			Charif (2021), Stirling and Siniff (1979), van Opzeeland et al. (2010)
Mirounga angustirostris	Northern elephant seal	Air	P	469	1579	246	1959	5.5	0.52	20	98	125	Bartholomew and Collias (1962), Casey et al. (2021), Insley (1992), Le Boeuf and Peterson (1969), Schusterman (1967)
Mirounga angustirostris	Northern elephant seal	Air	N	800			4500	1.1					Bartholomew and Collias (1962), Burgess et al. (1998), Casey et al. (2015), Holt et al. (2010), Le Boeuf and Petrinovich (1974), Mathevon et al. (2017), Petrinovich (1974), Sandegren (1976), Shipley et al. (1981), Southall (2002), Southall et al. (2019)
Mirounga angustirostris	Northern elephant seal	Water	P	290	30								Bartholomew and Collias (1962), Schusterman (1967)
Mirounga leonina	Southern elephant seal	Air	P	410	807	184	954	8.1	1	57	109		Burgess et al. (1998)
Monachus monachus	Mediterranean monk seal	Air	T	1976	1663	1534	3449	1.1	0.21	4.2			McCann (1981), Sanvito and Galimberti (2000a, b), Sanvito et al. (2007b), Sanvito et al. (2008)
Monachus monachus	Mediterranean monk seal	Air	BP	843	1635	347	1339	0.35	0.12	0.59			Charrier et al. (2017), Muñoz et al. (2011)
Monachus monachus	Mediterranean monk seal	Air	P	1261	2887	1176	4081	1.2	0.11	4.5			Charrier et al. (2017)
Monachus monachus	Mediterranean monk seal	Air	N		686	1003	1688	2.2	0.6	8.7			Charrier et al. (2017), Muñoz et al. (2011)
Monachus monachus	Mediterranean monk seal	Water	BP	385	326	222	548	0.5	0.07	2.5			Muñoz et al. (2011)
Monachus monachus	Mediterranean monk seal	Water	P	452	506	199	705	2.7	0.09	6.7			Charrier et al. (2023)
Neomonachus schauinslandi	Hawaiian monk seal	Air	T										Parnell (2018)

Species	Common name	Medium	Call type										References
Neomonachus schauinslandi	Hawaiian monk seal	Air	BP	693	200	567	767	0.62	0.13	1			Eliason et al. (1990), Miller and Job (1992)
Neomonachus schauinslandi	Hawaiian monk seal	Air	P		300	100	400	8.8	5	30			Miller and Job (1992), Sills et al. (2021)
Neomonachus schauinslandi	Hawaiian monk seal	Air	N					0.25					Miller and Job (1992)
Neomonachus schauinslandi	Hawaiian monk seal	Water	T	52	26	34	70	0.74	0.27	2.1	146		Parnell (2018), Sills et al. (2021)
Neomonachus schauinslandi	Hawaiian monk seal	Water	BP	153	146	17	289	1.3	0.37	2.1	137		Parnell (2018), Sills et al. (2021)
Neomonachus schauinslandi	Hawaiian monk seal	Water	P	169	96	94	244	1.1	0.1	4.4	146		Parnell (2018), Sills et al. (2021), Stirling and Thomas (2003)
Neomonachus schauinslandi	Hawaiian monk seal	Water	N	286	131	205	367	3.2	1.7	4.6	141		Parnell (2018), Sills et al. (2021)
Neophoca cinerea	Australian sea lion	Air	T	200				0.3					Gwilliam et al. (2008)
Neophoca cinerea	Australian sea lion	Air	BP	819	707	573	1280	0.94	0.54	1.4			Charrier and Harcourt (2006), Charrier et al. (2009), Gwilliam et al. (2008), Pitcher et al. (2009, 2012), Stirling (1972)
Neophoca cinerea	Australian sea lion	Air	P	860	1196	463	1659	0.06	0.02	0.11			Ahonen et al. (2014), Attard et al. (2010), Charrier et al. (2011a), Gwilliam et al. (2008), Pitcher et al. (2009), Stirling (1972)
Neophoca cinerea	Australian sea lion	Air	N		10,767	400	11,167	0.6					Stirling (1972)
Odobenus rosmarus	Walrus	Air	T	945	94	1251	1345	1.8	1.6	2.8	111		Charrier et al. (2011b), Miller (1985), Verboom and Kastelein (1995)
Odobenus rosmarus	Walrus	Air	BP	255	197	183	380	0.66	0.09	3.6			Charrier et al. (2010), Kastelein et al. (1995), Miller (1985)
Odobenus rosmarus	Walrus	Air	P	271	715	585	1300	2.4	0.1	0.56			Charrier et al. (2011b), Miller (1985), Stirling et al. (1987)
Odobenus rosmarus	Walrus	Air	N	100				1	0.5	1			Miller (1985)
Odobenus rosmarus	Walrus	Water	T	1451	954	462	1416	1.8	0.78	2		147	Jézéquel et al. (2022), Mouy et al. (2012), Ray and Watkins (1975), Reichmuth et al. (2009), Schusterman and Reichmuth (2008), Stirling et al. (1983)
Odobenus rosmarus	Walrus	Water	BP	250				0.5					Mouy et al. (2012), Schusterman and Reichmuth (2008)
Odobenus rosmarus	Walrus	Water	P	750	2066	279	2239	1.4	0.02	0.71	177	179	

(continued)

Table 5.1 (continued)

Species	Common	Medium	Sound type	Fpeak [Hz]	BW [Hz]	Fmin [Hz]	Fmax [Hz]	Dur [s]	Durmin [s]	Durmax [s]	SL_SPLrms [dB re 1 (water) or 20 (air) µPa m]	SL_SPLpk-pk [dB re 1 (water) or 20 (air) µPa m]	References
Odobenus rosmarus													Denes (2014), Denes et al. (2015), Hughes et al. (2011), Jézéquel et al. (2022), Larsen and Reichmuth (2021), Madan et al. (2020), Mouy et al. (2012), Ray and Watkins (1975), Reichmuth et al. (2009), Reichmuth and Quihuis (2022), Rideout et al. (2013), Schevill et al. (1966), Schusterman and Reichmuth (2008), Stirling et al. (1983, 1987)
Odobenus rosmarus	Walrus	Water	N	300									Schusterman and Reichmuth (2008)
Ommatophoca rossii	Ross seal	Air	T		700	100	800						Ray (1981)
Ommatophoca rossii	Ross seal	Air	BP		700	100	800	1.3	1	1.5			Ray (1981), Watkins and Ray (1985)
Ommatophoca rossii	Ross seal	Air	P		900	100	1000	0.38	0.25	0.5			Watkins and Ray (1985)
Ommatophoca rossii	Ross seal	Water	T	417	901	330	1231	2.2	1.5	2.8			Gedamke and Robinson (2010), Kindermann et al. (2008), Schwenke (2013), Seibert et al. (2007, 2008), Stacey (2006), van Opzeeland et al. (2010)
Ommatophoca rossii	Ross seal	Water	BP	1225	4162	460	4578	2.9	1	4.1			Gedamke and Robinson (2010), Kindermann et al. (2008), Ray 1981, Seibert et al. (2007, 2008), van Opzeeland et al. (2010), Watkins and Ray 1985
Ommatophoca rossii	Ross seal	Water	P		9100	1400	10,500						Kindermann et al. (2008)
Ommatophoca rossii	Ross seal	Water	N	2000	9859	1292	11,151	2.5	2.2	2.8			Schwenke (2013), Seibert et al. (2007, 2008), van Opzeeland et al. (2010)

Species	Common name	Medium	Type									References
Otaria byronia	South American sea lion	Air	T	444	490	199	689	0.45	0.15	0.84		Fernández-Juricic et al. (1999)
Otaria byronia	South American sea lion	Air	BP	683	770	298	1068	1.1	0.34	1.8		Fernández-Juricic et al. (1999), Trimble and Charrier (2011)
Otaria byronia	South American sea lion	Air	P	479	160	399	559	0.21	0.16	0.25		Fernández-Juricic et al. (1999)
Otaria byronia	South American sea lion	Air	N	628	1082	87	1169	0.49	0.31	0.66		Fernández-Juricic et al. (1999)
Pagophilus groenlandicus	Harp seal	Air	T	700	600	400	1000					Miller and Murray (1995)
Pagophilus groenlandicus	Harp seal	Air	BP		1492	772	2264	0.87	0.26	1.5		Miller and Murray (1995), van Opzeeland et al. (2009), van Opzeeland and Van Parijs (2004)
Pagophilus groenlandicus	Harp seal	Air	N	206	0	206	206	4.9	4.2	5.5		Serrano (2001), Serrano-Solis (1998)
Pagophilus groenlandicus	Harp seal	Water	T	1855	727	1446	2173	1	1.2	3.2	143	Møhl et al. (1975), Rossong and Terhune (2009), Serrano (2001), Serrano-Solis (1998), Terhune (1994)
Pagophilus groenlandicus	Harp seal	Water	BP	734	402	637	1038	1.3	0.24	2.6	145	Møhl et al. (1975), Rossong and Terhune (2009), Serrano (2001), Serrano-Solis (1998), Watkins and Schevill (1979)
Pagophilus groenlandicus	Harp seal	Water	P	4183	718	256	974	1.2	0.46	2.1	150	Møhl et al. (1975), Moors and Terhune (2003), Rossong and Terhune (2009), Serrano (2001), Serrano-Solis (1998), Terhune (1994), Watkins and Schevill (1979)
Pagophilus groenlandicus	Harp seal	Water	N	500								Møhl et al. (1975)
Phoca largha	Spotted seal	Air	T	281	213	174	387	0.73	0.42	0.91	106	Zhang et al. (2016)
Phoca largha	Spotted seal	Air	BP	1018	749	598	1346	0.84	0.39	0.74	103	Zhang et al. (2016, 2022)
Phoca largha	Spotted seal	Air	P	770	315	596	910	0.67	0.11	1.7	117	Beier and Wartzok (1979), Sills and Reichmuth (2022), Zhang et al. (2016, 2022)
Phoca largha	Spotted seal	Water	T	251	618	819	1415	0.75	0.27	2.1	111	Beier and Wartzok (1979), Gailey-Phipps (1984), Sills and Reichmuth (2022)

(continued)

Table 5.1 (continued)

Species	Common	Medium	Sound type	Fpeak [Hz]	BW [Hz]	Fmin [Hz]	Fmax [Hz]	Dur [s]	Durmin [s]	Durmax [s]	SL_SPLrms [dB re 1 (water) or 20 (air) µPa m]	SL_SPLpk-pk [dB re 1 (water) or 20 (air) µPa m]	References
Phoca largha	Spotted seal	Water	BP	232	245	30	434	5.3	3.2	9.4	142		Sills and Reichmuth (2022)
Phoca largha	Spotted seal	Water	P	245	279	172	442	0.24	0	1.5	143		Sills and Reichmuth (2022), Yang et al. (2017, 2022)
Phoca sibirica	Baikal seal	Water	BP	435	1700	250	1950	0.8					Baranov et al. (2014), Grigoriev et al. (2020), Pastukhov (1993), Petrov (2009)
Phoca vitulina	Harbor seal	Air	T	1522	2107	1130	3236	0.91	0.41	1.9			Ralls et al. (1985), Sauve et al. (2015a), Van Parijs and Kovacs (2002)
Phoca vitulina	Harbor seal	Air	BP	874	1763	791	2243	0.71	0.12	2.4			Khan et al. (2006), MacRae (2018), Perry and Renouf (1988), Ralls et al. (1985), Scheffer and Slipp (1944), Van Parijs and Kovacs (2002)
Phoca vitulina	Harbor seal	Air	P	2533	1398	1126	2524	1.6	0.12	7.8			Khan et al. (2006), Kocsis et al. (2024), Ralls et al. (1985), Scheffer and Slipp (1944), Van Parijs and Kovacs (2002)
Phoca vitulina	Harbor seal	Air	N	4050	1837	1338	3175	2.8	0.41	11			Ralls et al. (1985), Van Parijs and Kovacs (2002)
Phoca vitulina	Harbor seal	Water	T	168	1420	352	1772	2.5	0.5	7			Bjørgesæter et al. (2004), Hanggi and Schusterman (1992, 1994), Pozo Galván et al. (2024), Sabinsky et al. (2017)
Phoca vitulina	Harbor seal	Water	BP	540	1396	63	1428	4.3	0.17	13			Hanggi and Schusterman (1994), Nikolich et al. (2016), Perry and Renouf (1988), Rößler et al. (2021)
Phoca vitulina	Harbor seal	Water	P	1762	1488	119	1861	3.2	0	19	149	193	Bjørgesæter et al. (2004), Casey et al. (2021), Hanggi and Schusterman (1992, 1994), Matthews et al. (2017b, 2020), Renouf and Davis (1982), Sabinsky et al. (2017), Schusterman et al. (1970), Van Parijs et al. (1997), Wahlberg et al. (2002)
Phoca vitulina	Harbor seal	Water	N	521	719	355	961	4.8	0.68	15	141		Casey et al. (2016), Hanggi and Schusterman (1992, 1994), Hayes et al. (2004), Matthews (2017), Matthews et al. (2017a, 2018), Nicholson (2000), Nikolich et al. (2016, 2018), Sabinsky et al. (2017), Van Parijs et al. (2000a, b), Van Parijs and Kovacs (2002)

Species	Common name	Medium	Code									References
Pusa hispida	Ringed seal	Air	T	700					2.5			Sipilä et al. (1996)
Pusa hispida	Ringed seal	Air	P									Hyvärinen (1989)
Pusa hispida	Ringed seal	Water	T	663	917	338	1256	0.35	0.04	1.3		Jones et al. (2014), Lomac-MacNair et al. (2019), Mizuguchi et al. (2016b), Prawirasasra et al. (2021), Stirling (1973), Stirling et al. (1983)
Pusa hispida	Ringed seal	Water	BP	615	800	218	1018	0.93	0.04	10	112	Cummings et al. (1984), Jones et al. (2014), Kunnasranta et al. (1996), Lomac-MacNair et al. (2019), Mizuguchi et al. (2016b), Prawirasasra et al. (2021), Stirling (1973), Stirling et al. (1983)
Pusa hispida	Ringed seal	Water	P	747	1220	271	1491	0.44	0	1.3	130	Cummings et al. (1984), Kunnasranta et al. (1996), Mizuguchi et al. (2016b), Prawirasasra et al. (2021), Rautio et al. (2009), Stirling (1973), Stirling et al. (1983)
Pusa hispida	Ringed seal	Water	C	5500	3333	4000	7333	0.31	0.02	0.6		Hyvärinen (1989), Kunnasranta et al. (1996)
Pusa hispida	Ringed seal	Water	N	388	1100	200	1300	0.2				Halliday et al. (2020b), Stirling (1973), Stirling et al. (1983)
Zalophus californianus	California sea lion	Air	BP	670	775	313	1088	2.5	0.44	13		Peterson and Bartholomew (1969), Schusterman et al. (1992), Schusterman and Feinstein (1965), Schwalm (2014)
Zalophus californianus	California sea lion	Air	P									Peterson and Bartholomew (1969)
Zalophus californianus	California sea lion	Air	N	1000				2				Peterson and Bartholomew (1969)
Zalophus californianus	California sea lion	Water	T									Poulter (1968)
Zalophus californianus	California sea lion	Water	BP	720			3500	0.87	0.2	0.3		Poulter (1968), Schusterman (1967), Schusterman and Feinstein (1965), Schusterman et al. (1967)
Zalophus californianus	California sea lion	Water	P	836	1200	533	2175	0.76	0.1	23		Poulter (1968), Schevill et al. (1963), Schusterman (1967), Schusterman et al. (1967), Thompson (1965)
Zalophus californianus	California sea lion	Water	C		10,400	3000	16,700					Poulter 1963, Poulter (1968)
Zalophus wollebaeki	Galápagos sea lion	Air	BP	500	300	300	600					Trillmich (1981)
Zalophus wollebaeki	Galápagos sea lion	Air	P									Trillmich (1981)

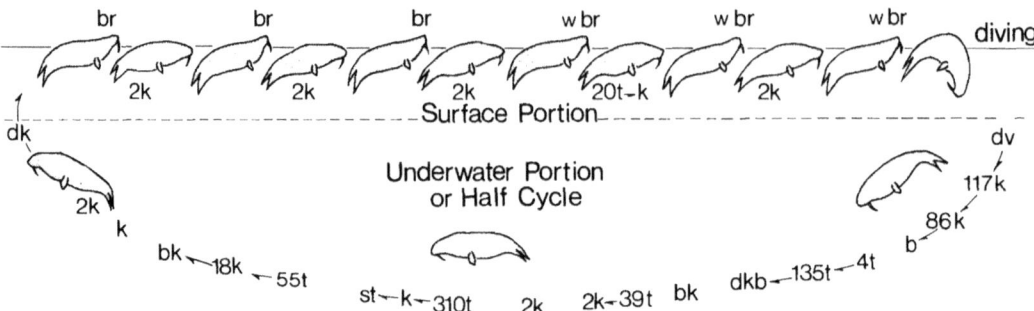

Fig. 5.9 Sketch of a walrus diving vocalization cycle. *B* bell, *bk* bell-knock, *br* breath, *dk* double-knock, *dkb* double-knock bell, *dv* diving vocalization, *k* knock, preceded by the number of knocks in a sequence; *st* strum, *t* tap, preceded by the number of taps in a sequence, *w* whistle; →, separable vocalizations that run together. Used with permission of Canadian Science Publishing, from Fig. 1 in Stirling I, Calvert W, Spencer C (1987) Evidence of stereotyped underwater vocalizations of male Atlantic walruses (*Odobenus rosmarus rosmarus*). Can J Zool 65 (9):2311–2321; https://doi.org/10.1139/z87-348. © Canadian Science Publishing 1987; permission conveyed through Copyright Clearance Center, Inc. All rights reserved

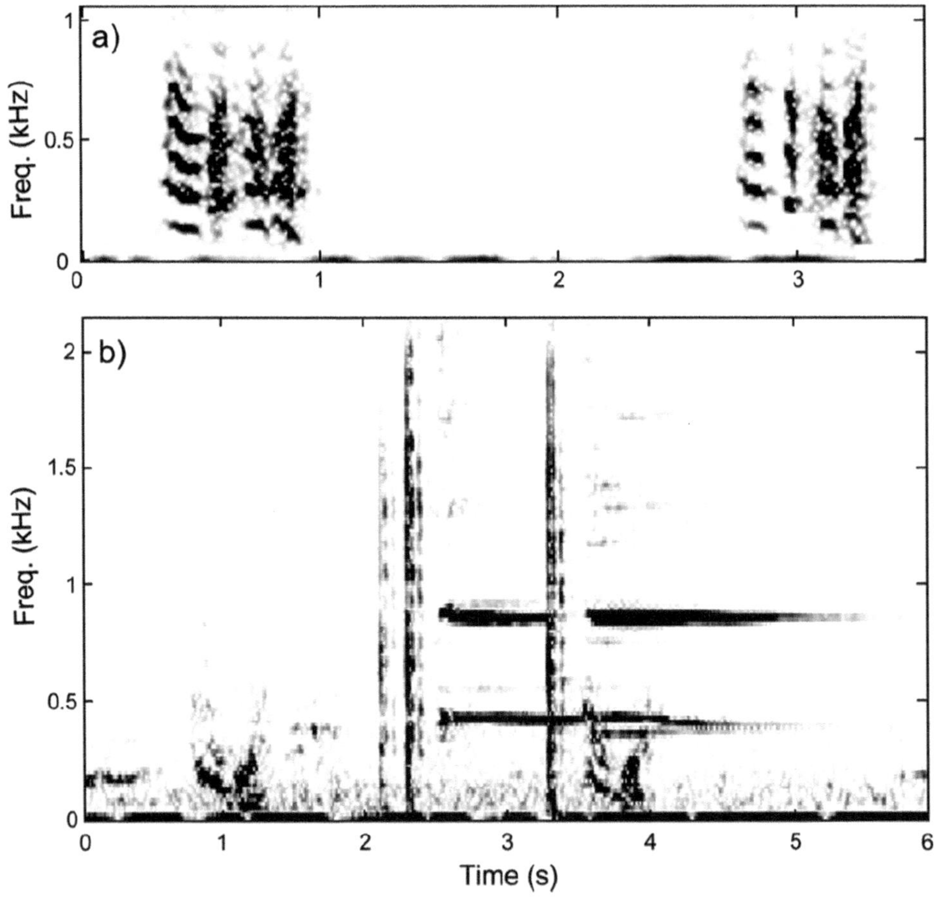

Fig. 5.10 Spectrograms of Pacific walrus sounds under water. (**a**) Two grunts, (**b**) one grunt, two knocks, one bell, one knock, one grunt, and one bell. Reproduced with permission by the ASA; from Mouy et al. 2012; https://doi.org/10.1121/1.3675008. © Acoustical Society of America, 2012. All rights reserved

Grunts and snorts may be used as mild threat sounds (e.g., toward an approaching conspecific), whereas guttural sounds were heard associated with fights; high-intensity threats and roars may have an alerting function (Miller 1985).

In captivity, walruses have been shown to learn new sounds or sound combinations (Schusterman and Reichmuth 2008). A male walrus came up with an attention-getting sound, which was later copied by female conspecifics, by using a rubber tug toy to produce a sharp trumpeting sound under water when "buzzing" it in a specific way against a small hole in the window frame. The hole was present because a bolt was removed by at least one of the walruses (Reichmuth and Quihuis 2022).

Walruses produce other nonvocal sounds. Knocks may be produced using their teeth (Fay 1982) and, like roars, can be heard kilometers away (Born and Wiig 2021). Whistles are produced by blowing air through curled lips while lifting their heads above water and bending their backs. Males may be able to discriminate foreign males' whistles. However, this has only been tested on two captive individuals (Charrier et al. 2011b).

During the breeding season, mature male walruses produce loud rhythmic clapping sounds under water with peak energy around 500 Hz (but a maximum frequency of 48 kHz), for a duration of 0.015 s and sound levels between 195 and 210 dB re 1 μPa pk-pk (Larsen and Reichmuth 2021; Reichmuth et al. 2009). Clapping the left wrist joint into the right palm creates small cavitation bubbles that implode loudly, producing the clapping sound (Fig. 5.11). This behavior has so far only been observed in captive individuals.

Feeding Atlantic walruses have been recorded producing two types of sound, with peak energy around 480 Hz (type 1) and 520 Hz (type 2) and a maximum frequency of <25 kHz, both with a duration of 0.1 s. Feeding sound type 1 has been compared to sounds "produced when opening a wine bottle" and corresponds to a walrus using its tongue to suck the soft tissues out of a bivalve's shell when it is buried in sand. Feeding sound type 2 was assumed to be the sound of crunching bivalve shells before chewing their soft tissues. This suggests that walruses consume bivalves living on the seafloor, which would be a new foraging method (Jézéquel et al. 2022).

Some differences between the Atlantic and Pacific subspecies have been described. Barking sounds showed less frequency modulation by Atlantic than Pacific walruses (Miller 1985). Compared to female Atlantic walruses in the wild, female Pacific walruses in captivity barked at higher frequencies and for shorter durations (Charrier et al. 2011b).

5.5.2 Family Otariidae: Eared Seals

5.5.2.1 *Arctocephalus australis*: South American Fur Seal

To our knowledge, South American fur seals vocalize in air but have not been recorded under water. Males and females emit sounds between 290 and 1590 Hz, with peak energy of 720 to 1500 Hz, lasting for 0.08 to 2.5 s, while pup calls are higher pitched, between 850 and 2260 Hz, with peak energy around 1100 to 1360 Hz, for 0.6 to 0.7 s (Phillips and Stirling 2000, 2001; Trillmich and Majluf 1981).

Philipps and Stirling (2001) grouped the 11 sound types they found in South American fur seals into four groups according to the associated behaviors: "investigative, threat, submissive, and affiliative calls" (Fig. 5.12). Barks were mostly used to approach conspecifics non-agonistically and were therefore classed as an investigation call. Threat calls include snorts, growls, chuffs, puffs, full-threat calls, low-intensity threat calls, and guttural threats that sound like "guug-guug-guug." Subordinate males and females use submissive or appeasement calls, often after an agonistic interaction, while displaying a submissive posture. Affiliative calls comprise FACs and PACs, which are individually distinctive and used to recognize each other (Phillips 2003; Phillips and Stirling 2000, 2001; Trillmich and Majluf 1981). Fundamental frequency varied most between females' calls, while pup calls could mostly be discriminated by their formant-like frequency ranges. Females

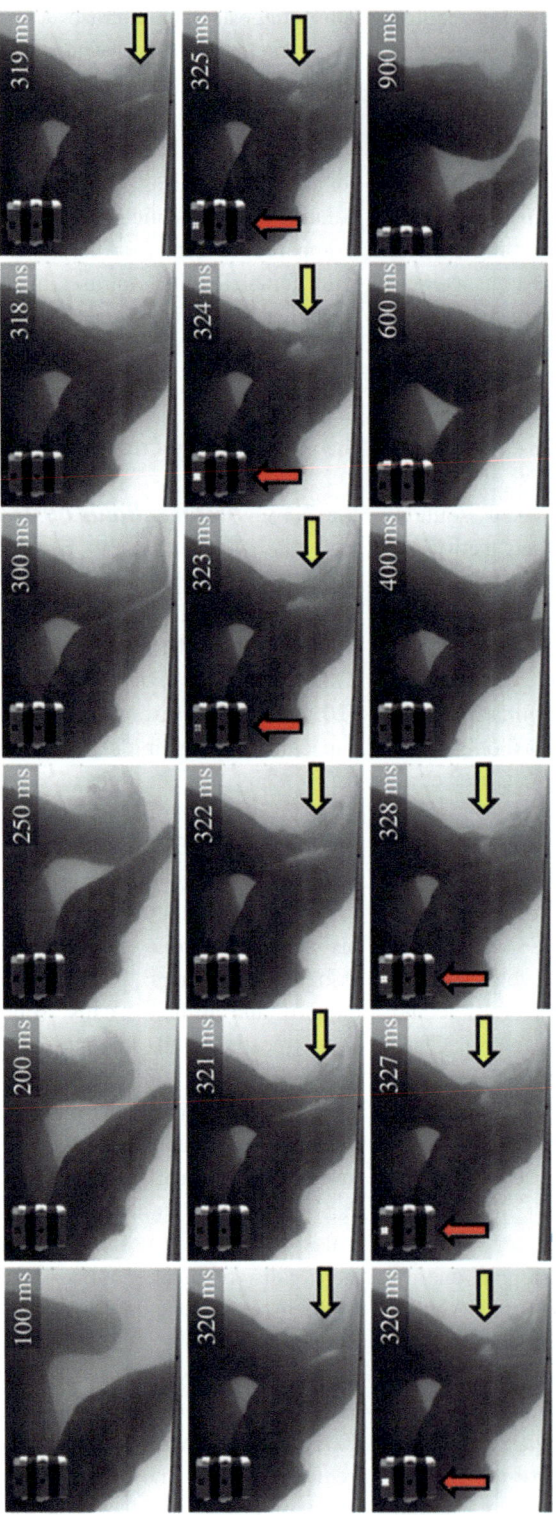

Fig. 5.11 Video frames showing walrus clap-induced cavitation (1 frame/ms). The clapping walrus is turned slightly to the left relative to the camera placed outside the pool. The right flipper of the walrus (the "anvil") is shown on the left portion of each image. Horizontal arrows (yellow) show cavitation cloud formation and collapse, and vertical arrows (red) show the onset and duration of the associated impulse sound. The brightest LED activation is associated with the cavitation cloud "budding" and shedding. Reprinted from (and full video in electronic supplementary material of) Larsen ON, Reichmuth C (2021) Walruses produce intense impulse sounds by clap-induced cavitation during breeding displays. R Soc Open Sci 8 (6):210197; https://doi.org/10.1098/rsos.210197. © Larsen and Reichmuth 2021. Published CC BY 4.0; https://creativecommons.org/licenses/by/4.0/

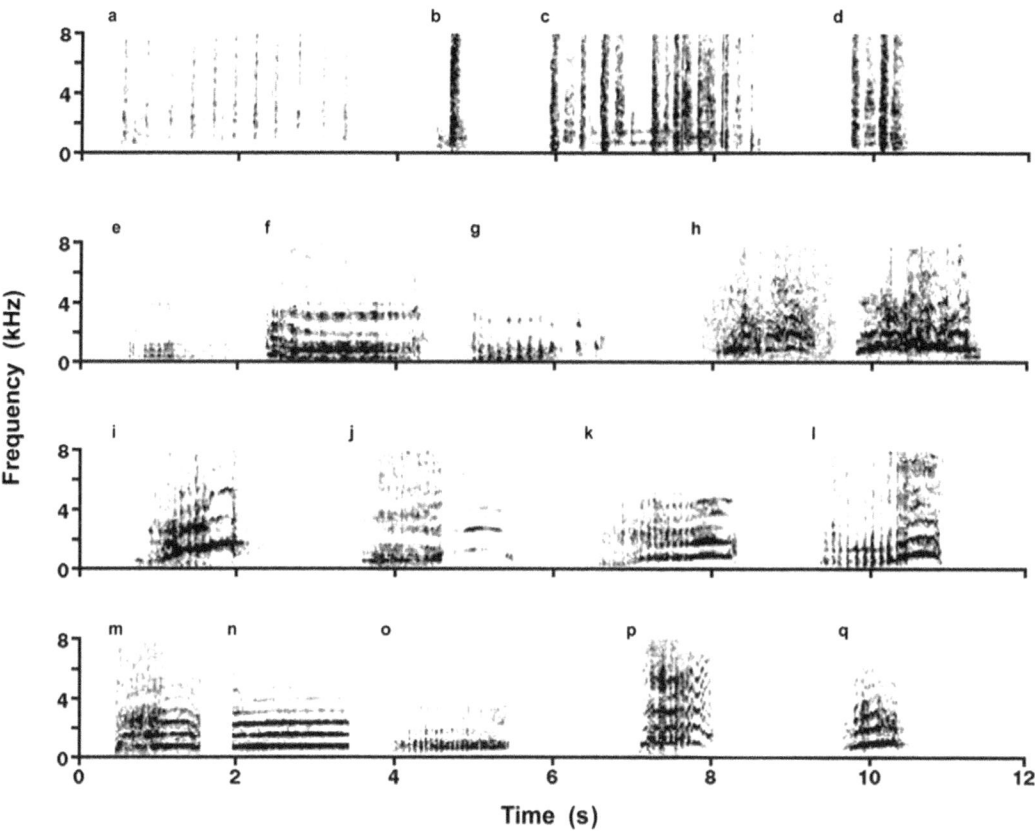

Fig. 5.12 Spectrograms of South American fur seal calls. (*a*) Bark. (*b*) Snort. (*c*) Puff. (*d*) Two chuffs. (*e*) Growl. (*f*) Low-intensity threat call. (*g*) Guttural threat call. (*h*) Two submissive calls, made by a retreating adult male after losing a fight with a territorial male. (*i, j, k*) Full-threat calls (FTCs) made by three different adult males. (*l*) FTC made by an adult female to a human observer. (*m, n, o*) PACs made by three different mothers. (*p*) FAC made by a pup. (*q*) FAC made by a yearling. Reprinted with permission from Canadian Science Publishing from Fig. 1 in Phillips AV, Stirling I (2001) Vocal repertoire of South American fur seals, *Arctocephalus australis*: structure, function, and context. Can J Zool 79 (3):420–437. https://doi.org/10.1139/z00-219. © Canadian Science Publishing 2001. All rights reserved. Permission conveyed through Copyright Clearance Center, Inc

appear to show more individual variation compared to the calls emitted by their pups (Phillips 2003; Phillips and Stirling 2000).

5.5.2.2 *Arctocephalus forsteri*: New Zealand Fur Seal, Long-Nosed Fur Seal

Sound types emitted by New Zealand fur seals have mostly been characterized descriptively, and only a few studies measured their acoustic features in air. Underwater vocalizations remain unknown for this species. In-air sounds fall between 160 and 7060 Hz, with peak energy between 440 and 2200 Hz, lasting 0.1 to 2 s (Page et al. 2001, 2002a, b; Stirling 1971a; Stirling and Warneke 1971). Females, pups, and one individual male have been observed producing a "fawp" sound by jaw clapping during agonistic interactions (Miller 1971).

Various threat calls have been described, including whimpering, snarling, (repetitive) barking, growls, high-intensity guttural threat calls (also called a choke call or full-threat call), and low- and high-intensity threat calls or roars, with the trumpeted roar being a long-range vocal threat call. These roars, emitted in varying intensities, are mostly used by territorial males. The trumpeted roars or male full-threat calls vary between individuals (Crawley and Wilson 1976; Miller 1971, 1974; Page et al. 2001, 2002a, b; Stirling 1970, 1971a; Stirling and Warneke 1971). These threat calls, and potentially also the male low-intensity threat call, male barking sounds, and the male guttural challenge call, also referred to as snorting, may be used for individual recognition among males (Stirling 1971a).

Other sound types used by New Zealand fur seals include moans and highly pitched submissive screeches or squeals, also referred to as a whine resembling "the whine of a dog" (Crawley and Wilson 1976; Miller 1971, 1974; Stirling 1970). For example, the latter is used in submissive situations by a retreating, defeated seal (Stirling 1970).

Females predominantly produce two sound types to address their pups: a lowing sound straight after birth and a loud and penetrating PAC, also compared to a cow's whine. Pups respond to their mothers' calls by emitting FACs—a bawling sound that becomes deeper as the pup grows. When in contact with its mother or to greet it, a pup may be "whimpering like a puppy" (Acevedo-Gutierrez et al. 2011; Crawley and Wilson 1976; McNab and Crawley 1975; Miller 1971; Page et al. 2002a; Stirling 1970, 1971a). The stereotyped FACs and PACs vary individually and may be used to recognize each other (Page et al. 2002a).

Three species of fur seals—New Zealand fur seals (*A. forsteri*), subantarctic fur seal (*A. tropicalis*), and Antarctic fur seals (*A. gazella*)—hybridize on Macquarie Island, Australia, with most hybrids probably being a cross between Antarctic and subantarctic fur seals (Goldsworthy et al. 1999). Male barking calls and PACs emitted by hybrid fur seals appeared to be intermediate sounds between the pure species calls, but no such pattern was found for full-threat calls and FACs. All hybrid calls, however, differ from the calls emitted by their parental species. Hybrid fur seals produced calls between 200 and 1740 Hz, with peak energy between 520 and 1200 Hz, lasting for 0.14 to 3.6 s (Page et al. 2001, 2002b; St Clair Hill et al. 2001; Fig. 5.13).

5.5.2.3 *Arctocephalus galapagoensis*: Galapagos Fur Seal

Very little information is available on the sound types produced by Galapagos fur seals under water and in air. Under water, they emit 100–1500 Hz drawn-out growls and snaps (or knocks) for 0.04–0.06 s, respectively (Merlen 2000). The snaps or knocks appear important for foraging (Merlen 2000).

Only females and pups have been recorded in air, producing calls that can be distinguished among individuals. Females call to their pups (PAC) between 650 and 1500 Hz from afar or emit pup contact calls when next to their pups (Trillmich 1981). FACs have been reported as pups bleating when searching for their mother and making greeting calls in staccato once they get in contact. The peak energy of pup sounds lies between 1000 and 1800 Hz (Trillmich 1981).

5.5.2.4 *Arctocephalus gazella*: Antarctic Fur Seal

To our knowledge, the first report of Antarctic fur seals producing sound under water was only recently published. Chevallay et al. (2024) reported on sounds from two females (i.e., no males) at Kerguelen Islands. Under water, females emitted series of pulses, with a mean peak frequency of 435 Hz and mean duration of 0.03 s, and inter-series sounds between each pulse series that lasted 0.1 s on average, with a mean peak frequency of 386 Hz. These vocalizations were produced during the bottom time of deep and long dives, assumed to be foraging dives, i.e., these underwater sounds are most likely associated with foraging (Chevallay et al. 2024).

In air, Antarctic fur seal calls last for 0.3–1.5 s, except one study measured barking sounds that lasted 5.6 s, on average. Their sounds were emitted between 220 and 7580 Hz, with peak frequencies of 710–5720 Hz (Aubin et al. 2015; Page et al. 2001, 2002a, b; St Clair Hill et al. 2001; Stirling 1971a). Males produce a range of threat calls, mostly during the breeding period, such as high-intensity guttural threat calls that are

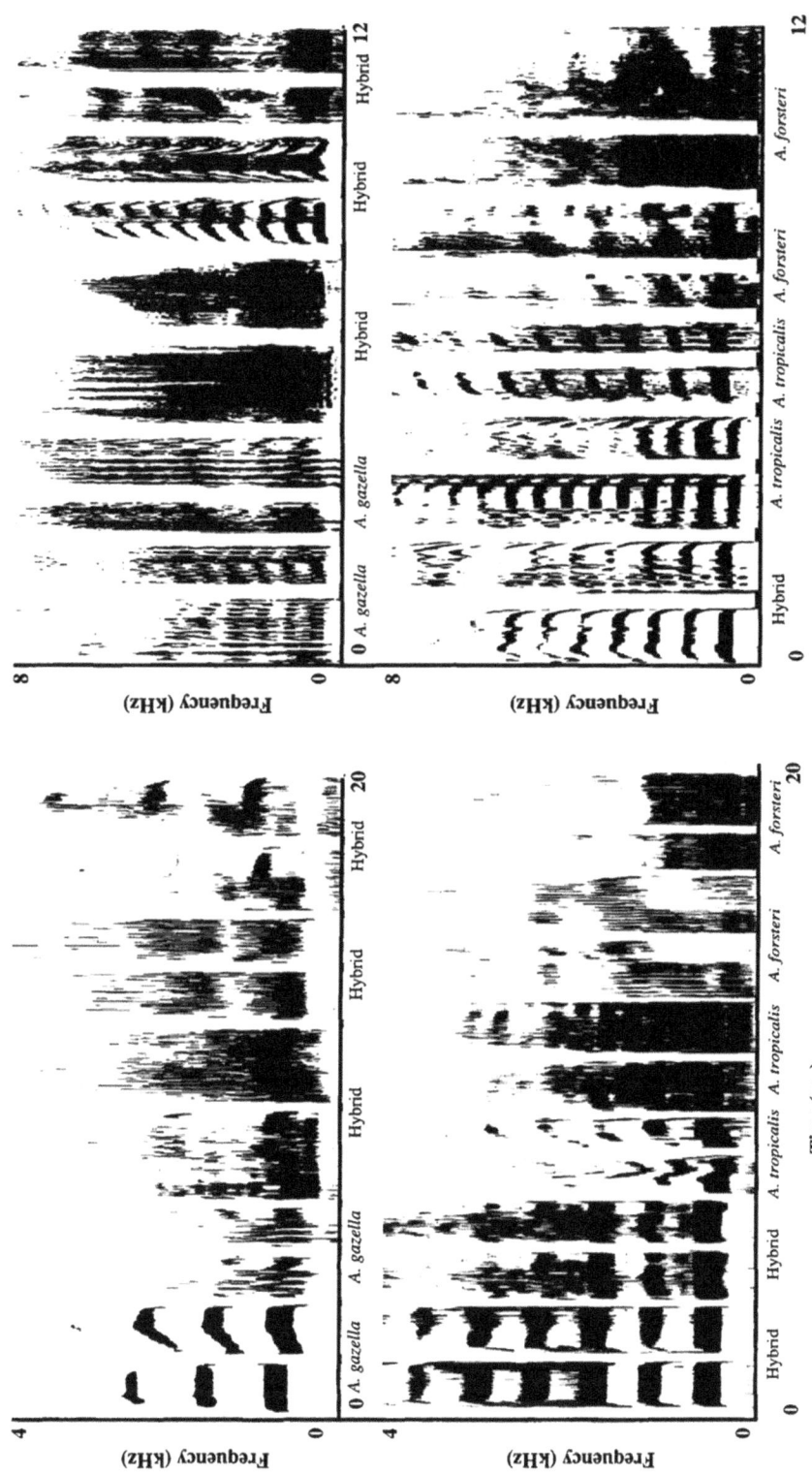

Fig. 5.13 Spectrograms of PACs (left panels) and FACs (right panels) of *Arctocephalus* spp. and hybrids (two calls/individual). Reprinted with permission; Fig. 1 of Page B, Goldsworthy SD, Hindell MA (2002) Individual vocal traits of mother and pup fur seals. Bioacoustics 13 (2):121–143. https://doi.org/10.1080/09524622.2002.9753491. © AB Academic Publishers, 2002. All rights reserved

either preceded by a growl or consist of a growl followed by a puff. Their calls include barks (e.g., to advertise territorial status) and low-intensity threat calls, referred to as short roars (Page et al. 2001, 2002b; St Clair Hill et al. 2001; Stirling and Warneke 1971). Females hiss toward foreign pups and emit PACs when searching for their own pup. Pups respond with FACs that sound like a bleat. These calls are also emitted by pups searching for their mothers or when hungry (Aubin et al. 2015; Dobson and Jouventin 2003; Page et al. 2002a; St Clair Hill et al. 2001; Stirling and Warneke 1971).

FACs vary among individuals and are important for individual recognition. Pups use the frequency-modulation pattern for long-distance recognition, and at closer range, they also use variations in amplitude and energy spectrum to identify their mothers (Aubin et al. 2015).

Although interspecies variations of male calls of Antarctic and subantarctic fur seals are considerable and discernable by the human ear, these fur seal species hybridize on some islands, suggesting that species-typical male vocalizations do not prevent interbreeding (St Clair Hill et al. 2001). See Sect. 5.5.2.2 and Fig. 5.13 for spectrograms of Antarctic fur seals and hybrids.

5.5.2.5 *Arctocephalus philippii*: Juan Fernández Fur Seal

Previous species lists have classified the Juan Fernández and Guadalupe fur seals as subspecies of *A. philippii*. We have followed the IUCN classification system that lists them both as individual species. Little is known about the vocal repertoire of the Juan Fernández fur seal. They moan and snarl in air and produce low-frequency pulse series under water. The latter carries most energy between 120 and 200 Hz, with a peak frequency of 150 Hz (Fig. 5.14, Norris and Watkins 1971). Individual pulses last, on average, 0.07 s (Norris and Watkins 1971).

5.5.2.6 *Arctocephalus townsendi*: Guadalupe Fur Seal

To our knowledge, sounds of Guadalupe fur seals have been recorded in air but not under water. Vocalizations lasted between 0.2 and 2 s, with peak frequencies of 500–800 Hz (Peterson et al. 1968; Pierson 1978). Threat calls used in situations of varying intensity include roars, barks also referred to as whickers, boundary puffs, growls, and full-threat calls. The latter vary considerably between individual males and may be used for individual recognition. Females and pups emit PACs and FACs, respectively, when interacting. The FAC sounds like a lamb-like bleat. Among other sounds, females may also bawl at their pups (PAC, Peterson et al. 1968; Pierson 1978; Reynoso 1994).

5.5.2.7 *Arctocephalus pusillus*: Cape Fur Seal and Australian Fur Seal

There are two subspecies of *A. pusillus*: *A. p. doriferus*, the Australian fur seal, and *A. p. pusillus*, the Cape fur seal.

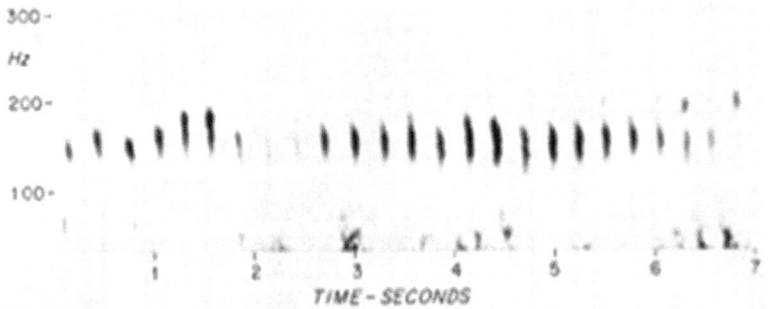

Fig. 5.14 Spectrogram of a pulse series emitted under water by a Juan Fernández fur seal. Reprinted with permission from Fig. 1 in Norris KS, Watkins WA (1971) Underwater sounds of *Arctocephalus philippii*, the Juan Fernandez fur seal. In: Burt WH (ed) Antarctic Pinnipedia, Antarctic Research Series, vol 18. John Wiley & Sons, Washington DC, pp 169–172. https://doi.org/10.1029/AR018p0169. © John Wiley & Sons, 2013. All rights reserved

5.5.2.7.1 *Arctocephalus pusillus doriferus*: Australian Fur Seal

To date, there are no records of Australian fur seals vocalizing under water. In air, Australian fur seals are relatively vocal, especially during the breeding season. Adult fur seals emit calls between 90 and 3180 Hz, with a peak energy of 210–830 Hz, lasting 0.1–1.8 s (Stirling and Warneke 1971; Tripovich et al. 2005, 2006, 2008a, b, 2009b, 2012). Male bark series are produced for 3.6 s on average lasting up to 9.9 s (Fig. 5.15, Tripovich et al. 2005). These barks are individually different, and males can distinguish barking calls between neighbors and strangers (Tripovich et al. 2008b). Further, males change the barking rate depending on the circumstances; for example, confrontations usually result in faster barking than when advertising their territory (Tripovich et al. 2008a). Females do not often emit barking sounds (Tripovich et al. 2008a).

Male and female Australian fur seals produce guttural threat sounds during aggressive interactions and submissive calls when retreating from another conspecific (Stirling and Warneke 1971; Tripovich et al. 2008a, 2009b). Growls are only emitted by females during aggressive interactions with conspecifics (Tripovich et al. 2008a).

In Australian fur seals, as in other pinnipeds, PACs emitted by females to call their pups and FACs emitted by pups responding to or searching for their mothers show individual variability and appear to be used to identify each other (Stirling and Warneke 1971; Tripovich et al. 2006, 2008a, 2009b). In FACs, the frequency decreases, and the duration and number of sections within a call change within the first year of a pup's life. Fundamental frequency, minimum frequency of the first harmonic band, and duration appear to be individually specific acoustic features. Pups and yearlings emit sounds between 410 and 1770 Hz, with peak energy from 800 to 1300 Hz, lasting 0.4–1.1 s (Stirling and Warneke 1971; Tripovich et al. 2006, 2008a, 2009b).

5.5.2.7.2 *Arctocephalus pusillus pusillus*: Cape Fur Seal

To our knowledge, there are no underwater recordings of the Cape fur seal subspecies, and in-air vocalizations have only recently been recorded and described. Cape fur seals produce

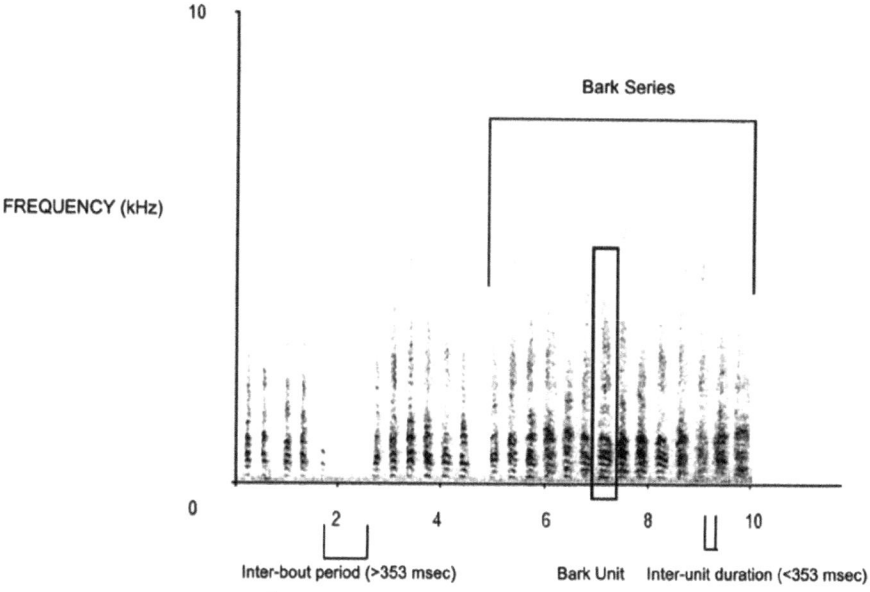

Fig. 5.15 Bark series spectrogram of a male Australian fur seal. Reprinted with permission from Fig. 1 in Tripovich JS, Rogers TL, Arnould JPY (2005) Species-specific characteristics and individual variation of the bark call produced by male Australian fur seals, *Arctocephalus pusillus doriferus*. Bioacoustics 15 (1):79–96. https://doi.org/10.1080/09524622.2005.9753539. © Taylor & Francis, 2012. All rights reserved

sound in air ranging from 60 to 3510 Hz, with peak energy between 290 and 1180 Hz and a duration ranging from 0.03 to 2.3 s, on average 0.1–1.4 s (Fig. 5.16; Martin et al. 2021b; Osiecka et al. 2022). Individually distinct attraction calls are emitted by females and pups when searching for each other and are used for both long- and short-distance communication. Males and females use soft and long growls and barks in aggressive and agonistic encounters. Barks vary between age and sex classes and are also produced in association with mating and territorial behaviors and by pups playing with each other (Martin et al. 2021a, b). Confronting bulls in a high arousal state increased their barking rate and bark fundamental energy by 17% and 20%, respectively, concurrently raising alertness levels in nearby subadults (Martin et al. 2022b).

Cape fur seal call types may vary geographically. Most have been shown to be individually distinctive, with attraction calls and bulls' territorial barks being more individually distinctive than barks by subadults or females (Martin et al. 2021a, b). Females learn to recognize their

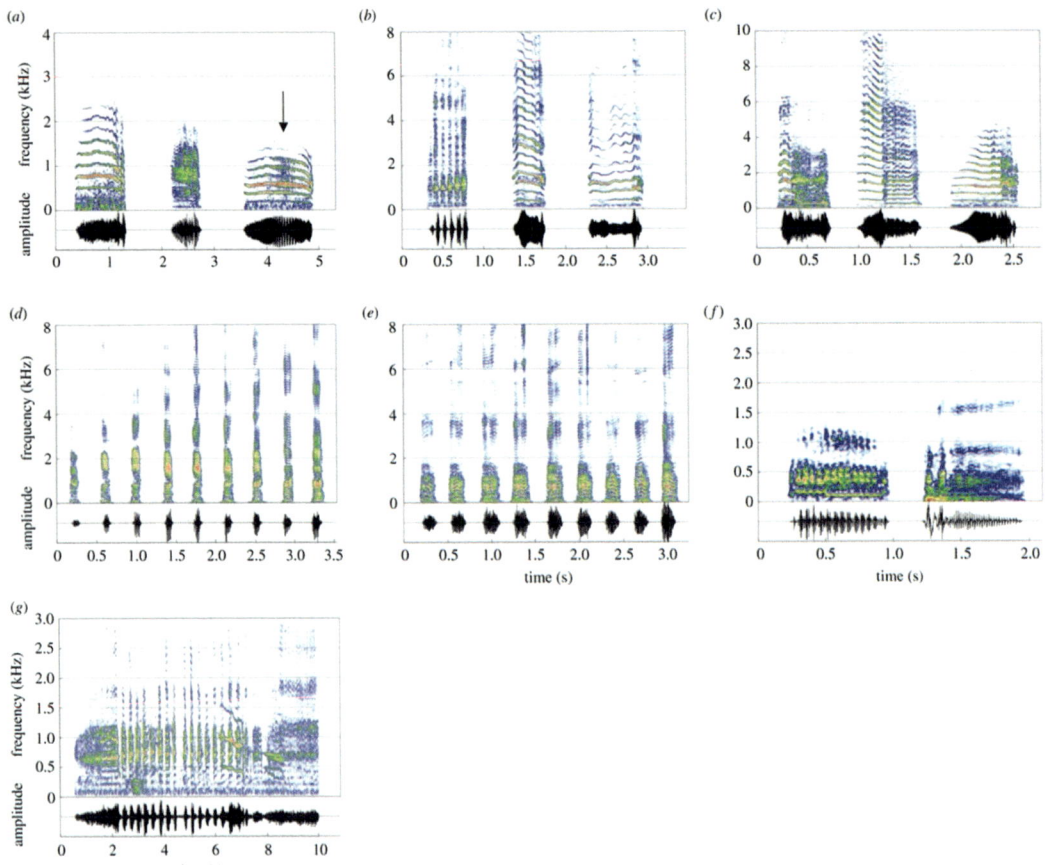

Fig. 5.16 Spectrograms of Cape fur seal aerial vocalizations. (**a**) PAC from three different females: clear harmonic structure, noisy call, and harmonic structure with a fast amplitude-modulated part (indicated by the arrow). (**b**) FAC from three pups at three different ages: (from left to right) <2 weeks, 1 month, and 2–4 months old. (**c**) FAC from one pup at different ages: <2 weeks, 1 month, 2–4 months old. (**d**) Female barks. (**e**) Male barks. (**f**) Soft and loud growls. (**g**) The sequence of male barks starting and ending with a long bark. fs = 44.1 kHz, 1024 NFFT, Hamming window, 90% overlap. Reprinted Fig. 1 from Martin M, Gridley T, Harvey Elwen S, Charrier I (2021) Vocal repertoire, microgeographical variation and within-species acoustic partitioning in a highly colonial pinniped, the Cape fur seal. R Soc Open Sci 8 (10):202241; https://doi.org/10.1098/rsos.202241. © Martin et al. 2021. Published CC BY 4.0; https://creativecommons.org/licenses/by/4.0/

pups' call 2–4 h postpartum, and pups have been shown to distinguish their mothers' call 4–6 h after birth (Martin et al. 2022a). After a female returns ashore, for example, from a foraging trip or swim, pups appear to be more active in the initial search for their mothers compared to females looking for their pups. Although the females' calls were longer and the time spent vocalizing was similar, pups produced more calls, called at a higher rate, and had shorter breaks between vocalizations (Osiecka et al. 2022).

5.5.2.8 *Arctocephalus tropicalis*: Subantarctic Fur Seal

In air, subantarctic fur seals produce sounds containing energy from 180 to 4490 Hz with peak frequencies between 390 and 3620 Hz, lasting for 0.07 to 8.3 s (Charrier et al. 2002, 2003b, c; Gwilliam et al. 2008; Page et al. 2001, 2002a, b; Roux and Jouventin 1987; St Clair Hill et al. 2001). To date, subantarctic fur seals have not been recorded under water.

Male subantarctic fur seals produce a range of threat calls, especially during their territorial period. Sound types include barks (also referred to as whickers), full-threat calls, territorial calls used for long-distance communication, and guttural challenges comprising two parts: a growl followed by a puff. Threat calls vary between individual males and are used to recognize each other and distinguish the calls of their neighbors from unknown males (Gwilliam et al. 2008; Page et al. 2001, 2002b; Roux and Jouventin 1987; St Clair Hill et al. 2001).

Females emit long and loud PACs, for example, when returning from a foraging trip and searching for their pup, and pups answer with FACs that resemble a high-frequency bleat (Charrier et al. 2002, 2003b, c; Page et al. 2002a; Roux and Jouventin 1987; St Clair Hill et al. 2001). The latter is also used when the pup is searching for its mother, for example, when hungry (Roux and Jouventin 1987; St Clair Hill et al. 2001). PACs and FACs carry individual-identifying information (Fig. 5.17). Pups mainly

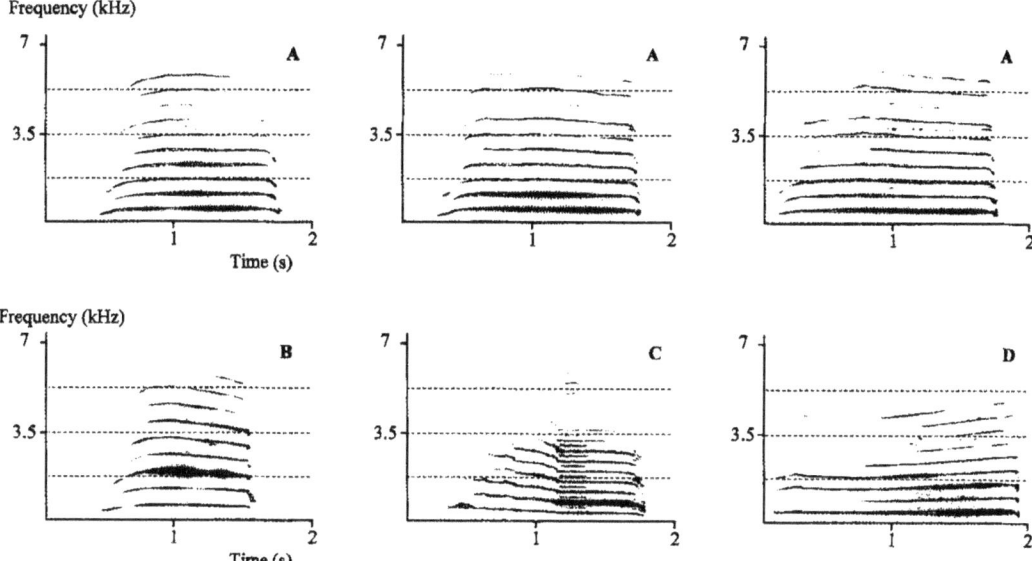

Fig. 5.17 Spectrograms of aerial calls from four female subantarctic fur seals (**a–d**). Intra-individual variability is small (top three panels of female **a**); inter-individual variability is great (bottom three panels of females **b**, **c**, and **d**). Reproduced with permission from Fig. 2 in Charrier I, Mathevon N, Jouventin P (2003) Individuality in the voice of fur seal females: An analysis study of the pup attraction call in *Arctocephalus tropicalis*. Mar Mamm Sci 19 (1):161–172; https://doi.org/10.1111/j.1748-7692.2003.tb01099.x. © John Wiley & Sons, 2006. All rights reserved

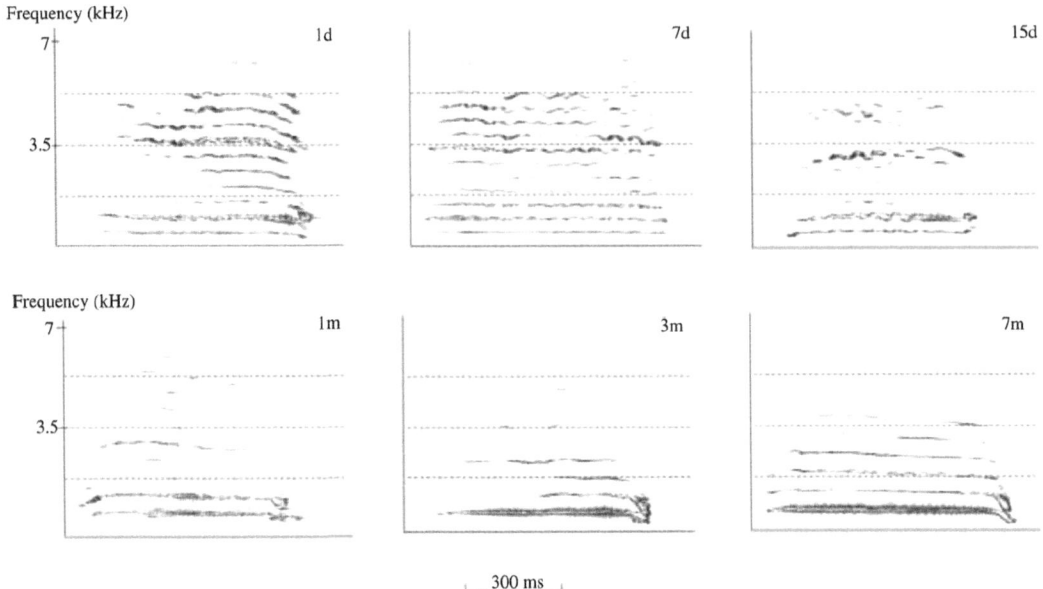

Fig. 5.18 Spectrograms of an individual subantarctic fur seal pup's calls at six different ages, showing a reduction in quavering and a decrease in peak frequency. Reproduced with permission from Oxford University Press. Fig. 2 in Charrier I, Mathevon N, Jouventin P (2003) Fur seal mothers memorize subsequent versions of developing pups' calls: adaptation to long-term recognition or evolutionary by-product? Biol J Linn Soc 80 (2): 305–312. https://doi.org/10.1046/j.1095-8312.2003. 00239.x. © The Linnean Society of London, 2003. All rights reserved

recognize their mother's calls by the energy spectrum and frequency modulation pattern. Mothers, conversely, rely mainly on the frequency modulation pattern to distinguish their pups (Charrier et al. 2002, 2003b, c).

Although FACs change with the age of the pup (e.g., pups do not display a tremble or jitter, which is called "quavering" after 15 days of age; Fig. 5.18), females are still able to distinguish their young and do this in the long term (Charrier et al. 2001, 2003a). Females seem to recognize their pups' call within the first hours after birth, while it takes the pups ~2–5 days to learn their mother's individual call. Females have also been shown to adjust the timing of their first foraging trip departure to when their pups have learned their specific acoustic communication (Charrier et al. 2001, 2003a).

5.5.2.9 *Callorhinus ursinus*: Northern Fur Seal

Northern fur seal sounds have mainly been recorded in air. In air, sounds last 0.02–5 s and are emitted between 30 and 3330 Hz, with peak energy of 300 to 1800 Hz (Antonelis and York 1983; Insley 1992, 2001; Lisitsyna 1973; Peterson 1965, 1968; Takemura et al. 1983; Winn and Schneider 1978). However, Poulter (1968) described sounds recorded in air and under water and found that some of the calls made in air strongly resembled those made under water and were often made up of a series of clicks.

Territorial males use trumpeted roars as threat signals, often directed at other males. These roars varied individually when tested among a small number of fur seals (Antonelis and York 1983). Males also produce sound types classified as guttural snoring sounds, low roar (also called a woofing sound or harem cry), and whicker, described as the "bark of a dog after laryngectomy," in association with threatening or courting behaviors (Gentry 2014; Insley 1992, 2001; Kenyon 1960; Lisitsyna 1973; Peterson 1965, 1968; Poulter 1968; Takemura et al. 1983; Winn and Schneider 1978). Boundary puffing calls are used in boundary displays, while males whine, sounding like a high-pitched squeal when being submissive (Peterson 1965, 1968; Winn and Schneider 1978). Males and females of all ages produce bleating sounds, which are lowest in frequency in bulls and highest in pups (Poulter 1968).

Females emit PACs and so-called estrus calls during the breeding period. Pups emit FACs that can be a short, loud call when scared. FACs and PACs are individually distinctive. Females and pups have been shown to remember and recognize each other for at least 4 years (Gentry 2014; Insley 1992, 2000, 2001; Lisitsyna 1973; Takemura et al. 1983).

5.5.2.10 *Eumetopias jubatus*: Steller Sea Lion

This species comprises the subspecies Western Steller (*E. j. jubatus*) and Loughlin's Steller (*E. j. monteriensis*) sea lions, which have been collectively described as "Steller sea lions" in this section.

Steller sea lions communicate in air and under water. However, their vocal repertoire, especially under water, has not been recorded often. Under water, they produce clicks (also called belching sounds) and other signals similar to their sounds in air but not further specified (Orr and Poulter 1967; Schusterman et al. 1970). In captivity, the underwater clicks had a maximum frequency of 1600 Hz with peak energy around 400 Hz (Schusterman et al. 1970).

The sounds emitted in air can be categorized into four classes based on the behavioral context in which the vocalizations are used (Orr and Poulter 1967; Park et al. 2006). "Communication" sounds are emitted to interact with conspecifics. "Threat" calls (e.g., roars and growls) aim to warn others and defend territories. "Wheedling" sounds are produced during socializing activities or eating prey. When responding to other sea lions, "acknowledgment" calls are used (Orr and Poulter 1967; Park et al. 2006). Sounds in the wild and captivity were emitted between 260 and 480 Hz with peak frequencies between 290 and 470 Hz for 0.25–1.7 s (Park et al. 2006; Sandegren 1970; Schusterman et al. 1970). Males produced sounds at lower frequencies and longer duration than did females, and sounds emitted in the wild were generally lower in frequency and shorter in duration than those in captivity (Park et al. 2006).

Females and their pups are very vocal with each other. Within a few days of birth, females and their pups learn to recognize each other's calls (Sandegren 1970). The pup's FAC cry was compared with "the bleat of a small lamb," while the female's call was described as sounding like the "low bellowing of cattle" or the "bleating of sheep." Females produce sounds between 150 and 1000 Hz for 1–1.5 s to call their pups (Campbell et al. 2000, 2002; Orr and Poulter 1967; Sandegren 1970).

Apart from emitting vocal sounds, some female Steller sea lions were observed clapping their front flippers onto their wet belly (called belly-clapping) to produce a loud threat sound that travels long distances. Similarly, some young bulls generated loud sounds by slapping their front flippers against a wet rock, causing nearby sea lions to flee (Sandegren 1970).

5.5.2.11 *Neophoca cinerea*: Australian Sea Lion

Australian sea lions are yet to be recorded under water, but they vocalize in air. Females and males bark, click, huff, bleat, roar, and hiss (Ahonen et al. 2014; Attard et al. 2010; Gwilliam et al. 2008; Marlow 1975; Stirling 1972). Females with pups emit PACs similar to the moo of a cow, and pups respond with FACs that may sound like a harsh lamb-like bleat, squeal, or a deep bleat in stressful situations (Charrier and Harcourt 2006; Charrier et al. 2009; Marlow 1975; Pitcher et al. 2009, 2012; Stirling 1972). Sounds are produced between 370 and 4500 Hz (guttural threat and hiss sounds up to 14,000 and 15,000 Hz, respectively) with peak energy ranging from 400 to 1300 Hz, lasting 0.03–1.5 s (Ahonen et al. 2014; Attard et al. 2010; Charrier et al.

2009, 2011a; Charrier and Harcourt 2006; Gwilliam et al. 2008; Pitcher et al. 2009, 2012; Stirling 1972).

Male barks differ among colonies, varying more between South and Western Australian colonies, and they can distinguish between conspecifics from their own and different colonies (Ahonen et al. 2018; Attard et al. 2010). Temporal spacing between barks gives information on the potential threat to opposing males, eliciting stronger responses with bark series of "accelerated rhythmicity" (Charrier et al. 2011a). As in other pinnipeds, males' barks show individual differences and may be used to discriminate between individuals (Ahonen et al. 2014; Gwilliam et al. 2008). On rare occasions, male Australian sea lions produce female-like calls, of which the function is unclear; they resemble the calls emitted by females and pups searching for each other (Gwilliam et al. 2008; McIntosh and Pitcher 2021).

Females and pups emit individually distinct calls, for example, with differences in amplitude and frequency modulations (Charrier and Harcourt 2006; Charrier et al. 2009; Pitcher et al. 2012). Females learn to distinguish their pup's call within 48 h after birth, but it takes the pup some time before recognizing its mother's calls, achieving this at ages of between 10 days and 2 months, though their recall of this information is long-lasting. Juvenile Australian sea lions could still recognize their mother's call 2 years after weaning (Marlow 1975; Pitcher et al. 2009, 2010a, b, 2012). As females and pups search for and approach each other, vocalizations are the first cue, with visual and olfactory (i.e., sniffing) cues becoming effective at reduced distances. For pups, females' calls are more important for recognizing their mothers than visual cues (Charrier et al. 2022; Marlow 1975; Pitcher et al. 2009, 2010a, 2012; Wierucka et al. 2018).

The acoustic characteristics (i.e., sound propagation conditions and ambient noise) of the environment in which calls are emitted affect the distance over which calls propagate and remain individually distinguishable. Over a sandy beach or an area with shrubs, calls could still be reliably distinguished at 16 m. However, herbaceous vegetation limited this distance to less than 8 m (Fig. 5.19; Charrier et al. 2009; Pitcher et al. 2012).

5.5.2.12 *Otaria byronia*: South American Sea Lion

South American sea lions communicate in air, but to date they have not been recorded under water. High-pitched calls and barks are produced by adult males during agonistic interactions, growls while interacting with a female, and exhalations follow their agonistic confrontations (Fig. 5.20; Fernández-Juricic et al. 1999, 2003). Females emit grunts during aggressive interactions with other females and mother primary calls (PACs) when attending their pups, which respond with pup primary calls (FACs). Yearlings call for their mothers to nurse or seek protection using a yearling primary call (Fernández-Juricic et al. 1999; Trimble and Insley 2010, 2011). South American sea lions produce sounds between 90 and 1810 Hz with a peak frequency of 240 to 1090 Hz, lasting 0.16 to 1.8 s (Fernández-Juricic et al. 1999; Trimble and Charrier 2011).

According to Fernández-Juricic et al. (1999), the following sound types vary among individuals: high-pitched calls and barks, as well as females' and pups' primary calls. Furthermore, female and pup primary calls tend to differ geographically between distant sea lion colonies (Trimble and Charrier 2011).

Vocalizing is important for females to reunite with pups after returning from foraging trips at sea. Long-range calls are usually undirected calls without additional sensory cues (e.g., olfactory) that are followed by short-range, directed calls in closer proximity and usually emitted by both females and pups while approaching each other (Kunc and Wolf 2008; Phillips and Stirling 2001; Trimble and Insley 2010). Young pups, up to 1 month of age, did not always respond to their mother's call; however, by the time pups were 2 months old, they all responded to returning females when they called. The success rate of females finding their pups was positively associated with the rate at which the females called (i.e., they had better chances of finding

Fig. 5.19 A female Australian sea lion call played back in three different environments (MB, main beach; SD, shrubby dunes; HD, herbaceous dunes) and recorded at four ranges, showing how the spectral content degrades differently in these environments. Reprinted with permission. Fig. 5 in Charrier I, Pitcher BJ, Harcourt RG (2009) Vocal recognition of mothers by Australian sea lion pups: individual signature and environmental constraints. Anim Behav 78 (5): 1127–1134. https://doi.org/10.1016/j.anbehav.2009.07.032. © The Association for the Study of Animal Behaviour, 2009. All rights reserved

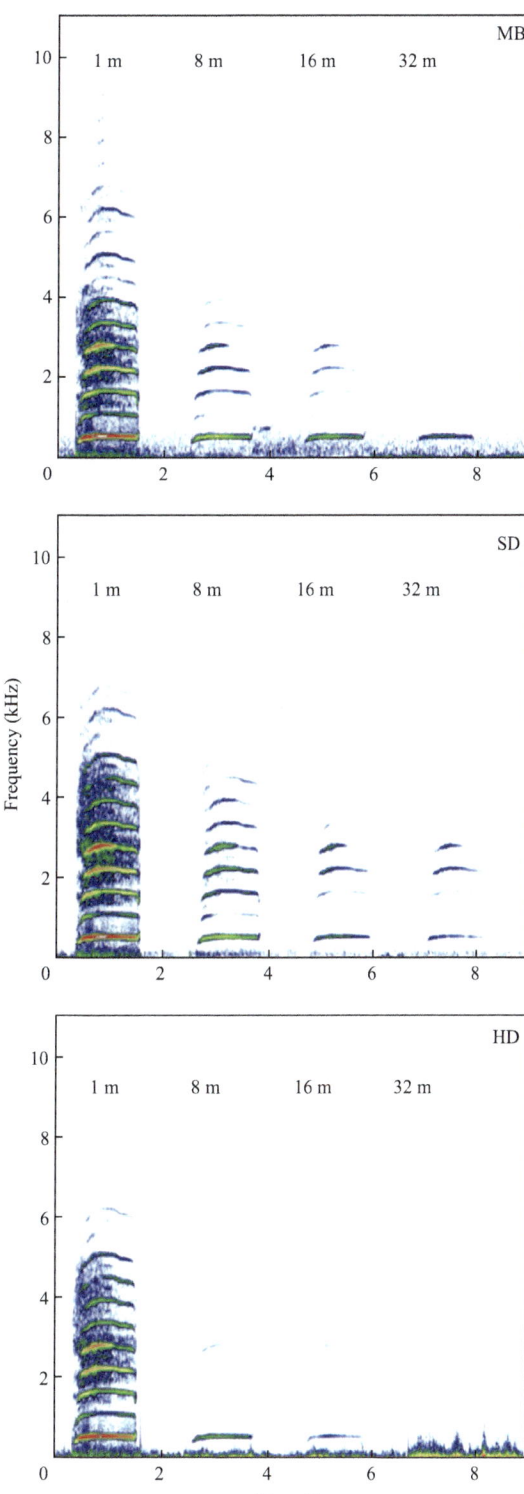

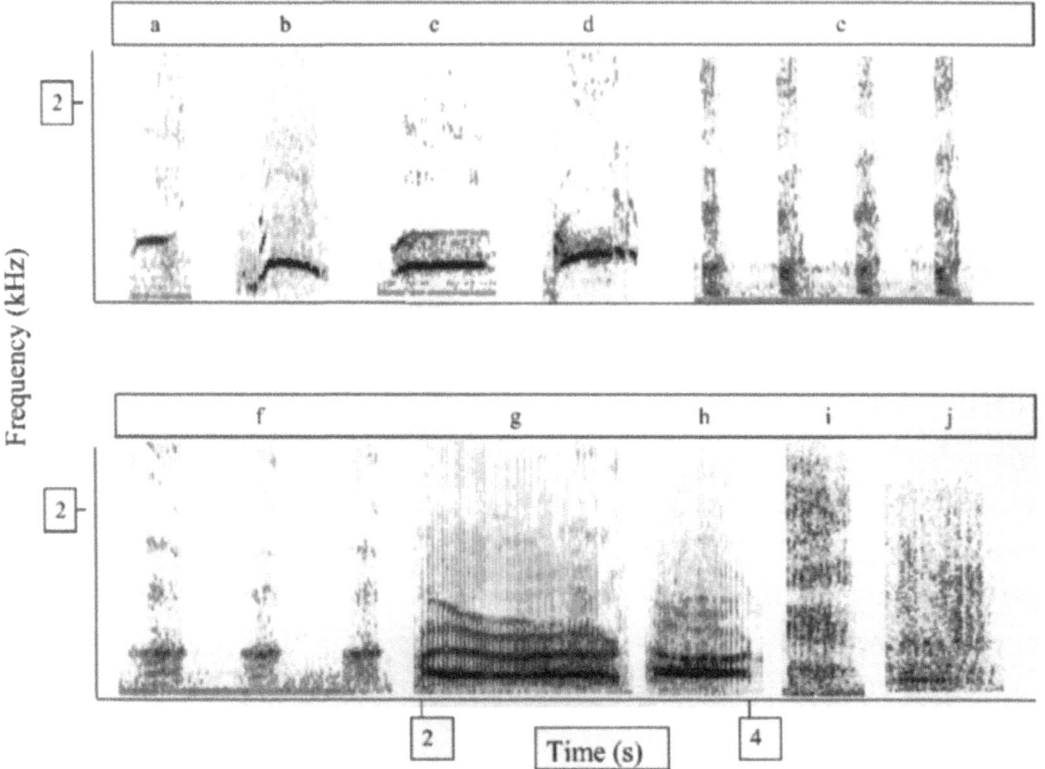

Fig. 5.20 Spectrograms of male South American sea lion calls. (**a–d**) High-pitched calls of four different males. (**e, f**) Barks of two males. (**g, h**) Growls of two males. (**i, j**) Two exhalations of the same male. Reprinted with permission. Fig. 2 in Fernández-Juricic E, Campagna C, Enriquez V, Ortiz CL (1999) Vocal communication and individual variation in breeding South American sea lions. Behavior 136 (4):495–517; https://www.jstor.org/stable/4535623. © Brill, 1947. All rights reserved

their pups when they called more; Trimble and Insley 2010).

5.5.2.13 *Phocarctos hookeri*: Hooker's Sea Lion, New Zealand Sea Lion

We were not able to find any recordings of underwater or in-air vocalizations by New Zealand sea lions, previously known as Hooker's sea lions. Marlow (1975) described pups' FACs as lamb-like bleats and females' PACs as guttural rolling moo, the latter being individually distinguishable. Males grunt at fighting females and emit guttural warning barks when fighting each other (Marlow 1975).

5.5.2.14 *Zalophus californianus*: California Sea Lion

California sea lions emit sounds in air and under water and have been recorded more often in captivity than in the wild. California sea lions bark in the wild, especially males during the breeding season, in air and under water, and emit clicks under water. In addition to PACs, females use short-range threat calls, including a high-pitched squeal, prolonged hoarse belch, and harsh, aspirate growl. Pups produce lamb-like bleats, FACs or mother-response calls, and barking sounds while playing. Sounds produced in the wild displayed energy between 500 and 1640 Hz with peak frequencies of 500–1200 Hz, lasting 0.1–2 s (Peterson and Bartholomew 1969;

Poulter 1968; Schusterman et al. 1992; Thompson 1965).

California sea lions in captivity were also recorded emitting snorts, chuckles, growls, whinny sounds resembling a neighing horse, popping sounds before clicks, buzzes, and bang or crack sounds under water. These included energy between 400 and 4000 Hz and peak frequencies of 500–1250 Hz for 0.2–23 s. In air, sounds were between 180 and 3500 Hz, with a peak frequency of 400–1000 Hz, lasting 0.3–4.1 s. Further sound types of captive sea lions in air were classified as goat, growl/grumble, bark/grumble, bark/growl, grumble/moan, snarl, and roar (Poulter 1968; Schevill et al. 1963; Schusterman 1967; Schusterman and Balliet 1969; Schusterman and Feinstein 1965; Schusterman et al. 1967; Schwalm 2014). Broadband pulses, up to 20 kHz, were hypothesized to serve biosonar functions, but the actual performance was never sufficiently demonstrated (Poulter 1963, 1966; Schusterman 1966).

Males change the number and repetition rates of their barking sounds according to the social context. Their barking also inhibits smaller males from barking and restricts their movement (Schusterman 1970, 1977; Schusterman and Balliet 1969; Schusterman and Dawson 1968).

As in other pinnipeds, PACs are emitted by females to reunite with their pups. Pups respond by producing FACs while they orient and move toward their mother. While pups increase moving toward their mother with age, females reduce approaching them, ultimately letting pups move the complete distance to join the mother (Schusterman et al. 1992). Pups recognize their mother's call; PACs vary individually, mostly in their fundamental frequency and the slope of the frequency change at the start and end of each call (Hanggi 1992; Peterson and Bartholomew 1969; Schusterman et al. 1992).

It has been shown in captive sea lions that related individuals, such as females and their offspring (including older offspring) and siblings, treat each other quite differently compared to their usually aggressive behaviors toward nonaffiliated sea lions (Schusterman et al. 1992). In captivity, mother–pup bonds lasted for at least 11 years, and females and pups recognized each other, even after extended periods of separation (Schusterman et al. 1992).

5.5.2.15 *Zalophus wollebaeki*: Galapagos Sea Lion

To date, Galapagos sea lions have only been recorded producing sound in air, focusing on calls by females and pups. When females are in close contact with their pups, they produce pup contact calls. To seek the attention of a pup from a distance, the female emits PACs, which range between 300 and 600 Hz with peak frequency around 500 Hz (Trillmich 1981). Pups' FACs sound like a bleat with peak frequencies around 1000 Hz. They bleat when searching for their mothers and emit a staccato greeting call once they make contact (Trillmich 1981).

Females and pups have individually distinct calls. Females learn to distinguish their pup's call within a few hours after birth, but it takes the pup a little longer to recognize its mother's call (Trillmich 1981).

Males have been heard emitting barking sounds while holding a territory (Merlen 2000) and have been described as producing long-range, undirected signals in air (Kunc and Wolf 2008). Males vocalize more while holding a territory, especially at the start of the territorial period and when more females are present (Kunc and Wolf 2008).

5.5.3 Family Phocidae: True, Earless, Hair Seals

5.5.3.1 *Cystophora cristata*: Hooded Seal

Hooded seals produce sounds in air and under water, mostly categorized into three types, usually described as A, B, and C, including variations of these main sound types designated with numbers and letters. Type A sounds (moaning growls, roars, "grungs," "blaat," and whooping sounds) last for 0.08–6.9 s at 50–6000 Hz in air and for 0.5–1 s at 140–1830 Hz under water. Females and males may use type A sounds in low- and high-threat situations.

Type B sounds (frequency-modulated growls, snorts, and alternating moaning–growling sounds) are mostly used by females fighting males but may also be emitted by males fighting each other. Type B sounds last 0.6–5.3 s in air and 0.5–1 s under water, at 500–5000 Hz and 100–1000 Hz, respectively (Ballard and Kovacs 1995; Frouin-Mouy and Hammill 2021; Terhune and Ronald 1973). Weaned pups used type A1iibI wailing moans at around 250 Hz in calm settings on land but B1 wailing moans/frequency-modulated growls during agonistic encounters, peaking at around 1500 Hz (Ballard and Kovacs 1995; Frouin-Mouy and Hammill 2021; Terhune and Ronald 1973). B1 growls may also be used by females in high-threat situations, especially with their pups in proximity (Frouin-Mouy and Hammill 2021).

Visual mating displays by males on land are accompanied by type C sounds that include bloops, wooshes, and pings, produced by deflating and inflating their proboscis and septum, and last 0.08–0.7 s (Ballard and Kovacs 1995; Frouin-Mouy and Hammill 2021; Terhune and Ronald 1973). Proboscis and septum-generated sounds were the dominant sound types recorded on ice

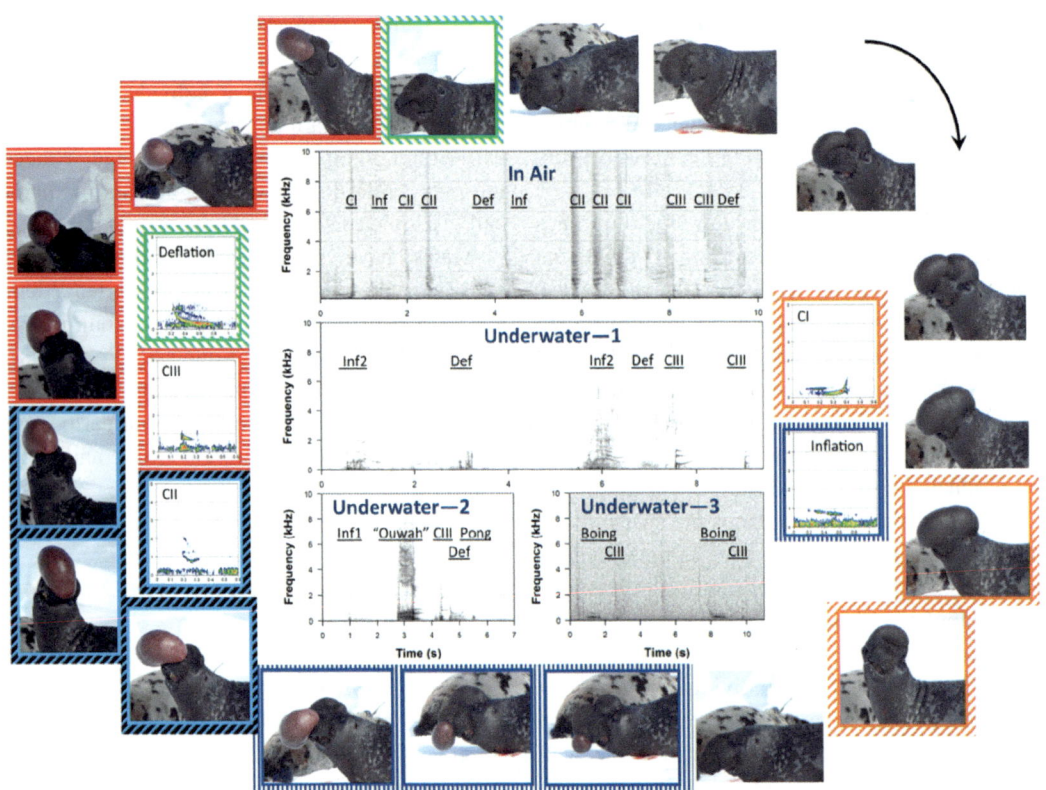

Fig. 5.21 Stages of proboscis inflation in a male hooded seal (clockwise from the top-right): inflated proboscis (hood), extruded nasal septum, and deflation. Photo frames are colored to match the in-air sound spectrograms. A full sound sequence in air is shown in the central spectrogram ("In Air"). Three example underwater sound sequences are also shown, containing proboscis and septum-generated sounds as well as vocal ouwahs and boings. The figure was constructed using *Seewave* (Sueur et al. 2008). Reprinted from Frouin-Mouy HC, Hammill MO (2021) In-air and underwater sounds of hooded seals during the breeding season in the Gulf of St. Lawrence. J Acoust Soc Am 150 (1):281–293; https://doi.org/10.1121/10.0005478. With permission from the ASA. © Acoustical Society of America, 2021. All rights reserved

(Fig. 5.21; Frouin-Mouy and Hammill 2021) and accounted for five of 12 in-air sound types between 50 and 1400 Hz and five of 22 underwater sound types between 30 and 5280 Hz. Type C sounds under water include beats, bloops, buzzes, chucks, clicks, knocks, pings, and trills and are emitted for 0.5–1 s, with peak energy around 1200 Hz (Ballard and Kovacs 1995; Terhune and Ronald 1973).

Additional underwater sounds, assumed to be mostly emitted by males, last up to 1.4 s and have been described as boing, downsweep, groan, howl, moan, "moo," "ouwah," roar, and three trill types. These sounds have peak energy between 100 and 900 Hz; the "moo" reaches as high as 15 kHz (Ballard and Kovacs 1995; Frouin-Mouy and Hammill 2021; Terhune and Ronald 1973). A pup in captivity was recorded producing two types of click sounds, with peak energy at 3000 Hz and a maximum frequency of 16 kHz (Schevill et al. 1963).

5.5.3.2 *Erignathus barbatus*: Bearded Seal

This species comprises the subspecies Atlantic (*E. b. barbatus*) and Pacific (*E. b. nauticus*) bearded seals, treated together in the following paragraphs. In-air vocalizations have not been recorded for this species, and females are not known to vocalize under water (Davies et al. 2006; Ray et al. 1969).

Male bearded seals produce a variety of trills under water that differ individually and geographically and are thought to be associated with breeding. Males are believed to either hold a territory or keep roaming through a larger area during the breeding season; trills emitted during the latter behavior were shorter than those of territorial males (Van Parijs et al. 2003b). Males appear to be able to distinguish between local calls and calls from other regions, reacting more strongly to sounds emitted by local than foreign males (Charrier et al. 2013; Cleator et al. 1989; Risch et al. 2007; Van Parijs et al. 2003b).

Different variations of trills exist: downsweeping, slow-dropping, fast-dropping, up-sweeping or ascending, high-pitched, hooked, linear, short, and long trills, as well as "trills with a plume," which can last up to 88 s (averages ranged from 0.3 to 65 s; Fig. 5.22). Trills were produced between 65 and 12,600 Hz, with peak energy from 300 to 5000 Hz (Charrier et al. 2013; Cleator et al. 1989; Davies et al. 2006; Frouin-Mouy et al. 2016; Heimrich et al. 2021; Jones et al. 2014; MacIntyre et al. 2013, 2015; Madan et al. 2020; Parisi et al. 2017; Ray et al. 1969; Risch et al. 2007; Stirling et al. 1983; Terhune 1999; Van Parijs et al. 2001). At some locations, over 300 trills are emitted per hour during the peak. The seasonal timing of the peak varies geographically, and trill rate may be influenced by ice coverage (greater number of acoustic detections with higher sea ice concentration) (de Vincenzi et al. 2019; Halliday et al. 2018, 2019; Llobet et al. 2021; Mattmüller et al. 2022).

Other underwater sounds emitted by males are moans and groans, which last for 0.5–13.5 s at 110–13,600 Hz, with peak frequencies between 150 and 3200 Hz (Cleator et al. 1989; Davies et al. 2006; Frouin-Mouy et al. 2016; Heimrich et al. 2021; Jones et al. 2014; Parisi et al. 2017; Risch et al. 2007; Van Parijs et al. 2001). Bearded seals increase the amplitude of their underwater calls with increasing ambient noise. This Lombard effect could be seen up to an ambient noise threshold of 100–105 dB rms, plateauing beyond this threshold (Fournet et al. 2021).

5.5.3.3 *Halichoerus grypus*: Gray Seal

This species comprises the subspecies Atlantic gray seal (*H. g. atlantica*) and Baltic gray seal (*H. g. grypus*), which are treated together in this section.

Male, female, and pup gray seals vocalize in air and under water, with some sound types occurring in both media. McCulloch (2000) recorded six sound types (types A–F) in air and ten sound types (types 1–10) under water. Others identified nine underwater sound types (types 1–9) and claps in their recordings (Fig. 5.23; Pérez Tadeo et al. 2023; Pozo Galván et al. 2024). Asselin et al. (1993) described seven underwater sounds. Nowak (2021) categorized gray seal calls, in air and under water, into three types (S1, S2, and S3), based on their tonal versus pulsed character and duration.

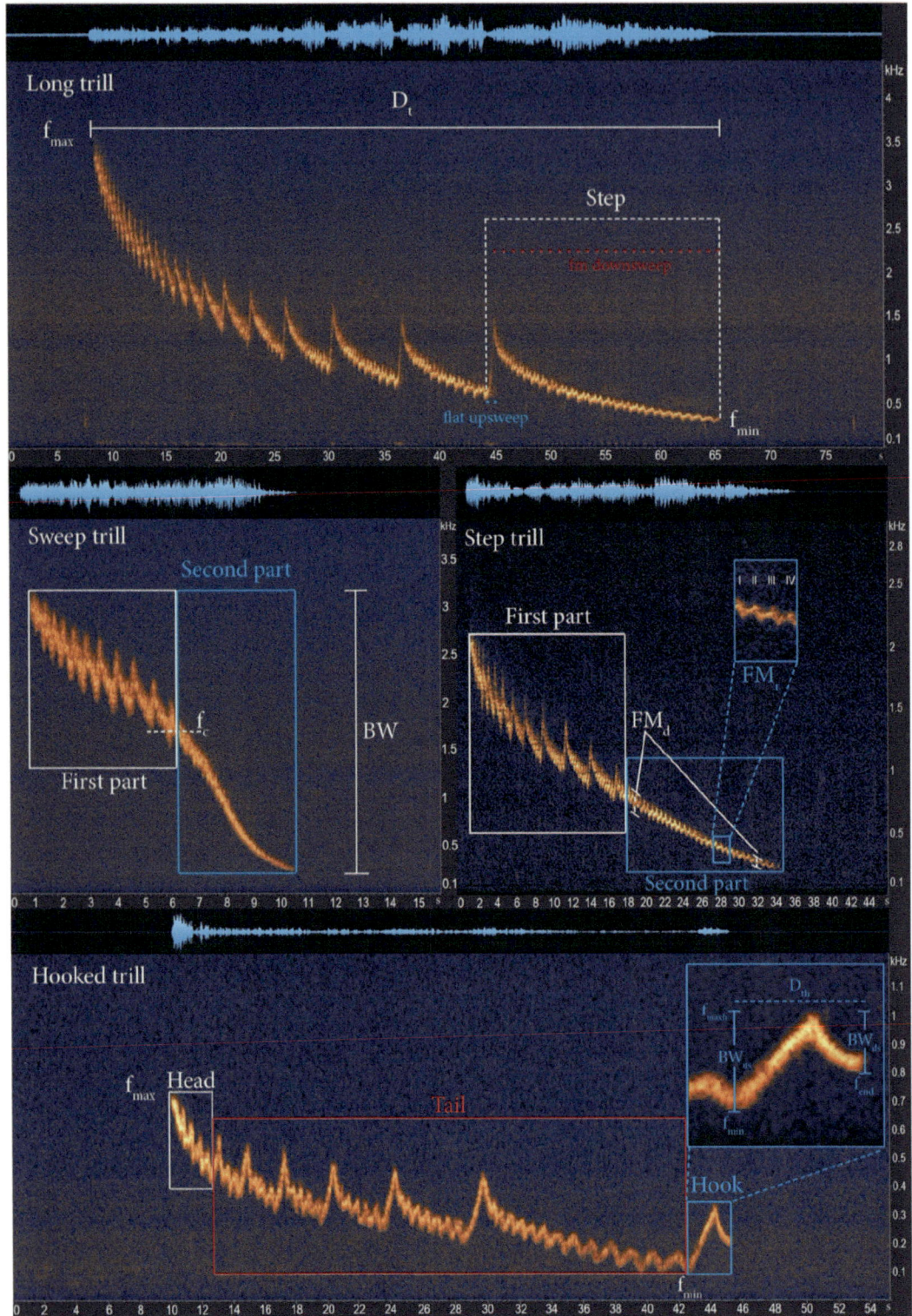

Fig. 5.22 Trill variations of bearded seals. *Dt* total duration, *BW* bandwidth, *fmin* minimum frequency, *fmax* maximum frequency, *fc* central frequency, *FMr* frequency modulation rate, *FMd* frequency modulation range, *fmid* frequency at Dt/2, *Harm* harmonics, *Head* first portion of vocalization, *Tail* portion of vocalization after head, *Step* up flat-sweep followed by FM downsweep. Reproduced with ASA permission from Parisi I, de Vincenzi G, Torri M, Papale E, Mazzola S, Bonanno A, Buscaino G (2017) Underwater vocal complexity of Arctic seal *Erignathus barbatus* in Kongsfjorden (Svalbard). J Acoust Soc Am 142 (5):3104–3115. https://doi.org/10.1121/1.5010887. © Acoustical Society of America. All rights reserved

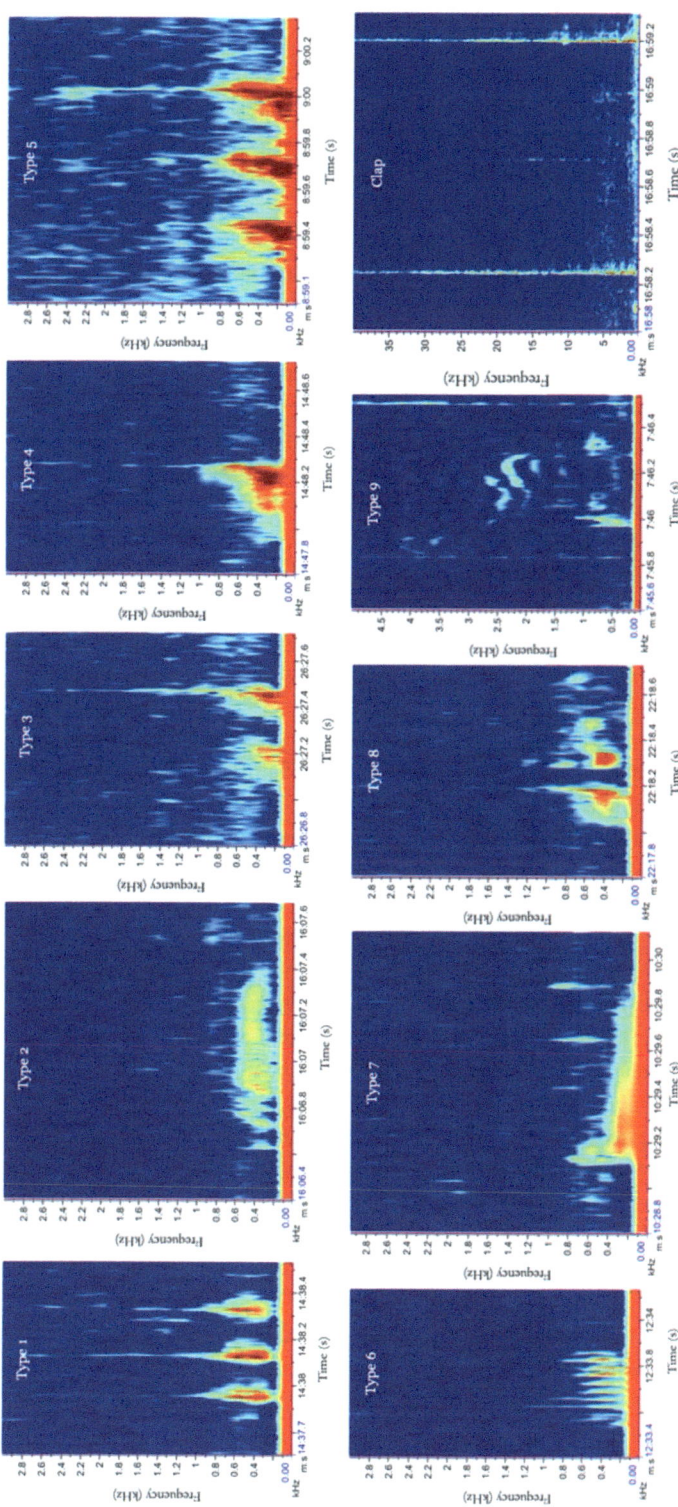

Fig. 5.23 Spectrograms of ten gray seal underwater sound types recorded off Ireland. Reprinted from Pérez Tadeo M, Gammell M, O'Brien J (2023) First steps towards the automated detection of underwater vocalisations of grey seals (*Halichoerus grypus*) in the Blasket Islands, Southwest Ireland. J Mar Sci Eng 11 (2):351; https://doi.org/10.3390/jmse11020351. © Pérez Tadeo et al. 2023. Published CC BY 4.0; https://creativecommons.org/licenses/by/4.0/

Vocalizations in air include "kataro," growls, gurgles, moans also described as a hoot, "moo," and yodel between 80 and 1170 Hz, lasting 0.8–5.9 s, besides Nowak's S2 sounds that were recorded for nearly 19 s (Hewer and Backhouse 1960; McCulloch 2000; Nowak 2021; Schusterman et al. 1970; Stansbury et al. 2015). Captive gray seals have been shown to discriminate between growls and moans and could also recognize each other; the frequency differs between individuals (Shapiro et al. 2004; Stansbury et al. 2015).

Sounds under water have been described as buzzes, clicks that are comparable to a creaking door, growls, guttural rup, guttural rupe, knocks that sound "like a hammer hitting metal," moans or roars, "moo," and "trrot" comparable to a jackhammer. Vocalizations lasted, on average, 0.1–3.6 s, ranging from 0.04 to 18.9 s, between 40 and 6800 Hz, excluding knocks that were measured at frequencies up to 14,300 Hz. Peak energy was measured from 200 to 3500 Hz and around 8600 Hz for knocks (Asselin et al. 1993; McCulloch 2000; Nowak 2021; Pérez Tadeo et al. 2023; Prawirasasra et al. 2021; Schevill et al. 1963; Schusterman et al. 1970). The guttural rup (Asselin et al. 1993) or type 1 vocalization (McCulloch 2000; Pérez Tadeo et al. 2023) was the most frequently recorded call, accounting for about 31% of underwater sounds (Pérez Tadeo et al. 2023). However, the previously described underwater knocks were most likely misclassified claps, produced by clapping the flippers together. These percussive sounds may be used by males during social or agonistic interactions, with peak energy recently measured up to >50 kHz and clap duration of <1 s. In contrast to walrus claps, gray seals only clap a few times in a row (Hocking et al. 2020; Larsen and Reichmuth 2021; Pérez Tadeo et al. 2023).

Pups were recorded producing pup calls, clicks, and hums, lasting 0.4–2.6 s, ranging from 200 to 1200 Hz, at peak energy of 370 Hz and 1000 Hz for clicks and hums, respectively (Caudron et al. 1998; McCulloch and Boness 2000; McCulloch et al. 1999; Schusterman et al. 1970). Captive pups appear to time their vocalizations to minimize masking by other pups' calls, meaning that they can produce their calls in a rhythm that limits acoustic interference (de Heer Kloots et al. 2020). At Isle of May, Scotland, pups displayed individually different calls. However, the females appeared to not discriminate between pups and allowed allosuckling at a relatively high rate. In contrast, females did distinguish between pups on Sable Island, Canada, and so allo-suckling by non-filial pups did not occur (McCulloch and Boness 2000; McCulloch et al. 1999).

Furthermore, like northern elephant seals, male gray seals may communicate through seismic vibrations, which can reportedly travel up to 126 m from a body slap and indicate the size of the male (Bishop et al. 2015).

5.5.3.4 *Histriophoca fasciata*: Ribbon Seal

Seven sound types emitted by ribbon seals under water have been reported: downsweep (Watkins (1977) distinguished between short, medium, and long sweeps), growl, grunt, hiss or broadband puffing sound, roar, shuffle, and yowl (Fig. 5.24). Depending on the call, these are produced between 10 and 7840 Hz, with peak frequencies between 400 and 4000 Hz and source levels of 170–178 dB re 1 µPa m. Vocalizations last between 0.2 and 14.9 s (Frouin-Mouy et al. 2019; Jones et al. 2014; Miksis-Olds and Parks 2011; Mizuguchi et al. 2016a; Otsuki et al. 2018; Watkins and Ray 1977).

Mizuguchi et al. (2016a) noted geographical variations in the downsweeps between different populations. Ribbon seals produce calls (songs) with stereotyped sequences linking up to ten calls together (Frouin-Mouy et al. 2019; Jones et al. 2014; Miksis-Olds and Parks 2011; Mizuguchi et al. 2016a; Otsuki et al. 2018; Watkins and Ray 1977).

Some of the calls emitted by ribbon seals may be related to breeding behaviors. However, calls are also produced outside the breeding season, and their behavioral contexts have not been linked to the different sound types (Mizuguchi et al. 2016a; Otsuki et al. 2018; Watkins and

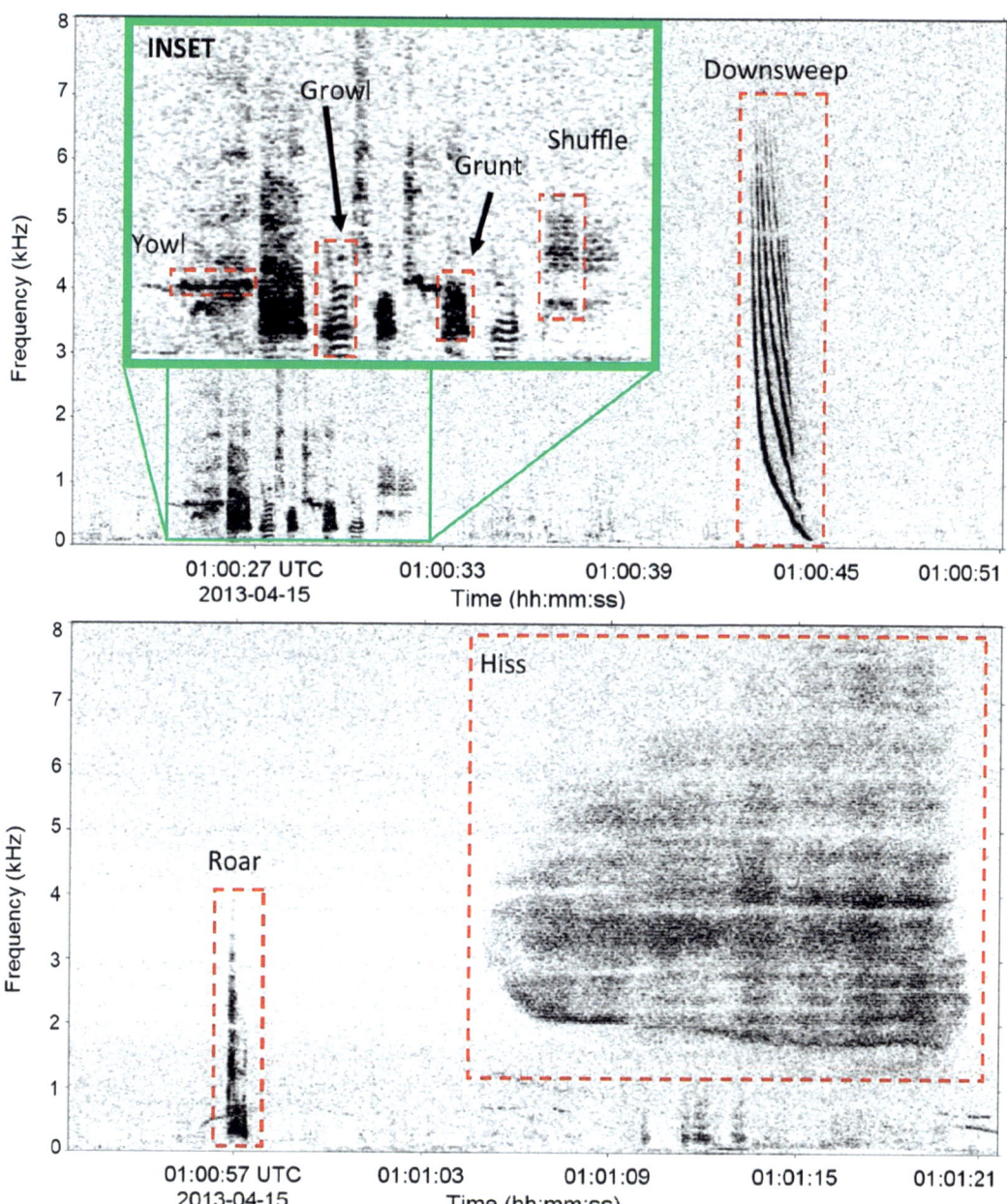

Fig. 5.24 Spectrograms of ribbon seal sound types in the Beaufort, Bering, and Chukchi seas. Reprinted with permission Fig. 2 in Frouin-Mouy H, Mouy X, Berchok CL, Blackwell SB, Stafford KM (2019) Acoustic occurrence and behavior of ribbon seals (*Histriophoca fasciata*) in the Bering, Chukchi, and Beaufort seas. Polar Biol 42 (4): 657–674. https://doi.org/10.1007/s00300-019-02462-y. © Springer Nature, 2019. All rights reserved

Ray 1977). Sounds produced under water can be loud and may be heard in air (Watkins and Ray 1977), but no sounds emitted in air have been reported—to our knowledge.

5.5.3.5 *Hydrurga leptonyx*: Leopard Seal

Underwater vocalizations of leopard seals range from 35 to 7160 Hz, with spectral peak frequencies between 70 and 3500 Hz, durations of 0.3–10 s, and source levels of 153–177 dB re

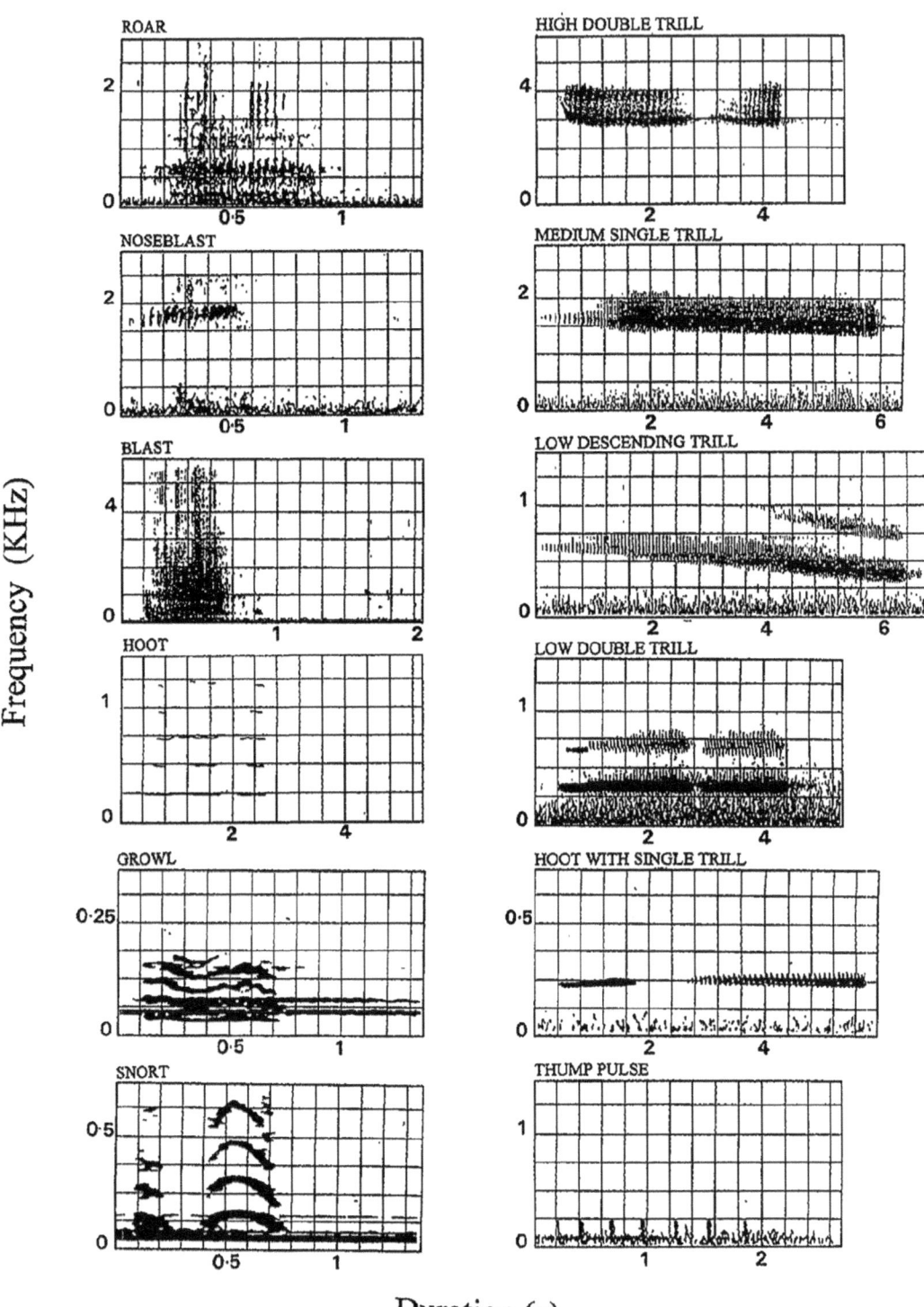

Fig. 5.25 Underwater sound types recorded from a male and a female captive leopard seal. Reprinted with permission from John Wiley and Sons. From Rogers TL, Cato DH, Bryden MM (1996) Behavioral significance of underwater vocalizations of captive leopard seals, *Hydrurga leptonyx*. Mar Mamm Sci 12 (3):414–427. https://doi.org/10.1111/j.1748-7692.1996.tb00593.x. © John Wiley & Sons, 2006. All rights reserved

1 μPa m (Awbrey et al. 2003; Gedamke and Robinson 2010; Klinck 2008; Kreiss et al. 2008, 2013; Meister 2017; Mossbridge and Thomas 1999; Rogers 2007, 2014, 2017; Rogers et al. 1995, 1996; Schwenke 2013; Stirling and Siniff 1979; Thomas and Golladay 1995; van Opzeeland et al. 2010).

Leopard seals produce a number of trills (Fig. 5.25), consisting of rapid pulse trains (e.g., low, medium, high single or double trills); some may be ascending or descending, and a trill may also follow hoots. Subadult males produce variants of trills, differing in their frequency content. Rogers (2007) classed the different types of trills as "broadcast calls" as they may be used for long-distance communication during the breeding season by males and females during estrus. "Local calls" are mostly used for agonistic interactions in close range and include growls, nose blasts that result in air bubbles coming out from the mouth, roars, snorts, and thump pulses (Gedamke and Robinson 2010; Klinck 2008; Rogers 2007; Rogers et al. 1995, 1996; Schwenke 2013; Stirling and Siniff 1979; Thomas and Golladay 1995; van Opzeeland et al. 2010).

In a captive individual, ultrasonic buzzes, chirps, clicks, and sweeps were recorded under water between 5 and 160 kHz, with an ultrasonic frequency peak around 55 kHz and duration of 0.009–0.1 s (Awbrey et al. 2003; Thomas et al. 1982).

In air, Stirling and Siniff (1979) mention a low trill being emitted. When disturbed, leopard seals gave a "throaty k-k-k sound" (Dearborn 1962). Although only few records describe sounds in air, the leopard seal's underwater calls can be heard well in air in certain circumstances (Rogers 2014; Stirling and Siniff 1979).

Leopard seal vocalizations have been shown to vary individually, by size and age, as well as geographically (Kreiss et al. 2013; Rogers 2007, 2017; Rogers and Cato 2002; Rogers et al. 1995; Thomas and Golladay 1995).

5.5.3.6 *Leptonychotes weddellii*: Weddell Seal

Weddell seals have a well-documented vocal repertoire under water and in air. Some underwater sound types may also be emitted in air. Several underwater calls are audible through the ice in air and may be used for communication in this manner (Terhune 2017).

Weddell seals produce sounds under water mostly between 10 and 13,000 Hz (except ultrasonic vocalizations up to 50 kHz), with peak energy ranging from 200 to 3720 Hz. Vocalizations last for 0.01–109 s, on average between 0.03 and 76 s, with source levels of 153–193 dB re 1 μPa m (Cziko et al. 2020; Doiron et al. 2012; Evans et al. 2004; Moors and Terhune 2004, 2005; Morrice et al. 1994; Pahl et al. 1997; Ray 1967; Russell et al. 2016; Schevill and Watkins 1965, 1971; Terhune 2016; Terhune 2017; Terhune and Dell'Apa 2006; Terhune et al. 2001; Thomas 1979; Thomas and Kuechle 1982; Thomas and Stirling 1983; van Opzeeland et al. 2010). Underwater ultrasonic sounds have peak energy between 11 and 36 kHz and last 1.2–21 s (Cziko et al. 2020; Russell et al. 2016; Schevill and Watkins 1971).

Several authors have grouped the range of sound types into categories, dividing them according to their call characteristics, such as duration, frequencies, and contour (Fig. 5.26). For example, Thomas and Kuechle (1982), Pahl et al. (1997), and Abgrall et al. (2003) found 34, 50, and 33 underwater sound types, which were grouped into 12, 13, and 13 call categories, respectively. Recorded and reported underwater vocalizations seem, however, to be dominated by only a few sound types in most of the studies: trills, whistles, and chugs (Doiron et al. 2012; Moors and Terhune 2004, 2005; Pahl et al. 1997). The underwater ultrasonic sounds were categorized into nine different call types, including downsweeping chirps, trills, and whistles (Cziko et al. 2020). Regardless of the sound type, most underwater calls (71%) were emitted between a water depth of 10 and 35 m in waters over 300 m deep (Moors and Terhune 2005).

Sex-related differences in sound production have been observed. Oetelaar et al. (2003) reported an ascending whistle only produced by females under water, while only males mewed, roared, and emitted all the variations of trills. Only male Weddell seals have been observed

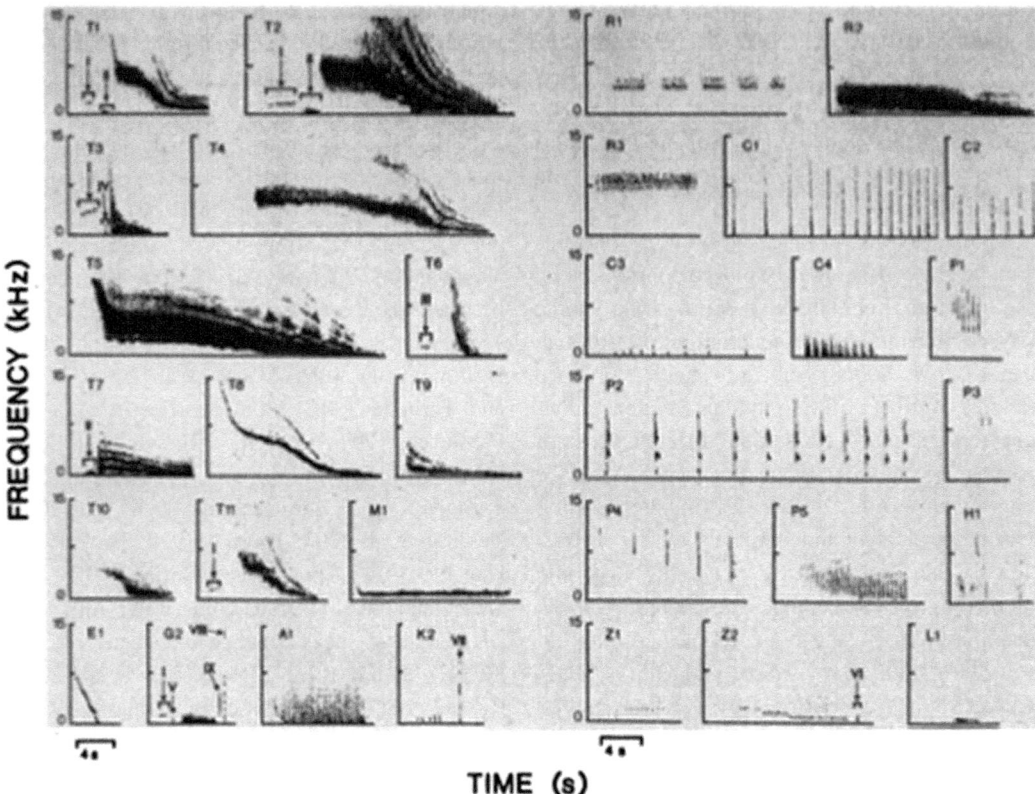

Fig. 5.26 Spectrograms of Weddell seal sound types. *T* trill, *M* mew, *E* eeyoo, *G* guttural thump or glug, *A* click, *K* pulse, *R* cricket call, *C* chunk, *P* chink, *chirp* chirrup, and chi-chi-chi, *H* jaw clap and snap, *Z* seitz, *L* growl. Reproduced with ASA permission from Thomas JA, Kuechle VB (1982) Quantitative analysis of Weddell seal (*Leptonychotes weddelli*) underwater vocalizations at McMurdo Sound, Antarctica. J Acoust Soc Am 72 (6): 1730–1738. https://doi.org/10.1121/1.388667. © Acoustical Society of America, 1982. All rights reserved

emitting clicks, cricket calls, and low-intensity threat sounds (guttural thump and glug; Thomas 1979; Thomas and Kuechle 1982). Males appear more vocal and produce a wider range of sound types than females (Russell et al. 2016). However, some complex underwater calls previously thought to only be produced by males may also be emitted by females as they have been recorded from tagged females with pups (Charrier and Casey 2022). Underwater call sequences were recorded from a breeding male; the most common sequence followed the pattern of low-frequency roar, high-frequency roar, middle-frequency roar, trill, and, finally, whistle-ascending grunt (Terhune and Dell'Apa 2006).

In air, sounds are emitted between 50 and 7200 Hz, with peak energy from 280 to 460 Hz, lasting 0.1–3.3 s (Collins et al. 2005, 2006; Collins and Terhune 2007; Schevill and Watkins 1965; Terhune et al. 1994; van Opzeeland et al. 2010). Aerial sounds include barks, chirps, grunts, howls, moans, snorts, descending and rising tones, whistles, whoops, and "teeth chatter." Chirps were emitted by a male Weddell seal in the presence of conspecifics, especially other males, in his territory, while growls, mews, trills, and "trills + knocks" were mainly emitted when females were located within his territory. Females produced chirps and jaw claps in a social context when other male or female seals were present

(Russell et al. 2016; Terhune et al. 1994; Thomas 1979).

Females and pups communicate with each other through mother–pup contact calls (comparable to FACs and PACs described for other species) or primary calls. Pups produce "aa" sounds and bawls, in addition to "ma" crying sounds. The latter may also be emitted under water, for example, when pups are learning to swim but are trying to leave the water (Terhune et al. 1994; Thomas 1979). From about 2 weeks of age, pups show some individual variation in their calls, mainly varying their fundamental frequency and distribution of the spectral energy. In playback experiments, however, females did not recognize (or at least respond differently to) their pups' calls from other familiar or unfamiliar pups (Collins et al. 2006; van Opzeeland et al. 2012). Pups play the main role when reuniting with their mothers. They mostly produce initiating calls, while females mostly give responding calls. The pup contact calls emitted by females searching for their pups vary individually, enabling pups to recognize their mother's call (Collins et al. 2005). Acoustic features of these contact calls (fundamental frequency, energy spectrum, and duration) also varied depending on the social context (e.g., if in search for or in contact with each other; Collins et al. 2011).

At different geographic locations, Weddell seals have been recorded emitting different sound types and using them to differing extents while also displaying variation in call features (frequency and duration; Abgrall et al. 2003; Morrice et al. 1994; Terhune et al. 2001; Thomas and Stirling 1983). Trills emitted by adult males showed variation between six locations around Antarctica, suggesting that males display breeding site fidelity (Terhune et al. 2008). Geographic differences have been documented for mother and pup contact calls in two Weddell seal populations, which mainly varied in their fundamental frequency and energy spectra among the harmonics (Collins and Terhune 2007).

Rouget et al. (2007) observed a diel variation in calling rates that varied seasonally. Calling rates in July (i.e., in midwinter) were low during daytime but high at night, nearly reaching the daytime calling rate during the breeding season in November (summer; 24 h daylight).

5.5.3.7 *Lobodon carcinophagus*: Crabeater Seal

In air, crabeater seals have been described to bark harshly and hiss when approached and disturbed by humans, and pups loudly bawl when separated from their mothers (Nel 1966; Siniff et al. 1979). Three juvenile crabeater seals in rehabilitation have recently been recorded in air, emitting brief, intermediate, and long moans, as well as hissing sounds created by exhaling air through the nostrils when aroused. The female juvenile also emitted croaks. Moans and croaks are burst-pulsed sounds, with a dominant frequency of up to 1240 Hz in moans and up to 2890 Hz in croaks, compared to 20,910 Hz in hisses. Moans lasted 0.4–2.3 s, croaks 0.09–0.25 s, and hisses 1–2.5 s (Shabangu et al. 2022).

Under water, crabeater seals produce high and low moans during the breeding period, a recently described short moan that occurred before the breeding season when sea ice formed, and long and short groans. Whistles, screeches, grunts, and short groans may be linked to foraging behaviors. These underwater sounds range from 10 to 2550 Hz, with peak energy between 600 and 1310 Hz. The shortest vocalization, a grunt, lasts for 0.13 s, while a long groan can last up to 6.01 s (Klinck et al. 2010; McCreery and Thomas 2009; Nagaraj et al. 2021; Shabangu and Charif 2021; Stirling and Siniff 1979; van Opzeeland et al. 2010).

5.5.3.8 *Mirounga angustirostris*: Northern Elephant Seal

In-air vocalizations by northern elephant seals contain energy between 10 and 12,000 Hz, with peak frequencies between 120 and 1060 Hz, and last 0.2–20 s. Source levels of calls emitted by females were 72–116 dB re 20 µPa m, depending on age, sex, and body size, compared to 120–130 dB re 20 µPa m for males. Older and larger animals appear to vocalize at lower frequencies and higher source levels (Bartholomew and Collias 1962; Burgess et al. 1998; Casey et al. 2015; Holt et al. 2010; Insley

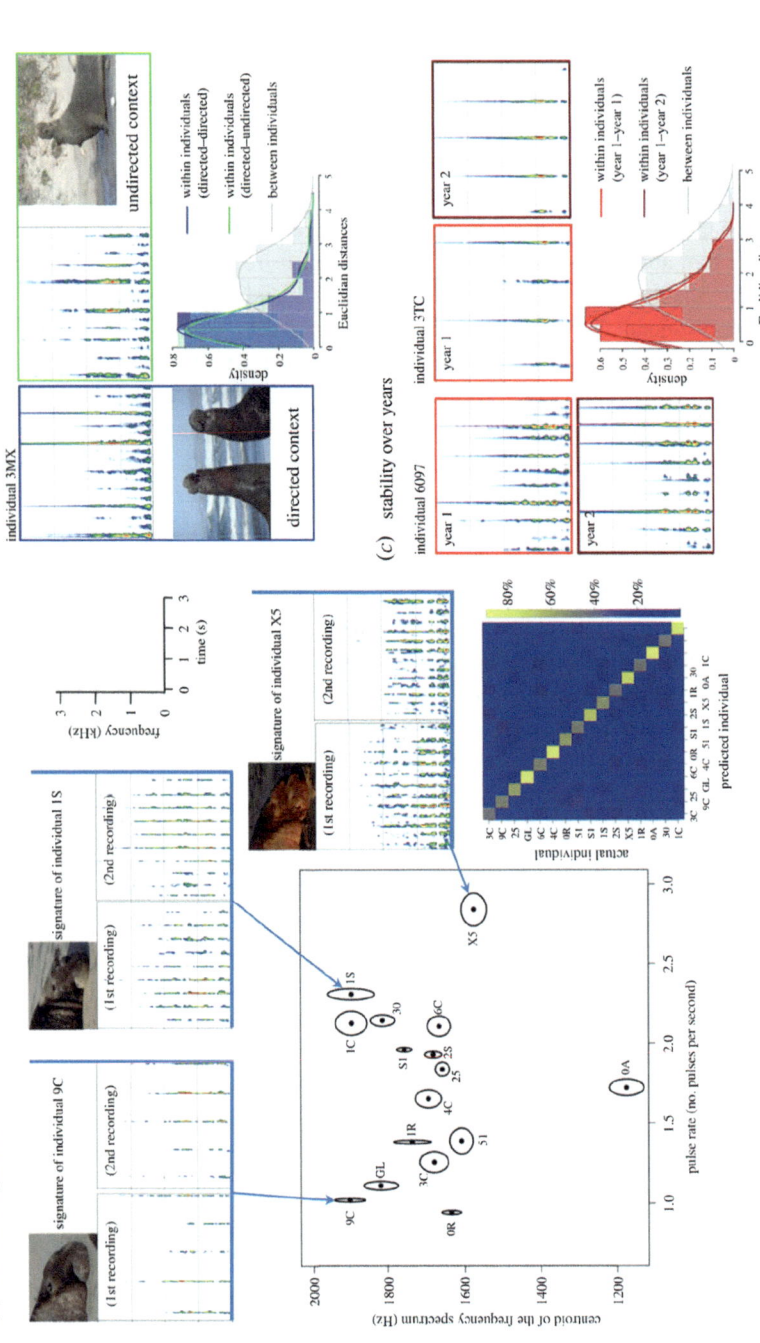

Fig. 5.27 Individual calls (signatures) of male northern elephant seals. (**a**) Spectrograms of two calls from each of the three individuals showing differences in frequency and pulse rate. The central plot shows how these two features cluster for additional individuals. The confusion matrix resulted from validated discriminant function analysis and shows by the color of cell(i,j) the conditional probability of guessing that the test call came from individual j when in fact it was emitted by individual i. The yellow diagonal of the matrix underscores the high probability of correct classification (average = 61.3% versus chance =6.3%), highlighting the strength of the individual signatures. (**b**) Both spectrograms illustrate the stereotypy of an individual's calls in two different social contexts (calling alone and calling to a rival). The distribution of the Euclidian distances (density curves) underscores the similarity of calls within and between contexts (in 2D space defined by the calls' frequency centroid and pulse rate; n = eight individuals, two to six calls per individual). (**c**) Both pairs of spectrograms illustrate the consistency of an individual's calls over successive years. The distribution of Euclidean distances (density curves) shows the remarkable proximity of calls within and between years (n = ten individuals, five to six calls per individual per year). Corresponding audio recordings are available in the supplementary material of: Casey C, Charrier I, Mathevon N, Reichmuth C (2015) Rival assessment among northern elephant seals: evidence of associative learning during male–male contests. R Soc Open Sci 2 (8): 150228. https://doi.org/10.1098/rsos.150228. © Casey et al. 2015. Published CC BY 4.0; https://creativecommons.org/licenses/by/4.0/

1992; Le Boeuf and Petrinovich 1974; Mathevon et al. 2017; Sanvito and Galimberti 2003; Shipley et al. 1981; Southall 2002). Elephant seals may also communicate via seismic vibrations, which can travel for 20 m (Shipley et al. 1992).

Northern elephant seals have been reported to emit individually distinct calls, varying in pulse rate and spectra (Fig. 5.27). Females and pups recognize each other's vocalizations, with females recognizing their pups' calls within 1–2 days after birth. In younger subadult males, vocalizations vary (mostly in frequency modulation) while maturing, decreasing in frequency with increasing size. However, bulls keep their calls stable over multiple seasons, and temporal and spectral components allow conspecifics to distinguish among rivals (Casey 2018; Casey et al. 2015, 2020; Insley and Holt 2012; Klopfer and Gilbert 1966; Petrinovich 1974; Shipley et al. 1986).

Fletcher et al. (1996) were unable to record any sounds under water in tagged juveniles. However, Poulter (1968) mentions underwater clicks in captive and wild northern elephant seals. Burgess et al. (1998) recorded sounds with a spectral peak at 290 Hz and source levels of 83–100 dB re 1 µPa m on a tagged female diving between 220 and 440 m deep.

Regarding sound types, pups bawl with a shrill yapping cry (a FAC or offspring primary call), to which females answer with a high-frequency warbling soft call (PAC or mother primary call). Pups also bark and squawk if a bull rolls or steps onto them. Female threat calls, in response to disturbance, have been called belch roar, described as a "loud, prolonged rasping noise" and "low-pitched staccato growl," whereas they emit croaks when a bull approaches to mount them. According to Poulter (1968), "cows and pups produce a raucous array of pulsating honks, whines, whimpers, squeaks, squeals, roars, moans, snorts, grunts, croaks, chuckles, snores, snarls, moos, bellows, barks, cries, cackles, and about anything that one can imagine." Yearlings and juvenile males may belch roar, similar to the female threat call, or hiss and emit high-pitched whimpering sounds (e.g., when bitten).

The use of the proboscis by male northern elephant seals in displays is described by Sandegren (1976), but its contribution to the acoustic characteristics of male calls is best described for the southern elephant seal (Sect. 5.5.3.9). Four call types have been described for bulls; the first three are used as a threat to others: VO (vocalization) 0 is a low-pitched sound produced when inhaling air through the mouth; VO 1 is a low-pitched sound, often when in water (i.e., with the nostrils under water), also referred to as a snort; VO 2 is a low-pitched, loud roar also called the clap threat; and VO 3 is used to show submission (Bartholomew and Collias 1962; Christenson and Le Boeuf 1978; Cox 1981; Insley 1992; Klopfer and Gilbert 1966; Le Boeuf 1972; Le Boeuf et al. 1972; Linossier et al. 2021; Petrinovich 1974; Poulter 1968; Sandegren 1976; Shipley et al. 1981, 1986). Shipley and Strecker (1986) and Shipley et al. (1981) further describe "clap threats" (threat 1 call), "clap burst threats" (threat 2 call), and "patterned clap threats."

Males of some populations have previously been reported as vocalizing in dialects, using different pulse rates. These dialects have disappeared, probably due to the near depletion of northern elephant seals, followed by migration in a recovering and growing population, and cultural mutation. Males now show higher variability in their vocal communication instead (Casey et al. 2018; Le Boeuf and Petrinovich 1974).

5.5.3.9 *Mirounga leonina*: Southern Elephant Seal

Southern elephant seals have been recorded in air, producing sounds between 100 and 1620 Hz, with peak frequencies between 380 and 500 Hz, durations of 1–50 s, and source levels of 103–120 dB re 20 µPa m (Sanvito and Galimberti 2000a, b, 2003; Sanvito et al. 2008). Bulls produce a threat-related high-pitched sound, also called vocalization 1 (V1), by exhaling air through their proboscis, on land and in water. If the seal is in the water, the nostrils are usually under water when this sound is produced. Bulls roar (sometimes called V2) to show their dominance; they emit short, high-pitched sounds or cries when displaying submissive behaviors (Carrick et al. 1962; Laws 1956; Matthews 1929; McCann 1981). Females produce harsh grunts and barks, also described as a harsh roaring noise (e.g., when disturbed), and high-pitched whining sounds (yodel) when attending to their pups (Carrick et al. 1962; Laws 1956; Matthews 1929). Pups have been reported to make sharp or high-pitched barking sounds (Carrick et al. 1962; Laws 1956).

Populations use different call types and numbers of "syllables," evidence of dialects in male communications (Sanvito and Galimberti 2000b; Sanvito et al. 2007b). This observation strengthens the hypothesis that dialects in northern elephant seals disappeared as a result of rapid population recovery, which has been associated with high immigration/emigration. Male southern elephant seal sound production also changes with age as their proboscis grows and develops, lengthening the vocal tract, thereby reducing the peak frequency of aggressive sounds (Sanvito et al. 2007a, 2008).

5.5.3.10 *Monachus monachus*: Mediterranean Monk Seal

In air, Mediterranean monk seals produce barks, chirps, grunts, screams, and snorts in agonistic interactions (except for grunts, for which the function is unclear). These sounds range in duration from 0.12 to 8.7 s and in frequency from 100 to 8010 Hz, with peak frequencies from 100 to 2420 Hz (Charrier et al. 2017; Muñoz et al. 2011). Although wild pups were only recorded producing barks from 120 to 1240 Hz, rehabilitating pups also emitted gaggle and squawk sounds from 1010 to 1270 Hz across all three call types (Muñoz et al. 2011).

Charrier et al. (2023) described 18 underwater calls classified into three categories, emitted between 60 and 1600 Hz, with peak energy between 100 and 1205 Hz and duration of 0.07–6.7 s. Harmonic calls include bark, croak, cry, "gloo," "gloogloo," groan, moan, scream, whines, "whoo," "wop," and "wom" sounds (Fig. 5.28). Growls, hiccups, and squeaks were classified as noisy calls and claps, knocks, and rumbles as pulsative calls.

5.5.3.11 *Neomonachus schauinslandi*: Hawaiian Monk Seal

In air, Hawaiian monk seals produce 0.13–1 s long sounds between 100 and 1300 Hz, with peak frequencies between 500 and 1200 Hz (Kenyon and Rice 1959; Miller and Job 1992). Hawaiian monk seal pups bleat to respond to females and use grunting bawls when disturbed, and both females and pups use loud broadcast calls when seeking each other (Kenyon and Rice 1959; Miller and Job 1992). While nursing, pups "utter medium to soft calls," returned by females by "huh-huhs," also called humming sounds, which are also heard when they give birth (Eliason et al. 1990; Miller and Job 1992; Parnell 2018). Although calls produced by pups vary individually in spectral content, their mothers do not necessarily recognize their own pup's calls, and other females may respond as well, contributing to the relatively high level of allo-nursing in this species (Boness 1990; Job et al. 1995).

The most frequent sounds produced by Hawaiian monk seals are not strictly vocalizations but sneezes (i.e., nasal snorts and sneeze snorts) and coughs (Kenyon and Rice 1959; Miller and Job 1992). The remaining sounds in the repertoire are associated with agonistic encounters. The bellowing sound, also called rolling bellows or a grunting bawl, is equivalent to groans emitted under water and is used by females to defend their pups and by males when pursuing females (Kenyon and Rice 1959). Roars, guttural expirations, and belch-coughs (probably equivalent to the coughing snorts) are threat sounds, usually produced by males. For example, they are used as agonistic calls while patrolling beaches or chasing females (Miller and Job 1992). Other reported sounds are bubbling sounds and loud snarling calls.

Underwater recordings were conducted with one captive animal, which emitted sounds for 0.1–4.6 s between 10 and 370 Hz, with most energy between 50 and 290 Hz, depending on its vocalization type. In ascending order, when listed by their average peak frequency, these vocalization types were moans, rumbles, groans,

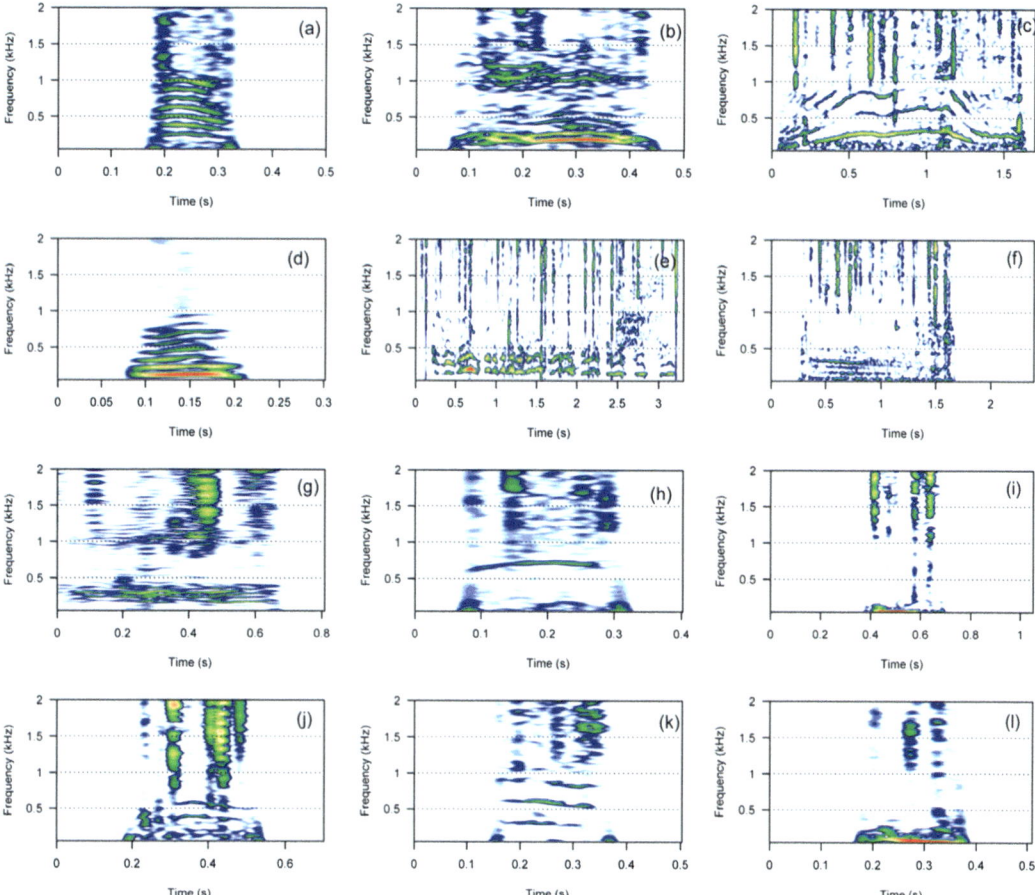

Fig. 5.28 Spectrograms of Mediterranean monk seal underwater harmonic calls. Bark (**a**), croak (**b**), cry (**c**), gloo (**d**), gloogloo (**e**), groan (**f**), moan (**g**), scream (**h**), whine (**i**), whoo (**j**), wop (**k**), and wom (**l**). Reprinted from Charrier I, Huetz C, Prevost L, Dendrinos P, Karamanlidis AA (2023) First description of the underwater sounds in the Mediterranean monk seal *Monachus monachus* in Greece: towards establishing a vocal repertoire. Animals 13 (6):1048. https://doi.org/10.3390/ani13061048. © Charrier et al. 2023. Published CC BY 4.0; https://creativecommons.org/licenses/by/4.0/

whoops, croaks, and growls. The seal was also able to produce these sounds in air (Parnell 2018; Sills et al. 2021). Foghorn-like sounds and barks have also been reported as part of their underwater repertoire (Stirling and Thomas 2003). Opportunistic dive videos have also recorded sounds of monk seals in the wild (e.g., http://www.jonathanbird.net/cgi-bin/video.asp?id=1041; accessed Jan 24, 2024).

5.5.3.12 *Ommatophoca rossii*: Ross Seal

Ross seals emit similar sounds in air and under water with some variation in frequency (Watkins and Ray 1985). They are known for their siren sounds, which are tones that alternate between downward and upward sweeping frequencies (Watkins and Ray 1985). In air, sirens, as well as pulses and downsweeps, which are also referred to as chugs, are produced between 100 and 1000 Hz and last 0.25–1.5 s (Ray 1970, 1981; Watkins and Ray 1985).

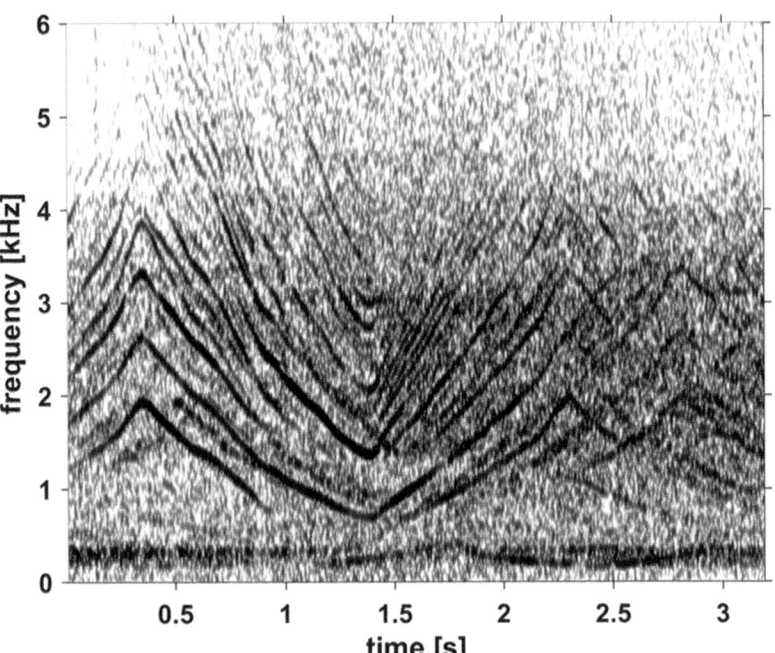

Fig. 5.29 Spectrogram of a Ross seal underwater siren recorded in the Davis Sea (Eastern Antarctica) in October 1996 (fs = 22 kHz, NFFT = 440, 50% overlap, Hann window); courtesy of Tracey Rogers. Reprinted from Erbe C, Dunlop R, Jenner KCS, Jenner M-NM, McCauley RD, Parnum I, Parsons M, Rogers T, Salgado-Kent C (2017) Review of underwater and in-air sounds emitted by Australian and Antarctic marine mammals. Acoust Aust 45:179–241. https://doi.org/10.1007/s40857-017-0101-z. © Erbe et al. 2017. Published CC BY 4.0; https://creativecommons.org/licenses/by/4.0/

Underwater vocalizations range from 110 to 12,300 Hz with spectral peak frequencies from 200 to 2000 Hz, lasting 2.5–4 s. Apart from high, mid, and low siren calls (Fig. 5.29), Ross seals also emit whoosh sounds, which consist of a tonal and a broadband element and which are sometimes classed as two sound types (Gedamke and Robinson 2010; Kindermann et al. 2008; Ray 1981; Schwenke 2013; Seibert et al. 2007, 2008; van Opzeeland et al. 2010; Watkins and Ray 1985).

5.5.3.13 *Pagophilus groenlandicus*: Harp Seal

Harp seals vocalize in air and under water; they have a large repertoire of narrowband, broadband, frequency-modulated, and amplitude-modulated sounds. Møhl et al. (1975) and Terhune (1994) described sound types 1–19 from underwater recordings in the wild (Fig. 5.30). Serrano (2001) and Serrano-Solis (1998) added types 20–27 to the list from underwater recordings in captivity as well as aerial type 1 from recordings in air of captive males. Most publications refer to these sound types with descriptions that include chirp, click, distressed blackbird, dove cooing, frequency shift keying, grunt, double grunt, gull's cry, morse call, knocking sound, passerine call, sine wave, single growl, squeak, "tink" (which was also described as a "hammer hitting a small metallic object"), "tjok," trill, warble, and whistle. Alternatively, sounds are described as patterns 1–3 (short–short calls, long–long calls, or short–long calls, respectively; Møhl et al. 1975; Rossong and Terhune 2009; Terhune 1994; Terhune and Ronald 1986).

Under water, harp seals vocalize between 100 and 6000 Hz with peak energy from 200 to 3320 Hz, excluding type 16 (click sounds), which can exhibit peak energy up to 30,000 Hz. Vocalizations last between 0.2 and 1 s. Source levels as high as 180 dB re 1 µPa m have been reported (Miller and Murray 1995; Møhl et al. 1975; Moors and Terhune 2003, 2005; Rossong and Terhune 2009; Terhune 1994; Terhune and Ronald 1986; Watkins and Schevill 1979). In low levels of ambient noise, the ensemble of calls from a colony could be heard up to 48 km away (Terhune 2008).

In air, including type 26, females and males vocalize between 400 and 3500 Hz with peak energy from 200 to 700 Hz for around 5 s.

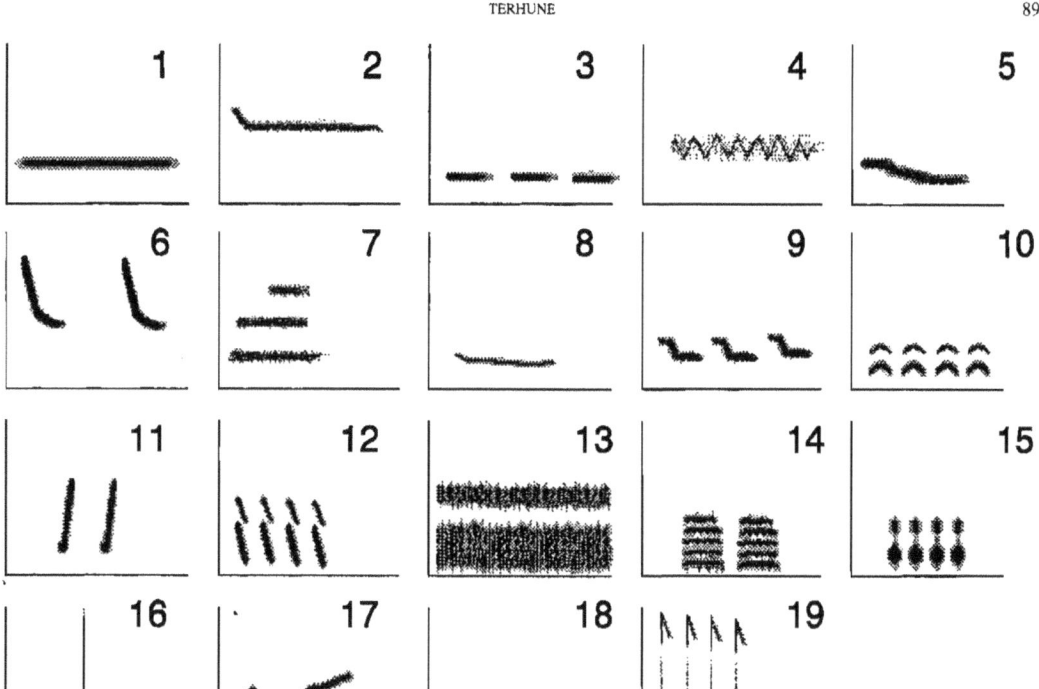

Fig. 5.30 Sketches of spectrograms of underwater vocalizations of harp seals (x-axis 0–2 s, y-axis 0–5 kHz). Used with permission of Canadian Science Publishing, from Terhune JM (1994) Geographical variation of harp seal underwater vocalizations. Can J Zool 72 (5):892–897; https://doi.org/10.1139/z94-121. © Canadian Science Publishing 1951. Permission conveyed through Copyright Clearance Center, Inc. All rights reserved

Females used two types of sound: one to attend to their pup, and the second was a long, loud, harsh call threatening foreign pups, females, or humans (Miller and Murray 1995; Serrano 2001; Serrano-Solis 1998).

Harp seals haul out in large groups and have developed strategies to avoid masking each other's vocalizations. Apart from using relatively short calls, individuals in groups use different call types; with higher calling rates from conspecifics, the calls had a larger number of elements (Serrano and Terhune 2001, 2002).

Harp seal pups emit various sounds between 140 and 2040 Hz, lasting for 0.3 to 1.5 s. In the Greenland Sea, female pups appeared to produce more individually distinctive calls than male pups, and pups vocalized for longer, compared with a population in Newfoundland. The latter also showed no calling difference between male and female pups (Miller and Murray 1995; van Opzeeland et al. 2009; van Opzeeland and Van Parijs 2004).

5.5.3.14 *Phoca largha*: Spotted Seal, Largha Seal

Most in-air and underwater recordings of sounds emitted by spotted seals are from captive individuals. In the wild, spotted seals have only been recorded under water, producing drum, growl, knock, and sweep sounds lasting 0.002–1.5 s, at 90–660 Hz, with peak energy between 260 and 500 Hz. Knocks are often emitted repetitively as knock trains (Yang et al. 2017, 2022).

In captivity, in air and under water, male and female spotted seals emit barks, buzzes, chirps, "creaky door" sounds, drums, growls, grunts, "gurgle grunts," "moo" sounds, and snorts, with

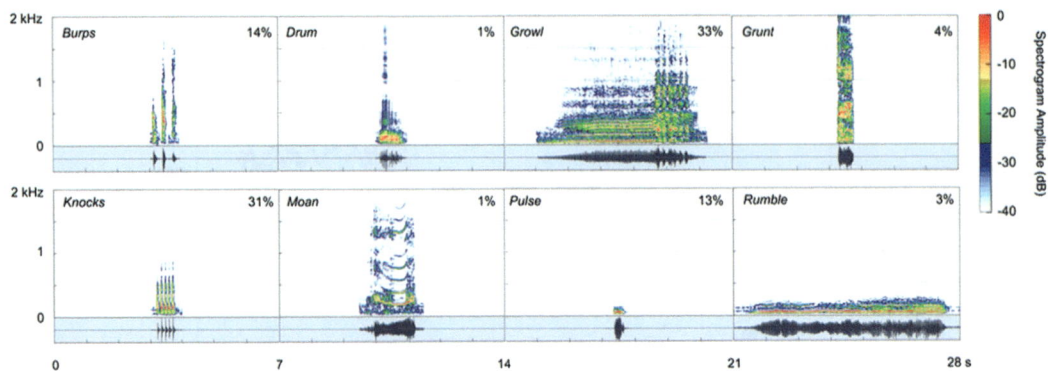

Fig. 5.31 Spectrograms and waveforms of underwater calls of male spotted seals (fs = 48 kHz, NFFT = 4096, 90% overlap, Hann window). Percentage shows how often each sound was recorded. Reprinted from Sills JM, Reichmuth C (2022) Vocal behavior in spotted seals (*Phoca largha*) and implications for passive acoustic monitoring. Frontiers in Remote Sensing 3:862435; https://doi.org/10.3389/frsen.2022.862435. © Sills and Reichmuth, 2022. Published CC BY 4.0; https://creativecommons.org/licenses/by/4.0/

males vocalizing 2.5 times more often than females. Males have also been recorded emitting burps, knocks/knock trains, moans, pulses, and rumbles. Growls were their most common sound type, followed by knocks/knock trains. Vocalizations lasted between 0.02 and 9.4 s at 200–3500 Hz, peak energy between 60 and 1300 Hz (Fig. 5.31; Beier and Wartzok 1979; Gailey-Phipps 1984; Sills and Reichmuth 2022; Zhang et al. 2016). Peak-to-peak sound levels have been measured in air between 100 and 130 dB re 20 µPa (Zhang et al. 2016). Drums, buzzes, and "gurgle grunts" may be used for advertisement displays, while growls and grunts were mostly observed accompanying territorial behaviors, and grunts, snorts, and barks were used to show aggression (Gailey-Phipps 1984).

In air, pup and yearling calls lasted 0.4–0.7 s, between 600 and 1420 Hz, peak energy between 970 and 1110 Hz, with sound levels of 103–119 dB re 20 µPa (Zhang et al. 2016). Peak frequencies, durations, and source levels measured in up to 1-year-old pups varied between females and males and between age classes, with all measurements being lowest for 1–3-month-old pups. Source level and duration increased with age. Usually, male pups emitted sounds at higher source levels and peak frequencies than female pups, whereas female pups called for longer (Zhang et al. 2022).

5.5.3.15 *Phoca vitulina*: Harbor Seal, Common Seal

This species comprises the subspecies Ungava (*P. v. mellonae*), Pacific (*P. v. richardii*), and Atlantic (*P. v. vitulina*) harbor seals, which are treated together in this section. Male harbor seals and pups vocalize in air and under water, while females older than pup age have only been recorded in air.

In air, males and females emit growls that sound like "rwoor rarrwrr," honks, moans, and "descending roars," and males also bark and grunt, the former resembling a dog-like "barf warf." These calls are produced between 780 and 4900 Hz, with peak energy ranging from 1100 to 4300 Hz, and last between 0.5 and 2.7 s (0.9–1.6 s on average), excluding the longer grunts and roars that last, on average, 5.7 and 8.3 s, respectively. In-air vocalizations appear to be used for male competition, as they were generally observed being emitted by males during agonistic interactions (Scheffer and Slipp 1944; Van Parijs and Kovacs 2002).

Under water, males produce a creak described "as if a rusty hinged door is being opened," groans, growls, grunts, roars, sweeps, and

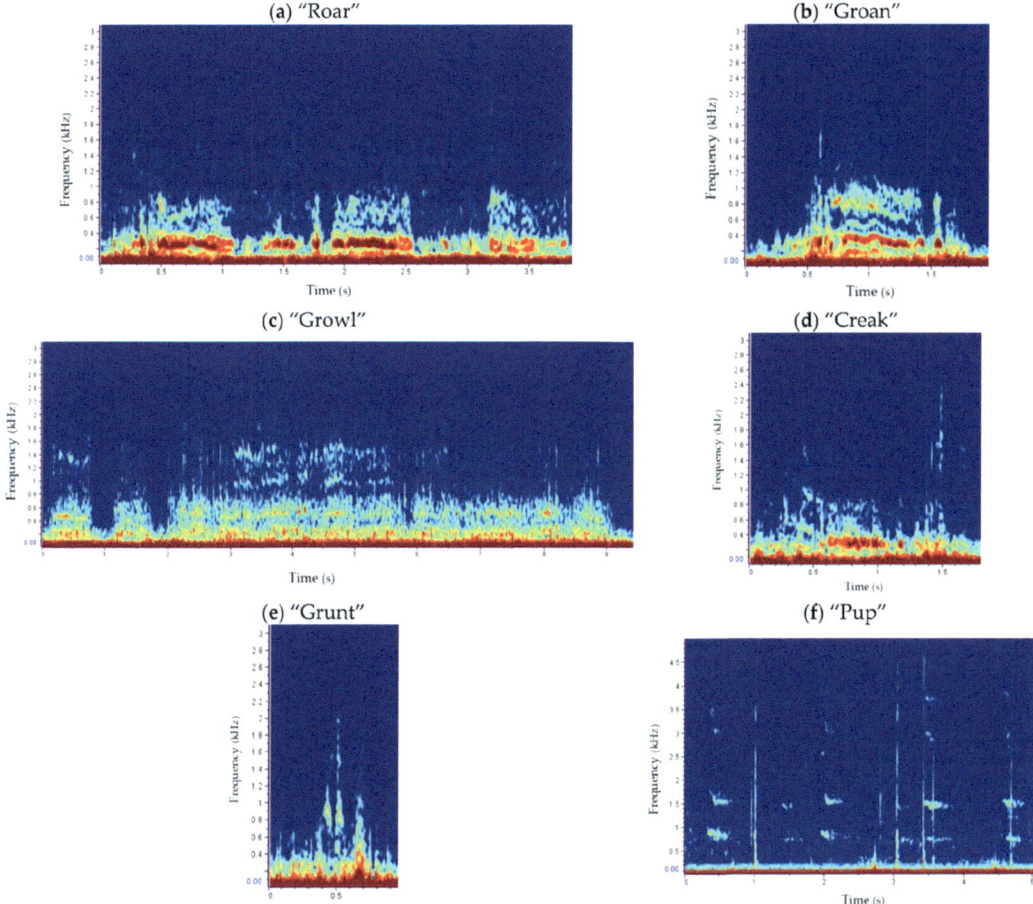

Fig. 5.32 Example spectrograms of underwater sounds of harbor seals; (a-e) by adults, (f) by a pup. Reprinted from Pozo Galván YP, Pérez Tadeo M, Pommier M, O'Brien J (2024) Static acoustic monitoring of harbor (*Phoca vitulina*) and gray seals (*Halichoerus grypus*) in the Malin Sea: a revolutionary approach in pinniped conservation. J Mar Sci Eng 12 (1):118; https://doi.org/10.3390/jmse12010118. © Pozo Galván et al. 2024. Published CC BY 4.0; https://creativecommons.org/licenses/by/4.0/

warbles (Fig. 5.32). Nikolich et al. (2016) suggested that harbor seals produce graded sounds on a vocal continuum rather than distinct call types during the breeding season. Their statistical approach to call-type classification has not been used in many studies of pinniped repertoire, so it is not clear whether it would have produced similar results in studies of other species.

Besides clicks (or click trains), which are emitted at peak frequencies of ~7500 Hz and last for 0.01 s, underwater calls were recorded between 28 and 4000 Hz, with peak energy ranging from 90 to 1530 Hz. Vocalizations under water last 0.1–19 s (average 0.9–15 s). Source levels have been measured up to 153 dB re 1 µPa m. Male roars may be heard over ~2100 m in slough environments, 630 m at the coast, and 400 m in a harbor (Bjorgesaeter et al. 2004; Casey et al. 2016; Hanggi and Schusterman 1992, 1994; Hayes et al. 2004; Matthews 2017; Matthews et al. 2017a, b, 2020; Nicholson 2000; Nikolich et al. 2016, 2018; Pozo Galván et al. 2024; Renouf and Davis 1982; Rößler et al. 2021; Sabinsky et al. 2017; Scheffer and Slipp 1944; Van Parijs et al. 1997, 2000a, b; Van Parijs and Kovacs 2002). The estimated propagation distance in a slough environment was estimated rather than measured; it was based on measurements of background noise without reference to other conditions affecting shallow-water propagation, such as water depth, topography, and bottom.

Males show individual differences in their acoustic displays, with varying use of sound types between individuals. Frequency features, such as minimum, mean, and maximum frequencies, vary individually in male roars (Hanggi and Schusterman 1994; Van Parijs et al. 2000a). Furthermore, geographical differences have been observed regarding the use of different sound types and differences in the frequency contour of common calls (Bjorgesaeter et al. 2004). Male roars varied between locations and in consecutive years at the same location (Rößler et al. 2021; Sabinsky et al. 2017; Van Parijs et al. 2000a, 2003a). Environmental variables, such as tidal variations or tidal and diel patterns combined, may also influence vocalizations differently at separate locations (Van Parijs et al. 1999).

Males cover a vocal underwater display area of 40–1910 m^2 (Matthews 2017; Van Parijs et al. 2000b) and predominantly call at night (Matthews 2017; Matthews et al. 2017a; Nikolich et al. 2018; Renouf and Davis 1982; Sabinsky et al. 2017). If other seals are in the vicinity, males increase the duration of their vocalizations compared to when alone. Also, more dominant males produce longer roars at lower frequencies than less dominant conspecifics (Nicholson 2000). In captivity, females appeared to prefer calls by more dominant males (Matthews 2017; Matthews et al. 2018).

Harbor seals have been observed to alter vocalizations when predators are present and as a response to anthropogenic disturbance. Male harbor seals may cease to vocalize after killer whales pass through an area; in one experiment, calling did not return to pre-killer-whale rates until 48 h later (Nikolich et al. 2018). Vessel noise also impacts harbor seal vocalizations. Males roar louder, but for a shorter period and with an increased minimum frequency, while vessels are in the vicinity (Matthews 2017); however, they do not vary their vocalizations sufficiently to prevent acoustic masking entirely (Matthews et al. 2020).

Hoover was an orphaned harbor seal pup that was hand-raised and started mimicking human sounds once he reached adulthood. His repertoire included "hello," "hello there," "how are you," "hey," "Hoover," "come over here," "hurry," and "get out of there," in addition to chest slapping, strange cries, raspy growling sounds, and a "blood-curdling scream" (Duengen et al. 2023; Ralls et al. 1985).

Pups vocalize in air and under water. They may call for their mothers in either medium by emitting mother attraction calls (termed FACs in earlier sections), mother–pup contact calls sounding like "kroo-roo-uh" or "kroo-roo." Nursing pups may produce a "sheep-like m-a-a-a," while, during agonistic encounters, pups growl, resembling harsh, broadband, staccato calls. Pup calls are emitted between 180 and 3106 Hz, with a peak frequency of 350–1700 Hz, at sound levels of 94 dB re 20 µPa, and last 0.1–2.4 s (Khan et al. 2006; Newby 1973; Perry and Renouf 1988; Ralls et al. 1985; Reiman and Terhune 1993; Renouf 1984; Sauve et al. 2015a; Scheffer and Slipp 1944; Van Parijs and Kovacs 2002).

Pups also produce so-called amphibious sounds when calling for their mothers. These calls are emitted while the pups are in the water with their heads above the surface. This mother–pup contact call can be heard in air and under water, with different acoustic features between the two media. In both media, however, the calls are individually distinctive, mostly varying in their fundamental frequency. Pup calls appear to be individually distinctive within the first few days of birth, and females can recognize their pup's call from about 3 days of age (Perry and Renouf 1988; Renouf 1984, 1989; Sauve et al. 2015a, b). While the pup matures, the structure of its mother attraction call changes, with, for example, a decrease in frequency range and fundamental frequency, as well as changes in call duration and frequency modulation (Khan et al. 2006; Sauve et al. 2015a).

Calls by female pups are emitted at higher frequencies and amplitudes, while those of male pups decrease in mean frequency and amplitude and increase in duration as they grow (Khan et al. 2006; MacRae 2018; Sauve et al. 2015a). Rhythmic features (i.e., temporal patterns and inter-onset and inter-peak intervals) develop in pups' individual calls while they mature (de Heer Kloots et al. 2020; Ravignani et al. 2018). Harbor seal pups can determine rhythmic patterns in calls (i.e., tempo, regularity, and call length) and can produce calls asynchronous with those of conspecifics, probably to minimize acoustic masking (de Heer Kloots et al. 2020; Ravignani 2019; Verga et al. 2022).

Under good conditions, some pup cries may be heard over 1 km and recognized from 140 m away (Reiman and Terhune 1993). If the pup is in the water, the mother is more likely to approach the pup than when hauled out on land (Renouf 1984). One- to 3-month-old pups decrease the fundamental frequency of their calls to adjust to noisy environments. Some pups may also change their call amplitude and lower call frequency in response to noise (Ravignani et al. 2022; Torres Borda et al. 2021).

Harbor seal pups also display pain in their calls, with higher call rates and peak frequencies, concomitant with increased orbital tightening and higher eye temperature, measured right after tagging and micro-chipping procedures (MacRae 2018).

Nonvocal sounds by harbor seals include a "bubbly growl" recorded between 100 and 200 Hz lasting 1–8 s (Hanggi 1992). They may also produce a "penetrating rasping noise" with their lips when blowing bubbles (Venables and Venables 1957). Bubble blowing was observed in self-advertisement displays of both females and males (Ralls et al. 1985; Venables and Venables 1957). Harbor seals may also convey information using flipper slaps, which last for 0.002 s at sound levels between 186 and 199 dB re 1 µPa pk-pk, reaching up to 5000 Hz, and slaps 2 and 3 probably exceeding 20,000 Hz (Wahlberg et al. 2002). Apart from emitting yelps and snarls, harbor seals also slapped the water's surface with their flippers during mating behaviors, while Newby (1973) observed mothers using their fore flippers to slap the water surface to signal danger. Flipper slapping was also a response of male harbor seals to playback of roaring conspecific males (Hayes et al. 2004).

5.5.3.16 *Pusa caspica*: Caspian Seal

We could not find any reports of the acoustic characteristics of sounds made by Caspian seals in the English peer-reviewed literature. The only description of sound production in this species relates to pups emitting distress calls when disturbed by vessels or losing their mother, such as when the female is moving too quickly for the pup to follow while moving to a new location (Wilson et al. 2017).

5.5.3.17 *Pusa hispida*: Ringed Seal

Most reports of ringed seal sounds describe underwater vocalizations with only a mention of the occurrence of in-air vocalizations. Although there are five recognized subspecies of ringed seals—Baltic (*P. h. botnica*), Arctic (*P. h. hispida*), Lake Ladoga (*P. h. ladogensis*), Okhotsk (*P. h. ochotensis*), and Saima (*P. h. saimensis*) ringed seals—we are summarizing their sound types together, due to the relatively sparse data for the species.

Frequencies of underwater barks, chirps, growls, knocks (described "like someone knocking on a door"), squeaks, woofs, and yelps ranged from 30 to 1500 Hz (20–4420 Hz in captivity) with peak frequencies from 150 to 1500 Hz and source levels of 112–131 dB re 1 µPa m. Durations are 0.03–1.3 s (Fig. 5.33; Cummings et al. 1984; Halliday et al. 2020a; Hyvärinen 1989; Jones et al. 2014; Kunnasranta et al. 1996; Lomac-MacNair et al. 2019; Mizuguchi et al. 2016b; Prawirasasra et al. 2021; Rautio et al. 2009; Stirling 1973; Stirling et al. 1983). Clicks were reported from 2000 to 10,000 Hz, with spectral peak frequencies around 5500 Hz, lasting 0.02–0.6 s (Blackwell et al. 2004; Hyvärinen 1989; Kunnasranta et al. 1996). In air, ringed seals roar or yowl/howl for ~2.5 s, at peak frequencies around 700 Hz (Hyvärinen 1989; Sipilä et al. 1996).

Rautio et al. (2009) reported pups producing sounds with nonlinear characteristics (see Ch. 7),

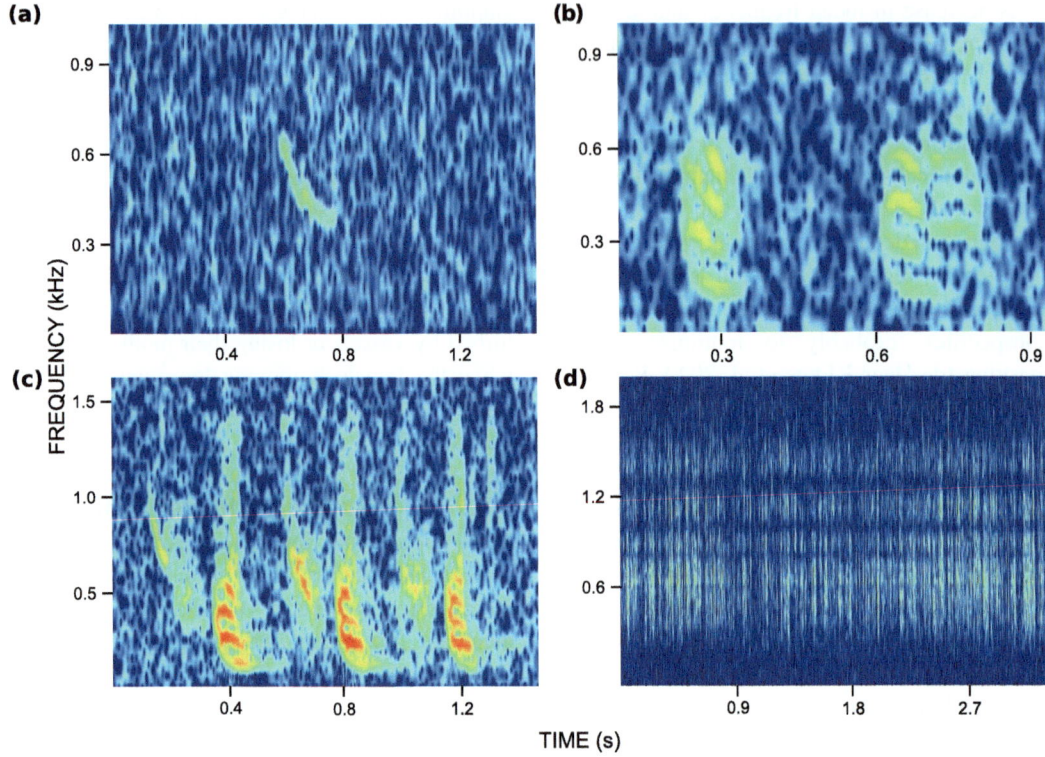

Fig. 5.33 Underwater sounds of ringed seals: (**a**) yelp, (**b**) bark, (**c**) alternating series of yelps and barks, and (**d**) digging of a hole in the ice. Reprinted from Prawirasasra MS, Mustonen M, Klauson A (2021) The underwater soundscape at Gulf of Riga Marine-Protected Areas. J Mar Sci Eng 9 (8):915; https://doi.org/10.3390/jmse9080915. © Prawirasasra et al. 2021. Published CC BY 4.0; https://creativecommons.org/licenses/by/4.0/

termed "harmonic vocalization–chaos–subharmonics," subharmonics, "harmonic vocalization–subharmonics," and chaos sound types, with peak frequencies ranging from 500 to 1700 Hz for 0.9–1.5 s. Overall, pups vocalized in air between 460 and 800 Hz for ~1 s and under water from 460 to 960 Hz for ~0.8 s.

While the behavioral context of ringed seal vocalizations is not fully understood, roars and barks usually accompany threat or aggressive behaviors, yelps may serve in submissive situations, and knocks may be used for long-distance communications (Hyvärinen 1989; Kunnasranta et al. 1996; Stirling 1973).

When ringed seals dig breathing holes into the ice, they produce scratching sounds between 300 and 1900 Hz with a peak frequency of ~620 Hz (Prawirasasra et al. 2021). Detection of underwater sounds of ringed seals was generally better with increasing broken sea ice concentration (Halliday et al. 2018).

5.5.3.18 *Pusa sibirica*: Baikal Seal

Although Baikal seals have been heard vocalizing in air and under water, and males and females have different sound types, recordings are scarce or have not been published in the English peer-reviewed literature (Martínková et al. 2001; Petrov et al. 2021).

The only description of aerial 0.02 s long bursts at peak energy of 2500 Hz and 0.2 s long growls at peak energy of 3500 Hz originated from recordings by the BBC (Glotin et al. 2017), which are no longer available. According to Petrov et al. (2021), Baikal seals also emit groans, "roaroink" (a sound between a bull's roar and a frightened pig's oink), snorts, and whistles when defending their haul-out spot. Threatening sounds are produced when smaller seals move too close to larger conspecifics; sniffing sounds are emitted while thrusting at a conspecific to attempt taking over a haul-out spot (Baranov et al. 2014). An alternative way of gaining haul-out space is to splash water with the fore flippers onto the seal on the desired location; Baikal seals appear to despise this "spa service" or "hydrotherapeutic procedure" and move away with snorts and head shakes (Ivanov 1938; Petrov et al. 2022). Females and pups communicate with each other with mooing sounds (Petrov et al. 2021), but pups may also whimper and weep to call their mothers or when hungry.

One spectrogram of a "quasi-harmonic signal with overtones" in underwater recordings from Lake Baikal has been published, depicting peak frequencies around 120 Hz, ranging from 100 to 1400 Hz, with a duration of ~1 s (Grigoriev et al. 2020). Captive Baikal seals also emit loud noises in air and under water by forcefully exhaling through tightly closed nostrils as attraction calls and to express emotions like resentment, protest, or impatience. Baranov et al. (2014) described hearing the sound from two stories above within a brick building. In the wild, females appear to produce burst-pulsed sounds under water to attract males and exhale air at a peak frequency of 750 Hz (~1000 Hz in captivity), ranging from 400 to 2500 Hz (250–2500 Hz in captivity) for ~0.6 s (Baranov et al. 2014; Petrov et al. 2021).

Baikal seals may signal alarm to conspecifics by a loud, splashing fall into the water when fleeing danger (Petrov et al. 2022).

5.6 Summary

Pinnipeds are a highly vocal clade of semiaquatic marine mammals, producing sounds above and below water. Of 34 extant species, 32 species have reports of sound production in air and 22 under water originating from 13 of the 18 International Union for Conservation of Nature (IUCN) global regions (exceptions being the Caribbean, Central Indian, Arabian, East Africa, and East Asian regions, regions 7 and 10–13, respectively). Like other marine fauna, their call characteristics and cue rates are often related to biological traits, learned and innate acoustic behavior, the associated life functions, and the physical environment in which sounds are produced, which vary to different degrees among species and families. For example, for all species, feeding mostly occurs in the water, while rearing pups occurs in air (i.e., on land or ice). In contrast, courtship, a highly vocal activity, is conducted primarily on land by otariids (eared seals) but mostly at sea by phocids (earless seals) and odobenids (walruses). The physical limitations of producing sound under water and the difference between propagation in water and air mean that call characteristics and ecology for sound production in these two media should be considered separately. Inter- and intraspecies variation is significant and not necessarily related to phylogeny. Some closely related species display vastly different repertoires, while distantly related species can be similar in complexity and repertoire size, suggesting that acoustic cues are often derived traits shaped by learning, natural selection, or genetic drift.

Descriptions of pinniped repertoires indicate a range of complexities, from tonal sounds, broadband grunts, roars, and barks to complex amplitude-modulated trills and chirps, repeated sequences, and songs. Numerous species have displayed evidence of vocal plasticity, call pattern sequencing and recognition, discrimination of physical traits, anti-masking techniques, dialects, and niche partitioning, whether by pups, juveniles, or adults. Here, we collate data published in the English language to provide an overview of the call characteristics, repertoire, and associated functions of the pinniped families.

5.7 Data Availability

Table S1 with pinniped call features that we compiled from the literature may be requested from the first author.

References

Abgrall P, Terhune JM, Burton HR (2003) Variation of Weddell seal (*Leptonychotes weddellii*) underwater vocalizations over mesogeographic ranges. Aquat Mamm 29(2):268–277. https://doi.org/10.1578/016754203101024202

Acevedo-Gutierrez A, Acevedo L, Belonvich O, Boren L (2011) How effective are posted signs to regulate tourism? An example with New Zealand fur seals. Tour Mar Environ 7(1):39–41. https://doi.org/10.3727/154427310X12826772784874

Ahonen H, Stow AJ, Harcourt RG, Charrier I (2014) Adult male Australian sea lion barking calls reveal clear geographical variations. Anim Behav 97:229–239. https://doi.org/10.1016/j.anbehav.2014.09.010

Ahonen H, Harcourt RG, Stow AJ, Charrier I (2018) Geographic vocal variation and perceptual discrimination abilities in male Australian sea lions. Anim Cogn 21(2):235–243. https://doi.org/10.1007/s10071-017-1158-7

Antonelis GA, York AE (1983) Identification of individual male northern fur seal, *Callorhinus ursinus* from their vocalizations. National Marine Mammal Laboratory, Northwest and Alaska Fisheries Center, National Marine Fisheries Service, NOAA, Seattle/Washington, DC

Asselin S, Hammill MO, Barrette C (1993) Underwater vocalizations of ice breeding grey seals. Can J Zool 71(11):2211–2219. https://doi.org/10.1139/z93-310

Attard MRG, Pitcher BJ, Charrier I, Ahonen H, Harcourt RG (2010) Vocal discrimination in mate guarding male Australian sea lions: familiarity breeds contempt. Ethology 116(8):704–712. https://doi.org/10.1111/j.1439-0310.2010.01786.x

Au WWL, Hastings M (2008) Principles of marine bioacoustics. Springer, New York, 679 pp. https://doi.org/10.1007/978-0-387-78365-9

Aubin T, Jouventin P, Charrier I (2015) Mother vocal recognition in Antarctic fur seal *Arctocephalus gazella* pups: a two-step process. PLoS One 10(9): e0134513. https://doi.org/10.1371/journal.pone.0134513

Awbrey FT, Thomas JA, Evans WE (2003) Ultrasonic underwater sounds from a captive leopard seal (*Hydrurga leptonyx*). In: Thomas JA, Moss CF, Vater M (eds) Echolocation in Bats and Dolphins. University of Chicago Press, Chicago, pp 535–541

Ballard KA, Kovacs KM (1995) The acoustic repertoire of hooded seals (*Cystophora cristata*). Can J Zool 73(7): 1362–1374. https://doi.org/10.1139/z95-159

Baranov EA, Granin NG, Kucher KM, Makarov MM (2014) On the possibility of acoustical long-range underwater communication in Baikal seal (*Pusa sibirica* Gm.) in the period of under-ice habitation. Paper presented at the marine mammals of the Holarctic; VIII international conference by the Marine Mammal Council (Russia), St. Petersburg, Russia, September 22–27, 2014

Barklow WE (1997) Some underwater sounds of the hippopotamus (*Hippopotamus amphibius*). Mar Freshw Behav Physiol 29(4):237–249. https://doi.org/10.1080/10236249709379008

Bartholomew GA, Collias NE (1962) The role of vocalization in the social behaviour of the northern elephant seal. Anim Behav 10(1):7–14. https://doi.org/10.1016/0003-3472(62)90124-0

Bartsh SS, Johnston SD, Siniff DB (1992) Territorial behavior and breeding frequency of male Weddell seals (*Leptonychotes weddellii*) in relation to age, size, and concentrations of serum testosterone and cortisol. Can J Zool 70(4):680–692. https://doi.org/10.1139/z92-102

Beier JC, Wartzok D (1979) Mating behaviour of captive spotted seals (*Phoca largha*). Anim Behav 27:772–781. https://doi.org/10.1016/0003-3472(79)90013-7

Berland B (1958) The Hood of the hooded seal, *Cystophora cristata* Erxl. Nature 182(4632):408–409. https://doi.org/10.1038/182408a0

Berta A, Sumich JL, Kovacs KM (2015) Marine mammals: evolutionary biology, 3rd edn. Academic, San Diego. https://doi.org/10.1016/C2011-0-07338-6

Berta A, Churchill M, Boessenecker RW (2022) The origin of phocid seals and evolution of key behavioral character traits. In: Costa DP, McHuron EA (eds) Ethology and behavioral ecology of phocids. Springer, Cham, pp 3–30. https://doi.org/10.1007/978-3-030-88923-4_1

Bishop AM, Lidstone-Scott R, Pomeroy P, Twiss SD (2014) Body slap: an innovative aggressive display by breeding male gray seals (*Halichoerus grypus*). Mar Mamm Sci 30(2):579–593. https://doi.org/10.1111/mms.12059

Bishop AM, Denton P, Pomeroy P, Twiss S (2015) Good vibrations by the beach boys: magnitude of substrate vibrations is a reliable indicator of male grey seal size. Anim Behav 100:74–82. https://doi.org/10.1016/j.anbehav.2014.11.008

Bjorgesaeter A, Ugland KI, Bjorge A (2004) Geographic variation and acoustic structure of the underwater vocalization of harbour seal (*Phoca vitulina*) in Norway, Sweden and Scotland. J Acoust Soc Am 116(4):2459–2468. https://doi.org/10.1121/1.1782933

Blackwell SB, Greene CR, Richardson WJ (2004) Drilling and operational sounds from an oil production Island in the ice-covered Beaufort Sea. J Acoust Soc Am 116(5): 3199–3211. https://doi.org/10.1121/1.1806147

Boness DJ (1990) Fostering behavior in Hawaiian monk seals: is there a reproductive cost? Behav Ecol Sociobiol 27(2):113–122. https://doi.org/10.1007/BF00168454

Born EW, Wiig Ø (2021) Ecology and behavior of Atlantic walruses. In: Keighley X, Tange Olsen M, Jordan P, Desjardins J (eds) . Academic, The Atlantic Walrus, pp 39–76. https://doi.org/10.1016/B978-0-12-817430-2.00001-7

Boye TK, Simon MJ, Laidre KL, Rigét F, Stafford KM (2020) Seasonal detections of bearded seal (*Erignathus barbatus*) vocalizations in Baffin Bay and Davis Strait in relation to sea ice concentration. Polar Biol 43: 1493–1502. https://doi.org/10.1007/s00300-020-02723-1

Brady B, Hedwig D, Trygonis V, Gerstein E (2020) Classification of Florida manatee (*Trichechus manatus latirostris*) vocalizations. J Acoust Soc Am 147(3): 1597–1606. https://doi.org/10.1121/10.0000849

Brauer RW, Jennings R, Poulter T (1966) The effect of substituting helium and oxygen for air on the vocalization of the California sea lion, *Zalophus californianus*. In: Proceedings of the Third Annual Conference on Biological Sonar and Diving Mammals, pp 68–73

Burgess WC, Tyack PL, Le Boeuf BJ, Costa DP (1998) A programmable acoustic recording tag and first results from free-ranging northern elephant seals. Deep-Sea Res II 45(7):1327–1351. https://doi.org/10.1016/S0967-0645(98)00032-0

Burns JJ (1981) Bearded seals *Erignathus barbatus* Erxleben, 1777. In: Ridgeway SH, Harrison RJ (eds) Handbook of marine mammals, vol 2. Academic, London, pp 145–170

Campagna C, Harcourt R (2021) Ethology and behavioral ecology of otariids and the odobenid. Springer, Cham, 672 pp. https://doi.org/10.1007/978-3-030-59184-7

Campbell GS, Gisner R, Helweg DA (2000) Acoustic identification of female Steller sea lions. J Acoust Soc Am 108(5):2541. https://doi.org/10.1121/1.4743416

Campbell GS, Gisner RC, Helweg DA, Milette LL (2002) Acoustic identification of female Steller Sea lions (*Eumetopias jubatus*). J Acoust Soc Am 111(6): 2920–2928. https://doi.org/10.1121/1.1474443

Carrick R, Csordas SE, Ingham SE (1962) Studies on the southern elephant seal, *Mirounga leonina* (L.). IV. Breeding and development. CSIRO Wildl Res 7 (2):161–197. https://doi.org/10.1071/CWR9620161

Casey C (2018) The vocal behavior of the male northern elephant seal. University of California, Santa Cruz, 171 pp

Casey C, Charrier I, Mathevon N, Reichmuth C (2015) Rival assessment among northern elephant seals: evidence of associative learning during male–male contests. R Soc Open Sci 2(8). https://doi.org/10.1098/rsos.150228

Casey C, Sills J, Reichmuth C (2016) Source level measurements for harbor seals and implications for estimating communication space. In: The effects of noise on aquatic life, Dublin, Ireland, 10–16 July 2016, vol 1. Proceedings of Meetings on Acoustics, p 010034. https://doi.org/10.1121/2.0000353

Casey C, Reichmuth C, Costa DP, Le Boeuf B (2018) The rise and fall of dialects in northern elephant seals. Proc Royal Soc B 285(1892):2176. https://doi.org/10.1098/rspb.2018.2176

Casey C, Charrier I, Mathevon N, Nasr C, Forman P, Reichmuth C (2020) The genesis of giants: behavioural ontogeny of male northern elephant seals. Anim Behav 166:247–259. https://doi.org/10.1016/j.anbehav.2020.06.014

Casey C, Sills JM, Knaub S, Sotolotto K, Reichmuth C (2021) Lifelong patterns of sound production in two seals. Aquat Mamm 47(5):499–514. https://doi.org/10.1578/AM.47.5.2021.499

Caudron AK, Kondakov AA, Siryanov SV (1998) Acoustic structure and individual variation of grey seal (*Halichoerus grypus*) pup calls. J Mar Biol Assoc 78(2):651–658. https://doi.org/10.1017/S0025315400041680

Charlton BD (2015) The acoustic structure and information content of female koala vocal signals. PLoS One 10(10):e0138670. https://doi.org/10.1371/journal.pone.0138670

Charrier I (2020) Mother–offspring vocal recognition and social system in pinnipeds. In: Aubin T, Mathevon N (eds) Coding strategies in vertebrate acoustic communication, vol 7. Springer, pp 231–246. https://doi.org/10.1007/978-3-030-39200-0_9

Charrier I (2021) Vocal communication in otariids and odobenids. In: Campagna C, Harcourt R (eds) Ethology and behavioral ecology of otariids and the odobenid. Springer, Cham, pp 265–289. https://doi.org/10.1007/978-3-030-59184-7_14

Charrier I, Casey C (2022) Social communication in phocids. In: Costa DP, McHuron EA (eds) Ethology and behavioral ecology of marine mammals. Springer, Cham, pp 69–100. https://doi.org/10.1007/978-3-030-88923-4_3

Charrier I, Harcourt RG (2006) Individual vocal identity in mother and pup Australian sea lions (*Neophoca cinerea*). J Mammal 87(5):929–938. https://doi.org/10.1644/05-MAMM-A-344R3.1

Charrier I, Mathevon N, Jouventin P (2001) Mother's voice recognition by seal pups. Nature 412(6850): 873. https://doi.org/10.1038/35091136

Charrier I, Mathevon N, Jouventin P (2002) How does a fur seal mother recognize the voice of her pup? An experimental study of *Arctocephalus tropicalis*. J Exp Biol 205(5):603–612. https://doi.org/10.1242/jeb.205.5.603

Charrier I, Mathevon N, Jouventin P (2003a) Fur seal mothers memorize subsequent versions of developing pups' calls: adaptation to long-term recognition or evolutionary by-product? Biol J Linn Soc 80(2): 305–312. https://doi.org/10.1046/j.1095-8312.2003.00239.x

Charrier I, Mathevon N, Jouventin P (2003b) Individuality in the voice of fur seal females: an analysis study of the pup attraction call in *Arctocephalus tropicalis*. Mar Mamm Sci 19(1):161–172. https://doi.org/10.1111/j.1748-7692.2003.tb01099.x

Charrier I, Mathevon N, Jouventin P (2003c) Vocal signature recognition of mothers by fur seal pups. Anim Behav 65(3):543–550. https://doi.org/10.1006/anbe.2003.2073

Charrier I, Pitcher BJ, Harcourt RG (2009) Vocal recognition of mothers by Australian sea lion pups: individual

signature and environmental constraints. Anim Behav 78(5):1127–1134. https://doi.org/10.1016/j.anbehav.2009.07.032
Charrier I, Aubin T, Mathevon N (2010) Mother-Calf vocal communication in Atlantic walrus: a first field experimental study. Anim Cogn 13(3):471–482. https://doi.org/10.1007/s10071-009-0298-9
Charrier I, Ahonen H, Harcourt RG (2011a) What makes an Australian sea lion (*Neophoca cinerea*) male's bark threatening? J Comp Psychol 125(4):385–392. https://doi.org/10.1037/a0024513
Charrier I, Brulet A, Aubin T (2011b) Social vocal communication in captive Pacific walruses *Odobenus rosmarus divergens*. Mamm Biol 76(5):622–627. https://doi.org/10.1016/j.mambio.2010.10.006
Charrier I, Mathevon N, Aubin T (2013) Bearded seal males perceive geographic variation in their trills. Behav Ecol Sociobiol 67(10):1679–1689. https://doi.org/10.1007/s00265-013-1578-6
Charrier I, Marchesseau S, Dendrinos P, Tounta E, Karamanlidis AA (2017) Individual signatures in the vocal repertoire of the endangered Mediterranean monk seal: new perspectives for population monitoring. Endang Species Res 32:459–470. https://doi.org/10.3354/esr00829
Charrier I, Pitcher BJ, Harcourt RG (2022) Mother–pup recognition mechanisms in Australia Sea lion (*Neophoca cinerea*) using uni- and multi-modal approaches. Anim Cogn 25:1019–1028. https://doi.org/10.1007/s10071-022-01641-5
Charrier I, Huetz C, Prevost L, Dendrinos P, Karamanlidis AA (2023) First description of the underwater sounds in the Mediterranean monk seal *Monachus monachus* in Greece: towards establishing a vocal repertoire. Animals 13(6):1048. https://doi.org/10.3390/ani13061048
Chevallay M, Guinet C, Jeanniard du Dot T (2024) Underwater vocalizations in foraging female Antarctic fur seals (*Arctocephalus gazella*) in the Kerguelen Islands. Mar Mamm Sci e13118. https://doi.org/10.1111/mms.13118
Chou E, Antunes R, Sardelis S, Stafford KM, West L, Spagnoli C, Southall BL, Robards M, Rosenbaum HC (2020) Seasonal variation in Arctic marine mammal acoustic detection in the northern Bering Sea. Mar Mamm Sci 36(2):522–547. https://doi.org/10.1111/mms.12658
Christenson TE, Le Boeuf BJ (1978) Aggression in the female northern elephant seal, *Mirounga angustirostris*. Behaviour 64(1–2):158–171. https://doi.org/10.1163/156853978X00495
Churchill M, Clementz MT (2016) The evolution of aquatic feeding in seals: insights from Enaliarctos (Carnivora: Pinnipedimorpha), the oldest known seal. J Evol Biol 29(2):319–334. https://doi.org/10.1111/jeb.12783
Cleator HJ, Stirling I, Smith TG (1989) Underwater vocalizations of the bearded seal (*Erignathus barbatus*). Can J Zool 67(8):1900–1910. https://doi.org/10.1139/z89-272
Collins KT, Terhune JM (2007) Geographic variation of Weddell seal (*Leptonychotes weddellii*) airborne mother-pup vocalisations. Polar Biol 30(11):1373–1380. https://doi.org/10.1007/s00300-007-0297-8
Collins KT, Rogers TL, Terhune JM, McGreevy PD, Wheatley KE, Harcourt RG (2005) Individual variation of in-air female 'pup contact' calls in Weddell seals, *Leptonychotes weddellii*. Behaviour 142(2):167–189. https://doi.org/10.1163/1568539053627668
Collins KT, Terhune JM, Rogers TL, Wheatley KE, Harcourt RG (2006) Vocal individuality of in-air Weddell seal (*Leptonychotes weddellii*) pup "primary" calls. Mar Mamm Sci 22(4):933–951. https://doi.org/10.1111/j.1748-7692.2006.00074.x
Collins KT, McGreevy PD, Wheatley KE, Harcourt RG (2011) The influence of behavioural context on Weddell seal (*Leptonychotes weddellii*) airborne mother-pup vocalisation. Behav Process 87(3):286–290. https://doi.org/10.1016/j.beproc.2011.06.005
Committee on Taxonomy (2023) List of marine mammal species and subspecies. Society for Marine Mammalogy. Accessed 26/03/2024
Costa DP, McHuron EA (2022) Ethology and behavioral ecology of phocids. Springer, Cham. https://doi.org/10.1007/978-3-030-88923-4
Cox CR (1981) Agonistic encounters among male elephant seals: frequency, context, and the role of female preference. Am Zool 21(1):197–209. https://doi.org/10.1093/icb/21.1.197
Crawley MC, Wilson GJ (1976) The natural history and behaviour of the New Zealand fur seal (*Arctocephalus forsteri*). Tuatara J Biol Soc 22(1):1–28
Cummings WC, Holliday DV, Lee BJ (1984) Potential impacts of man-made noise on ringed seals: vocalizations and reactions (NTIS PB87-107546). National Oceanic and Atmospheric Administration, Anchorage, pp 95–230
Cziko PA, Munger LM, Santos NR, Terhune JM (2020) Weddell seals produce ultrasonic vocalizations. J Acoust Soc Am 148(6):3784–3796. https://doi.org/10.1121/10.0002867
Davies CE, Kovacs KM, Lydersen C, Van Parijs SM (2006) Development of display behavior in young captive bearded seals. Mar Mamm Sci 22(4):952–965. https://doi.org/10.1111/j.1748-7692.2006.00075.x
de Heer Kloots M, Carlson D, Garcia M, Kotz SA, Lowry A, Poli-Nardi L, de Reus K, Rubio-Garcia A, Sroka M, Varola M, Ravignani A (2020) Rhythmic perception, production and interactivity in harbour and grey seals. In: EvoLang XIII, Brussels, 14 April 2020–17 April 2020. EvoLang, pp 59–62. https://doi.org/10.5167/uzh-192382
de Reus K, Carlson D, Lowry A, Gross S, Garcia M, Rubio-Garcia A, Salazar-Casals A, Ravignani A

(2022) Vocal tract allometry in a mammalian vocal learner. J Exp Biol 225(8). https://doi.org/10.1242/jeb.243766

de Vincenzi G, Parisi I, Torri M, Papale E, Mazzola S, Nuth C, Buscaino G (2019) Influence of environmental parameters on the use and spatiotemporal distribution of the vocalizations of bearded seals (*Erignathus barbatus*) in Kongsfjorden, Spitsbergen. Polar Biol. https://doi.org/10.1007/s00300-019-02514-3

De Vreese S, van der Schaar M, Weissenberger J, Erbs F, Kosecka M, Solé M, André M (2018) Marine mammal acoustic detections in the Greenland and Barents Sea, 2013–2014 seasons. Sci Rep 8(16882):1–14. https://doi.org/10.1038/s41598-018-34624-z

Dearborn JH (1962) An unusual occurrence of the leopard seal at McMurdo Sound, Antarctica. J Mammal 43(2):273–275. https://doi.org/10.2307/1377114

Denes SL (2014) Ocean environmental effects on walrus communication. PhD thesis, Pennsylvania State University, University Park, 172 pp

Denes SL, Miksis-Olds JL, Mellinger D, Otjen E, Bowles AE (2015) Comparison of managed care and wild walrus source characteristics. J Acoust Soc Am 137(4):2196. https://doi.org/10.1121/1.4919985

Dobson FS, Jouventin P (2003) How mothers find their pups in a colony of Antarctic fur seals. Behav Process 61(1–2):77–85. https://doi.org/10.1016/S0376-6357(02)00164-X

Doiron E, Rouget P, Terhune J (2012) Proportional underwater call type usage by Weddell seals (*Leptonychotes weddellii*) in breeding and nonbreeding situations. Can J Zool 90(2):237–247. https://doi.org/10.1139/z11-131

Duengen D, Fitch WT, Ravignani A (2023) Hoover the talking seal. Curr Biol 33(2):R50–R52. https://doi.org/10.1016/j.cub.2022.12.023

Dziak RP, Hong J, Kang S-G, Lau T-K, Haxel JH, Matsumoto H (2017) The Balleny Island hydrophone array: hydro-acoustic records of sea-ice dynamics, seafloor volcano-tectonic activity, and marine mammal vocalizations off Antarctica. In: OCEANS 2017-Aberdeen. IEEE, pp 1–8

Eliason JJ, Johanos TC, Webber MA (1990) Parturition in the Hawaiian monk seal (*Monachus schauinslandi*). Mar Mamm Sci 6(2):146–151

Evans WE, Thomas JA, Davis RW (2004) Vocalizations from Weddell seals (*Leptonychotes weddellii*) during diving and foraging. In: Thomas JA, Moss CF, Vatek M (eds) Echolocation in bats and dolphins. The University of Chicago Press, Chicago, pp 541–547

Fay FH (1982) Ecology and biology of the Pacific walrus, *Odobenus rosmarus divergens* Illiger. Fish Wildlife Serv Bul North American Fauna 74:1–279. https://doi.org/10.3996/nafa.74.0001

Fay FH, Ray CG, Kibal'chich AA (1984) Time and location of mating and associated behavior of the Pacific walrus, *Odobenus rosmarus divergens* Illiger. Soviet-Am Cooperat Res Marine Mamm 1:89–99

Fernández-Juricic E, Campagna C, Enriquez V, Ortiz CL (1999) Vocal communication and individual variation in breeding South American sea lions. Behaviour 136(4):495–517. https://doi.org/10.1163/156853999501441

Fernández-Juricic E, Campagna C, Mauro DS (2003) Variations in the arrangement of South American sea lion (*Otaria flavescens*) male vocalizations during the breeding season: patterns and contexts. Aquat Mamm 29(2):289–296. https://doi.org/10.1578/016754203101024112

Fischer J, Wadewitz P, Hammerschmidt K (2017) Structural variability and communicative complexity in acoustic communication. Anim Behav 134:229–237. https://doi.org/10.1016/j.anbehav.2016.06.012

Fitch WT (2006) The biology and evolution of music: a comparative perspective. Cognition 100(1):173–215. https://doi.org/10.1016/j.cognition.2005.11.009

Fitch WT, Neubauer J, Herzel H (2002) Calls out of chaos: the adaptive significance of nonlinear phenomena in mammalian vocal production. Anim Behav 63(3):407–418. https://doi.org/10.1006/anbe.2001.1912

Fletcher S, LeBoeuf BJ, Costa DP, Tyack PL, Blackwell SB (1996) Onboard acoustic recording from diving northern elephant seals. J Acoust Soc Am 100(4):2531–2539. https://doi.org/10.1121/1.417361

Fournet MEH, Silvestri M, Clark CW, Klinck H, Rice AN (2021) Limited vocal compensation for elevated ambient noise in bearded seals: implications for an industrializing Arctic Ocean. Proc R Soc B 288:20202712. https://doi.org/10.1098/rspb.2020.2712

Frankel AS (2009) Sound production. In: Perrin WF, Würsig B, Thewissen JGM (eds) Encyclopedia of marine mammals, 2nd edn. Academic, London, pp 1056–1071. https://doi.org/10.1016/B978-0-12-373553-9.00242-X

Frouin-Mouy HC, Hammill MO (2021) In-air and underwater sounds of hooded seals during the breeding season in the Gulf of St. Lawrence. J Acoust Soc Am 150(1):281–293. https://doi.org/10.1121/10.0005478

Frouin-Mouy H, Mouy X, Martin B, Hannay D (2016) Underwater acoustic behavior of bearded seals (*Erignathus barbatus*) in the northeastern Chukchi Sea, 2007–2010. Mar Mamm Sci 32(1):141–160. https://doi.org/10.1111/mms.12246

Frouin-Mouy H, Mouy X, Berchok CL, Blackwell SB, Stafford KM (2019) Acoustic occurrence and behavior of ribbon seals (*Histriophoca fasciata*) in the Bering, Chukchi, and Beaufort seas. Polar Biol 42(4):657–674. https://doi.org/10.1007/s00300-019-02462-y

Gailey-Phipps JJ (1984) Acoustic communication and behavior of the spotted seal (*Phoca largha*). Johns Hopkins University, Baltimore

Gedamke J, Robinson SM (2010) Acoustic survey for marine mammal occurrence and distribution off East Antarctica (30–80°E) in January-February 2006. Deep Sea Res Part II 57(9):968–981. https://doi.org/10.1016/j.dsr2.2008.10.042

Gentry RL (2014) Behavior and ecology of the northern fur seal, vol 382. Princeton University Press

Glotin H, Poupard M, Marxer R, Ferrari M, Ricard J, Roger V, Patris J, Malige F, Giraudet P, Prevot J-M, Komarov M (2017) Big data passive acoustic for Baikal Lake (Озеро Байкал) soundscape & ecosystem observatory. In: Kornilov VV (ed) Russian-French workshop in big data and applications, Moscow. Higher School of Economic Publishing House, pp 22–43

Goldsworthy SD, Boness DJ, Fleischer RC (1999) Mate choice among sympatric fur seals: female preference for conphenotypic males. Behav Ecol Sociobiol 45: 253–267. https://doi.org/10.1007/s002650050560

Grigoriev VA, Kucher KM, Lunkov AA, Makarov MM, Petnikov VG (2020) Acoustic parameters of the bottom in Lake Baikal. Acoust Phys 66(5):508–516. https://doi.org/10.1134/S1063771020050048

Gwilliam J, Charrier I, Harcourt RG (2008) Vocal identity and species recognition in male Australian sea lions, *Neophoca cinerea*. J Exp Biol 211(14):2288–2295. https://doi.org/10.1242/jeb.013185

Halliday WD, Insley SJ, de Jong T, Mouy X (2018) Seasonal patterns in acoustic detections of marine mammals near Sachs Harbour, Northwest Territories. Arctic Science 4:259–278. https://doi.org/10.1139/as-2017-0021

Halliday WD, Pine MK, Insley SJ, Soares RN, Kortsalo P, Mouy X (2019) Acoustic detections of Arctic marine mammals near Ulukhaktok, Northwest Territories, Canada. Can J Zool 97:72–80. https://doi.org/10.1139/cjz-2018-0077

Halliday WD, Pine MK, Insley SJ (2020a) Underwater noise and Arctic marine mammals: review and policy recommendations. Environ Rev 28(4):438–448. https://doi.org/10.1139/er-2019-0033

Halliday WD, Pine MK, Mouy X, Kortsalo P, Hilliard RC, Insley SJ (2020b) The coastal Arctic marine soundscape near Ulukhaktok, Northwest Territories, Canada. Polar Biol 43:623–636. https://doi.org/10.1007/s00300-020-02665-8

Hanggi EB (1992) The importance of vocal cues in mother-pup recognition in California sea lion. Mar Mamm Sci 8(4):430–432. https://doi.org/10.1111/j.1748-7692.1992.tb00061.x

Hanggi EB, Schusterman RJ (1992) Underwater acoustic displays by male harbour seals (*Phoca vitulina*): initial results. In: Thomas JA, Kastelein RA, Supin AY (eds) Marine mammal sensory systems. Plenum Press, New York, pp 449–457. https://doi.org/10.1007/978-1-4615-3406-8_31

Hanggi EB, Schusterman RJ (1994) Underwater acoustic displays and individual variation in male harbour seals, *Phoca vitulina*. Anim Behav 48(6):1275–1283. https://doi.org/10.1006/anbe.1994.1363

Hayes SA, Kumar A, Costa DP, Mellinger DK, Harvey JT, Southall BL, Le Boeuf BJ (2004) Evaluating the function of the male harbour seal, *Phoca vitulina*, roar through playback experiments. Anim Behav 67: 1133–1139. https://doi.org/10.1016/j.anbehav.2003.06.019

Heimrich AF, Halliday WD, Frouin-Mouy H, Pine MK, Juanes F, Insley SJ (2021) Vocalizations of bearded seals (*Erignathus barbatus*) and their influence on the soundscape of the western Canadian Arctic. Mar Mamm Sci 37:173–192. https://doi.org/10.1111/mms.12732

Hewer HR, Backhouse KM(1960) A preliminary account of a colony of grey seals *Halichoerus grypus* (Fab.) in the southern Inner Hebrides. In: Proceedings of the Zoological Society of London, vol 2. Wiley Online Library, pp 157–195

Higgins LV (1984) Maternal behavior and attendance patterns of the Stellar sea lion in California. MSc thesis, University of California, Santa Cruz

Hocking DP, Burville B, Parker WM, Evans AR, Park T, Marx FG (2020) Percussive underwater signaling in wild gray seals. Mar Mamm Sci 36(2):728–732. https://doi.org/10.1111/mms.12666

Holt MM, Southall BL, Insley SJ, Schusterman RJ (2010) Call directionality and its behavioural significance in male northern elephant seals, *Mirounga angustirostris*. Anim Behav 80(3):351–361. https://doi.org/10.1016/j.anbehav.2010.06.013

Hughes WR, Reichmuth C, Mulsow JL, Næsbye Larsen O (2011) Source characteristics of the underwater knocking displays of a male Pacific walrus (*Odobenus rosmarus divergens*). J Acoust Soc Am 129(4):2506. https://doi.org/10.1121/1.3588276

Hyvärinen H (1989) Diving in darkness: whiskers as sense organs of the ringed seal (*Phoca hispida saimensis*). J Zool 218(4):663–678. https://doi.org/10.1111/j.1469-7998.1989.tb05008.x

Insley SJ (1992) Mother-offspring separation and acoustic stereotypy: a comparison of call morphology in two species of pinnipeds. Behaviour 120 (1–2):103–122. https://doi.org/10.1163/156853992X00237

Insley SJ (2000) Long-term vocal recognition in the northern fur seal. Nature 406(6794):404–405. https://doi.org/10.1038/35019064

Insley SJ (2001) Mother–offspring vocal recognition in northern fur seals is mutual but asymmetrical. Anim Behav 61(1):129–137. https://doi.org/10.1006/anbe.2000.1569

Insley S, Holt M (2012) Do male northern elephant seals recognize individuals or merely relative dominance rank? J Acoust Soc Am 131 (1):EL35–EL41. https://doi.org/10.1121/1.3665259

Insley SJ, Phillips AV, Charrier I (2003) A review of social recognition in pinnipeds. Aquat Mamm 29(2): 181–201

Ivanov TM (1938) Baikal seal, its biology and harvest. Izv Biol-Geogr Nauchno-Issled Inst Vost-Sib Gos Univ, Irkutsk 8(1–2):5–119

Janik VM, Knörnschild M (2021) Vocal production learning in mammals revisited. Philos Trans R Soc B 376(1836):20200244. https://doi.org/10.1098/rstb.2020.0244

Janik VM, Slater PJB (1997) Vocal learning in mammals. Adv Study Behav 26:59–99

Jenkins W, Johnson H, Vakhutinsky S, Helmberger MN, Storheim E, Sagen H, Sandven S (2022) Analysis of underwater acoustic data collected under sea ice during the Useful Arctic Knowledge 2021 cruise. In: International Conference on Underwater Acoustics; Underwater Acoustics: Polar Acoustics, Southampton, UK, 20–23 June 2022. vol 070003. Proceedings of Meetings on Acoustics, pp 1–14. https://doi.org/10.1121/2.0001574

Jézéquel Y, Mathias D, Olivier F, Amice E, Chauvaud S, Jolivet A, Bonnel J, Sejr MK, Chauvaud L (2022) Passive acoustics suggest two different feeding mechanisms in the Atlantic walrus (*Odobenus rosmarus rosmarus*). Polar Biol 45:1157–1162. https://doi.org/10.1007/s00300-022-03055-y

Job DA, Boness DJ, Francis JM (1995) Individual variation in nursing vocalizations of Hawaiian monk seal pups, *Monachus schauinslandi* (Phocidae, Pinnipedia), and lack of maternal recognition. Can J Zool 73(5): 975–983. https://doi.org/10.1139/z95-114

Jones JM, Thayre BJ, Roth EH, Mahoney M, Sia I, Merculief K, Jackson C, Zeller C, Clare M, Bacon A, Weaver S, Gentes Z, Small RJ, Stirling I, Wiggins SW, Hildebrand JA (2014) Ringed, bearded, and ribbon seal vocalizations north of Barrow, Alaska: seasonal presence and relationship with sea ice. Arctic 67(2): 203–222. https://doi.org/10.14430/arctic4388

Kastelein RA, Postma J, Verboom WC (1995) Airborne vocalizations of Pacific walrus pups (*Odobenus rosmarus divergens*). In: Kastelein RA, Thomas JA, Nachtigall PE (eds), Sensory systems of aquatic mammals. De Spil Publishers, Woerden, The Netherlands, pp. 265–285

Kastelein R, Dubbeldam J, de Bakker M (1997) The anatomy of the walrus head (*Odobenus rosmarus*). Part 5: the tongue and its function in walrus ecology. Aquat Mamm 23 (1):29–47

Kenyon KW (1960) Territorial behavior and homing in the Alaska fur seal. Mammalia 24:431–444

Kenyon KW, Rice DW (1959) Life history of the Hawaiian monk seal. Pac Sci 13(3):215–252

Khan CB, Markowitz H, McCowan B (2006) Vocal development in captive harbor seal pups, *Phoca vitulina richardii*: age, sex, and individual differences. J Acoust Soc Am 120(3):1684–1694. https://doi.org/10.1121/1.2226530

Kindermann L, Boebel O, Bornemann H, Burkhardt E, Klinck H, Opzeeland IC, Plötz J, Seibert AM (2008) A perennial acoustic observatory in the Antarctic Ocean. In: Frommolt K-H, Bardeli R, Clausen M (eds) Computational bioacoustics for assessing biodiversity: proceedings of the international expert meeting on IT-based detection of bioacoustical patterns, Bundesamt Naturschutz Skr, pp 15–28

Klinck H (2008) Automated passive acoustic detection, localization and identification of leopard seals: from hydro-acoustic technology to leopard seal ecology. University of Trier

Klinck H, Mellinger DK, Klinck K, Hager J, Kindermann L, Boebel O (2010) Long-range underwater vocalizations of the crabeater seal (*Lobodon carcinophaga*). J Acoust Soc Am 128(1):474–479. https://doi.org/10.1121/1.3442362

Klopfer PH, Gilbert BK (1966) A note on retrieval and recognition of young in the elephant seal, *Mirounga angustirostris*. Zeitschrift für Tierpsychologie 23(6): 757–760. https://doi.org/10.1111/j.1439-0310.1966.tb01708.x

Kocsis K, Duengen D, Jadoul Y, Ravignani A (2024) Harbour seals use rhythmic percussive signalling in interaction and display. Anim Behav 207:223–234. https://doi.org/10.1016/j.anbehav.2023.09.014

Kovacs KM (1995) Mother–pup reunions in harp seals, *Phoca groenlandica*: cues for the relocation of pups. Can J Zool 73(5):843–849. https://doi.org/10.1139/z95-099

Kreiss C, Klinck H, Kindermann L, Rogers T, Bornemann H, van Opzeeland I, Boebel O (2008) PALAOA: a study on the leopard seal (*Hydrurga leptonyx*) acoustic repertoire and its geographic variation. AWI

Kreiss CM, Boebel O, Bornemann H, Kindermann L, Klinck H, Klinck K, Plötz J, Rogers TL, van Opzeeland I (2013) Call characteristics of high-double trill leopard seal (*Hydrurga leptonyx*) vocalizations from three Antarctic locations. Polarforschung 83(2): 63–71. https://doi.org/10.2312/polarforschung.83.2.63

Kunc HP, Wolf JBW (2008) Seasonal changes of vocal rates and their relation to territorial status in male Galápagos sea lions (*Zalophus wollebaeki*). Ethology 114(4):381–388. https://doi.org/10.1111/j.1439-0310.2008.01484.x

Kunnasranta M, Hyvarinen H, Sorjonen J (1996) Underwater vocalizations of Ladoga ringed seals (*Phoca hispida ladogensis* Nordq.) in summertime. Mar Mamm Sci 12 (4):611–618. https://doi.org/10.1111/j.1748-7692.1996.tb00076.x

Larsen ON, Reichmuth C (2021) Walruses produce intense impulse sounds by clap-induced cavitation during breeding displays. R Soc Open Sci 8(6):210197. https://doi.org/10.1098/rsos.210197

Laws RM (1956) The elephant seal (*Mirounga leonina*, Linn.) II. General, social and reproductive behaviour. Falkland Islands Dependecies Survey Scientific Reports 13:1–88

Le Boeuf BJ (1972) Sexual behavior in the northern elephant seal *Mirounga angustirostris*. Behaviour 41(1–2):1–26. https://doi.org/10.1163/156853972X00167

Le Boeuf BJ, Peterson RS (1969) Dialects in elephant seals. Science 166(3913):1654–1656. https://doi.org/10.1126/science.166.3913.1654

Le Boeuf BJ, Petrinovich LF (1974) Dialects of northern elephant seals, *Mirounga angustirostris*: origin and reliability. Anim Behav 22(3):656–663. https://doi.org/10.1016/S0003-3472(74)80013-8

Le Boeuf BJ, Whiting RJ, Gantt RF (1972) Perinatal behavior of northern elephant seal females and their young. Behaviour 43(1):121–156. https://doi.org/10.1163/156853973x00508

Linossier J, Casey C, Charrier I, Mathevon N, Reichmuth C (2021) Maternal responses to pup calls in a high-cost lactation species. Biol Lett 17(12):20210469. https://doi.org/10.1098/rsbl.2021.0469

Lisitsyna TY (1973) Povedenie i zvukovaya signalizatsiya severnogo morskogo kotika (*Callorhinus ursinus*) na lezhbishchakh (The behavior and acoustical signals of the northern fur seal, *Callorhinus ursinus*, on the rookery). Zoologicheskii Zhurnal 52(8):1220–1228

Llobet SM, Ahonen H, Lydersen C, Berge J, Ims R, Kovacs KM (2021) Bearded seal (*Erignathus barbatus*) vocalizations across seasons and habitat types in Svalbard, Norway. Polar Biol 44:1273–1278. https://doi.org/10.1007/s00300-021-02874-9

Lomac-MacNair KS, Smultea MA, Yack T, Lammers M, Norris T, Green G, Dunleavey K, Steckler D, James V (2019) Marine mammal visual and acoustic surveys near the Alaskan Colville River Delta. Polar Biol 42:441–448. https://doi.org/10.1007/s00300-018-2434-y

Lombard É (1911) Le signe de l'élévation de la voix. Annales des Maladies de L'Oreille et du Larynx XXXVII 2:101–109

Looby A, Erbe C, Bravo S, Cox K, Davies HL, Di Iorio L, Jézéquel Y, Juanes F, Martin CW, Mooney TA, Radford C, Reynolds LK, Rice AN, Riera A, Rountree R, Spriel B, Stanley J, Vela S, Parsons MJG (2023) Global inventory of species categorized by known underwater sonifery. Scientif Data 10(1):892. https://doi.org/10.1038/s41597-023-02745-4

MacIntyre KQ, Stafford KM, Berchok CL, Boveng PL (2013) Year-round acoustic detection of bearded seals (*Erignathus barbatus*) in the Beaufort Sea relative to changing environmental conditions, 2008–2010. Polar Biol 36(8):1161–1173. https://doi.org/10.1007/s00300-013-1337-1

MacIntyre K, Stafford KM, Conn PB, Laidre KL, Boveng PL (2015) The relationship between sea ice concentration and the spatio-temporal distribution of vocalizing bearded seals (*Erignathus barbatus*) in the Bering, Chukchi, and Beaufort Seas from 2008 to 2011. Prog Oceanogr 136:241–249. https://doi.org/10.1016/j.pocean.2015.05.008

MacRae AM (2018) Facial expression, vocalizations and eye temperature as potential indicators of pain in harbour seals (*Phoca vitulina*). University of British Columbia, Vancouver

Madan MM, Latha G, Ashokan M, Raguraman G, Thirunavukkarasu A (2020) Identification of soundscape components and their temporal patterns in Kongsfjorden, Svalbard Archipelago. Polar Biol 26:100604. https://doi.org/10.1016/j.polar.2020.100604

Marlow BJ (1975) The comparative behaviour of the Australasian sea lions *Neophoca cinerea* and *Phocarctos hookeri* (Pinnipedia: Otariidae). Mammalia 39(2):159–230. https://doi.org/10.1515/mamm.1975.39.2.159

Martin K, Tucker MA, Rogers TL (2017) Does size matter? Examining the drivers of mammalian vocalizations. Evolution 71(2):249–260. https://doi.org/10.1111/evo.13128

Martin M, Gridley T, Elwen SH, Charrier I (2021a) Extreme ecological constraints lead to high degree of individual stereotypy in the vocal repertoire of the Cape fur seal (*Arctocephalus pusillus pusillus*). Behav Ecol Sociobiol 75(104):1–16. https://doi.org/10.1007/s00265-021-03043-y

Martin M, Gridley T, Harvey Elwen S, Charrier I (2021b) Vocal repertoire, microgeographical variation and within-species acoustic partitioning in a highly colonial pinniped, the Cape fur seal. R Soc Open Sci 8(10):202241. https://doi.org/10.1098/rsos.202241

Martin M, Gridley T, Elwen S, Charrier I (2022a) Early onset of postnatal individual vocal recognition in a highly colonial mammal species. Proc R Soc B 289(1988):20221769. https://doi.org/10.1098/rspb.2022.1769

Martin M, Gridley T, Elwen SH, Charrier I (2022b) Feel the beat: cape fur seal males encode their arousal state in their bark rate. Sci Nat 109:5. https://doi.org/10.1007/s00114-021-01778-2

Martin M, Gridley T, Elwen SH, Charrier I (2023a) Good fences make good neighbours: territorial male Cape fur seals use spatial acoustic map of neighbours. Behaviour 160(6):499–514. https://doi.org/10.1163/1568539X-bja10218

Martin M, Stow J, Gridley T, Elwen S, Charrier I (2023b) Geographical variation in Cape fur seals' in-air vocalizations across Southern Africa (Namibia and South Africa). Mar Mamm Sci 40:1–16. https://doi.org/10.1111/mms.13084

Martínková N, Zahradníková A, Budeev JA, Vrsanský P (2001) Surface home ranges of the Baikal seal (*Phoca sibirica*). Biologia Bratislava 56(2):219–224

Mathevon N, Casey C, Reichmuth C, Charrier I (2017) Northern elephant seals memorize the rhythm and timbre of their rivals' voices. Curr Biol 27(15):2352–2356. https://doi.org/10.1016/j.cub.2017.06.035

Matthews LH (1929) The natural history of the elephant seal, with notes on other seals found at South Georgia. Discovery Rep 1:235–255

Matthews L (2017) Harbor seal (*Phoca vitulina*) reproductive advertisement behavior and the effects of vessel noise. Syracuse University

Matthews LP, Gabriele CM, Parks SE (2017a) The role of season, tide, and diel period in the presence of harbor seal (*Phoca vitulina*) breeding vocalizations in Glacier Bay National Park and Preserve. Alaska Aquat Mamm 43(5):537–546. https://doi.org/10.1578/AM.43.5.2017.537

Matthews LP, Parks SE, Fournet MEH, Gabriele CM, Womble JN, Klinck H (2017b) Source levels and call parameters of harbor seal breeding vocalizations near a terrestrial haulout site in Glacier Bay National Park and

Preserve. J Acoust Soc Am 141 (3):EL274-EL280. https://doi.org/10.1121/1.4978299

Matthews LP, Blades B, Parks SE (2018) Female harbor seal (*Phoca vitulina*) behavioral response to playbacks of underwater male acoustic advertisement displays. PeerJ 6:e4547. https://doi.org/10.7717/peerj.4547

Matthews LP, Fournet ME, Gabriele C, Klinck H, Parks SE (2020) Acoustically advertising male harbour seals in southeast Alaska do not make biologically relevant acoustic adjustments in the presence of vessel noise. Biol Lett 16(4):20190795. https://doi.org/10.1098/rsbl.2019.0795

Mattmüller RM, Thomisch K, Van Opzeeland I, Laidre K, Simon M (2022) Passive acoustic monitoring reveals year-round marine mammal community composition off Tasiilaq, Southeast Greenland. J Acoust Soc Am 151:1380. https://doi.org/10.1121/10.0009429

McCann TS (1981) Aggression and sexual activity of male Southern elephant seals, *Mirounga leonina*. J Zool 195(3):295–310. https://doi.org/10.1111/j.1469-7998.1981.tb03467.x

McCarthy R (2023) The seasonal variation of ultrasonic vocalizations produced by Weddell seals (*Leptonychotes weddellii*) in McMurdo Sound, Antarctica with ambient illumination levels. University of Oregon

McCreery L, Thomas JA (2009) Acoustic analysis of underwater vocalizations from crabeater seals (*Lobodon carcinophagus*): not so monotonous. Aquat Mamm 35(4):490–501. https://doi.org/10.1578/AM.35.4.2009.490

McCulloch S (2000) The vocal behaviour of the grey seal (*Halichoerus grypus*). PhD thesis, University of St Andrews, Scotland

McCulloch S, Boness DJ (2000) Mother–pup vocal recognition in the grey seal (*Halichoerus grypus*) of Sable Island, Nova Scotia, Canada. Journal of Zoology 251(4):449–455. https://doi.org/10.1111/j.1469-7998.2000.tb00800.x

McCulloch S, Pomeroy PP, Slater PJB (1999) Individually distinctive pup vocalizations fail to prevent allosuckling in grey seals. Can J Zool 77(5):716–723. https://doi.org/10.1139/z99-023

McIntosh RR, Pitcher BJ (2021) The enigmatic life history of the Australian sea lion. In: Campagna C, Harcourt R (eds) Ethology and behavioral ecology of otariids and the odobenid. Springer, Cham, pp 557–585. https://doi.org/10.1007/978-3-030-59184-7_26

McNab AG, Crawley MC (1975) Mother and pup behaviour of the New Zealand fur seal, *Arctocephalus forsteri* (lesson). Mauri Ora 3:77–88

Meister M (2017) Temporal patterns in the acoustic presence of marine mammals off Elephant Island, Antarctica. Universität Bremen, Bremen

Merlen G (2000) Nocturnal acoustic location of the Galapagos fur seal *Arctocephalus galapagoensis*. Mar Mamm Sci 16(1):248–253. https://doi.org/10.1111/j.1748-7692.2000.tb00917.x

Miksis-Olds J, Parks S (2011) Seasonal trends in acoustic detection of ribbon seal (*Histriophoca fasciata*) vocalizations in the Bering Sea. Aquat Mamm 37(4):464–471. https://doi.org/10.1578/AM.37.4.2011.464

Miksis-Olds JL, Opzeeland IC, Parijs SM, Jones J (2016) Pinniped sounds in the polar oceans. In: Au LWW, Lammers OM (eds) Listening in the ocean. Springer, New York, pp 257–308. https://doi.org/10.1007/978-1-4939-3176-7_11

Miller EH (1971) Social and thermoregulatory behaviour of the New Zealand fur seal, *Arctocephalus forsteri* (Lesson, 1828). MSc thesis, University of Canterbury, Christchurch, New Zealand

Miller EH (1974) Social behaviuor between adult male and female New Zealand fur seals, *Arctocephalus forsteri* (Lesson) during the breeding season. Aust J Zool 22(2):155–173. https://doi.org/10.1071/ZO9740155

Miller EH (1985) Airborne acoustic communication in the walrus *Odobenus rosmarus*. Natl Geogr Res 1(1):124–145

Miller EH, Job DA (1992) Airborne acoustic communication in the Hawaiian monk seal, *Monachus schauinslandi*. In: Thomas JA, Kastelein RA, Supin AY (eds) Marine mammal sensory systems. Plenum, New York, pp 485–451. https://doi.org/10.1007/978-1-4615-3406-8_33

Miller EH, Murray AV (1995) Structure, complexity, and organization of vocalizations in harp seal (*Phoca groenlandica*) pups. In: Kastelein R, Thomas JA, Nachtigall PE (eds) Sensory abilities of aquatic animals. De Spil, Woerden, pp 237–264

Mizuguchi D, Mitani Y, Kohshima S (2016a) Geographically specific underwater vocalizations of ribbon seals (*Histriophoca fasciata*) in the Okhotsk Sea suggest a discrete population. Mar Mamm Sci 32(3):1138–1151. https://doi.org/10.1111/mms.12301

Mizuguchi D, Tsunokawa M, Kawamoto M, Kohshima S (2016b) Underwater vocalizations and associated behavior in captive ringed seals (*Pusa hispida*). Polar Biol 39(4):659–669. https://doi.org/10.1007/s00300-015-1821-x

Møhl B, Terhune JM, Ronald K (1975) Underwater calls of the harp seal, *Pagophilus groenlandicus*. Rapports et Proces-verbaux des Réunions Conseil International pour l'Éxploration de la Mer 169:533–543

Moors HB, Terhune JM (2003) Repetition patterns within harp seal (*Pagophilus groenlandicus*) underwater calls. Aquat Mamm 29(2):278–288. https://doi.org/10.1578/016754203101024211

Moors HB, Terhune JM (2004) Repetition patterns in Weddell seal (*Leptonychotes weddellii*) underwater multiple element calls. J Acoust Soc Am 116(2):1261–1270. https://doi.org/10.1121/1.1763956

Moors HB, Terhune JM (2005) Calling depth and time and frequency attributes of harp (*Pagophilus groenlandicus*) and Weddell (*Leptonychotes weddellii*) seal underwater vocalizations. Can J Zool 83(11):1438–1452. https://doi.org/10.1139/z05-135

Morrice MG, Burton HR, Green K (1994) Microgeographic variation and songs in the underwater vocalisation repertoire of the Weddell seal (*Leptonychotes weddellii*) from the Vestfold Hills, Antarctica. Polar Biol 14:441–446. https://doi.org/10.1007/BF00239046

Mossbridge JA, Thomas JA (1999) An "acoustic niche" for antarctic killer whale and leopard seal sounds. Mar Mamm Sci 15(4):1351–1357. https://doi.org/10.1111/j.1748-7692.1999.tb00897.x

Mouy X, Hannay D, Zykov M, Martin B (2012) Tracking of Pacific walruses in the Chukchi Sea using a single hydrophone. J Acoust Soc Am 131(2):1349–1358. https://doi.org/10.1121/1.3675008

Muñoz G, Karamanlidis AA, Dendrinos P, Thomas JA (2011) Aerial vocalizations by wild and rehabilitating Mediterranean monk seals (*Monachus monachus*) in Greece. Aquat Mamm 37(3):262–279. https://doi.org/10.1578/AM.37.3.2011.262

Nagaraj H, Owen K, Lea MA, Miller BS (2021) Acoustic analysis of crabeater seal (*Lobodon carcinophaga*) vocalizations in the Southern Kerguelen Plateau region of East Antarctica. J Acoust Soc Am 150(5):3353–3361. https://doi.org/10.1121/10.0006789

Nel JAJ (1966) On the behaviour of the crabeater seal *Lobodon carcinophagus* (Hombron & Jacquinot). Afr Zool 2(1):91–93. https://doi.org/10.1080/00445096.1966.11447335

Newby TC (1973) Observations on the breeding behavior of the harbor seal in the state of Washington. J Mammal 54(2):540–543. https://doi.org/10.2307/1379151

Nicholson TE (2000) Social structure and underwater behavior of harbor seals in southern Monterey Bay, California. San Francisco State University,

Nikolich K, Frouin-Mouy H, Acevedo-Gutiérrez A (2016) Quantitative classification of harbor seal breeding calls in Georgia Strait, Canada. J Acoust Soc Am 140(2):1300–1308. https://doi.org/10.1121/1.4961008

Nikolich K, Frouin-Mouy H, Acevedo-Gutiérrez A (2018) Clear diel patterns in breeding calls of harbor seals (*Phoca vitulina*) at Hornby Island, British Columbia, Canada. Can J Zool 96(11):1236–1243. https://doi.org/10.1139/cjz-2018-0018

Norris KS, Watkins WA (1971) Underwater sounds of *Arctocephalus philippii*, the Juan Fernandez fur seal. In: Burt WH (ed) Antarctic Pinnipedia, Antarctic research series, vol 18. Wiley, Washington, DC, pp 169–172. https://doi.org/10.1029/AR018p0169

Nowak LJ (2021) Observations on mechanisms and phenomena underlying underwater and surface vocalisations of grey seals. Bioaccoustics 30(6):696–715. https://doi.org/10.1080/09524622.2020.1851298

Oetelaar ML, Terhune JM, Burton HR (2003) Can the sex of a Weddell seal (*Leptonychotes weddellii*) be identified by its surface call? Aquat Mamm 29(2):261–267. https://doi.org/10.1578/016754203101024194

Orr RT, Poulter TC (1967) Some observations on reproduction, growth, and social behavior in the Steller Sea lion. Proc Calif Acad Sci 35(10):193–226

Osiecka AN, Gridley T, Fearey J (2022) Temporal patterns in Cape fur seal (*Arctocephalus pusillus pusillus*) mother and pup attraction calls. Belg J Zool 152:117–138. https://doi.org/10.26496/bjz.2022.103

Osiecka AN, Fearey J, Ravignani A, Burchardt L (2024) Isochrony in barks of Cape fur seal (*Arctocephalus pusillus pusillus*) pups and adults. Ecol Evol 14(3):e11085. https://doi.org/10.1002/ece3.11085

Otsuki M, Akamatsu T, Nobetsu T, Mizuguchi D, Mitani Y (2018) Diel changes in ribbon seal *Histriophoca fasciata* vocalizations during sea ice presence in the Nemuro Strait, Sea of Okhotsk. Polar Biol 41(3):451–456. https://doi.org/10.1007/s00300-017-2203-3

Page B, Goldsworthy SD, Hindell MA (2001) Vocal traits of hybrid fur seals: intermediate to their parental species. Anim Behav 61(5):959–967. https://doi.org/10.1006/anbe.2000.1663

Page B, Goldsworthy SD, Hindell MA (2002a) Individual vocal traits of mother and pup fur seals. Bioacoustics 13(2):121–143. https://doi.org/10.1080/09524622.2002.9753491

Page B, Goldsworthy SD, Hindell MA, Mckenzie J (2002b) Interspecific differences in male vocalizations of three sympatric fur seals (*Arctocephalus* spp.). J Zool 258 (1):49–56. https://doi.org/10.1017/S095283690200119X

Pahl BC, Terhune JM, Burton HR (1997) Repertoire and geographic variation in underwater vocalisations of Weddell seals (*Leptonychotes weddellii*, Pinnipedia: Phocidae) at the Vestfold Hills, Antarctica. Aust J Zool 45(2):171–187. https://doi.org/10.1071/ZO95044

Parisi I, de Vincenzi G, Torri M, Papale E, Mazzola S, Bonanno A, Buscaino G (2017) Underwater vocal complexity of Arctic seal *Erignathus barbatus* in Kongsfjorden (Svalbard). J Acoust Soc Am 142(5):3104–3115. https://doi.org/10.1121/1.5010887

Park T-G, Iida K, Mukai T (2006) Characteristics of vocalizations in Steller Sea lions. In: Trites AW, Atkinson SK, DeMaster DP et al. (eds) Sea lions of the world, vol AK-SG-06-01. Alaska Sea Grant College Program, University of Alaska Fairbanks, pp 549–560. https://doi.org/10.4027/slw.2006.34

Parnell K (2018) Underwater vocal repertoire of the endangered Hawaiian monk seal, *Neomonachus schauinslandi*. MSc thesis. University of California, Santa Cruz

Parsons MJ, Looby A, Chanda K, Di Iorio L, Erbe C, Frazao F, Havlik M, Juanes F, Lammers M, Li S, Liffers M, Lin T-H, Linke S, Mooney TA, Radford CN, Rice AN, Rountree R, Sayigh L, Sousa-Lima R, Stanley J, Thomisch K, Urban E, van Zeeland L, Vela S, Zuffi S, Nedelec SL (2024) A global library of underwater biological sounds (GLUBS): an online platform with multiple passive acoustic monitoring applications. In: Popper AN, Sisneros J, Hawkins AD, Thomsen F (eds) The effects of noise on aquatic life: principles and practical considerations. Springer, Cham. https://doi.org/10.1007/978-3-031-10417-6_123-1

Pastukhov VD (1993) Nerpa Baikala (Baikal Seal), Novosibirsk: Nauka

Pérez Tadeo M, Gammell M, O'Brien J (2023) First steps towards the automated detection of underwater vocalisations of grey seals (*Halichoerus grypus*) in the Blasket Islands, Southwest Ireland. J Mar Sci Eng 11(2):351. https://doi.org/10.3390/jmse11020351

Perry EA, Renouf D (1988) Further studies of the role of harbour ceal (*Phoca vitulina*) pup vocalizations in preventing separation of mother-pup pairs. Can J Zool 66(4):934–938. https://doi.org/10.1139/z88-138

Peterson RS (1965) Behavior of the northern fur seal. John Hopkins University

Peterson RS (1968) Social behavior in pinnipeds with particular reference to the northern fur seal. In: Harrison RJ (ed) The behavior and physiology of pinnipeds. Appleton-Century Croft, NY, pp 3–53

Peterson RS, Bartholomew GA (1969) Airborne vocal communication in the California Sea lion, *Zalophus californianus*. Anim Behav 17(1):17–24. https://doi.org/10.1016/0003-3472(69)90108-0

Peterson RS, Hubbs CL, Gentry RL, DeLong RL (1968) The Guadalupe fur seal: habitat, behavior, population size, and field identification. J Mammal 49(4):665–675. https://doi.org/10.2307/1378727

Petrinovich L (1974) Individual recognition of pup vocalization by northern elephant seal mothers. Z Tierpsychol 34(3):308–312. https://doi.org/10.1111/j.1439-0310.1974.tb01803.x

Petrov EA (2009) Baikal'skaya nerpa (Baikal seal), Ulan-Ude: ID Ekos

Petrov EA, Kupchinsky AB, Fialkov VA, Badardinov AA (2021) The importance of hauling grounds in the life of the Baikal seals (*Pusa sibirica* Gmelin 1788, Pinnipedia): 2. Behavior on hauling grounds. Biol Bull 48(9):1715–1728. https://doi.org/10.1134/S1062359021090193

Petrov EA, Kupchinsky AB, Fialkov VA, Badardinov AA (2022) The importance of coastal hauling grounds in the life of the Baikal seal (*Pusa sibirica* Gmelin 1788, Pinnipedia): 4. Behavior of seals on coastal hauling grounds of Tonkii Ushkan islet (Ushkan Islands, Lake Baikal), based on video observations. Biol Bull 49(7):992–1010. https://doi.org/10.1134/S1062359022070160

Phillips AV (2003) Behavioral cues used in reunions between mother and pup South American fur seals (*Arctocephalus australis*). J Mammal 84(2):524–535. https://doi.org/10.1644/1545-1542(2003)084<0524:BCUIRB>2.0.CO;2

Phillips AV, Stirling I (2000) Vocal individuality in mother and pup South American fur seals, *Arctocephalus australis*. Mar Mamm Sci 16(3):592–616. https://doi.org/10.1111/j.1748-7692.2000.tb00954.x

Phillips AV, Stirling I (2001) Vocal repertoire of South American fur seals, *Arctocephalus australis*: structure, function, and context. Can J Zool 79(3):420–437. https://doi.org/10.1139/cjz-79-3-420

Piérard J (1969) Le larynx du phoque de Weddell (*Leptonychotes weddelli*, Lesson, 1826). Can J Zool 47(1):77–87. https://doi.org/10.1139/z69-015

Pierson MO (1978) A study of the population dynamics and breeding behavior of the Guadalupe fur seal, *Arctocephalus townsendi*. University of California, Santa Cruz

Pierson MO (1987) Breeding behavior of the Guadalupe fur seal, *Arctocephalus townsendi*. In: Croxall JP, Gentry RL (eds) Status, biology and ecology of fur seals. NOAA Technical Report, NMFS, vol 51, pp 83–94

Pitcher BJ, Ahonen H, Harcourt RG, Charrier I (2009) Delayed onset of vocal recognition in Australian sea lion pups (*Neophoca cinerea*). Naturwissenschaften 96(8):901–909. https://doi.org/10.1007/s00114-009-0546-5

Pitcher BJ, Harcourt RG, Charrier I (2010a) The memory remains: long-term vocal recognition in Australian sea lions. Anim Cogn 13(5):771–776. https://doi.org/10.1007/s10071-010-0322-0

Pitcher BJ, Harcourt RG, Charrier I (2010b) Rapid onset of maternal vocal recognition in a colonially breeding mammal, the Australian sea lion. PLoS One 5(8):e12195. https://doi.org/10.1371/journal.pone.0012195

Pitcher B, Harcourt R, Charrier I (2012) Individual identity encoding and environmental constraints in vocal recognition of pups by Australian sea lion mothers. Anim Behav 83(3):681–690. https://doi.org/10.1016/j.anbehav.2011.12.012

Porov LA (1961) Data on the general morphology of the Greenland hooded seal (*Cystophora cristata* Erxl.). Tr Soveshch Ikhtiol Kom 12:180–191

Poulter TC (1963) Sonar signals of the sea lion. Science 139(3556):753–755. https://doi.org/10.1016/0011-7471(64)90135-4

Poulter TC (1966) The use of active sonar by the California sea lion *Zalophus californianus* (Lesson). J Audit Res 6:165–173

Poulter TC (1968) Marine mammals. In: Sebeok TA (ed) Animal communication: techniques of study and results of research. Indiana University Press, Bloomington, pp 405–465

Pozo Galván YP, Pérez Tadeo M, Pommier M, O'Brien J (2024) Static acoustic monitoring of harbour (*Phoca vitulina*) and grey seals (*Halichoerus grypus*) in the Malin Sea: a revolutionary approach in pinniped conservation. J Mar Sci Eng 12(1):118. https://doi.org/10.3390/jmse12010118

Prawirasasra MS, Mustonen M, Klauson A (2021) The underwater soundscape at gulf of Riga marine-protected areas. J Mar Sci Eng 9(8):915. https://doi.org/10.3390/jmse9080915

Ralls K, Fiorelli P, Gish S (1985) Vocalizations and vocal mimicry in captive harbor seals, *Phoca vitulina*. Can J

Zool 63(5):1050–1056. https://doi.org/10.1139/z85-157

Rautio A, Niemi M, Kunnasranta M, Holopainen IJ, Hyvärinen H (2009) Vocal repertoire of the Saimaa ringed seal (*Phoca hispida saimensis*) during the breeding season. Mar Mamm Sci 25(4):920–930. https://doi.org/10.1111/j.1748-7692.2009.00299.x

Ravignani A (2019) Timing of antisynchronous calling: a case study in a harbor seal pup (*Phoca vitulina*). J Comp Psychol 133(2):272–277. https://doi.org/10.1037/com0000160

Ravignani A, Kello CT, de Reus K, Kotz SA, Dalla Bella S, Méndez-Aróstegui M, Rapado-Tamarit B, Rubio-Garcia A, de Boer B (2018) Ontogeny of vocal rhythms in harbor seal pups: an exploratory study. Curr Zool 65(1):107–120. https://doi.org/10.1093/cz/zoy055

Ravignani A, Torres Borda L, Rasilo H, Salazar Casals A, Jadoul Y (2022) Parselmouth for bioacoustics: analysis pipelines for seal vocalizations. J Acoust Soc Am 151(4):A29. https://doi.org/10.1121/10.0010551

Ray C (1967) Social behavior and acoustics of the Weddell seal. Antarct J US 2(4):105–106

Ray C (1970) Population ecology of Antarctic seals. Antarctic Ecol 1:398–414

Ray GC (1981) Ross seal *Ommatophoca rossii* Gray, 1844. In: Ridgeway SH, Harrison RJ (eds) Handbook of marine mammals, vol 2, pp 237–260

Ray GC, Watkins WA (1975) Social function of underwater sounds in the walrus *Odobenus rosmarus*. Rapports et Proces-Verbaux des Reunions 169:524–526

Ray C, Watkins WA, Burns JJ (1969) The underwater song of *Erignathus* (bearded seal). Zoologica 54(2):79–83

Reichmuth C, Casey C (2014) Vocal learning in seals, sea lions, and walruses. Curr Opin Neurobiol 28:66–71. https://doi.org/10.1016/j.conb.2014.06.011

Reichmuth C, Quihuis D (2022) Social transmission of innovative sound production in walruses (*Odobenus rosmarus*). Aquat Mamm 48(6):720–723. https://doi.org/10.1578/AM.48.6.2022.720

Reichmuth C, Mulsow J, Schusterman R (2009) Underwater acoustic displays of a Pacific walrus (*Odobenus rosmarus divergens*): source level estimates and temporal patterning. In: 18th biennial conference on the biology of marine mammals, Quebec City, Canada, p. 210

Reidenberg JS (2017) Terrestrial, semiaquatic, and fully aquatic mammal sound production mechanisms. Acoustics Today 13(2):35–43

Reidenberg JS, Laitman JT (2010) Chapter 10.4 – generation of sound in marine mammals. In: Brudzynski SM (ed) Handbook of behavioral neuroscience, vol 19. Elsevier, pp 451–465. https://doi.org/10.1016/B978-0-12-374593-4.00041-3

Reiman AJ, Terhune JM (1993) The maximum range of vocal communication in air between a harbor seal (*Phoca vitulina*) pup and its mother. Mar Mamm Sci 9(2):182–189. https://doi.org/10.1111/j.1748-7692.1993.tb00442.x

Renouf D (1984) The vocalization of the harbour seal pup (*Phoca vitulina*) and its role in the maintenance of contact with the mother. J Zool 202(4):583–590. https://doi.org/10.1111/j.1469-7998.1984.tb05055.x

Renouf D (1989) Sensory function in the harbor seal. Sci Am 260(4):90–95. https://doi.org/10.1038/scientificamerican0489-90

Renouf D, Davis MB (1982) Evidence that seals may use echolocation. Nature 300(5893):635–637. https://doi.org/10.1038/300635a0

Reynoso JPG (1994) Factors affecting the population status of Guadalupe fur seal, *Arctocephalus townsendi* (Merriam, 1897), at Isla de Guadalupe, Baja California, México. University of California, Santa Cruz

Rideout BP, Dosso SE, Hannay DE (2013) Underwater passive acoustic localization of Pacific walruses in the northeastern Chukchi Sea. J Acoust Soc Am 134(3):2534–2545. https://doi.org/10.1121/1.4816580

Risch D, Clark CW, Corkeron PJ, Elepfandt A, Kovacs KM, Lydersen C, Stirling I, Van Parijs SM (2007) Vocalizations of male bearded seals, *Erignathus barbatus*: classification and geographical variation. Anim Behav 73(5):747–762. https://doi.org/10.1016/j.anbehav.2006.06.012

Rogers TL (2003) Factors influencing the acoustic behaviour of male phocid seals. Aquat Mamm 29(2):247–260

Rogers TL (2007) Age-related differences in the acoustic characteristics of male leopard seals, *Hydrurga leptonyx*. J Acoust Soc Am 122(1):596–605. https://doi.org/10.1121/1.2736976

Rogers TL (2014) Source levels of the underwater calls of a male leopard seal. J Acoust Soc Am 136(4):1495–1498. https://doi.org/10.1121/1.4895685

Rogers TL (2017) Calling underwater is a costly signal: size-related differences in the call rates of Antarctic leopard seals. Cur Zool 63(4):433–443. https://doi.org/10.1093/cz/zox028

Rogers TL, Cato DH (2002) Individual variation in the acoustic behaviour of the adult male leopard seal, *Hydrurga leptonyx*. Behaviour 139(10):1267–1286. https://doi.org/10.1163/156853902321104154

Rogers TL, Cato DH, Bryden MM (1995) Underwater vocal repertoire of the leopard seal, *Hydrurga leptonyx*, in Prydz Bay, Antarctica. In: Kastelein RA, Thomas JA, Nachtigal PE (eds) Sensory abilities of aquatic animals. DeSpil Publishers, Amsterdam, pp 223–236

Rogers TL, Cato DH, Bryden MM (1996) Behavioral significance of underwater vocalizations of captive leopard seals, *Hydrurga leptonyx*. Mar Mamm Sci 12(3):414–427. https://doi.org/10.1111/j.1748-7692.1996.tb00593.x

Rößler H, Tougaard J, Sabinsky PF, Rasmussen MH, Granquist SM, Wahlberg M (2021) Are Icelandic harbor seals acoustically cryptic to avoid predation? JASA

Express Lett 1(3):031201. https://doi.org/10.1121/10.0003782

Rossong MA, Terhune JM (2009) Source levels and communication-range models for harp seal (*Pagophilus groenlandicus*) underwater calls in the Gulf of St. Lawrence, Canada. Can J Zool 87 (7): 609–617. https://doi.org/10.1139/Z09-048

Rouget PA, Terhune JM, Burton HR (2007) Weddell seal underwater calling rates during the winter and spring near Mawson Station, Antarctica. Mar Mamm Sci 23(3):508–523. https://doi.org/10.1111/j.1748-7692.2007.00129.x

Roux JP, Jouventin P (1987) Behavioral cues to individual recognition in the subantarctic fur seal, *Arctocephalus tropicalis* (NOAA technical report NMFS 51). National Marine Fisheries Service, USA, pp 95–102

Russell LR, Purdy JE, Davis RW (2016) Social context predicts vocalization use in the courtship behaviors of Weddell seals (*Leptonychotes weddellii*): a case study. Anim Behav Cogn 3(2):95–119. https://doi.org/10.12966/abc.04.05.2016

Sabinsky PF, Larsen ON, Wahlberg M, Tougaard J (2017) Temporal and spatial variation in harbor seal (*Phoca vitulina* L.) roar calls from southern Scandinavia. J Acoust Soc Am 141(3):1824–1834. https://doi.org/10.1121/1.4977999

Sandegren FE (1970) Breeding and maternal behavior of the Steller sea lion (*Eumetopias jubata*) in Alaska. Masters Thesis. University of Alaska Fairbanks, 149 pp

Sandegren FE (1976) Agonistic behavior in the male northern elephant seal. Behaviour 57(1–2):136–157. https://doi.org/10.1163/156853976X00145

Sanvito S, Galimberti F (2000a) Bioacoustics of southern elephant seals. I. Acoustic structure of male aggressive vocalisations. Bioacoustics 10(4):259–285. https://doi.org/10.1080/09524622.2000.9753438

Sanvito S, Galimberti F (2000b) Bioacoustics of southern elephant seals. II. Individual and geographical variation in male aggressive vocalisations. Bioacoustics 10 (4):287–307. https://doi.org/10.1080/09524622.2000.9753439

Sanvito S, Galimberti F (2003) Source level of male vocalisations in the genus *Mirounga*: repeatability and correlates. Bioacoustics 14(1):47–59. https://doi.org/10.1080/09524622.2003.9753512

Sanvito S, Galimberti F, Miller EH (2007a) Having a big nose: structure, ontogeny, and function of the elephant seal proboscis. Can J Zool 85(2):207–220. https://doi.org/10.1139/z06-193

Sanvito S, Galimberti F, Miller EH (2007b) Observational evidences of vocal learning in southern elephant seals: a longitudinal study. Ethology 113(2):137–146. https://doi.org/10.1111/j.1439-0310.2006.01306.x

Sanvito S, Galimberti F, Miller EH (2007c) Vocal signalling of male southern elephant seals is honest but imprecise. Anim Behav 73(2):287–299. https://doi.org/10.1016/j.anbehav.2006.08.005

Sanvito S, Galimberti F, Miller EH (2008) Development of aggressive vocalizations in male southern elephant seals (*Mirounga leonina*): maturation or learning? Behaviour 145(2):137–170. https://doi.org/10.1163/156853907783244729

Sauve C, Beauplet G, Hammill MO, Charrier I (2015a) Acoustic analysis of airborne, underwater, and amphibious mother attraction calls by wild harbor seal pups (*Phoca vitulina*). J Mammal 96(3):591–602. https://doi.org/10.1093/jmammal/gyv064

Sauve CC, Beauplet G, Hammill MO, Charrier I (2015b) Mother-pup vocal recognition in harbour seals: influence of maternal behaviour, pup voice and habitat sound properties. Anim Behav 105:109–120. https://doi.org/10.1016/j.anbehav.2015.04.011

Scheffer VB, Slipp JW (1944) The harbor seal in Washington State. Am Midl Nat 32(2):373–416. https://doi.org/10.2307/2421307

Schevill WE, Watkins WA (1965) Underwater calls of *Leptonychotes* (Weddell seal). Zoologica 50(1):45–46

Schevill WE, Watkins WA (1971) Directionality of the sound beam in *Leptonychotes weddelli* (Mammalia: Pinnipedia). Antarctic Res Ser 18:163–168. https://doi.org/10.1029/AR018p0163

Schevill WE, Watkins WA, Ray C (1963) Underwater sounds of pinnipeds. Science 141(3575):50–53. https://doi.org/10.1126/science.141.3575.50

Schevill WE, Watkins WA, Ray C (1966) Analysis of underwater *Odobenus* calls with remarks on the development and function of the pharyngeal pouches. Zoologica 51(3):103–110

Schusterman RJ (1966) Underwater click vocalizations by a California Sea lion: effects of visibility. Psychol Rec 16(2):129–136

Schusterman RJ (1967) Perception and determinants of underwater vocalization in the California sea lion. Les Systemes Sonars Animaux, Biologie et Bionique 1:535–617

Schusterman RJ (1970) Determinants and control of underwater vocalizations in the California sea lion. Stanford Research Institute, Menlo Park

Schusterman RJ (1977) Temporal patterning in sea lion barking (*Zalophus californianus*). Behav Biol 20(3): 404–408. https://doi.org/10.1016/S0091-6773(77)90964-6

Schusterman RJ (2008) Vocal learning in mammals with special emphasis on pinnipeds. In: Oller DK, Gribel U (eds) The evolution of communicative flexibility: complexity, creativity, and adaptability in human and animal communication. MIT Press, Cambridge, MA, pp 41–70

Schusterman RJ, Balliet RF (1969) Underwater barking by male sea lions (*Zalophus californianus*). Nature 222(5199):1179–1181. https://doi.org/10.1038/2221179a0

Schusterman RJ, Dawson RG (1968) Barking, dominance, and territoriality in male sea lions. Science 160(3826): 434–436. https://doi.org/10.1126/science.160.3826.434

Schusterman RJ, Feinstein SH (1965) Shaping and discriminative control of underwater click vocalizations in a California sea lion. Science 150(3704):1743–1744. https://doi.org/10.1126/science.150.3704.1743

Schusterman RJ, Reichmuth C (2008) Novel sound production through contingency learning in the Pacific walrus (*Odobenus rosmarus divergens*). Anim Cogn 11(2):319–327. https://doi.org/10.1007/s10071-007-0120-5

Schusterman RJ, Gentry R, Schmook J (1967) Underwater sound production by captive California sea lions, *Zalophus californianus*. Zoologica 52(1):21–24

Schusterman RJ, Balliet RF, St. John S (1970) Vocal displays under water by the gray seal, the harbor seal, and the Steller sea lion. Psychon Sci 18(5):303–305. https://doi.org/10.3758/BF03331839

Schusterman R, Hanggi E, Gisiner R (1992) Acoustic signalling inn mother-pup reunions, interspecies bonding, and affiliation by kinship in California Sea lions (*Zalophus californianus*). In: Thomas J (ed) Marine mammal sensory systems. Plenum Press, New York, pp 533–551. https://doi.org/10.1007/978-1-4615-3406-8_34

Schusterman RJ, Kastak D, Levenson DH, Reichmuth CJ, Southall BL (2000) Why pinnipeds don't echolocate. J Acoust Soc Am 107(4):2256–2264. https://doi.org/10.1121/1.428506

Schwalm AL (2014) Determination of sound types and source levels of airborne vocalizations by California sea lions, *Zalophus californianus*, in rehabilitation at the Marine Mammal Center in Sausalito, California. Western Illinois University

Schwenke T (2013) Underwater acoustic behaviour of leopard seals (*Hydrurga leptonyx*) and Ross seals (*Ommatophoca rossii*) in Antarctic coastal and offshore habitats. MSc thesis, Bremen University, Germany

Scronce BL, Ridgway SH (1980) Grey seal, *Halichoerus*: echolocation not demonstrated. In: Busnel R-G, Fish JF (eds) Animal sonar systems. Plenum, New York, pp 991–993

Seibert A-M, Klinck H, Kindermann L, Bornemann H, van Opzeeland I, Plötz J, Boebel O (2007) Characteristics of underwater calls of the Ross seal. In: 17th biennial conference on the biology of marine mammals, Cape Town, South Africa

Seibert A-M, Kindermann L, Klinck H, Boebel O (2008) PALAOA: ross seal presence and calling patterns. In: 22nd conference of the European Cetacean Society, Egmond aan Zee

Serrano A (2001) New underwater and aerial vocalizations of captive harp seals (*Pagophilus groenlandicus*). Can J Zool 79(1):75–81. https://doi.org/10.1139/z00-182

Serrano A, Terhune JM (2001) Within-call repetition may be an anti-masking strategy in underwater calls of harp seals. Can J Zool 79(8):1410–1413. https://doi.org/10.1139/z01-107

Serrano A, Terhune JM (2002) Antimasking aspects of harp seal (*Pagophilus groenlandicus*) underwater vocalizations. J Acoust Soc Am 112(6):3083–3090. https://doi.org/10.1121/1.1518987

Serrano-Solis A (1998) Underwater vocalizations and vocal activity of captive harp seals (*Pagophilus groenlandicus*). Masters Thesis. Memorial University, Newfoundland

Shabangu FW, Charif RA (2021) Short moan call reveals seasonal occurrence and diel-calling pattern of crabeater seals in the Weddell Sea, Antarctica. Bioaccoustics 30(5):543–563. https://doi.org/10.1080/09524622.2020.1819877

Shabangu FW, Rogers TL (2021) Summer circumpolar acoustic occurrence and call rates of Ross, *Ommatophoca rossii*, and leopard, *Hydrurga leptonyx*, seals in the Southern Ocean. Polar Biol 44(2):433–450. https://doi.org/10.1007/s00300-021-02804-9

Shabangu FW, Hofmeyr GG, Probert R, Connan M, Buhrmann CA, Gridley T (2022) In-air acoustic repertoire and associated behaviour of wild juvenile crabeater seals during rehabilitation. Bioaccoustics:1–23 https://doi.org/10.1080/09524622.2022.2108145

Shapiro AD, Slater PJ, Janik VM (2004) Call usage learning in gray seals (*Halichoerus grypus*). J Comp Psychol 118(4):447–454. https://doi.org/10.1037/0735-7036.118.4.447

Shipley C, Strecker G (1986) Day and night patterns of vocal activity of northern elephant seal bulls. J Mammal 67(4):775–778. https://doi.org/10.2307/1381150

Shipley C, Hines M, Buchwald JS (1981) Individual differences in threat calls of northern elephant seal bulls. Anim Behav 29(1):12–19. https://doi.org/10.1016/S0003-3472(81)80147-9

Shipley C, Hines M, Buchwald JS (1986) Vocalizations of nothern elephant seal bulls: development of adult call characteristics during puberty. J Mammal 67(3):526–536. https://doi.org/10.2307/1381284

Shipley C, Stewart BS, Bass J (1992) Seismic communication in northern elephant seals. In: Thomas JA, Kastelein RA, Supin AY (eds) Marine mammal sensory systems. Plenum, New York, pp 553–562. https://doi.org/10.1007/978-1-4615-3406-8_35

Sills JM, Reichmuth C (2022) Vocal behavior in spotted seals (*Phoca largha*) and implications for passive acoustic monitoring. Front Remote Sens 3:862435. https://doi.org/10.3389/frsen.2022.862435

Sills JM, Parnell K, Ruscher B, Lew C, Kendall TL, Reichmuth C (2021) Underwater hearing and communication in the endangered Hawaiian monk seal *Neomonachus schauinslandi*. Endang Species Res 44:61–78. https://doi.org/10.3354/esr01092

Siniff DB, Stirling I, Bengston JL, Reichle RA (1979) Social and reproductive behavior of crabeater seals (*Lobodon carcinophagus*) during the austral spring. Can J Zool 57(11):2243–2255

Sipilä T, Medvedev NV, Hyvärinen H (1996) The Ladoga seal (*Phoca hispida ladogensis* Nordq.). Hydrobiologia 322 (1):193–198. https://doi.org/10.1007/BF00031827

Sjare B, Stirling I (1996) The breeding behavior of Atlantic walruses, *Odobenus rosmarus rosmarus*, in the Canadian High Arctic. Can J Zool 74(5):897–911. https://doi.org/10.1139/z96-103

Sjare B, Stirling I, Spencer C (2003) Structural variation in the songs of Atlantic walruses breeding in the Canadian high Arctic. Aquat Mamm 29(2):297–318. https://doi.org/10.1578/016754203101024121

Southall BL (2002) Northern elephant seal field bioacoustics and aerial auditory masked hearing thresholds in three pinnipeds. University of California, Santa Cruz

Southall BL, Casey C, Holt M, Insley S, Reichmuth C (2019) High-amplitude vocalizations of male northern elephant seals and associated ambient noise on a breeding rookery. J Acoust Soc Am 146(6):4514–4524. https://doi.org/10.1121/1.5139422

Southall BL, Southall H, Antunes R, Nichols R, Rouse A, Stafford KM, Robards M, Rosenbaum HC (2020) Seasonal trends in underwater ambient noise near St. Lawrence Island and the Bering Strait. Mar Pollut Bull 157:111283. https://doi.org/10.1016/j.marpolbul.2020.111283

St Clair Hill M, Ferguson JWH, Bester MN, Kerley GIH (2001) Preliminary comparison of calls of the hybridizing fur seals *Arctocephalus tropicalis* and *A. gazella*. Afr Zool 36 (1):45–53. https://doi.org/10.1080/15627020.2001.11657113

Stacey RM (2006) Airborne and underwater vocalizations of the Antarctic Ross seal (*Ommatophoca rossii*). Western Illinois University

Stansbury AL, Janik VM (2019) Formant modification through vocal production learning in gray seals. Curr Biol 29(13):2244–2249. https://doi.org/10.1016/j.cub.2019.05.071

Stansbury AL, Janik VM (2021) The role of vocal learning in call acquisition of wild grey seal pups. Philos Trans R Soc B 376(1836):20200251. https://doi.org/10.1098/rstb.2020.0251

Stansbury AL, de Freitas M, Wu G-M, Janik VM (2015) Can a gray seal (*Halichoerus grypus*) generalize call classes? J Comp Psychol 129(4):412–420. https://doi.org/10.1037/a0039756

Stirling I (1970) Observations on the behavior of the New Zealand fur seal (*Arctocephalus forsteri*). J Mammal 51(4):766–778. https://doi.org/10.2307/1378300

Stirling I (1971a) Studies on the behaviour of the South Australian fur seal, *Arctocephalus forsteri* (Lesson). I. Annual cycle, postures and calls, and adult males during the breeding season. Aust J Zool 19(3):243–266. https://doi.org/10.1071/ZO9710243

Stirling I (1971b) Studies on the behaviour of the South Australian fur seal, *Arctocephalus forsteri* (Lesson). II. Adult females and pups. Austral J Zool 19:167–273

Stirling I (1972) Observations on the Australian sea lion, *Neophoca cinerea* (Peron). Aust J Zool 20(3):271–279. https://doi.org/10.1071/ZO9720271

Stirling I (1973) Vocalization in the ringed seal (*Phoca hispida*). J Fish Res Board Can 30(10):1592–1595. https://doi.org/10.1139/f73-253

Stirling I, Siniff DB (1979) Underwater vocalizations of leopard seals (*Hydrurga leptonyx*) and crabeater seals (*Lobodon carcinophagus*) near the South Shetland Islands, Antarctica. Can J Zool 57(6):1244–1248. https://doi.org/10.1139/z79-160

Stirling I, Thomas JA (2003) Relationships between underwater vocalizations and mating systems in phocid seals. Aquat Mamm 29(2):227–246. https://doi.org/10.1578/016754203101024176

Stirling I, Warneke RM (1971) Implications of a comparison of the airborne vocalizations and some aspects of the behaviour of the two Australian fur seals, *Arctocephalus* spp. on the evolution and present taxonomy of the genus. Aust J Zool 19 (3):227–241. https://doi.org/10.1071/ZO9710227

Stirling I, Calvert W, Cleator HJ (1983) Underwater vocalizations as a tool for studying the distribution and relative abundance of wintering pinnipeds in the high arctic. Arctic 36(3):262–274. https://doi.org/10.14430/arctic2275

Stirling I, Calvert W, Spencer C (1987) Evidence of stereotyped underwater vocalizations of male Atlantic walruses (*Odobenus rosmarus rosmarus*). Can J Zool 65(9):2311–2321. https://doi.org/10.1139/z87-348

Swallow AD (2001) Hoover the seal, and George. Freeport Village Press

Takemura A, Yoshida K, Baba N (1983) Distinction of individual Northern fur seal pups, *Callorhinus ursinus*, through their call. Bull Facult Fisher Nagasaki University 54:29–34

Terhune JM (1994) Geographical variation of harp seal underwater vocalizations. Can J Zool 72(5):892–897. https://doi.org/10.1139/z94-121

Terhune JM (1999) Pitch separation as a possible jamming-avoidance mechanism in underwater calls of bearded seals (*Erignathus barbatus*). Can J Zool 77(7):1025–1034. https://doi.org/10.1139/z99-067

Terhune JM (2008) Variation in harp seal underwater acoustic communication range estimations. Bioacoustics 17(1–3):50–52. https://doi.org/10.1080/09524622.2008.9753761

Terhune JM (2016) Weddell seals do not lengthen calls in response to conspecific masking. Bioacoustics 25(1):75–88. https://doi.org/10.1080/09524622.2015.1089791

Terhune JM (2017) Through-ice communication by Weddell seals (*Leptonychotes weddellii*) is possible. Polar Biol 40(10):2133–2136. https://doi.org/10.1007/s00300-017-2124-1

Terhune JM (2019) The underwater vocal complexity of seals (Phocidae) is not related to their phylogeny. Can J Zool 97(3):232–240. https://doi.org/10.1139/cjz-2018-0190

Terhune JM, Dell'Apa A (2006) Stereotyped calling patterns of a male Weddell seal (*Leptonychotes weddellii*). Aquat Mamm 32(2):175–181. https://doi.org/10.1578/AM.32.2.2006.175

Terhune JM, Ronald K (1973) Some hooded seal (*Cystophora cristata*) sounds in March. Can J Zool 51(3):319–321. https://doi.org/10.1139/z73-045

Terhune JM, Ronald K (1986) Distant and near-range functions of harp seal underwater calls. Can J Zool 64(5):1065–1070. https://doi.org/10.1139/z86-159

Terhune JM, Stewart REA, Ronald K (1979) Influence of vessel noises on underwater vocal activity of harp seals. Can J Zool 57(6):1337–1338. https://doi.org/10.1139/z79-170

Terhune JM, Burton H, Green K (1994) Weddell seal in-air call sequences made with closed mouths. Polar Biol 14(2):117–122. https://doi.org/10.1007/BF00234973

Terhune JM, Healey SR, Burton HR (2001) Easily measured call attributes can detect vocal differences between Weddell seals from two areas. Bioacoustics 11(3):211–222. https://doi.org/10.1080/09524622.2001.9753463

Terhune JM, Quin D, Dell'Apa A, Mirhaj M, Plötz J, Kindermann L, Bornemann H (2008) Geographic variations in underwater male Weddell seal trills suggest breeding area fidelity. Polar Biol 31(6):671–680. https://doi.org/10.1007/s00300-008-0405-4

Thomas JA (1979) Quantitative analysis of the vocal repertoire of Weddell seals (*Leptonychotes weddelii*) in McMurdo Sound, Antarctica. PhD thesis, University of Minnesota

Thomas JA, Demaster DP (1982) An acoustic technique for determining diurnal activities in leopard (*Hydrurga leptonyx*) and crabeater (*Lobodon carcinophagus*) seal. Can J Zool 60(8):2028–2031. https://doi.org/10.1139/z82-260

Thomas JA, Golladay CL (1995) Geographic variation in leopard seal (*Hydrurga leptonyx*) underwater vocalizations. In: Kastelein RA, Thomas JA, Nachtigall PE (eds) Sensory abilities of aquatic animals. De Spil Publishers, Woerden, pp 201–221

Thomas JA, Kuechle VB (1982) Quantitative analysis of Weddell seal (*Leptonychotes weddelli*) underwater vocalizations at McMurdo Sound, Antarctica. J Acoust Soc Am 72(6):1730–1738. https://doi.org/10.1121/1.388667

Thomas JA, Stirling I (1983) Geographic variation in the underwater vocalizations of weddell seals (*Leptonychotes weddelli*) from Palmer Peninsula and McMurdo Sound, Antarctica. Can J Zool 61(10):2203–2212. https://doi.org/10.1139/z83-291

Thomas JA, Fisher SR, Evans WE, Awbrey FT (1982) Ultrasonic vocalizations of leopard seals (*Hydrurga leptonyx*). Antarct J US 17(5):186

Thomas JA, Awbrey FT, Fisher SR (1983) Incidental evidence for echolocation in polar pinnipeds. J Acoust Soc Am 74(S1):S75. https://doi.org/10.1121/1.2021129

Thomas JA, Puddicombe RA, George M, Lewis D (1988) Variations in underwater vocalizations of Weddell seals (*Leptonychotes weddelli*) at the Vestfold Hills as a measure of breeding population discreteness. In: Ferris JM, Burton HR, Johnstone GW, Bayly IAE (eds) Hydrobiologia. Biology of the Vestfold Hills, Antarctica, vol 1. Springer, Dordrecht, pp 279–284. https://doi.org/10.1007/BF00025597

Thompson PO (1965) Deep water recordings of pinniped sounds. Navy Electronics Lab, San Diego

Torres Borda L, Jadoul Y, Rasilo H, Salazar Casals A, Ravignani A (2021) Vocal plasticity in harbour seal pups. Philos Trans R Soc B 376:20200456. https://doi.org/10.1098/rstb.2020.0456

Trillmich F (1981) Mutual mother-pup recognition in Galapagos fur seals and sea lions: cues used and functional significance. Behaviour 78(1–2):21–42. https://doi.org/10.1163/156853981X00248

Trillmich F, Majluf P (1981) First observations on colony structure, behavior, and vocal repertoire of the South American fur seal (*Arctocephalus australis* Zimmermann, 1783) in Peru. Zeitschrift für Säugetierkunde 46(5):310–322

Trimble M, Charrier I (2011) Individuality in South American sea lion (*Otaria flavescens*) mother-pup vocalizations: implications of ecological constraints and geographical variations? Mammal Biol-Zeitschrift für Säugetierkunde 76(2):208–216. https://doi.org/10.1016/j.mambio.2010.10.009

Trimble M, Insley SJ (2010) Mother–offspring Reunion in the South American sea lion *Otaria flavescens* at Isla de Lobos (Uruguay): use of spatial, acoustic and olfactory cues. Ethol Ecol Evol 22(3):233–246. https://doi.org/10.1080/03949370.2010.502318

Tripovich JS, Rogers TL, Arnould JPY (2005) Species-specific characteristics and individual variation of the bark call produced by male Australian fur seals, *Arctocephalus pusillus doriferus*. Bioacoustics 15(1):79–96. https://doi.org/10.1080/09524622.2005.9753539

Tripovich JS, Rogers TL, Canfield R, Arnould JPY (2006) Individual variation in the pup attraction call produced by female Australian fur seals during early lactation. J Acoust Soc Am 120(1):502–509. https://doi.org/10.1121/1.2202864

Tripovich JS, Canfield R, Rogers TL, Arnould JPY (2008a) Characterization of Australian fur seal vocalizations during the breeding season. Mar Mamm Sci 24(4):913–928. https://doi.org/10.1111/j.1748-7692.2008.00229.x

Tripovich JS, Charrier I, Rogers TL, Canfield R, Arnould JPY (2008b) Acoustic features involved in the neighbour–stranger vocal recognition process in male Australian fur seals. Behav Process 79(1):74–80. https://doi.org/10.1016/j.beproc.2008.04.007

Tripovich JS, Canfield R, Rogers TL, Arnould JPY (2009a) Individual variation of the female attraction call produced by Australian fur seal pups throughout the maternal dependence period. Bioacoustics 18(3):259–276. https://doi.org/10.1080/09524622.2009.9753605

Tripovich JS, Rogers TL, Dutton G (2009b) Faecal testosterone concentrations and the acoustic behaviour of two captive male Australian fur seals. Aust Mammal 31(2):117–122. https://doi.org/10.1071/AM09009

Tripovich JS, Hall-Aspland S, Charrier I, Arnould JPY (2012) The behavioural response of Australian fur seals to motor boat noise. PLoS One 7(5):e37228. https://doi.org/10.1371/journal.pone.0037228

Turnbull SD, Terhune JM (1994) Descending frequency swept tones have lower thresholds than ascending frequency swept tones for a harbor seal (*Phoca vitulina*) and human listeners. J Acoust Soc Am 96(5): 2631–2636

Tyack PL, Miller EH (2002) Vocal anatomy, acoustic communication and echolocation. In: Hoelzel AR (ed) Marine mammal biology: an evolutionary approach. Blackwell Science, Oxford, pp 142–184

van Opzeeland IC, Van Parijs SM (2004) Individuality in harp seal, *Phoca groenlandica*, pup vocalizations. Anim Behav 68(5):1115–1123. https://doi.org/10.1016/j.anbehav.2004.07.005

van Opzeeland IC, Corkeron PJ, Risch D, Stenson G, Van Parijs SM (2009) Geographic variation in vocalizations of pups and mother-pup behavior of harp seals *Pagophilus groenlandicus*. Aquat Biol 6(1–3): 109–120. https://doi.org/10.3354/ab00170

van Opzeeland I, Van Parijs S, Bornemann H, Frickenhaus S, Kindermann L, Klinck H, Plotz J, Boebel O (2010) Acoustic ecology of Antarctic pinnipeds. Mar Ecol Prog Ser 414:267–291. https://doi.org/10.3354/Meps08683

van Opzeeland I, Van Parijs S, Frickenhaus S, Kreiss C, Boebel O (2012) Individual variation in pup vocalizations and absence of behavioral signs of maternal vocal recognition in Weddell seals (*Leptonychotes weddellii*). Mar Mamm Sci 28(2):E158–E172. https://doi.org/10.1111/j.1748-7692.2011.00505.x

Van Parijs SM (2003) Aquatic mating in pinnipeds: a review. Aquat Mamm 29(2):214–226

Van Parijs SM, Kovacs KM (2002) In-air and underwater vocalizations of eastern Canadian harbour seals, *Phoca vitulina*. Can J Zool 80(7):1173–1179. https://doi.org/10.1139/z02-088

Van Parijs SM, Thompson PM, Tollit DJ, Mackay A (1997) Distribution and activity of male harbour seals during the mating season. Anim Behav 54:35–43. https://doi.org/10.1006/anbe.1996.0426

Van Parijs SM, Hastie GD, Thompson PM (1999) Geographical variation in temporal and spatial vocalization patterns of male harbour seals in the mating season. Anim Behav 58(6):1231–1239. https://doi.org/10.1006/anbe.1999.1258

Van Parijs SM, Hastie GD, Thompson PM (2000a) Individual and geographical variation in display behaviour of male harbour seals in Scotland. Anim Behav 59(3): 559–568. https://doi.org/10.1006/anbe.1999.1307

Van Parijs SM, Janik VM, Thompson PM (2000b) Display-area size, tenure length, and site fidelity in the aquatically mating male harbour seal, *Phoca vitulina*. Can J Zool 78(12):2209–2217. https://doi.org/10.1139/z00-165

Van Parijs SM, Kovacs KM, Lydersen C (2001) Spatial and temporal distribution of vocalising male bearded seals – implications for male mating strategies. Behaviour 138(7):905–922. https://doi.org/10.1163/156853901753172719

Van Parijs SM, Corkeron PJ, Harvey J, Hayes SA, Mellinger DK, Rouget PA, Thompson PM, Wahlberg M, Kovacs KM (2003a) Patterns in the vocalizations of male harbor seals. J Acoust Soc Am 113(6):3403–3410. https://doi.org/10.1121/1.1568943

Van Parijs SM, Lydersen C, Kovacs KM (2003b) Vocalizations and movements suggest alternative mating tactics in male bearded seals. Anim Behav 65(2): 273–283. https://doi.org/10.1006/anbe.2003.2048

Venables UM, Venables LSV (1957) Mating behaviour of the seal *Phoca vitulina* in Shetland. J Zool 128:387–396. https://doi.org/10.1111/j.1096-3642.1957.tb00332.x

Verboom WC, Kastelein RA (1995) Rutting whistles of a male Pacific walrus (*Odobenus rosmarus divergens*). In: Kastelein RA, Thomas JA, Nachtigall PE (eds) Sensory abilities of aquatic animals. De Spil Publ, Woerden, pp 287–298

Verga L, Ravignani A (2021) Strange seal sounds: claps, slaps, and multimodal pinniped rhythms. Front Ecol Evol 9(644497):1–5. https://doi.org/10.3389/fevo.2021.644497

Verga L, Sroka MG, Varola M, Villanueva S, Ravignani A (2022) Spontaneous rhythm discrimination in a mammalian vocal learner. Biol Lett 18:20220316. https://doi.org/10.1098/rsbl.2022.0316

Wahlberg M, Lunneryd S-G, Westerberg H (2002) The source level of harbour seal flipper slaps. Aquat Mamm 28(1):90–92

Watkins WA, Ray GC (1977) Underwater sounds from ribbon seal, *Phoca (Histriophoca) fasciata*. Fish Bull 75(2):450–453

Watkins WA, Ray GC (1985) In air and underwater sounds of the Ross seal, *Ommatophoca rossi*. J Acoust Soc Am 77(4):1598–1600. https://doi.org/10.1121/1.392003

Watkins WA, Schevill WE (1979) Distinctive characteristics of underwater calls of the harp seal, *Phoca groenlandica*, during the breeding season. J Acoust Soc Am 66(4):983–988. https://doi.org/10.1121/1.383375

Wierucka K, Pitcher BJ, Harcourt R, Charrier I (2018) Multimodal mother–offspring recognition: the relative importance of sensory cues in a colonial mammal. Anim Behav 146:135–142. https://doi.org/10.1016/j.anbehav.2018.10.019

Wilson SC, Dolgova E, Trukhanova I, Dmitrieva L, Crawford I, Baimukanov M, Goodman SJ (2017) Breeding behavior and pup development of the Caspian seal, *Pusa caspica*. J Mammal 98(1): 143–153. https://doi.org/10.1093/jmammal/gyw176

Winn HE, Schneider J (1978) Communication in sireniens, sea otters, and pinnipeds. In: Sebeok TA (ed) How animals communicate. Indiana University Press, Bloomington & London, pp 809–840

Wirthlin M, Chang EF, Knörnschild M, Krubitzer LA, Mello CV, Miller CT, Pfenning AR, Vernes SC, Tchernichovski O, Yartsev MM (2019) A modular approach to vocal learning: disentangling the diversity of a complex behavioral trait. Neuron 104(1):87–99. https://doi.org/10.1016/j.neuron.2019.09.036

Yang TC, Giellis GR, Votaw CW, Diachok OI (1987) Acoustic properties of ice edge noise in the Greenland Sea. J Acoust Soc Am 82(3):1034–1038. https://doi.org/10.1121/1.395377

Yang L, Xu X, Zhang P, Han J, Li B, Berggren P (2017) Classification of underwater vocalizations of wild spotted seals (*Phoca largha*) in Liaodong Bay. China J Acoust Soc Am 141(3):2256–2262. https://doi.org/10.1121/1.4979056

Yang L, Xu X, Berggren P (2022) Spotted seal *Phoca largha* underwater vocalisations in relation to ambient noise. Mar Ecol Prog Ser 683:209–220. https://doi.org/10.3354/meps13951

Zhang P, Lu J, Li S, Han J, Wang Q, Yang L (2016) In-air vocal repertoires of spotted seals, *Phoca largha*. J Acoust Soc Am 140(2):1101–1107. https://doi.org/10.1121/1.4961048

Zhang P, Yang L, Han J, Yang Y, Lu Z, Li S (2022) Age and sex differences in in-air vocalization characteristics of spotted seal pups from newborn to 1 year old in captivity. Front Mar Sci 9:943030. https://doi.org/10.3389/fmars.2022.943030

Open Access This chapter is licensed under the terms of the Creative Commons Attribution 4.0 International License (http://creativecommons.org/licenses/by/4.0/), which permits use, sharing, adaptation, distribution and reproduction in any medium or format, as long as you give appropriate credit to the original author(s) and the source, provide a link to the Creative Commons license and indicate if changes were made.

The images or other third party material in this chapter are included in the chapter's Creative Commons license, unless indicated otherwise in a credit line to the material. If material is not included in the chapter's Creative Commons license and your intended use is not permitted by statutory regulation or exceeds the permitted use, you will need to obtain permission directly from the copyright holder.

Otter Sounds

Renata S. Sousa-Lima, Samara Almeida, Izabela Laurentino, and Caroline Leuchtenberger

Contents

6.1	Introduction	441
6.2	Sound Production in Otters	441
6.3	Functions of Otter Sounds	442
6.4	Overview of Otter Sounds	451
6.4.1	Small-Clawed Otter	454
6.4.2	Eurasian Otter	454
6.4.3	Sea Otter	455
6.4.4	North American River Otter	455
6.4.5	Neotropical Otter	455
6.4.6	Giant Otter	455
6.5	Summary	456
	References	456

R. S. Sousa-Lima (✉) · I. Laurentino
Laboratory of Bioacoustics, EcoAcoustic Research Hub, Universidade Federal do Rio Grande do Norte, Natal, RN, Brazil
e-mail: sousalima.renata@gmail.com; izabelaurentino@gmail.com

S. Almeida
Laboratory of Bioacoustics, EcoAcoustic Research Hub, Universidade Federal do Rio Grande do Norte, Natal, RN, Brazil

Projeto Ariranhas, Instituto Natureza do Tocantins—Naturatins, Palmas, Brazil
e-mail: samaraalmeida@gmail.com

C. Leuchtenberger
Instituto Federal Farroupilha, Santa Maria, RS, Brazil

Projeto Ariranhas, Arroio do Meio, RS, Brazil

IUCN Species Survival Commission, Otter Specialist Group, Gland, Switzerland
e-mail: caroleucht@gmail.com

© The Author(s) 2025
C. Erbe et al. (eds.), *Marine Mammal Acoustics in a Noisy Ocean*,
https://doi.org/10.1007/978-3-031-77022-7_6

6.1 Introduction

Otters are semiaquatic mammals in the subfamily Lutrinae. There are 13 extant species. Only the sea otter is classed as a marine mammal; the others occupy rivers or estuaries. Sound production has been studied in six otter species.

6.2 Sound Production in Otters

There are still many gaps in our understanding of the anatomical and physiological characteristics related to sound production in the subfamily Lutrinae. A recent review of mustelid communication suggests that sound signals are favored in habitats where chemical cues are less efficient

(Mumm and Knörnschild 2022). Otters are semi-aquatic mammals, and most of their vocal communication happens at the water's surface. Their airborne sounds are produced by the vibration of vocal cords, as in other terrestrial mammals. Despite recording efforts, underwater acoustic communication has not been well documented for most species, with a few underwater sounds described only for the social giant otter (Schenck et al. 1995; Mumm and Knörnschild 2014). The otters' aerial vocalization system may present adaptations for propagation in aquatic environments, such as the emission of loud and high-pitched sounds that suffer less attenuation at the water surface, as suggested by Bettoni et al. (2021).

Vocal communication in otters is well developed (Leuchtenberger et al. 2014; Lélias et al. 2021). The 13 species of otters vary considerably in their social systems, which range from relatively solitary (e.g., Eurasian otters, *Lutra lutra*, and neotropical otters, *Lontra longicaudis*) to highly social and gregarious (e.g., sea otters, *Enhydra lutris*, and giant otters, *Pteronura brasiliensis*) (Fig. 6.1). Thus, the acoustic communication patterns of otters are a possible model for understanding how mammalian vocal behavior has evolved from terrestrial to aquatic habitats and among different social systems since it has been shown that vocal communication in otters is strongly correlated with sociability (Leuchtenberger et al. 2014, Table 6.1). However, even for the most solitary otter species, sounds play an important role in social interactions (e.g., in neotropical otters; Laurentino 2020; Laurentino et al. unpublished).

6.3 Functions of Otter Sounds

Despite the differences in the social systems and, consequently, in the complexity of the vocal repertoire (Leuchtenberger et al. 2014), otters share similar vocalization types, which present some consistency in acoustic structure and usage in certain behavioral contexts (Bettoni et al. 2021). Affiliative sounds, such as "chirps," "twitters," and "coos," are usually emitted in close contact, such as between mothers and their offspring. Other long-distance contact calls, such as "chirps" and "adult calls," may be emitted by individuals isolated from their social group. Aggressive and agonistic sounds, such as "screams" and "growls," are used during conflicts, such as in defense of food or during territorial fights.

Some sounds may encode the identity of senders, suggesting that receivers could use them to identify familiar conspecifics (McShane et al. 1995). Almonte (2014) observed individual variation in almost all sound types described for North American river otters. Vocal signatures were identified in contact calls of the social small-clawed otters (Lemasson et al. 2013) and giant otters (Mumm et al. 2014), supporting the role of these sounds in individual recognition and maintenance of group cohesion. McShane et al. (1995) observed significant acoustic differences in the "screams" of sea otters, particularly among mothers and young, suggesting that this sound may function in individual recognition to ensure correct direction of maternal care (as in other marine mammals such as manatees; Sousa-Lima et al. 2002; see Chap. 7).

Acoustic traits that are determined by vocal tract filtering, such as formants (Fitch 1997), can aid receivers in estimating the sender's sex and size as well as recognizing individuals (Fitch 1997; Fitch and Fritz 2006; Vannoni and McElligott 2007). Nonlinear phenomena (e.g., deterministic chaos, biphonations, and subharmonics), transitions between call types (e.g., calls which transition from pulsed to tonal characteristics over the duration of the call), and gradations between call types (i.e., calls which display features intermediate of different call types) might add complexity to calls and repertoires and encode information about the sender (Leuchtenberger et al. 2014). Acoustic recognition of individual sex can help receivers find or assess potential mating partners or avoid conflict with rivals (Briefer et al. 2008; Leuchtenberger et al. 2016a). In particular, the frequency and dispersion of formants in the harsh alarm calls of giant otters might encode the sex of senders (Leuchtenberger et al. 2016a).

Fig. 6.1 Six species of otters for which the vocal repertoire has been described: (**a**) small-clawed otter (*Aonyx cinereus*; Photo: N. Duplaix with permission); (**b**) Eurasian otter (*Lutra lutra*; Photo: N. Duplaix with permission); (**c**) sea otter (*Enhydra lutris*; Photo: John Stewart with permission); (**d**) North American river otter (*Lontra canadensis*; Photo: N. Duplaix with permission); (**e**) neotropical otter (*Lontra longicaudis*; Photo: N. Duplaix with permission); (**f**) giant otter (*Pteronura brasiliensis*; Photo: C. Leuchtenberger with permission)

Although sexual dimorphism in giant otters is not adequately documented, some authors suggest that giant otter adult males have larger body sizes and wider necks than adult females (Duplaix 1980; Carter and Rosas 1997). Their "snorts" could provide the means for identifying and selecting suitable mating partners in the species (Leuchtenberger et al. 2016a).

Giant otters are the most social among otter species and are highly territorial (Leuchtenberger et al. 2015). Agonistic encounters are very costly and may lead to the death of individuals, and the death of key individuals in the group can lead to loss of offspring and disruption of the group structure as a whole (Ribas and Mourão 2004; Gonchoroskiz et al. 2025). Therefore, giant otters are expected to employ strategies to avoid conflict, such as patrolling, scent-marking (Leuchtenberger and Mourão 2009), and acoustic communication (Leuchtenberger et al. 2014; Mumm and Knörnschild 2014, 2022). Loud and long-range "screams" are commonly used when giant otters are alarmed, excited, or isolated (Leuchtenberger et al. 2014; Mumm and Knörnschild 2014) and, in threatening situations, when giant otters from a particular group vocalize in chorus simultaneously (Mumm and Knörnschild 2017). Mumm and Knörnschild (2017) found evidence for acoustic signatures in group choruses, which could be encoding not just the presence of the resident group to an intruder but also information on group composition, size, strength, and willingness to fight (Mumm and Knörnschild 2017). The combination of giant otter calls in a chorus during mobbing events may deter potential predators, such as jaguars (Leuchtenberger et al. 2016b).

Although sociability is hypothesized to explain the complexity of acoustic communication in otters (Leuchtenberger et al. 2014), there

Table 6.1 Call type characteristics of otter sounds

Call type	Frequency range (Hz)[a]	Duration (s)[b]	Comments[c]	Source level (dB)[d]	Location[e]	References
Small-clawed otter (*Aonyx cinereus*)						
U1	f_p 2429–2797	0.19–0.21	Captive; atonal	–	Planete Sauvage Animal Park (FRA)	Lemasson et al. (2014)
U2	f_p 2143–2895	0.26–0.35	Captive; many harmonics; FM	–	Planete Sauvage Animal Park (FRA)	Lemasson et al. (2014)
U3	f_p 3954–4946	0.12–0.15	Captive; many harmonics; FM	–	Planete Sauvage Animal Park (FRA)	Lemasson et al. (2014)
U4	f_p 251–301	2.25–2.71	Captive; no harmonics; no FM	–	Planete Sauvage Animal Park (FRA)	Lemasson et al. (2014)
Eurasian otter (*Lutra lutra*)						
Blow also called "hah!"	1–13,000	0.3–0.41	Captive; atonal	–	Otter Breeding Centre of the Ticino Valley Nature Park, (ITA);	Gnoli and Prigioni (1995)
					Zoo Tierpark–Bern Dählhölzli (CHE)	Castillo (2019)
Mewing	~1–16,000 estimated from printed spectrogram	1.1–4	Captive	–	Otter Breeding Centre of the Ticino Valley Nature Park, (ITA);	Gnoli and Prigioni (1995)
					Zoo Tierpark–Bern Dählhölzli (CHE)	Castillo (2019)
Cries (subcategories: staccato and scream)	2–>16,000	1.1 (variable)	Captive; tonal; agonistic	–	Otter Breeding Centre of the Ticino Valley Nature Park, (ITA);	Gnoli and Prigioni (1995)
					Zoo Tierpark–Bern Dählhölzli (CHE)	Castillo (2019)
Food call	f_0 666–846	0.1–0.19	Captive	–	Zoo Tierpark–Bern Dählhölzli (CHE)	Castillo (2019)
Murmur similar to "coo"	LF	–	Captive; mothers and cubs; contact call	–	Otter Breeding Centre of the Ticino Valley Nature Park, (ITA);	Gnoli and Prigioni (1995)
					Zoo Tierpark–Bern Dählhölzli (CHE)	Castillo (2019)
Loud whistle	5000–7000	Longer than the feeble whistle	Captive; distance call	–	Otter Breeding Centre of the Ticino Valley Nature Park, (ITA);	Gnoli and Prigioni (1995)
					Zoo Tierpark–Bern Dählhölzli (CHE)	Castillo (2019)
Feeble whistle	–	–	Captive	–	Otter Breeding Centre of the Ticino Valley Nature Park, (ITA);	Gnoli and Prigioni (1995)
					Zoo Tierpark–Bern Dählhölzli (CHE)	Castillo (2019)
Twitter	4–16,000	~0.5 estimated from printed spectrogram	Captive; FM; produced by newborn pups toward their mother inside the den	–	Zoo Wildnispark, Zürich (CHE)	Gnoli and Prigioni (1995)
						Castillo (2019)

(continued)

Table 6.1 (continued)

Call type	Frequency range (Hz)[a]	Duration (s)[b]	Comments[c]	Source level (dB)[d]	Location[e]	References
Sea otter (*Enhydra lutris*)						
Scream young	f_0 804–814	0.60–0.66	Captive and wild; complex	–	Monterey Bay Aquarium, California (USA)	McShane et al. (1995)
Scream female	f_0 840–870	0.59–0.63	Captive and wild; complex	–	Monterey Bay Aquarium, California (USA)	McShane et al. (1995)
Whine	f_0 301–319	0–1.25	Captive and wild; several harmonics	–	Monterey Bay Aquarium, California (USA)	McShane et al. (1995)
Whistle	f_0 2253–2575	0.53–0.59	Captive and wild; young; tonal with 3–4 harmonics	–	Monterey Bay Aquarium, California (USA)	McShane et al. (1995)
Squeal–whine	f_0 697–729	0.65–0.73	Captive and wild; harmonic	–	Monterey Bay Aquarium, California (USA)	McShane et al. (1995)
Squeal–scream	f_0 763–817	0.69–0.73	Captive and wild; harmonic	–	Monterey Bay Aquarium, California (USA)	McShane et al. (1995)
Whimper	f_0 438–508	0.20–0.24	Captive	–	Monterey Bay Aquarium, California (USA)	McShane et al. (1995)
Squeak type 1	f_0 2044–2676	0.25–0.33	Captive and wild; 1–2 harmonics	–	Monterey Bay Aquarium, California (USA)	McShane et al. (1995)
Squeak type 2	f_0 615–1239	0.21–0.29	Captive and wild; 1–2 harmonics	–	Monterey Bay Aquarium, California (USA)	McShane et al. (1995)
Hiss	1596–15,166	0.44–0.52	Wild	–	Moss Landing Harbor (USA, during tagging and translocation)	McShane et al. (1995)
Growl	f_0 347–423	0.43–0.55	Wild; complex	–	Moss Landing Harbor (USA, during tagging and translocation)	McShane et al. (1995)
Coo	f_0 266–278	0.69–0.97	Captive	–	Monterey Bay Aquarium, California (USA)	McShane et al. (1995)
Grunt	f_0 253–293	0–0.48	Young	–	–	McShane et al. (1995)
North American river otter (*Lontra canadensis*)						
Whine adult	108–7714	0.3–2.5	Captive; 1–2 harmonics	59–99	Atlantis Aquarium, NY; Turtle Back Zoo, NJ; Beardsley Zoo, CT; Palisades Park Conservancy, NY; Stamford Farm and Museum, CT (USA);	Almonte (2014)
	703–10,258	0.77			The Maritime Aquarium and Stamford Museum (USA)	Walkley et al. (2018)
Whine pup	715–5606	0.3–0.7	Captive; 1–2 harmonics	51–74	Atlantis Aquarium, NY; Turtle Back Zoo, NJ; Beardsley Zoo, CT; Palisades Park Conservancy, NY; Stamford Farm and Museum, CT (USA)	Almonte (2014)

(continued)

Table 6.1 (continued)

Call type	Frequency range (Hz)[a]	Duration (s)[b]	Comments[c]	Source level (dB)[d]	Location[e]	References
Chirp adult	615–12,159	0.2–0.6	Captive; 1–3 harmonics	66–99	Atlantis Aquarium, NY; Turtle Back Zoo, NJ; Beardsley Zoo, CT; Palisades Park Conservancy, NY; Stamford Farm and Museum, CT (USA);	Almonte (2014)
	2082–11,084	0.08			The Maritime Aquarium and Stamford Museum (USA)	Walkley et al. (2018)
Chirp pup	1206–5329	0.1–0.3	Captive; 1–2 harmonics	61–87	Atlantis Aquarium, NY; Turtle Back Zoo, NJ; Beardsley Zoo, CT; Palisades Park Conservancy, NY; Stamford Farm and Museum, CT (USA)	Almonte (2014)
Chatter adult also called chatter chirp	189–7644	0.5–3.1	Captive	66–103	Atlantis Aquarium, NY; Turtle Back Zoo, NJ; Beardsley Zoo, CT; Palisades Park Conservancy, NY; Stamford Farm and Museum, CT (USA);	Almonte (2014)
	1253–15,412	0.7		64–91	The Maritime Aquarium and Stamford Museum (USA)	Walkley et al. (2018)
Chatter pup	2206–5555	0.5–0.7	Captive	44–60	Atlantis Aquarium, NY; Turtle Back Zoo, NJ; Beardsley Zoo, CT; Palisades Park Conservancy, NY; Stamford Farm and Museum, CT (USA)	Almonte (2014)
Creek	400–7697	0.2–2.2	Captive	57–96	Atlantis Aquarium, NY; Turtle Back Zoo, NJ; Beardsley Zoo, CT; Palisades Park Conservancy, NY; Stamford Farm and Museum, CT (USA)	Almonte (2014)
Squeak	112–12,077	0.6–3.6	Captive; transition, gradation; 2–4 harmonics	82–104	Atlantis Aquarium, NY; Turtle Back Zoo, NJ; Beardsley Zoo, CT; Palisades Park Conservancy, NY; Stamford Farm and Museum, CT (USA);	Almonte (2014)
	416–17,426	2.4		76–104	The Maritime Aquarium and Stamford Museum (USA)	Walkley et al. (2018)
Scream	61–6925	0.3–2.7	Captive	86–103	Atlantis Aquarium, NY; Turtle Back Zoo, NJ; Beardsley Zoo, CT; Palisades Park Conservancy, NY; Stamford Farm and Museum, CT (USA)	Almonte (2014)
Grunt	621–5099	0.3–1.1	Captive	70–82	Atlantis Aquarium, NY; Turtle Back Zoo, NJ; Beardsley Zoo, CT; Palisades Park Conservancy, NY; Stamford Farm and Museum, CT (USA)	Almonte (2014)

(continued)

Table 6.1 (continued)

Call type	Frequency range (Hz)[a]	Duration (s)[b]	Comments[c]	Source level (dB)[d]	Location[e]	References
Swish	581–4527	0.4–2.4	Captive	64–81	Atlantis Aquarium, NY; Turtle Back Zoo, NJ; Beardsley Zoo, CT; Palisades Park Conservancy, NY; Stamford Farm and Museum, CT (USA)	Almonte (2014)
Hiss	1137–6488	0.6–1.4	Captive	65–80	Atlantis Aquarium, NY; Turtle Back Zoo, NJ; Beardsley Zoo, CT; Palisades Park Conservancy, NY; Stamford Farm and Museum, CT (USA)	Almonte (2014)
Blow	290–15,810	0.2–0.4	Captive	75–94	Atlantis Aquarium, NY; Turtle Back Zoo, NJ; Beardsley Zoo, CT; Palisades Park Conservancy, NY; Stamford Farm and Museum, CT (USA);	Almonte (2014)
	2362–20,124	0.16		52–70	The Maritime Aquarium and Stamford Museum (USA)	Walkley et al. (2018)
Hiccup	98–6472	0.1–0.3	Captive; 1–2 harmonics	60–97	Atlantis Aquarium, NY; Turtle Back Zoo, NJ; Beardsley Zoo, CT; Palisades Park Conservancy, NY; Stamford Farm and Museum, CT (USA)	Almonte (2014)
Whistle	724–5815	0.3–0.5	Captive; 1–2 harmonics	58–74	Atlantis Aquarium, NY; Turtle Back Zoo, NJ; Beardsley Zoo, CT; Palisades Park Conservancy, NY; Stamford Farm and Museum, CT (USA)	Almonte (2014)
Call A	268–3797	0.1	Captive	69–82	The Maritime Aquarium and Stamford Museum (USA)	Walkley et al. (2018)
Call B	268–3797	0.1	Captive	69–82	The Maritime Aquarium and Stamford Museum (USA)	Walkley et al. (2018)
Call C	408–3021	0.1	Captive	78–93	The Maritime Aquarium and Stamford Museum (USA)	Walkley et al. (2018)
Neotropical otter (*Lontra longicaudis*)						
Chirp also called type 1	f_0 479–1410	0.02–0.60	Captive and wild; tonal	–	Rio Grande do Sul (BRA)	Bettoni et al. (2021)
	247–5502	0.08–0.16			Rio Grande do Norte (BRA)	Laurentino et al. (unpublished)
Squeak	f_0 302–836	0.02–0.06	Captive; tonal	–	Rio Grande do Sul (BRA)	Bettoni et al. (2021)
Chuckle	f_0 137–301	0.09–0.18	Captive; tonal	–	Rio Grande do Sul (BRA)	Bettoni et al. (2021)
Growl	f_0 89–135	0.19–0.88	Captive; tonal	–	Rio Grande do Sul (BRA)	Bettoni et al. (2021)
Hah also called type 6	f_p 750–3450	0.3–0.9	Captive; irregular	–	Rio Grande do Sul (BRA)	Bettoni et al. (2021)
	730–4088	0.3			Rio Grande do Norte (BRA)	Laurentino et al. (unpublished)

(continued)

Table 6.1 (continued)

Call type	Frequency range (Hz)[a]	Duration (s)[b]	Comments[c]	Source level (dB)[d]	Location[e]	References
Scream also called type 5	f_p 550–4950	0.08–2.57	Captive and wild; irregular	–	Rio Grande do Sul (BRA)	Bettoni et al. (2021)
	1733–4889	0.8–1.4			Rio Grande do Norte (BRA)	Laurentino et al. (unpublished)
Type 2	493–4935	0.08–0.1	Wild; mother and pup contact call	–	Rio Grande do Norte (BRA)	Laurentino et al. (unpublished)
Type 3	1127–5102	1.02–1.6	Wild; male adult agonistic encounter	–	Rio Grande do Norte (BRA)	Laurentino et al. (unpublished)
Type 4 exhalation sound	1421–4313	0.25–0.39	Wild	–	Rio Grande do Norte (BRA)	Laurentino et al. (unpublished)
Giant otter (*Pteronura brasiliensis*)						
Adult call also called contact call	f_0 110–7410	0.2–0.4	Captive and wild; all ages	–	Pantanal (BRA)	Leuchtenberger et al. (2014)
	f_0 120–26,210	0.23–0.48			Manu National Park (PER); Zoo Tierpark Hagenbeck; Zoo Duisburg and Zoo Dortmund (GER)	Mumm and Knörnschild (2014)
Ascending scream	f_0 154–30,020	0.3–0.9	Captive and wild; young, subadults, adults	–	Manu National Park (PER); Zoo Tierpark Hagenbeck; Zoo Duisburg and Zoo Dortmund (GER)	Mumm and Knörnschild (2014)
Bark	f_0 60–10,076	0.1–0.19	Captive and wild; young, subadults, adults	–	Manu National Park (PER); Zoo Tierpark Hagenbeck; Zoo Duisburg and Zoo Dortmund (GER)	Mumm and Knörnschild (2014)
Begging call	f_0 37–36,172	0.15–0.25	Captive and wild; all ages	–	Manu National Park (PER); Zoo Tierpark Hagenbeck; Zoo Duisburg and Zoo Dortmund (GER)	Mumm and Knörnschild (2014)
Begging scream also called begging scream gradation	f_0 20–8430	0.18–2.24	Captive and wild; young, subadults, adults; gradation	–	Pantanal (BRA)	Leuchtenberger et al. (2014)
	f_0 3–37,216	0.4–0.8			Manu National Park (PER); Zoo Tierpark Hagenbeck; Zoo Duisburg and Zoo Dortmund (GER)	Mumm and Knörnschild (2014)
Contact call gradation	f_0 48–29,536	0.12–0.36	Captive and wild; all ages	–	Manu National Park (PER); Zoo Tierpark Hagenbeck; Zoo Duisburg and Zoo Dortmund (GER)	Mumm and Knörnschild (2014)
Coo also called close call	f_0 130–6230	0.25–0.47	Captive and wild; subadults, adults; discrete harmonic sound	–	Pantanal (BRA)	Leuchtenberger et al. (2014)
	f_0 113–16,053	0.15–0.29	Transition, gradation		Manu National Park (PER); Zoo Tierpark Hagenbeck; Zoo Duisburg and Zoo Dortmund (GER)	Mumm and Knörnschild (2014)

(continued)

Table 6.1 (continued)

Call type	Frequency range (Hz)[a]	Duration (s)[b]	Comments[c]	Source level (dB)[d]	Location[e]	References
Coo–hum	f_0 140–6060	0.11–0.29	Wild; subadults, adults; < 3 harmonics	–	Pantanal (BRA)	Leuchtenberger et al. (2014)
Cub call	f_0 950–21,720	0.2–0.5	Wild; cub; FM	–	Pantanal (BRA)	Leuchtenberger et al. (2014)
Growl	f_0 580–4550	1.64–4.06	Captive and wild; all ages	–	Pantanal (BRA)	Leuchtenberger et al. (2014)
	f_0 71–11,536	0.53–2.57			Manu National Park (PER); Zoo Tierpark Hagenbeck; Zoo Duisburg and Zoo Dortmund (GER)	Mumm and Knörnschild (2014)
Hah	f_0 0–4930	0.09–0.23	Captive and wild; subadults, adults; atonal	–	Pantanal (BRA)	Leuchtenberger et al. (2014)
	f_0 96–31,014	0.13–0.25			Manu National Park (PER); Zoo Tierpark Hagenbeck; Zoo Duisburg and Zoo Dortmund (GER)	Mumm and Knörnschild (2014)
Hum gradation	f_0 60–20,592	0.66–1.54	Captive and wild; subadults, adults; transition, gradation		Manu National Park (PER); Zoo Tierpark Hagenbeck; Zoo Duisburg and Zoo Dortmund (GER)	Mumm and Knörnschild (2014)
High scream	f_0 150–7780	1.89–2.61	Wild; subadults, adults; nonlinear phenomena, biphonation; transition, gradation	–	Pantanal (BRA)	Leuchtenberger et al. (2014)
Hum also called hum short	f_0 20–6020	0.14–0.52	Wild; all ages; transition, gradation; <5 harmonics	–	Pantanal (BRA)	Leuchtenberger et al. (2014)
	f_0 41–11,656	0.19–0.37			Manu National Park (PER); Zoo Tierpark Hagenbeck; Zoo Duisburg and Zoo Dortmund (GER)	Mumm and Knörnschild (2014)
Isolation call	f_0 121–30,126	0.83–121	Wild; young, subadults, adults		Manu National Park (PER)	Mumm and Knörnschild (2014)
Mating call	200–4000 estimated from printed spectrogram	0.6 estimated from printed spectrogram	Captive, wild; adults	–	Manu National Park (PER); Zoo Tierpark Hagenbeck; Zoo Duisburg and Zoo Dortmund (GER)	Mumm and Knörnschild (2014)
Purr	f_0 20–4110	0.27–0.81	Wild; subadults, adults; pulsed	–	Pantanal (BRA)	Leuchtenberger et al. (2014)
	f_0 130–8790	0.52–1.34		–	Pantanal (BRA)	

(continued)

Table 6.1 (continued)

Call type	Frequency range (Hz)[a]	Duration (s)[b]	Comments[c]	Source level (dB)[d]	Location[e]	References
Scream also called wavering scream	f_0 113–13,066	0.68–2.06	Captive and wild; subadults, adults; amplitude modulation; < 11 harmonics; transition, gradation		Manu National Park (PER); Zoo Tierpark Hagenbeck; Zoo Duisburg and Zoo Dortmund (GER)	Leuchtenberger et al. (2014) Mumm and Knörnschild (2014)
Scream gurgle also called suckling call	f_0 130–9910	0.91–1.11	Captive and wild; cub; transition, gradation	–	Pantanal (BRA)	Leuchtenberger et al. (2014)
	f_0 91–25,420	0.02–2.02			Manu National Park (PER); Zoo Tierpark Hagenbeck; Zoo Duisburg and Zoo Dortmund (GER)	Mumm and Knörnschild (2014)
Squeak also called distress call	f_0 0–19,230	0.34–054	Captive and wild; cub	–	Pantanal (BRA)	Leuchtenberger et al. (2014)
	2500–18,000 estimated from printed spectrogram	0.6 estimated from printed spectrogram			Manu National Park (PER); Zoo Tierpark Hagenbeck; Zoo Duisburg and Zoo Dortmund (GER)	Mumm and Knörnschild (2014)
Snort	f_0 120–9990	0.17–0.37	Captive and wild; all ages; pulsed	–	Pantanal (BRA)	Leuchtenberger et al. (2014)
	f_0 59–39,900	0.1–0.22			Manu National Park (PER); Zoo Tierpark Hagenbeck; Zoo Duisburg and Zoo Dortmund (GER)	Mumm and Knörnschild (2014)
Whine	f_0 157–20,772	0.33–1.04	Captive and wild; young, subadults, adults		Manu National Park (PER); Zoo Tierpark Hagenbeck; Zoo Duisburg and Zoo Dortmund (GER)	Mumm and Knörnschild (2014)
Whistle	f_0 1765–34,224	0.14–0.24	Captive and wild; all ages; tonal		Manu National Park (PER); Zoo Tierpark Hagenbeck; Zoo Duisburg and Zoo Dortmund (GER)	Mumm and Knörnschild (2014)
Whistle double	f_0 1012–41,590	0.28–0.68	Captive and wild; cubs, young, subadults; tonal		Manu National Park (PER); Zoo Tierpark Hagenbeck; Zoo Duisburg and Zoo Dortmund (GER)	Mumm and Knörnschild (2014)

Note that sound type classifications present the original descriptions from different authors, and similar sound types might be listed with different names. Abbreviations: *LF* low frequency, *FM* frequency modulation, f_0 fundamental frequency, f_p frequency of peak energy

[a]The overall frequency range is a combination of ranges given (minimum–maximum). When authors gave minimum and maximum frequencies separately, with a standard deviation (SD) for each, the range listed in this column is from the minimum minus its SD to the maximum plus its SD

[b]Authors measured duration in varying resolutions. A maximum of two significant digits is reported

[c]Information about the origin of the animals from which the sound was recorded: captive or wild animals, by which age group the sound was vocalized (e.g., pups, juveniles, or adults), as well as other relevant information about the acoustic characteristics of the sound, such as tonal/atonal, pulsed, or harmonic

[d]Source level is in dB re 20 μPa m. Authors measured amplitudes, but information about the procedure for estimating source level often was not given

[e]ISO alpha-3 country codes: http://www.nationsonline.org/oneworld/country_code_list.htm

are still many gaps in understanding the various factors that correlate with their sound communication system. The North American otter, with its fluid social structure that varies according to habitat and certain ecological conditions, deserves further study to better understand the flexibility of otters' communication systems (Almonte 2014). Another important factor influencing which and how many sound types are produced by otters is the recording context. Studies of sounds produced by otters in captivity and the wild have detected different sounds. For example, studies of neotropical otters recorded in captivity (Bettoni et al. 2021) and in communal latrines in the wild (Laurentino et al. unpublished) resulted in three sounds that are emitted in both contexts (see Table 6.1) but also different sound types in each context, yielding a larger total repertoire size known for the species—nine sound types considering both recording contexts, as opposed to six found in captivity alone (Bettoni et al. 2021; Laurentino et al. unpublished).

6.4 Overview of Otter Sounds

Vocal repertoires have been described for six species: Eurasian otter (Gnoli and Prigioni 1995), sea otter (McShane et al. 1995), North American river otter (Almonte 2014), small-clawed otter (Lemasson et al. 2014), giant otter (Leuchtenberger et al. 2014; Mumm and Knörnschild 2014), and neotropical otter (Bettoni et al. 2021). All repertoires have a unique set of vocalizations emphasizing particular acoustic traits and complexities, usually related to different behavioral contexts. The complexity of the repertoires may be correlated with sociability (Lemasson 2011; Leuchtenberger et al. 2014). The social, group-living giant otter has up to 22 call types (Leuchtenberger et al. 2014; Mumm and Knörnschild 2014, 2022), while solitary species, such as the North American river otter and the neotropical otter, have smaller repertoires of up to nine sounds (Almonte 2014; Bettoni et al. 2021; Laurentino et al. unpublished) (Table 6.1).

The terminology applied to otter calls is not consistent across species, but all species share similarities among several call types within their vocal repertoires, including tonal "chirps," atonal "hahs," harmonic "screams," and harmonic/pulsed "growls." The loud, high-pitched "chirp" (Fig. 6.2) seems to be homologous in all otter species (see "whistles" in Eurasian otters, "squeaks" in sea otters, "adult calls" and "contact calls" in giant otters, and "U3 call" in small-clawed otter). Commonality across species suggests that this call type might have been present in common ancestors (Bettoni et al. 2021).

All species' vocal repertoires include noisy and atonal sounds emitted by a forceful exhalation of air. These sounds are used in inquiry, alarm, and aggressive contexts, such as the "hah" in giant otters and neotropical otters (Duplaix 1980; Leuchtenberger et al. 2014; Mumm and Knörnschild 2014; Bettoni et al. 2021; Laurentino et al. unpublished), the "hiss" in sea otters (McShane et al. 1995), the "blow" in river otters (Almonte 2014) and Eurasian otters (Gnoli and Prigioni 1995), and the "U1 call" in small-clawed otters (Lemasson et al. 2014) (Fig. 6.3).

Otter repertoires also include close-contact calls, such as the "twitter" or "chuckle" shared by neotropical otters (Bettoni et al. 2021) and Eurasian otters (Gnoli and Prigioni 1995), while "coos" are emitted by sea otters (McShane et al. 1995) and giant otters (Leuchtenberger et al. 2014; Mumm and Knörnschild 2014) in friendly close-contact contexts (Fig. 6.4).

Low-frequency, noisy, and pulsed sounds have been described in agonistic contexts. These include "growls" in the vocal repertoires of giant otters (Leuchtenberger et al. 2014; Mumm and Knörnschild 2014), neotropical otters (Bettoni et al. 2021), and sea otters (McShane et al. 1995) (Fig. 6.5).

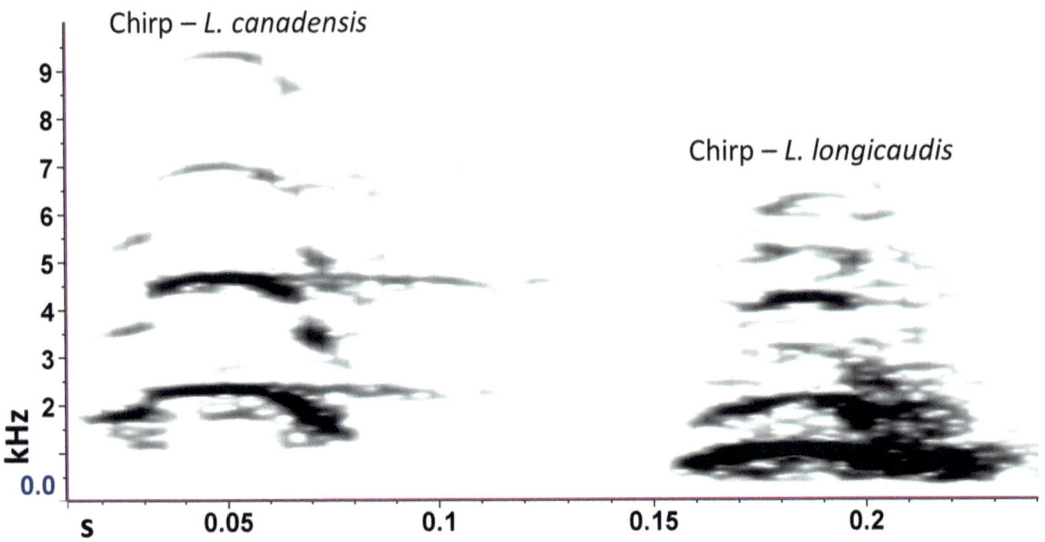

Fig. 6.2 Spectrograms (44,100 Hz sampling frequency, Hann window, 1024 samples DFT size, 85.2% overlap) of "chirps" from North American otters (*Lontra canadensis*) and neotropical otters (*Lontra longicaudis*)

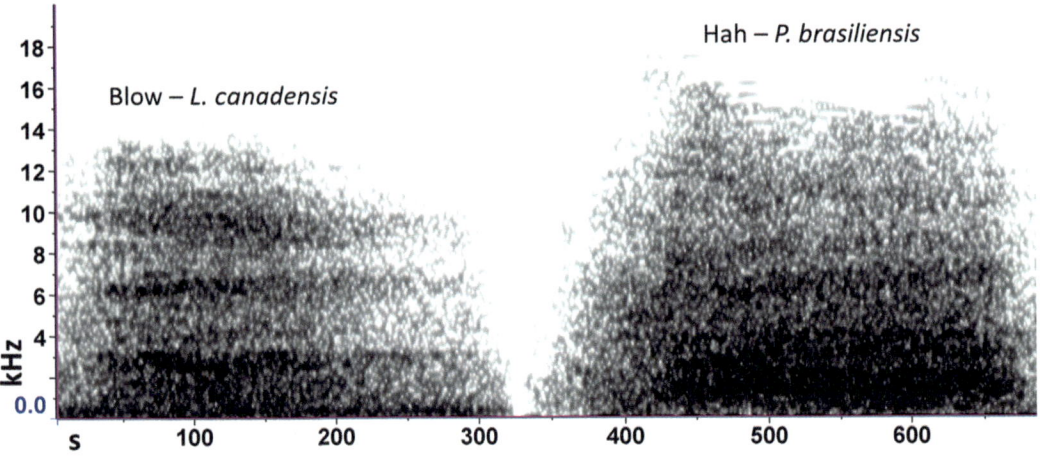

Fig. 6.3 Spectrograms (44,100 Hz sampling frequency, Hann window, 512 samples DFT size, 85.2% overlap) of noisy and atonal sounds from North American river otters (*Lontra canadensis*) and giant otters (*Pteronura brasiliensis*)

Amplitude modulation was reported in some calls, such as the "screams" in giant otters (Leuchtenberger et al. 2014) and neotropical otters (Bettoni et al. 2021) and "grunts" in sea otters (McShane et al. 1995). Frequency modulation was reported in several harmonic calls from all species (Table 6.1). Transitions and gradations were reported in "squeaks" emitted by North American river otters (Almonte 2014) and in several combined sounds emitted by Eurasian otters (Castillo 2019), sea otters (e.g., "growl," "whine," and "squeal"; McShane et al. 1995), and giant otters (e.g., "hum," "purr," "coo," "scream," and "high scream"; Leuchtenberger et al. 2014). Nonlinear phenomena, such as deterministic chaos, biphonations, and subharmonics, were observed in giant otters' "adult screams" and "high screams" (Fig. 6.6; Leuchtenberger et al. 2014) and neotropical otters' "hahs" and "screams" (Bettoni et al. 2021). The ending of

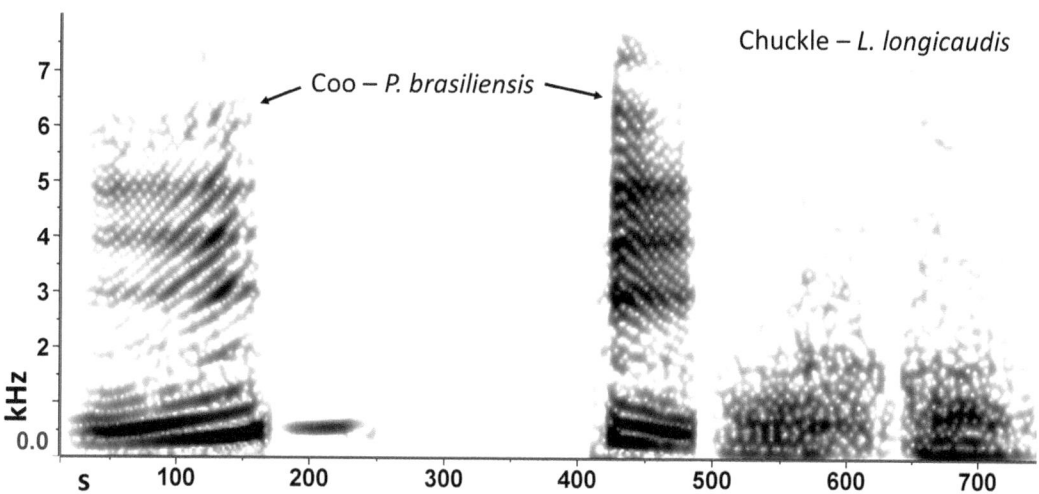

Fig. 6.4 Spectrograms (44,100 Hz sampling frequency, Hann window, 512 samples DFT size, 85.2% overlap) of close-contact calls: "coo" from giant otters (*Pteronura brasiliensis*) and "chuckle" from neotropical otters (*Lontra longicaudis*)

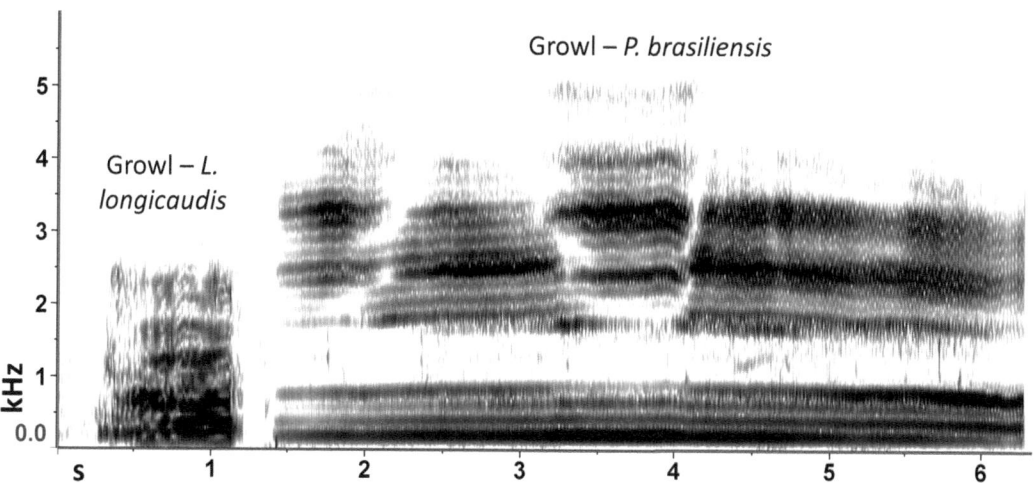

Fig. 6.5 Spectrograms (44,100 Hz sampling frequency, Hann window, 1024 samples DFT size, 85.2% overlap) of agonistic "growls" from neotropical otters (*Lontra longicaudis*; low frequency and noisy) and giant otters (*Pteronura brasiliensis*; pulsed quality of this sound not visible due to set spectrogram parameters)

some giant otters' sounds (e.g., "screams") becomes harsher during more excited contexts, which may reflect the arousal state of senders, as it does in sea otters (McShane et al. 1995). The presence of formants in giant otters' "snorts" and the fundamental frequency of harmonic sounds may be considered an honest indicator of body size and may be individually distinct (Leuchtenberger et al. 2014).

Knowledge about the vocal ontogeny of otters is still limited for all species. Some studies have reported unique calls emitted by young otters, such as the "whistle" by pups under 8 weeks of age in North American river otters (Almonte 2014), "twitter" by newborn Eurasian otters (Gnoli and Prigioni 1995), and "scream gurgle," "squeak," "distress call," and "high whistle" of giant otter cubs (Leuchtenberger et al. 2014;

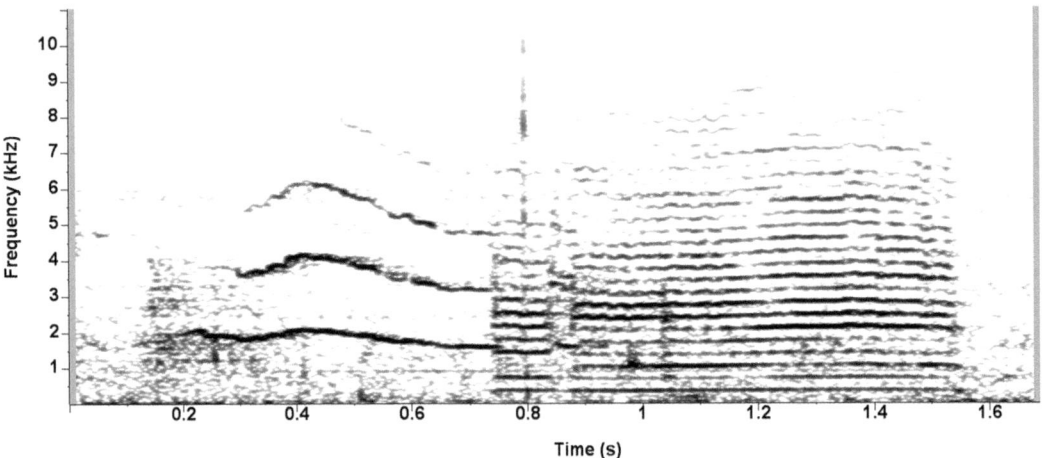

Fig. 6.6 Spectrogram (44,100 Hz sampling frequency, Hann window, 1024 samples DFT size, 87.5% overlap) of a giant otter high scream showing nonlinear phenomena of chaos (at the beginning of the signal) and three frequency jumps at 0.75, 0.85, and 0.9 s

Mumm and Knörnschild 2014). Some calls seem to be comparable to those of adults from an early age, such as the "growl" in giant otters (Leuchtenberger et al. 2014) and the "whine," "chirp," and "chatter" in North American river otters (Almonte 2014). However, ontogenetic studies are needed to improve our understanding of vocal development in otters.

Despite all acoustic similarities among the vocal repertoires described for six otter species, giant otters have the most distinct repertoire. Bettoni et al. (2021) suggested that this distinctiveness could be a phylogenetic signal within the Lutrinae because giant otters are more basal within the family evolutionary tree (de Ferran et al. 2022). Differences in otter species' vocal communication may reflect the greater phylogenetic distance between giant otters and other otters (Bettoni et al. 2021), as well as ecological, morphological, and sociobiological differences. There are still seven species for which vocalizations have not been described, and there are additional gaps in our understanding of the acoustic communication system of otters, such as geographic variation in those species with wide distribution ranges (e.g., neotropical, giant, and Eurasian otters).

Commonalities in otter repertoires are described above to the extent that they could be inferred from available literature. The following sections describe repertoires of individual species, including the terminology adopted and what is known about the behavioral context of each call (Table 6.1).

6.4.1 Small-Clawed Otter

The vocal repertoire of this species has been described from a captive group of ten small-clawed otters; it is composed of four vocal units (U1, U2, U3, U4), emitted in different behavioral contexts. Some repertoire elements are formed from vocal units in combination (CO = U2 + U3) or by repeating single units (RE1 = U2 repeated; RE2 = U3 repeated), for a total of seven types of vocalizations (Lemasson et al. 2014). The U3 call was the most frequently uttered and presented a high potential for acoustic identity coding in adult male.

6.4.2 Eurasian Otter

Captive Eurasian otters use seven types of sounds ("loud whistle," "feeble whistle," "cry," "murmur," "meow," "chirp," "blow," and "twitter"; Gnoli and Prignoli 1995), but the vocal repertoire of this species may include at least one more vocalization (food call) and additional graded calls (Castillo 2019). The "whistle" is the most frequently uttered call (Gnoli and Prignoli 1995; Castillo 2019) and can be statistically discriminated among individuals, based on mean fundamental frequency and call duration parameters (Castillo 2019).

6.4.3 Sea Otter

The vocal repertoire of the California sea otter of both captive and wild populations comprises 13 distinct calls (McShane et al. 1995): "scream," "whistle young," "whistle female," "whine," "hiss," "growl," "coo," "grunt," "squeal–whine," "squeal–scream," "squeak type 1," "squeak type 2," and "whimper." "Screams" vary significantly among individuals in both mothers and young.

6.4.4 North American River Otter

The vocal repertoire of captive North American river otters includes 18 call types (Almonte 2014; Walkley et al. 2018). Eleven adult calls (i.e., "blow," "chatter," "chirp," "creek," "grunt," "hiccup," "hiss," "scream," "squeak," "swish," and "whine") are described by Almonte (2014), and three calls (i.e., A, B, and C) are described by Walkley et al. (2018). Almonte (2014) also described four pup calls ("chatter," "chirp," "whine," and "whistle"). Different call types are associated with different behavioral contexts and arousal states (Almonte 2014). The "squeak" is a combination of "whine" and "chirp." Most sound types showed significant individual variation, except "scream" and "hiss" calls (Almonte 2014).

6.4.5 Neotropical Otter

Vocalizations from neotropical otters maintained in captivity in Southern Brazil can be divided into six call types (Bettoni et al. 2021): "chirp," "chuckle," "growl," "hah," "scream," and "squeak," associated with different behavioral contexts (see Table 1 in Bettoni et al. 2021 for specific contexts of calls). A study on free-living individuals in communal latrines in Northeastern Brazil also identified six sound types (Laurentino et al. unpublished). Nonetheless, the sound types found in captivity and latrine recordings were not identical. Three sound types were similar ("chirp," type 1; "scream," type 5; and "hah," type 6), but types 2, 3, and 4 were only recorded at communal latrines in the wild (Laurentino et al. unpublished). The most frequently uttered call in captivity was the "chuckle," followed by the "hah" (Bettoni et al. 2021), as opposed to the most frequent call recorded in latrines, which was the contact call "chirp," also called type 1 (Laurentino et al. unpublished). The differences found in these repertoires could arise from geographical variation or different environmental contexts of individuals recorded (captive versus wild).

6.4.6 Giant Otter

The vocal repertoire of giant otters is the most complex among otters. Giant otters are also the most social of all otter species. The repertoire has been classified by several authors, resulting in between 6 and 22 call types (see Table 6.1; Duplaix 1980; Leuchtenberger et al. 2014; Mumm and Knörnschild 2014). Mumm and Knörnschild (2014) further distinguished 11 vocalizations emitted exclusively by newborn cubs. The repertoire is enlarged by subtle gradations between call types (i.e., where a call exhibits features intermediate of two types) and combinations of call types (e.g., "scream" and "high scream"; Leuchtenberger et al. 2014). Call types and combinations are associated with different behavioral contexts, and many of them can also reflect the arousal state of senders, such as the "scream" (Leuchtenberger et al. 2014). The "purr" was the most frequently uttered by adults in the Brazilian Pantanal, followed by the "snort," while the "cub call" was the most frequently emitted vocalization by cubs (Leuchtenberger et al. 2014).

Giant otter vocalizations can serve several important functions. Variations in "contact calls" and the affiliative "hum" encode individual differences, with the potential for vocal discrimination and recognition (Mumm et al. 2014). Group choruses play an important role in territorial defense; the identity of groups can be determined from acoustic characteristics of these choruses and may facilitate defense (Mumm and Knörnschild 2017). "Snort" calls encode the sex of senders and reflect physical traits, which may be important for sexual selection (Leuchtenberger et al. 2016a).

6.5 Summary

Otters are aquatic/semiaquatic mammals, and most of their vocal communication happens at the water's surface. The vocal repertoire of six of the 13 species of otters has been studied and seems to be correlated with sociality. There is still considerable uncertainty in the terminology applied to calls, assessment of gradation among call types, and association of calls with behavioral contexts that should be resolved in the future. Nevertheless, because the vocal behavior of the subfamily Lutrinae is relatively well known and because there is considerable variation in the degree of sociality within and between otter species, the Lutrinae represent an important opportunity to clarify relationships among vocal complexity and sociality. This chapter summarized what is known about otter vocal communication, addressing aspects of their sound production and function. It provided an overview of the described species-specific vocal repertoires, with insights into similarities and differences among them.

References

Almonte C (2014) Classification of captive North American river otters (*Lontra canadensis*) vocal repertoires: individual variations, and age class comparisons. Anim Behav Cogn 1(4):502–517. https://doi.org/10.12966/abc.11.07.2014

Bettoni S, Stoeger A, Rodriguez C, Fitch WT (2021) Airborne vocal communication in adult neotropical otters (*Lontra longicaudis*). PLoS One 16(5): e0251974. https://doi.org/10.1371/journal.pone.0251974

Briefer E, Aubin T, Lehongre K, Rybak F (2008) How to identify dear enemies: the group signature in the complex song of the skylark *Alauda arvensis*. J Exp Biol 211(3):317–326. https://doi.org/10.1242/jeb.013359

Carter SK, Rosas FC (1997) Biology and conservation of the giant otter *Pteronura brasiliensis*. Mammal Rev 27(1):1–26. https://doi.org/10.1111/j.1365-2907.1997.tb00370.x

Castillo D (2019) Who is present?—individuality in the call structure of the Eurasian otter (*Lutra lutra*) whistle. Master's thesis, University of Zürich

de Ferran V, Figueiró HV, de Jesus Trindade F, Smith O, Sinding M-HS, Trinca CS, Lazzari GZ, Veron G, Vianna JA, Barbanera F, Kliver S, Serdyukova N, Bulyonkova T, Ryder OA, Gilbert MTP, Koepfli K-P, Eizirik E (2022) Phylogenomics of the world's otters. Curr Biol 32(16):3650–3658. https://doi.org/10.1016/j.cub.2022.06.036

Duplaix N (1980) Observations on the ecology and behavior of the giant river otter *Pteronura brasiliensis* in Suriname. Rev Ecol (Terre Vie) 34(4):495–620

Fitch WT (1997) Vocal tract length and formant frequency dispersion correlate with body size in rhesus macaques. J Acoust Soc Am 102(2):1213–1222. https://doi.org/10.1121/1.421048

Fitch WT, Fritz JB (2006) Rhesus macaques spontaneously perceive formants in conspecific vocalizations. J Acoust Soc Am 120(4):2132–2141. https://doi.org/10.1121/1.2258499

Gnoli C, Prigioni C (1995) Preliminary study on the acoustic communication of captive otters (*Lutra lutra*). Hystrix 7(1–2). https://doi.org/10.4404/hystrix-7.1-2-4083

Gonchoroski GZ, Martin A, Raphael G, Rodrigues LA, Furtado MG, Mourão GM, Leuchtenberger C (2025) Intergroup conflict and myiasis-induced mortality in a giant otter from the Brazilian Pantanal: implications for population conservation. IUCN Otter Spec. Group Bull 42(2):63–70

Laurentino IC (2020) Comportamento e análise bioacústica do repertório vocal da *Lontra longicaudis* (Olfers, 1818), Rio Grande do Norte, Brasil. Master's thesis, Universidade Federal do Rio Grande do Norte

Laurentino I, Sousa R, Corso G, Sousa-Lima RS (unpublished). Vocal repertoire of free-ranging neotropical otters in communal latrines

Lélias ML, Lemasson A, Lodé T (2021) Social organization of otters in relation to their ecology. Biol J Linn Soc Lond 133(1):1–27. https://doi.org/10.1093/biolinnean/blab016

Lemasson A (2011) What can forest guenons "tell" us about the origin of language? In: Vilain A, Schwartz

J-L, Abry C, Vauclair J (eds) Primate communication and human language: vocalisation, gestures, imitation and deixis in humans and non-humans. John Benjamins Publishing Company, pp 39–70. https://doi.org/10.1075/ais.1.04lem

Lemasson A, Mikus MA, Blois-Heulin C, Lodé T (2013) Social partner discrimination based on sounds and scents in Asian small-clawed otters (*Aonyx cinereus*). Sci Nat 100:275–279. https://doi.org/10.1007/s00114-013-1022-9

Lemasson A, Mikus MA, Blois-Heulin C, Lodé T (2014) Vocal repertoire, individual acoustic distinctiveness, and social networks in a group of captive Asian small-clawed otters (*Aonyx cinerea*). J Mammal 95(1):128–139. https://doi.org/10.1644/12-MAMM-A-313.1

Leuchtenberger C, Mourao G (2009) Scent-marking of giant otter in the southern Pantanal, Brazil. Ethology 115(3):210–216. https://doi.org/10.1111/j.1439-0310.2008.01607.x

Leuchtenberger C, Sousa-Lima R, Duplaix N, Magnusson WE, Mourão G (2014) Vocal repertoire of the social giant otter. J Acoust Soc Am 136(5):2861–2875. https://doi.org/10.1121/1.4896518

Leuchtenberger C, Magnusson WE, Mourão G (2015) Territoriality of giant otter groups in an area with seasonal flooding. PLoS One 10(5):e0126073. https://doi.org/10.1371/journal.pone.0126073

Leuchtenberger C, Sousa-Lima R, Ribas C, Magnusson WE, Mourão G (2016a) Giant otter alarm calls as potential mechanisms for individual discrimination and sexual selection. Bioacoustics 25(3):279–291. https://doi.org/10.1080/09524622.2016.1157704

Leuchtenberger C, Almeida SB, Andriolo A, Crawshaw PG (2016b) Jaguar mobbing by giant otter groups. Acta Ethol 19:143–146. https://doi.org/10.1007/s10211-016-0233-4

McShane LJ, Estes JA, Riedman ML, Staedler MM (1995) Repertoire, structure, and individual variation of vocalizations in the sea otter. J Mammal 76(2):414–427. https://doi.org/10.2307/1382352

Mumm CA, Knörnschild M (2014) The vocal repertoire of adult and neonate giant otters (*Pteronura brasiliensis*). PLoS One 9(11):e112562. https://doi.org/10.1371/journal.pone.0112562

Mumm CA, Knörnschild M (2017) Territorial choruses of giant otter groups (*Pteronura brasiliensis*) encode information on group identity. PLoS One 12(10):e0185733. https://doi.org/10.1371/journal.pone.0185733

Mumm CA, Knörnschild M (2022) Mustelid communication. In: Vonk J, Shackelford TK (eds) Encyclopedia of animal cognition and behavior. Springer, Cham. https://doi.org/10.1007/978-3-319-55065-7_1191

Mumm CA, Urrutia MC, Knörnschild M (2014) Vocal individuality in cohesion calls of giant otters, *Pteronura brasiliensis*. Anim Behav 88:243–252. https://doi.org/10.1016/j.anbehav.2013.12.005

Ribas C, Mourão G (2004) Intraspecific agonism between giant otter groups. IUCN Otter Spec Group Bull 21(2):89–93

Schenck C, Staib E, Yasseri AM (1995) Unterwasserlaute bei Riesenottern (*Pteronura brasiliensis*). Mamm Biol Z Saugetierkd 60(5):310–313

Sousa-Lima RS, Paglia AP, da Fonseca GAB (2002) Signature information and individual recognition in the isolation calls of Amazonian manatees, *Trichechus inunguis* (Mammalia: Sirenia). Anim Behav 63(2):301–310. https://doi.org/10.1006/anbe.2001.1873

Vannoni E, McElligott AG (2007) Individual acoustic variation in fallow deer (*Dama dama*) common and harsh groans: a source-filter theory perspective. Ethology 113(3):223–234. https://doi.org/10.1111/j.1439-0310.2006.01323.x

Walkley S, Zapetis M, Lyn H (2018) Vocalizations of North American river otters (*Lontra canadensis*) in two human care populations. J Acoust Soc Am 144(3):1954. https://doi.org/10.1121/1.5068543

Open Access This chapter is licensed under the terms of the Creative Commons Attribution 4.0 International License (http://creativecommons.org/licenses/by/4.0/), which permits use, sharing, adaptation, distribution and reproduction in any medium or format, as long as you give appropriate credit to the original author(s) and the source, provide a link to the Creative Commons license and indicate if changes were made.

The images or other third party material in this chapter are included in the chapter's Creative Commons license, unless indicated otherwise in a credit line to the material. If material is not included in the chapter's Creative Commons license and your intended use is not permitted by statutory regulation or exceeds the permitted use, you will need to obtain permission directly from the copyright holder.

Sirenian Sounds

7

Renata S. Sousa-Lima, Juan Carlos Azofeifa-Solano,
Vera M. F. da Silva, Giovanna A. Dantas, Isadora M. Carletti,
Ann Bowles, Rodney Rountree, and Christine Erbe ⓘ

Contents

7.1	**Introduction**	460
7.2	**Sound Production Mechanisms**	461
7.3	**Biological and Ecological Functions of Sirenian Sounds**	462
7.3.1	Social, Behavioral, and Motivational Contexts	464
7.3.2	Diel Patterns	464
7.3.3	Individual Identity	465
7.3.4	Mother-Calf Contact	465
7.3.5	Vocal Ontogeny	465
7.3.6	Geographic Variability	467
7.4	**Overview of Sirenian Sounds**	467
7.4.1	*Dugong dugon*: Dugong	468
7.4.2	*Trichechus inunguis*: Amazonian Manatee	471

R. S. Sousa-Lima (✉) · I. M. Carletti
Laboratory of Bioacoustics, EcoAcoustic Research Hub,
Universidade Federal do Rio Grande do Norte, Natal, RN,
Brazil
e-mail: sousa.lima@ufrn.br

J. C. Azofeifa-Solano · C. Erbe
Centre for Marine Science and Technology, Curtin
University, Perth, WA, Australia
e-mail: juan.solano@curtin.edu.au; c.erbe@curtin.edu.au

V. M. F. da Silva · G. A. Dantas
Aquatic Mammals Laboratory, Instituto Nacional de
Pesquisas da Amazônia, Manaus, AM, Brazil
e-mail: tucuxi@inpa.gov.br

A. Bowles
Hubbs-Sea World Research Institute, San Diego, CA, USA
e-mail: abowles@hswri.org

R. Rountree
Biology Department, University of Victoria, Victoria, BC,
Canada

The Fish Listener, Waquoit, MA, USA
e-mail: rrountree@fishecology.org

© The Author(s) 2025
C. Erbe et al. (eds.), *Marine Mammal Acoustics in a Noisy Ocean*,
https://doi.org/10.1007/978-3-031-77022-7_7

7.4.3	*Trichechus manatus*: West Indian Manatee	477
7.4.4	*Trichechus senegalensis*: West African Manatee	480
7.4.5	The Case of *Hydrodamalis gigas*: Steller's Sea Cow	483
7.5	**Sirenians in the Anthropocene**	483
7.6	**Sirenian Sounds and Conservation Opportunities**	483
7.7	**Summary**	484
	References	485

7.1 Introduction

"The Siren's song drew me like honey… for the Sirens kept on singing most beautifully, and as one spoke, they all chimed in, singing in unison." The *Odyssey* Book 12 is one of the earliest known references to sirens and their sound. However, in this epic poem, Homer did not describe the physical appearance of these creatures, as it was expected that the seventh-century BCE audience knew what the sirens looked like (Schur 2014). Later artistic representations of sirens underwent a transformation from hybrid creatures with both human and bird elements to women with fish-like traits, and historically in some languages, the terms "siren" and "mermaid" have been interchangeable (Medeiros 2021). It is not surprising that Christopher Columbus, influenced by the beliefs of his time, found "sirens" during his first voyage to America (Flint 2017): "On the previous day, when the Admiral went to the Gold River, he said he quite distinctly saw three mermaids, which rose well out of the sea; but they are not so beautiful as they are said to be, for their faces had some masculine traits," reads an entry dated January 9, 1493 CE, in the compilation of Bartolomé de Las Casas. Today, we believe Columbus probably saw manatees in the Caribbean Sea. However, this is not the only cultural and historical example that links sirenians (manatees and dugongs) with "sirens" or a similar form of mythological feminine creature. The word dugong is borrowed from the Tagalog and derived from the Bahasa word *duyung*, meaning "lady of the sea" in both languages (Marsh 2018). The word manatee is borrowed from the Spanish word *manatí*, which according to Simpson (1941) is believed to be derived from the Carib word *manati*, meaning "a (woman's) breasts," although no recorded language from the Carib or Arawak families uses this word to refer to the manatee but a variation of the words *kuyumuru, koyumolu*, and *koimuru* (Simpson 1941). Other language and cultural similarities include *Mami Wata* and its variations (a water goddess with a mermaid figure) possibly associated with manatees throughout the African Atlantic regions (Drewal 2008; Chang 2016), "queen of the sea" in Kenya, and *peixe mulher* ("womanfish") in Brazil (also called *peixe boi* or "ox fish"). In the eighteenth century CE, the naturalist Georges-Louis Leclerc, Comte de Buffon, mentioned that the *lamantin* (manatee) or *vache marine* (sea cow) was called *sirènes* by numerous travelers, and in fact this animal might have been the "siren" (or "mermaid") of the ancients (Buffon 1799). In the early nineteenth century CE, Sirenia was coined to include both manatees and dugongs (Illiger 1811).

The sirenians, generally known as sea cows, are the only fully aquatic and obligate herbivorous mammals. They inhabit coastal waters, estuaries, and salt and freshwater wetlands, usually in shallow waters that can support their diet (O'Shea et al. 2022). There are two extant families: Dugongidae and Trichechidae. In the eighteenth century CE, there were five known species of sea cows (Committee on Taxonomy 2023): the dugong (*Dugong dugon*), the Amazonian manatee (*Trichechus inunguis*), the West Indian manatee (*Trichechus manatus*, which presently has two recognized subspecies: the Florida manatee, *T. m. latirostris*, and the Antillean manatee, *T. m. manatus*), the West African manatee or African manatee (*Trichechus senegalensis*), and the now extinct Steller's sea cow (*Hydrodamalis gigas*). Sirenia is the living sister taxon of the order Proboscidea (elephants) within the clade Paenungulata (clade Afrotheria). The taxon

Sirenia has had a rich evolutionary and biogeographic history for >48 million years, with more than 60 recorded species (living and fossil) and with records from every continent but Antarctica (Marsh et al. 2011; Domning 2022; Heritage and Seiffert 2022).

Sirenians have an important role in ecosystems as their foraging activities can directly influence the biomass, productivity, composition, and carbon storage of macrophytes and sediments through herbivory and disturbance, and thus indirectly influence other components of the community (Wirsing et al. 2022). They have life history traits that make them susceptible to disturbances: long lives, slow breeding, and low metabolic rates (Marsh et al. 2011). Presently, these species are facing anthropogenic pressures derived from direct human-wildlife interactions (poaching, hunting, incidental mortality in fisheries, and watercraft collisions), indirect human-derived pressures (avoidance of humans, habitat loss from cattle impacts on shores, and modification of waterways), natural threats (predation, severe climatic events, infectious diseases, harmful algal blooms, and macroparasites), and climate change, which is expected to become an increasingly influential threat (Jiménez 2002; Marsh et al. 2011, 2022; Broadwater et al. 2018; Hieb et al. 2021).

Presently, all living sirenian species are at risk according to the IUCN Red List: the dugong is vulnerable (Marsh and Sobtzick 2019), but the subpopulations in Eastern Africa and the Ryukyu Islands are critically endangered (Brownell et al. 2019; Trotzuk et al. 2022); the Amazonian manatee (Marmontel et al. 2016) and the West Indian manatee are vulnerable (Deutsch et al. 2008); the Florida manatee has varied from threatened to endangered in US listings but is listed as endangered by the IUCN (Deutsch 2008); and the Antillean manatee is endangered (Self-Sullivan and Mignucci-Giannoni 2008; Deutsch 2008), with the Antillean manatee population of Brazil critically endangered (Meirelles et al. 2022); finally, the West African manatee is vulnerable (Keith-Diagne 2015).

Early studies investigated sirenian sound production in captivity or, on rare occasions, when free-ranging animals were restrained for a limited time. Although captive studies of sirenian acoustic communication continue to date, technological advances have made investigations in free-ranging individuals practical, revealing a great deal more about the usage patterns and functions of sirenian sounds. Present studies are advancing toward the use of sound as a method to study and develop management strategies to protect these endangered species (Rycyk et al. 2022; Factheu et al. 2023). The vocal behaviors of the dugong and the Amazonian and West Indian manatees have been relatively well described, in contrast to those of the West African manatee. Recently, O'Shea et al. (2022) and Henaut et al. (2022) reviewed most of the information about sirenian vocal communication.

Not all our readers might consider sirenian vocalizations as spellbinding as the legendary songs of Homer's sirens; however, we think that sirenian bioacoustics and ecology are as mesmerizing as any epic poem. The following sections present an overview of sound production, usage, and features.

7.2 Sound Production Mechanisms

Sirenian sound production is hypothesized to involve vocal-fold-like oscillators, situated above the trachea, as sound sources and nasal cavities that act as filters. Air forced through the excised larynx of Florida manatees produced sounds similar to those of other living mammals (Grossman et al. 2014). Although the mechanisms are still hypothetical because manatee sound production has not been measured directly using imaging techniques, it seems clear that sound is generated when air vibrates the bilateral strips of tissue located in the lateral laryngeal walls (vocal ligaments in Fig. 7.1). These strips of tissue are thought to be modified vocal folds containing ligaments that are connected anteriorly to the posterior side of the thyroid cartilage and posteriorly to the arytenoid cartilages (Fig. 7.1; Grossman et al. 2014). Movements of these modified vocal folds (Vf in Fig. 7.2) are likely to modulate the airflow from

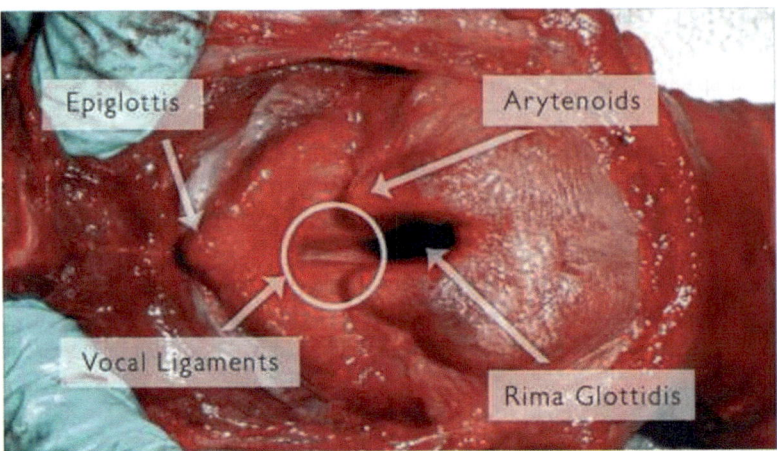

Fig. 7.1 Front view of a Florida manatee larynx showing the glottal opening (*rima glottidis*) surrounded by the two protruding arytenoid cartilages. The leaf-shaped epiglottis can be seen as a projecting structure in the middle left. The circled region marked as "vocal ligaments" indicates the location of two bilateral strips of cartilaginous tissue embedded in the lateral laryngeal walls along the margins of the glottis, which are in close approximation anteriorly but diverge posteriorly. These presumably modified vocal ligaments are connected anteriorly to the posterior side of the thyroid cartilage and posteriorly with arytenoid cartilage. Reprinted Fig. 6 from Grossman et al. (2014). The vocalization mechanism of the Florida manatee (*Trichechus manatus latirostris*). Online J Biol Sci 14 (2):127–149. https://doi.org/10.3844/ojbsci.2014.127.149. © Grossman et al. (2014). Published CC BY 4.0; https://creativecommons.org/licenses/by/4.0/

the trachea under the muscular control of the arytenoid cartilages (A in Fig. 7.2; Landrau-Giovannetti et al. 2014). Measurements of sound intensity in living beached West Indian (probably Antillean based on the recording locations) and Amazonian manatees support the anatomical localization of the sound source (Landrau-Giovannetti et al. 2014) indicated in the study of the excised larynx of Florida manatees (Grossman et al. 2014).

Sounds generated in manatee vocal folds are subsequently filtered and modified within the resonance cavities in the frontal area of the head. Air flux within this region causes a transient deformation behind the nostrils before returning to the lungs (Fig. 7.3; Grossman et al. 2014). Usually, no bubbles are released from the nostrils when manatees vocalize under water (Hartman 1979) indicating that there is no airflow through the nostrils. Visible movement contracts and wrinkles the region dorsal to the maxilla and caudal to the closed nostrils (Fig. 7.3) in conjunction with observed call production in dugong calves (Nair and Lal Mohan 1975; Anderson 1981; Anderson and Barclay 1995) and Amazonian and Antillean manatees (Sousa-Lima 1999; Sousa-Lima et al. 2008; Grossman et al. 2014; Landrau-Giovannetti et al. 2014; Brady et al. 2022).

7.3 Biological and Ecological Functions of Sirenian Sounds

The evolution of complex vocal communication systems in mammals has been associated with high sociability or group hunting (Freeberg et al. 2012; Marino 2022). Extant sirenians, however, have solitary habits and are the only obligate herbivorous aquatic mammals, which might limit the development of complex vocal communication systems seen in other marine mammals (Sousa-Lima 1999; Umeed et al. 2017; Brady et al. 2020). Nonetheless, slow reproductive rates and extended maternal care may have promoted selection for some level of communication complexity in dugongs and manatees (Hedwig et al. 2021).

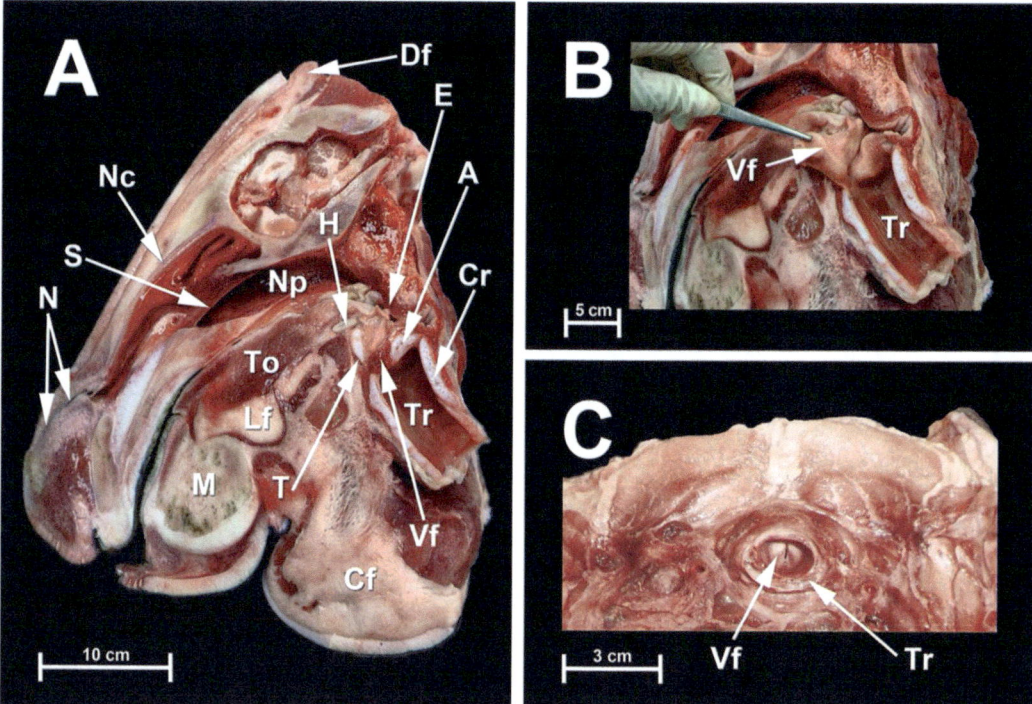

Fig. 7.2 (**a**) Adult Florida manatee head cut along the midsagittal plane, right side. The trachea (Tr) would normally be positioned rotated ~90° dorsal from the position in this figure, running approximately parallel to the vertebral column. In the natural position, the large region of cervical fat (Cf) would lie ventral to the larynx. *A* arytenoid cartilage, *Nc* nasal cartilage, *Cr* cricoid cartilage, *Df* dorsal fat, *E* epiglottis, *H* hyoid bone, *LF* lingual fat, *M* mandible, *N* naris, *Np* nasal passage, *S* septum, *T* thyroid cartilage, *To* tongue, *Tr* trachea, and *Vf* vocal fold. (**b**) Detail of the vocal folds in the larynx. Retraction with the forceps shows the left vocal fold laterally. This reveals the entirety of the right vocal fold traversing the laryngeal lumen. (**c**) Caudal view of a whole manatee male calf head showing a close-up of the cut trachea (Tr) and the lumen that leads to the larynx. The vocal folds (Vf) can be seen as two opposing masses of tissue that obstruct the lumen. Reprinted with permission; Fig. 3 from Landrau-Giovannetti et al. (2014). Acoustical and anatomical determination of sound production and transmission in West Indian (*Trichechus manatus*) and Amazonian (*T. inunguis*) manatees. Anat Rec 297 (10):1896–1907; https://doi.org/10.1002/ar.22993 © Wiley Periodicals, 2014. All rights reserved

Sirenians produce sounds in diverse social and ecological contexts. These sounds mediate interactions among individuals within groups; they transmit the sender's location and encode information about individual identity, gender, age, motivational state, and reproductive status and receptivity (O'Shea et al. 2022). For example, call types and structures and calling rates have been reported to change in the presence of calves in a group (Hartman 1979; Bengston and Fitzgerald 1985; Dantas 2009; Carletti 2016; Tanaka et al. 2017, 2023; Brady et al. 2022). Also, sounds have been associated with functional behaviors, such as territorial defense, foraging, mating, and calf rearing (Henaut et al. 2022).

As with other aquatic animals, environmental conditions can influence the evolution of communication systems and vocal characteristics (van der Sluijs et al. 2011; Frommen 2020). Sirenian sound energy is mainly distributed within 0.5–20 kHz, a frequency range known to propagate efficiently in freshwater, marine, and brackish habitats (Rivera-Chavarría et al. 2015), as well as in shallow seagrass beds (Miksis-Olds and Tyack 2009).

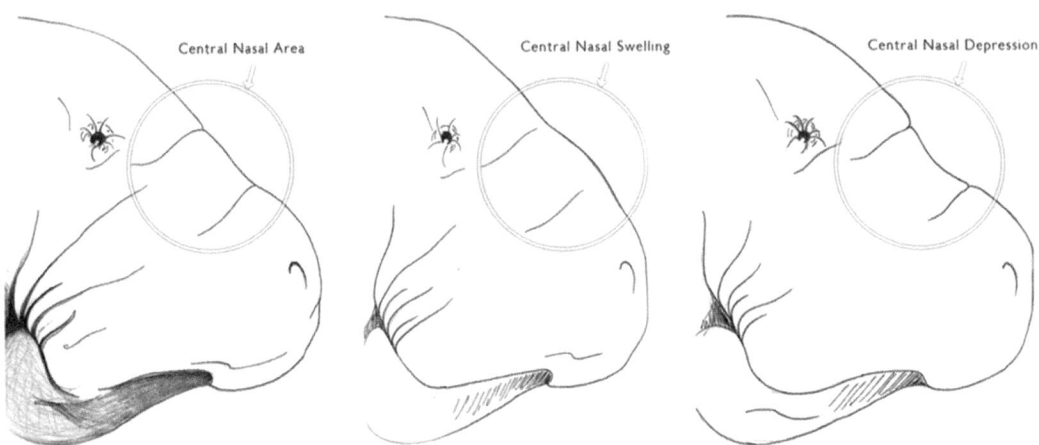

Fig. 7.3 Illustration of the sequential changes observed in the nasal region of a vocalizing Florida manatee at the Cincinnati Zoo and Botanical Garden. Note that the central nasal region swells and then rebounds during an active vocalization. Reprinted Fig. 16 from Grossman et al. (2014). © Grossman et al., 2014. The vocalization mechanism of the Florida manatee (*Trichechus manatus latirostris*). Online J Biol Sci 14 (2):127–149. https://doi.org/10.3844/ojbsci.2014.127.149. Published CC BY 4.0; https://creativecommons.org/licenses/by/4.0/

7.3.1 Social, Behavioral, and Motivational Contexts

There is extensive evidence that sirenians use sound for communication among individuals. For example, dugongs responded to conspecific chirp playbacks; the source level and duration of these response chirps increased significantly with range to the playback speaker, indicating that dugongs compensate for propagation loss in two-way communication (Ichikawa et al. 2011). Antiphonal calling (i.e., alternate calling between two individuals) is typically most pronounced between mothers and their calves (O'Shea and Poché 2006; Vanderhoff and Bernal Hoverud 2022).

Certain call types have been associated with specific behaviors. For example, dugongs use chirp-squeaks while roaming their mutually exclusive "activity zones" in the wild, suggesting a ranging function of these sounds (Anderson and Barclay 1995). These individually exclusive zones and the irregularly distributed calling in space (Ichikawa et al. 2012b; Tanaka et al. 2023) are evidence that dugong acoustic behavior is associated with territorial defense and/or a resource-holding display during bottom feeding and patrolling (Anderson and Barclay 1995; O'Shea et al. 2022). In manatees, calling rates have been reported to increase during traveling and prior to rejoining other individuals (Hartman 1979; Reynolds 1981; Sousa-Lima et al. 2002, 2008; O'Shea and Poché 2006).

Particular call types likely communicate some motivational states. For example, complex call types have been associated with a heightened state of arousal during social milling, while longer and frequency-modulated calls were emitted during stress in Florida manatees (Brady et al. 2022). Furthermore, manatees have been reported to call more frequently if in distress or when frightened (Sousa-Lima et al. 2002; O'Shea and Poché 2006). Captive manatees vocalized during exploratory behaviors, with changes in the number of squeaks and squeals according to the kind of stimuli, and they showed greater interest and vocalized more with submerged stimuli that they were able to manipulate (Charles et al. 2024).

7.3.2 Diel Patterns

Higher nocturnal calling activity was found in dugongs (Ichikawa et al. 2006) and African manatees (Rycyk et al. 2022). Authors hypothesized that this was related to diel feeding patterns, perhaps shaped by human hunting pressure in the case of the African manatee (Factheu et al. 2023). Brady et al. (2023) recently studied the diel vocalization patterns of the Antillean manatee in Belize. They found no significant differences in the number of calls per hour between sites, but the proportion of tonal calls decreased with hours after sunset and increased in boat presence (Brady et al. 2023).

7.3.3 Individual Identity

While context determines the choice of call types, studies have shown that variation in call structure results from differences in the anatomical sound production apparatus related to individuality, sex, and age (Sousa-Lima 1999; Sousa-Lima et al. 2002, 2008). The caller's individual identity information is coded in spectral features, nonlinear phenomena, and duration (Steel 1982; Anderson and Barclay 1995; Sousa-Lima et al. 2002, 2008; Williams 2005; Mann et al. 2006; Umeed et al. 2017; Dietrich et al. 2022). Moreover, vocalization rates vary across individuals (Sousa-Lima 1999) while also depending on social context and motivation (Williams 2005).

The stability of vocal identity over time has been studied in Amazonian (Sousa-Lima and da Silva 2001; Dantas 2009), Florida (Williams 2005), and Antillean manatees (Merchan et al. 2019; Dietrich et al. 2022). Stability was individual-specific, meaning that some individuals maintained their vocal features over time, while others did not. Despite changes in individual features over time, it appears that, within captive groups, individuals retain vocal discriminability (Fig. 7.4; Dietrich et al. 2022). This indicates that multiple animals can be individually discriminated (and possibly recognized) based on their call characteristics at any point in time.

7.3.4 Mother-Calf Contact

The recognition of individual differences facilitates one of the most important social functions of sirenian sounds: the maintenance of contact and proximity between mothers and their dependent calves (Fig. 7.5). The evolution of an acoustic mother-calf recognition mechanism ensures correct direction of prolonged maternal care in the often murky underwater habitats of sirenians (Sousa-Lima et al. 2002, 2008). Calf calling rate tends to increase while isolated from their mothers and prior to (sometimes during) suckling (Sousa-Lima pers. obs.). Sousa-Lima et al. (2002), using playbacks of different calf voices, found that a mother recognizes her own calf among calls from other Amazonian manatee calves of the same age and sex and changes her behavior to approach the sound source. Therefore, isolation calls indeed carry individual information that is recognized by another individual and mediates proximity maintenance between a mother and her calf.

7.3.5 Vocal Ontogeny

Dantas (2009) studied the vocalizations of 38 individual Amazonian manatees living in captivity at the Instituto Nacional de Pesquisas da Amazônia (Brazil), which had their vocalizations recorded over 11 years. Calling rates were compared across age classes, sexes, and reproductive status. Lactating females and calves had higher calling rates. Twenty-three calves were recorded to investigate vocal ontogeny. Cross-correlation of spectrograms over time showed that call acoustic features were fairly stable. Comparisons among all age classes showed that calls tended to become longer with age. In Antillean and Florida manatees, frequency modulation of the call decreased (i.e., the contour flattened) with animal age (Fig. 7.6; Sousa-Lima et al. 2008; Brady et al. 2022). Additionally, fundamental frequency decreased with age in captive Antillean manatees (Sousa-Lima et al. 2008), and larger Florida manatees had lower fundamental frequencies than smaller animals (O'Shea and Poché 2006).

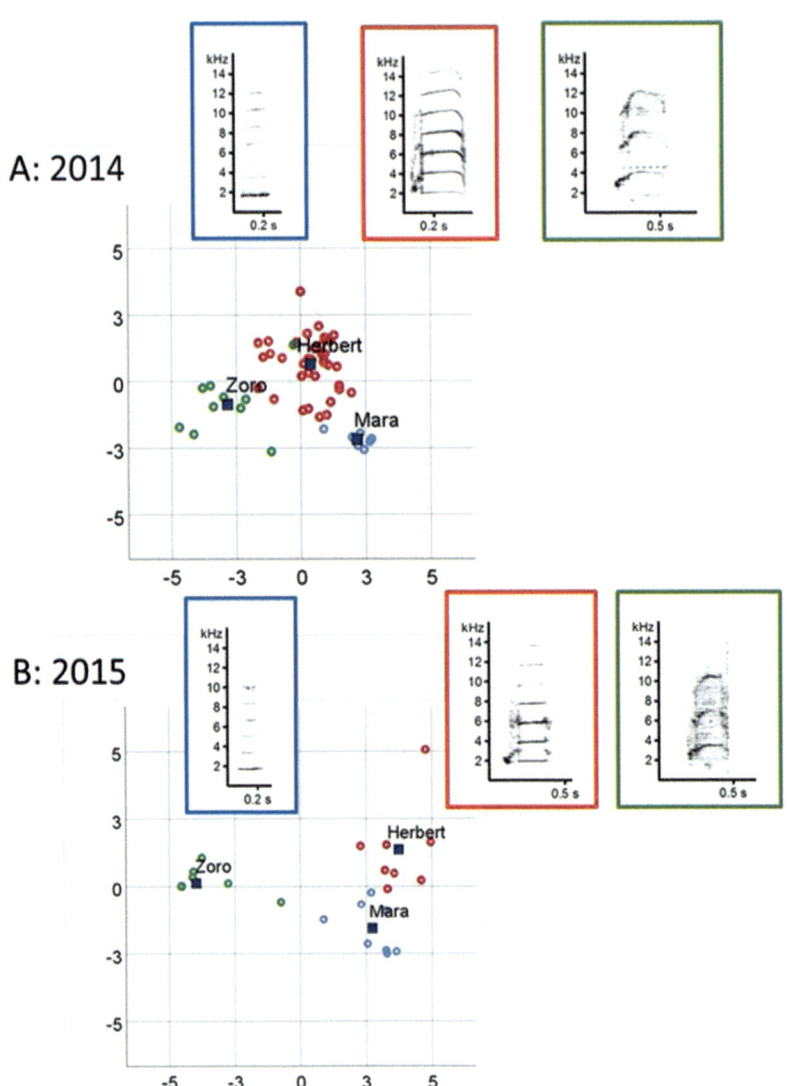

Fig. 7.4 Results of discriminant function analysis of calls from three individual West Indian manatees housed at the Nuremberg Zoo, Germany, in 2014 and 2015. Spectrograms of representative calls for each manatee are discriminated by name and color. Reprinted with permission from Fig. 2 in Dietrich et al. (2022). Signature calls in West Indian Manatee (*Trichechus manatus manatus*)? Aquat Mamm 48 (4):349–354. https://doi.org/10.1578/AM.48.4.2022.349. ⓒAquatic Mammals, 2014. All rights reserved

Mother-calf pair recordings in captivity (Sousa-Lima et al. 2002, 2008) and in the wild (Umeed et al. 2023) show that offspring vocal patterns are similar to those of their mothers and twin siblings. Inherited morphological structures and social learning might explain such similarities (Sousa-Lima et al. 2002, 2008; Umeed et al. 2023). Evolutionary similarities between the closely related Sirenia and Proboscidea (elephants) suggest that a shared ability for vocal learning is worth investigating (Baotic et al. 2022). For example, in both groups, the strongest social bond is between a mother and her calf with a rearing period extending beyond feeding independency; the modulation of sounds occurs where the nasal passages enter the skull; and other anatomical similarities in the highly flexible and multifunctional orofacial structures result in sounds that are structurally similar. However, the only evidence for vocal learning in Sirenia to date comes from an orphaned Amazonian manatee that adjusted its isolation call frequency gradually to match that of its foster mother (Sousa-Lima 2001).

7 Sirenian Sounds

Fig. 7.5 Top: Dugong mother and her calf traveling across the reef lagoon in Nyinggulu (Ningaloo) off Western Australia (photo by JC Azofeifa-Solano). Bottom: Amazonian manatee mother and her calf, which was conceived and born under human care at the Robin Best Aquatic Center, Instituto Nacional de Pesquisas da Amazônia, Manaus, Amazonas (photo by Anselmo d'Affonseca, with permission)

7.3.6 Geographic Variability

Differences in vocalizations across populations of the same species may result from genetic, environmental, or ecological differences driven by geographic separation. Geographic variations in sound characteristics have been documented in dugongs (Ichikawa et al. 2006, 2012a; Tanaka et al. 2017, 2023; Kotaro et al. 2023), Amazonian manatees (Carletti 2016; Sousa-Lima et al. 2002), and West Indian manatees (Nowacek et al. 2003; Brady et al. 2022; Reyes-Arias et al. 2023), yet the behavioral, population genetics, and conservation implications of these differences remain to be studied.

Sample recordings of Amazonian manatees were taken in captivity and in nature along the Amazon River Basin in Brazil and Peru (Fig. 7.7). The acoustic parameters of vocalizations recorded in nature were similar to those in captivity (Sousa-Lima et al. 2013; Carletti 2016), despite significant differences in fundamental frequency and signal duration between the sample groups. Reyes-Arias et al. (2023) found differences in three types of calls of the West Indian manatee among geographic locations at Belize, Panama, and Florida (Fig. 7.8). Such differences could be related to an accumulation of changes in vocalizations between the separated populations, but the authors acknowledged these might as well be the result of individual differences (Sousa-Lima et al. 2002; Carletti 2016; Reyes-Arias et al. 2023). Tanaka et al. (2017, 2023) and Kotaro et al. (2023) related differences in dugong vocalization rates between localities to environmental attributes such as seagrass distribution, presence of other soniferous species, temperature, and oceanographic currents.

7.4 Overview of Sirenian Sounds

Sirenian sounds have acoustically similar characteristics among species; however, species-specific acoustic features can be discriminated statistically (Sousa-Lima et al. 1999; Brady et al. 2022). Call types fall within a continuum along the vocal repertoire, ranging from tonal, harmonic series with little frequency modulation to amplitude-modulated, pulsed, and ultimately noisy sounds that are characterized by high entropy (i.e., disorder).

Nonlinear phenomena (e.g., deterministic chaos, biphonation, and frequency jumps) add more complexity to calls, often encoding metacommunicative information such as state of arousal (Fitch et al. 2002; Mann et al. 2006; Fig. 7.9). These phenomena might be produced

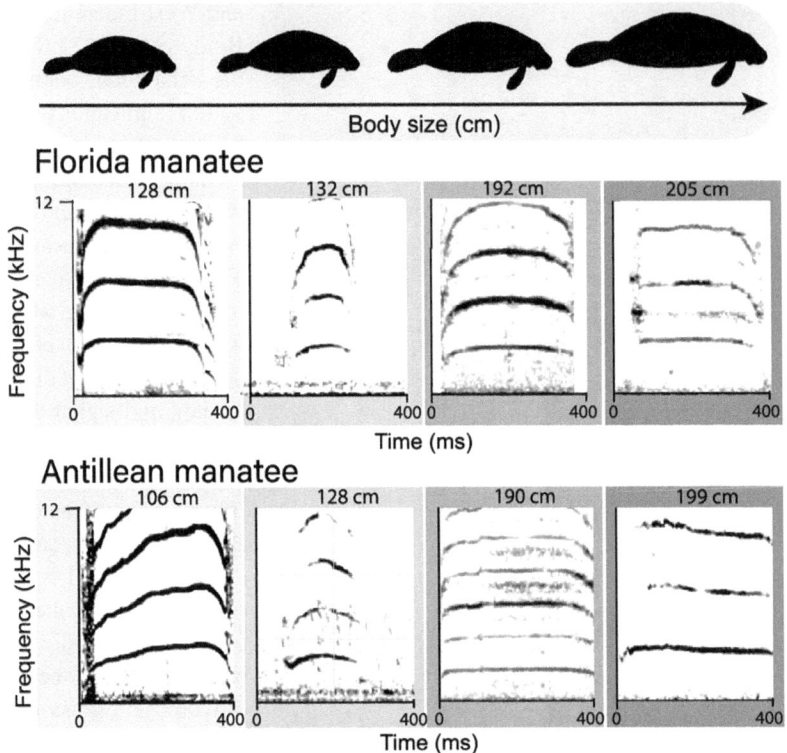

Fig. 7.6 Ontogenetic changes in call structure associated with the size (age class) of different individuals in the two subspecies of the West Indian manatees: the Florida manatee and the Antillean manatee. Reprinted from Brady et al. (2022). Manatee calf call contour and acoustic structure varies by species and body size. Sci Rep 12 (1):19597. https://doi.org/10.1038/s41598-022-23321-7. © Brady et al. (2022). Published CC BY 4.0; https://creativecommons.org/licenses/by/4.0/

in the nasal cavities when sounds originating from the vocal folds are modified. Turbulence at the start and end of the airflow from the lungs can cause chaotic frequency content in sirenian calls. The prevalent nonlinear phenomenon in West Indian manatee calls is deterministic chaos (Mann et al. 2006) and often appears at the beginning or end of the call (also described as asynchronous frequency bands; Schevill and Watkins 1965; Sonoda and Takemura 1973; Sousa-Lima 1999; Mann et al. 2006). In African manatees, deterministic chaos is less common, but a fifth of the calls contained subharmonics (Rycyk et al. 2021).

Sirenians further produce gastrointestinal sounds, grinding sounds while chewing, and mechanical, clicking sounds that may be associated with tooth movement (manatee teeth are constantly being replaced; Domning and Hayek 1984). These sounds are not vocal and may or may not have a communicative function (Sousa-Lima et al. 2002). The following sections present an overview of the sounds produced by each of the four extant sirenian species; detailed characteristics of vocalizations are given by call type in Table 7.1.

7.4.1 *Dugong dugon*: Dugong

The dugong is widespread in the Indo-Pacific region in tropical, inshore waters up to 40 m depth, from Madagascar to the Indo-Malay archipelago, Australia, New Guinea, islands of the Solomon and Coral Seas, and north to the Philippines. A remnant, critically endangered population is found off Okinawa Island, Japan—the northern limit of its range (Kayanne et al. 2022). Dugongs are occasional visitors to New South Wales coastal and estuarine waters (Allen et al. 2004). They often cluster in "hot spots" in shallow, inshore waters, where it is easier to detect their calling activity (Ichikawa et al. 2006, 2012b; Tanaka et al. 2023).

Airborne sounds of a captive young male dugong were described as short (0.1–0.3 s) chirp-squeaks, with frequencies in the band 3–8 kHz (Nair and Lal Mohan 1975). Dugong underwater sounds (Fig. 7.10) have been recorded at a few locations, with quantitative data from Shark Bay, Australia, and Talibong Island, Thailand (off the Malay Peninsula). The majority of their repertoire is composed of amplitude- and

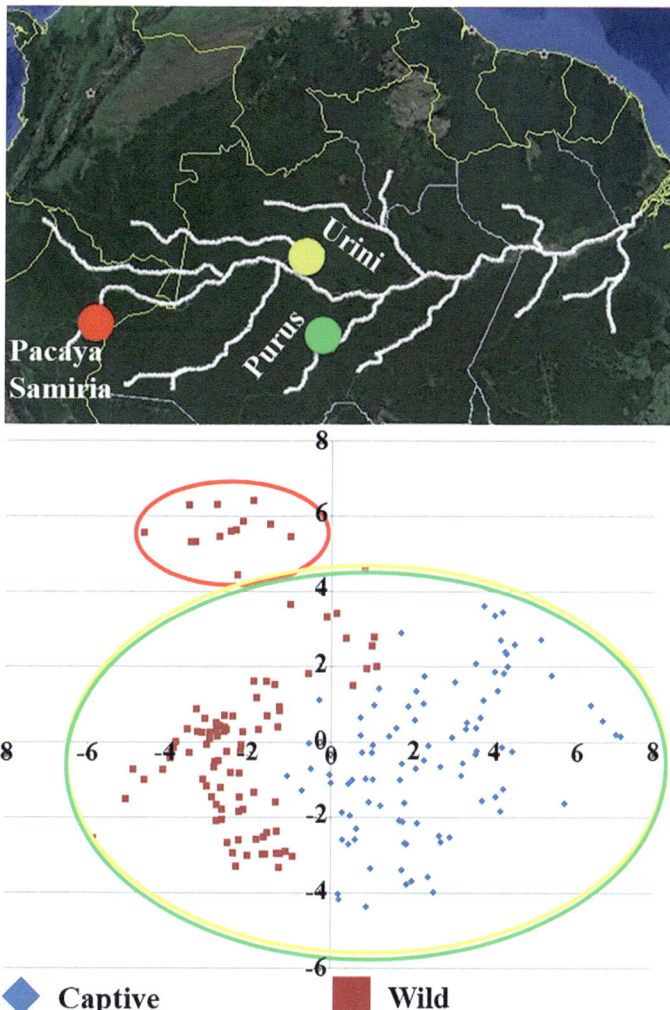

Fig. 7.7 Principal component analysis plot of calls from captive (blue diamonds) and wild (red squares) manatees showing two main groups: Points circled in red are from calls recorded in Peru; points circled in yellow and green are from calls recorded in Brazil. The map shows the recording sites in the Amazon River Basin: Lake Urini in Brazil (yellow, S2° 43′60″ W64°30′00″), Purus River in Brazil (green, S3° 41′35″ W61°28′12″), and Pacaya-Samiria National Reserve in Peru (red, S5° 15′00″ W74°40′00″). Modified from Figs. 2 and 5 in Carletti (2016). Invisíveis porém detectáveis: a utilização do monitoramento acústico passivo em peixes-boi da Amazônia (*Trichechus inunguis*). B.Sc. Thesis, Universidade Federal do Rio Grande do Norte, Natal, Rio Grande do Norte, Brazil. © Carletti (2016). All rights reserved

frequency-modulated chirp-squeaks (or chirps) with fundamental frequencies of 2.7–8 kHz and up to five harmonics (note that the number of harmonics can be affected by recording distance and characteristics of recording systems). The chirp frequency-modulation pattern varies from convex-shaped sounds to downsweeps (Parsons et al. 2013). These short calls last 0.05–0.37 s with source levels of 131–161 dB re 1 µPa m (Anderson and Barclay 1995; Ichikawa et al. 2003, 2006, 2011; Parsons et al. 2013; Tanaka et al. 2023).

Dugongs also produce longer (0.1–4 s) amplitude- and frequency-modulated calls described as trills, long calls, or squeaks (Anderson and Barclay 1995; Ichikawa et al. 2003, 2006, 2011; Parsons et al. 2013). The fundamental frequency of these calls varies between 1.8 and 3.5 kHz, overlapping with the frequency range of chirps. Trills contain repeated modulations and/or multiple elements. They have been suggested as affiliative sounds produced during visual displays (Anderson and Barclay 1995). Dugong pulsed sounds (bursts) cover the range 300 Hz–22 kHz, last 0.1–1.2 s, and have source levels of 129–163 dB re 1 µPa m (Fig. 7.10b; Anderson and Barclay 1995; Parsons et al. 2013).

Dugong barks are harsh broadband sounds with energy between 500 Hz and 2.2 kHz and durations of 0.03–0.12 s (Fig. 7.10c; Anderson

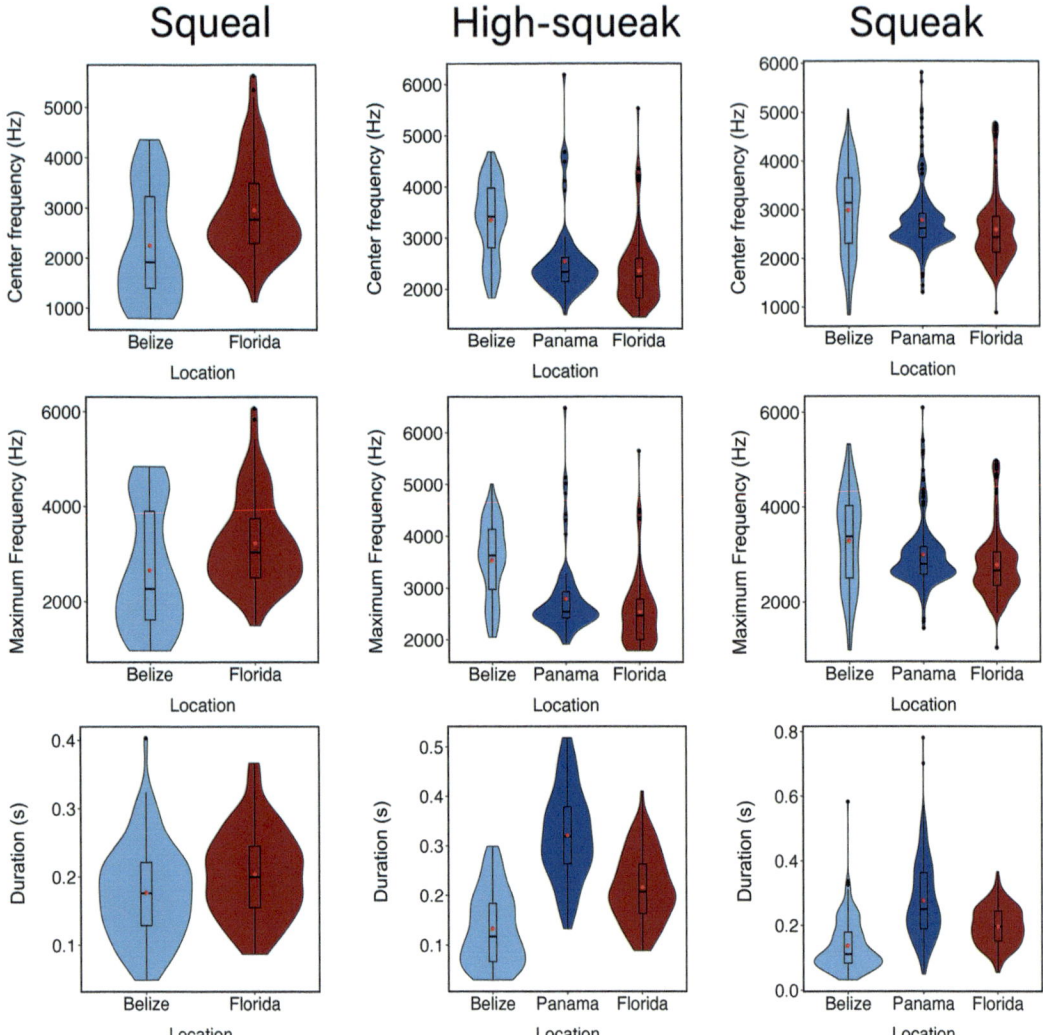

Fig. 7.8 Violin plots and boxplots showing differences in center frequency, maximum frequency, and duration of three call types among three geographically isolated manatee populations. Antillean manatees in blue, Florida manatees in red. Reprinted Fig. 4 from Reyes-Arias et al. (2023). Vocalizations of wild West Indian manatee vary across subspecies and geographic location. Sci Rep 13 (1): 11028. https://doi.org/10.1038/s41598-023-37882-8. © Reyes-Arias et al. (2023). Published CC BY 4.0; https://creativecommons.org/licenses/by/4.0/

and Barclay 1995). Among territorial males, barks emitted by an unhabituated dugong approached by a vessel were associated with aggressive behavior (Anderson and Barclay 1995). Anderson and Barclay (1995) identified "prebarks" and "pretrills" as "transitional sounds between chirp-squeaks and barks and trills" in free-ranging individuals from Australia. Prebarks are modified lower-frequency and gradually noisier chirp-squeaks that are emitted before barks.

Pretrills precede trills, and their acoustic features also gradually change: Their frequency-modulation pattern tends to flatten with increasing duration (Anderson and Barclay 1995).

Dugong calls are usually emitted in sequences—mainly chirp and trill combinations. These transitional sequences and gradual changes in pretrills and prebarks suggest a graded repertoire with sound types that are intermediate between chirps, barks, and trills (Anderson and

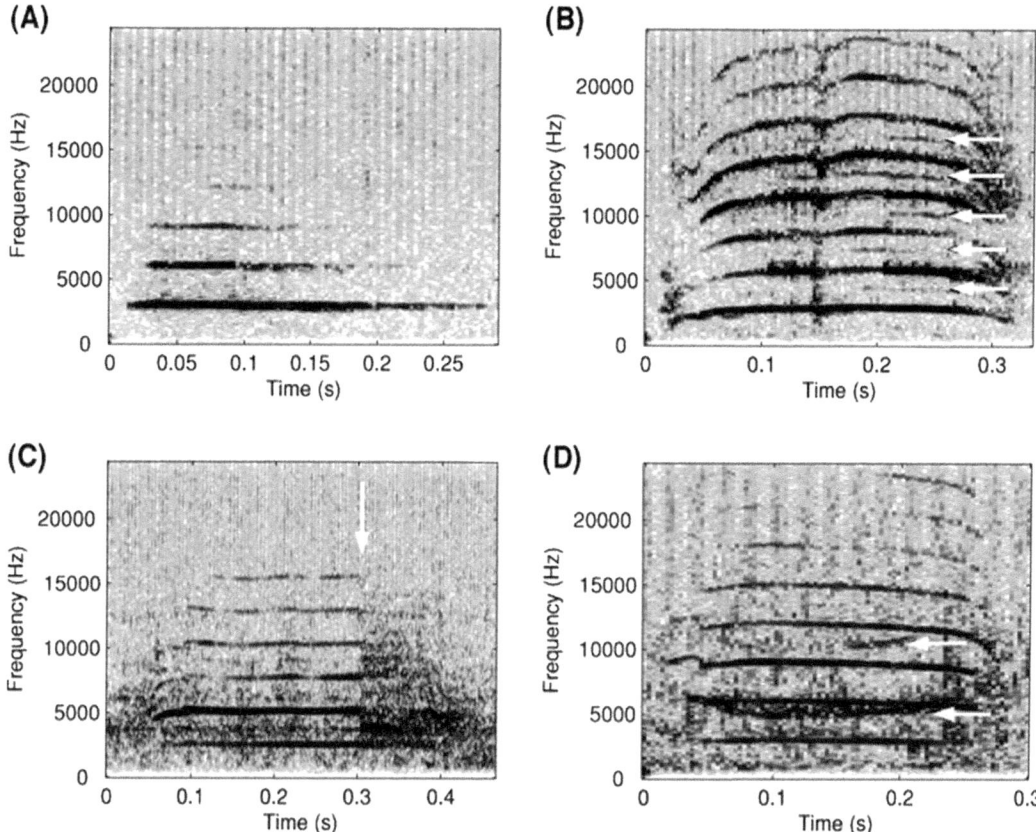

Fig. 7.9 Spectrograms of representative West Indian manatee vocalizations illustrating nonlinear phenomena. The illustrated features are indicated with white arrows. (a) Vocalization showing only harmonics. (b) Vocalization with subharmonics. (c) Bifurcation from tonal harmonic signal to chaotic signal. (d) Biphonation. Reprinted with permission from John Wiley and Sons; Fig. 1 from Mann et al. (2006). Nonlinear dynamics in manatee vocalizations. Mar Mamm Sci 22 (3):548–555. https://doi.org/10.1111/j.1748-7692.2006.00036.x. © Society for Marine Mammalogy, 2006. All rights reserved

Barclay 1995; Okumura et al. 2007). This pattern of emission can complicate classification of their vocalizations. No differences in captive call sequences have been observed between individuals from the separate populations of the Philippines and Thailand (Okumura et al. 2007). Chirp-to-chirp transitions were most frequent, followed by chirp-to-trill and trill-to-trill transitions; trills mostly occurred in the middle and at the end of a call sequence (Okumura et al. 2007). Dugong broadband feeding (mastication) sounds ranging from 1 to over 22 kHz were described from Thailand (Tsutsumi et al. 2006).

7.4.2 *Trichechus inunguis*: Amazonian Manatee

The Amazonian manatee is the smallest of the three manatee species, confined to the Amazon River Basin, including tributaries and appended lakes in Brazil, Guyana, Colombia, Ecuador, and Peru (Best 1984; Amaral et al. 2023). It is generally solitary but may be found in groups while foraging and mating. Individuals within groups do not appear to socialize except for mother-calf pairs or during the mating season. Amazonian manatees avoid humans and are difficult to detect

Table 7.1 Catalog of sirenian sounds by species: description/type of call, frequency range, duration, source level, and recording location. Recordings were taken under water unless otherwise noted

Species/description/type of call	Frequency range[c] [Hz] (dominant)	Duration [s]	Comments[d]	Source level[e] [dB]	Recording location[f]	References
Dugong (*Dugong dugon*)						
Tonal; FM and AM whistles; chirp-squeak	f₀: 3.0–5.9 k; 2–5 harmonics	0.06	IVI: 0.26–0.46 s	—	Shark Bay, AUS	Anderson and Barclay (1995)
Tonal; FM whistles	f_p: 2–6 k	0.1–2.7	72% chirps	131–161 rms; 145–180 pk-pk; 110–146 SEL	Shark Bay, AUS	Parsons et al. (2013)
Tonal; FM whistles; short calls	2.9–6.2 k	0.03–0.2	92% of calls	104–158 rms	Talibong, THA	Ichikawa et al. (2003, 2006, 2011)
Pulse burst; AM bark	0.58–2.2 k; ≤5 harmonics	0.03–0.1	IVI: 0.05–0.14 s	—	Shark Bay, AUS	Anderson and Barclay (1995)
Pulse burst; AM bark, "quack"	0.3–22 k	0.3–1.2	23% of calls	129–163 rms; 148–172 pk-pk; 110–135 SEL	Shark Bay, AUS	Parsons et al. (2013)
Complex: FM whistle, "trill"	3–18 k[a]; 2–4 harmonics	2.2		—	Shark Bay, AUS	Anderson and Barclay (1995)
Complex: FM, "squeak"	0.5–5.5 k	0.9	One call	158 rms; 171 pk-pk; 145 SEL	Shark Bay, AUS	Parsons et al. (2013)
Complex: FM long call, multiple segments	3.0–5.3 k	0.7–4	8% of calls	—	Talibong, THA	Ichikawa et al. (2006, 2011)
Amazonian manatee (*Trichechus inunguis*)						
Tonal; FM (notes and calls)	1–16 k[b]; f₀: 1–8 k	x̄: 0.18	One to four notes/call; Call duration x̄: 0.23 s		Amazon basin, BRA; captive	Sousa-Lima (1999) and Sousa-Lima et al. (2002)
Tonal; FM calls; variable FM; "squeaky"	6–16 k[b]; f₀: 6–8 k	0.1–0.2	One young male		Amazon basin; Leticia, COL; captive	Evans and Herald (1970)
Tonal; FM calls, some noisy or breathy	0.70–17 k; f₀: 1.8–3.9 k; f_p: 0.92–12 k	0.07–0.52	In air, contact hydrophone		Amazon basin; PER; captive	Landrau-Giovannetti et al. (2014)

Call description	Freq range (kHz)	f_0 / f_p	Notes	dB	Location	Reference
Tonal: FM whistle, "frog-like," "chirp of birds"	2–10k[b] (2–3 k)	0.1–0.2	Immature at capture	—	Amazon basin; SUR; captive	Sonoda and Takemura (1973)
Pulses; combined with squeaky calls	5–16 k[b]	<0.02	Pulses occurred during feeding	—	Amazon basin; Leticia, COL; captive	Evans and Herald (1970)
West Indian manatee (*Trichechus manatus*)						
Tonal; FM; squeaky chirp	0.6–16 k (2.5–5 k)	0.15–0.5	TRML "Not loud"	—	Ft. Lauderdale (FL), USA	Schevill and Watkins (1965)
Tonal; FM calls	1.1–5.0 k	0.08–0.47	TRML Identified five call subtypes	—	Homosassa Springs (FL), USA	Steel (1982)
Tonal; FM call	0.6–18 k[a] (2–5 k)	0.2–0.5	TRML Rate: up to 1.8/min	100–134 rms	Homosassa Springs (FL), USA	Phillips et al. (2004) Gur and Niezrecki (2007)
Tonal; FM call	f_0: 1.8–3.9 k; f_p: 3.3–7.2 k	0.15–0.30	TRML Call types pooled[1] Rate: 1.3/min	—	Crystal River (FL), USA	Nowacek et al. (2003)
Tonal; FM call	0.6–18 k; f_0: 1.7–4.2 k	0.12–0.64	TRML 0–5 harmonics Geometric means ±95%CI 12 of 14 manatees: first and second harmonics dominant	—	Blue Springs and St. Johns River (FL), USA	O'Shea and Poché (2006) and Niezrecki et al. (2003)
Tonal; FM chirp	1.8–18 k; f_p: 5.1 k	0.13–0.31	TRML 76% of sample Rate: 0.2–0.8/min	116–128 rms	Sarasota Bay (FL), USA	Miksis-Olds and Tyack (2009)
Tonal; squeak	1.4–14.1 k; f_p: 3.3 k	0.13–0.27	TRML 24% of sample	113–127 rms	Sarasota Bay (FL), USA	Miksis-Olds and Tyack (2009)
Tonal; FM call, contact call	f_0: 1.1–5.0 k (2.0–3.0 k); f_p: 3.7–5.7 k (0.28–0.43); Δf_0: 0.97	0.18–0.48	TRMM Rate: 6/min (calves) 1/min (adults) Nonharmonic content	—	Captive, bred in captivity, kept in CMA (PE), and released in Mundaú Lagoon (AL), BRA	Sousa-Lima (1999) Sousa-Lima et al. (2008)
Tonal; FM call	f_0: 2.1–5.2 k; f_p: 2.4–9.3 k	0.15–0.32	TRMM 0.09–0.75/min Call types pooled[1]	86–105 rms	Southern lagoon, BLZ	Nowacek et al. (2003)

(continued)

Table 7.1 (continued)

Species/description/type of call	Frequency range[c] [Hz] (dominant)	Duration [s]	Comments[d]	Source level[e] [dB]	Recording location[f]	References
Tonal; FM calls, some noisy or breathy	0.45–22 k f_0: 1.8–3.9 k f_p: 0.52–20 k	0.04–0.50	TRMM Males, two calves, one adult In air, contact hydrophone	–	Captive (CUB; PRI; COL)	Landrau-Giovannetti et al. (2014)
Tonal; FM call	f_0: 0.72–8.1 k (1.2–2.9 k) f_p: 2.0–15.4 k (3.5–8.5 k)	0.25–0.48	TRMM	–	San San River, PAN	Rivera-Chavarría et al. (2015)
Pulses; clicks, "click-like"	2–7 k (2–4 k)	<0.2	TRMM	–	Captive, SUR	Sonoda and Takemura (1973)
Pulses; clicks	1.0–20 k[a] (1.0–4.0 k)	–	TRMM Possibly produced with teeth	–	Captive, bred in captivity, kept in CMA (PE), and released in Mundaú Lagoon (AL), BRA	Sousa-Lima et al. (2008)
Squeak	1.65–3.10 k p_0: 2.06–2.86 k	0.138–0.254	20 calls from males and females	–	Captive, kept in CMA (PE), BRA	Umeed et al. (2017)
Screech	1.25–4.12 k p_0: 2.05–3.58 k	0.126–0.265	30 calls from males, females, and juveniles	–	Captive, kept in CMA (PE), BRA	Umeed et al. (2017)
Trill	2.56–4.61 k p_0: 2.42–3.87 k	0.212–0.273	20 calls from females and juveniles	–	Captive, kept in CMA (PE), BRA	Umeed et al. (2017)
Creak	1.9–5.7 k p_0: 2.77–4.33 k	0.162–0.241	20 calls from males and females	–	Captive, kept in CMA (PE), BRA	Umeed et al. (2017)
Whine	4.11–5.76 k p_0: 4.46–5.42 k	0.137–0.170	Ten calls from males	–	Captive, kept in CMA (PE), BRA	Umeed et al. (2017)
Rubbing	0.55–3.62 k p_0: 1.29–1.49 k	0.66–0.91	Ten calls from juveniles	–	Captive, kept in CMA (PE), BRA	Umeed et al. (2017)
Chirp	2.01–3.13 k f_p: 1.94–2.84 k	0.94–1.02	Seven calls from mother-calf pairs	–	Mamanguape River (PB) Riacho Tabatinga (AL), BRA	Umeed et al. (2023)
Pulse	–	0.125–0.230	17 calls from mother-calf pairs	–	Riacho Tabatinga (AL), BRA	Umeed et al. (2023)
Rubbing	1.03–1.66 k f_p: 0.96–1.14 k	0.45–0.61	Ten calls from mother-calf pairs	–	Riacho Tabatinga (AL), BRA	Umeed et al. (2023)

Squeak	1.74–4.28 k f_p: 1.73–3.46 k	0.199–0.326	241 calls from mother-calf pairs	—	Mamanguape River (PB) Riacho Tabatinga (AL), BRA	Umeed et al. (2023)
Trill	1.84–4.17 k f_p: 1.61–3.15 k	0.211–0.324	65 calls from mother-calf pairs	—	Mamanguape River (PB) Riacho Tabatinga (AL), BRA	Umeed et al. (2023)
Squeak	1.82–4.29 k f_p: 2.06–3.90 k	0.066–0.206	203 calls	—	Belize	Reyes-Arias et al. (2023)
Squeak	1.88–3.69 k f_p: 2.07–3.46 k	0.267–0.283	203 calls	—	Panama	Reyes-Arias et al. (2023)
Squeak	1.72–3.50 k f_p: 1.90–3.31 k	0.194–0.202	203 calls	—	Florida (FL), USA	Reyes-Arias et al. (2023)
High squeak	2.34–4.27 k f_p: 2.59–4.11 k	0.063–0.209	81 calls	—	Belize	Reyes-Arias et al. (2023)
High squeak	1.19–3.26 k f_p: 1.64–3.07 k	0.148–0.284	81 calls	—	(FL), USA	Reyes-Arias et al. (2023)
High squeak	1.76–3.54 k f_p: 1.72–3.27 k	0.231–0.411	81 calls	—	Panama	Reyes-Arias et al. (2023)
Squeal	0.87–3.90 k f_p: 1.16–3.40 k	0.108–0.244	87 calls	—	Belize	Reyes-Arias et al. (2023)
Squeal	1.81–4.17 k f_p: 2.03–3.87 k	0.142–0.266	87 calls	—	(FL), USA	Reyes-Arias et al. (2023)
West African manatee (*Trichechus senegalensis*)						
FM and AM tonal vocalizations	f_0: 3.95 k–5.35 k	0.11–0.25	Varied FM patterns	—	Lake Ossa, CMR	Rycyk et al. (2021)

Abbreviations: *IVI* inter-vocalization interval, *FM* frequency modulation, *AM* amplitude modulation, f_0 fundamental or frequency based on cycle period, f_p frequency of peak energy, *TRML Trichechus manatus latirostris*, *TRMM Trichechus manatus manatus*, *CMA* Centro Mamíferos Aquáticos (PE), Brazil

[a]Limits of the recording system to the reported range

[b]Frequency range taken from spectrograms

[c]The overall frequency range is a combination of ranges given (minimum – maximum). When authors gave minimum and maximum frequencies separately, with a standard deviation (SD) for each, then the range listed in this column is from the minimum minus its SD to the maximum plus its SD. When available, dominant ranges are listed separately in parentheses. Authors used different methods for specifying the dominant range. Frequency ranges reported to two significant digits. k: ×1000

[d]Where subspecies is specified in the references, it is indicated in this column

[e]Source level in terms of rms, root-mean-square sound pressure level (SPL) [dB re 1 μPa m]; pk-pk, peak-to-peak SPL [dB re 1 μPa m]; and SEL, sound exposure level [dB re 1 μPa²m²s]

[f]ISO alpha-3 country codes: http://www.nationsonline.org/oneworld/country_code_list.htm

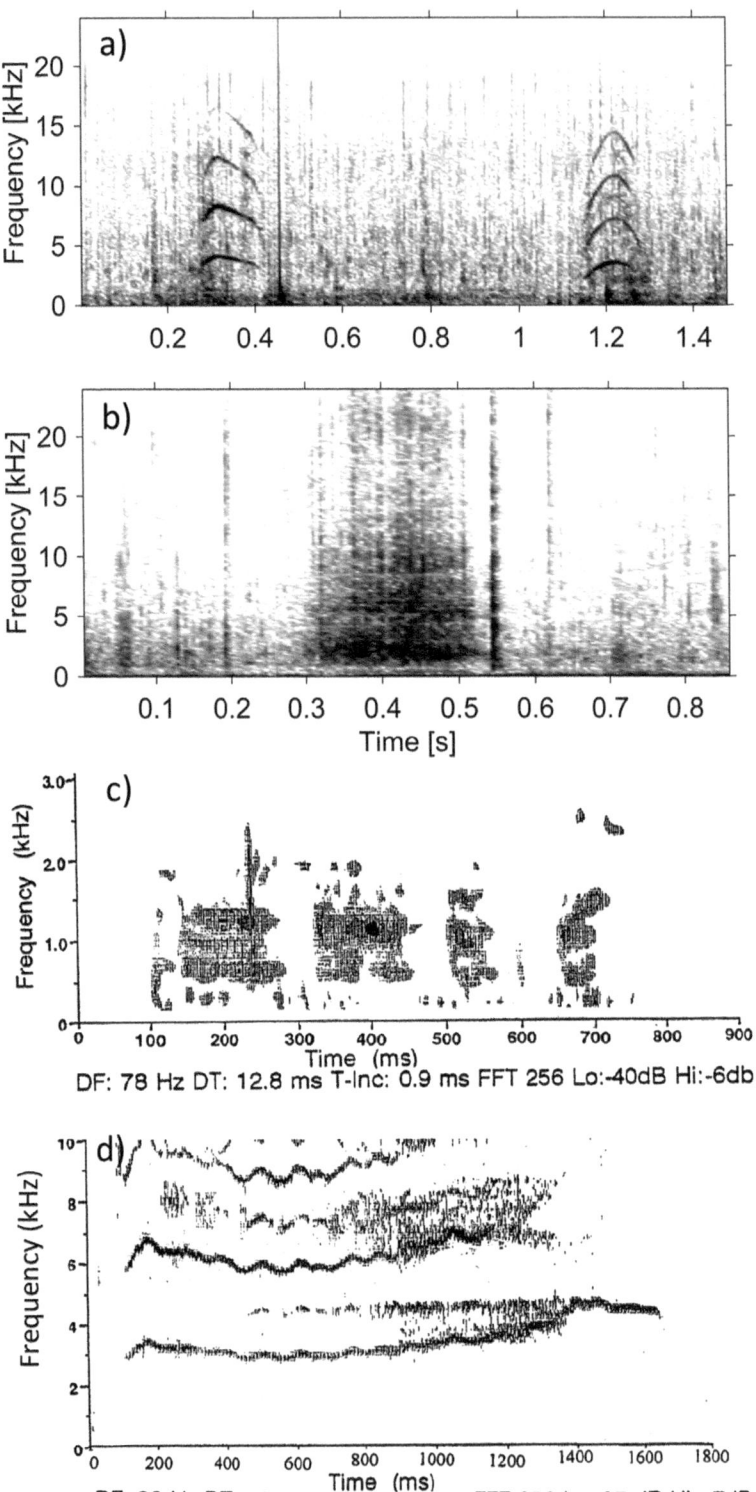

Fig. 7.10 Dugong sounds: (**a**) FM chirps and (**b**) burst-pulse sound (bark) (both from Erbe et al. 2017; fs = 48 kHz, NFFT = 512, 50% overlap, Hann window); (**c**) four barks and (**d**) trill. Both from Anderson and Barclay (1995) Acoustic signals of solitary dugongs: physical characteristics and behavioural correlates. J Mammal 76 (4): 1226–1237. https://doi.org/10.2307/1382616. Reproduced with permission from Oxford University Press. © The American Society of Mammalogists, 1995. All rights reserved

in their dark, turbid, vegetation-choked habitats (Best 1984; Amaral et al. 2023).

Nunes Pereira (1944) reported that traditional inhabitants of the Brazilian Amazon mentioned whistles emitted (probably in air) by Amazonian manatees. Later, the underwater sounds of captive Amazonian manatees were described (Evans and Herald 1970; Sonoda and Takemura 1973) as loud (15–22 dB above the background at 1 m or less from the hydrophone) with the fundamental frequency less intense than the second harmonic (Sousa-Lima et al. 2002) and showing higher complexity (i.e., nonlinear phenomena, e.g., Landrau-Giovannetti et al. 2014; see Fig. 7.11) at the start and end of the signal.

To date, the most comprehensive studies of the acoustic behavior of Amazonian manatees have been conducted in Brazil, both in the wild (Rosas et al. 2003; Carletti 2016) and in captivity (Sousa-Lima 1999; Sousa-Lima et al. 2002; Dantas 2009). Amazonian manatees predominantly produce short frequency-modulated tonal calls (chirps and squeaks) and click-like signals associated with some calls. The fundamental frequency of tonal calls ranges from 1.07 to 8 kHz (Evans and Herald 1970; Sousa-Lima et al. 2002). The Amazonian manatee is the only species that produces frequency-modulated calls with multiple notes per vocalization (1–4; Sousa-Lima et al. 2002) lasting 50–500 ms and usually produced in sequences (with intervals between notes ~10 ms; Fig. 7.11). Amazonian manatees produce click-like, short (~20 ms) broadband pulses (1–17 kHz) when feeding (Evans and Herald 1970; Sousa-Lima 1999). Limited evidence supports the hypothesis that these sounds are produced with the teeth. Kikuchi et al. (2014) equipped Amazonian manatees in captivity with tail-mounted recorders, describing broadband mastication sounds ~200 ms in duration.

Sousa-Lima et al. (2002) found evidence that Amazonian manatee chirps are individually distinctive (Fig. 7.11). Females had higher fundamental frequency, and there was a significant relationship between body length and the range of fundamental frequency (Sousa-Lima et al. 2002; Brady et al. 2022).

7.4.3 *Trichechus manatus*: West Indian Manatee

Schevill and Watkins (1965) first described manatee vocalizations as "squeaky and rather ragged." Later studies found that manatee vocalizations also included tonal calls with harmonic overtones (Reynolds 1981; Steel 1982; O'Shea and Poché 2006; Sousa-Lima et al. 2008) and atonal calls lacking harmonic structures (Steel 1982; O'Shea and Poché 2006; Miksis-Olds and Tyack 2009). The tonal calls can be frequency-modulated, including nonlinear elements like deterministic chaos, bifurcations, biphonation, and subharmonics (Mann et al. 2006; Sousa-Lima et al. 2008).

The West Indian manatee has two recognized subspecies: the Florida manatee, ranging from the US southeast coast to Florida and into the Gulf of Mexico as far as Texas (Self-Sullivan and Mignucci-Giannoni 2008), and the Antillean manatee, ranging with discontinuities from northern Mexico to Alagoas State, northeastern Brazil, and eastward to islands of the Caribbean (Deutsch et al. 2008). The vocal behaviors of both subspecies have been relatively well studied. However, it is important to note that differences in the published acoustic behavior and call descriptions may be methodological, as a larger proportion of the studies are based on data collected under controlled conditions from captive manatees (often rehabilitated or orphaned animals) and in one case were collected in air with a contact hydrophone (Landrau-Giovannetti et al. 2014).

7.4.3.1 *Trichechus manatus latirostris*: Florida Manatee

Florida manatees can be seen solitary or in small groups of less than six animals, but they can occur in aggregations of hundreds in warm water refugia during the winter, especially at the northern limit of their range. Vocal behavior appears to be predominantly social, primarily mediating the mother-calf bond, but also present when adults travel in small groups that engage in intense social activity, particularly during breeding (Hartman 1979).

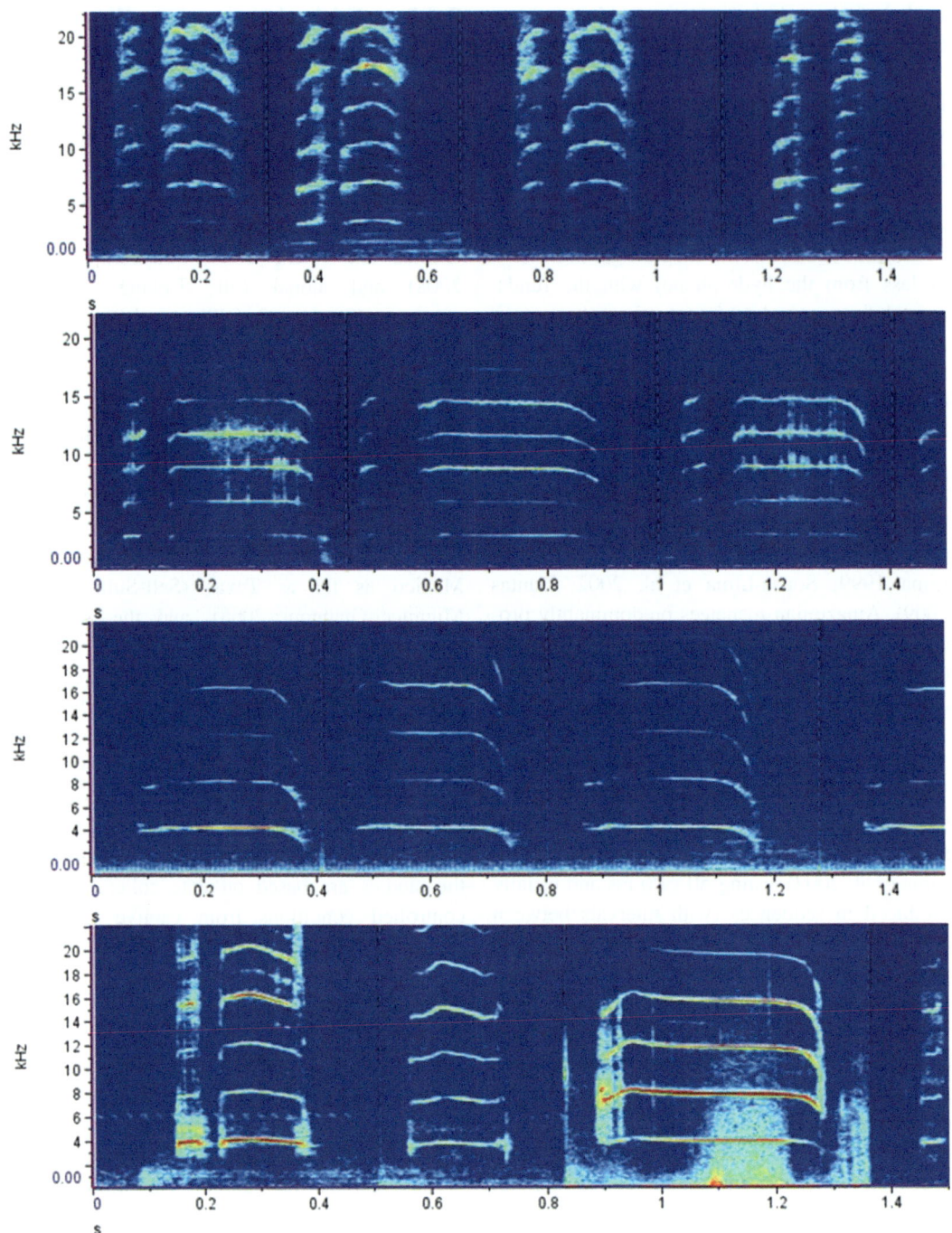

Fig. 7.11 Example spectrograms (fs = 44.1 kHz, NFFT = 1024, 75% overlap, Hann window) of chirps of four individual Amazonian manatees with one and two notes showing intra- and interindividual variation in acoustic parameters

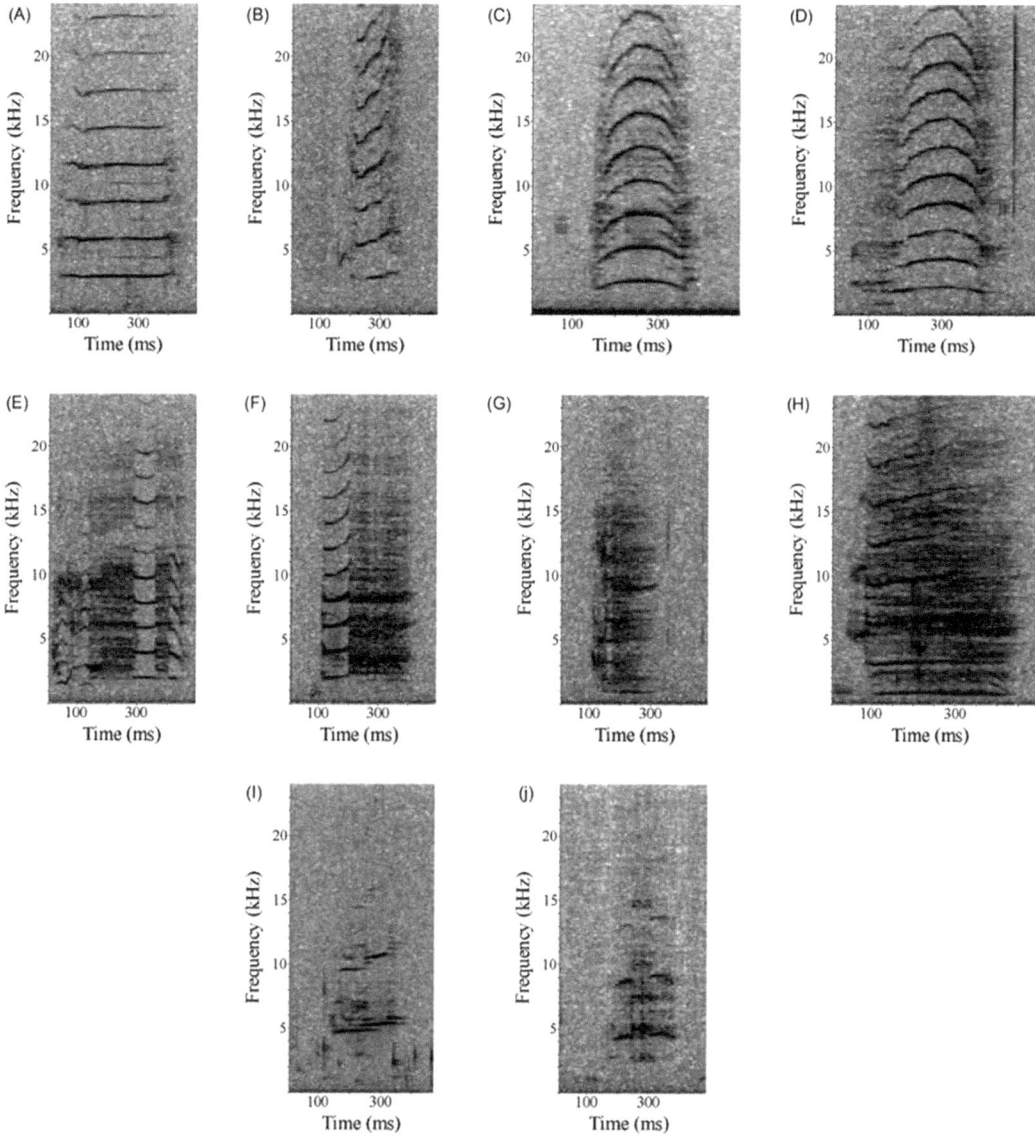

Fig. 7.12 Spectrograms of call types produced by Florida manatees. Two spectrograms are shown for each call type to represent variability. (**a, b**) Squeak, (**c, d**) high squeak, (**e, f**) squeak-squeal, (**g, h**) squeal, and (**i, j**) chirp. Reprinted with ASA permission from Brady et al. (2020). Classification of Florida manatee (*Trichechus manatus latirostris*) vocalizations. J Acoust Soc Am 147 (3):1597–1606. https://doi.org/10.1121/10.0000849. © Acoustical Society of America, 2020. All rights reserved

The primary vocalization of the Florida manatee is a short single note, an up-down frequency-modulated tone typically characterized as two discrete types: chirps and squeaks (Miksis-Olds and Tyack 2009). Brady et al. (2020) further classified these calls as squeaks, high squeaks, squeak-squeals, squeals, and chirps (Fig. 7.12). The figure illustrates some of the challenges of classifying these call types given their variability, intergrading, and nonlinear characteristics.

Chirps vary substantially in details, such as the shape of the contour, number of harmonics, presence of nonlinear acoustic energy, abrupt steps in frequency, and duration (Mann et al. 2006). Fundamental frequency ranges from 1.7 to 4.2 kHz (Nowacek et al. 2003; O'Shea and Poché 2006), with energy in the range 0.6–18 kHz, peak energy at 3.3–7.2 kHz (Nowacek et al. 2003), and duration of 0.08–0.64 s. Phillips et al. (2004) estimated a mean source level of 113 dB re 1 µPa m using triangulation across multiple hydrophones and cylindrical spreading loss. The latter might have been an underestimation of the spreading loss (and hence an underestimation of source level) in shallow water and with the animal near the sea surface, possibly creating a dipole radiation pattern and thus greater horizontal propagation loss (see Chap. 1).

Steel (1982) proposed the first categorization for the Florida manatee calls with six discrete types, mostly from captive animals. Hartman (1979) and O'Shea and Poché (2006) broke the frequency-modulated tonal repertoire into five to nine subtypes, providing evidence that there are probably categories within the general class (particularly begging vocalizations of calves). Miksis-Olds and Tyack (2009) categorized Florida manatee calls into two broad categories: chirps and squeaks. Brady et al. (2020) categorized wild Florida manatee vocalizations into five broadly defined call categories. O'Shea and Poché (2006) identified frequency-modulated tonal calls with complex modulations, suggesting some similarity with the dugong repertoire. However, no studies to date have identified or measured burst-pulse sounds or clicks from the Florida manatee.

7.4.3.2 *Trichechus manatus manatus*: Antillean Manatee

Antillean manatee vocalizations are short (0.03–0.5 s) with a fundamental frequency ranging from 1.07 to 5.9 kHz (Sousa-Lima et al. 2008; Landrau-Giovannetti et al. 2014; Nowacek et al. 2003; Rivera-Chavarría et al. 2015) and mean dominant frequencies between 3.18 and 7.08 kHz (Nowacek et al. 2003; Phillips et al. 2004; Sousa-Lima et al. 2008). Females emit longer calls; although they have lower mean fundamental frequencies, the energy is concentrated in higher harmonics, resulting in higher-pitched signals (Fig. 7.13; Sousa-Lima et al. 2008). Sousa-Lima et al. (2008) suggested that Antillean manatee vocalizations do not fall into distinct classes but rather are graded along a continuum from simple, tonal harmonic series similar to the squeaks of Florida manatees (Fig. 7.13a) to more complex sounds with a harsh noisy quality, chaos, and broader frequency bands similar to squeals of Florida manatees (Fig. 7.13m).

Antillean manatees produce click-like pulses (Sonoda and Takemura 1973; Sousa-Lima et al. 2008), which have been hypothesized to be produced with the teeth (Sousa-Lima et al. 2008). The behavioral context of these sounds remains unknown. Click frequency ranges up to at least 20 kHz, but the dominant energy is 1–4 kHz, and the duration is <200 ms.

Captive, temporarily isolated Antillean manatees call on average three times per minute (Sousa-Lima et al. 2008). Call emission rates tend to increase when free-ranging manatees are socializing in groups but are usually lower than when individuals are isolated, varying between 0.25 and 4.75 vocalizations per individual every 5 min (Bengtson and Fitzgerald 1985; Phillips et al. 2004; Miksis-Olds and Tyack 2009).

Based on structure, emphasized frequency, frequency modulation, and duration, Alicea-Pou (2001) found five different call types of the Antillean manatee in Puerto Rico and captive Florida manatee (Fig. 7.14). Otherwise, Umeed (2017) found six types from captive Antillean manatees, also noting that the use of the different types varied among sex and age classes. Ramos et al. (2020) recorded Antillean manatee calls' harmonics with energy extending to ultrasonic ranges (20–150 kHz) in captivity and in the wild (Belize).

7.4.4 *Trichechus senegalensis*: West African Manatee

The West African manatee is found in shallow, tropical coastal waters and enclosed lagoons from

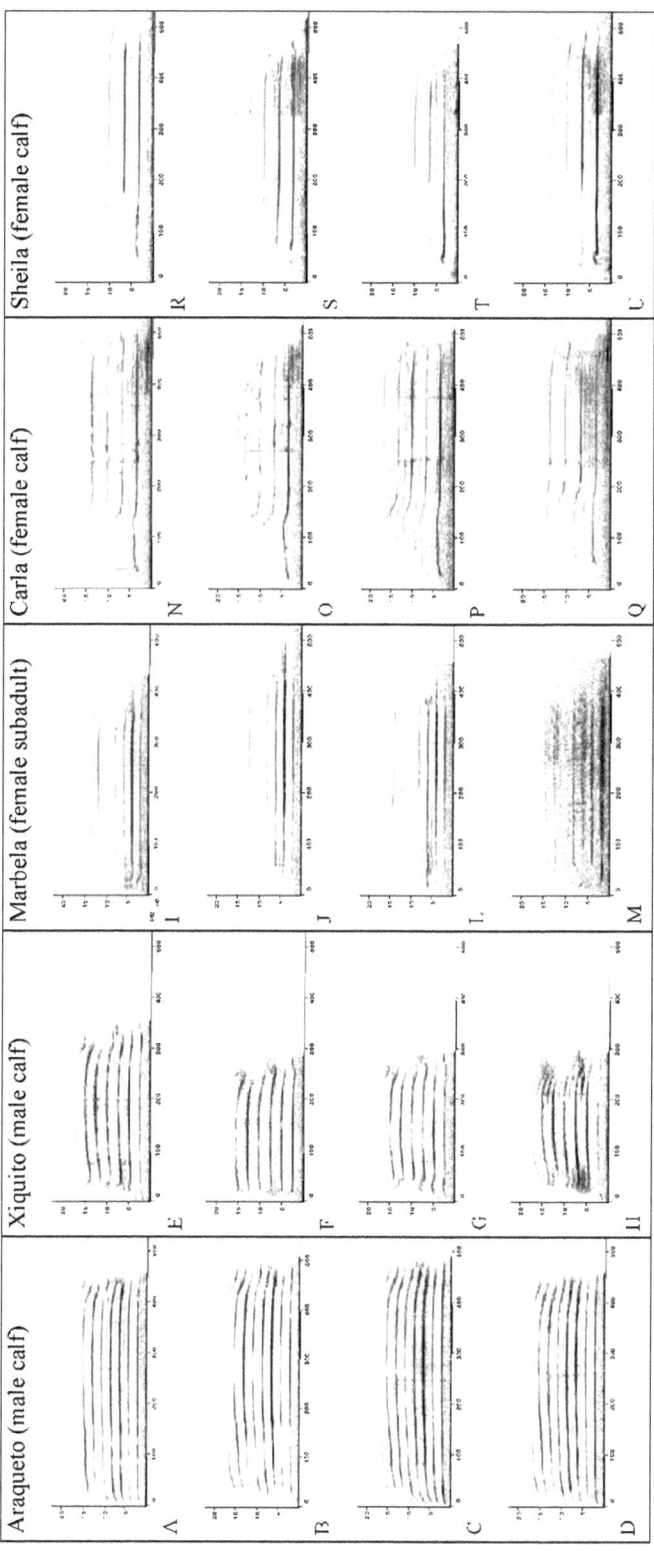

Fig. 7.13 Spectrogram examples of FM vocalizations ("isolation calls") from five different Antillean manatees (*Trichechus manatus manatus*). Time on the x-axis (0–500 ms) and frequency on the y-axis (0–20 kHz). The calves in N–Q and R–U were twin siblings. Gender, age, and identity in the isolation calls of Antillean manatees (*Trichechus manatus manatus*). Aquat Mamm 34 (1):109–122. https://doi.org/10.1578/AM.34.1.2008.109. Reproduced with permission. © Aquatic Mammals, 2008. All rights reserved

Senegal to Angola (Keith-Diagne 2015). It may also reach into rivers flowing into the Gulf of Guinea (Marsh and Lefebvre 1994). These animals are subject to hunting throughout their range (Reeves et al. 1988) and as a result, are declining in many areas and difficult to approach. Little is known of their life history or behavior (Marsh and Lefebvre 1994; Laudisoit et al. 2017).

West African manatee vocalizations were first recorded from an isolated captive calf by Thomas O'Shea in the 1980s in Côte d'Ivoire (see O'Shea and Powell pers. comm. in Rycyk et al. 2021). This small sample of sounds was described as similar to the calls of Florida manatees. In 2020, over 3000 tonal vocalizations were collected by autonomous acoustic recorders deployed at Lake Ossa, Cameroon (Rycyk et al. 2021). Vocalizations were structurally similar to those of other manatee species (overlapping values of fundamental frequency range and duration) with varying frequency- and amplitude-modulation patterns (Fig. 7.15). After the highest-quality vocalizations had been selected, all but nine vocalizations presented harmonics (typically three or more harmonics above the fundamental); the frequency band with most energy was predominantly the fundamental, but 27% of the detected vocalizations had more energy in the harmonics (Rycyk et al. 2021), similar to Amazonian and West Indian manatees (Figs. 7.11, 7.12, and 7.13). West African manatee calls tend to have higher mean fundamental frequency and shorter mean duration compared to other manatee species calls (Rycyk et al. 2021). Passive acoustic monitoring in Africa resulted in the highest call detections during the evening and at night (Rycyk et al. 2021), consistent with avoidance of humans.

Fig. 7.14 Juvenile Antillean manatee released in the Tatuamunha River estuary, state of Alagoas, Brazil. Photo courtesy of Fundação Mamíferos Aquáticos

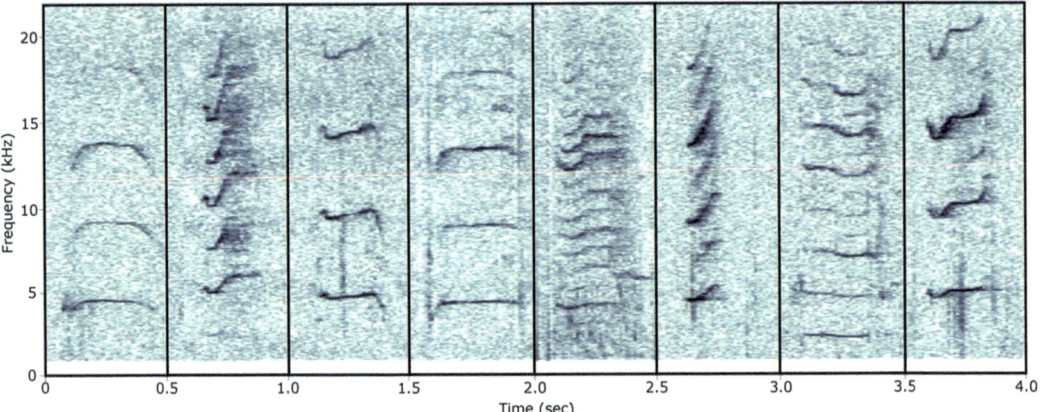

Fig. 7.15 Spectrogram examples of frequency- and amplitude-modulated tonal vocalization of West African manatees (fs = 44.1 kHz, 1024-point DFT, Hann window, 50% overlap, 1 kHz high-pass filter). Reprinted with ASA permission; Fig. 3 from Rycyk et al. (2021). First characterization of vocalizations and passive acoustic monitoring of the vulnerable African manatee (*Trichechus senegalensis*). J Acoust Soc Am 150 (4):3028–3037. https://doi.org/10.1121/10.0006734. © Acoustical Society of America. All rights reserved

7.4.5 The Case of *Hydrodamalis gigas*: Steller's Sea Cow

Information about Steller's sea cow is mostly limited to Steller's accounts during the Great Northern Expedition or Second Kamchatka Expedition. The size of this extinct sirenian reached a length of 4–5 fathoms (7–9 m) length (Steller 1793). The last known population lived in the *Komandorskiye Ostrova* (Commander Islands) and became extinct ca. 1768 CE (Stejneger 1887; Crerar et al. 2014), 27 years after its discovery by Europeans during the winter of 1741–42 CE (Steller 1793). The last population was restricted to Bering and Copper Islands (Crerar et al. 2014), but there is evidence suggesting the species inhabited most of the North Pacific in the Late Pleistocene, including the Kuril Islands, the Kamchatka Peninsula, the Chukotka and Bering Seas, and the Eastern North Pacific, as far south as today's border between the USA and Mexico (Sharko et al. 2021). The Steller's sea cow was driven to extinction mainly by overhunting (Turvey and Risley 2006), but some studies suggest that loss of feeding habitat was a factor because kelp (i.e., the main diet of the Steller's sea cow) declined due to an explosion of sea urchin populations after their natural predator, the sea otter, was overhunted (Estes et al. 2016).

Steller included anatomical descriptions of the Steller's sea cow ears; however, there is no mention of sounds produced by these animals. Steller's accounts suggest strong social bonds, complex mating behaviors, and tight mother-calf bonds (Steller 1793). One can only speculate about the possible sounds produced by the Steller's sea cow during the mating season, during group interactions, and within mother-calf pairs, considering what we have learned about the other sirenian species. It remains to be found whether any indigenous human populations that might have hunted and potentially interacted with the Steller's sea cow in the Bering Sea region have stories about sounds produced by the Steller's sea cow, similar to the stories of Aboriginal Australians about dugongs (Smith 1987) and Amazonian tribes about manatees (Nunes Pereira 1944). Based on indigenous accounts of extant species, their putative distress calling behavior could have facilitated their extinction by calling in conspecifics during capture and poaching by humans.

7.5 Sirenians in the Anthropocene

The Anthropocene ocean is noisier than the historic ocean (Duarte et al. 2021). Noise may influence sirenian acoustic communication (Buckingham et al. 1999; Nowacek et al. 2004; Rycyk et al. 2018; see Chaps. 9 and 10). Noise that can affect sirenians is mostly produced by motorized boats, which often overlaps with signals from these animals (e.g., dugongs; Tanaka et al. 2023). Dugongs exposed to boat noise increased the number of harmonics in their calls and tended to change call rate, duration, and frequency (Sakamoto et al. 2006; Ando-Mizobata et al. 2014). Florida manatees used fewer call types during feeding and social behaviors in elevated noise and decreased call rate and duration (Miksis-Olds and Tyack 2009). Evidence suggests that ambient noise might reveal a vocal plasticity indicative of undocumented vocal learning in these species (Baotic et al. 2022).

7.6 Sirenian Sounds and Conservation Opportunities

All sirenian species are threatened (Deutsch 2008; Deutsch et al. 2008; Self-Sullivan and Mignucci-Giannoni 2008; Keith-Diagne 2015; Marmontel et al. 2016; Brownell et al. 2019; Marsh and Sobtzick 2019; Trotzuk et al. 2022), and some populations are presently rare and under severe threat (Moore et al. 2017; Alvarez-Alemán et al. 2021; Panyawai and Prathep 2022; Meirelles et al. 2022). Understanding the biological and ecological importance of sound for these species and the impacts of noise pollution could aid in the conservation of sirenians.

There are important efforts toward a positive coexistence between humans and sirenians. Some examples include establishment and signposting of go-slow areas in places of known manatee or dugong occurrence, better fishing practices and waterway control, efforts to reduce hunting and exploitation, and rehabilitation of calves and adults (Adimey et al. 2012; Bonde and Flint 2017). These actions could have a positive impact on wild populations of dugongs and manatees.

Passive acoustic monitoring is a useful tool for conservation management; in particular, species that exhibit individuality in their calls (i.e., individual signatures) lend themselves to acoustic monitoring of population size and structure (Sousa-Lima 1999; Sousa-Lima et al. 2002, 2008, 2013; Umeed et al. 2017; Tanaka et al. 2023). However, understanding the social behavioral contexts and the environmental variables that may affect sirenian vocalizations is key to using passive acoustics as a monitoring tool (Tanaka et al. 2023). For example, population density estimation from passive acoustic monitoring might not be simple. Tests with captive Amazonian manatees have shown that the number of acoustic detections does not increase with the number of individuals (Carletti 2016). Not all individuals vocalize at the same rate (Sousa-Lima 1999; Sousa-Lima et al. 2002; Dantas 2009). However, a significant relationship between the number of detections and the presence of calves was found in captivity (Carletti 2016): Calves and lactating females vocalize more than other individuals in captivity (Dantas 2009). Assuming this pattern holds in the wild, we should expect a higher number of vocalization detections during the calving season, as well as sounds typical of calves in some species.

Recent studies have targeted passive acoustic detectability (LaCommare et al. 2008; Kikuchi et al. 2014; Rivera-Chavarria et al. 2015; Merchan et al. 2019; Rycyk et al. 2021, 2022; Tanaka et al. 2023; Jiang et al. 2024). In a review of methods for detecting wild West African manatees, passive acoustics was assessed as more effective compared to visual scans (Factheu et al. 2023). Expanding passive acoustic monitoring efforts of African manatees into Nigeria, Benin, and Senegal might capitalize on the high success achieved on manatees elsewhere to document occurrence and distribution of the species around the African continent (Factheu et al. 2023). Furthermore, the fact that sirenians respond vocally to acoustic playback (with increased vocalization rates and detectability) enhances passive acoustic monitoring efficiency (Sousa-Lima 1999; Sousa-Lima et al. 2002; Rosas et al. 2003; Phillips et al. 2004; Ichikawa et al. 2011). However, monitoring in habitats with high boat noise could underestimate population density, as boat noise can mask sirenian vocalizations (Tanaka et al. 2023).

7.7 Summary

Despite poems about singing "sirens" in ancient Greek poetry, Columbus' descriptions of sirenians on his voyage to America, and Aboriginal Australian and Amazonian tribe knowledge on sirenian sounds, these animals (in the scientific literature) were thought not to vocalize for a long time. The sirenian communication system might not be as complex as that of other aquatic mammals, which has been attributed to their predominantly solitary habits and herbivorous diet. However, their repertoires are not unusually small among mammals; their slow reproductive rate and extended maternal care are thought to be the main drivers for the evolution of their communication system.

Sirenian sounds are produced when air makes the vocal ligaments vibrate, and it is modulated in nasal cavities. They are mainly produced under water without expelling air. These vocalizations can have frequency and amplitude modulation and range from 0.5 to 22 kHz (some species reach ultrasonic ranges) in frequency and 0.02–1 s in duration. Several studies have recorded a variety of calls, including barks, chirps, clicks, quacks, squeaks, squeals, trills, and whistles. These calls are usually emitted in sequences—mainly chirps and trills in combination. Sirenian vocalizations are produced in different social and environmental contexts. The vocal repertoire can be statistically differentiated between species, and some intraspecific geographical variations have been observed, suggesting processes of isolation between populations. Calls can be used to identify individuals, age, sex, and motivational state. Pairs of mothers and calves are prolific in sound production, and these sounds are possibly important to mediate interactions during rearing.

Vocal repertoires, behavior, and ecology of the dugong, the Amazonian, and the West Indian manatee have been relatively well studied, in contrast to those of the West African manatee. Early publications focused on captive or temporarily restrained wild animals, but advances in recording technology and analysis have extended sirenian acoustic research to free-ranging individuals and more complex ecological and behavioral questions. Presently, passive acoustic monitoring is used to improve our understanding of dugong and manatee behaviors, population numbers, and ecology. These advances will contribute to the conservation of these threatened aquatic mammals.

References

Adimey NM, Mignucci-Giannoni AA, Auil-Gomez NE, Da Silva VM, Alvite CMC, Morales-Vela B, Rosas FC (2012) Manatee rescue, rehabilitation, and release efforts as a tool for species conservation. In: Hines EM, Reynolds JE III, Aragones L, Mignucci-Giannoni AA, Marmontel M (eds) Sirenian conservation: issues and strategies in developing countries. University Press of Florida, Gainesville, pp 204–217

Alicea-Pou JA (2001) Vocalizations and behavior of the Antillean and Florida manatee (*Trichechus manatus*): individual variability and geographical comparison. MSc thesis, San Francisco State University, Moss Landing, California, USA

Allen SG, Marsh H, Hodgson AJ (2004) Occurrence and conservation of the dugong (Sirenia: Dugongidae) in New South Wales. Proc Linnean Soc NSW 125:211–216

Alvarez-Alemán A, García Alfonso E, Powell JA, Jacoby CA, Austin JD, Frazer TK (2021) Causes of mortality for endangered Antillean manatees in Cuba. Front Mar Sci 8:646021. https://doi.org/10.3389/fmars.2021.646021

Amaral RS, Marmontel M, Souza DA, Carvalho CC, Valdevino GCM, Guterres-Pazin MG, Mello DMD, Lima DS, Chávez-Pérez HI, da Silva VMF (2023) Advances in the knowledge of the biology and conservation of the Amazonian manatee (*Trichechus inunguis*). Lat Am J Aquat Mamm 18(1):125–138. https://doi.org/10.5597/lajam00296

Anderson PK (1981) Dugong behaviour: observations, extrapolations, and speculations. In: Marsh H (ed) The Dugong. Proceedings of a seminar/workshop held at James Cook University 8–13 May 1979, Townsville, pp 91–111

Anderson PK, Barclay RMR (1995) Acoustic signals of solitary dugongs: physical characteristics and behavioral correlates. J Mammal 76(4):1226–1237. https://doi.org/10.2307/1382616

Ando-Mizobata N, Ichikawa K, Arai N, Kato H (2014) Does boat noise affect dugong (*Dugong dugon*) vocalization? Mamm Study 39(2):121–127. https://doi.org/10.3106/041.039.0208

Baotic A, Brady B, Ramos EA, Stoeger A (2022) Elephants and sirenians: a comparative review across related taxa in regard to learned vocal behavior. Comp Cogn Behav Rev 17:89–108. https://doi.org/10.3819/CCBR.2022.170004

Bengtson JL, Fitzgerald SM (1985) Potential role of vocalizations in West Indian manatees. J Mammal 66(4):816–819. https://doi.org/10.2307/1380821

Best RC (1984) The aquatic mammals and reptiles of the Amazon. In: Sioli H (ed) The Amazon: limnology and landscape ecology of a mighty tropical river and its basin. Springer, Dordrecht, pp 371–412. https://doi.org/10.1007/978-94-009-6542-3

Bonde RK, Flint M (2017) Human interactions with sirenians (manatees and dugongs). In: Butterworth A (ed) Marine mammal welfare: human induced change in the marine environment and its impacts on marine mammal welfare. Springer, Cham, pp 299–314. https://doi.org/10.1007/978-3-319-46994-2_17

Brady B, Hedwig D, Trygonis V, Gerstein E (2020) Classification of Florida manatee (*Trichechus manatus latirostris*) vocalizations. J Acoust Soc Am 147(3):1597–1606. https://doi.org/10.1121/10.0000849

Brady B, Ramos EA, May-Collado L, Landrau-Giovannetti N, Lace N, Arreola MR, Santos GM, da Silva VMF, Sousa-Lima RS (2022) Manatee calf call contour and acoustic structure varies by species and body size. Sci Rep 12:19597. https://doi.org/10.1038/s41598-022-23321-7

Brady B, Sarbacker C, Lasala JA, Maust-Mohl M, Collom KA, Searle L, May-Collado LJ, Ramos EA (2023) Manatees display diel trends in acoustic activity at two microhabitats in Belize. PLoS One 18(11): e0294600. https://doi.org/10.1371/journal.pone.0294600

Broadwater MH, van Dolah FM, Fire SE (2018) Vulnerabilities of marine mammals to harmful algal blooms. In: Shumway SE, Burkholder JM, Morton SL (eds) Harmful algal blooms: a compendium desk reference. Wiley, West Sussex, pp 191–222

Brownell Jr RL, Kasuya T, Marsh H (2019) *Dugong dugon* (Nansei subpopulation). The IUCN Red List of Threatened Species. https://doi.org/10.2305/IUCN.UK.2019-3.RLTS.T157011948A157011982.en. Accessed 12 Jan 2024

Buckingham CA, Lefebvre LW, Schaefer JM, Kochman HI (1999) Manatee response to boating activity in a thermal refuge. Wildl Soc Bull 27(2):514–522. https://www.jstor.org/stable/3783921

Buffon GLLcd (1799) Quadrupedes t.9. In: Histoire naturelle t.33. P. Didot l'ainé et Firmin Didot, Paris. https://doi.org/10.5962/bhl.title.64910

Carletti IM (2016) Invisíveis porém detectáveis: a utilização do monitoramento acústico passivo em peixes-boi da Amazônia (*Trichechus inunguis*). BSc thesis, Universidade Federal do Rio Grande do Norte, Natal, Rio Grande do Norte, Brazil

Chang Y (2016) Making of Mami Wata: diasporic encounter of African, European and Asian spirits. J Asian Afr Stud 39:53–70

Charles A, Henaut Y, Saint-Jalme M, Mulot B, Lecu A, Delfour F (2024) Visual and acoustic exploratory behaviors toward novel stimuli in Antillean manatees (*Trichechus manatus manatus*) under human care. J Comp Psychol 138(2):118–129. https://doi.org/10.1037/com0000360

Committee on Taxonomy (2023) List of marine mammal species and subspecies. Society for Marine Mammalogy. https://www.marinemammalscience.org. Accessed 12 Jan 2024

Crerar LD, Crerar AP, Domning DP, Parsons EC (2014) Rewriting the history of an extinction—was a population of Steller's sea cows (*Hydrodamalis gigas*) at St. Lawrence Island also driven to extinction? Biol Lett 10(11):20140878. https://doi.org/10.1098/rsbl.2014.0878

Dantas GA (2009) Ontogenia do padrão vocal individual do peixe-boi da Amazônia *Trichechus inunguis* (Sirenia, Trichechidae). MSc thesis, Instituto Nacional de Pesquisas da Amazônia-Universidade Federal do Amazonas, Manaus, Amazonas, Brazil

Deutsch CJ (2008) *Trichechus manatus* ssp. *latirostris*. The IUCN Red List of Threatened Species. https://doi.org/10.2305/IUCN.UK.2008.RLTS.T22106A9359881.en. Accessed 12 Jan 2024

Deutsch CJ, Self-Sullivan C, Mignucci-Giannoni A (2008) *Trichechus manatus*. The IUCN Red List of Threatened Species. https://doi.org/10.2305/IUCN.UK.2008.RLTS.T22103A9356917.en. Accessed 12 Jan 2024

Dietrich A, von Fersen L, Hammerschmidt K (2022) Signature calls in West Indian manatee (*Trichechus manatus manatus*)? Aquat Mamm 48(4):349–354. https://doi.org/10.1578/AM.48.4.2022.349

Domning DP (2022) What can we infer about the behavior of extinct sirenians? In: Marsh H (ed) Ethology and behavioral ecology of Sirenia. Springer, Cham, pp 1–17. https://doi.org/10.1007/978-3-030-90742-6_1

Domning DP, Hayek LAC (1984) Horizontal tooth replacement in the Amazonian manatee (*Trichechus inunguis*). Mammalia 48:105–128. https://doi.org/10.1515/mamm.1984.48.1.105

Drewal HJ (2008) Mami Wata: arts for water spirits in Africa and its diasporas. Afr Arts 41(2):60–83. https://www.jstor.org/stable/20447886

Duarte CM, Chapuis L, Collin SP, Costa DP, Devassy RP, Eguiluz VM, Erbe C, Gordon TAC, Halpern BS, Harding HR, Havlik MN, Meekan M, Merchant ND, Miksis-Olds JL, Parsons M, Predragovic M, Radford AN, Radford CA, Simpson SD, Slabbekoorn H, Staaterman E, Van Opzeeland IC, Winderen J, Zhang X, Juanes F (2021) The soundscape of the anthropocene ocean. Science 371(6529):eaba4658. https://doi.org/10.1126/science.aba4658

Erbe C, Dunlop R, Jenner KCS, Jenner M-NM, McCauley RD, Parnum I, Parsons M, Rogers T, Salgado-Kent C (2017) Review of underwater and in-air sounds emitted by Australian and Antarctic marine mammals. Acoust Aust 45:179–241. https://doi.org/10.1007/s40857-017-0101-z

Estes JA, Burdin A, Doak DF (2016) Sea otters, kelp forests, and the extinction of Steller's sea cow. Proc Natl Acad Sci USA 113(4):880–885. https://doi.org/10.1073/pnas.1502552112

Evans WE, Herald ES (1970) Underwater calls of a captive Amazon manatee, *Trichechus inunguis*. J Mammal 51(4):820–823. https://doi.org/10.2307/1378319

Factheu C, Rycyk AM, Kekeunou S, Keith-Diagne LW, Ramos EA, Kikuchi M, Kamla AT (2023) Acoustic methods improve the detection of the endangered African manatee. Front Mar Sci 9:1032464. https://doi.org/10.3389/fmars.2022.1032464

Fitch WT, Neubauer J, Herzel H (2002) Calls out of chaos: The adaptive significance of nonlinear phenomena in mammalian vocal production. Anim Behav 63:407–418. https://doi.org/10.1006/anbe.2001.1912

Flint VIJ (2017) The imaginative landscape of Christopher Columbus, vol 4849. Princeton University Press, Princeton

Freeberg TM, Dunbar RIM, Ord TJ (2012) Social complexity as a proximate and ultimate factor in communicative complexity. Philos Trans R Soc Lond Ser B Biol Sci 367(1597):1785–1801. https://doi.org/10.1098/rstb.2011.0213

Frommen JG (2020) Aggressive communication in aquatic environments. Funct Ecol 34(2):364–380. https://doi.org/10.1111/1365-2435.13482

Grossman CJ, Hamilton RE, Wit MD, Johnson J, Faul R, Herbert S, Tierney D, Buot M, Latham ML, Boivin G, Boivin GP (2014) The vocalization mechanism of the Florida manatee (*Trichechus manatus latirostris*). Online J Biol Sci 14(2):127–149. https://doi.org/10.3844/ojbsci.2014.127.149

Gur BM, Niezrecki C (2007) Autocorrelation based denoising of manatee vocalizations using the undecimated discrete wavelet transform. J Acoust Soc Am 122(1):188–199. https://doi.org/10.1121/1.2735111

Hartman DS (1979) Ecology and behavior of the manatee (*Trichechus manatus*) in Florida. Am Soc Mamm Spec Pub 5:1–153

Hedwig D, Poole J, Granil P (2021) Does social complexity drive vocal complexity? Insights from the two African elephant species. Animals 11(11):3071. https://doi.org/10.3390/ani11113071

Henaut Y, Charles A, Delfour F (2022) Cognition of the manatee: past research and future developments. Anim Cogn 25:1049–1058. https://doi.org/10.1007/s10071-022-01676-8

Heritage S, Seiffert ER (2022) Total evidence time-scaled phylogenetic and biogeographic models for the evolution of sea cows (Sirenia, Afrotheria). PeerJ 10: e13886. https://doi.org/10.7717/peerj.13886

Hieb EE, Eniang EA, Keith-Diagne LW, Carmichael RH (2021) In-water bridge construction effects on manatees with implications for marine megafauna species. J Wildl Manag 85(4):674–685. https://doi.org/10.1002/jwmg.22030

Ichikawa K, Arai N, Akamatsu T, Shinke T, Hara T, Adulyanukosol K (2003) Acoustical analyses on the calls of dugong. In: Proceedings of the 4th SEASTAR 2000 workshop, Kyoto, Japan, pp 72–76

Ichikawa K, Tsutsumi C, Arai N, Akamatsu T, Shinke T, Hara T, Adulyanukosol K (2006) Dugong (*Dugong dugon*) vocalization patterns recorded by automatic underwater sound monitoring systems. J Acoust Soc Am 119(6):3726–3733. https://doi.org/10.1121/1.2201468

Ichikawa K, Akamatsu T, Shinke T, Adulyanukosol K, Arai N (2011) Callback response of dugongs to conspecific chirp playbacks. J Acoust Soc Am 129(6): 3623–3629. https://doi.org/10.1121/1.3586791

Ichikawa K, Akamatsu T, Adulyanukosol K, Lanyon J, Nawata H (2012a) Intraspecific variation in vocal repertoire among dugong populations. J Acoust Soc Am 131(4):3456–3456. https://doi.org/10.1121/1.4709024

Ichikawa K, Akamatsu T, Shinke T, Arai N, Adulyanukosol K (2012b) Clumped distribution of vocalising dugongs (*Dugong dugon*) monitored by passive acoustic and visual observations in Thai waters. In: Proceedings of Acoustics 2012 Fremantle: acoustics, development and the environment, Australian Acoustical Society, Fremantle, Fremantle, Australia, pp 130–134

Illiger JKW (1811) Caroli Illigeri. Prodromus systematis mammalium et avium additis terminis zoographicis utriusque classis, eorumque versione germanica. C. Salfeld, Berlin. https://doi.org/10.5962/bhl.title.106965

Jiang Y, Liu Z, Yang C, White P, Wang X, Lü LG, Xia T, Zhang X, Kittiwattanawon K (2024) Dugong chirp type classification based on fundamental contour extraction and hierarchical cluster analysis. Appl Acoust 217:109812. https://doi.org/10.1016/j.apacoust.2023.109812

Jiménez I (2002) Heavy poaching in prime habitat: the conservation status of the West Indian manatee in Nicaragua. Oryx 36(3):272–278. https://doi.org/10.1017/S0030605302000492

Kayanne H, Hara T, Arai N, Yamano H, Matsuda H (2022) Trajectory to local extinction of an isolated dugong population near Okinawa Island, Japan. Sci Rep 12:6151. https://doi.org/10.1038/s41598-022-09992-2

Keith-Diagne L (2015) *Trichechus senegalensis* (errata version published in 2016). The IUCN Red List of Threatened Species. https://doi.org/10.2305/IUCN.UK.2015-4.RLTS.T22104A81904980.en. Accessed 12 Jan 2024

Kikuchi M, Akamatsu T, Gonzalez-Socoloske D, De Souza DA, Olivera-Gomez LD, Da Silva VMF (2014) Detection of manatee feeding events by animal-borne underwater sound recorders. J Mar Biol Assoc UK 94(6):1139–1146. https://doi.org/10.1017/S0025315413001343

Kotaro T, Ichikawa K, Kittiwattanawong K, Arai N, Mitamura H (2023) Spatial variation of vocalising dugongs around Talibong Island, Thailand. Bioacoustics 32(1):33–47. https://doi.org/10.1080/09524622.2022.2058614

LaCommare KS, Self-Sullivan C, Brault S (2008) Distribution and habitat use of antillean manatees (*Trichechus manatus manatus*) in the Drowned Cayes area of Belize, central America. Aquat Mamm 34:35–43. https://doi.org/10.1578/AM.34.1.2008.35

Landrau-Giovannetti N, Mignucci-Giannoni AA, Reidenberg JS (2014) Acoustical and anatomical determination of sound production and transmission in West Indian (*Trichechus manatus*) and Amazonian (*T. inunguis*) manatees. Anat Rec 297(10): 1896–1907. https://doi.org/10.1002/ar.22993

Laudisoit A, Collet M, Muyaya B, Mauwa C, Ntadi S, Wendelen W, Guiet A, Baudouin M, Leirs H, Vanhoutte N, Micha JC, Verheyen E (2017) West African manatee *Trichechus senegalensis* (Link, 1795) in the Estuary of the Congo River (Democratic Republic of the Congo): review and update. J Biodivers Endanger Species 5(1):1000181. https://doi.org/10.4172/2332-2543.1000181

Mann DA, O'Shea TJ, Nowacek DP (2006) Nonlinear dynamics in manatee vocalizations. Mar Mamm Sci 22(3):548–555. https://doi.org/10.1111/j.1748-7692.2006.00036.x

Marino L (2022) Cetacean brain, cognition, and social complexity. In: di Sciara GN, Würsig B (eds) Marine mammals: the evolving human factor. Springer, Cham, pp 113–148. https://doi.org/10.1007/978-3-030-98100-6_4

Marmontel M, de Souza D, Kendall S (2016) *Trichechus inunguis*. The IUCN Red List of Threatened Species. https://doi.org/10.2305/IUCN.UK.2016-2.RLTS.T22102A43793736.en. Accessed 12 Jan 2024

Marsh H (2018) Dugong. In: Wursig B, Thewissen JGM, Kovacs KM (eds) Encyclopedia of marine mammals. Academic, London, pp 275–277. https://doi.org/10.1016/C2015-0-00820-6

Marsh H, Lefebvre LW (1994) Sirenian status and conservation efforts. Aquat Mamm 20:155–170

Marsh H, Sobtzick S (2019) *Dugong dugon* (amended version of 2015 assessment). The IUCN Red List of Threatened Species. https://doi.org/10.2305/IUCN.UK.2015-4.RLTS.T6909A160756767.en. Accessed 12 Jan 2024

Marsh H, O'Shea TJ, Reynolds JE (2011) Ecology and conservation of the Sirenia: dugongs and manatees (No. 18). Cambridge University Press

Marsh H, Albouy C, Arraut E, Castelblanco-Martínez DN, Collier C, Edwards H, James C, Keith-Diagne L (2022) How might climate change affect the ethology and behavioral ecology of dugongs and manatees? In: Marsh H (ed) Ethology and behavioral ecology of Sirenia. Springer, Cham, pp 351–406. https://doi.org/10.1007/978-3-030-90742-6_8

Medeiros JSD (2021) The iconographical diversity of the Sirens' physical forms in medieval bestiaries. Anastasis: Res Medieval Cult Art 8(1):51–64. https://doi.org/10.35218/armca.2021.1.03

Meirelles ACO, dos Santos Lima D, de Oliveira Alves MD, Borges JCG, Marmontel M, Carvalho VL, dos Santos FR (2022) Don't let me down: West Indian manatee, *Trichechus manatus*, is still critically endangered in Brazil. J Nat Conserv 67:126169. https://doi.org/10.1016/j.jnc.2022.126169

Merchan F, Echevers G, Poveda H, Sanchez-Galan JE, Guzman HM (2019) Detection and identification of manatee individual vocalizations in Panamanian wetlands using spectrogram clustering. J Acoust Soc Am 146(3):1745–1757. https://doi.org/10.1121/1.5126504

Miksis-Olds JL, Tyack PL (2009) Manatee (*Trichechus manatus*) vocalization usage in relation to environmental noise levels. J Acoust Soc Am 125(3):1806–1815. https://doi.org/10.1121/1.3068455

Moore AM, Ambo-Rappe R, Ali Y (2017) "The lost princess (putri duyung)" of the small islands: dugongs around Sulawesi in the anthropocene. Front Mar Sci 4:284. https://doi.org/10.3389/fmars.2017.00284

Nair RV, Lal Mohan RS (1975) Studies on the vocalisation of the sea cow *Dugong dugon* in captivity. Indian J Fish 22:277–278

Niezrecki C, Phillips R, Meyer M, Beusse DO (2003) Acoustic detection of manatee vocalizations. J Acoust Soc Am 114(3):1640–1647. https://doi.org/10.1121/1.1598196

Nowacek DP, Casper BM, Wells RS, Nowacek SM, Mann DA (2003) Intraspecific and geographic variation of West Indian manatee (*Trichechus manatus* spp.) vocalizations (L). J Acoust Soc Am 114(1):66–69. https://doi.org/10.1121/1.1582862

Nowacek SM, Wells RS, Owen EC, Speakman TR, Flamm RO, Nowacek DP (2004) Florida manatees, *Trichechus manatus latirostris*, respond to approaching vessels. Biol Conserv 119(4):517–523. https://doi.org/10.1016/j.biocon.2003.11.020

Nunes Pereira MN (1944) O peixe-boi da Amazônia. Bol Minist Agric Serv Florest Bras 33:21–95

Okumura N, Ichikawa K, Akamatsu T, Arai, Shinke T, Hara T, Adulyanukosol K (2007) Stability of call sequence in dugongs' vocalization. In: Proceeding of the OCEANS 2006-Asia Pacific, vol 19. IEEE, pp 12–15. https://doi.org/10.1109/OCEANSAP.2006.4393936

O'Shea TJ, Poché LB (2006) Aspects of underwater sound communication in Florida manatees (*Trichechus manatus latirostris*). J Mammal 87(6):1061–1071. https://doi.org/10.1644/06-MAMM-A-066R1.1

O'Shea TJ, Beck CA, Hodgson AJ, Keith-Diagne L, Marmontel M (2022) Social and reproductive behaviors. In: Marsh H (ed) Ethology and behavioral ecology of Sirenia. Springer, Cham, pp 101–154. https://doi.org/10.1007/978-3-030-90742-6_4

Panyawai J, Prathep A (2022) A systematic review of the status, knowledge, and research gaps of dugong in Southeast Asia. Aquat Mamm 48(3):203–222. https://doi.org/10.1578/AM.48.3.2022.203

Parsons MJG, Holley D, McCauley RD (2013) Source levels of dugong (*Dugong dugon*) vocalizations recorded in Shark Bay. J Acoust Soc Am 134(3):2582–2588. https://doi.org/10.1121/1.4816583

Phillips R, Niezrecki C, Beusse DO (2004) Determination of West Indian manatee vocalization levels and rate. J Acoust Soc Am 115(1):422–428. https://doi.org/10.1121/1.1635839

Ramos EA, Maust-Mohl M, Collom KA, Brady B, Gerstein ER, Magnasco MO, Reiss D (2020) The Antillean manatee produces broadband vocalizations with ultrasonic frequencies. J Acoust Soc Am 147(2):EL80–EL86. https://doi.org/10.1121/10.0000602

Reeves RR, Tuboku-Metzger D, Kapindi RA (1988) Distribution and exploitation of manatees in Sierra Leone. Oryx 22(2):75–84. https://doi.org/10.1017/S0030605300027538

Reyes-Arias JD, Brady B, Ramos EA, Henaut Y, Castelblanco-Martínez DN, Maust-Mohl M, Searle L, Pérez-Lachaud G, Guzmán HM, Poveda H, Merchan F, Contreras K, Sanchez-Galan JE, Collom KA, Magnasco MO (2023) Vocalizations of wild West Indian manatee vary across subspecies and geographic location. Sci Rep 13(1):11028. https://doi.org/10.1038/s41598-023-37882-8

Reynolds JE III (1981) Aspects of the social behaviour and herd structure of a semi-isolated colony of West Indian manatees, *Trichechus manatus*. Mammalia 45(4):

431–451. https://doi.org/10.1515/mamm.1981.45.4.431

Rivera-Chavarría M, Castro J, Camacho A (2015) The relationship between acoustic habitat, hearing and tonal vocalizations in the Antillean manatee (*Trichechus manatus manatus*, Linnaeus, 1758). Biol Open 4(10):1237–1242. https://doi.org/10.1242/bio.013631

Rosas FCW, Sousa-Lima RS, da Silva VMF (2003) Capítulo 6: Avaliação preliminar dos mamíferos do baixo Rio Purus. In: de Deus CP, da Silveira R, Py-Daniel LHR (eds) Piagaçu-Purus: Bases Científicas para a Criação de uma Reserva de Desenvolvimento Sustentável. Instituto de Desenvolvimento Sustentável Mamirauá, Manaus, pp 49–59

Rycyk AM, Deutsch CJ, Barlas ME, Hardy SK, Frisch K, Leone EH, Nowacek DP (2018) Manatee behavioral response to boats. Mar Mamm Sci 34(4):924–962. https://doi.org/10.1111/mms.12491

Rycyk AM, Factheu C, Ramos EA, Brady BA, Kikuchi M, Nations HF, Kapfer K, Hampton MC, Garcia ER, Kamla AT (2021) First characterization of vocalizations and passive acoustic monitoring of the vulnerable African manatee (*Trichechus senegalensis*). J Acoust Soc Am 150(4):3028–3037. https://doi.org/10.1121/10.0006734

Rycyk AM, Berchem C, Marques TA (2022) Estimating Florida manatee (*Trichechus manatus latirostris*) abundance using passive acoustic methods. JASA Express Lett 2(5):051202. https://doi.org/10.1121/10.0010495

Sakamoto S, Ichikawa K, Akamatsu T, Shinke T, Arai N, Hara T, Adulyanukosol K (2006) Effect of ship sound in the vocal behavior of dugongs. In: Proceedings of the 7th SEASTAR 2000 workshop, Bangkok, Thailand, pp 69–75

Schevill WE, Watkins WA (1965) Underwater calls of *Trichechus* (manatee). Nature 205(4969):373–374. https://doi.org/10.1038/205373a0

Schur D (2014) The silence of Homer's sirens. Arethusa 47(1):1–17. https://doi.org/10.1353/are.2014.0000

Self-Sullivan C, Mignucci-Giannoni A (2008) *Trichechus manatus* ssp. *manatus*. The IUCN Red List of Threatened Species. https://doi.org/10.2305/IUCN.UK.2008.RLTS.T22105A9359161.en. Accessed 12 Jan 2024

Sharko FS, Boulygina ES, Tsygankova SV, Slobodova NV, Alekseev DA, Krasivskaya AA, Rastorguev SM, Tikhonov AN, Nedoluzhko AV (2021) Steller's sea cow genome suggests this species began going extinct before the arrival of Paleolithic humans. Nat Commun 12(1):2215. https://doi.org/10.1038/s41467-021-22567-5

Simpson GG (1941) Vernacular names of South American mammals. J Mammal 22(1):1–17

Smith AJ (1987) An ethnobiological study of the usage of marine resources by two Aboriginal communities on the east coast of Cape York Peninsula, Australia. PhD thesis, James Cook University, Townsville, Queensland, Australia

Sonoda S, Takemura A (1973) Underwater sounds of the manatees, *Trichechus manatus manatus* and *T. inunguis* (Trichechidae). Rep Inst Breed Res Tokyo Univ Agric 4:19–24

Sousa-Lima RS (1999) Comunicação acústica em peixes-boi (Sirenia: Trichechidae): repertório, discriminação vocal e aplicações no manejo e conservação das espécies no Brasil. MSc thesis, Universidade Federal de Minas Gerais, Belo Horizonte, Minas Gerais, Brazil

Sousa-Lima RS (2001) Ontogeny of individually distinct vocal patterns in manatees. In: Culture in marine mammals workshop at the 14th biennial conference on the biology of marine mammals, Vancouver, Canada, p 27

Sousa-Lima RS, da Silva VMF (2001) Four-year consistency in individual vocal patterns of *Trichechus inunguis*. In: Abstracts of the 14th biennial conference on the biology of marine mammals, Vancouver, Canada, pp 201–202

Sousa-Lima RS, Paglia AP, da Fonseca GAB (1999) Vocal discrimination of two species of manatees (*Trichechus inunguis* and *Trichechus manatus manatus*) in Brazil. J Acoust Soc Am 106(4):2164. https://doi.org/10.1121/1.427204

Sousa-Lima RS, Paglia AP, da Fonseca GAB (2002) Signature information and individual recognition in the isolation calls of Amazonian manatees, *Trichechus inunguis* (Mammalia: Sirenia). Anim Behav 63(2):301–310. https://doi.org/10.1006/anbe.2001.1873

Sousa-Lima RS, Paglia AP, da Fonseca GAB (2008) Gender, age, and identity in the isolation calls of Antillean manatees (*Trichechus manatus manatus*). Aquat Mamm 34(1):109–122. https://doi.org/10.1578/AM.34.1.2008.109

Sousa-Lima RS, Rountree RA, Brito MRM, Carletti IM, da Silva VMF (2013) Invisible yet detectable: wild Amazonian Manatees (*Trichechus inunguis*) can be monitored by passive acoustics. In: XI Congresso de Ecologia do Brasil & I Congresso Internacional de Ecologia, Porto Seguro, Brasil, pp 1–5

Steel C (1982) Vocalization patterns and corresponding behavior of the West Indian manatee (*Trichechus manatus*). PhD thesis, Florida Institute of Technology, Melbourne, Florida, USA

Stejneger L (1887) How the great northern sea-cow (Rytina) became exterminated. Am Nat 21(12):1047–1054

Steller GW (1793) Reise von Kamtschatka nach Amerika mit dem Commandeur-Capitän Bering: ein Pendant zu dessen Beschreibung von Kamtschatka. Johann Zacharias Logan, Saint Petersburg

Tanaka K, Ichikawa K, Nishizawa H, Kittiwattanawong K, Arai N, Mitamura H (2017) Differences in vocalisation patterns of dugongs between fine-scale habitats around Talibong Island, Thailand. Acoust Aust 45:243–251. https://doi.org/10.1007/s40857-017-0094-7

Tanaka K, Ichikawa K, Kittiwattanawong K, Arai N, Mitamura H (2023) Spatial variation of vocalising dugongs around Talibong Island, Thailand. Bioacoustics 32(1):33–47. https://doi.org/10.1080/09524622.2022.2058614

Trotzuk E, Allen K, Cockcroft V, Findlay K, Gaylard A, Marsh H, Matos L, West L, Guissamulo A (2022) *Dugong dugon* (Eastern Africa subpopulation). The IUCN Red List of Threatened Species. https://doi.org/10.2305/IUCN.UK.2022-2.RLTS.T218582764A218589142.en. Accessed 12 Jan 2024

Tsutsumi C, Ichikawa K, Arai N, Akamatsu T, Shinke T, Hara T, Adulyanukosol K (2006) Feeding behavior of wild dugongs monitored by a passive acoustical method. J Acoust Soc Am 120(3):1356–1360. https://doi.org/10.1121/1.2221529

Turvey ST, Risley CL (2006) Modelling the extinction of Steller's sea cow. Biol Lett 2(1):94–97. https://doi.org/10.1098/rsbl.2005.0415

Umeed R, Attademo FLN, Bezerra B (2017) The influence of age and sex on the vocal repertoire of the Antillean manatee (*Trichechus manatus manatus*) and their responses to call playback. Mar Mamm Sci 34(3):577–594. https://doi.org/10.1111/mms.12467

Umeed R, Lucchini K, Coutinho PDF, dos Santos PJP, Borges JCG, Normade I, Attademo FLN, Luna F, Bezerra B (2023) Acoustic interactions between free-living mother–calf Antillean manatees, *Trichechus manatus manatus*. J Ethol 41:243–251. https://doi.org/10.1007/s10164-023-00788-z

van der Sluijs I, Gray SM, Amorim MCP, Barber I, Candolin U, Hendry AP, Krahe R, Maan ME, Utne-Palm AC, Wagner H-J, Wong BB (2011) Communication in troubled waters: responses of fish communication systems to changing environments. Evol Ecol 25:623–640. https://doi.org/10.1007/s10682-010-9450-x

Vanderhoff EN, Bernal Hoverud N (2022) Perspectives on antiphonal calling, duetting and counter-singing in non-primate mammals: an overview with notes on the coordinated vocalizations of bamboo rats (*Dactylomys* spp., Rodentia: Echimyidae). Front Ecol Evol 10:906546. https://doi.org/10.3389/fevo.2022.906546

Williams LE (2005) Individual distinctiveness, short- and long-term comparisons, and context specific rates of Florida manatee vocalizations. MSc thesis, University of North Carolina, Wilmington, North Carolina, USA

Wirsing AJ, Kiszka JJ, Allen AC, Heithaus MR (2022) Ecological roles and importance of sea cows (Order: Sirenia): a review and prospectus. Mar Ecol Prog Ser 689:191–215. https://doi.org/10.3354/meps14031

Open Access This chapter is licensed under the terms of the Creative Commons Attribution 4.0 International License (http://creativecommons.org/licenses/by/4.0/), which permits use, sharing, adaptation, distribution and reproduction in any medium or format, as long as you give appropriate credit to the original author(s) and the source, provide a link to the Creative Commons license and indicate if changes were made.

The images or other third party material in this chapter are included in the chapter's Creative Commons license, unless indicated otherwise in a credit line to the material. If material is not included in the chapter's Creative Commons license and your intended use is not permitted by statutory regulation or exceeds the permitted use, you will need to obtain permission directly from the copyright holder.

Marine Mammal Hearing

Dorian Houser

Contents

8.1	**Introduction**	492
8.2	**Measuring Hearing**	493
8.2.1	Terminology	493
8.2.2	Measurement Methods	496
8.2.3	Amphibious vs. Fully Aquatic Auditory Systems	501
8.2.4	Approaches to Hearing Studies	504
8.3	**Audiograms**	505
8.3.1	Behavioral Audiograms	509
8.3.2	Electrophysiological Audiograms	520
8.3.3	Baleen Whale Hearing	533
8.3.4	Population-Level Variability	535
8.3.5	Hearing Loss	537
8.3.6	Hearing at Depth	537
8.3.7	Control of Hearing Sensitivity	538
8.4	**Critical Ratio, Critical Band, and Auditory Filters**	539
8.4.1	Critical Ratio and Critical Band	539
8.4.2	Adaptations for the Reduction of Masking	542
8.5	**Frequency and Level Discrimination**	544
8.5.1	Frequency Discrimination	544
8.5.2	Level Discrimination	546
8.6	**Temporal Summation and Temporal Resolution**	546
8.6.1	Temporal Summation	546
8.6.2	Temporal Resolution	548
8.7	**Directional Hearing**	549
8.8	**Source Localization**	549
8.8.1	Odontocetes	550
8.8.2	Pinnipeds	552
8.8.3	Sirenians	554

D. Houser (✉)
National Marine Mammal Foundation, San Diego, CA, USA
e-mail: dorian.houser@nmmf.org

© The Author(s) 2025
C. Erbe et al. (eds.), *Marine Mammal Acoustics in a Noisy Ocean*,
https://doi.org/10.1007/978-3-031-77022-7_8

8.9	Auditory Weighting Functions	554
8.10	Summary	560
References		560

8.1 Introduction

The study of marine mammal hearing is necessary to understand the role of sound in marine mammal behavioral ecology and the potential for anthropogenic noise to impact marine mammals. Hearing directly affects the ability of marine mammals to detect and localize acoustic communication signals, predators, or other signals of interest, as well as to navigate and forage. Hearing characteristics also affect the magnitude and type of potential impacts that a marine mammal might incur from anthropogenic noise exposure. For example, noise outside the hearing range of an animal is unlikely to be perceived (except at very high levels) and is unlikely to result in a behavioral reaction. Conversely, acute sensitivity across a range of frequencies might correlate with an increase in responsiveness to noise, greater potential for masking, and the potential for physiological auditory effects [e.g., noise-induced hearing loss (NIHL)] at the same or higher frequencies than the noise exposure (see Sect. 8.4).

Sound travels more efficiently through water than air, whereas light penetrates poorly through water. It is therefore no surprise that most marine mammals evolved proficient sound production and reception mechanisms. Some marine mammals (e.g., odontocetes) might depend primarily on sound, as opposed to most terrestrial mammals, which are primarily visual. The evolution of hearing in marine mammals has resulted in specializations for underwater hearing (e.g., cetaceans) and for balancing hearing in air and under water (e.g., pinnipeds). As a result, studies of sound perception across species have focused not only on underwater hearing, but also on how hearing in air differs from hearing under water in amphibious species.

Hearing studies have been completed on odontocetes, otariids, phocids, walruses, sirenians, sea otters, and polar bears. Information about hearing sensitivity in previously unstudied species is now published more frequently, largely because of the maturity and availability of electrophysiological methods that allow studies to be conducted without the time and resources required of psychoacoustic studies (see below). However, direct measurements of hearing in mysticetes have been elusive with only a couple attempts using electrophysiological methods to test hearing in these large whales.

Most information on hearing in marine mammals is collected at frequencies above 1 kHz, and because of the ultrasonic range of hearing in odontocetes, much of this occurs above 10 kHz. Concern about the impact of noise on marine mammals has historically focused on low frequencies (<1 kHz), where the ability to project sound and provide adequate test environments is more difficult. Concern for the impact of low-frequency sound is greatest for mysticetes, which themselves produce low-frequency sounds and have an auditory system anatomy believed to support low-frequency hearing (Ketten 1998, 2000). However, cetacean strandings in association with mid-frequency sonar systems (1–10 kHz range) and discoveries about the sensitivity of some marine mammals to high-frequency sound (e.g., behavioral responses and susceptibility to NIHL) has broadened the scope of potentially impacting sound sources and frequencies of concern (e.g., Evans and England 2001).

Hearing ability is not explained by the range of hearing and hearing sensitivity alone; to comprehensively understand hearing ability, and therefore the effects of noise, the sensitivity, frequency and intensity discrimination, integration time, potential for masking, and the ability to localize sound sources should be studied. Studies should be conducted across marine mammal species to look for trends and commonalities among closely related species while also identifying species' specializations. All of this must be performed with an understanding of the caveats of the experimental approaches used, particularly when comparing results from behavioral and electrophysiological studies.

8.2 Measuring Hearing

Before discussing the measurement of hearing in marine mammals, it is important to define some terms relevant to hearing studies. These terms form the basis for characterizing marine mammal hearing, and in some instances, the application of hearing information to estimating the impact of noise on marine mammals.

8.2.1 Terminology

The following section discusses terminology relevant to marine mammal hearing and its measurement. Where standard definitions exist (e.g., American National Institute of Standards (ANSI) documents), the reference to the standard containing the definition is provided. The section does not repeat verbatim the definition from the standard.

> *Threshold of hearing (hearing threshold; ANSI/ASA S3.20)*—The threshold of hearing is the lowest level of sound perceivable by an animal and is determined by the minimum sound pressure level (SPL; root-mean-square sound pressure in dB re 1 µPa) necessary to evoke an auditory sensation in a specified proportion of test trials. The threshold of hearing is dependent upon whether background noise is present that can interfere with the perception of the sound. In the presence of background noise that affects the threshold, the threshold is termed a *masked threshold* (ANSI/ASA S1.1). Although not part of its standard definition (ANSI/ASA S3.20), the *absolute threshold* is generally accepted as the threshold measured when noise levels are too low to affect the threshold. The difference is important because many studies on marine mammal hearing occur in conditions where some degree of masking is likely. The threshold is determined as a function of frequency and thresholds change with frequency; in mammals, thresholds are lowest within some mid-frequency region of the hearing range, and increase toward lower and higher frequencies. The frequency range across which thresholds are the lowest defines the frequency range of *best hearing*. (For purposes of the discussion that follows, the region of best hearing is defined as the range of frequencies within which the thresholds are no more than 20 dB above the lowest measured threshold.) A plot of hearing thresholds as a function of frequency is called the *audiogram* (ANSI/ASA S3.20).
>
> Hearing thresholds are not constant across individuals, can demonstrate variability within individuals over time, and differ as a function of hearing test conditions and methods. It is therefore important that the hearing test methods and the statistical methods used to determine thresholds are defined. Hearing thresholds can be determined behaviorally or through electrophysiological methods, but in either case some statistical definition of threshold must be provided.
>
> *Masking* (ANSI/ASA S1.1)—Interference in the perception of one sound by the presence of other sounds. Masking affects the ability of animals to detect and recognize sound in the natural environment. For marine mammals, masking can affect communication, navigation, foraging, and predator/prey detection.

(continued)

Marine mammals contend with masking from natural noise sources, such as wind, rain, ice breakup, breaking waves, natural seismic activity, and the noise produced by other animals. Anthropogenic noise can also cause masking. Masking is most often characterized with respect to signal detection, but it can also influence signal recognition or interpretation.

Under experimental conditions, the ability of an animal to resolve a signal in noise is determined by measuring masked thresholds. The signal-to-noise ratio (SNR; ANSI/ASA S3.20) for a tone at the threshold of detection in broadband background noise of constant spectral density is known as the *critical ratio* (also called the "critical ratio for masking" or the "Fletcher critical band," ANSI/ASA S3.20). It is calculated as the difference between the SPL of the signal at threshold and the pressure spectral density level of the background noise. *Critical bands* (ANSI/ASA S3.20) are determined by characterizing how the bandwidth of noise centered on a signal affects the signal's detectability. For bandwidths wider than the critical band, the threshold of signal detection does not change. As the bandwidth becomes smaller than the critical band, the threshold of detection decreases. The critical band is one approximation of the bandwidth of *auditory filters*. The auditory system can be represented by a continuum of bandpass filters distributed across the basilar membrane of the inner ear. The shape of the auditory filter plays a large role in determining the ability of an animal to differentiate signal from noise. Critical ratios, critical bands, and auditory filters all change as a function of frequency, and though they are related, the critical band and the critical ratio are not the same and the critical ratio cannot be reliably used to predict the critical band (e.g., see Yost and Shofner 2009); see Sect. 8.4.

Frequency and level discrimination—Information can be encoded in the frequencies and levels of communication signals (e.g., dolphin whistles) and the frequencies and levels of a sound can indicate its source (e.g., boat or sonar). Sound frequency and level are important considerations in the ability to distinguish a signal from background noise. Frequency discrimination is typically determined by comparing a test tone to a reference tone and asking a subject to determine when the tones are different. By this method, the difference in frequency required for the discrimination of two tones can be determined. The minimum discriminable difference between the frequencies of two tones is referred to as the *difference limen* (*DL*; ANSI/ASA S3.20). Similar types of methods can be used for determining the discrimination limits in level differences. Note that *level discrimination* is also sometimes called *intensity discrimination*.

Temporal resolution—Limit of the ability of the auditory system to differentiate the time-separation of two sequential sounds, or avoid interpreting two sequential sounds as a single sound. Temporal resolving power is important in the ability of marine mammals to differentiate the components of communication signals, in processing sequential echoes from objects insonified during echolocation, and in differentiating sources of signals in complex acoustic fields (e.g., noisy environments). In marine mammals, temporal resolution has been studied by both behavioral and electrophysiological methods. Below the limits of temporal resolution,

(continued)

distinct sounds may be integrated as a single event within which discrimination cues may rely more on the spectra of the overall signal. This period of time is sometimes referred to as the *temporal integration time*.

Temporal summation (ANSI/ASA S3.20)—The relationship between the duration of a tone and its detectability, often set equal to the duration above which detectability does not change; that is, the hearing threshold remains the same. It is also sometimes called *auditory summation*, the *auditory integration time*, or the *temporal integration time*, although the latter terms are also used with respect to the temporal window within which an animal processes two distinct sounds as one. This can obviously create confusion since similar terms can be used to describe different processes. Temporal summation is important for understanding signal detection in relation to the threshold of hearing. Below a certain summation time, which will be called the critical summation time, the detectability of the sound decreases (i.e., the hearing threshold increases).

Source localization (or *auditory localization*; ANSI/ASA S3.20)—The ability to locate, in space, the origin of a sound. Source localization is critical for marine mammals to localize soniferous prey, conspecifics (members of the same species), and predators based upon the sounds they make. For echolocating odontocetes, it is also critical to determining where in space an echo originates, thus allowing odontocetes to forage and navigate at range and under conditions of poor visibility.

The directionality of hearing, sometimes described as the *receiving beam pattern*, is an important aspect of sound source localization. The orientation of an animal relative to a sound source is important in source localization, both because of the directionality of hearing and differences in the time of arrival of sound at the two ears; sounds to one side or the other of an animal are generally easier to localize than those directly in front or behind the animal. Similarly, the relationship between the location of a sound source and the location of other masking sounds can affect both detectability and localization. Separation between a sound source and the source of masking noise can facilitate detection and localization, a phenomenon called the *spatial release of masking* (or spatial masking release).

Auditory weighting function—A mathematical function that accounts for the frequency-dependent sensitivity of animals across the range of hearing, and which is applied to a sound to obtain a single, *weighted sound level*. The weighting function is a filter that emphasizes frequencies to which animals are most sensitive and de-emphasizes frequencies to which they are less sensitive. Weighting functions are used to address species differences in hearing sensitivity. In other words, the same sound will not be perceived in the same way by both a mysticete and a dolphin because of differences in their hearing abilities for each frequency. Auditory weighting functions can be used to predict the potential impact of sound on an organism. They have a long history of use in predicting noise impacts on humans, such as annoyance and hearing loss. In more recent years, they have primarily been adopted for use in predicting NIHL in marine mammals.

8.2.2 Measurement Methods

8.2.2.1 Behavioral Methods

The study of hearing in marine mammals is rooted in psychoacoustics, which is the study of the relationship between the physical characteristics of sound and the perception of sound (Fechner 1860). Psychoacoustic approaches to studying hearing rely on behavioral methods in which the subject (animal) is required to respond behaviorally to a presentation of one or more sounds. Various psychoacoustic procedures have been used in the study of marine mammal hearing, but nearly all ultimately produce a statistical description of threshold (e.g., either a 50% or 75% probability of detection or discrimination). Psychoacoustic methods have been used with marine mammals to study hearing thresholds, the directionality of hearing, frequency discrimination capabilities, auditory filter shapes, the onset of noise-induced hearing loss, and many other aspects of hearing.

Two common psychoacoustic approaches used in the study of marine mammal hearing are the "go/no-go" procedure and the "two-alternative forced choice (2AFC)" procedure (Levitt 1971; Stebbins 1970). In both procedures, the position of the animal during the presentation of a sound, or sounds, is generally consistent; an animal may touch its nose to an object, position its head in a hoop, or bite onto an object (e.g., a biteplate) so its position is maintained. The go/no-go procedure requires the animal to perform an action when a sound is detected or perceived as different from a comparison sound ("go" condition). The animal is to do nothing otherwise ("no-go" condition). The "go" response varies across studies, but typically involves either the animal leaving its station and touching a response paddle or producing a sound (e.g., dolphin whistle) to make a response. The 2AFC is similar, but during the stimulus presentation the subject is presented with two alternatives and must decide following the presentation corresponding to one of two possible decisions. The most common 2AFC procedure requires the animal to decide whether two sounds are the same or different.

An important aspect of behavioral studies of signal detection is determining the bias of the subject being studied (for examples with sea lions see Schusterman and Johnson 1975). Bias can be defined as the tendency of an animal to be either conservative or liberal in its decision making. An animal's bias is determined by several factors, such as its temperament, expectation of reward, motivation, and probabilities of signal presentation. If an animal has an expectation that a reward will be great for a correct response and the consequences of being incorrect are minor, it may adopt a liberal response strategy. Conversely, if the consequences of being incorrect are severe, the strategy might be more conservative. Motivationally, an animal's desire to participate in psychoacoustic studies may vary across time as a function of external and internal factors (e.g., distractions in the test environment, reproductive condition or not feeling well, time since the last meal). The real probabilities of stimulus presentation, or an a priori expectation of the animal for the probabilities of stimulus presentation, may also affect the response strategy. For example, if the probability of presenting a stimulus is 75% on one day, but 50% on the next, the strategy on the second day may be influenced by the animal's prior experience with the test procedure. Collectively, these factors can influence how an animal responds under conditions of uncertainty. The bias of marine mammal subjects is typically addressed in signal detection studies by measuring false alarm rates over time or by establishing a relative operating characteristic, or ROC curve. The ROC curve allows the bias of an individual subject to be determined (Swets 2003).

An advantage of behavioral methods is that they provide an "integrated" response of the animal. The process not only accounts for the physical and physiological aspects of sound reception, such as the anatomy and physiology of the ear, but also higher order cognitive processes (e.g., decision making) that cannot be adequately monitored at a physiological level. In short, behavioral methods provide a measure of perception, including both sensory operations and decision operations. A drawback to behavioral methods is animal access. The ability to implement behavioral methods in the study of marine mammal hearing requires time to condition animals on the behavioral procedures and additional time to collect data. This necessitates long-term access to animals and a commitment to house, feed, and provide veterinary care. Because of the experience required and expense associated with the long-term care of marine mammals, relatively few subjects have been and are available for psychoacoustic studies. The result is a double-edged sword; a small number of marine mammals become expert at psychoacoustic studies, thus increasing the efficiency with which information can be obtained, but representatives for the broad number of marine mammal species groups remains limited (often only one or a few subjects are available for a species, if any).

8.2.2.2 Electrophysiological Methods

Electrophysiological methods are an alternative to behavioral methods in which some aspect of a sensory system pathway (e.g., auditory nerve, brain stem, and cortex) is studied by looking at its neural response to a stimulus. The methods used to study hearing in marine mammals commonly record the voltages, or potentials, produced by the auditory nerve, brain stem, or cortex in response to an acoustic stimulus. Voltages produced by the nervous system in response to a sound are called auditory evoked potentials (AEPs). Because the nervous system is naturally noisy, and because the neural recordings are subject to other electrical noise sources (e.g., muscle activity, external electrical sources), the technique generally involves presenting the acoustic stimulus many times so that the AEP waveform (i.e., the change in recorded voltage over time) can be averaged and the noise reduced (i.e., random noise, when averaged, tends to approach a zero-value).

The study of marine mammal hearing through electrophysiological means has been around for decades, dating back to the pioneering work of Bullock and colleagues in the 1960s (Bullock et al. 1968). Historically, electrophysiological methods were invasive and potentially required the sacrifice of the animal subject. However, in the latter part of the last century, methods using noninvasive (surface electrodes) or relatively noninvasive approaches (e.g., small subcutaneous electrodes) became more common (Fig. 8.1). This contributed to growth in the use of electrophysiological methods to study marine mammal hearing (e.g., see Supin et al. 2001).

Marine mammal hearing studies utilize a variety of AEPs and look for stimulus-dependent changes in the voltage-amplitude and/or time-latency of the AEP waveform, or changes in the spectrum of the waveform. The auditory brainstem response (ABR) is an AEP that is characterized by a series of waves produced within the auditory nerve and brain stem within several milliseconds after receiving a sound (Fig. 8.2b). Broadband clicks and tone-bursts are commonly used to produce the ABR (Fig. 8.2a). Each wave of the ABR corresponds to a different level of the auditory system, with the earliest wave corresponding to the auditory nerve. Characterizing the relationship between ABR latencies and amplitudes and the amplitude, spectrum, and timing of the sound that produced the ABR permits certain aspects of auditory system function and design to be characterized, such as the hearing threshold, auditory filter shape, directionality of hearing, and recovery from noise exposure.

One technique that has become prominent in determining marine mammal hearing thresholds is the use of the auditory steady-state response, or ASSR (e.g., Lins et al. 1995). The ASSR is produced when a sound is delivered at a sufficiently rapid rate that the corresponding evoked response approaches a quasi-periodic steady state. Stimuli commonly used to obtain the ASSR include series of tone-bursts or amplitude-modulated tones (Fig. 8.2c). Tone-bursts and amplitude-modulated tones have a narrower spectrum than the broadband clicks often used in ABR studies. Their narrower spectrum makes them more amenable to determining frequency-specific hearing thresholds and audiograms. The ASSR to repetitive tone-bursts or amplitude-modulated tones follows the envelope of sound modulation (Fig. 8.2d). For this reason, it is also sometimes

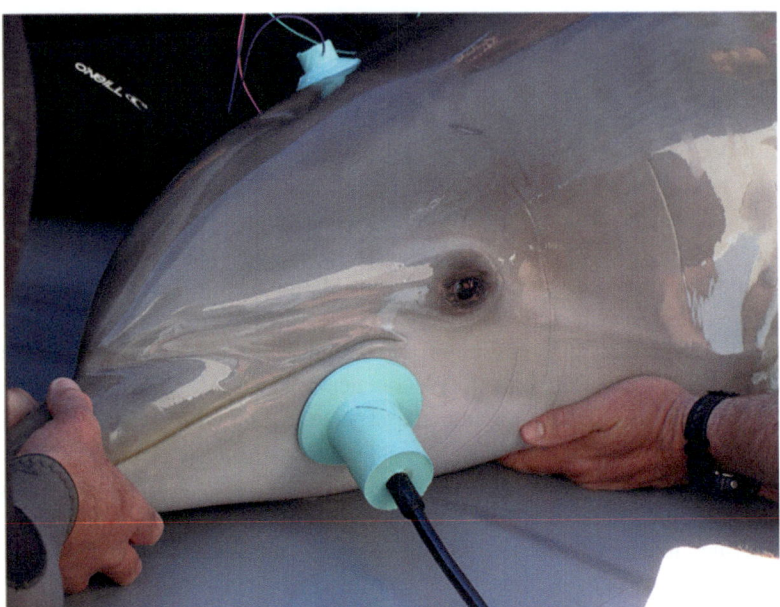

Fig. 8.1 A dolphin undergoing an electrophysiological hearing test while resting out of water. The suction cups on the dorsal surface hold the recording electrodes to the skin. The suction cup on the jaw contains the transducer that delivers sound to the dolphin. Used with permission of Taylor & Francis LLC, from Houser D, Mulsow J (2016) Chapter 11 Acoustics, in Castellini MA, Mellish J, eds, Marine mammal physiology: Requisites for ocean living. © Taylor & Francis, 2016; permission conveyed through Copyright Clearance Center, Inc., https://doi.org/10.1201/b19614. All rights reserved

called the *envelope following response* (e.g., Popov et al. 2005). Whereas the ABR is generally described by the latencies and amplitudes of the ABR waves, the periodicity of the ASSR allows it to be analyzed in the frequency domain (i.e., using spectral analysis). This, in turn, allows the response to be analyzed by additional statistical means.

Because of the broad use of AEPs in determining marine mammal hearing thresholds, it is important to understand the differences in methodological approaches and how they potentially influence the threshold estimate. Different laboratories and researchers use different techniques, and the results obtained are not necessarily equivalent. All approaches have one thing in common: They rely on the ability to distinguish between when a particular evoked response is present in a recording of brain activity, and when it is absent. To obtain a hearing threshold, this transition from presence to absence must somehow be related to a sound level. A traditional approach often employed by researchers using tone-bursts or clicks as acoustic stimuli determines the hearing threshold by "looking" for the disappearance of the ABR as the level of the sound stimulus is decreased. In other words, presence of an evoked response is determined by visual inspection of the voltage-waveform recorded from the animal. Similar approaches can be used with ASSR recordings; once converted to the frequency domain, researchers can observe the frequency peak in the spectrum of the ASSR that corresponds to the stimulus modulation rate, as well as the voltage time-waveform. The peak height varies with stimulus level and the disappearance of the peak as stimulus level declines can be determined visually. Visual determination of the presence or absence of an evoked response is straightforward and intuitive, but it also has problems. The presence of a response can be difficult to determine near threshold and subjective decisions may be biased (e.g., by prior experience or expectation of

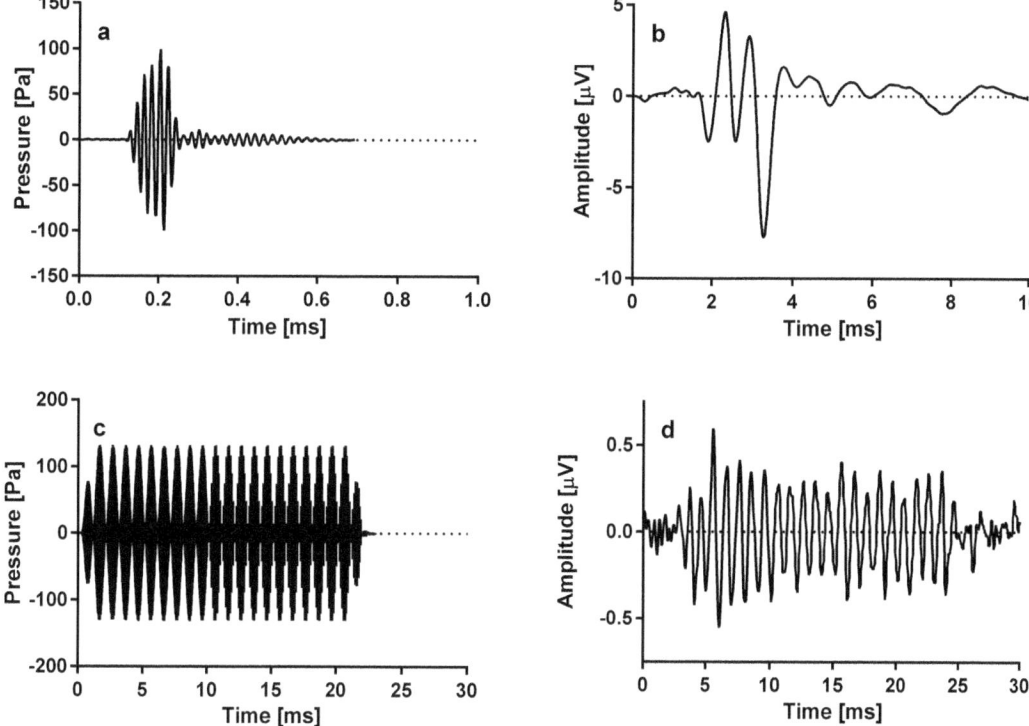

Fig. 8.2 Acoustic waveforms and evoked responses to those waveforms. (**a**) Transient tone-burst (note the slight ringing of the transducer following production of the tone-burst). (**b**) Auditory brainstem response (ABR) to the tone-burst in (**a**). (**c**) A sinusoidal amplitude-modulated (SAM) tone. (**d**) The evoked response to the SAM tone in (**c**). Note that the evoked response follows the envelope of the stimulus amplitude modulation

what the threshold should be). Because of these potential confounds, statistical approaches have been developed for the objective detection of evoked responses. To date, the most used statistical approaches in electrophysiological marine mammal hearing studies consist of the F-test, magnitude-squared coherence test, and the modified variance ratio (Finneran 2008, 2009; Finneran et al. 2007a).

Multiple methods for determining the hearing threshold through AEP methods exist, just as they do in behavioral studies. The "linear extrapolation" method relies upon measuring the amplitude of AEPs to various levels of a sound presentation, ensuring that the range of stimulus levels encompasses both robust evoked responses and levels where no response is observed. A plot of the AEP amplitude vs. stimulus level is then created and a linear regression applied to the most linear set of continuous data points just above the highest stimulus level where no evoked response is observed (Fig. 8.3). Once established, the linear regression is extrapolated to a 0-volt crossing and the SPL at the crossing is used as the hearing threshold. An alternative threshold estimation method is to set the hearing threshold equal to the lowest SPL associated with a detected AEP, or the midpoint between this level and the highest stimulus level at which no AEP is detected. The approach can be modified by setting an electrophysiological noise floor (i.e., from external and internal (e.g., myogenic) electrical noise sources) below which no evoked response is expected to be detected.

The various approaches to estimating hearing thresholds have varying degrees of bias and affect the threshold estimate differently. For example, the selection of "the most linear" series of data

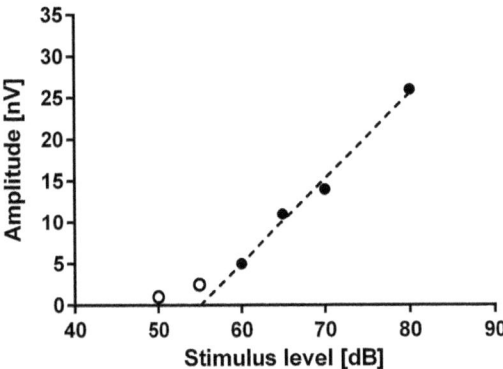

Fig. 8.3 Comparison of the different methods by which threshold can be estimated from the relationship of ASSR amplitude to the stimulus presentation level. The plot of the evoked response amplitude as a function of stimulus presentation level creates the IO function. Filled circles correspond to positively identified ASSRs, either by objective or subjective methods, and open circles correspond to when no ASSR is detected in the electrophysiological recording. By using the lowest stimulus at which an ASSR is detected, the threshold estimate is 60 dB. By using the linear extrapolation method (indicated by the dashed line), in which the threshold is determined by extrapolating the linear regression line to the 0-voltage crossing, the threshold estimate is 55 dB

points used to create the linear regression is subjectively determined and may vary from one analyst to the next, whereas the use of the lowest stimulus level where a response is objectively detected limits experimenter bias to the selection of the objective response parameters. For a given set of data, a threshold obtained via linear extrapolation should nearly always be lower than that obtained with the lowest-detected response because of the slope of the regression line (Fig. 8.3). Because there are different approaches to estimating threshold, and each affects the threshold differently, an American National Standard was created to standardize approaches for AEP hearing tests in odontocetes (American National Standards Institute (ANSI) 2018), which is the marine mammal group within which the most marine mammal AEP hearing tests have been completed.

Electrophysiological hearing tests can be performed on animals that are untrained, asleep, or anesthetized, enabling investigations of hearing in wild-caught animals, stranded marine mammals, and those brought into rehabilitation centers. AEP systems have become portable (e.g., Finneran 2009), making these tests more practical. Species have been tested that are not typically kept in marine parks or aquaria, including beaked whales (Cook et al. 2006; Finneran et al. 2009; Pacini et al. 2011), a group of whales that are of concern due to historical relationships between mass strandings and naval sonar operations. Relative to behavioral hearing tests, AEP hearing tests are rapid, typically taking less than an hour to complete. For these reasons, AEP hearing tests are now performed in more subjects than are behavioral hearing tests. The AEP hearing test is how estimates of population-level hearing variability will likely be determined. Enough animals have already been tested in some species so that population-level variability in AEP audiograms can begin to be assessed (e.g., Castellote et al. 2014; Houser and Finneran 2006b; Houser et al. 2008b; Mulsow et al. 2011a, 2014b; Popov et al. 2007; Ruser et al. 2016). Furthermore, AEP methods provide the possibility of directly measuring hearing in small mysticetes.

Although there are many benefits to using electrophysiological methods to study hearing in marine mammals, the results are not equivalent to those obtained from behavioral studies. Hearing thresholds obtained with AEPs are generally higher than those obtained with behavioral methods, although they might sometimes be lower when linear extrapolation is used to estimate threshold (see Sect. 8.3.2). Behavioral and AEP thresholds may differ by 20 dB or more, and the agreement between the two approaches often worsens at the high- and low-frequency ends of the hearing range (Finneran and Houser 2006; Houser and Finneran 2006a; Yuen et al. 2005). Behavioral methods provide a whole-animal response, capturing cognitive aspects of hearing and other intrinsic factors such as animal motivation and attention. Electrophysiological approaches generally measure averaged neural responses corresponding to one or more parts of the auditory system, and typically do not account for higher-order signal processing and cognitive factors, motivation, or attention. In addition, the determination of AEP thresholds often uses signals much shorter in duration (μs to tens of ms) than those used in most behavioral hearing studies. Because the signal durations are below the temporal summation time (see Sect. 8.6), additional confounds in relating results from AEP and behavioral hearing studies occur. Thus, although there are several advantages to using electrophysiological methods to study hearing in marine mammals, the results must be interpreted in context of the approach's shortcomings relative to behavioral hearing studies.

8.2.2.3 Behavioral Response As an Indirect Method

It is possible to monitor whether a marine mammal provides an unconditioned response to the presentation of a noise to determine that it heard the sound. This approach has been tried with several marine mammals under water and in air (Dahlheim and Ljungblad 1990; Ghoul and Reichmuth 2014b; Kastelein et al. 1993). If an animal responds to a sound, in the absence of other factors (e.g., approach of a ship), then it can reasonably be assumed that the animal heard the sound. This can be useful information, particularly as it pertains to the frequency range of hearing (see Sect. 8.3). However, there are serious limitations to the approach. First, if an animal does not respond to a sound presentation it does not mean that it did not hear it. Animals may not respond for many reasons, including the motivation of the animal at the time of the exposure, the context of the exposure, the personality of the individual (e.g., risk-averse vs. risk-tolerant), prior experience with the sound source, the biological meaning of the sound (if any), the context of the exposure, etc. Second, although the approach might provide information about the frequencies an animal can hear, it cannot provide reliable data on the threshold of hearing. It is possible under certain conditions that the probability of response within an individual marine mammal declines with the level of received sound, but unconditioned response thresholds are probably well above the threshold of hearing in most cases.

8.2.2.4 Anatomical Modeling

Anatomical modeling is not truly a method of hearing measurement, but it is briefly included here because the use of anatomical modeling in predicting hearing abilities in marine mammals has grown over the last several decades. Utilizing advanced laboratory analyses and physical modeling techniques, such as finite-element modeling (FEM), the anatomy and morphology of marine mammals has been used to predict the range of hearing, hearing sensitivity, and sound reception and production pathways of different species, predominantly cetaceans (e.g., Cranford and Krysl 2015; Houser et al. 2001; Parks et al. 2007b; Tubelli et al. 2018). Because no empirical measures of hearing have been made in any baleen whale, and as it has never been practical to conduct direct behavioral hearing studies with them, anatomical models have become the primary means by which the range of hearing and hearing sensitivity have been inferred, albeit in only a few baleen whale species (see Sect. 8.3.3).

8.2.3 Amphibious vs. Fully Aquatic Auditory Systems

The obligate aquatic existence of certain marine mammals has resulted in a greater modification to the ancestral terrestrial mammal ear than is observed in amphibious species, in part due to the need of amphibious species to detect and localize sound in support of life-history functions both in air and in water (e.g., breeding, foraging, predator detection/avoidance). Profound differences in the auditory system design between the obligate aquatic and amphibious marine mammal species are also influenced by factors other than hearing, such as the need to reduce drag and improve swimming efficiency. Design and function are coupled and differences in auditory system design influence hearing capabilities, the scientific questions addressed in hearing studies, and the way some hearing studies are performed.

The amphibious marine mammals include the pinnipeds, sea otters, and polar bear. Each of these carnivorous species performs certain life functions both in air and water. The degree to which the auditory system has been modified in these groups is driven by the relative importance of activities supporting life-history functions in each medium, as well as its evolutionary history. How this has been balanced across evolutionary scales is of particular interest to scientists because of the selective pressures driven by the physical characteristics of the medium. An effective pathway from the environment to the ear is going to be strongly influenced by whether sound is transmitted through water or through air.

Hearing abilities within the pinnipeds vary by family: the otariids generally show similar hearing characteristics across species; the phocids show a greater diversity in hearing abilities by species, probably reflecting their greater evolutionary and ecological diversity; the odobenids (walrus) demonstrate hearing characteristics somewhat intermediate of the otariids and phocids. The otariids have ears most similar to their terrestrial ancestors, the fissiped carnivores (Nummela 2008). The pinna is greatly diminished, but still exists, and the middle ear bones are similar to that of terrestrial mammals. In phocids, the pinna is absent and a pinhole-sized meatus exists. A muscular sphincter surrounds the meatus, presumably to exclude water during diving. Some phocids (e.g., harbor seal) have demonstrated the ability to manipulate the meatal opening to improve hearing sensitivity in air (Kastak et al. 2005a), while others (e.g., monk seal) have a permanently closed meatal opening (Ruscher et al. 2021). The middle ear bones and auditory bullae may be hypertrophied, depending on species, which likely contributes to the bone conduction of sound. The walrus lacks a pinna, but appears to have a functional acoustic meatus and ossicles, that though hypertrophied, are similar in design to that of the otariids (Kastelein et al. 1996a; Nummela 2008; Repenning 1972).

The frequency range of hearing in the otariids is generally similar in air and under water (Moore and Schusterman 1987; Reichmuth et al. 2013), whereas that of the phocids is broader under water than in air (Kastak and Schusterman 1999; Kastelein et al. 2009c; Møhl 1968a; Reichmuth et al. 2013; Sills et al. 2014, 2015). Underwater high-frequency hearing limits in phocids and otariids are likely dictated by cochlear structure, but it has been suggested that high-frequency hearing limits of phocids in air may be limited by the inertial constraints imposed by their more massive middle ear ossicles (Hemilä et al. 2006). Comparative tests of in-air and underwater hearing in the walrus are more limited and prevent any conclusive statements about whether the frequency range of hearing substantially changes when walruses are submerged (Kastelein et al. 1996b, 2002b; Reichmuth et al. 2020). It should also be noted that work with otariids and phocids indicates the ability to detect ultrahigh-frequency signals, above the frequency range typically associated with the upper limit of functional hearing (Cunningham and Reichmuth 2016). The findings challenge the presumption that cochlear constraints limit high-frequency hearing in air, and question whether the constraints are dictated by conductive mechanisms. However, it should be noted that the thresholds at ultrahigh frequencies are high, between 122 and 148 dB re 1 µPa for frequencies of 140 and 180 kHz, and the functional biological implications for signal detection at these frequencies might be questionable.

Debate about whether pinniped ears are better adapted for in-air or underwater hearing, and how this varies by species, has been around for many years. In recent years, research has shown that most prior aerial hearing measurements in pinnipeds had been masked and underestimated sensitivity. Contemporary utilization of sound-attenuating chambers has demonstrated that the unmasked hearing sensitivity of many pinnipeds (excluding the northern elephant seal and Hawaiian monk seal) is similar to that of terrestrial carnivores, suggesting that the ears of most pinnipeds are acutely adapted to hearing in both media (Reichmuth et al. 2013).

Sea otters and polar bears both have ears more similar to their fully terrestrial relatives and less derived for an aquatic existence. Underwater hearing in the sea otter is much less sensitive

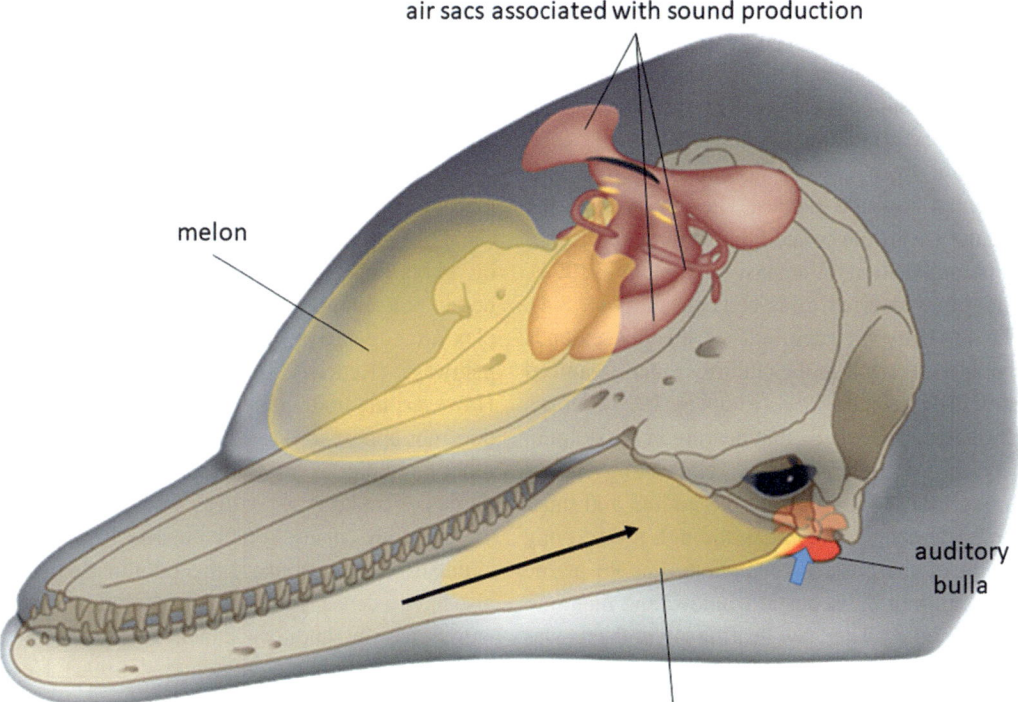

Fig. 8.4 Depiction of the dolphin cranial anatomy highlighting the auditory pathway for high-frequency sound reception. (Modified from Fig. 5.4 of Cozzi et al. 2017). The thick black arrow depicts the sound reception pathway for high-frequency echoes; sound is received through the lower jaw and is transmitted to the auditory bulla (red) through a fatty channel connecting the acoustic fats of the jaw to the bulla (blue arrow). Lower frequency sound has a more omnidirectional path to the auditory bullae. Other anatomical structures of interest include the fatty melon and air sacs utilized in sound production. Used with permission of Elsevier Inc., from Cozzi B, Huggenberger S, Oelschläger H (2017) Anatomy of dolphins: Insights into body structure and function. © Elsevier, 2017; permission conveyed through Copyright Clearance Center, Inc. https://doi.org/10.1016/B978-0-12-407229-9.00001-4. All rights reserved

than in pinnipeds and cetaceans suggesting their ears are primarily adapted for aerial hearing (Ghoul and Reichmuth 2014b). Only aerial hearing measurements have been made in polar bears and nothing is known of their frequency range of hearing and hearing sensitivity under water (Nachtigall et al. 2007; Owen and Bowles 2011).

The fully aquatic marine mammals are the cetaceans and sirenians. Cetaceans have a telescoped skull and no pinnae; the external ear (meatus) resembles a pin-sized hole on each side of the cetacean head. In odontocetes (Fig. 8.4), the bony structures containing the middle and inner ear form a complex of exceptionally dense bone (auditory bulla or tympanoperiotic complex) that is separate and lateral to the skull (Oelschläger 1990). The complex is covered dorsomedially by an air-filled peribullar sinus (Houser et al. 2004), which helps to acoustically isolate the ears and potentially contributes to sound source localization. The tympanic membrane does not connect to the external auditory canal and it has been argued that the outer ear of the dolphin contributes relatively little to sound reception (McCormick et al. 1970; Ridgway 2000); however, a possible role for the outer ear in pressure reception has been suggested more recently (De Vreese 2014).

Many odontocete species have upper limits of hearing >100 kHz. High-frequency hearing enables echolocation, which is directional. Directionality is supported in part by modifications to the peripheral auditory system that enable echo reception from objects insonified by forward-projected biosonar signals. The jaw bone of the odontocete is hollow and filled with fat of distinctive composition (Varanasi and Malins 1970), sometimes referred to as acoustic fat (Fig. 8.4). The fat body on each side of the jaw extends caudally to attach to the auditory bulla (Cranford et al. 2010; Ketten 1994; Ridgway 1999). Norris (1968) first speculated that hearing in the dolphin may be accomplished by channeling sound through these fat bodies to the middle and inner ears. The sound first enters through a thin bony region of the posterior mandible, known as the pan region or the acoustic window, and then into the fat bodies. The "jaw hearing" hypothesis was confirmed experimentally years later by engaging a dolphin in an echolocation task with the lower jaw shielded by an acoustically opaque material (Brill 1988; Brill and Harder 1991). Subsequent work further showed that dolphins were acoustically sensitive to frequencies of sound in the echolocation band at a point on the jaw corresponding to the "pan," a thin region of bone below the gape of the mouth (e.g., Møhl et al. 1999). However, more recent work has supported a dual-mode of hearing hypothesis. Evidence of the acoustic delays between the location of sound presentation on a dolphin's head and electrophysiological indications of when sound reception occurs demonstrate that two anatomical regions may be optimized for sound reception (Popov and Supin 1990a; Popov et al. 1992, 2008): one over the pan region corresponding to frequencies used in echolocation, and one over the meatus corresponding to lower frequencies typically associated with communication.

Mysticetes lack the jaw specializations of the echolocating odontocetes. However, the auditory bullae are enlarged and separated from the skull, although some bony connections between the bullae and the skull remain. The ossicles of mysticete whales are massive and the basilar membrane thicknesses and widths are flaccid and broad, all suggestive of low-frequency specialization (Ketten 2000). Experimental evidence for sound reception pathways in mysticete whales does not exist. However, anatomy is providing some clues on possible sound conduction mechanisms. Connections of the external ear canal to the highly derived tympanic membrane, or "glove finger," and recent discoveries of fatty bodies connecting with the tympanoperiotic complex are suggestive of lateral sound conduction pathways (Ketten 2000; Yamato et al. 2012). However, the lipid composition of the fat bodies does not appear to have the distinctiveness from other fat bodies as observed in odontocete acoustic fats (Yamato et al. 2014).

As with many other marine mammals, sirenians lack a pinna. Like odontocetes, they also have a dense, bony tympanoperiotic complex and a narrow auditory canal, the function of which is questionable (Ketten et al. 1992). However, the tympanoperiotic complex is not isolated from the skull, as in odontocetes. The ossicles are massive, loosely connected to one another, and with respect to the stapes, somewhat derived (Ketten et al. 1992). Relatively little is known about sound reception pathways in this fully aquatic group.

8.2.4 Approaches to Hearing Studies

The above brief overview of differences between the auditory systems of the fully aquatic and amphibious marine mammals is presented to discuss how these differences affect the design of hearing studies. Experiments for marine mammal hearing studies can be set up in many ways. The use of "direct field" studies, in which the marine mammal is stationed such that near-field nonlinearities in the sound field and potential reflections are avoided, are commonly used in both amphibious and fully aquatic species (both in air and water). Direct field hearing tests must consider the potential for the environment to affect the hearing test measurements, for example, reflections within small pools might require the stimulus to be changed (for instance using frequency sweeps and narrow noise bands instead of pure tones) to mitigate the occurrence of reflections and standing waves (Finneran and Schlundt 2007), or baffle boards may be installed. Direct field hearing tests are easier to conduct in open-water situations, but such studies typically occur in the natural environment where masking by biological noise sources may occur. Conversely, performance of hearing studies in pools potentially allows for better control of background noise, but at the possible cost of violating direct field assumptions.

For in-air hearing studies in amphibious marine mammals, both direct field acoustic stimulations and the delivery of acoustic stimuli through headphones have been used. For direct field studies, all the aforementioned considerations of the direct field in underwater testing conditions must be taken into account (e.g., reflections and masking). In some instances, the use of a hemi-anechoic chamber has been useful to substantially reduce background noise so that absolute aerial thresholds could be obtained (Reichmuth et al. 2013). Headphones, which have been used in pinnipeds, can also reduce background noise, but thresholds using headphones for sound delivery have not produced the same absolute thresholds as observed in hemi-anechoic chamber studies (Reichmuth et al. 2013).

In some instances, testing of odontocete species can also be conducted in air with results predicting sensitivity to underwater sound. This occurs either when an animal is stranded, or in situations where human care is provided, when the animal voluntarily beaches itself or is pulled out of the water for a procedure. Under these circumstances, sound delivery can be accomplished through a contact transducer, or "jawphone," which is a sound projector embedded in a suction cup that has properties similar to water (Brill et al. 2001; Møhl et al. 1999). The jawphone is not used to test aerial hearing capabilities, but rather relies on the coupling of the jawphone to the skin to emulate the transfer of sound energy from the water to the animal. In many delphinids, jawphones are typically attached to the pan region of the jaw. However, there are species differences in the optimal placement of jawphones (e.g., compare Møhl et al. 1999; Mooney et al. 2008). Jawphones are particularly suited for studying hearing at frequencies where echolocation occurs; their attachment capitalizes on the acoustic pathway used for echolocation, and the size of the transducers used in the jawphones is often optimal for the projection of high frequencies. Jawphones are not restricted to studies of odontocete hearing in air; a number of studies have used jawphones on submerged/partially submerged odontocetes (Brill et al. 2001; Cook et al. 2006; Finneran et al. 2009; Houser and Finneran 2006a; Kastelein et al. 1997).

8.3 Audiograms

The audiogram is a graphical representation of an animal's hearing sensitivity at specific frequencies. Audiograms can be obtained through behavioral and electrophysiological methods, but as previously noted, the two approaches are not equivalent. Behavioral methods can be used to determine absolute detection thresholds. Electrophysiological methods typically overestimate the absolute threshold (underestimate sensitivity), and the degree to which thresholds are overestimated can vary by species and animal size. However, electrophysiological results can correlate well with the frequency range of hearing and the hearing sensitivity curve as determined behaviorally (e.g., Finneran and Houser 2006; Schlundt et al. 2007; Yuen et al. 2005).

In this section, behavioral audiograms are presented first as they provide the most accurate information on hearing sensitivity. Within the discussion of behavioral audiograms, the "region of best hearing" is defined as the range of frequencies within which the thresholds are no more than 20 dB above the lowest measured threshold (Table 8.1). For species in which multiple subjects were available for hearing tests, the region of best hearing is calculated as the average across individuals. In instances where animals had suspected hearing loss (e.g., Sills et al. 2015), significant masking was expected (e.g.,

Table 8.1 Region of best hearing and bandwidth of best hearing for various species of marine mammals

Species	Region of best hearing [kHz]		Bandwidth of best hearing [octaves]				Low-frequency roll-off [dB/octave]			
	Under water	In air	Under water	In air	N	References	Under water	In air	N	References
HF odontocetes										
Bottlenose dolphin	5.3–125.5		4.6		3	Johnson (1967), Schlundt et al. (2008) and Finneran et al. (2010)	10.4		1	Johnson (1967)
Pacific white-sided dolphin	2.4–130.5		5.8		1	Tremel et al. (1998)	11.0		1	Tremel et al. (1998)
Killer whale	5.3–69.6		3.7		8	Szymanski et al. (1999) and Branstetter et al. (2017a)	9.5		6	Branstetter et al. (2017a)
Risso's dolphin	3.4–83.8		4.6		1	Nachtigall et al. (1995)				
False killer whale	10.4–75.5		2.9		1	Thomas et al. (1988)				
Beluga	1.5–89.3		5.9		4	White et al. (1978), Ridgway et al. (2001) and Finneran et al. (2005)	9.0		3	Awbrey et al. (1988) and Johnson et al. (1989)
Striped dolphin	19.6–135.0		2.8		1	Kastelein et al. (2003)	8.0		1	Kastelein et al. (2003)
Tucuxi	14.3–126.0		3.1		1	Sauerland and Denhardt (1998)				
VHF odontocetes										
Harbor porpoise	3.6–142.3[e]		5.3		3	Andersen (1970a, b) and Kastelein et al. (2002a, 2010)				
Baiji	9.4–69.0		2.9		1	Wang et al. (1992)				
Boto	10.0–96.9		3.3		1	Jacobs and Hall (1972)				
Phocids										
Harbor seal	0.4–47.6	0.7–13.9	6.9	4.3	3	Reichmuth et al. (2013) and Kastelein et al. (2013)	6.5	11.8	3	Kastelein et al. (2009a, c, 2013) and Reichmuth et al. (2013)
Ringed seal	0.3–52.5	0.7–11.7	7.5	4.1	2	Sills et al. (2015)	9.5	10.9	2	Sills et al. (2015)
Spotted seal	0.3–56.0	0.7–11.7	7.5	4.1	2	Sills et al. (2014)	8.7	12.0	2	Sills et al. (2014)
Bearded seal	0.3–44.9		7.2		2	Sills et al. (2020)	8.8		2	Sills et al. (2020)
Northern elephant seal	0.2–48.9	0.2–26.9	7.9	7.1	1	Kastak and Schusterman (1999)	[a]			
Hawaiian monk seal	0.3–33.2	0.1–32.4	6.8	8.1	1	Sills et al. (2021) and Ruscher et al. (2021)	8.3	7.3	1	Sills et al. (2021) and Ruscher et al. (2021)
Odobenids										
Walrus	0.2–13.3		6.1		1	Kastelein et al. (2002b)				

8 Marine Mammal Hearing

Species									References
Otariids									
California sea lion	0.5–33.7	1.2–23.8	6.1	4.3	4/3	13.5	14.1	3	Schusterman et al. (1972), Mulsow et al. (2011a, 2012), Reichmuth and Southall (2012) and Reichmuth et al. (2013, 2017)
Norther fur seal	1.7–32.1	1.1–25.5	4.2	4.5	2/3[b]		12.9	1	Moore et al. (1987) and Babushina et al. (1991)
Steller sea lion		0.9–19.2					14.7	1	Mulsow et al. (2010)
Sirenians									
Manatee	2.4–28.6		3.6		4	7.8		3[c]	Gerstein et al. (1999) and Mann et al. (2009)
Ursids									
Polar bear		2.0–15.5		3.0	5[d]		9.3	5[d]	Owen and Bowles (2011)
Mustellids									
Sea otter	2.3–26.6	1.2–27.1	3.5	4.5	1	4.6	16.9	1	Ghoul et al. (2014)

Where sufficient low-frequency sensitivity has been determined, the low-frequency roll-off is also calculated. For species with multiple individuals tested, the average value of the individuals is presented

[a]The elephant seal's underwater sensitivity is relatively flat from 200 Hz to 3.2 kHz

[b]Babushina et al. (1991) excluded for calculation of region of best hearing under water (see text)

[c]Insufficient data below 1 kHz for one animal (Gerstein et al. 1999); calculation from the other animal in the same study excludes thresholds at 400 and 500 Hz due to elevated thresholds

[d]Data used for calculation were averages of two animals from one facility and three animals from another

[e]For one animal the highest frequency tested was imputed as the upper limit of the best region of hearing because the threshold did not exceed 20 dB above the lowest threshold obtained

Ljungblad et al. 1982), or there was insufficient data to calculate the region of best hearing, the animals were excluded from the calculation. Occasional subjects were reported with thresholds at the frequency of best hearing as low as ~30 dB re 1 μPa under water. However, the broader collection of data suggests that these values are probably erroneous and that thresholds at the frequency of best hearing in normal hearing animals are likely 40–45 dB re 1 μPa when measured under water. These outlier data were excluded from the analysis as they disproportionately affected the determination of the region of best hearing.

Where sufficient behavioral data exist to describe the decline in hearing sensitivity below 1 kHz, the change in sensitivity with frequency, or the low-frequency "roll-off," is defined (Table 8.1). A high-frequency roll-off is not presented as hearing sensitivity declines rapidly near the "upper limit of hearing," which is defined here as the frequency at which the underwater behavioral threshold of hearing equals or exceeds 100 dB re 1 μPa.

Electrophysiological audiograms are presented following the discussion of behavioral audiograms. The region of best hearing and low-frequency roll-off are not noted in this discussion. There is a known divergence between behavioral and electrophysiological (AEP) estimates of hearing sensitivity at the lower and higher frequencies where sensitivity declines (see Sect. 8.2.2.2). Electrophysiological methods therefore tend to provide a narrower frequency range of best hearing and a steeper low-frequency roll-off than is observed behaviorally, and would likely provide a narrower estimate of the hearing range. The upper limit of hearing is still feasible to quantify, but because of the differences between AEP and behavioral thresholds, the upper limit of hearing for AEP studies is defined here as the frequency at which the underwater AEP threshold of hearing equals or exceeds 120 dB re 1 μPa (Houser and Finneran 2006b).

Reflecting on a suggestion from Terhune (1981) that sufficient information existed to generically talk about odontocete, phocid and otariid hearing, Richardson et al. (1995) warned that the few individuals studied and the potential for intra- and interspecies differences should be cautiously considered. Over the last two decades, the idea of classifying marine mammals according to shared hearing characteristics has gained momentum. Ketten (2000) proposed that cetaceans could be classified into acoustic groups according to their inner ear anatomy: Type M, infrasonic to low-frequency mysticetes; and Type I and Type II high- and low-range ultrasonic odontocetes. Southall et al. (2007) subsequently grouped all marine mammal species into hearing groups during the development of marine mammal noise exposure criteria. The functional hearing groups were based largely upon species hearing ranges, as well as auditory system anatomy, and whether species were amphibious. The proposed hearing groups (and their associated auditory bandwidths) consisted of the low-frequency (LF) cetaceans, mid-frequency (MF) cetaceans, high-frequency (HF) cetaceans, pinnipeds in water, and pinnipeds in air. The LF cetaceans were mysticetes, and the MF and HF cetaceans were odontocetes. In subsequent guidance on how to estimate the impact of anthropogenic sound on the hearing of marine mammal species under their purview (National Marine Fisheries Service 2016, 2018), the National Marine Fisheries Service (NMFS) modified the hearing groups proposed by Southall et al. (2007). Another proposed modification to the groupings was made in 2019, based on consideration of new scientific findings related to species-typical vocalizations, echolocation characteristics (for odontocetes), auditory system anatomy, audiograms, and differences between aerial and underwater audiograms in amphibious marine mammals (Southall et al. 2019b). The proposed groupings included the LF cetaceans, HF cetaceans, very high-frequency (VHF) cetaceans, sirenians (SI), phocid carnivores in water (PCW), phocid carnivores in air (PCA), other marine carnivores in water (OCW), and other marine carnivores in air (OCA). As is apparent, three major changes were made: pinnipeds were split into phocid and "other carnivore" groups, the latter consisting of the otariids and other marine carnivores, including the sea otter and polar bear;

both the phocid and other carnivore groups were split into in-air and underwater groups; and the odontocetes were reorganized into HF and VHF groups to address differences in the regions of best hearing sensitivity across species. It should be noted that an additional group was proposed for future consideration—the very low frequency (VLF) group—as well as possibly treating sperm whales, beaked whales, and killer whales differently that other members of the HF cetaceans because of a better sensitivity to lower frequency sound. The VLF group would consist of the largest baleen whales (e.g., blue, fin) which produce very low frequency vocalizations and have larger basilar membranes and cochlear radii ratios than those (often smaller) mysticetes that vocalize at higher frequencies (e.g., minkes, humpbacks).

8.3.1 Behavioral Audiograms

8.3.1.1 Odontocetes

Complete or partial behavioral audiograms have been collected for 11 species of odontocete, all small to moderate size and excluding sperm and beaked whales. The audiograms represent delphinid, monodont, and phocoenid species and cover both oceanic and riverine ecotypes. In the following sections, hearing abilities are discussed in reference to animals that appear to have "normal" hearing ranges. Animals with suspected hearing loss are not included in the review of audiograms that follow (both behavioral and electrophysiological), but are discussed in a later section. Nevertheless, audiogram variability within a species is likely explained by variation in noise masking during measurements and individual differences in hearing sensitivity.

The bandwidth of best hearing in HF and VHF cetaceans ranges from 2.8 to 5.9 octaves (Table 8.1). The upper limit to the region of best hearing is somewhat lower in larger odontocetes, although this does not necessarily correspond to a narrower bandwidth of best hearing when considered on an octave spacing. For example, the beluga has the widest measured region of best hearing among cetaceans, but the upper limit of the region is 89.3 kHz, which is 30–40 kHz lower than that of other odontocetes such as the bottlenose dolphin, Pacific white-sided dolphin, striped dolphin, and tucuxi.

Hearing sensitivity has been tested down to 75 Hz in the bottlenose dolphin (Fig. 8.5a), 40 Hz in the beluga (Fig. 8.5b), 100 Hz in the killer whale and Pacific white-sided dolphin (Fig. 8.6a, b), and 250 Hz in the harbor porpoise (Fig. 8.7). In general, hearing data from odontocetes collected to-date suggest that odontocetes are relatively insensitive to sound below 1 kHz. These findings are consistent with additional results of hearing tests with a false killer whale and a Risso's dolphin trained to detect a 75 Hz signal; mean thresholds for both

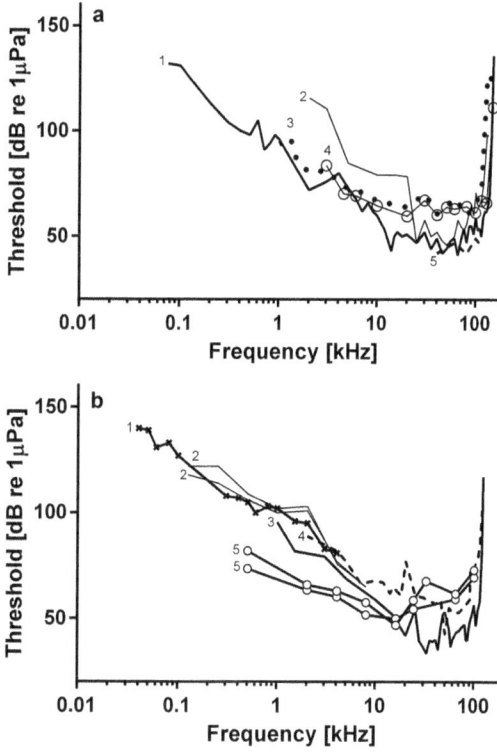

Fig. 8.5 Underwater behavioral audiograms of (**a**) bottlenose dolphin: 1 (thick line) (Johnson 1967), 2 (thin line) (Ljungblad et al. 1982), 3 (dotted line) (Finneran et al. 2010), 4 (thin line, open circles) (Schlundt et al. 2008), 5 (dashed line) (Lemonds 1999). (**b**) beluga: 1 (thick line, x's) (Johnson et al. 1989), 2 (thin lines, $n = 2$) (Awbrey et al. 1988), 3 (thick line) (White et al. 1978), 4 (dashed line) (Finneran et al. 2005), 5 (thick line and open circles, $n = 2$) (Ridgway et al. 2001)

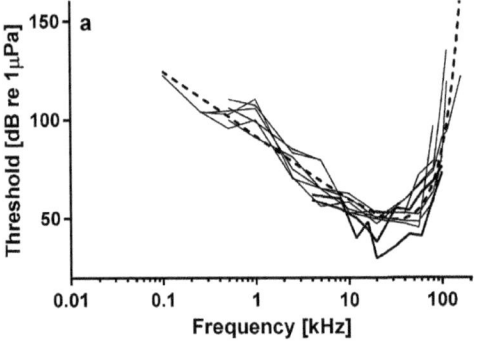

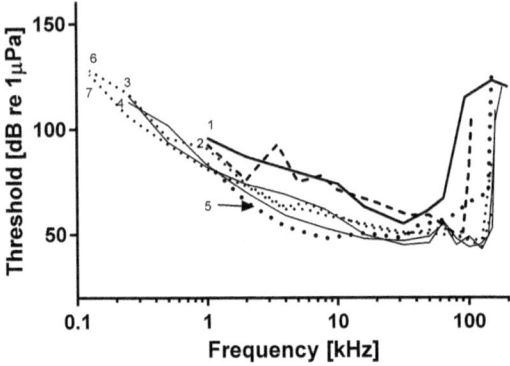

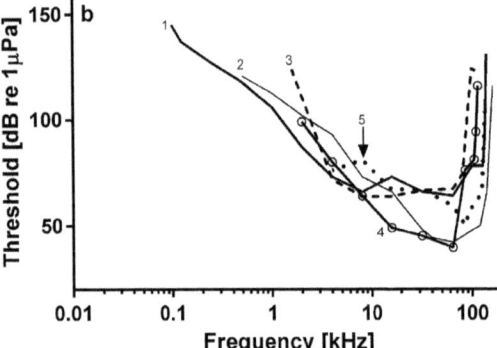

Fig. 8.6 Underwater behavioral audiograms of (**a**) killer whales: thin lines represent individuals from Branstetter et al. (2017a) and thick lines represent individuals from Szymanski et al. (1999). The dashed line represents the composite audiogram of the killer whale from Branstetter et al. (2017a). (**b**) Other HF cetaceans: Pacific white-sided dolphin, 1 (thick line) (Tremel et al. 1998); striped dolphin, 2 (thin line) (Kastelein et al. 2003); Risso's dolphin, 3 (dashed line) (Nachtigall et al. 1995); false killer whale, 4 (thick line, open circles) (Thomas et al. 1988); tucuxi, 5 (dotted line) (Sauerland and Dehnhardt 1998)

Fig. 8.7 Underwater behavioral audiograms of VHF cetaceans: Chinese river dolphin, 1 (thick line) (Wang et al. 1992); Amazon river dolphin, 2 (dashed line) (Jacobs and Hall 1972); harbor porpoise, 3 and 4 (thin lines) (Kastelein et al. 2002a, 2010); harbor porpoise, 5 (dotted line, large dots) (Andersen 1970a); harbor porpoise, 6 and 7 (dotted line, small dots) (Kastelein et al. 2017)

animals were >140 dB re 1 μPa (Au et al. 1997). Utilizing a linear best-fit to threshold data from species that have been tested below 1 kHz (Awbrey et al. 1988; Branstetter et al. 2017a; Johnson 1967; Johnson et al. 1989; Kastelein et al. 2003; Tremel et al. 1998), the average low-frequency roll-off below 1 kHz is 9.5 dB/octave (range of 8–11 dB/octave) in odontocetes. At low frequencies, it has been suggested that bottlenose dolphins (and possibly other odontocetes) might be more sensitive to a combination of acoustic pressure and low-frequency particle motion (Turl 1993). However, experimental evidence in the beluga and bottlenose dolphin have since suggested that acoustic pressure is the primary contributor to low-frequency, underwater hearing (Finneran et al. 2002a).

The upper limit to hearing in bottlenose dolphins, the striped dolphin, Pacific white-sided dolphin, and the tucuxi extends above 120 kHz. In the harbor porpoise it can extend to 150 kHz or more. This limit appears slightly lower in the killer whale, Risso's dolphin, beluga, and false killer whale (~95 to 115 kHz), which may be an indicator of a relationship between the size of the whale species and the upper limit of hearing. Electrophysiological audiograms collected on other odontocetes demonstrate upper limits of hearing consistent with this idea (Sect. 8.3.2). The frequency range of hearing in all odontocetes supports echolocation, which is dependent upon the production of high-frequency pulses, and the upper limit of hearing matches well with the bandwidth of echolocation signals across species.

Not surprisingly, the highest frequency hearing limits have been recorded in the youngest animals (e.g., Branstetter et al. 2017a; Johnson 1967). The age of animals likely also affects hearing threshold levels. For example, the lowest thresholds for bottlenose dolphin obtained within the region of best hearing were obtained with animals that were 12 years (Ljungblad et al. 1982) and 8–9 years (Johnson 1967) of age. By comparison, the highest thresholds obtained within the same frequency range were from animals 25 years (Finneran et al. 2010) and 21 years of age (Schlundt et al. 2008). Although this is by no means conclusive, reductions in hearing sensitivity with age are well known in mammals and should be considered in the interpretation of audiograms and the potential for both natural and anthropogenic ocean noise to affect marine mammals.

Extrinsic factors can also contribute to variability in hearing thresholds and these should be considered in the interpretation of audiograms. For example, masking is a concern for wild animals subject to natural and anthropogenic background noise, and is also a critical concern in behavioral audiometry. Hall and Johnson (1972) and Wang et al. (1992) both performed audiometric studies under conditions where background noise was a masking concern. By comparison to more recent studies of porpoises in very controlled, low-level noise conditions (Kastelein et al. 2002a, 2010), the potential impact of masking can be observed (e.g., note differences in thresholds in Fig. 8.7). Although species differences cannot be ruled out, the higher hearing thresholds in the Chinese river dolphin (Wang et al. 1992) and Amazon river dolphin, or boto (Jacobs and Hall 1972), must be considered as possibly masked thresholds. Similarly, Ljungblad et al. (1982) speculated that signal detection in their bottlenose dolphin was masked at frequencies <5 kHz, and a comparison to other measurements in bottlenose dolphins suggests this might very well be the case. Conversely, Ridgway et al. (2001) measured thresholds lower than previously measured in the beluga at frequencies <10 kHz (Fig. 8.5b). Although no background noise measurements were presented, the measurements made by Ridgway and his colleagues were in deep ocean water under calm conditions. It is quite possible that low-frequency noise under these conditions was less than that observed in the presence of tank noise (Awbrey et al. 1988) or the noise in bays and other coastal waters where anthropogenic and biological noise might be prevalent (Johnson et al. 1989).

Taken in whole, the audiograms of belugas and dolphins seem similar (Fig. 8.5a, b), displaying similar thresholds and frequency ranges of hearing with possible small differences (<1/2 octave) in the upper limit of hearing. In the context of predicting impacts of ocean noise upon marine mammals, this is important. Audiograms from these animals are, by necessity, used as surrogates for many phylogenetically related species for which audiograms have not been obtained (i.e., untested members of the HF cetacean hearing group). The other HF cetaceans for which behavioral audiograms exist generally follow the same pattern, although many species are only represented by a single individual and it is uncertain as to how representative each audiogram is for the species. In contrast, Branstetter et al. (2017a) collected audiograms from eight killer whales under similar experimental conditions. All but two of these individuals, which were suspected of having hearing loss, are presented in Fig. 8.6a. Given that the animals in this study represented both sexes ranging in age from 12 to 52 years of age, the distribution of audiograms enables a composite audiogram for the species to be generated (dashed line of Fig. 8.6a; taken from Branstetter et al. 2017a). Although various methods for generating composite audiograms can be used (e.g., mean, median, and other model approaches), the generation of composite audiograms (with associated threshold variances) will be useful in characterizing hearing abilities from a "population-level" perspective (see Sect. 8.3.4).

The VHF cetaceans are less represented than their HF counterparts. The collective data from five different harbor porpoises suggest that the frequency at which harbor porpoise hearing thresholds exceed 100 dB re 1μ Pa is higher than in HF cetaceans. The two other VHF cetaceans (Amazon and Chinese river dolphin) for which behavioral audiograms exist demonstrate lower frequencies for the upper roll-off in sensitivity than in some HF cetaceans. As noted with the less-studied HF cetaceans, single individuals need to be considered cautiously as representative of species hearing capabilities. Indeed, classification of the Chinese river dolphin with VHF cetaceans is probably weighted more by its echolocation signal than by its hearing range (Li et al. 2005), although similar evidence for the Amazon river dolphin is debatable (Evans 1973; Kamminga 1994; Ladegaard et al. 2015). No hearing data are available for *Cephalorhynchus* dolphins, another VHF cetacean group, although their echolocation and vocal repertoire are suggestive as well (e.g., Götz et al. 2010; Kyhn et al. 2010; Morisaka et al. 2011).

8.3.1.2 Pinnipeds

The pinnipeds are composed of species that vary broadly in their ecology, ranging from those that regularly haul out on land to the primarily aquatic, from shallow divers to deep divers, and from capital breeders (those that acquire body reserves prior to pupping and fast during lactation) to income breeders (those that continue feeding throughout lactation). However, they are all amphibious and they call and hear both under water and on land. Knowledge of the hearing abilities of the amphibious marine mammals in air and under water is important, not only to better understand the ecology of the species and the adaptations that allow them to exploit terrestrial and aquatic environments, but also because they are potentially susceptible to the impact of anthropogenic noise in both environments.

8.3.1.2.1 Underwater Hearing

Underwater hearing has been behaviorally measured in three species of otariids and nine species of phocids. Additionally, some data are available for the only extant odobenid species, the walrus. However, not all the pinniped behavioral audiometry studies represent absolute thresholds and issues with either masking or individual hearing problems may exist. Studies of aerial hearing conducted in recent years have challenged the notion that the pinnipeds are more adapted to hearing under water than in air (Reichmuth et al. 2013; Schusterman 1981), but this will be addressed later in the discussion of aerial audiograms. In certain instances, some subject animals have been tested more than once. In these instances, the data that are reported here were determined by verifying (1) which study exhibited the greatest amount of experimental control, (2) covered the greatest range of hearing, and (3) reflected normal hearing without indication of hearing loss (e.g., as might occur with age or NIHL).

Phocids

The phocids can be broken into the Phocinae and Monachinae, or the northern and southern seals, as their respective subfamilies are commonly known. (These are also sometimes referred to as the *phocinid* and *monachid* seals, which is the form that will be used in this chapter.) Within the phocids, hearing abilities appear to vary in accordance with both species phylogeny and ecology. In contrast to otariids, the frequency range of hearing in phocids is greater under water than in air.

The harbor seal has been subject to more hearing studies than any other phocid seal (Kastelein et al. 2009a, c; Møhl 1968a; Reichmuth et al. 2013; Terhune 1988). To a lesser degree, studies have also been conducted with the Caspian, harp, spotted, ringed and bearded seals (Babushina 1997; Sills et al. 2014, 2015, 2020; Terhune and Ronald 1972, 1975b). Audiograms for the harbor seal are presented in Fig. 8.8a and the audiograms for all other phocinid seals are presented in Fig. 8.8b. There are large differences in the thresholds obtained in older hearing studies and those made in phocinid seals in recent years (Babushina 1997; Sills et al. 2014, 2015; Terhune and Ronald 1972, 1975b). Masking seems a likely explanation for the threshold differences. Comparisons of critical ratios measured in the seals to levels of background noise led to the conclusion that masking may have occurred, particularly in the youngest animals with the best hearing (Sills et al. 2015).

Insufficient data are available to calculate the region of best hearing in the Caspian and harp seal, but the region of best hearing in harbor, spotted, bearded, and ringed seals appears remarkably similar, spanning between 6.9 and 7.5 octaves. The region of best hearing is broader

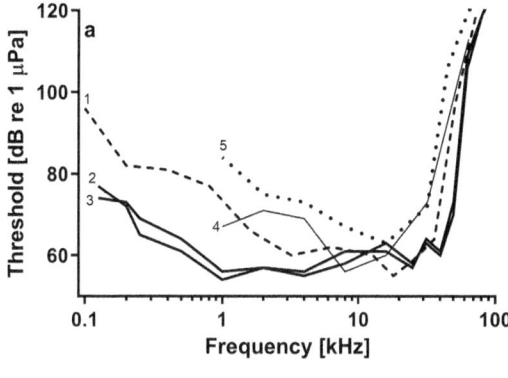

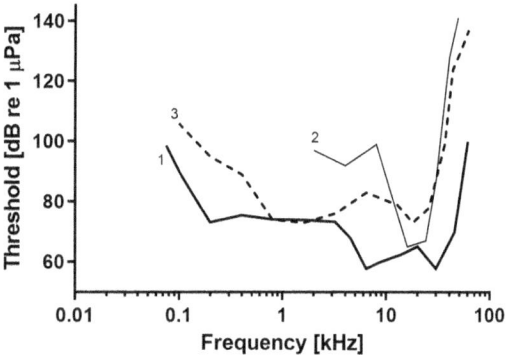

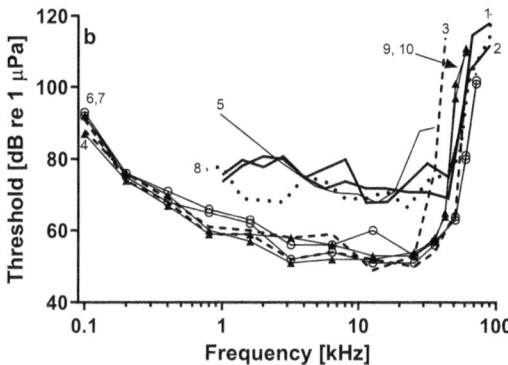

Fig. 8.8 Underwater behavioral audiograms of phocinid seals. (**a**) Audiograms from the harbor seal: 1 (dashed line) (Reichmuth et al. 2013); 2 and 3 (thick lines) (Kastelein et al. 2009a); 4 (thin line) (Terhune 1988); 5 (dotted line) (Møhl 1968a). (**b**) Other northern seals: ringed seal, 1 and 2 (thick lines, no symbols) (Terhune and Ronald 1975b); ringed seal, 3 and 4 (dashed lines) (Sills et al. 2015); Caspian seal, 5 (thin line) (Babushina 1997); spotted seal, 6 and 7 (thin lines, open circles) (Sills et al. 2014); harp seal, 8 (dotted line) (Terhune and Ronald 1972); bearded seal, 9 and 10 (thin line, filled triangles) (Sills et al. 2020)

Fig. 8.9 Underwater behavioral audiograms of monachid seals: Northern elephant seal, 1 (thick line) (Kastak and Schusterman 1999); Hawaiian monk seal, 2 (thick line) (Thomas et al. 1990a); 3 (dashed line) (Sills et al. 2021)

than that observed in odontocetes, which is due to a greater sensitivity to low frequency sound. Hearing sensitivity has been tested to as low as 100 Hz in many phocinid seal species (Fig. 8.8). Thresholds below 1 kHz are lower than in the odontocetes (e.g., thresholds at 100 Hz are <100 dB re 1 μPa).

Phocinid seals exhibit similar upper limits of hearing (Fig. 8.8). Thresholds exceeding 100 dB re 1 μPa generally occur at frequencies between 50 and 60 kHz, although one ringed seal showed a lower, upper limit of hearing that was possibly due to age (he was 16 years old) or prior exposure to ototoxic antibiotics (Sills et al. 2015). Although the harbor seal has been shown capable of hearing at frequencies >60 kHz, it has been suggested that the ability to discriminate between frequencies above 60 kHz is limited (Møhl 1967). If this is true, it potentially holds true for other phocinid seals. Underwater hearing tests with harbor seals just sub-surface (i.e., with ears 6 cm below the water surface) found little difference in thresholds compared to tests performed in the same seals at 1 m of depth, suggesting little impact to thresholds due to surface reflections, at least in laboratory situations (Kastelein et al. 2018).

The monachid seals are less represented in seal hearing studies than are phocinid species. To date, behavioral audiograms have been obtained from two monachid species, the northern elephant seal and the Hawaiian monk seal (Fig. 8.9). The elephant seal demonstrates hearing characteristics that appear to be most adapted to underwater hearing relative to their ability to hear in air. The elephant seal is the most aquatic and deep diving of the seal species, spending much of its life at sea. The region of best underwater hearing in the elephant seal is broader than that of other phocids that have been tested. However, the lowest thresholds measured in the elephant seal are higher than observed in the phocinid seals and the upper limit of hearing also is slightly lower. The only other monachid seal for which behavioral audiograms exist is the Hawaiian monk seal.

The earliest Hawaiian monk seal audiogram does not fit expectation based on other phocid audiograms and there is a question as to whether this single individual was representative of the species (Thomas et al. 1990a). A more recent Hawaiian monk seal audiogram is like that of the elephant seal (Sills et al. 2021), although the Hawaiian monk seal appears to be less sensitive to sound across most of the range of hearing. Both monk seal audiograms show a similar upper frequency limit of hearing, which suggests that like the elephant seal, Hawaiian monk seals have a constrained upper frequency limit to hearing relative to the phocinids. Because there are so few monachid audiograms, caution should be taken when extrapolating to their respective species until there are more samples and greater consistency in the measurements.

Otariids

The underwater hearing of the otariids is represented by three species; the California sea lion (Fig. 8.10a), the Steller sea lion (Fig. 8.11), and the northern fur seal (Fig. 8.10b). Taken as a whole, the otariids demonstrate a sensitivity within the region of best hearing that is comparable to that observed in the phocids (Table 8.1). Note that the audiogram from Babushina et al. (1991) was not used in determining the region of best hearing for the northern fur seal because the 20 dB sensitivity bandwidth was substantially different from that measured in two other subjects (Moore and Schusterman 1987) and the California sea lion. The frequency range of best hearing is possibly from 1 to 30 kHz in the Steller sea lion (Kastelein et al. 2005b). Though consistent with the range of best hearing in the California sea lion, the range of best hearing in the Steller sea lion is questionable because the upper and lower limits of hearing were not well defined.

The lowest thresholds in the otariids ranged from 56 to 60 dB re 1 μPa. The work of Schusterman was not considered in this analysis because the thresholds reported by Schusterman (1972) were much higher across the bandwidth of

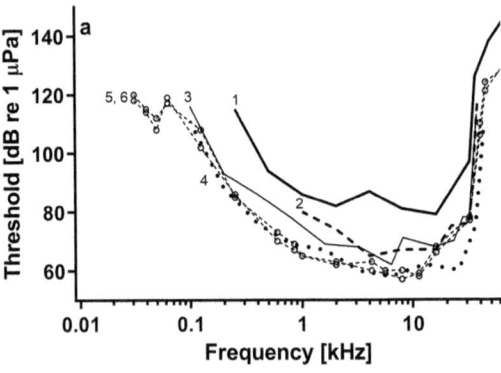

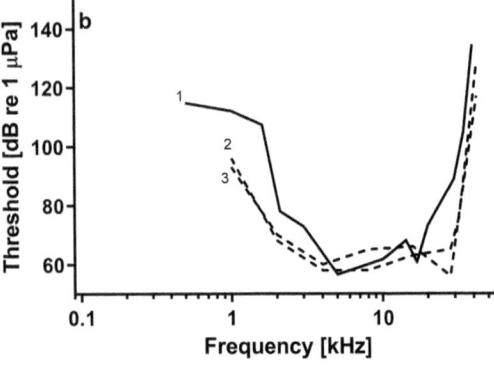

Fig. 8.10 Underwater behavioral audiograms of (**a**) California sea lion: 1 (thick line) (Schusterman et al. 1972), 2 (dashed line) (Mulsow et al. 2012), 3 (thin line) (Reichmuth and Southall 2012), 4 (dotted line) (Reichmuth et al. 2013), 5 and 6 (thin dashed line, open circles) (Kastelein et al. 2023b). (**b**) Northern fur seal: 1 (solid line) (Babushina et al. 1991), 2 and 3 (dashed lines) (Moore and Schusterman 1987)

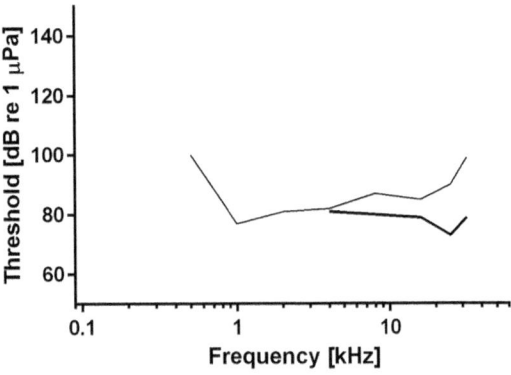

Fig. 8.11 Underwater behavioral audiograms of two Steller sea lions (Kastelein et al. 2005b). Thick and thin lines denote the audiograms of the two individuals. Note that frequencies below 4 kHz were only tested in one animal

hearing than in other studies. As noted by Reichmuth and Southall (2012), this was due in part to a psychophysical approach in which a 75% correct detection rate was used in determining thresholds, as opposed to the more commonly used 50% detection limit. The method used by Schusterman results in a more conservative audiogram by roughly 5–10 dB. The lowest thresholds in the northern fur seal were in the same range as in the California sea lion suggesting that sensitivity within the otariids measured to date is similar, albeit slightly poorer than in some odontocetes.

The upper limit of hearing in otariids is notably lower than in phocids with steep reductions in sensitivity occurring around 32 kHz (Fig. 8.10). Sensitivity to frequencies <1 kHz primarily has been studied in the California sea lion. Hearing sensitivity declines at a slightly steeper rate than that observed in the odontocetes and greater than that observed in the phocids (Table 8.1). Thresholds at 1 kHz are generally lower in the otariids (and other pinnipeds) than in the odontocetes tested to date, supporting the observation that pinnipeds appear relatively more sensitive to low-frequency sound (even if the roll-off below 1 kHz is greater). However, comparable thresholds found in belugas tested in potentially quieter environments (Ridgway et al. 2001) may require this assertion to be revisited as masking may be a confounding issue.

Odobenids
Underwater hearing has been tested in only one walrus (Kastelein et al. 2002b). The walrus showed a much narrower range of hearing than did other pinnipeds and sensitivity declined rapidly above 12 kHz (Fig. 8.12; Table 8.1). The lowest thresholds within the range of best hearing (67–77 dB re 1 µPa) were like that of the northern elephant seal and Hawaiian monk seal between ~500 Hz and 4 kHz, but higher than observed in other pinnipeds within their best range of hearing. Kastelein et al. (2002b) suspected a combination of the methodological approach and possible masking may have resulted in a conservative audiogram. A peculiar notch in the audiogram at 2 kHz was not believed to be an artifact of the

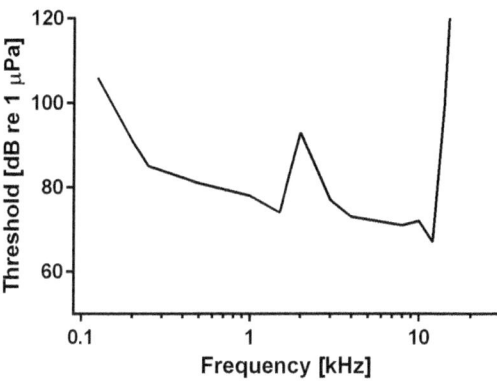

Fig. 8.12 Underwater behavioral audiogram of a Pacific walrus (Kastelein et al. 2002b). Note that the notch at 2 kHz may not be representative of the species

experimental approach, but no explanation for the notch could be confirmed. As with other pinnipeds, the walrus seems better adapted to low-frequency underwater sound reception than odontocetes.

8.3.1.2.2 In-Air Hearing

In-air behavioral audiograms have been determined for the California sea lion, Steller sea lion, northern fur seal, harbor seal, Caspian seal, spotted seal, ringed seal, northern elephant seal, and Hawaiian monk seal. In recent years, improved ambient noise conditions in test enclosures have resulted in much lower in-air hearing thresholds than previously reported, both with direct field studies and those in which headphones were used (Reichmuth et al. 2013). Prior arguments that pinnipeds are better adapted to underwater hearing now need to be reconsidered as the more recent audiometric data indicate that many pinnipeds have aerial hearing on par with that of terrestrial carnivores (Kastak and Schusterman 1995; Møhl 1968a; Moore and Schusterman 1987; Reichmuth et al. 2013; Sills et al. 2014, 2015; Terhune 1991).

Audiometric studies that likely experienced significant masking during in-air testing are not presented as part of this review. These include Kastak and Schusterman (1998) and Schusterman (1974) for the California sea lion; Kastak and Schusterman (1998), Wolski et al. (2003), Terhune and Turnbull (1995), and Møhl (1968a) for the harbor seal; Babushina (1997) for the Caspian seal; and Kastak and Schusterman (1998, 1999) for the northern elephant seal. The results of some of these studies raised suspicion about the potential issues of masking. Several of the pinniped subjects in these papers were later tested in lower ambient noise conditions and the results confirmed the suspicions. Another audiogram not included is that of a harp seal (Terhune and Ronald 1971), which showed relatively high aerial hearing thresholds. In this study, the seal would swim submerged before each trial. Although replication of this study's experimental design with a ringed seal obtained similar results, the noise and critical ratios in Terhune and Ronald (1971) suggest that masking was not the explanation for the increased thresholds (Sills et al. 2015). A physiological response of the auditory system to being in water (i.e., in preparation for diving) could possibly explain some of the differences.

Phocids

Aerial audiograms for phocinid and monachid seals are presented in Fig. 8.13. The three phocinid species for which aerial audiograms have been recorded show remarkable similarity, possibly because all the subjects were tested in the same low-noise environment (Reichmuth et al. 2013; Sills et al. 2014, 2015). From 1.6 to 6.4 kHz, many of the tested seals had thresholds less than 0 dB re 20 µPa, which demonstrates acute sensitivity comparable to other terrestrial carnivores at their lowest thresholds. Below 1 kHz, the low-frequency roll-off ranges from 10.9 to 12.0 dB/octave, which is steeper than observed in the underwater hearing of the same seals (Table 8.1). The decline in sensitivity near the upper limit of hearing is less steep in air than in water, and the upper frequency limit of hearing is more similar to that of terrestrial carnivores.

The only monachid seals for which behavioral measures of in-air hearing exist are the northern elephant seal and the Hawaiian monk seal. The region of best hearing in air is broader than observed in the phocinid seals. However, the audiogram is flatter in both species and the lowest thresholds are more than 25 dB higher than in other phocids. The relative differences between the audiograms of the elephant seal and Hawaiian monk seal collected in air and under water, as well as comparisons to terrestrial carnivores, other phocids, and fully aquatic marine mammals, suggest that the elephant seal and Hawaiian monk seal are adapted for underwater hearing. Modifications to the auditory system, such as an occluded external ear canal, enlarged auditory bullae, etc., are consistent with the deep-diving behavior of the elephant seal and the requirements for pressure compensation at depth (Kastak and Schusterman 1998; Nummela 2008; Repenning 1972). However, these features seem to be shared among the monachid seals regardless of the depth profiles they typically exploit. The anatomical features might similarly reduce hearing sensitivity in aerial environments for all monachid seals, although additional elephant seals and Hawaiian monk seals need to be tested and other southern seal species need to be assessed.

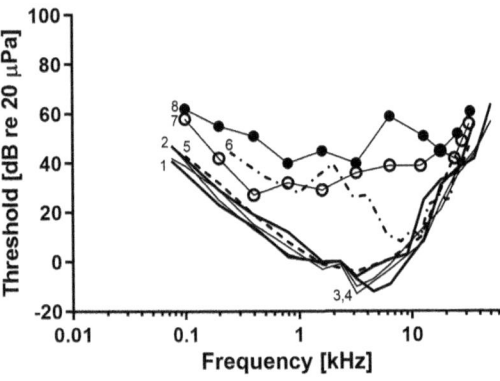

Fig. 8.13 Aerial behavioral audiograms of the phocids: ringed seal, 1 and 2 (thick lines) (Sills et al. 2015); spotted seal, 3 and 4 (thin lines) (Sills et al. 2014); harbor seal, 5 (dashed line) (Reichmuth et al. 2013); 6 (dotted-dashed line) (Wolski et al. 2003); Northern elephant seal, 7 (thin line, open circles) (Reichmuth et al. 2013); Hawaiian monk seal, 8 (thin line, filled circles) (Ruscher et al. 2021)

Otariids

In-air audiograms for the otariids are presented in Fig. 8.14. Aerial audiograms exist for three species: the northern fur seal (Fig. 8.14a), the California sea lion (Fig. 8.14b), and the Steller sea lion (Fig. 8.14b). The region of best hearing

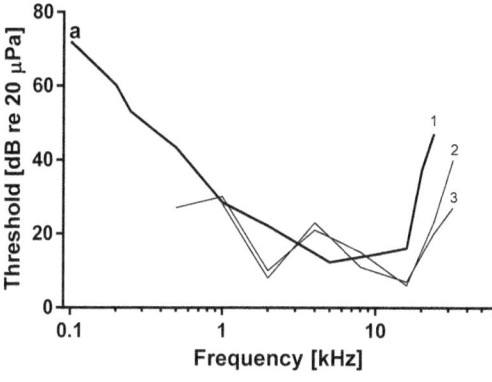

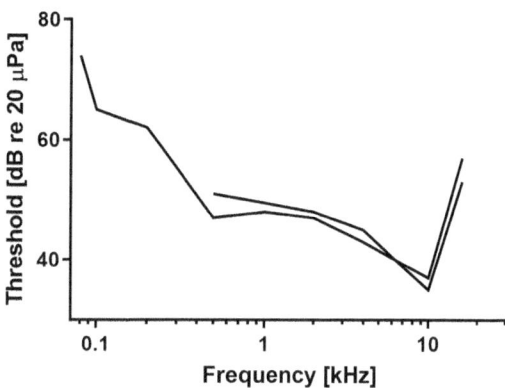

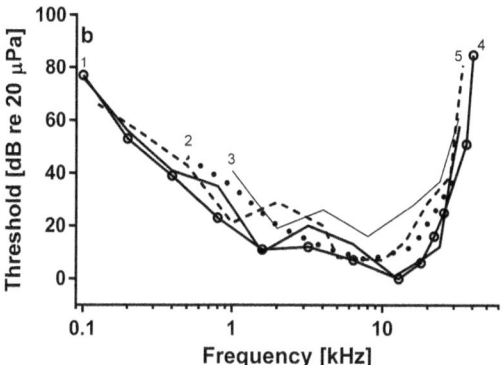

Fig. 8.14 Aerial behavioral audiograms of otariids. (**a**) Northern fur seal: 1 (thick line) (Babushina et al. 1991), 2 and 3 (thin lines) (Moore and Schusterman 1987). (**b**) California sea lion: 1 (thick line) (Reichmuth et al. 2013), 2 (dotted line) (Mulsow et al. 2011b), 3 (thin line) (Moore and Schusterman 1987), 4 (solid line, open circles) (Reichmuth et al. 2017); Steller's sea lion: 5 (dashed line) (Mulsow and Reichmuth 2010)

and low-frequency roll-offs in hearing sensitivity are provided in Table 8.1. For all otariids, a steep increase in thresholds occurs at high frequencies, similar to that observed under water (i.e., the upper limit of hearing is about the same). The low-frequency roll-off appears similar in air and under water, although a direct comparison can only be made in the California sea lion.

Odobenids

Playbacks of sound to walrus colonies were first used to estimate the ability of the walrus to detect aerial sound in natural background noise (Kastelein et al. 1993). A subsequent attempt at measuring aerial hearing thresholds in a walrus found thresholds at frequencies ranging from

Fig. 8.15 Aerial behavioral audiograms of two Pacific walruses (Jones et al. 2023)

125 Hz to 8 kHz were higher than expected (>40 dB re 20 µPa) based on comparisons to other pinnipeds, and varied considerably between direct field tests and testing using headphones (Kastelein et al. 1996b). The thresholds were likely affected by many factors, including masking, calibration issues, direct field assumptions (e.g., reflective surfaces), background noise, and choice of psychophysical approaches, and are not presented here. Reichmuth et al. (2020) not only confirmed that walruses could hear at the frequencies measured by Kastelein et al. (1996b), but that the frequency range of suprathreshold sensitivity extended from 60 Hz to 23 kHz. Jones et al. (2023) later measured the audiograms of two walruses in outdoor ambient noise (Fig. 8.15). The best hearing sensitivity was measured at 10 kHz in both animals, but declined rapidly at higher frequencies suggesting a narrower bandwidth of hearing than in other pinnipeds. The rapid decline in sensitivity above 10 kHz was consistent with the underwater audiogram previously measured (Kastelein et al. 2002b).

Notes on Pinniped Hearing

As previously noted, pinniped aerial audiograms recently collected in controlled noise conditions challenge prior hypotheses that pinnipeds are better adapted to underwater hearing than they are to hearing in air (Reichmuth et al. 2013). Although some seals may truly be better adapted to underwater hearing (e.g., elephant seal), audiograms from the California sea lion and several phocinid seals suggest that aerial thresholds are comparable to those of terrestrial carnivores at the most sensitive frequencies of hearing. How this is achieved is not well understood. The ears of the phocinid seals and otariids probably operate like a terrestrial mammal ear in air, but modifications to sound reception pathways must occur under water. Bone and tissue conduction, the cavernous tissues in the auditory canal and middle ear, and other anatomical adaptations likely have important roles in underwater sound reception (Møhl 1967, 1968b; Møhl and Ronald 1975; Nummela 2008; Ramprashad 1975; Repenning 1972), but details of underwater sound reception mechanisms remain unresolved.

One intriguing observation that must be considered in the study of phocid in-air hearing is the observation that the harbor seal can open and close the auditory meatus (Kastak et al. 2005a; Møhl 1968b). Kastak et al. (2005a) observed a harbor seal mitigate noise-induced hearing loss during aerial noise exposures by closing the meatus. Although this apparently learned behavior was observed during sound exposures designed to induce temporary threshold shift (TTS), and not for the purpose of determining sensitivity (when sounds are just audible), it is a potential confound to aerial hearing experiments. Møhl (1968b) was able to demonstrate that irritation of the skin around the external ear opening can induce closure in the harbor seal. Concern over the potential for the same effect in the walrus was voiced by Kastelein et al. (1996b) who obtained relatively high hearing thresholds in a walrus tested with headphones in air. In instances where seals were not conditioned with positive rewards (fish) for detecting sound but were tested via electrophysiological means while chemically immobilized, ear inserts were used to prevent the potential closure of the auditory meatus (Ruser et al. 2014).

8.3.1.3 Polar Bears

The polar bear is the most terrestrial of the marine mammals, but spends much of its life in

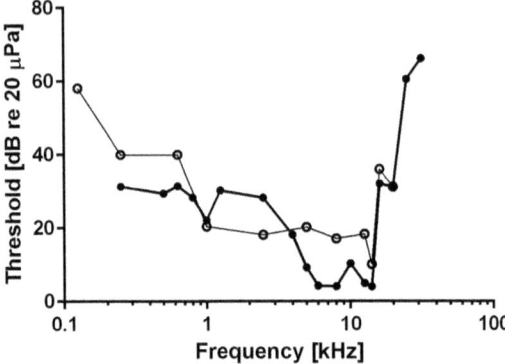

Fig. 8.16 Mean aerial behavioral audiograms of polar bears tested at two facilities (Owen and Bowles 2011): bears from the San Diego Zoo (thick line, filled circles; $n = 2$) and bears from SeaWorld San Diego (thin lines, open circles; $n = 3$)

association with the ocean. Polar bears utilize sea ice as hunting grounds for seals and for breeding and maternal denning. Polar bears are capable of diving and will sometimes stalk prey from the water. However, the amount of time polar bears spend submerged in sea water is short relative to the time they spend in air, which is where the greatest potential for noise impacts would occur. Owen and Bowles (2011) measured the in-air audiograms of five polar bears from zoos. The measurements were averaged by facility to create a mean audiogram for the animals at each facility (Fig. 8.16). Depending upon the facility, the region of best hearing varied; the region of best hearing determined from the mean audiogram of San Diego Zoo bears was ~3.1–15.4 kHz, and that of the SeaWorld bears was ~800 Hz–15.5 kHz. The region of best hearing in the mean SeaWorld audiogram hinged largely on the threshold at 14 kHz, which was 8 dB lower than any other threshold in the audiogram. Within the San Diego Zoo bears, considerable variation in thresholds between 1 and 16 kHz contributed to a narrower region of best hearing. Differences in the thresholds of the mean audiograms within the overlapping frequencies of best hearing are difficult to explain as the reported background noise at the two facilities was similar. Thresholds at the lowest and highest frequencies fall within the range observed in terrestrial carnivores and suggest that the polar bear has hearing abilities typical of terrestrial carnivores.

8.3.1.4 Sea Otter

The sea otter is a member of the family Mustelidae, which includes animals such as the weasel and skunk. The sea otter is amphibious but spends a significant portion of its time in air while floating at the sea surface. These animals are awkward on land, but occasionally haul out on shore. The time spent submerged, when underwater hearing would be important, is primarily spent foraging (Bodkin et al. 2007).

8.3.1.4.1 Underwater Hearing

There has been only one behavioral audiogram collected from the southern sea otter (Ghoul and Reichmuth 2014b). Fortunately, as with several pinnipeds, both aerial and underwater audiograms were collected in the same animal, allowing a comparison of hearing within the two media and development of hypotheses about the evolution of the sea otter's amphibious ear. Under water, hearing sensitivity declines rapidly above ~23 kHz (Fig. 8.17). The low-frequency roll-off is shallow but no thresholds less than 100 dB re 1μ Pa were detected below 1 kHz. Compared to other marine carnivores, the sea otter has a slightly narrower region of best hearing under water and has a best threshold of 69 dB re 1μ Pa, which is more than 10 dB higher than that of the other marine carnivores (Ghoul and Reichmuth 2014b). Sensitivity is also worse than in the pinnipeds below and above the best range of hearing.

8.3.1.4.2 In-Air Hearing

The in-air audiogram of the sea otter has a typical mammalian shape and a steep decline in sensitivity above 22.6 kHz, like what is observed under water (Fig. 8.17). The low-frequency roll-off is much steeper than that observed for the sea otter's hearing under water. The in-air hearing sensitivity is similar to that of some terrestrial carnivores with respect to the lowest thresholds and upper limit of hearing, but it is most similar to that of the California sea lion, which shares a similar overall bandwidth of hearing and similar thresholds both within the region of best hearing and at high and

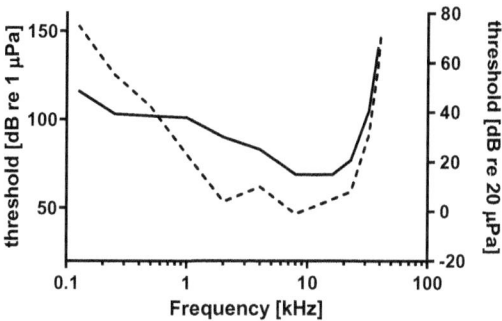

Fig. 8.17 Behavioral audiograms of the sea otter (Ghoul and Reichmuth 2014b): underwater audiogram (solid line referenced to the left y-axis); in-air audiogram (dashed line referenced to the right y-axis)

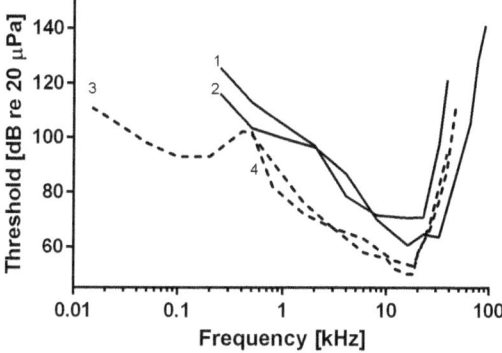

Fig. 8.18 Underwater behavioral audiograms of West Indian manatees: 1 and 2 (solid lines) (Gaspard et al. 2012); 3 and 4 (dashed lines) (Gerstein et al. 1999)

low frequencies (Ghoul and Reichmuth 2014b). Collectively, comparison of the underwater and in-air hearing of the sea otter suggests that the sea otter ear is better adapted to aerial hearing, which likely reflects both its amphibious nature and its relatively recent divergence from a fully terrestrial species (i.e., the in-air audiogram is similar to that of terrestrial mustelids (Fay 1988; Ghoul and Reichmuth 2014b)).

8.3.1.5 Sirenians

Behavioral audiograms have been obtained on four West Indian manatees conducted by two different laboratories (Gaspard et al. 2012; Gerstein et al. 1999). The two hearing studies produced generally similar audiometric shapes (Fig. 8.18), although the low-frequency roll-offs differed between studies. The frequency range of best hearing is slightly narrower than in odontocetes and pinnipeds, but has a similar steep decline in sensitivity near the upper limit of hearing. The low-frequency roll-off was 4.8 dB/octave in the Gerstein et al. (1999) study (calculated for the single animal with sufficient data below 1 kHz) and 8.2 and 10.3 dB/octave in the Gaspard et al. (2012) study. Sirenians do not appear to be as sensitive to low-frequency sound as are the pinnipeds and Gerstein et al. (1999) suggested that below 400 Hz sirenians might detect signals via vibrotactile means. Subsequent work by Gaspard et al. (2013) demonstrated acute vibrotactile sensitivity in the manatee.

8.3.2 Electrophysiological Audiograms

The use of electrophysiological methods to test hearing, particularly approaches in which the ASSR is used for threshold estimation, has grown rapidly in the last two decades. A considerable amount of information on species hearing abilities has been obtained and numerous audiograms have been acquired from species that would probably not otherwise have been tested, particularly within the odontocetes. The information obtained with AEP methods is, however, not without constraint and the methods used for data collection and threshold estimation warrant careful evaluation. Furthermore, as previously discussed, AEP audiograms should not be considered the final word on hearing sensitivity; behavioral audiograms provide the only means for obtaining absolute hearing thresholds and AEP audiograms must be considered in context of how they differ from behavioral audiograms.

The upper limit of hearing is often comparable across studies that used different AEP methods, but hearing thresholds can vary widely within a species and even within the same individual tested with different methods (e.g., Houser and Finneran 2006b; Popov et al. 2007; Schlundt et al. 2007). Differences in AEP methods and the impact it has on hearing test results was of sufficient concern that an effort to standardize testing methods within medium-sized odontocetes was made in 2018 (American National Standards Institute (ANSI) 2018). Nevertheless, not all studies conducted in odontocetes and published since 2018 followed the guidelines of the standard. Therefore, before accepting the results of different AEP hearing studies as comparable and reflective of species or individual differences, aspects of each study that should be considered include:

1. Method of threshold estimation: Thresholds may be determined by linear extrapolation of a regression line through a linear region of the AEP amplitude vs. stimulus level function, otherwise called the input-output (IO) function (e.g., Klishin et al. 2000; Popov et al. 2005). Alternatively, thresholds may be determined in relation to the lowest level at which an AEP is detected (e.g., Houser et al. 2008b; Klishin et al. 2000). Of the two approaches, the linear regression approach should typically produce the lower of the two thresholds.

2. Calibration methods: Transient signals (e.g., tone-bursts) and sinusoidal amplitude-modulated (SAM) tones are often used for AEP hearing tests. Historically, there was little consensus as to how signals should be calibrated, particularly for transients. For example, some approaches calibrate the SPL_{rms} of the sound stimulus, some the peak or peak-peak SPL, and some the peak-equivalent SPL (e.g., Finneran and Houser 2006; Popov and Supin 1998; Reichmuth et al. 2007; Supin and Popov 2007). The location of calibration hydrophones for direct field approaches, and the distance between source and calibration hydrophone when jawphones are used, are also variable between studies (e.g., Finneran and Houser 2006; Mann et al. 2010). Differences in calibration measurements can contribute to variability in thresholds obtained between studies.

3. Stimuli used: The stimulus bandwidth impacts the magnitude of the evoked response; the broader the bandwidth of the stimulus, the

broader the bandwidth of stimulation along the cochlear partition and the more robust the evoked response (Supin and Popov 2007). The steepness of IO functions can therefore be impacted by the stimulus bandwidth, with potential impact for subsequent threshold estimation, depending on the threshold estimation used. In addition, the greater the bandwidth of the stimulus, the more likely that the true threshold will be underestimated (sensitivity overestimated) at frequencies where the roll-off in sensitivity is steep, as occurs near the upper frequency limit of hearing (Coffinger et al. 2018). This occurs due to spectral "splatter" at lower frequencies where hearing sensitivity is better than the intended test frequency.

4. Test environment: AEP methods may be used in air or under water, in the direct field, with headphones, and in cetaceans with jawphones (both in air and under water). Although comparisons between test situations have generally shown good agreement within odontocetes, differences in test environments contribute some degree of variability to threshold estimates.

It has been noted that hearing thresholds determined with AEPs are generally higher in larger animals. This occurs because the voltages produced by the brain attenuate with distance between the source of the voltage and the location of the recording [i.e., the brain-to-body mass ratio is an important consideration for subcutaneous recordings or those made at the surface of the animal (Popov and Supin 1990c, d; Schlundt et al. 2011; Supin et al. 2001)]. Matters are further complicated by the depth of the blubber layer since fat is a poor conductor of electricity. These issues are apparent in attempts to determine hearing thresholds in large odontocetes, such as killer whales, pilot whales, beaked whales, and adult northern elephant seals (Finneran et al. 2009; Houser et al. 2008a, 2019; Lucke et al. 2016; Schlundt et al. 2011). It appears that for AEPs to be useful in studying the hearing of the largest mysticetes, some ingenuity in recording evoked responses closer to the source of the response will be required.

There is a need to reevaluate the frequency range over which evoked potential methods are most useful. Studies in bottlenose dolphins using various bandwidths of masking noise and broad-band stimuli indicate that, below certain frequencies, there is a danger of exciting activity in the more basal (higher frequency) regions of the cochlea that do not correspond to the frequency of sound being delivered to the animal (Finneran et al. 2016). For this reason, it has been recommended that threshold testing in small odontocetes at frequencies <10 kHz be conducted and interpreted with caution (Finneran et al. 2016). This issue is a concern for studies of low-frequency sound and its impact on marine mammals. Concern about the effects of ocean noise on marine mammals is largely focused on low-frequency sound and there is probably a limit to the frequency-specific information that AEP methods can provide. However, the lower limit for which frequency-specific AEP methods will be informative is species-specific and related to the hearing range and inner ear anatomy of the species. AEP methods can be measured at frequencies <1 kHz in pinnipeds (Mulsow et al. 2014b), which have better sensitivity to low-frequency sound than do the odontocetes. Thus, AEP methods hold promise for studying hearing in mysticetes, which are presumed to have sensitive low-frequency hearing and functional hearing ranges that extend below 1 kHz.

Electrophysiological audiograms, specifically AEP audiograms, are presented in the following section. Where hearing loss was believed to exist, animals have been excluded from the presented data (hearing loss is addressed later in this chapter).

8.3.2.1 Odontocetes

Figure 8.19a shows AEP audiograms for presumably normal hearing bottlenose dolphins from two different studies (Houser et al. 2008b; Popov et al. 2007). (Note that all published AEP audiograms for bottlenose dolphins appear in the section on Population-Level Variability.) These studies used two different approaches to testing and determining hearing thresholds via AEPs. Popov et al. (2007) tested dolphins in the direct field, utilized

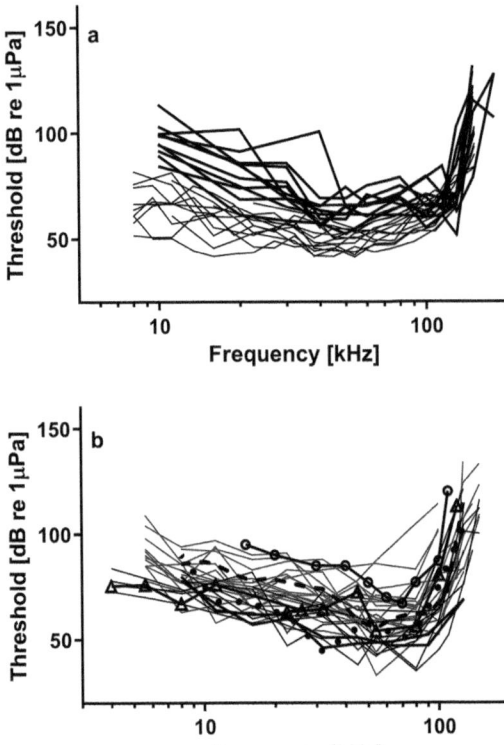

Fig. 8.19 AEP audiograms obtained from (**a**) bottlenose dolphins: thin lines (Popov et al. 2007), thick lines (Houser et al. 2008b); (**b**) belugas: thin lines (Castellote et al. 2014; Mooney et al. 2018), thick lines (Popov et al. 2013), dotted line (Mooney et al. 2008), dashed line (Klishin et al. 2000), thick line with open circles (Popov and Supin 1990b), thick line with open triangles (Mooney et al. 2020)

trains of tone-bursts as stimuli, and used a linear extrapolation of a regression line fit to the AEP amplitude vs. stimulus level to estimate threshold. In contrast, Houser et al. (2008b) used a jawphone for stimulus delivery, SAM tones as stimuli, and estimated threshold based upon the lowest detected evoked response. Although variability in thresholds between animals within a study is similar, differences in the threshold levels across the populations are apparent and are likely due to the differences in study approaches. Also note, based on previously discussed findings in bottlenose dolphins, that threshold levels obtained at frequencies ≤10 kHz were suspect because of differences between the frequency of acoustic stimulation and site of cochlear excitation (Finneran et al.

2016). These considerations are relevant to the audiograms of other species that follow and a summary of the methods used in the reported studies is provided in Table 8.2.

Figure 8.19b shows the electrophysiological audiograms of the beluga from seven different studies (Castellote et al. 2014; Klishin et al. 2000; Mooney et al. 2008, 2018, 2020; Popov and Supin 1990b; Popov et al. 2013). Most of the audiograms come from Castellote et al. (2014) and Mooney et al. (2018), who measured audiograms in wild belugas captured and released in Bristol Bay, Alaska. Best thresholds between studies differed by as much as 50 dB, even at the frequencies of best hearing, which probably reflects both actual differences between animals and methodological differences. Nevertheless, the upper limits of hearing and frequency range of hearing were generally consistent with that observed in behavioral audiograms.

Prior to the year 2000, only one AEP audiogram had been collected from a common dolphin (Popov and Klishin 1998). In more recent years, the number of common dolphin audiograms, both short-beaked and long-beaked, has increased because of the implementation of AEP hearing tests during stranding responses and rehabilitation efforts (Houser et al. 2018, 2022). Common dolphin AEP audiograms are presented in Fig. 8.20a. The audiogram from Popov and Klishin (1998) is in good agreement with that of the other common dolphin measurements. The upper limit of hearing is well above 100 kHz, although the sloping decrease in sensitivity with increasing frequency is shallower in some individuals. Several subjects were excluded from Houser et al. (2018) because of apparent hearing loss.

Nachtigall et al. (2008) captured and tested wild white-beaked dolphins. A full AEP audiogram was obtained from a male dolphin and AEP thresholds were obtained at two frequencies in a female (Fig. 8.20b). The best AEP thresholds obtained were ~45 dB re 1 μPa. The steep decline in sensitivity associated with the upper limit of hearing was not observed until frequencies exceeded 150 kHz in the male; AEP thresholds >120 dB re 1 μPa were not observed until testing occurred at 180 kHz. In contrast, AEP thresholds

Table 8.2 Test conditions, subjects, and test methods for AEP hearing tests conducted in various odontocete species

References	Species	Test condition	Subject(s)[b]	Stimulus	Stimuls delivery	Threshold estimation method based on…[d]
Popov et al. (2007)	Bottlenose dolphin	Water bath	Research facility (c-r)	Tone-burst trains	Direct field[a]	Linear extrapolation method
Houser et al. (2008b)	Bottlenose dolphin	Out of water	Display facility (wc, fb)	SAM tones	Jawphone	Lowest detected response
Houser and Finneran (2006a)	Bottlenose dolphin	Bay	Research facility (wc)	SAM tones	Jawphone	Linear extrapolation method
Houser and Finneran (2006b)	Bottlenose dolphin	Out of water	Research facility (wc, fb)	SAM tones	Jawphone	Linear extrapolation method
Finneran and Houser (2006)	Bottlenose dolphin	Out of water	Research facility (wc)	SAM tones	Jawphone	Linear extrapolation method
Klishin et al. (2000)	Beluga	Shallow pool	Research facility (wc)	SAM tones	Direct field[a]	Linear extrapolation method
Mooney et al. (2008)	Beluga	Pool	Display facility[a]	SAM tones	Direct field	Linear extrapolation method
Popov and Supin (1990b)	Beluga	Bay	Research facility (c)[a]	Tone-bursts	Direct field	Lowest detected response
Popov et al. (2013)	Beluga	Water bath	Research facility[a]	Tone-burst trains	Direct field[a]	Linear extrapolation method
Castellote et al. (2014)	Beluga	Shallow bay water	Wild (c-r)	SAM tones	Jawphone	Linear extrapolation method
Mooney et al. (2020)	Beluga	Shallow pool	Rehab facility (s)	SAM tones	Jawphone	Linear extrapolation method, lowest detected response
Popov and Klishin (1998)	Common dolphin	Water bath	Research facility (s)	Tone-bursts	Direct field[a]	Lowest detected response
Houser et al. (2018)	Common dolphin	Out of water	Wild (s)	SAM tones	Jawphone	Lowest detected response
Nachtigall et al. (2008)	White-beaked dolphin	Water bath	Wild (c-r)	SAM tones	Direct field[a]	Linear extrapolation method
Houser et al. (2018)	Atlantic white-sided dolphin	Out of water	Wild (s)	SAM tones	Jawphone	Lowest detected response
Montie et al. (2011)	Pygmy killer whale	Pool	Rehab facility (s)	SAM tones	Jawphone	Lowest detected response
Houser et al. (2022)	Pygmy killer whale	Pool	Rehab facility (s)	SAM tones	Jawphone	Lowest detected response
Pacini et al. (2010)	Long-finned pilot whale	Pool	Display facility (s)	SAM tones	Direct field	Linear extrapolation method
Greenow et al. (2014)	Short-finned pilot whale	Shallow bay water	Rehabilitation facility (s)	SAM tones	Jawphone	Lowest detected response
Schlundt et al. (2011)	Short-finned pilot whale	Pool	Display facility (wc, s)	SAM tones	Direct field	Lowest detected response
Popov et al. (2005)	Finless porpoise	Water bath	Research facility[a]	SAM tones	Direct field[a]	Linear extrapolation method
Popov et al. (2011)	Finless porpoise	Water bath	Research facility[a]	Tone-burst trains	Direct field[a]	Lowest detected response (reversal in ASSR detections)

(continued)

Table 8.2 (continued)

References	Species	Test condition	Subject(s)[b]	Stimulus	Stimulus delivery	Threshold estimation method based on...[d]
Lucke et al. (2009)	Harbor porpoise	Bay water	Display facility (s)	SAM tones	Direct field	Lowest detected response
Ruser et al. (2016)	Harbor porpoise	Pool, open water	Rehab facility (s), wild (s)	SAM tones	Direct field	Lowest detected response
Cook et al. (2006)	Gervais' beaked whale	Pool	Rehab facility (s)	SAM tones	Jawphone	Lowest detected response
Finneran et al. (2009)	Gervais' beaked whale	Shallow bay water	Rehab facility (s)	SAM tones	Jawphone	Lowest detected response
Pacini et al. (2011)	Blainville's beaked whale	Pool	Rehab facility (s)	SAM tones	Direct field	Linear extrapolation method
Nachtigall et al. (2005)	Risso's dolphin	Pool	Rehab facility (s)	SAM tones	Direct field	Linear extrapolation method
Popov and Supin (1990b)	Tucuxi	Water bath	Wild (c)[a]	Tone-bursts	Direct field[a]	Lowest detected response
Popov and Supin (1990b)	Boutu	Water bath	Wild (c)[a]	Tone-bursts	Direct field[a]	Lowest detected response
Li et al. (2012)	Indo-Pacific humpback dolphin	Pool	Display facility (s)	Tone-burst trains	Direct field	Linear extrapolation method
Mann et al. (2010)[c]	Rough-toothed dolphin	–	–	SAM tones	Jawphone	Lowest detected response
Szymanski et al. (1999)	Killer whale	Pool	Display facility (wc)	Tone-bursts	Direct field	Lowest detected response

[a]Information is either unknown, or is not explicitly stated but is assumed from the description in the manuscript
[b]Letter designations correspond to the following: *c-r* capture-release, *fb* facility born, *s* stranded, *wc* wild caught
[c]Multiple species of animals were tested at different facilities and conditions. The exact test condition for the audiogram of the rough-toothed dolphin was not reported
[d]The method of threshold estimation does not state how the evoked response was detected. Studies used various approaches, including visual determination and the application of objective tests, such as the F-test and magnitude squared coherence test

obtained from the related Atlantic white-sided dolphin were higher at all frequencies tested (Fig. 8.20b). Three dolphins were tested after they stranded in the Cape Cod, Massachusetts region (Houser et al. 2018). The upper limit of hearing appeared to be ~160 kHz. One of the dolphins had higher thresholds and had a lower, upper limit of hearing, so it is possible that this animal had hearing loss. However, the sample size for this species remains small and it is therefore difficult to exclude this animal.

Four AEP audiograms have been obtained from the deep-ocean pygmy killer whale. All individuals were tested after stranding and during rehabilitation (Houser et al. 2022; Montie et al. 2011). Test conditions for these animals were relatively consistent across studies and the thresholds across individuals also were relatively consistent (Fig. 8.21a). The lowest thresholds obtained were 48–49 dB re 1 µPa and the upper limit of hearing was between 110 and 120 kHz.

Both long-finned and short-finned pilot whales have been subject to AEP threshold audiometry (Fig. 8.21b). A young (~2 years old) long-finned pilot whale that stranded and was subsequently rehabilitated was tested at the Lisbon Zoo in

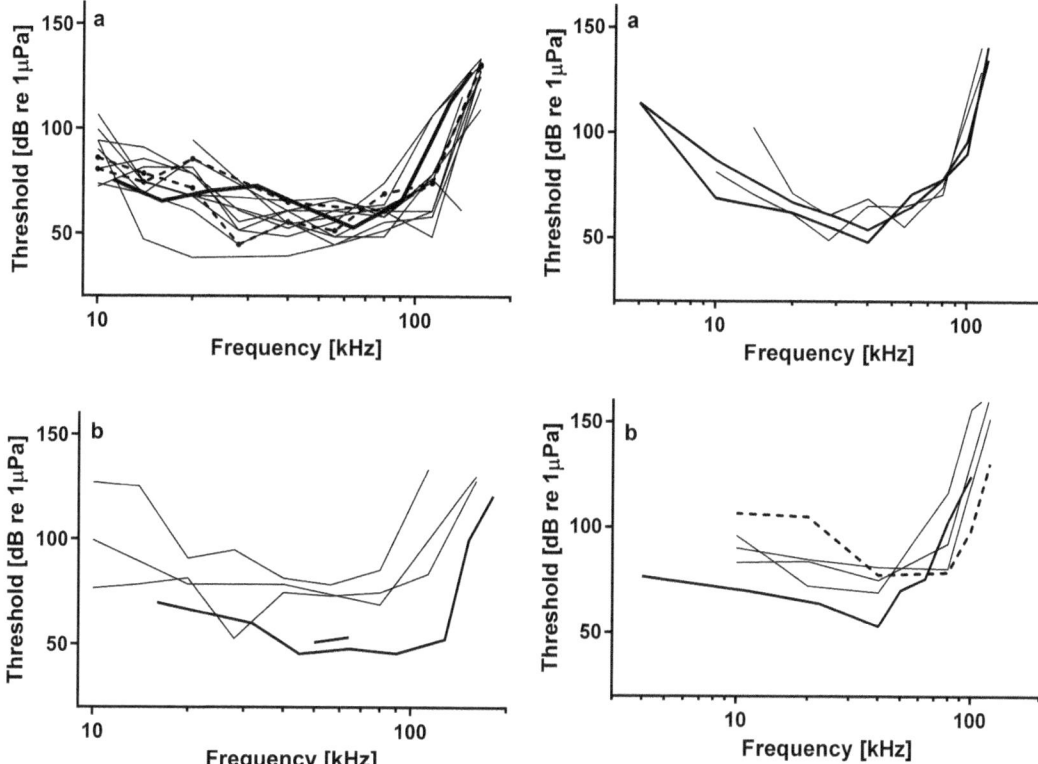

Fig. 8.20 AEP audiograms of the (**a**) short-beaked common dolphins: thick line ($n = 1$) (Popov and Klishin 1998), thin lines ($n = 10$) (Houser et al. 2018); long-beaked common dolphins: dashed lines, filled circles ($n = 2$) (Houser et al. 2022). (**b**) White-beaked dolphin (thick lines, $n = 2$) (Nachtigall et al. 2008); Atlantic white-sided dolphin (thin lines, $n = 3$) (Houser et al. 2018)

Fig. 8.21 AEP audiograms of (**a**) Pygmy killer whales: thin lines ($n = 2$) (Houser et al. 2022); thick lines ($n = 2$) (Montie et al. 2011). (**b**) Long-finned pilot whale, thick line (Pacini et al. 2010); short-finned pilot whales, thin lines ($n = 3$) (Greenhow et al. 2014); short-finned pilot whale, dashed line (Schlundt et al. 2011)

Portugal (Pacini et al. 2010). The animal showed good hearing sensitivity up to ~50 kHz and had a lowest AEP threshold of 53 dB re 1 µPa. Greenhow et al. (2014) capitalized on a stranding/rehabilitation effort to test four short-finned pilot whales. Only three of the subjects are presented in Fig. 8.21b, as one of the subjects had an apparent hearing loss. Schlundt et al. (2011) tested the hearing of a 30+ year-old short-finned pilot whale at SeaWorld of San Diego. For the most part, good sensitivity was observed in all the short-finned pilot whales up to about 80 kHz. In all the pilot whales tested, sensitivity began to rapidly roll-off before 100 kHz, which indicates a pattern that is distinct from the smaller odontocetes.

Popov et al. (2005, 2011) tested multiple finless porpoises that were held at aquaria. Wang et al. (2020) subsequently tested additional individuals in the wild. The audiograms were similar across studies and showed best sensitivities of ~47–48 dB re 1 µPa (both from the earliest study; Fig. 8.22a). A broad frequency range of good hearing sensitivity was observed and all animals had upper limits of hearing >139 kHz.

Twenty-nine AEP audiograms have been collected and reported for the harbor porpoise, with most ($n = 27$) being reported recently (Lucke et al. 2009; Popov and Supin 1990b; Ruser et al. 2016). Figure 8.22b shows the collective audiograms for the harbor porpoise, including two that have not been previously published (A. Ruser unpublished data). An additional

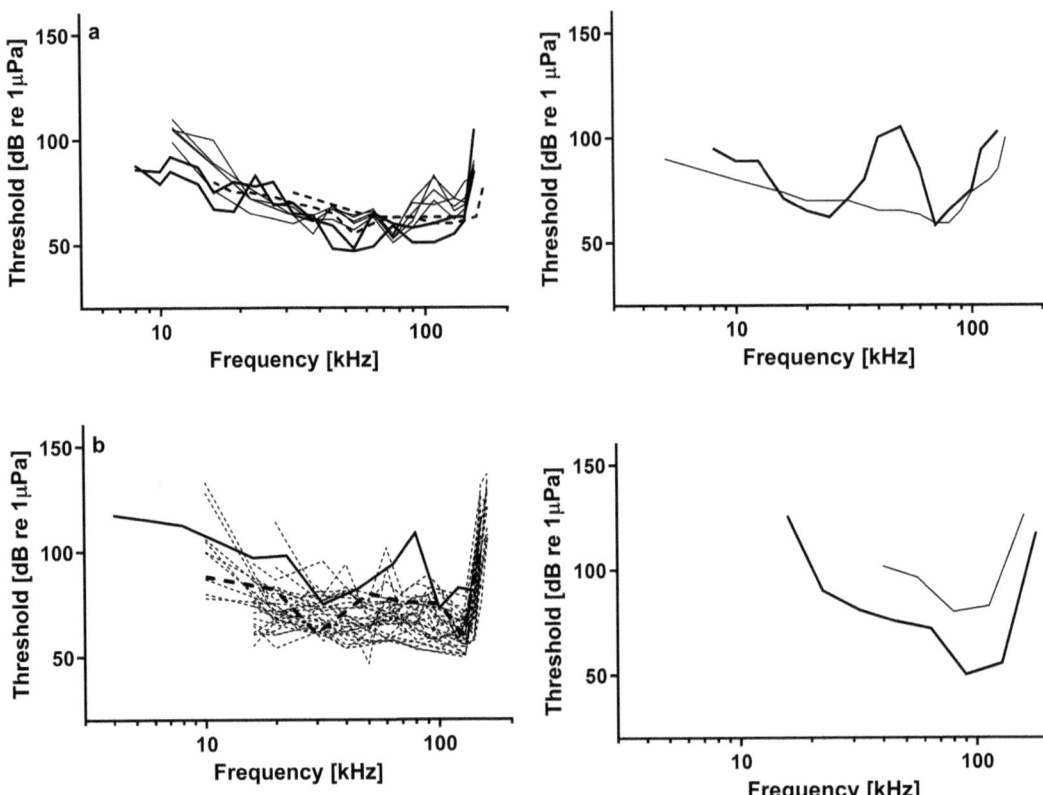

Fig. 8.22 AEP audiograms of (**a**) Finless porpoises: dashed lines (Popov et al. 2011); thick lines (Popov et al. 2005); thin lines (Wang et al. 2020). (**b**) Harbor porpoises: thick line (Lucke et al. 2009); dashed line (Popov and Supin 1990b); thin, dashed lines (Ruser et al. 2016; Ruser unpublished)

Fig. 8.23 AEP audiograms of fresh water dolphins and *Kogia* spp. (**a**) Amazon river dolphin (thin line, average of four animals) (Popov and Supin 1990b); tucuxi (thin line, average of two subjects) (Popov and Supin 1990b). (**b**) Pygmy sperm whale (thin line) (Houser et al. 2022); dwarf sperm whale (thick line) (Houser et al. 2022)

harbor porpoise audiogram was obtained but is not included in the figure because the thresholds reported were well below expectation (Bibikov 1992)—a best threshold of ~9 dB re 1 μPa was obtained, which is an unbelievably low SPL. Most of the thresholds were obtained with animals in the direct field but the lowest thresholds were obtained on two animals resting out of water and using a jawphone. Overall, the frequency range of hearing is consistent with that determined from the highest quality behavioral audiograms and shows upper limits of hearing ≥160 kHz. Thresholds were also relatively close between the AEP and behavioral measurements, which might reflect the greater ease of obtaining evoked responses in smaller odontocetes.

Besides the Yangtze finless porpoise, there are a handful of other odontocetes that inhabit freshwater habitats. Two species that have been tested inhabit the Amazon River (Fig. 8.23a), the Amazon river dolphin and the tucuxi (Popov and Supin 1990b). The Amazon river dolphin demonstrated a rapid decline in sensitivity at frequencies >100 kHz, which is similar to where the high frequency roll-off begins in the one individual for which a behavioral audiogram has been obtained (90–100 kHz). However, it should be noted that the Amazon river dolphin audiogram is an average of four individuals and is potentially skewed by one or more animals with compromised hearing. The large "notch" observed in the mid-frequency range of the audiogram was not observed in the behavioral audiogram. Although notches in hearing sensitivity are occasionally observed, it is difficult to assess whether they are artefactual or represent true hearing loss without repeated testing. The tucuxi also had a high-frequency roll-off that began ~90 kHz, but the audiogram was otherwise similar in form to that of oceanic dolphins.

The kogiids consist of two extant species of small, deep-diving odontocetes. Ridgway and Carder (2001) made the first attempt at recording AEPs in a pygmy sperm whale. Robust AEPs to 90 kHz and 150 kHz tone bursts were evident in the pygmy sperm whale, but responses at lower frequencies were difficult to obtain. It is only recently that AEP audiograms have been reported for both extant kogiid species, the dwarf sperm whale and the pygmy sperm whale (Houser et al. 2022). Audiograms from individuals of these two species are found in Fig. 8.23b. The audiograms deviate from the typical oceanic dolphin audiogram in that the region of best sensitivity is roughly 80–130 kHz (both species considered). This narrower region of best sensitivity aligns with the fact that these species are known to be narrow-band high-frequency echolocators, with echolocation clicks having center frequencies of 120–130 kHz (Malinka et al. 2021; Marten 2000). Based on the limited information from the individuals presented in Fig. 8.23b, it might be assumed that the dwarf sperm whale has less sensitive hearing than the pygmy sperm whale. However, as noted by Houser et al. (2022), these were the first of these species to be tested with AEP methods and the placement of the jawphone may not have been optimal for the dwarf sperm whale, the first one tested. Alternatively, it is feasible that the dwarf sperm whale had a hearing deficit, although it was a neonate, so age-related hearing loss seems unlikely.

The hearing of beaked whales is of particular interest with respect to the impact of noise on marine mammals because of stranding incidents associated with naval sonar operations (Cox et al. 2006; Evans and England 2001; Fernández et al. 2005; Frantzis 1998; Jepson et al. 2003; Rommel et al. 2006). The hearing of beaked whales has only been tested with AEP methods and in stranded animals (Fig. 8.24a). A Blainville's beaked whale had its lowest thresholds between 40 and 50 kHz (~49 dB re 1 µPa), beyond which there was a steep decrease in sensitivity with increasing frequency (Pacini et al. 2011). Oddly, the slope of the audiogram became shallow once again above 100 kHz. Two Gervais' beaked whales were also tested following stranding (Cook et al. 2006; Finneran et al. 2009). In the first test (Cook et al. 2006), frequencies were only tested up to 80 kHz because of equipment limitations. The AEP thresholds from these studies were much higher than that previously determined in the Blainville's beaked whale except at frequencies ≥80 kHz, which is also the frequency at which the steep decline in sensitivity begins in the Gervais' beaked whale. Given the broad variability in audiograms between the studies, it is apparent that additional audiograms should be collected.

AEP audiograms have been collected for multiple other odontocete species that are represented by fewer than two individuals (Fig. 8.24b). A stranded Risso's dolphin was tested during rehabilitation (Nachtigall et al. 2005). Thresholds were lower than those behaviorally obtained in an older animal and the upper limit of hearing was higher. Additional studies, with varying stimulus types and methods of stimulus presentation and threshold estimation have been made on the Indo-Pacific humpback dolphin (Li et al. 2012), spinner dolphin (Pacini et al. 2016), Atlantic spotted dolphin (Houser et al. 2022), melon-headed whale (Houser et al. 2022), northern right whale dolphin (Houser et al. 2022), and rough-toothed dolphin (Mann et al. 2010). The rough-toothed dolphin was not tested across the full-range of hearing, but the Indo-Pacific humpback dolphin showed similar frequencies at which high-frequency sensitivity began to steeply decline (~130 kHz). The spinner dolphin had the lowest frequency at which the high-frequency sensitivity began to steeply decline, which might reflect hearing loss, although this is uncertain with only one individual tested. For the Atlantic spotted dolphin and melon-headed whale, the high-frequency sensitivity began to roll-off between 80 and 113 kHz, but otherwise appeared typically delphinid. Similarly, the northern right whale dolphin had excellent hearing across the range of frequencies tested but otherwise looked typically delphinid in form.

Relatively few investigators have attempted to record ABRs and AEP audiograms in killer whales (Houser et al. 2019; Lucke et al. 2016; Szymanski 1996; Szymanski et al. 1995, 1999, 1998). The only AEP audiograms published for killer whales have been questioned because of methodological concerns and the fact that a few thresholds were lower than the behavioral thresholds measured in the same animal (Lucke et al. 2016). Additional work needs to be conducted in killer whales to determine the best AEP methods for performing hearing tests, specifically addressing factors that challenge AEP utility, such as the brain-to-body mass ratio and the depth of the blubber layer (Sect. 8.3.2).

Ridgway and Carder (2001) capitalized upon the rehabilitation of a stranded sperm whale calf to obtain AEPs to tone-burst stimuli. The highest amplitude AEP responses in the sperm whale calf occurred at 5, 10 and 20 kHz, with some responsiveness noted up to 60 kHz. From these results, Ridgway and Carder concluded that the sperm whale calf likely had its best hearing between 5 to 20 kHz and a hearing bandwidth that extended up to 60 kHz. Unfortunately, no audiogram was obtained.

8.3.2.1.1 Relationship Between AEP and Behavioral Audiograms in Odontocetes

Across all the odontocetes tested, a big question remains: How do AEP thresholds compare to behavioral hearing thresholds? Studies quantifying differences between AEP and behavioral approaches are necessary to make AEP data more applicable to ocean noise issues. Comparisons between hearing thresholds obtained with AEP methods and behavioral methods have been made in the bottlenose dolphin and false killer whale (Finneran and Houser 2006; Houser and Finneran 2006a; Yuen et al. 2005). In these studies, the stimuli differed. Either short duration, intermittent SAM tones ($\leq$32 ms for all but one test at 5 kHz) were used (Finneran and Houser 2006; Yuen et al. 2005), or a constant SAM tone of 10 s was used in the AEP hearing tests (Finneran and Houser 2006). These stimuli differ greatly from tones used in behavioral hearing tests and which range in duration from 0.5 to 3 s. All three studies used extrapolation of the linear regression between ASSR amplitude and stimulus level to estimate hearing sensitivity. Most of the studies were conducted in ocean water with natural background noise and were undoubtedly masked to some degree. Most of the animals tested also were older and were known to have high-frequency hearing loss. Nevertheless, across these conditions, relatively good agreement between AEP thresholds and behavior thresholds existed, particularly within the mid-frequency region of the hearing range. AEP thresholds generally exceeded behavioral thresholds, but not always. AEP thresholds were as much as 27 dB higher, although mean differences tended to be less than 20 dB in most comparisons. At the low- and high-frequency tails of the audiograms the differences between behavioral and AEP thresholds tended to increase. A subsequent study in which a dolphin performed a behavioral signal detection of a continuous SAM tone stimulus while the ASSR was recorded attempted to determine how differences between AEP and behavioral thresholds might change when the acoustic stimulus used was the same (Schlundt et al. 2007). The trends in differences were the same, for example, differences were greatest at the lowest and highest frequencies tested; however, the overall magnitude of differences between thresholds declined (Fig. 8.25). Differences were within ±5 dB at all frequencies but 10 kHz (12 dB), 20 kHz (8 dB), 30 kHz (7 dB), and 150 kHz (24 dB), suggesting that differences in stimulus types between behavioral and AEP hearing tests can contribute significantly to differences in the thresholds obtained.

8.3.2.2 Pinnipeds

Electrophysiological studies of the seal brain extend back to the early part of the twentieth century, when electrical stimulation of the motor cortex of harbor seals was used to study the linkage between motor cortex and movement (Langworthy et al. 1938). Brain responses to

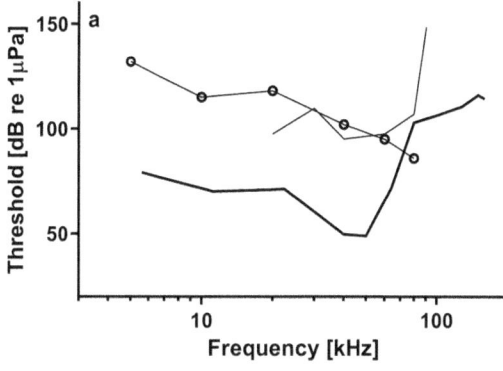

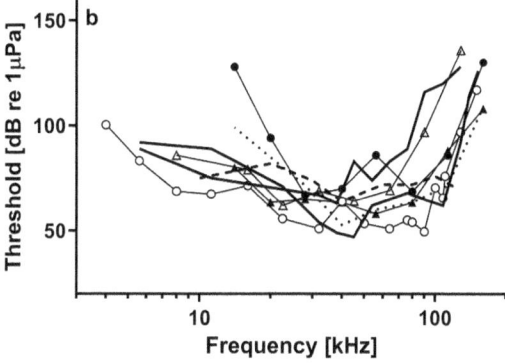

Fig. 8.24 AEP audiograms of (**a**) Beaked whales: Blainville's beaked whale (thick line) (Pacini et al. 2011); Gervais' beaked whale (thin line) (Finneran et al. 2009); Gervais' beaked whale (thin line, open circles) (Cook et al. 2006). (**b**) Other odontocete species: Risso's dolphin (thin line, open circles) (Nachtigall et al. 2005); Indo-Pacific humpback dolphin (thick lines, $n = 2$) (Li et al. 2012); rough-toothed dolphin (dashed line) (Mann et al. 2010); spinner dolphin (thin line, open triangles) (Pacini et al. 2016); Atlantic spotted dolphin (thin line, filled circles) (Houser et al. 2022); melon-headed whale (thin line, filled triangles) (Houser et al. 2022); northern right whale dolphin (dotted line) (Houser et al. 2022)

click stimuli were later mapped to the harbor seal cortex (Alderson et al. 1960). Auditory evoked potentials in the midbrain structures of the California sea lion were measured using clicks and frequency modulated stimuli (Bullock et al. 1971). This early work was used to estimate an in-air and underwater upper frequency limit of hearing in the California sea lion (30–35 kHz), which was close to that measured behaviorally, as well as a region of best hearing (4–5 kHz). Ridgway and Joyce (1975) subsequently attempted to determine an audiogram by recording cortical auditory evoked responses in the gray seal. Attempts to collect AEP audiograms in pinnipeds declined in the years following the work of Ridgway and Joyce, but have increased in the last two decades.

The use of AEPs to study aerial hearing in phocids has some unique issues related to the potential for seals to close their external meatus. This may be overcome in the trained seal through behavioral conditioning in which a positive reward is provided for responding to the presence of a tone. In the unconditioned or sedate animal, which makes up most pinniped AEP hearing test subjects, there is no guarantee that the meatus will remain open. In addition, unlike odontocete AEP studies that typically use surface electrodes attached with suction cups, AEP studies in pinnipeds generally use subcutaneous needle electrodes. The use of subcutaneous electrodes is challenging for studies of underwater hearing and all but one AEP hearing study in pinnipeds have been performed in air (Ridgway and Joyce 1975).

The greatest number of in-air AEP audiograms collected for a species of pinniped is in the gray seal (Fig. 8.26a). The early work of Ridgway and Joyce (1975) measured audiograms from changes in cortical-evoked responses. Thresholds obtained with the cortical-evoked responses were higher than those later obtained with ASSRs originating in the brain stem (Ruser et al. 2014). The differences were likely due to the evoked responses monitored (cortical-evoked responses are affected by attentiveness, brainstem-evoked responses are not), as well as other methodological differences (e.g., calibration, stimulus bandwidth, stimulus delivery). Ruser et al. (2014) delivered acoustic stimuli to their animals through ear inserts placed within the external auditory canal to prevent the seals from closing the auditory meatus during testing. The ear inserts also likely reduced the degree of masking that the seals of Ridgway and Joyce (1975) experienced, but created a new challenge in determining how to best calibrate signals; the impact of the ear inserts on signal fidelity varied with signal type (e.g., tone-bursts or SAM tones) and frequency. The ear inserts, however, showed

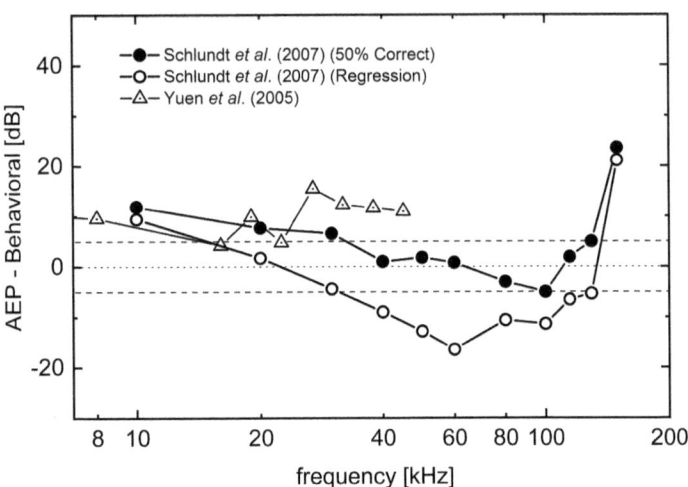

Fig. 8.25 Differences between AEP and behavioral thresholds in a dolphin in which the same stimulus is used for both threshold measurements. Data from Yuen et al. (2005), which compared thresholds obtained with different stimuli, are included for comparison. The graph is modified from that of Schlundt et al. (2007). The dashed lines are at ±5 dB

promise for future hearing studies in phocids provided methodological issues can be addressed. The lowest thresholds obtained in both studies were between 4 and 10 kHz, which is generally consistent with behavioral auditory test results in other phocids tested in noise-attenuated environments. Unfortunately, except for one animal, sensitivity in the gray seal has not been tested via AEP methods above 20 kHz and comparisons to the upper limit of hearing obtained behaviorally cannot be made.

As the gray seal is the only pinniped for which underwater AEP audiograms have been obtained (Ridgway and Joyce 1975), the audiograms are provided here for sake of completeness (Fig. 8.26b). The audiograms, based on cortical AEPs and collected with tone-bursts as stimuli, show relatively good sensitivity in both animals tested up to ~20–24 kHz. Beyond this point, there is a gradual decline in hearing sensitivity with increasing frequency. The sharp decline in sensitivity observed in other phocid behavioral audiograms is not observed. The reason for the difference is not immediately apparent, but might have something to do with test and calibration conditions and the attention-dependence of cortical response amplitudes.

Reichmuth et al. (2007) compared ABRs to click and tone-burst stimuli and made recommendations for optimal stimulus and electrode configurations for ABR recordings in the harbor seal. However, only one AEP audiogram for a harbor seal currently exists (Fig. 8.26a, Wolski et al. 2003). Thresholds obtained in that study are generally consistent with those of the gray seal, although a direct comparison is difficult because the reference values and method of calibration were different (see figure legend).

No complete in-air AEP audiograms exist for other phocids, although several electrophysiological studies have been attempted. Tone-burst evoked ABRs were used to study hearing in the leopard seal at three frequencies from 1 to 4 kHz (Tripovich et al. 2011). Thresholds measured in reference to the peak-equivalent sound pressure level (peSPL) were high, but suggested best sensitivity at 4 kHz. Efforts in elephant seals have also shown elevated thresholds to SAM, tone-burst, and click stimuli, which is consistent with behavioral hearing thresholds in air (Houser et al. 2007, 2008a; Reichmuth et al. 2007). Work with different age classes of elephant seal, which can vary an order of magnitude in mass between yearlings and adults, also demonstrated the need for increased AEP averaging and an expectation of decreased AEP amplitude and increased AEP latency with increasing size (Houser et al. 2008a).

The vast majority of in-air AEP audiograms in otariids have been obtained with California sea lions. Reichmuth et al. (2007) performed some preliminary investigations into optimal stimulus presentations and recording parameters in stranded sea lions and detailed descriptions of the ABR to clicks have also been provided

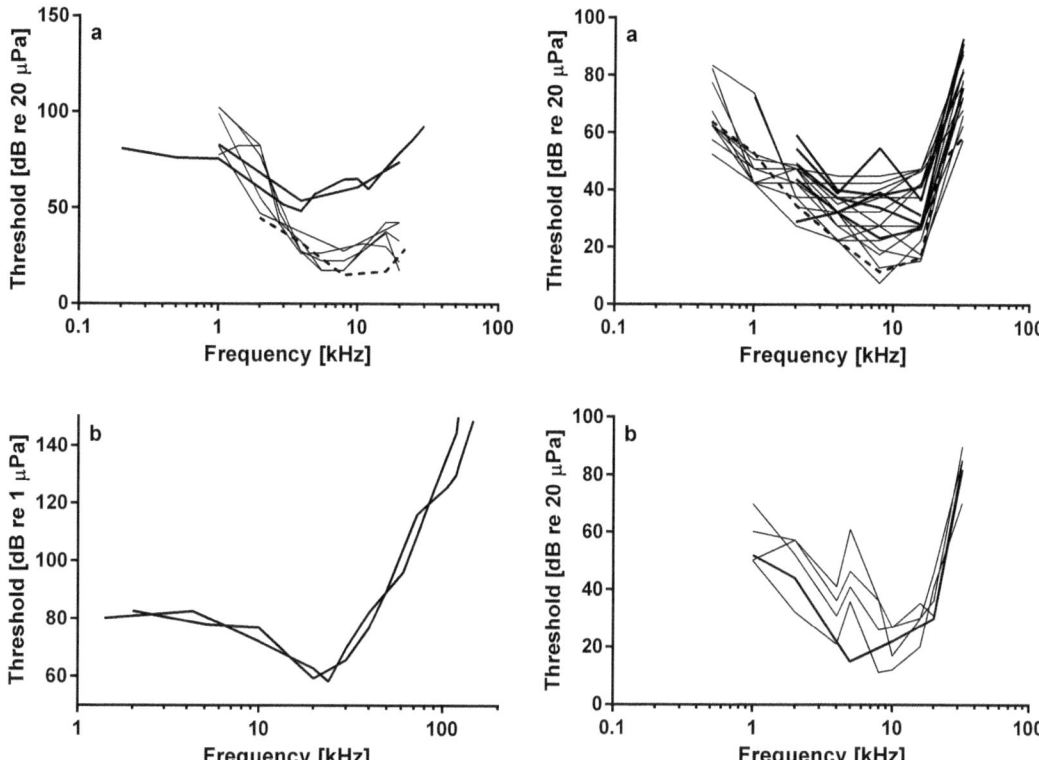

Fig. 8.26 AEP audiograms for the gray seal and harbor seal. (**a**) In-air AEP audiograms for the gray seal: thick lines (Ridgway and Joyce 1975), thin lines (Ruser et al. 2014). In-air AEP audiogram for the harbor seal: dashed line (Wolski et al. 2003). Note that the y-axis is presented in dB re 20 μPa but that each study used different threshold metrics: Ridgway and Joyce did not describe calibration (assumed RMS SPL), Ruser et al. used peSPL, and Wolski et al. normalized thresholds to a reference of 1 μPa²·s. (**b**) Underwater AEP audiograms of the gray seal (Ridgway and Joyce 1975)

Fig. 8.27 AEP audiograms for the California sea lion and the Steller sea lion in air. (**a**) California sea lions: thick lines (Mulsow et al. 2011a); thin lines (Mulsow et al. 2014b); dashed line-average for one individual tested twice (Mulsow et al. 2011b, 2014b). (**b**) Steller sea lions: thick line (Mulsow and Reichmuth 2010); thin lines (Mulsow et al. 2011a)

(Mulsow and Reichmuth 2013). Audiograms have been collected from stranded sea lions and those kept under human care (Mulsow et al. 2011a, 2014b), as these provide the most accessible animals for testing. The in-air AEP audiograms for the California sea lion are presented in Fig. 8.27a. Note that one animal is represented as the average of two AEP audiograms collected on this animal at different times (Mulsow et al. 2011b, 2014b). The audiograms are consistent with respect to the upper limit of hearing (limits between 16 and 32 kHz) and an apparent region of best sensitivity (~2–16 kHz), and comparable in these regards to what is observed in behavioral audiograms. Differences in AEP thresholds, however, differed by as much as 40 dB, even though methods used were similar. There was a wide range of ages tested within the group under human care (1–12 years, Mulsow et al. 2014b). Ages were not estimated for the stranded animals (Mulsow et al. 2011a) and these animals also had symptoms of domoic acid toxicity, which can affect the nervous system. Given the combination of subject conditions and ages, a broad variability in thresholds might be expected.

Mulsow and colleagues also obtained AEP audiograms from Steller sea lions (Mulsow et al. 2011a; Mulsow and Reichmuth 2010). The range of hearing and AEP thresholds (including variability) were similar to that observed in

California sea lion AEP audiograms (Fig. 8.27b). A peculiar notch at 5 kHz in the animals tested by Mulsow et al. (2011a) was not easily explained, although the methods were very similar to those from the earlier study (Mulsow and Reichmuth 2010). It seems unlikely that all the subjects had a hearing deficit at the same frequency and a consistent (yet unidentified) methodological issue is a reasonable alternative explanation.

Differences between AEP and behavioral hearing thresholds within the pinnipeds are likely similar to patterns observed in odontocetes. The specifics, such as the magnitude of deviations and where thresholds begin to show increasing divergence, might be specific to the species tested. Comparisons between AEP and behavioral thresholds within the same animal have been conducted in a California sea lion (Mulsow et al. 2011b) and a Steller sea lion (Mulsow and Reichmuth 2010). In the Steller sea lion, AEP thresholds determined from ASSR methods reliably demonstrated the upper limit of hearing and reasonably predicted the frequencies at which the animal had greatest hearing sensitivity, but were less reliable at frequencies ≤2 kHz. Differences between AEP and behavioral thresholds ranged from +1 dB at 20 kHz to as high as +31 dB at 1 kHz in the Steller sea lion. In the California sea lion, differences between AEP and behavioral thresholds differed from +3 to +20 dB. Differences were smallest where hearing was most sensitive, but increased at the low- and high-frequency ends of the audiogram. For animals which had multiple tests conducted at the same frequency, the standard deviation of the measurements were <10 dB.

8.3.2.3 Polar Bear

Only one effort has been made to measure polar bear hearing with AEPs (Nachtigall et al. 2007). The raw measures from bears tested are presented in Fig. 8.28b. Thresholds were higher than those measured behaviorally, but Nachtigall and colleagues made a correction for temporal summation (see Sect. 8.6.1) to compensate for the short duration of the tone-bursts used in the AEP tests. This reduced the thresholds by 26–35 dB, depending on the frequency of the

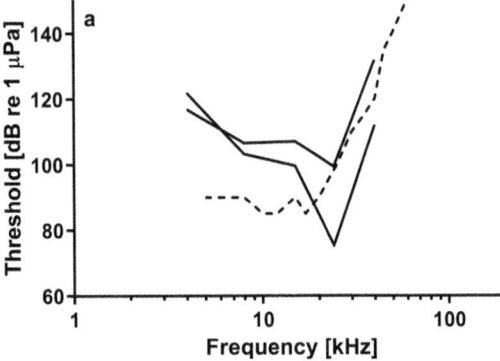

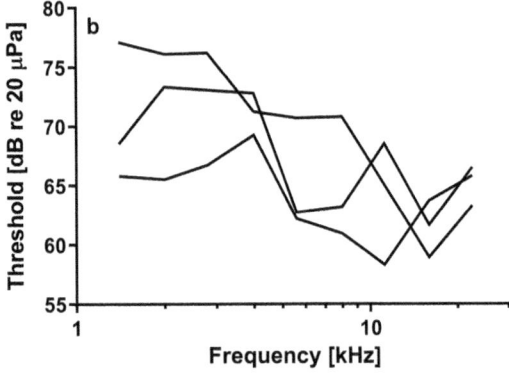

Fig. 8.28 AEP audiograms of (**a**) manatees: thick lines (underwater, Mann et al. 2005), see text for a description of how thresholds were determined; dashed lines (Klishin et al. 1990). (**b**) Aerial AEP audiograms of three bears from Kolmården Zoo, Sweden (Nachtigall et al. 2007)

tone-bursts. The trend of sensitivity with frequency was similar between behavioral and AEP audiograms, but the upper limit of hearing observed behaviorally (~16 kHz) was not observed in the AEP audiograms.

8.3.2.4 Sirenians

Bullock et al. (1980) were the first to investigate manatee hearing with AEP methods. They tested an Amazonian manatee in air and monitored cortical-evoked responses to click and tone-burst stimuli. They concluded that the manatee was most sensitive to aerial sound at ~3 kHz with possible sensitivity to ultrasonic frequencies. Follow-on studies in four West Indian manatees verified some degree of auditory sensitivity at frequencies up to 35 kHz and suggested greatest responsiveness at 1–1.5 kHz (Bullock et al.

1982). Klishin et al. (1990) utilized tone-bursts to estimate ABR thresholds and found a relatively flat sensitivity curve from ~5 to 20 kHz and then a rapid decrease in sensitivity (Fig. 8.28a). The use of ASSR methods was later applied to two West Indian manatees in an effort to determine AEP thresholds (Mann et al. 2005). Two different approaches were used to estimate AEP thresholds and differences between the approaches varied considerably depending on frequency: differences were more than 40 dB at 24 kHz in one individual, but were generally ~20 dB or less at other frequencies tested. The AEP audiograms were different from the behavioral audiogram previously collected in the manatee (Gerstein et al. 1999), but the upper limits of hearing were somewhat similar between the studies of Mann et al. (2005), using the first method, and Klishin et al. (1990). Mann and colleagues argued that the distance between the source of the AEPs (the brain stem) and the location of the ASSR recording some distance away was a factor in causing the differences between the AEP audiogram (s) and the behavioral audiogram.

8.3.3 Baleen Whale Hearing

The collection of behavioral audiograms requires access to trained animals; so, no behavioral audiogram has ever been collected from a baleen whale. However, an attempt was made to obtain electrophysiological measurements of hearing on a gray whale calf (Ridgway and Carder 2001). The calf stranded in California in 1997 and, in a rare event requiring substantial commitment, was rescued and transferred to SeaWorld San Diego for rehabilitation. Throughout its rehabilitation, several attempts to acquire AEP data from the gray whale were made. The effort had limited success and no frequency-specific threshold information was obtained. However, the experience was invaluable for planning future efforts to collect AEPs in mysticetes.

Considerable effort has gone into modeling hearing as a function of the tissue properties and the anatomical arrangement of auditory system components. A middle ear transfer function (METF) was created for the middle ear of the minke whale through a combination of finite element modeling and measurements of other mechanical properties of the middle ear (Tubelli et al. 2012). The frequencies predicted by the METF at which hearing was sensitive depended on the location of the acoustic input to the modeled system: stimulation at the point of the "glove finger" suggested best sensitivity from 30 Hz to 7.5 kHz; stimulation at the tympanic bone, which is assumed to be a sound conduction mechanism through fat bodies similar to that in odontocetes, suggested best sensitivity between 100 Hz and 25 kHz. As noted by the authors, a predictive audiogram could not be created without including information on the biomechanics of the basilar membrane.

Parks et al. (2007b) utilized measurements from 18 North Atlantic right whale inner ears to estimate the frequency range of hearing for this species. The study utilized measurements of basilar membrane dimensions, pitch, and turn ratio in a model previously applied to odontocetes and mysticetes (Ketten 1994). Based upon the model, the researchers concluded that the hearing of the North Atlantic right whale potentially ranged from 10 Hz to 20 kHz. Utilizing similar methods for the humpback whale, Houser et al. (2001) predicted a frequency range of hearing for the humpback from 100 Hz to 18 kHz. By combining the frequency-by-position map derived from the basilar membrane, and incorporating the dynamic range observed in the bandwidth of hearing in the cat and human, Houser and colleagues derived an audiometric function and a prediction of best hearing range from 700 Hz to 10 kHz. The function, however, was not truly an audiogram and only provided an estimate of relative sensitivity. Tubelli et al. (2018) subsequently created a finite-element modeling of the humpback whale middle ear, like what had been done in the minke whale. The model predicted a frequency range of best sensitivity between 15 Hz and 3 kHz when acoustic stimulation was modeled at the tympanic membrane, and between 200 Hz and 9 kHz when modeled at the thin region of the tympanic bone next to the tympanic membrane.

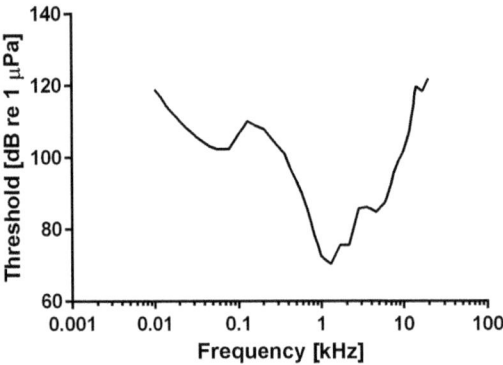

Fig. 8.29 Predicted fin whale "audiogram" based upon finite element modeling (Cranford and Krysl 2015)

Cranford and Krysl (2015) utilized computed tomography and finite element modeling to make a frequency-specific transfer function prediction for the fin whale. The model assumed both a bone conduction mechanism (through the skull and into the bulla) and what was termed a pressure mechanism, which is analogous to soft tissue conduction of the pressure wave to the tympanic bulla (as believed to occur for odontocetes). Cranford and Krysl concluded that, depending on the mechanism used, the velocity at the stapes footplate for a given pressure excitation varied. The stapes-velocity transfer function was then used to predict an audiogram based on the bone conduction mechanism, the pressure mechanism, and the sum of the effects of the two mechanisms. As no model of the basilar membrane was included, the frequency-specific predictions were referenced to an arbitrary minimum threshold of 70 dB re 1 μPa. The "audiogram" predicted by the combined effects of the pressure and bone conduction mechanisms is presented in Fig. 8.29. Overall, the model suggested a bandwidth of hearing of ~10 Hz–20 kHz. The shallow slope to the upper limit of hearing was argued to be controlled by the sensory epithelium of the inner ear, which was not incorporated in the model. The audiogram has a narrower region of best sensitivity than might be expected based upon the broader U-shape of typical audiograms, due primarily to the jagged nature of the model prediction.

The various model predictions of the frequency bandwidth of hearing for the fin, humpback, and right whale are reasonably consistent with one another. The predicted hearing ranges encompass the frequencies where the animals produce sound (see Sect. 6.5), but it may not be overly important that mysticetes hear "well" at these frequencies. The audiogram prediction of the fin whale suggests that the animals are relatively insensitive at the frequencies at which they commonly vocalize (~20 Hz). At these low frequencies ambient ocean noise possibly limits detectability, not the absolute hearing sensitivity (Richardson et al. 1995). Even in quiet ocean conditions, the ambient noise in 1/3-octave bands below 1 kHz can exceed 75 dB re 1 μPa (e.g., Chapman and Price 2011). If critical bands are larger than 1/3-octave (see Sect. 8.4), then the contribution of ambient ocean noise to detectability is even greater. Thus, there may have been no great selective pressure for acute hearing at low frequencies and it should not be assumed that baleen whale vocalizations necessarily overlap with the frequency range of best hearing. Conversely, if ambient noise played a role in the evolution of hearing sensitivity, at least within the region of best hearing, then the threshold might result from the sum of historical ambient noise levels and the critical ratio (see Sect. 8.4; Clark and Ellison 2004).

Numerous behavioral response studies and observations of baleen whale responses to sound from conspecifics and anthropogenic sources point to sensitivity at frequencies within the vocal range (e.g., Castellote et al. 2012; Frankel and Clark 2000, 2002; Frankel et al. 1995; Greene 1987; Richardson et al. 1986, 1990, 1999; Robertson et al. 2013; Watkins 1981). However, responses to anthropogenic sources have demonstrated that at least some mysticetes also hear higher frequencies. Blue, minke, and humpback whales, have responded to sources with frequencies of up to 2–4 kHz (e.g., Goldbogen et al. 2013; Lien et al. 1992; Maybaum 1989; Todd et al. 1992; Dunlop et al. 2013; Kvadsheim et al. 2011). Such observations give credence to models that suggest higher frequency hearing. Absolute sensitivity cannot be obtained through such methods as whales may not show a behavioral response to a sound they hear (see Sect. 8.2.2.3), but a confirmation of a behavioral response can provide a conservative estimate of hearing sensitivity.

Empirical measures of hearing are still needed in mysticetes and would be particularly useful in anatomical model validation. Electrophysiological approaches to testing baleen whale hearing hold promise, but might be limited to the smallest mysticete species or calves of larger species. Recent AEP measurements in young minke whales have demonstrated that the ABR can be recorded to transient stimuli in this species (Houser et al. 2024). There are inherent limitations in this approach, but the information obtained would go a long way to understanding the hearing abilities of mysticetes.

8.3.4 Population-Level Variability

AEP audiogram sample sizes are now sufficient for some species to begin determining population-level variability in hearing. Determining species-representative audiograms and variability in threshold estimates based on what are presumably normal hearing subjects should improve predictions regarding noise impacts at the population level, albeit with the caveats regarding increased variation in thresholds using the AEP technique (see below). Modeling the representative audiogram can be achieved by different modeling and statistical approaches (Branstetter et al. 2017a; Castellote et al. 2014; Finneran 2016). To date, however, there is still no consensus on the approach for establishing representative audiograms for groupings of marine mammals (e.g., species, populations, or hearing groups).

It would be advantageous if population estimates of hearing could be achieved through the collection of behavioral audiograms. Behavioral audiograms directly relate the perception of the signal to a physical measure of the signal, whereas AEPs do not provide a direct measure of perception. The use of behavioral audiograms also would allow species-representative audiograms to be referenced to absolute pressures, provided issues with masking are resolved across studies. Unfortunately, amassing behavioral audiograms for any marine mammal is a daunting task. The species for which the most behavioral audiograms have been published is the killer whale (Branstetter et al. 2017a; Szymanski et al. 1999). A species-representative audiogram based upon animals with presumed normal hearing has been created for this species (Branstetter et al. 2017a). The species-representative audiogram is presented in Fig. 8.6a as a dashed line. Estimates of threshold variance could also be included, although specifics of how the variance estimates would relate to the modeled audiogram have not been broadly addressed.

The remaining species for which sufficient sample sizes are available to create composite audiograms and determine variability in thresholds for normal hearing subjects have been tested with AEP methods. Unfortunately, AEP thresholds have a greater inherent variability than do behavioral thresholds; repeat measures in the same individuals show that threshold estimates are typically <10 dB in California sea lions and dolphins, but can be as high as 19 dB in the latter (Mulsow et al. 2011a; Houser et al. unpublished). Thus, there is a trade-off in the number of individuals that can be tested and the uncertainty in the accuracy of the threshold measurements. Figure 8.30 presents the AEP audiograms available for the bottlenose dolphin and considering only animals with presumably normal hearing (Finneran et al. 2008; Houser and Finneran 2006b; Houser et al. 2008b; Mann et al. 2010; Popov and Supin 1990c; Popov et al. 2007). Utilizing the approach proposed by Finneran (2016), a composite audiogram was created (thick, dashed line). The use of AEP thresholds to model a species-representative audiogram brings with it a set of problems. In theory, the distribution of thresholds at any particular frequency could be used to create the composite audiogram and build confidence intervals and error predictions around it. Unfortunately, much of the variation in the AEP threshold data is probably due to the impact of different testing

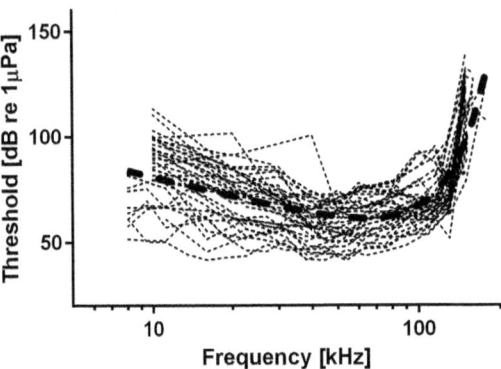 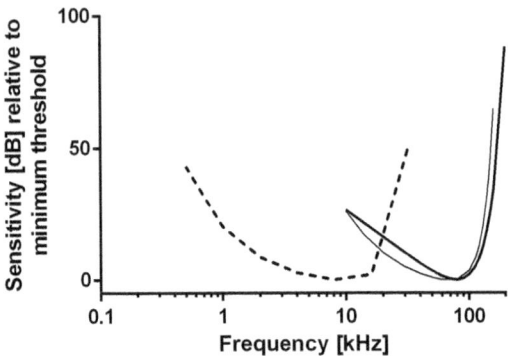

Fig. 8.30 Composite AEP audiogram of the bottlenose dolphin based upon the collection of published AEP audiograms (Finneran et al. 2008; Houser and Finneran 2006b; Houser et al. 2008b; Mann et al. 2010; Popov and Supin 1990c; Popov et al. 2007). (Note that the audiogram from Popov and Supin (1990c) is the average taken from four individuals.) Only animals with presumably normal hearing are included. The thick-dashed line is the composite audiogram for the species utilizing the method proposed by Finneran (2016)

Fig. 8.31 Representative underwater audiograms modeled for the bottlenose dolphin (thin line) and harbor porpoise (thick line) and in-air audiogram of the California sea lion (dashed line) based on AEP data in which each contributing audiogram has been normalized to the lowest threshold, as described in Finneran (2016). Note that the y-axis is in relative dB due to the normalization

methods, configurations, and threshold estimation techniques. As noted previously, differences in methods of predicting threshold can have a large impact on the distribution of AEP threshold data. Across studies, the data included in Fig. 8.30 were collected using different stimuli (e.g., tone-bursts, SAM tones), different means of projecting stimuli (e.g., direct field, jawphone), different methods of threshold estimation (e.g., extrapolation of IO functions, using lowest detectable evoked response), different test environments (e.g., submerged in pools or the ocean, tested in air using jawphones), and different means of calibrating. Without some type of correction, the study with the most data will inherently bias the modeled audiogram based on the methods it used for data collection. It will also be difficult to obtain a robust estimate of threshold variability using AEP methods until the variability arising from different test methods is accounted for, as well as the variance in repeated measures of threshold in the same individual. In this regard, standardization of AEP methods specifically for the purpose of determining audiograms, as proposed for odontocetes (American National Standards Institute (ANSI) 2018), would be beneficial to increasing the collective utility of AEP audiograms.

One potential means of addressing the variability in AEP threshold data is to model the audiogram after all individual audiograms are normalized to their lowest threshold. This reduces the variance around any measure of central tendency and at least provides some greater confidence in the shape and upper limit of hearing defined by the audiogram. The resultant modeled audiogram can then be shifted such that its minimum threshold is equal to zero, which in effect creates an audiogram that could be loosely related to a "pseudo-sensation level." This would be an inherently inaccurate approach because sensation level relies on the perception of a signal, which is something that AEP measures do not provide. Although the approach has shortcomings, it does provide a common ground for incorporating AEP data in the creation of species-representative functions. As an example, this procedure has been applied to the bottlenose dolphin data, as well as available California sea lion (Mulsow et al. 2011a, b, 2014b) and porpoise data (Lucke et al. 2009; Popov and Supin 1990b; Ruser et al. 2016; Ruser unpublished), and the resultant audiometric functions are presented in Fig. 8.31. These species are well-represented in the behavioral literature and are not the species for which such a procedure would be of greatest benefit.

8.3.5 Hearing Loss

Consideration of the potential impact that sound can have on marine mammal hearing must account for other factors affecting hearing. It is known that high-frequency hearing declines with age in mammals (e.g., Willott 2001), and many marine mammals are long-lived. Other factors that can cause hearing loss outside of noise exposure include disease and parasites, traumatic injuries, and congenital conditions. Captive or rehabilitated stranded marine mammals might also be affected by exposure to ototoxic drugs. Hearing loss has repercussions to marine mammals and potentially affects predator avoidance, foraging, mating behavior, navigation, etc. In odontocetes, the impact of hearing loss is observed in the changing of echolocation signals to compensate for the loss of sensitivity at high frequencies (Houser et al. 1999, 2005; Kloepper et al. 2010a, b).

Possibly the earliest measured hearing loss in a marine mammal was made in the killer whale (Hall and Johnson 1972), although the hearing loss was not verified until audiograms were collected in additional killer whales (Szymanski et al. 1999). Ridgway and Carder (1997) assessed the behavioral thresholds of a group of bottlenose dolphins to frequencies ranging from 5 to 120 kHz. They identified hearing deficits in several animals with a bias toward high frequency hearing loss in older males (≥23 years of age). They also identified a dolphin that appeared both deaf and mute. Some years later, a broadscale assessment of hearing in bottlenose dolphins conducted with AEP methods demonstrated that bottlenose dolphins typically begin to lose their high-frequency hearing after the age of 20 and that hearing loss begins to occur at a younger age in males than in females (Houser and Finneran 2006b). Work with known-age bottlenose dolphins at other marine mammal facilities has been consistent with these findings (e.g., Houser et al. 2008b).

Hearing loss has now been observed in several odontocete species (Finneran et al. 2005; Mann et al. 2010; Nachtigall et al. 2005; Popov et al. 2007; Schlundt et al. 2011; Yuen et al. 2005), California sea lions (Mulsow et al. 2014b), and possibly sea otters (Ghoul and Reichmuth 2014b). Across species, the trend of hearing loss with age seems to hold. Within the odontocetes a peculiar progression of hearing loss has also been observed. Hearing loss typically (but not always) occurs at high frequencies and then stops such that ~1–2 octaves of hearing range is lost. The hearing range may then stabilize for many years; for example, upper limits of hearing settle at 40–80 kHz in many delphinids, with no further loss in range apparent. The reason for this pattern of hearing loss is likely related to cochlear structure, although no specific mechanism has been identified that explains it.

Stranded odontocetes have been found with hearing loss, even though they were young (Andre et al. 2003; Mann et al. 2010; Schlundt et al. 2011; Strahan et al. 2020). Given the reliance of odontocetes on echolocation, it would not be surprising if hearing loss contributed to strandings. The cause of hearing loss generally cannot be proven, although in at least one instance hearing loss in a beluga was believed to result from aggressive treatment with ototoxic antibiotics (Finneran et al. 2005). Ototoxicity is a concern with the use of some antibiotics and known ototoxic antibiotics are limited in use and monitored closely by veterinarians in the treatment of stranded and rehabilitating odontocetes.

8.3.6 Hearing at Depth

Questions exist as to whether hearing in marine mammals changes with depth. As the depth of diving increases, so does the hydrostatic pressure. It is unknown how and whether the volume of intrameatal (pinnipeds) and peribullar sinus (odontocetes) air effects sound reception, or whether special vascularized (cavernous) tissues associated with the ear and sinus spaces regulate hearing at depth. Given the anatomical differences between the cetaceans and amphibious pinnipeds, it might be expected that the impact of hydrostatic pressure on hearing varies between the groups. The impact of depth on hearing has been largely overlooked, in part due to the difficulties in conducting such studies. However, two studies have addressed this issue.

Ridgway et al. (2001) investigated the impact of depth on the hearing of two belugas to depths of 300 m. They found no impact of depth on hearing sensitivity up to frequencies of 100 kHz and argued that the highly specialized middle ear of the beluga (and other odontocetes) contributed to a sound reception mechanism that did not rely on the ossicular chain conduction of sound common to terrestrial mammals. Different results were obtained in California sea lions that had their hearing tested at depths up to 100 m (Kastak and Schusterman 2002). Hearing sensitivity changed in a frequency-dependent manner with depth. At 50 m of depth, sensitivity to frequencies $\leq$10 kHz was worse than sensitivity measured at 10 m of depth. However, at 35 kHz, hearing sensitivity improved. The authors concluded that changes in the resonant characteristics of the middle ear cavity and expansion of the cavernous tissues were plausible mechanisms that explained the observed changes in hearing sensitivity at depth in the California sea lion.

8.3.7 Control of Hearing Sensitivity

Supin et al. (2008) were the first to demonstrate that the hearing sensitivity of a false killer whale changed during an echolocation task depending on whether a target was presented to the animal or whether it was absent. Using AEP methods, they demonstrated that hearing sensitivity was greater when no target was presented. They subsequently concluded that the whale used a form of hearing "gain control" during echolocation. Subsequent work suggested that knowledge about the returning echoes (e.g., in a stream of echoes) allowed the whale to consciously adjust the sensitivity of its receiving system. Nachtigall and Supin (2013) later demonstrated with the same animal that this change in hearing response could be conditioned. This was achieved by using series of tone-bursts as both a means to obtain AEP thresholds and a means to warn of an impending exposure to a 170 dB re 1 µPa (rms) signal. Once conditioned to relate the test signals to the impending high-level exposure, the AEP thresholds obtained with the test signals increased. Thresholds increased by as much as 12.7 dB and decreased as the duration of the test (warning) signal increased. Later work demonstrated that changes in hearing sensitivity could be conditioned in other odontocetes and with similar degrees of hearing suppression (Nachtigall and Supin 2014; Nachtigall et al. 2016a). Work with a beluga, which demonstrated that the response could be extinguished by running multiple sessions without a loud sound after the response was established, led Nachtigall and colleagues to the conclusion that the phenomenon was truly a conditioned response and not a reflexive or artefactual, temporary noise-induced hearing loss (Nachtigall et al. 2016a). The studies of Nachtigall and his colleagues were later repeated with a modified experimental design in two bottlenose dolphins (Finneran 2018). The results of the earlier studies were confirmed along with new findings: the magnitude of the threshold increase was found to be as high as 40 dB, and the duration over which the effect was observed was much longer than previously reported. A subsequent study in bottlenose dolphins indicated that the animals could learn the timing of repetitive noise exposure and attenuate their hearing sensitivity if the received level of the acoustic stimulus is sufficiently high (Finneran et al. 2023).

These findings suggest a shared capability for controlling hearing sensitivity within the odontocetes, as well as a potential limit to the magnitude that hearing sensitivity can be reduced (possibly up to 40 dB). The mechanism remains unknown and both muscular contraction of middle ear muscles (e.g., like the acoustic reflex) and inhibitory neural processes have been suggested (Nachtigall et al. 2016b). However, Finneran (2018, 2020) argued that the occurrence of reduced sensitivity at and above that of the intense sound exposure, and the presence of bilateral attenuation of sensitivity under monaural stimulation, suggest the phenomenon is most likely a neural mechanism operating at the level of the cochlea or auditory nerve. The finding of conditioned changes in hearing might have implications for both experimental tests of TTS and mitigation of loud sounds experienced in the wild (Finneran 2018; Nachtigall and Supin 2013). Animals used in TTS studies might protect themselves from the fatiguing effects of high-level sound exposures if they can predict when they will be exposed. This could result in elevated exposure estimates for the onset of TTS. Conversely, animals in the wild might be able to protect themselves from high-level sounds if the sounds are predictable in time (e.g., sonar emissions at a set duty cycle, air guns firing at regular intervals, or offshore pile driving). The impact that this "self-mitigation" in odontocetes has on our understanding of the effects of noise has yet to be fully realized.

8.4 Critical Ratio, Critical Band, and Auditory Filters

When a sound interferes with the detection of another sound *masking* occurs. Measurements of the threshold of hearing strive to achieve quiet test conditions where unwanted sounds (i.e., noise) cannot interfere with the measurement. However, constructing a test environment that is sufficiently quiet that threshold measurements are not masked is difficult to achieve, particularly when little is known about the auditory capabilities of a species in advance. A large number of marine mammal audiograms have been affected by masking (e.g., see Reichmuth et al. 2013), that is, they include masked (elevated) hearing thresholds. However, in the ocean, masking noise arises from physical processes (e.g., waves, storms, earthquakes), biological sources of noise (e.g., marine mammals, snapping shrimp, fish), and human activity. Noise levels are typically higher than a sea state of 0, and higher than noise levels present in some laboratory testing facilities. Knowledge of a marine mammal's ability to detect and comprehend signals in noise is therefore important to understanding the potential impact of human-made sound.

The ear is often modeled as a bank of overlapping auditory filters that describe the frequency selectivity of the auditory system. Knowledge of auditory filter shapes is useful in modeling the potential for masking and information about auditory filters (e.g., critical bandwidths, critical ratios) can be acquired through behavioral and electrophysiological means. Masking, critical ratios, critical bands, auditory filters, and mechanisms used to mitigate masking are covered in the following section. A couple of reviews of masking and its relevance to marine mammals are available (Branstetter and Sills 2022; Erbe et al. 2016).

8.4.1 Critical Ratio and Critical Band

Ambient noise can be measured as a broadband signal and characterized as a SPL_{rms}, or it can be characterized according to its spectrum level. The spectrum level represents the time-mean-square pressure within a narrow frequency band, typically 1 Hz. It is also called the pressure spectral density level or the noise spectral density level, expressed in water as dB re 1 $\mu Pa^2/Hz$.

At the masked threshold of detection, the power of a signal (S) is assumed equal to the noise power (N_o) within some critical frequency band that overlaps the frequency of the signal (equal-power assumption, Fletcher 1940). Beyond this *critical bandwidth (CBW)*, the addition of noise at other frequencies does not contribute to additional masking. The concept can be understood if the CBW is thought of as the upper and lower frequency limits of a filter. Noise within the frequency limits of the filter contributes to masking, whereas noise outside of the filter, above or below the frequency limits, does not.

An indirect estimate of the ability of an animal to detect signals in overlapping noise is the *critical ratio (CR)*, which is typically determined by testing an animal's ability to detect a signal in the presence of *white noise*. White noise is a broadband sound in which all frequencies are of equal level within the noise spectrum. As long as the bandwidth of the white noise is greater than the CBW, then measures of the masked threshold are unaffected and the CR can be calculated as the degree to which the signal pressure level exceeds the noise spectrum level at threshold (assuming a continuous noise of constant spectral density). For example, if an animal were able to "just" detect a tonal signal with a 100 dB re 1 µPa level in broadband white noise with a spectrum level of 80 dB re 1 µPa^2/Hz, then the CR would be 20 dB.

Critical ratios have been measured in many odontocetes (Fig. 8.32a). They increase with frequency above ~5 kHz and are generally similar for all species measured to date, even though the methodologies for obtaining CRs differed somewhat between studies. The CRs are ~18–20 dB near 5 kHz and increase to ~40 dB near 150 kHz. The trend changes at lower frequencies with little change in the CR at frequencies below 5 kHz; however, CRs at frequencies <5 kHz have only been tested in the beluga, killer whale, and the harbor porpoise (Branstetter et al. 2021; Johnson et al. 1989; Kastelein and Wensveen 2008; Kastelein et al. 2009b).

Underwater CRs in phocids (Fig. 8.33a) are comparable to those observed in odontocetes across the frequency ranges measured and increase with frequency in a similar manner. The CRs range from 10 to 17 dB from 100 to 200 Hz and increase to between 25 and 37 dB from 16 to 32 kHz. Critical ratios have only been measured under water in two otariid species, but show a similar trend to the phocids (Fig. 8.33b). In-air measurements of the CR appear similar between the California sea lion and walrus (Fig. 8.33b), and from 200 Hz to 3 kHz, CRs appear to be similar when measured in air and under water in the California sea lion. The latter observation is indicative of the CR as an inner ear phenomenon (i.e., it is the structure of the filters along the basilar membrane that determine the ability to detect a signal in noise). The filtering of sound by head tissues or the external and middle ear, prior to reaching the cochlea, should affect the signal and white noise similarly

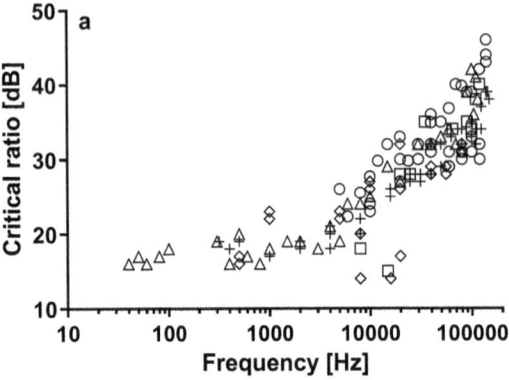

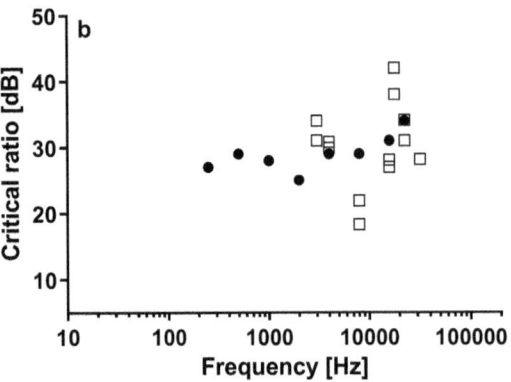

Fig. 8.32 Critical ratios measured in (**a**) false killer whale: square (Thomas et al. 1990b); bottlenose dolphin: circle (Au and Moore 1990; Branstetter et al. 2017b; Johnson 1968; Lemonds et al. 2011); beluga: triangle (Johnson et al. 1989); harbor porpoise: plus sign (Kastelein and Wensveen 2008; Kastelein et al. 2009b); killer whale: diamond-(Bain and Dahlheim 1994; Branstetter et al. 2021). (**b**) West Indian manatee: squares (Gaspard et al. 2012; Gerstein 1999); sea otter (in air): filled circles (Ghoul and Reichmuth 2014a)

prior to reaching the inner ear and have little effect on the ability of the animal to detect signals in noise (Renouf 1980; Southall et al. 2003; Turnbull and Terhune 1990).

The critical ratios of marine mammals are somewhat lower than those measured in terrestrial mammals but increase at a similar rate with increasing frequency (Fay 1988). This suggests that marine mammals have no particular cochlear specializations for extracting certain frequencies from noise (Southall et al. 2003). Trends are more difficult to determine in the sea otter and manatee, the only other marine mammals for which CRs

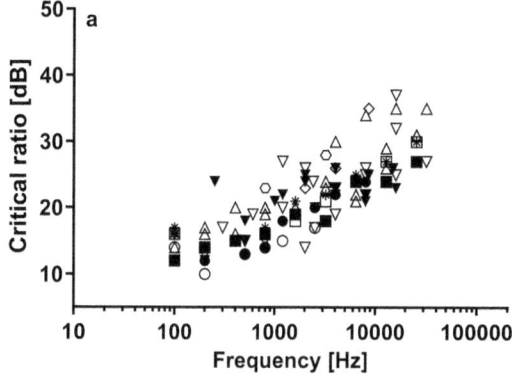

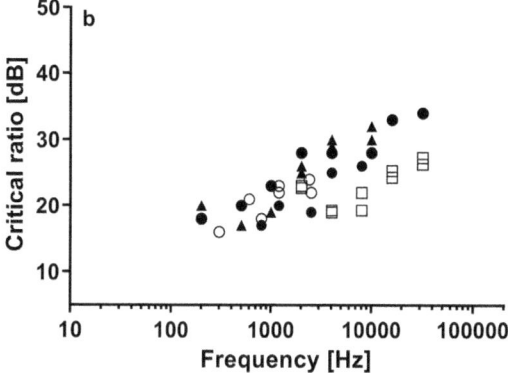

Fig. 8.33 Critical ratios measured in pinnipeds. Open symbols and symbols that are not enclosed are CRs measured to underwater sound. Closed symbols are CRs measured to sound presented in air. (**a**) Phocids. Spotted seal: square (Sills et al. 2014); northern elephant seal: circle (Southall et al. 2000, 2003); harbor seal: inverted triangle (Renouf 1980; Southall et al. 2000, 2003; Terhune 1991; Turnbull 1994; Turnbull and Terhune 1990, 1993); ringed seal: triangle (Sills et al. 2015; Terhune and Ronald 1975a); harp seal: diamond (Terhune and Ronald 1971); Hawaiian monk seal: hexagon-(Ruscher et al. 2021); bearded seal: asterisk (Sills et al. 2020). (**b**) Otariids and odobenids. Northern fur seal: square (Moore and Schusterman 1987); California sea lion: circles (Jones et al. 2023; Southall et al. 2000, 2003); walrus: triangles (Jones et al. 2023)

have been measured (Gaspard et al. 2012; Gerstein 1999; Ghoul and Reichmuth 2014a). Above 2 kHz, CRs in the sea otter increase predictably (~2.9 dB/octave) but are more variable below 2 kHz (Fig. 8.32b). The CRs are slightly higher than observed in other terrestrial carnivores (Ghoul and Reichmuth 2014a). The collective CR data in the manatee are quite variable with frequency-specific CRs being both lower and higher than observed in other marine mammals. However, if trends are considered from individual studies (Gaspard et al. 2012; Gerstein 1999), then a direct relationship between CR and frequency is observed.

Critical ratios have been used as an indirect measure of the CBW largely because directly measuring the CBW is challenging and time consuming. Under the equal-power assumption (Fletcher 1940), the bandwidth of the effective masker could be determined as:

$$CBW = 10^{CR/10}$$

Indirect estimates of the CBW from the CR vary considerably from direct measures of CBW, in part, due to violations of assumptions about the equal-power method (Yost and Shofner 2009). A few direct measures of CBW have been made behaviorally in marine mammals using notched-noise, band-widening, or tone-on-tone techniques. Relationships between direct measures of CBW and indirect measures derived from CRs vary within these studies. In a tone-on-tone masking study with a harbor seal, direct CBWs were found to be narrower than indirect estimates, ranging from 1% to 2.5% of stimulus center frequency (Turnbull and Terhune 1990). In contrast, band-widening experiments with a dolphin and some pinnipeds, including another harbor seal, found direct CBWs to be wider than those found in the tone-on-tone experiment (Au and Moore 1990; Southall et al. 2003), ranging from 13% to 57% of stimulus center frequency. Notch-noise studies have been used to estimate the shape of auditory filters in bottlenose dolphins and a beluga at frequencies ranging from 20 to 100 kHz (Finneran et al. 2002b; Lemonds et al. 2012). The filters were modeled as rounded exponential (*roex*) functions and were subsequently used to estimate the equivalent rectangular bandwidth (ERB), which is a simple way of modeling the auditory filter as a rectangular bandpass filter. Filter widths measured in this manner tended to be narrower than estimates obtained from band-widening experiments and ranged from ~12% to 17% of the filter center frequency.

A number of AEP approaches have also been used in odontocetes to investigate tuning curves, the function describing the neuronal response to a continuous stimulus attribute, and auditory filter shapes (e.g., Bibikov 1992; Klishin et al. 2000; Popov and Klishin 1998; Popov et al. 1997, 2001, 2006b; Sysueva et al. 2014). The ERBs derived from these studies are variable and likely influenced by both the source of the evoked potential (e.g., brain stem vs. cortical) and the methods used (e.g., tone-on-tone vs. rippled spectra). However, for the most part, ERBs determined from AEP studies provide the narrowest estimates of the critical bandwidth, up to an order of magnitude lower than obtained with some behavioral approaches.

Collectively, the ERBs from electrophysiological and behavioral approaches suggest different things about the structure of auditory filters in odontocetes. Much of the electrophysiological work is suggestive of constant bandwidth filters across the frequencies used in echolocation, where behavioral work is more suggestive of constant-Q filters (i.e., filters with a bandwidth that is a constant proportion of the center frequency). Comparisons to measures of critical bands obtained behaviorally and electrophysiologically should be made cautiously as the approaches do not measure the same thing. Behavioral approaches measure an integrated animal response (including cognitive aspects of signal detection), whereas electrophysiological approaches measure some neural component of the auditory system.

8.4.2 Adaptations for the Reduction of Masking

The ocean is not a laboratory environment and ocean noise is typically quite different from the noise used in masking studies under controlled laboratory conditions (e.g., white noise). Many sources of ocean noise are dynamic in time, frequency, and space, and will provide variable degrees of masking. Examples of how types of anthropogenic and natural noise differ in the masking of marine mammal communication signals have been demonstrated in the laboratory (Branstetter et al. 2013b; Erbe and Farmer 1998). Marine mammals benefit from auditory phenomena and behavioral adaptations that enable them to better extract signals from background noise, even at SNRs that are near-zero or negative.

When the source of a signal and noise are spatially separated, the signal is easier to detect than if the sources of both noise and signal are co-located—a phenomenon referred to as the spatial release from masking (SRM). This occurs because of binaural phase and level cues available to the animal that may improve the perception of the signal (see Sect. 8.5). Laboratory demonstrations of SRM in air performed in a harbor seal and sea lion demonstrated that movement of a sound source to a location 45° and 90° in the horizontal from a masking source improved (i.e., lowered) masked thresholds by as much as 19 and 12 dB, respectively (Holt and Schusterman 2007). Under water, SRM has been observed in the harbor seal and killer whale to frequencies less than 10 kHz (Bain and Dahlheim 1994; Turnbull 1994). Relatively little other work has been done at frequencies below 20 kHz, or in other species, but it is certain that other marine mammals also utilize binaural cues to resolve spatially separated noise and signal sources.

Odontocetes have highly directional echolocation beams and receiving (hearing) beams that support echolocation (Au et al. 1986, 1993, 1995, 1999, 2012; Au and Moore 1984; Finneran et al. 2014; Kastelein et al. 2005a; Popov and Supin 2009; Zimmer et al. 2005). The directionality of the echolocation beam and the directionality of hearing at high frequencies potentially provides a substantial SRM for echolocating odontocetes. Spatial separation of masking noise and signal sources resulted in up to ~24 dB improvement in the CR in a bottlenose dolphin listening to sound at echolocation frequencies (Zaĭtseva et al. 1975, 1980). Work with a beluga also suggests that odontocetes may learn to utilize surface reflected echolocation signals to avoid point sources of noise in-line with a target of interest, thus producing echoic returns from an angle different from that of the noise source (Penner et al. 1986). However, AEP studies of how the echolocation beam affects SRM suggest that the improvement in SRM is modest (Popov et al. 2020). The authors however noted that aspects of the testing conditions (e.g., testing performed in a small tank) could have affected the outcome of the study.

Noise in the ocean is seldom tonal, can cover a broad range of frequencies, or contain a substantial number of harmonics. As the noise varies in amplitude, the amplitude sometimes changes across frequencies in a correlated manner (i.e., the noise is comodulated with respect to its level). Many natural and anthropogenic sounds are comodulated, including ship noise, rain, snapping shrimp, ice creaks, and animal vocalizations (Branstetter et al. 2013b; Cunningham et al. 2014; Nelken et al. 1999). A signal that is not comodulated with the background noise will be easier to detect than one embedded in uncorrelated noise (e.g., Gaussian noise), a phenomenon called comodulation masking release (CMR). Experiments in dolphins suggest that this is probably accomplished by across-channel listening (Branstetter et al. 2013b). Across-channel listening occurs when amplitude modulation patterns are compared across a bank of auditory filters. If the amplitude of a signal of interest is not correlated with the pattern of amplitude modulation across other auditory filters, then a temporal cue to the signal's presence may be detected. Within-valley listening, or "dip" listening, occurs when the SNR improves because of a drop in the amplitude (the valley) of the noise, that is, the noise is deeply amplitude-modulated. It is most effective for "deep" valleys and possibly for slowly varying amplitude-modulated noise wherein more "looks" at the signal are permitted in short time windows (see Chap. 2 on underwater noise).

Comodulation masking release and within-valley listening have been experimentally demonstrated in bottlenose dolphins, belugas, harbor porpoises, harbor seals, spotted seals, ringed seals and California sea lions (Branstetter and Finneran 2008; Branstetter et al. 2013b; Cunningham et al. 2014; Erbe and Farmer 1998; Kastelein et al. 2021, 2023a; Sills et al. 2017; Trickey et al. 2010). The degree of CMR that occurs is specific to the masking noise, but as much as 17 dB of masking release has been observed in a bottlenose dolphin (Branstetter and Finneran 2008). Thresholds of detection and comparisons of CRs measured in various types of comodulated noise relative to those measured in the presence of white noise, particularly when complex signals are used, further suggest that CRs measured in white noise are conservative estimates, underestimate the ability of marine mammals to detect signals in the presence of many noise sources, and should be used cautiously (Branstetter and Finneran 2008; Cunningham et al. 2014). Interestingly, comodulated noise may increase thresholds for odontocetes to broadband clicks, similar to the echoes received during echolocation (Branstetter et al. 2016). In such situations, the directionality of hearing and of the outgoing echolocation beam may be of critical importance to signal detection.

Marine mammals might also mitigate masking by varying the level, frequency, duration or repetition of a communication signal; this is called the Lombard Effect (Lombard 2011). Increases in the level of communication signals have been observed in humpback whales, minke whales, North Atlantic right whales, killer whales, and belugas under conditions of elevated natural and human-made noise (Dunlop et al. 2014; Fournet et al. 2021; Guazzo et al. 2020; Helble et al. 2020; Holt et al. 2008, 2011; Parks et al. 2011; Scheifele et al. 2005; Southall et al. 2019a), although this is not always the case. For example, humpback whales off the east coast of Australia have been observed to increase calling levels in the presence of wind noise but not to boating noise, likely because they can use other mechanisms (e.g., SRM or CMR) to mitigate masking that are not as suitable for the diffuse nature of wind noise (Girola et al. 2023). Bottlenose dolphins and belugas have been observed to increase the rate of communication calls in the presence of boat noise (Buckstaff 2004; Lesage et al. 1999), and killer whales and humpback whales have been observed to increase the duration of their calls or songs in the presence of boat noise or low frequency sonar, respectively (Foote et al. 2004; Miller et al. 2000). Increases in signal duration or repetition should increase the number of looks, or opportunities, for conspecifics to detect a call within background noise. Northern right whales, belugas, and common dolphins have been observed to change the frequency of their calls in the presence of noise, presumably to reduce masking from overlapping or nearby frequencies (Ansmann et al. 2007; Lesage et al. 1999; Parks et al. 2007a). In addition, many marine mammal calls are complex, have broad bandwidth, or distinctive structures (e.g., down-sweeps), which presumably improve their detectability in noise (Cunningham et al. 2014; Turnbull and Terhune 1994).

Echolocation provides a special situation in which the signaler is also the receiver. Odontocetes produce echolocation clicks to generate echoes off targets that then return to the whale. Odontocetes are capable of increasing echolocation signal source levels and modifying the peak frequency and bandwidth of echolocation signals in response to environmental noise and experimental conditions (Au et al. 1974, 1985; Moore and Pawloski 1990; Romanenko and Kitain 1992; Thomas and Turl 1990). Bottlenose dolphins and harbor porpoises are capable of steering and varying the width of the echolocation beam (Koblitz et al. 2012; Moore et al. 2008; Wisniewska et al. 2012, 2015), which should be useful for optimizing the SNR of echoes generated from elusive prey, while simultaneously optimizing the SRM.

Additional studies of masking are needed, particularly at low frequencies. Masking studies should, however, not be limited to detection; recognition or comprehension of a signal is equally, if not more, important to an animal (Branstetter and Sills 2022). Experiments with a bottlenose dolphin have demonstrated that thresholds for signal recognition are higher than detection thresholds under the same noise conditions (Branstetter et al. 2016), as expected. Furthermore, noise is often characterized with respect to frequency, whereas temporal aspects of noise, such as its amplitude modulation, are often unreported. Temporal characteristics of noise likely play an important part in signal detection in noise (e.g., CMR) and should be considered in future studies of masking and characterizations of noise (Branstetter et al. 2013a).

8.5 Frequency and Level Discrimination

8.5.1 Frequency Discrimination

A marine mammal's ability to detect signals in background noise and to correctly discriminate them depends in part on their ability to differentiate between frequencies and detect differences in the level of sounds. This is important to animal communication (e.g., mating displays, individual identification), foraging, and possibly, predator detection and avoidance. Fine frequency and level discrimination is also required for the discrimination of echoes by echolocating odontocetes.

The frequency difference limen (DL, see Sect. 8.2.1) is calculated by subtracting the frequency of a just-discriminable comparison tone from the frequency of a reference tone (in Hz). To facilitate comparisons across frequencies, since the DL can change with frequency, the DL is divided by the frequency of the reference tone (F). The resulting ratio (DL/F) is known as the *Weber ratio* and it is often changed to a percentage value to describe the DL as a percentage difference from the reference frequency. The smaller the Weber ratio, the better the resolution of the sensory system.

Several studies demonstrated that odontocetes have a high degree of frequency discrimination. A partial study of frequency DLs in the bottlenose dolphin demonstrated Weber ratios ranging from 0.003 to 0.012 across a frequency range of 0.9–90 kHz (Jacobs 1972). Research by Thompson and Herman (1975) and Herman and Arbeit (1972) later confirmed low Weber ratios in the bottlenose dolphin. Frequency discrimination abilities of the dolphin are shown in Fig. 8.34a. The trends between the studies are similar with best frequency discrimination between ~2 and 56 kHz. All the studies showed a decline in frequency discrimination at 1 kHz, but disagreed with respect to the high frequency at which a decline in frequency discrimination occurred.

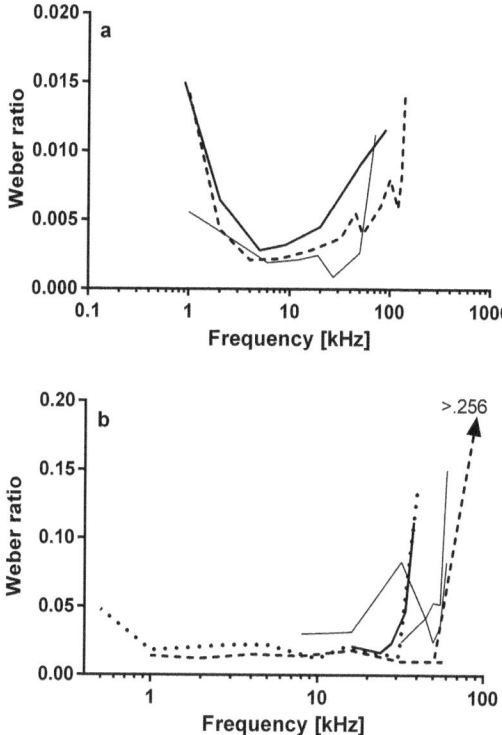

Fig. 8.34 Frequency difference limens (presented as Weber ratios) (**a**) Bottlenose dolphin: the thick line corresponds to Jacobs (1972), the thin line to Herman and Arbeit (1972), and the dashed line to Thompson and Herman (1975). For Herman and Arbeit (1972), the 36 kHz value is an average from two different experimental setups. For Jacobs (1972), values are a mix of single measurements and mean values from replicate measurements. (**b**) California sea lion, thick line (Schusterman and Moore 1978); northern fur seal, dotted line (Babushina et al. 1991); ringed seal ($n = 2$), thin lines (Terhune and Ronald 1976); harbor seal, dashed line (Møhl 1967)

The latter might seem somewhat surprising given that Thompson and Herman (1975) and Herman and Arbeit (1972) used the same dolphin subject, but the difference potentially exists because of differences in experimental methodologies between the three studies and the experience of the subjects. Nevertheless, dolphins appear to have frequency discrimination capabilities similar to those of humans (in air) across the frequency range at which both species hear well, and also show a dependence of the DL upon other signal features, such as amplitude (Herman and Arbeit 1972; Moore 1973). Frequency discrimination appears several times better than other terrestrial mammals (as summarized in Fay 1988; Heffner et al. 1971).

Soviet researchers utilized a conditioned (respiratory, heart rate, and galvanic) reflex to measure frequency discrimination in the harbor porpoise and found Weber ratios less than 0.005 across the range of frequencies tested (3–190 kHz) and from 0.001 to 0.002 for most of the hearing range (Sukhoruchenko 1973; Supin and Sukhoruchenko 1974). In view of the behavioral measurements made in the bottlenose dolphin and the methods employed here, the reliability of these results has been questioned (Bullock and Gurevich 1979). However, Popov et al. (1986) found that cortical-evoked response amplitudes changed in response to changing frequencies with a threshold of detection occurring at a ~0.1% change in frequency (a corresponding Weber ratio of 0.001). The lowest thresholds were observed between 120 and 125 kHz, which is approximately the frequency region at which harbor porpoises produce echolocation pulses (Au et al. 1999).

Pinniped frequency discrimination is at least three times coarser than that of the echolocating odontocetes (Fig. 8.34b). A harbor seal was found to have Weber ratios ranging from 0.01 to 0.014 across the frequency range of 1–57 kHz (Møhl 1967). The seal was unable to discriminate frequencies that deviated by as much as 25.6% at frequencies above 60 kHz, suggesting that detection of sound was not accompanied by an ability to discriminate sounds at frequencies >60 kHz. Similar findings of an upper limit to frequency discrimination (>60 kHz) were made in the ringed seal (Terhune and Ronald 1976). Weber ratios measured under water in the northern fur seal range from 0.01 to 0.02 at frequencies ranging from 1 to 30 kHz, and from 0.016 to 0.018 in air when measured across a much narrower frequency range of 3–5 kHz (Babushina et al. 1991). From 16 to 28 kHz the Weber ratio of the California sea lion listening under water averaged 0.02, but the ratio doubled at 34 kHz and deteriorated further at 38 kHz (Schusterman and Moore 1978). At 38 kHz, the California sea lion was only able to discriminate frequency differences of 11.2% (Weber ratio = 0.112). Unfortunately, few frequency discrimination studies have been conducted with pinnipeds over the last 20 years. What is known continues to be based on a few animal subjects and more investigations would be beneficial.

8.5.2 Level Discrimination

The ability to discriminate changes in level is important when detecting signals in noise, localizing sound (see Sect. 8.8), and classifying sound sources (e.g., as in dolphin echo classification). Humans discriminate level differences by as little as 1 dB within the range of frequencies over which they hear well (e.g., Jesteadt et al. 1977). Little direct information exists on the abilities of marine mammals to differentiate sound level differences, with more information available for the odontocetes than for the pinnipeds.

Level discrimination is critical to the odontocete ability to differentiate complex targets and find and pursue prey, which produce species-specific and aspect-dependent echoic cues (Au et al. 2009; Helweg et al. 1996). Based on an echolocation target discrimination experiment, Nakahara et al. (1997) concluded that the finless porpoise might be able to discriminate level differences as little as 1 dB if it utilized differences in target strength (TS) as a means of discriminating targets. In a review of early echolocation discrimination studies, Evans (1973) arrived at a similar discrimination ability for the bottlenose dolphin and boto, based on TS differences in the targets employed. By modulating the amplitude envelope of a stream of electronically-produced echoes, the ability to correctly classify amplitude-modulated streams in a bottlenose dolphin was found to be as fine as 0.8 dB (Dankiewicz et al. 2002). This was similar to Dubrovskiy's finding that a dolphin could discriminate between two sets of echo streams that differed in level by 1 dB when presented 40 dB above threshold, and 2 dB when presented near threshold (Dubrovskiy 1990). Johnson estimated that level discrimination based on threshold audiometry and tone masking studies in the dolphin was also about ~1 dB, with modeled estimates ranging from 0.35 to 2 dB (Johnson 1967, 1971). An alternative estimate based on interaural intensity differences using jawphones suggested that the bottlenose dolphin could discriminate levels that differ by as little as 0.6–1.0 dB (Moore et al. 1995). Behavioral findings are supported by electrophysiological studies that demonstrate the amplitude of evoked potentials change with stimulus level differences between 0.5 and 3 dB (Bullock et al. 1968; Popov et al. 1986). Thus, multiple lines of research with tonal and pulsed stimuli point to an excellent level discrimination ability in the odontocetes, which is comparable to that in humans (~1 dB), and better than that of many other vertebrates (>2 dB; Fay 1988).

Unfortunately, little is known about level discrimination in pinnipeds. A study with a California sea lion conducted nearly 50 years ago remains the only published information on pinniped level discrimination (Moore and Schusterman 1976). In that study, the sea lion was able to approach a level discrimination threshold of ~3 dB when comparing underwater 16 kHz test tones presented at different levels. The sea lion's performance improved with repeated testing and the threshold of discrimination showed some dependence on the method of testing employed. Nevertheless, the authors concluded that the discrimination ability was probably sufficient for underwater sound source localization.

8.6 Temporal Summation and Temporal Resolution

8.6.1 Temporal Summation

The detectability of a signal increases with the duration of the signal, at least to a point (Yost 2000). The signal duration at which detectability ceases to increase ranges from 100 to 1000 ms in mammals and can show frequency dependency (Gelfand 1990). The phenomenon is possibly explained as the ear acting as an integrator with a long time constant, or through multiple, brief observations, or "looks," which are stored in memory and selectively processed (Buus 1999; Viemeister and Wakefield 1991). Most behavioral audiograms collected in marine mammals have been obtained with stimuli that are 500 ms or longer in duration and are therefore unlikely to have been affected by the duration of the stimulus. In contrast, most AEP audiograms have been collected with much shorter stimuli, which probably contributes, in part, to differences in AEP and behavioral hearing thresholds.

Early measures of temporal summation in cetaceans were performed on the bottlenose dolphin (Johnson 1968). Across a range of frequencies, Johnson found that tone durations longer than 100–200 ms resulted in similar detection thresholds for the dolphin and that there was some variability based upon the frequency tested. The shortest duration at which detection probability became constant was 36 ms at 40 kHz. Johnson later found the summation time for a 60 kHz tone in the beluga was similarly ~100 ms (Johnson 1991). He also found that as shorter duration pulses were transmitted to the beluga in series (i.e., a pulse train), the threshold of detection approached the behavioral threshold for the long-duration tone as the pulse interval declined below 10 ms. In the harbor porpoise, the signal duration at which detectability was found to be constant was similar to that of the dolphin and beluga, at least across a comparable range of frequencies tested (Kastelein et al. 2010). However, Wang et al. (1992) suggested an overall longer duration in the Chinese river dolphin. Thresholds improved as the signal duration was changed from 500 to 5000 ms, but no intermediate durations were tested so the point at which detection performance came to an asymptote is not known.

Electrophysiological studies in the harbor porpoise demonstrated that evoked responses recorded from the cortex of the brain increase in amplitude as tone duration increases (Popov and Supin 1976). It should be noted that the durations of the signals used (tens of ms) were much shorter than behaviorally measured temporal summation times (hundreds of ms). Although the trend is informative, particularly as it relates to potential nervous system processing of longer duration signals, it is not possible to directly relate the findings to those from behavioral studies. In contrast to cortical studies, studies of the dolphin midbrain show that it is primarily responsive to the rapid onset of acoustic stimuli and much less influenced by signal duration (Bullock and Ridgway 1972). The responsiveness to rapid onset stimuli is undoubtedly an adaptation for the processing of echolocation signals, and the responses in the midbrain appear maximal at echolocation frequencies.

Terhune (1988) found that a harbor seal's thresholds for frequencies spanning the range of hearing did not change significantly when the test tone duration varied from 50 to 500 ms (i.e., thresholds were generally within 6 dB of one another). Below a duration of 50 ms, thresholds increased with decreasing duration. Thresholds for 16 kHz tone bursts presented at durations ranging from 0.625 ms (1 cycle) to 100 ms (1600 cycles) and broadband clicks were 30–40 dB higher than thresholds collected with longer duration tones (Terhune 1989). Kastelein et al. (2010) investigated the dependence of threshold on signal duration in two other harbor seals and found that (except at 200 Hz) thresholds did not improve when signal durations exceeded 200–500 ms. Furthermore, independent of frequency, thresholds increased as the number of cycles in the stimulus fell below ~780 cycles. Oddly, for a 200 Hz signal the detection threshold continued to improve for signal durations of up to 5 s. The result, however, is potentially artefactual and might have occurred because of fluctuations in low-frequency noise or nearfield effects resulting from the proximity of the seals to the sound projector (Kastelein et al. 2010). A follow-up study with another harbor seal investigated the finding at 200 Hz but found no difference in thresholds for 200 Hz tones presented at 500 ms and 2.5 s durations, consistent with findings in other mammals (Reichmuth et al. 2012).

Holt et al. (2012) conducted the only study of temporal summation in an otariid. Using tones presented in air at frequencies of 2.5, 5, and 10 kHz, they determined the signal duration beyond which detectability no longer improved to be ~300 ms across this frequency range. As with the harbor seal, the results were also consistent with findings in other marine and terrestrial mammals.

Understanding temporal summation is important to understanding basic auditory processes in marine mammals. Determining the potential impact of noise on marine mammals must consider that many sounds that animals encounter under water might be shorter in duration than that at which signal detectability is optimal. An understanding of how the threshold of detection changes as a function of signal duration is therefore necessary. Further investigations would help identify whether the effect of temporal summation is relatively consistent across frequencies and marine mammal groups.

8.6.2 Temporal Resolution

The auditory system must differentiate the time-separation of sequential sounds to segregate and localize sound sources, particularly if the sources are co-located relative to the receiver. Temporal resolution is particularly important to odontocetes because of their need to process sequential echoes at a high rate of return and to resolve targets at various ranges. Temporal resolution also plays a role in forward and backward masking, where signal and noise are separated by only a brief gap in time. If the temporal gap between signal and noise is too short, then the noise can mask the signal, even though it does not coincide exactly in time with it. Behavioral estimates of temporal resolution in dolphins have been remarkably consistent showing they can resolve a temporal separation between 250 and 300 µs (Au 1988, 1990; Dubrovskiy 1990; Moore et al. 1984).

Electrophysiological estimates of temporal resolution can be achieved by monitoring the amplitude of evoked responses to series of clicks, tone bursts, or amplitude-modulated tones. By assessing the amplitude of the evoked response across a range of amplitude modulation or stimulus presentation rates, a modulation rate transfer function (MRTF) can be developed (e.g., Dolphin et al. 1995). The upper limit of the function (i.e., its bandwidth) is argued to be related to the temporal resolution of the auditory system and an estimate of temporal resolution can be calculated from it (Dolphin 2000). Utilizing this method, a temporal resolution of ~350 µs or less has been estimated (or is apparent from the MRTF) for the bottlenose dolphin (Dolphin et al. 1995; Finneran et al. 2007b; Popov and Supin 1998), harbor porpoise (Linnenschmidt et al. 2013), false killer whale (Dolphin et al. 1995), white-beaked dolphin (Mooney et al. 2009), common dolphin (Popov and Klishin 1998), spinner dolphin (Smith et al. 2018), beluga (Dolphin et al. 1995), beaked whales (Cook et al. 2006; Finneran et al. 2009; Smith et al. 2018), pygmy killer whale (Montie et al. 2011; Smith et al. 2018), long-finned pilot whale (Smith et al. 2018), and Yangtze finless porpoise (Mooney et al. 2011). In the killer whale and Risso's dolphin, the estimate is somewhat higher at ~420 µs (estimated from Szymanski et al. 1998) and ~430 µs (Mooney et al. 2006), respectively. Other electrophysiological approaches have provided similar (or finer) estimates of temporal resolution in the odontocetes (e.g., Popov and Supin 1997; Supin and Popov 1995). Across multiple methods, estimates of the temporal resolution of odontocetes show remarkable consistency with a potential relationship of decreasing temporal resolving capability with increasing body size (Linnenschmidt et al. 2013; Smith et al. 2018).

Relatively little work has been conducted into the temporal resolution of other marine mammals. Mulsow and Reichmuth (2007) obtained MRTFs from a group of California sea lions. An estimate of the limit of temporal resolution based on these MRTFs was ~500 µs. Using the same process, similar estimates were found for the elephant seal and harbor seal (Houser et al. 2007; Mulsow and Reichmuth 2007). An estimate of temporal resolution from MRTFs in manatees was ~700 µs, suggesting their ability to resolve the separation of sounds is slightly worse than the odontocetes and pinnipeds (Mann et al. 2005).

It has been suggested that the temporal resolution ability of the marine mammals is an adaptation accommodating the increased speed of sound under water (Mann et al. 2005). However, a comparison to the temporal resolution of the dog (*Canis familiaris*) suggests that it has similar capabilities to that of the pinnipeds (Mulsow and Reichmuth 2007). Mulsow and Reichmuth concluded from their direct comparison to the temporal resolution of the dog that selective pressures related to underwater hearing cannot entirely explain the resolving ability of non-echolocating marine mammals.

8.7 Directional Hearing

The ability to localize sound requires directional hearing, which is important to a marine mammal's ability to find prey, avoid predators, and locate conspecifics. Directional hearing also aids signal detection in noise. Most of the work on the directionality of hearing has been conducted with odontocetes (Accomando et al. 2020; Au and Moore 1984; Kastelein et al. 2005a; Popov and Supin 2009; Popov et al. 2006a), largely because of the interest in the directionality of hearing in relation to echolocation.

By recording ABRs to tone bursts with frequencies ranging from 16 to 128 kHz and varying the location of the sound source, Popov and Supin (2008) calculated the distance to the primary sound receiving area on the bottlenose dolphin head by accounting for the acoustic delay from the sound source position and the latency of the ABR waves. They concluded that at least two primary sound receiving areas exist; one for frequencies <22.5 kHz and located near the auditory meatus, and one for higher frequencies associated with echolocation and corresponding to the pan region of the lower jaw. The results supported prior work in which monaural and binaural directivity indices were calculated from AEP response amplitudes (Popov et al. 2006a). The best sensitivity azimuth for monaural measurements was frequency dependent with lower frequencies having a best sensitivity azimuth at 22.5° and high frequencies at 0°.

Au and Moore (1984) used psychophysical procedures to measure the receiving beam pattern of the bottlenose dolphin in the vertical and horizontal planes. Maximum sensitivity in the vertical plane was found to be 5–10° above the midline of the mouth and somewhat asymmetrical. Beam patterns in the horizontal plane were more symmetrical. In both planes, the beam pattern became narrower with increasing frequency. Beamwidths were defined by the points on the beam pattern that corresponded to an amplitude 3 dB below the amplitude maximum. In the vertical plane, beamwidths were 30.4°, 22.7°, and 17.0° for frequencies of 30, 60, and 120 kHz, respectively. In the horizontal plane, they were 59.1°, 32.0°, and 13.7°, respectively. Accomando et al. (2020) later studied the directionality of hearing in two bottlenose dolphins at frequencies of 2, 10, 20 and 30 kHz. The relationship between stimulus frequency and directionality of hearing was consistent with the trend observed by Au and Moore (1984), mainly that directivity decreased with decreasing frequency. Additionally, results similar to Au and Moore (1984) have been found in another bottlenose dolphin, as well as a harbor porpoise and a beluga (Kastelein et al. 2005a; Popov and Supin 2009). In absolute terms, the receiving beam appears broader at all frequencies measured in the beluga and harbor porpoise than in the bottlenose dolphin, although this might be logically explained by differences in the measurement methods used (e.g., see discussion in Popov and Supin 2009). Although the beam pattern describes the directionality of hearing and is related to binaural reception, the narrowness of the beam pattern is particularly critical to echolocation. A target being insonified by an odontocete will show changes in echo amplitude dependent upon the narrowness of the receiving beam and the target's change in position relative to the echolocating dolphin, thus providing amplitude cues to target localization.

8.8 Source Localization

Mammals rely on several different cues to determine the location of a sound source (Branstetter and Mercado III 2006). These include binaural cues such as interaural level differences (ILD), which are differences in the received level of sound at the two ears [also called interaural intensity differences (IID)], and interaural time differences (ITD), which are differences in the time of arrival of sound at the two ears. The ILD and ITD are particularly critical to the localization of sound in azimuth. Spectral cues can also contribute to azimuthal localization and potentially contribute to localization in the vertical plane and at distance (i.e., high-frequency sound attenuates more rapidly with distance than does low-frequency sound). A fourth potential cue for signals of sufficient duration is the interaural phase difference (IPD). The IPD, which is more important at low frequencies, uses phase differences between signals arriving at the two ears to estimate the ITD.

Sound travels ~5 times faster under water than in air resulting in shorter ITDs for a given interaural distance. The path that sound travels under water is also different. Many of the tissues of marine mammals have impedances that are like water, and sound will pass through these tissues from one ear to the other. The ITD is therefore affected mostly by distance between the auditory bullae. In terrestrial mammals, sound must travel around the outside of the head because of the impedance mismatch between the air and the skull/skin; so, the circumference of the head becomes more prominent in the ITD. When marine mammals are under water, differences in tissue impedances (e.g., between skull and fat) and the presence of air spaces (e.g., peribullar sinuses) likely contribute to spectral cues and ILD. The precise role of tissue properties and air spaces on the spectral shading of a sound received at spatially separated ears (e.g., through low-pass filtering) is unknown.

The minimum audible angle (MAA) is the smallest discriminable angular difference between two sound sources located at an equal distance from a receiver. It is commonly used to describe source localization abilities. Tests of MAA are typically performed with the sound sources located in front of the subject. Closely related to the MAA is angular difference discrimination, which utilizes a slightly different paradigm to investigate angular discrimination abilities (Branstetter et al. 2022).

8.8.1 Odontocetes

Studies of the MAA in odontocetes began with the harbor porpoise (Andersen 1970b; Dudok van Heel 1962). These early studies exercised relatively little control over the test subject's orientation relative to the sound source as the animals were allowed to swim a "constrained" course while swimming toward a sound source. Estimates of the MAA from these studies ranged from 3° to 8° but varied depending on the frequency tested and on the method of threshold determination (i.e., 50% vs. 75% correct). Because of the uncontrolled nature of the studies, the MAA estimates are possibly too high (Nachtigall et al. 2000). To appropriately determine the MAA, the animal's head should be in a fixed orientation relative to the sound source. This prevents the animal from moving its head and potentially capitalizing on other cues. Nevertheless, using a similar paradigm with a free-swimming harbor porpoise, Kastelein et al. (2007) demonstrated the importance of the source signal characteristics in the localization process. The harbor porpoise improved its source localization as the duration (from 600 to 1000 ms) and the sound pressure level of the sound source increased. The porpoise located sound sources up to 124° to the left or right more easily than when the sound source was behind it.

Renaud and Popper (1975) trained a bottlenose dolphin to maintain a stable position relative to a buzzer (the reference point) placed in front of it by biting onto an acoustically transparent bar. Two movable sound transducers were placed equidistant from the dolphin with one on each side of the dolphin's midline. The dolphin was trained to station on the bar as sound was presented to it from one of the speakers and then select which side the sound came from by touching a response paddle. Localization thresholds were obtained for tonal sounds ranging from 6 to 100 kHz and clicks, like those produced by dolphins during echolocation. The bottlenose dolphin demonstrated relatively constant MAAs of 2°–3° for tonal sounds ranging from 10 to 80 kHz. The MAAs at the high (90 and 100 kHz) and low (6 kHz) end of the tested frequency range varied from 3.2° to 3.8° and were significantly greater than those measured from 10 to 80 kHz. The dolphin was subsequently trained to position itself so the reference point for the sounds was at azimuthal angles of 15°, 30°, 345°, or 330°. The study was then repeated but only for a test sound of 40 kHz. At azimuthal reference points of 15° and 345°, the MAA of the dolphin improved to 1.7° and 1.4°, respectively. At azimuthal reference points of 30° and 330°, the MAA of the dolphin worsened to 5.3° and 5.2°, respectively. The finding differed from humans and other terrestrial mammals that have the best MAAs when sounds are displaced relative to a 0° azimuth. The finding, however, makes sense with respect to what is known about dolphin sound production and reception. A 40 kHz signal is within the bandwidth of echolocation frequencies used by dolphins, and dolphins are known to have a forward directed receiving beam pattern. The decline in sensitivity with azimuth increasing/decreasing relative to 0° may contribute to angular resolution, at least to a point.

Renaud and Popper also trained the dolphin to perform the same procedure while stationing sideways on the bite bar. This allowed them to test the dolphin's MAA in the vertical plane. For the frequencies tested (30, 60, and 90 kHz), they found no difference from the MAAs to the same frequencies measured in the horizontal plane. It is unlikely that binaural cues were used to make this distinction as the inner ears would be in the same position relative to the sound source regardless of angular displacement. Later measurements of ABRs to clicks delivered to a dolphin via a jawphone and placed at different locations on the dolphin's head demonstrated increased responsiveness to sounds delivered to the ventral surface of the dolphin relative to those presented on the melon (Møhl et al. 1999). Similar findings are found in the measurements of the receiving beam pattern in the vertical plane (Au and Moore 1984). Thus, for a given stimulus level, the dolphin of Renaud and Popper potentially capitalized on pathway-dependent amplitude differences to make its determination of sound directionality.

Renaud and Popper also determined the MAA of the dolphin to clicks. The clicks were 35 μs in duration and had a center frequency of ~64 kHz. The MAAs to the clicks relative to the vertical and horizontal planes were 0.7° and 0.9°, respectively. The smaller MAA is not surprising given that the broader bandwidth of the clicks provides a greater amount of acoustic information to the dolphin.

Branstetter et al. (2003) investigated angular difference discrimination in the bottlenose dolphin. They performed a study in which a dolphin stationed in a hoop was required to echolocate on two arrays, one containing four rods (the S+, or positive stimulus) and one containing two rods (the S−, or negative stimulus). The dolphin was trained to echolocate on the arrays while wearing eyecups to prevent the dolphin from using visual cues. The dolphin then reported the side where the S+ array was located by touching an appropriate response paddle. The arrays were moved closer together over several trials covering angular separations of the arrays from 0.25° to 4°. Based on the dolphin's performance, Branstetter and colleagues calculated angular resolution for the dolphin of 1.6°, which is between the MAA values of 0.9° and 2.1° found by Renaud and Popper (1975) for clicks and tones, respectively, in the horizontal plane. Branstetter et al. (2007) repeated the echolocation task with a two-rod (S+) and single rod (S−) configuration in order to determine if performance improved when the number of echoes produced by each insonifying click was reduced. They found angular discrimination thresholds of 0.7° and 1.5°, depending on the spacing between rods. Branstetter et al. (2022) later examined true angular resolution by having two dolphins echoically determine the location (left or right) of an S+ target relative to a closely spaced S− target. Consistent with prior findings, the angular resolution of the best-performing dolphin was found to be 1.5°.

Moore et al. (1995) investigated the ILD and ITD resolving capabilities of the bottlenose dolphin to clicks with center frequencies ranging from 5 to 90 kHz. Clicks were presented as 2 s sequences with individual clicks separated in time by 40 ms. Clicks were presented through jawphones attached to the pan region (acoustic window) on each side of the lower jaw. Click trains were presented simultaneously, but with a time delay or difference in the amplitude presented between the two sides. The dolphin was trained to report which click train had higher amplitude or which started first by touching a response paddle associated with either the right or left jawphone. Calculated ITD thresholds ranged from 7.0 to 10.6 μs for clicks with center frequencies ranging from 5 to 30 kHz, with the lowest ITD threshold at 30 kHz. At 60, 80, and 90 kHz the ITD threshold increased to 17–18 μs. The ILD thresholds were found to be ≤1 dB and independent of the click center frequency. These values, though quite low, are on par with ITD and ILD thresholds in humans across the audible range of hearing. However, as noted by Branstetter et al. (2003), the thresholds seem too high to explain the superior angular discrimination acuity observed in their study. They suggested that mechanisms other than differences in arrival time and level (e.g., spectral cues) might be involved.

Earlier work by Popov and Supin (1990b) measured ABR thresholds with a sound source that varied in angular location about a bottlenose dolphin, an Amazon river dolphin, and a tucuxi. The results showed greatest sensitivity to broadband clicks when the clicks were presented directly in front of the animals. Subsequent work using band-limited clicks of different frequency and covering the range of dolphin hearing showed that the relationship between ABR threshold and azimuth became narrower with increasing frequency (Popov et al. 1992). Supin and Popov (1993) subsequently used subcutaneous electrodes to record the amplitude of the auditory nerve response (ANR) of each ear to estimate monaural receiving patterns and calculate the ILD between ears as a function of sound source azimuth in an Amazon river dolphin. They concluded that monaural azimuthal sensitivity at most frequencies was greatest at ~5–10° for sources presented ipsilateral to the ear (to the same side) from which the ANR was recorded.

This was similar to electrophysiological findings in the harbor porpoise, which suggested greatest receiving sensitivity was 15–30° for sources presented ipsilateral to the site of recording (Voronov and Stosman 1983). Supin and Popov (1993) used differences in ANR amplitude as a means of assessing ILD and concluded that ILD resolution was greatest when a sound source was presented at an azimuth of ~15° for frequencies above 55 kHz. At the most sensitive azimuths, the ILDs corresponded to differences in ANR amplitude between the two ears by as much as 25 dB. Mulsow et al. (2014a) measured the ANR at both ears of a dolphin simultaneously in response to 25 μs clicks presented via a jawphone placed over the pan region. From the latencies and amplitudes of the ANRs, ITDs and the attenuation of the ANR due to propagation through the dolphin's interaural tissues were calculated. Two dolphins were used to make this assessment and the ITD was calculated as 70 μs and 118 μs, with the larger dolphin having the greater ITD. The measured ITD corresponded well to a simple propagation model through the dolphin's head using a computed tomographic (CT) scan of one of the dolphin subjects. Differences in ANR amplitude also showed a reduction of ~20 dB between the two ears, an interaural level difference that was consistent with prior ANR studies in the dolphin.

8.8.2 Pinnipeds

The MAA in pinnipeds has been tested in both air and water (Table 8.3). The earliest studies of sound source localization in pinnipeds were performed on the harbor seal (Møhl 1964, 1968a). Terhune (1974) later found that a harbor seal was better at localizing broadband sources and 1/3-octave band sources (center frequencies >3 kHz) in air than it was at localizing low frequency and tonal sources. Conducting studies in a hemi-anechoic chamber, Holt et al. (2004) confirmed the harbor seal's aerial MAA to clicks to be similar to previous findings. Subsequent work with pure tones demonstrated that the harbor seal had relatively small aerial MAAs for

Table 8.3 Minimum audible angles for three species of pinniped collected in air and under water

Species	Stimulus	References	Frequency (kHz)									Clicks
			0.2	0.5	0.8	1	2	4	6	8	16	20	
			In air										
Harbor seal	Clicks, tones	Holt et al. (2004, 2005)			4.3	4.1	5.4	7.9		12.7	3.8	6.9	3.6
Harbor seal	Tones	Møhl (1964)		4.8									
Harbor seal	Clicks	Terhune (1974)											~3 and ~2[a]
	1/3 octave band noise	Terhune (1974)				–		6.6	0.7	1.7	1.2		
	Tones	Terhune (1974)				4.6	–	–		5.5			
Northern elephant seal	Clicks, tones	Holt et al. (2004, 2005)				3.3	4.8	3.7	3.9	15	14.2		4.7
California sea lion	Clicks, tones	Holt et al. (2004, 2005)			4.7	6	11.9	8		11.1	3.9	8.7	4.2
			Under water										
Harbor seal	Tones	Bodson et al. (2007)[b]	3.2	3.8		4.7	10.3	9.9	9.3	13.2	15.5		
Harbor seal	Tones	Møhl (1968b)					3.1						~4.5
Harbor seal	Clicks	Terhune (1974)											9
California sea lion	Clicks	Moore (1975)											9
California sea lion	Tones (in pulse sequences)	Moore and Au (1975)		12		4	[c]		42.4		13.5	18	

Studies reported were all conducted with the head of the subject fixed in one location
[a]Results are from two experiments with different methodologies
[b]Reported MAA's are the average from two subjects
[c]Indicates that a MAA could not be determined, although the frequency was tested

frequencies ≤2 kHz and ≥8 kHz (Holt et al. 2005), but that MAAs were larger at 4 kHz and 8 kHz (Table 8.3). The results supported the duplex theory of sound localization—that sounds are predominantly localized through ITDs at low frequencies and ILDs at high frequencies (Strutt 1909). Bodson et al. (2006) found that the underwater MAA of a harbor seal was 9.8° for a 6 kHz tone at 0° azimuth and 9.7° at 180° azimuth. However, the ability to localize the same sound source improved significantly if the animal was allowed to freely swim; the seal was able to determine the location of the source within 2.8° of the actual source location. Subsequent work with two harbor seals determined the dependence of underwater MAAs on frequency (Bodson et al. 2007). MAAs were smallest at the lowest frequencies tested and increased with increasing frequency (Table 8.3). By comparison, the work of Holt et al. (2005), pointed to a decreased importance of ILDs at high frequencies under water relative to sounds received in air.

Holt et al. (2004, 2005) similarly studied the aerial MAA of the elephant seal to clicks and tones (Table 8.3). The aerial MAA for pure tones was like that of the harbor seal at the lower test frequencies but became progressively worse at the higher frequencies. The worsening of MAAs at the high frequencies suggests that the elephant seal has a poorer ability to capitalize on ILDs in air than other tested pinnipeds.

Holt et al. (2004, 2005) also studied the aerial MAA of the California sea lion to clicks and tones (Table 8.3). The MAAs associated with pure tones presented in air followed a pattern somewhat like that of the harbor seal, although more variable within the mid-frequency range of tones (Holt et al. 2005). At frequencies ≥16 kHz, the MAA for in-air tones improved. Moore (1975) found the underwater MAA of the sea lion to be 9°, and using a series of pulsed tones (0.25–4.0 kHz), he determined that the ability of the sea lion to localize sound cues declined sharply above 1 kHz. Moore and Au (1975) extended this work by characterizing the underwater MAAs of the sea lion to pulsed tones ranging in frequency from 0.5 to 16 kHz in 1-octave steps. The smallest MAA of 4° was found to occur at 1 kHz. At 2 kHz and 4 kHz, the sea lion had great difficulty in localizing the sound source, but improved again at frequencies of 8 kHz and 16 kHz. By comparison, Gentry (1967) found that for a given set of source azimuths, a sea lion performed better at localizing 6 kHz tones than 3.5 kHz tones. With the sound source orientation set at an azimuth of 25°, the percentage of correct source localizations achieved by the sea lion increased with frequency (1.5–6.5 kHz), except at 2.5 kHz, where the performance was no better than chance. Collectively, the results from sea lion studies of MAA are more variable and firm conclusions more difficult to make. However, the evidence suggests that sea lions have difficulty localizing signals within the approximate range of 2–4 kHz. As with harbor seals, their ability to localize sound is also poorer than that observed in dolphins, particularly when considering the sound localization acuity of the dolphin within the range of echolocation frequencies.

8.8.3 Sirenians

Mann et al. (2005) studied underwater sound localization in two manatees. Tests used three broadband stimuli (0.2–2 kHz, 0.2–20 kHz, and 6–20 kHz) and two tonal signals at 14 kHz and 16 kHz (Colbert-Luke et al. 2015). The manatees were trained to station within an array of underwater speakers arranged at angular placements relative to the animal's midline of 270°, 315°, 45°, and 90°. Upon listening to the playback of the test signal, the manatee swam to and touched the appropriate speaker. One manatee was found to localize broadband stimuli fairly well (>87%), regardless of signal duration or signal bandwidth. The other manatee had lower performance and performed better with longer signal durations and those with the greatest bandwidths. Neither animal performed particularly well with tonal signals. Both animals showed low accuracy and behaviors consistent with frustration when shorter duration tones were used. Collectively, the available data suggest that sound localization in manatees is much less refined than in the odontocetes, but should be sufficient for anthropogenic broadband signal localization likely to be encountered in their environment (such as boat noise).

8.9 Auditory Weighting Functions

Auditory weighting functions are mathematical functions applied to sound to account for frequency-specific changes in sensitivity within the auditory system. Weighting functions emphasize frequencies where animals hear well and de-emphasize frequencies where they do not hear well. Weighting is achieved by adding the auditory weighting function value to each frequency of the noise spectral amplitude of the sound to be weighted. Weighting is done in dB and the resultant weighted noise spectral density is converted to linear units, integrated across frequency, and subsequently converted back into logarithmic units. The result is a weighted sound level that represents the impact of frequency-specific hearing sensitivity across the range of hearing. In regulatory practice, the weighted sound level is compared to a weighted exposure threshold (e.g., SPL or SEL) above which an adverse impact is assumed.

In the United States, weighting functions have become part of the process by which the potential for NIHL caused by anthropogenic sound is estimated (National Marine Fisheries Service 2016, 2018). The focus on NIHL is partly based on the historical development of weighting functions in relation to an emerging understanding of the potential impact of sound on marine mammals. In the mid-1990s, as awareness of noise impacts on marine mammals was increasing, impacts (from a regulatory perspective) were primarily based on the potential for anthropogenic sound to cause either temporary or permanent threshold shifts (i.e., changes in hearing sensitivity; TTS and PTS, respectively—see Chap. 9). A substantial amount of research since then has investigated the potential for sound to cause NIHL, including the collection of data to inform weighting function development. This differs from the history and development of weighting functions in humans, which has its roots in the telecommunications industry. Readers are recommended to read Houser et al. (2017) for a review of the similarities and differences in development and application of auditory weighting functions in humans and marine mammals. A brief synopsis is provided here.

Research contributing to the development of auditory weighting functions for humans began at Bell Laboratories in the early part of the twentieth century (Gertner 2012; Yost 2015). The goal of research at that time was to improve human speech intelligibility across telephone lines. Early work focused on the loudness of tones. Loudness is the "perception" or "auditory sensation" related to a signal's amplitude, which progresses from quiet (lower signal strength) to loud (higher signal strength). Fletcher and Munson (1933) found that the loudness of tones as a function of tone level was frequency dependent. By having listeners compare the loudness of tones to a standard tone at 1 kHz, they developed equal-loudness functions for humans. A characteristic of the functions was that changes in perceived loudness for low- and high-frequency signals changed more rapidly than did mid-frequency signals closer to the standard tone. The "phon," a metric used as the measure of loudness, was based on the equal-loudness contours.

Since the work of Fletcher and Munson, several studies have refined the human equal-loudness contours (e.g., Robinson and Dadson 1956). These contours were subsequently used in the creation of weighting functions: A-, B-, C-, and D-weighting. A-weighting and C-weighting are the curves typically encountered. The use of A-weighting has achieved international acceptance and broad use in the estimation of human noise impacts, including NIHL, annoyance, etc. (Houser et al. 2017). The A-weighting function (based on the 40-phon equal-loudness contour) is most similar to the human audiogram, whereas the C-weighting function (based on the 100-phon equal-loudness contour) is flatter across most of the hearing range. The flatter 100-phon equal-loudness contour likely results from the widening of auditory filters at higher sound exposures; thus, C-weighting is applicable to situations where humans may be exposed to high-amplitude sounds, such as sonic booms or explosions. In practice, C-weighting is seldom used.

Compared to humans, the data available for the development of auditory weighting functions in marine mammals is severely limited. The problem is exacerbated by the fact that there are well over one hundred species of marine mammal and it is unlikely that data relevant to weighting function development will be available for most. It has been proposed that a simple starting point for the development of marine mammal weighting functions would be to weight a sound exposure by a marine mammal's audiogram (Nedwell and Turnpenny 1998; Nedwell et al. 2007). This is, in effect, a form of weighting by sensation level that does not account for the scaling of perception according to noise level (i.e., loudness and sensation level are not equivalent). This approach has been applied in several situations, including the prediction of low-frequency noise impacts for harbor porpoises (Terhune 2013), and has been advocated for broader use (Tougaard et al. 2015).

The first major use of a marine mammal weighting function for federal regulatory purposes was in noise impact predictions made as part of two U.S. Navy ship-shock trials (Department of the Navy 1998; Department of the Navy 2001). The ship-shock trials tested the integrity of new vessels by detonating 10,000-pound explosive charges at varying proximities and orientations to the vessel. For purposes of predicting injury to marine mammals in the exposure area, only frequencies of sound ≥100 Hz for odontocetes and ≥10 Hz for mysticetes were considered. The result was a "brick-wall" function in which all sound below the frequency limit was discounted and all sound above the frequency limit was considered equally hazardous. The output of the weighted sound (essentially an application of a rectangular filter) was then compared to weighted thresholds for TTS based on TTS studies previously conducted on the bottlenose dolphin (Ridgway et al. 1997). A second threshold for impact, a peak unweighted SPL, was also applied such that exceedance of either impact threshold resulted in a predicted TTS (i.e., there was a dual criterion for impact). No weighting or thresholds were established for pinnipeds or sirenians as none were expected to be in the area of impact where the shock-trials occurred.

Southall et al. (2007) proposed the first set of marine mammal auditory weighting functions for broad use (i.e., with respect to species and to the types of sound filtered). The functions, collectively referred to as the "M-weighting functions," were based largely on human C-weighting. Parameter values of the C-weighting function were modified so as to better fit the known or predicted range of hearing in marine mammals. The form of the function was:

$$W(f) = k + 20 \log_{10} \left[\frac{b^2 f^2}{(a^2 + f^2)(b^2 + f^2)} \right]$$

where f is frequency (Hz), $W(f)$ is the amplitude (dB) of the weighting function as a function of frequency, k is a constant that normalizes the equation at a particular frequency, and a and b are constants related to the upper and lower limits of hearing, respectively. Five simple weighting functions were created using this approach; these included a function for the low-frequency (LF) cetaceans (mysticetes), one for the mid-frequency (MF) cetaceans (e.g., delphinids and beaked whales), one for the high-frequency (HF) cetaceans (e.g., porpoises), and two for the pinnipeds (one for in-air and one for underwater exposures).

Along with the proposed weighting functions, Southall et al. (2007) proposed new weighted thresholds for TTS based on measured TTS thresholds and TTS growth functions. Data underlying these thresholds were obtained in studies on dolphins, belugas, and pinnipeds (Finneran et al. 2002c; Kastak et al. 1999, 2004, 2005b; Schlundt et al. 2000). Thresholds for the onset of PTS were determined as any exposure that resulted in TTS ≥40 dB. For both pulsed and non-pulsed sounds, the dual-criteria approach was applied—one criterion based on the unweighted peak SPL of the exposure, and one based on the M-weighted SEL.

The M-weighting functions reflected the broadly flattened nature of the C-weighting function upon which they were based. This was believed to be a conservative approach to weighting as the contributions of sound were only discounted at the tail ends of the frequency range of hearing assumed for the species groups. However, the application of M-weighting to the prediction of TTS was based on TTS-thresholds derived from exposures that were predominantly made at frequencies ≤10 kHz. As more TTS studies were conducted and it became apparent that TTS was a frequency-dependent phenomenon, it also became apparent that the assumption about M-weighting conservativeness was incorrect because it did not reflect frequency-dependent differences in the vulnerability to NIHL.

Finneran and Schlundt (2011) trained a bottlenose dolphin to perform a loudness comparison test between two sequentially presented tones. Most comparisons were performed between tones of similar (1/2-octave spacing) or the same frequency and were presented at large differences in SPL (10–30 dB) so that the loudness relationship was known. Approximately 30% of the trials were used as "probes." In these trials, one of the tones was a 10 kHz reference presented at one of three different SPLs (90, 105, and 115 dB), and the other was a tone of different frequency and level to which the loudness relationship was unknown. The data acquired through this study were used to create dolphin equal-loudness functions, similar to what had been done in humans. Three different equal-loudness functions were determined based upon the level of the reference tone. The equal-loudness functions were subsequently used in the creation of a new auditory weighting function. The function was more similar to the design of the human weighting functions because of the underlying loudness relationships.

The equal-loudness functions measured by Finneran and Schlundt (2011) for the bottlenose dolphin are unique among animals and required considerable time and resources to collect. Because the study was complex and demanding, alternative strategies at investigating loudness relationships were sought. Some evidence suggests that reaction time is related to loudness in humans and that equal-latency functions might have utility as surrogates for equal-loudness functions (e.g., May et al. 2009; Pfingst et al. 1975). To develop equal-latency functions, an animal's reaction time in a signal detection task is related to the level of the test signal. In theory, the louder the signal, the shorter the reaction time required to report the detection.

Equal-latency studies have been completed in the harbor seal (Kastelein et al. 2011; Reichmuth 2013), harbor porpoise (Wensveen et al. 2014), California sea lion (Mulsow et al. 2015; Reichmuth 2013), and bottlenose dolphins (Mulsow et al. 2015). The studies varied in many ways, but all derived equal-latency functions from reaction time and sound pressure data. Because of a limited dynamic range of reaction times (i.e., beyond some stimulus level there is no improvement in reaction time), Reichmuth (2013) and Mulsow et al. (2015) reasoned that applying the equal-latency functions to the creation of auditory weighting functions would be limited to lower level sound exposures. However, Wensveen et al. (2014), who created equal-loudness contours for reaction times ranging from 150 to 200 ms, subsequently developed a family of harbor porpoise weighting functions through a process of smoothing, inversion and normalization. The functions were effectively flat above 10 kHz, but declined in amplitude at rates of 10–16 dB/octave at lower frequencies (depending on the reaction time).

As part of the U.S. Navy's Tactical Theater Training Assessment and Planning (TAP) program, a new set of auditory weighting functions were proposed that integrated aspects of M-weighting with the weighting functions based on dolphin equal-loudness functions (Finneran and Jenkins 2012; Finneran and Schlundt 2011). Two types of weighting functions were proposed: Type I, which was used for auditory effects in non-cetaceans and for behavioral effects from exposure to non-impulsive sounds in all marine mammals (except sensitive species such as harbor porpoises or beaked whales); and Type II, which was developed for the prediction of auditory effects in cetaceans. The Type I functions were similar to the M-weighting functions proposed by Southall et al. (2007) but the Type II function modified the M-weighting function to account for the increased susceptibility to TTS at specific frequencies, which was better approximated by the dolphin equal-loudness functions (Finneran and Schlundt 2010, 2011). Because frequency-specific TTS data only existed for bottlenose dolphins at the time of its development, the use of the Type II function was constrained to use on cetaceans. The Type II function was mathematically defined as

$$W_{II}(f) = \text{maximum}\{G_1(f), G_2(f)\}$$

where $W_{II}(f)$ is the weighting function amplitude (dB) at frequency f (Hz), and $G_1(f)$ and $G_2(f)$ are the components dictated by the M-weighting function and the equal-loudness weighting function, respectively.

$$G_1(f) = k_1 + 20\log_{10}\left[\frac{b_1^2 f^2}{(a_1^2+f^2)(b_1^2+f^2)}\right]$$

$$G_2(f) = k_2 + 20\log_{10}\left[\frac{b_2^2 f^2}{(a_2^2+f^2)(b_2^2+f^2)}\right]$$

The parameters k, a, and b are as previously defined for the M-weighting function.

For regulatory purposes, weighted sound levels from the Type I and Type II weighting functions were compared to updated TTS thresholds. The new weighted thresholds were based on TTS studies to impulsive and non-impulsive sound in a number of species, including the bottlenose dolphin (Schlundt et al. 2000), beluga (Finneran et al. 2002c; Schlundt et al. 2000), harbor porpoise (Lucke et al. 2009), California sea lion (Kastak et al. 2005b, 2007), and harbor seal (Kastak et al. 2004, 2005b).

Marine mammal weighting functions were modified once again in 2016 and 2018 (National Marine Fisheries Service 2016, 2018). NMFS proposed the new weighting functions to account for recent data on marine mammal TTS (e.g., Finneran and Schlundt 2013; Finneran et al. 2015; Kastelein et al. 2012a, b, 2013, 2014a, b, 2015a, b; Popov et al. 2013, 2014, 2015), hearing sensitivity (e.g., Ghoul and Reichmuth 2014a; Sills et al. 2014, 2015), masking studies, and equal-latency contours (e.g., Mulsow et al. 2015; Wensveen et al. 2014). The process also considered input from the public, an expert peer-review panel, and other governmental agencies. The updated weighting functions were restricted to underwater exposures and for application to animals for which NMFS had oversight (i.e., they did not apply the functions to species under the authority of the U.S. Fish and Wildlife Service, although some of those species were included in the weighting function development). Species groupings consisted of low-frequency cetaceans, high-frequency cetaceans, sirenians, otariids, and phocids. For each of these groups a species-representative audiogram, or normalized composite audiogram, was created to inform the shape of the auditory weighting function.

The mathematical form of the weighting function was

$$W(f) = C + 10\log_{10}\left\{\frac{f/f_1^{2a}}{\left[1+(f/f_1)^2\right]^a\left[1+(f/f_2)^2\right]^b}\right\}$$

As noted earlier in the chapter, Southall et al. (2019b) proposed updating marine mammal hearing groups and derivation of new categories based on scientific findings related to species-typical vocalizations, echolocation characteristics (for odontocetes), auditory system anatomy, masking studies, and more recently acquired audiograms. The groups included the low-frequency (LF) cetaceans, high-frequency (HF) cetaceans, very high-frequency (VHF) cetaceans, sirenians (SI), phocid carnivores in water (PCW), phocid carnivores in air (PCA), other marine carnivores in water (OCW), and other marine carnivores in air (OCA). The changes necessitated the creation of new auditory weighting functions but utilized the same functional form and process implemented by NMFS. The functions were based on audiometric data and data on TTS, when available. For groups that had no data supporting the form of the weighting function, e.g., mysticete whales, extrapolations from other groups were made. The marine mammal weighting functions proposed by Southall et al. (2019b) are shown in Fig. 8.35 and the parameters for the different group functions provided in Table 8.4. Associated weighted thresholds for the onset of TTS and PTS are shown in Table 8.5.

Auditory weighting functions are one means by which to estimate the impact of noise on marine mammals by taking into account measures of an animal's hearing ability. The approach does, however, have drawbacks. For example, there are no empirically determined audiograms for mysticetes upon which to base auditory weighting functions, so the relationship between the

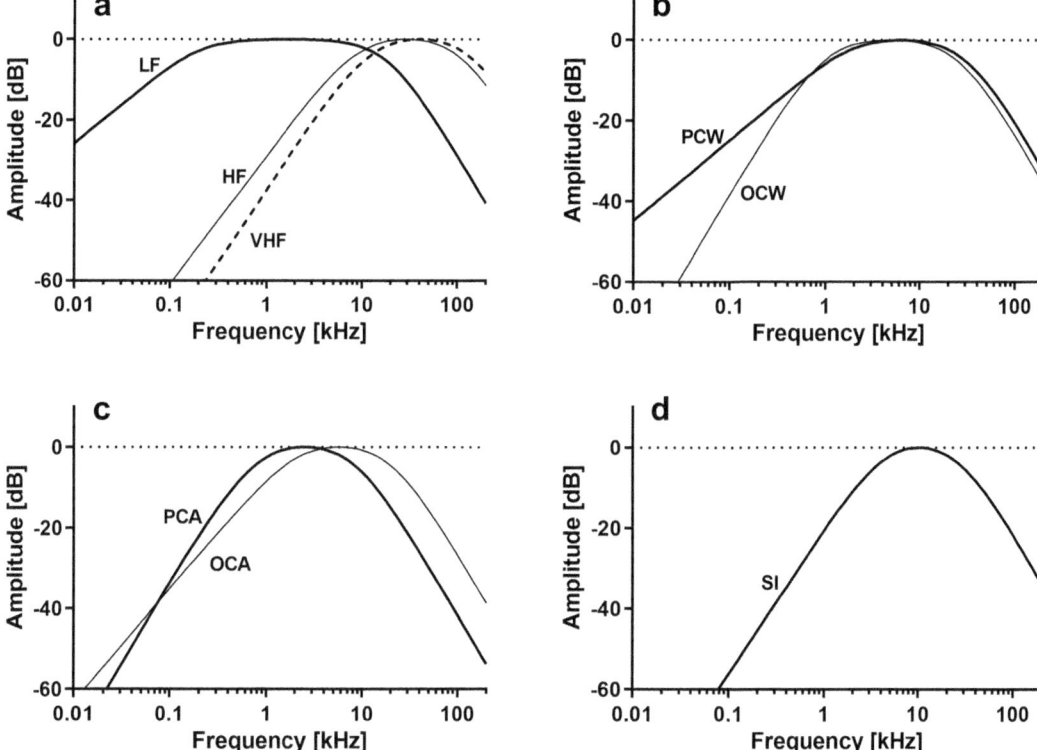

Fig. 8.35 Auditory weighting functions as presented by Southall et al. (2019b) (**a**) Weighting functions for the cetaceans; low-frequency cetaceans (LF, thick line), high-frequency cetaceans (HF, thin line), and very high-frequency cetaceans (VHF, dashed line). (**b**) Weighting functions for the pinnipeds and other carnivores in water; otariids and other carnivores (OCW, thin line), and phocids (PCW, thick line). (**c**) Weighting functions for the pinnipeds and other carnivores in air; otariids and other carnivores (OCA, thin line), and phocids (PCA, thick line). (**d**) Weighting function for the sirenians

Table 8.4 Weighting function parameters from Southall et al. (2019b)

Group	a	b	f_1 (kHz)	f_2 (kHz)	C
LF	1	2	0.2	19	0.13
HF	1.6	2	8.8	110	1.2
VHF	1.8	2	12	140	1.36
SI	1.8	2	4.3	25	2.62
PCW	1	2	1.9	30	0.75
OCW	2	2	0.94	25	0.64
PCA	2	2	0.75	8.3	1.5
OCA	1.4	2	2	20	1.39

proposed functions and baleen whale hearing is uncertain. The weighting functions also have largely discounted AEP hearing data, which account for an increasing proportion of hearing-related research in marine mammals. This has occurred because the relationships between AEP and behavioral measures of hearing are not well-defined and there is considerable variability in AEP threshold data (see Sects. 8.2.2.2 and 8.3.2). Nevertheless, auditory weighting functions have moved into the regulatory world and will likely continue to be used in predicting

Table 8.5 TTS and PTS-onset thresholds for impulsive and nonimpulsive sound exposure types

Group	Nonimpulsive		Impulsive			
	TTS onset: SEL (weighted)	PTS onset: SEL threshold (weighted)	TTS onset: SEL (weighted)	PTS onset: SEL (weighted)	TTS onset: Peak SPL (unweighted)	PTS onset: Peak SPL (unweighted)
LF	179	199	168	183	213	219
HF	178	198	170	185	224	230
VHF	153	173	140	155	196	202
SI	186	206	175	190	220	226
PCW	181	201	170	185	212	218
OCW	199	219	188	203	226	232
PCA	134	154	123	138	138	144
OCA	157	177	146	161	161	167

Thresholds for underwater exposures are either in SEL (dB re 1 $\mu Pa^2 s$) or peak SPL (dB re 1 μPa). Thresholds for in-air exposures are either SEL (dB re (20 $\mu Pa)^2 s$) or peak pressure (dB re 20 μPa). Thresholds are used in conjunction with the auditory weighting functions proposed by Southall et al. (2019b)

noise impacts to marine mammals. Whether sound weighting will expand in regular use beyond predictions of NIHL in marine mammals has yet to be determined; for example, weighting functions have been proposed for the prediction of behavioral responses (Kastelein et al. 2022). Even in the case of NIHL, continued use might depend on the performance of weighting functions at predicting TTS in marine mammals prior to the collection of TTS data (i.e., their prediction accuracy needs to be assessed). It is likely that these functions will continue to be updated as new information about marine mammal hearing and the impact of noise on marine mammal hearing is acquired. A significant amount of literature on TTS in marine mammals has been published since 2019, and will likely be an integral part of any future recommendation updates.

8.10 Summary

This chapter provided an overview of the hearing abilities of marine mammals. Audiograms were presented for cetaceans in water and for marine carnivores in both air and water. Audiograms collected with behavioral methods and electrophysiological methods were presented and differences in the audiograms due to the methods used were discussed. Additional concepts specifically relevant to hearing in noise were introduced (e.g., critical ratio, critical bandwidth), as were adaptations in marine mammals for reducing masking and facilitating sound detection and sound source localization in noise. Despite steady progress, our understanding of marine mammal hearing remains limited; the majority of marine mammal species remain untested, and for species that have been tested, measurements have been made on only one or a few individuals and often with limited stimulus conditions. Direct measures of baleen whale hearing are absent and will likely have to rely on electrophysiological methods and chance opportunities (e.g., strandings). Low-frequency hearing measurements are difficult for both behavioral and electrophysiological approaches, yet it is at low frequencies where the greatest impacts of ocean noise might occur. Predictions of the impacts of sound on marine mammals have been improved through the use of auditory weighting functions, but the applicability of weighting functions beyond the prediction of NIHL in marine mammals remains to be seen. Continued research into basic hearing abilities and physiology are required to better understand, model, predict, and assess the potential impacts of noise on marine mammals.

References

Accomando AW, Mulsow J, Branstetter BK, Schlundt CE, Finneran JJ (2020) Directional hearing sensitivity for

2-30 kHz sounds in the bottlenose dolphin, *Tursiops truncatus*. J Acoust Soc Am 147(1):388–398. https://doi.org/10.1121/10.0000557

Alderson AM, Diamantopoulos E, Downman CBB (1960) Auditory cortex of the seal (*Phoca vitulina*). J Anat 94: 506–511

American National Standards Institute (ANSI) (2018) Procedure for determining audiograms in toothed whales through evoked potential methods, Accredited Standards Committee S3/SC 1, Animal Bioacoustics, vol ANSI/ASA S3/SC1.6–2018. Acoustical Society of America, New York

Andersen S (1970a) Auditory sensitivity of the harbor porpoise, *Phocoena phocoena*. In: Pilleri G (ed) Investigations on cetacea, vol 2. Hirnanatomisches Institut, Berne, pp 255–259

Andersen S (1970b) Directional hearing in the harbour porpoise *Phocoena phocoena*. In: Pilleri G (ed) Investigations on cetacea, vol II. Hirnanatomisches Institute, Berne, pp 260–263

Andre M, Supin A, Delory E, Kamminga C, Degollada E, Alonso JM (2003) Evidence of deafness in a striped dolphin, *Stenella coeruleoalba*. Aquat Mamm 29(1): 3–8

Ansmann IC, Goold JC, Evans PGH, Keith SG, Simmonds M (2007) Variation in the whistle characteristics of short-beaked common dolphins, *Delphinus delphis*, at two locations around the British Isles. J Mar Biol Assoc UK 87(1):19–26. https://doi.org/10.1017/S0025315407054963

Au WWL (1990) Target detection in noise by echolocating dolphins. In: Thomas JA, Kastelein RA (eds) Sensory abilities of cetaceans: laboratory and field evidence. Plenum Press, New York, pp 203–216. https://doi.org/10.1007/978-1-4899-0858-2_12

Au WWL, Moore PWB (1984) Receiving beam patterns and directivity indices of the Atlantic bottlenosed dolphin (*Tursiops truncatus*). J Acoust Soc Am 75(1): 255–262. https://doi.org/10.1121/1.390403

Au WWL, Moore PWB (1990) Critical ratio and critical bandwidth for the Atlantic bottlenose dolphin. J Acoust Soc Am 88(3):1635–1638. https://doi.org/10.1121/1.400323

Au WWL, Floyd RW, Penner RH, Murchison AE (1974) Measurement of echolocation signals of the Atlantic bottlenose dolphin, *Tursiops truncatus* Montagu, in open waters. J Acoust Soc Am 56(4):1280–1290. https://doi.org/10.1121/1.1903419

Au WWL, Carder DA, Penner RH, Scronce BL (1985) Demonstration of adaptation in beluga whale echolocation signals. J Acoust Soc Am 77(2):726–730. https://doi.org/10.1121/1.392341

Au WWL, Moore PWB, Pawloski D (1986) Echolocation transmitting beam of the Atlantic bottlenose dolphin. J Acoust Soc Am 80(2):688–694. https://doi.org/10.1121/1.394012

Au WWL, Moore PWB, Pawloski DA (1988) Detection of complex echoes in noise by an echolocating dolphin. J Acoust Soc Am 83(2):662–668. https://doi.org/10.1121/1.396161

Au WWL, Pawloski JL, Cranford TW, Gisiner RC, Nachtigall PE (1993) Transmission beam pattern of a false killer whale. J Acoust Soc Am 93(4):2358. https://doi.org/10.1121/1.406207

Au WWL, Pawloski J, Nachtigall PE, Blonz M, Gisner RC (1995) Echolocation signals and transmission beam pattern of a false killer whale (*Pseudorca crassidens*). J Acoust Soc Am 98(1):51–59. https://doi.org/10.1121/1.413643

Au WWL, Nachtigall PE, Pawloski JL (1997) Acoustic effects of the ATOC signal (75 Hz, 195 dB) on dolphins and whales. J Acoust Soc Am 101(5): 2973–2977. https://doi.org/10.1121/1.419304

Au WWL, Kastelein RA, Rippe T, Schooneman NM (1999) Transmission beam pattern and echolocation signals of a harbor porpoise (*Phocoena phocoena*). J Acoust Soc Am 106(6):3699–3705. https://doi.org/10.1121/1.428221

Au WWL, Branstetter BK, Benoit-Bird KJ, Kastelein RA (2009) Acoustic basis for fish prey discrimination by echolocating dolphins and porpoises. J Acoust Soc Am 126(1):460–467. https://doi.org/10.1121/1.3147497

Au WWL, Branstetter B, Moore PW, Finneran JJ (2012) The biosonar field around an Atlantic bottlenose dolphin (*Tursiops truncatus*). J Acoust Soc Am 131(1): 569–576. https://doi.org/10.1121/1.3662077

Awbrey FT, Thomas JA, Kastelein RA (1988) Low-frequency underwater hearing sensitivity in belugas, *Delphinapterus leucas*. J Acoust Soc Am 84(6):2273–2275. https://doi.org/10.1121/1.397022

Babushina ES (1997) Audiograms of the Caspian seal under water and in air. Sens Syst 11(2):67–71

Babushina YS, Zaslavskii GL, Yurkevich LI (1991) Air and underwater hearing characteristics of the northern fur seal: audiograms, frequency and differential thresholds. Biophysics 36(5):909–913

Bain DE, Dahlheim ME (1994) Effects of masking noise on detection thresholds of killer whales. In: Loughlin TR (ed) Marine mammals and the Exxon Valdez. Academic, San Diego, pp 243–256. https://doi.org/10.1016/b978-0-12-456160-1.50021-7

Bibikov NG (1992) Auditory brainstem responses in the Harbor Porpoise (*Phocoena phocoena*). In: Thomas JA, Kastelein RA, Supin AY (eds) Marine mammal sensory systems. Plenum Press, New York, pp 197–211. https://doi.org/10.1007/978-1-4615-3406-8_11

Bodkin JL, Monson DH, Esslinger GG (2007) Activity budgets derived from time–depth recorders in a diving mammal. J Wildl Manag 71(6):2034–2044. https://doi.org/10.2193/2006-258

Bodson A, Miersch L, Mauck B, Dehnhardt G (2006) Underwater auditory localization by a swimming harbor seal (*Phoca vitulina*). J Acoust Soc Am 120(3): 1550–1557. https://doi.org/10.1121/1.2221532

Bodson A, Miersch L, Dehnhardt G (2007) Underwater localization of pure tones by harbor seals (*Phoca vitulina*). J Acoust Soc Am 122(4):2263–2269. https://doi.org/10.1121/1.2775424

Branstetter BK, Finneran JJ (2008) Comodulation masking release in bottlenose dolphins (*Tursiops truncatus*). J Acoust Soc Am 124(1):625–633. https://doi.org/10.1121/1.2918545

Branstetter B, Mercado E III (2006) Sound localization by cetaceans. Int J Comp Psychol 19:26–61. https://doi.org/10.46867/ijcp.2006.19.01.05

Branstetter BK, Sills JM (2022) Mechanisms of auditory masking in marine mammals. Anim Cogn 25:1029–1047. https://doi.org/10.1007/s10071-022-01671-z

Branstetter BK, Mevissen SJ, Herman LM, Pack AA, Roberts SP (2003) Horizontal angular discrimination by an echolocating bottlenose dolphin *Tursiops truncatus*. Bioacoustics 14:15–34. https://doi.org/10.1080/09524622.2003.9753510

Branstetter B, Mevissen SJ, Pack AA, Herman LM, Roberts SR, Carsrud LK (2007) Dolphin (*Tursiops truncatus*) echoic angular discrimination: effects of object separation and complexity. J Acoust Soc Am 121(1):626–635. https://doi.org/10.1121/1.2400664

Branstetter BK, Trickey JS, Aihara H, Finneran JJ, Liberman TR (2013a) Time and frequency metrics related to auditory masking of a 10kHz tone in bottlenose dolphins (*Tursiops truncatus*). J Acoust Soc Am 134(6):4556–4565. https://doi.org/10.1121/1.4824680

Branstetter BK, Trickey JS, Bahktiari K, Black A, Aihara H, Finneran JJ (2013b) Auditory masking patterns in bottlenose dolphins (*Tursiops truncatus*) with natural, anthropogenic, and controlled noise. J Acoust Soc Am 133(3):1811–1818. https://doi.org/10.1121/1.4789939

Branstetter BK, Bakhtiari K, Black A, Trickey JS, Finneran JJ, Aihara H (2016) Energetic and informational masking of complex sounds by a bottlenose dolphin (*Tursiops truncatus*). J Acoust Soc Am 140(3):1904–1917. https://doi.org/10.1121/1.4962530

Branstetter BK, St. Leger J, Acton D, Stewart J, Houser D, Finneran JJ, Jenkins K (2017a) Killer whale (*Orcinus orca*) behavioral audiograms. J Acoust Soc Am 141(4):2387–2398. https://doi.org/10.1121/1.4979116

Branstetter BK, Van Alstyne KR, Wu TA, Simmons RA, Curtis LD, Xitco MJ Jr (2017b) Composite critical ratio functions for odontocete cetaceans (L). J Acoust Soc Am 142(4):1897–1900. https://doi.org/10.1121/1.5006186

Branstetter BK, Felice M, Robeck T (2021) Auditory masking in killer whales (*Orcinus orca*): critical ratios for tonal signals in Gaussian noise. J Acoust Soc Am 149(3):2109–2115. https://doi.org/10.1121/10.0003923

Branstetter BK, Brientenstein R, Goya G, Tormey T, Wu T, Finneran JJ (2022) Spatial acuity of the bottlenose dolphin (*Tursiops truncatus*) biosonar system with a bat and human comparison. J Acoust Soc Am 151(6):3847–3857. https://doi.org/10.1121/10.0011676

Brill RL (1988) The jaw-hearing dolphin: preliminary behavioral and acoustical evidence. In: Nachtigall PE, Moore PWB (eds) Animal sonar processes and performance. Plenum Press, New York, pp 281–287. https://doi.org/10.1007/978-1-4684-7493-0_28

Brill RL, Harder PJ (1991) The effects of attenuating returning echolocation signals at the lower jaw of a dolphin (*Tursiops truncatus*). J Acoust Soc Am 89(6):2851–2857. https://doi.org/10.1121/1.400723

Brill RL, Moore PWB, Dankiewicz LA (2001) Assessment of dolphin (*Tursiops truncatus*) auditory sensitivity and hearing loss using jawphones. J Acoust Soc Am 109(4):1717–1722. https://doi.org/10.1121/1.1356704

Buckstaff KC (2004) Effects of watercraft noise on the acoustic behavior of bottlenose dolphins, *Tursiops truncatus*, in Sarasota Bay, Florida. Mar Mamm Sci 20(4):709–725. https://doi.org/10.1111/j.1748-7692.2004.tb01189.x

Bullock TH, Gurevich V (1979) Soviet literature on the nervous system and psychobiology of Cetacea. Int Rev Neurobiol 21:48–127. https://doi.org/10.1016/s0074-7742(08)60637-6

Bullock TH, Ridgway SH (1972) Evoked potentials in the central auditory system of alert porpoises to their own and artificial sounds. J Neurobiol 3(1):79–99. https://doi.org/10.1002/neu.480030107

Bullock TH, Grinnell AD, Ikezono E, Kameda K, Katsuki K, Nomoto M, Sato O, Suga N, Yanagisawa K (1968) Electrophysiological studies of central auditory mechanisms in cetaceans. Z Vgl Physiol 59:117–156. https://doi.org/10.1007/BF00339347

Bullock TH, Ridgway SH, Nobuo S (1971) Acoustically evoked potentials in midbrain auditory structures in sea lions (Pinnipedia). Z Vgl Physiol 74:372–387. https://doi.org/10.1007/BF00341402

Bullock TH, Domning DP, Best RC (1980) Evoked brain potentials demonstrate hearing in a manatee (*Trichechus inunguis*). J Mammal 61(1):130–133. https://doi.org/10.2307/1379969

Bullock TH, O'Shea TJ, McClune MC (1982) Auditory evoked potentials in the West Indian manatee (Sirenia: *Trichechus manatus*). J Comp Physiol A 148(4):547–554. https://doi.org/10.1007/BF00619792

Buus S (1999) Temporal integration and multiple looks, revisited: weights as a function of time. J Acoust Soc Am 105(4):2466–2475. https://doi.org/10.1121/1.426859

Castellote M, Clark CW, Lammers MO (2012) Acoustic and behavioural changes by fin whales (*Balaenoptera physalus*) in response to shipping and airgun noise. Biol Conserv 147:115–122. https://doi.org/10.1016/j.biocon.2011.12.021

Castellote M, Mooney TA, Quakenbush L, Hobbs R, Goertz C, Gaglione E (2014) Baseline hearing abilities and variability in wild beluga whales (*Delphinapterus leucas*). J Exp Biol 217:1682–1691. https://doi.org/10.1242/jeb.093252

Chapman NR, Price A (2011) Low frequency deep ocean ambient noise trend in the Northeast Pacific Ocean. J

Acoust Soc Am 129(5):EL161–EL165. https://doi.org/10.1121/1.3567084

Clark CW, Ellison WT (2004) Potential use of low-frequency sounds by baleen whales for probing the environment: evidence from models and empirical measurements. In: Thomas JA, Moss CF, Vater M (eds) Echolocation in bats and dolphins. University of Chicago Press, Chicago, pp 564–582

Coffinger S, Houser D, Finneran JJ, Mulsow J, Gentner TQ, Burkard R (2018) Stimulus bandwidth impact on auditory evoked potential thresholds and estimated upper-frequency limits of hearing in dolphins. J Acoust Soc Am 144(6):3575–3581. https://doi.org/10.1121/1.5084043

Colbert-Luke DE, Gaspard JC, Reep RL, Bauer GB, Dziuk K, Cardwell A, Mann DA (2015) Eight-choice sound localization by manatees: performance abilities and head related transfer functions. J Comp Physiol A 201(2):249–259. https://doi.org/10.1007/s00359-014-0973-4

Cook MLH, Varela RA, Goldstein JD, McCulloch SD, Bossart GD, Finneran JJ, Houser DS, Mann DA (2006) Beaked whale auditory evoked potential hearing measurements. J Comp Physiol A 192(5):489–495. https://doi.org/10.1007/s00359-005-0086-1

Cox TM, Ragen TJ, Read AJ, Vos E, Baird RW, Balcomb K, Barlow J, Caldwell J, Cranford TW, Crum L, D'Amico A, D'Spain G, Fernández A, Finneran JJ, Gentry R, Gerth W, Gulland FM, Hildebrand J, Houser DS, Hullar T, Jepson PD, Ketten D, Macleod CD, Miller P, Moore S, Mountain DC, Palka DL, Ponganis PJ, Rommel SA, Rowles T, Taylor BL, Tyack P, Wartzok D, Gisiner R, Meads J, Benner L (2006) Understanding the impacts of anthropogenic sound on beaked whales. J Cetacean Res Manag 7(3):177–187. https://doi.org/10.47536/jcrm.v7i3.729

Cranford TW, Krysl P (2015) Fin whale sound reception mechanisms: skull vibration enables low-frequency hearing. PLoS One 10(1):1–17. https://doi.org/10.1371/journal.pone.0116222

Cranford TW, Krysl P, Amundin M (2010) A new acoustic portal into the odontocete ear and vibrational analysis of the tympanoperiotic complex. PLoS One 5(8):e11927. https://doi.org/10.1371/journal.pone.0011927

Cunningham KA, Reichmuth C (2016) High-frequency hearing in seals and sea lions. Hear Res 331:83–91. https://doi.org/10.1016/j.heares.2015.10.002

Cunningham KA, Southall BL, Reichmuth C (2014) Auditory sensitivity of seals and sea lions in complex listening scenarios. J Acoust Soc Am 136(6):3410. https://doi.org/10.1121/1.4900568

Dahlheim ME, Ljungblad DK (1990) Preliminary hearing study on gray whales (*Eschrichtius robustus*) in the field. In: Thomas JA, Kastelein RA (eds) Sensory abilities of cetaceans: laboratory and field evidence. Springer US, Boston, pp 335–346. https://doi.org/10.1007/978-1-4899-0858-2_22

Dankiewicz LA, Helweg DA, Moore PW, Zafran JM (2002) Discrimination of amplitude-modulated synthetic echo trains by an echolocating bottlenose dolphin. J Acoust Soc Am 112(4):1702–1708. https://doi.org/10.1121/1.1504856

De Vreese S (2014) The external ear canal of cetaceans: vestigial or not? Vlaams Diergeneeskundig Tijdschrift 83(6):284–292. https://doi.org/10.21825/vdt.v83i6.16625

Department of the Navy (1998) Final environmental impact statement, shock testing the SEAWOLF submarine. Department of the Navy, Washington, DC, 563 pp

Department of the Navy (2001) Final environmental impact statement, shock trial of the WINSTON S. CHURCHILL (DDG81). Department of the Navy, Washington, DC

Dolphin WF (2000) Electrophysiological measures of auditory processing in odontocetes. In: Au WWL, Popper AN, Fay RR (eds) Hearing by whales and dolphins. Springer, New York, pp 294–329. https://doi.org/10.1007/978-1-4612-1150-1_7

Dolphin WF, Au WWL, Nachtigall PE, Pawloski J (1995) Modulation rate transfer functions to low-frequency carriers in three species of cetaceans. J Comp Physiol A 177(2):235–245. https://doi.org/10.1007/BF00225102

Dubrovskiy NA (1990) On the two auditory subsystems in dolphins. In: Thomas JA, Kastelein RA (eds) Sensory abilities of cetaceans: laboratory and field evidence. Plenum Press, New York, pp 233–254. https://doi.org/10.1007/978-1-4899-0858-2_14

Dudok van Heel WHD (1962) Sound and Cetacea. Neth J Sea Res 1(4):407–507. https://doi.org/10.1016/0077-7579(62)90001-7

Dunlop RA, Noad MJ, Cato DH, Kniest E, Miller PJO, Smith JN, Stokes MD (2013) Multivariate analysis of behavioural response experiments in humpback whales (*Megaptera novaeangliae*). J Exp Biol 216(5):759–7701. https://doi.org/10.1242/jeb.071498

Dunlop RA, Cato DH, Noad MJ (2014) Evidence of a Lombard response in migrating humpback whales (*Megaptera novaeangliae*). J Acoust Soc Am 136(1):430. https://doi.org/10.1121/1.4883598

Erbe C, Farmer DM (1998) Masked hearing thresholds of a beluga whale (*Delphinapterus leucas*) in icebreaker noise. Deep-Sea Res 45:1373–1378

Erbe C, Reichmuth C, Cunningham K, Lucke K, Dooling R (2016) Communication masking in marine mammals: a review and research strategy. Mar Pollut Bull 103(1–2):15–38. https://doi.org/10.1016/j.marpolbul.2015.12.007

Evans WE (1973) Echolocation by marine delphinids and one species of fresh-water dolphin. J Acoust Soc Am 54(1):191–199. https://doi.org/10.1121/1.1913562

Evans DL, England GR (2001) Joint interim report Bahamas marine mammal stranding event of 14–16 March 2000. National Oceanic and Atmospheric Administration, Washington, DC

Fay RR (1988) Hearing in vertebrates: a psychophysics handbook. Hill-Fay Associates, Winnetka

Fechner GT (1860) Elemente der Psychophysik, vol 1. Breitkopf und Härtel

Fernández A, Edwards J, Martín V, Rodríguez F, Espinosa de los Monteros A, Herráez P, Castro P, Jaber JR, Arbelo M (2005) "Gas and fat embolic syndrome" involving a mass stranding of beaked whales (family Ziphiidae) exposed to anthropogenic sonar signals. J Vet Pathol 42:446–457. https://doi.org/10.1354/vp.42-4-446

Finneran JJ (2008) Modified variance ratio for objective detection of transient evoked potentials in bottlenose dolphins (Tursiops truncatus). J Acoust Soc Am 124(6):4069–4082. https://doi.org/10.1121/1.2996320

Finneran JJ (2009) Evoked response study tool: a portable, rugged system for single and multiple auditory evoked potential measurements. J Acoust Soc Am 126(1):491–500. https://doi.org/10.1121/1.3148214

Finneran JJ (2016) Auditory weighting functions and TTS/PTS exposure functions for marine mammals exposed to underwater noise (SSC Pacific Technical Report SSC Pacific TR 3026). SSC Pacific, San Diego, 58 pp

Finneran JJ (2018) Conditioned attenuation of auditory brainstem responses in dolphins warned of an intense noise exposure: temporal and spectral patterns. J Acoust Soc Am 143(2):795–810. https://doi.org/10.1121/1.5022784

Finneran JJ (2020) Conditioned attenuation of dolphin monaural and binaural auditory evoked potentials after preferential stimulation of one ear. J Acoust Soc Am 147(4):2302–2313. https://doi.org/10.1121/10.0001033

Finneran JJ, Houser DS (2006) Comparison of in-air evoked potential and underwater behavioral hearing thresholds in four bottlenose dolphins (Tursiops truncatus). J Acoust Soc Am 119(5):3181–3192. https://doi.org/10.1121/1.2180208

Finneran JJ, Jenkins AK (2012) Criteria and thresholds for U.S. navy acoustic and explosive effects analysis. SSC Pacific, San Diego, 60 pp

Finneran JJ, Schlundt CE (2007) Underwater sound pressure variation and bottlenose dolphin (Tursiops truncatus) hearing thresholds in a small pool. J Acoust Soc Am 122(1):606–614. https://doi.org/10.1121/1.2743158

Finneran JJ, Schlundt CE (2010) Frequency-dependent and longitudinal changes in noise-induced hearing loss in a bottlenose dolphin (Tursiops truncatus). J Acoust Soc Am 128(2):567–570. https://doi.org/10.1121/1.3458814

Finneran JJ, Schlundt CE (2011) Subjective loudness level measurements and equal loudness contours in a bottlenose dolphin (Tursiops truncatus). J Acoust Soc Am 130(5):3124–3136. https://doi.org/10.1121/1.3641449

Finneran JJ, Schlundt CE (2013) Effects of fatiguing tone frequency on temporary threshold shift in bottlenose dolphins (Tursiops truncatus). J Acoust Soc Am 133(3):1819–1826. https://doi.org/10.1121/1.4776211

Finneran JJ, Carder DA, Ridgway SH (2002a) Low-frequency acoustic pressure, velocity, and intensity thresholds in a bottlenose dolphin (Tursiops truncatus) and white whale (Delphinapterus leucas). J Acoust Soc Am 111(1):447–456. https://doi.org/10.1121/1.1423925

Finneran JJ, Schlundt CE, Carder DA, Ridgway SH (2002b) Auditory filter shapes for the bottlenose dolphin (Tursiops truncatus) and the white whale (Delphinapterus leucas) derived with notched noise. J Acoust Soc Am 112(1):322–328. https://doi.org/10.1121/1.1488652

Finneran JJ, Schlundt CE, Dear R, Carder DA, Ridgway SH (2002c) Temporary shift in masked hearing thresholds (MTTS) in odontocetes after exposure to single underwater impulses from a seismic watergun. J Acoust Soc Am 111(6):2929–2940. https://doi.org/10.1121/1.1479150

Finneran JJ, Carder DA, Dear R, Belting T, McBain J, Dalton L, Ridgway SH (2005) Pure tone audiograms and possible aminoglycoside-induced hearing loss in belugas (Delphinapterus leucas). J Acoust Soc Am 117(6):3936–3943. https://doi.org/10.1121/1.1893354

Finneran JJ, Houser DS, Schlundt CE (2007a) Objective detection of bottlenose dolphin (Tursiops truncatus) steady-state auditory evoked potentials in response to AM/FM tones. Aquat Mamm 33(1):43–54. https://doi.org/10.1578/AM.33.1.2007.43

Finneran JJ, London HR, Houser DS (2007b) Modulation rate transfer functions in bottlenose dolphins (Tursiops truncatus) with normal hearing and high-frequency hearing loss. J Comp Physiol A 193:835–843. https://doi.org/10.1007/s00359-007-0238-6

Finneran JJ, Houser DS, Blasko D, Hicks C, Hudson J, Osborn M (2008) Estimating bottlenose dolphin (Tursiops truncatus) hearing thresholds from single and multiple simultaneous auditory evoked potentials. J Acoust Soc Am 123(1):542–551. https://doi.org/10.1121/1.2812595

Finneran JJ, Houser DS, Mase-Guthrie B, Ewing RY, Lingenfelser RG (2009) Auditory evoked potentials in a stranded Gervais' beaked whale (Mesoplodon europaeus). J Acoust Soc Am 126(1):484–490. https://doi.org/10.1121/1.3133241

Finneran JJ, Carder DA, Schlundt CE, Dear RL (2010) Growth and recovery of temporary threshold shift (TTS) at 3 kHz in bottlenose dolphins (Tursiops truncatus). J Acoust Soc Am 127(5):3256–3266. https://doi.org/10.1121/1.3372710

Finneran JJ, Branstetter BK, Houser DS, Moore PW, Mulsow J, Martin C, Perisho S (2014) High-resolution measurement of a bottlenose dolphin's (Tursiops truncatus) biosonar transmission beam pattern in the horizontal plane. J Acoust Soc Am 136(4):2025–2038. https://doi.org/10.1121/1.4895682

Finneran JJ, Schlundt CE, Branstetter BK, Trickey J, Bowman V, Jenkins K (2015) Effects of multiple impulses from a seismic air gun on bottlenose dolphin hearing and behavior. J Acoust Soc Am 137(4): 1634–1646. https://doi.org/10.1121/1.4916591

Finneran JJ, Mulsow J, Houser DS, Burkard RF (2016) Place specificity of the click-evoked auditory brainstem response in the bottlenose dolphin (*Tursiops truncatus*). J Acoust Soc Am 140(4):2593–2602. https://doi.org/10.1121/1.4964274

Finneran JJ, Lally K, Strahan MG, Donohoe K, Mulsow J, Houser DS (2023) Dolphin conditioned hearing attenuation in response to repetitive tones with increasing level. J Acoust Soc Am 153:496–504. https://doi.org/10.1121/10.0016868

Fletcher H (1940) Auditory patterns. Rev Mod Phys 12: 47–65. https://doi.org/10.1103/RevModPhys.12.47

Fletcher H, Munson WA (1933) Loudness, its definition, measurement and calculation. J Acoust Soc Am 5:82–108. https://doi.org/10.1121/1.1915637

Foote AD, Osborne RW, Hoelzel AR (2004) Whale-call response to masking boat noise. Nature 428:910. https://doi.org/10.1038/428910a

Fournet MEH, Silvestri M, Clark CW, Klinck H, Rice AN (2021) Limited vocal compensation for elevated ambient noise in bearded seals: implications for an industrializing Arctic Ocean. Proc R Soc Lond B 288(1945):20202712. https://doi.org/10.1098/rspb.2020.2712

Frankel AS, Clark CW (2000) Behavioral responses of humpback whales (*Megaptera novaeangliae*) to full-scale ATOC signals. J Acoust Soc Am 108(4): 1930–1937. https://doi.org/10.1121/1.1289668

Frankel AS, Clark CW (2002) ATOC and other factors affecting the distribution and abundance of humpback whales (*Megaptera novaeangliae*) off the north shore of Kauai. Mar Mamm Sci 18(3):644–662. https://doi.org/10.1111/j.1748-7692.2002.tb01064.x

Frankel AS, Mobley J, Herman LM (1995) Estimation of auditory response thresholds in humpback whales using biologically meaningful sounds. In: Kastelein RA, Thomas JA, Nachtigall PE (eds) Sensory systems of aquatic mammals, vol 55. De Spil Publishers, Woerden, p 70

Frantzis A (1998) Does acoustic testing strand whales? Nature 392(6671):29. https://doi.org/10.1038/32068

Gaspard JC III, Bauer GB, Reep RL, Dziuk K, Cardwell A, Read L, Mann DA (2012) Audiogram and auditory critical ratios of two Florida manatees (*Trichechus manatus latirostris*). J Exp Biol 215: 1442–1447. https://doi.org/10.1242/jeb.065649

Gaspard JC III, Bauer GB, Reep RL, Dziuk K, Read L, Mann DA (2013) Detection of hydrodynamic stimuli by the Florida manatee (*Trichechus manatus latirostris*). J Comp Physiol A Neuroethol Sens Neural Behav Physiol 199(6):441–450. https://doi.org/10.1007/s00359-013-0822-x

Gelfand SA (1990) Hearing: an introduction to psychological and physiological acoustics, 2nd edn. Marcel Dekker, Inc, New York. https://doi.org/10.1201/9781315154718

Gentry RL (1967) Underwater auditory localization in the California sea lion (*Zalophus californianus*). J Aud Res 7:187–193. https://doi.org/10.1121/1.380720

Gerstein ER (1999) Psychoacoustic evaluation of the West Indian Manatee (*Trichechus manatus latirosris*). PhD, Florida Atlantic University, Boca Raton, FLorida

Gerstein ER, Gerstein L, Forsythe SE, Blue JE (1999) The underwater audiogram of the West Indian manatee (*Trichechus manatus*). J Acoust Soc Am 105(6): 3575–3583. https://doi.org/10.1121/1.424681

Gertner J (2012) The idea factory: bells labs and the great age of American innovation. Penguin Press, New York

Ghoul A, Reichmuth C (2014a) Hearing in the sea otter (*Enhydra lutris*): auditory profiles for an amphibious marine carnivore. J Comp Physiol A 200(11):967–981. https://doi.org/10.1007/s00359-014-0943-x

Ghoul A, Reichmuth CJ (2014b) Hearing in the sea otter (*Enhydra lutris*): auditory profiles for an amphibious marine carnivore. J Comp Physiol A 200:967–981. https://doi.org/10.1007/s00359-014-0943-x

Girola E, Dunlop RA, Noad MJ (2023) Singing humpback whales respond to wind noise, but not to vessel noise. Proc R Soc Lond B 290(1998):20230204. https://doi.org/10.1098/rspb.2023.0204

Goldbogen JA, Southall BL, DeRuiter SL, Calambokidis J, Friedlaender AS, Hazen EL, Falcone EA, Schorr GS, Douglas A, Moretti DJ, Kyburg C, McKenna MF, Tyack PL (2013) Blue whales respond to simulated mid-frequency military sonar. Proc R Soc Lond B 280(1765):20130657. https://doi.org/10.1098/rspb.2013.0657

Götz T, Antunes R, Heinrich S (2010) Echolocation clicks of free-ranging Chilean dolphins (*Cephalorhynchus eutropia*) (L). J Acoust Soc Am 128(2):563–566. https://doi.org/10.1121/1.3353078

Greene CR Jr (1987) Characteristics of oil industry dredge and drilling sounds in the Beaufort Sea. J Acoust Soc Am 82(4):1315–1324. https://doi.org/10.1121/1.395265

Greenhow DR, Brodsky MC, Lingenfelser RG, Mann DA (2014) Hearing threshold measurements of five stranded short-finned pilot whales (*Globicephala macrorhynchus*). J Acoust Soc Am 135(1):531–536. https://doi.org/10.1121/1.4829662

Guazzo RA, Helble TA, Alongi GC, Durbach IN, Martin CR, Martin SW, Henderson EE (2020) The Lombard effect in singing humpback whales: source levels increase as ambient ocean noise levels increase. J Acoust Soc Am 148(2):542–555. https://doi.org/10.1121/10.0001669

Hall JD, Johnson CS (1972) Auditory thresholds of a killer whale *Orcinus orca* Linnaeus. J Acoust Soc Am 51(2B):515–517. https://doi.org/10.1121/1.1912871

Heffner R, Heffner H, Masterton B (1971) Behavioral measurements of absolute and frequency-difference thresholds in guinea pig. J Acoust Soc Am 49(6B):1888–1895. https://doi.org/10.1121/1.1912596

Helble TA, Guazzo RA, Martin CR, Durbach IN, Alongi GC, Martin SW, Boyle JK, Henderson EE (2020)

Lombard effect: minke whale boing call source levels vary with natural variations in ocean noise. J Acoust Soc Am 147(2):698–712. https://doi.org/10.1121/10.0000596

Helweg DA, Au WWL, Roitblat HL, Nachtigall PE (1996) Acoustic basis for recognition of aspect-dependent three-dimensional targets by an echolocating bottlenose dolphin. J Acoust Soc Am 99(4): 2409–2420. https://doi.org/10.1121/1.415429

Hemilä S, Nummela S, Berta A, Reuter T (2006) High-frequency hearing in phocid and otariid pinnipeds: an interpretation based on inertial and cochlear constraints (L). J Acoust Soc Am 120(6):3463–3466. https://doi.org/10.1121/1.2372712

Herman LM, Arbeit WR (1972) Frequency difference limens in the bottlenosed dolphin: 1-70 KC/S. J Aud Res 2:109–120

Holt MM, Schusterman RJ (2007) Spatial release from masking of aerial tones in pinnipeds. J Acoust Soc Am 121(2):1219–1225. https://doi.org/10.1121/1.2404929

Holt MM, Schusterman RJ, Southall BL, Kastak D (2004) Localization of aerial broadband noise by pinnipeds. J Acoust Soc Am 115(5):2339–2345. https://doi.org/10.1121/1.1694995

Holt MM, Schusterman RJ, Kastak D, Southall BL (2005) Localization of aerial pure tones by pinnipeds. J Acoust Soc Am 118(6):3921–3926. https://doi.org/10.1121/1.2126931

Holt MM, Noren DP, Veirs V, Emmons CK, Veirs S (2008) Speaking up: killer whales (*Orcinus orca*) increase their call amplitude in response to vessel noise. J Acoust Soc Am 125(1):EL27–EL32. https://doi.org/10.1121/1.3040028

Holt MM, Noren DP, Emmons CK (2011) Effects of noise levels and call types on the source levels of killer whale calls. J Acoust Soc Am 130(5):3100–3106. https://doi.org/10.1121/1.3641446

Holt MM, Ghoul A, Reichmuth C (2012) Temporal summation of airborne tones in a California sea lion (*Zalophus californianus*). J Acoust Soc Am 132(5): 3569–3575. https://doi.org/10.1121/1.4757733

Houser DS, Finneran JJ (2006a) A comparison of underwater hearing sensitivity in bottlenose dolphins (*Tursiops truncatus*) determined by electrophysiological and behavioral methods. J Acoust Soc Am 120(3): 1713–1722. https://doi.org/10.1121/1.2229286

Houser DS, Finneran JJ (2006b) Variation in the hearing sensitivity of a dolphin population obtained through the use of evoked potential audiometry. J Acoust Soc Am 120(6):4090–4099. https://doi.org/10.1121/1.2357993

Houser DS, Helweg DA, Moore PWB (1999) Classification of dolphin echolocation clicks by energy and frequency distributions. J Acoust Soc Am 106(3): 1579–1585. https://doi.org/10.1121/1.427153

Houser DS, Helweg DA, Moore PWB (2001) A bandpass filter-bank model of auditory sensitivity in the humpback whale. Aquat Mamm 27(2):82–91

Houser DS, Finneran JJ, Carder DA, Van Bonn W, Smith C, Hoh C, Mattrey R, Ridgway SH (2004) Structural and functional imaging of bottlenose dolphin (*Tursiops truncatus*) cranial anatomy. J Exp Biol 207:3657–3665. https://doi.org/10.1242/Jeb.01207

Houser DS, Martin SW, Bauer EJ, Phillips M, Herrin T, Cross M, Vidal A, Moore PW (2005) Echolocation characteristics of free-swimming bottlenose dolphins during object detection and identification. J Acoust Soc Am 117(4):2308–2317. https://doi.org/10.1121/1.1867912

Houser DS, Crocker DE, Kastak C, Mulsow J, Finneran JJ (2007) Auditory evoked potentials in northern elephant seals (*Mirounga angustirostris*). Aquat Mamm 33(1): 110–121. https://doi.org/10.1578/AM.33.1.2007.110

Houser DS, Crocker DE, Finneran JJ (2008a) Click-evoked potentials in a large marine mammal, the adult male northern elephant seal (*Mirounga angustirostris*). J Acoust Soc Am 124(1):44–47. https://doi.org/10.1121/1.2932063

Houser DS, Gomez-Rubio A, Finneran JJ (2008b) Evoked potential audiometry of 13 Pacific bottlenose dolphins (*Tursiops truncatus gilli*). Mar Mamm Sci 24(1): 28–41. https://doi.org/10.1111/j.1748-7692.2007.00148.x

Houser DS, Yost W, Burkard R, Finneran JJ, Reichmuth C, Mulsow J (2017) A review of the history, development and application of auditory weighting functions in humans and marine mammals. J Acoust Soc Am 141(3):1371–1413. https://doi.org/10.1121/1.4976086

Houser DS, Moore KMT, Sharp SM, Hoppe JM, Finneran JJ (2018) Cetacean evoked potential audiometry by stranding networks enables more rapid accumulation of hearing information in stranded odontocetes. J Cetacean Res Manag 18:93–101. https://doi.org/10.47536/jcrm.v18i1.436

Houser DS, Mulsow J, Almunia J, Finneran JJ (2019) Frequency-modulated up-chirp stimuli enhance the auditory brainstem response of the killer whale (*Orcinus orca*). J Acoust Soc Am 146(1):289–296. https://doi.org/10.1121/1.5116141

Houser DS, Noble L, Fougeres E, Mulsow J, Finneran JJ (2022) Audiograms and click spectra of seven novel and seldom-tested odontocetes. Front Mar Sci 9: 984333. https://doi.org/10.3389/fmars.2022.984333

Houser DS, Kvadsheim PH, Kleivane L, Mulsow J, Ølberg RA, Harms CA, Teilmann J, Finneran JJ (2024) Direct hearing measurements in a baleen whale suggest ultrasonic sensitivity. Science 386 (6724):902–906. https://doi.org/10.1126/science.ado7580

Jacobs DW (1972) Auditory frequency discrimination in the Atlantic bottenosed dolphin, *Tursiops truncatus* Montague: a preliminary report. J Acoust Soc Am 52(2B):696–698. https://doi.org/10.1121/1.1913160

Jacobs DW, Hall JD (1972) Auditory thresholds of a fresh water dolphin, *Inia geoffrensis* Blainville. J Acoust Soc

Am 51(2B):530–533. https://doi.org/10.1121/1.1912874

Jepson PD, Arbelo M, Deaville R, Patterson IAR, Castro P, Baker JR, Degollada E, Ross HM, Herráez P, Pocknell AM, Rodriguez E, Howie FE, Espinosa A, Reid RJ, Jaber JR, Martin V, Cunningham AA, Fernández A (2003) Gas-bubble lesions in stranded cetaceans: was sonar responsible for a spate of whale deaths after an Atlantic military exercise? Nature 425:575–576. https://doi.org/10.1038/425575a

Jesteadt W, Wier CC, Green DM (1977) Intensity discrimination as a function of frequency and sensation level. J Acoust Soc Am 61(1):169–177. https://doi.org/10.1121/1.381278

Johnson CS (1967) Sound detection thresholds in marine mammals. In: Tavolga WN (ed) Marine bioacoustics. Pergamon Press, Oxford, pp 247–260

Johnson CS (1968) Relation between absolute threshold and duration of tone pulses in the bottlenosed porpoise. J Acoust Soc Am 43(4):757–763. https://doi.org/10.1121/1.1910893

Johnson CS (1971) Auditory masking of one pure tone by another in the bottlenosed porpoise. J Acoust Soc Am 49(4 part 2):1317–1318. https://doi.org/10.1121/1.1912496

Johnson CS (1991) Hearing thresholds for periodic 60-kHz tone pulses in the beluga whale. J Acoust Soc Am 89(6):2996–3001. https://doi.org/10.1121/1.400736

Johnson CS, McManus MW, Skaar D (1989) Masked tonal hearing thresholds in the beluga whale. J Acoust Soc Am 85:2651–2654. https://doi.org/10.1121/1.397759

Jones RA, Sills JM, Synnott M, Mulsow J, Williams R, Reichmuth C (2023) Auditory masking in odobenid and otariid carnivores. J Acoust Soc Am 154(3):1746–1756. https://doi.org/10.1121/10.0020911

Kamminga C (1994) Research on dolphin sounds. Technische Hogeschool, Delft

Kastak D, Schusterman RJ (1995) Aerial and underwater hearing thresholds for 100 Hz pure tones in two species of pinnipeds. In: Kastelein RA, Thomas JA, Nachtigall PE (eds) Sensory systems of aquatic mammals. De Spil Publishers, Woerden, pp 71–79

Kastak D, Schusterman RJ (1998) Low-frequency amphibious hearing in pinnipeds: methods, measurements, noise, and ecology. J Acoust Soc Am 103(4):2216–2228. https://doi.org/10.1121/1.421367

Kastak D, Schusterman RJ (1999) In-air and underwater hearing sensitivity of a northern elephant seal (*Mirounga angustirostris*). Can J Zool 77(11):1751–1758. https://doi.org/10.1139/z99-151

Kastak D, Schusterman RJ (2002) Changes in auditory sensitivity with depth in a free-diving California sea lion (*Zalophus californianus*). J Acoust Soc Am 112(1):329–333. https://doi.org/10.1121/1.1489438

Kastak D, Schusterman RJ, Southall BL, Reichmuth CJ (1999) Underwater temporary threshold shift induced by octave-band noise in three species of pinniped. J Acoust Soc Am 106(2):1142–1148. https://doi.org/10.1121/1.427122

Kastak D, Southall B, Holt M, Kastak CR, Schusterman R (2004) Noise-induced temporary threshold shifts in pinnipeds: effects of noise energy. J Acoust Soc Am 116(4):2531–2532(A). https://doi.org/10.1121/1.4785103

Kastak D, Holt MM, Kastak CJR, Southall BL, Mulsow J, Schusterman RJ (2005a) A voluntary mechanism of protection from airborne noise in a harbor seal. Paper presented at the 16th biennial conference on the biology of marine mammals, San Diego, CA, December 12–16, 2005

Kastak D, Southall BL, Schusterman RJ, Kastak CR (2005b) Underwater temporary threshold shift in pinnipeds: effects of noise level and duration. J Acoust Soc Am 118(5):3154–3163. https://doi.org/10.1121/1.2047128

Kastak D, Reichmuth C, Holt MM, Mulsow J, Southall BL, Schusterman RJ (2007) Onset, growth, and recovery of in-air temporary threshold shift in a California sea lion (*Zalophus californianus*). J Acoust Soc Am 122(5):2916–2924. https://doi.org/10.1121/1.2783111

Kastelein RA, Wensveen PJ (2008) Effect of two levels of masking noise on the hearing threshold of a harbor porpoise (*Phocoena phocoena*) for a 4.0 kHz signal. Aquat Mamm 34(4):420–425. https://doi.org/10.1578/AM.34.4.2008.420

Kastelein RA, van Ligtenberg CL, Gjertz I, Verboom WC (1993) Free field hearing tests on wild Atlantic walruses (*Odobenus rosmarus rosmarus*) in air. Aquat Mamm 19(3):143–148

Kastelein RA, Dubbeldam JL, De Bakker MAG, Gerrits NM (1996a) The anatomy of the walrus head (*Odobenus rosmarus*). Part 4: the ears and their function in aerial and underwater hearing. Aquat Mamm 22(2):95–125

Kastelein RA, Mosterd P, van Ligtenberg CL, Verboom WC (1996b) Aerial hearing sensitivity tests with a male Pacific walrus (*Odobenus rosmarus divergens*), in the free field and with headphones. Aquat Mamm 22(2):81–93

Kastelein RA, van der Kaay LSP, Staal C, Schooneman NM, van Ligtenberg CL (1997) Detection of bone conductor signals by a harbour porpoise (*Phocoena phocoena*): a pilot study. In: Read AJ, Wiepkema PR, Nachtigall PE (eds) The biology of the harbour porpoise. De Spil Publishers, Woerden, pp 313–327

Kastelein RA, Bunskoek P, Hagedoorn M, Au WWL, de Haan D (2002a) Audiogram of a harbor porpoise (*Phocoena phocoena*) measured with narrow-band frequency-modulated signals. J Acoust Soc Am 112(1):334–344. https://doi.org/10.1121/1.1480835

Kastelein RA, Mosterd P, van Santen B, Hagedoorn M, de Haan D (2002b) Underwater audiogram of a Pacific walrus (*Odobenus rosmarus divergens*) measured with narrow-band frequency-modulated signals. J Acoust Soc Am 112(5):2173–2182. https://doi.org/10.1121/1.1508783

Kastelein RA, Hagedoorn M, Au WWL, de Haan D (2003) Audiogram of a striped dolphin (*Stenella coeruleoalba*). J Acoust Soc Am 113(2):1130–1137. https://doi.org/10.1121/1.1532310

Kastelein RA, Janssen M, Verboom WC, Haan D (2005a) Receiving beam patterns in the horizontal plane of a harbor porpoise (*Phocoena phocoena*). J Acoust Soc Am 118(2):1172–1179. https://doi.org/10.1121/1.1945565

Kastelein RA, van Schie R, Verboom WC, de Haan D (2005b) Underwater hearing sensitivity of a male and a female Steller sea lion (*Eumetopias jubatus*). J Acoust Soc Am 118(3):1820–1829. https://doi.org/10.1121/1.1992650

Kastelein RA, de Haan D, Verboom WC (2007) The influence of signal parameters on the sound source localization ability of a harbor porpoise (*Phocoena phocoena*). J Acoust Soc Am 122(2):1238–1248. https://doi.org/10.1121/1.2747202

Kastelein RA, Wensveen P, Hoek L, Terhune JM (2009a) Underwater hearing sensitivity of harbor seals (*Phoca vitulina*) for narrow noise bands between 0.2 and 80 kHz. J Acoust Soc Am 126(1):476–483. https://doi.org/10.1121/1.3132522

Kastelein RA, Wensveen PJ, Hoek L, Au WWL, Terhune JM, de Jong CAF (2009b) Critical ratios in harbor porpoises (*Phocoena phocoena*) for tonal signals between 0.315 and 150 kHz in random Gaussian white noise. J Acoust Soc Am 126(3):1588–1597. https://doi.org/10.1121/1.3177274

Kastelein RA, Wensveen PJ, Hoek L, Verboom WC, Terhune JM (2009c) Underwater detection of tonal signals between 0.125 and 100 kHz by harbor seals (*Phoca vitulina*). J Acoust Soc Am 125(2):1222–1229. https://doi.org/10.1121/1.3050283

Kastelein RA, Hoek L, de Jong CAF, Wensveen PJ (2010) The effect of signal duration on the underwater detection thresholds of a harbor porpoise (*Phocoena phocoena*) for single frequency-modulated tonal signals between 0.25 and 160 kHz. J Acoust Soc Am 128(5):3211–3222. https://doi.org/10.1121/1.3493435

Kastelein RA, Wensveen PJ, Terhune JM, de Jong CAF (2011) Near-threshold equal-loudness contours for harbor seals (*Phoca vitulina*) derived from reaction times during underwater audiometry: a preliminary study. J Acoust Soc Am 129(1):488–495. https://doi.org/10.1121/1.3518779

Kastelein RA, Gransier R, Hoek L, Macleod A, Terhune JM (2012a) Hearing threshold shifts and recovery in harbor seals (*Phoca vitulina*) after octave-band noise exposure at 4 kHz. J Acoust Soc Am 132(4):2745–2761. https://doi.org/10.1121/1.4747013

Kastelein RA, Gransier R, Hoek L, Olthuis J (2012b) Temporary threshold shifts and recovery in a harbor porpoise (*Phocoena phocoena*) after octave-band noise at 4 kHz. J Acoust Soc Am 132(5):3525–3537. https://doi.org/10.1121/1.4757641

Kastelein RA, Gransier R, Hoek L (2013) Comparative temporary threshold shifts in a harbor porpoise and harbor seal, and severe shift in a seal. J Acoust Soc Am 134(1):13–16. https://doi.org/10.1121/1.4808078

Kastelein RA, Hoek L, Gransier R, Rambags M, Claeys N (2014a) Effect of level, duration, and inter-pulse interval of 1-2 kHz sonar signal exposures on harbor porpoise hearing. J Acoust Soc Am 136(1):412–422. https://doi.org/10.1121/1.4883596

Kastelein RA, Schop J, Gransier R, Hoek L (2014b) Frequency of greatest temporary hearing threshold shift in harbor porpoises (*Phocoena phocoena*) depends on the noise level. J Acoust Soc Am 136(3):1410–1418. https://doi.org/10.1121/1.4892794

Kastelein RA, Gransier R, Marijt MAT, Hoek L (2015a) Hearing frequency thresholds of harbor porpoises (*Phocoena phocoena*) temporarily affected by played back offshore pile driving sounds. J Acoust Soc Am 137(2):556–564. https://doi.org/10.1121/1.4906261

Kastelein RA, Gransier R, Schop J, Hoek L (2015b) Effects of exposure to intermittent and continuous 6–7 kHz sonar sweeps on harbor porpoise (*Phocoena phocoena*) hearing. J Acoust Soc Am 137(4):1623–1633. https://doi.org/10.1121/1.4916590

Kastelein RA, Helder-Hoek L, Voorde SV (2017) Hearing thresholds of a male and a female harbor porpoise (*Phocoena phocoena*). J Acoust Soc Am 142(2):1006–1010. https://doi.org/10.1121/1.4997907

Kastelein RA, Helder-Hoek L, Terhune JM (2018) Hearing thresholds, for underwater sounds, of harbor seals (*Phoca vitulina*) at the water surface. J Acoust Soc Am 143(4):2554–2563. https://doi.org/10.1121/1.5034173

Kastelein RA, Helder-Hoek L, Covi J, Terhune JM, Klump G (2021) Masking release at 4 kHz in harbor porpoises (*Phocoena phocoena*) associated with sinusoidal amplitude-modulated masking noise. J Acoust Soc Am 150(3):1721–1732. https://doi.org/10.1121/10.0006103

Kastelein RA, de Jong CAF, Tougaard J, Helder-Hoek L, Defillet LN (2022) Behavioral responses of a harbor porpoise (*Phocoena phocoena*) depend on the frequency content of pile-driving sounds. Aquat Mamm 48(2):97–109. https://doi.org/10.1578/AM.48.2.2022.97

Kastelein RA, Helder-Hoek L, Defillet LN, Terhune JM, Beutelmann R, Klump GM (2023a) Masking release at 4 and 32 kHz in harbor seals associated with sinusoidal amplitude-modulated masking noise. J Acoust Soc Am 154(1):81–94. https://doi.org/10.1121/10.0019631

Kastelein RA, Helder-Hoek L, Van Acoleyen L, Defillet LN, Huijser LAE, Terhune JM (2023b) Underwater sound detection thresholds (0.031-80 kHz) of two California sea lions (*Zalophus californianus*) and a revised generic audiogram for the species. Aquat Mamm 49(5):422–435. https://doi.org/10.1578/AM.49.5.2023.422

Ketten DR (1994) Functional analyses of whale ears: adaptations for underwater hearing. In: IEEE proceedings in underwater acoustics, 14626, pp 264–270

Ketten DR (1998) Marine mammal auditory systems: a summary of audiometric and anatomical data and its implications for underwater acoustic impacts (NOAA-TM-NMFS-SWFSC-256). Dolphin-Safe Research Program, Southwest Fisheries Science Center, LA Jolla, 74 pp

Ketten DR (2000) Cetacean ears. In: Au W, Popper AN, Fay RR (eds) Hearing by whales and dolphins. Springer handbook of auditory research, 1st edn. Springer, New York, pp 43–108. https://doi.org/10.1007/978-1-4612-1150-1_2

Ketten DR, Odell DK, Domning DP (1992) Structure, function, and adaptation of the manatee ear. In: Thomas JA, Kastelein RA, Supin AY (eds) Marine mammal sensory systems. Plenum Press, New York, pp 77–95. https://doi.org/10.1007/978-1-4615-3406-8_4

Klishin VO, Pezo Diaz R, Popov VV, Supin AY (1990) Some characteristics of hearing of the Brazilian manatee, *Trichechus inunguis*. Aquat Mamm 16(3):139–144

Klishin VO, Popov VV, Supin AY (2000) Hearing capabilities of a beluga whale, *Delphinapterus leucas*. Aquat Mamm 26(3):212–228

Kloepper LN, Nachtigall PE, Breese M (2010a) Change in echolocation signals with hearing loss in a false killer whale (*Pseudorca crassidens*). J Acoust Soc Am 128(4):2233–2237. https://doi.org/10.1121/1.3478851

Kloepper LN, Nachtigall PE, Gisiner R, Breese M (2010b) Decreased echolocation performance following high-frequency hearing loss in the false killer whale (*Pseudorca crassidens*). J Exp Biol 213:3717–3722. https://doi.org/10.1242/jeb.042788

Koblitz JC, Wahlberg M, Stilz P, Madsen PT, Beedholm K, Schnitzler H-U (2012) Asymmetry and dynamics of a narrow sonar beam in an echolocating harbor porpoise. J Acoust Soc Am 131(3):2315–2324. https://doi.org/10.1121/1.3683254

Kvadsheim P, Lam F-P, Miller P, Doksæter L, Visser F, Kleivane L, van Ijsselmuide S, Samarra F, Wensveen P, Curé C, Hickmott L, Dekeling R (2011) Behavioural response studies of cetaceans to naval sonar signals in Norwegian waters—3S-2011 cruise report (FFI-rapport 2011/01289)

Kyhn LA, Jensen FH, Beedholm K, Tougaard J, Hansen M, Madsen PT (2010) Echolocation in sympatric Peale's dolphins (*Lagenorhynchus australis*) and Commerson's dolphins (*Cephalorhynchus commersonii*) producing narrow-band high-frequency clicks. J Exp Biol 213(11):1940–1949. https://doi.org/10.1242/jeb.042440

Ladegaard M, Jensen FH, de Freitas M, Ferreira da Silva VM, Madsen PT (2015) Amazon river dolphins (*Inia geoffrensis*) use a high-frequency short-range biosonar. J Exp Biol 218(19):3091–3101. https://doi.org/10.1242/jeb.120501

Langworthy OR, Hesser FH, Kolb LC (1938) A physiological study of the cerebral cortex of the hair seal (*Phoca vitulina*). J Comp Neurol 69:351–369

Lemonds DW (1999) Auditory filter shapes in an Atlantic bottlenose dolphin (*Tursiops truncatus*). PhD thesis, University of Hawaii

Lemonds DW, Kloepper LN, Nachtigall PE, Au WWL, Vlachos SA, Branstetter BK (2011) A re-evaluation of auditory filter shape in delphinid odontocetes: evidence of constant-bandwidth filters. J Acoust Soc Am 130(5):3107–3114. https://doi.org/10.1121/1.3644912

Lemonds DW, Au WWL, Vlachos SA, Nachtigall PE (2012) High-frequency auditory filter shape for the Atlantic bottlenose dolphin. J Acoust Soc Am 132(2):1222–1228. https://doi.org/10.1121/1.4731212

Lesage V, Barrette C, Kingsley MCS, Sjare B (1999) The effect of vessel noise on the vocal behavior of belugas in the St. Lawrence River estuary, Canada. Mar Mamm Sci 15(1):65–84. https://doi.org/10.1111/j.1748-7692.1999.tb00782.x

Levitt H (1971) Transformed up-down methods in psychoacoustics. J Acoust Soc Am 49:467–477

Li S, Wang K, Wang D, Akamatsu T (2005) Echolocation signals of the free-ranging Yangtze finless porpoise (*Neophocaena phocaenoides asiaeorientialis*). J Acoust Soc Am 117(5):3288–3296. https://doi.org/10.1121/1.1882945

Li S, Wang D, Wang K, Taylor EA, Cros E, Shi W, Wang Z, Fang L, Chen Y, Kong F (2012) Evoked-potential audiogram of an Indo-Pacific humpback dolphin (*Sousa chinensis*). J Exp Biol 215(17). https://doi.org/10.1242/jeb.070904

Lien J, Barney W, Todd S, Seton R, Guzzwell J (1992) Effects of adding sounds to cod traps on the probability of collisions by humpback whales. In: Thomas JA, Kastelein RA, Supin AY (eds) Marine mammal sensory systems. Plenum Press, New York, pp 701–708. https://doi.org/10.1007/978-1-4615-3406-8_43

Linnenschmidt M, Wahlberg M, Damsgaard Hansen J (2013) The modulation rate transfer function of a harbour porpoise (*Phocoena phocoena*). J Comp Physiol A 199(2):115–126. https://doi.org/10.1007/s00359-012-0772-8

Lins OG, Picton PE, Picton TW, Champagne SC, Durieux-Smith A (1995) Auditory steady-state responses to tones amplitude-modulated at 80-110 Hz. J Acoust Soc Am 97(5):3051–3063. https://doi.org/10.1121/1.411869

Ljungblad DK, Scroggins PD, Gilmartin WG (1982) Auditory thresholds of a captive Eastern Pacific bottle-nosed dolphin, *Tursiops* spp. J Acoust Soc Am 72(6):1726–1729. https://doi.org/10.1121/1.388666

Lombard E (2011) Le signe de l'élévation de la voix. Annals des Maladies de L'Oreille et du Larynx 37(2):101–109

Lucke K, Siebert U, Lepper PA, Blanchet M-A (2009) Temporary shift in masked hearing thresholds in a harbor porpoise (*Phocoena phocoena*) after exposure to seismic airgun stimuli. J Acoust Soc Am 125(6):4060–4070. https://doi.org/10.1121/1.3117443

Lucke K, Finneran JJ, Almunia J, Houser DS (2016) Variability in click-evoked potentials in killer whales

(*Orcinus orca*) and identification of a hearing impaired whale. Aquat Mamm 42(2):184–192. https://doi.org/10.1578/AM.42.2.2016.184

Malinka CE, Tønnesen P, Dunn CA, Claridge DE, Gridley T, Elwen SH, Teglberg Madsen P (2021) Echolocation click parameters and biosonar behaviour of the dwarf sperm whale (*Kogia sima*). J Exp Biol 224(6). https://doi.org/10.1242/jeb.240689

Mann DA, Colbert DE, Gaspard JC, Casper BM, Cook MLH, Reep R, Bauer GB (2005) Temporal resolution of the Florida manatee *(Trichechus manatus latirostris)* auditory system. J Comp Physiol A 191(10):903–908. https://doi.org/10.1007/s00359-005-0016-2

Mann D, Hill-Cook M, Manire C, Greenhow D, Montie E, Powell J, Wells R, Bauer G, Cunningham-Smith P, Lingenfelser R, DiGiovanni R Jr, Stone A, Brodsky M, Stevens R, Kieffer G, Hoetjes P (2010) Hearing loss in stranded odontocete dolphins and whales. PLoS One 5(11):1–5. https://doi.org/10.1371/journal.pone.0013824

Marten K (2000) Ultrasonic analysis of pygmy sperm whale (*Kogia breviceps*) and Hubbs' beaked whale (*Mesoplodon carlhubbsi*) clicks. Aquat Mamm 26(1):45–48

May BJ, Little N, Saylor S (2009) Loudness perception in the domestic cat: reaction time estimates of equal loudness contours and recruitment effects. J Assoc Res Otolaryngol 10:295–308. https://doi.org/10.1007/s10162-009-0157-z

Maybaum HL (1989) Effects of 3.3 kHz sonar system on humpback whales, *Megaptera novaeangliae*, in Hawaiian waters. MS, Univeristy of Hawaii at Manoa

McCormick JG, Wever EG, Palin J, Ridgway SH (1970) Sound conduction in the dolphin ear. J Acoust Soc Am 48(6):1418–1428. https://doi.org/10.1121/1.1912302

Miller PJO, Biassoni N, Samuels A, Tyack PL (2000) Whale songs lengthen in response to sonar. Nature 405(6789):903. https://doi.org/10.1038/35016148

Møhl B (1964) Preliminary studies on hearing in seals. Vidensk Medd fra Dansk naturh Foren 127:283–294

Møhl B (1967) Frequency discrimination in the common seal and a discussion of the concept of upper hearing limit. In: Underwater acoustics, vol 4. Plenum Press, New York, pp 43–54

Møhl B (1968a) Auditory sensitivity of the common seal in air and water. J Aud Res 8:27–38

Møhl B (1968b) Hearing in seals. In: Harrison RJ, Hubbard R, Rice C, Schusterman RJ (eds) Behavior and physiology of pinnipeds, vol 5. Appleton-Century, New York, pp 172–195

Møhl B, Ronald K (1975) The peripheral auditory system of the harp seal, *Pagophilus groenlandicus*, (Erxleben, 1777). In: Ronald K, Mansfield AW (eds) Biology of the seal. Conseil International Pour L'exploration, Denmark, pp 516–523

Møhl B, Au WWL, Pawloski J, Nachtigall PE (1999) Dolphin hearing: relative sensitivity as a function of point of application of a contact sound source in the jaw and head region. J Acoust Soc Am 105(6):3421–3424. https://doi.org/10.1121/1.426959

Montie EW, Manire CA, Mann DA (2011) Live CT imaging of sound reception anatomy and hearing measurements in the pygmy killer whale, *Feresa attenuata*. J Exp Biol 214:945–955. https://doi.org/10.1242/jeb.051599

Mooney TA, Nachtigall PE, Yuen MML (2006) Temporal resolution of the Risso's dolphin, *Grampus griseus*, auditory system. J Comp Physiol A 192:373–380. https://doi.org/10.1007/s00359-005-0075-4

Mooney TA, Nachtigall PE, Castellote M, Taylor KA, Pacini AF, Esteban J-A (2008) Hearing pathways and directional sensitivity of the beluga whale, *Delphinapterus leucas*. J Exp Mar Biol Ecol 362:108–116. https://doi.org/10.1016/j.jembe.2008.06.004

Mooney TA, Nachtigall PE, Taylor KA, Rasmussen MH, Miller LA (2009) Auditory temporal resolution of a wild white-beaked dolphin (*Lagenorhynchus albirostris*). J Comp Physiol A 195:375–384. https://doi.org/10.1007/s00359-009-0415-x

Mooney TA, Li S, Ketten DR, Wang K, Wang D (2011) Auditory temporal resolution and evoked responses to pulsed sounds for the Yangtze finless porpoises (*Neophocaena phocaenoides asiaeorientalis*). J Comp Physiol A 197:1149–1158. https://doi.org/10.1007/s00359-011-0677-y

Mooney TA, Castellote M, Quakenbush L, Hobbs R, Gaglione E, Goertz C (2018) Variation in hearing within a wild population of beluga whales (*Delphinapterus leucas*). J Exp Biol 221(9):1–13. https://doi.org/10.1242/jeb.171959

Mooney TA, Castellote M, Jones I, Rouse N, Rowles T, Mahoney B, Goertz CEC (2020) Audiogram of a Cook Inlet beluga whale (*Delphinapterus leucas*). J Acoust Soc Am 148(5):3141–3148. https://doi.org/10.1121/10.0002351

Moore BCJ (1973) Frequency difference limens for short-duration tones. J Acoust Soc Am 54(3):610–619. https://doi.org/10.1121/1.1913640

Moore PWB (1975) Underwater localization of click and pulsed pure-tone signals by the California sea lion (*Zalophus californianus*). J Acoust Soc Am 57(2):406–410. https://doi.org/10.1121/1.380456

Moore PWB, Au WWL (1975) Underwater localization of pulsed pure tones by the California sea lion (*Zalophus californianus*). J Acoust Soc Am 58(3):721–727. https://doi.org/10.1121/1.380720

Moore PWB, Pawloski DA (1990) Investigations on the control of echolocation pulses in the dolphin (*Tursiops truncatus*). In: Thomas JA, Kastelein RA (eds) Sensory abilities of cetaceans: laboratory and field evidence. Plenum, New York, pp 305–316. https://doi.org/10.1007/978-1-4899-0858-2_19

Moore PWB, Schusterman RJ (1976) Discrimination of pure-tone intensities by the California sea lion. J Acoust Soc Am 60(6):1405–1407. https://doi.org/10.1121/1.381234

Moore PWB, Schusterman RJ (1987) Audiometric assessment of northern fur seals, *Callorhinus ursinus*. Mar Mamm Sci 3(1):31–53. https://doi.org/10.1111/j.1748-7692.1987.tb00150.x

Moore PWB, Hall RW, Friedl WA, Nachtigall PE (1984) The critical interval in dolphin echolocation: what is it? J Acoust Soc Am 76:314–317. https://doi.org/10.1121/1.391016

Moore PWB, Pawloski DA, Dankiewicz L (1995) Interaural time and intensity difference thresholds in the bottlenose dolphin (*Tursiops truncatus*). In: Kastelein RA, Thomas JA, Nachtigall PE (eds) Sensory systems of aquatic mammals. De Spil Publishers, Woerden, pp 11–23. https://doi.org/10.1121/1.4892795

Moore PW, Dankiewicz LA, Houser DS (2008) Beamwidth control and angular target detection in an echolocating bottlenose dolphin (*Tursiops truncatus*). J Acoust Soc Am 124(5):3324–3332. https://doi.org/10.1121/1.2980453

Morisaka T, Karczmarski L, Akamatsu T, Sakai M, Dawson S, Thornton M (2011) Echolocation signals of Heaviside's dolphins (*Cephalorhynchus heavisidii*). J Acoust Soc Am 129(1):449–457. https://doi.org/10.1121/1.3519401

Mulsow J, Reichmuth C (2007) Electrophysiological assessment of temporal resolution in pinnipeds. Aquat Mamm 33(1):122–131. https://doi.org/10.1578/AM.33.1.2007.122

Mulsow JL, Reichmuth C (2010) Psychophysical and electrophysiological aerial audiograms of a Steller sea lion (*Eumetopias jubatus*). J Acoust Soc Am 127(4):2692–2701. https://doi.org/10.1121/1.3327662

Mulsow J, Reichmuth C (2013) The binaural click-evoked auditory brainstem response of the California sea lion (*Zalophus californianus*). J Acoust Soc Am 133(1):579–586. https://doi.org/10.1121/1.4770253

Mulsow J, Reichmuth C, Gulland F, Rosen DAS, Finneran JJ (2011a) Aerial audiograms of several California sea lions (*Zalophus californianus*) and Steller sea lions (*Eumetopias jubatus*) measured using single and multiple simultaneous auditory steady-state response methods. J Exp Biol 214:1138–1147. https://doi.org/10.1242/jeb.052837

Mulsow JL, Finneran JJ, Houser DS (2011b) California sea lion (*Zalophus californianus*) aerial hearing sensitivity measured using auditory steady-state response and psychophysical methods. J Acoust Soc Am 129(4):2298–2306. https://doi.org/10.1121/1.3552882

Mulsow J, Houser DS, Finneran JJ (2012) Underwater psychophysical audiogram of a young male California sea lion (*Zalophus californianus*). J Acoust Soc Am 131(5):4182–4187. https://doi.org/10.1121/1.3699195

Mulsow J, Finneran JJ, Houser DS (2014a) Interaural differences in the bottlenose dolphin (*Tursiops truncatus*) auditory nerve response to jawphone stimuli. J Acoust Soc Am 136(3):1402–1409. https://doi.org/10.1121/1.4892795

Mulsow J, Houser DS, Finneran JJ (2014b) Aerial hearing thresholds and detection of hearing loss in male California sea lions (*Zalophus californianus*) using auditory evoked potentials. Mar Mamm Sci 30(4):1383–1400. https://doi.org/10.1111/mms.12123

Mulsow J, Schlundt CE, Brandt L, Finneran JJ (2015) Equal latency contours for bottlenose dolphins (*Tursiops truncatus*) and California sea lions (*Zalophus californianus*). J Acoust Soc Am 138(5):2678–2691. https://doi.org/10.1121/1.4932015

Nachtigall PE, Supin AY (2013) A false killer whale reduces its hearing sensitivity when a loud sound is preceded by a warning. J Exp Biol 216(Pt 16):3062–3070. https://doi.org/10.1242/jeb.085068

Nachtigall PE, Supin AY (2014) Conditioned hearing sensitivity reduction in a bottlenose dolphin (*Tursiops truncatus*). J Exp Biol 217(Pt 15):2806–2813. https://doi.org/10.1242/jeb.104091

Nachtigall PE, Au WWL, Pawloski J, Moore PWB (1995) Risso's dolphin (*Grampus griseus*) hearing thresholds in Kaneohe Bay, Hawaii. In: Kastelein RA, Thomas JA, Nachtigall PE (eds) Sensory systems of aquatic mammals. DeSpil, Woerden, pp 49–53

Nachtigall PE, Lemonds DW, Roitblat HL (2000) Psychoacoustic studies of dolphin and whale hearing. In: Au WWL, Popper AN, Fay RR (eds) Hearing by whales and dolphins. Springer, New York, pp 330–363. https://doi.org/10.1007/978-1-4612-1150-1_8

Nachtigall PE, Yuen MML, Mooney TA, Taylor KA (2005) Hearing measurements from a stranded infant Risso's dolphin, *Grampus griseus*. J Exp Biol 208:4181–4188. https://doi.org/10.1242/jeb.01876

Nachtigall PE, Supin AY, Amundin M, Roken B, Møller T, Mooney TA, Taylor KA, Yuen M (2007) Polar bear *Ursus maritimus* hearing measured with auditory evoked potentials. J Exp Biol 210(7):1116–1122. https://doi.org/10.1242/jeb.02734

Nachtigall PE, Mooney TA, Taylor KA, Miller LA, Rasmussen MH, Akamatsu T, Teilmann J, Linnenschmidt M, Vikingsson GA (2008) Shipboard measurements of the hearing of the white-beaked dolphin *Lagenorhynchus albirostris*. J Exp Biol 211:642–647. https://doi.org/10.1242/Jeb.014118

Nachtigall PE, Supin AY, Estaban JA, Pacini AF (2016a) Learning and extinction of conditioned hearing sensation change in the beluga whale (*Delphinapterus leucas*). J Comp Physiol A 202(2):105–113. https://doi.org/10.1007/s00359-015-1056-x

Nachtigall PE, Supin AY, Pacini AF, Kastelein RA (2016b) Conditioned hearing sensitivity change in the harbor porpoise (*Phocoena phocoena*). J Acoust Soc Am 140(2):960–967. https://doi.org/10.1121/1.4960783

Nakahara F, Takemura A, Koido T, Hiruda H (1997) Target discrimination by an echolocating finless porpoise, *Neophocaena phocaenoides*. Mar Mamm Sci 13(4):639–649. https://doi.org/10.1111/j.1748-7692.1997.tb00088.x

National Marine Fisheries Service (2016) Technical guidance for assessing the effects of anthropogenic sound on marine mammal hearing—underwater acoustic thresholds for onset of permanent and temporary threshold shifts. National Oceanic and Atmospheric Administration, Silver Springs, 178 pp

National Marine Fisheries Service (2018) Revision to: technical guidance for assessing the effects of anthropogenic sound on marine mammal hearing (Version 2.0)—underwater acoustic thresholds for onset of permanent and temporary threshold shifts. National Oceanic and Atmospheric Administration, Silver Springs, 167 pp

Nedwell JR, Turnpenny AWH (1998) The use of a generic frequency weighting scale in estimating environmental effect. In: Proceedings of the workshop on seismics and marine mammals, London, 23–25th June 1998

Nedwell JR, Turnpenny AWH, Lovell J, Parvin SJ, Workman R, Spinks JAL, Howell D (2007) A validation of the dBht as a measure of the behavioural and auditory effects of underwater noise (534R1231). Subacoustech Acoustic Research Consultancy, pp 1–74

Nelken I, Rotman Y, Bar Yosef O (1999) Response of auditory-cortex neurons to structural features of natural sounds. Nature 397:154–157. https://doi.org/10.1038/16456

Norris KS (1968) The evolution of acoustic mechanisms in odontocete cetaceans. In: Drake ET (ed) Evolution and environment, vol 10. Yale University Press, New Haven, pp 297–324

Nummela S (2008) Hearing in aquatic mammals. In: Thewissen JGM, Nummela S (eds) Sensory evolution on the threshold. University of California Press, Berkeley, pp 211–224. https://doi.org/10.1525/california/9780520252783.003.0013

Oelschläger HA (1990) Evolutionary morphology and acoustics in the dolphin skull. In: Thomas JA, Kastelein RA (eds) Sensory abilities of cetaceans: laboratory and field evidence. Plenum Press, New York, pp 137–162. https://doi.org/10.1007/978-1-4899-0858-2_8

Owen MA, Bowles AE (2011) In-air auditory psychophysics and the management of a threatened carnivore, the polar bear (*Ursus maritimus*). Int J Comp Psychol 24:244–254. https://doi.org/10.46867/ijcp.2011.24.03.05

Pacini AF, Nachtigall PE, Kloepper LN, Linnenschmidt M, Sogorb A, Matias S (2010) Audiogram of a formerly stranded long-finned pilot whale (*Globicephala melas*) measured using auditory evoked potentials. J Exp Biol 213(Pt 18):3138–3143. https://doi.org/10.1242/jeb.044636

Pacini AF, Nachtigall PE, Quintos CT, Schofield TD, Look DA, Levine GA, Turner JP (2011) Audiogram of a stranded Blainville's beaked whale (*Mesoplodon densirostris*) measured during auditory evoked potentials. J Exp Biol 214:2409–2415. https://doi.org/10.1242/jeb.054338

Pacini AF, Nachtigall PE, Smith AB, Suarez LJA, Magno C, Laule GE, Aragones LV, Braun R (2016) Evidence of hearing loss due to dynamite fishing in two species of odontocetes. Proc Meet Acoust 27(1):010043. https://doi.org/10.1121/2.0000393

Parks SE, Clark CW, Tyack PL (2007a) Short- and long-term changes in right whale calling behavior: the potential effects of noise on acoustic communication. J Acoust Soc Am 122(6):3725–3731. https://doi.org/10.1121/1.2799904

Parks SE, Ketten DR, O'Malley JT, Arruda J (2007b) Anatomical predictions of hearing in the North Atlantic right whale. Anat Rec 290:734–744. https://doi.org/10.1002/Ar.20527

Parks SE, Johnson M, Nowacek D, Tyack PL (2011) Individual right whales call louder in increased environmental noise. Biol Lett 7:33–35. https://doi.org/10.1098/rsbl.2010.0451

Penner RH, Turl CW, Au WWL (1986) Target detection by the beluga using a surface-reflected path. J Acoust Soc Am 80:1842–1843. https://doi.org/10.1121/1.394301

Pfingst BE, Hienz R, Kimm J, Miller J (1975) Reaction-time procedure for measurement of hearing. I. Suprathreshold functions. J Acoust Soc Am 57(2):421–430. https://doi.org/10.1121/1.380465

Popov VV, Klishin VO (1998) EEG study of hearing in the common dolphin, *Delphinus delphis*. Aquat Mamm 24(1):13–20

Popov VV, Supin AY (1976) Determination of the hearing characteristics of dolphins by measuring induced potentials. Fiziol Zh 62(4):550–557

Popov V, Supin A (1990a) Localization of the acoustic window at the dolphin's head. In: Thomas JA, Kastelein RA (eds) Sensory abilities of cetaceans: laboratory and field evidence. Plenum Press, New York, pp 417–426

Popov VV, Supin AY (1990b) Electrophysiological studies of hearing in some cetaceans and a manatee. In: Thomas JA, Kastelein RA (eds) Sensory abilities in cetaceans. Plenum Press, New York, pp 405–415. https://doi.org/10.1007/978-1-4899-0858-2_27

Popov VV, Supin AY (1990c) Auditory brain stem responses in characterization of dolphin hearing. J Comp Physiol A 166(3):385–393. https://doi.org/10.1007/BF00204811

Popov VV, Supin AY (1990d) Electrophysiological investigation of hearing in the freshwater dolphin *Inia geoffrensis*. Dokl Biol Sci 313(1):238–241

Popov VV, Supin A (1997) Detection of temporal gaps in noise in dolphins: evoked-potential study. J Acoust Soc Am 102(2):1169–1176. https://doi.org/10.1121/1.419935

Popov VV, Supin AY (1998) Auditory evoked responses to rhythmic sound pulses in dolphins. J Comp Physiol A 183(4):519–524. https://doi.org/10.1007/s003590050277

Popov VV, Supin AY (2009) Comparison of directional selectivity of hearing in a beluga whale and a bottlenose dolphin. J Acoust Soc Am 126(3):1581. https://doi.org/10.1121/1.3177273

Popov VV, Ladygina TF, Supin AY (1986) Evoked potentials of the auditory cortex of the porpoise, *Phocoena phocoena*. J Comp Physiol A 158(5): 705–711. https://doi.org/10.1007/bf00603828

Popov VV, Supin AY, Klishin VO (1992) Electrophysiological study of sound conduction in dolphins. In: Thomas JA, Kastelein RA, Supin AY (eds) Marine mammal sensory systems. Plenum Press, New York, pp 269–276. https://doi.org/10.1007/978-1-4615-3406-8_18

Popov VV, Supin AY, Klishin VO (1997) Frequency tuning of the dolphin's hearing as revealed by auditory brain-stem response with notch-noise masking. J Acoust Soc Am 102(6):3795–3801. https://doi.org/10.1121/1.420142

Popov VV, Supin AY, Klishin VO (2001) Auditory brainstem response recovery in the dolphin as revealed by double sound pulses of different frequencies. J Acoust Soc Am 110(4):2227–2233. https://doi.org/10.1121/1.1404382

Popov VV, Supin AY, Wang D, Wang K, Xiao J, Li S (2005) Evoked-potential audiogram of the Yangtze finless porpoise *Neophocaena phocaenoides asiaeorientalis* (L). J Acoust Soc Am 117(5): 2728–2731. https://doi.org/10.1121/1.1880712

Popov VV, Supin AY, Klishin VO, Bulgakova TN (2006a) Monaural and binaural hearing directivity in the bottlenose dolphin: evoked-potential study. J Acoust Soc Am 119(1):636–644. https://doi.org/10.1121/1.2141093

Popov VV, Supin AY, Wang D, Wang K (2006b) Nonconstant quality of auditory filters in the porpoises, *Phocoena phocoena* and *Neophocaena phocaenoides* (Cetacea, Phocoenidae). J Acoust Soc Am 119(5): 3173–3180. https://doi.org/10.1121/1.2184290

Popov VV, Supin AY, Pletenko MG, Tarakanov MB, Klishin VO, Bulgakova TN, Rosanova EI (2007) Audiogram variability in normal bottlenose dolphins (*Tursiops truncatus*). Aquat Mamm 33(1):24–33. https://doi.org/10.1578/AM.33.1.2007.24

Popov VV, Supin AY, Klishin VO, Tarakanov MB, Pletenko MG (2008) Evidence for double acoustic windows in the dolphin, *Tursiops truncatus*. J Acoust Soc Am 123(1):552–560. https://doi.org/10.1121/1.2816564

Popov VV, Supin AY, Wang D, Wang K, Dong L, Wang S (2011) Noise-induced temporary threshold shift and recovery in Yangtze finless porpoises *Neophocaena phocaenoides asiaeorientalis*. J Acoust Soc Am 130(1):574–584. https://doi.org/10.1121/1.3596470

Popov VV, Supin AY, Rozhnov VV, Nechaev DI, Sysueyva EV, Klishin VO, Pletenko MG, Tarakanov MB (2013) Hearing threshold shifts and recovery after noise exposure in beluga whales *Delphinapterus leucas*. J Exp Biol 216:1587–1596. https://doi.org/10.1242/jeb.078345

Popov VV, Supin AY, Rozhnov VV, Nechaev DI, Sysueva EV (2014) The limits of applicability of the sound exposure level (SEL) metric to temporal threshold shifts (TTS) in beluga whales, *Delphinapterus leucas*. J Exp Biol 217(Pt 10):1804–1810. https://doi.org/10.1242/jeb.098814

Popov VV, Nechaev DI, Sysueva EV, Rozhnov VV, Supin AY (2015) Spectrum pattern resolution after noise exposure in a beluga whale, *Delphinapterus leucas*: evoked potential study. J Acoust Soc Am 138(1):377–388. https://doi.org/10.1121/1.4923157

Popov VV, Supin AY, Gvozdeva AP, Nechaev DI, Tarakanov MB, Sysueva EV (2020) Spatial release from masking in a bottlenose dolphin *Tursiops truncatus*. J Acoust Soc Am 147(3):1719–1726. https://doi.org/10.1121/10.0000909

Ramprashad F (1975) Aquatic adaptations in the ear of the Harp seal *Pagophilus groenlandicus* (Erxleben, 1777). In: Ronald K, Mansfield AW (eds) Biology of the seal. Conseil International Pour L'exploration, Denmark, pp 102–111

Reichmuth C (2013) Equal loudness contours and possible weighting functions for pinnipeds. J Acoust Soc Am 134(5):4210(A). https://doi.org/10.1121/1.4831454

Reichmuth C, Southall BL (2012) Underwater hearing in California sea lions (*Zalophus californianus*): expansion and interpretation of existing data. Mar Mamm Sci 28(2):358–363. https://doi.org/10.1111/j.1748-7692.2011.00473.x

Reichmuth C, Mulsow J, Finneran JJ, Houser DS, Supin AY (2007) Measurement and response characteristics of auditory brainstem responses in pinnipeds. Aquat Mamm 33(1):132–150. https://doi.org/10.1578/AM.33.1.2007.132

Reichmuth C, Ghoul A, Southall BL (2012) Temporal processing of low-frequency sounds by seals (L). J Acoust Soc Am 132(4):2147–2150. https://doi.org/10.1121/1.4746030

Reichmuth C, Holt MM, Mulsow J, Sills JM, Southall BL (2013) Comparative assessment of amphibious hearing in pinnipeds. J Comp Physiol A 199(6):491–507. https://doi.org/10.1007/s00359-013-0813-y

Reichmuth C, Sills JM, Ghoul A (2017) Psychophysical audiogram of a California sea lion listening for airborne tonal sounds in an acoustic chamber. Proc Meet Acoust 30(1):010001. https://doi.org/10.1121/2.0000525

Reichmuth C, Sills JM, Brewer A, Triggs L, Ferguson R, Ashe E, Williams R (2020) Behavioral assessment of in-air hearing range for the Pacific walrus (*Odobenus rosmarus divergens*). Polar Biol 14(3):e4433. https://doi.org/10.1007/s00300-020-02667-6

Renaud DL, Popper AN (1975) Sound localization by the bottlenose porpoise *Tursiops truncatus*. J Exp Biol 63(3):569–585. https://doi.org/10.1242/jeb.63.3.569

Renouf D (1980) Masked hearing thresholds of harbour seals (*Phoca vitulina*) in air. J Aud Res 20:263–269

Repenning CA (1972) Underwater hearing in seals: functional morphology. In: Harrison RJ (ed) Functional anatomy of marine mammals, vol 1. Academic, London, pp 307–331

Richardson WJ, Würsig B, Greene CR Jr (1986) Reactions of bowhead whales, *Balaena mysticetus*, to seismic exploration in the Canadian Beaufort Sea. J Acoust Soc Am 79(4):1117–1128. https://doi.org/10.1121/1.393384

Richardson JW, Würsig B, Greene CR Jr (1990) Reactions of bowhead whales, *Balaena mysticetus*, to drilling and dredging noise in the Canadian Sea. Mar Envion Res 29:135–160. https://doi.org/10.1016/0141-1136(90)90032-J

Richardson WJ, Greene CR, Malme CI, Thomson DH (1995) Marine mammals and noise. Academic. https://doi.org/10.1016/C2009-0-02253-3

Richardson WJ, Miller GW, Greene CR Jr (1999) Displacement of migrating bowhead whales by sounds from seismic surveys in shallow waters of the Beaufort Sea. J Acoust Soc Am 106(4_Suppl):2281. https://doi.org/10.1121/1.427801

Ridgway SH (1999) An illustration of Norris' acoustic window. Mar Mamm Sci 15(4):926–930. https://doi.org/10.1111/j.1748-7692.1999.tb00861.x

Ridgway SH (2000) The auditory central nervous system of dolphins. In: Au WWL, Fay RR, Popper AN (eds) Hearing by whales and dolphins, Springer handbook of auditory research, vol 12. Springer, New York, pp 273–293. https://doi.org/10.1007/978-1-4612-1150-1_6

Ridgway SH, Carder DA (1997) Hearing deficits measured in some *Tursiops truncatus*, and discovery of a deaf/mute dolphin. J Acoust Soc Am 101(1):590–594. https://doi.org/10.1121/1.418122

Ridgway SH, Carder DA (2001) Assessing hearing and sound production in cetaceans not available for behavioral audiograms: experiences with sperm, pygmy sperm, and gray whales. Aquat Mamm 27(3):267–276

Ridgway SH, Joyce PL (1975) Studies on seal brain by radiotelemetry. Rap P-v Reun Cons Int Explor Mer 169:81–91

Ridgway SH, Carder DA, Smith RR, Kamolnick T, Schlundt CE, Elsberry WR (1997) Behavioral responses and temporary shift in masked hearing thresholds of bottlenose dolphins, *Tursiops truncatus*, to 1-second tones of 141–201 dB re 1 μPa (Technical Report 1751). Naval Command, Control, and Ocean Surveillance Center, RDT&E Division, San Diego, 31 pp

Ridgway SH, Carder DA, Kamolnick T, Smith RR, Schlundt CE, Elsberry WR (2001) Hearing and whistling in the deep sea: depth influences whistle spectra but does not attenuate hearing by white whales (*Delphinapterus leucas*) (Odontoceti, Cetacea). J Exp Biol 204:3829–3841. https://doi.org/10.1242/jeb.204.22.3829

Robertson FC, Koski WR, Thomas TA, Richardson WJ, Würsig B, Trites AW (2013) Seismic operations have variable effects on dive-cycle behavior of bowhead whales in the Beaufort Sea. Endanger Species Res 21:143–160. https://doi.org/10.3354/esr00515

Robinson DW, Dadson RS (1956) A re-determination of the equal-loudness relations for pure tones. Br J Appl Phys 7(5):166–181. https://doi.org/10.1088/0508-3443/7/5/302

Romanenko EV, Kitain VY (1992) The functioning of the echolocation system of *Tursiops truncatus* during noise masking. In: Thomas JA, Kastelein RA, Supin AY (eds) Marine mammal sensory systems. Plenum Press, New York, pp 415–419. https://doi.org/10.1007/978-1-4615-3406-8_27

Rommel SA, Costidis AM, Fernández A, Jepson PD, Pabst DA, McLellan WA, Houser DS, Cranford TW, Van Helden AL, Aleen DM, Barros NB (2006) Elements of beaked whale anatomy and diving physiology and some hypothetical causes of sonar-related stranding. J Cetacean Res Manag 7(3):189–209. https://doi.org/10.47536/jcrm.v7i3.730

Ruscher B, Sills JM, Richter BP, Reichmuth C (2021) In-air hearing in Hawaiian monk seals: implications for understanding the auditory biology of Monachinae seals. J Comp Physiol A 207(4):561–573. https://doi.org/10.1007/s00359-021-01498-y

Ruser A, Dähne M, Sundermeyer J, Lucke K, Houser DS, Finneran JJ, Driver J, Pawliczka I, Rosenberger T, Siebert U (2014) In-air evoked potential audiometry of grey seals (*Halichoerus grypus*) from the North and Baltic Sea. PLoS One 9(3):e90824. https://doi.org/10.1371/journal.pone.0090824.g001

Ruser A, Dähne M, Av N, Lucke K, Sundermeyer J, Siebert U, Houser DS, Finneran JJ, Everaarts E, Meerbeek J, Dietz R, Sveegaard S, Teilmann J (2016) Assessing auditory evoked potentials of wild harbor porpoises (*Phocoena phocoena*). J Acoust Soc Am 140(1):442–452. https://doi.org/10.1121/1.4955306

Sauerland M, Dehnhardt G (1998) Underwater audiogram of a tucuxi (*Sotalia fluviatilis guianensis*). J Acoust Soc Am 103(2):1199–1204. https://doi.org/10.1121/1.421228

Scheifele PM, Andrew S, Cooper RA, Darre M, Musiek FE, Max L (2005) Indication of a Lombard vocal response in the St. Lawrence River beluga. J Acoust Soc Am 117(3):1486–1492. https://doi.org/10.1121/1.1835508

Schlundt CE, Finneran JJ, Carder DA, Ridgway SH (2000) Temporary shift in masked hearing thresholds of bottlenose dolphins, *Tursiops truncatus*, and white whales, *Delphinapterus leucas*, after exposure to intense tones. J Acoust Soc Am 107(6):3496–3508. https://doi.org/10.1121/1.429420

Schlundt CE, Dear RL, Green L, Houser DS, Finneran JJ (2007) Simultaneously measured behavioral and electrophysiological hearing thresholds in a bottlenose dolphin (*Tursiops truncatus*). J Acoust Soc Am 122(1):615–622. https://doi.org/10.1121/1.2737982

Schlundt CE, Finneran JJ, Branstetter BK, Dear RL, Houser DS, Hernandez E (2008) Evoked potential and behavioral hearing thresholds in nine bottlenose dolphins (*Tursiops truncatus*). J Acoust Soc Am 123: 3506(A)

Schlundt CE, Dear RL, Houser DS, Bowles AE, Reidarson T, Finneran JJ (2011) Auditory evoked potentials in two short-finned pilot whales (*Globicephala macrorhynchus*). J Acoust Soc Am 129(2):1111–1116. https://doi.org/10.1121/1.3531875

Schusterman RJ (1974) Auditory sensitivity of a California sea lion to airborne sound. J Acoust Soc Am 756:1248–1251. https://doi.org/10.1121/1.1903415

Schusterman RJ (1981) Behavioral capabilities of seals and sea lions: a review of their hearing, visual, learning and diving skills. Psychol Rec 31:125–143. https://doi.org/10.1007/BF03394729

Schusterman RJ, Johnson BW (1975) Signal probability and response bias in California sea lions. Psychol Rec 25:39–45. https://doi.org/10.1007/BF03394287

Schusterman RJ, Moore PW (1978) The upper limit of underwater auditory frequency discrimination in the California sea lion. J Acoust Soc Am 63(5): 1591–1595. https://doi.org/10.1121/1.381853

Schusterman RJ, Balliet RF, Nixon J (1972) Underwater audiogram of the California sea lion by the conditioned vocalization technique. J Exp Anal Behav 17:339–350. https://doi.org/10.1901/jeab.1972.17-339

Sills JM, Southall BL, Reichmuth C (2014) Amphibious hearing in spotted seals (*Phoca largha*): underwater audiograms, aerial audiograms and critical ratio measurements. J Exp Biol 217(Pt 5):726–734. https://doi.org/10.1242/jeb.097469

Sills JM, Southall BL, Reichmuth C (2015) Amphibious hearing in ringed seals (*Pusa hispida*): underwater audiograms, aerial audiograms and critical ratio measurements. J Exp Biol 218(Pt 14):2250–2259. https://doi.org/10.1242/jeb.120972

Sills JM, Southall BL, Reichmuth C (2017) The influence of temporally varying noise from seismic air guns on the detection of underwater sounds by seals. J Acoust Soc Am 141(2):996. https://doi.org/10.1121/1.4976079

Sills JM, Reichmuth C, Southall BL, Whiting A, Goodwin J (2020) Auditory biology of bearded seals (*Erignathus barbatus*). Polar Biol 43(11):1681–1691. https://doi.org/10.1007/s00300-020-02736-w

Sills JM, Parnell K, Ruscher-Hill B, Lew C, Kendall TL, Reichmuth C (2021) Underwater hearing and communication in the endangered Hawaiian monk seal, *Neomonachus schauinslandi*. Endanger Species Res 44:61–78. https://doi.org/10.3354/esr01092

Smith AB, Pacini AF, Nachtigall PE (2018) Modulation rate transfer functions from four species of stranded odontocete (*Stenella longirostris*, *Feresa attenuata*, *Globicephala melas*, and *Mesoplodon densirostris*). J Comp Physiol A 204(4):377–389. https://doi.org/10.1007/s00359-018-1246-4

Southall BL, Schusterman RJ, Kastak D (2000) Masking in three pinnipeds: underwater, low-frequency critical ratios. J Acoust Soc Am 108(3):1322–1326. https://doi.org/10.1121/1.1288409

Southall BL, Schusterman RJ, Kastak D (2003) Auditory masking in three pinnipeds: aerial critical ratios and direct critical bandwidth measurements. J Acoust Soc Am 114(3):1660–1666. https://doi.org/10.1121/1.1587733

Southall BL, Bowles AE, Ellison WT, Finneran JJ, Gentry RL, Greene CR Jr, Kastak D, Ketten DR, Miller JH, Nachtigall PE, Richardson WJ, Thomas JA, Tyack PL (2007) Marine mammal noise exposure criteria: initial scientific recommendations. Aquat Mamm 33(4): 411–521. https://doi.org/10.1578/AM.45.2.2019.125

Southall BL, Casey C, Holt M, Insley S, Reichmuth C (2019a) High-amplitude vocalizations of male northern elephant seals and associated ambient noise on a breeding rookery. J Acoust Soc Am 146(6):4514–4524. https://doi.org/10.1121/1.5139422

Southall BL, Finneran JJ, Reichmuth C, Nachtigall PE, Ketten DR, Bowles AE, Ellison WT, Nowacek DP, Tyack PL (2019b) Marine mammal noise exposure criteria: auditory weighting functions and TTS/PTS onset. Aquat Mamm 45(2):125–232. https://doi.org/10.1578/AM.45.2.2019.125

Stebbins WC (1970) Animal psychophysics: the design and conduct of sensory experiments. Appleton, New York

Strahan MG, Finneran JJ, Mulsow J, Houser D (2020) Effects of dolphin hearing bandwidth on biosonar click emissions. J Acoust Soc Am 148(1):243–252. https://doi.org/10.1121/10.0001497

Strutt JW (1909) On the perception of the direction of sound. Proc R Soc A 83(559):61–64. https://doi.org/10.1098/rspa.1909.0073

Sukhoruchenko MN (1973) Frequency discrimination in dolphins (*Phocoena phocoena*). Sechenov Physiol J USSR 59(8):1205–1209

Supin AY, Nachtigall PE (2013) Gain control in the sonar of odontocetes. J Comp Physiol A 199(6):471–478. https://doi.org/10.1007/s00359-012-0773-7

Supin AY, Popov VV (1993) Direction-dependent spectral sensitivity and interaural spectral difference in a dolphin: evoked potential study. J Acoust Soc Am 93(6): 3490–3495. https://doi.org/10.1121/1.405679

Supin AY, Popov VV (1995) Temporal resolution in the dolphin's auditory system revealed by double-click evoked potential study. J Acoust Soc Am 97(4): 2586–2593. https://doi.org/10.1121/1.411913

Supin AY, Popov VV (2007) Improved techniques of evoked-potential audiometry in odontocetes. Aquat Mamm 33(1):14–23. https://doi.org/10.1578/AM.33.1.2007.14

Supin AY, Sukhoruchenko MN (1974) Characteristics of acoustic analyzer of the *Phocoena phocoena* L. dolphin. Morfologiya, Fiziologiya I Akustinka Morskikh Mlekopitayushchikh:129–136

Supin AY, Popov VV, Mass AM (2001) The sensory physiology of aquatic mammals. Kluwer Academic

Publishers, Boston. https://doi.org/10.1007/978-1-4615-1647-7

Supin AY, Nachtigall PE, Breese M (2008) Hearing sensitivity during target presence and absence while a whale echolocates. J Acoust Soc Am 123(1):534–541. https://doi.org/10.1121/1.2812593

Swets JA (2003) The relative operating characteristic in psychology. Science 182(2):990–1000. https://doi.org/10.1126/science.182.4116.990

Sysueva EV, Nechaev DI, Popov VV, Supin AY (2014) Frequency tuning of hearing in the beluga whale: discrimination of rippled spectra. J Acoust Soc Am 135(2):963–974. https://doi.org/10.1121/1.4823846

Szymanski MD (1996) Auditory evoked potentials of killer whales (Orcinus orca). PhD thesis, University of California Davis

Szymanski MD, Bain DE, Henry KR (1995) Auditory evoked potentials of a killer whale (Orcinus orca). In: Kastelein RA, Thomas JA, Nachtigall PE (eds) Sensory systems of aquatic mammals. De Spil Publishers, Woerden, pp 1–10

Szymanski MD, Supin AY, Bain DE, Henry KR (1998) Killer whale (Orcinus orca) auditory evoked potentials to rhythmic clicks. Mar Mamm Sci 14(4):676–691. https://doi.org/10.1111/j.1748-7692.1998.tb00756.x

Szymanski MD, Bain DE, Kiehl K, Pennington S, Wong S, Henry KR (1999) Killer whale (Orcinus orca) hearing: auditory brainstem response and behavioral audiograms. J Acoust Soc Am 106(2):1134–1141. https://doi.org/10.1121/1.427121

Terhune JM (1974) Directional hearing of a harbor seal in air and water. J Acoust Soc Am 56:1862–1865. https://doi.org/10.1121/1.1903523

Terhune, JM (1981) Influence of loud vessel noises on marine mammal hearing and vocal communication. In: Peterson, NM (ed), The question of sound from icebreaker operations: The proceedings of a workshop. Arctic Pilot Project, Petro-Canada, Calgary, p 350

Terhune JM (1988) Detection thresholds of a harbour seal to repeated underwater high-frequency, short-duration sinusoidal pulses. Can J Zool 66:1578–1582. https://doi.org/10.1139/z88-230

Terhune JM (1989) Underwater click hearing thresholds of a harbour seal, Phoca vitulina. Aquat Mamm 15(1):22–26

Terhune JM (1991) Masked and unmasked pure tone thresholds of a harbour seal listening in air. Can J Zool 69:2059–2066. https://doi.org/10.1139/z91-287

Terhune JM (2013) A practical weighting function for harbor porpoise underwater sound level measurements. J Acoust Soc Am 134(3):2405–2408. https://doi.org/10.1121/1.4816556

Terhune JM, Ronald K (1971) The harp seal, Pagophilus groenlandicus (Erxleben, 1777) X. The air audiogram. Can J Zool 49:385–390. https://doi.org/10.1139/z71-057

Terhune JM, Ronald K (1972) The harp seal, Pagophilus groenlandicus (Erxleben, 1777) III. The underwater audiogram. Can J Zool 50:565–569. https://doi.org/10.1139/z72-077

Terhune JM, Ronald K (1975a) Masked hearing thresholds of ringed seals. J Acoust Soc Am 58:515–516

Terhune JM, Ronald K (1975b) Underwater hearing sensitivity of two ringed seals (Pusa hispida). Can J Zool 53:227–231. https://doi.org/10.1139/z75-028

Terhune JM, Ronald K (1976) The upper frequency limit of ringed seal hearing. Can J Zool 54:1226–1229. https://doi.org/10.1139/z76-139

Terhune JM, Turnbull S (1995) Variation in the psychometric functions and hearing thresholds of a harbor seal. In: Kastelein RA, Thomas JA, Nachtigall PE (eds) Sensory systems of aquatic mammals. De Spil Publishing, Woerden

Thomas JA, Turl CW (1990) Echolocation characteristics and range detection threshold of a false killer whale (Pseudorca crassidens). In: Thomas JA, Kastelein RA (eds) Sensory abilities of cetaceans: laboratory and field evidence. Plenum Press, New York, pp 321–334. https://doi.org/10.1007/978-1-4899-0858-2_21

Thomas J, Chun N, Au W, Pugh K (1988) Underwater audiogram of a false killer whale (Pseudorca crassidens). J Acoust Soc Am 84:936–940. https://doi.org/10.1121/1.396662

Thomas J, Moore P, Withrow R, Stoermer M (1990a) Underwater audiogram of a Hawaiian monk seal (Monachus schauinslandi). J Acoust Soc Am 87(1):417–420. https://doi.org/10.1121/1.399263

Thomas JA, Pawloski JL, Au WWL (1990b) Masked hearing abilities in a false killer whale (Pseudorca crassidens). In: Thomas JA, Kastelein RA (eds) Sensory abilities of cetaceans: laboratory and field evidence. Plenum Press, New York, pp 395–404. https://doi.org/10.1007/978-1-4899-0858-2_26

Thompson RKR, Herman LM (1975) Underwater frequency discrimination in the bottlenosed dolphin (1-140 kHz) and the human (1-8 kHz). J Acoust Soc Am 57(4):943–948. https://doi.org/10.1121/1.380513

Todd S, Lien J, Verhulst A (1992) Orientation of humpback whales (Megaptera novaeangliae) and minke whales (Balaenoptera acutorostrata) to acoustic alarm devices designed to reduce entrapment in fishing gear. In: Thomas JA, Kastelein RA, Supin AY (eds) Marine mammal sensory systems. Plenum Press, New York, pp 727–739. https://doi.org/10.1007/978-1-4615-3406-8_45

Tougaard J, Wright AJ, Madsen PT (2015) Cetacean noise criteria revisited in the light of proposed exposure limits for harbour porpoises. Mar Pollut Bull 90:196–208. https://doi.org/10.1016/j.marpolbul.2014.10.051

Tremel DP, Thomas JA, Ramierez KT, Dye GS, Bachman WA, Orban AN, Grimm KK (1998) Underwater hearing sensitivity of a Pacific white-sided dolphin, *Lagenorhynchus obliquidens*. Aquat Mamm 24(2): 63–69

Trickey JS, Branstetter BK, Finneran JJ (2010) Auditory masking of a 10 kHz tone with environmental, comodulated, and Gaussian noise in bottlenose dolphins (*Tursiops truncatus*). J Acoust Soc Am 128(6):3799–3804. https://doi.org/10.1121/1.3506367

Tripovich JS, Purdy SC, Hogg C, Rogers LT (2011) Toneburst-evoked auditory brainstem response in a leopard seal, *Hydrurga leptonyx*. J Acoust Soc Am 129(1):483–487. https://doi.org/10.1121/1.3514370

Tubelli AA, Zosuls A, Ketten DR, Yamato M, Mountain DC (2012) A prediction of the minke whale (*Balaenoptera acutorostrata*) middle-ear transfer function. J Acoust Soc Am 132(5):3263–3272. https://doi.org/10.1121/1.4756950

Tubelli AA, Zosuls A, Ketten DR, Mountain DC (2018) A model and experimental approach to the middle ear transfer function related to hearing in the humpback whale (*Megaptera novaeangliae*). J Acoust Soc Am 144(2):525–535. https://doi.org/10.1121/1.5048421

Turl CW (1993) Low-frequency sound detection by a bottlenose dolphin. J Acoust Soc Am 94(5): 3006–3008. https://doi.org/10.1121/1.407333

Turnbull SD (1994) Changes in masked thresholds of a harbour seal (*Phoca vitulina*) associated with angular separation of signal and noise sources. Can J Zool 72(11):1863–1866. https://doi.org/10.1139/z94-253

Turnbull SD, Terhune JM (1990) White noise and pure tone masking of pure tone thresholds of a harbour seal listening in air and under water. Can J Zool 68:2090–2097. https://doi.org/10.1139/z90-291

Turnbull SD, Terhune JM (1993) Repetition enhances hearing detection thresholds in a harbour seal (*Phoca vitulina*). Can J Zool 71(5):926–932. https://doi.org/10.1139/z93-120

Turnbull SD, Terhune JM (1994) Descending frequency swept tones have lower thresholds than ascending frequency swept tones for a harbor seal (*Phoca vitulina*) and human listeners. J Acoust Soc Am 96(5): 2631–2636. https://doi.org/10.1121/1.411296

Varanasi U, Malins DC (1970) Unusual wax esters from the mandibular canal of the porpoise (*Tursiops gilli*). Biochemistry 9(18):3629–3631. https://doi.org/10.1021/bi00820a020

Viemeister NF, Wakefield GH (1991) Temporal integration and multiple looks. J Acoust Soc Am 90(2): 858–865. https://doi.org/10.1121/1.401953

Voronov VA, Stosman IM (1983) On sound perception in the dolphin *Phocoena phocoena*. J Evol Biochem Physiol 18(5):499–506

Wang D, Wang K, Xiao Y, Sheng G (1992) Auditory sensitivity of a Chinese River dolphin, *Lipotes vexillifer*. In: Thomas JA, Kastelein RA, Supin AY (eds) Marine mammal sensory systems. Plenum Press, New York, pp 213–221. https://doi.org/10.1007/978-1-4615-3406-8_12

Wang Z-T, Li J, Duan P-X, Mei Z-G, Niu F-Q, Akamatsu T, Lei P-Y, Zhou L, Yuan J, Chen Y-W, Supin AY, Wang D, Wang K-X (2020) Evoked-potential audiogram variability in a group of wild Yangtze finless porpoises (*Neophocaena asiaeorientalis asiaeorientalis*). J Comp Physiol A 206(4):527–541. https://doi.org/10.1007/s00359-020-01426-6

Watkins WA (1981) Activities and underwater sounds of fin whales. Sci Rep Whales Res Inst 33:83–117

Wensveen PJ, Huijser LAE, Hoek L, Kastelein RA (2014) Equal latency contours and auditory weighting functions for the harbour porpoise (*Phocoena phocoena*). J Exp Biol 217(Pt 3):359–369. https://doi.org/10.1242/jeb.091983

White MJ, Norris J, Ljungblad DK, Baron K, di Sciara GN (1978) Auditory thresholds of two beluga whales (*Delphinapterus leucas*). Technical Report 78-108, Hubbs Sea World Research Institute, San Diego, 36 pp

Willott JF (2001) Animal models of presbycusis and the aging auditory system. In: Hof PR, Mobbs CV (eds) Functional neurobiology of aging. Academic, San Diego, pp 605–621. https://doi.org/10.1016/B978-012351830-9/50044-5

Wisniewska DM, Johnson M, Beedholm K, Wahlberg M, Madsen PT (2012) Acoustic gaze adjustments during active target selection in echolocating porpoises. J Exp Biol 215:4358–4373. https://doi.org/10.1242/jeb.074013

Wisniewska DM, Ratcliffe JM, Beedholm K, Christensen CB, Johnson M, Koblitz JC, Wahlberg M, Madsen PT (2015) Range-dependent flexibility in the acoustic field of view of echolocating porpoises (*Phocoena phocoena*). elife 4:e05651. https://doi.org/10.7554/eLife.05651

Wolski LF, Anderson RC, Bowles AE, Yochem PK (2003) Measuring hearing in the harbor seal (*Phoca vitulina*): comparison of behavioral and auditory brainstem response techniques. J Acoust Soc Am 113(1):629–637. https://doi.org/10.1121/1.1527961

Yamato M, Ketten DR, Arruda J, Cramer S, Moore K (2012) The auditory anatomy of the minke whale (*Balaenoptera acutorostrata*): a potential fatty sound reception pathway in a baleen whale. Anat Rec 295(6): 991–998. https://doi.org/10.1002/ar.22459

Yamato M, Koopman H, Niemeyer M, Ketten D (2014) Characterization of lipids in adipose depots associated with minke and fin whale ears: comparison with "acoustic fats" of toothed whales. Mar Mamm Sci 30(4):1549–1563. https://doi.org/10.1111/mms.12120

Yost WA (2000) Fundamentals of hearing: an introduction, 4th edn. Academic, New York

Yost WA (2015) Psychoacoustics: a brief historical overview. Acoust Today 11(3):46–53

Yost WA, Shofner WP (2009) Critical bands and critical ratios in animal psychoacoustics: an example using chinchilla data. J Acoust Soc Am 125(1):315–323. https://doi.org/10.1121/1.3037232

Yuen MML, Nachtigall PE, Breese M, Supin AY (2005) Behavioral and auditory evoked potential audiograms of a false killer whale (*Pseudorca crassidens*). J Acoust Soc Am 118(4):2688–2695. https://doi.org/10.1121/1.2010350

Zaĭtseva KA, Akopian AI, Morozov VP (1975) Noise resistance of the dolphin auditory analyzer as a function of the directional angle of the noise. Biofizika 20(3):519–521

Zaĭtseva KA, Morozov VP, Akopian AI (1980) Comparative characteristics of spatial hearing in the dolphin *Tursiops truncatus* and man. Neurosci Behav Physiol 10(2):180–182. https://doi.org/10.1007/BF01148460

Zimmer WMX, Tyack PL, Johnson MP, Madsen PT (2005) Three-dimensional beam pattern of regular sperm whale clicks confirms bent-horn hypothesis. J Acoust Soc Am 117(3):1473–1485. https://doi.org/10.1121/1.1828501

Open Access This chapter is licensed under the terms of the Creative Commons Attribution 4.0 International License (http://creativecommons.org/licenses/by/4.0/), which permits use, sharing, adaptation, distribution and reproduction in any medium or format, as long as you give appropriate credit to the original author(s) and the source, provide a link to the Creative Commons license and indicate if changes were made.

The images or other third party material in this chapter are included in the chapter's Creative Commons license, unless indicated otherwise in a credit line to the material. If material is not included in the chapter's Creative Commons license and your intended use is not permitted by statutory regulation or exceeds the permitted use, you will need to obtain permission directly from the copyright holder.

Physiological Effects of Sound on Marine Mammals

Dorian Houser

Contents

9.1 **Effects on Auditory Physiology** 579
9.1.1 Temporary Threshold Shift (TTS) 580
9.1.2 Permanent Threshold Shift (PTS) 588

9.2 **Effects on Nonauditory Physiology** 589
9.2.1 Acoustic Resonance Hypothesis 589
9.2.2 Acoustically Mediated Bubble Formation Hypotheses 590
9.2.3 Stress Responses 593

9.3 **Summary** 600

References 600

9.1 Effects on Auditory Physiology

A noise exposure can be of sufficiently high amplitude and/or long in duration that the sensitivity of the ear decreases (i.e., the hearing threshold increases) due to fatigue or injury. The change in sensitivity is termed a noise-induced hearing loss (NIHL). If the threshold of hearing returns to normal after some time, the change in threshold is termed a temporary threshold shift (TTS). If the ear does not fully recover, then the permanent change in the hearing sensitivity is referred to as a permanent threshold shift (PTS; Fig. 9.1), which usually only affects a part of the animal's hearing frequency range.

Human noise exposure guidelines have been developed over decades of study on the noise factors that result in NIHL with the goal of preventing or mitigating PTS in the human workforce (see Houser et al. 2017). In accordance with harassment definitions included in the 1994 amendments to the U.S. Marine Mammal Protection Act, the U.S. Navy proposed to use TTS as the sole criterion for impact to marine mammals in its Final Environmental Impact Statement (FEIS) prepared for shock tests of the SEAWOLF submarine (Department of the Navy 1998). However, at the time the FEIS was published, the only data available on the levels of noise required to induce TTS in a marine mammal were from studies conducted on the bottlenose dolphin (Ridgway et al. 1997). In the years since, studies of TTS have expanded to other species and encompassed myriad sound types. The thresholds of noise exposure assumed to cause TTS have

D. Houser (✉)
National Marine Mammal Foundation, San Diego, CA, USA
e-mail: dorian.houser@nmmf.org

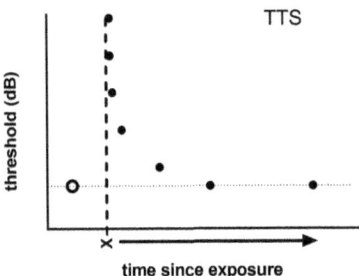 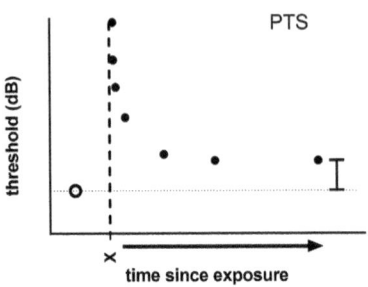

Fig. 9.1 (Left) The baseline threshold of an animal is measured before a noise exposure (open circle). Upon receiving a noise exposure (X and vertical dashed line), an increase in the threshold occurs. Because the threshold returns to baseline over time (horizontal dotted line), it is a TTS. (Right) If the threshold does not return to the baseline threshold over time, it is a PTS. At any point in time, the difference between the shifted threshold and the baseline threshold is the magnitude of the threshold shift (demonstrated as a PTS by the vertical bar)

naturally morphed in response to new data. Thresholds now exist for various species groups and sound source types (National Marine Fisheries Service 2018; Southall et al. 2019).

9.1.1 Temporary Threshold Shift (TTS)

A considerable body of literature has been established on acoustic factors that result in TTS in humans and laboratory animals. A substantially smaller number of studies have occurred in marine mammals, largely because of limited access to subjects. Most TTS studies have been conducted with trained marine mammals. Although this has allowed detailed information about the relationship between noise exposure and the onset and growth of TTS, it has also resulted in a limited number of subjects and a limited number of species that can be tested. From a regulatory perspective, this has required that information from a particular species be extrapolated to related species when predicting the levels of noise exposure that might result in a TTS. This has generally been perceived as a more reasonable approach than extrapolating from terrestrial species.

Table 9.1 lists studies of TTS resulting from underwater exposures in various species of marine mammals, as well as the few TTS studies in marine mammals conducted in air. Marine mammal TTS studies have utilized numerous sound types as fatiguing stimuli, including continuous and intermittent sounds, tones and bandpass noise (typically octave to 1/6-octave), and both frequency- and amplitude-modulated sounds. Studies have also investigated the impact of noise exposure duration and the quiet time between noise exposures on TTS onset and growth, and have used both behavioral and auditory evoked potential (AEP) methods to quantify threshold shifts. A review of NIHL in marine mammals provides an excellent summary of the TTS experiments conducted until the year 2015, and provides a more detailed description of the methods used (see Table 1 of Finneran 2015).

Temporary threshold shift is quantified as the difference between a post-noise exposure (hereafter, simply post-exposure) hearing threshold and a pre-exposure hearing threshold at the same frequency, which is considered the baseline or normal hearing condition. The onset and growth of TTS varies as a function of the level and frequency of the fatiguing stimulus, its bandwidth, exposure duration, species tested, and frequency of hearing that is tested. These are the factors that are of greatest interest in determining the potential for TTS in a marine mammal, particularly as it relates to anthropogenic noise. However, there are other factors that can affect the measurement of an NIHL and it is important to understand these factors when comparing results across studies. Two of these are the ambient noise and acoustic

9 Physiological Effects of Sound on Marine Mammals

Table 9.1 List of TTS studies conducted with different species for underwater and in-air noise exposures

Species	Underwater studies of TTS
Bottlenose dolphin	Finneran et al. (2000, 2002, 2005, 2007, 2010a, b, 2015), Finneran and Schlundt (2010, 2013), Mooney et al. (2009a, b), Nachtigall et al. (2003, 2004) and Schlundt et al. (2000)
Beluga	Finneran et al. (2000, 2002), Popov et al. (2011a, 2013, 2014, 2015) and Schlundt et al. (2000)
Harbor porpoise	Kastelein et al. (2012b, 2013a, b, 2014a, b, 2015b, 2016a, b, 2017a, b, 2019a, d, 2020a, b, d, f, 2021a) and Lucke et al. (2009)
Yangtze finless porpoise	Popov et al. (2011b)
Harbor seal	Kastak et al. (1999, 2005) and Kastelein et al. (2012a, 2013a, 2018, 2019b, c, 2020c, e, g)
California sea lion	Finneran et al. (2003), Kastak et al. (1999, 2005) and Kastelein et al. (2021b, 2022a, b)
Northern elephant seal	Kastak et al. (1999, 2005)
Ringed seal	Reichmuth et al. (2016) and Sills et al. (2020)
Spotted seal	Reichmuth et al. (2016)
Bearded seal	Sills et al. (2020)
	In-air studies of TTS
California sea lion	Kastak et al. (2007)
Harbor seal	Kastak and Schusterman (1996)

field of the test environment, and whether behavioral or AEP methods are used for threshold testing. In the following sections, these factors are discussed in relation to what is known about TTS in marine mammals.

9.1.1.1 SPL, Duration, and the Equal-Energy Rule

There is a maximum noise SPL that an animal can be exposed to at which no TTS occurs, regardless of the duration of the exposure, and which does not affect recovery from TTS. This is called "effective quiet." However, there is a direct but nonlinear relationship between the SPL of a fatiguing noise of fixed duration and the magnitude of the TTS, once the SPL at which TTS is first observed is exceeded (the onset of TTS) (Fig. 9.2a, d, g). This phenomenon is well-known in terrestrial mammals and has also been found in both odontocetes and pinnipeds (e.g., Kastak et al. 2007; Kastelein et al. 2012b; Popov et al. 2014). Similarly, TTS increases near-linearly with the logarithm of signal duration for a fatiguing stimulus of constant SPL, at least for moderate levels of TTS and exposure duration (Fig. 9.2b, e, h) (Popov et al. 2014). Since TTS is a function of both noise exposure duration and SPL, it is common to describe TTS as a function of the SEL, which integrates the instantaneous squared sound pressure across the duration of the exposure. Because the SEL increases by 3 dB for each doubling of the noise exposure duration, it has been termed an "equal-energy" rule based on the assumption that equal-energy sound exposures should cause the same level of NIHL. For single, continuous exposures of similar duration, the equal-energy rule seems to apply in marine mammals. However, it becomes a poor predictor of TTS for longer duration exposures, intermittent exposures, or high SPL exposures (Kastak et al. 2005). Studies with both pinnipeds and odontocetes suggest that duration is potentially a greater contributor to TTS than SPL when the energy of the noise exposure is held constant (Finneran et al. 2010a; Kastak et al. 2007; Mooney et al. 2009a; Popov et al. 2014), but it has also been argued that the equal-energy rule might break down in the opposite direction when SPLs are decreased for long duration exposures in order to keep an equivalent SEL (Finneran et al. 2015; Popov et al. 2014).

The growth of TTS in marine mammals as a function of the SEL increases in a curvilinear manner (Fig. 9.2). Because TTS does not scale simply as a function of the SEL, but rather as a complex interaction between noise SPL and exposure duration, multiple "models" of TTS as a function of both SPL and duration have been

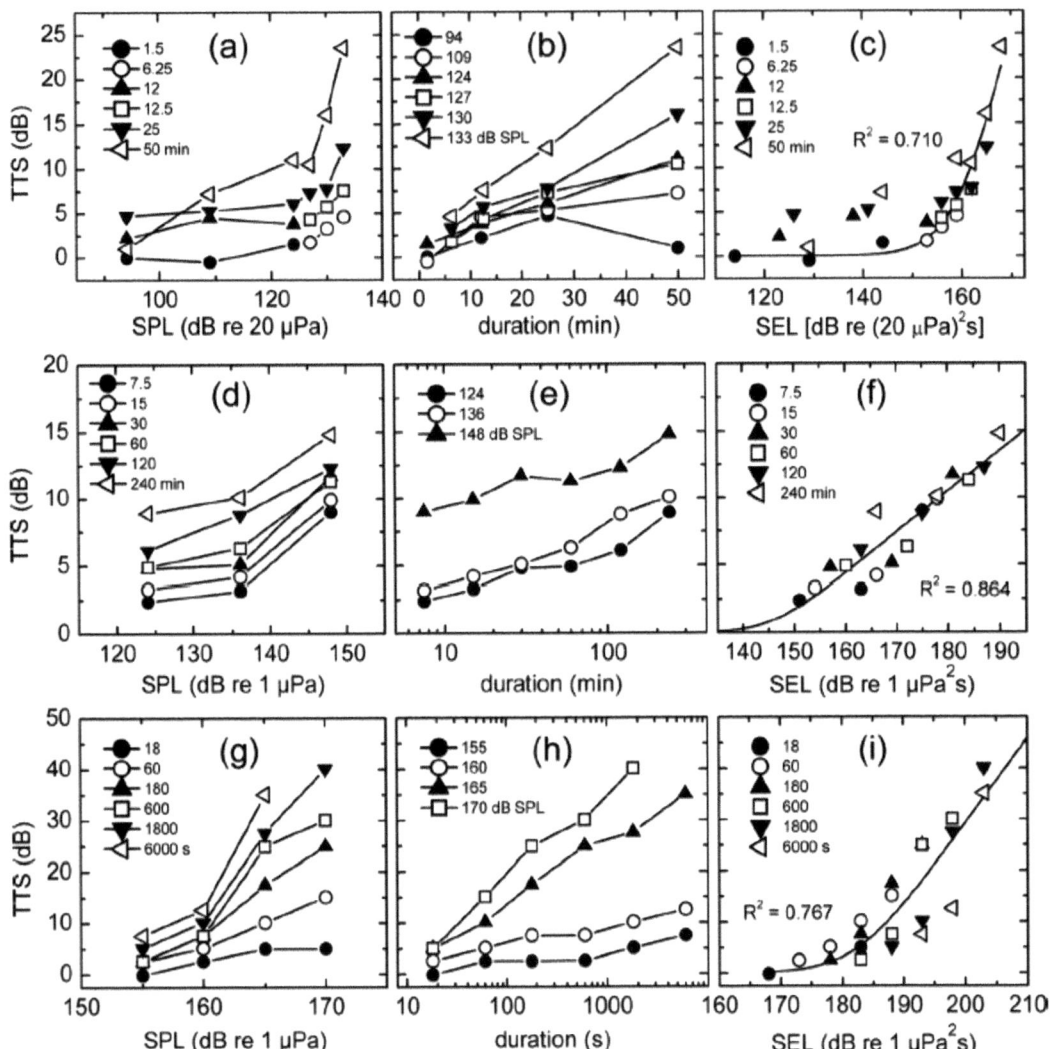

Fig. 9.2 TTS as a function of SPL (left column), duration (middle column), and SEL (right column) of the fatiguing stimulus. Values within the panels correspond to duration of the stimulus (left and right columns) or SPL (center column). The solid lines in the panels of the right column are nonlinear fits of a model demonstrating growth in TTS as a function of the SEL. Panels (**a**–**c**) are from a California sea lion exposed to octave band noise centered at 2.5 kHz (Kastak et al. 2007). Panels (**d**–**f**) are from a harbor porpoise exposed to octave band noise centered at 4 kHz (Kastelein et al. 2012b). Panels (**g**–**i**) are from a beluga exposed to half-octave band noise centered at 22.5 kHz (Popov et al. 2014). See Finneran (2015) or original papers for more detail on exposure and hearing test conditions. Reproduced from Finneran JJ (2015) Noise-induced hearing loss in marine mammals: a review of temporary threshold shift studies from 1996 to 2015. J Acoust Soc Am 138(3): 1702–1726. https://doi.org/10.1121/1.4927418. © Acoustical Society of America, 2015. All rights reserved

offered (Finneran et al. 2010a; Mooney et al. 2009a; Popov et al. 2014). Linear portions of the models of TTS growth have been used to predict TTS growth rates in marine mammals (i.e., the dB increase in TTS per dB increase in the exposure level). Using this approach, TTS growth rates in marine mammals have been observed to range from as low as 0.2 dB TTS/dB SEL to as high as 4.5 dB TTS/dB SEL with significant variation between species and as a function of the characteristics (e.g., bandwidth, duty cycle, frequency) of the fatiguing noise (e.g., Finneran 2015; Finneran and Schlundt 2013; Kastelein et al. 2014a). The function can be complex and can show a significant change in the TTS growth rate at some critical combination of duration and exposure (Fig. 9.2a). The growth rate of TTS as a function of exposure level has been proposed as one means by which the onset of PTS might be predicted in marine mammals exposed to anthropogenic noise (Southall et al. 2007, 2019).

9.1.1.2 Intermittent Noise

Quiet periods between multiple noise exposures allow the ear to recover from acoustic insult. For this reason, two noise exposures with the same SEL and which are identical in every way except for whether they are intermittent or continuous will result in different levels of TTS (Ward 1997). Even though the SEL between the signals is the same, the intermittent signal will produce a lower level of TTS because of the recovery of hearing that occurs during the intervening quiet periods. As the duty cycle decreases (i.e., the quiet period between noises increases), the amount of TTS that occurs will also decrease (Fig. 9.3).

Intermittent signals are some of the most likely anthropogenic sounds to be encountered by marine mammals in the ocean. Sonar signals and air-gun impulses are used by navies, academia, industry, and the public (e.g., fish finders) for military, scientific, commercial, and recreational purposes. It is therefore a little surprising that there have only been a handful of marine mammal studies that investigated the impact of signal intermittency on TTS; Mooney et al. (2009b) and Finneran et al. (2010b) performed studies in bottlenose dolphins, and Kastelein et al. performed studies in the harbor porpoise (2014a, 2015b) and in California sea lions (2021b, 2022b). Across these studies, the impact of intermittent exposures on TTS was found to be in qualitative agreement with what has been observed in terrestrial mammals.

The lack of sufficient data on the impact of signal intermittency on TTS has hampered the

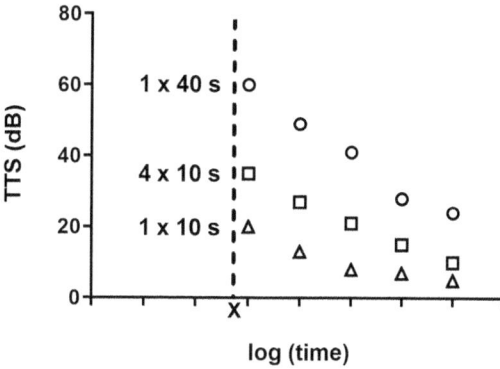

Fig. 9.3 Depiction of the change in the level of TTS as a function of duty cycle and SEL. X denotes the time of the noise exposure. For a noise of constant SPL, TTS will increase as SEL increases (compare the single 10-s exposure to the single 40-s exposure, where the SEL would be 6 dB higher). In contrast, if the SEL of two noise exposures is the same, but one exposure is interrupted by periods of quiet, the resulting TTS will be lower after the exposure with the quiet intervals (compare single 40-s exposure to the four intermittent exposures of 10 s each)

ability to develop methods of predicting the TTS that might occur following exposure to intermittent underwater noise. Multiple predictive models of TTS resulting from repetitive/intermittent noise exposures have been proposed (Finneran 2015): exposure summation models that do not account for the recovery of the ear between exposures (Department of the Navy 2008; Southall et al. 2007); the single-ping equivalent model, which attempted to account for ear recovery between noise exposures by transforming the cumulative exposure into a summed sound intensity value (Department of the Navy 2001); and various forms of TTS summation models that account for the summation of the NIHL from individual exposures, such as the modified power law (Finneran et al. 2010b). Unfortunately, because of the paucity of data on the impact of signal intermittency on TTS, there has been limited model validation. Only gross conclusions can be made about model predictions to date, mainly that those models that do not account for the recovery of the ear show the greatest overestimates of TTS due to intermittent noise exposures.

9.1.1.3 Stimulus (Fatiguing Sound) Frequency and Hearing Test Frequency

The onset and growth of TTS varies with the frequency of the fatiguing stimulus, but does not necessarily grow more quickly or occur at lower exposure levels at the frequencies where a species has its best hearing. For example, Finneran and Schlundt (2013) exposed bottlenose dolphins to 16-s tones ranging in frequency from 3 to 80 kHz and measured TTS onset and growth rate through behavioral methods. Growth rates of TTS were greatest from 14.1 to 28 kHz, whereas the best hearing sensitivity is typically between 40 and 80 kHz in a bottlenose dolphin with normal hearing. Conversely, the exposure levels necessary for the onset of TTS were highest at the lowest frequencies tested, but were lower and within 5 dB of one another at frequencies from 10 to 56 kHz. No TTS was observed at 80 kHz. Across studies and subjects within bottlenose dolphins, a trend of higher SEL being required for the onset of TTS at lower noise exposure frequencies (<10 kHz) exists (Finneran et al. 2005; Finneran and Schlundt 2010, 2013). A similar trend has emerged from studies on harbor porpoises; data suggest the SEL required for onset TTS are lowest for noise exposure frequencies greater than 6.5 kHz, and the SEL required for TTS onset increases toward lower frequencies (Kastelein et al. 2019a, d, 2020a, b, c, d, 2021a). Frequency-dependent differences in TTS growth rates and onset have also been measured in belugas and the finless porpoise using electrophysiological methods (Popov et al. 2011b, 2013). Collectively, TTS data on odontocetes suggests that they might be more susceptible to TTS within a region of ~10–40 kHz (Finneran 2015), but with a higher frequency limit for phocoenids.

Less information on the impact of fatiguing noise frequency on TTS onset and growth exists for the pinnipeds (specifically the California sea lion, elephant seal, and harbor seal), which have primarily had TTS growth rates measured following exposure to fatiguing noises containing center frequencies <4 kHz (also, see Fig. 5(j–l) of Finneran 2015; Kastak et al. 2004, 2005, 2007; Kastelein et al. 2012a, 2020c, 2021b, 2022a). However, in more recent years, Kastelein et al. (2020e, g, 2022b) explored the onset and growth of TTS across a broader frequency range in harbor seals and California sea lions. For harbor seals, the potential for TTS to occur did not significantly vary from 2 to 40 kHz, but higher SELs were required to induce TTS at frequencies <2 kHz. Similarly, a higher SEL was required to induce TTS in California sea lions at frequencies <1 kHz, whereas SELs required to induce TTS at frequencies >1 kHz were roughly similar (except possibly at 8 kHz, where the lowest SEL required for TTS was determined).

The level of TTS measured in marine mammals is typically greatest at frequencies ranging from the center frequency of an exposure up to an octave above the frequency of the fatiguing noise (Finneran et al. 2007; Kastak et al. 2005; Kastelein et al. 2012a, 2013b; Mooney et al. 2009a; Nachtigall et al. 2004; Popov et al. 2011b, 2013). There are possibly species differences in the frequency of maximum TTS, relative to the center frequency of the fatiguing noise, as well as dependencies on the spectral and temporal characteristics of the noise used as a fatiguing noise. Evidence exists in terrestrial mammals that an upward spread in the frequencies most affected by TTS occurs as the fatiguing noise SPL increases (McFadden and Plattsmier 1983). Figure 9.4 depicts this relationship. This occurs due to broadening of the cochlear filters in response to increasing fatiguing noise level, that is, there is a spread in the excitation patterns along the cochlear partition. Some evidence for this phenomenon has been found in the harbor porpoise (Kastelein et al. 2014b, 2019a, d, 2020b, d), harbor seal (Kastelein et al. 2019b, c, 2020c, e, g), and California sea lion (Kastelein et al. 2021b, 2022a, b), but the phenomenon has not been observed at all exposure frequencies (e.g., Kastelein et al. 2020a, 2021a). Additional investigation is needed to determine the degree that this trend holds true for other marine mammal species and at what frequencies of noise exposure the pattern holds.

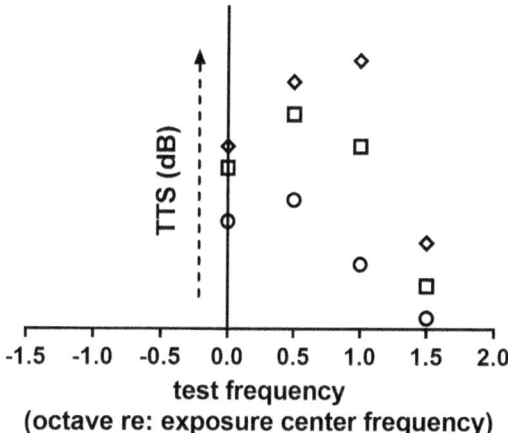

Fig. 9.4 The degree of TTS is generally greatest up to an octave above the center frequency of the fatiguing noise. An upward spread in the frequencies affected by TTS can occur as the SPL of the noise exposure increases. (open circles = lower SPL exposure; open squares = moderate SPL exposure; open diamond = higher SPL exposure)

9.1.1.4 Impulsive Noise

Impulsive noise is characterized by short durations, very fast changes in pressure (i.e., very short pressure rise time), and high peak pressures. Because of these characteristics and the frequency response of the sound sources and projectors employed, impulsive noise contains energy in a broad spectrum of frequencies. These same characteristics have been observed to make impulsive noise exposure particularly hazardous to hearing in terrestrial mammals (Henderson and Hamernik 1986). Because it is difficult to determine the exact physical factors of the impulsive noise that results in TTS, it is common to see multiple acoustic metrics associated with the reporting of TTS. In marine mammals, these are most commonly the SEL and the peak or peak-peak SPL of the exposure.

Anthropogenic impulse noise in the ocean consists of underwater explosions (e.g., as used in demolitions, ship-shock trials, or detonation of unexploded ordnance), echo-sounding, sounds produced in marine seismic surveys (e.g., as used for geological purposes or petroleum exploration; see Chap. 2), and offshore pile driving. Because real-world, high-level impulsive signals are not easy to produce or control in laboratory environments, relatively few TTS studies with impulsive noise as the fatiguing stimulus have been conducted. Of these, no studies have been conducted with explosive charges, most have been conducted with real-world sources (Finneran et al. 2002, 2003, 2015; Lucke et al. 2009; Reichmuth et al. 2016; Sills et al. 2020), and some have been performed with either playbacks of impulsive recordings or with surrogate sources (Finneran et al. 2000; Kastelein et al. 2015a, 2017b, 2018, 2020f).

Observations of the effect of impulse noise exposure on hearing were most pronounced in a beluga exposed to a seismic water gun (Finneran et al. 2002) and a harbor porpoise exposed to a seismic air gun (Lucke et al. 2009). The beluga was tested through behavioral methods and the harbor porpoise via AEP methods, but the different methods still provide the most pronounced measures of TTS resulting from impulse exposure to date. Finneran et al. (2002) were able to measure a TTS of 6 and 7 dB at 400 Hz and 30 kHz, respectively, to a seismic water gun impulse with an unweighted SEL of 186 dB re 1 $\mu Pa^2 s$ and a peak SPL of 224 dB re 1 μPa. Lucke et al. (2009) found threshold shifts as high as 15 dB at 4 kHz following an exposure to a single impulse from an air gun with peak-peak SPL of 202 dB re 1 μPa and an unweighted SEL of 166 dB re 1 $\mu Pa^2 s$. In a related study, unweighted SELs of 188 and 191 dB re 1 $\mu Pa^2 s$ resulting from 10 and 20 discharges from two small air guns resulted in a shift of ~4.4 dB in a harbor porpoise tested at 4 kHz (Kastelein et al. 2017b). No shift was noted at lower frequencies. A repeat study conducted with the same animal but higher exposure levels was unable to reproduce a threshold shift (Kastelein et al. 2020f). The authors speculated that the inability to obtain a shift might have been due to the self-attenuation of hearing sensitivity (see Chap. 8), since the animal might have learned to predict the timing of the air-gun pulses (e.g., due to the sound of the solenoid valves, which were noisier in the second experiment). Reflecting the ability of AEP measures to potentially record impacts to the auditory system that are not observed behaviorally, Finneran et al.

(2015) found no behavioral shift in hearing threshold following exposure to 10 impulses from an air gun transduced at one impulse every 10 s (maximum cumulative, unweighted SEL ranged from 193 to 195 dB re 1 $\mu Pa^2 s$), but did observe a 9 dB shift in the AEP threshold of one dolphin tested at 8 kHz. Two dolphins were also observed to have changes in their AEP IO functions (see Chap. 8), but without a concomitant threshold shift, suggesting a potential impact to the auditory system that was not reflected in a change in threshold. In one of the few studies involving pinnipeds exposed to air-gun impulses, a TTS was observed in a bearded seal after it was exposed to between 4 and 10 shots from an air gun (Sills et al. 2020). The cumulative SEL from the shots ranged from 191 to 194 dB re 1 $\mu Pa^2 s$ and had a peak SPL of ~203 dB re 1 μPa. The median TTS measured at a frequency of 400 Hz was 7.5 dB.

Several TTS studies using impulsive sources have not been able to produce a measurable TTS, even at the maximum level of SEL and SPL produced by the sound sources (Finneran et al. 2000, 2003; Reichmuth et al. 2016). Whether this speaks to the resilience of marine mammals to impulsive underwater signals, the inability to accurately recreate signals that exist in the ocean, the ability of the animals to mitigate the noise exposure, or a combination of the three is uncertain. Until this is resolved, it seems that the real-world scenarios where marine mammals would be exposed to impulsive signals causing TTS would be rather uncommon, unless individuals were very near such sources. Impulsive signals lose their impulsive "nature" as they travel away from their source. In instances where pile driving strikes resembled signals tens of kilometers from the source (Kastelein et al. 2016b), playbacks ranging from 15 to 360 min (duty cycle of ~9.5% and cumulative SEL of 173–187 dB re 1 $\mu Pa^2 s$) resulted in only minor TTS (<6 dB). Thus, the magnitude of the impact of threshold shifts on marine mammal hearing from impulsive seismic or pile-driving sources might not be large, unless animals are near the sources.

9.1.1.5 Recovery from TTS

The recovery of TTS resulting from exposure to narrowband noise or tones has been reviewed by Finneran (2015). When restricted to these types of fatiguing stimuli, a pattern of recovery from TTS emerges in which recovery appears to occur as a linear function of the logarithm of recovery time (Fig. 9.3), but with recovery rates becoming more variable as the duration of recovery increases. Recovery rates have been observed ranging from 4 to 23 dB/decade of time (e.g., Finneran and Schlundt 2013; Finneran et al. 2007; Kastelein et al. 2012a, b, 2022b; Mooney et al. 2009a; Nachtigall et al. 2004; Popov et al. 2014). Recovery rates generally increase as the initial TTS increases, but larger initial shifts still typically take longer to recover. Beyond these constraints, patterns of TTS recovery become difficult to predict, likely because of differences in fatiguing stimuli, test methods, intersubject variability, and interspecies variability. As found in humans, the recovery from TTS is not solely a function of the magnitude of the TTS, but is dependent on the noise conditions that lead to the TTS (Melnick 1991). Patterns of recovery from TTS can be complex and involve nonlinear recovery patterns and delayed TTS growth or recovery (see Fig. 9 of Finneran 2015), all of which makes predicting the duration of impacts to marine mammals that have experienced TTS difficult (e.g., Finneran and Schlundt 2013; Popov et al. 2013). Furthermore, as noted by Finneran (2015), the utility of extrapolating models to predict recovery to untested species and exposure scenarios is uncertain and should be evaluated cautiously. Nevertheless, as the duration of TTS might be a critical factor affecting marine mammals (e.g., in determining the time over which the range of acoustic signal detection might be reduced), further effort in elucidating patterns of recovery as a function of fatiguing noise characteristics should be pursued.

9.1.1.6 Ambient Noise and the Acoustic Field of the Test Environment

Baseline conditions can be affected by ambient noise with higher noise conditions potentially elevating the baseline hearing threshold. Complications in synthesizing TTS data across studies arise because studies have been conducted in different environments with different noise profiles. These include ocean sites (e.g., Finneran et al. 2000, 2002, 2003, 2015; Lucke et al. 2009; Mooney et al. 2009a, b; Nachtigall et al. 2003, 2004; Ridgway et al. 1997; Schlundt et al. 2000), where snapping shrimp and anthropogenic noise can substantially contribute to the ambient noise, small water-filled tanks within which animals were suspended and supported by a sling (e.g., Popov et al. 2011a, b, 2013, 2014, 2015, 2017), and pools (e.g., Finneran et al. 2005, 2007, 2010a, b; Finneran and Schlundt 2010; Kastak et al. 1999, 2005; Kastelein et al. 2019c, 2021a; Reichmuth et al. 2016; Sills et al. 2020), some of which may be purposely built to minimize noise (e.g., Kastelein et al. 2019c, 2021a). In-air TTS measurements with pinnipeds have been conducted both in the presence of background noise and in negligible background noise conditions inside a hemi-anechoic chamber (Kastak et al. 2007; Kastak and Schusterman 1996).

Ocean noise conditions vary depending on location. All marine mammal TTS studies performed in the ocean have been conducted in shallow water environments where biological (e.g., snapping shrimp) and anthropogenic (e.g., motorboats) noise sources might be prevalent. The ambient noise conditions of these environments are an important consideration in the interpretation of the results of TTS studies. Prior research has demonstrated that the introduction of Gaussian masking noise for the purpose of elevating the pre-exposure threshold can lower the measured TTS to a fatiguing stimulus. For a constant level of the fatiguing stimulus, the magnitude of the TTS decreases as the masked hearing threshold increases (Humes 1980). Nearshore environments, however, do not have ambient noise fields characterized by Gaussian noise. These environments are typically dominated by comodulated noise sources, within which marine mammals can capitalize on the comodulated nature of noise to better detect acoustic signals (see Chap. 8, as well as Branstetter and Finneran 2008).

Pools and water-filled tanks sometimes have the advantage that background noise is reduced (unless machinery or filtration equipment are nearby) relative to that observed in nearshore, ocean environments. However, the acoustic fields in pools and tanks can be complicated because of the small volume of water in the pools (relative to the ocean) and the reflective surfaces (i.e., surface, bottom, and pool sides). This requires careful consideration of stimulus waveforms for the hearing tests, as well as a good understanding of the fatiguing stimulus and hearing test stimuli as they are received at the animal.

According to Finneran (2015), a synthesis of results across marine mammal TTS studies involving Gaussian masking noise, ambient ocean noise, or pool noise conditions has not produced a good understanding of the relationship between masking noise and its effect on TTS measurements. The synthesis is further hampered by the fact that no single individual animal has ever been used in a methodical assessment of the impact of masking noise on TTS.

9.1.1.7 Behavioral and AEP Methods

Differences in hearing thresholds determined with behavioral and AEP measures, and the causes for them (e.g., differences in stimulus calibration), have been discussed previously and are not repeated here (see Chap. 8). Nevertheless, these factors deserve consideration in the comparison of TTS studies that utilize AEPs and those that use behavioral methods. As of the writing of this chapter, only two studies have directly compared AEP and behavioral TTS measurements in the same subjects exposed to the same fatiguing stimuli (Finneran et al. 2007, 2015), and the results of those studies suggest that AEPs might provide a more sensitive means of detecting impact to the auditory system than is provided by behavioral thresholds. Finneran et al. (2007) compared behavioral and AEP threshold measurements within the same bottlenose dolphin subject. TTS measured by AEPs was 40–45 dB and was 19–33 dB greater than that measured behaviorally. AEP threshold shifts were found (~10 dB) even when no behavioral shift was observed and AEP threshold shifts always took longer to recover. In a comparison of AEP and behavioral thresholds obtained in three bottlenose dolphins exposed to multiple impulses from a seismic air gun, no changes in the behavioral thresholds were observed following the noise exposure but a small amount of TTS was observed using AEP methods (Finneran et al. 2015). In one of the dolphins, a 9 dB TTS was observed using AEPs and testing at 8 kHz. The TTS was accompanied by a shift in the IO function created from the ASSR amplitude vs. stimulus level data (see Chap. 8 for an explanation of IO functions and the ASSR). Differences in the IO function of one of the other dolphins was also observed, but without a change in the post-exposure behavioral threshold.

Previous discussion (Chap. 8) highlighted the issues with directly comparing AEP and behavioral hearing thresholds, and the same caveats exist when considering how to compare TTS data obtained with AEP and behavioral methods. The same types of studies needed to better relate the two approaches to studying baseline hearing are also needed for TTS studies. Nevertheless, the use of AEPs might have some advantage in that the measurement method might indicate an impact to the auditory system that is not observed with behavioral measures: the magnitude of measured TTS is usually greater than that measured behaviorally; TTS is measured for longer durations post-exposure; and impacts to the auditory system can be observed in suprathreshold measurements, for example through changes in the IO function, when no behavioral shift occurs. It is difficult to interpret exactly what this means for the animal. If the behavioral threshold returns to normal then this might imply that the animal can function normally for activities that involve sound detection, localization, etc. Conversely, the persistence of AEP threshold shifts and changes in the suprathreshold IO functions suggest that some form of impact to the auditory system persists. How or whether this impacts the animal's ability to thrive is unknown; for example, whether this predisposes the animal to greater TTS from a subsequent exposure, but prior to full recovery, has not been studied.

9.1.2 Permanent Threshold Shift (PTS)

There are essentially no studies of PTS in marine mammals because of the ethical issue of causing permanent injury to the hearing of a marine mammal. However, an accidental PTS was caused in a harbor seal participating in a study of TTS induced by a tonal signal at 4.1 kHz (Reichmuth et al. 2019). Over a period of 3 months, the exposure level to which the seal was exposed was gradually increased. Following an underwater exposure with an SEL of 199 dB re 1 $\mu Pa^2 s$, the seal showed an initial estimated shift of 47 dB one-half octave above the stimulus frequency (5.8 kHz). Although hearing at the frequency of noise exposure (4.1 kHz) fully recovered within 48 h of the exposure, recovery at 5.8 kHz was not complete and a threshold shift of ~8 dB persisted for over 10 years after the exposure. It is important to reiterate that the goal of this study was the induction of TTS, not PTS. It is unlikely that whatever led to the PTS will be fully understood, and it is also unlikely that studies will be conducted to intentionally cause PTS in marine mammals in the future. For this reason, other approaches to predicting PTS have been proposed, such as the use of the TTS growth rate to predict the level of exposure required to produce a PTS (Southall et al. 2007, 2019).

A few extra words are warranted for the discussion of PTS. The focus on PTS as a form of injury increased throughout the late 1990s. From a regulatory perspective, PTS was first agreed to be injury by the U.S. Navy based on the assumption that a PTS could not occur without the destruction of tissue (Department of the Navy 2008). Within the United States, noise exposures that correspond to the onset of PTS are now often used as the thresholds for Level A harassment under the MMPA, which is defined as an injury or the potential for injury. In more recent years, arguments have been made that TTS might also be considered injurious, based primarily on work by Kujawa and Liberman (2009). Kujawa and Liberman demonstrated that a significant TTS in mice (~40 dB measured 24 h after the fatiguing noise exposure) measured with ABRs (see Chap. 8) resulted in a delayed, but progressive, neural damage to the hearing apparatus that was apparent for weeks and months after the exposure, even though thresholds returned to normal over several days. Based on these findings, and others (e.g., Lin et al. 2011), it has been questioned as to whether injury, as defined by the destruction of tissue, might also be associated with TTS. In a review of this topic, Houser et al. (2021) noted that TTS is almost always measured within minutes of the cessation of a fatiguing noise exposure in experimental measures of TTS in marine mammals, not 24 h later. In addition, even though levels of TTS might rise as high as 40 dB (using both behavioral and AEP methods), recovery typically occurs within 24 h for shifts of up to 30 dB; no shifts have been noted to have levels comparable to those observed by Kujawa and Liberman (2009) 24 h after exposure, except for the single incident of PTS previously noted (Reichmuth et al. 2019). Furthermore, reconciling the observations of Kujawa and Liberman (2009) with what happens in the marine environment is difficult, as the potential for marine mammals to be exposed to noise sources that would result in 40 dB of TTS measured 24 h after the exposure is likely very small. Based upon the transient use of most high-level noise sources, most TTS occurrences are likely to be lower in magnitude and recover much more quickly. This may change, however, if high-level sound sources move to more continuous active operations (e.g., Continuous Active Sonar).

9.2 Effects on Nonauditory Physiology

Not only hearing and behavior can be affected in animals exposed to noise. It has long been known from terrestrial studies that noise exposure can impact wildlife by inducing physiological changes to the neuroendocrine system, cardiovascular system, metabolism, reproduction and development, immune responses, and cognitive ability (Halfwerk et al. 2011; Kight and Swaddle 2011; Münzel et al. 2014; Tennessen et al. 2014). Anthropogenic noise can also disrupt sleep and cause annoyance (e.g., Hume et al. 2012), a factor which cannot be easily determined in nonverbal animals. These "nonauditory" physiological responses may not in all cases be deleterious to animals. The degree of impact is likely mediated by the duration and magnitude of the noise event and the experience of the animal with human activity producing sound (e.g., Payne et al. 2012).

Marine mammals are fundamentally different from terrestrial mammals in that nearly all marine mammals must dive to forage and perform activities such as traveling and mating displays. Because of the diving lifestyle of marine mammals, several rather unique potentials for noise to either directly or indirectly induce nonauditory physiological responses exist (see below). One of these is related to the interaction of sound fields with the changing air spaces within marine mammals that occur while diving; that is, as the depth of a dive and the hydrostatic pressure change, certain compressible air spaces (e.g., the lung) change in volume in inverse relation to the hydrostatic pressure. Some tissues of marine mammals can potentially become supersaturated with nitrogen gas while diving, which establishes the potential for nitrogen bubble formation either through direct interaction with high-level acoustic fields or through altered dive behavior. The following sections describe the potential nonauditory physiological responses to noise exposure that have been observed or have been hypothesized to occur in marine mammals. The section begins with the more unusual possibilities.

9.2.1 Acoustic Resonance Hypothesis

Every physical body, including a gas-filled cavity (such as a lung or air sinus), has a particular frequency at which it preferentially vibrates. This is called its "resonant frequency" and is dictated by the size, shape, and material properties of the body. Given sufficient stimulation at the resonant frequency, some bodies can vibrate violently and become disrupted, as is observed in the ever-familiar shattering of the wine glass by the opera singer singing at the resonant frequency of the wine glass. Such dramatic responses are uncommon because of the damping properties associated with an object's material composition and shape, which helps to attenuate vibrations at the resonant frequency.

Early in the millennium, following the mass stranding of beaked whales in the Bahamas (Balcomb III and Claridge 2001; Department of Commerce and Department of the Navy 2001), speculation arose that one of the potential contributors to the stranding might have been the destructive resonance of tissues and tissue-lined air spaces in whales exposed to high levels of mid-frequency (1–10 kHz) active sonar. A panel of scientists was convened in 2002 to investigate the potential for acoustic resonance to have led to the beaked whale mass strandings in 2000, although it was also offered as a potential explanation of other whale strandings (Evans 2002). Based upon the analyses, laboratory animal experiments, and modeling provided by workshop participants, a consensus agreement that acoustic resonance of whale tissues or air spaces was an unlikely cause of the strandings was obtained. This was based upon multiple findings, including:

- Tissue displacements were likely too small to result in serious harm, even under the unrealistic situation in which air spaces were undamped by surrounding tissues.
- Air spaces in marine mammals were generally too large, such that their resonant frequencies did not correspond to the frequencies of mid-frequency active sonar systems (resonant lung frequencies were predicted to be less than 100 Hz).
- Diving decreased tissue displacements with depth, thus off-setting the increased resonant frequency of smaller air spaces that occur at higher hydrostatic pressures.

- Exposure levels were too low to have caused significant resonance for most marine mammals under real-world exposure conditions.
- The time required for resonant vibrations to cause damage to tissues was longer than the typical exposure of mid-frequency active sonar systems (~1 s).
- There was no specific evidence of lung damage pointing to resonance effects in the necropsies of whales stranded coincident with sonar activity.

One other hypothesis that was proposed at the workshop was that nitrogen bubble formation and/or growth might be activated, and driven by, the exposure of microbubbles in whale tissue to high levels of sound—a process termed "rectified diffusion" (Crum and Mao 1996).

9.2.2 Acoustically Mediated Bubble Formation Hypotheses

9.2.2.1 Rectified Diffusion

Rectified diffusion is the process by which an acoustic field causes a bubble to oscillate and grow. When a gas bubble is insonified, it alternately increases and decreases in volume with the negative and positive pressure cycles of the sound wave, respectively (Crum and Mao 1996). Over each acoustic cycle, the expansion of the bubble (negative pressure phase) forces gas into the bubble via diffusion, whereas gas diffuses out of the bubble during compression (positive pressure phase). Diffusion of gas into and out of the bubble is typically unequal due to surface area and boundary effects producing gradual bubble growth over the progression of the acoustic exposure. The hypothesis that this might explain the strandings of certain whale species exposed to naval sonar assumed that marine mammals have "microbubbles" in their tissues that are stabilized by the structural heterogeneity of tissues (i.e., stabilized by microscopic crevices) and/or caused by the mechanical movement of tissues (e.g., shear forces created by the bending of joints). It was proposed that these microbubbles, which are stabilized under normal dive conditions, could become destabilized by high levels of sound exposure. Once destabilized, the bubbles would then continue to grow by rectified diffusion during the period of insonification. Whether or not bubbles become destabilized and grow is a function of the received SPL at the site of the microbubble and the duration of the sound exposure. If bubbles could grow in this manner in situ, then it was argued that bubbles might grow to the point that they become emboli, rupture tissue, and enter the bloodstream. They might even cause an immune response (i.e., the bubbles are treated as foreign bodies), although recent work suggests the immune response to nitrogen bubbles is less severe than observed in humans (Thompson et al. 2020). The result could possibly be symptoms like that experienced by divers suffering from "the bends," otherwise known as decompression sickness (DCS). The symptoms would include pain and neurological dysfunction resulting from vascular blockage and tissue damage.

Rectified diffusion is enhanced if the tissue/fluid around the bubble is supersaturated with gas. Crum and Mao (1996) argued that once destabilized by sound, microbubbles could continue to grow through the static diffusion of gas into the bubble provided the tissue/fluid around the bubble was supersaturated with gas. Crum et al. (2005) subsequently exposed supersaturated kidney and liver tissues from cows to a 37-kHz source at 214 dB re 1 µPa and were able to demonstrate a substantial increase in bubble growth resulting from the insonification under conditions of tissue gas supersaturation. Modeling of nitrogen gas saturation in diving marine mammals conducted by Houser et al. (2001) first suggested that if rectified diffusion or static bubble growth (once activated) could occur in marine mammals, it would likely be most prevalent in the deepest-diving species with the slowest ascent/descent times and shortest surface intervals, since these would likely produce the greatest nitrogen tissue saturations. Repetitive and deep diving has been shown to cause supersaturation of the blood and some tissues of marine mammals (Falke et al. 1985; Ridgway and Howard 1979), although the data are limited. Nevertheless, subsequent, more sophisticated modeling has agreed with some of the initial conclusions about dive behavior and the degree of risk for nitrogen gas supersaturation.

Although rectified diffusion was demonstrated to be feasible (Crum et al. 2005), the tissue saturation conditions under which it was observed seemed unlikely to occur in marine mammals given their typical dive behavior (Fahlman et al. 2009; Saunders et al. 2008). Furthermore, the levels of sound exposure used in the experiments of Crum et al. (2005) seemed unlikely for real-world exposure scenarios; it is doubtful that marine mammals would be exposed to the high levels of sound used to demonstrate rectified diffusion unless they were very close to a powerful sound source during operation (e.g., mid-frequency active sonar). Unfortunately, little other empirical testing of the rectified diffusion hypothesis has been conducted and it is unknown as to whether lower received levels of sound and lower degrees of supersaturation could still result in bubble formation and/or growth. Soon after the proposition that rectified diffusion might be possible in marine mammals, the finding of gas emboli and bubble-induced tissue separations in stranded whales (some associated with naval sonar operations) increased the fervor over the potential for sonar to be related to a "DCS-like" condition in marine mammals (Fernández et al. 2005; Jepson et al. 2003, 2005).

9.2.2.2 Decompression Risk and Changes in Dive Behavior

Up until the turn of the century, it was essentially dogma that marine mammals did not suffer from DCS. Findings of bubble emboli in stranded marine mammals around the same period began to challenge this idea, which was further fueled by the finding of fat emboli in beaked whales that stranded in association with naval sonar operations (Fernández et al. 2005; Jepson et al. 2003). The presence of the fat emboli was argued to be consistent with a decompression insult since nitrogen bubbles were believed to appear in fatty tissue (nitrogen is highly soluble in lipid) and the rupture of fatty tissue would presumably result in the delivery of fat droplets into the bloodstream. Disagreement among scientists about the potential for bubble formation grew (e.g., Piantadosi and Thalmann 2004), prompting further investigation and modeling of nitrogen gas accumulation in marine mammals. At the same time, a new hypothesis regarding bubble formation began to evolve as an explanation for whale strandings. Multiple variants of the hypothesis exist, but all proposed that a change in dive behavior and/or physiology (e.g., dramatic increase in ascent rate or increase in exercise) following a noise exposure was sufficient to cause nitrogen bubble formation (Blix et al. 2013; Cox et al. 2006; Fahlman et al. 2014; Hooker et al. 2012; Jepson et al. 2003; Rommel et al. 2006; Talpalar and Grossman 2005). The requirement for a noise exposure to directly activate or drive bubble growth, which presumably required high levels or long durations of received noise, was no longer necessary. The implications of the hypothesis were also greater as the received levels of noise required to trigger a change in dive behavior and/or physiology could presumably be much lower than those required for rectified diffusion or the direct activation of bubble growth.

Numerous models have been developed to predict nitrogen tensions in the tissues of diving marine mammals since direct measurements are extremely difficult to make (e.g., Fahlman et al. 2006, 2009; Hooker et al. 2009; Kvadsheim et al. 2012; Saunders et al. 2008; Zimmer and Tyack 2007). Models have addressed different marine mammal species, though the focus has been on beaked whales because of the association between beaked whale mass strandings and naval sonar operations (e.g., Hooker et al. 2009). Models vary in their predictions of the risk of a DCS-like response, which reflects both the structure of the different models and their assumptions (empirical data on gas kinetics in diving marine mammals is sorely lacking). However, several models have suggested that deep-diving species, including beaked whales, might be more prone to bubble formation because of their dive behavior (Fahlman et al. 2014; Houser et al. 2001; Kvadsheim et al. 2012). The impact of lung compression and alveolar collapse, circulatory adjustments, and the kinetics of gas exchange between tissues are critical components of these models. The uptake of nitrogen gas is greatly affected by the depth of lung compression and alveolar collapse, which occurs as diving depth (and hydrostatic pressure) increases and which serves to limit gas exchange between the lung and the blood. Lung compression and alveolar collapse are the primary mechanisms by which gas exchange is reduced while diving. In odontocetes, it has been hypothesized that gas exchange also might occur across venous networks that are closely associated with air-filled sinus spaces, thus allowing nitrogen gas uptake to continue below the depth at which complete lung and alveolar collapse occurs (Costidis and Rommel 2012). No empirical investigation of this hypothesis has been completed.

The post-mortem presence of bubbles does not necessarily mean that a bubble pathology exists. Bubbles might form due to tissue decomposition, the off-gassing of tissues in situ of animals that have died at depth and been brought to the surface, or through the introduction of bubbles into the vascular system during dissection (Bernaldo de Quirós et al. 2012, 2013a, b; Moore et al. 2009). Care must be taken during postmortem analyses so as not to introduce bubble artifacts or incorrectly identify bubbles as arising from nitrogen gas. Great efforts have been made to minimize artifacts and identify the gas composition of bubbles found in stranded marine mammals, which has helped to demonstrate the origin of bubbles in postmortem samples (Bernaldo de Quirós et al. 2011, 2012, 2013a, b; Dennison et al. 2012). The prevalence of bubbles with predominantly a nitrogen gas composition appears to be higher in deep-diving species (Bernaldo de Quirós et al. 2012), which is consistent with some modeling predictions.

The direct measurement of bubbles in the blood of any free-diving marine mammal has only been tried once with a bottlenose dolphin (Houser et al. 2009). The dolphin was trained to dive repetitively to depths of 100 m and remain there for a short period of time (90 s) during each dive. At the end of a series of 10 dives, the dolphin underwent ultrasound inspection to determine if vascular bubbles were present. No vascular bubbles were observed, but other observations have either directly or indirectly supported the notion that bubbles can form in marine mammals. Moore et al. (2009) demonstrated that marine mammals which died in gillnet fisheries had a greater prevalence of bubbles in their tissues after being brought to the surface than did stranded marine mammals. This demonstrated that marine mammals had tissues that were supersaturated with gas once they were brought to the surface—as the hydrostatic pressure decreased with surfacing, the gas solubility of the tissues also decreased such that they became supersaturated. Similar findings were obtained by Bernaldo de Quirós et al. (2013b). Moore and Early (2004) argued that a progressive osteonecrosis observed in sperm whale bones might have resulted from an accumulation of decompression insults experienced throughout the whales' lives. However, the most convincing evidence for bubbles to occur in live marine mammals was through an investigation of stranded common dolphins in which ultrasound was used to identify bubbles in the kidneys and portal vein (Dennison et al. 2011). Conversely, no bubbles were found in wild-caught bottlenose dolphins or in one trained to dive repetitively to depth (Dennison et al. 2011; Houser et al. 2009). The difference in the presence of bubbles might be due to the shallow diving nature of wild-caught or trained bottlenose dolphins (Dennison et al. 2012), but might also be related to the stranding event itself. Dennison et al. (2012) suggested the presence of the bubbles might reflect the inability of the stranded animal to dive and recompress the bubbles. Similarly, Houser et al. (2009) hypothesized that beaked whales that rapidly stranded in association with naval sonar might not have been able to effectively off-gas supersaturated tissues with long half-lives (i.e., the time it takes for the concentration of gas to decline by 50%) because of the cardiovascular collapse that develops with time on the beach (i.e., the gas cannot effectively circulate to the lung to be cleared). The persistence of bubbles in marine mammals sufficient to cause cavitary lesions, whether of nitrogen gas origin or not, has also been observed in several pinniped and cetacean species (Jepson et al. 2003, 2005; Van Bonn et al. 2011).

The link between noise exposure and the peculiar bubble and fat emboli observed in beaked whales stranded in association with high-intensity naval sonar activity has yet to be fully explained. Nevertheless, the accumulated evidence of bubble formation in marine mammals has resulted in some interesting hypotheses regarding how noise might impact marine mammals. Whether a behavioral or physiological change following noise exposure results in bubble formation remains to be observed, but recent findings in another deep diver, the Risso's dolphin, suggests that other factors besides noise might also trigger a behavioral response that can contribute to the formation of bubble and fat emboli (Fernández et al. 2017). Whatever the causative mechanism, the appearance of bubble and fat emboli appears to be much more extensive in beaked whales that have stranded in association with high-intensity sonar operations than in other observed instances of bubble and fat emboli (e.g., Fernández et al. 2012). Whether this argues for beaked whales being more sensitive to high levels of anthropogenic noise, whether there is a particular susceptibility due to their dive behavior or physiology, or whether some ecological factor is at play (e.g., the occurrence is higher due to a higher lipid content of blood due to feeding), remains to be determined.

9.2.3 Stress Responses

It is possible that marine mammals exposed to anthropogenic noise can be "stressed" by the exposure. Numerous studies have identified a relationship between noise exposure (the stressor) and a stress response in vertebrates (e.g., Kight and Swaddle 2011). The classic stress response, which consists of the rapidly occurring "fight-or-flight" response and the more prolonged "general adaptation syndrome" (Cannon 1932; Selye 1936), is characterized in large part by changes in hormones that subsequently alter an animal's physiology to help it deal with the stressor. The "fight-or-flight" response is typified by the release of catecholamines from the adrenal glands. The main catecholamines are epinephrine and norepinephrine, which are also called adrenaline and noradrenaline. Other catecholamines, such as dopamine, also exist. The catecholamines affect the availability of blood glucose, heart rate, blood pressure, oxygen consumption, and the responsiveness of the sympathetic nervous system. These occur to prepare an animal physically to either fight or flee from a stressful situation. The "general adaptation syndrome" is primarily characterized by the release of cortisol from the adrenal gland. The increase in cortisol is slower than the increase in catecholamines and occurs over a period of minutes following the introduction of a stressor. Cortisol has both genetic and nongenetic effects, with the nongenetic effects occurring earlier and influencing both metabolism and immune responses. Cortisol belongs to a group of hormones known as the glucocorticoids, which are so named because of their influence on the regulation of glucose.

If the stress response to one or more stressors is too great in magnitude or duration, the action of the same hormones can cause the animal to experience physiological dysfunction that potentially affects reproduction, growth, disease susceptibility, etc. Various models have been proposed (e.g., allostasis and the reactive scope model) to describe the relationship between the stress response and its benefit or detriment in animals (McEwen and Wingfield 2010; Romero et al. 2009). These models provide a framework within which an understanding of the stress response can potentially be evaluated in marine mammals exposed to anthropogenic sound, should a sufficient understanding of the stress response in marine mammals be obtained.

An important component in evaluating the impact of a stressor on an animal is understanding how markers of stress change as part of an animal's life history and how they change in response to a specific stressor. Unfortunately, relatively little is known about how stress markers vary as a function of life history in most marine mammals, and evidence suggests that some hormones associated with the stress response might function differently in marine mammals than in terrestrial mammals because of their secondary adaptation to a fully or semi-aquatic existence (reviewed in Atkinson et al. 2015). Fasting periods, breeding cycles, conspecific social interactions, and molting (for pinnipeds) are stressors with associated changes in hormones that help the animal cope with the stress, but all are predictable components of an animal's life history. Other natural stressors to marine mammals include exposure to disease and naturally occurring toxins, encounters with predators, changes in food availability, and changes in weather, ocean temperature, and other ocean conditions (Atkinson et al. 2015). Additional stress due to anthropogenic exposure can be additive to that which occurs naturally (Fair et al. 2014, 2017; Rolland et al. 2012). These stressors might be acute, occurring over short periods of time, or chronic, being persistent in the environment or at least existing for extended periods of time. Both acute and chronic stressors can result in immediate and long-term potential consequences to the animal depending on the characteristics of the stressor, the animal's proclivity for responding to the stressor (see below), and the animals physical and psychological condition during the stress event.

Little information on the relationship between anthropogenic sound and stress in marine mammals exists, and of the relatively few studies conducted to date, most have focused on short-term stress responses. Our knowledge about the long-term consequences of sound-induced stress is severely limited. Studies focused on components of the fight-or-flight response have primarily measured the catecholamine hormones (Romano et al. 2004; Thomas et al. 1990) or measured the heart rate of the sound-exposed animal (Bakhchina et al. 2017; Kvadsheim et al. 2010; Lyamin et al. 2011; Miksis et al. 2001), which is affected by catecholamine release. Measures of cortisol, the classical steroid hormone associated with the "general adaptation syndrome," and aldosterone, an adrenal hormone that has been suggested as being a stronger indicator of a stress response than cortisol in marine mammals (Atkinson et al. 2015; St. Aubin and Dierauf 2001), have also been made.

9.2.3.1 Hormones as Indicators of Acute Stress

Marine mammals have been shown to follow the typical mammalian stress response when trapped, captured, transported, or otherwise manipulated by humans. In these limited situations where observations of stress hormones could be made, increases in the catecholamines, glucocorticoids, and other stress-related physiological responses were observed in belugas (Schmitt et al. 2002, 2010; Spoon and Romano 2012; Trana et al. 2015), dolphins (Champagne et al. 2018; St. Aubin et al. 2013; Thomson and Geraci 1986), killer whales (Steinman et al. 2020), Guadalupe fur seals (DeRango et al. 2019), and elephant seals (Champagne et al. 2012). Specific linkages to sound exposure are similarly limited.

Thomas et al. (1990) exposed four belugas to playbacks of a semisubmersible drilling platform to assess whether the exposure would result in changes in behavior or increases in the catecholamines. The playback occurred in a pool complex; the source level of the playback, which was located on one side of the pool complex, was 153 dB re 1 µPa while sound pressure levels at a monitoring hydrophone on the opposite side of the pool complex ranged from 134 to 137 dB re 1 µPa. No changes in behavior during the playbacks were noted and no elevations in catecholamines were noted in blood samples taken immediately after the playbacks.

Romano et al. (2004) subjected a beluga and bottlenose dolphin to impulses from a seismic water gun with peak pressures of 198–226 dB re 1 µPa and 213–226 dB re 1 µPa, respectively. For the beluga, at least one of the exposures was also associated with a measurable TTS. The beluga showed significant increases in catecholamines (including dopamine) for exposures with peak sound pressures greater than 220 dB re 1 µPa, but not for lower-level exposures. Conversely, the bottlenose dolphin showed no change in catecholamines, but did show an increase in aldosterone following the exposure. However, changes in aldosterone were much lower than the magnitude of increase observed with handling stress or chemical stimulation of the adrenal glands in dolphins (Champagne et al. 2018; Thomson and Geraci 1986) and were within the normal variation observed in this species (Houser et al. 2021; St. Aubin et al. 1996). The same dolphin was also exposed to 1 s, 3 kHz tones with SPLs as high as 201 dB re 1 µPa. No changes in the stress hormones were observed in response to the tonal stimuli. In both the dolphin and beluga exposures of Romano et al. (2004), the timing of sampling following the sound exposure complicates interpretation of the stress response. Samples were collected 1 h after the exposures. Catecholamines have very short biological half-lives, typically no more than a few minutes, and it is uncertain as to how samples taken an hour after an exposure are reflective of the acute stress response. A similar concern exists for aldosterone, which rapidly declines following the cessation of exposure to a stressor (Champagne et al. 2018).

Houser et al. (2020) exposed 30 bottlenose dolphins performing a trained behavior to simulated sonar signals and collected voluntary blood samples 1 week prior to, immediately after (within 5 min), and 1 week after the exposure. Six groups of five animals each were trained to swim across an enclosure, touch a paddle, and return to their starting location. The animals were requested to perform the behavior 10 times and were exposed to a simulated sonar signal (3250–3450 Hz) as they crossed the middle of the enclosure. Each group of five animals was exposed to received SPLs of either 115, 130, 145, 160, 175, or 185 dB re 1 µPa. The blood samples were analyzed for the hormones aldosterone, cortisol, and epinephrine. No significant changes in any hormone were observed nor was any relationship observed between hormone levels and the received SPL, even though animals showed abandonment of the trained behavior at high received levels and the severity of responses scaled with increasing SPL. The findings suggested that changes in dolphin behavior might not reflect physiological changes in circulating stress hormones, and that behavior alone may not be adequate to assess an animal's state of stress.

Yang et al. (2021) exposed three dolphins to 800-Hz tones for 30 min. Unfortunately, the exposure is difficult to evaluate in this study as the authors described the noise as 150-ms tones presented as 40 strikes/min. It is further unclear whether the levels reported corresponded to a root-mean-square measure or a peak measure. Nevertheless, the authors reported exposure conditions consisting of controls (no sound), and low and high-level exposures with the latter having estimated received levels of 120 and 140 dB re 1 µPa. Exposures occurred every 3 days and each condition was tested five times. Animals provided voluntary blood samples prior to and immediately following (exact time not given) the exposure sessions and the blood samples were analyzed for cortisol and mRNA expression of several immunological factors. Although no absolute values were provided for the blood parameters measured, ratios of before and after values were analyzed to determine if changes occurred under the various exposure conditions. The authors reported a significantly higher ratio for cortisol and immune factors in the high-level exposure condition than in the low-level and control conditions, suggesting a stress response occurred due to the exposure. Unfortunately, because concentrations of cortisol were not reported, it is not possible to determine whether the variation observed in any condition was outside the range of normal variation in the bottlenose dolphin (Houser et al. 2021).

Collectively, the studies mentioned above do not suggest a strong hormonal response to acute sound exposures, or at least leave it questionable as to whether one exists. However, as all the animals in these studies had been under human care for an extended period, it cannot be ruled out that they had become more tolerant of anthropogenic noise. In the case of Romano et al. (2004), the animals were also conditioned to tolerate the exposures. How this affects the hormonal response to the acute noise exposure cannot be stated with certainty, but it seems quite possible that the experience and conditioning with the noise sources could have dampened any hormonal response. It should also be noted that all the animals involved were either bottlenose dolphins or belugas. Information on the endocrine response to acute noise exposure in other marine mammal species is sorely needed.

9.2.3.2 Hormones as Markers of Chronic Stress

Very little is known about how elevations in stress hormones might be related to chronic anthropogenic noise exposures, and the impact to animals in such situations is even less known. Limited information exists as to how chronic stressors unrelated to life-history changes impact stress hormones (e.g., see Houser et al. 2011, for an example of how cold stress elevates cortisol and aldosterone in bottlenose dolphins). Rolland et al. (2012) measured the levels of glucocorticoid metabolites in the feces of northern Atlantic right whales, a critically endangered species, and compared them between periods of ocean noise when shipping activity was normal and when it was substantially reduced following the terrorist attacks of September 11, 2001. Fecal glucocorticoid metabolites (fGC), which are believed to arise primarily from cortisol and/or corticosterone, presumably reflect the levels of cortisol in the blood over some period and can therefore be used as an indirect measure of circulating glucocorticoid levels (Wasser et al. 2000). The authors found that fGC levels were significantly lower during the period of reduced ship traffic. Ocean noise during that period was reduced by ~6 dB, particularly at frequencies <150 Hz. Unfortunately, no direct noise exposures were made on the animals (e.g., acoustic dosimeter) and fecal collections were not related to individuals. This makes any direct relationship between ocean noise and fecal glucocorticoids difficult to ascertain.

Fecal hormone metabolites have also been used to study the impact of vessel traffic on killer whales (Ayres et al. 2012) and gray whales (Lemos et al. 2022). Lemos et al. (2022) reported regional changes in ambient SPLs and its relationship to vessel traffic, but no sound measurements were reported by Ayres et al. (2012); however, there was an inherent assumption that the magnitude of vessel traffic should be related to noise levels within the frequency bands of noise generated by vessel engines. The goal of Ayres et al. (2012) was to determine whether vessel traffic or the lack of prey was the stronger impact to killer whales. Feces collected from the water surface were assessed for both thyroid hormone and fGC. Because fGC were most strongly influenced by the presence or absence of prey, the authors concluded that the impact of prey availability was a much stronger stressor to the killer whales than was the presence of vessel traffic. Conversely, by assessing fGC in gray whales exposed to vessel traffic that varied across years, Lemos et al. (2022) concluded that there was a direct relationship between the degree of vessel traffic and fGC levels, even though noise exposure levels were likely low. Unfortunately, like in the work of Rolland et al. (2012), no direct measures of noise exposure on any individuals were made and other confounds, both natural and anthropogenic, could be related to the results.

Stress-linked hormones can be measured in blubber samples and at least one study has attempted to link cortisol measured in humpback whale blubber biopsies to whale-watching vessel activity off the coast of Alaska (Teerlink et al. 2018). The researchers compared cortisol in blubber biopsies collected from a high vessel activity region with two regions with low vessel activity. They hoped to determine whether cortisol would increase in the blubber of animals exposed to extensive whale watching during the 3 to 4-month whale watching season. No relationship between blubber cortisol and the degree of whale watching vessel activity was found. However, as with the fGC studies, no direct measures of noise exposure were obtained and it is difficult to assess what an individual whale might have been exposed to across the duration of the whale watching season.

Measuring stress hormones in a wild animal concomitant with a long-term noise exposure estimate is a daunting task. However, the potential to assess large numbers of animals and more generally monitor noise conditions in an affected habitat has potential for elucidating low-level impacts that might pervade populations but not be apparent at the individual level. Collectively, research studies utilizing fGC and blubber hormone measurements have a unique potential for assessing chronic noise impacts provided experimental designs can be refined to address the study shortcomings observed to date.

9.2.3.3 Heart Rate as an Indicator of Acute Stress

Heart rate can rapidly change in response to a stressful disturbance. Because of the response, heart rate is used to monitor the stress of terrestrial animals in the wild and in domestic and agricultural settings (Baldock and Sibly 1990; Ditmer et al. 2015; Hopster and Blokhuis 1994; Schmidt et al. 2010a, b). Heart rate might decrease or increase during a stressful disturbance and is possibly linked to whether the animal's defense strategy in the face of predators is to freeze or to flee (e.g., Espmark and Langvatn 1985). In some animals, heart rate has been shown to elevate to variable degrees depending on the stressor (Chabot et al. 1996; Nephew et al. 2003; Weisenberger et al. 1996); for example, the presence of predators might elicit a maximal heart rate whereas the presence of predictable human activity might only slightly elevate the heart rate (MacArthur et al. 1979, 1982; Viblanc et al. 2012). The heart rate response has also been observed to be scaled by the distance to the threat (Ackerman et al. 2004; MacArthur et al. 1982). An organism often rapidly habituates (decreased magnitude of the heart rate response) to repetitive exposures of a stressor (Thompson et al. 1968; Weisenberger et al. 1996) and can vary as a function of age (Espmark and Langvatn 1985).

Heart rate as an indicator of stress in marine mammals is less understood than in terrestrial mammals. Capture stress produced variable heart rate responses in harbor porpoises with one individual showing an increased heart rate, another showing a decreased heart rate, and the remaining subjects demonstrating no significant difference in heart rate (Eskesen et al. 2009). Only relatively minor differences were observed in the electrocardiogram and heart rate of wild-caught bottlenose dolphins and those trained to be held out of the water (Harms et al. 2013). In contrast, pinnipeds subjected to forced dives show maximal reductions in heart rate, which is likely a combination of the dive response and the stress caused by the forced dive (Elsner et al. 1966; Grinnell et al. 1942; Jobsis et al. 2001; Scholander 1940). The response was attenuated in harbor seals that were trained for the procedure, thus showing a form of cardiac habituation as the animals learned that the procedure was nonthreatening (Jobsis et al. 2001).

Only a handful of studies exist that have related acoustic exposures to heart rate as a measure of stress. Miksis et al. (2001) compared the heart rate before an acoustic stimulus to the heart rate following the stimulus for different types of playbacks. The measurement period spanned the 10 beats just prior to stimulus presentation and the 20 beats immediately following the stimulus presentation. The authors found that the heart rate of a bottlenose dolphin initially increased when presented with playbacks of tank noise, known conspecific whistles, or jawclaps. For playbacks of tank noise, heart rate quickly returned to normal, but remained elevated for the duration of the measurement period for conspecific whistles and jawclaps. The authors suggested the initial response to all playbacks was potentially related to a startle response, where the more persistent response to conspecific whistles and jawclaps was potentially reflective of a stress response (although it is unclear why conspecific whistles would be stressful). No changes in heart rate were observed in a second animal exposed to the same acoustic stimuli. The authors concluded that the heart rate response observed in the singular dolphin was consistent with that observed in terrestrial mammals experiencing threatening stimuli.

Lyamin et al. (2011) subjected a young beluga, captured 2 months prior to testing, to a series of noise exposures. The beluga was placed into a small (4 m × 0.8 m × 0.8 m) bath and was exposed to bands of noise with various bandwidths for variable durations (1, 3, and 10 min) and exposure levels (140, 150, and 160 dB re 1 µPa). The beluga showed increases in heart rate to noise exposures that were dependent upon the frequency band of the exposure and the level of noise. The lowest frequency band of 19–27 kHz produced the greatest increase in heart rate. The magnitude of the heart rate change decreased as the bandpass noise increased in frequency, and for higher frequency bands of noise, the change in the heart rate was only significant within the first minute of the noise exposure. Heart rate also showed a direct relationship to the noise sound pressure level. A subset of the data consisting of 10-min playbacks with noise bands of 19–27 kHz and 27–38 kHz and noise levels of 150 and 160 dB re 1 µPa were subsequently analyzed via a spectral analysis (Bakhchina et al. 2017). The results were consistent with prior direct measurements. However, a repeat of the experiment 1 year later, but using only 19–27 kHz bandpass noise and a sound pressure level of 165 dB re 1 µPa, showed no change in the heart rate of the same animal. The authors concluded that the beluga likely became habituated to the noise exposures, but that the initial findings were indicative of a heart rate response like that observed in terrestrial mammals exposed to a startling or acute (fight-or-flight) stressor. It is also possible that the age of the animal affected the heart rate response and changed as the beluga matured.

The heart rate of two captive harbor porpoises declined below the normal bradycardia that occurs with diving when exposed to various sounds with frequencies between 100 and 140 kHz (Teilmann et al. 2006). All signals were 200 ms in duration and had maximum source levels of 153 dB re 1 µPa, although some signals were ramped up from lower starting amplitudes. Received levels at the animal were not obtained. The sounds were presented to the harbor porpoises every 4 s and sound exposure sessions were 5 min in duration (and compared to a 10-min quiet period prior to the exposure). Multiple sessions were conducted with the porpoises with the sound types varying across sessions. The decline in heart rate was notable only in the early presentations of the signals and the porpoises appeared to habituate rapidly over successive presentations of the noise. In another study, two harbor porpoises were exposed to a 500 ms, 6–9 kHz frequency sweep with a

~50–100 ms rise time (Elmegaard et al. 2021). These animals showed a dramatic bradycardia upon the first few noise exposures, but also habituated quickly across repeated exposures. Exposure to higher frequency noise bursts (center frequency of 40 kHz) consistently produced a startle response, but no change in heart rate was observed.

Kvadsheim et al. (2010) measured the heart rate of captive hooded seals exposed to linear frequency-modulated upsweeps ranging from 1.3 to 1.7 kHz, 3.7 to 4.3 kHz, and 6.0 to 7.0 kHz. The order of signal presentation was randomized but consisted of a "soft start," in which the source level was gradually increased from 134 to 194 dB re 1 µPa over a 10-min period. A second experiment was performed with the same signals and at a source level of 194 dB re 1 µPa but with an increasing duty cycle (from 1% to 10%). Heart rate measured in the exposed seals demonstrated an increase only when the animals were at the surface. During diving, when bradycardia occurs, no change in heart rate was noticed. Neither gray seals nor harbor seals exposed to sonic booms during the gray seal breeding period showed any significant change in heart rate. No exposure levels were recorded at the animals, but peak SPLs of ~122–133 dB re 20 µPa were recorded near the rookery (Perry et al. 2002).

In one of the only studies to look at changes in heart rate in wild cetaceans exposed to a known level of noise, Williams et al. (2022) found that two narwhals exposed to three different seismic events ($n = 1$ and $n = 2$ for events, respectively) did not show a significant change in bradycardia with diving when compared to controls (no noise exposure). However, the lack of change co-occurred with an increase in exercise effort as measured through stroke rate. The authors concluded that the animals likely had a fear response to the exposures, but that the physiological response was reflected in increased heart rate variability; this was due to changes in dive/stroke effort and increased surface respiration rate, but not a consistent reduction in the diving heart rate.

The interaction between the acute stress response's impact on heart rate and that of the bradycardia that occurs during diving must be resolved, particularly regarding how it might differ between cetaceans and pinnipeds, and considering how exercise effort might be altered in response to the exposure (Williams et al. 2022). Seals seem to demonstrate a reduction in heart rate when startled while at depth, but an increase in heart rate when acutely stressed by noise while at the surface. The response in cetaceans is not as clear. Slight increases in heart rate were noted in dolphins exposed to jawclaps and whistles of conspecifics while submerged at shallow depth (Miksis et al. 2001), as well as in a beluga exposed to noise while constrained in a bath (Lyamin et al. 2011). However, freely diving harbor porpoises demonstrated a decrease in heart rate following initial exposures to noise (Teilmann et al. 2006), and free-ranging narwhals demonstrated a complex interaction that resulted in increased heart rate variability relative to controls (Williams et al. 2022). As recently observed in a study of the heart rate of captured narwhals (Williams et al. 2017), a profound bradycardia persisted when diving immediately after the animals were exposed to the stress of capture and handling, even while exercise effort increased. The response was more graded when the animals were free-ranging and exposed to noise from seismic air guns. An understanding of the impact that exposure context might have on heart rate is also needed. It cannot be determined whether the signals played to the dolphins in the study of Miksis et al. (2001) are truly reflective of a stress response or the anticipation of being reunited with conspecifics from which they were recently separated. There is also substantial variation in whether an individual exhibits a change in heart rate following noise exposure (e.g., one dolphin showed a heart rate increase and one did not when exposed to the same stimuli; Miksis et al. 2001) and that habituation to noise exposure is rapid in many cases. Whether other aspects of the stress response (e.g., increase in cortisol) are attenuated with repeated sound exposures remains to be resolved.

9.2.3.4 Future Work

The characteristics of sound that correlate with specific stress responses in marine mammals remain largely unknown. Current evidence suggests that acute responses to sound exposure might be mitigated by habituation. Experience with a particular type of sound exposure undoubtedly plays a role in the stress response (i.e., animals might rapidly habituate or become sensitized to repetitive sound exposures of a given type). Other factors also likely impact the stress response of marine mammals to anthropogenic sound exposure, including the sex, age (e.g., Houser et al. 2013), reproductive status, and behavioral phenotype of the animal (e.g., whether it is a risk-taker or is risk-averse), as well as the context of the exposure. Generally, the impact of chronic stress is likely much more important to the health and well-being of a marine mammal than an acute stressor (St. Aubin and Dierauf 2001), provided the acute response does not lead to death, debilitation, or the interruption of breeding. Studies on humans and other land mammals has shown that chronic stress can lead to impaired growth and reproduction, and increased susceptibility to disease, all of which can be profoundly impactful if occurring across populations.

Much of the information summarized in this section has focused on stress hormones and heart rate as an indicator of stress. It should be noted that advances have occurred in recent years that should help to shed light on the stress response of marine mammals. These include the ability to look at stress hormones in other "matrices" than blood, blubber, or feces. For example, assessing stress hormones in the exhalant (blows) noninvasively collected from whales is actively being researched (e.g., Burgess et al. 2016). Beyond stress hormones, new techniques in the world of "-omics" also hold promise. Work on stress-related transcriptomics, which is the collection of RNA transcripts expressed under specific situations, has shown changes in the transcriptome following an acute stressor in elephant seals (Khudyakov et al. 2017). With further development, these and similar techniques might be used to better understand the nonauditory physiological responses of marine mammals to sound exposure.

9.3 Summary

Noise exposure can affect the auditory pathway of marine mammals by causing auditory fatigue or damage that results in hearing loss, which is an increase in the hearing threshold. The hearing loss can be either temporary or permanent, with the latter usually related to more extreme noise exposures. The growth and onset of NIHL is a function of the duration, level, and frequency of the noise exposure, and appears to vary by species. Nonauditory physiological effects can also occur in response to noise exposure. Hypothesized effects include nitrogen and fat emboli formation and the potential development of symptoms related to decompression sickness. Other effects include cardiac and endocrine responses, specifically the release of hormones that mitigate the effects of the stressor (noise exposure). Interpreting cardiac responses to noise exposures are complicated due to the dive-related bradycardia that is common to marine mammals. Marine mammals have shown rapid habituation in the cardiac response to noise exposure when exposed repeatedly. Although cortisol is known to increase when marine mammals are stressed, little conclusive evidence exists that noise predictably elevates cortisol in marine mammals under either acute or chronic noise exposure conditions. Nevertheless, studies of cardiac and endocrine responses to noise are in their infancy and will likely yield more conclusive results as experimental procedures evolve. New techniques being developed in the field of "-omics" also hold promise for helping to understand nonauditory physiological responses to noise exposure.

References

Ackerman JT, Takekawa JY, Kruse KL, Orthmeyer DL, Yee JL, Ely CR, Ward DH, Bollinger KS, Mulcahy DM (2004) Using radiotelemetry to monitor cardiac response of free-living tule greater white-fronted geese

(*Anser albifrons elgasi*) to human disturbance. Wilson Bull 116(2):146–151. https://doi.org/10.1676/03-110

Atkinson S, Crocker D, Houser D, Mashburn K (2015) Stress physiology in marine mammals: how well do they fit the terrestrial model? J Comp Physiol B 185(5): 463–486. https://doi.org/10.1007/s00360-015-0901-0

Ayres KL, Booth RK, Hempelmann JA, Koski KL, Emmons CK, Baird RW, Balcomb-Bartok K, Hanson MB, Ford MJ, Wasser SK (2012) Distinguishing the impacts of inadequate prey and vessel traffic on an endangered killer whale (*Orcinus orca*) population. PLoS One 7(6). https://doi.org/10.1371/journal.pone.0036842

Bakhchina AV, Mukhametov LM, Rozhnov VV, Lyamin OI (2017) Spectral analysis of heart rate variability in the beluga (*Delphinapterus leucas*) during exposure to acoustic noise. J Evol Biochem Physiol 53(1):60–65. https://doi.org/10.1134/s0022093017010070

Balcomb KC III, Claridge DE (2001) A mass stranding of cetaceans caused by naval sonar in the Bahamas. Bahamas J Sci 8(2):2–12

Baldock NM, Sibly RM (1990) Effects of handling and transportation on the heart rate and behaviour of sheep. Appl Anim Behav Sci 28(1):15–39. https://doi.org/10.1016/0168-1591(90)90044-E

Bernaldo de Quirós Y, González-Díaz OS, Saavedra P, Arbelo M, Sierra E, Sacchini S, Jepson PD, Mazzariol S, Di Guardo G, Fernández A (2011) Methodology for *in situ* gas sampling, transport and laboratory analysis of gases from stranded cetaceans. Sci Rep 1(193). https://doi.org/10.1038/srep00193

Bernaldo de Quirós Y, González-Díaz O, Arbelo M, Sierra E, Sacchini S, Fernández A (2012) Decompression vs. decomposition: distribution, amount, and gas composition of bubbles in stranded marine mammals. Front Physiol 3(177). https://doi.org/10.3389/fphys.2012.00177

Bernaldo de Quirós Y, González-Díaz O, Møllerløkken A, Brubakk AO, Hjelde A, Saavedra P, Fernández A (2013a) Differentiation at autopsy between in vivo gas embolism and putrefaction using gas composition analysis. Int J Legal Med 127(2):437–445. https://doi.org/10.1007/s00414-012-0783-6

Bernaldo de Quirós Y, Seewald JS, Sylva SP, Greer B, Niemeyer M, Bogomolni AL, Moore MJ (2013b) Compositional discrimination of decompression and decomposition gas bubbles in bycaught seals and dolphins. PLoS One 8(12). https://doi.org/10.1371/journal.pone.0083994

Blix AS, Walloe L, Messelt EB (2013) On how whales avoid decompression sickness and why they sometimes strand. J Exp Biol 216(18):3385–3387. https://doi.org/10.1242/jeb.087577

Branstetter BK, Finneran JJ (2008) Comodulation masking release in bottlenose dolphins (*Tursiops truncatus*). J Acoust Soc Am 124(1):625–633. https://doi.org/10.1121/1.2918545

Burgess EA, Hunt KE, Kraus SD, Rolland RM (2016) Get the most out of blow hormones: validation of sampling materials, field storage and extraction techniques for whale respiratory vapour samples. Conserv Physiol 4(1):cow024. https://doi.org/10.1093/conphys/cow024

Cannon WB (1932) The wisdom of the body. W. W. Norton and Company, Inc., New York

Chabot D, Gagnon P, Dixon EA (1996) Effect of predator odors on heart rate and metabolic rate of wapiti (*Cervus elaphus canadensis*). J Chem Ecol 22(4):839–868. https://doi.org/10.1007/bf02033590

Champagne CD, Houser DS, Costa DP, Crocker DE (2012) The effects of handling and anesthetic agents on the stress response and carbohydrate metabolism in northern elephant seals. PLoS One 7(5):e38442. https://doi.org/10.1371/journal.pone.0038442

Champagne CD, Kellar NM, Trego ML, Delehanty B, Boonstra R, Wasser SK, Booth RK, Crocker DE, Houser DS (2018) Comprehensive endocrine response to acute stress in the bottlenose dolphin from serum, blubber, and feces. Gen Comp Endocrinol 266:178–193. https://doi.org/10.1016/j.ygcen.2018.05.015

Costidis A, Rommel SA (2012) Vascularization of air sinuses and fat bodies in the head of the bottlenose dolphin (*Tursiops truncatus*): morphological implications on physiology. Front Physiol 3(243): 1–23. https://doi.org/10.3389/fphys.2012.00243

Cox TM, Ragen TJ, Read AJ, Vos E, Baird RW, Balcomb K, Barlow J, Caldwell J, Cranford TW, Crum L, D'Amico A, D'Spain G, Fernández A, Finneran JJ, Gentry R, Gerth W, Gulland FM, Hildebrand J, Houser DS, Hullar T, Jepson PD, Ketten D, Macleod CD, Miller P, Moore S, Mountain DC, Palka DL, Ponganis PJ, Rommel SA, Rowles T, Taylor BL, Tyack P, Wartzok D, Gisiner R, Meads J, Benner L (2006) Understanding the impacts of anthropogenic sound on beaked whales. J Cetacean Res Manag 7(3):177–187. https://doi.org/10.47536/jcrm.v7i3.729

Crum LA, Mao Y (1996) Acoustically enhanced bubble growth at low frequencies and its implications for human diver and marine mammal safety. J Acoust Soc Am 99(5):2898–2907. https://doi.org/10.1121/1.414859

Crum LA, Bailey MR, Jingfeng G, Hilmo PR, Kargl SG, Matula TJ (2005) Monitoring bubble growth in supersaturated blood and tissue *ex vivo* and the relevance to marine mammal bioeffects. Acoust Res Lett Online 6(3):214–220. https://doi.org/10.1121/1.1930987

Dennison S, Moore MJ, Fahlman A, Moore K, Sharp S, Harry CT, Hoppe J, Niemeyer M, Lentell B, Wells RS (2011) Bubbles in live-stranded dolphins. Proc R Soc Lond B 279:1396–1404. https://doi.org/10.1098/rspb.2011.1754

Dennison S, Fahlman A, Moore M (2012) The use of diagnostic imaging for identifying abnormal gas accumulations in cetaceans and pinnipeds. Front

Physiol 3(181):1–8. https://doi.org/10.3389/fphys.2012.00181

Department of Commerce (DoC), Department of the Navy (DoN) (2001) Joint interim report bahamas marine mammal stranding event of 15-16 March 2000. Department of Commerce, Washington, DC, 66 pp

Department of the Navy (1998) Final environmental impact statement, shock testing the SEAWOLF submarine. Department of the Navy, Washington, DC, 563 pp

Department of the Navy (DoN) (2001) Final overseas environmental impact statement and environmental impact statement for surveillance towed array sensor system low frequency active (SURTASS LFA) sonar. Department of the Navy, Washington, DC, 688 pp

Department of the Navy (DoN) (2008) Atlantic fleet active sonar training: final environmental impact statement/overseas environmental impact statement (FEIS/OEIS). Department of the Navy, Washington, DC, 590 pp

DeRango EJ, Greig DJ, Gálvez C, Norris TA, Barbosa L, Elorriaga-Verplancken FR, Crocker DE (2019) Response to capture stress involves multiple corticosteroids and is associated with serum thyroid hormone concentrations in Guadalupe fur seals (*Arctocephalus philippii townsendi*). Mar Mamm Sci 35(1):72–92. https://doi.org/10.1111/mms.12517

Ditmer MA, Vincent JB, Werden LK, Tanner JC, Laske TG, Iaizzo PA, Garshelis DL, Fieberg JR (2015) Bears show a physiological but limited behavioral response to unmanned aerial vehicles. Curr Biol 25(17): 2278–2283. https://doi.org/10.1016/j.cub.2015.07.024

Elmegaard SL, McDonald BI, Teilmann J, Madsen PT (2021) Heart rate and startle responses in diving, captive harbour porpoises (*Phocoena phocoena*) exposed to transient noise and sound. Biol Open 10(6). https://doi.org/10.1242/bio.058679

Elsner R, Franklin DL, Van Citters RL, Kenney DW (1966) Cardiovascular defense against asphyxia. Science 153:941–949. https://doi.org/10.1126/science.153.3739.941

Eskesen IG, Teilmann J, Geertsen BM, Desportes G, Riget F, Dietz R, Larsen F, Siebert U (2009) Stress level in wild harbour porpoises (*Phocoena phocoena*) during satellite tagging measured by respiration, heart rate and cortisol. J Mar Biol Assoc UK 89(5):885–892. https://doi.org/10.1017/S0025315408003159

Espmark Y, Langvatn R (1985) Development and habituation of cardiac and behavioral responses in young red deer calves (*Cervus elaphus*) exposed to alarm stimuli. J Mammal 66(4):702–711. https://doi.org/10.2307/1380796

Evans DL (2002) Report of the workshop on acoustic resonance as a source of tissue trauma in cetaceans. NOAA, Silver Spring

Fahlman A, Olszowka A, Bostrom B, Jones DR (2006) Deep diving mammals: dive behavior and circulatory adjustments contribute to bends avoidance. Respir Physiol Neurobiol 153(1):66–77. https://doi.org/10.1016/j.resp.2005.09.014

Fahlman A, Hooker SK, Olszowka A, Bostrom BL, Jones DR (2009) Estimating the effect of lung collapse, pulmonary shunt on gas exchange during breath-hold diving: the Scholander and Kooyman legacy. Respir Physiol Neurobiol 165(1):28–39. https://doi.org/10.1016/j.resp.2008.09.013

Fahlman A, Tyack PL, Miller PJ, Kvadsheim PH (2014) How man-made interference might cause gas bubble emboli in deep diving whales. Front Physiol 5. https://doi.org/10.3389/fphys.2014.00013

Fair PA, Schaefer AM, Romano TA, Bossart GD, Lamb SV, Reif JS (2014) Stress response of wild bottlenose dolphins (*Tursiops truncatus*) during capture-release health assessment studies. Gen Comp Endocrinol 206:203–212. https://doi.org/10.1016/j.ygcen.2014.07.002

Fair P, Schaefer A, Houser D, Bossart G, Romano T, Champagne C, Stott J, Rice C, White N, Reif J (2017) The environment as a driver of immune and endocrine responses in dolphins (*Tursiops truncatus*). PLoS One 12(5):e0176202. https://doi.org/10.1371/journal.pone.0176202

Falke KJ, Hill RD, Qvist J, Schneider RC, Guppy M, Liggins GC, Hochachka PW, Elliott RE, Zapol WM (1985) Seal lungs collapse during free diving: evidence from arterial nitrogen tensions. Science 229:556–558. https://doi.org/10.1126/science.4023700

Fernández A, Edwards J, Martín V, Rodríguez F, Espinosa de los Monteros A, Herráez P, Castro P, Jaber JR, Arbelo M (2005) "Gas and fat embolic syndrome" involving a mass stranding of beaked whales (family Ziphiidae) exposed to anthropogenic sonar signals. J Vet Pathol 42:446–457. https://doi.org/10.1354/vp.42-4-446

Fernández A, Sierra E, Martin V, Mendez M, Sacchinni S, Bernaldo de Quirós Y, Andrada M, Rivero M, Quesada O, Tejedor M, Arbelo M (2012) Last "atypical" beaked whales mass stranding in the Canary Islands (July, 2004). J Mar Sci Res Dev 2(2). https://doi.org/10.4172/2155-9910.1000107

Fernández A, Sierra E, Díaz-Delgado J, Sacchini S, Sánchez-Paz Y, Suárez-Santana C, Arregui M, Arbelo M, Bernaldo de Quirós Y (2017) Deadly acute decompression sickness in Risso's dolphins. Sci Rep 7(1):13621. https://doi.org/10.1038/s41598-017-14038-z

Finneran JJ (2015) Noise-induced hearing loss in marine mammals: a review of temporary threshold shift studies from 1996 to 2015. J Acoust Soc Am 138(3): 1702–1726. https://doi.org/10.1121/1.4927418

Finneran JJ, Schlundt CE (2010) Frequency-dependent and longitudinal changes in noise-induced hearing loss in a bottlenose dolphin (*Tursiops truncatus*). J Acoust Soc Am 128(2):567–570. https://doi.org/10.1121/1.3458814

Finneran JJ, Schlundt CE (2013) Effects of fatiguing tone frequency on temporary threshold shift in bottlenose dolphins (*Tursiops truncatus*). J Acoust Soc Am 133(3):1819–1826. https://doi.org/10.1121/1.4776211

Finneran JJ, Schlundt CE, Carder DA, Clark JA, Young JA, Gaspin JB, Ridgway SH (2000) Auditory and behavioral responses of bottlenose dolphins (*Tursiops truncatus*) and a beluga whale (*Delphinapterus leucas*) to impulsive sounds resembling distant signatures of underwater explosions. J Acoust Soc Am 108(1): 417–431. https://doi.org/10.1121/1.429475

Finneran JJ, Schlundt CE, Dear R, Carder DA, Ridgway SH (2002) Temporary shift in masked hearing thresholds (MTTS) in odontocetes after exposure to single underwater impulses from a seismic watergun. J Acoust Soc Am 111(6):2929–2940. https://doi.org/10.1121/1.1479150

Finneran JJ, Dear R, Carder DA, Ridgway SH (2003) Auditory and behavioral responses of California sea lions (*Zalophus californianus*) to single underwater impulses from an arc-gap transducer. J Acoust Soc Am 114(3):1667–1677. https://doi.org/10.1121/1.1598194

Finneran JJ, Carder DA, Schlundt CE, Ridgway SH (2005) Temporary threshold shift (TTS) in bottlenose dolphins (*Tursiops truncatus*) exposed to mid-frequency tones. J Acoust Soc Am 118(4): 2696–2705. https://doi.org/10.1121/1.2032087

Finneran JJ, Schlundt CE, Branstetter B, Dear RL (2007) Assessing temporary threshold shift in a bottlenose dolphin (*Tursiops truncatus*) using multiple simultaneous auditory evoked potentials. J Acoust Soc Am 122(2):1249–1264. https://doi.org/10.1121/1.2749447

Finneran JJ, Carder DA, Schlundt CE, Dear RL (2010a) Growth and recovery of temporary threshold shift (TTS) at 3 kHz in bottlenose dolphins (*Tursiops truncatus*). J Acoust Soc Am 127(5):3256–3266. https://doi.org/10.1121/1.3372710

Finneran JJ, Carder DA, Schlundt CE, Dear RL (2010b) Temporary threshold shift in a bottlenose dolphin (*Tursiops truncatus*) exposed to intermittent tones. J Acoust Soc Am 127(5):3267–3272. https://doi.org/10.1121/1.3377052

Finneran JJ, Schlundt CE, Branstetter BK, Trickey J, Bowman V, Jenkins K (2015) Effects of multiple impulses from a seismic air gun on bottlenose dolphin hearing and behavior. J Acoust Soc Am 137(4): 1634–1646. https://doi.org/10.1121/1.4916591

Grinnell SW, Irving L, Scholander PF (1942) Experiments on the relation between blood flow and heart rate in the diving seal. J Cell Comp Physiol 19:341–346

Halfwerk W, Holleman LJM, Lessells CM, Slabbekoorn H (2011) Negative impact of traffic noise on avian reproductive success. J Appl Ecol 48(1):210–219. https://doi.org/10.1111/j.1365-2664.2010.01914.x

Harms CA, Jensen ED, Townsend FI, Hansen LJ, Schwacke LH, Rowles TK (2013) Electrocardiograms of bottlenose dolphins (*Tursiops truncatus*) out of water: habituated collection versus wild postcapture animals. J Zoo Wildl Med 44:972–981. https://doi.org/10.1638/2013-0093.1

Henderson D, Hamernik RP (1986) Impulse noise: critical review. J Acoust Soc Am 80(2):569–584. https://doi.org/10.1121/1.394052

Hooker SK, Baird RW, Fahlman A (2009) Could beaked whales get the bends? Effect of diving behaviour and physiology on modelled gas exchange for three species: *Ziphius cavirostris*, *Mesoplodon densirostris* and *Hyperoodon ampullatus*. Respir Physiol Neurobiol 167:235–246. https://doi.org/10.1016/j.resp.2009.04.023

Hooker SK, Fahlman A, Moore MJ, Aguilar de Soto N, Bernaldo de Quirós Y, Brubakk AO, Costa DP, Costidis AM, Dennison S, Falke KJ, Fernández A, Ferrigno M, Fitz-Clarke JR, Garner MM, Houser DH, Jepson PD, Ketten DR, Kvadsheim PH, Madsen PT, Pollock NW, Rostein DS, Rowles TK, Simmons SE, Van Bonn W, Weathersby PK, Weise MJ, Williams TM, Tyack PL (2012) Deadly diving? Physiological and behavioural management of decompression stress in diving mammals. Proc R Soc Lond B 279:1041–1050. https://doi.org/10.1098/rspb.2011.2088

Hopster H, Blokhuis HJ (1994) Validation of a heart-rate monitor for measuring a stress response in dairy cows. Can J Anim Sci 74(3):465–474. https://doi.org/10.4141/cjas94-066

Houser DS (2021) When is temporary threshold shift injurious to marine mammals? J Mar Sci Eng 9(7): 757. https://doi.org/10.3390/jmse9070757

Houser DS, Howard R, Ridgway SH (2001) Can diving-induced tissue nitrogen supersaturation increase the chance of acoustically driven bubble growth in marine mammals? J Theor Biol 213(2):183–195. https://doi.org/10.1006/jtbi.2001.2415

Houser DS, Dankiewicz-Talmadge LA, Stockard TK, Ponganis PJ (2009) Investigation of the potential for vascular bubble formation in a repetitively diving dolphin. J Exp Biol 213:52–62. https://doi.org/10.1242/jeb.028365

Houser DS, Yeates LC, Crocker DE (2011) Cold stress induces an adrenocortical response in bottlenose dolphins (*Tursiops truncatus*). J Zoo Wildl Med 42(4):565–571. https://doi.org/10.1638/2010-0121.1

Houser DS, Martin SW, Finneran JJ (2013) Behavioral responses of California sea lions to mid-frequency (3250-3450 Hz) sonar signals. Mar Envion Res 92: 268–278. https://doi.org/10.1016/j.marenvres.2013.10.007

Houser DS, Yost W, Burkard R, Finneran JJ, Reichmuth C, Mulsow J (2017) A review of the history, development and application of auditory weighting functions in humans and marine mammals. J Acoust Soc Am 141(3):1371–1413. https://doi.org/10.1121/1.4976086

Houser DS, Martin S, Crocker DE, Finneran JJ (2020) Endocrine response to simulated U.S. Navy mid-frequency sonar exposures in the bottlenose dolphin (*Tursiops truncatus*). J Acoust Soc Am 147(3): 1681–1687. https://doi.org/10.1121/10.0000924

Houser DS, Champagne CD, Wasser SK, Booth RK, Romano T, Crocker DE (2021) Influence of season, age, sex, and time of day on the endocrine profile of the common bottlenose dolphin (*Tursiops truncatus*). Gen Comp Endocrinol 313:113889. https://doi.org/10.1016/j.ygcen.2021.113889

Hume K, Brink M, Basner M (2012) Effects of environmental noise on sleep. Noise Health 14(61):297–302. https://doi.org/10.4103/1463-1741.104897

Humes LE (1980) Temporary threshold shift for masked pure tones. Audiology 19:335–345. https://doi.org/10.3109/00206098009072674

Jepson PD, Arbelo M, Deaville R, Patterson IAR, Castro P, Baker JR, Degollada E, Ross HM, Herráez P, Pocknell AM, Rodriguez E, Howie FE, Espinosa A, Reid RJ, Jaber JR, Martin V, Cunningham AA, Fernández A (2003) Gas-bubble lesions in stranded cetaceans: was sonar responsible for a spate of whale deaths after an Atlantic military exercise? Nature 425:575–576. https://doi.org/10.1038/425575a

Jepson PD, Deaville R, Patterson IAP, Pocknell AM, Ross HM, Baker JR, Howie FE, Reid RJ, Colloff A, Cunningham AA (2005) Acute and chronic gas bubble lesions in cetaceans stranded in the United Kingdom. Vet Pathol 42(3):291–305. https://doi.org/10.1354/vp.42-3-291

Jobsis PD, Ponganis PJ, Kooyman GL (2001) Effects of training on forced submersion responses in harbor seals. J Exp Biol 204(22):3877–3885. https://doi.org/10.1242/jeb.204.22.3877

Kastak D, Schusterman RJ (1996) Temporary threshold shift in a harbor seal (*Phoca vitulina*). J Acoust Soc Am 100(3):1905–1908. https://doi.org/10.1121/1.416010

Kastak D, Schusterman RJ, Southall BL, Reichmuth CJ (1999) Underwater temporary threshold shift induced by octave-band noise in three species of pinniped. J Acoust Soc Am 106(2):1142–1148. https://doi.org/10.1121/1.427122

Kastak D, Southall B, Holt M, Kastak CR, Schusterman R (2004) Noise-induced temporary threshold shifts in pinnipeds: effects of noise energy. J Acoust Soc Am 116(4):2531–2532(A). https://doi.org/10.1121/1.4785103

Kastak D, Southall BL, Schusterman RJ, Kastak CR (2005) Underwater temporary threshold shift in pinnipeds: effects of noise level and duration. J Acoust Soc Am 118(5):3154–3163. https://doi.org/10.1121/1.2047128

Kastak D, Reichmuth C, Holt MM, Mulsow J, Southall BL, Schusterman RJ (2007) Onset, growth, and recovery of in-air temporary threshold shift in a California sea lion (*Zalophus californianus*). J Acoust Soc Am 122(5):2916–2924. https://doi.org/10.1121/1.2783111

Kastelein RA, Gransier R, Hoek L, Macleod A, Terhune JM (2012a) Hearing threshold shifts and recovery in harbor seals (*Phoca vitulina*) after octave-band noise exposure at 4 kHz. J Acoust Soc Am 132(4): 2745–2761. https://doi.org/10.1121/1.4747013

Kastelein RA, Gransier R, Hoek L, Olthuis J (2012b) Temporary threshold shifts and recovery in a harbor porpoise (*Phocoena phocoena*) after octave-band noise at 4 kHz. J Acoust Soc Am 132(5):3525–3537. https://doi.org/10.1121/1.4757641

Kastelein RA, Gransier R, Hoek L (2013a) Comparative temporary threshold shifts in a harbor porpoise and harbor seal, and severe shift in a seal. J Acoust Soc Am 134(1):13–16. https://doi.org/10.1121/1.4808078

Kastelein RA, Gransier R, Hoek L, Rambags M (2013b) Hearing frequency thresholds of a harbor porpoise (*Phocoena phocoena*) temporarily affected by a continuous 1.5 kHz tone. J Acoust Soc Am 134(3): 2286–2292. https://doi.org/10.1121/1.4816405

Kastelein RA, Hoek L, Gransier R, Rambags M, Claeys N (2014a) Effect of level, duration, and inter-pulse interval of 1-2 kHz sonar signal exposures on harbor porpoise hearing. J Acoust Soc Am 136(1):412–422. https://doi.org/10.1121/1.4883596

Kastelein RA, Schop J, Gransier R, Hoek L (2014b) Frequency of greatest temporary hearing threshold shift in harbor porpoises (*Phocoena phocoena*) depends on the noise level. J Acoust Soc Am 136(3):1410–1418. https://doi.org/10.1121/1.4892794

Kastelein RA, Gransier R, Marijt MAT, Hoek L (2015a) Hearing frequency thresholds of harbor porpoises (*Phocoena phocoena*) temporarily affected by played back offshore pile driving sounds. J Acoust Soc Am 137(2):556–564. https://doi.org/10.1121/1.4906261

Kastelein RA, Gransier R, Schop J, Hoek L (2015b) Effects of exposure to intermittent and continuous 6–7 kHz sonar sweeps on harbor porpoise (*Phocoena phocoena*) hearing. J Acoust Soc Am 137(4): 1623–1633. https://doi.org/10.1121/1.4916590

Kastelein RA, Gransier R, Hoek L (2016a) Cumulative effects of exposure to continuous and intermittent sounds on temporary hearing threshold shifts induced in a harbor porpoise (*Phocoena phocoena*). In: The effects of noise on aquatic life II. Springer, New York, pp 523–528

Kastelein RA, Helder-Hoek L, Covi J, Gransier R (2016b) Pile driving playback sounds and temporary threshold shift in harbor porpoises (*Phocoena phocoena*): effect of exposure duration. J Acoust Soc Am 139(5): 2842–2851. https://doi.org/10.1121/1.4948571

Kastelein RA, Helder-Hoek L, Voorde SVD (2017a) Effects of exposure to sonar playback sounds (3.5–4.1 kHz) on harbor porpoise (*Phocoena phocoena*) hearing. J Acoust Soc Am 142(4): 1965–1975. https://doi.org/10.1121/1.5005613

Kastelein RA, Helder-Hoek L, Voorde SVD, Benda-Beckmann AMV, Lam F-PA, Jansen E, de Jong CAF, Ainslie MA (2017b) Temporary hearing threshold shift in a harbor porpoise (*Phocoena phocoena*) after exposure to multiple airgun sounds. J Acoust Soc Am 142(4):2430–2442. https://doi.org/10.1121/1.5007720

Kastelein RA, Helder-Hoek L, Kommeren A, Covi J, Gransier R (2018) Effect of pile-driving sounds on harbor seal (*Phoca vitulina*) hearing. J Acoust Soc Am 143(6):3583–3594. https://doi.org/10.1121/1.5040493

Kastelein RA, Helder-Hoek L, Cornelisse S, Huijser LAE, Gransier R (2019a) Temporary hearing threshold shift in harbor porpoises (*Phocoena phocoena*) due to one-sixth-octave noise band at 32 kHz. Aquat Mamm 45(5):549–562. https://doi.org/10.1578/AM.45.5.2019.549

Kastelein RA, Helder-Hoek L, Cornelisse S, Huijser LAE, Terhune JM (2019b) Temporary hearing threshold shift in harbor seals (*Phoca vitulina*) due to a one-sixth-octave noise band centered at 16 kHz. J Acoust Soc Am 146(5):3113–3122. https://doi.org/10.1121/1.5130385

Kastelein RA, Helder-Hoek L, Gransier R (2019c) Frequency of greatest temporary hearing threshold shift in harbor seals (*Phoca vitulina*) depends on fatiguing sound level. J Acoust Soc Am 145(3):1353–1362. https://doi.org/10.1121/1.5092608

Kastelein RA, Helder-Hoek L, van Kester R, Huisman R, Gransier R (2019d) Temporary hearing threshold shift in harbor porpoises (*Phocoena phocoena*) due to one-sixth octave noise band at 16 kHz. Aquat Mamm 45(3):280–292. https://doi.org/10.1578/am.45.3.2019.280

Kastelein RA, Cornelisse SA, Huijser LA, Helder-Hoek L (2020a) Temporary hearing threshold shift in harbor porpoises (*Phocoena phocoena*) due to one-sixth-octave noise bands at 63 kHz. Aquat Mamm 46(2):167–182. https://doi.org/10.1578/AM.46.2.2020.167

Kastelein RA, Helder-Hoek L, Cornelisse SA, Defiller LN, Huijser LAE (2020b) Temporary threshold shift in a second harbor porpoise (*Phocoena phocoena*) after exposure to a one-sixth-octave noise band at 1.5 kHz and a 6.5 kHz continuous wave. Aquat Mamm 46(5):431–443. https://doi.org/10.1578/AM.46.5.2020.431

Kastelein RA, Helder-Hoek L, Cornelisse SA, Defillet LN, Huijser LAE, Terhune JM (2020c) Temporary hearing threshold shift in harbor seals (*Phoca vitulina*) due to one-sixth-octave noise bands centered at 0.5, 1, and 2 kHz. J Acoust Soc Am 148(6):3873–3885. https://doi.org/10.1121/10.0002781

Kastelein RA, Helder-Hoek L, Cornelisse SA, Huijser LAE, Gransier R (2020d) Temporary hearing threshold shift at ecolglcally relevant frequencies in a harbor porpoise (*Phocoena phocoena*) due to exposure to a noise band centered at 88.4 kHz. Aquat Mamm 46(5):444–453. https://doi.org/10.1578/AM.46.5.2020.444

Kastelein RA, Helder-Hoek L, Cornelisse SA, Huijser LAE, Terhune JM (2020e) Temporary hearing threshold shift in harbor seals (*Phoca vitulina*) due to a one-sixth-octave noise band centered at 32 kHz. J Acoust Soc Am 147(3):1885–1896. https://doi.org/10.1121/10.0000889

Kastelein RA, Helder-Hoek L, Cornelisse SA, von Benda-Beckmann AM, Lam F-PA, de Jong CAF, Ketten DR (2020f) Lack of reproducibility of temporary hearing threshold shifts in a harbor porpoise after exposure to repeated airgun sounds. J Acoust Soc Am 148(2):556–565. https://doi.org/10.1121/10.0001668

Kastelein RA, Parlog C, Helder-Hoek L, Cornelisse SA, Huijser LAE, Terhune JM (2020g) Temporary hearing threshold shift in harbor seals (*Phoca vitulina*) due to a one-sixth-octave noise band centered at 40 kHz. J Acoust Soc Am 147(3):1966–1976. https://doi.org/10.1121/10.0000908

Kastelein RA, Helder-Hoek L, Cornelisse SA, Defillet LN, Huijser LAE, Gransier R (2021a) Temporary hearing threshold shift in a harbor porpoise (*Phocoena phocoena*) due to exposure to a continuous one-sixth-octave noise band centered at 0.5 kHz. Aquat Mamm 47(2):135–145. https://doi.org/10.1578/AM.47.2.2021.135

Kastelein RA, Helder-Hoek L, Cornelisse SA, Defillet LN, Huijser LAE, Terhune JM, Gransier R (2021b) Temporary hearing threshold shift in California sea lions (*Zalophus californianus*) due to one-sixth-octave noise bands centered at 2 and 4 kHz: effect of duty cycle and testing the equal-energy hypothesis. J Acoust Soc Am 47(4):394–418. https://doi.org/10.1578/AM.47.4.2021.394

Kastelein RA, Helder-Hoek L, Defillet LN, Kuiphof F, Huijser LAE, Terhune JM (2022a) Temporary hearing threshold shift in California sea lions (*Zalophus californianus*) due to one-sixth-octave noise bands centered at 0.6 and 1 kHz. Aquat Mamm 48(3):248–265. https://doi.org/10.1578/AM.48.3.2022.248

Kastelein RA, Helder-Hoek L, Defillet LN, Kuiphof F, Huijser LAE, Terhune JM (2022b) Temporary hearing threshold shift in California sea lions (*Zalophus californianus*) due to one-sixth-octave noise bands centered at 8 and 16 kHz: Effect of duty cycle and testing the equal-energy hypothesis. Aquat Mamm 48(1):36–58. https://doi.org/10.1578/am.48.1.2022.36

Khudyakov JI, Champagne CD, Meneghetti LM, Crocker DE (2017) Blubber transcriptome response to acute stress axis activation involves transient changes in adipogenesis and lipolysis in a fasting-adapted marine mammal. Sci Rep 7:42110. https://doi.org/10.1038/srep42110

Kight CR, Swaddle JP (2011) How and why environmental noise impacts animals: an integrative, mechanistic review. Ecol Lett 14(10):1052–1061. https://doi.org/10.1111/j.1461-0248.2011.01664.x

Kujawa SG, Liberman MC (2009) Adding insult to injury: cochlear nerve degeneration after "temporary" noise-induced hearing loss. J Neurosci 29(45):14077–14085. https://doi.org/10.1523/JNEUROSCI.2845-09.2009

Kvadsheim PH, Sevaldsen EM, Folkow LP, Blix AS (2010) Behavioural and physiological responses of hooded seals (*Cystophora cristata*) to 1 to 7 kHz sonar signals. Aquat Mamm 36(3):239–247. https://doi.org/10.1578/AM.36.3.2010.239

Kvadsheim PH, Miller PJO, Tyack PL, Sivle LD, Lam FPA, Fahlman A (2012) Estimated tissue and blood N2 levels and risk of decompression sickness in deep-, intermediate-, and shallow-diving toothed whales during exposure to naval sonar. Front Physiol 3(125). https://doi.org/10.3389/fphys.2012.00125

Lemos LS, Haxel JH, Olsen A, Burnett JD, Smith A, Chandler TE, Nieukirk SL, Larson SE, Hunt KE, Torres LG (2022) Effects of vessel traffic and ocean noise on gray whale stress hormones. Sci Rep 12(1):18580. https://doi.org/10.1038/s41598-022-14510-5

Lin HW, Furman AC, Kujawa SG, Liberman MC (2011) Primary neural degeneration in the guinea pig cochlea after reversible noise-induced threshold shift. J Assoc Res Otolaryngol. https://doi.org/10.1007/s10162-011-0277-0

Lucke K, Siebert U, Lepper PA, Blanchet M-A (2009) Temporary shift in masked hearing thresholds in a harbor porpoise (*Phocoena phocoena*) after exposure to seismic airgun stimuli. J Acoust Soc Am 125(6):4060–4070. https://doi.org/10.1121/1.3117443

Lyamin OI, Korneva SM, Rozhnov VV, Mukhametov LM (2011) Cardiorespiratory changes in beluga in response to acoustic noise. Dokl Biol Sci 440:275–278. https://doi.org/10.1134/S0012496611050218

MacArthur RA, Johnston RH, Geist V (1979) Factors influencing heart rate in free-ranging bighorn sheep: a physiological approach to the study of wildlife harassment. Can J Zool 57(10):2010–2021. https://doi.org/10.1139/z79-265

MacArthur RA, Geist V, Johnston RH (1982) Cardiac and behavioral responses of mountain sheep to human disturbance. J Wildl Manag 46(2):351–358. https://doi.org/10.2307/3808646

McEwen BS, Wingfield JC (2010) What is in a name? Integrating homeostasis, allostasis and stress. Horm Behav 57:105–111. https://doi.org/10.1016/j.yhbeh.2009.09.011

McFadden D, Plattsmier HS (1983) Frequency patterns of TTS for different exposure intensities. J Acoust Soc Am 74(4):1178–1184. https://doi.org/10.1121/1.390041

Melnick W (1991) Human temporary threshold shift (TTS) and damage risk. J Acoust Soc Am 90(1):147–154. https://doi.org/10.1121/1.401308

Miksis JL, Connor RC, Grund MD, Nowacek DP, Solow AR, Tyack PL (2001) Cardiac responses to acoustic playback experiments in the captive bottlenose dolphin (*Tursiops truncatus*). J Comp Psychol 115(3):227–232. https://doi.org/10.1037/0735-7036.115.3.227

Mooney TA, Nachtigall PE, Breese M, Vlachos S, Au WWL (2009a) Predicting temporary threshold shifts in a bottlenose dolphin (*Tursiops truncatus*): the effects of noise level and duration. J Acoust Soc Am 125(3):1816–1826. https://doi.org/10.1121/1.3068456

Mooney TA, Nachtigall PE, Vlachos S (2009b) Sonar-induced temporary hearing loss in dolphins. Biol Lett 5(4):565–567. https://doi.org/10.1098/rsbl.2009.0099

Moore MJ, Early GA (2004) Cumulative sperm whale bone damage and the bends. Science 306:2215. https://doi.org/10.1126/science.1105452

Moore MJ, Bogomolni AL, Dennison SE, Early G, Garner MM, Hayward BA, Lentell BJ, Rotstein DS (2009) Gas bubbles in seals, dolphins, and porpoises entangled and drowned at depth in gillnets. Vet Pathol 46:536–547. https://doi.org/10.1354/vp.08-VP-0065-M-FL

Münzel T, Gori T, Babisch W, Basner M (2014) Cardiovascular effects of environmental noise exposure. Eur Heart J 35(13):829–836. https://doi.org/10.1093/eurheartj/ehu030

Nachtigall PE, Pawloski J, Au WWL (2003) Temporary threshold shifts and recovery following noise exposure in the Atlantic bottlenosed dolphin (*Tursiops truncatus*). J Acoust Soc Am 113(6):3425–3429. https://doi.org/10.1121/1.1570438

Nachtigall PE, Supin AY, Pawloski J, Au WWL (2004) Temporary threshold shifts after noise exposure in the bottlenose dolphin (*Tursiops truncatus*) measured using evoked auditory potentials. Mar Mamm Sci 20(4):673–687. https://doi.org/10.1111/j.1748-7692.2004.tb01187.x

National Marine Fisheries Service (2018) Revision to: technical guidance for assessing the effects of anthropogenic sound on marine mammal hearing (Version 2.0)—underwater acoustic thresholds for onset of permanent and temporary threshold shifts. National Oceanic and Atmospheric Administration, Silver Springs, 167 pp

Nephew BC, Kahn SA, Michael Romero L (2003) Heart rate and behavior are regulated independently of corticosterone following diverse acute stressors. Gen Comp Endocrinol 133(2):173–180. https://doi.org/10.1016/S0016-6480(03)00165-5

Payne CJ, Jessop TS, Guay P-J, Johnstone M, Feore M, Mulder RA, Iwaniuk A (2012) Population, behavioural and physiological responses of an urban population of black swans to an intense annual noise event. PLoS One 7(9):1–9. https://doi.org/10.1371/journal.pone.0045014

Perry EA, Boness DJ, Insley SJ (2002) Effects of sonic booms on breeding gray seals and harbor seals on Sable Island, Canada. J Acoust Soc Am 111(1 Pt 2):599–609. https://doi.org/10.1121/1.1349538

Piantadosi CA, Thalmann ED (2004) Whales, sonar and decompression sickness. Nature 15. https://doi.org/10.1038/nature02527a

Popov VV, Klishin VO, Nechaev DI, Pletenko MG, Rozhnov VV, Supin AY, Sysueva EV, Tarakanov MB (2011a) Influence of acoustic noises on the white whale hearing thresholds. Dokl Biol Sci 440:332–334. https://doi.org/10.1134/S001249661105019X

Popov VV, Supin AY, Wang D, Wang K, Dong L, Wang S (2011b) Noise-induced temporary threshold shift and recovery in Yangtze finless porpoises *Neophocaena phocaenoides asiaeorientalis*. J Acoust Soc Am 130(1):574–584. https://doi.org/10.1121/1.3596470

Popov VV, Supin AY, Rozhnov VV, Nechaev DI, Sysuyeva EV, Klishin VO, Pletenko MG, Tarakanov MB (2013) Hearing threshold shifts and recovery after noise exposure in beluga whales *Delphinapterus leucas*. J Exp Biol 216:1587–1596. https://doi.org/10.1242/jeb.078345

Popov VV, Supin AY, Rozhnov VV, Nechaev DI, Sysueva EV (2014) The limits of applicability of the sound exposure level (SEL) metric to temporal threshold shifts (TTS) in beluga whales, *Delphinapterus leucas*. J Exp Biol 217(Pt 10):1804–1810. https://doi.org/10.1242/jeb.098814

Popov VV, Nechaev DI, Sysueva EV, Rozhnov VV, Supin AY (2015) Spectrum pattern resolution after noise exposure in a beluga whale, *Delphinapterus leucas*: evoked potential study. J Acoust Soc Am 138(1):377–388. https://doi.org/10.1121/1.4923157

Popov VV, Sysueva EV, Necharev DI, Rozhnov VV, Supin AY (2017) Influence of fatiguing noise on auditory evoked responses to stimuli of various levels in a beluga whale, *Delphinapterus leucas*. J Exp Biol 220: 1090–1096. https://doi.org/10.1242/jeb.149294

Reichmuth C, Ghoul A, Rouse A, Sills J, Southall B (2016) Low-frequency temporary threshold shift not observed in spotted or ringed seals exposed to single air gun impulses. J Acoust Soc Am 140(4):2646–2658. https://doi.org/10.1121/1.4964470

Reichmuth C, Sills JM, Mulsow J, Ghoul A (2019) Long-term evidence of noise-induced permanent threshold shift in a harbor seal (*Phoca vitulina*). J Acoust Soc Am 146(4):2552–2561. https://doi.org/10.1121/1.5129379

Ridgway SH, Howard R (1979) Dolphin lung collapse and intramuscular circulation during free diving: Evidence from nitrogen washout. Science 206:1182–1183. https://doi.org/10.1126/science.505001

Ridgway SH, Carder DA, Smith RR, Kamolnick T, Schlundt CE, Elsberry WR (1997) Behavioral responses and temporary shift in masked hearing thresholds of bottlenose dolphins, *Tursiops truncatus*, to 1-second tones of 141-201 dB re 1 μPa (Technical Report 1751). Naval Command, Control, and Ocean Surveillance Center, RDT&E Division, San Diego, 31 pp

Rolland RM, Parks SE, Hunt KE, Castellote M, Corkeron PJ, Nowacek DP, Wasser SK, Kraus SD (2012) Evidence that ship noise increases stress in right whales. Proc R Soc Lond B 279(1737):2363–2368. https://doi.org/10.1098/rspb.2011.2429

Romano T, Keogh M, Kelly C, Feng P, Berk L, Schlundt CE, Carder DA, Finneran JJ (2004) Anthropogenic sound and marine mammal health: measures of the nervous and immune systems before and after intense sound exposures. Can J Fish Aquat Sci 61:1124–1134. https://doi.org/10.1139/f04-055

Romero LM, Dickens MJ, Cyr NE (2009) The reactive scope model—a new model integrating homeostasis, allostasis, and stress. Horm Behav 55:375–389. https://doi.org/10.1016/j.yhbeh.2008.12.009

Rommel SA, Costidis AM, Fernández A, Jepson PD, Pabst DA, McLellan WA, Houser DS, Cranford TW, Van Helden AL, Aleen DM, Barros NB (2006) Elements of beaked whale anatomy and diving physiology and some hypothetical causes of sonar-related stranding. J Cetacean Res Manag 7(3):189–209. https://doi.org/10.47536/jcrm.v7i3.730

Saunders KJ, White PR, Leighton TG (2008) Models for predicting nitrogen tensions and decompression sickness risk in diving beaked whales. In: Institute of Acoustics: Underwater Noise Measurement, Impact and Mitigation, 14–15 October 2008, pp 82–89

Schlundt CE, Finneran JJ, Carder DA, Ridgway SH (2000) Temporary shift in masked hearing thresholds of bottlenose dolphins, *Tursiops truncatus*, and white whales, *Delphinapterus leucas*, after exposure to intense tones. J Acoust Soc Am 107(6):3496–3508. https://doi.org/10.1121/1.429420

Schmidt A, Aurich J, Möstl E, Müller J, Aurich C (2010a) Changes in cortisol release and heart rate and heart rate variability during the initial training of 3-year-old sport horses. Horm Behav 58(4):628–636. https://doi.org/10.1016/j.yhbeh.2010.06.011

Schmidt A, Möstl E, Wehnert C, Aurich J, Müller J, Aurich C (2010b) Cortisol release and heart rate variability in horses during road transport. Horm Behav 57(2):209–215. https://doi.org/10.1016/j.yhbeh.2009.11.003

Schmitt TL, Dunn JL, St. Aubin DJ (2002) Physiologic stress response of long-term captive belugas, *Delphinapterus leucas*, to routine blood collection, out-of-water physical examination, and wading-contact programs. In: Smith SA (ed) IAAAM proceedings, Albufeira, Portugal. Proceedings of the 33rd conference of the IAAAM. 14517, pp 52–53

Schmitt TL, St. Aubin DJ, Schaefer AM, Dunn JL (2010) Baseline, diurnal variations, and stress-induced changes of stress hormones in three captive beluga whales, *Delphinapterus leucas*. Mar Mamm Sci 26(3):635–647. https://doi.org/10.1111/j.1748-7692.2009.00366.x

Scholander PF (1940) Experimental investigations on the respiratory function in diving mammals and birds. Hvalradets Skrifter, Scientific Results of Marine Biological Research, Oslo

Selye H (1936) A syndrome produced by diverse nocuous agents. Nature 138:32. https://doi.org/10.1038/138032a0

Sills JM, Ruscher B, Nichols R, Southall BL, Reichmuth C (2020) Evaluating temporary threshold shift onset levels for impulsive noise in seals. J Acoust Soc Am 148(5):2973–2986. https://doi.org/10.1121/10.0002649

Southall BL, Bowles AE, Ellison WT, Finneran JJ, Gentry RL, Greene CR Jr, Kastak D, Ketten DR, Miller JH, Nachtigall PE, Richardson WJ, Thomas JA, Tyack PL (2007) Marine mammal noise exposure criteria: initial scientific recommendations. Aquat Mamm 33(4):411–521. https://doi.org/10.1578/AM.45.2.2019.125

Southall BL, Finneran JJ, Reichmuth C, Nachtigall PE, Ketten DR, Bowles AE, Ellison WT, Nowacek DP, Tyack PL (2019) Marine mammal noise exposure criteria: auditory weighting functions and TTS/PTS onset. Aquat Mamm 45(2):125–232. https://doi.org/10.1578/AM.45.2.2019.125

Spoon TR, Romano TA (2012) Neuroimmunological response of beluga whales (*Delphinapterus leucas*) to translocation and a novel social environment. Brain Behav Immun 26:122–131. https://doi.org/10.1016/j.bbi.2011.08.003

St. Aubin DJ, Dierauf LA (2001) Stress and marine mammals. In: Dierauf LA, Gulland FMD (eds) Marine mammal medicine, vol 13, 2nd edn. CRC Press, Boca Raton, pp 253–269

St. Aubin DJ, Ridgway SH, Wells RS, Rhinehart H (1996) Dolphin thyroid and adrenal hormones: circulating levels in wild and semidomesticated *Tursiops truncatus*, and influence of sex, age, and season. Mar Mamm Sci 12(1):1–13. https://doi.org/10.1111/j.1748-7692.1996.tb00301.x

St. Aubin DJ, Forney KA, Chivers SJ, Scott MD, Danil K, Romano TA, Wells RS, Gulland FMD (2013) Hematological, serum, and plasma chemical constituents in pantropical spotted dolphins (*Stenella attenuata*) following chase, encirclement, and tagging. Mar Mamm Sci 29(1):14–35. https://doi.org/10.1111/j.1748-7692.2011.00536.x

Steinman KJ, Robeck TR, Fetter GA, Schmitt TL, Osborn S, DiRocco S, Nollens HH, O'Brien JK (2020) Circulating and excreted corticosteroids and metabolites, hematological, and serum chemistry parameters in the killer whale (*Orcinus orca*) before and after a stress response. Front Mar Sci 6(830). https://doi.org/10.3389/fmars.2019.00830

Talpalar AE, Grossman Y (2005) Sonar versus whales: noise may disrupt neural activity in deep-diving cetaceans. Undersea Hyperb Med 32(2):135–139

Teerlink S, Hostmann L, Witteveen B (2018) Humpback whale (*Megaptera novaeangliae*) blubber steroid hormone concentration to evaluate chronic stress response from whale-watching vessels. Aquat Mamm 44(4):411–425. https://doi.org/10.1578/AM.44.4.2018.411

Teilmann J, Tougaard J, Miller LA, Kirketerp T, Hansen K, Brando S (2006) Reactions of captive harbor porpoises (*Phocoena phocoena*) to pinger-like sounds. Mar Mamm Sci 22(2):240–260. https://doi.org/10.1111/j.1748-7692.2006.00031.x

Tennessen JB, Parks SE, Langkilde T (2014) Traffic noise causes physiological stress and impairs breeding migration behaviour in frogs. Conserv Physiol 2(1):cou032. https://doi.org/10.1093/conphys/cou032

Thomas JA, Kastelein RA, Awbrey FT (1990) Behavior and blood catecholamines of captive belugas during playbacks of noise from an oil drilling platform. Zoo Biol 9(5):393–402. https://doi.org/10.1002/zoo.1430090507

Thompson RD, Grant CV, Pearson EW, Corner GW (1968) Differential heart rate response of starlings to sound stimuli of biological origin. J Wildl Manag 32(4):888–893. https://doi.org/10.2307/3799566

Thompson LA, Hindle AG, Black SR, Romano TA (2020) Variation in the hemostatic complement (C5a) responses to in vitro nitrogen bubbles in monodontids and phocids. J Comp Physiol B 190(6):811–822. https://doi.org/10.1007/s00360-020-01297-y

Thomson CA, Geraci JR (1986) Cortisol, aldosterone, and leukocytes in the stress response of bottlenose dolphins, *Tursiops truncatus*. Can J Fish Aquat Sci 43(5):1010–1016. https://doi.org/10.1139/f86-125

Trana MR, Roth JD, Tomy GT, Anderson GW, Ferguson SH (2015) Increased blubber cortisol in ice-entrapped beluga whales (*Delphinapterus leucas*). Polar Biol:1–7. https://doi.org/10.1007/s00300-015-1881-y

Van Bonn WG, Montie E, Dennison S, Pussini N, Cook P, Greig DJ, Barakos J, Colegrove KM, Gulland FM (2011) Evidence of injury caused by gas bubbles in a live marine mammal: barotrauma in a California sea lion *Zalophus californianus*. Dis Aquat Org 96:89–96. https://doi.org/10.3354/dao02376

Viblanc VA, Smith AD, Gineste B, Groscolas R (2012) Coping with continuous human disturbance in the wild: insights from penguin heart rate response to various stressors. BMC Ecol 12(1):10. https://doi.org/10.1186/1472-6785-12-10

Ward WD (1997) Effects of high-intensity sound. In: Crocker MJ (ed) Encyclopedia of acoustics. Wiley, New York, pp 1497–1507

Wasser SK, Hunt KE, Brown JL, Cooper K, Crockett CM, Bechert U, Millspaugh JJ, Larson S, Monfort SL (2000) A generalized fecal glucocorticoid assay for use in a diverse array of nondomestic mammalian and avian species. Gen Comp Endocrinol 120:260–275. https://doi.org/10.1006/gcen.2000.7557

Weisenberger ME, Krausman PR, Wallace MC, De Young DW, Maughan OE (1996) Effects of simulated jet aircraft noise on heart rate and behavior of desert ungulates. J Wildl Manag 60(1):52–61. https://doi.org/10.2307/3802039

Williams TM, Blackwell SB, Richter B, Sinding M-HS, Heide-Jørgensen MP (2017) Paradoxical escape responses by narwhals (*Monodon monoceros*). Science 358(6368):1328–1331. https://doi.org/10.1126/science.aao2740

Williams TM, Blackwell SB, Tervo O, Garde E, Sinding M-HS, Richter B, Heide-Jørgensen MP (2022) Physiological responses of narwhals to anthropogenic noise: a case study with seismic airguns and vessel traffic in the Arctic. Funct Ecol https://doi.org/10.1111/1365-2435.14119

Yang W-C, Chen C-F, Chuah Y-C, Zhuang C-R, Chen I-H, Mooney TA, Stott J, Blanchard M, Jen I-F, Chou L-S (2021) Anthropogenic sound exposure-induced stress in captive dolphins and implications for cetacean health. Front Mar Sci 8. https://doi.org/10.3389/fmars.2021.606736

Zimmer WMX, Tyack PL (2007) Repetitive shallow dives pose decompression risk in deep-diving beaked whales. Mar Mamm Sci 23(4):888–925. https://doi.org/10.1111/j.1748-7692.2007.00152.x

Open Access This chapter is licensed under the terms of the Creative Commons Attribution 4.0 International License (http://creativecommons.org/licenses/by/4.0/), which permits use, sharing, adaptation, distribution and reproduction in any medium or format, as long as you give appropriate credit to the original author(s) and the source, provide a link to the Creative Commons license and indicate if changes were made.

The images or other third party material in this chapter are included in the chapter's Creative Commons license, unless indicated otherwise in a credit line to the material. If material is not included in the chapter's Creative Commons license and your intended use is not permitted by statutory regulation or exceeds the permitted use, you will need to obtain permission directly from the copyright holder.

Behavioral Responses to Underwater Noise

10

Christine Erbe ⓘ, Ann Bowles, Dorian Houser, Capri Jolliffe, Shyam Madhusudhana, Sarah A. Marley, Angela Recalde Salas, Chandra Salgado-Kent, Renee Schoeman, Valeria Senigaglia, Cristina Tollefsen, Leah Trigg, and Rebecca Wellard

---

C. Erbe (✉) · C. Jolliffe · R. Schoeman · V. Senigaglia ·
C. Tollefsen · R. Wellard
Centre for Marine Science and Technology, Curtin
University, Perth, WA, Australia
e-mail: c.erbe@curtin.edu.au; cristina.tollefsen@curtin.edu.au

A. Bowles
Hubbs-Sea World Research Institute, San Diego, CA,
USA
e-mail: ABowles@hswri.org

D. Houser
National Marine Mammal Foundation, San Diego, CA,
USA
e-mail: dorian.houser@nmmf.org

S. Madhusudhana
Centre of Ocean and Earth Science and Technology,
Curtin University, Telfair, Moka, Mauritius
e-mail: shyam.m@curtin.edu.au

S. A. Marley
Scotland's Rural College (SRUC), Aberdeen, UK
e-mail: sarah.marley@sruc.ac.uk

A. Recalde Salas
Bush Heritage Australia, Perth, WA, Australia
e-mail: angela.recaldesalas@bushheritage.org.au

C. Salgado-Kent
Centre for Marine Ecosystems Research, School of
Science, Edith Cowan University, Joondalup, WA,
Australia
e-mail: c.salgadokent@ecu.edu.au

L. Trigg
Department of Oceanography, Dalhousie University,
Halifax, NS, Canada
e-mail: leah.trigg@dal.ca

© The Author(s) 2025
C. Erbe et al. (eds.), *Marine Mammal Acoustics in a Noisy Ocean*,
https://doi.org/10.1007/978-3-031-77022-7_10

Contents

10.1	Introduction	612
10.2	**Behavioral Responses of Mysticetes**	616
10.2.1	Bowhead Whales	616
10.2.2	Right Whales	618
10.2.3	Rorquals	619
10.2.4	Gray Whales	624
10.3	**Behavioral Responses of Odontocetes**	625
10.3.1	Bottlenose Dolphins	625
10.3.2	Killer Whales	629
10.3.3	Pilot Whales	633
10.3.4	Other Delphinids	635
10.3.5	River Dolphins	638
10.3.6	Porpoises	638
10.3.7	Beluga and Narwhal	642
10.3.8	Sperm Whales	643
10.3.9	*Kogia* spp.	646
10.3.10	Beaked Whales	646
10.4	**Behavioral Responses of Sirenians**	649
10.4.1	Ships and Boats	650
10.4.2	Acoustic Deterrent Devices	651
10.5	**Behavioral Responses of Pinnipeds**	651
10.5.1	Otariids	651
10.5.2	Phocids	653
10.5.3	Walrus	658
10.6	**Behavioral Responses of Polar Bears and Otters**	659
10.7	**Compilation of Quantitative Data**	660
10.8	**Discussion**	666
10.8.1	Characterizing Received Sound at the Animal	666
10.8.2	Individual Variability in Responsiveness	669
10.8.3	The Challenge of Context	672
10.9	**Summary**	673
10.10	**Data Availability**	674
	References	674

10.1 Introduction

In the late 1960s, growing environmental awareness led to the authorization of the US National Environmental Policy Act (signed January 1, 1970) and the Marine Mammal Protection Act (signed October 21, 1972), which in turn led to an increase in research on the effects of anthropogenic noise on marine mammal behavior. The earliest published hypotheses about the potential effects of anthropogenic noise on marine mammals concerned the noise from heavy shipping and its impact on fin whale communication space (Payne and Webb 1971). By the early 1980s, the scope of concerns about the effects of noise had grown, particularly as oil and gas industry activities penetrated the Bering Sea and Arctic Ocean (Mansfield 1983; Stirling and Calvert 1983). By the time the book *Marine Mammals and Noise* (Richardson et al. 1995b) was published, concerns had been extended to many marine mammal species and any anthropogenic sound source, including seismic surveys and other oil and gas industry activities, missile launches and space shuttles, aircraft overflights, sonic booms, motorized vessels, explosives, construction noise, navy tactical sonars, etc. (Chap. 2).

Even though the earliest published concerns about the potential effects of noise on marine mammals related to masking and were based on theoretical calculations of fin whale communication ranges in ship noise (Payne and Webb 1971), efforts to directly measure the consequences of masking were not performed under real-world conditions for several decades (e.g., Holt et al. 2011). Instead, the earliest detailed experimental studies measured short-term behavioral responses to relatively intense transient or localized noise. Marine mammal behavior was observed from platforms in air, such as aircraft, boats, and ships, or shore-based stations, with supplemental acoustic recordings collected using single hydrophones deployed from vessels, sonobuoys, or bottom-mounted emplacements. It was not unusual for data returned from a field season to contain recordings of actively vocalizing animals lasting minutes to a few hours (e.g., Ljungblad et al. 1982), little to none of which could be directly connected with the behaviors of individuals observed from above the water. Surface-based observations of behavior were limited to the time that animals could be seen. Research predominantly focused on a few mysticete species but was invaluable for determining whether individuals detected sounds, whether the sounds aroused behaviors indicating aversion or safety-seeking response (hereafter, defensive response), and whether varying properties of the sounds had greater or lesser effects on defensive responses. Even in the case of mysticetes, however, individuals often could not be followed reliably between surfacings, their behavior below the surface could not be monitored, and sounds emitted were not associated with visual observations reliably. In the case of smaller cetacean species and pinnipeds at sea, the limitations were even greater.

Initially, any response to an exposure (i.e., any change in behavior that could be associated with an anthropogenic sound) had the potential to be interpreted as a disturbance. So little was known about the consequences of behavioral responses that it was difficult to justify distinctions between behaviors that potentially caused biologically significant effects (see Chap. 11) and those that did not. Responses were to be expected when an animal perceived a novel or surprising sound, such as orienting (e.g., changing orientation to focus on a stimulus), investigation (e.g., stopping vocalizations to listen), and species-typical defensive behaviors. At high stimulus amplitudes, particularly when the rise time of the stimulus was rapid, some behaviors could be easily interpreted as startle reflexes, abrupt dives, altering direction, swimming away from the source, fast swimming, congregating, recalling young, and visual or vocal alarms. Changes in activity states could also indicate defense, such as changes from feeding to traveling. These behaviors made intuitive sense as negative responses because they were similar to defensive behaviors of other mammals. However, only avoidance was sufficiently consistent as a response and interpretable as a significant effect (e.g., abandoning an area) that was deemed appropriate for predictive modeling (e.g., Malme et al. 1983; Richardson and Malme 1993).

The research leading to behavioral response models consisted of a series of controlled exposure experiments (CEEs), conducted by placing noise sources in the path of animals engaged in activities such as migrating or feeding and quantifying both the sound exposure and response (Richardson et al. 1995b). Typically, exposures and responses lasted a short time: minutes to hours. Although the case for defensive responses could be made, the short-term exposures and responses in these experiments were difficult to link with biologically significant effects (National Research Council 2005). When short-term behaviors were extrapolated to longer periods, effects could be inferred, such as changes in activity state extrapolated to degraded opportunities for foraging.

However, the extrapolations assumed that responses could continue unchanged over time with prolonged or repeated exposure. CEEs did not last long enough to confirm such extrapolations or alternatively show whether animals had compensatory mechanisms that would lead to reduced responses when costs of response outweighed benefits. Animals also might react differently if responses were limited by conditions, such as less avoidance when a narrow lead left migrants with no alternative route (Richardson et al. 1991). The assumption that strong defensive responses were a good measure of risk was questioned. For example, absence of response might indicate a greater impact than overt startled flight if animals had no choice but to stay in the presence of a disturbance (Beale and Monaghan 2004). Such absence of response would be different from the strict sense of habituation (i.e., a decline in the intensity of response because a stimulus proves to be neutral and therefore safe to ignore; Bejder et al. 2009). If animals cease to respond to an aversive stimulus because the cost of the response is higher than the payoff for avoiding it (e.g., because it is a manageable risk; Watkins 1986), the implications are different. Alternatively, animals might become sensitized (i.e., increase responsiveness). When such scenarios were considered, the experience of an animal with a stimulus over time and its strategic alternatives became relevant to modeling effects.

In the 1980s and 1990s, when the context of the experiment was relatively consistent (i.e., a source in the path of migrating mysticetes), multiple reports found a fairly consistent association between level and the probability of avoidance (Malme et al. 1986), allowing dose-response models to be calculated (Richardson and Malme 1993). However, evidence began to accumulate that context (i.e., the ensemble of environmental, social, and internal conditions animals experience at the time of exposure) plays an important role in determining the probability of responses (Ellison et al. 2012). For example, Tyack (2008) sought experimental evidence for a dosage threshold for avoidance in migrating gray whales exposed to low-frequency sonar noise but found that the dosage relationship was different when the source was outside versus inside the nearshore migratory corridor. When the source was placed offshore, the whales did not alter their path at comparable received levels (RLs).

A decade later, when a larger literature on behavioral responses to anthropogenic noise was analyzed in an effort to develop generalized dose-response relationships (Southall et al. 2007), no simple dose-response relationship could be formulated by marine mammal species grouping. A more recent, broad, cross-grouping look at the cumulative body of literature came to similar conclusions (Gomez et al. 2016). Variability by species, exposure type, and several types of context appeared to swamp any general relationship between received noise level (i.e., the dosage) and response.

Within the US regulatory context, debate about the interpretation of behavioral effects resulted in a regulatory separation of disturbance effects into Level A and Level B categories (1994 Amendments to the US Marine Mammal Protection Act), with Level B recognized as of lower concern, requiring simpler regulatory review. In the late 1990s, the US National Oceanic and Atmospheric Administration (NOAA) published the first acoustic thresholds for behavioral effects, but based on the older dose-response literature (Scholik-Schlomer 2015). Although not explicitly discussed in most cases, these thresholds did not represent the exposure level at which the most sensitive individual was likely to respond, but rather the level at which a specific proportion of the population (50%) was assumed to respond with avoidance. These thresholds have become the standard that is typically applied in environmental impact assessments around the globe where jurisdictions do not have their own guidance on noise effect thresholds.

Over the years, models for behavioral change have become more sophisticated. Attempts to improve models began with a theoretical framework for connecting behavior with biologically significant outcomes, developed in a series of workshops sponsored by the Ocean Studies Board of the National Academy of Sciences (National Research Council 1994, 2005, 2017). The European Union initiated a similar effort (Fleishman et al. 2016). There is now a substantial body of additional modeling research stimulated by these efforts (e.g., New et al. 2013a), including methods for modeling cumulative exposure as animals move through sound fields (e.g., Frankel et al. 2002; Houser 2006) and statistical tools for documenting relationships between noise and response (e.g., DeRuiter et al. 2017).

In parallel with modeling efforts, methods for collecting response data have become more diverse, making it possible to address questions about effects on behavior over greater timescales and distances. The technologies applied to marine mammal noise studies include large strategic hydrophone arrays originally developed to monitor submarine activity during the Cold War (e.g., Watkins et al. 2000), hydrophone arrays deployed from shorelines or shore-fast ice (e.g., Clark and Ellison 2000), towed hydrophone arrays (e.g., Barlow et al. 2013), arrays of bottom-mounted autonomous recording systems (e.g., Blackwell et al. 2012; Clark et al. 2007), and recording devices attached to free-ranging animals (e.g., Costa et al. 2003; Johnson et al. 2009; Kimura et al. 2013), which also collect other data, such as depth and three-dimensional acceleration (e.g., Goldbogen et al. 2017). In some cases, individuals can be tracked over long periods and large areas (e.g., Širović et al. 2007). In some conditions, they can even be localized using single hydrophones (Tiemann et al. 2006) or small, sparse arrays (Thode et al. 2006). Arrays of bottom-mounted recorders covering large areas make it possible to track animals and document changes in behavior over long periods in response to sources that ensonify large areas, such as heavy shipping (Clark et al. 2010) or seismic surveying (Blackwell et al. 2015). Acoustic sensors have been added to oceanographic gliders, which can actively survey large areas without human guidance (Baumgartner et al. 2013). Passive acoustic monitoring (PAM) now represents the most effective tool for monitoring status or movements of some cryptic, oceanic, or deep-diving species over periods of years (Clark et al. 2010; Jaramillo-Legorreta et al. 2017; Kerosky et al. 2012; Warren et al. 2017).

Although PAM has made long-term observations possible, it has several limitations. Vocal intermittency presents some of the same difficulties as visual behaviors detected at the surface. Behavior can only be inferred when animals are vocally active; the observations are not continuous streams of states or events (e.g., Warren et al. 2017). When animals are not making sounds, little can be said about what they are doing. The limitation is minimal when species vocalize almost constantly during active periods, but some species can only be monitored seasonally, such as during brief reproductive periods (many pinnipeds). Moreover, if animals become silent (a potential defensive behavior in cetaceans), avoidance cannot be differentiated from defensive silence (e.g., Bowles et al. 1994). If an effect on vocalization rate is observed, it may be informative, but the consequences of absence of effect are much less interpretable. Thus, although PAM data can be collected over the long term, it generally provides little information useful for predicting effects of noise on individuals. Animal-attached tags, on the other hand, provide reliable, detailed information about individual movement, foraging patterns, and vocal behavior, but currently cannot be deployed for long periods.

In summary, although new technology to monitor marine mammal behavior in context of noise exposure has had a large effect on both the behaviors observed and the timescales at which they can be monitored, individuals still cannot be followed nor noise exposures measured over broad timescales (seasons, years) except perhaps through short serial observations of well-known individuals. Thus, there is still a large gap in knowledge of effects that play out over long time periods, such as the impact of prolonged and reduced foraging efficiency or prolonged social effects of noise exposure. Cumulative noise effects are still difficult to measure and poorly understood (see Chap. 11), and the progression of technological adoption has varied by region and species, adding another layer of variability to broadscale assessments of marine mammal responses to noise.

10.2 Behavioral Responses of Mysticetes

10.2.1 Bowhead Whales

Bowhead whales are the only endemic mysticete species in the Arctic. Increasing anthropogenic noise in the marine soundscape across their habitats (Halliday et al. 2021) is of concern. Increasing commercial shipping (Pizzolato et al. 2016) and vessel traffic (Dawson et al. 2018) in the Arctic pose increased risk of auditory masking as vessel noise is dominant in the frequency range (10–1000 Hz) of bowhead vocalizations. Petroleum companies have been undertaking airgun surveys in the region during ice-free periods. In a warming climate, a decrease in polar ice is expected to increase the presence of anthropogenic noise sources in both numbers and in seasonal duration.

10.2.1.1 Marine Seismic Surveys and Vessels

There has been a significant amount of research on the potential effects of seismic surveying on bowhead whales. Some of these studies looked at the effects when the survey vessel was operating airguns versus not. But it can be hard to disentangle the effects of airgun noise, vessel noise, and vessel proximity; therefore, these sources are presented together here.

Using CEEs, bowhead whale responses to approaching seismic vessels were investigated (Ljungblad et al. 1988; Richardson et al. 1986, 1999). Whales were observed from light aircraft; RLs were measured with sonobuoys dropped in the vicinity of the whales. Collectively, these studies assessed numerous hours of focal group data, collected under both undisturbed and potentially disturbed conditions. Offering the ability to contrast and compare with prior observations of undisturbed behaviors (Ljungblad et al. 1988; Richardson et al. 1995a, b; Richardson and Malme 1993; Richardson and Würsig 1997), these works provided some of the best evidence that whales exhibit a cascade of behaviors when exposed to noise with increasing RL and proximity, with stages in the cascade roughly corresponding to the intensity of exposure. Experiments using a single airgun deployed from a moored or moving boat offered evidence of avoidance response to airgun pulses as opposed to approaching vessels (Richardson et al. 1986). These experiments solidified the evidence for an avoidance threshold of a root-mean-square (rms) sound pressure level (SPL) of ~160 dB re 1 µPa rms for airgun pulses, while exposures at lower amplitudes elicited no or more subtle responses. In the Ljungblad et al. experiments, as the airgun arrays were activated on the approaching vessels, the whales exhibited abrupt diving that created a splash at the surface and surface-active gestures such as fluke slapping. As approach continued, grouping behavior and synchrony increased: group members sometimes moved within a few whale lengths of one another and began surfacing more synchronously. Richardson et al. (1986) reported changes in the numbers of whales in echelon formations when exposed to pulses from a single airgun. At close range, Ljungblad et al. (1988) observed a consistent and significant increase in blow interval and dispersal, as the whales began turning out of the path of the oncoming vessel, traveling away from it at medium to fast swim speeds. Following shutdown of the airgun array, experimental vessels moved away from the whales' position, and surfacing/dive cycles and activity began to return to baseline conditions (within 30–60 min), with decreasing swim speeds and a return to milling and socializing. In all trials, whales did not react to the lowest pulse levels (RL 131–142 dB re 1 µPa rms). Subtle behavioral responses and partial avoidance occurred at 142–158 dB re 1 µPa rms and ranges of 3.5–7.6 km, while complete avoidance was seen at 152–165 dB re 1 µPa rms and 1.3–2.9 km. The highest exposure was 178 dB re 1 µPa rms.

Miller (2005) conducted visual (aerial and shipboard) surveys in conjunction with noise measurements during the passage of active and inactive seismic vessels. At close range (within visual range of the seismic vessel), sighting rates were halved and initial sighting distance was nearly twice as great when the vessel was operating airguns, but over the much wider area of the aerial survey, there was no evidence that the density of whales decreased. Thus, whale avoidance responses appeared to be local to the active vessel. A study of non-seismic vessels found no bowhead responses within 8–50 km; observations were not made at shorter ranges (Martin et al. 2023b).

In PAM-only studies, Blackwell et al. (2013, 2015) investigated how bowhead whales adjust their vocalization rates in response to pulsed noise from seismic vessels. Using arrays of bottom-mounted Directional Autonomous Seafloor Acoustic Recorders (DASARs), they recorded hundreds of thousands of bowhead calls in the presence and absence of seismic pulses in the Alaskan Beaufort Sea. They localized bowhead calls and estimated the RL of seismic pulses at the callers' positions. They found that as the RL increased to 94 dB re 1 $\mu Pa^2 s$, whales increased calling rate (calls detected per 10-min period). The calling rate remained constant in the range of 94–127 dB re 1 $\mu Pa^2 s$, possibly because the whales had reached a limit. Above this range, the whales began to reduce or stop calling, and when RL reached 160 dB re 1 $\mu Pa^2 s$, all calling ceased. The use of sound exposure level (SEL) as a metric was important because received seismic signals were composed of a short pulse followed by a period of reverberant decay. The study also showed that the noise was present at a higher duty cycle than might have been expected given the short duration of the pulses (Guerra et al. 2011). In a follow-up study, with abundant additional DASAR recordings constituting a 7-year (2008–2014) dataset from across the bowhead fall migration route in the Alaskan Beaufort Sea, Thode et al. (2020) found that the whales also adjusted source levels (SLs), besides altering calling rates. Considering detected calls that were localized to within a range threshold, they found that call density (calls per km^2 per minute) increased in the presence of weak airgun pulses. At higher RLs of airgun noise, however, individual call rates decreased, and call density reduced. On the other hand, the mean SL of the calls increased only by a few dB when the airgun activity increased noise levels by 40 dB cumulative SEL. The authors suggested that, given the impulsive nature of the noise, the whales may be calling more often during the time between airgun pulses, allowing them to continue communicating using their normal SLs.

10.2.1.2 Drilling

In open-water playback experiments, Richardson et al. (1991) exposed migrating bowhead whales to drilling noise in narrow leads off the north coast of Alaska, at ranges as low as 35 m and RLs as high as 140 dB re 1 μPa, with signal-to-noise ratios (SNR) up to 46 dB. Aside from altering their behavior (e.g., varying their swim speed) as they approached the playback source (at RL > 107 dB re 1 μPa and <4 km distance) and turning more frequently (at RL > 118 dB re 1 μPa and <2 km distance), the whales did not halt migration. These responses were consistent with the 120 dB re 1 μPa rms level later adopted as a threshold of disturbance. The experiments also showed that whales did not avoid the source at the expense of migrating when exposed to RLs 20 dB higher than this threshold. The whales turned and changed swim path to hug the edge of the lead, suggesting wariness while passing a source they might have given a wider berth in more open water. This experiment provided some of the earliest evidence that context could affect the threshold of avoidance.

In a PAM-only study using recordings from arrays of DASARs, Blackwell et al. (2017) assessed calling rates as a function of tonal machinery noise that included those from offshore drilling. Accurately determining RL of non-playback tonal sounds in the presence of other noise is tricky. Instead, the authors defined and used a "tonal quality" metric akin to SNR (the stronger the tonals, the higher the metric's value). The study found that bowheads initially increased their calling rates with increasing tonal levels and reached a peak rate, but then decreased with any further increases in tonal levels. This pattern of response was similar to that observed in increasing levels of airgun pulses (Thode et al. 2020). Additionally, Blackwell et al. (2017) also found that, in the presence of both machinery tonals and airgun pulses, the calling rates increased to a higher peak (than when only tonals were present), before decreasing.

10.2.1.3 Aircraft

Low-flying light aircraft and helicopters have been found to elicit behavioral responses from bowhead whales. Use of helicopters (by offshore oil industry) and fixed-wing aircraft (for coastal transport) along the whales' migratory routes can be of concern when operations scale up. In studies using aerial surveys for monitoring responses to other stimuli, response to overflights might introduce undesirable biases. Patenaude et al. (2002) assessed responses to overflights of a Bell 212 helicopter and a Twin Otter fixed-wing aircraft. Using controlled flights (at different altitudes), they measured RLs at different depths under water. Though the authors did not present noise level thresholds for responses, they observed short-term reactions (abrupt dives, short surfacing, turning, etc.), with less pronounced responses to the aircraft than to the helicopter. Pooling all behavior classes, Richardson et al. (1985) showed that responses (including avoidance) to a twin-engine Britten-Norman Islander were common at aircraft altitudes ≤ 305 m, but infrequent at aircraft altitudes ≥ 457 m.

10.2.2 Right Whales

There have been few controlled studies targeting right whale behavioral responses to noise. Their vocal behavior has been the subject of PAM-only studies in environments differing in ambient noise, in both hemispheres, with little else on nonvocal behavior. No studies have quantified acoustic or non-acoustic behavioral responses of the pygmy right whale.

10.2.2.1 Ships and Boats

Based on decades of sighting logs from Cape Cod waters, Watkins (1986) found that North Atlantic right whales were rarely affected by low-amplitude vessel noise, potentially leading to vessel collisions with these slow-moving whales. In fact, the vessel's presence might be a stronger stimulus than its noise, but often too late (Kraus 1990). In playback experiments, Nowacek et al. (2004a) found that the whales neither responded to playbacks of container ship noise nor to transiting vessels in the vicinity of the experimental exposures (all exposures at levels ≤ 173 dB re 1 µPa). Whale-watching boats reduced resting behavior in southern right whale mothers and calves; the response was stronger if boats departed at high speed (Sprogis et al. 2023).

In their study assessing vocal response, Parks et al. (2007) followed photo-identified female North Atlantic right whales in the Bay of Fundy from a vessel, recording and attributing screams to the focal whales and upcalls to the associated group. The noise exposure metric was the SNR of ship noise, collected from a single receiver deployed from the tracking boat. Parks et al. were able to collect data in both the presence and absence of ship noise; thus, single individuals (female screams) and groups (upcalls) served as their own controls. Parks et al. reported that at higher levels of shipping noise, the whales called less often, increased minimum (scream) and start (upcall) frequency, and increased call duration.

10.2.2.2 Ambient Noise

Parks et al. (2007) related frequency characteristics of upcalls to ambient noise in the Bay of Fundy over time (1956–2004) and at three sites along the eastern seaboard. They reported that, at higher noise levels, while call rates decreased, the starting and ending frequencies and bandwidth of calls increased. The change in starting and ending frequency between recordings from 1956 and 2000–2004 was on the order of 20–30 Hz, significant for the low-frequency sounds in their analyses. The Bay of Fundy dataset was compared with recordings of southern right whales from an area with substantially less vessel traffic, Golfo San Jose on the coast of Argentina. Southern right whales had lower starting and ending frequencies than North Atlantic right whales and also higher frequencies in 2000 than in 1977—a period during which vessel traffic increased substantially worldwide.

The period over which changes in call frequency were observed is within the lifespan of individual right whales and thus indicative of a behavioral shift in calling (Lombard response). The approach had the advantage of making comparisons on temporal and spatial scales appropriate for large, migratory, long-lived balaenid whales. However, different groups of whales were observed at different sites. Factors other than or in addition to noise could have explained the trends observed.

10.2.2.3 Alerting Devices

In a playback study, North Atlantic right whales were exposed to playbacks (SL ≤ 173 dB re 1 µPa) of conspecific social sounds and tonal signals (alert signals intended to prevent vessel collisions; Nowacek et al. 2004a). The tonal alerts consisted of three 2-min signals played three times, including tones at 500 and 850 Hz, downsweeps from 4500 to 500 Hz, and two tonals (1.5 and 2 kHz) modulated at 120 Hz. Whale responses were monitored with attached DTAGs and from a vessel. The whales oriented in response to right whale vocalizations. Five of six subjects responded to tonals with a rapid ("high-power") ascent to the surface, followed by prolonged surfacing intervals.

10.2.2.4 Unmanned Aerial Vehicles

Operational sounds from unmanned aerial vehicles typically manifest as low-intensity noise under water (Christiansen et al. 2016; Erbe et al. 2017). Sounds noticeable at sufficient horizontal distances and altitudes away from an underwater receiver (Laute et al. 2023) become barely noticeable in nearshore regions having high ambient levels (Nielsen et al. 2019). In a study assessing potential disturbance to southern right whales in Australian waters, Christiansen et al. (2020) compared pre-approach controls (30-min focal follows) with during-approach focal-follow data on mother-calf pairs and found that these aircraft did not elicit significant behavioral responses (swim speed, turning, surfacing).

10.2.3 Rorquals

The rorquals include the genus *Balaenoptera* and the humpback whale. The humpback whale is one of the most intensively studied marine mammal species in and outside the context of noise effects. The species appears to differ substantially from other rorquals in feeding behavior, social and mating system, and acoustic communication system. In addition, it is known to use attack as an antipredator strategy (Pitman et al. 2017), which has not been reported in other rorquals. Therefore, its responses to noise might differ from those of whales in the genus *Balaenoptera*.

10.2.3.1 Ships and Boats

Ship noise overlaps in frequency with rorqual vocalizations, and so several studies have investigated and reported changes in acoustic behavior in response to ship noise. For example, a higher shipping noise RL was correlated with a decrease in fin whale frequency measures, as well as note duration (Castellote et al. 2012). Humpback whales changed their singing behavior at RLs as low as 103 dB re 1 µPa (Tsujii et al. 2018). During the study, distance to the noise source was measured, and at distances of between 235 and 1200 m, humpback whales decreased their singing rate or ceased singing altogether (Tsujii et al. 2018). The Lombard effect was reported in humpback whales, consistent with wind noise increase (Girola et al. 2023b), but additional vessel noise (up to ~23 dB on top of prevalent wind noise) did not elicit a Lombard response. Subject whales continued to compensate only for wind noise even in the presence of vessel noise (Girola et al. 2023a). Conversely, an increase in blue whale acoustic detections (measured as the number of hours with D-calls present) was found in shipping noise >125 dB re 1 µPa (Melcón et al. 2012).

Whale-watching vessels target whales and hence behave differently from other vessels that move independently from animals. Whale-watching vessels approach in a directed manner, often following animals and exposing them to noise at relatively close range. Also, they may be present for large portions of the day. A CEE study of 42 humpback whale mother-calf pairs on a resting ground found significant levels of disturbance from playback of whale-watching boats at an RL of 135 dB re 1 µPa (Sprogis et al. 2020). Mothers' responses included a doubling of respiration rate, a 37% increase in swim speed, and a 30% reduction in resting. Similarly, theodolite-tracked fin whales altered their movement patterns in the presence of whale-watching vessels (Santos-Carvallo et al. 2021). Fin whales engaged in both resting and traveling behavior were found to reorient significantly more, swim faster, and decrease directness of travel, consistent with an avoidance response. Interestingly, as the number of whale-watching boats increased, resting fin whales' direction of travel became more consistent, indicating they were less likely to employ directional avoidance responses when there were multiple vessels. Responses to whale-watching vessels increased as the vessels moved away, though it was not clear if this was a result of an increase in noise RL (due to vessel orientation and increased departure speed) or other factors (Santos-Carvallo et al. 2021).

Shipping has been found to impact foraging behavior across several rorqual species. The presence of ships reduced the foraging opportunities of blue whales, with the proximity of vessels having a significant influence on the severity of the behavioral response (Lesage et al. 2017). In humpback whales, the passage of shipping traffic has been associated with a reduction in foraging behavior at relatively low RL (77–104 dB re 1 µPa; Blair et al. 2016). Specifically, in the presence of ships, ascent and descent rates slowed and humpbacks engaged in fewer side rolls, indicating a reduction in feeding effort or success when ships were close (Blair et al. 2016). An abrupt behavioral change consistent with avoidance was observed in a blue whale in response to a near vessel strike (Szesciorka et al. 2019).

10.2.3.2 Seismic Surveys

Blue whale call rate decreased in the presence of noise from a seismic sparker (bandwidth 30–450 Hz) at an RL of 131 dB re 1 µPa pk-pk and 114 dB re 1 $\mu Pa^2 s$ (di Iorio and Clark 2010). Fin whales moved away from seismic survey pulses and decreased their vocalization bandwidth and peak frequency in relation to a 15 dB increase in background noise in the presence of a seismic survey (Castellote et al. 2012). Displacement persisted for more than 10 days. Humpback whale singer count decreased with increasing seismic survey noise with RL of 111–157 dB re 1 µPa pk-pk and 66–133 dB re 1 $\mu Pa^2/Hz$ (in the band 15–1000 Hz), but these measures of RL explained only a small proportion of the variance (1–2%), which was affected more strongly by seasonality (i.e., seasonal variation in whale song; Cerchio et al. 2014). Such decreases in vocal activity in the presence of airgun noise render PAM as a mitigation tool inefficient (Van Parijs et al. 2021; Verfuss et al. 2018).

The behavioral responses of humpback whales to seismic survey noise were studied over two decades off Australia. The earlier studies exposed southbound migrants to pulses from a 0.33 l Bolt airgun at 8–15 s intervals (McCauley et al. 1998, 2000). Groups avoided the active vessel at 2–5 km (RL 157–164 dB re 1 µPa rms), but there were differences in responses by age and sex. Resting pods with cows began avoidance at 140 dB re 1 µPa rms (126 dB re 1 $\mu Pa^2 s$). Resting cows avoided the source at 7 km, while individual males approached the source (in nine trials), coming as close as 100 m, with RLs up to 179 dB re 1 µPa rms and 195 dB re 1 µPa pk-pk. Avoidance close to the vessel was the only response observed, and no difference in whale density could be detected across depth contours over a larger area. The Behavioral Response of Australian Humpback whales to Seismic Surveys (BRAHSS) experiments continued these investigations and provided some of the most detailed information on humpback whale behavior and response strategies when exposed to impulsive (seismic survey) and continuous (vessel and wind-driven) noise. Humpbacks were more likely to show an avoidance response to airguns when the RL was >130 dB re 1 $\mu Pa^2 s$ and whales were within 4 km of the source (Dunlop et al. 2018). At 150–155 dB re 1 $\mu Pa^2 s$ (within 2.5 km range), the probability of response was 50%. Some individuals did not avoid the source vessel at the highest exposures of 160–170 dB re 1 µPa²s. The extensive studies of humpback response to seismic sound have highlighted the importance of proximity, as well as RL, in driving behavioral responses.

10.2.3.3 Sonar

The effect of naval sonar on rorquals has been the focus of considerable research, largely owing to the overlap of naval training areas with key foraging areas for blue whales in the Southern California Bight. One of the earliest studies (Croll et al. 2001) measured effects of low-frequency (<1000 Hz) active sonar (LFAS) on the density of blue and fin whales detected from the surface, supplemented with call rates recorded on bottom-mounted pop-up recorders. This study was one of the few to measure prey density concurrently, using echosounder sampling ground-truthed with net tows. The density of rorquals in the study area was low, so sightings of blue and fin whales were pooled. There appeared to be a decline in whale density during the period of exposure, with RL < 150 dB re 1 µPa rms. The effect of sonar exposure on density was, however, less significant when prey availability was taken into consideration, leading to the conclusion that the density of blue and fin whales was better described by prey distribution than exposure to LFAS (Croll et al. 2001). Further studies of the behavioral response of blue whales to simulated mid-frequency active sonar (MFAS) and pseudorandom noise also found that prey density influenced results, with variability in whale diving behavior attributed to all three factors (Friedlaender et al. 2016).

The effects of sonar exposure on caller density, which is often used as a surrogate measure of animal density, are difficult to disentangle from effects on calling rate and singing behavior. The effects of MFAS and associated military operations on the call density of common minke whales were studied on a large spatial scale (54 km × 70 km; Martin et al. 2015). During MFAS transmissions and subsequent operations with military vessels, the 95% confidence interval of calls per km² dropped from 3.3–4.0/km² to 2.3–3.4/km² (MFAS alone) to 0.3–1.8/km² (during operations). After the operations, call density rebounded to 4.0–4.9/km². Movements of the whales in and out of the area could not be distinguished from changes in call rate. Unlike Blackwell et al. (2015), the authors did not estimate RL at the call locations, so it is unclear whether minke whale call rates were correlated with RL. Blue whale D-call counts declined in the presence of MFAS with RL above 105 dB re 1 µPa rms (maximum RL was 150 dB re 1 µPa; Melcón et al. 2012). Exposure to LFAS has also been shown to influence acoustic behavior of singing humpback whales on breeding grounds off the Hawaiian Islands (Biassoni et al. 2000; Fristrup et al. 2003; Miller et al. 2000). While 30% of focal whales ceased singing, there was a significant increase in song duration during the LFAS transmissions, consistent with what would be expected under the Lombard effect. RL at the singing whales ranged from 120 to 155 dB re 1 µPa rms; however, there was no clear correlation between RL and increase in singing bout duration.

Changes in the movement patterns of animals, including changes in dive profile, foraging, or resting behavior in response to sonar exposure, have been documented across a variety of species. Blue whale responses to MFAS have been investigated using CEEs (DeRuiter et al. 2017; Friedlaender et al. 2016; Goldbogen et al. 2013; Southall et al. 2012). Whales exposed to 3.3–4.2 kHz MFAS at a rate of 1.5 s every 25 s for 30 min showed highly variable responses influenced by initial behavioral state and environmental conditions. While the most common response was a reduction in deep foraging dives (DeRuiter et al. 2017; Goldbogen et al. 2013), abrupt changes in movement patterns and movement away from the source at RLs between 102 and 165 dB re 1 µPa rms were reported (Southall et al. 2019b). Only two individual common minke whales have been subjects in experimental studies (Kvadsheim et al. 2017; Sivle et al. 2015). One subject off Spitsbergen was approached by a vessel projecting 1.3–2.0 kHz Socrates II sonar signals (1 s every 20 s) with an RL between 83 and 158 dB re 1 µPa rms. The authors measured responses in two ways: (1) using a response severity code (after the method of Southall et al. 2007) and (2) documenting avoidance from surfacing sequences and the pattern of dives measured with an attached tag. Severity scoring indicated a progressive increase in response intensity during the approach. The whale's initial response was a change in dive depth, increasing to a threshold of avoidance at 146 dB re 1 µPa rms. At maximum RL, it had transitioned to fast, shallow swimming away from the source vessel. The same whale was also exposed to MFAS (3.5–4.05 kHz), with a similar exposure conducted on a second common minke whale in the Southern California Bight (Kvadsheim et al. 2017). Procedures for the two minke whales differed, with a lower maximum RL and shorter range of exposure for the California whale. Off Norway, the minke whale threshold for avoidance was 156 dB re 1 µPa rms (166 dB re 1 $\mu Pa^2 s$) and off California 146 dB re 1 µPa rms (149 dB re 1 $\mu Pa^2 s$). In a CEE experiment exposing tagged whales to MFAS, blue and fin whales did not change their lunge rate relative to baseline, while humpback whales were found to be more responsive during and after sonar exposure (Harris et al. 2019). Five out of 15 fin whales responded behaviorally to MFAS and pseudorandom noise; responses were lower in occurrence, duration, and severity compared to those of blue whales (Southall et al. 2023). Although there was great individual variability, an MFAS effect on the dive profile (i.e., increase in steep dives) and swim speed (i.e., slowing) of humpback whales was reported at RL up to 158 dB re 1 µPa rms (Henderson et al. 2019). A study of the behavioral response of feeding humpback whales to MFAS transmissions from an approaching source was undertaken using DTAGs. All six tagged feeding whales reduced lunge rates during exposure to the Socrates II sonar, indicating a reduction in foraging. The strongest responses were observed in groups, including mother-calf pairs (Sivle et al. 2015, 2016). Most of these responses occurred at RL greater than 140 dB re 1 µPa rms, with a 50% response threshold of 179 dB re 1 $\mu Pa^2 s$. Confidence intervals on this measure were wide, suggesting substantial variability in responsiveness among subjects. The authors included severity scaling to categorize observed behaviors. The highest response severity was observed at an RL of 164 dB re 1 µPa rms (Sivle et al. 2015). Not a sonar source but a tonal research source, the ATOC M-sequence signals led to variable responses in humpback whales in both simulated and real exposures, with some groups moving towards the source, some moving away, and others continuing on an unaltered path (Frankel and Clark 1998, 2000). While there was no consistent avoidance response observed, Frankel and Clark (1998) did find a significant effect of sonar exposure on the duration and distances traveled between surfacing in both simulated and real exposure studies for sound exposures extending from local ambient (~105 dB re 1 µPa) to 130 dB re 1 µPa rms.

Ramp-up is a mitigation procedure using a slow increase in stimulus level designed to give subjects an opportunity to avoid full power transmissions at close range. Although current permitting requirements often include a ramp-up, it is not clear that the procedure arouses consistent avoidance responses. The efficacy of ramp-up protocols was assessed during a behavioral response study of humpback whales to sonar. While initially 9 of 12 instrumented whales exposed to the ramp-up procedure during active vessel approach avoided the sound, the limited 5-min duration of the ramp-up meant that, on average, the range to full exposure was increased by less than 300 m equating to around a 5 dB reduction in sound exposure. In addition, during a second exposure to the ramp-up, the whales became less responsive (Wensveen et al. 2017). Consequently, while ramp-ups may provide some benefit, particularly in areas where sounds are novel, they should be of sufficient duration for whales to move a significant distance from the source. Furthermore, they may be of limited efficacy if operating in the same area over an extended period of time.

10.2.3.4 Acoustic Deterrent Devices

The effect of acoustic deterrent devices (ADDs) on whales continues to be debated, despite their common use as deterrents for aquaculture and fishing operations and increasing use as a mitigation measure for underwater noise impacts. Given the contrasting acoustic characteristics of different types of ADDs (see Chap. 2), it is difficult to generalize the findings of these studies. Responses appear to vary between species and ADD type.

Some species, such as minke whales, appear to be rather sensitive to ADD exposure. During all experimental deployments, focal animals exhibited an immediate avoidance response, moving directly away from the source with an increase in horizontal speed and extended dive duration (Boisseau et al. 2021). Minke whales continued to move away from the source even after it had been deactivated, consistent with a flee response. Moreover, source proximity had a greater influence on response severity than the signal magnitude had, indicating the flee response of minke whales was likely driven by perceived threat. Humpback whales also exhibited an avoidance response with a significant increase in swim speed and decrease in time spent feeding when exposed to an ADD. The consistency of humpback whale responses appeared to vary between ADD sources, with higher-frequency sources yielding a less consistent response, suggesting humpback whales may be less sensitive to sounds at the expected upper limit of their hearing capability (Basran et al. 2020).

While acoustic deterrent devices may be useful in some applications as a mitigation measure to decrease the interactions between marine mammals and anthropogenic gear and operations, they might carry their own risk of acoustic impacts from long-term exposure.

10.2.3.5 Predator Sounds

It is often speculated that marine mammals' responses to anthropogenic noise stem from antipredator responses. Despite their size, rorqual whales and their young are targeted by killer whales as a prey source (Jefferson et al. 1991). If noise is perceived as a threat (as are predators), behavioral responses might be similar. For example, exposure of humpback whales to the playback of killer whale sounds resulted in avoidance (increase in swim speed, longer and deeper dives, and behavioral change from foraging to traveling); humpback whale mother-calf pairs made alternating horizontal turns as they swam, which have been interpreted as specialized defensive behavior to protect the calf from predation (Curé et al. 2015). Antipredator responses, including changes in dive profiles, cessation of foraging, and fleeing, all carry energetic and opportunity costs and yet have evolved because they presumably reduce mortality risk from predation, increasing individual fitness (Miller et al. 2022). The ability to detect and avoid predators inevitably comes with a fitness benefit, and it is likely that responsiveness to anthropogenic noise is correlated with responsiveness to the threat of predators: Species that are more sensitive to the presence of predators appear to exhibit stronger responses to anthropogenic noise (Miller et al. 2022). Miller and colleagues evaluated the responsiveness of a small number of species, including humpback whales, to both killer whale signals and 1–4 kHz naval sonar. The authors found a strong correlation between the reduction in the average foraging intensity in response to both killer whales and naval sonar playback, indicating that the sensitivity of a species to predator presence is a strong indicator of their sensitivity to anthropogenic noise. Despite variability in individual responses, comparison of average responsiveness by species confirmed that those species whose life history characteristics increase their predation risk are the most sensitive to both predatory cues and anthropogenic noise. In the Miller et al. study, northern bottlenose whales and humpback whales had the highest sensitivity to both killer whales and sonar, likely as a result of their limited capacity to fight off killer whales and limited refugia opportunities. On the contrary, sperm and long-finned pilot whales, which are not a common diet of killer whales (likely due to their large pod sizes, which offer some level of protection), had a lower level of sensitivity to both predatory cues and anthropogenic noise (Miller et al. 2022).

10.2.3.6 Wind Noise

Humpback whales increased their SLs in increasing wind-driven noise (Lombard response; Dunlop et al. 2014). As RLs increase, humpbacks have further been found to cease vocalizing and shift to surface-generated sound production, such as pectoral slapping and breaching (Dunlop et al. 2010).

10.2.3.7 Long-Term and Population-Level Effects

Pirotta et al. (2022) utilized empirical data on vital rates to fully parametrize a Population Consequences of Acoustic Disturbance (PCAD) model to assess population effects (specifically, female reproductive success and survival) of behavioral responses and environmental climate change. They found no effects of the current level of active sonar exposure on vital rates, compared to severe effects from climate change. Simulated disturbances had a stronger effect on blue whale reproductive success than on survival.

10.2.4 Gray Whales

The literature on gray whale responses encompasses all phases of the gray whale seasonal cycle and both Eastern Pacific and Western Pacific populations. It includes some of the most intensive efforts to measure population-level effects and dose-response relations. Gray whales were studied in the calving lagoons off Baja California (Dahlheim 1993; Dahlheim and Castellote 2016; Richardson et al. 1995a), off Soberanes Point (CA) while on southbound (Malme et al. 1983, 1984; Tyack 2009; Tyack and Clark 1998) and northbound (Malme et al. 1983, 1984) migrations, and off Sakhalin Island during feeding seasons in 2001 (Gailey et al. 2007; Johnson et al. 2007; Rutenko et al. 2007), 2010 (Gailey et al. 2016; Muir et al. 2016), and 2015 (Gailey et al. 2022a, b). The works of Malme and team Malme et al. 1983, 1984) were reviewed by Richardson et al. (1995b) and significantly informed the development of NOAA's exposure thresholds for behavioral disturbance at the time. The long-term programs off Sakhalin have been particularly important from a management perspective, because they have focused on the behaviors and distribution of a small population that has been at risk for over a decade—monitoring the whales while feeding in a limited area of critical habitat close to long-term seismic exploration and oil and gas extraction. Although never planned as an experiment, these programs have come as close as any to a population-level experiment on responses of a mysticete to anthropogenic noise.

10.2.4.1 Seismic Surveys, Drilling, and Vessels

In controlled experiments exposing gray whales in the northern Bering Sea to sounds from a single airgun, Malme et al. (1986, 1988) found that ~50% of the feeding and migrating whales interrupted activity at RLs of 173 dB re 1 μPa rms and 170 dB re 1 μPa rms, respectively. About 10% of the animals were observed interrupting feeding and migration at RLs of 163 dB re 1 μPa rms and 164 dB re 1 μPa rms, respectively. They found that most whales returned and resumed feeding after the seismic noise ended.

The Sakhalin programs had mitigation and monitoring plans in place with the aim to keep exposure levels to within the 10% response threshold of 163 dB re 1 μPa rms. In the 2010 Sakhalin program (Gailey et al. 2016), gray whales were estimated to be exposed to per-pulse SEL (from the seismic survey) in the range 99–156 dB re 1 μPa²s and SEL_{1s} in the range 75–137 dB re 1 μPa²s for continuous sounds (vessel activity). During the 2015 program, a 30-s exposure level (i.e., SEL_{30s}) was considered for assessing behavioral responses (Gailey et al. 2022a), while cumulative SEL values (over periods of 2 h, 8 h, 1 day, 3 days, and 7 days) were considered for assessing the effects of short- and long-term exposures on population density (Gailey et al. 2022b). As vessels approached whales or sound exposures increased,

Gailey et al. (2022a) observed whales exhibiting increased breathing rates and higher swim speeds. There was a significant correlation between the direction and approach distance of the closest vessel and the direction of movement of whales (Gailey et al. 2016, 2022a). Density was found to decrease with increasing cumulative SEL during shorter exposure periods, and additional exposure during longer periods decreased density further (Gailey et al. 2022b).

Dahlheim and colleagues (Dahlheim et al. 1984, Dahlheim and Castellote 2016) conducted playback experiments in San Ignacio Lagoon, Baja California, exposing gray whales to recordings of oil drilling and outboard engine noise using a before/during/after protocol. With exposure levels ≥130 dB re 1 μPa rms, they observed a Lombard response in calling rate, call level, frequency modulation, and pulses per call. The range of call frequencies (including peak frequency) and call durations were not affected as much as the other call parameters.

10.2.4.2 Sonar

Tyack and Clark conducted controlled dose-response experiments along migration paths by placing a moored playback source at inshore and offshore locations (Tyack and Clark 1998). The transmissions were Surveillance Towed Array Sensor System (SURTASS) LFAS in the 160–330 Hz band. The playbacks, 42 s in duration and repeated every 6 min, were at SLs of 170–185 dB re 1 μPa m for the inshore experiments and 185–200 dB re 1 μPa m for the offshore experiments. Tyack and Clark (Tyack 2009; Tyack and Clark 1998) found that the whales exhibit greatly reduced avoidance response to offshore playbacks relative to inshore playbacks. Buck and Tyack analyzed data from these experiments using different models and reported avoidance thresholds of 141 dB re 1 μPa rms (for 50% probability) and 135 dB re 1 μPa rms (median) for the inshore playbacks (Buck and Tyack 2000, 2003).

10.2.4.3 Long-Term and Population-Level Effects

Villegas-Amtmann et al. (2015) used a bio-energetic model to predict the consequences of energetic imbalances under different levels of anthropogenic disturbance and estimated that a mere 4% energetic loss during pregnancy would lead to unsuccessful reproduction, possibly resulting in population effects. Different results were obtained by McHuron et al. (2021) who simulated acoustic disturbance in a foraging area of an endangered population of gray whales off northeast Sakhalin Island but found no evidence of impact on reproduction or survival; however, the study highlighted a higher individual heterogeneity in disturbance impact due to differences in size.

10.3 Behavioral Responses of Odontocetes

As there has been a great amount of research on some species, these are presented separately from their family groups.

10.3.1 Bottlenose Dolphins

As a result of their cosmopolitan and primarily coastal distribution, bottlenose dolphins experience a range of anthropogenic sound sources. They are also the most common cetacean species in captivity. Consequently, they have been the focus of many field and captive studies, which have documented a range of physical, acoustical, and biological responses.

10.3.1.1 Ships and Boats

The effects of vessel traffic on bottlenose dolphins have been of particular concern, in terms of both direct, targeted interactions via tourism operations and general areas of high vessel density. Although many studies have the intention of investigating the behavioral responses of dolphins to vessel traffic, relatively few consider underwater noise levels in their analyses. Instead, responses are generally related to visual observations of vessel presence, number, and/or behavior. While these may provide a proxy of noise levels, it remains unresolved whether dolphins are responding to the physical presence of vessels or to vessel noise. Many of these studies are further confounded by the fact that they are vessel-based, introducing the possibility that the research vessel itself may influence dolphin behavior (although some attempt to account for this; e.g., Heiler et al. 2016; Lusseau 2003). Indeed, dolphins have shown differential behavioral responses between vessel types (e.g., tourism versus transport vessels; Perez-Ortega et al. 2021).

Studies that considered noise levels typically recorded ambient conditions in varying levels of vessel traffic and predominantly focused on dolphin acoustic responses. This was particularly true of studies comparing COVID-19 lockdown periods with unrestricted periods, as visual observations were generally not possible under lockdown restrictions, and so inferences were made acoustically about both dolphin and vessel behaviors at such times (e.g., Gagne et al. 2022; Longden et al. 2022). Studies of dolphin acoustical responses have typically found that whistle duration and frequency characteristics vary as noise levels increase (Antichi et al. 2022; Fouda et al. 2018; Gagne et al. 2022; La Manna et al. 2019; Marley et al. 2017; Morisaka et al. 2005; Rako Gospić and Picciulin 2016; Sobreira et al. 2024; van Ginkel et al. 2018). Bottlenose dolphins are also known to partially compensate for increased noise levels by adjusting whistle amplitude, with signature whistles demonstrating less adjustments than non-signature whistles, likely as a reflection of differing signal functions (Kragh et al. 2019). However, acoustic-based studies often struggle with two main complications. Firstly, only "high-quality" whistles (i.e., those with a high SNR) can be included in analyses; thus, whistles that are partially or completely masked by noise (or indeed other dolphin sounds) cannot be examined. Secondly, additional contextual factors, such as group size, activity state, and calf presence, can also influence whistle characteristics (Heiler et al. 2016; La Manna et al. 2019; Marley et al. 2017; Rako-Gospić et al. 2021); yet not all studies were able to consider these variables.

Some studies have linked acoustic behavior with functional behavior. For example, a reduction in the recording of feeding buzzes likely indicates a disruption of foraging. Pirotta et al. (2015) showed that boat presence can result in a 49% reduction in foraging activity, but did not find a correlation with the noise level.

Other studies have focused on non-acoustic behavioral responses to increased ambient noise. Rako et al. (2013) found seasonal displacements of dolphins from noisy areas and correlated increased ambient noise levels with intense vessel activity. The lowest total number of dolphins was recorded within areas of highest anthropogenic impact. Yet, other studies have found dolphins to continually utilize noisy habitats, likely due to their importance as key foraging sites (Marley et al. 2017; Pirotta et al. 2015).

This is not to say that there are no subtler consequences of noise. A captive study demonstrated that dolphins had reduced success achieving a cooperative task with increasing noise levels, from 85% success rate in 115 dB re 1 µPa rms to 62.5% in 150 dB re 1 µPa rms (Sørensen et al. 2023). Despite a range of behavioral and acoustical compensatory mechanisms (e.g., orienting toward their partner, increasing whistle duration and amplitude), these were insufficient to overcome the effects of noise. This aligns with observations of bottlenose dolphins engaging in cooperative fishing with artisanal fishers in Brazil, where dolphin acoustic behavior during this task alters in the presence of boats (Pellegrini et al. 2021). Unfortunately, underwater noise levels and cooperative-foraging success rates were not reported in this study. Further research would be beneficial to better equate vessel-noise impacts to cooperative activities (e.g., group foraging) and other subtler behaviors in wild dolphins.

10.3.1.2 Seismic Surveys

Captive dolphins participating in trials involving airguns have shown clear avoidance behaviors. Finneran et al. (2015) measured hearing thresholds in bottlenose dolphins before and after exposure to ten airgun impulses. Two of the three dolphins exhibited anticipatory avoidance behavior at the highest exposure condition (196–210 dB re 1 µPa pk), altering their orientation away from the sound source.

In the wild, differences in bottlenose dolphin sightings have been reported during seismic surveys, with encounter rates lowest during seismic operations (Holst et al. 2017). Fernandez-Betelu et al. (2021) report that offshore seismic surveys (20–30 km away, predicted maximum RL of 139 dB re µPa^{2}s) did not cause dolphin displacement over longer temporal scales (years); however, over short temporal scales (days), there was an increase in dolphin acoustic detections.

10.3.1.3 Pile Driving

After the ubiquitous noise from vessel traffic, the most likely sound source to be encountered by coastal bottlenose dolphins is that from construction activities, including pile driving. Branstetter et al. (2018) examined the effects of vibratory pile driver noise on dolphin echolocation, requiring animals to scan their enclosure and indicate the occurrence of phantom echoes during five different SLs of playback sound. Three of the five dolphins demonstrated a significant decrease in echolocation performance at the highest SPL of 140 dB re 1 µPa, at which clicks were also produced.

Field studies investigating behavioral responses to construction noise all report some degree of avoidance in the immediate area. While no studies report an overall exclusion of animals from sites in the vicinity of pile driving and construction, there is typically a reduction in dolphin occurrence and the time spent in those areas (Buckstaff et al. 2013; Graham et al. 2017; Paiva et al. 2015). Culloch et al. (2020) detected no evidence of construction activity causing short- or long-term reductions in dolphin abundance within a 10-km-wide bay, but note that fine-scale effects (e.g., movement within the bay) may have occurred and were simply not captured by the available data. At distances of 40–70 km away from piling activities (predicted maximum RLs of 141 dB re µPa^{2}s), Fernandez-Betelu et al. (2021) found that acoustic detections of dolphins remained constant over longer temporal scales (years), but short-term increases existed over smaller temporal scales (days). This may reflect changes in acoustic behavior (as seen in Branstetter et al. 2018), rather than increased animal abundance.

One of the challenges with field studies of construction-related disturbance is that activities rarely occur in isolation; there may be significant temporal and spatial overlap in pile-driving, multibeam surveys, dredging, rock manipulation, etc. This can make untangling potential consequences challenging, especially if using "coarse" measurements like species presence/absence or changes in abundance, which are the metrics frequently used for environmental impact assessments (Culloch et al. 2020). Such measurements overlook subtler effects, particularly if animals place high value on the impacted habitat for key life functions (e.g., foraging, mating), forcing them to tolerate disturbance.

10.3.1.4 Explosions

Explosives are often used in demolition (e.g., prior to new construction) as well as for artisanal fishing practices. During demolitions, Buckstaff et al. (2013) found that wild dolphins increased group sizes, decreased inter-animal distances, and displayed heading changes within a kilometer of the underwater explosion (peak amplitude 119 dB re 1 µPa^2/Hz at 1.6 kHz). Broader-scale spatiotemporal studies investigating dolphin occurrence in relation to underwater explosions suggest that the radius of physical responses may be up to 6 km (Lammers et al. 2017).

Finneran (2000) exposed two bottlenose dolphins in captivity to waveforms resembling those from distant explosions. While the objective of the experiment was to measure TTS, the two animals started to show behavioral responses (i.e., ignoring the start cue, delaying or refusing to station, swimming around the pool, and vocalizing after exposures) at levels corresponding to a 5 kg charge at 9.3 km and 1.5 km range, respectively.

10.3.1.5 Sonar

Behavioral responses to sonar have been contradictory. Some field studies have reported that tagged bottlenose dolphins remain in the area during MFAS exercises (modeled RLs 149–168 dB re 1 µPa rms; Baird et al. 2014), while other studies of non-tagged bottlenose dolphins reported animals leaving the area at moderate speed, changing their behavioral state, and increasing intensity of vocalizations after sonar exposure (mean peak SPL 122 dB re 1 µPa when a behavioral response was observed; Henderson et al. 2014).

Houser et al. (2013b) played a simulated sonar signal (3250–3450 Hz) to 30 bottlenose dolphins of the US Navy Marine Mammal Program. The dolphins were divided into six groups of five individuals each, and each group received exposures to the signal at one of six received SPLs (115, 130, 145, 160, 175, or 185 dB re 1 µPa rms). Dolphins showed changes in respiration rate at the lowest exposure condition of 115 dB re 1 µPa. At the second-lowest exposure of 130 dB re 1 µPa, the first refusal to participate in some of the trials was observed. The percentage of trials in which dolphins refused to participate increased with increasing exposure level, and by the highest exposure condition (185 dB re 1 µPa rms), all five dolphins in the group refused to participate in at least some of the trials. Conversely, as the trial sequence progressed (and repetitive exposures increased), rapid habituation to the exposure occurred for received levels $\leq$160 dB re 1 µPa rms.

10.3.1.6 Acoustic Deterrent Devices

Acoustic deterrent and harassment devices (ADDs and AHDs) have been trialed in several coastal sites to mitigate dolphin bycatch and/or depredation on fishing gear. Avoidance responses have been reported for active devices with mean SLs of 120–145 dB re 1 µPa m, along with increased surfacing rates, aggressive behavior, and echolocation behavior (Bowles and Anderson 2012; Bruno et al. 2021; Buscaino et al. 2021; Cox et al. 2003; Gazo et al. 2008; Niu et al. 2012; Waples et al. 2013). Yet trials with devices at higher SLs (<194 dB re 1 µPa) have reported no effect on dolphin presence, distance from device, group size, or time spent in the area (Díaz López and Marino 2011). There is also evidence of habituation over time (Buscaino et al. 2021; Niu et al. 2020).

Modeling of different ADDs suggests potential audible ranges of up to 90 km for bottlenose dolphins (Todd et al. 2019). Increasingly, there are concerns regarding the risk of broader habitat displacement and masking if multiple devices are deployed simultaneously (Todd et al. 2021), which is also translating into animal welfare concerns (Dolman et al. 2022). For example, on the west coast of Scotland, 52% of salmon farms reported using ADDs (Quick et al. 2004), with acoustic point surveys detecting a significant increase in ADD presence over time (median 6.78% increase over 10 years; Findlay et al. 2018). Yet field studies of potential broadscale and/or cumulative effects on bottlenose dolphins are limited.

10.3.1.7 Aircraft

Unmanned aerial vehicles (UAVs) are being increasingly used to study cetacean behavior as a cheaper, more convenient, and potentially more accurate option to conventional vessel- or aerial-based cetacean surveys (Brown et al. 2023; Fettermann et al. 2022). However, UAV noise can exceed underwater ambient levels by up to 30 dB between 100 and 10,000 Hz (Erbe et al. 2017). Fettermann et al. (2019) assessed the short-term behavioral responses of resting bottlenose dolphins to a UAV flown at different altitudes. Changes in group swim direction and the frequencies of specific behavioral events were recorded at the lowest altitude of 10 m, at which the mean noise levels were 95 dB re 1 µPa rms at 1 m depth. Similarly, bottlenose dolphins have been recorded to orient upward and turn toward the UAV (likely to observe it) during flights at altitudes of 11–30 m (Ramos et al. 2018).

10.3.1.8 Long-Term and Population-Level Effects

While particularly well studied around the globe, there is a paucity of long-term observations of bottlenose dolphins, and so assessment of population-level effects of noise largely relies on modeling. New et al. (2013a) used a simulation approach to investigate the spatial and social behaviors of coastal bottlenose dolphins in relation to a modeled increase in vessel traffic in Moray Firth, Scotland. The study reported that even a sixfold increase in the number of vessels per year did not impact the dolphins in a biologically significant way, leading to the conclusion that bottlenose dolphins in this area are able to compensate for increasing vessel presence. While it is fair to assume that an increase in commercial vessel traffic might correspond to an increase in noise disturbance, the study was unable to differentiate between the effects of vessel presence and vessel noise and to disentangle the possible compounding effects of multiple stressors. Reed et al. (2020) compared the forecasted population trajectories of two Indo-Pacific bottlenose dolphin populations under different scenarios, which also included various levels of acoustic pollution. While climate change and epizootic outbreaks were ultimately the biggest contributors to lower abundance and fecundity, the two populations responded differently to variations in disturbance intensity. Similarly, New et al. (2020) found that four populations of bottlenose dolphins (from Doubtful Sound, New Zealand; Sarasota Bay, USA; Durban Bay, South Africa; and Jervis Bay, Australia) had different degrees of tolerance to whale-watching disturbance depending on their size, habitat use, and prey availability. These studies highlight the intra-specific variability in bottlenose dolphin behavioral responses and population dynamics, which is relevant for noise disturbance as well.

10.3.2 Killer Whales

Killer whales are the most widely ranging marine mammal, but the species is broken into populations and ecotypes with distinctive predatory lifestyles and stratified in a hierarchy of social units based on matrilines. Sympatric ecotypes can be reproductively isolated to the point of being classed as distinct population segments; some eventually could be recognized as separate (sub-)species (Morin et al. 2015). As a result, killer whale responses to anthropogenic noise must be interpreted against a backdrop of considerable environmental, genetic, lifestyle, and social complexity.

Most of what is known about killer whale behavior comes from work on small populations that prey on fishes in the Northeast Pacific and Norway. The research on the Northeast Pacific populations has been remarkable for intensive efforts to measure acoustic characteristics, behavioral responses, and demography associated with exposures. Two distinct population segments, northern and southern resident killer whales, enter inshore waters of Washington State and British Columbia to exploit Chinook salmon runs, bringing whales into contact with heavy shipping and whale-watching activity. Concerns about the viability of the southern resident killer whale have intensified in recent years because it has declined with Chinook salmon populations (Ford and Ellis 2010). Measures to reduce exposure to both heavy shipping (e.g., Joy et al. 2019; Williams et al. 2019) and whale-watching (e.g., Ferrara et al. 2017) have been instituted in an effort to help the population recover. These measures constitute an experimental treatment that may ultimately show whether population-level change resulting from exposure to anthropogenic noise can be quantified and managed in a long-lived marine mammal (e.g., Lacy et al. 2017; Williams et al. 2019).

10.3.2.1 Ships and Boats

The southern resident killer whale population is certainly exposed intensively to vessels in inshore waters, in particular during spring and summer. In addition to the potential for masking by shipping and small-vessel noise, which could affect both social communication and predatory efficiency, the whales must manage all the activities of their daily lives in the presence of repeated approaches by numerous whale-watching vessels. Foote et al. (2004) reported a fivefold increase in the number of vessels around whales over the decade 1990–2000, reaching an average of 22 whale-watching vessels following pods during daylight hours. Lusseau et al. (2009) reported that each group was within 400 m of a vessel most of the time between May and September 2003–2005, leading to a reduction in time spent foraging. The southern residents alter their behavior to cope with repeated close vessel approaches as measured by changes in path directness, travel speed, surfacing patterns, and surface activity (Williams et al. 2009). Given the high proportion of their time spent in close proximity to vessels, there is a case for energetic expenditure and lost foraging opportunity sufficient for biologically significant effects (Lusseau et al. 2009; Williams et al. 2006). Indeed, Holt et al. (2021) assessed foraging success using tags on southern resident killer whales and found the probability of prey capture decreased as vessel speeds increased.

There has been considerable effort to measure and model ship-noise propagation and predict possible impact of shipping noise on the southern residents (e.g., Cominelli et al. 2018; Erbe 2002; Erbe et al. 2012, 2014; Joy et al. 2019; Williams et al. 2015). Measured levels at one long-term monitoring site were found to be elevated chronically at frequencies that the whales hear well and depend on for echolocation (5–13 dB increase in the 10–96 kHz range; Veirs et al. 2016).

Holt and colleagues quantified the masking of killer whale communication and echolocation by ship noise. Using data from a 4-hydrophone fixed array, Holt et al. (2008) measured SLs of killer whale calls (128–162 dB re 1 μPa) in varying ambient noise (95–120 dB re 1 μPa in the range 1–40 kHz). Counts of nearby vessels were correlated with increases in ambient noise. They reported an increase of 0.5 dB in call SL for every 1 dB increase in ambient noise. Later measurements showed that the response may be even greater, as much as 1 dB in call SL for every 1 dB increase in ambient noise (Holt et al. 2009). Based on recordings made directly on the whales using DTAGs, the most important predictor of RL was vessel speed (Houghton et al. 2015).

The whales may use multiple strategies to compensate for vessel masking noise. Killer whales produce stereotyped burst-pulse calls, and one call in the repertoire typically predominates, the primary matriline call (Hoelzel and Osborne 1986). Foote et al. (2004) found that southern resident killer whales increased duration of the primary call by 15% in 2001–2003 compared with earlier periods (1977–1981, 1989–1992). Foote et al. suggested that the change had occurred because noise had exceeded a response threshold in the last period of their study relative to the first two, coincident with a sharp increase in the number of whale-watching vessels at close range. However, RL was not assessed at the whales, nor call absolute level. There are other possible explanations for the shift, such as a generation-scale shift in the features of the call or identities of callers. Interestingly, call rate did not seem to be affected. Conversely, De Clerck et al. (2019) analyzed killer whale vocalizations using 2.5 months of autonomous recordings acquired near Iceland, observing increased frequency and decreased duration of their calls as broadband noise levels increased.

Behavioral responses of northern resident killer whales to vessels followed a dose-response relationship. Williams et al. (2014) quantified behaviors of northern resident killer whales traveling a shipping lane in Johnstone Strait during transits by cargo vessels, cruise ships, marine tugboats, and a range of smaller craft. The authors took advantage of a long-term theodolite tracking dataset collected from a high point, totaling 6 years of July–August summering periods, and paired the tracks with modeled acoustic data estimated from ship tracks. RL in the 10–50 kHz band ranged from 130 to 150 dB re 1 μPa. Because observations were not made from a vessel, there was no possibility of confounding effects of self-noise from the observation platform. The authors classed killer whale behavior into before and during periods and compared response parameters statistically, including surfacing/diving parameters, swim speed, path directness and smoothness, and surface activity. Behavioral responses were quantified on a severity scale recommended by Southall et al. (2007). Covariates in the analysis included killer whale identity, age, and sex; vessel information (number of vessels, ship speed, and closest point of approach); and RL estimated from propagation modeling (Erbe et al. 2012), including both weighted and unweighted levels. Responses were best explained by models incorporating the number of ships in each category, concurrent counts of smaller vessels, time of year, whale age and sex, and RL. The 50% response threshold for response severity ≥2 was estimated to be at 130 dB re 1 µPa RL and >150 dB re 1 µPa for higher response severities (≥3), spanning the range of RL in the dataset.

Several studies (Noren et al. 2009; Williams et al. 2009) have addressed killer whale surface activity, considered to be a form of social communication, as a response to close approach by vessels. None tied the behavior to acoustic characteristics such as RL. Noren et al. found that the highest frequencies of surface-active behaviors occurred when the nearest vessel was within 75–149 m of the focal observational subject, with 70% of surface-active behaviors occurring when the nearest vessel was within 224 m of the whale. Surface activity was most likely near the time and place of vessel closest approach.

10.3.2.2 Sonar

Responses of killer whales to military tactical sonars have been explored in a time-area study of distribution in the presence and absence of sonar exercises as well as in short-term CEEs. Kuningas et al. (2013) reported the results of surveys for killer whales before, during, and after FLOTEX naval exercises in a Norwegian fjord in 2002–2007. They had sonar fishery monitoring data on herring abundance and sightings from whale-watching effort during exercises, as well as experimental exposures with tactical sonars in 1 year. SLs up to 209 dB re 1 µPa rms and frequencies of 1–2 kHz LFAS or 6–7 kHz MFAS were used. The authors did not collect sonar RL or data on whale-watching vessel approach. They found that the primary explanation for killer whale use of the fjords was herring presence; they reported a marginally significant change in killer whale abundance in the presence of sonar when herring resources were poor.

Miller et al. (2012) reported the results of CEEs on killer whales (3S experiments; see Lam et al. 2016) with a Socrates sonar source producing hyperbolic 1–2 kHz LFAS sweeps and 6–7 kHz MFAS upsweeps, as well as projected killer whale sounds (sympatric fish-eating whales). Subjects were instrumented with DTAGs and observed visually from surface vessels, with vocalizations tracked using passive line arrays. A total of 14 experimental exposures were conducted on 6 tagged killer whales monitored for 15–18 h. The whales were approached from a range of 6–7 km by a vessel towing an active or silent source. Individual tagged whales received one to three experimental exposures. Onset of sonar exposure included a ramp-up (SL 152–158 dB re 1 µPa m to a maximum of 198–214 dB re 1 µPa m over 10 min), with a total of 30 min of exposure. The killer whale sounds were projected at 140–155 dB re 1 µPa m. Killer whales were more responsive than sperm whales or pilot whales in the same study (Harris et al. 2015). Before exposure to sonar signals, the whales were traveling slowly and with evidence of feeding such as circular travel paths, echolocation, tail slap sounds, and stunned herring at the surface. Few calls were recorded while the killer whales were engaged in this activity. Onset of sonar signals was associated with a transition in behavior. Killer whales called more, ceased behaviors associated with feeding, changed diving patterns, and moved directly away from the approaching source. The most common behavior was avoidance. In several cases, there were increases in vocal behavior that were interpreted as social signaling (contact calling) or counter calling with the source. Changes could consist of more than one transition in behavior, but these tended to be consistent with the broader change in behavioral state.

In a single experiment, a calf was separated from its group during an avoidance response, one of the few examples of mother-offspring separation during experimental exposures in the literature on cetaceans (Miller et al. 2012). It was the most severe response in the experimental series. The calf became separated during the ramp-up phase of the experiment (RL ~152 dB re 1 µPa rms) when its group began to move away from the source. It successfully tracked the group in association with a change in vocal behavior to include high-frequency whistles and rejoined successfully after 86 min.

Behavioral changes were significantly more severe during sonar exposures than during silent vessel passes, with more changes associated with higher severity scoring (Miller et al. 2012). However, the relationship between RL and severity was highly variable. These observations imply that there was an overall tendency for increased risk of a severe behavioral response above 120–130 dB re 1 µPa rms RL. One whale showed indications of response at low RL (<100 dB re 1 µPa rms), at an RL expected to be close to the detection threshold. The authors considered the proportion of sessions with a maximum severity ≥4 to indicate possible effects on vital biological functions (feeding, resting, or caring for young) and severities of 7–9 to indicate likely effects. Killer whales had relatively more severe-scored responses to LFAS than MFAS despite better hearing sensitivity in the MFAS range. Sivle et al. (2012) reported that they altered the diving pattern from deep to shallow dives during LFAS but not MFAS experiments. Prolonged avoidance of the sound source continued for at least 5 h after the end of the sonar transmissions, accounting for two severity 7 responses, both associated with prolonged cessation of feeding as well as avoidance.

10.3.2.3 Acoustic Harassment Devices

Killer whales appear to avoid tonal signals. Morton and Symonds (2002) compared the distribution of killer whales in two adjacent areas with (Broughton Archipelago) and without (Johnstone Strait) AHDs from 1985 through 2000. The Airmar AHD tested in these experiments was designed to deter pinnipeds attempting to depredate salmon farms, emitting a 10 kHz signal at 194 dB re 1 µPa m under water that could have been detected at very long ranges under conducive conditions. Salmon farms tested AHDs in the Broughton Archipelago from 1993 to 1999. Whale occurrence was relatively stable in both areas until the installation, after which killer whale sightings declined in the Broughton Archipelago, returning to baseline levels only after the experimental AHDs were removed. Both local ecotypes (northern residents and mammal-eating transients) responded with avoidance. Morton and Symonds did not estimate RL in the exposed area, but Southall et al. (2007) used the characteristics of AHDs to estimate levels of 140–150 dB re 1 µPa rms in the area avoided.

10.3.2.4 Long-Term and Population-Level Effects

Southern resident killer whales in the Salish Sea suffer a decrease of >20% in daily foraging time due to vessel presence, through a combination of acoustic masking and behavioral responses (Tollit et al. 2017). Lacy et al. (2017) examined the effects of multiple stressors, including reduced availability of Chinook salmon and reduced foraging efficacy due to anthropogenic noise and disturbance. Results from this population viability analysis support the concern of Tollit et al. (2017) and identify the need to halve noise disturbance and increase Chinook salmon availability by 15% to allow the target recovery goal of 2.3% killer whale population growth (Lacy et al. 2017). Murray et al. (2021) also modeled population dynamics of killer whales under different types of disturbance. Results from this study echo previous findings from Lacy et al. (2017) that cumulative effects of multiple stressors, including reduced foraging efficiency attributed to acoustic disturbance, are the likely cause of southern resident killer whale population decline. However, similar cumulative disturbances in population viability models of northern resident killer whales did not cause a decline in population growth (Murray et al. 2021).

10.3.3 Pilot Whales

Pilot whales are a gregarious, largely offshore species less likely to be impacted by coastal anthropogenic activities and generally considered to be resilient to disturbance from anthropogenic sound. However, recent studies describe a further level of complexity in pilot whale behavioral reactions to sound, mirroring high plasticity in responding to predation risk.

10.3.3.1 Ships and Boats

Pilot whales typically occur in large groups. The role of synchronized swimming was studied in long-finned pilot whales in the presence of boats. This behavior allows for close proximity and rapid coordinated response, with the dual function of showing affiliation and reacting to threats. In fact, synchronization increased in the presence of boats, confirming that also in this species, antipredator behavior occurs during anthropogenic disturbance (Senigaglia et al. 2012).

The effects of whale-watching vessels on short-finned pilot whales were the topic of a comparative CEE study between boats with petrol engines and those with electric engines (Arranz et al. 2021). Exposed animals were resting mothers and calves in small groups. Resting time decreased by 29% and nursing time by 81% in the presence of petrol-powered boats, with no significant behavioral changes in the presence of electric boats, which tend to be quieter (Parsons et al. 2020).

10.3.3.2 Seismic Surveys

Pilot whales inhabit environments where exposure to airgun operations is potentially common, yet little information on RLs correlated with behavioral responses exists. Long-finned pilot whales were observed to approach within 300 m of operational seismic airgun arrays off Nova Scotia, potentially being exposed to RL > 190 dB re 1 µPa (Moulton and Miller 2005). Limited movement of short-finned pilot whales away from an airgun array was observed during a ramp-up procedure off Gabon in 2008 (Weir 2008b). The movement was noted to deviate approximately 90° from the path of the source vessel but included multiple bouts of milling, logging, and orienting toward the airgun array. Neither persistent, directed movement away from the source nor high-speed swimming was observed. Although one group of long-finned pilot whales was presumed to show a startle response to a ramp-up of an airgun array operating in UK waters, the majority of pilot whales (67%) observed in the vicinity of an airgun array during the ramp-up period approached the source (Stone and Tasker 2006). While these results seem to support a certain resilience of pilot whales against noise disturbance, McGeady et al. (2016) found a significant positive correlation between offshore seismic surveys and mass strandings of pilot whales in the North Atlantic.

10.3.3.3 Sonar

Pilot whale responses to sonar have been the topic of several CEE studies, most notably the Sea mammals and Sonar Safety (3S) project. In the early phases of this project, a research vessel approached groups of long-finned pilot whales while gradually increasing the SL of its sonar system; 1–2 kHz LFAS and 6–7 kHz MFAS were used. Pilot whale responses included avoidance of the approaching vessel, cessation of foraging, and changes in locomotion, orientation, dive profiles, vocal behavior, individual spacing, and synchrony (Miller et al. 2012). The initiation of responses as a function of RL varied broadly with response type, ranging from 80 to 189 dB re 1 µPa rms. The most common form of response to the escalation of the sonar source was avoidance, occasionally across the approach path of the vessel, and with little evidence of avoidance when the vessel approached without sonar (Antunes et al. 2014). Onset of avoidance in long-finned pilot whales appeared at higher RL than in other species, in that 50% of long-finned pilot whales were modeled to show avoidance at a maximum RL of 170 dB re 1 µPa rms and a cumulative SEL of 173 dB re 1 µPa^{2}s, based on 3S data (Antunes et al. 2014). Avoidance typically ceased following the cessation of sonar (Antunes et al. 2014). Furthermore, pilot whales were more likely to change from deep foraging dives to shallow dives upon exposure to LFAS but generally maintained deep diving behavior during MFAS (Sivle et al. 2012). Subsequent time allocation analysis and modeling suggested that long-finned pilot whales reduced their foraging time by 83% (95% CI of 29–96%) when first exposed to the sonar signals (Isojunno et al. 2017). However, an increase in foraging time over subsequent sonar exposures suggested that the whales might have habituated with repeated exposures. Moreover, long-finned pilot whales took fewer breaths during sonar exposure than expected based on baseline relationships between diving locomotion and post-dive breathing behavior (Isojunno et al. 2018). A study on short-finned pilot whales found that their responses to sonar occurred at lower RL (147 dB re 1 µPa rms) than previously reported for closely related long-finned pilot whales in the 3S experiments, but decreased with time (Baird et al. 2019). In fact, temporary displacement of the tagged animals from the area lasted only 24 h before pilot whales returned to closer proximity to the noise source, suggesting that exposure history can affect animal responses to MFAS.

Based on data from the 3S experiments, Wensveen et al. (2015) built a model to simulate the efficacy of pilot whales' antipredator behavioral responses to sonar. Vertical avoidance proved effective at reducing RL during exposure; however, empirical observations report that long-finned pilot whales tend to increase or maintain shallow dive patterns (which, in turn, might result in increased RL) and favor grouping behavior and social coordination over RL reduction tactics (Visser et al. 2016). A dose-response probability function based on the 3S data found that behavioral state of the animals prior to exposure influenced the onset and severity of the behavioral response (Harris et al. 2015). Feeding long-finned pilot whales exposed to LFAS responded to a lower cumulative SEL than nonfeeding pilot whales, while the model predicted the opposite for MFAS (Harris et al. 2015). While the low sample size does not allow for extrapolation nor generalizations, the developed dose-response functions predicted a steady increase in the probability of recording any type of behavioral reaction (from mild to severe) with RL above 87 dB re 1 µPa^{2}s reaching 100% probability of response at an RL of 168 dB re 1 µPa^{2}s.

Observations of vocal responses to noise in pilot whales are variable. Observations of pilot whale vocal patterns at the time of the low-frequency sound exposures during the Heard Island Feasibility Test suggested that pilot whales reduced vocal activity during transmissions (Bowles et al. 1994); yet they increased vocal activity immediately following exposure to MFAS (Rendell and Gordon 1999). In the 3S studies, the application of a dissimilarity metric between the spectrogram of pilot whale calls and those of the sonar signals suggested that, in some instances, long-finned pilot whales altered their vocalizations to be more similar to the sonar signal (Alves et al. 2014). In contrast, short-finned pilot whales showed no change in vocal behavior following exposure to MFAS elsewhere (DeRuiter et al. 2013a).

Finally, short-finned pilot whales demonstrated a less notable response to a scientific echosounder with a center frequency of 38 kHz; an RL of 125 dB re 1 µPa rms elicited variance of animal headings during the exposure period, but no other behavioral changes were observed (Quick et al. 2017).

10.3.3.4 Predator Sounds

Interactions between pilot whales and killer whales are complex. The two species live in sympatry (Selbmann et al. 2022). Pilot whales are likely preyed upon by killer whales (Jefferson et al. 1991); however, pilot whales have been observed mobbing killer whales (de Stephanis et al. 2015), and killer whales have been observed fleeing pilot whales (Selbmann et al. 2022). Pilot whales share call characteristics with sympatric killer whales (Courts et al. 2020). Playbacks of killer whale sounds to pilot whales have provided insights into their relationship.

Four out of five groups of long-finned pilot whales exposed to playbacks of the sounds of killer whales foraging on herring approached the source (Curé et al. 2012). The approach response was coupled with an increase in group size, as individuals in the area that were not targeted for the playback joined the exposed group. The authors of the study noted that the pilot whales likely had experience with the specific killer whale sounds, which were recorded from killer whales inhabiting the same region. This behavior could represent a response to the presence of competitors for prey items. Bowers et al. (2018) reported attraction of pilot whales exposed to killer whale playbacks, in particular when the calls contained multiple nonlinearities (e.g., biphonation, subharmonics, and frequency jumps), possibly indicating arousal. Curé et al. (2019) exposed long-finned pilot whales to sounds of sympatric fish-eating killer whales (a possible foraging competitor) and unfamiliar mammal-eating killer whales (a potential predator) and saw different reactions to the two stimuli. Pilot whales ceased foraging and reduced time spent at the surface only when exposed to mammal-eating killer whale sounds but increased spy-hopping and social group cohesion in response to fish-eating killer whale sounds. Thus, pilot whales seem to be able to discern between different ecotypes of killer whales and employ a potential intimidation strategy toward the familiar fish-eating ecotype of low perceived risk while acting more warily when exposed to the predation risk posed by mammal-eating killer whales.

Isojunno et al. (2018) discovered that 17 tagged long-finned pilot whales increased their breathing rate prior to and immediately after deep dives. However, during playbacks of mammal-eating killer whale sounds (and during initial sonar exposure), breathing rate decreased by 13–16%. Visser et al. (2016) compared the responses of tagged long-finned pilot whales across three types of disturbance: killer whale playback, sonar exposure, and tagging effort. All treatments resulted in an increase in grouping behavior and social cohesion. In addition, during killer whale playback, pilot whales approached the source and increased their vocal activity (indicative of a mobbing response); during sonar exposure, surface resting increased; and tagging increased synchrony, reduced surface logging, and reduced vocal activity. Similarly, results from Miller et al. (2022) on pilot whale sensitivity to noise highlight the influence of context-specific antipredator adaptations on cetacean responses to anthropogenic noise.

10.3.3.5 Long-Term and Population-Level Effects

The link between pilot whale disturbance, foraging disruption, and vital rates was explored by applying state-dependent life history theory to a dynamic energy budget model (Pirotta et al. 2020). Disturbance was simulated as a cessation of foraging for 5–50 days. The model showed that unpredictable and sparse resource availability leads to the evolution of a precautionary approach toward risk (as additional energy resource to compensate for predation risk might not be available), which, in turn, represents a more robust response to disturbance and is linked to higher resilience of reproductive behavior. On the contrary, species experiencing low variation in resource availability are left unprepared for situations in which disturbance might prevent feeding opportunities. Pilot whales tend to live in seasonal environments or within large habitat ranges and thus likely experience highly heterogenous resources. These factors likely make them more prone to evolve resilient reproductive strategies, creating a buffer against the potential effects of natural and anthropogenic disturbance. Pilot whales exhibit a propensity for risk in their reproductive strategy, which affects their susceptibility to anthropogenic disturbance (Pirotta et al. 2020). Pilot whales appear to be quite resilient to acoustic disturbance; however, responses are context-dependent, and repeated disturbance might cause more significant disruption of life functions (e.g., nursing), potentially resulting in population-level consequences.

10.3.4 Other Delphinids

10.3.4.1 Ships and Boats

Hawaiian spinner dolphins have been observed to increase both aerial and vocal activity in the presence of vessels, particularly when there are targeted vessel interactions with the dolphins (Courbis and Timmel 2009; Heenehan et al. 2016). Unfortunately, it is not feasible with currently published data to determine the relative impact to behavior resulting from vessel presence and human interaction versus that of vessel noise alone. Tyne and colleagues (Tyne et al. 2018) reported that a population of Hawaiian spinner dolphins was exposed to human activities within 100 m for 82.7% of their daytime, raising concerns over potential population impact due to chronic exposure. Poupard et al. (2019) also suggest behavioral and acoustic variations of pantropical spotted dolphins during encounters with motorized tourism vessels. In New Zealand, vessel traffic affected the acoustic detections and thus likely distribution of Hector's dolphins (Carome et al. 2023).

10.3.4.2 Sonar

In 2008, the UK experienced its largest common dolphin mass stranding event. A minimum of 26 individuals died in Falmouth Bay, Cornwall. A similar number of individuals were returned to the sea. Necropsied dolphins were without obvious disease but had microscopic hemorrhages. The conclusion was that these normally pelagic animals entered the bay and were stranded, possibly after exposure to naval activities (Jepson et al. 2013).

Henderson et al. (2014) examined the behavioral responses of free-ranging dolphins (including short- and long-beaked common dolphins, Pacific white-sided dolphins, Risso's dolphins, and bottlenose dolphins) to incidental MFAS in the Southern California Bight. Using visual focal follows, static hydrophones, and autonomous recorders, they observed changes in behavioral state, direction of travel, vocalization rate, and call level in roughly half of the focal groups observed; the other half did not respond. Moreover, roughly half of the unexposed groups also changed their behavior. Mean peak (over a 5-s window) SPL of MFAS when changes were observed was ~122 dB re 1 µPa (2–8 kHz).

Tagged rough-toothed dolphins moved away from areas of MFAS exercises (143–147 dB re 1 µPa rms) for 24 h before returning to regions of comparable RL (146 and 151 dB re 1 µPa rms), while melon-headed whales had moved 100 km away before the exercises began and continued to move away during the sonar exposure (Baird et al. 2019).

Tagged false killer whales whistled more frequently following reception of an MFAS pulse and changed their whistles to sound more like the test pulses—an effect that decreased as time increased since pulse reception (RL 116–161 dB re 1 µPa rms; DeRuiter et al. 2013a).

10.3.4.3 Seismic Surveys

Gray and Van Waerebeek (2011) reported that a pantropical spotted dolphin showed erratic movement and aberrant behavior in close spatial and temporal association with an operational seismic airgun array. The dolphin appeared to lose the ability to move before going beneath the surface ~300 m ahead of the moving array. Although no examination of the individual could be conducted, the close association with the airgun array was used to argue a likely linkage between airgun exposure and the animal's condition.

10.3.4.4 Pile Driving

Würsig et al. (2000) examined the effects of noise from pile driving on Indo-Pacific humpback dolphins in Hong Kong and the efficiency of bubble screening as a mitigation measure. By tracking dolphins via a theodolite, they noted no overt behavioral changes with and without pile driving. However, the average speed of dolphin groups tracked by theodolite during periods of active piling was twice as high as during other periods (a significant difference) with RL of ~160 dB re 1 µPa at a distance of 250 m from piling activity. Line-transect estimates of abundance indicated potential evidence of a decline in dolphins in the area during the piling and on-jetty construction phases, with a lower abundance observed following the industrial activity.

10.3.4.5 Synthetic Signals and Acoustic Deterrent Devices

Berrow et al. (2008) studied the potential responses of wild common dolphins to different types of ADDs off Ireland. While initial tests with bottlenose dolphins showed evasive responses to some ADDs, these ADDs failed to elicit any similar response in common dolphins. Two other commercially available ADDs, however, did elicit an occasional mild evasive response. Fu et al. (2022) reported a decrease in click count and amplitude (91% and 84%, respectively), following exposure of pantropical spotted dolphins to ADDs with a pk-pk SL of 182 dB re 1 µPa m.

Captive Commerson's dolphins showed avoidance and defensive behaviors when a pinger (SL = 131 dB re 1 µPa m) was active. There was a dramatic difference in response to the presence versus absence of the pinger, with animals demonstrating agonistic behaviors and a change in surface behavior such as rooster tailing and fluke slaps. The response of a Pacific white-sided dolphin was less dramatic, but the individual exhibited avoidance behavior when the pinger was active and spent more time in a refuge pool during trials (Bowles and Anderson 2012).

Rasmussen et al. (2016) investigated how wild Icelandic white-beaked dolphins responded to playback of novel anthropogenic sounds: synthetic, amplitude-modulated tones (100, 200, or 250 kHz, 2 s duration) and burst-pulse sounds (prerecorded white-beaked dolphin clicks, played at a rate of 300 clicks/2 s). Boat-based survey playback experiments consisted of a playback portion and recording portion, during which video and audio responses were recorded. Behavioral responses included changes in swim direction, turnaround and camera approach, tail slaps, jumps, circling the hydrophone array, emitting burst-pulse sounds at the speaker, and bubble production at RL within 110–160 dB re 1 µPa for tones and 152–166 dB re 1 µPa pk-pk for burst-pulse sounds.

Most work on false killer whales has focused on their reaction to anthropogenic noise for deterrence purposes (e.g., Akamatsu et al. 1993; Mooney et al. 2009). Trials with two captive whales exposed to a variety of pulsed sounds, including pure tones, modulated tones, tone combinations, and impulsive sounds, demonstrated that single pure tones and rudimentary impulsive sounds (banging on iron plates and rods) had no effect on their behavior (Akamatsu et al. 1993). However modulated and combined pulses in the frequency range 24–75 kHz were "somewhat effective" to "effective" in changing the whales' swimming behavior at RLs >170 dB re 1 µPa, though there was also some evidence of habituation. In Mooney's study, a very broadband (1–250 kHz) ADD affected false killer whale echolocation performance. In subsequent sessions, performance recovered somewhat; however, it was unclear whether the recovery was caused by adaptation or the decrease in ADD SL during the experiment (Mooney et al. 2009).

10.3.4.6 Explosions

In 2011, long-beaked common dolphins entered a mine countermeasure training zone unexpectedly, resulting in at least four deaths (Danil and St. Leger 2011). A 3.97 kg charge was detonated in 15 m of water, 0.5–0.75 nm from shore. Minutes before the scheduled explosion, a group of 100–150 long-beaked common dolphins entered the 640 m mitigation zone and could not be deterred. Three dolphins died instantly, while the others continued their travel. A fourth individual was found on the beach, 68 km north, 3 days later. All four dolphins were necropsied showing blast injuries.

Lara et al. (2023) reported a 2.5 times increase of whistle vocalizations of delphinids immediately after an underwater explosion and significant changes in minimum frequency and duration of vocalizations before and after detonation. RLs at the acoustic recorder were above 150 dB re 1 µPa.

Akamatsu et al. (1993) found that the explosion of a "water bomb" (SL 213 dB 1 µPa m) did not cause any behavioral reaction in captive false killer whales.

10.3.4.7 Predator Sounds

Playback of killer whale calls resulted in significant behavioral changes (fleeing) in four tagged Risso's dolphins, when the calls contained multiple nonlinearities, possibly indicating predatory intentions (Bowers et al. 2018).

10.3.5 River Dolphins

Studying the responses of river dolphins to noise can be difficult. Water is shallow and often murky, multiple noises occur simultaneously in the river and on the banks, and sound propagation mechanisms are complex, making it difficult to determine RL. However, because of these murky environments, dolphins likely rely on acoustics for communication and environmental sensing. Dey et al. (2019) conducted field-based acoustic recordings and modeling to assess responses of the endangered Ganges River dolphins to vessels in India. An increase in underwater noise due to motorized vessels resulted in changes in acoustic behavior, masking of echolocation clicks, and high metabolic energy expenditure. Although no RLs were estimated, these dolphins showed enhanced activity during acute noise exposure and suppressed activity during chronic exposure. The authors suggested that continuous noise exposure might compel dolphins to substantially alter baseline acoustic activity, leading to missed foraging and altered social behaviors.

10.3.6 Porpoises

Out of all porpoise species, the harbor porpoise is by far the most studied species with regard to the effect of noise. The species has a wide distribution range within the cold and temperate regions of the northern hemisphere, particularly inhabiting continental shelf areas that overlap with anthropogenic activities (Hammond et al. 2008). Their cold-water habitat and small size present a thermoregulation challenge, demanding high metabolic rates, which may drive high feeding rates in this species (Rojano-Doñate et al. 2018; Wisniewska et al. 2016). However, foraging rates and feeding success can be expected to be highly variable over time and between age classes in this generalist species (Booth 2020; Hoekendijk et al. 2018; Kastelein et al. 2019c). Disturbance of harbor porpoises by noise could therefore have energetic impacts that may reduce individual fitness.

10.3.6.1 Ships and Boats

Harbor porpoise hearing sensitivity is relatively low below 1 kHz, where vessel noise peaks (Kastelein et al. 2017a). Impacts of vessel noise on harbor porpoise behavior have therefore not been extensively studied. Dyndo et al. (2015) observed the behavior of two captive harbor porpoises in a seminatural net pen during uncontrolled vessel passes. Thirty percent of vessel passes elicited porpoising behavior at M-weighted broadband maximum RLs of 113–138 dB re 1 µPa, mean 123 dB re 1 µPa. Acoustic tags on wild harbor porpoises showed that most individuals exposed to vessel noise >96 dB re 1 µPa within the 16 kHz one-third-octave band decreased echolocation—a measure for foraging activity (Wisniewska et al. 2018). Two individuals additionally dove to deeper depths at RLs of ~80 and ~100 dB re 1 µPa, respectively. Similarly, Frankish et al. (2023) found that ship noise led to changes in swim and dive behavior in tagged harbor porpoises. Elsewhere, harbor porpoise displacement correlated with levels of vessel activity prior to pile driving (Benhemma-Le Gall et al. 2023).

10.3.6.2 Seismic Surveys

Harbor porpoise behavioral responses to airgun noise have been found to vary with airgun and survey configuration. Two captive individuals were exposed to ten discharges (4 s intervals, 2–4 bar pressure) of a 10 in^3 airgun. One animal displayed an increase in respiration rate, respiration force, swimming speed, and the number of aerial jumps in response to single-pulse SEL of 145–149 dB re 1 µPa^{2}s (Kastelein et al. 2019b). The second individual did not respond to single-pulse SEL < 156 dB re 1 µPa^{2}s but showed an increased respiration rate, swimming speed, and distance from the source to single-strike SEL of 156–160 dB re 1 µPa^2 s. Exposure to a maximum of four airguns (i.e., total volume 40 in^3, 8 bar pressure) resulted in similar behavioral responses at single-pulse SEL of 164 dB re 1 µPa^{2}s. No behavioral response was observed for either animal when kept behind a bubble curtain (single-pulse SEL 143–157 dB re 1 µPa^{2}s), which reduced noise levels above 250 Hz by 20–40 dB. The lack of response in the presence of a bubble curtain suggests that the captive harbor porpoises responded to the higher-frequency content of the airgun signal (Kastelein et al. 2019b). A wild harbor porpoise equipped with an acoustic tag responded to lower RLs produced by a 10 in^3 single airgun (single-pulse SEL ~141 dB re 1 µPa^{2}s, SPL ~171 dB re 1 µPa pk-pk), with an increase in swimming speed and distance to the source while decreasing its dive depth, dive duration, and time at the surface (van Beest et al. 2018).

Marine seismic surveys for hydrocarbon exploration are conducted with large airgun arrays, which have been shown to elicit harbor porpoise responses over areas where single-pulse SELs exceeded ~130 dB re 1 µPa^2 s (Pirotta et al. 2014; Sarnocińska et al. 2020; Thompson et al. 2013a). Sarnocińska et al. found that click-detection rates decreased within 8–12 km from an active 3D seismic survey (single-pulse SEL 132–157 dB re 1 µPa^{2}s). Thompson et al. (2013a) similarly recorded a decrease in harbor porpoise click-train detections up to 10 km from an active 2D seismic survey. Aerial surveys confirmed that the decrease in acoustic detections resembled a decrease in harbor porpoise density, translating to avoidance of received single-pulse SELs of 145–162 dB re 1 µPa^{2}s (165–172 dB re 1 µPa pk-pk). A second study on the same 2D seismic survey reported a decrease in the detection of high-speed click-trains (i.e., buzzes) up to 35 km (single-pulse SEL 130–165 dB re 1 µPa^{2}s; Pirotta et al. 2014). Buzzes are presumed to indicate foraging behavior (Wisniewska et al. 2018), and so while animals did not avoid the area >10 km from the source, results suggested that foraging behavior was reduced up to 35 km (i.e., minimum single-pulse SEL 130 dB re 1 µPa^{2}s; Pirotta et al. 2014).

10.3.6.3 Pile Driving

The potential effects of pile driving noise associated with offshore wind farm construction have been of particular concern with regard to harbor porpoises. Percussive pile driving can generate high-intensity sounds that are, depending on background noise levels and sound propagation conditions, audible to porpoises over tens of kilometers in range (Kastelein et al. 2013a). Europe has been a frontier in developing offshore wind farms in areas that overlap with harbor porpoise habitat, leading to research in captivity and the wild that has shown adverse behavioral responses to various levels of pile driving noise.

Captive animals have shown a startle response to broadband single-strike SEL of 65–107 dB re 1 µPa^{2}s, with a 50% response rate at single-strike SEL of 92 dB re 1 µPa^{2}s (Kastelein et al. 2013b). Increased swimming speeds, respiration rates, and leaps out of the water have been observed from single-strike SEL of 121, 127, and 145 dB re 1 µPa^{2}s, respectively (Kastelein et al. 2013c, 2015a, 2018b). For one individual, feeding success decreased at received single-strike SEL > 134 dB re 1 µPa^{2}s (Kastelein et al. 2019d). However, the behavioral response at a particular single-strike SEL depended on the frequency content of the acoustic signal, which is better captured by frequency-weighted sound levels (Kastelein et al. 2022). In addition, habituation was observed in respiration rate changes within 20 min of exposure to single-strike SEL of 127–145 dB re 1 µPa^{2}s (Kastelein et al. 2013c), while effects of similar single-strike SEL on behavioral changes and foraging efficiency showed strong inter-individual differences (Kastelein et al. 2019d). These results highlight the limitations of captive studies that are mostly conducted with only one or two animals, making it difficult to infer a population response. In addition, exposure duration in captive and wild CEEs is short in comparison with the time it can take to construct a wind farm; so, these fail to describe any long-term effects.

Wild harbor porpoises avoided underwater speakers up to 200 m during playback of pile driving noise, which corresponded with RL ~140 dB re 1 µPa pk-pk (Tougaard et al. 2012). Haelters et al. (2015) subsequently used the same pk-pk SPL as the assumed level of discomfort to model the area of avoidance around pile driving events off Belgium. Model results indicated an area of avoidance with a 21 km radius, which corresponded to observations made during aerial surveys in the same area (Haelters et al. 2015) and in the German North Sea (Dähne et al. 2013). Using an array of automated click detectors on two Scottish offshore wind farms before and during construction, the probability of detecting porpoises and buzzing activity increased with distance from pile driving activities and decreased with levels of vessel activity and background noise (Benhemma-Le Gall et al. 2021). The duration of displacement appeared positively correlated with the duration of piling (of one pile: Dähne et al. 2013) and was observed at ranges of up to 12 km for piling activities (Benhemma-Le Gall et al. 2021). PAM studies similarly reported decreased acoustic detection rates up to ~18 km (Brandt et al. 2011), with a 50% decrease in acoustic detections at 7.4 km (single-strike SEL 144 dB re 1 µPa^{2}s; Graham et al. 2019). Acoustic detections remained low up to 24 h after pile driving had ceased (Brandt et al. 2011). However, overall, the severity of response decreased with distance to the pile driving site and time since the start of construction (Brandt et al. 2011, 2012; Graham et al. 2019). In China, acoustic detection rates of finless porpoises decreased during pile driving, indicating a reduced use of habitat (Fang et al. 2023).

While the above studies demonstrate that percussive pile driving has the potential to affect the behavior of harbor porpoises, piling is often not the only source of noise at marine construction sites. Graham et al. (2019), for example, found that ADDs used during piling events and the number of vessels within 1 km from the acoustic receiver significantly affected harbor porpoise acoustic detection rates in addition to pile driving alone; yet, studies generally do not account for these additional noise sources and contextual covariates. In a recent PAM-based study, Benhemma-Le Gall et al. (2023) reported up to 33% decline in detected clicks prior to piling, associating the displacement with increased levels of vessel-based preparation activity and broadband noise.

10.3.6.4 Sonar

The use of low-, mid-, and high-frequency sonar during naval activities is of concern in harbor porpoise habitat (Kastelein et al. 2018a). The potential effects of low- and mid-frequency sonar sweeps have been studied in captive porpoises (Elmegaard et al. 2021; Kastelein et al. 2012b, 2013b, 2014a, b, 2019a, e). Generally, individuals were startled, avoided areas near the transducer, increased swimming speed and respiration rate, and were more likely to leap out of the water. Kastelein et al. (2012b) found that 50% startle response threshold levels were lower for low-frequency sonar up- and downsweeps (i.e., 1–2 kHz) with harmonics (99 dB re 1 µPa) than for the same signal without harmonics (133 dB re 1 µPa) and lower for mid-frequency sweeps (i.e., 6–7 kHz: 101 dB re 1 µPa) than for low-frequency sweeps (133 dB re 1 µPa). Responses did not differ between up- and downsweeps when presented at low duty cycles (i.e., 82–106 dB re 1 µPa, ~1% duty cycle; Kastelein et al. 2012b), but a stronger response was observed to upsweeps when signals were presented at higher RLs and duty cycles (i.e., 123 dB re 1 µPa, ~19% duty cycle; Kastelein et al. 2014b).

Responses to tones, signals that combine tonal and sweep components, and high-frequency sonar have been less studied; yet similar patterns have been observed. For example, Kastelein et al. (2013b) found that the 50% startle response threshold was lower for a 1380 Hz tone with harmonics (124 dB re 1 µPa) than without harmonics (131 dB re 1 µPa). In addition, one harbor porpoise increased his respiration rate in response to a 25 kHz tone with harmonics at RLs of 77 dB re 1 µPa, which was only observed at 131 dB re 1 µPa when exposed to the same signal without harmonics (Kastelein et al. 2015d). In both signals, higher RLs resulted in a more severe response, including leaping out of the water (~118 dB re 1 µPa) and an increase in swimming speed (131–153 dB re 1 µPa; Kastelein et al. 2015d). Exposure of two harbor porpoises to AN/SQS-53C sonar playbacks (i.e., both tone and sweep components) highlighted the importance of duty cycle settings; while no response was observed in response to playbacks at a 2.7% duty cycle (119–143 dB re 1 µPa), an increased respiration rate was observed under a 96% duty cycle at RLs of 119–131 dB re 1 µPa (Kastelein et al. 2018a). Additional changes in respiration rate and leaping behavior occurred at RLs of 143 dB re 1 µPa (Kastelein et al. 2018a).

These results indicate that harbor porpoises are sensitive to disturbance from sonar signals at all frequencies, with the severity of response depending on the signal frequency spectrum and duty cycle. However, there is a paucity of knowledge on how these captive study results translate to open-ocean settings.

10.3.6.5 Acoustic Deterrent Devices

A significant threat to harbor porpoises is incidental bycatch in fishing gear, with the highest mean annual number of 5591 individuals recorded in the Danish North Sea in 1987 and 2001 (Vinther and Larsen 2004). As a result, numerous studies have assessed the efficacy of ADDs to deter this species from fishing nets. Deterrent signals differ in their fundamental frequency, signal type (i.e., pulsed, sweeps, complex, as well as with or without harmonics), and SL, leading to variable behavioral responses and response thresholds.

Kastelein et al. (1995) exposed two captive harbor porpoises to either of two acoustic alarms: a single-frequency 2.5 kHz tone and the same tone with a strong harmonic overtone at 17.5 kHz. The second signal elicited increased swimming speed and schooling behavior, while individuals were rather inquisitive toward the 2.5 kHz pure tone despite the latter being emitted at a slightly higher SL (119 versus 115 dB re 1 µPa). Harbor porpoise optimal hearing falls between 4 and 140 kHz, likely explaining the adverse effect of the second alarm, although the first alarm should have been audible at the SL used (Kastelein et al. 1995). Wild harbor porpoises generally avoided ADDs emitting signals with fundamental frequencies of ~10–11 kHz over areas where RLs exceed 123 dB re 1 µPa (Cox et al. 2001; Johnston 2002). Similar avoidance responses were observed for seal scarers with a fundamental frequency of ~14 kHz, albeit at lower RL (113 dB re 1 µPa; Brandt et al. 2013a, b). Harbor porpoises showed a significant avoidance response to an ADD at 10.5 kHz peak frequency (5.5–20 kHz range), leaving the area within 1 km of the sound source where the estimated RL was 131–155 dB re 1 µPa (Hiley et al. 2021). Studies on wild harbor porpoises suggest that ADD signals can elicit a physical response over distances of 12 km (Dähne et al. 2017). There is, however, a lack of research on how this distance differs between, for example, locations with different propagation loss properties and ambient noise levels (Brandt et al. 2013a, b). In addition, very little research has looked at potential habituation effects. Yet, the probability of harbor porpoises around Grand Manan Island (Canada) approaching a pinger within audible range increased again after 10–11 days of exposure, suggesting some habituation (Cox et al. 2001).

Captive studies on the deterrent potential of very high-frequency signals found them all aversive to harbor porpoises, but signal composition and duty cycle affected tolerance threshold levels. Porpoises avoided a 50 kHz continuous tone at RL > 107 dB re 1 µPa (Kastelein et al. 2008a) but tolerated higher levels of a continuous 120 kHz signal (120 dB 1 µPa) and 70 kHz signal (130 dB re 1 µPa; Kastelein et al. 2008b). Tolerance threshold values for 70 kHz tonal signals decreased to 117 dB re 1 µPa when using an 8% or 25% duty cycle (Kastelein et al. 2008b). Higher sensitivities were reported for exposure to a collection of predominantly upsweeps with variable fundamental frequencies between 60 and 150 kHz played back at a 65% duty cycle (mean RL 104–110 dB re 1 µPa; Kastelein et al. 2017c).

10.3.6.6 Other Sounds

A few captive studies have focused on other signals. Kastelein et al. (2005) exposed two harbor porpoises to four signals (centered at ~12 kHz) from the Acoustic Communication network for Monitoring the underwater Environment (ACME). Avoidance and increased respiration rates were observed in response to all signals, but discomfort thresholds varied from 99 dB re 1 µPa for chirps and sweeps to 112 dB re 1 µPa for frequency-modulated shift-key signals. Avoidance of white noise and brown noise has also been reported (148 dB re 1 μPa and 83 dB re 1 μPa, respectively; Kastelein et al. 2012a; Kok et al. 2017).

10.3.6.7 Long-Term and Population-Level Effects

The consequences of seismic surveys on the population viability of harbor porpoises were investigated using an energy budget model to emphasize the direct link between disturbance and vital rates (Gallagher et al. 2021). Results from this study highlight the importance of the spatiotemporal context of the disturbance as the greatest effects of seismic surveying were predicted to occur in late summer and fall, corresponding to higher energy requirements due to water temperature, lactation cost, and relative body fat storage. Population-consequences models of harbor porpoise disturbance from wind farm construction (King et al. 2015) and from naval exercises (Wilson et al. 2020) have been constructed.

10.3.7 Beluga and Narwhal

10.3.7.1 Icebreakers, Ships, and Boats

Research on the potential effects of underwater noise on belugas and narwhals was driven by concerns over increasing industrial presence in the Arctic, specifically increased icebreaker traffic and growing oil and gas development. Approached by icebreakers, beluga whales moved rapidly, dove in asynchrony, changed their vocalization types, and lost their pod integrity at RLs of 94–105 dB re 1 μPa (20–1000 Hz) (Cosens and Dueck 1988; Finley et al. 1990; LGL Ltd. 1986). Playback of icebreaker noise elicited avoidance behavior at RLs of 78–84 dB re 1 μPa and 78–84 dB SNR at 5 kHz (Richardson et al. 1995c). In a 6-month study of tagged beluga whales in the Beaufort, Chukchi, and Bering Seas, beluga swim speed was negatively correlated with ship distance when ships were up to 79 km away (Martin et al. 2023c). Changes in lateral and vertical movements were observed when the whales were within 50 km of ships, indicating disruption of behavior and avoidance responses at RLs of 98–133 dB re 1 μPa.

Narwhals traveled more slowly along straighter paths or became motionless and sank and stopped vocalizing at RLs of 124 dB re 1 μPa (20–1000 Hz) (Cosens and Dueck 1988; Finley et al. 1990; LGL Ltd. 1986). Behavioral responses can become fatal, as in the case of delayed migration out of rapidly freezing regions. Offshore seismic surveys were hypothesized to have caused the entrapments and deaths of over 1000 narwhals (Heide-Jørgensen et al. 2013).

Outside of the Arctic, in the St. Lawrence Estuary, the effects of ship noise (i.e., commercial ships and whale-watching boats) were investigated, documenting the Lombard effect in beluga whales (Scheifele et al. 2005). A linear correlation was found, whereby the level of beluga whale vocalizations increased by 0.9 dB for every 1 dB increase in vessel noise, over a range of 60–120 dB re 1 μPa in vocalization level.

10.3.7.2 Seismic Surveys

Recent work has explored the effects of airguns and airgun arrays on narwhal physical and acoustic behaviors. A group of 16 tagged narwhals was observed moving away from vessels and remaining closer to shore when a single airgun was firing, compared to baseline measurements (Heide-Jørgensen et al. 2021). Narwhal swimming speed and direction changed up to 2 h before the vessels were in their line-of-sight, when SELs were below ambient noise levels. Buzzing rates were halved when the narwhals were 12 km from the ship (SEL < 134 dB re 1 μPa^{2}s or HF-weighted SEL of 107 dB re 1 μPa^{2}s) and foraging clicks stopped entirely 7–8 km from the ship, despite an SEL less than 135 dB re 1 μPa^{2}s (Tervo et al. 2021).

10.3.7.3 Drilling

Playback of drilling noise resulted in temporary behavioral changes (i.e., slowing, reversing swim direction, and milling); however, belugas continued their route past the noise projector within 15 min. RLs at the locations of behavioral change were 112–134 dB re 1 μPa (20–1000 Hz)

(Richardson et al. 1990, 1991). In contrast, captive belugas showed no statistically significant change in behavior (or stress hormones) during 30-min playbacks of drilling noise at RLs of 134–153 dB re 1 µPa (40–20,000 Hz) (Thomas et al. 1990).

10.3.7.4 Aircraft

As offshore drilling operations are typically accompanied by service flights, beluga whale reactions to helicopter and fixed-wing aircraft overflight (as well as takeoff and landing nearby) were also studied in the Arctic. Abrupt behavioral changes, avoidance, and alteration of surface and respiration patterns were noted at RLs of 95–125 dB re 1 µPa (10–500 Hz) (Patenaude et al. 2002). Evasive responses (sudden diving) to airborne drones were observed in beluga whales (Aubin et al. 2023).

10.3.7.5 Predator Sounds

Beluga whale responses to playbacks of killer whale vocalizations were examined as a means of repelling them from rivers important to fisheries management. The instantaneous effects were dramatic. Sound was played as beluga whales approached and they turned immediately, leaving the river against strong current. Their blow rate increased, as did time spent at the surface. Some beluga whales swam to the opposite bank. There was no behavioral change during control trials. Beluga whales also vocalized less when killer whale sounds were played. At the onset of behavioral response, RLs were 110–132 dB re 1 µPa (500–5000 Hz) and 25–47 dB SNR (Fish and Vania 1971). These strong antipredator responses are consistent with behavioral responses and displacement reported for narwhals in the presence of actual killer whales (from concurrent tracking; Breed et al. 2017).

10.3.7.6 Explosions

Noteworthy is an observation by Finneran et al. (2000), who had trained a captive beluga whale for behavioral hearing tests before and after exposure to playbacks of noise from distant underwater explosions. While the goal of the study was to determine a potential threshold shift in hearing, the beluga whale diverted from its trained behavior at RLs of 220–221 dB re 1 µPa pk-pk or 177–179 dB re 1 µPa^{2}s SEL (900–20,000 Hz), equivalent to a 500 kg charge at 1.9 km range.

10.3.7.7 Long-Term and Population-Level Effects

Williams et al. (2017) modeled population consequences to beluga whales when exposed to multiple stressors including noise—as a potential influence on prey detection and, in turn, demographic parameters (i.e., calf mortality). While the study reported a decline in population mean size when exposed to a wide range of acoustic disturbances, the effects of noise-compromised prey detection were not as extensive as scarcity of prey availability itself. More recent modeling showed synergistic effects of reduced prey availability and anthropogenic disturbance on beluga whales (McHuron et al. 2023).

10.3.8 Sperm Whales

Sperm whale responses to noise have been measured predominantly using passive acoustic tracking, not surprising given the high amplitude of their click vocalizations and prolonged dive times.

10.3.8.1 Ships and Boats

Sperm whale responses to ship noise are variable. Azzara et al. (2013) counted sperm whale clicks in 36 days of acoustic recordings from the Gulf of Mexico and found fewer clicks during vessel passes, but could not distinguish nonvocal periods from avoidance. Conversely, André et al. (2017) found no decrease in sperm whale acoustic detections in the presence of ship noise in a 2-year PAM dataset from the Ligurian Sea, a sperm whale foraging area within the Mediterranean Sea. In the presence of whale-watching vessels, sperm whales decreased surface time, respiration interval, and blows per surfacing (Gordon et al. 1992). In other contexts, sperm whales have been reported to actively approach vessels. In a population that commonly depredates sablefish longlining vessels, an increase in sperm whale acoustic and visual detections near longline vessels was documented[1]; the pattern of vessel speed changes when retrieving longlines creates a distinctive pattern of propeller cavitation noise, likely acting as the "dinner bell" attracting sperm whales to depredate (Thode et al. 2007). This hypothesis was supported when the Southeast Alaska Sperm Whale Avoidance Project (SEASWAP) demonstrated that an acoustic decoy broadcasting vessel longline hauling noises successfully led sperm whales to swap sources, moving away from true fishing vessels at ranges >9 km (Wild et al. 2017).

10.3.8.2 Seismic Surveys

The effects of pulsed sounds on sperm whale behavior are equally variable. CEEs with eight tagged whales found no evidence for horizontal avoidance of seismic airguns with RLs of 111–147 dB re 1 µPa at ranges of 1.4–12.6 km (Miller et al. 2009). None of the eight whales changed behavioral state (seven were foraging; one was resting) during ramp-up and full exposures. A statistically insignificant 20% decrease in buzz rate might indicate decreased foraging success in some animals. No evidence for changes in vertical movement was found.

The available time-area studies of sperm whale distribution during seismic surveys have reported heterogeneous responses, but they are limited and did not report RL. An abstract of a time-area study in the Gulf of Mexico found evidence for a change in distribution during exposure to seismic surveys. The whales were monitored visually from a survey vessel and disappeared during the surveys except at the periphery of the study area (Mate et al. 1994). The observation was opportunistic (the seismic survey was not part of a planned experiment) and not substantiated with more lengthy analysis or further research. Weir (2008a) compared marine mammal sightings from an active seismic vessel off Angola during periods when the airgun array was firing (51% of 2769 h of effort) versus not firing. The survey lasted 10 months. Mean distance to sperm whale sightings was greater when the array was active, but not significantly different, and the number of sightings was significantly higher.

10.3.8.3 Explosions

Small detonations of 1 g TNT at 3–15 m depth for the purpose of calibrating hydrophone positions in a large-aperture array did not change click rates and surfacing behavior of five male sperm whales at RLs estimated at 173–179 dB re 1 µPa (pulse-equivalent rms; Madsen and Møhl 2000). Inter-pulse intervals and amplitudes of clicks before and after discharge were also unchanged.

10.3.8.4 Sonar

Exposure to sonar might affect functional behavior (i.e., foraging). Stanistreet et al. (2022) reported a decrease in the number of hours per day with sperm whale echolocation clicks exposed to active sonar with a median and maximum RL of 120 dB and 164 dB re 1 µPa rms (200 ms window), respectively. DTAG-instrumented sperm whales exposed to LFAS transmissions (1–2 kHz, SL 214 dB re 1 µPa m) from an approaching vessel changed from foraging to non-foraging, non-resting behaviors at RLs of 131–165 dB re 1 µPa rms (Isojunno et al. 2016). They did not respond to vessel approach without active sonar, a broadband noise control stimulus, or MFAS exposures (6–7 kHz, SL 199 dB re 1 µPa m) with RLs of 73–158 dB re 1 µPa. Little change in foraging behavior was observed during incidental 4.7–5.1 kHz sonar with RLs of 89–133 dB re 1 µPa rms (Isojunno et al. 2016). Another study compared the effects of continuous active sonar (CAS) and pulsed active sonar (PAS) on tagged sperm whales that were tracked both visually and acoustically (Isojunno et al. 2020). No impact on foraging was observed during medium-level PAS matching the low-amplitude CAS exposures. Foraging was reduced to a similar degree during CAS and higher-amplitude PAS exposures, which matched in source level energy (14 individuals switched from foraging to non-foraging, non-resting at an individual-average, maximum

---

[1] https://www.youtube.com/watch?v=5panVo_3oVQ&t=9s accessed on 13 April 2023

SEL of 143 dB re 1 µPa²s). Response thresholds agreed across the two types of sonar if compared on the basis of sound energy, rather than peak pressure (Isojunno et al. 2020). Masking of echolocation clicks was modeled at high levels of 160 dB re 1 µPa²s for PAS and 173 dB re 1 µPa²s for CAS (Isojunno et al. 2022; von Benda-Beckmann et al. 2021).

Most of these observational data on sperm whale behavior in the presence of sonar were pooled and scored in terms of severity from 0 (no response), 1 (brief orientation), 2 (brief change in dive profile), 3 (brief shift in group distribution), 4 (brief cessation of feeding or resting), up to 7 (prolonged cessation of feeding, prolonged avoidance) (see Curé et al. 2021 for more detailed descriptions). Responses of severity 4 or greater (i.e., potentially impacting vital rates) commenced at cumulative SEL of 137–177 dB re 1 µPa²s for CAS and 143–181 dB re 1 µPa²s for PAS. The probability of response increased in sperm whales previously exposed to predatory killer whale or competing long-finned pilot whale sounds (Curé et al. 2021).

10.3.8.5 Narrowband Sounds

Sperm whales responded to acoustic deterrent device (pinger) tone pips in the 6–13 kHz range by interrupting click production for periods of minutes (Watkins and Schevill 1975). The authors did not report RL, but the SLs of the pingers were low (110–130 dB re 1 µPa m) and exposure ranges were short. Others reported a decrease in click production in response to simulated sperm whale codas, while 10 kHz pings produced startle responses in whales at the surface (André et al. 1997). Bowles et al. (1994) reported silence as a sperm whale response to an intense low-frequency sound, the M-sequence transmissions of the Heard Island Feasibility Test (HIFT). Systematic vessel-based visual surveys were conducted before, during, and after HIFT transmissions from a large accompanying vessel, and sonobuoys were deployed concurrently along the vessel track line. Sperm whales were not sighted frequently at the surface, but clicks were frequently detected in acoustic data during the period before transmissions and beginning a day after the end of the transmission experiment. However, the whales were no longer acoustically detected during the transmissions. No RLs could be estimated for these whales, and it was unclear whether they left the area during transmissions or simply ceased clicking.

10.3.8.6 Aircraft

Some studies attempted to qualify behavioral responses of sperm whales to aircraft and UAVs and recorded a variety of reactions from mild changes in respiration rate to abrupt diving, horizontal avoidance, and grouping behavior (Smultea et al. 2008). Similar findings, including sudden diving and delayed clicking after diving, were reported elsewhere (Richter et al. 2006; Würsig et al. 1998). An indication of habituation is the observation that resident animals reacted less strongly than transients. However, animal sensitivity to aircraft seems to be variable and dependent on the exposure context (including behavioral state at the time of exposure and altitude and lateral distance of the aircraft; Luksenburg and Parsons 2009). So far, studies have been unable to determine on-animal RL from aircraft noise.

10.3.8.7 Predator Sounds

Five tagged male sperm whales demonstrated horizontal avoidance, cessation of foraging, interruption of resting dives and returning to the surface, and grouping behaviors during playback of mammal-eating killer whale calls (Curé et al. 2016; Isojunno et al. 2016), similar to their responses to naval sonar. Killer whales are predators of sperm whales (Pitman et al. 2001). Curé et al. (2013) highlighted that sperm whales would not "simply" out-dive their predators. Also, grouping behavior is otherwise rare in these solitary male individuals. Therefore, predator vocalizations disrupted functional behaviors of sperm whales and provoked previously unrecognized antipredator responses. The authors suggested looking for antipredator type responses in studies on the effects of noise, to assess perceived level of threat and disturbance.

10.3.8.8 Long-Term and Population-Level Effects

More recently, the focus of impact studies has shifted from stimulus-response toward the biological significance of noise disturbance using severity scoring, contextual variables, and statistical, spatiotemporal modeling (e.g., Ellison et al. 2012; Isojunno et al. 2022; Southall et al. 2021). Farmer and colleagues modeled the resilience and population viability of sperm whales in the Gulf of Mexico using a bioenergetic approach and found that individual resilience to foraging disruptions was mostly dependent on animal size (and thus energetic reserves available) and energetic needs (i.e., reproductive status; Farmer et al. 2018a, b). Recent simulation work with sperm whales showed that if animal responses depend on their body condition (as might be expected based on foraging-risk trade-offs), then these can further exaggerate or dampen consequences to individuals and populations (Burslem et al. 2022). The simulation paper also showed that results are affected by assumptions made about the bioavailability of different fat types. Population modeling of sperm whales in the Gulf of Mexico under two stressors (oil exposure from the Deepwater Horizon oil spill and seismic surveys) predicted a stock decline of 26% by 2025 as a result of oil exposure and a population proportion of 4.4% reaching terminal starvation by 2025 as a result of noise exposure. Additionally, 11% of unsuccessful gestations were predicted. Behavioral disturbance was modeled as a step function at 160 dB re 1 µPa (other dose-response relationships were also explored), resulting in reduced foraging success. Sound exposure estimation involved individual agent-based modeling. Bioenergetic models tracked the impacts of lost foraging opportunities through the population (Farmer et al. 2018a).

The diverse responses of sperm whales to noise (from no response to changes in behavior, avoidance, and silence) suggest that whales do not react solely on the basis of exposure level but highlight the importance of context (e.g., noise type, behavioral state, age and sex, environmental parameters, resource availability, etc.).

10.3.9 *Kogia* spp.

Kogiidae (pygmy and dwarf sperm whales) are deep-diving odontocetes with widespread presence in temperate and tropical ocean basins. However, little information is available on these elusive species (Malinka et al. 2021; Staudinger et al. 2014). Pygmy and dwarf sperm whales are almost indiscernible at sea, which further complicates our ability to study species-specific traits and behaviors. Moreover, most of the data available for both species are derived from stranded animals, which may or may not be representative of the species at large. There are no specific studies in the literature that characterize and quantify *Kogia* spp. responses to sound, although strong avoidance of survey ships and airplanes has been documented (Würsig et al. 1998). Live stranding of one individual dwarf sperm whale occurred shortly after the start of a US Navy sonar exercise in Hawaii (Baird 2016). Overlap of *Kogia* spp. with areas of known acoustic disturbance (including sonar exposure), coupled with reported strandings of these species (McCullough et al. 2021; McGeady et al. 2016), are a source of concern about the potential impact of sonar and other sources (Hohn et al. 2006; Parsons 2017).

10.3.10 Beaked Whales

Even though beaked whales are often referred to as "cryptic" because they live away from the coasts, in the deep ocean, their responses to anthropogenic noise have been the subject of a significant amount of research, mostly since a stranding incident in the Bahamas (Balcomb and Claridge 2001) in 2000 raised concerns over their potentially lethal behavioral responses to noise.

10.3.10.1 Ships and Boats

Beaked whale response to ship noise is not consistent across studies. Aguilar Soto et al. (2006) recorded shortened foraging dives and fewer successful prey captures by a DTAG-instrumented Cuvier's beaked whale exposed to motorized vessels at a maximum RL of 136 dB re 1 µPa (356 Hz–44.8 kHz). However, the authors cautioned against generalization given the small sample size. Pirotta et al. (2012) found that foraging duration of Blainville's beaked whale was not affected by vessel noise, even when controlling for distance from the source, but an RL of 135 dB re 1 µPa reduced the horizontal area in which animals foraged, potentially leading to a decrease in foraging efficiency.

10.3.10.2 Sonar

Most of the literature on beaked whale responses to noise deals with ship-based sonar transmissions (e.g., DeRuiter et al. 2013b; Goldbogen et al. 2013; Kvadsheim et al. 2017; Sivle et al. 2015; Stanistreet et al. 2022). The types of behavioral responses observed were similar across species, including foraging disruption, changes in dive-surface cycles, and temporary displacement, consistent with previously observed antipredator defensive strategies.

There are several examples in the literature on the likely disruption of foraging during sonar exercises. A bottlenose whale exposed to 1–2 kHz sonar in Jan Mayen, Norway, ceased foraging and left the location at an RL of 130 dB re 1 µPa rms (Sivle et al. 2015). In detail, this whale made regular foraging dives to 2000 m in the 5 h prior to exposure, with resting dives to 100–200 m. Feeding activity was detected as clicks and buzzes indicating prey capture. After the start of transmissions, at an RL of 98 dB re 1 µPa rms, the whale approached the oncoming source, but above 107 dB re 1 µPa, it turned away from the source, began straight-line swimming, and began diving deeply. It made an unusually deep dive initially (2340 m) and then shallow dives to about 250 m throughout the rest of the tagging period. This tagged whale remained out of the area and silent for the 7 h that its tag functioned. The same study reported avoidance and extended silent periods of another 12 tagged bottlenose whales exposed to sonar activities with RLs ranging from 107 to 141 dB re 1 µPa rms (Miller et al. 2015). Sixteen satellite-tagged Cuvier's beaked whales documented longer dives and longer surface intervals during MFAS exposure, hinting at a disruption of foraging (Falcone et al. 2017). A multiyear study recorded Blainville's beaked whales using a bottom-mounted array at the US Navy's Atlantic Undersea Test and Evaluation Center (AUTEC) in the Bahamas during multiship MFAS operations (McCarthy et al. 2011). Clicking declined during sonar exercises and increased after the transmissions ceased, but was not restored to pre-exposure levels (in 2007) or remained low for a protracted period (35 h in 2008). The whales continued foraging at RLs up to 157 dB re 1 µPa rms, averaging 128 dB re 1 µPa rms. Another publication on Blainville's beaked whales at AUTEC reported continued foraging during 3–8 kHz MFAS transmissions but a reduction in time spent on deep dives (Joyce et al. 2020). An empirical risk function predicted a 50% probability that clicking during foraging dives ceased at an RL of 150 dB re 1 µPa rms (95% confidence interval [CI] 144–155 dB re 1 µPa rms; Moretti et al. 2014). Even more significant results were reported in a recent study at the US Navy Pacific Missile Range Facility (PMRF) in Hawaii, where the authors found a 77% decrease in detection probability of Blainville's beaked whales exposed to MFAS at an RL of 150 dB re 1 µPa rms (Jacobson et al. 2022). Based on passive acoustic recordings alone, it is not possible to distinguish between silence and avoidance, although both imply a reduction or cessation of foraging in the area.

Actual vertical or horizontal avoidance has been demonstrated in a few tracking studies. A tagged Baird's beaked whale exposed to simulated MFAS (3.5–4 kHz) for 30 min (8 min ramp-up, 22 min full) doubled its swim speed within 3 min of transmission onset, as indicated by increased fluke rate. It also altered its dive pattern. The response persisted until 1.6 min after the end of exposure. RL closest to the onset of response was 100 dB re 1 µPa rms. Since the tagged animal was traveling in a group of seven, individual vocal pattern could not be distinguished from the DTAG data, but there was evidence that acoustic foraging ceased during exposure (Stimpert et al. 2014). A satellite-tagged, male Blainville's beaked whale was tracked before, during, and after a 5-day sonar exercise (Tyack et al. 2011). During the exercise, the whale was temporarily displaced from the center of the AUTEC range (up to 54 km). The highest RL was ~146 dB re 1 µPa in the highest one-third-octave band (3111–3920 Hz). The whale moved back within the range over a period of ~3 days following the cessation of the exercise, further supporting the hypothesis that exposed whales left the range rather than merely ceased to forage as recorded in previous studies. Moreover, five of seven satellite-tagged Blainville's beaked whales moved 28–68 km away from AUTEC during 3–8 kHz sonar exercises and returned 2–4 days after sonar transmissions ceased, reducing their RL from 145–172 dB re 1 µPa to 70–150 dB re 1 µPa (Joyce et al. 2020).

Shipboard echosounders have had a similar effect on beaked whales as military sonar. A set of singlebeam echosounders operating simultaneously at multiple frequencies between 18 and 300 kHz reduced beaked whale acoustic detections to 3% (Cholewiak et al. 2017), but the study was unable to differentiate between species and between displacement and cessation of foraging. Finally, a negative correlation between Cuvier's beaked whale detection rate and shipboard echosounders and ultrasonic antifouling devices (transmitting pulses with energy in between 19 and 166 kHz at short ranges) was reported (Trickey et al. 2022).

More and more studies model a suite of predictor variables for behavioral responses and are showing that RL is not necessarily the only or strongest predictor. Studies on satellite-tagged Cuvier's beaked whales reported foraging interruption (rapid, silent avoidance; extended dive duration and surface interval) by two individuals during simulated MFAS exposure at short range with RLs of 89–127 dB re 1 µPa rms (DeRuiter et al. 2013b; Southall et al. 2016); however, a lack of response of one individual to an incidental, long-range sonar exposure from over 118 km at an RL of 78–106 dB re 1 µPa rms was also noted (DeRuiter et al. 2013b). Falcone et al. (2017) studied the effect of MFAS on dive and surfacing patterns of Cuvier's beaked whales and reported distance-mediated effects with longer dive duration and longer surface intervals at closer distance (and thus likely higher RL). However, individuals exposed to mid-power MFAS deployed by helicopters (4.1 kHz; SL = 217 dB re 1 µPa m) showed more pronounced behavioral responses than individuals exposed to high-power MFAS from a surface-based ship (3.5 kHz; SL = 235 dB re 1 µPa m) at similar distance from the source (~50 km; Falcone et al. 2017), hinting at the importance of proximity or other contextual factors. Conversely, in a CEE conducted on 12 northern bottlenose whales in an area with few human activities, the onset and intensity of the behavioral response were independent of the distance of the noise source, with an avoidance threshold ranging from 117 to 126 dB re 1 µPa rms (Wensveen et al. 2019).

The importance of context, source type, individual variability, and inter-annual variability was highlighted in some studies. A DTAG-instrumented female Blainville's beaked whale responded to simulated MFAS by ceasing foraging at lower RL (138 dB re 1 µPa rms) compared to a male individual exposed to pseudorandom noise with timing and bandwidth similar to MFAS (RL 142 dB re 1 µPa rms; Tyack et al. 2011). Passive acoustic recordings of Blainville's beaked whales at the Pacific Missile Range Facility in Hawaii were compared over 3 years; while more group vocal periods were detected before exercises as opposed to during or after, great interannual differences were discovered (Manzano-Roth et al. 2016). Varghese et al. (2020) found no indication of reduced group vocal periods nor an abandonment of the area by Cuvier's beaked whales during a 12 kHz multibeam echosounder survey. While the study did not report RL, the authors hypothesized that the exposure context (including number of vessels, duration, and predictable transects of the survey) might explain the different responses of beaked whales to multibeam echosounders versus MFAS.

10.3.10.3 Aircraft

During an aerial cetacean survey in the Gulf of Mexico with a DeHavilland twin otter aircraft flying at 229 m altitude, beaked whales changed their behavior in response to the survey airplane during 89% of their sightings (Würsig et al. 1998).

10.3.10.4 Predator Sounds

Blainville's beaked whales reacted to the playback of killer whale calls at an RL of 98 dB re 1 µPa, marginally above the background noise (Tyack et al. 2011). The responses included a cessation of foraging (indicated by the number of buzzes), shorter foraging duration, changed rate of ascent, and changed dive interval. Similar responses were reported in a playback experiment in which a DTAGged Blainville's beaked whale ceased foraging when exposed to recorded killer whale calls at an RL of 138 dB re 1 µPa (Allen et al. 2014); a single northern bottlenose whale also ceased foraging completely during killer whale playbacks (Miller et al. 2022). Beaked whales are preyed upon by killer whales (Wellard et al. 2016), and so, this type of predator avoidance likely represents an evolutionary adaptation to predation risk. It has been suggested that predator avoidance might also be involved in sonar-induced mass strandings (Aguilar de Soto et al. 2020).

10.3.10.5 Long-Term and Population-Level Effects

Bioenergetic modeling approaches have been used to evaluate the consequences of disturbance for beaked whales. These models linked the estimated energy loss due to lost foraging opportunities and additional energy exertion during noise avoidance to their effects on life history traits (Czapanskiy et al. 2021; New et al. 2013b). Results showed that prey quality and/or availability affect inter-calving intervals and that the displacement of whales during and after noise exposure (which can, in turn, affect prey quality and availability) can significantly impact the reproductive parameters and life history traits of beaked whales (New et al. 2013b). Recent observations of the nursing period of beaked whales being up to three to four times longer than previously thought raise concern about low resilience of beaked whales to anthropogenic disturbance (Feyrer et al. 2020). However, Czapanskiy et al. (2021) compared the energetic cost of disturbance response relative to baseline metabolic rates across cetacean species and found that even extreme reactions displayed by deep divers with slow life history traits (including sperm whales, beaked whales, and pilot whales) would pose a lesser threat to life history outcomes than a mild response by large baleen species. However, as the authors point out, their analysis focused on relative energetic costs and did not account for inter-specific differences in ability to compensate for lost feeding opportunities. Interestingly, northern bottlenose whales in lower blubber story body condition, expected to have a greater motivation to feed, did not prioritize foraging and instead maintained high levels of antipredator behaviors during baseline states (Siegal et al. 2022).

10.4 Behavioral Responses of Sirenians

The order Sirenia includes four extant species of aquatic herbivorous marine mammals, which fall within two families: the Dugongidae and the Trichechidae. The Dugongidae includes only one living species, the dugong. The family Trichechidae includes three species of manatees: the Amazonian, the West Indian, and the West African manatee. While various studies have been conducted to evaluate the response of Sirenians to human activities, only a handful have reported associated RLs. Of those that have, the subjects under study have been limited to West Indian manatees, and most of the studies focused on behavior associated with vessel noise or noise in environments with varying levels of vessel usage. The focus on vessels has been motivated by the significant number of manatee mortalities documented over past decades in the USA resulting from vessel collisions (Ackerman et al. 1995; Calleson and Kipp Frohlich 2007) and an attempt to understand whether collisions are due to manatees' inabilities to hear approaching vessels (Calleson and Kipp Frohlich 2007; Gerstein et al. 1999).

10.4.1 Ships and Boats

A wide range of changes in behaviors have been reported in association with noise. Sleep disruption and movement away from the source were reported in a vessel-noise playback experiment with manatees at a zoo at RLs of 85–97 dB re 1 µPa (700 Hz–14 kHz; Lace 2016). While the speaker was stationary, vessel passes were simulated by fading the sound in and out. Interestingly, male manatees were also exposed to conspecific female vocalizations, and while sleep disruption was also observed for these stimuli, manatees moved toward rather than away from the source as they did for vessel noise. In another study, higher probability in occurrence of behavioral change (roll, heading, depth, or fluking behavior) in 18 tagged manatees was associated with vessels' closest point of approach (CPA) of <10 m and RLs of 94–152 dB re 1 µPa (500 Hz–30 kHz; Rycyk 2013; Rycyk et al. 2018). Pass duration was found to be important, with longer boat passes associated with a greater likelihood of manatees making a change in behavior. The authors suggested that longer passes would provide more time for a manatee to detect, locate, make a decision on how to respond to avoid collision, and undertake the response. While the RL at CPA itself was not important in determining the response magnitude or the latency between the start of the boat pass and the first behavioral change, the rate in rise of the RL was (which is related to either boat speed or proximity to the boat trajectory). Changes in heading and pitch (followed by depth) of tagged manatees were documented during experimental approaches by a vessel at two speeds (12–13 km/h and 25 km/h) at RLs of 108 and 114 dB re 1 µPa (125 Hz–15.5 kHz; see figures in Nowacek et al. 2002). The response was stronger in the case of the faster approach. In a different study conducting visual observations of wild manatees during 170 vessel approaches and 187 control periods, manatees exhibited a consistent flight response into deeper water. If they were in shallow water at the edge of the deeper boat channel, they swam toward deep water at increased speed. The response was stronger for closer boat approaches and in shallower water. However, movement toward the deeper waters of the boat channel could result in the manatee moving directly into the trajectory of an approaching vessel before reaching sufficiently deep water to move below the vessel (Nowacek et al. 2004b). Similarly, changes in heading, pitch, swim speed (which were described as consistent with fast swimming away from the source), and an increase in respiration rate variability were reported in a vessel-noise playback experiment at RLs of 150–166 dB re 1 µPa (240 Hz–20 kHz), which corresponded to vessels at ~10 m range (Miksis-Olds et al. 2007b).

Changes in habitat usage and behavioral states associated with environmental noise levels have also been reported. One study combining visual surveys and PAM discovered that manatees forage on seagrass plains with lower low-frequency (<1 kHz) ambient noise levels. The number of boats was negatively correlated with the number of manatees in the mornings, but not at other times of day (Miksis-Olds et al. 2007a). Noise levels were positively correlated with an increase in the average number of vessels passing per 5 min. Manatee responses further depended on their behavioral state. In seagrass beds, an increase in the proportion of time foraging and a decrease in the proportion of time milling were correlated with an increase in ambient noise levels between 39 and 83 dB re 1 µPa in the one-third-octave band at 4 kHz (Miksis-Olds and Wagner 2011). The authors suggested several possible explanations for these behavioral changes. For instance, increased feeding may maximize energy consumption under more threatening conditions (i.e., elevated noise levels), suggesting that elevated noise conditions in grass beds may spur manatees to engage in feeding to meet energetic demands and enhance fitness.

Changes in manatee vocalizations have been reported in association with environmental noise conditions (Miksis-Olds and Tyack 2009). Changes were complex, and a few are described here. In general, reduced vocalization rate during feeding and social behaviors were reported in elevated noise conditions. In high low-frequency noise (LFN; measured as the one-third-octave band level centered at 500 Hz) and low mid-frequency noise (MFN; one-third-octave band centered at 4 kHz), a longer duration of chirp vocalizations was reported when calves were present. In contrast, in high LFN conditions, the duration of squeaks was shorter when calves were present. In elevated MFN conditions, similarly nuanced changes in vocal behavior were observed. The peak frequency of chirp vocalizations was lower in low MFN conditions when calves were present than when they were absent. Changes in vocalization SL in responses to high-frequency noise (HFN; one-third-octave band centered at 32 kHz) were also reported, which depended upon behavioral state. Squeaks during feeding were louder than during milling when HFN was low, with the reverse true for high HFN.

10.4.2 Acoustic Deterrent Devices

Behavioral responses of dugongs to ADDs were evaluated in Moreton Bay, Queensland. Pingers (4 and 10 kHz, SL 133 dB re 1 µPa m) were deployed 50–55 m apart (Hodgson 2004; Hodgson et al. 2007). No change was observed in the number of dugongs when pingers were active versus inactive Dugongs did not seem to pay attention to the pingers and swam between them. While this study extended our knowledge to include pingers and dugongs, there is an urgent need to evaluate behavioral responses to sound from a wider range of noise sources and sirenian species.

10.5 Behavioral Responses of Pinnipeds

Pinnipeds are a group of semiaquatic marine mammals comprising seals (Phocidae), sea lions and fur seals (Otariidae), and walrus (Odobenidae). Because of their aquatic and land-based behavior, they are susceptible to both underwater and airborne noise. Furthermore, this group is vulnerable to noise generated by human activities prevalent around their habitats (e.g., port operations; water, air, and road traffic; and construction) in coastal areas close to human settlements.

10.5.1 Otariids

10.5.1.1 Boats

A few studies have looked at the impacts of watercraft on otariids. At Kanowna Island in Bass Strait, CEEs were conducted on hauled-out, breeding Australian fur seals using playbacks of in-air motor boat noise. Seals showed changes in body position, increased alertness, increased aggression toward each other, avoidance of the sound source, and differences in call types (including increased bark rate) at RLs of 75–85 dBC (Tripovich et al. 2012). At the same site, experimental boat approaches showed that reductions in colony attendance (i.e., reductions in the number of individuals remaining on land) and resting behavior depended on approach distance and time of day. Resting behavior was disrupted at 75-m distance, and colony attendance declined at 25 m, with a greater number of animals responding in the morning than afternoon (Back et al. 2018). A study on the effects of seal-swim tourism observed Australian fur seals near 61 tour boats and 74 recreational boats. Boats within 200-m range provoked aggressive behaviors; swim-with-seals swimmers provoked haul-outs, at least initially; and hauled-out seals entered the water during boat approaches (Stafford-Bell et al. 2012). South American sea lions within a breeding colony in Chile were affected by tour vessels. Subadult males were most affected at breeding sites and adult females at nonbreeding sites (Pavez et al. 2015). Longer visitation times and shorter approach distances generated a stronger negative response. Variability in Cape fur seal behavioral response to boats was documented across study sites (Martin et al. 2023a).

10.5.1.2 Seismic Surveys

New Zealand fur seals were observed offshore during a marine seismic survey over the continental slope, where they engage in nighttime foraging dives and daytime resting at the sea surface. The number of seals in the vicinity and their responses as a function of distance from the source vessel were measured. Responses were observed at distances <200 m. Results were inconclusive as to whether the source vessel and seismic gear presented obstacles leading to behavioral responses or whether seals reacted to noise (Lalas and Mcconnell 2016).

10.5.1.3 Navy Activities

The effects of sonar exposure were investigated in California sea lions. In a CEE, 15 individuals were exposed to simulated MFAS (3.3–3.5 kHz, 1-s duration) while swimming across a pool. A dose-response relationship was found, with stronger responses at higher RL. Responses included an increase in time spent submerged and an increase in respiration rate at RL < 155 dB re 1 µPa rms and hauling out and refusal to participate at RL > 170 dB re 1 µPa rms (Houser et al. 2013a). Younger sea lions exhibited stronger reactions at lower RLs. Unlike in bottlenose dolphins where rapid habituation was observed across repeated exposures to low or moderate RLs (Houser et al. 2013b), there was no sign of habituation after repeated exposures in the sea lions.

Two captive California sea lions exposed to pulsed sounds from an arc-gap transducer did not show a masked hearing threshold shift even at the highest exposures (183 dB re 1 µPa rms, 163 dB re 1 µPa^{2}s). Besides, exposure did not cause appetite changes, health effects, or injuries. However, changes in trained behavior were observed with increasing levels of exposure (Finneran et al. 2003).

In the wild, the majority of California sea lions startled and became more vigilant after missile launches and moved toward the water. Responses were related to RL and distance to the missiles at closest point of approach (Holst et al. 2011).

Ordnance disposal and on-land demolition training with explosives went on for decades and temporarily, albeit repeatedly, disturbed Steller sea lions at Race Rocks Ecological Reserve, British Columbia. Avoidance occurred at RL > 101 dB re 20 µPa rms (RL > 78 dBA; SEL > 98 dB re (20 µPa)^{2}s) and lasted only a few minutes with individuals returning to haul-outs (Demarchi et al. 2012).

10.5.1.4 Acoustic Deterrent Devices

The influence on pinnipeds of acoustic deterrents like specific devices, firecrackers, rifle gunshots, and playbacks of killer whale calls is of interest to fisheries and other industries to prevent depredation, damage to fishing nets, entanglement, bycatch, and other issues. The efficacy of these devices was studied with wild Cape fur seals depredating the purse-seine fishery off southern Africa (Shaughnessy et al. 1981). Firecrackers were thrown into the group of seals (at times comprising hundreds of individuals), and more than half of seals left by diving but returned within 5 min, requiring repeated lighting of firecrackers. Playback of killer whale sounds (115 dB re 1 µPa at 1 m from the underwater speaker) and artificial tones and banging steel shackles on the ship's hull repelled seals, but they returned within a minute. Adding dummy orca dorsal fins in the water did not improve results. Rifle shots fired into the water (142 dB re 1 µPa as the bullet hit the water) resulted in temporary (15–20 s) displacement; rifle shots fired above their heads had no effect.

Captive Steller sea lions were tempted by fish in a net in a pool while a variety of sounds were played through an underwater speaker at the net, as well as a boomer. Tonal sweeps and killer whale calls were transmitted with an SL of 145–165 dB re 1 µPa m, and the impulsive boomer was fired at a fixed SL of 210 dB re

1 µPa m. The animals hauled out of the pool during boomer firing and playbacks >165 dB re 1 µPa m. The study found differences in the cohorts affected by each sound type. For example, tonal sounds affected only juveniles, while boomer sounds generated responses in adults and juveniles. The results also suggested that the animals might habituate after repeated exposures (Akamatsu et al. 1996).

10.5.1.5 Urban Noises

Northern fur seals on rookeries in Alaska abruptly moved away when noise from aircraft, land vehicles, or construction was less than 300 m from their locations (Insley 1990). Adult females and pups at a rookery near St. George Island airport, Alaska, showed behavioral changes (activity levels increased) for 5 min after overflights (Williams 1997). Similarly, Cape fur seal females and pups at the Pelican Point breeding colony in Namibia exhibited higher levels of vigilance and locomotion during playbacks of boat and car engine noise. These energetically costly behaviors replaced time spent in vital activities such as nursing or resting. In this study, the animals were exposed to different noise levels (from 60 to 80 dB re 20 µPa), and the results were similar regardless of the level of exposure. Stronger responses occurred during boat noise playbacks from near shore than car noise playbacks from land. Individuals recovered pre-exposure behaviors within 2 min (Martin et al. 2022). The impact of New Year's fireworks was studied in South American sea lions during their breeding season (Pedreros et al. 2016). A cessation of vocalization and departures from the site were noted.

10.5.1.6 Long-Term and Population-Level Effects

Costa et al. (2016) modeled the potential consequences of reduced foraging efficiency due to acoustic disturbance on the reproductive rate and success of elephant seals and California sea lions. Elephant seals were more resilient, thanks to their wider foraging range; California sea lions were more vulnerable to noise disturbance because of their more coastal ranges, closer to noise sources. This study highlighted the importance of considering species-specific ecological traits when assessing the biological significance of behavioral responses to stressors.

10.5.2 Phocids

Phocid seals, like otariids, are amphibious, and their need to haul out on land or ice places many species within a rapidly industrializing coastal zone and exposes them to a variety of anthropogenic noises. Harbor seals have been the subject of half of the noise impact literature. This could, in part, be due to their widespread distribution in the northern hemisphere, concern for their declining population in some areas, and their protection under the EU habitats directive. The literature on other phocids is sparse, and several species have never been the subject of noise impact research.

10.5.2.1 Ships and Boats

Phocid seals are exposed to significant levels of underwater noise from ships, and the distributions of many species overlap with high-density zones of shipping (Jones et al. 2017; Prawirasasra et al. 2021; Trigg et al. 2020). Animal-borne tag data have shown potential changes in the diving behavior of gray seals (Chen et al. 2017; Mikkelsen et al. 2019), ringed seals (Prawirasasra et al. 2022), northern elephant seals (Burgess et al. 1998; Fletcher et al. 1996), and harbor seals (Mikkelsen et al. 2019) in relation to shipping in the Celtic, Baltic, Pacific, and North Seas, respectively. Observed responses included a cessation of deep dives (Chen et al. 2017), an interruption of resting and traveling (Mikkelsen et al. 2019), an occurrence of a deep dive (Fletcher et al. 1996; Prawirasasra et al. 2022), a change in the length of time at the surface (Prawirasasra et al. 2022), and a reduction in swim speed (Burgess et al. 1998) at broadband (10–1000 Hz) RL > 95 dB re 1 µPa (see Suppl. Material). There was considerable variation in observed responses, which might be attributable to species-specific behaviors, variable RL throughout a dive profile, and contextual factors, such as age. Moreover, most studies were limited to a qualitative rather than quantitative inspection of dive profiles.

Whether seals exhibit a behavioral response to vessels or not, sound exposures even in dense shipping zones, such as the Celtic Sea and English Channel, are not expected to be high enough to cause hearing threshold shifts. This was modeled based on predicted exposure metrics, weighted using early functions based on seal audiograms (24 h cumulative, weighted SEL 124–170 dB re 1 µPa^{2}s) and later updated functions (24 h cumulative, weighted SEL 106–152 dB re 1 µPa^{2}s; Southall et al. 2007, 2019c; Trigg et al. 2020). However, a threshold shift was assessed to be possible in the study by Jones et al. (2017).

Phocid seals have a relatively simple vocal repertoire when compared to cetaceans. This may contribute, in part, to the low number of studies that have investigated the impact of shipping noise on phocid vocalizations despite the overlap of call frequencies with those of shipping noise (Bagočius 2014; Matthews et al. 2020; Prawirasasra et al. 2021). While one study found the amplitude of harp seal calls reduced in the presence of a vessel (Terhune et al. 1979), another documented that the amplitude of harbor seal calls increased by 0.16 dB for every 1 dB increase in ambient noise (87–107 dB re 1 µPa; 40–500 Hz) (Matthews et al. 2020). However, with every 1 dB increase in ambient noise, the SNR decreased by 0.84 dB, and so, call level compensation was not enough to account for the masking effect of higher ambient noise levels. Under water, spotted seals and bearded seals increased call amplitude with ambient noise levels. The same was observed only for growls in spotted seals (Yang et al. 2022) and until call amplitude reached a threshold (~100–105 dB re 1 µPa) for bearded seals (Fournet et al. 2021). Similarly, no evidence for a Lombard effect was found in northern elephant seals on their breeding rookery in a range of in-air ambient noise conditions (Southall et al. 2019a).

There has been a particular focus on the response of hauled-out seals to approaching vessels. These studies do not often measure in-air or underwater RLs of sound at the animal. Consequently, it is difficult to disentangle the visual and audio stimuli from ships that may result in a behavioral response. Small vessels and ships have elicited alert and head-raising behaviors in harbor seals (Henry and Hammill 2001) and Saimaa ringed seals (Niemi et al. 2013) and the flushing of hauled-out harbor seals into the sea (Andersen et al. 2012; Blundell and Pendleton 2015; Jansen et al. 2010; Lomac-MacNair et al. 2019; Paterson et al. 2019). While many return to haul-out sites after disturbance, this is not the case for all seals. Paterson et al. (2019) found only 52% of harbor seals had returned to a haul-out 30 min after disturbance by a small vessel. It could take over 5 h for seal numbers on a haul-out to recover after a flushing event with small boats (Andersen et al. 2012).

10.5.2.2 Seismic Surveys

Polar species (ringed seals, spotted seals, and bearded seals) have been the focus of studies investigating the impacts of seismic survey activity. Studies have primarily used visual observations with a small number supplementing these with animal-borne tags to measure behavior. Consequently, there is little information about the underwater behavior of pinnipeds in response to seismic surveys.

Ringed seals, spotted seals, and bearded seals have been observed in the water at similar rates during no airgun, one airgun, and full seismic array operation. However, seals were observed farther away from the seismic vessel during full array operation, suggesting avoidance at RLs up to 190 dB re 1 µPa (10–4000 Hz) (Harris et al. 2001). There was no significant difference in the occurrence of behavioral categories dive, swim, mill, thrash, and look (Harris et al. 2001). Ringed seals avoided terrestrial seismic surveys at 69 dB re 20 µPa (Kelly et al. 1986). They abandoned subnivean lairs within 150 m of a seismic survey at a greater rate than lairs further away from the survey, with one observation of abandonment 644 m away from the survey location (Kelly et al. 1986). Marine seismic airguns caused ringed seals to flee from ice (Stemland et al. 2019). In contrast, captive bearded seals that had been trained to conduct a hearing test only exhibited small flinches during test exposures, suggesting the contextual motivation for reward or lack of space could extinguish the avoidance response seen in wild seals (SEL 166–185 dB re 1 µPa^{2}s; 0.01–20 kHz) (Sills et al. 2020).

10.5.2.3 Pile Driving

Research related to pile driving and pinnipeds has primarily been driven by the large populations of gray and harbor seals in European waters as well as a burgeoning offshore wind industry in these regions (e.g., Hastie et al. 2015; Thompson et al. 2013b). However, other studies have investigated general construction and pipe driving noise in Arctic oil fields and found ringed seals showed no or curious responses (Blackwell et al. 2004). The most reported response to pile driving during wind farm construction has been a reduction in the numbers of seals around pile driving activities based on haul-out counts (Edrén et al. 2010; Skeate et al. 2012) or at-sea distribution (Aarts et al. 2018; Brasseur et al. 2022; Russell et al. 2016; Whyte et al. 2020). However, only a small number of these have measured or estimated RLs experienced by the seals. Some authors investigated the density of tagged seals during piling events and periods of no piling for wind farm construction in the North Sea and found a significant decrease in seal abundance within 25 km of the site during piling events (Russell et al. 2016). There was a 19–83% decrease in usage of the area around the wind farm during piling events at RL > 166–178 dB re 1 µPa pk-pk. Density also decreased with increasing RLs. There was a significant decrease in seal density at single-strike SEL > 145 dB re 1 µPa^{2}s; the SEL varied with both depth and the specific pile installation (Whyte et al. 2020).

It must be noted that many of the responses reported were not considered long term, with declines in numbers only occurring during pile driving, not construction as a whole (Edrén et al. 2010). Within 2 h of piling at the Lincs wind farm in the North Sea, the number of seals had recovered (Russell et al. 2016). In contrast, Skeate et al. (2012) reported a long-term reduction in harbor seal numbers at a haul-out during a 5-year aerial survey at Scroby Sands in the North Sea. However, while that was partially attributed to pile driving, it was also accompanied by an increase in vessel traffic and an increase in gray seal numbers, suggesting other factors may be contributing to the long-term decline of harbor seals in the area.

There is great variation in individual diving behavior during pile driving, and only two studies in a small geographic area have investigated this. Responses included altered surfacing, change in swim direction, stopping, going to shore, and change in descent speed, but for many individuals, there was no response observed (Aarts et al. 2018; Brasseur et al. 2022). The closer seals were to pile driving events, the more likely it was for a response to occur (Aarts et al. 2018; Brasseur et al. 2022). During the construction of Luchterduinen and Gemini wind farms in the Wadden Sea (the Netherlands), responses were more likely within a 36-km range (Aarts et al. 2018), but during Borssele wind farm construction in the same region, the distance was only 14 km, with the reduction attributed to the use of mitigation bubble curtains (Brasseur et al. 2022). The most common response observed was a decrease in dive descent speed, interpreted by authors as a switch from foraging dives to horizontal movement at single-strike SELs between 130 and 150 dB re 1 µPa^{2}s (Aarts et al. 2018; Brasseur et al. 2022).

In captive settings, a study assessing the impact of prey availability on seal responses found shorter foraging duration during playbacks of pile driving strikes. At low density prey patches, seals spent less time foraging and were less successful when foraging during sound exposure than at high-density prey patches. These patterns were seen during playbacks of both pile driving and tidal turbine noise and indicate foraging context is important when interpreting the variable responses shown by seals (Hastie et al. 2021).

10.5.2.4 Sonar

We know very little about the behavioral impact of sonar on wild phocid seals. However, a study with captive hooded seals of 1–7 kHz MFAS reported active avoidance, increased heart rate, reduced diving, rapid swimming, and floating at the surface (160–170 dB re 1 µPa rms, 10% duty cycle; Kvadsheim et al. 2010). At lower RL (125 dB re 1 µPa rms, 2.4% duty cycle), captive harbor seals reacted to frequency-modulated 25.5–24.5 kHz hyperbolic downsweeps by spending increased time at the surface, hauling out, and increasing swim speed and number of jumps (Kastelein et al. 2015c). Not all seals responded, and a continuous-waveform signal showed fewer responses, highlighting the possible importance of signal characteristics in mediating responses (Kastelein et al. 2015c). Gray seals exposed to 200 kHz high-frequency sonar in captivity spent more time hauled out (Hastie et al. 2014). At 375 kHz, seals stayed in the water but were farther away from the sonar source even though the frequency exceeded their hearing range (Hastie et al. 2014). These captive studies suffered from a low sample size, common in all captive studies discussed, not just those relating to sonar, and the conclusions require corroboration with wild individuals in sufficient numbers. Wild harbor seals and southern elephant seals have also been exposed to very-high-frequency echosounders (720 and 1500 kHz) to track seal movement around tidal turbines and study mid-trophic level organisms (Hastie et al. 2019a, b; Tournier et al. 2021). These studies did not consider the possible impact of this noise on seals due to the high frequency.

Although not used for sonar but rather climate research, the ATOC signal changed the dive patterns of wild northern elephant seals (descent rate) at RLs of 118–137 dB re 1 µPa rms (60–90 Hz; Costa et al. 2003). Previous studies had found no effect at similar RLs but only studied the responses of a single northern elephant seal (Burgess et al. 1998). Finally, underwater acoustic communication involving the use of tonal signals (12 kHz) resulted in avoidance by harbor seals in captivity (Kastelein et al. 2006b).

10.5.2.5 Acoustic Deterrent Devices

In contrast to incidental noise discussed in the sections above, ADDs aim to introduce noise into the environment to deter animals from the location of the device. For phocid seals, they are primarily used to prevent predation on commercial fisheries and aquaculture (Götz and Janik 2013), and in some regions they form a significant component of the soundscape (Findlay et al. 2018, 2022; Todd et al. 2021). A number of studies demonstrated a reduction in the predation of fish by seals and a decrease in the number of seals sighted near aquaculture enclosures when ADDs were in operation (Götz and Janik 2015, 2016; Graham et al. 2009; Harris et al. 2014; Jacobs and Terhune 2002; Kastelein et al. 2015b, 2017b; Lehtonen et al. 2022; Vetemaa et al. 2021; Yurk and Trites 2000). Similarly, captive seals increased the amount of time spent hauled out (Bowles and Anderson 2012; Kastelein et al. 2010). However, there was no change in surfacing distance, which might have been due to the limited space available in captivity (Kastelein et al. 2010). These studies used a range of ADD devices, which differ in their frequency, inter-pulse interval, and SL (which were as high as 195 dB re 1 µPa m, peak frequency 10 kHz; Jacobs and Terhune 2002). However, Mikkelsen et al. (2017) reported an increase in seal numbers within 100 m of simulated ADD sounds with pk-pk SLs of 165 dB re 1 µPa m (peak frequency 12 kHz), and Kastelein et al. (2015b) only observed the haul-out of captive seals when RLs were above 124 dB re 1 µPa rms (10–20 kHz). Furthermore, captive seals habituated to commercial ADDs when food was present, and in wild seals, initial success in the reduction of predation can diminish over time (Fjälling et al. 2006; Götz and Janik 2010; Mate and Harvey 1987).

ADDs are also used to deter seals from areas where potentially harmful anthropogenic activities (such as pile driving and tidal turbines) are located (Sparling et al. 2015). Controlled exposure of harbor seals to ADDs in this context reported seals moved away from the sound source, traveled further inshore when near the coast, and deviated around the source when traveling (Gordon et al. 2019). Tag data also suggested net swim speed, distance between surfacing points, and duration between two surfacing points increased during ADD exposure. All exposures at less than 1 km (estimated

RL = 135 dB re 1 μPa rms) resulted in a detected response (Gordon et al. 2019). This is one of only a few studies that focus on the detailed vertical and horizontal movement of seals in response to ADDs, with the majority of studies measuring seal counts and fish predation to infer avoidance. The literature relating to the efficacy of ADDs and impacts on target and nontarget species has been of much interest to regulatory bodies who have commissioned several reports reviewing the available ADDs and their effectiveness (Coram et al. 2014; Lepper et al. 2014; McGarry et al. 2022; Sparling et al. 2015; Thompson et al. 2021).

Phocid seals can be held in captivity with relative ease and, hence, have been a model species for studies that address fundamental questions regarding habituation, learning, startle responses, and the aversiveness of different sound types. Götz and Janik (2011) demonstrated that sudden onset sounds with quick rise times elicited startle responses, which resulted in sensitization over progressive trials and avoidance of the noise source. In contrast, under novel non-startling stimuli with food available, the first presentation of an aversive stimulus led to fleeing and avoidance by seals (Götz and Janik 2010). However, for these non-startling stimuli, there was a reduction in the intensity of behavioral responses, and seals demonstrated habituation to the sound source (Götz and Janik 2010). This is reflected in studies of ADDs that report a diminishing response by seals (Götz and Janik 2013). However, harbor seals exposed to pulses (8–45 kHz, pulse duration 250 ms, pulse interval 5 s) for 45 min over 40 days showed consistent displacement and increased number of surfacing events across all days (Kastelein et al. 2006a). This suggests that short use periods may be less likely to lead to habituation than continuous ADD operations (Kastelein et al. 2006a). Although studies in captivity have allowed carefully controlled exposures to sounds, a greater emphasis on at-sea exposure could elucidate potential changes in behavior not available to captive seals due to their restricted environment.

10.5.2.6 Tidal Turbine Noise

The possible disturbance effect from tidal turbines is considered a beneficial side effect of their operation to ensure animals are not harmed in the turbines themselves. Tidal turbine noise had no impact on the number of seals sighted in the water, but there was a significant increase in the distance of seals from the location of tidal turbine playbacks (Hastie et al. 2018). As described above for pile driving, during playbacks to captive gray seals at low density prey patches, the animals spent less time foraging and were less successful at foraging when exposed to tidal turbine playbacks than at high-density prey patches (Hastie et al. 2021). This suggests that while tidal turbine noise may be beneficial in reducing the presence of seals near harmful infrastructure, it may impact foraging success. However, these studies form a limited evidence base with respect to the impact of underwater noise from tidal turbines on phocid seal species.

10.5.2.7 Airborne Noises

Due to their amphibious lifestyle, phocid seals are potentially vulnerable to in-air noise sources. No response or only minor changes in behavior have been reported in response to a sonic boom from Concorde above harbor and gray seals (Perry et al. 2002), unmanned aerial vehicles near Weddell seals (Laborie et al. 2021), explosions near northern elephant seals (Fletcher et al. 1996), and snow mobiles passing Weddell seals (van Polanen Petel et al. 2006). In contrast, missile launches in California caused 68% of hauled-out harbor seals to enter the water for 34 of 48 observed launches (Holst et al. 2011). Less pronounced but similar movement responses were seen in the same seals in a later study (Ugoretz and Greene 2012). Elsewhere, unmanned aerial vehicles significantly increased vigilance behavior and decreased resting in hauled-out harbor seals (Pérez Tadeo et al. 2023). A possible link between in-air noise from city construction and higher numbers of harbor seals hauled out at night was reported; however, time of day and tidal level were also major influences on seal haul-out patterns (Acevedo-Gutiérrez and Cendejas-Zarelli 2011). The difficulty in assessing the effects of multiple variables among urban sites was also highlighted by observations that numbers of hauled-out harbor seals depended on noise level at a site distant from human activities, but not at a site close to human activities, suggesting probable habituation (Bankhead et al. 2023). Ringed seals were seen at similar densities on landfast ice during aerial surveys before and after construction at an oil development site; construction included the building of roads, transport of gravel by trucks, and pipeline installation (Moulton et al. 2003).

10.5.2.8 Predator Sounds

Controlled exposures of tagged harbor seals to playbacks of killer whale calls found avoidance responses similar to those to other sound sources, but also great individual variation (Gordon et al. 2019). Avoidance in response to noise may reflect the natural predator-avoidance behavior of seals.

10.5.2.9 Long-Term and Population-Level Effects

As air-breathing mammals that forage at various depths within the water column, diving is central to the life strategy of phocid seals, and the disturbance of diving behavior could have implications for foraging success, energetic balance, and ultimately vital rates (Booth and Heinis 2018; Costa et al. 2016; Schwarz et al. 2016). As with other marine mammals, this has been explored through frameworks looking at population-level consequences.

New et al. (2014) modeled the effect of reduced foraging time due to disturbance on vital rates and population trajectories of southern elephant seals. The consequences of a lower lipid mass manifested in reduced pup survival and thus a decline in abundance in the year following the disturbance. Further, the study forecasted, assuming no behavioral adaptations from the elephant seals, a decline in population size of ~10% in 30 years. While "disturbance" was described by its consequences (loss of foraging time) and not by its source, the authors included sound, pollution, and environmental change as examples of stressors.

Models of disturbance for northern elephant seals that experience reduced foraging in a 100-km patch resulted in no change to birth rate and pup survival (Costa et al. 2016). However, this did not consider the breeding strategy. Such models that looked at capital versus income breeders found pup recruitment was sensitive to foraging patch productivity, probability of disturbance, and individual behavioral choices in the face of disturbance (McHuron et al. 2017).

10.5.3 Walrus

Available information for the walrus is scarce and mostly consists of opportunistic observations without acoustic measurements. For example, an omnidirectional hydrophone was used to record the acoustic reaction of Pacific walruses to vessels arriving at Round Island, Alaska. While vocal rates were lower when vessels approached the area, acoustical analysis was not performed (Quin 1997). Behavioral reaction to seismic surveys was observed in a 7-year study of Pacific walruses in the Chukchi Sea, in which 23% of animals exhibited an "attention" response; 7% approached, avoided, or fled from the ship; and the remaining animals did not respond (Born et al. 2021). The theoretical detection range for a ship in summer and an icebreaker in winter by walruses in Baffin Bay was calculated by comparing RLs to ambient noise and the walrus audiogram, concluding that ships could be detected at ranges of 37–64 km in summer and 75–154 km in winter (Schack and Haapaniemi 2017). Noise from small boats near walrus colonies has not been measured; however, regulations intended to reduce the potential impact of encounters with human activity are in place for both seismic and tourism operations (Born et al. 2021).

Hauled-out walruses were observed fleeing areas where icebreaking activity was taking place, but later returned, and showed little response to other drilling operations (Brueggeman 1993). Born et al. (2021), summarizing an extensive set of observations by Fay et al. (1984), concluded that walrus reaction to icebreaking activities may be due more to the icebreaking noise than to the ship noise. Various investigators have observed behavioral responses including changes in swim speed for walrus-to-ship ranges of 100–800 m, while at 4.8-km range, no reaction was observed (Born et al. 2021). Pups may theoretically be abandoned during a flee response, but it was assessed that the temporary separation is unlikely to be permanent, since walrus females have been shown to recognize their pups' voices (Born et al. 2021).

Airborne noise disturbance potentially affecting hauled-out groups of walruses has been studied more frequently than that from underwater noise, with reactions to icebreakers, ships, aircraft, and small boats comprehensively summarized in Born et al. (2021). However, the only study that included estimated RLs was one that explored the reactions of walruses and other marine mammals to UAV overflights (Palomino-González et al. 2021). In their first trial of exposure of wild walruses to UAV noise (42–70 dB re 20 μPa), the animals moved toward the water, while in a second trial, they only raised their heads. Aircraft overflights are another potential area of concern, with hauled-out groups of walruses seen startling when airliners were at high altitudes (10,000 m), low altitudes (150–1500 m), and ranges less than 5 km (Born et al. 2021). Helicopters were more likely to startle walruses than fixed-wing aircraft were, with walrus reaction ranging from head lifts to escape into the water when the aircraft was within 1.3–2.5 km range depending on altitude (Born et al. 2021).

10.6 Behavioral Responses of Polar Bears and Otters

Information on noise impacts on polar bears is scarce. Some studies have measured behavioral responses to aircraft, vehicles, oil developments, or construction activities in or close to polar bear dens during winter (Amstrup 1993; Linnell et al. 2000). Linnell et al. (2000) did a systematic literature review and found that bears tolerate disturbances caused by human activities. However, this was dependent on the distance at which the activity occurred. For example, when a disturbance occurred at <1 km, bears left the den and did not return. The observations by Amstrup (1993) support these results. This study reported that polar bears left the den after exposure to aircraft flights in Alaska. However, polar bears did not react to seismic surveys and oil field operations. Andersen and Aars (2007) documented behavioral responses of polar bears to snowmobile traffic in Norway and found that bears moved away when vehicles were present, and females with small cubs showed stronger reactions than males. They also reported that the reactions were related to the distance of the vehicle as displacement was observed at distances of 3 km or less. However, polar bears increased their alertness when the vehicle was 5 km away. Palomino-González et al. (2021) observed polar bears during an experiment assessing marine mammal reactions to UAV overflights. A mother bear with two cubs was observed moving toward the cubs, checking the cubs, and watching the UAV during approaches at RLs of 47–65 dB re 20 μPa. Lomac-MacNair et al. (2019) reported on three incidental observations of individual polar bears reacting to icebreaker noise. Animal reactions depended on whether the ship was engaged in icebreaking activities at the time. A polar bear approached a ship drifting in ice, directly placing its forepaws on the stern, sniffing, and otherwise investigating the ship for ~12 min. On two separate occasions while icebreaking was underway, an individual polar bear looked at the ship but remained at distances greater than 800 m, watching the ship. In a study covering 7 years and 56,901 km of observation effort in the Chukchi and Southern Beaufort Seas, 55% of the 42 groups of bears exhibited a behavioral response to vessel presence (vigilance or flee), including all 5 groups of mothers with cubs. The mean distance at which a behavioral response was observed was 805 m, and the probability of response was higher for bears in the water than on ice (Lomac-MacNair et al. 2021).

The available information on noise impacts on sea otters comes from acoustic measurements on three launches of a Delta II aircraft in the

Vandenberg Air Force Base (Berg et al. 2001). The authors measured the noise levels from the Delta II and counted the number of adults and pups in an area close to the base (San Miguel Island, California), before and after each launch. They found that the number of adults remained the same but the number of pups was lower afterward. However, this result was perhaps caused by poor observation conditions and not a direct consequence of the launch. The authors did not report injuries, mortality, or abnormal behavior caused by the launch. The results are based on unweighted SELs ranging from 127 to 129 dB re 1 $\mu Pa^2 s$.

Curland (1997) investigated how kayaking and boating activities influenced the behavior of sea otters in different areas in Monterey Bay, California. This research found that kayaks were the main source of disturbance and sea otters spent longer times traveling in areas with higher levels of human disturbance.

10.7 Compilation of Quantitative Data

While undertaking a literature review for this chapter, we produced an overview of quantitative data in the form of a spreadsheet of sound levels at which behavioral responses were (or were not) observed (see Suppl. Material). The literature review started from the marine mammal species listed on the Society for Marine Mammalogy website.[2] Extinct and assumed extinct species were excluded, and subspecies were grouped at the species level, resulting in 133 marine mammal species. Attempts were made to identify all studies that included an estimate of the noise level experienced by the animals. In such studies, noise levels were typically measured using hydrophones in the area or acoustic tags on the animal or modeled using knowledge of the source level coupled with a sound propagation model. There were no studies of behavioral reaction to noise for 83 of the 133 marine mammal species; the remaining 50 species and numbers of references for each are listed in Table 10.1. Of the 50 species with references, 39 had studies that measured or modeled RLs. The remaining 11 species were included in Table 10.1 when a related variable was measured, in particular for species with sparse results such as walrus and polar bear, where the best available studies did not include measurements of airborne noise exposure but other quantitative values (such as distance from a noise source resulting in a behavioral response) were rigorously measured.

Experiment types were standardized as being one of combinations of before, during, and after exposures; controlled exposure experiments; exposures of varying dosage; observations in the presence and absence of uncontrolled exposure; and opportunistic observations.

Observation types fell into three main categories: visual observations, acoustic observations, and tags. Visual observations included those from aerial platforms, vessels, or land; underwater visual observations (through a viewing window); video recordings; and photographic identification by trained personnel. Acoustic observations included passive acoustic monitoring and fish-detecting sonar. Tag types and capabilities varied widely, with different data available in different versions of some tags, and not all authors fully specified tag versions and capabilities. Therefore, the specific make or model of tag was not listed in the summary spreadsheet; the reader should refer to the original references. Tag capabilities could include almost any combination of acoustic recording, time-depth records, and position at the surface (via GPS, satellite, and radio frequency).

Behavioral responses were categorized into three main groups: physical responses (involving changes in individual location, body movement, or locomotion), acoustical responses (involving changes in acoustic output), and other biological responses (Table 10.2). If an animal showed no response to an estimated sound level, the study was still included in the table and coded as "no response." If a study was quantitative, the anthropogenic, environmental, and biological covariates

---

[2] https://marinemammalscience.org/science-and-publications/list-marine-mammal-species-subspecies/ accessed 8 Aug 2023

10 Behavioral Responses to Underwater Noise

Table 10.1 Species included in literature review for which references were found and the number of references by species

Species	Common names	Number of studies
Balaena mysticetus	Bowhead whale, Greenland whale	11
Eubalaena australis	Southern right whale	3
Eubalaena glacialis	North Atlantic right whale	4
Eschrichtius robustus	Gray whale	5
Balaenoptera acutorostrata	Minke whale	4
Balaenoptera musculus	Blue whale	9
Balaenoptera physalus	Fin whale	3
Megaptera novaeangliae	Humpback whale	16
Cephalorhynchus commersonii	Commerson's dolphin	1
Delphinus capensis	Long-beaked common dolphin	1
Delphinus delphis	Common dolphin	1
Globicephala macrorhynchus	Short-finned pilot whale	4
Globicephala melas	Long-finned pilot whale	6
Grampus griseus	Risso's dolphin, grampus	1
Lagenorhynchus albirostris	White-beaked dolphin	1
Lagenorhynchus obliquidens	Pacific white-sided dolphin	2
Orcinus orca	Killer whale, orca	6
Peponocephala electra	Melon-headed whale, electra dolphin	2
Pseudorca crassidens	False killer whale	3
Sousa chinensis	Indo-Pacific humpback dolphin	1
Stenella attenuata	Pantropical spotted dolphin	1
Stenella coeruleoalba	Striped dolphin	1
Stenella frontalis	Atlantic spotted dolphin	1
Steno bredanensis	Rough-toothed dolphin	2
Tursiops aduncus	Indian Ocean bottlenose dolphin	3
Tursiops truncatus	Common bottlenose dolphin	32
Delphinapterus leucas	Beluga, white whale	11
Monodon monoceros	Narwhal	3
Phocoena phocoena	Harbor porpoise	57
Physeter macrocephalus	Sperm whale	10
Berardius bairdii	Baird's beaked whale	1
Hyperoodon ampullatus	Northern bottlenose whale	3
Mesoplodon densirostris	Blainville's beaked whale, dense-beaked whale	7
Ziphius cavirostris	Cuvier's beaked whale, goose-beaked whale	3
Trichechus manatus	West Indian manatee, Florida manatee	6
Arctocephalus pusillus	Cape fur seal, Australian fur seal	3
Callorhinus ursinus	Northern fur seal	1
Eumetopias jubatus	Steller sea lion	2
Zalophus californianus	California sea lion	6
Odobenus rosmarus	Pacific walrus, Atlantic walrus	5
Cystophora cristata	Hooded seal	1
Erignathus barbatus	Bearded seal	3
Halichoerus grypus	Gray seal	8
Leptonychotes weddellii	Weddell seal	2
Mirounga angustirostris	Northern elephant seal	7
Phoca largha	Spotted seal, largha seal	2
Phoca vitulina	Harbor seal, common seal	24
Pusa hispida	Ringed seal	5
Ursus maritimus	Polar bear	4
Enhydra lutris	Sea otter	1

Table 10.2 List of possible behavioral responses

Response category	Response
Physical response	Orienting toward or away from noise
	Surface-respiration-dive response
	Abrupt movement
	Avoidance: any avoidance of the vicinity of a noise source
	Approach noise source
	Change in surface activity
	Movement: change in swim speed or walking speed
	Head raising, vigilance behaviors
	Trained animal still completes task but hesitates in noise
	Entered water
Acoustic response	Change in call RL
	Change in call count or rate of calling
	Change in call duration
	Change in call frequency
	Change in call type usage
	Change in echolocation behavior
	Change in source level
Biological response	Change in aggregation
	Change in social interactions other than vocalizations
	Change in activity state
	Change in foraging dive pattern
	Change in social association
	Change in species or group density
	Change in detection rate
	Change in habitat use

Table 10.3 Covariates used and cataloged in the spreadsheet

Covariate category	Covariates
Anthropogenic	Azimuth to noise source, distance to noise source, duration, land vehicles (presence/absence), missile type, previous exposure, rise rate, session number, sound type, sonar (in use or off), sonic boom (presence/absence), vessels (distance, number of vessels, noise, operational mode, orientation, presence/absence, speed, type)
Environmental	Air temperature, ambient noise, cloud cover, distance to boating/traffic channel, distance to shore, food availability, habitat type, ice cover, prey density, prey depth, sea state, sound speed profile, sea surface temperature, tidal level, time of day, time of year, water depth, wind speed
Biological	Behavioral state during exposure, behavioral state before exposure, cow-calf pairs, demographic (age/sex) composition, dive depth, group size, heart rate, individuals, nearest neighbor group (presence/distance), predator presence, season during exposure (e.g., pupping, breeding), social state during exposure, song presence/level, vocalization rate/type

evaluated were standardized as given in Table 10.3.

Noise sources were described using a list of 47 possible noise sources that fell into 5 different categories: impulsive, continuous, natural, tonals/sweeps, and sonar. The noise descriptions and their corresponding categories are summarized in Table 10.4. RL estimation method was selected from a list including geometric spreading, interpolation, measurement, modeling, or source level only.

Acoustic units and weighting always present a particular problem of standardization, due to the different reference levels in air and under water and the variety of metrics used for different types of sounds over the years. Abbreviations were

Table 10.4 Noise source categorization scheme

Noise category	Noise sources
Impulsive	Airgun, airgun array, boomer, explosion, firecracker, impact piling, rifle, terrestrial seismic survey, seismic survey vessel, sparker
Continuous	Construction, dredging, drilling, icebreaking, fixed-wing aircraft, rotary-wing aircraft, land vehicle, naval missile, outboard motor, vibratory piling, whale-watching vessel, ship(s) (nearby), shipping (distant), snow vehicle, tidal turbine, unmanned aerial vehicle, vessel(s)
Natural	Ambient noise, brown noise, killer whale sounds, psychophysiological model, social vocalizations, white noise, wind
Tonals/sweeps	Acoustic Communication network for Monitoring of underwater Environment (ACME), acoustic deterrent devices (ADDs), M-sequences, pulses (continuous wave), sweeps (frequency-modulated), tonals
Sonar	Continuous low-frequency active sonar (CLFA), high-frequency sonar (HFS), echosounders, low-frequency active sonar (LFA) (US: <1 kHz; UK: 1–2 kHz), mid-frequency sonar (MFS) (US: 1–10 kHz; UK: >2 kHz)

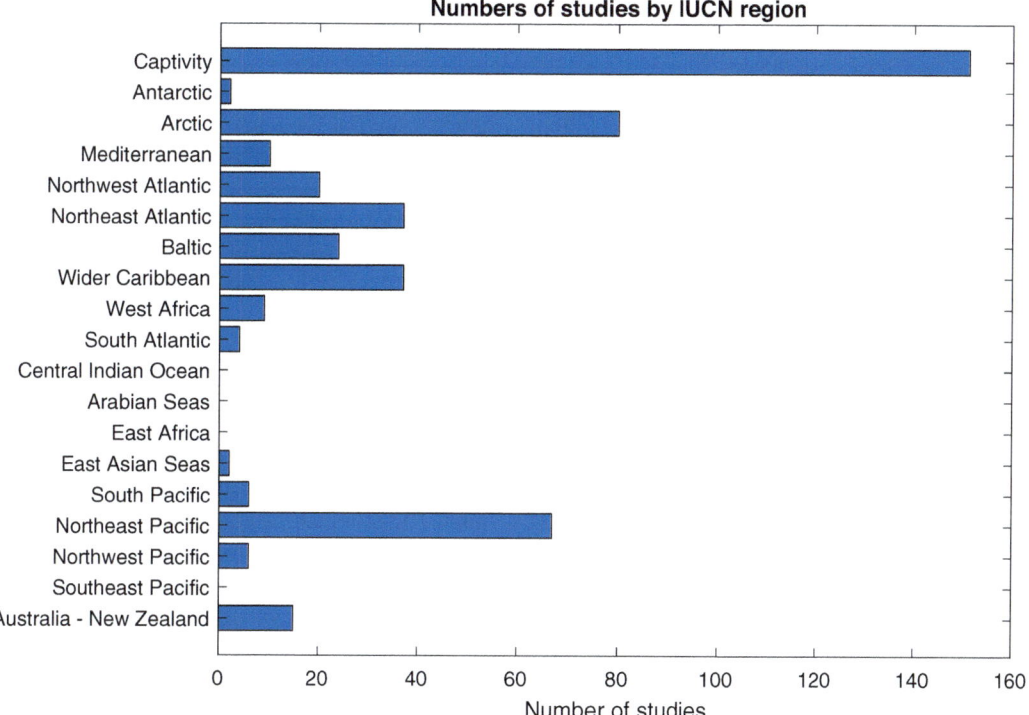

Fig. 10.1 Histogram of the number of studies linking physical or acoustic behavioral responses to received levels of sound by geographic region. (Based on the data in Suppl. Material)

standardized to remove ambiguity between in-air and underwater measurements, using appropriate standards and reference values for each: in air, a reference pressure of 20 µPa (IEEE 2019); under water, a reference pressure of 1 µPa (International Organization for Standardization 2017). The most common units used under water were the root-mean-square (rms) sound pressure level (SPL) for continuous sounds and the sound exposure level (SEL) for impulsive or intermittent sounds. However, other units included zero-to-peak or peak-to-peak sound pressure level, one-third-octave band level, single-strike SEL, and cumulative SEL. Various weightings were used,

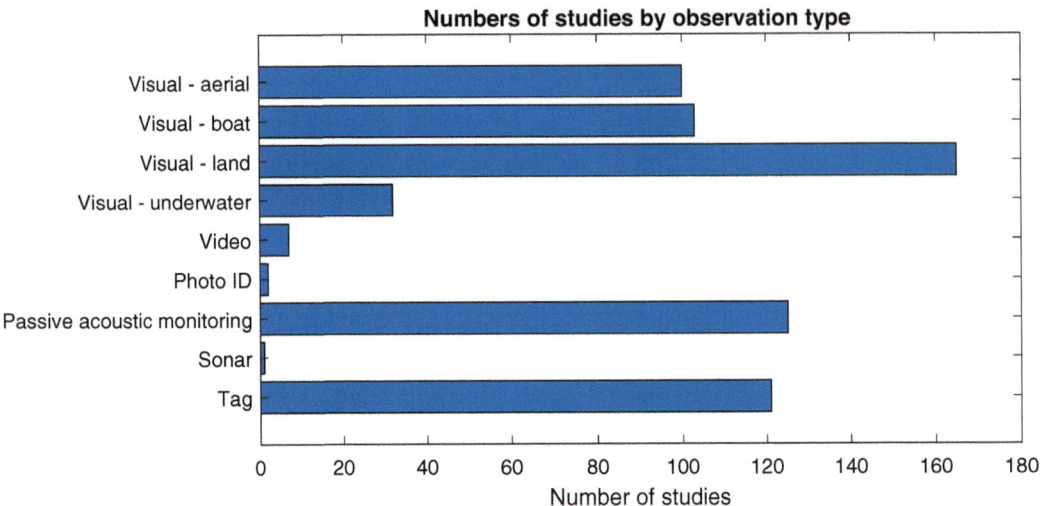

Fig. 10.2 Histogram of the observational methods used

depending on species, typically M-weighting and occasionally audiogram weighting. For in-air measurements, there was more variation in reported units, including A-, C-, M-, and linear weighted SPL, peak unweighted sound level, and A-, M-, and Z- (linear) weighted SEL.

The majority of studies linking physical or acoustic behavioral responses to RL were done under controlled experimental conditions in captivity (Fig. 10.1). In the wild, most of these studies occurred in the Arctic and then the Northeast Pacific. Partly driven by the great number of studies in captivity, the most common observation method was land-based visual (Fig. 10.2). Aerial observations, boat-based observations, PAM, and tag data collection were roughly equally utilized. MFAS and ADDs were the most commonly studied noise sources (Fig. 10.3). Avoidance, no response, changes in movement speed, and changes in surface-respiration-dive behavior were most commonly observed, in that order (Fig. 10.4).

Scatter plots were derived from the data for underwater sound (Suppl. Material) as follows. The acoustic data derived from individual studies were standardized as much as possible into three values (mean, maximum, and minimum) as follows. If an animal showed any response (of the 27 possible responses in Table 10.2), it was included in the plot; sound levels at which animals showed no response are recorded in the spreadsheet, but were not plotted. Depending on the study design, mean sound levels at which a response was observed were either stated explicitly by the original author or were recorded in the spreadsheet as the level at which 50% of a study population showed a behavioral response. If the author provided a range of sound levels at which responses were observed, those were used to populate the "minimum" and "maximum" fields, while "mean" was left blank (unless additional information was provided in the original study). To clarify, the maximum value was thus the maximum RL at which response was observed, and not a maximum level of exposure. The quantities plotted are SEL for the "impulsive," "tonals/sweeps," and "sonar" sound categories and SPLrms for the "continuous," "natural," "tonals/sweeps," and "sonar" categories. The "sonar" and "tonals/sweeps" categories were determined by sound context: if the experiment was intended to replicate sonar signals, but did not involve an operational sonar or echosounder mounted on a boat or ship operating in its usual environment, it was categorized as tonals/sweeps. Some authors measured additional levels (e.g., M-weighted SEL or in-air measurements), which were captured in the spreadsheet but not plotted, in general because there were not enough data of a single

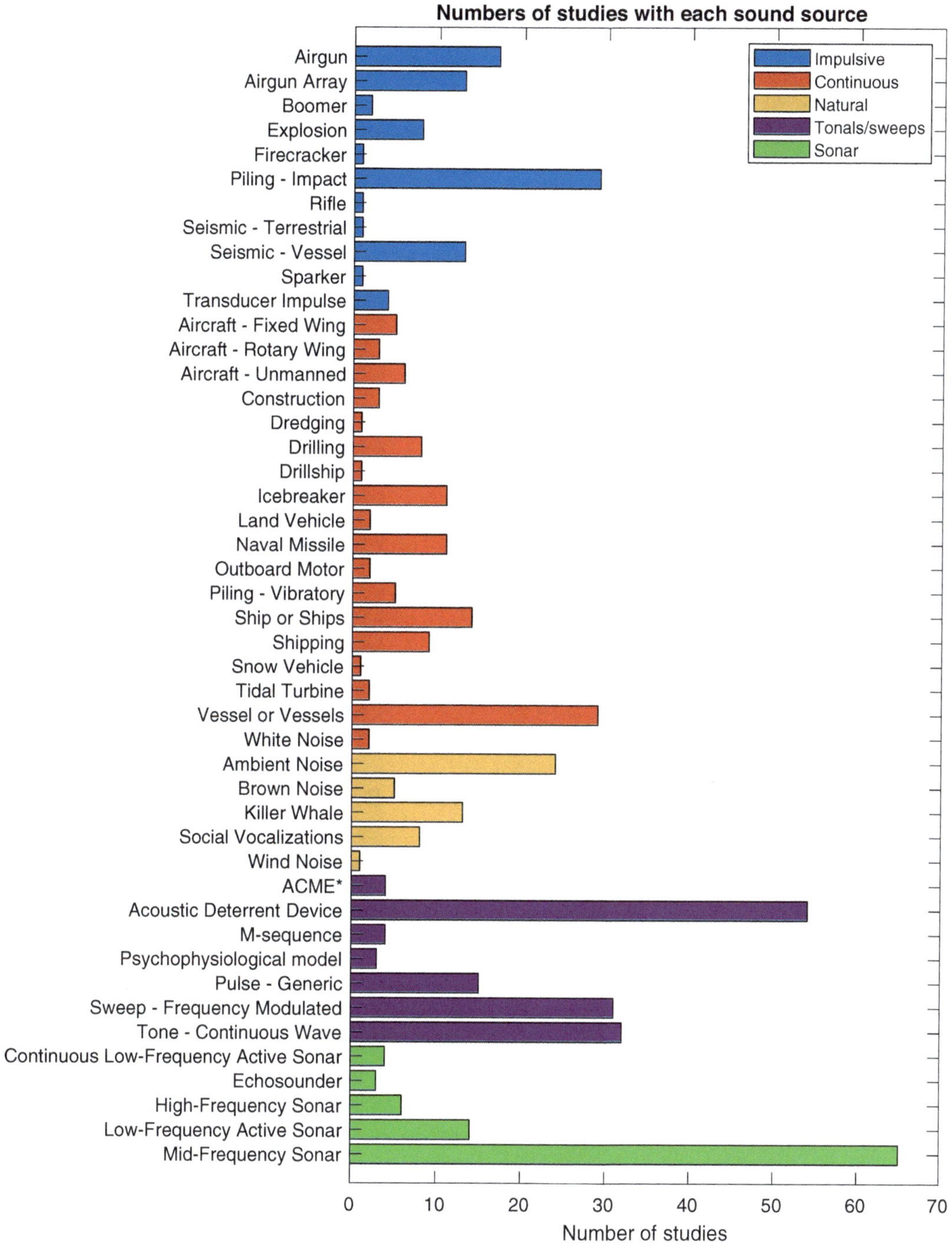

Fig. 10.3 Histogram showing how often each sound source was studied

type (beyond SPLrms and SEL) to produce meaningful plots.

Bearing in mind that every study design was different, and individual animals may respond differently to noise in the same or different contexts, a few general statements may perhaps be made from Figs. 10.5, 10.6, and 10.7. First, minimum RLs at which behavioral reactions of

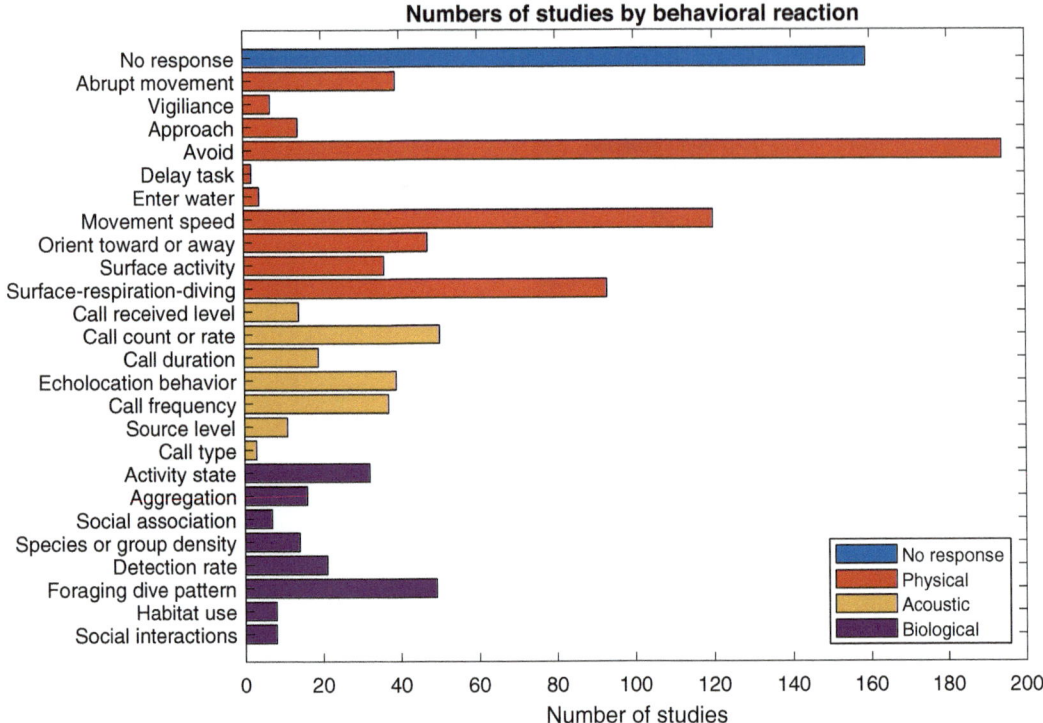

Fig. 10.4 Histogram of the types of behavioral responses observed, as summarized in Suppl. Material

any kind have been observed for a given species to a given sound category may vary by >50 dB. Second, responses to natural noise occur at similar RL ranges as responses to anthropogenic noise. That's partly because the variability is so great for all noise types. Third, no particular species stands out as particularly sensitive or particularly tolerant. Having said that, three species (harbor porpoises, bottlenose dolphins, and humpback whales) have been proportionately well studied, while quantitative studies on other species are sparse or nonexistent. To some degree, this reflects the difficulty of study design (Booth et al. 2022) with remote, cryptic, endangered, or evasive species, compared to studying species that are relatively common, occur in easily accessible coastal waters, or are available in captivity for structured studies.

10.8 Discussion

10.8.1 Characterizing Received Sound at the Animal

Many reports of marine mammals responding to anthropogenic sound are not included in this review or are only mentioned topically because there was no characterization of the noise received by the animal. There are many complexities in the relationship between the probability and magnitude of a marine mammal's behavioral response and an exposure to noise; and to understand this relationship, some knowledge of the level of noise exposure is required. At a minimum, every effort should be made to report the level of sound received by an animal in intentional exposure studies and incidental exposure observations. If a modicum of the details of the event are known, such as the distance between the noise source and the marine mammal and the

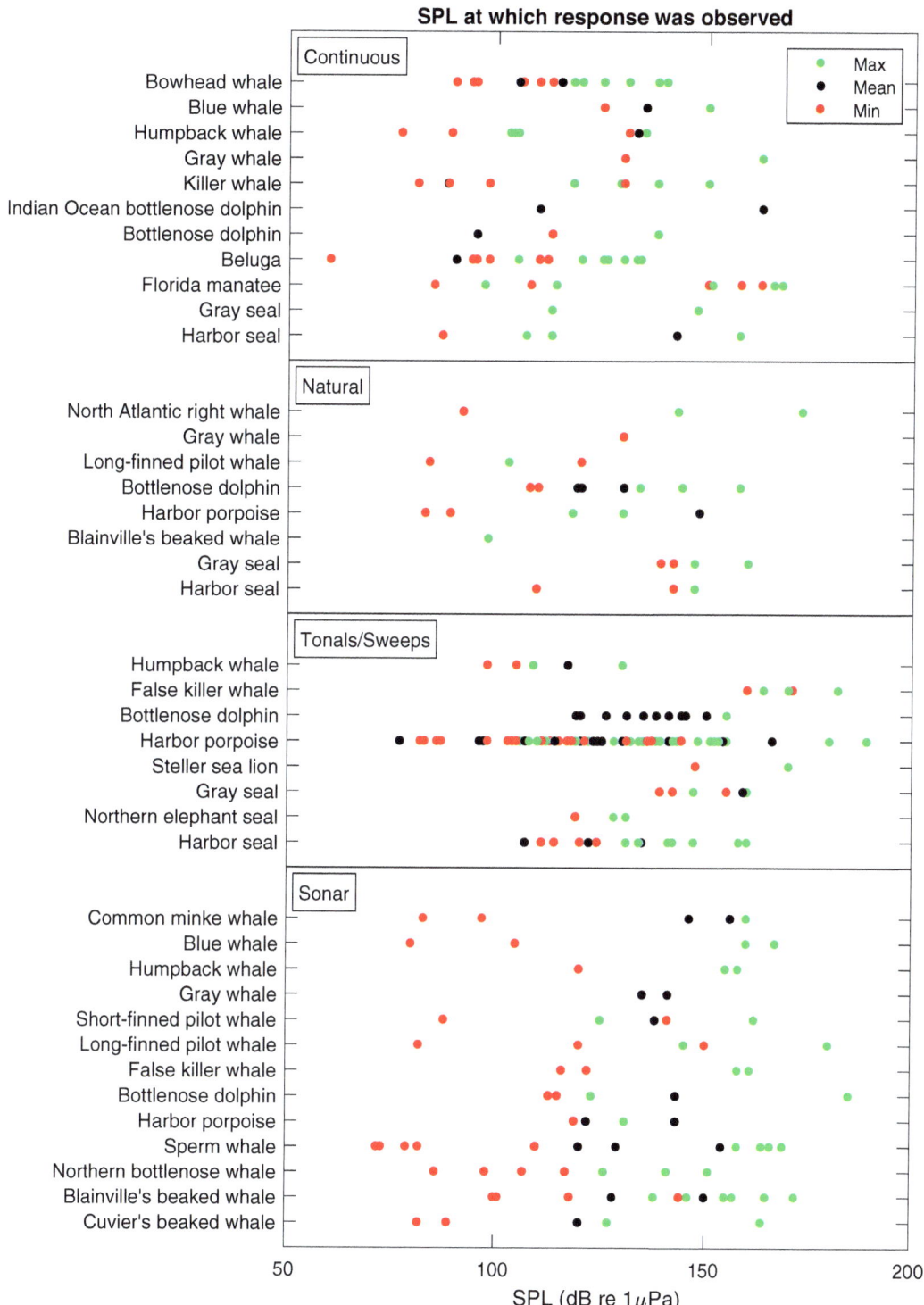

Fig. 10.5 Minimum, mean, and maximum received sound pressure levels at which behavioral responses were observed

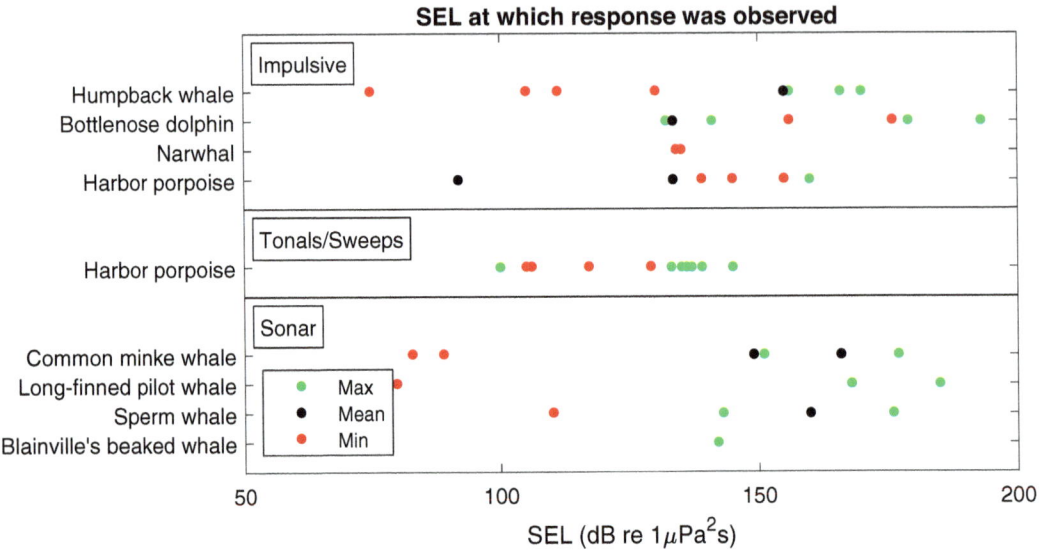

Fig. 10.6 Minimum, mean, and maximum sound exposure levels at which behavioral responses were observed

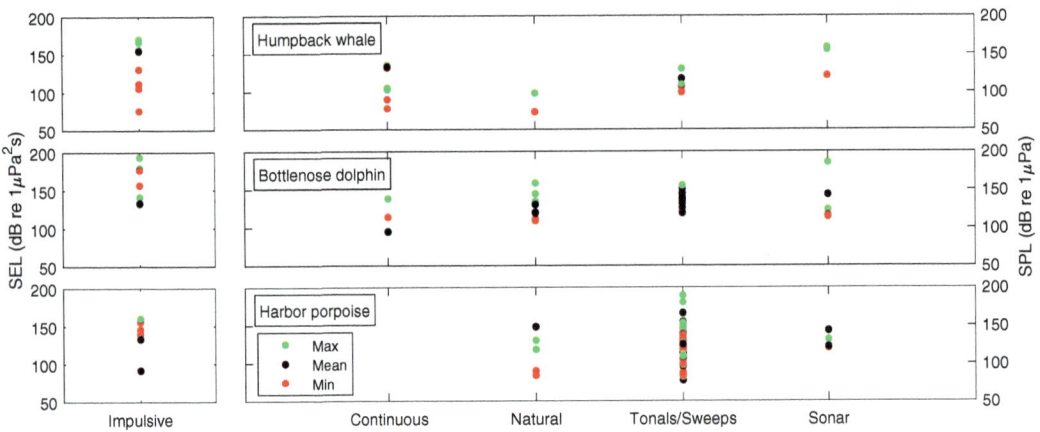

Fig. 10.7 Minimum, mean, and maximum received sound levels at which behavioral responses have been observed in the three most studied underwater species

noise SL, then even a crude model might be employed to estimate the RL. This approach at least provides some exposure information that can add to our growing knowledge on the subject. Sadly, many reports of the interaction between marine mammals and anthropogenic sound sources do not meet this bar.

The characteristics of the received noise are also a critical component of the exposure context. Beyond only reporting the RL of the noise exposure, studies and observations will be more informative if they report additional characteristics of the noise and exposure context. These include (but are not limited to) the duty cycle, information on frequency and amplitude modulation, distance between the source and receiver, etc. All of these factors could impact the probability that a noise-exposed marine mammal might elicit a behavioral response to the exposure. Over the last 30 years, marine mammal noise exposure studies have increased the amount and types of data that are collected, largely because of advances in monitoring technology. This has enabled study designs to become more sophisticated and reporting of exposure details more comprehensive. Nevertheless, there is considerable variation in the quality of reporting related to noise exposures. A high standard for reporting the acoustic details of a noise exposure should be maintained in order to facilitate our understanding of the relationship between noise and marine mammal behavior.

10.8.2 Individual Variability in Responsiveness

Not all members of a given species will react in the same way to noise exposure. For example, three harbor porpoises exposed to the same acoustic alarms showed differences in their degree of response to the alarms (Kastelein et al. 1995, 1997), and singing humpback whales were found to both increase song length and stop singing in the presence of low-frequency active sonar playbacks (Miller et al. 2000). These are just two examples of what is often a difficult aspect of understanding the behavioral response of animals to disturbance—individual variability. Numerous factors, both intrinsic (within an animal) and extrinsic (factors outside the animal), can affect whether a marine mammal responds to a noise exposure (Southall et al. 2016; Tablado and Jenni 2017; Wartzok et al. 2003). In this section, some of the intrinsic and extrinsic factors that potentially affect an animal's response to noise are discussed. These are factors that should be considered when judging both the response of a marine mammal to noise and the potential biological impact of the response.

10.8.2.1 Intrinsic Factors

Intrinsic factors are those characteristics of an individual or species that are a result of its biological makeup and experience. Intrinsic factors that have been observed to affect responsiveness to disturbance in animals include an animal's age, sex, reproductive status, behavioral state at the time of disturbance, and prior experience with the noise source (Tablado and Jenni 2017). In addition, species have different propensities for responsiveness due to differences in sensory abilities, degree of vigilance, and antipredator defense mechanisms. Sensory abilities and the degree of vigilance both directly impact when and how far away an animal can detect noise, which subsequently impacts a species' responsiveness to acoustic stimuli (Tablado and Jenni 2017). Since sensory capabilities of marine mammals differ across species, it should not be assumed that all marine mammal species have the same ability to equivalently detect anthropogenic sound (e.g., compare the hearing sensitivity of a dolphin to that of an elephant seal under water; see Chap. 8). Some terrestrial species have a propensity to freeze in response to disturbance, whereas others flee or fight. The response arises from antipredatory defense mechanisms but can be modulated by prior experience. It is quite likely that the same patterns exist in marine mammals (e.g., for discussion of the risk-disturbance hypothesis, see Harris et al. 2017).

Differences in responsiveness due to sensory abilities and antipredator defense mechanisms are also observed in marine mammal species. For example, differences in responsiveness to the same missile launches have been observed in elephant seals (least responsive), California sea lions, and harbor seals (most responsive; Holst et al. 2011). Responses to repetitive exposures of a gillnet pinger attached to benign objects and simulated fisheries equipment elicited quite variable responses from California sea lions, harbor seals, bottlenose dolphins, and Commerson's dolphins, with the latter species being the most reactive (Bowles and Anderson 2012). For similar types of sonar signals, long-finned pilot whales demonstrated a higher threshold of responsiveness (measured as the RL) and shorter response period than did killer whales (Antunes et al. 2014; Miller et al. 2014). In contrast, beaked whales appeared particularly responsive to sonar signals, chiefly with regard to cessation of foraging behaviors (Southall et al. 2016). Long-finned pilot whales have been observed to increase in group size and approach the source of playbacks of killer whale vocalizations obtained while the killer whales were feeding on herring (Curé et al. 2012). In contrast, avoidance responses have been observed in belugas, humpback whales, and gray whales exposed to playbacks of killer whale vocalizations (Cummings and Thompson 1971; Curé et al. 2015; Fish and Vania 1971). Male sperm whales exposed to the vocalizations of mammal-eating killer whales interrupted foraging and resting behavior and initiated social responses (Curé et al. 2013). The differences in the responses might point to a difference in the species' vulnerability to predation; belugas and gray whales are known prey of mammal-eating killer whales, but long-finned pilot whales have been observed to approach and chase killer whales (as reported in Curé et al. 2012; de Stephanis et al. 2015). Humpback whales have been observed to defend themselves once attacked and are considered "fight" strategists (Ford and Reeves 2008), but they demonstrate evasiveness upon initial detection of killer whales at range (Curé et al. 2015). Differences in the types of killer whale vocalizations used in the playback studies (mammal-eating vs. fish-eating) could modulate the response of the individual animals, complicating interpretation of the observations. However, the results overall are consistent with variability in species responsiveness to the presence of potential predators, which is likely further modulated by other intrinsic factors.

Age is inherently associated with hearing ability and experience and is often correlated with responsiveness to novel or threatening stimuli. In general, younger animals are expected to be more responsive than older, more experienced animals because they can have greater hearing sensitivity and do not have the same experience in categorizing signals as threatening or non-threatening. Houser et al. (2013a) found that behavioral dose-response curves of the California sea lion exposed to MFAS changed dramatically once young (<2 years old) sea lions were removed from the analysis (i.e., the probability of responsiveness decreased at low to moderate RL). In contrast, an older California sea lion completed most of the trials in which sonar signals were received at the highest levels of 185 dB re 1 µPa. The sea lions were different in their overall experience with anthropogenic sound sources; the naïve sea lions were more responsive to the novel stimulus, but the older, more experienced male was much more tolerant of the acoustic stimulus, possibly because of his much longer history of encountering and living around anthropogenic sound sources. In a similar study with bottlenose dolphins, the dose-response functions to simulated sonar signals also appeared to be affected by age, although the age distribution of animals involved in the study was quite different (Houser et al. 2013b). Although resulting in wide confidence intervals, the modeling of killer whale responsiveness to ship noise also suggested age as a factor affecting responsiveness (Williams et al. 2014). For at-sea animals, an accurate age assessment will likely be difficult under many circumstances. The experience of an animal with a noise source is related to an animal's age and also to the processes of habituation and sensitization. There will likely be distinct differences between the marine mammal exposed to a novel noise exposure and the marine mammal that has learned the consequence of a noise by repeated exposure through time. Several studies have demonstrated that marine mammals with prior experience with a disturbance are often less reactive on subsequent disturbances provided there is no immediate negative impact (e.g., Suryan and Harvey 1998) (but also see Chap. 11).

Sex and reproductive status have also been observed to affect the responsiveness of terrestrial mammals exposed to noise or other disturbance; females and females with offspring generally were the most responsive (e.g., Loehr et al. 2005; Saïd et al. 2012; Stankowich 2008). Similar trends have been observed in some marine mammals exposed to human (tourist) disturbance, such as the New Zealand fur seal (Boren et al. 2002), harp seal (Kovacs and Innes 1990), and possibly bottlenose dolphin (Symons et al. 2014). Evidence for sex and reproductive status as modulators of responses in marine mammals exposed specifically to noise are much less developed. Male Stellar sea lions holding territory for breeding are more difficult to displace due to noise or other disturbance, and California sea lions have been observed to be less responsive to rocket launches during the breeding season (Calkins 1979 [published 1983]; Holst et al. 2011). Conversely, harbor seals were more responsive to rocket launches during the breeding/pupping period (Holst et al. 2011). The presence of cetacean calves has also been observed to affect responses to noise exposure or disturbance, possibly as a function of the differences in dive capacity and behavior of the less developed calves (e.g., Robertson et al. 2013). However, considerably more research is needed to differentiate differences in responsiveness due to sex and reproductive state in marine mammals, particularly in cetaceans.

The behavior of an animal can modulate responsiveness to disturbance, particularly if the behavior during the disturbance is critical to survival, growth, or reproduction or if it requires a lot of attentional resources. It has previously been noted how pinnipeds change their responsiveness to acoustic disturbance during breeding periods. Belugas, bowhead whales, and right whales also seem less affected by aircraft noise when feeding, mating, or socializing than when they are resting (as summarized in Richardson and Würsig 1997). Broad variation in responsiveness has been observed in acute behavioral response studies performed with sonar-like signals across both mysticete and odontocete species (Harris and Thomas 2015; Southall et al. 2016). For example, blue whales off of the coast of Southern California have been found to more readily abandon deep foraging dives to dispersed prey when exposed to sonar signals than when conducting shallow dives on more condensed prey (Goldbogen et al. 2013). Using a recurrent survival analysis, Harris et al. (2015) found that the behavioral state and the species of the noise-exposed animal (long-finned pilot whales, sperm whales, and killer whales) modulated responsiveness to sonar signal exposure, with killer whales being the most responsive species. Because of the potential for the behavioral state of the animal to modulate responsiveness to a noise exposure, it is critical that future research efforts continue to document the behavior of noise-exposed marine mammals at the time of the noise exposure.

The temperament (or "personality") of an animal, defined by consistent individual differences in traits such as boldness, docility, curiosity, reactiveness, sociability, aggressiveness, etc., is an intrinsic factor that is not easily characterized in short-term observations of animals but which affects an animal's response to disturbance and novelty (Carrete and Tella 2010; Dingemanse et al. 2010; Martin and Réale 2008; May et al. 2016). Within a given population, individuals might be extremely responsive to disturbance or indifferent (e.g., Storch 2013), reflecting (at least partially) the temperament of the individual. The temperament of individuals can affect the makeup of a previously disturbed population of animals (Lowry et al. 2013); for example, animals that are tolerant of disturbance might remain in a disturbed area, whereas the more docile or risk-averse members of the population might relocate. As has been seen in comparisons of some urban and rural individuals of the same species, this can subsequently result in differences in the behaviors employed by the separate populations (Lowry et al. 2013).

10.8.2.2 Extrinsic Factors

Extrinsic factors are those factors external to an animal that contribute to the context of a noise exposure and the animal's responsiveness to the exposure. The noise characteristics (e.g., SL, frequency, duty cycle); the location, proximity, and bearing of the sound source; and its changing distance relative to the receiving marine mammal are important factors affecting the probability (and likely the magnitude) of a behavioral response (Ellison et al. 2012; Southall et al. 2016). These factors are under human control, and modifications to the use of underwater sound sources have been a subject of ongoing and evolving mitigation practices. Other extrinsic factors are those related to the habitat (e.g., prey availability), sound propagation conditions, weather, etc. These are factors that are not under human control and might not be known at the time the sound exposure occurs.

Terrestrial wildlife are generally more responsive to disturbance when the source of the disturbance is close and are less responsive when it is farther away (Holmes et al. 1993). Limited evidence suggests that the same pattern is possibly true in Cuvier's beaked whales exposed to sonar signals because responsiveness to the sonar exposure was less dependent on RL than it was on proximity of the sound source (DeRuiter et al. 2013b). Although data are too limited at this time to disentangle the relationship between RL, distance to the sound source, and the probability of a behavioral response, determining this relationship is a high priority (Harris and Thomas 2015; Southall et al. 2016).

Distance is not the only spatial factor affecting an animal's response to noise. Whether the sound source is approaching or moving away, and how quickly it is doing so, is also a consideration (Stankowich 2008). It is certain that wildlife species have the ability to capitalize on acoustic cues that tell the direction a sound source is moving (either approaching or moving away). The responses of long-finned pilot whales and killer whales to sonar playbacks showed a particular avoidance pattern if the vessel approached more quickly than the exposed individual's horizontal rate of movement (Antunes et al. 2014; Miller et al. 2014); animals moved away from (or at an acute angle from) the trajectory of the vessel playing sonar signals and generally moved 180° from the track of the vessel. Efforts have attempted to model the benefit of marine mammal movement strategies in response to noise and to understand why animals might choose to move in a particular way relative to a moving sound source (Wensveen et al. 2015). Although much yet remains to be discovered with respect to the dynamic spatial and acoustic relationship between a noise source and marine mammals, it is apparent that the trajectory of an active noise source can impact a marine mammal's decision to respond.

One final extrinsic factor worth noting is the degree to which a noise exposure is unexpected. In areas where noise sources are chronic, there is likely an anticipation of the noise and possibly some learned responses by animals due to prior experience with the noise. Unpredictable and sudden noise, on the other hand, is generally more likely to produce a behavioral response the magnitude of which is likely to be modulated by other factors (e.g., presence of ships). For this reason, ramp-up procedures in which the SL of a sound source is gradually increased have been adopted by or promoted for many noise producers (e.g., Johnson et al. 2007). The effectiveness of ramp-up procedures has been modeled or investigated for some scenarios (Cato et al. 2013; Miller et al. 2009; Robinson et al. 2007; von Benda-Beckmann et al. 2014; Wensveen et al. 2017), but the effectiveness of ramp-up procedures remains to be fully investigated (Barlow and Gisiner 2006; Cato et al. 2013).

10.8.3 The Challenge of Context

It is now well-accepted that context is critically important to a decision by an animal to behaviorally respond to a noise exposure and probably often affects the magnitude and duration of any response that occurs. Adequately determining and truly understanding the context of a noise exposure are a daunting task as there are certain aspects of the context that will almost always be uncertain (e.g., prior experience with a particular noise source or motivation for being in a particular area or behavioral state). Furthermore, contextual variability inherent to animals involved in noise exposure studies is extremely high, making it difficult to integrate the results of studies that may vary according to location (e.g., northern vs. southern hemisphere), time of year, social composition, or even between wild animals and animals held under human care. For example, both wild and captive belugas have been exposed to playbacks of drilling noise. While some wild belugas slowed, reversed, or milled at the onset of exposure at RLs of 112–134 dB (20–1000 Hz) (Richardson et al. 1990, 1991), captive belugas exhibited no statistically significant behavioral change at RLs of 134–153 dB (40–20,000 Hz) (Thomas et al. 1990). Behavior in captivity was averaged over 30-min observation windows, whereas instantaneous measures of behavior were taken in the wild. The behavior of the wild belugas returned to baseline shortly after initial responses were observed (within 15 min), and not all of the wild belugas responded as they swam past the noise projector. So then, how does one determine how the context of the exposures affected the results and how should the results be integrated? Did the belugas under human care have greater tolerance of the exposure due to prior exposures with industrial sounds at their facility? Did prior experience with trainer reinforcement change the threshold of responsiveness? How might the social composition of groups have affected the response dynamics of wild belugas? Why did some wild belugas respond while others did not; was it due to prior experience, or possibly differences in hearing sensitivities? How might the sound exposure have affected the response since the bandwidths of the two signals were different? Contextual complexities may never be completely or even adequately resolved, but the identification of certain contexts (e.g., exposures during deep vs. shallow foraging dives) has been useful in bracketing when behavioral responses to a particular noise source might be more likely in some species.

A major challenge exists between integrating results from captive and wild studies, largely because the contexts are so fundamentally different. Wild animals exposed to noise are the animals of primary concern, and exposures in the wild encompass the natural and anthropogenic contexts, the dynamics of which we wish to understand. However, it is nearly impossible to have complete knowledge of the overall context of a noise exposure in the wild, as inherent sensitivities and prior experiences may never be known. Conversely, much of the context of a noise exposure can be controlled by the experimenter when working with animals under human care, and a comprehensive history of the animal's experiences with noise could theoretically be obtained, but the context is unnatural and undoubtedly influenced by the animal-human relationship. Thus, whereas wild studies provide the most relevant information but with the greatest level of contextual uncertainty, captive studies provide increased contextual certainty that can address targeted questions (e.g., how does age affect responsiveness) but with a degree of artificiality that can be difficult to relate to wild counterparts.

Wild noise exposure studies have dramatically improved in defining and characterizing contextual covariates over the last two decades. However, few studies manage to fully incorporate all measured covariates into their analyses. This is likely due, in part, to the complexity of considering such inherently variable and potentially interactive features, particularly when unknown influences (e.g., animal temperament) are also factors. Consequently, while many studies report collecting contextual data, these data are partially (or even fully) discarded when reporting the results of the study. Recently, studies have improved in their ability to consider a greater number of contextual covariates in behavioral response research, but overall, contextual data continue to be underutilized in the majority of noise exposure studies.

10.9 Summary

This chapter presented an overview of physical and acoustic behavioral responses of marine mammals to underwater sounds. A literature review was undertaken, and data on received levels at the animal when certain types of responses were observed were compiled in an online supplementary spreadsheet. Based on this, an overview of responses was written, organized first by species and then by sound type. In-air and underwater sound sources were considered. The most studied sound types were mid-frequency sonar and acoustic deterrent devices and then impact pile driving and vessels. The most observed response was avoidance, followed by changes in swim speed and surface-respiration-dive behavior. However, no response was the second most common observation, after avoidance. Easily accessible coastal or captive species (i.e., harbor porpoises, bottlenose dolphins, and humpback whales) have been comparatively well studied. There has been great variability in study design, response observation and classification, received level derivation and unit, as well as (statistical) analyses. Given the different environments, contexts, populations, and individuals that have been studied, it is not surprising that great variability has been reported in minimum received levels at which behavioral responses were observed: >50 dB variability for the most studied species and sound type combinations.

10.10 Data Availability

Supplementary material may be found at https://cmst.curtin.edu.au/research/marine-mammal-bioacoustics/ or requested from the first author.

References

Aarts G, Brasseur S, Kirkwood R (2018) Behavioural response of grey seals to pile-driving (Report C006/18). Wageningen Marine Research, Den Helder

Acevedo-Gutiérrez A, Cendejas-Zarelli S (2011) Nocturnal haul-out patterns of harbor seals (*Phoca vitulina*) related to airborne noise levels in Bellingham, Washington, USA. Aquat Mamm 37(2):167–174. https://doi.org/10.1578/AM.37.2.2011.167

Ackerman BB, Wright SD, Bonde RK, Odell DK, Banowetz DJ (1995) Trends and patterns in mortality of manatees in Florida, 1974-1992. In: O'Shea TJ, Ackerman BB, Percival HF (eds) Population biology of the Florida manatee (*Trichechus manatus latirostris*). National Biological Service, Washington, DC, pp 223–258

Aguilar de Soto N, Visser F, Tyack PL, Alcazar J, Ruxton G, Arranz P, Madsen PT, Johnson M (2020) Fear of killer whales drives extreme synchrony in deep diving beaked whales. Sci Rep 10(1):13. https://doi.org/10.1038/s41598-019-55911-3

Aguilar Soto N, Johnson M, Madsen PT, Tyack PL, Bocconcelli A, Borsani JF (2006) Does intense ship noise disrupt foraging in deep-diving Cuvier's beaked whales (*Ziphius cavirostris*)? Mar Mamm Sci 22(3):690–699. https://doi.org/10.1111/j.1748-7692.2006.00044.x

Akamatsu T, Hatakeyama Y, Takatsu N (1993) Effects of pulse sounds on escape behavior of false killer whales. Bull Jpn Soc Sci Fish 59(8):1297–1303. https://doi.org/10.2331/suisan.59.1297

Akamatsu T, Nakamura K, Nitto H, Watabe M (1996) Effects of underwater sounds on escape behaviour of Steller sea lions. Fish Sci 62(4):503–510. https://doi.org/10.2331/fishsci.62.503

Allen AN, Schanze JJ, Solow AR, Tyack PL (2014) Analysis of a Blainville's beaked whale's movement response to playback of killer whale vocalizations. Mar Mamm Sci 30(1):154–168. https://doi.org/10.1111/mms.12028

Alves A, Antunes R, Tyack PL, Miller PJO, Lam F-PA, Kvadsheim PH (2014) Vocal matching of naval sonar signals by long-finned pilot whales (*Globicephala melas*). Mar Mamm Sci 30(3):1248–1257. https://doi.org/10.1111/mms.12099

Amstrup SC (1993) Human disturbances of denning polar bears in Alaska. Arctic 46(3):246–250. https://www.jstor.org/stable/40511412

Andersen M, Aars J (2007) Short-term behavioural response of polar bears (*Ursus maritimus*) to snowmobile disturbance. Polar Biol 31(4):501-507. https://doi.org/10.1007/s00300-007-0376-x

Andersen S, Teilmann J, Dietz R, Schmidt N, Miller L (2012) Behavioural responses of harbour seals to human-induced disturbances. Aquat Conserv 22(1):113–121. https://doi.org/10.1002/aqc.1244

André M, Terada M, Watanabe Y (1997) Sperm whale (*Physeter macrocephalus*) behavioural response after the playback of artificial sounds. Rep Int Whal Comm 47:499–504

André M, Caballe A, Van der Schaar M, Solsona A, Houegnigan L, Zaugg S, Sanchez AM, Castell JV, Sole M, Vila F, Djokic D, Adrian-Martinez S, Albert A, Anghinolfi M, Anton G, Ardid M, Aubert JJ, Avgitas T, Baret B, Barrios-Marti J, Basa S, Bertin V, Biagi S, Bormuth R, Bouwhuis MC, Bruijn R, Brunner J, Busto J, Capone A, Caramete L, Carr J, Celli S, Chiarusi T, Circella M, Coleiro A, Coniglione R, Costantini H, Coyle P, Creusot A, Deschamps A, De Bonis G, Distefano C, Di Palma I, Donzaud C, Dornic D, Drouhin D, Eberl T, El Bojaddaini I, Elsasser D, Enzenhofer A, Fehn K, Felis I, Fusco LA, Galata S, Gay P, Geisselsoder S, Geyer K, Giordano V, Gleixner A, Glotin H, Gracia-Ruiz R, Graf K, Hallmann S, van Haren H, Heijboer AJ, Hello Y, Hernandez-Rey JJ, Hossl J, Hofestadt J, Hugon C, Illuminati G, James CW, de Jong M, Jongen M, Kadler M, Kalekin O, Katz U, Kiessling D, Kouchner A, Kreter M, Kreykenbohm I, Kulikovskiy V, Lachaud C, Lahmann R, Lefevre D, Leonora E, Loucatos S, Marcelin M, Margiotta A, Marinelli A, Martinez-Mora JA, Mathieu A, Melis K, Michael T, Migliozzi P, Moussa A, Mueller C, Nezri E, Pavalas GE, Pellegrino C, Perrina C, Piattelli P, Popa V, Pradier T, Racca C, Riccobene G, Roensch K, Saldana M, Samtleben DFE, Sanguineti M, Sapienza P, Schnabel J, Schussler F, Seitz T, Sieger C, Spurio M, Stolarczyk T, Sanchez-

Losa A, Taiuti M, Trovato A, Tselengidou M, Turpin D, Tonnis C, Vallage B, Vallee C, Van Elewyck V, Vivolo D, Wagner S, Wilms J, Zornoza JD, Zuniga J (2017) Sperm whale long-range echolocation sounds revealed by ANTARES, a deep-sea neutrino telescope. Sci Rep 7:45517. https://doi.org/10.1038/srep45517

Antichi S, Jaramillo-Legorreta AM, Urbán RJ, Martínez-Aguilar S, Viloria-Gómora L (2022) Small vessel impact on the whistle parameters of two ecotypes of common bottlenose dolphin (*Tursiops truncatus*) in La Paz Bay, Mexico. Diversity 14(9):712. https://doi.org/10.3390/d14090712

Antunes R, Kvadsheim PH, Lam FPA, Tyack PL, Thomas L, Wensveen PJ, Miller PJO (2014) High thresholds for avoidance of sonar by free-ranging long-finned pilot whales (*Globicephala melas*). Mar Pollut Bull 83(1):165–180. https://doi.org/10.1016/j.marpolbul.2014.03.056

Arranz P, Glarou M, Sprogis KR (2021) Decreased resting and nursing in short-finned pilot whales when exposed to louder petrol engine noise of a hybrid whale-watch vessel. Sci Rep 11(1):21195. https://doi.org/10.1038/s41598-021-00487-0

Aubin JA, Mikus M-A, Michaud R, Mennill D, Vergara V (2023) Fly with care: belugas show evasive responses to low altitude drone flights. Mar Mamm Sci 39(3): 718–739. https://doi.org/10.1111/mms.12997

Azzara AJ, Zharen WMV, Newcomb JJ (2013) Mixed-methods analytic approach for determining potential impacts of vessel noise on sperm whale click behavior. J Acoust Soc Am 134(6):4566–4574. https://doi.org/10.1121/1.4828819

Back JJ, Hoskins AJ, Kirkwood R, Arnould JPY (2018) Behavioral responses of Australian fur seals to boat approaches at a breeding colony. Nat Conserv 31:35–52. https://doi.org/10.3897/natureconservation.31.26263

Bagočius D (2014) Potential masking of the Baltic grey seal vocalisations by underwater shipping noise in the Lithuanian area of the Baltic Sea. Environ Res Eng Manag 70(4):66–72. https://doi.org/10.5755/j01.erem.70.4.6913

Baird RW (2016) The lives of Hawaii's dolphins and whales: natural history and conservation. University of Hawaii Press, Honolulu

Baird RW, Martin SW, Webster DL, Southall BL (2014) Assessment of modelled received sound pressure levels and movements of satellite-tagged odontocetes exposed to mid-frequency active sonar at the Pacific Missile Range Facility: February 2011 through February 2013. NAVFAC PAC, U.S. Pacific Fleet, Report by HDR Environmental, Operations and Construction Inc

Baird RW, Henderson E, Martin SW, Southall BL (2019) Assessing odontocete exposure and response to mid-frequency active sonar during submarine command courses at the Pacific Missile Range Facility: 2016 through 2018 (Report submitted to Naval Facilities Engineering Command Pacific under HDR Environmental, Operations and Construction, Inc. Contract No. N62470-15-D-8006, TO KB16). HDR Inc., Honolulu

Balcomb KC, Claridge DE (2001) A mass stranding of cetaceans caused by naval sonar in the Bahamas. Bahamas J Sci 01(05):2–12

Bankhead KR, Freeman G, Goebel WH, Acevedo-Gutiérrez A (2023) Effects of anthropogenic noise on haul-out numbers of harbor seals (*Phoca vitulina*). Can J Zool 101(9):720-728. https://doi.org/10.1139/cjz-2023-0053

Barlow J, Gisiner R (2006) Mitigating, monitoring and assessing the effects of anthropogenic sound on beaked whales. J Cetacean Res Manag 7(3):239–249. https://doi.org/10.47536/jcrm.v7i3.734

Barlow J, Tyack PL, Johnson MP, Baird RW, Schorr GS, Andrews RD, Aguilar de Soto N (2013) Trackline and point detection probabilities for acoustic surveys of Cuvier's and Blainville's beaked whales. J Acoust Soc Am 134(3):2486–2496. https://doi.org/10.1121/1.4816573

Basran CJ, Woelfing B, Neumann C, Rasmussen MH (2020) Behavioural responses of humpback whales (*Megaptera novaeangliae*) to two acoustic deterrent devices in a northern feeding ground off Iceland. Aquat Mamm 46(6):584–602. https://doi.org/10.1578/AM.46.6.2020.584

Baumgartner MF, Fratantoni DM, Hurst TP, Brown MW, Cole TVN, Van Parijs SM, Johnson M (2013) Real-time reporting of baleen whale passive acoustic detections from ocean gliders. J Acoust Soc Am 134(3):1814–1823. https://doi.org/10.1121/1.4816406

Beale CM, Monaghan P (2004) Behavioural responses to human disturbance: a matter of choice? Anim Behav 68(5):1065–1069. https://doi.org/10.1016/j.anbehav.2004.07.002

Bejder L, Samuels A, Whitehead H, Finn H, Allen S (2009) Impact assessment research: use and misuse of habituation, sensitisation and tolerance in describing wildlife responses to anthropogenic stimuli. Mar Ecol Prog Ser 395:177–185. https://doi.org/10.3354/meps07979

Benhemma-Le Gall A, Graham IM, Merchant ND, Thompson PM (2021) Broad-scale responses of harbor porpoises to pile-driving and vessel activities during offshore windfarm construction. Front Mar Sci 8: 664724. https://doi.org/10.3389/fmars.2021.664724

Benhemma-Le Gall A, Thompson P, Merchant N, Graham I (2023) Vessel noise prior to pile driving at offshore windfarm sites deters harbour porpoises from potential injury zones. Environ Impact Assess Rev 103:107271. https://doi.org/10.1016/j.eiar.2023.107271

Berg EA, Nieto MP, Thorson PH, Francine JK, Oliver G (2001) Acoustic measurements of the 21 November 2000 Delta II EO-1 launch and quantitative analysis of behavioral responses of Pacific Harbor Seals, Brown Pelicans and Southern Sea Otters on Vandenberg Air Force Base and Selected Pinnipeds on San Miguel

Island, CA (Report for United States Air Force). SRS Technologies, Manhattan Beach

Berrow S, Cosgrove R, Leeney RH, O'Brien J, McGrath D, Dalgard J, Gall YL (2008) Effect of acoustic deterrents on the behaviour of common dolphins (*Delphinus delphis*). J Cetacean Res Manag 10(3):227–233. https://doi.org/10.47536/jcrm.v10i3.639

Biassoni N, Miller PJO, Tyack PL (2000) Preliminary results of the effects of SURTASS-LFA sonar on singing humpback whales (Report WHOI-2000-06, ADA378666). Woods Hole Oceanographic Institution

Blackwell SB, Lawson JW, Williams MT (2004) Tolerance by ringed seals (*Phoca hispida*) to impact pipe-driving and construction sounds at an oil production island. J Acoust Soc Am 115(5):2346–2357. https://doi.org/10.1121/1.1701899

Blackwell S, McDonald T, Kim K, Aerts L, Richardson WJ, Greene CJ, Streever B (2012) Directionality of bowhead whale calls measured with multiple sensors. Mar Mamm Sci 28(1):200–212. https://doi.org/10.1111/j.1748-7692.2011.00464.x

Blackwell SB, Nations CS, McDonald TL, Greene CRJ, Thode AM, Guerra M, Macradner AM (2013) Effects of airgun sounds on bowhead whale calling rates in the Alaskan Beaufort Sea. Mar Mamm Sci 29(4):E342–E365. https://doi.org/10.1111/mms.12001

Blackwell SB, Nations CS, McDonald TL, Thode AM, Mathias D, Kim KH, Green CR, Macrander AM (2015) Effects of airgun sounds on bowhead whale calling rates: evidence for two behavioral thresholds. PLoS One 10(6):e0125720. https://doi.org/10.1371/journal.pone.0125720

Blackwell SB, Nations CS, Thode AM, Kauffman ME, Conrad AS, Norman RG, Kim KH (2017) Effects of tones associated with drilling activities on bowhead whale calling rates. PLoS One 12(11):e0188459. https://doi.org/10.1371/journal.pone.0188459

Blair HB, Merchant ND, Friedlaender AS, Wiley DN, Parks SE (2016) Evidence for ship noise impacts on humpback whale foraging behaviour. Biol Lett 12(8): 20160005. https://doi.org/10.1098/rsbl.2016.0005

Blundell GM, Pendleton GW (2015) Factors affecting haul-out behavior of harbor seals (*Phoca vitulina*) in tidewater glacier inlets in Alaska: can tourism vessels and seals coexist? PLoS One 10(5):e0125486. https://doi.org/10.1371/journal.pone.0125486

Boisseau O, McGarry T, Stephenson S, Compton R, Cucknell AC, Ryan C, McLanaghan R, Moscrop A (2021) Minke whales *Balaenoptera acutorostrata* avoid a 15 kHz acoustic deterrent device (ADD). Mar Ecol Prog Ser 667:191–206. https://doi.org/10.3354/meps13690

Booth CG (2020) Food for thought: harbor porpoise foraging behavior and diet inform vulnerability to disturbance. Mar Mamm Sci 36(1):195–208. https://doi.org/10.1111/mms.12632

Booth C, Heinis F (2018) Updating the Interim PCoD Model: workshop report—new transfer functions for the effects of permanent threshold shifts on vital rates in marine mammal species (Report for the University of Aberdeen and Department for Business, Energy and Industrial Strategy (BEIS) SMRUC-UOA-2018-006). SMRU Consulting, St. Andrews

Booth CG, Brannan N, Dunlop R, Friedlander A, Isojunno S, Miller P, Quick N, Southall B, Pirotta E (2022) A sampling, exposure and receptor framework for identifying factors that modulate behavioural responses to disturbance in cetaceans. J Anim Ecol 91(10):1948–1960. https://doi.org/10.1111/1365-2656.13787

Boren L, Gemmell N, Barton K (2002) Tourist disturbance on New Zealand fur seals (*Arctocephalus forsteri*). Aust Mammal 24(1):85–96. https://doi.org/10.1071/AM02085

Born EW, Wiig Ø, Olsen MT (2021) Chapter 12—anthropogenic impacts on the Atlantic walrus. In: Keighley X, Olsen MT, Jordan P, Desjardins S (eds) The Atlantic walrus. Academic, pp 263–308. https://doi.org/10.1016/B978-0-12-817430-2.00013-3

Bowers MT, Friedlaender AS, Janik VM, Nowacek DP, Quick NJ, Southall BL, Read AJ (2018) Selective reactions to different killer whale call categories in two delphinid species. J Exp Biol 221(11). https://doi.org/10.1242/jeb.162479

Bowles A, Anderson R (2012) Behavioural responses and habituation of pinnipeds and small cetaceans to novel objects and simulated fishing gear with and without a pinger. Aquat Mamm 38(2):161–188. https://doi.org/10.1578/AM.38.2.2012.161

Bowles AE, Smultea M, Würsig B, Demaster DP, Palka D (1994) Relative abundance and behavior of marine mammals exposed to transmissions from the Heard Island Feasibility Test. J Acoust Soc Am 96(4): 2469–2484. https://doi.org/10.1121/1.410120

Brandt MJ, Diederichs A, Betke K, Nehls G (2011) Responses of harbour porpoises to pile driving at the Horns Rev II offshore wind farm in the Danish North Sea. Mar Ecol Prog Ser 421:205–216. https://doi.org/10.3354/meps08888

Brandt MJ, Diederichs A, Betke K, Nehls G (2012) Effects of offshore pile driving on harbor porpoises (*Phocoena phocoena*). In: Popper AN, Hawkins A (eds) The effects of noise on aquatic life. Springer, New York, pp 281–284. https://doi.org/10.1007/978-1-4419-7311-5_62

Brandt MJ, Hoschle C, Diederichs A, Betke K, Matuschek R, Nehls G (2013a) Seal scarers as a tool to deter harbour porpoises from offshore construction sites. Mar Ecol Prog Ser 475:291–302. https://doi.org/10.3354/meps10100

Brandt MJ, Hoschle C, Diederichs A, Betke K, Matuschek R, Witte S, Nehls G (2013b) Far-reaching effects of a seal scarer on harbour porpoises, *Phocoena phocoena*. Aquat Conserv 23(2):222–232. https://doi.org/10.1002/aqc.2311

Branstetter BK, Bowman VF, Houser DS, Tormey M, Banks P, Finneran JJ, Jenkins K (2018) Effects of vibratory pile driver noise on echolocation and vigilance in bottlenose dolphins (*Tursiops truncatus*). J Acoust Soc Am 143(1):429–439. https://doi.org/10.1121/1.5021555

Brasseur S, Aarts G, Schop J (2022) Measurement of effects of piledriving in the Borssele wind farm zone on the seals in the Dutch Delta area: changes in dive behaviour, haul-out and stranding of harbour and grey seals (Report C055/22). Wageningen Marine Research, Den Helder

Breed GA, Matthews CJD, Marcoux M, Higdon JW, LeBlanc B, Petersen SD, Orr J, Reinhart NR, Ferguson SH (2017) Sustained disruption of narwhal habitat use and behavior in the presence of Arctic killer whales. Proc Natl Acad Sci 114(10):2628–2633. https://doi.org/10.1073/pnas.1611707114

Brown AM, Allen SJ, Kelly N, Hodgson AJ (2023) Using unoccupied aerial vehicles to estimate availability and group size error for aerial surveys of coastal dolphins. Remote Sens Ecol Conserv 9(3):340–353. https://doi.org/10.1002/rse2.313

Brueggeman J (1993) Walrus response to offshore drilling operations. J Acoust Soc Am 94(3_Suppl):1828. https://doi.org/10.1121/1.407788

Bruno C, Caserta V, Salzeri P, Bonanno Ferraro G, Pecoraro F, Lucchetti A, Boitani L, Blasi M (2021) Acoustic deterrent devices as mitigation tool to prevent dolphin-fishery interactions in the Aeolian Archipelago (Southern Tyrrhenian Sea, Italy). Mediterr Mar Sci 22:408–421. https://doi.org/10.12681/mms.23129

Buck JR, Tyack PL (2000) Response of gray whales to low-frequency sounds (Abstract). J Acoust Soc Am 107(5_Suppl):2774. https://doi.org/10.1121/1.428908

Buck JR, Tyack PL (2003) An avoidance behavior model for migrating whale populations (Abstract). J Acoust Soc Am 113(4_Suppl):2326. https://doi.org/10.1121/1.4780811

Buckstaff KC, Wells RS, Gannon JG, Nowacek DP (2013) Responses of bottlenose dolphins (*Tursiops truncatus*) to construction and demolition of coastal marine structures. Aquat Mamm 39(2):174–186. https://doi.org/10.1578/AM.39.2.2013.174

Burgess WC, Tyack PL, Le Boeuf BJ, Costa DP (1998) A programmable acoustic recording tag and first results from free-ranging northern elephant seals. Deep-Sea Res II Top Stud Oceanogr 45(7):1327–1351. https://doi.org/10.1016/S0967-0645(98)00032-0

Burslem A, Isojunno S, Pirotta E, Miller PJO (2022) Modelling the impact of condition-dependent responses and lipid-store availability on the consequences of disturbance in a cetacean. Conserv Physiol 10(1):coac069. https://doi.org/10.1093/conphys/coac069

Buscaino G, Ceraulo M, Alonge G, Pace DS, Grammauta R, Maccarrone V, Bonanno A, Mazzola S, Papale E (2021) Artisanal fishing, dolphins, and interactive pinger: a study from a passive acoustic perspective. Aquat Conserv 31(8):2241–2256. https://doi.org/10.1002/aqc.3588

Calkins DG (1979 [published 1983]) Marine mammals of Lower Cook Inlet and the potential for impact from outer continental shelf oil and gas exploration, development, and transport (NTIS: PB85-201226). National Oceanic and Atmospheric Administration, Juneau, AK

Calleson CS, Kipp Frohlich R (2007) Slower boat speeds reduce risks to manatees. Endanger Species Res 3(3):295–304. https://doi.org/10.3354/esr00056

Carome W, Rayment W, Slooten E, Bowman MH, Dawson SM (2023) Vessel traffic influences distribution of Aotearoa New Zealand's endemic dolphin (*Cephalorhynchus hectori*). Mar Mamm Sci 39(2):626–647. https://doi.org/10.1111/mms.12995

Carrete M, Tella JL (2010) Individual consistency in flight initiation distances in burrowing owls: a new hypothesis on disturbance-induced habitat selection. Biol Lett 6(2):167–170. https://doi.org/10.1098/rsbl.2009.0739

Castellote M, Clark C, Lammers M (2012) Acoustic and behavioural changes by fin whales (*Balaenoptera physalus*) in response to shipping and airgun noise. Biol Conserv 147(1):115–122. https://doi.org/10.1016/j.biocon.2011.12.021

Cato DH, Noad MJ, Dunlop RA, McCauley RD, Gales NJ, Salgado Kent CP, Kniest H, Paton D, Jenner KCS, Noad J, Maggi AL, Parnum IM, Duncan AJ (2013) A study of the behavioural response of whales to the noise of seismic air guns: design, methods and progress. Acoust Aust 41(1):88–97

Cerchio S, Strindberg S, Collins T, Bennett C, Rodenbaum H (2014) Seismic surveys negatively affect humpback whale singing activity off northern Angola. PLoS One 9(3):e86464. https://doi.org/10.1371/journal.pone.0086464

Chen F, Shapiro GI, Bennett KA, Ingram SN, Thompson D, Vincent C, Russell DJF, Embling CB (2017) Shipping noise in a dynamic sea: a case study of grey seals in the Celtic Sea. Mar Pollut Bull 114(1):372–383. https://doi.org/10.1016/j.marpolbul.2016.09.054

Cholewiak D, DeAngelis AI, Palka D, Corkeron PJ, Van Parijs SM (2017) Beaked whales demonstrate a marked acoustic response to the use of shipboard echosounders. R Soc Open Sci 4:170940. https://doi.org/10.1098/rsos.170940

Christiansen F, Rojano-Doñate L, Madsen PT, Bejder L (2016) Noise levels of multi-rotor unmanned aerial vehicles with implications for potential underwater impacts on marine mammals. Front Mar Sci 3:277. https://doi.org/10.3389/fmars.2016.00277

Christiansen F, Nielsen MLK, Charlton C, Bejder L, Madsen PT (2020) Southern right whales show no behavioral response to low noise levels from a nearby unmanned aerial vehicle. Mar Mamm Sci 36(3):953–963. https://doi.org/10.1111/mms.12699

Clark CW, Ellison WT (2000) Calibration and comparison of the acoustic location methods used during the spring migration of the bowhead whale, *Balaena mysticetus*, off Pt. Barrow, Alaska, 1984-1993. J Acoust Soc Am 107(6):3509–3517. https://doi.org/10.1121/1.429421

Clark CW, Gillespie D, Nowacek DP, Parks SE (2007) Listening to their world: acoustics for monitoring and protecting right whales in an urbanized ocean. In: Kraus SD, Rolland RM (eds) The urban whale: North Atlantic right whales at the crossroads. Harvard University Press, Cambridge, MA, pp 333–357

Clark CW, Brown MW, Corkeron P (2010) Visual and acoustic surveys for North Atlantic right whales, *Eubalaena glacialis*, in Cape Cod Bay, Massachusetts, 2001-2005: management implications. Mar Mamm Sci 26(4):837–854. https://doi.org/10.1111/j.1748-7692.2010.00376.x

Cominelli S, Devillers R, Yurk H, MacGillivray A, McWhinnie L, Canessa R (2018) Noise exposure from commercial shipping for the southern resident killer whale population. Mar Pollut Bull 136:177–200. https://doi.org/10.1016/j.marpolbul.2018.08.050

Coram A, Gordon J, Thompson D, Northridge SP (2014) Evaluating and assessing the relative effectiveness of acoustic deterrent devices and other non-lethal measures on marine mammals. University of St Andrews, Sea Mammal Research Unit, St Andrews

Cosens SE, Dueck LP (1988) Responses of migrating narwhal and beluga to icebreaker traffic at the Admiralty Inlet ice-edge, N.W.T. in 1986. In: Sackinger WM, Jeffries MO (eds) Port and ocean engineering under Arctic conditions. Geophysical Institute, University of Alaska, Fairbanks, pp 39–54

Costa DP, Crocker DE, Gedamke J, Webb PM, Houser DS, Blackwell SB, Waples D, Hayes SA, Le Boeuf BJ (2003) The effect of a low-frequency sound source (acoustic thermometry of the ocean climate) on the diving behavior of juvenile northern elephant seals, *Mirounga angustirostris*. J Acoust Soc Am 113(2):1155–1165. https://doi.org/10.1121/1.1538248

Costa DP, Schwarz L, Robinson P, Schick RS, Morris PA, Condit R, Crocker DE, Kilpatrick AM (2016) A bioenergetics approach to understanding the population consequences of disturbance: elephant seals as a model system. In: Popper AN, Hawkins T (eds) The effects of noise on aquatic life II. Springer, New York, pp 161–169. https://doi.org/10.1007/978-1-4939-2981-8_19

Courbis S, Timmel G (2009) Effects of vessels and swimmers on behaviour of Hawaiian spinner dolphins (*Stenella longirostris*) in Kealake'akua, Honaunau, and Kauhako bays, Hawai'i. Mar Mamm Sci 25(2):430–440. https://doi.org/10.1111/j.1748-7692.2008.00254.x

Courts R, Erbe C, Wellard R, Boisseau O, Jenner KC, Jenner M-N (2020) Australian long-finned pilot whales (*Globicephala melas*) emit stereotypical, variable, biphonic, multi-component, and sequenced vocalisations, similar to those recorded in the northern hemisphere. Sci Rep 10(1):20609. https://doi.org/10.1038/s41598-020-74111-y

Cox TM, Read AJ, Solow A, Tregenza N (2001) Will harbour porpoises (*Phocoena phocoena*) habituate to pingers? J Cetacean Res Manag 3(1):81–86. https://doi.org/10.1111/mms.12880

Cox TM, Read AJ, Swanner D, Urian K, Waples D (2003) Behavioral responses of bottlenose dolphins, *Tursiops truncatus*, to gillnets and acoustic alarms. Biol Conserv 115:203–212. https://doi.org/10.1016/S0006-3207(03)00108-3

Croll DA, Clark CW, Calambokidis J, Ellison WT, Tershy BR (2001) Effect of anthropogenic low-frequency noise on the foraging ecology of *Balaenoptera* whales. Anim Conserv 4(1):13–27. https://doi.org/10.1017/s1367943001001020

Culloch RM, Foley A, Haberlin D, McGovern B, Pinfield R, Jessopp M, Cronin M (2020) Occurrence, site-fidelity and abundance of bottlenose dolphins (*Tursiops truncatus*) in Broadhaven bay, northwest Ireland during long-term construction of a gas pipeline. Reg Stud Mar Sci 34:100983. https://doi.org/10.1016/j.rsma.2019.100983

Cummings WC, Thompson PO (1971) Gray whales, *Eschrichtius robustus*, avoid the underwater sounds of killer whales, *Orcinus orca*. Fish Bull 69(3):525–530. https://doi.org/10.1121/1.407772

Curé C, Antunes R, Samarra F, Alves AC, Visser F, Kvadsheim P, Miller PJO (2012) Pilot whales attracted to killer whale sounds: acoustically-mediated interspecific interactions in cetaceans. PLoS One 7(12):e52201. https://doi.org/10.1371/journal.pone.0052201

Curé C, Antunes R, Alves AC, Visser F, Kvadsheim PH, Miller PJO (2013) Responses of male sperm whales (*Physeter macrocephalus*) to killer whale sounds: implications for anti-predator strategies. Sci Rep 3:1579. https://doi.org/10.1038/srep01579

Curé C, Doksaeter Sivle L, Visser F, Wensveen PJ, Isojunno S, Harris CM, Kvadsheim PH, Lam F-PA, Miller PJO (2015) Predator sound playbacks reveal strong avoidance responses in a fight strategist baleen whale. Mar Ecol Prog Ser 526:267–282. https://doi.org/10.3354/meps11231

Curé C, Isojunno S, Visser F, Wensveen PJ, Sivle LD, Kvadsheim PH, Lam FPA, Miller PJO (2016) Biological significance of sperm whale responses to sonar: comparison with anti-predator responses. Endanger Species Res 31:89–102. https://doi.org/10.3354/esr00748

Curé C, Isojunno SI, Vester H, Visser F, Oudejans M, Biassoni N, Massenet M, Barluet de Beauchesne LJ, Wensveen P, Sivle LD, Tyack PL, Miller PJO (2019) Evidence for discrimination between feeding sounds of familiar fish and unfamiliar mammal-eating killer whale ecotypes by long-finned pilot whales. Anim Cogn 22(5):863–882. https://doi.org/10.1007/s10071-019-01282-1

Curé C, Isojunno S, Siemensma ML, Wensveen PJ, Buisson C, Sivle LD, Benti B, Roland R, Kvadsheim PH, Lam F-PA, Miller PJO (2021) Severity scoring of behavioural responses of sperm whales (*Physeter macrocephalus*) to novel continuous versus conventional pulsed active sonar. J Mar Sci Eng 9(4): 444. https://doi.org/10.3390/jmse9040444

Curland JM (1997) Effects of disturbances on sea otters (*Enhydra lutris*) near Monterey, California. MSc thesis, San Jose State University, CA, USA

Czapanskiy MF, Savoca MS, Gough WT, Segre PS, Wisniewska DM, Cade DE, Goldbogen JA (2021) Modelling short-term energetic costs of sonar disturbance to cetaceans using high-resolution foraging data. J Appl Ecol 58:1643–1657. https://doi.org/10.1111/1365-2664.13903

Dahlheim M (1993) Responses of gray whales, *Eschrichtius robustus*, to noise (Abstract). J Acoust Soc Am 94(3):1830

Dahlheim M, Castellote M (2016) Changes in the acoustic behavior of gray whales *Eschrichtius robustus* in response to noise. Endanger Species Res 31:227–242. https://doi.org/10.3354/esr00759

Dahlheim ME, Fisher HD, Schempp JD (1984) Sound production by the gray whale and ambient noise levels in Laguna San Ignacio, Baja California Sur, Mexico. In: Jones ML, Swartz SL, Leatherwood S (eds) The gray whale *Eschrichtius robustus*. Academic, New York, pp 511–541

Dähne M, Gilles A, Lucke K, Peschko V, Adler S, Krügel K, Sundermeyer J, Siebert U (2013) Effects of pile-driving on harbour porpoises (*Phocoena phocoena*) at the first offshore wind farm in Germany. Environ Res Lett 8(2):025002. https://doi.org/10.1088/1748-9326/8/2/025002

Dähne M, Tougaard J, Carstensen J, Rose A, Nabe-Nielsen J (2017) Bubble curtains attenuate noise from offshore wind farm construction and reduce temporary habitat loss for harbour porpoises. Mar Ecol Prog Ser 580:221–237. https://doi.org/10.3354/meps12257

Danil K, St. Leger JA (2011) Seabird and dolphin mortality associated with underwater detonation exercises. Mar Technol Soc J 45(6):89–95. https://doi.org/10.4031/MTSJ.45.6.5

Dawson J, Pizzolato L, Howell SEL, Copland L, Johnston ME (2018) Temporal and spatial patterns of ship traffic in the Canadian Arctic from 1990 to 2015. Arctic 71(1):15–26. https://doi.org/10.14430/arctic4698

De Clerck S, Samarra FIP, Svavarsson J, Mouy X, Wensveen P (2019) Noise influences the acoustic behavior of killer whales, *Orcinus orca*, in Iceland. Proc Meet Acoust 37(1):040003. https://doi.org/10.1121/2.0001219

de Stephanis R, Giménez J, Esteban R, Gauffier P, García-Tiscar S, Sinding M-HS, Verborgh P (2015) Mobbing-like behavior by pilot whales towards killer whales: a response to resource competition or perceived predation risk? Acta Ethol 18(1):69–78. https://doi.org/10.1007/s10211-014-0189-1

Demarchi M, Holst M, Robichaud D, Waters M, MacGillivray A (2012) Responses of Stellar Sea Lions (*Eumetopias jubatus*) to in-air blast noise from military explosions. Aquat Mamm 38(3):279–289. https://doi.org/10.1578/AM.38.3.2012.279

DeRuiter SL, Boyd IL, Claridge DE, Clark CW, Gagnon C, Southall BL, Tyack PL (2013a) Delphinid whistle production and call matching during playback of simulated military sonar. Mar Mamm Sci 29(2): E46–E59. https://doi.org/10.1111/j.1748-7692.2012.00587.x

DeRuiter SL, Southall BL, Calambokidis J, Zimmer WMX, Sadykova D, Falcone EA, Friedlaender AS, Joseph JE, Moretti D, Schorr GS, Thomas L, Tyack PL (2013b) First direct measurements of behavioural responses by Cuvier's beaked whales to mid-frequency active sonar. Biol Lett 9(4):20130223. https://doi.org/10.1098/rsbl.2013.0223

DeRuiter SL, Langrock R, Skirbutas T, Goldbogen JA, Calambokidis J, Friedlaender AS, Southall BL (2017) A multivariate mixed hidden Markov model for blue whale behaviour and responses to sound exposure. Ann Appl Stat 11(1):362–392. https://doi.org/10.1214/16-AOAS1008

Dey M, Krishnaswamy J, Morisaka T, Kelkar N (2019) Interacting effects of vessel noise and shallow river depth elevate metabolic stress in Ganges river dolphins. Sci Rep 9(1):15426. https://doi.org/10.1038/s41598-019-51664-1

di Iorio L, Clark CW (2010) Exposure to seismic survey alters blue whale acoustic communication. Biol Lett 6: 334–335. https://doi.org/10.1098/rsbl.2009.0967

Díaz López B, Marino F (2011) A trial of acoustic harassment device efficacy on free-ranging bottlenose dolphins in Sardinia, Italy. Mar Freshw Behav Physiol 44(4):197–208. https://doi.org/10.1080/10236244.2011.618216

Dingemanse NJ, Kazem AJN, Réale D, Wright J (2010) Behavioural reaction norms: animal personality meets individual plasticity. Trends Ecol Evol 25(2):81–89. https://doi.org/10.1016/j.tree.2009.07.013

Dolman SJ, Breen CN, Brakes P, Butterworth A, Allen SJ (2022) The individual welfare concerns for small cetaceans from two bycatch mitigation techniques. Mar Policy 143:105126. https://doi.org/10.1016/j.marpol.2022.105126

Dunlop RA, Cato DH, Noad MJ (2010) Your attention please: increasing ambient noise levels elicits a change in communication behaviour in humpback whales (*Megaptera novaeangliae*). Proc R Soc B 277(1693): 2521–2529. https://doi.org/10.1098/rspb.2009.2319

Dunlop RA, Cato DH, Noad MJ (2014) Evidence of a Lombard response in migrating humpback whales (*Megaptera novaeangliae*). J Acoust Soc Am 136(1): 430–437. https://doi.org/10.1121/1.4883598

Dunlop RA, Noad MJ, McCauley RD, Kniest E, Slade R, Paton D, Cato DH (2018) A behavioural dose-response model for migrating humpback whales and seismic air gun noise. Mar Pollut Bull 133:506–516. https://doi.org/10.1016/j.marpolbul.2018.06.009

Dyndo M, Wiśniewska DM, Rojano-Doñate L, Madsen PT (2015) Harbour porpoises react to low levels of high frequency vessel noise. Sci Rep 5:11083. https://doi.org/10.1038/srep11083

Edrén SMC, Andersen SM, Teilmann J, Carstensen J, Harders PB, Dietz R, Miller LA (2010) The effect of a large Danish offshore wind farm on harbor and gray seal haul-out behavior. Mar Mamm Sci 26(3): 614–634. https://doi.org/10.1111/j.1748-7692.2009.00364.x

Ellison W, Southall B, Clark C, Frankel A (2012) A new context-based approach to assess marine mammal behavioral responses to anthropogenic sounds. Conserv Biol 26(1):21–28. https://doi.org/10.1111/j.1523-1739.2011.01803.x

Elmegaard SL, McDonald BI, Teilmann J, Madsen PT (2021) Heart rate and startle responses in diving, captive harbour porpoises (*Phocoena phocoena*) exposed to transient noise and sonar. Biol Open 10(6): bio058679. https://doi.org/10.1242/bio.058679

Erbe C (2002) Underwater noise of whale-watching boats and its effects on killer whales (*Orcinus orca*). Mar Mamm Sci 18(2):394–418. https://doi.org/10.1111/j.1748-7692.2002.tb01045.x

Erbe C, MacGillivray AO, Williams R (2012) Mapping cumulative noise from shipping to inform marine spatial planning. J Acoust Soc Am 132(5):EL 423–EL 428. https://doi.org/10.1121/1.4758779

Erbe C, Williams R, Sandilands D, Ashe E (2014) Identifying modelled ship noise hotspots for marine mammals of Canada's Pacific region. PLoS One 9(3): e89820. https://doi.org/10.1371/journal.pone.0089820

Erbe C, Parsons M, Duncan AJ, Osterrieder S, Allen K (2017) Aerial and underwater sound of unmanned aerial vehicles (UAV, drones). J Unmanned Veh Syst 5(3):92–101. https://doi.org/10.1139/juvs-2016-0018

Falcone EA, Schorr GS, Watwood SL, DeRuiter SL, Zerbini AN, Andrews RD, Morrissey RP, Moretti DJ (2017) Diving behaviour of Cuvier's beaked whales exposed to two types of military sonar. R Soc Open Sci 4:170629. https://doi.org/10.1098/rsos.170629

Fang L, Li M, Wang X, Chen Y, Chen T (2023) Indo-Pacific finless porpoises presence in response to pile driving on the Jinwan Offshore Wind Farm, China. Front Mar Sci 10:1005374. https://doi.org/10.3389/fmars.2023.1005374

Farmer NA, Baker K, Zeddies DG, Denes SL, Noren DP, Garrison LP, Machernis A, Fougères EM, Zykov M (2018a) Population consequences of disturbance by offshore oil and gas activity for endangered sperm whales (*Physeter macrocephalus*). Biol Conserv 227: 189–204. https://doi.org/10.1016/j.biocon.2018.09.006

Farmer NA, Noren DP, Fougères EM, Machernis A, Baker K (2018b) Resilience of the endangered sperm whale *Physeter macrocephalus* to foraging disturbance in the Gulf of Mexico, USA: a bioenergetic approach. Mar Ecol Prog Ser 589:241–261. https://doi.org/10.3354/meps12457

Fay FH, Ray GC, Kibal'chich AA (1984) Time and location of mating and associated behavior of the Pacific walrus, *Odobenus rosmarus divergens* illiger (NOAA Technical Report NMFS 12). U.S. Department of Commerce, National Oceanic and Atmospheric Administration, National Marine Fisheries Service

Fernandez-Betelu O, Graham IM, Brookes KL, Cheney BJ, Barton TR, Thompson PM (2021) Far-field effects of impulsive noise on coastal bottlenose dolphins. Front Mar Sci 8:664230. https://doi.org/10.3389/fmars.2021.664230

Ferrara GA, Mongillo TM, Barre LM (2017) Reducing disturbance from vessels to Southern Resident killer whales: assessing the effectiveness of the 2011 federal regulations in advancing recovery goals (NOAA Technical Memorandum NMFS-OPR-58). National Oceanic and Atmospheric Administration (NOAA)

Fettermann T, Fiori L, Bader M, Doshi A, Breen D, Stockin KA, Bollard B (2019) Behaviour reactions of bottlenose dolphins (*Tursiops truncatus*) to multirotor Unmanned Aerial Vehicles (UAVs). Sci Rep 9(1): 8558. https://doi.org/10.1038/s41598-019-44976-9

Fettermann T, Fiori L, Gillman L, Stockin KA, Bollard B (2022) Drone surveys are more accurate than boat-based surveys of bottlenose dolphins (*Tursiops truncatus*). Drones 6(4):82. https://doi.org/10.3390/drones6040082

Feyrer LJ, Zhao S, Whitehead H, Matthews CJD (2020) Prolonged maternal investment in northern bottlenose whales alters our understanding of beaked whale reproductive life history. PLoS One 15(6):e0235114. https://doi.org/10.1371/journal.pone.0235114

Findlay CR, Ripple HD, Coomber F, Froud K, Harries O, van Geel NCF, Calderan SV, Benjamins S, Risch D, Wilson B (2018) Mapping widespread and increasing underwater noise pollution from acoustic deterrent devices. Mar Pollut Bull 135:1042–1050. https://doi.org/10.1016/j.marpolbul.2018.08.042

Findlay CR, Hastie GD, Farcas A, Merchant ND, Risch D, Wilson B (2022) Exposure of individual harbour seals (*Phoca vitulina*) and waters surrounding protected habitats to acoustic deterrent noise from aquaculture. Aquat Conserv 32(5):766–780. https://doi.org/10.1002/aqc.3800

Finley KJ, Miller GW, Davis RA, Greene CR (1990) Reactions of belugas, *Delphinapterus leucas*, and narwhals, *Monodon monoceros*, to ice-breaking ships in the Canadian high Arctic. Can Bull Fish Aquat Sci 224:97–117

Finneran JJ, Schlundt CE, Carder DA, Clark JA, Young JA, Gaspin JB, Ridgway SH (2000) Auditory and behavioural responses of bottlenose dolphins (*Tursiops truncatus*) and a beluga whale (*Delphinapterus leucas*) to impulsive sounds resembling distant signatures of underwater explosions. J Acoust Soc Am 108(1): 417–431. https://doi.org/10.1121/1.429475

Finneran JJ, Dear R, Carder DA, Ridgway SH (2003) Auditory and behavioral responses of California sea lions (*Zalophus californianus*) to single underwater impulses from an arc-gap transducer. J Acoust Soc Am 114(3):1667–1677. https://doi.org/10.1121/1.1598194

Finneran JJ, Schlundt CE, Branstetter BK, Trickey JS, Bowman V, Jenkins K (2015) Effects of multiple impulses from a seismic air gun on bottlenose dolphin hearing and behavior. J Acoust Soc Am 137(4): 1634–1646. https://doi.org/10.1121/1.4916591

Fish JF, Vania JS (1971) Killer whale, *Orcinus orca*, sounds repel white whales, *Delphinapterus leucas*. Fish Bull 69(3):531–535

Fjälling A, Wahlberg M, Westerberg H (2006) Acoustic harassment devices reduce seal interaction in the Baltic salmon-trap, net fishery. ICES J Mar Sci 63(9): 1751–1758. https://doi.org/10.1016/j.icesjms.2006.06.015

Fleishman E, Costa DP, Harwood J, Kraus S, Moretti D, New LF, Schick RS, Schwarz LK, Simmons SE, Thomas L, Wells RS (2016) Monitoring population-level responses of marine mammals to human activities. Mar Mamm Sci 32(3):1004–1021. https://doi.org/10.1111/mms.12310

Fletcher S, LeBoeuf BJ, Costa DP, Tyack PL, Blackwell SB (1996) Onboard acoustic recording from diving northern elephant seals. J Acoust Soc Am 100(4): 2531–2539. https://doi.org/10.1121/1.417361

Foote AD, Osborne RW, Hoelzel AR (2004) Whale-call response to masking boat noise. Nature 428:910. https://doi.org/10.1038/428910a

Ford JKB, Ellis GM (2010) Linking killer whale survival and prey abundance: food limitation in the oceans' apex predator? Biol Lett 6:139–142. https://doi.org/10.1098/rsbl.2009.0468

Ford JKB, Reeves RR (2008) Fight or flight: antipredator strategies of baleen whales. Mammal Rev 38(1):50–86. https://doi.org/10.1111/j.1365-2907.2008.00118.x

Fouda L, Wingfield JE, Fandel AD, Garrod A, Hodge KB, Rice AN, Bailey H (2018) Dolphins simplify their vocal calls in response to increased ambient noise. Biol Lett 14(10):20180484. https://doi.org/10.1098/rsbl.2018.0484

Fournet MEH, Silvestri M, Clark CW, Klinck H, Rice AN (2021) Limited vocal compensation for elevated ambient noise in bearded seals: implications for an industrializing Arctic Ocean. Proc R Soc B 288(1945):20202712. https://doi.org/10.1098/rspb.2020.2712

Frankel AS, Clark CW (1998) Results of low-frequency playback of M-sequence noise to humpback whales, *Megaptera novaeangliae*, in Hawaii. Can J Zool 76(3): 521–535. https://doi.org/10.1139/z97-223

Frankel AS, Clark CW (2000) Behavioural responses of humpback whales (*Megaptera novaeangliae*) to full-scale ATOC signals. J Acoust Soc Am 108(4): 1930–1937. https://doi.org/10.1121/1.1289668

Frankel AS, Ellison WT, Buchanan J (2002) Application of the Acoustic Integration Model (AIM) to predict and minimize environmental impacts. Oceans 2002 MTS/IEEE conference & exhibition, vols 1–4, Conference proceedings, pp 1438–1443. https://doi.org/10.1109/OCEANS.2002.1191849

Frankish CK, von Benda-Beckmann AM, Teilmann J, Tougaard J, Dietz R, Sveegaard S, Binnerts B, de Jong CAF, Nabe-Nielsen J (2023) Ship noise causes tagged harbour porpoises to change direction or dive deeper. Mar Pollut Bull 197:115755. https://doi.org/10.1016/j.marpolbul.2023.115755

Friedlaender AS, Hazen EL, Goldbogen JA, Stimpert AK, Calambokidis J, Southall BL (2016) Prey-mediated behavioral responses of feeding blue whales in controlled sound exposure experiments. Ecol Appl 26(4): 1075–1085. https://doi.org/10.1002/15-0783

Fristrup KM, Hatch LT, Clark CW (2003) Variation in humpback whale (*Megaptera novaeangliae*) song length in relation to low-frequency sound broadcasts. J Acoust Soc Am 113(6):3411–3424. https://doi.org/10.1121/1.1573637

Fu W, Song Z, Wang T, Gao Z, Li J, Zhang P, Zhang Y (2022) Acoustic deterrence to facilitate the conservation of pantropical spotted dolphins (*Stenella attenuata*) in the Western Pacific Ocean. Front Mar Sci 9:1023860. https://doi.org/10.3389/fmars.2022.1023860

Gagne E, Perez-Ortega B, Hendry AP, Melo-Santos G, Walmsley SF, Rege-Colt M, Austin M, May-Collado LJ (2022) Dolphin communication during widespread systematic noise reduction—a natural experiment amid COVID-19 lockdowns. Front Remote Sens 3:934608. https://doi.org/10.3389/frsen.2022.934608

Gailey G, Würsig B, McDonald TL (2007) Adundance, behavior, and movement patterns of western gray whales in relation to a 3-D seismic survey, Sakhalin Island, Russia. Environ Monit Assess 134:75–91. https://doi.org/10.1007/s10661-007-9812-1

Gailey G, Sychenko O, McDonald T, Racca R, Rutenko A, Bröker K (2016) Behavioural responses of western gray whales to a 4-D seismic survey off northeastern Sakhalin Island, Russia. Endanger Species Res 30:53–71. https://doi.org/10.3354/esr00713

Gailey G, Sychenko O, Zykov M, Rutenko A, Blanchard A, Melton RH (2022a) Western gray whale behavioral response to seismic surveys during their foraging season. Environ Monit Assess 194(1): 740. https://doi.org/10.1007/s10661-022-10023-w

Gailey G, Zykov M, Sychenko O, Rutenko A, Blanchard AL, Aerts L, Melton RH (2022b) Gray whale density during seismic surveys near their Sakhalin feeding ground. Environ Monit Assess 194(1):739. https://doi.org/10.1007/s10661-022-10025-8

Gallagher CA, Grimm V, Kyhn LA, Kinze CC, Nabe-Nielsen J (2021) Movement and seasonal energetics mediate vulnerability to disturbance in marine mammal populations. Am Nat 197(3):296–311. https://doi.org/10.1086/712798

Gazo M, Gonzalvo J, Aguilar A (2008) Pingers as deterrents of bottlenose dolphins interacting with trammel nets. Fish Res 92(1):70–75. https://doi.org/10.1016/j.fishres.2007.12.016

Gerstein ER, Gerstein L, Forsythe SE, Blue JE (1999) The underwater audiogram of the West Indian manatee

(*Trichechus manatus*). J Acoust Soc Am 105(6):3575–3583. https://doi.org/10.1121/1.424681

Girola E, Dunlop RA, Noad MJ (2023a) Singing humpback whales respond to wind noise, but not to vessel noise. Proc R Soc B 290(1998):20230204. https://doi.org/10.1098/rspb.2023.0204

Girola E, Dunlop RA, Noad MJ (2023b) Singing in a noisy ocean: vocal plasticity in male humpback whales. Bioacoustics 32(3):301–324. https://doi.org/10.1080/09524622.2022.2122560

Goldbogen JA, Southall BL, DeRuiter SL, Calambokidis J, Friedlaender AS, Hazen EL, Falcone EA, Schorr GS, Douglas A, Moretti DJ, Kyburg C, McKenna MF, Tyack PL (2013) Blue whales respond to simulated mid-frequency military sonar. Proc R Soc B 280(1765):20130657. https://doi.org/10.1098/rspb.2013.0657

Goldbogen JA, Cade DE, Boersma AT, Calambokidis J, Kahane-Rapport SR, Segre PS, Stimpert AK, Friedlaender AS (2017) Using digital tags with integrated video and inertial sensors to study moving morphology and associated function in large aquatic vertebrates. Anat Rec 300:1935–1941. https://doi.org/10.1002/ar.23650

Gomez C, Lawson J, Wright AJ, Buren A, Tollit D, Lesage V (2016) A systematic review on the behavioural responses of wild marine mammals to noise: the disparity between science and policy. Can J Zool 94(12):801–819. https://doi.org/10.1139/cjz-2016-0098

Gordon J, Leaper R, Hartley FG, Chappell O (1992) Effects of whale-watching vessels on the surface and underwater acoustic behaviour of sperm whales off Kaikoura, New Zealand. Conserv Te Papa Atawhai 48(6):1–64

Gordon J, Blight C, Bryant E, Thompson D (2019) Measuring responses of harbour seals to potential aversive acoustic mitigation signals using controlled exposure behavioural response studies. Aquat Conserv 29(S1):157–177. https://doi.org/10.1002/aqc.3150

Götz T, Janik VM (2010) Aversiveness of sounds in phocid seals: psycho-physiological factors, learning processes and motivation. J Exp Biol 213(9):1536–1548. https://doi.org/10.1242/Jeb.035535

Götz T, Janik VM (2011) Repeated elicitation of the acoustic startle reflex leads to sensitisation in subsequent avoidance behaviour and induces fear conditioning. BMC Neurosci 12(1):30. https://doi.org/10.1186/1471-2202-12-30

Götz T, Janik VM (2013) Acoustic deterrent devices to prevent pinniped depredation: efficiency, conservation concerns and possible solutions. Mar Ecol Prog Ser 492:285–302. https://doi.org/10.3354/meps10482

Götz T, Janik VM (2015) Target-specific acoustic predator deterrence in the marine environment. Anim Conserv 18(1):102–111. https://doi.org/10.1111/acv.12141

Götz T, Janik VM (2016) Non-lethal management of carnivore predation: long-term tests with a startle reflex-based deterrence system on a fish farm. Anim Conserv 19(3):212–221. https://doi.org/10.1111/acv.12248

Graham IM, Harris RN, Denny B, Fowden D, Pullan D (2009) Testing the effectiveness of an acoustic deterrent device for excluding seals from Atlantic salmon rivers in Scotland. ICES J Mar Sci 66(5):860–864. https://doi.org/10.1093/icesjms/fsp111

Graham IM, Pirotta E, Merchant ND, Farcas A, Barton TR, Cheney B, Hastie GD, Thompson PM (2017) Responses of bottlenose dolphins and harbor porpoises to impact and vibration piling noise during harbor construction. Ecosphere 8(5):e01793. https://doi.org/10.1002/ecs2.1793

Graham IM, Merchant ND, Farcas A, Barton TR, Cheney B, Bono S, Thompson PM (2019) Harbour porpoise responses to pile-driving diminish over time. R Soc Open Sci 6(6):190335. https://doi.org/10.1098/rsos.190335

Gray H, Van Waerebeek K (2011) Postural instability and akinesia in a pantropical spotted dolphin, *Stenella attenuata*, in proximity to operating airguns of a geophysical seismic vessel. J Nat Conserv 19:363–367. https://doi.org/10.1016/j.jnc.2011.06.005

Guerra M, Thode A, Blackwell S, Macrander M (2011) Quantifying seismic survey reverberation off the Alaskan North Slope. J Acoust Soc Am 130(5):3046–3058. https://doi.org/10.1121/1.3628326

Haelters J, Dulière V, Vigin L, Degraer S (2015) Towards a numerical model to simulate the observed displacement of harbour porpoises *Phocoena phocoena* due to pile driving in Belgian waters. Hydrobiologia 756(1):105–116. https://doi.org/10.1007/s10750-014-2138-4

Halliday WD, Barclay D, Barkley AN, Cook E, Dawson J, Hilliard RC, Hussey NE, Jones JM, Juanes F, Marcoux M, Niemi A, Nudds S, Pine MK, Richards C, Scharffenberg K, Westdal K, Insley SJ (2021) Underwater sound levels in the Canadian Arctic, 2014–2019. Mar Pollut Bull 168:112437. https://doi.org/10.1016/j.marpolbul.2021.112437

Hammond PS, Bearzi G, Bjørge A, Forney K, Karczmarski L, Kasuya T, Perrin WF, Scott MD, Wang JY, Wells RS, Wilson B (2008) *Phocoena phocoena*. www.iucnredlist.org. Accessed 20 Sept 2019

Harris CM, Thomas LT (2015) Status and future of research on the behavioral responses of marine mammals to U.S. Navy sonar. CREEM Technical Report 2015-3, University of St. Andrews

Harris RE, Miller GW, Richardson WJ (2001) Seal responses to airgun sounds during summer seismic surveys in the Alaskan Beaufort Sea. Mar Mamm Sci 17(4):795–812. https://doi.org/10.1111/j.1748-7692.2001.tb01299.x

Harris RN, Harris CM, Duck CD, Boyd IL (2014) The effectiveness of a seal scarer at a wild salmon net fishery. ICES J Mar Sci 71(7):1913–1920. https://doi.org/10.1093/icesjms/fst216

Harris CM, Sadykova D, DeRuiter SL, Tyack PL, Miller PJO, Kvadsheim PH, Lam FPA, Thomas L (2015) Dose response severity functions for acoustic disturbance in cetaceans using recurrent event survival analysis. Ecosphere 6(11):1–14. https://doi.org/10.1890/ES15-00242.1

Harris CM, Thomas L, Falcone EA, Hildebrand J, Houser D, Kvadsheim PH, Lam F-PA, Miller PJO, Moretti DJ, Read AJ, Slabbekoorn H, Southall BL, Tyack PL, Wartzok D, Janik VM (2017) Marine mammals and sonar: dose-response studies, the risk-disturbance hypothesis and the role of exposure context. J Appl Ecol 55(1):396–404. https://doi.org/10.1111/1365-2664.12955

Harris CM, Burt ML, Allen AN, Wensveen PJ, Miller PJO, Sivle LD (2019) Foraging behavior and disruption in blue, fin, and humpback whales in relation to sonar exposure: the challenges of generalizing responsiveness in species with high individual variability. Aquat Mamm 45(6):646–660. https://doi.org/10.1578/AM.45.6.2019.646

Hastie GD, Donovan C, Götz T, Janik VM (2014) Behavioural responses by grey seals (*Halichoerus grypus*) to high frequency sonar. Mar Pollut Bull 79(1-2):205–210. https://doi.org/10.1016/j.marpolbul.2013.12.013

Hastie GD, Russell DJF, McConnell B, Moss S, Thompson D, Janik VM (2015) Sound exposure in harbour seals during the installation of an offshore wind farm: predictions of auditory damage. J Appl Ecol 52(3):631–640. https://doi.org/10.1111/1365-2664.12403

Hastie GD, Russell DJF, Lepper P, Elliott J, Wilson B, Benjamins S, Thompson D (2018) Harbour seals avoid tidal turbine noise: implications for collision risk. J Appl Ecol 55(2):684–693. https://doi.org/10.1111/1365-2664.12981

Hastie GD, Bivins M, Coram A, Gordon J, Jepp P, MacAulay J, Sparling C, Gillespie D (2019a) Three-dimensional movements of harbour seals in a tidally energetic channel: application of a novel sonar tracking system. Aquat Conserv 29(4):564–575. https://doi.org/10.1002/aqc.3017

Hastie GD, Wu G-M, Moss S, Jepp P, MacAulay J, Lee A, Sparling CE, Evers C, Gillespie D (2019b) Automated detection and tracking of marine mammals: a novel sonar tool for monitoring effects of marine industry. Aquat Conserv 29(S1):119–130. https://doi.org/10.1002/aqc.3103

Hastie GD, Lepper P, McKnight JC, Milne R, Russell DJF, Thompson D (2021) Acoustic risk balancing by marine mammals: anthropogenic noise can influence the foraging decisions by seals. J Appl Ecol 58(9):1854–1863. https://doi.org/10.1111/1365-2664.13931

Heenehan HL, Johnston DW, Van Parijs SM, Bejder L, Tyne JA (2016) Acoustic response of Hawaiian spinner dolphins to human disturbance. Proc Meet Acoust 27:010001. https://doi.org/10.1121/2.0000232

Heide-Jørgensen MP, Hansen R, Westdal K, Reeves R, Mosbech A (2013) Narwhals and seismic exploration: is seismic noise increasing the risk of ice entrapments? Biol Conserv 158:50–54. https://doi.org/10.1016/j.biocon.2012.08.005

Heide-Jørgensen MP, Blackwell SB, Tervo OM, Samson AL, Garde E, Hansen RG, Ngô MCòn, Conrad AS, Trinhammer P, Schmidt HC, Sinding MHS, Williams TM, Ditlevsen S (2021) Behavioral response study on seismic airgun and vessel exposures in narwhals. Front Mar Sci 8(665):658173. https://doi.org/10.3389/fmars.2021.658173

Heiler J, Elwen SH, Kriesell HJ, Gridley T (2016) Changes in bottlenose dolphin whistle parameters related to vessel presence, surface behaviour and group composition. Anim Behav 117:167–177. https://doi.org/10.1016/j.anbehav.2016.04.014

Henderson EE, Smith MH, Gassmann M, Wiggins SM, Douglas AB, Hildebrand JA (2014) Delphinid behavioral responses to incidental mid-frequency active sonar. J Acoust Soc Am 136(4):2003–2014. https://doi.org/10.1121/1.4895681

Henderson EE, Aschettino J, Deakos M, Alongi G, Leota T (2019) Quantifying the behavior of humpback whales (*Megaptera novaeangliae*) and potential responses to sonar. Aquat Mamm 45(6):612–631. https://doi.org/10.1578/AM.45.6.2019.612

Henry E, Hammill MO (2001) Impact of small boats on the haulout activity of harbour seals (*Phoca vitulina*) in Metis Bay, Saint Lawrence Estuary, Quebec, Canada. Aquat Mamm 27(2):140–148

Hiley HM, Janik VM, Götz T (2021) Behavioural reactions of harbour porpoises *Phocoena phocoena* to startle-eliciting stimuli: movement responses and practical applications. Mar Ecol Prog Ser 672:223–241. https://doi.org/10.3354/meps13757

Hodgson AJ (2004) Dugong behaviour and responses to human influences. PhD dissertation, James Cook University, Townsville, QLD, Australia

Hodgson AJ, Marsh H, Delean S, Marcus L (2007) Is attempting to change marine mammal behaviour a generic solution to the bycatch problem? A dugong case study. Anim Conserv 10(2):263–273. https://doi.org/10.1111/j.1469-1795.2007.00104.x

Hoekendijk JPA, Spitz J, Read AJ, Leopold MF, Fontaine MC (2018) Resilience of harbor porpoises to anthropogenic disturbance: must they really feed continuously? Mar Mamm Sci 34(1):258–264. https://doi.org/10.1111/mms.12446

Hoelzel AR, Osborne RW (1986) Killer whale call characteristics: implications for cooperative foraging strategies. In: Behavioural biology of killer whales. Alan R. Liss, New York, pp 373–403

Hohn AA, Rotstein DS, Harms CA, Southall BL (2006) Multispecies mass stranding of pilot whales (*Globicephala macrorhynchus*), minke whale (*Balaenoptera acutorostrata*), and dwarf sperm whales (*Kogia sima*) in North Carolina on 15-16 January 2005 (NOAA Technical Memorandum NMFS-SEFSC-537). Department of Commerce, Beaufort, 222 pp

Holmes TL, Knight RL, Stegall L, Craig GR (1993) Responses of wintering grassland raptors to human disturbance. Wildl Soc Bull (1973-2006) 21(4):461–468. https://doi.org/10.2307/3783420

Holst M, Greene CR, Richardson WJ, McDonald TL, Bay K, Schwartz SJ, Smith G (2011) Responses of pinnipeds to navy missile launches at San Nicolas Island, California. Aquat Mamm 37(2):139–150. https://doi.org/10.1578/AM.37.2.2011.139

Holst M, Smultea MA, Koski WR, Sayegh AJ, Pavan G, Beland J, Goldstein HH (2017) Cetacean sightings and acoustic detections during a seismic survey off Nicaragua and Costa Rica, November-December 2004. Rev Biol Trop 65(2):599–611. https://doi.org/10.15517/rbt.v65i2.25477

Holt MM, Veirs V, Veirs S (2008) Noise effects on the call amplitude of southern resident killer whales (Orcinus orca). Bioacoustics:164–166. https://doi.org/10.1080/09524622.2008.9753802

Holt MM, Noren DP, Veirs V, Emmons CK, Veirs S (2009) Speaking up: killer whales (Orcinus orca) increase their call amplitude in response to vessel noise. J Acoust Soc Am 125(1):El27–El32. https://doi.org/10.1121/1.3040028

Holt M, Noren D, Emmons C (2011) Effects of noise levels and call types on the source levels of killer whale calls. J Acoust Soc Am 130(5):3100–3106. https://doi.org/10.1121/1.3641446

Holt MM, Tennessen JB, Hanson MB, Emmons CK, Giles DA, Hogan JT, Ford MJ (2021) Vessels and their sounds reduce prey capture effort by endangered killer whales (Orcinus orca). Mar Environ Res:105429. https://doi.org/10.1016/j.marenvres.2021.105429

Houghton J, Holt MM, Giles DA, Hanson MB, Emmons CK, Hogan JT, Branch TA, VanBlaricom GR (2015) The relationship between vessel traffic and noise levels received by killer whales (Orcinus orca). PLoS One. https://doi.org/10.1371/journal.pone.0140119

Houser DS (2006) A method for modeling marine mammal movement and behavior for environmental impact assessment. IEEE J Ocean Eng 31(1):76–81. https://doi.org/10.1109/JOE.2006.872204

Houser DS, Martin SW, Finneran JJ (2013a) Behavioural responses of California sea lions to mid-frequency (3250-3450 Hz) sonar signals. Mar Environ Res 92:268–278. https://doi.org/10.1016/j.marenvres.2013.10.007

Houser DS, Martin SW, Finneran JJ (2013b) Exposure amplitude and repetition affect bottlenose dolphin behavioural responses to simulated mid-frequency sonar signals. J Exp Mar Biol Ecol 443:123–133. https://doi.org/10.1016/j.jembe.2013.02.043

IEEE (2019) Standard for letter symbols and abbreviations for quantities used in acoustics. IEEE Std 2604-2018 (Revision of IEEE Std 2604-1996):1–71. https://doi.org/10.1109/IEEESTD.2019.8654216

Insley SJ (1990) Impact of airborne noise on northern fur seals in the Pribilof Islands, Alaska: a preliminary assessment. In: Kajimura H, Sinclair E (eds) Fur seal investigations, NOAA Technical Memorandum NMFS-AFSC-2. Alaska Fisheries Science Centre, Seattle, pp 27–47

International Organization for Standardization (2017) Underwater acoustics—terminology (ISO 18405). Geneva, Switzerland

Isojunno S, Curé C, Kvadsheim PH, Lam F-PA, Tyack PL, Wensveen PJ, Miller PJOM (2016) Sperm whales reduce foraging effort during exposure to 1–2 kHz sonar and killer whale sounds. Ecol Appl 26(1):77–93. https://doi.org/10.1890/15-0040

Isojunno S, Sadykova D, DeRuiter S, Cure C, Visser F, Thomas L, Miller PJO, Harris CM (2017) Individual, ecological, and anthropogenic influences on activity budgets of long-finned pilot whales. Ecosphere 8(12):e02044. https://doi.org/10.1002/ecs2.2044

Isojunno S, Aoki K, Curé C, Kvadsheim PH, Miller PJOM (2018) Breathing patterns indicate cost of exercise during diving and response to experimental sound exposures in long-finned pilot whales. Front Physiol 9:1462. https://doi.org/10.3389/fphys.2018.01462

Isojunno S, Wensveen PJ, Lam F-PA, Kvadsheim PH, von Benda-Beckmann AM, Martín López LM, Kleivane L, Siegal EM, Miller PJO (2020) When the noise goes on: received sound energy predicts sperm whale responses to both intermittent and continuous navy sonar. J Exp Biol 223(7):jeb.219741. https://doi.org/10.1242/jeb.219741

Isojunno S, von Benda-Beckmann AM, Wensveen PJ, Kvadsheim PH, Lam F-PA, Gkikopoulou KC, Pöyhönen V, Tyack PL, Benti B, Foskolos I, Bort J, Neves M, Biassoni N, Miller PJO (2022) Sperm whales exhibit variation in echolocation tactics with depth and sea state but not naval sonar exposures. Mar Mamm Sci 38(2):682–704. https://doi.org/10.1111/mms.12890

Jacobs SR, Terhune JM (2002) The effectiveness of acoustic harassment devices in the Bay of Fundy, Canada: seal reactions and a noise exposure model. Aquat Mamm 28:147–158

Jacobson EK, Henderson EE, Miller DL, Oedekoven CS, Moretti DJ, Thomas L (2022) Quantifying the response of Blainville's beaked whales to US naval sonar exercises in Hawaii. Mar Mamm Sci 38(4):1549–1565. https://doi.org/10.1111/mms.12944

Jansen JK, Boveng PL, Dahle SP, Bengtson JL (2010) Reaction of harbor seals to cruise ships. J Wildl Manag 74(6):1186–1194. https://doi.org/10.1111/j.1937-2817.2010.tb01239.x

Jaramillo-Legorreta A, Cardenas-Hinojosa G, Nieto-Garcia E, Rojas-Bracho L, Ver Hoef J, Moore J, Tregenza N, Barlow J, Gerrodette T, Thomas L, Taylor B (2017) Passive acoustic monitoring of the decline of Mexico's critically endangered vaquita. Conserv Biol 31(1):183–191. https://doi.org/10.1111/cobi.12789

Jefferson TA, Stacey PJ, Baird RW (1991) A review of killer whale interactions with other marine mammals: predation to co-existence. Mammal Rev 21(4):151–180. https://doi.org/10.1111/j.1365-2907.1991.tb00291.x

Jepson PD, Deaville R, Acevedo-Whitehouse K, Barnett J, Brownlow A, Brownell RL Jr, Clare FC, Davison N,

Law RJ, Loveridge J, Macgregor SK, Morris S, Murphy S, Penrose R, Perkins MW, Pinn E, Seibel H, Siebert U, Sierra E, Simpson V, Tasker ML, Tregenza N, Cunningham AA, Fernández A (2013) What caused the UK's largest common dolphin (*Delphinus delphis*) mass stranding event? PLoS One 8(4):e60953. https://doi.org/10.1371/journal.pone.0060953

Johnson SR, Richardson WJ, Yazvenko SB, Blokhin SA, Gailey G, Jenkerson MR, Meier SK, Melton HR, Newcomer MW, Perlov AS, Rutenko SA, Würsig B, Martin CR, Egging DE (2007) A Western gray whale mitigation and monitoring program for a 3-D seismic survey, Sakhalin Island, Russia. Environ Monit Assess 134(1-3):1–19. https://doi.org/10.1007/s10661-007-9813-0

Johnson M, de Soto NA, Madsen PT (2009) Studying the behaviour and sensory ecology of marine mammals using acoustic recording tags: a review. Mar Ecol Prog Ser 395:55–73. https://doi.org/10.3354/Meps08255

Johnston DW (2002) The effect of acoustic harassment devices on harbour porpoises (*Phocoena phocoena*) in the Bay of Fundy, Canada. Biol Conserv 108(1):113–118. https://doi.org/10.1016/s0006-3207(02)00099-x

Jones EL, Hastie GD, Smout S, Onoufriou J, Merchant ND, Brookes KL, Thompson D (2017) Seals and shipping: quantifying population risk and individual exposure to vessel noise. J Appl Ecol 54(6):1930–1940. https://doi.org/10.1111/1365-2664.12911

Joy R, Tollit D, Wood J, MacGillivray A, Li Z, Trounce K, Robinson O (2019) Potential benefits of vessel slowdowns on endangered southern resident killer whales. Front Mar Sci 6:344. https://doi.org/10.3389/fmars.2019.00344

Joyce TW, Durban JW, Claridge DE, Dunn CA, Hickmott LS, Fearnbach H, Dolan K, Moretti D (2020) Behavioral responses of satellite tracked Blainville's beaked whales (*Mesoplodon densirostris*) to mid-frequency active sonar. Mar Mamm Sci 36(1):29–46. https://doi.org/10.1111/mms.12624

Kastelein RA, Goodson AD, Lien J, de Haan D (1995) The effects of acoustic alarms on harbour porpoise (*Phocoena phocoena*) behaviour. In: Nachtigall PE, Lien J, Au WWL, Read AJ (eds) Harbour porpoises—laboratory studies to reduce bycatch. De Spil Publishers, Woerden, pp 157–168

Kastelein RA, de Haan D, Goodson AD, Staal C, Vaughan N (1997) The effects of various sounds on harbor porpoise. In: Read AJ, Wiepkema PR, Nachtigall PE (eds) The biology of the harbor porpoise. De Spil Publishers, Woerden, pp 367–383

Kastelein RA, Verboom WC, Muijsers M, Jennings NV, van der Heul S (2005) The influence of acoustic emissions for underwater data transmission on the behaviour of harbour porpoises (*Phocoena phocoena*) in a floating pen. Mar Environ Res 59(4):287–307. https://doi.org/10.1016/j.marenvres.2004.05.005

Kastelein RA, van der Heul S, Terhune JM, Verboom WC, Triesscheijn RJV (2006a) Deterring effects of 8-45 kHz tone pulses on harbour seals (*Phoca vitulina*) in a large pool. Mar Environ Res 62(5):356–373. https://doi.org/10.1016/j.marenvres.2006.05.004

Kastelein RA, van der Heul S, Verboom WC, Triesscheijn RJV, Jennings NV (2006b) The influence of underwater data transmission sounds on the displacement behaviour of captive harbour seals (*Phoca vitulina*). Mar Environ Res 61(1):19–39. https://doi.org/10.1016/j.marenvres.2005.04.001

Kastelein RA, Verboom WC, Jennings N, de Haan D (2008a) Behavioural avoidance threshold level of a harbour porpoise (*Phocoena phocoena*) for a continuous 50 kHz pure tone (L). J Acoust Soc Am 123(4):1858–1861. https://doi.org/10.1121/1.2874557

Kastelein RA, Verboom WC, Jennings N, de Haan D, van der Heul S (2008b) The influence of 70 and 120 kHz tonal signals on the behaviour of harbour porpoises (*Phocoena phocoena*) in a floating pen. Mar Environ Res 66(3):319–326. https://doi.org/10.1016/j.marenvres.2008.05.005

Kastelein R, Hoek L, Jennings N, Jong CD, Terhune J, Dieleman M (2010) Acoustic Mitigation Devices (AMDs) to deter marine mammals from pile driving areas at sea: audibility & behavioural response of a harbour porpoise & harbour seals (Technical Report for COWRIE Ltd. SEAMAMD-09). Sea Mammal Research Company SEAMARCO, Harderwijk, 68 pp

Kastelein R, Gransier R, Hoek L, Olthuis J (2012a) Temporary threshold shifts and recovery in a harbour porpoise (*Phocoena phocoena*) after octave-band noise at 4 kHz. J Acoust Soc Am 132(5):3525–3537. https://doi.org/10.1121/1.4757641

Kastelein RA, Steen N, Gransier R, Wensveen PJ, Jong CAF (2012b) Threshold received sound pressure levels of single 1–2 kHz and 6–7 kHz up-sweeps and down-sweeps causing startle responses in a harbor porpoise (*Phocoena phocoena*). J Acoust Soc Am 131(3):2325–2333. https://doi.org/10.1121/1.3682032

Kastelein RA, Gransier R, Hoek L, Rambags M (2013a) Hearing frequency thresholds of a harbour porpoise (*Phocoena phocoena*) temporarily affected by a continuous 1.5 kHz tone. J Acoust Soc Am 134(3):2286–2292. https://doi.org/10.1121/1.4816405

Kastelein RA, Gransier R, van den Hoogen M, Hoek L (2013b) Brief behavioural response threshold levels of a harbour porpoise (*Phocoena phocoena*) to five helicopter dipping sonar signals (1.33 to 1.43 kHz). Aquat Mamm 39(2):162–173. https://doi.org/10.1578/AM.39.4.2013.315

Kastelein RA, Van Heerden D, Gransier R, Hoek L (2013c) Behavioural responses of a harbour porpoise (*Phocoena phocoena*) to playbacks of broadband pile driving sounds. Mar Environ Res 92:206–214. https://doi.org/10.1016/j.marenvres.2013.09.020

Kastelein RA, Hoek L, Gransier R, Rambags M, Claeys N (2014a) Effect of level, duration, and inter-pulse interval of 1-2kHz sonar signal exposures on harbor porpoise hearing. J Acoust Soc Am 136(1):412–422. https://doi.org/10.1121/1.4883596

Kastelein RA, Schop J, Gransier R, Steen N, Jennings N (2014b) Effect of series of 1 to 2 kHz and 6 to 7 kHz up-sweeps and down-sweeps on the behavior of a harbor porpoise (*Phocoena phocoena*). Aquat Mamm 40(3):232–242. https://doi.org/10.1578/AM.40.3.2014.232

Kastelein RA, Gransier R, Marijt MAT, Hoek L (2015a) Hearing frequency thresholds of harbor porpoises (*Phocoena phocoena*) temporarily affected by played back offshore pile driving sounds. J Acoust Soc Am 137(2):556-564. https://doi.org/10.1121/1.4906261

Kastelein RA, Helder-Hoek L, Gransier R, Terhune JM, Jennings N, De Jong CAF (2015b) Hearing thresholds of harbor seals (*Phoca vitulina*) for playbacks of seal scarer signals, and effects of the signals on behavior. Hydrobiologia 756(1):75–88. https://doi.org/10.1007/s10750-014-2152-6

Kastelein RA, Helder-Hoek L, Janssens G, Gransier R, Johansson T (2015c) Behavioral responses of Harbor Seals (*Phoca vitulina*) to sonar signals in the 25-kHz range. Aquat Mamm 41(4):388–399. https://doi.org/10.1578/AM.41.4.2015.388

Kastelein RA, van den Belt I, Gransier R, Johansson T (2015d) Behavioral responses of a harbor porpoise (*Phocoena phocoena*) to 25.5- to 24.5-kHz sonar down-sweeps with and without side bands. Aquat Mamm 41(4):400–411. https://doi.org/10.1578/AM.41.4.2015.400

Kastelein RA, Helder-Hoek L, Voorde SV (2017a) Hearing thresholds of a male and a female harbor porpoise (*Phocoena phocoena*). J Acoust Soc Am 142(2):1006–1010. https://doi.org/10.1121/1.4997907

Kastelein RA, Horvers M, Helder-Hoek L, Voorde SV, Hofstede RT, Meij HVD (2017b) Behavioral responses of harbor seals (*Phoca vitulina*) to FaunaGuard Seal Module sounds at two background noise levels. Aquat Mamm 43(4):347–363. https://doi.org/10.1578/AM.43.4.2017.347

Kastelein RA, Huybrechts J, Covi J, Helder-Hoek L (2017c) Behavioral responses of a harbor porpoise (*Phocoena phocoena*) to sounds from an acoustic porpoise deterrent. Aquat Mamm 43(3):233–244. https://doi.org/10.1578/AM.43.3.2017.233

Kastelein RA, Helder-Hoek L, Van de Voorde S, de Winter S, Janssen S, Ainslie MA (2018a) Behavioral responses of harbor porpoises (*Phocoena phocoena*) to sonar playback sequences of sweeps and tones (3.5-4.1 kHz). Aquat Mamm 44(4):389–404. https://doi.org/10.1578/AM.44.4.2018.389

Kastelein RA, Van de Voorde S, Jennings N (2018b) Swimming speed of a harbor porpoise (*Phocoena phocoena*) during playbacks of offshore pile driving sounds. Aquat Mamm 44(1):92–99. https://doi.org/10.1578/AM.44.1.2018.92

Kastelein RA, Ainslie MA, van Kester R (2019a) Behavioral responses of harbor porpoises (*Phocoena phocoena*) to US Navy 53C sonar signals in noise. Aquat Mamm 45(4):359–366. https://doi.org/10.1578/AM.45.4.2019.359

Kastelein RA, Benda-Beckmann AMV, Lam FPA, Jansen E, Jong CAFD (2019b) Effect of a bubble screen on the behavioral responses of captive harbor porpoises (*Phocoena phocoena*) exposed to airgun sounds. Aquat Mamm 45(6):703–716. https://doi.org/10.1578/AM.45.6.2019.706

Kastelein RA, Helder-Hoek L, Booth C, Jennings N, Leopold M (2019c) High levels of food intake in harbor porpoises (*Phocoena phocoena*): insight into recovery from disturbance. Aquat Mamm 45(4):380–388. https://doi.org/10.1578/AM.45.4.2019.380

Kastelein RA, Huijser LA, Cornelisse S, Helder-Hoek L, Jennings N, de Jong CA (2019d) Effect of pile-driving playback sound level on fish-catching efficiency in harbor porpoises (*Phocoena phocoena*). Aquat Mamm 45(4):398–410. https://doi.org/10.1578/AM.45.4.2019.398

Kastelein RA, Verhoeven A, Helder-Hoek L (2019e) Behavioral responses of a harbor porpoise (*Phocoena phocoena*) to a series of four different simulated low-frequency sonar sounds (1.33-1.43 kHz). Aquat Mamm 45(6):632–645. https://doi.org/10.1578/AM.45.6.2019.632

Kastelein RA, de Jong CAF, Tougaard J, Helder-Hoek L, Defillet LN (2022) Behavioral responses of a harbor porpoise (*Phocoena phocoena*) depend on the frequency content of pile-driving sounds. Aquat Mamm 48(2):97–109. https://doi.org/10.1578/AM.48.2.2022.97

Kelly BP, Quakenbush LT, Rose JR (1986) Ringed-seal winter ecology and effects of noise disturbance (Final Report, Contract NA81RAC00045). Institute of Marine Science, University of Alaska, Fairbanks

Kerosky SM, Širović A, Roche LK, Baumann-Pickering S, Wiggins SM, Hildebrand JA (2012) Bryde's whale seasonal range expansion and increasing presence in the Southern California Bight from 2000 to 2010. Deep-Sea Res I Oceanogr Res Pap 65:125–132. https://doi.org/10.1016/j.dsr.2012.03.013

Kimura S, Akamatsu T, Wang D, Li S, Wang K, Yoda K (2013) Variation in the production rate of biosonar signals in freshwater porpoises. J Acoust Soc Am 133(5):3128–3134. https://doi.org/10.1121/1.4796129

King SL, Schick RS, Donovan C, Booth CG, Burgman M, Thomas L, Harwood J (2015) An interim framework for assessing the population consequences of disturbance. Methods Ecol Evol 6:1150–1158. https://doi.org/10.1111/2041-210X.12411

Kok ACM, Engelberts JP, Kastelein RA, Helder-Hoek L, Van de Voorde S, Visser F, Slabbekoorn H (2017) Spatial avoidance to experimental increase of intermittent and continuous sound in two captive harbour porpoises. Environ Pollut 233:1024–1036. https://doi.org/10.1016/j.envpol.2017.10.001

Kovacs K, Innes S (1990) The impact of tourism on harp seals (*Phoca groenlandica*) in the Gulf of

St. Lawrence, Canada. Appl Anim Behav Sci 26(1): 15–26. https://doi.org/10.1016/0168-1591(90)90083-P

Kragh IM, McHugh K, Wells RS, Sayigh LS, Janik VM, Tyack PL, Jensen FH (2019) Signal-specific amplitude adjustment to noise in common bottlenose dolphins (*Tursiops truncatus*). J Exp Biol 222(Pt 23):216606. https://doi.org/10.1242/jeb.216606

Kraus SD (1990) Rates and potential causes of mortality in North Atlantic right whales (*Eubalaena glacialis*). Mar Mamm Sci 6(4):278–291. https://doi.org/10.1111/j.1748-7692.1990.tb00358.x

Kuningas S, Kvadsheim PH, Lam F-PA, Miller PJO (2013) Killer whale presence in relation to naval sonar activity and prey abundance in northern Norway. ICES J Mar Sci 70(7):1287–1293. https://doi.org/10.1093/icesjms/fst127

Kvadsheim PH, Sevaldsen EM, Folkow LP, Blix AS (2010) Behavioural and physiological responses of hooded seals (*Cystophora cristata*) to 1 to 7 kHz sonar signals. Aquat Mamm 36(3):239–247. https://doi.org/10.1578/AM.36.3.2010.239

Kvadsheim PH, DeRuiter S, Sivle LD, Goldbogen J, Roland-Hansen R, Miller PJO, Lam F-PA, Calambokidis J, Friedlaender A, Visser F, Tyack PL, Kleivane L, Southall BL (2017) Avoidance responses of minke whales to 1-4 kHz naval sonar. Mar Pollut Bull 121:60–68. https://doi.org/10.1016/j.marpolbul.2017.05.037

La Manna G, Rako-Gòspic N, Manghi M, Ceccherelli G (2019) Influence of environmental, social and behavioural variables on the whistling of the common bottlenose dolphin (*Tursiops truncatus*). Behav Ecol Sociobiol 73(9):121. https://doi.org/10.1007/s00265-019-2736-2

Laborie J, Christiansen F, Beedholm K, Madsen PT, Heerah K (2021) Behavioural impact assessment of unmanned aerial vehicles on Weddell seals (*Leptonychotes weddellii*). J Exp Mar Biol Ecol 536:151509. https://doi.org/10.1016/j.jembe.2020.151509

Lace N (2016) The effect of auditory stimulation on sleep disruption in West Indian Manatee (*Trichechus manatus latirostris*). PhD thesis, University of Southern Mississippi, Hattiesburg, MS, USA

Lacy RC, Williams R, Ashe E, Balcomb Iii KC, Brent LJN, Clark CW, Croft DP, Giles DA, MacDuffee M, Paquet PC (2017) Evaluating anthropogenic threats to endangered killer whales to inform effective recovery plans. Sci Rep 7(1):14119. https://doi.org/10.1038/s41598-017-14471-0

Lalas C, McConnell H (2016) Effects of seismic surveys on New Zealand fur seals during daylight hours: do fur seals respond to obstacles rather than airgun noise? Mar Mamm Sci 32(2):643–663. https://doi.org/10.1111/mms.12293

Lam FPA, Kvadsheim PH, Miller PJO, Tyack PL, Ainslie MA, Curé C, Kleivane L, Sivle LD, van Ijsselmuide SP, Visser F, von Benda-Beckmann AM, Wensveen PJ, Dekeling RPA (2016) Controlled sonar exposure experiments on cetaceans in Norwegian waters: overview of the 3S-Project, Ch 71. In: Hawkins A, Popper AN (eds) The effects of noise on aquatic life II, Advances in experimental medicine and biology, pp 590–598. https://doi.org/10.1007/978-1-4939-2981-8_71

Lammers MO, Howe M, Zang E, McElligott M, Engelhaupt A, Munger L (2017) Acoustic monitoring of coastal dolphins and their response to naval mine neutralization exercises. R Soc Open Sci 4(12):170558. https://doi.org/10.1098/rsos.170558

Lara G, Bou-Cabo M, Llorens S, Miralles R, Espinosa V (2023) Acoustical behavior of delphinid whistles in the presence of an underwater explosion event in the Mediterranean coastal waters of Spain. J Mar Sci Eng 11(4):780. https://doi.org/10.3390/jmse11040780

Laute A, Glarou M, Dodds F, Gomez Røsand SC, Grove TJ, Stoller A, Rasmussen MH, Fournet MEH (2023) Underwater sound of three unoccupied aerial vehicles at varying altitudes and horizontal distances. J Acoust Soc Am 153(6):3419–3427. https://doi.org/10.1121/10.0019805

Lehtonen E, Lehmonen R, Kostensalo J, Kurkilahti M, Suuronen P (2022) Feasibility and effectiveness of seal deterrent in coastal trap-net fishing—development of a novel mobile deterrent. Fish Res 252:106328. https://doi.org/10.1016/j.fishres.2022.106328

Lepper PA, Gordon J, Booth C, Theobald P, Robinson SP, Northridge S, Wang L (2014) Establishing the sensitivity of cetaceans and seals to acoustic deterrent devices in Scotland (Report for Scottish Natural Heritage No. 517). Loughborough University, Inverness

Lesage V, Omrane A, Doniol-Valcroze T, Mosnier A (2017) Increased proximity of vessels reduces feeding opportunities of blue whales in the St. Lawrence Estuary, Canada. Endanger Species Res 32:351–361. https://doi.org/10.3354/esr00825

LGL Ltd (1986) Reactions of beluga whales and narwhals to ship traffic and icebreaking along ice edges in the eastern Canadian High Arctic: 1982-1984. Indian and Northern Affairs Canada, Ottawa, 301 pp

Linnell JDC, Swenson JE, Andersen R, Barnes B (2000) How vulnerable are denning bears to disturbance? Wildl Soc Bull (1973-2006) 28(2):400–413

Ljungblad DK, Thompson PO, Moore SE (1982) Underwater sounds recorded from migrating bowhead whales, *Balaena mysticetus*, in 1979. J Acoust Soc Am 71(2):477–482. https://doi.org/10.1121/1.387419

Ljungblad DK, Würsig B, Swartz SL, Keene JM (1988) Observations on the behavioural responses of bowhead whales (*Balaena mysticetus*) to active geophysical vessels in the Alaskan Beaufort Sea. Arctic 41(3):183–194. https://doi.org/10.14430/arctic1717

Loehr J, Kovanen M, Carey J, Högmander H, Jurasz C, Kärkkäinen S, Suhonen J, Ylönen H (2005) Gender- and age-class-specific reactions to human disturbance in a sexually dimorphic ungulate. Can J Zool 83(12):1602–1607. https://doi.org/10.1139/z05-162

Lomac-MacNair K, Pedro J, Esteves E (2019) Seal and polar bear behavioral response to an icebreaker vessel in northwest Greenland. Human-Wildl Interact 13(2): 277–289. https://doi.org/10.26077/pxn3-h858

Lomac-MacNair K, Wisdom S, Pedro De Andrade J, Stepanuk JE, Esteves E (2021) Polar bear behavioral response to vessel surveys in northeastern Chukchi Sea, 2008–2014. Ursus 2021(32e8):1–14. https://doi.org/10.2192/ursus-d-20-00023.2

Longden EG, Gillespie D, Mann DA, McHugh KA, Rycyk AM, Wells RS, Tyack PL (2022) Comparison of the marine soundscape before and during the COVID-19 pandemic in dolphin habitat in Sarasota Bay, FL. J Acoust Soc Am 152(6):3170–3185. https://doi.org/10.1121/10.0015366

Lowry H, Lill A, Wong BBM (2013) Behavioural responses of wildlife to urban environments. Biol Rev 88(3):537–549. https://doi.org/10.1111/brv.12012

Luksenburg J, Parsons E (2009) The effects of aircraft on cetaceans: implications for aerial whalewatching. In: Proceedings of the 61st Meeting of the International Whaling Commission, Funchal, Portugal, 31, May–26, June 6. SC/61/WW2.

Lusseau D (2003) Male and female bottlenose dolphins *Tursiops* spp. have different strategies to avoid interactions with tour boats in Doubtful Sound, New Zealand. Mar Ecol Prog Ser 257:267–274. https://doi.org/10.3354/meps257267

Lusseau D, Bain DE, Williams R, Smith JC (2009) Vessel traffic disrupts the foraging behavior of southern resident killer whales *Orcinus orca*. Endanger Species Res 6(3):211–221. https://doi.org/10.3354/esr006211

Madsen PT, Møhl B (2000) Sperm whales (*Physeter catodon* L. 1758) do not react to sounds from detonators. J Acoust Soc Am 107(1):668–671. https://doi.org/10.1121/1.428568

Malinka CE, Tønnesen P, Dunn CA, Claridge DE, Gridley T, Elwen SH, Teglberg Madsen P (2021) Echolocation click parameters and biosonar behaviour of the dwarf sperm whale (*Kogia sima*). J Exp Biol 224(6):jeb240689. https://doi.org/10.1242/jeb.240689

Malme CI, Miles PR, Clark CW, Tyack P, Bird JE (1983) Investigations of the potential effects of underwater noise from petroleum industry activities on migrating gray whale behavior (Report for U.S. Minerals Management Service, BBN Report 5366, NTIS PB86-174174). Bolt Beranek and Newman Inc., Cambridge, MA

Malme CI, Miles PR, Clark CW, Tyack P, Bird JE (1984) Investigations of the potential effects of underwater noise from petroleum industry activities on migrating gray whale behavior. Phase II: January 1984 migration. (BBN Report 5586, NTIS PB86-218377). Bolt Beranek and Newman Inc., Anchorage, 357 pp

Malme CI, Würsig B, Bird JE, Tyack PL (1986) Behavioral responses of gray whales to industrial noise: feeding observations and predictive modeling (Final Report for NOAA Outer Continental Shelf Environmental Assessment Program, BBN Report 6265, OCS Study MMS 88-0048, NTIS PB88-249008). BBN Laboratories, Inc., Cambridge, MA

Malme CI, Miles Würsig B, Bird JE, Tyack P (1988) Observations of feeding gray whale responses to controlled industrial noise exposure. In: Sackinger WM, Jeffries MO, Imm JL, Treacy SD (eds) Port and ocean engineering under Arctic conditions, vol III. University of Alaska, Fairbanks, pp 55–73

Mansfield AW (1983) The effects of vessel traffic in the Arctic on marine mammals and recommendations for research. Can Tech Rep of Fisheries and Ocean Sciences No. 1186, 97 pp

Manzano-Roth R, Henderson EE, Martin SW, Martin C, Matsuyama BM (2016) Impacts of US Navy training events on Blainville's beaked whale (*Mesoplodon densirostris*) foraging dives in Hawaiian waters. Aquat Mamm 42(4):507-518. https://doi.org/10.1578/AM.42.4.2016.507

Marley SA, Salgado-Kent CP, Erbe C, Parnum I (2017) Effects of vessel traffic and underwater noise on the movement, behaviour and vocalisations of bottlenose dolphins in an urbanised estuary. Sci Rep 7:13437. https://doi.org/10.1038/s41598-017-13252-z

Martin JGA, Réale D (2008) Animal temperament and human disturbance: implications for the response of wildlife to tourism. Behav Process 77(1):66–72. https://doi.org/10.1016/j.beproc.2007.06.004

Martin SW, Martin CR, Matsuyama BM, Henderson EE (2015) Minke whales (*Balaenoptera acutorostrata*) respond to navy training. J Acoust Soc Am 137(5): 2533–2541. https://doi.org/10.1121/1.4919319

Martin M, Gridley T, Elwen SH, Charrier I (2022) Assessment of the impact of anthropogenic airborne noise on the behaviour of Cape fur seals during the breeding season in Namibia. J Exp Mar Biol Ecol 550:151721. https://doi.org/10.1016/j.jembe.2022.151721

Martin M, Gridley T, Elwen S, Charrier I (2023a) Inter-site variability in the Cape fur seal's behavioural response to boat noise exposure. Mar Pollut Bull 196:115589. https://doi.org/10.1016/j.marpolbul.2023.115589

Martin MJ, Halliday WD, Citta JJ, Quakenbush L, Harwood L, Lea EV, Juanes F, Dawson J, Nicoll A, Insley SJ (2023b) Exposure and behavioural responses of tagged bowhead whales (*Balaena mysticetus*) to vessels in the Pacific Arctic. Arctic Sci 9(3):600–615. https://doi.org/10.1139/as-2022-0052

Martin MJ, Halliday WD, Storrie L, Citta JJ, Dawson J, Hussey NE, Juanes F, Loseto LL, MacPhee SA, Moore L, Nicoll A, O'Corry-Crowe G, Insley SJ (2023c) Exposure and behavioral responses of tagged beluga whales (*Delphinapterus leucas*) to ships in the Pacific Arctic. Mar Mamm Sci 39(2):387–421. https://doi.org/10.1111/mms.12978

Mate BR, Harvey JT (1987) Acoustical deterrents in marine mammal conflicts with fisheries. Oregon State University Sea Grant, ORESU-W-86-001, Corvallis, p 116

Mate BR, Stafford KM, Ljungblad DK (1994) A change in sperm whale (*Physeter macroephalus*) distribution correlated to seismic surveys in the Gulf of Mexico

(Abstract). J Acoust Soc Am 96(5, pt. 2):3268. https://doi.org/10.1121/1.410971

Matthews LP, Fournet MEH, Gabriele C, Klinck H, Parks SE (2020) Acoustically advertising male harbour seals in southeast Alaska do not make biologically relevant acoustic adjustments in the presence of vessel noise. Biol Lett 16(4):20190795. https://doi.org/10.1098/rsbl.2019.0795

May TM, Page MJ, Fleming PA (2016) Predicting survivors: animal temperament and translocation. Behav Ecol 27(4):969–977. https://doi.org/10.1093/beheco/arv242

McCarthy E, Moretti D, Thomas L, DiMarzio N, Morrissey R, Jarvis S, Ward J, Izzi A, Dilley A (2011) Changes in spatial and temporal distribution and vocal behavior of Blainville's beaked whales (*Mesoplodon densirostris*) during multiship exercises with mid-frequency sonar. Mar Mamm Sci 27(3):E206–E226. https://doi.org/10.1111/j.1748-7692.2010.00457.x

McCauley RD, Jenner M-N, Jenner C, McCabe KA, Murdoch J (1998) The response of humpback whales (*Megaptera novaeangliae*) to offshore seismic survey noise: preliminary results of observations about a working seismic vessel and experimental exposures. Aust Petrol Prod Explor Assoc J 38:692–707. https://doi.org/10.1071/AJ97045

McCauley RD, Fewtrell J, Duncan AJ, Jenner C, Jenner M-N, Penrose JD, Prince RIT, Adihyta A, Murdoch J, McCabe K (2000) Marine seismic surveys: analysis and propagation of air gun signals, and effects of exposure on humpback whales, sea turtles, fishes and squid (Report for the Australian Petroleum Exploration and Production Association, CMST R99-15). Centre for Marine Science and Technology, Curtin University, Perth

McCullough JLK, Wren JLK, Oleson EM, Allen AN, Siders ZA, Norris ES (2021) An acoustic survey of beaked whales and *Kogia* spp. in the Mariana Archipelago using drifting recorders. Front Mar Sci 8:664292. https://doi.org/10.3389/fmars.2021.664292

McGarry T, De Silva R, Canning S, Mendes S, Prior A, Stephenson S, Wilson J (2022) Evidence base for application of Acoustic Deterrent Devices (ADDs) as marine mammal mitigation (Version 4) (JNCC Report No. 615). Joint Nature Conservation Committee, Peterborough

McGeady R, McMahon BJ, Berrow S (2016) The effects of seismic surveying and environmental variables on deep diving odontocete stranding rates along Ireland's coast. Proc Meet Acoust 27(1):040006. https://doi.org/10.1121/2.0000281

McHuron EA, Costa DP, Schwarz L, Mangel M, Matthiopoulos J (2017) State-dependent behavioural theory for assessing the fitness consequences of anthropogenic disturbance on capital and income breeders. Methods Ecol Evol 8(5):552–560. https://doi.org/10.1111/2041-210X.12701

McHuron EA, Aerts L, Gailey G, Sychenko O, Costa DP, Mangel M, Schwarz LK (2021) Predicting the population consequences of acoustic disturbance, with application to an endangered gray whale population. Ecol Appl 31(8):e02440. https://doi.org/10.1002/eap.2440

McHuron EA, Castellote M, Himes Boor GK, Shelden KEW, Warlick AJ, McGuire TL, Wade PR, Goetz KT (2023) Modeling the impacts of a changing and disturbed environment on an endangered beluga whale population. Ecol Model 483:110417. https://doi.org/10.1016/j.ecolmodel.2023.110417

Melcón M, Cummins A, Kerosky S, Roche L, Wiggins S, Hildebrand J (2012) Blue whales respond to anthropogenic noise. PLoS One 7(2):e32681. https://doi.org/10.1371/journal.pone.0032681

Mikkelsen L, Hermannsen L, Beedholm K, Madsen PT, Tougaard J (2017) Simulated seal scarer sounds scare porpoises, but not seals: species-specific responses to 12 kHz deterrence sounds. R Soc Open Sci 4(7):170286. https://doi.org/10.1098/rsos.170286

Mikkelsen L, Johnson M, Wisniewska DM, van Neer A, Siebert U, Madsen PT, Teilmann J (2019) Long-term sound and movement recording tags to study natural behavior and reaction to ship noise of seals. Ecol Evol 9:2588–2601. https://doi.org/10.1002/ece3.4923

Miksis-Olds JL, Tyack PL (2009) Manatee (*Trichechus manatus*) vocalization usage in relation to environmental noise levels. J Acoust Soc Am 125(3):1806–1815. https://doi.org/10.1121/1.3068455

Miksis-Olds JL, Wagner T (2011) Behavioral response of manatees to variations in environmental sound levels. Mar Mamm Sci 27(1):130–148. https://doi.org/10.1111/j.1748-7692.2010.00381.x

Miksis-Olds JL, Donaghay PL, Miller JH, Tyack PL, Nystuen JA (2007a) Noise level correlates with manatee use of foraging habitats. J Acoust Soc Am 121(5):3011–3020. https://doi.org/10.1121/1.2713555

Miksis-Olds JL, Donaghay PL, Miller JH, Tyack PL, Reynolds JE (2007b) Simulated vessel approaches elicit differential responses from manatees. Mar Mamm Sci 23(3):629–649. https://doi.org/10.1111/j.1748-7692.2007.00133.x

Miller PJO, Biassoni N, Samuels A, Tyack PL (2000) Whale songs lengthen in response to sonar. Nature 405(6789):903. https://doi.org/10.1038/35016148

Miller GW, Moulton VD, Davis RA, Holst M, Millman P, MacGillivray A, Hannay D (2005) Monitoring seismic effects on marine mammals—Southeastern Beaufort Sea, 2001-2002. In: Armsworthy SL, Cranford PJ, Lee K (eds) Offshore oil and gas environmental effects monitoring: approaches and technologies. Battelle Press, Columbus, pp 511–542

Miller PJO, Johnson MP, Madsen PT, Biassoni N, Quero M, Tyack PL (2009) Using at-sea experiments to study the effects of airguns on the foraging behavior of sperm whales in the Gulf of Mexico. Deep-Sea Res I Oceanogr Res Pap 56(7):1168–1181. https://doi.org/10.1016/j.dsr.2009.02.008

Miller PJ, Kvadsheim PH, Lam F-PA, Wensveen PJ, Antunes R, Alves AC, Visser F, Kleivane L, Tyack PL, Sivle LD (2012) The severity of behavioral changes observed during experimental exposures of killer (*Orcinus orca*), long-finned pilot (*Globicephala melas*), and sperm (*Physeter macrocephalus*) whales to naval sonar. Aquat Mamm 38(4):362–401. https://doi.org/10.1578/AM.38.4.2012.362

Miller PJO, Antunes RN, Wensveen PJ, Samarra FIP, Alves AC, Tyack PL, Kvadsheim PH, Kleivane L, Lam F-PA, Ainslie MA, Thomas L (2014) Dose-response relationships for the onset of avoidance of sonar by free-ranging killer whales. J Acoust Soc Am 135(1):975-993. https://doi.org/10.1121/1.4861346

Miller PJO, Kvadsheim PH, Lam F-PA, Tyack PL, Cure C, Deruiter SL, Kleivane L, Sivle LD, Van Ijsselmuide SP, Visser F, Wensveen PJ, Von Benda-Beckmann AM, Martin Lopez LM, Narazaki T, Hooker SK (2015) First indications that northern bottlenose whales are sensitive to behavioural disturbance from anthropogenic noise. R Soc Open Sci 2: 140484. https://doi.org/10.1098/rsos.140484

Miller PJO, Isojunno S, Siegal E, Lam F-PA, Kvadsheim PH, Curé C (2022) Behavioral responses to predatory sounds predict sensitivity of cetaceans to anthropogenic noise within a soundscape of fear. Proc Natl Acad Sci 119(13):e2114932119. https://doi.org/10.1073/pnas.2114932119

Mooney TA, Pacini AF, Nachtigall PE (2009) False killer whale (*Pseudorca crassidens*) echolocation and acoustic disruption: implications for longline bycatch and depredation. Can J Zool 87:726–733. https://doi.org/10.1139/Z09-061

Moretti D, Thomas L, Marques T, Harwood J, Dilley A, Neales B, Shaffer J, McCarthy E, New L, Jarvis S, Morrissey R (2014) A risk function for behavioral disruption of Blainville's beaked whales (*Mesoplodon densirostris*) from mid-frequency active sonar. PLoS One 9(1):e85064. https://doi.org/10.1371/journal.pone.0085064

Morin PA, Parsons KM, Archer FI, Ávila-Arcos M, Barrett-Lennard LG, Dalla Rosa L, Duchéne S, Durban JW, Ellis GM, Ferguson SM, Ford JKB, Ford MJ, Garilao C, Gilbert MTP, Kashner K, Matkin CO, Petersen SD, Roberts KM, Visser IN, Wade PR, Ho SYW, Foote AD (2015) Geographic and temporal dynamics of a global radiation and diversification in the killer whale. Mol Ecol 24:3964–3979. https://doi.org/10.1111/mec.13284

Morisaka T, Shinohara M, Nakahara F, Akamatsu T (2005) Effects of ambient noise on the whistles of Indo-Pacific bottlenose dolphin populations. J Mammal 86:541–546. https://doi.org/10.1644/1545-1542(2005)86[541:EOANOT]2.0.CO;2

Morton AB, Symonds HK (2002) Displacement of *Orcinus orca* (L.) by high amplitude sound in British Columbia, Canada. ICES J Mar Sci 59(1):71–80. https://doi.org/10.1006/jmsc.2001.1136

Moulton VD, Miller GW (2005) Marine mammal monitoring of a seismic survey on the Scotian Slope, 2003. In: Lee K, Bain H, Hurley GV (eds) Acoustic monitoring and marine mammal surveys in The Gully and outer Scotian Shelf before and during active seismic programs, Environmental studies research funds report no. 151. Fisheries and Oceans Canada, Dartmouth, pp 29–40

Moulton VD, Richardson WJ, Williams MT (2003) Ringed seal densities and noise near an icebound artificial island with construction and drilling. Acoust Res Lett Online 4(4):112–117. https://doi.org/10.1121/1.1605091

Muir JE, Ainsworth L, Racca R, Bychkov Y, Gailey G, Vladimirov V, Starodymov S, Bröker K (2016) Gray whale densities during a seismic survey off Sakhalin Island, Russia. Endanger Species Res 29(3):211–227. https://doi.org/10.3354/esr00709

Murray CC, Hannah LC, Doniol-Valcroze T, Wright BM, Stredulinsky EH, Nelson JC, Locke A, Lacy RC (2021) A cumulative effects model for population trajectories of resident killer whales in the Northeast Pacific. Biol Conserv 257:109124. https://doi.org/10.1016/j.biocon.2021.109124

National Research Council (1994) Low-frequency sound and marine mammals: current knowledge and research needs. National Academies Press, Washington, DC. https://doi.org/10.17226/4557

National Research Council (2005) Marine mammal populations and ocean noise: determining when noise causes biologically significant effects. National Academies Press, Washington, DC. https://doi.org/10.17226/11147

National Research Council (2017) Approaches to understanding the cumulative effects of stressors on marine mammals. National Academies Press, Washington, DC. https://doi.org/10.17226/23479

New LF, Harwood J, Thomas L, Donovan C, Clark JS, Hastie G, Lusseau D (2013a) Modelling the biological significance of behavioural change in coastal bottlenose dolphins in response to disturbance. Funct Ecol 27:314–322. https://doi.org/10.1111/1365-2435.12052

New LF, Moretti DJ, Hooker SK, Costa DP, Simmons SE (2013b) Using energetic models to investigate the survival and reproduction of beaked whales (family Ziphiidae). PLoS One 8(7):e68725. https://doi.org/10.1371/journal.pone.0068725

New LF, Clark JS, Costa DP, Fleishman E, Hindell MA, Klanjcek T, Lusseau D, Kraus S, McMahon CR, Robinson PW, Schick RS, Schwarz LK, Simmons SE, Thomas L, Tyack PL, Harwood J (2014) Using short-term measures of behaviour to estimate long-term fitness of southern elephant seals. Mar Ecol Prog Ser 496:99–108. https://doi.org/10.3354/meps10547

New L, Lusseau D, Harcourt R (2020) Dolphins and boats: when is a disturbance, disturbing? Front Mar Sci 7: 353. https://doi.org/10.3389/fmars.2020.00353

Nielsen MLK, Bejder L, Videsen SKA, Christiansen F, Madsen PT (2019) Acoustic crypsis in southern right whale mother–calf pairs: infrequent, low-output calls to avoid predation? J Exp Biol 222(13):jeb190728. https://doi.org/10.1242/jeb.190728

Niemi M, Auttila M, Valtonen A, Viljanen M, Kunnasranta M (2013) Haulout patterns of Saimaa ringed seals and their response to boat traffic during the moulting season. Endanger Species Res 22(2): 115–124. https://doi.org/10.3354/esr00541

Niu F-g, Liu Z-w, Wen H-t, Zu D-w, Yang Y-m (2012) Behavioral responses of two captive bottlenose dolphins (*Tursiops truncatus*) to a continuous 50 kHz tone. J Acoust Soc Am 131(2):1643–1649. https://doi.org/10.1121/1.3675945

Niu F, Yang Y, Xue R, Zhou Z, Chen S (2020) Behavioral responses by captive bottlenose dolphins (*Tursiops truncatus*) to 15-to 50-kHz tonal signals. Aquat Mamm 46(1):1–10. https://doi.org/10.1578/AM.46.1.2020.1

Noren D, Johnson A, Rehder D, Larson A (2009) Close approaches by vessels elicit surface active behaviors by southern resident killer whales. Endanger Species Res 8(3):179–192. https://doi.org/10.3354/esr00205

Nowacek SM, Nowacek DP, Johnson MP, Shorter KA, Powell JA, Wells RS (2002) Manatee behavioral responses to vessel approaches: results of digital acoustic data logger tagging of manatees in Belize (Report for Florida Fish and Wildlife Conservation Commission No. 847). Mote Marine Laboratory, Sarasota, 61 pp

Nowacek DP, Johnson MP, Tyack PL (2004a) North Atlantic right whales (*Eubalaena glacialis*) ignore ships but respond to alerting stimuli. Proc R Soc Lond Ser B Biol Sci 271(1536):227–231. https://doi.org/10.1098/rspb.2003.2570

Nowacek SM, Wells RS, Owen ECG, Speakman TR, Flamm RO, Nowacek DP (2004b) Florida manatees, *Trichechus manatus latirostris*, respond to approaching vessels. Biol Conserv 119(4):517–523. https://doi.org/10.1016/j.biocon.2003.11.020

Paiva EG, Salgado Kent CP, Gagnon MM, McCauley R, Finn H (2015) Reduced detection of Indo-Pacific bottlenose dolphins (*Tursiops aduncus*) in an inner harbour channel during pile driving activities. Aquat Mamm 41(4):455–468. https://doi.org/10.1578/AM.41.4.2015.455

Palomino-González A, Kovacs KM, Lydersen C, Ims RA, Lowther AD (2021) Drones and marine mammals in Svalbard, Norway. Mar Mamm Sci 37(4):1212–1229. https://doi.org/10.1111/mms.12802

Parks SE, Clark CW, Tyack PL (2007) Short- and long-term changes in right whale calling behavior: the potential effects of noise on acoustic communication. J Acoust Soc Am 122(6):3725–3731. https://doi.org/10.1121/1.2799904

Parsons ECM (2017) Impacts of navy sonar on whales and dolphins: now beyond a smoking gun? Front Mar Sci 4:295. https://doi.org/10.3389/fmars.2017.00295

Parsons MJG, Duncan AJ, Parsons SK, Erbe C (2020) Reducing vessel noise: an example of a solar-electric passenger ferry. J Acoust Soc Am 147(5):3575–3583. https://doi.org/10.1121/10.0001264

Patenaude NJ, Richardson WJ, Smultea MA, Koski WR, Miller GW, Würsig B, Greene CR (2002) Aircraft sound and disturbance to bowhead and beluga whales during spring migration in the Alaskan Beaufort Sea. Mar Mamm Sci 18(2):309–335. https://doi.org/10.1111/j.1748-7692.2002.tb01040.x

Paterson WD, Russell DJF, Wu G-M, McConnell B, Currie JI, McCafferty DJ, Thompson D (2019) Post-disturbance haulout behaviour of harbour seals. Aquat Conserv 29(S1):144–156. https://doi.org/10.1002/aqc.3092

Pavez G, Muñoz L, Barilari F, Sepúlveda M (2015) Variation in behavioral responses of the South American sea lion to tourism disturbance: implications for tourism management. Mar Mamm Sci 31(2):427–439. https://doi.org/10.1111/mms.12159

Payne R, Webb D (1971) Orientation by means of long range acoustic signaling in baleen whales. Ann N Y Acad Sci 188:110–141. https://doi.org/10.1111/j.1749-6632.1971.tb13093.x

Pedreros E, Sepúlveda M, Gutierrez J, Carrasco P, Quiñones RA (2016) Observations of the effect of a New Year's fireworks display on the behavior of the South American sea lion (*Otaria flavescens*) in a colony of central-south Chile. Mar Freshw Behav Physiol 49(2):127–131. https://doi.org/10.1080/10236244.2015.1125099

Pellegrini AY, Romeu B, Ingram SN, Daura-Jorge FG (2021) Boat disturbance affects the acoustic behaviour of dolphins engaged in a rare foraging cooperation with fishers. Anim Conserv 24(4):613-625. https://doi.org/10.1111/acv.12667

Pérez Tadeo M, Gammell M, O'Callaghan SA, O'Connor I, O'Brien J (2023) Disturbances due to unmanned aerial vehicles (UAVs) on harbor seal (*Phoca vitulina*) colonies: recommendations on best practices. Mar Mamm Sci 39(3):757–779. https://doi.org/10.1111/mms.13002

Perez-Ortega B, Daw R, Paradee B, Gimbrere E, May-Collado LJ (2021) Dolphin-watching boats affect whistle frequency modulation in bottlenose dolphins. Front Mar Sci 8(102):618420. https://doi.org/10.3389/fmars.2021.618420

Perry EA, Boness DJ, Insley SJ (2002) Effects of sonic booms on breeding gray seals and harbor seals on Sable Island, Canada. J Acoust Soc Am 111(1): 599–609. https://doi.org/10.1121/1.1349538

Pirotta E, Milor R, Quick N, Moretti D, Di Marzio N, Tyack P, Boyd I, Hastie G (2012) Vessel noise affects beaked whale behavior: results of a dedicated acoustic response study. PLoS One 7(8):e42535. https://doi.org/10.1371/journal.pone.0042535

Pirotta E, Brookes KL, Graham IM, Thompson PM (2014) Variation in harbour porpoise activity in response to seismic survey noise. Biol Lett 10(5):20131090. https://doi.org/10.1098/rsbl.2013.1090

Pirotta E, Merchant ND, Thompson PM, Barton TR, Lusseau D (2015) Quantifying the effect of boat disturbance on bottlenose dolphin foraging activity. Biol Conserv 181:82–89. https://doi.org/10.1016/j.biocon.2014.11.003

Pirotta E, Hin V, Mangel M, New L, Costa DP, Roos AM, Harwood J (2020) Propensity for risk in reproductive strategy affects susceptibility to anthropogenic disturbance. Am Nat 196(4):E71–E87. https://doi.org/10.1086/710150

Pirotta E, Booth CG, Calambokidis J, Costa DP, Fahlbusch JA, Friedlaender AS, Goldbogen JA, Harwood J, Hazen EL, New L, Santora JA, Watwood SL, Wertman C, Southall BL (2022) From individual responses to population effects: integrating a decade of multidisciplinary research on blue whales and sonar. Anim Conserv 25(6):796–810. https://doi.org/10.1111/acv.12785

Pitman RL, Ballance LT, Mesnick SI, Chivers SJ (2001) Killer whale predation on sperm whales: observations and implications. Mar Mamm Sci 17(3):494–507. https://doi.org/10.1111/j.1748-7692.2001.tb01000.x

Pitman RL, Deecke VB, Gabriele CM, Srinivasan M, Black N, Denkinger J, Durban JW, Mathews EA, Matkin DR, Neilson JL, Schulman-Janiger A, Shearwater D, Stap P, Ternullo R (2017) Humpback whales interfering when mammal-eating killer whales attack other species: mobbing behavior and interspecific altruism? Mar Mamm Sci 33(1):7–58. https://doi.org/10.1111/mms.12343

Pizzolato L, Howell SEL, Dawson J, Laliberté F, Copland L (2016) The influence of declining sea ice on shipping activity in the Canadian Arctic. Geophys Res Lett 43(23):12,146–112,154. https://doi.org/10.1002/2016GL071489

Poupard M, Montgolfier BD, Glotin H (2019) Ethoacoustic by Bayesian non parametric and stochastic neighbor embedding to forecast anthropic pressure on dolphins. In: OCEANS 2019—Marseille, 17–20 June 2019, pp 1–5. https://doi.org/10.1109/OCEANSE.2019.8867126

Prawirasasra MS, Mustonen M, Klauson A (2021) The underwater soundscape at Gulf of Riga Marine-Protected Areas. J Mar Sci Eng 9(8):915. https://doi.org/10.3390/jmse9080915

Prawirasasra MS, Jüssi M, Mustonen M, Klauson A (2022) Underwater noise impact of a ferry route on dive patterns of transiting Baltic ringed seals. Est J Earth Sci 71(4):201–213. https://doi.org/10.3176/earth.2022.14

Quick NJ, Middlemas SJ, Armstrong JD (2004) A survey of antipredator controls at marine salmon farms in Scotland. Aquaculture 230(1):169–180. https://doi.org/10.1016/S0044-8486(03)00428-9

Quick N, Scott-Hayward L, Sadykova D, Nowacek D, Read AJ (2017) Effects of a scientific echo sounder on the behavior of short-finned pilot whales (*Globicephala macrorhynchus*). Can J Fish Aquat Sci. 74(5):716-726 https://doi.org/10.1139/cjfas-2016-0293

Quin D (1997) Report of an acoustic disturbance at the Round Island Walrus Sanctuary. Alaska Department of Fish and Game, Dillingham

Rako Gospić N, Picciulin M (2016) Changes in whistle structure of resident bottlenose dolphins in relation to underwater noise and boat traffic. Mar Pollut Bull 105(1):193–198. https://doi.org/10.1016/j.marpolbul.2016.02.030

Rako N, Fortuna CM, Holcer D, Mackelworth P, Nimak-Wood M, Pleslić G, Sebastianutto L, Vilibić I, Wiemann A, Picciulin M (2013) Leisure boating noise as a trigger for the displacement of the bottlenose dolphins of the Cres–Lošinj archipelago (northern Adriatic Sea, Croatia). Mar Pollut Bull 68(1):77–84. https://doi.org/10.1016/j.marpolbul.2012.12.019

Rako-Gospić N, La Manna G, Picciulin M, Ceccherelli G (2021) Influence of foraging context on the whistle structure of the common bottlenose dolphin. Behav Process 182:104281. https://doi.org/10.1016/j.beproc.2020.104281

Ramos EA, Maloney B, Magnasco MO, Reiss D (2018) Bottlenose dolphins and Antillean manatees respond to small multi-rotor unmanned aerial systems. Front Mar Sci 5:316. https://doi.org/10.3389/fmars.2018.00316

Rasmussen MH, Atem ACG, Miller LA (2016) Behavioral responses by Icelandic white-beaked dolphins (*Lagenorhynchus albirostris*) to playback sounds. Aquat Mamm 42(3):317–329. https://doi.org/10.1578/AM.42.3.2016.317

Reed J, Harcourt R, New L, Bilgmann K (2020) Extreme effects of extreme disturbances: a simulation approach to assess population specific responses. Front Mar Sci 7:519845. https://doi.org/10.3389/fmars.2020.519845

Rendell LE, Gordon JCD (1999) Vocal response of long-finned pilot whales (*Globicephala melas*) to military sonar in the Ligurian Sea. Mar Mamm Sci 15(1):198–204. https://doi.org/10.1111/j.1748-7692.1999.tb00790.x

Richardson WJ, Malme CI (1993) Man-made noise and behavioral response. In: Burns JJ, Montague JJ, Cowles CJ (eds) The bowhead whale, Special publication 2nd edn. Society for Marine Mammalogy, Lawrence, pp 631–700

Richardson WJ, Würsig B (1997) Influences of man-made noise and other human actions on cetacean behaviour. Mar Freshw Behav Physiol 29(1-4):183–209. https://doi.org/10.1080/10236249709379006

Richardson WJ, Fraker MA, Würsig B, Wells RS (1985) Behaviour of bowhead whales *Balaena mysticetus* summering in the Beaufort Sea: reactions to industrial activities. Biol Conserv 32(3):195–230. https://doi.org/10.1016/0006-3207(85)90111-9

Richardson WJ, Würsig B, Greene CR Jr (1986) Reactions of bowhead whales, *Balaena mysticetus*, to seismic exploration in the Canadian Beaufort Sea. J Acoust

Soc Am 79(4):1117–1128. https://doi.org/10.1121/1.393384

Richardson WJ, Greene CRJ, Koski WR, Malme CI, Miller GW, Smultea MA (1990) Acoustic effects of oil production activities on bowhead and white whales visible during spring migration near Pt. Barrow, Alaska—1989 phase: sound propagation and whale responses to playbacks of continuous drilling noise from an ice platform, as studied in pack ice conditions. Annual report for U.S. Minerals Management Service, LGL Report TA848-4, 284 pp

Richardson WJ, Greene CRJ, Koski WR, Smultea MA, Cameron G, Holdsworth C (1991) Acoustic effects of oil production activities on bowhead and white whales visible during spring migration near Pt. Barrow, Alaska—1990 phase: sound propagation and whale responses to playbacks of continuous drilling noise from an ice platform, as studied in pack ice conditions (Report for U.S. Minerals Management Service OCS Study MMS 91-0037, NTIS PB92-170430). LGL Ltd., Herndon., 311 pp

Richardson WJ, Finley KJ, Miller GW, Davis RA, Koski WR (1995a) Feeding, social and migration behavior of bowhead whales, *Balaena mysticetus*, in Baffin Bay vs. the Beaufort Sea—regions with different amounts of human activity. Mar Mamm Sci 11(1): 1–45. https://doi.org/10.1111/j.1748-7692.1995.tb00272.x

Richardson WJ, Greene CR, Malme CI, Thomson DH (1995b) Marine mammals and noise. Academic, San Diego. https://doi.org/10.1016/C2009-0-02253-3

Richardson WJ, Greene CRJ, Hanna JS, Koski WR, Miller GW, Patenaude NJ, Smultea MA, Blaylock R, Elliott R, Würsig B (1995c) Acoustic effects of oil production activities on bowhead and white whales visible during spring migration near Pt. Barrow, Alaska—1991 and 1994 phases: sound propagation and whale responses to playbacks of icebreaker noise. Report for U.S. Minerals Management Service OCS Study MMS 95-0051, 570 pp

Richardson WJ, Miller GW, Charles R, Greene J (1999) Displacement of migrating bowhead whales by sounds from seismic surveys in shallow waters of the Beaufort Sea (Abstract). J Acoust Soc Am 106(4):2281. https://doi.org/10.1121/1.427801

Richter C, Dawson S, Slooten E (2006) Impacts of commercial whale watching on male sperm whales at Kaikoura, New Zealand. Mar Mamm Sci 22(1): 46–63. https://doi.org/10.1111/j.1748-7692.2006.00005.x

Robertson FC, Koski WR, Thomas TA, Richardson WJ, Würsig B, Trites AW (2013) Seismic operations have variable effects on dive-cycle behavior of bowhead whales in the Beaufort Sea. Endanger Species Res 21(2):143–160. https://doi.org/10.3354/esr00515

Robinson ST, Lepper PA, Ablitt J (2007) The measurement of the underwater radiated noise from marine piling including characterisation of a "soft start" period. In: IEEE Oceans 2007 conference, Aberdeen. https://doi.org/10.1109/OCEANSE.2007.4302326

Rojano-Doñate L, McDonald BI, Wisniewska DM, Johnson M, Teilmann J, Wahlberg M, Højer-Kristensen J, Madsen PT (2018) High field metabolic rates of wild harbour porpoises. J Exp Biol 221(23): jeb185827. https://doi.org/10.1242/jeb.185827

Russell DJF, Hastie GD, Thompson D, Janik VM, Hammond PS, Scott-Hayward LAS, Matthiopoulos J, Jones EL, McConnell BJ (2016) Avoidance of wind farms by harbour seals is limited to pile driving activities. J Appl Ecol 53(6):1642–1652. https://doi.org/10.1111/1365-2664.12678

Rutenko AN, Borisov SV, Gritsenko AV, Jenkerson MR (2007) Calibrating and monitoring the western gray whale mitigation zone and estimating acoustic transmission during a 3D seismic survey, Sakhalin Island, Russia. Environ Monit Assess 134(1-3):21–44. https://doi.org/10.1007/s10661-007-9814-z

Rycyk A (2013) Manatee behavioral response to approaching boats. PhD thesis, Florida State University, Florida, USA

Rycyk AM, Deutsch CJ, Barlas ME, Hardy SK, Frisch K, Leone EH, Nowacek DP (2018) Manatee behavioral response to boats. Mar Mamm Sci 34(4):924–962. https://doi.org/10.1111/mms.12491

Saïd S, Tolon V, Brandt S, Baubet E (2012) Sex effect on habitat selection in response to hunting disturbance: the study of wild boar. Eur J Wildl Res 58(1): 107–115. https://doi.org/10.1007/s10344-011-0548-4

Santos-Carvallo M, Barilari F, Pérez-Alvarez MJ, Gutiérrez L, Pavez G, Araya H, Anguita C, Cerda C, Sepúlveda M (2021) Impacts of whale-watching on the short-term behavior of fin whales (*Balaenoptera physalus*) in a Marine Protected Area in the Southeastern Pacific. Front Mar Sci 8:623954. https://doi.org/10.3389/fmars.2021.623954

Sarnocińska J, Teilmann J, Balle JD, van Beest FM, Delefosse M, Tougaard J (2020) Harbor porpoise (*Phocoena phocoena*) reaction to a 3D seismic airgun survey in the North Sea. Front Mar Sci 6:824. https://doi.org/10.3389/fmars.2019.00824

Schack H, Haapaniemi J (2017) Potential impact of noise from shipping on key species of marine mammals in waters off Western Greenland-Case Baffinland (Report for WWF Danmark). H B Schack Consulting

Scheifele PM, Andrew S, Cooper RA, Darre M, Musiek FE, Max L (2005) Indication of a Lombard vocal response in the St. Lawrence River beluga. J Acoust Soc Am 117(3):1486–1492. https://doi.org/10.1121/1.1835508

Scholik-Schlomer A (2015) Where the decibels hit the water: perspectives on the application of science to real-world underwater noise and marine protected species issues. Acoust Today 11(3):36–44

Schwarz LK, McHuron E, Mangel M, Wells RS, Costa DP (2016) Stochastic dynamic programming: an approach for modelling the population consequences of disturbance due to lost foraging opportunities. Proc Meet

Acoust 27(1):040004. https://doi.org/10.1121/2.0000276

Selbmann A, Basran CJ, Bertulli CG, Hudson T, Mrusczok M-T, Rasmussen MH, Rempel JN, Scott J, Svavarsson J, Wensveen PJ, Whittaker M, Samarra FIP (2022) Occurrence of long-finned pilot whales (*Globicephala melas*) and killer whales (*Orcinus orca*) in Icelandic coastal waters and their interspecific interactions. Acta Ethol 25(3):141–154. https://doi.org/10.1007/s10211-022-00394-1

Senigaglia V, de Stephanis R, Verborgh P, Lusseau D (2012) The role of synchronized swimming as affiliative and anti-predatory behavior in long-finned pilot whales. Behav Process 91(1):8–14. https://doi.org/10.1016/j.beproc.2012.04.011

Shaughnessy PD, Semmelink A, Cooper J, Frost PGH (1981) Attempts to develop acoustic methods of keeping Cape fur seals *Arctocephalus pusillus* from fishing nets. Biol Conserv 21(2):141–158. https://doi.org/10.1016/0006-3207(81)90076-8

Siegal E, Hooker SK, Isojunno S, Miller PJO (2022) Beaked whales and state-dependent decision-making: how does body condition affect the trade-off between foraging and predator avoidance? Proc R Soc B 289(1967):20212539. https://doi.org/10.1098/rspb.2021.2539

Sills JM, Ruscher B, Nichols R, Southall BL, Reichmuth C (2020) Evaluating temporary threshold shift onset levels for impulsive noise in seals. J Acoust Soc Am 148(5):2973–2986. https://doi.org/10.1121/10.0002649

Širović A, Hildebrand JA, Wiggins SM (2007) Blue and fin whale call source levels and propagation range in the Southern Ocean. J Acoust Soc Am 122(2):1208–1215. https://doi.org/10.1121/1.2749452

Sivle LD, Kvadsheim PH, Fahlman A, Lam FPA, Tyack PL, Miller PJO (2012) Changes in dive behavior during naval sonar exposure in killer whales, long-finned pilot whales, and sperm whales. Front Physiol 3(400):1–11. https://doi.org/10.3389/fphys.2012.00400

Sivle LD, Kvadsheim PH, Cure C, Isojunno S, Wensveen PJ, Lam F-PA, Visser F, Kleivane L, Tyack PL, Harris CM, Miller PJO (2015) Severity of expert-identified behavioural responses of humpback whale, minke whale, and northern bottlenose whale to naval sonar. Aquat Mamm 41(4):469–502. https://doi.org/10.1578/AM.41.4.2015.469

Sivle LD, Wensveen PJ, Kvadsheim PH, Lam FPA, Visser F, Curé C, Harris CM, Tyack PL, Miller PJO (2016) Naval sonar disrupts foraging in humpback whales. Mar Ecol Prog Ser 562:211–220. https://doi.org/10.3354/meps11969

Skeate E, Perrow M, Gilroy J (2012) Likely effects of construction of Scroby Sands offshore wind farm on a mixed population of harbour *Phoca vitulina* and grey *Halichoerus grypus* seals. Mar Pollut Bull 64(4):872–881. https://doi.org/10.1016/j.marpolbul.2012.01.029

Smultea MA, Mobley JR Jr, Fertl D, Fulling GL (2008) An unusual reaction and other observations of sperm whales near fixed-wing aircraft. Gulf Caribb Res 20(1):75–80. https://doi.org/10.18785/gcr.2001.10

Sobreira FV, Luís AR, Alves IS, Couchinho MN, dos Santos ME (2024) Raise your pitch! Changes in the acoustic emissions of resident bottlenose dolphins in the proximity of vessels. Mar Mamm Sci. 40(2):e13090. https://doi.org/10.1111/mms.13090

Sørensen PM, Haddock A, Guarino E, Jaakkola K, McMullen C, Jensen FH, Tyack PL, King SL (2023) Anthropogenic noise impairs cooperation in bottlenose dolphins. Curr Biol 33(4):749–754. https://doi.org/10.1016/j.cub.2022.12.063

Southall BL, Bowles AE, Ellison WT, Finneran JJ, Gentry RL, Greene CRJ, Kastak D, Ketten DR, Miller JH, Nachtigall PE, Richardson WJ, Thomas JA, Tyack PL (2007) Marine mammal noise exposure criteria: initial scientific recommendations. Aquat Mamm 33(4):411–521. https://doi.org/10.1080/09524622.2008.9753846

Southall BL, Moretti D, Abraham B, Calambokidis J, DeRuiter SL, Tyack PL (2012) Marine mammal behavioral response studies in Southern California: advances in technology and experimental methods. Mar Technol Soc J 46(4):48–59. https://doi.org/10.4031/MTSJ.46.4.1

Southall BL, Nowacek DP, Miller PJO, Tyack PL (2016) Experimental field studies to measure behavioral responses of cetaceans to sonar. Endanger Species Res 31:293–315. https://doi.org/10.3354/esr00764

Southall BL, Casey C, Holt M, Insley S, Reichmuth C (2019a) High-amplitude vocalizations of male northern elephant seals and associated ambient noise on a breeding rookery. J Acoust Soc Am 146(6):4514–4524. https://doi.org/10.1121/1.5139422

Southall BL, DeRuiter SL, Friedlaender A, Stimpert AK, Goldbogen JA, Hazen E, Casey C, Fregosi S, Cade DE, Allen AN, Harris CM, Schorr G, Moretti D, Guan S, Calambokidis J (2019b) Behavioral responses of individual blue whales (*Balaenoptera musculus*) to mid-frequency military sonar. J Exp Biol 222(5):jeb190637. https://doi.org/10.1242/jeb.190637

Southall BL, Finneran JJ, Reichmuth C, Nachtigall PE, Ketten DR, Bowles AE, Ellison WT, Nowacek DP, Tyack PL (2019c) Marine mammal noise exposure criteria: updated scientific recommendations for residual hearing effects. Aquat Mamm 45(2):125–232. https://doi.org/10.1578/AM.45.2.2019.125

Southall BL, Nowacek DP, Bowles AE, Senigaglia V, Bejder L, Tyack PL (2021) Marine mammal noise exposure criteria: assessing the severity of marine mammal behavioral responses to human noise. Aquat Mamm 47(5):421–464. https://doi.org/10.1578/AM.47.5.2021.421

Southall BL, Allen AN, Calambokidis J, Casey C, DeRuiter SL, Fregosi S, Friedlaender AS, Goldbogen JA, Harris CM, Hazen EL, Popov V, Stimpert AK (2023) Behavioural responses of fin whales to military mid-frequency active sonar. R Soc Open Sci 10(12): 231775. https://doi.org/10.1098/rsos.231775

Sparling C, Sams C, Stephenson S, Joy R, Wood J, Gordon J, Thompson D, Plunkett R, Miller B, Götz T (2015) The use of acoustic deterrents for the mitigation of injury to marine mammals during pile driving for offshore wind farm construction (Report for Carbon Trust ORJIP Project 4, Stage 1 of Phase 2). SMRU Consulting, St Andrews, 152 pp

Sprogis KR, Videsen S, Madsen PT (2020) Vessel noise levels drive behavioural responses of humpback whales with implications for whale-watching. eLife 9: e56760. https://doi.org/10.7554/eLife.56760

Sprogis KR, Holman D, Arranz P, Christiansen F (2023) Effects of whale-watching activities on southern right whales in Encounter Bay, South Australia. Mar Policy 150:105525. https://doi.org/10.1016/j.marpol.2023.105525

Stafford-Bell R, Scarr M, Scarpaci C (2012) Behavioural responses of the Australian fur seal (*Arctocephalus pusillus doriferus*) to vessel traffic and presence of swimmers in Port Phillip Bay, Victoria, Australia. Aquat Mamm 38(3):241–249. https://doi.org/10.1578/AM.38.3.2012.241

Stanistreet JE, Beslin WAM, Kowarski K, Martin SB, Westell A, Moors-Murphy HB (2022) Changes in the acoustic activity of beaked whales and sperm whales recorded during a naval training exercise off eastern Canada. Sci Rep 12(1):1973. https://doi.org/10.1038/s41598-022-05930-4

Stankowich T (2008) Ungulate flight responses to human disturbance: a review and meta-analysis. Biol Conserv 141(9):2159–2173. https://doi.org/10.1016/j.biocon.2008.06.026

Staudinger MD, McAlarney RJ, McLellan WA, Ann Pabst D (2014) Foraging ecology and niche overlap in pygmy (*Kogia breviceps*) and dwarf (*Kogia sima*) sperm whales from waters of the U.S. mid-Atlantic coast. Mar Mamm Sci 30(2):626–655. https://doi.org/10.1111/mms.12064

Stemland HM, Johansen TA, Ruud BO, Aniceto AS (2019) Measured sound levels in ice-covered shallow water caused by seismic shooting on top of and below floating ice, reviewed for possible impacts on true seals. First Break 37(1):35–42. https://doi.org/10.3997/1365-2397.2018010

Stimpert AK, Deruiter SL, Southall BL, Moretti DJ, Falcone EA, Goldbogen JA, Friedlaender A, Schorr GS, Calambokidis J (2014) Acoustic and foraging behavior of a Baird's beaked whale, *Berardius bairdii*, exposed to simulated sonar. Sci Rep 4:7031. https://doi.org/10.1038/srep07031

Stirling I, Calvert W (1983) Environmental threats to marine mammals in the Canadian Arctic. Polar Rec 21(134):433–449. https://doi.org/10.1017/S003224740002163X

Stone CJ, Tasker ML (2006) The effects of seismic airguns on cetaceans in UK waters. J Cetacean Res Manag 8(3):255–263. https://doi.org/10.47536/jcrm.v8i3.721

Storch I (2013) Human disturbance of grouse—why and when? Wildl Biol 19(4):390–403. https://doi.org/10.2981/13-006

Suryan RM, Harvey JT (1998) Variability in reactions of Pacific harbor seals, *Phoca vitulina richardsi*, to disturbance. Fish Bull 97:332–339

Symons J, Pirotta E, Lusseau D (2014) Sex differences in risk perception in deep-diving bottlenose dolphins leads to decreased foraging efficiency when exposed to human disturbance. J Appl Ecol 51(6):1584–1592. https://doi.org/10.1111/1365-2664.12337

Szesciorka AR, Allen AN, Calambokidis J, Fahlbusch J, McKenna MF, Southall B (2019) A case study of a near vessel strike of a blue whale: perceptual cues and fine-scale aspects of behavioral avoidance. Front Mar Sci 6:761. https://doi.org/10.3389/fmars.2019.00761

Tablado Z, Jenni L (2017) Determinants of uncertainty in wildlife responses to human disturbance. Biol Rev 92(1):216–233. https://doi.org/10.1111/brv.12224

Terhune JM, Stewart REA, Ronald K (1979) Influence of vessel noises on underwater vocal activity of harp seals. Can J Zool 57(6):1337–1338. https://doi.org/10.1139/z79-170

Tervo OM, Blackwell SB, Ditlevsen S, Conrad AS, Samson AL, Garde E, Hansen RG, Mads Peter H-J (2021) Narwhals react to ship noise and airgun pulses embedded in background noise. Biol Lett 17(11):20210220. https://doi.org/10.1098/rsbl.2021.0220

Thode AM, Gerstoft P, Burgess WC, Sabra KG, Guerra M, Stokes MD, Noad M, Cato DH (2006) A portable matched-field processing system using passive acoustic time synchronization. IEEE J Ocean Eng 31(3): 696–710. https://doi.org/10.1109/Joe.2006.880431

Thode A, Straley J, Tiemann CO, Folkert K, O'Connell V (2007) Observations of potential acoustic cues that attract sperm whales to longline fishing in the Gulf of Alaska. J Acoust Soc Am 122(2):1265–1277. https://doi.org/10.1121/1.2749450

Thode AM, Blackwell SB, Conrad AS, Kim KH, Marques T, Thomas L, Oedekoven CS, Harris D, Bröker K (2020) Roaring and repetition: how bowhead whales adjust their call density and source level (Lombard effect) in the presence of natural and seismic airgun survey noise. J Acoust Soc Am 147(3): 2061–2080. https://doi.org/10.1121/10.0000935

Thomas JA, Kastelein RA, Awbrey FT (1990) Behavior and blood catecholamines of captive belugas during playbacks of noise from an oil drilling platform. Zoo Biol 9(5):393–402. https://doi.org/10.1002/zoo.1430090507

Thompson PM, Brookes KL, Graham IM, Barton TR, Needham K, Bradbury G, Merchant ND (2013a) Short-term disturbance by a commercial two-dimensional seismic survey does not lead to long-term displacement of harbour porpoises. Proc R Soc Lond Ser B Biol Sci 280(1771):20132001. https://doi.org/10.1098/rspb.2013.2001

Thompson PM, Hastie GD, Nedwell J, Barham R, Brookes KL, Cordes LS, Bailey H, McLean N

(2013b) Framework for assessing impacts of pile-driving noise from offshore wind farm construction on a harbour seal population. Environ Impact Assess Rev 43:73–85. https://doi.org/10.1016/j.eiar.2013.06.005

Thompson D, Coram AJ, Harris RN, Sparling CE (2021) Review of non-lethal seal control options to limit seal predation on salmonids in rivers and at finfish farms. Scottish Marine and Freshwater Science 12(6):1–137. https://doi.org/10.7489/12369-1

Tiemann CO, Thode AM, Straley J, O'Connell V, Folkert K (2006) Three-dimensional localization of sperm whales using a single hydrophone. J Acoust Soc Am 120(4):2355–2365. https://doi.org/10.1121/1.2335577

Todd VLG, Jiang J, Ruffert M (2019) Potential audibility of three acoustic harassment devices (AHDs) to marine mammals in Scotland, UK. Int J Acoust Vib 24(4):792–800. https://doi.org/10.20855/ijav.2019.24.41528

Todd VLG, Williamson LD, Jiang J, Cox SE, Todd IB, Ruffert M (2021) Prediction of marine mammal auditory-impact risk from Acoustic Deterrent Devices used in Scottish aquaculture. Mar Pollut Bull 165:112171. https://doi.org/10.1016/j.marpolbul.2021.112171

Tollit D, Joy R, Wood J (2017) Estimating the effects of noise from commercial vessels and whale watch boats on southern resident killer whales (Report for Vancouver Fraser Port Authority). SMRU Consulting, Friday Harbor

Tougaard J, Kyhn LA, Amundin M, Wennerberg D, Bordin C (2012) Behavioral reactions of harbor porpoise to pile-driving noise. In: Popper AN, Hawkins A (eds) The effects of noise on aquatic life. Springer, New York, pp 277–280. https://doi.org/10.1007/978-1-4419-7311-5_61

Tournier M, Goulet P, Fonvieille N, Nerini D, Johnson M, Guinet C (2021) A novel animal-borne miniature echosounder to observe the distribution and migration patterns of intermediate trophic levels in the Southern Ocean. J Mar Syst 223:103608. https://doi.org/10.1016/j.jmarsys.2021.103608

Trickey JS, Cárdenas-Hinojosa G, Rojas-Bracho L, Schorr GS, Rone BK, Hidalgo-Pla E, Rice A, Baumann-Pickering S (2022) Ultrasonic antifouling devices negatively impact Cuvier's beaked whales near Guadalupe Island, México. Commun Biol 5(1):1005. https://doi.org/10.1038/s42003-022-03959-9

Trigg LE, Chen F, Shapiro GI, Ingram SN, Vincent C, Thompson D, Russell DJF, Carter MID, Embling CB (2020) Predicting the exposure of diving grey seals to shipping noise. J Acoust Soc Am 148(2):1014–1029. https://doi.org/10.1121/10.0001727

Tripovich JS, Hall-Aspland S, Charrier I, Arnould JPY (2012) The behavioural response of Australian fur seals to motor boat noise. PLoS One 7(5):e37228. https://doi.org/10.1371/journal.pone.0037228

Tsujii K, Akamatsu T, Okamoto R, Mori K, Mitani Y, Umeda N (2018) Change in singing behavior of humpback whales caused by shipping noise. PLoS One 13(10):e0204112. https://doi.org/10.1371/journal.pone.0204112

Tyack PL (2008) Implications for marine mammals of large-scale changes in the marine acoustic environment. J Mammal 89(3):549–558. https://doi.org/10.1644/07-MAMM-S-307R.1

Tyack P (2009) Acoustic playback experiments to study behavioral responses of free-ranging marine animals to anthropogenic sound. Mar Ecol Prog Ser 395:187–200. https://doi.org/10.3354/Meps08363

Tyack P, Clark C (1998) Quick Look—playback of low frequency sound to gray whales migrating past the central California coast—January, 1998. Quick Look Report, Space and Warfare Systems Command (SPAWAR), pp 1–34

Tyack PL, Zimmer WMX, Moretti D, Southall BL, Claridge DE, Durban JW, Clark CW, D'Amico A, DiMarzio N, Jarvis S, McCarthy E, Morrissey R, Ward J, Boyd IL (2011) Beaked whales respond to simulated and actual navy sonar. PLoS One 6(3):e17009. https://doi.org/10.1371/journal.pone.0017009

Tyne JA, Christiansen F, Heenehan HL, Johnston DW, Bejder L (2018) Chronic exposure of Hawaii Island spinner dolphins (*Stenella longirostris*) to human activities. R Soc Open Sci 5(10):171506. https://doi.org/10.1098/rsos.171506

Ugoretz J, Greene CR (2012) Pinniped monitoring during missile launches on San Nicolas Island, California, September 2011–September 2012 (Report by Naval Air Warfare Center Weapons Division, Point Mugu, CA, USA). National Marine Fisheries Service, Silver Spring

van Beest FM, Teilmann J, Hermannsen L, Galatius A, Mikkelsen L, Sveegaard S, Balle JD, Dietz R, Nabe-Nielsen J (2018) Fine-scale movement responses of free-ranging harbour porpoises to capture, tagging and short-term noise pulses from a single airgun. R Soc Open Sci 5(1):170110. https://doi.org/10.1098/rsos.170110

van Ginkel C, Becker DM, Gowans S, Simard P (2018) Whistling in a noisy ocean: bottlenose dolphins adjust whistle frequencies in response to real-time ambient noise levels. Bioacoustics 27(4):391–405. https://doi.org/10.1080/09524622.2017.1359670

Van Parijs SM, Baker K, Carduner J, Daly J, Davis GE, Esch C, Guan S, Scholik-Schlomer A, Sisson NB, Staaterman E (2021) NOAA and BOEM minimum recommendations for use of passive acoustic listening systems in offshore wind energy development monitoring and mitigation programs. Front Mar Sci 8:760840. https://doi.org/10.3389/fmars.2021.760840

van Polanen Petel TD, Terhune JM, Hindell MA, Giese MA (2006) An assessment of the audibility of sound from human transport by breeding Weddell seals (*Leptonychotes weddellii*). Wildl Res 33(4):275–291. https://doi.org/10.1071/WR05001

Varghese HK, Miksis-Olds J, DiMarzio N, Lowell K, Linder E, Mayer L, Moretti D (2020) The effect of two 12 kHz multibeam mapping surveys on the foraging behavior of Cuvier's beaked whales off of southern California. J Acoust Soc Am 147(6): 3849–3858. https://doi.org/10.1121/10.0001385

Veirs S, Veirs V, Wood JD (2016) Ship noise extends to frequencies used for echolocation by endangered killer whales. PeerJ 4:e1657. https://doi.org/10.7717/peerj. 1657

Verfuss UK, Gillespie D, Gordon J, Marques TA, Miller B, Plunkett R, Theriault JA, Tollit DJ, Zitterbart DP, Hubert P, Thomas L (2018) Comparing methods suitable for monitoring marine mammals in low visibility conditions during seismic surveys. Mar Pollut Bull 126:1–18. https://doi.org/10.1016/j.marpolbul. 2017.10.034

Vetemaa M, Päädam U, Fjälling A, Rohtla M, Svirgsden R, Taal I, Verliin A, Eschbaum R, Saks L (2021) Seal-induced losses and successful mitigation using Acoustic Harassment Devices in Estonian Baltic trap-net fisheries. Proc Est Acad Sci 70(2):207–214. https://doi.org/10.3176/proc.2021.2.09

Villegas-Amtmann S, Schwarz LK, Sumich JL, Costa DP (2015) A bioenergetics model to evaluate demographic consequences of disturbance in marine mammals applied to gray whales. Ecosphere 6(10):art183. https://doi.org/10.1890/ES15-00146.1

Vinther M, Larsen F (2004) Updated estimates of harbour porpoise (*Phocoena phocoena*) bycatch in the Danish North Sea bottom-set gillnet fishery. J Cetacean Res Manag 6(1):19–24. https://doi.org/10.47536/jcrm. v6i1.785

Visser F, Curé C, Kvadsheim PH, Lam F-PA, Tyack PL, Miller PJO (2016) Disturbance-specific social responses in long-finned pilot whales, *Globicephala melas*. Sci Rep 6:28641. https://doi.org/10.1038/ srep28641

von Benda-Beckmann AM, Wensveen PJ, Kvadsheim PH, Lam F-PA, Miller PJO, Tyack PL, Ainslie MA (2014) Modeling effectiveness of gradual increases in source level to mitigate effects of sonar on marine mammals. Conserv Biol 28(1):119–128. https://doi.org/10.1111/ cobi.12162

von Benda-Beckmann AM, Isojunno S, Zandvliet M, Ainslie MA, Wensveen PJ, Tyack PL, Kvadsheim PH, Lam FPA, Miller PJO (2021) Modeling potential masking of echolocating sperm whales exposed to continuous 1–2 kHz naval sonar. J Acoust Soc Am 149(4):2908–2925. https://doi.org/10.1121/10. 0004769

Waples D, Thorne L, Hodgy L, Burke E, Urian K, Read A (2013) A field test of acoustic deterrent devices used to reduce interactions between bottlenose dolphins and a coastal gillnet fishery. Biol Conserv 157:163–171. https://doi.org/10.1016/j.biocon.2012.07.012

Warren VE, Marques TA, Harris D, Thomas L, Tyack PL, Soto NA, Hickmott LS, Johnson MP (2017) Spatiotemporal variation in click production rates of beaked whales: implications for passive acoustic density estimation. J Acoust Soc Am 141(3):1962–1974. https:// doi.org/10.1121/1.4978439

Wartzok D, Popper AN, Gordon J, Merrill J (2003) Factors affecting the responses of marine mammals to acoustic disturbance. Mar Technol Soc J 37(4):6–15. https:// doi.org/10.4031/002533203787537041

Watkins WA (1986) Whale reactions to human activities in Cape Cod waters. Mar Mamm Sci 2(4):251–262. https://doi.org/10.1111/j.1748-7692.1986.tb00134.x

Watkins WA, Schevill WE (1975) Sperm whales (*Physeter catodon*) react to pingers. Deep-Sea Res 22:123–129. https://doi.org/10.1016/0011-7471(75) 90052-2

Watkins WA, Daher MA, Reppucci GM, George JE, Martin DL, Dimarzio NA, Gannon DP (2000) Seasonality and distribution of whale calls in the North Pacific. Oceanography 13(1):62–67. https://doi.org/10.5670/ oceanog.2000.54

Weir CR (2008a) Overt responses of humpback whales (*Megaptera novaeangliae*), sperm whales (*Physeter macrocephalus*), and Atlantic spotted dolphins (*Stenella frontalis*) to seismic exploration off Angola. Aquat Mamm 34(1):71–83. https://doi.org/10.1578/ AM.34.1.2008.71

Weir CR (2008b) Short-finned pilot whales (*Globicephala macrorhynchus*) respond to an airgun ramp-up procedure off Gabon. Aquat Mamm 34(3):349–354. https:// doi.org/10.1578/AM.34.3.2008.349

Wellard R, Lightbody K, Fouda L, Blewitt M, Riggs D, Erbe C (2016) Killer whale (*Orcinus orca*) predation on beaked whales (*Mesoplodon* spp.) in the Bremer Sub-Basin, Western Australia. PLoS One 11(12): e0166670. https://doi.org/10.1371/journal.pone. 0166670

Wensveen PJ, Von Benda-Beckmann AM, Ainslie MA, Lam F-PA, Kvadsheim PH, Tyack PL, Miller JO (2015) How effectively do horizontal and vertical response strategies of long-finned pilot whales reduce sound exposure from naval sonar? Mar Environ Res 106:68–81. https://doi.org/10.1016/j.marenvres.2015. 02.005

Wensveen PJ, Kvadsheim PH, Lam F-PA, von Benda-Beckmann AM, Sivle LD, Visser F, Curé C, Tyack PL, Miller PJO (2017) Lack of behavioural responses of humpback whales (*Megaptera novaeangliae*) indicate limited effectiveness of sonar mitigation. J Exp Biol 220(22):4150–4161. https://doi.org/10.1242/jeb. 161232

Wensveen PJ, Isojunno S, Hansen RR, Benda-Beckmann AMV, Kleivane L, IJsselmuide SV, Lam F-PA, Kvadsheim PH, Deruiter SL, Curé C, Narazaki T, Tyack PL, Miller PJO (2019) Northern bottlenose whales in a pristine environment respond strongly to close and distant navy sonar signals. Proc R Soc B 286(1899):20182592. https://doi.org/10.1098/rspb. 2018.2592

Whyte KF, Russell DJF, Sparling CE, Binnerts B, Hastie GD (2020) Estimating the effects of pile driving sounds on seals: pitfalls and possibilities. J Acoust Soc Am 147(6):3948–3958. https://doi.org/10.1121/ 10.0001408

Wild L, Thode A, Straley J, Rhoads S, Falvey D, Liddle J (2017) Field trials of an acoustic decoy to attract sperm whales away from commercial longline fishing vessels in western Gulf of Alaska. Fish Res 196:141–150. https://doi.org/10.1016/j.fishres.2017.08.017

Williams MTS (1997) The impact of aircraft activity on the behavior and productivity of Northern Fur Seals (*Callorhinus ursinus*) on St. George Island, Alaska. MSc thesis, University of Alaska Fairbanks, Fairbanks, AK, USA

Williams R, Lusseau D, Hammond PS (2006) Estimating relative energetic costs of human disturbance to killer whales (*Orcinus orca*). Biol Conserv 133(3):301–311. https://doi.org/10.1016/j.biocon.2006.06.010

Williams R, Bain DE, Smith JC, Lusseau D (2009) Effects of vessels on behaviour patterns of individual southern resident killer whales *Orcinus orca*. Endanger Species Res 6:199–209. https://doi.org/10.3354/esr00150

Williams R, Erbe C, Ashe E, Beerman A, Smith J (2014) Severity of killer whale behavioural responses to ship noise: a dose-response study. Mar Pollut Bull 79:254–260. https://doi.org/10.1016/j.marpolbul.2013.12.004

Williams R, Erbe C, Ashe E, Clark CW (2015) Quiet(er) marine protected areas. Mar Pollut Bull 100(1):154–161. https://doi.org/10.1016/j.marpolbul.2015.09.012

Williams R, Lacy RC, Ashe E, Hall A, Lehoux C, Lesage V, McQuinn I, Pourde S (2017) Predicting responses of St. Lawrence beluga to environmental change and anthropogenic threats to orient effective management actions (Report for Fisheries and Oceans Canada). Canadian Science Advisory Secretariat, Ottawa

Williams R, Veirs S, Veirs V, Ashe E, Mastick N (2019) Approaches to reduce noise from ships operating in important killer whale habitats. Mar Pollut Bull 139:459–469. https://doi.org/10.1016/j.marpolbul.2018.05.015

Wilson LJ, Harwood J, Booth CG, Joy R, Harris CM (2020) A decision framework to identify populations that are most vulnerable to the population level effects of disturbance. Conserv Sci Pract 2(2):e149. https://doi.org/10.1111/csp2.149

Wisniewska DM, Johnson M, Teilmann J, Rojano-Doñate L, Shearer J, Sveegaard S, Miller Lee A, Siebert U, Madsen Peter T (2016) Ultra-high foraging rates of harbor porpoises make them vulnerable to anthropogenic disturbance. Curr Biol 26(11):1441–1446. https://doi.org/10.1016/j.cub.2016.03.069

Wisniewska DM, Johnson M, Teilmann J, Siebert U, Galatius A, Dietz R, Madsen PT (2018) High rates of vessel noise disrupt foraging in wild harbour porpoises (*Phocoena phocoena*). Proc R Soc B 285(1872):20172314. https://doi.org/10.1098/rspb.2017.2314

Würsig B, Lynn SK, Jefferson TA, Mullin KD (1998) Behaviour of cetaceans in the northern Gulf of Mexico relative to survey ships and aircraft. Aquat Mamm 24(1):41–50

Würsig B, Greene CR Jr, Jefferson TA (2000) Development of an air bubble curtain to reduce underwater noise of percussive piling. Mar Environ Res 49:79–93. https://doi.org/10.1016/S0141-1136(99)00050-1

Yang L, Xu X, Berggren P (2022) Spotted seal *Phoca largha* underwater vocalisations in relation to ambient noise. Mar Ecol Prog Ser 683:209–220. https://doi.org/10.3354/meps13951

Yurk H, Trites AW (2000) Experimental attempts to reduce predation by harbor seals on out-migrating juvenile salmonids. Trans Am Fish Soc 129(6):1360–1366. https://doi.org/10.1577/1548-8659(2000)1292.0.CO;2

Open Access This chapter is licensed under the terms of the Creative Commons Attribution 4.0 International License (http://creativecommons.org/licenses/by/4.0/), which permits use, sharing, adaptation, distribution and reproduction in any medium or format, as long as you give appropriate credit to the original author(s) and the source, provide a link to the Creative Commons license and indicate if changes were made.

The images or other third party material in this chapter are included in the chapter's Creative Commons license, unless indicated otherwise in a credit line to the material. If material is not included in the chapter's Creative Commons license and your intended use is not permitted by statutory regulation or exceeds the permitted use, you will need to obtain permission directly from the copyright holder.

Biological Significance of Responses to Noise

11

Valeria Senigaglia, Dorian Houser, Capri Jolliffe, and Christine Erbe

Contents

11.1	**Acute (Short-Term) and Chronic (Long-Term) Noise Effects**	699
11.1.1	Mortality and Injury	700
11.1.2	Displacement	702
11.1.3	Chronic Masking and Loss of Communication Space	705
11.2	**Habituation, Sensitization, and Tolerance**	706
11.2.1	Habituation as a Progressive Decrease in Responsiveness	706
11.2.2	Sensitization as a Progressive Increase in Responsiveness	708
11.2.3	Tolerance	709
11.3	**Biological Significance**	709
11.3.1	The Concept of Resilience and its Application in Impact Assessment	710
11.3.2	Challenges with Data Collection and Analysis	711
11.3.3	From Short-Term Responses to Impact on Vital Rates	712
11.3.4	Population Consequences of Acoustic Disturbance	713
11.4	**Cumulative Impacts**	716
11.4.1	Multiple Sources of Disturbance and the Birth of PCoMS	716

11.1 Acute (Short-Term) and Chronic (Long-Term) Noise Effects

The impact of sound exposure can be acute (in the short term) or chronic (over the long term). Except for extreme acute responses (e.g., death due to stranding), chronic (or repetitive) effects are generally of greater concern. This is particularly true when considering population impacts where the effect may span many individuals.

Acute behavioral effects resulting from sound exposure are generally characterized by the cessation of normal behaviors and possible replacement by other behaviors (e.g., fleeing or

V. Senigaglia (✉)
Securing Antarctica's Environmental Future, School of Mathematical Sciences, Queensland University of Technology, Brisbane, QLD, Australia
e-mail: valeria.senigaglia@qut.edu.au

D. Houser
National Marine Mammal Foundation, San Diego, CA, USA
e-mail: dorian.houser@nmmf.org

C. Jolliffe · C. Erbe
Centre for Marine Science and Technology, Curtin University, Perth, WA, Australia
e-mail: c.erbe@curtin.edu.au

© The Author(s) 2025
C. Erbe et al. (eds.), *Marine Mammal Acoustics in a Noisy Ocean*,
https://doi.org/10.1007/978-3-031-77022-7_11

stampeding). Provided the acoustic disturbance is short, the duration of the observable effect is generally short-lived, and behavior returns to normal within some time following the removal of the acoustic disturbance. Similarly, most acute physiological effects are likely to resolve over a short period of time following removal of an acoustic disturbance. For example, acute elevations in stress hormones and a temporary reduction in hearing sensitivity (temporary threshold shift, or TTS) are both likely to resolve in minutes to hours following the cessation of the sound exposure. Chronic behavioral effects include long-term displacement and long-term changes in vocal behavior (e.g., shifting call frequencies or source levels). Chronic behavioral effects can arise from the repetition of acute disturbances, highlighting the importance of long-term monitoring. Chronic effects can result in reduced foraging success (Senigaglia et al. 2022), increased susceptibility to disease, decreased reproductive effort or success (Foroughirad and Mann 2013; Senigaglia et al. 2019), and lower survival. However, the effects of acoustic disturbance to marine mammals should be interpreted as acute or chronic with caution as short-term responses, or lack of thereof, might mask more significant effects from chronic disturbance (Bejder et al. 2006).

In the following sections, various forms of acute and chronic effects are discussed. We explore how the severity and duration of an effect potentially relates to a marine mammal's ability to thrive.

11.1.1 Mortality and Injury

Death has on occasion been reported following intense underwater noise exposure. Specifically, strandings have been associated with active sonar operations. Mass strandings of beaked whales were reported after exposure to naval sonar operations or in areas where sonar operations have historically occurred with some frequency (D'Amico et al. 2009; Evans and England 2001; Fernández et al. 2005, 2012; Filadelfo et al. 2009; Frantzis 1998; Jepson et al. 2003). In some of the better-characterized events (e.g., Evans and England 2001), source levels of the operational sonars were identified, but only the nominal source level of 235 dB re 1 µPa was reported, if at all. Many of the whales that were involved in these strandings died from the stranding event, whereas some that were pushed backed to sea were later resighted (Claridge 2013). Instances of cetacean strandings in which species other than beaked whales have died are also potentially sonar-related (Hohn et al. 2006; Southall et al. 2006), but the dramatic manner in which beaked whales mass-stranded in association with sonar has not been seen in other species.

Aircraft overflights have a variable effect on pinnipeds, depending on the type of aircraft, its altitude and distance from the rookery, the species exposed to the overflight, and the frequency with which the overflight occurs (Born et al. 1999; Bowles and Stewart 1980; Calkins 1979 [published 1983]; Johnson et al. 1989; see Sect. 9.2 of Richardson et al. 1995; Salter 1979; Southwell 2005). Similar observations have been made for sonic booms, explosions, and rocket launches (e.g., Demarchi et al. 2012; Stewart 1993). Closer overflights, high visibility of the aircraft, and high levels of received sound can cause dramatic responses such as stampeding and mother-pup separations, which have the potential to cause mortality (e.g., Johnson 1977; Lewis 1987). Only a few instances of mortality to pups occurring after mother-pup separations due to sound exposure have been documented, although Johnson (1977, in Richardson et al. 1995) estimated that >10% of the harbor seal pups born on a rookery in Alaska might have died due to mother-pup separations following aircraft overflights at low altitude. Similarly, walrus deaths after overflights have been reported (D. Fisher as cited in Johnson et al. 1989).

Explosions have been reported on rare cases to result in the death of marine mammals. Although lethality is primarily attributed to exposure to a shock wave, not sound, the two are related to one another. The shock wave, which has substantial shear force, can cause serious damage to animals where boundaries between tissues of different density occur, particularly in gas-containing bodies such as the lung or gastrointestinal tract (Fetherston et al. 2019). Ears are also vulnerable because of their sensitivity to changes in pressure (Ketten 1995). Large explosive charges are used in military operations and to "shock-test" new ships, which is a means to determine the ability of the ship's hull to tolerate live fire (Department of the Navy 1998, 2001). Smaller charges are also commonly used in military operations (von Benda-Beckmann et al. 2015), marine demolition, and construction (Mate and Harvey 1987). California sea lions have been killed when exposed to explosives used in seismic exploration (Fitch and Young 1947), and northern fur seals within 23 m of 11.4 kg dynamite charges have been found dead (Trasky 1976). Three long-beaked common dolphins died in proximity of 3.4 kg explosives detonated at a depth of 14.6 m (Danil and St. Leger 2011). A fourth animal was later found stranded that had injuries consistent with blast damage and was hypothesized to have been injured by the same event. Distance between the dolphins that died and the explosive at the time of detonation is assumed to have been several tens of meters. In 2019, NATO cleared over 40 unexploded mines within the marine protected area of Fehmarn Belt in the Baltic Sea. Eight of 24 dead harbor porpoises found in the months after the event demonstrated trauma associated with blast injury (Siebert et al. 2022). Further concerns on the potential physiological and behavioral impact on harbor porpoises exposed to seal bombs were raised by Simonis et al. (2020). Two humpback whales that died and stranded in a region with underwater blasting showed significant damage consistent with blast exposure, including the fracturing of the periotic bones in one whale (Ketten et al. 1993). Unfortunately, there was no information on the distance between the explosions and the whales, so proximity could not be determined. Dugongs have been incidentally killed during dynamite fishing, as well as killed intentionally by dynamite for food (Silas and Bastian Fernando 1985). In the latter case, the dynamite is typically thrown directly beside the dugong. Cases of Irrawaddy River dolphin deaths incidental to dynamite fishing have also been reported, and harpoons with small explosive tips are sometimes intentionally used during whaling in order to cause rapid death through blast-induced neurotrauma (Baird and Mounsouphom 1994; Knudsen and Oen 2003).

It is apparent from these cases that explosives have the potential to cause mortality, even with relatively small charges. The size of the charge and the proximity of the animal at the time of detonation will be critical factors in determining whether death occurs. However, because there are obvious ethical issues with testing on marine mammals, trauma predictions based on knowledge of the charge (shape, weight, etc.) and its distance to a marine mammal are necessarily based on modeling and observations from experiments on terrestrial mammals or marine mammal carcasses (Ketten 1995, 2004; Yelverton et al. 1973). While the instantaneous death of marine mammals from noise (or blast) exposure is likely relatively rare, explosions can cause internal and external injuries and significant damage to the hearing apparatus. The injuries described above for humpback whales that died following exposure to underwater explosives (Ketten et al. 1993), as well as the injuries observed in long-beaked common dolphins that died following exposure to an explosion (Danil, pers. comm., Danil and St. Leger 2011), could occur in other marine mammals exposed to similar events. Unfortunately, in almost every case, it is impossible to determine what internal injuries might have been experienced by a marine mammal that was exposed but continued to live. Injured animals must commit resources to healing and injuries might be permanent, potentially hindering foraging, diving, the ability to escape predators, and the ability to compete for mates and successfully reproduce. Injured individuals might therefore be at a disadvantage relative to their conspecifics and might not contribute as much to the population through reproduction as if they had otherwise been healthy.

Marine mammal deaths from blast explosions are likely a minor contributor to population declines, particularly in relation to the number of animals killed due to fishery interactions (Read et al. 2006). Except in cases where populations are already threatened or challenged (see Sect. 11.3.3), the loss of a few individuals from an otherwise healthy population likely has little impact on population numbers. Similar arguments about injury due to sound exposure can also be made, although the distance from a sound source at which injury can occur is undoubtedly greater than that at which mortality will occur.

11.1.2 Displacement

11.1.2.1 Short-Term Displacement

Short-term displacement is one means by which marine mammals mitigate noise impacts; indeed, the logic behind using ramp-up procedures for noise sources (i.e., slowly increasing the noise level; see Chap. 10) is that animals will displace and thus reduce their noise exposure level. Although displacement potentially has negative consequences (e.g., increased energetic costs, relocation to less suitable habitat), the cost of displacement might be less than the cost of remaining in an area where noise exposure occurs. For example, displacing for a short period of time might mitigate the potential for noise-induced hearing loss (NIHL; see Chap. 9) or a potential increase in stress hormones. Short-term displacements due to sound exposure can be identified when observations of marine mammals are spatially and temporally located in time with the acoustic event. Many of the behavioral response studies conducted over the last decade have characterized short-term displacements in response to intentional exposures of marine mammals to sound (see Chap. 10).

11.1.2.2 Long-Term Displacement

Long-term displacements are generally more difficult to identify and characterize as they require regular and prolonged monitoring for marine mammals in the areas of interest. Additionally, the multiple factors that determine whether marine mammals return to utilize a particular habitat are difficult to disentangle from noise as a potential causative factor. Changing ocean temperature, redistribution of prey, an increased presence of predators, or increases in other anthropogenic activities (e.g., fishing or boating) are a few factors that can change over time and which make it difficult to isolate a single causative factor in long-term animal displacement. Because environmental changes can be slow, occurring over months to years, it is often inappropriate to make conclusions regarding habitat abandonment in regions where marine mammal surveys and observations are sporadic or separated by long periods of time. Thus, it can be difficult to identify with confidence whether abandonment of an area due to noise exposure occurs, and conversely, site abandonment might be more common than is supported by available observations.

11.1.2.3 Evidence of Displacement

11.1.2.3.1 Acoustic Harassment Devices

Documented instances of displacement due to acoustic harassment devices (AHDs) or acoustic alarms range from hundreds of meters to tens of square kilometers and can affect untargeted marine mammal species (e.g., as noted in Hodgson et al. 2007). Species displacement probability varies based on several factors, including the species, social grouping, individuals' prior experience with the AHD/alarm, and the specific characteristics of the AHD/alarm sound. Killer whales, displaced from the Broughton Archipelago following the introduction of AHDs on salmon farms, were observed to relocate to the nearby Johnstone Strait, which was absent of AHDs (Morton and Symonds 2002). The killer whales returned to the Broughton Archipelago once the AHDs were removed from the salmon farms. Acoustic alarms have been shown to reduce the fisheries bycatch of harbor porpoises (Dawson et al. 2013; Gearin et al. 2000; Kraus et al. 1997; Larsen and Eigaard 2014; Larsen et al. 2013) and to displace porpoises ~100–500 m from a pingered net (Culik et al. 2001; Laake et al. 1998). The degree of habitat displacement should logically increase with an increase in the area over which nets with acoustic alarms are deployed, which is a concern for regions where fish farms and/or gillnet fishing and harbor porpoise habitat overlap. Several dolphin species appear to be resilient to AHD or alarm exposures, typically showing either no displacement or little variation in their behavior because of their presence (Berg Soto et al. 2013; Bruno and Marino 2012; Cox et al. 2003). Certain types of alarms have been reported as effective at reducing interactions between dolphins and fishing nets (Brotons et al. 2008), but habitat displacement has not been reported. Moreover, harbor porpoises have been observed in some instances to either habituate or develop a tolerance to acoustic alarms with habitats being reoccupied after 10–11 days of exposure (Cox et al. 2001). Evidence for the displacement of mysticetes by acoustic alarms or deterrents is largely lacking, although limited evidence suggests that humpback whales of the southern ocean showed no displacement in the presence of 3 kHz alarms (Harcourt et al. 2014).

11.1.2.3.2 Boat Traffic

Bejder et al. (2006) found that a significant percentage of a bottlenose dolphin population were displaced in response to increased whale-watching activity. The trend was found by comparing periods encompassing no whale-watch operations to those when there were one and two whale-watch operations and to a control area where no whale-watch operations were present. Gray whales were displaced from a breeding lagoon for ~10 years by barges moving salt into and out of the lagoon (Bryant et al. 1984). During the time the saltworks were operational, whale presence declined until whales were no longer observed to use the lagoon. The whales returned to the lagoon over a period of years following the closure of the saltworks. More recently, dredging activity in Aberdeen harbor has been associated with temporary displacement of bottlenose dolphins that forage in the harbor (Pirotta et al. 2013). Although the impact of boat presence is difficult to disentangle from the impact of the sound they make (Pirotta et al. 2015), additional vessels will certainly increase the noise to which the animals are exposed. The impact of the long-term displacements of some or all of a localized population is unknown, but the repopulation of a region where noise exposure ceases argues for some desirable habitat quality that might be forfeit under certain noise exposure conditions.

11.1.2.3.3 Seismic Survey

Fin whales tracked via passive acoustics were observed leaving the Alboran basin in the Mediterranean during a 10-day seismic survey but not in response to shipping noise, suggesting the characteristics of the airgun noise were important to the whales' decision to leave the area (Castellote et al. 2012). Conversely, harbor porpoises were observed to displace in response to a seismic survey (Thompson et al. 2013a) only for a few hours, and the responsiveness of porpoises declined over the 10-day survey. The ecological importance of noise at the frequencies generated by seismic surveys to porpoises is likely quite different than that of fin whales, which produce communication signals that overlap in frequencies with those of the seismic surveys. Decisions to displace from a habitat for long durations might vary depending upon the ecological significance of the frequency bands within which the anthropogenic noise occurs.

11.1.2.3.4 Sonar

A possible long-term displacement of beaked whales occurred after a short but intense sonar training exercise in the Bahamas in 2000, where, following a dramatic sonar-related stranding event, Cuvier's beaked whales were reported to be largely absent from the site of the event for several years (Claridge 2006, 2013). To determine chronic displacement of marine mammals affected by ocean noise (particularly as ocean conditions vary and habitat conditions change), it is critical to have an understanding of habitat use prior to disturbance lest erroneous conclusions about animal movement and habitat abandonment are made. Even shorter-term observations are useful if occurring prior to and following noise exposure. Passive acoustic monitoring of beaked whales on the Atlantic Undersea Test and Evaluation Center (AUTEC), which is located in the Tongue of the Ocean in the Bahamas, has shown changes in group echolocation behavior during sonar exercises suggesting beaked whales move away from the habitat during sonar operations (McCarthy et al. 2011). The displaced Blainville's beaked whales appeared to return over a period of 1–3 days following the exercise. However, detection of echolocation behavior during the exercises suggests that some animals remained to forage despite the increased noise.

11.1.2.4 Cost of Displacement

In contrast to documentation of long-term displacement or habitat abandonment, reports of marine mammals that remain in a noisy area are common. The lack of displacement is often attributed to the tolerance or habituation of animals to the presence of noise. Unfortunately, in many instances, there is little information on the populations of marine mammals prior to the introduction of anthropogenic noise, and so it is difficult to ascertain whether any changes in the population occurred around the time the noise was introduced. When animals are displaced from a region used for important biological functions and individuals cannot readily recover from disruption of its use, the cost of displacement becomes more apparent. For example, the displacement of pinnipeds from a rookery (important areas for pupping and nursing) could have serious consequences for young animals that are not nutritionally independent nor capable of efficient foraging. The vaquita, which appears to be headed for extinction, is an excellent example of an odontocete species that has a constrained habitat and that has been progressively impacted by human activity, specifically gillnet fishing (D'Agrosa et al. 2000). Unable to leave the area due to a lack of other suitable habitats, the vaquita is in serious decline because of fisheries-related mortality (Armando et al. 2007).

When assessing the impact of displacement, consideration should be given to the following:

- What is the area affected by noise exposure relative to the area of the habitat used by the marine mammal?
- How critical is the habitat to the species (e.g., is it used for breeding, foraging, etc.)?
- Are suitable alternative habitats available within a reasonable distance of displacement?
- How far were the animals displaced and for how long?
- What is the number/density of animals affected and what percentage of the population does it represent?

Widely distributed individuals would be less impacted by an anthropogenic noise event when considered as a percentage of a population. Temporal usage is also a concern, as marine mammals might use certain habitats only during specific points in their life history. For example, gray whales have been observed to use certain narrow passages during the spring and fall migration through the Aleutian Islands. Some mysticete species are also known to congregate in specific areas for breeding, calving, or feeding (e.g., North Atlantic right whale calving grounds off the southeast coast of the USA).

Understanding the impact that displacement has on marine mammals requires well-designed studies with effective monitoring capabilities and appropriate monitoring durations. To this end, acoustic monitoring of habitat use has grown dramatically. For example, passive acoustic monitoring has been employed to characterize marine mammal habitat use and animal density in various vocal species, including dolphins (Castellote et al. 2015; Elliott et al. 2011; Rayment et al. 2009; Thompson et al. 2015), killer whales (Yurk et al. 2010), belugas (Castellote et al. 2013; Roy et al. 2010), harbor porpoises (Brookes et al. 2013; Sveegaard et al. 2011), beaked whales (Yack et al. 2013), blue whales (Širović and Hildebrand 2011; Širović et al. 2004), bowhead whales (Charif et al. 2013), fin whales (Širović and Hildebrand 2011; Širović et al. 2004), minke whales (Marques et al. 2012; Martin et al. 2015), right whales (Marques et al. 2011; Matthews et al. 2014; Munger et al. 2008), and seal species (e.g., Rogers et al. 2013). Areas such as AUTEC, which have numerous bottom-mounted hydrophones, provide unique opportunities to study whale ecology and assessments of habitat quality (DiMarzio et al. 2008; Hazen et al. 2011). Such assessments, along with investigations of beaked whale foraging behavior (e.g., Johnson et al. 2004, 2006, 2008; Ward et al. 2008), shed light on the importance of the Tongue of the Ocean as a beaked whale habitat. Other undersea training ranges are also capitalizing on the existence of bottom-mounted hydrophone arrays to study marine mammal habitat use and to estimate animal densities, as well as responses of animals to sonar training exercises (Marques et al. 2009, 2012; Martin et al. 2015; Tiemann et al. 2004, 2006). These sites provide some of the best opportunities for long-term monitoring of marine mammal habitat use and displacement, albeit none of them can be considered sites where marine mammals would be naïve to sound exposures.

The continuing development of remote tags for tracking and monitoring of marine mammals has also greatly assisted in obtaining a better understanding of the natural movements of animals over time, as well as understanding the duration and distance of displacements that are associated with noise exposure (e.g., Baird et al. 2014). Nevertheless, there remain significant logistical barriers to completing the types of studies necessary to characterize the distance and duration of displacements potentially associated with noise exposure, including small sample size, the potential for pseudo-replication, and effective tools for acoustically characterizing habitat (Merchant et al. 2015). Continuing improvements in remote tag longevity (battery life) and attachment duration should facilitate our understanding of the impact of displacement, but future studies will nevertheless require careful thought as to the appropriate methods used and durations over which the studies will occur.

11.1.3 Chronic Masking and Loss of Communication Space

Continuously operated and distributed noise sources are potentially the greatest masking issue for some species. The most obvious example is shipping noise. The increase in low-frequency noise associated with shipping overlaps with the communication frequencies of many of the mysticete whales and concerns over chronic masking have been raised on numerous occasions (Hatch et al. 2012; McDonald et al. 2009). Low-frequency noise associated with shipping is higher than in the preindustrial period; for example, over a span of four to five decades (1960s to the early 2000s), ocean noise levels at frequencies <80 Hz increased by as much as 6–12 dB in the North Pacific (Andrew et al. 2002; McDonald et al. 2006). Estimates of the loss of communication space for North Atlantic right whales due to shipping noise are high when comparing present and derived historical noise levels (Hatch et al. 2012). North Atlantic right whales have been observed to increase the fundamental frequency of their vocalizations from the 1950s through the 2000s, a time period corresponding to an individual lifespan and during which ocean noise also increased (Parks et al. 2007). Thus, it is possible that the long-term increase in regional shipping noise has had a chronic masking impact on North Atlantic right whales with costs related to altered calling behavior and reduced acoustic communication space. Delphinids also rely on vocal communication in both foraging and social contexts and can thus be affected by anthropogenic noise masking. Based on lower whistle and click rates in the presence of boats, masking was suggested by Pellegrini et al. (2021) as a possible explanation for reduced cooperative behavior among Lahille's bottlenose dolphin in southern Brazil.

Changes in the level of produced sounds in the presence of elevated noise have been observed in other marine mammals, such as the beluga (Scheifele et al. 2005) and manatee (Miksis-Olds et al. 2007), but more investigation is required to ascertain whether trends exist across a broader range of species. Furthermore, noise varies both temporally and spatially such that the potential for masking and reductions in communication space likely differ for marine mammals based on their sensory capabilities and habitat utilization (Williams et al. 2013). This is particularly important when considering animals that range widely or are constrained in the use of specific (critical) habitats. Modeling the impact of sound on the acoustic communication space of marine mammals will continue to be useful (e.g., Jensen et al. 2009), but incorporating information about the ability of marine mammals to mitigate masking (e.g., through comodulation release of masking and spatial release of masking; see Chap. 9) into the models will provide more accurate estimates of the impact that chronic masking noise might have.

11.2 Habituation, Sensitization, and Tolerance

11.2.1 Habituation as a Progressive Decrease in Responsiveness

Harris (1943) defined habituation as a "response decrement as a result of repeated stimulation," where the stimulus is often assumed to have no negative consequence for the individual. A more recent definition for habituation, offered by Rankin et al. (2009), is "a behavioral response decrement that results from repeated stimulation and that does not involve sensory adaptation/sensory fatigue or motor fatigue." The change in the definition is important since it excludes reductions in responsiveness resulting from an impact to an animal's sensory systems; for example, a reduction in an animal's behavioral responsiveness to noise because it suffered an NIHL would not be considered habituation. Sensitization is the opposite of habituation: It is an increase in responsiveness due to repeated stimulation, and it is typically associated with some cost to the animal. These terms have relevance to understanding the behavioral responses of marine mammals to noise as either can occur.

There are characteristics of habituation that need to be considered when interpreting marine mammal behavioral responses to noise. The characteristics of habituation presented below are derived from Rankin et al. (2009) and Thompson and Spencer (1966) but discussed in regard to marine mammals and noise. The discussion is not meant to be exhaustive with respect to the characteristics presented by Rankin et al. (2009) but addresses only those most pertinent to assessing noise impacts. In that regard, the characteristics of habituation include the following:

1. Repeated exposure to a noise results in a progressive decrease of behavioral responsiveness.

This characteristic is the hallmark of habituation, which is a decline in responsiveness due to a repetitive stimulus exposure. Habituation is a matter of degree, but usually the responsiveness declines to some asymptotic level. Note that the habituation is a process that occurs over time, which is important to remember when interpreting point measures of marine mammal responsiveness to noise (see below). Moreover, it is important to remember that a decline in behavioral responses might not be associated with a correspondent decline in physiological responses, thus not qualifying as habituation.

2. If the specific noise to which a marine mammal has been habituated (or has shown some decrease in responsiveness) is withheld, the response will recover (at least partially) over time.

Marine mammals that become habituated to noise will not maintain the habituation indefinitely in the absence of further exposures. If the noise exposure to which a marine mammal has become habituated ceases, then the responsiveness of the animal to the noise exposure, if it were to occur again at some point in the future, will return. It is possible that the behavioral response will never return to the magnitude observed upon the initial noise exposure, but some degree of recovery in responsiveness will occur. This is an important factor to consider when assessing the degree of habituation that far-ranging species might experience with spatially separated but similar noise sources or species to which the noise exposure is seasonal or sporadic. For example, an individual that has become habituated to industrial noise may partially recover its responsiveness between periods of frequent noise and periods without this stimulus.

Generally, a more frequent exposure will result in a speedier and/or more marked decrease in responsiveness. It is unlikely that marine mammals will ever become habituated to sporadic events such as ship-shock trials. Conversely, some degree of habituation is more likely, and probably occurs faster, in situations where marine mammals are regularly exposed to a specific noise source over time. Boating in coastal regions, particularly where recreational boating is common, is likely a noise source to which many coastal-dwelling marine mammals became rapidly habituated (Southall et al. 2021). Similarly, some whales exposed to industrial noise in arctic and subarctic regions during the summer likely habituate to the noise due to its chronic presence (e.g., Richardson et al. 1990). Ringed seals have also been observed to produce little to no reaction in response to pile driving noise, which would produce impulsive strikes at inter-strike intervals <2 s. This suggests habituation to frequent pile strikes over time, although the seals did not habituate to helicopter overflights during the same period (Blackwell et al. 2004). Rapid habituation of killer whales to AHDs in a longline fishery has also been observed (Tixier et al. 2015). Even though killer whales initially responded by displacing from AHDs when activated, they quickly returned to distances at which they could depredate from the fishery boats after a handful of exposures.

3. The lower the intensity of the noise exposure, the speedier and/or more marked the decrease in behavioral responsiveness will be. Conversely, if the noise exposure is intense, no significant reduction in the behavioral response may occur.

Houser et al. (2013) found that bottlenose dolphins in captivity exposed to a simulated sonar signal at the midpoint of performing a task quickly habituated to the signal at moderate to low received SPLs ($\leq$160 dB re 1 µPa). This was apparent in a change in the form of behavioral dose-response functions describing the received level-dependent probability of response for the dolphins on each trial; the response functions became very steep across the trial sequence because of a progressive reduction in responsiveness at low to moderate received levels. However, consistent with the impact of an intense acoustic signal, no habituation was observed when the received SPL was $\geq$175 dB re 1 µPa. Many marine mammals in the wild are likely to follow similar trends, habituating quickly if received SPLs are low, but never habituating to intense acoustic signals. Thus, if an animal shows a reduction in responsiveness to a particular noise, it cannot be assumed that the reduction in responsiveness will persist at all levels of the noise exposure.

4. Habituation to a specific noise source might exhibit generalization to other noise sources.

Different noise sources may have similar characteristics, or a noise source might be associated with an activity (e.g., construction) in which other disparate noise sources exist. Either due to similarities in acoustic features or through association of noise with certain sources or activities, the habituation to one noise may be generalized to other noise sources. Although no specific instances of generalizing habituation across noise sources have been documented in marine mammals, it is not difficult to fathom that such generalization occurs. For example, recreational boating in coastal regions can involve noise from many different types of boat motors. In these regions, it is highly likely that resident marine mammals have generalized across the different boat motor types and attend to them as one "class" of noise source, even if a particular boat motor might be novel. Conversely, animals might habituate just to the presence of the boat and generalize to other types of boats.

5. Exposure to a different noise source can result in dishabituation to the original noise source to which habituation occurred.

Even though a marine mammal has become habituated to a certain type of noise, exposure to another noise source might result in dishabituation to the original noise source. The process of dishabituation is more likely if the different/novel noise is of a high received SPL or is perceived as noxious by the animal. While this seems plausible conceptually, conclusive demonstrations of active dishabituation to noise are not available.

6. Repetitive noise exposures of the same type may result in long-term habituation.

Long-term habituation might be perceived as good or bad, depending on the context. Long-term habituation to AHDs and alarms associated with fisheries gear is unwanted as it defeats the purpose of the devices, i.e., reducing marine mammal and fisheries interactions. On the other hand, long-term habituation to an innocuous noise source (e.g., underwater acoustic modem) within the animal's foraging area might be desired if the noise source must necessarily remain there. In areas where animals have long coexisted with anthropogenic noise sources and show little response to their presence, long-term habituation is often assumed (e.g., the case of bottlenose dolphins having habituated to boats in Sarasota Bay).

There are many instances where marine mammals have been supposed to habituate to noise exposure (e.g., Mobley 2005; Richardson and Würsig 1997), however, for purposes of understanding noise impacts to marine mammals in the wild, the concept of habituation requires refinement and consideration beyond simply quantifying a response decrement resulting from repeated noise exposure (Bejder et al. 2009). The response of a marine mammal to noise exposure is complex. Not all individuals within a population will react in the same manner, as factors such as age, sex, species sensitivity, and experience can influence the responsiveness of an animal. Moreover, the response, as well as its magnitude and duration, might depend on habitat suitability, prey abundance, distance to conspecifics, and predation risk. A lack of responsiveness might well occur with repeated noise exposures, if enduring the consequences is considered less impactful than other forms of behavioral response. For example, dolphins in heavily boat-trafficked areas might remain in the area because there is no suitable habitat nearby to which to displace. However, by staying in the area, the dolphins might incur costs associated with noise exposure, stress responses, and possible harassment by recreational boaters.

11.2.2 Sensitization as a Progressive Increase in Responsiveness

Sensitization is an increase in responsiveness following exposure to a repetitive stimulus (noise). Sensitization is typically associated with a direct or indirect cost to the animal. Sounds might be a simple annoyance or could be associated with more severe costs, such as the loss of life (e.g., the sound of predators). When sensitization to noise occurs, the response of the animal is greater in magnitude or duration (e.g., distance and duration of fleeing) and might occur at lower received levels than observed on the initial noise exposure. Harbor seals off the coast of British Columbia are likely sensitized to the vocalizations of mammal-eating killer whales, which are probably a significant cause of harbor seal mortality, but habituated to the sounds of familiar fish-eating killer whales that pose no threat (Deecke et al. 2002). The habituation is selective, and it is interesting that the sensitivity to mammal-eating killer whales is extended to the vocalizations of unfamiliar fish-eating killer whales—a form of generalized sensitization based on uncertainty about the potential for the whale to be a predator. This could change, however, through repeated exposure to the new fish-eating killer whale sounds. Animals can show an initial increase in responsiveness followed by a decrease in responsiveness; in other words, sensitization can later be followed by habituation.

One variant of sensitization is the "dinner bell" effect (Richardson et al. 1995), which occurs when an animal learns to associate a sound with food and is thus attracted to the noise source. It is notable in sea lions that associate AHDs with the presence of prey (Bordino et al. 2002), although concerns about dolphins capitalizing on AHDs for depredation have also been voiced (Cox et al. 2003). While AHDs have shown promise in deterring some odontocetes from depredating fishery catches (Dawson et al. 2013), the use of AHDs has sometimes resulted in increased depredation and net damage by other species (Bordino et al. 2002). The effect is debatable; a study of depredation in the swordfish and thresher shark drift gillnet fishery off California assessed rates of California sea lion depredation over a 14-year period of acoustic pinger use and found no increase in depredation (Carretta and Barlow 2011). In fact, increased bycatch of California sea lions (which is a measure of depredation effort) was better predicted by other variables, such as the size of the catch and the location. Still, to avoid a potential increase in fisheries depredation, the use of pingers with frequencies above the range of pinniped hearing is sometimes considered (Kastelein et al. 2008). Researchers have also noted increased predation, presumably by pinnipeds, on fish implanted with high-frequency acoustic tags (Cunningham et al. 2014). Experimental work with gray seals supports the idea that pinnipeds can capitalize on the acoustic signatures of implanted pingers to find prey (Stansbury et al. 2014).

11.2.3 Tolerance

It should be emphasized that when animal responsiveness is measured at a point in time, it is the "tolerance" of the animal to the exposure that is investigated (Bejder et al. 2009). However, repeated noise exposures could have occurred long before any behavioral response study was conducted. In such instances, a lack of responsiveness might not indicate habituation but could instead indicate that the vulnerable, and thus more impacted, individuals are no longer represented in the population (e.g., see Bejder et al. 2006). Only the most tolerant of individuals might remain giving a misleading perspective of the impact of the noise exposure. It is imperative that future studies of marine mammal responses to noise plan for longer-term monitoring so that the processes of habituation/sensitization can be properly characterized through repeated measures. Realistically, even in these scenarios, habituation and sensitization will likely only be determined by the monitoring of the same individuals, which, given the lack of funding and long-term monitoring programs, is a barrier to future research in this area.

11.3 Biological Significance

Marine mammals respond to underwater noise in a variety of ways with varying levels of severity. The varying responses of different groups of marine mammals to sources of underwater noise were discussed in Chap. 10. However, not all disturbance translates to a biologically significant impact. The National Research Council (NRC) of the USA recommends that biological significance should be assessed with respect to changes in the ability of an individual animal to survive, grow, and reproduce and that this should be considered with respect to the cumulative impact of changes to all individuals that might result in a population effect (National Research Council 2005). When it comes to understanding the potential for disturbance to translate to a biologically significant impact at the individual or population level, there are numerous factors that must be considered. The biological significance of disturbance at the individual level is influenced by a suite of factors including the type of response to the disturbance, its severity and duration, and the context at the time of disturbance (see Chap. 10). The likelihood and severity of responses vary between individuals of the same species and are influenced by a variety of factors including prior experience and habituation/sensitization to a sound source, risk tolerance, body condition, and life stage. Whether or not biologically significant disturbance at the individual level will manifest at the population level depends largely on the status of the population, the proportion of the population affected, and the duration and severity of disturbances in concert with other (cumulative) stressors.

11.3.1 The Concept of Resilience and its Application in Impact Assessment

Resilience is the ability of an individual marine mammal to express the same functional behavior during or after a disturbance as prior to it. A resilient individual should cope with disturbance by quickly returning to its pre-disturbance behavioral state (Nattrass and Lusseau 2016). As with the likelihood of responding to a disturbance, the resilience of an individual can be influenced by species-specific factors, individual context, experience, body condition, and life stage. Resilience can also be expressed at the population level, with some species and populations of species being more resilient to disturbance than others. Resilience at the population level is influenced by population size, ecology, life history characteristics, and the combined pressure of other stressors on the population. A species' resilience to long-term disturbance is shaped by its capacity for behavioral and evolutionary adaptation. Species that have a high level of genetic diversity are presumably more resilient to change because there is greater opportunity for an adaptive response to the evolutionary force (Moore and Reeves 2018). Genes that convey a fitness benefit during a long-term disturbance should be favored by sexual selection and thus preferentially passed on, leading to a greater prevalence of the beneficial traits in the population. Typically, when species population sizes are reduced, genetic diversity declines, leading to less opportunity for population-level adaptation and lower resilience to disturbance.

Behavioral plasticity describes a species' ability to change behavior based on environmental or social factors. Changes in foraging strategy or vocal behavior that provide a fitness benefit (and which can be transmitted via social learning) are examples of behavioral plasticity that may improve a species' resilience to change (Ducatez et al. 2020; Moore and Reeves 2018). Animals adapted to heterogeneous environments and variable conditions are often characterized by higher levels of plasticity, which allow quicker adaptation and ultimately represent a lower sensitivity to disturbance at the population level (Pirotta et al. 2020).

Repeated and prolonged exposure to a disturbance can impact an individual's vital rates and, if enough individuals are similarly affected by the disturbance, the viability of the entire population. To assess marine mammal resilience to noise disturbance, it is important to interpret behavioral responses considering the physiology and ecology of the species (e.g., sex differences, reproductive status, capital vs. income breeders, fasting-adapted vs. non-fasting-adapted species, and social characteristics). For example, the impact of being displaced from foraging habitat for several days might be more severe to small animals with high energy demands (e.g., harbor porpoises) than to large animals with low mass-specific metabolic rates and large amounts of blubber upon which to draw (e.g., humpback whales). Similarly, duration and frequency of the disturbance can affect its impact. Short-term disturbance of pinnipeds on rookeries (ranging from mere minutes to over a full day) is often considered to be of little impact to the population (Demarchi et al. 2012; Holst et al. 2011). However, chronic human interactions, which include but are not limited to noise, have been observed to affect the reproduction and growth rate of California sea lions and the reproduction of Mediterranean monk seals (Dendrinos et al. 2008; French et al. 2011). Similarly, whale-watching activities can have significant impacts on odontocetes due to increased energetic costs, loss of foraging opportunities, and reduced reproductive success (Bain et al. 2006; Bejder and Samuels 2003; Lusseau 2003; Senigaglia et al. 2016; Williams et al. 2006). As whale-watching activities become more common, the short-term behavioral actions used by marine mammals to mitigate interactions may lead to impact on vital rates and loss of critical habitat from long-term displacement (Lusseau and Bejder 2007).

11.3.2 Challenges with Data Collection and Analysis

At the core of the analysis challenge is the fundamental difficulty in assessing whether a behavioral response has occurred following a noise exposure, with limited baseline data on "normal" animal behavior for the purpose of statistical comparison. Despite advances in the means available for collecting animal behavior (e.g., deployable sensors and drones), there remains a general lack of information regarding the typical behaviors of marine mammals and their expected variability. Somewhat basic biological knowledge, such as species ranges, diving abilities, and even the mere existence or presence of a species, is regularly reassessed considering new discoveries (e.g., Accardo et al. 2018; Cranford and Krysl 2015; Yamada et al. 2019). It is, perhaps, then no surprise that the intricacies of marine mammal behavior are not fully realized for any species or population—let alone the behavioral variability within an individual. Since behavioral responses to noise are also variable, having knowledge of baseline behavior is paramount to analyses determining whether a behavioral response has occurred following noise exposure. Even with extensive knowledge of the species or animal ethogram, clearly identifying to what the animal is responding remains a challenge. Discerning whether the animal is in fact reacting to an acoustic stimulus vs. a social or environmental stimulus or other perceived risk (or a combination thereof) is not always possible, especially when conducting group follows of either tagged or untagged animals. Observing animals prior to an exposure, either through tagging or remote observation, is suggested to characterize variation in "natural" behavior. However, caution must be exercised as observations conducted during this so-called "pre-exposure" period have the potential themselves to alter animal behavior (e.g., as the mere presence of the observation vessel). Overall, there is no perfect experimental design or statistical approach. Yet, continuing efforts to characterize the normal behaviors of marine mammals (including their variability) without affecting behavior over the pre-exposure observation period, as well as long-term behavioral characterization in differing environments and contexts, will help elucidate the degree to which the response to noise is impactful.

Much of the initial work on behavioral responses to noise was qualitative in nature. Simple descriptive statistics (e.g., means and encounter rates) and comparative statistics (e.g., t-tests and ANOVAs) to assess behavior in the presence or absence of a noise source were sometimes applied and are occasionally still utilized (Cox et al. 2005; Holst et al. 2017). However, as computational power has improved and statistical techniques have increased in sophistication, efforts to identify behavioral responses statistically and objectively have become more common. Increasingly complicated, spatial, temporal, and individual-based models have been developed with the goal of both incorporating more contextual covariates in the analyses and investigating the effects of noise at individual and population levels (e.g., Nabe-Nielsen et al. 2014; Pirotta et al. 2015). In many studies, data limitations continue to drive the choice of what research questions can be answered and thus the analyses needed; selection of the correct analytical method requires careful consideration of the experimental design, sample size, availability of contextual data, and the existence of observer or availability bias. Model development should be guided by an understanding of the acoustic environment and noise sources being considered. The ability to collect the types and quantity of data required for many of the modern-day multivariate models has prompted large-scale behavioral response studies with a multitude of scientific collaborators. At the core of these collaborative efforts is the development of techniques for analyzing complex animal behavior. It is certain that new statistical tools will continue to be developed, particularly with respect to moving beyond the identification of behavioral responses to the prediction of the impacts (both acute and cumulative) resulting from the noise exposure.

11.3.3 From Short-Term Responses to Impact on Vital Rates

While it is easy to imagine the consequences of mortality or injuries (lethal effects), the impact of short-term behavioral and physiological effects on vital rates can be less intuitive. Below, we provide a rough categorization of consequences of noise disturbance that potentially affect vital rates.

11.3.3.1 Mortality and Injury

The death of an individual impacts a population in terms of population abundance and loss of reproductive potential. The impact could be negligible or critical depending on the status of the population from which the individual is lost and whether the population is considered endangered or at risk of extinction. While mortality from direct exposure to underwater noise has never been documented, lethal effects may arise indirectly because of behavioral responses. Mass strandings and mortalities of beaked whales have been associated with proximate naval sonar operations (D'Amico et al. 2009; Filadelfo et al. 2009), while pinniped pups have been abandoned or trampled during noise-induced stampeding (Lewis 1987). The physical effects of noise, including some possible types of injury, are discussed in Chap. 9. While injuries from underwater noise are difficult to quantify, particularly in free-ranging marine mammals, they can have both immediate and future impacts on the ability of an individual to feed, reproduce, or survive. Even recoverable injuries, or a temporary loss of sensory abilities, may impact an animal's ability to find resources, avoid predators, or meet energetic needs. However, reports of mortality resulting from noise exposure remain much rarer than those from fishery interactions (Read et al. 2006).

11.3.3.2 Energetics

Ideally, an animal will acquire sufficient energy to survive, grow, and reproduce. A noise disturbance can affect an animal's energy balance in different ways. For instance, the disruption of foraging behavior can reduce the rate of energy acquisition. Various forms of noise exposure have been documented to disrupt feeding behavior in marine mammals (e.g., humpback whales, (Sivle et al. 2015), blue whales (Goldbogen et al. 2013), and beaked whales (Tyack et al. 2011)). Factors such as prey density and social grouping can further modulate the behavioral responsiveness of animals to noise by modifying the trade-off between leaving a fertile feeding ground and perceived risk (e.g., for blue whales, see Friedlaender 2016). As an example, adult manatees responded to higher ambient noise by initially increasing foraging activity. This was, presumably, a response to a perceived risk whereby the animals increased foraging rate as a compensatory measure to a complete cessation of foraging in case the threat would escalate and lead to fleeing. This behavioral response was not observed if calves were present (Miksis-Olds and Wagner 2011). Animals that exhibit displacement or avoidance from a noise exposure might incur an increased energetic cost associated with the cost of transport, particularly if moving at high speed or for extended periods of time (DeRuiter et al. 2013). Measures of energy requirements and expenditure can be quantified using a bioenergetic approach (Farmer et al. 2018; Pirotta et al. 2018b) or by assigning an energy metric to different motivational states (Pirotta et al. 2018a).

11.3.3.3 Reproduction

Individual reproductive status and resource availability influence the decision of an individual to reproduce and/or nurture offspring, and reproductive strategies emerge from trade-offs between fitness maximization and risk avoidance. Pirotta et al. (2020) explored this trade-off in long-finned pilot whales. They compared a dynamic energy budget model to state-dependent life history theory in the presence and absence of simulated disturbance, in this case represented by variation in resource availability due to climate change and the disturbance of feeding. Results from this study suggest that pilot whales favor a cautionary approach when resources are patchy or heterogeneously available, demonstrating how a noise disturbance can have the potential to affect behaviors related to reproduction as well as nurturing required to successfully rear offspring (Pirotta et al. 2020). Changes in behavioral budget and nursing behaviors caused by disturbance may limit breeding opportunities and impact nurturing behaviors, which, in turn, can affect the ability of females to successfully raise their progeny (Currey et al. 2009; Foroughirad and Mann 2013; Senigaglia et al. 2019).

11.3.3.4 Stress Response

Noise disturbance can cause changes to the neural immune and endocrine systems. These can subsequently affect immune system function, possibly increasing vulnerability to disease (Lafferty and Holt 2003), as well as energy metabolism and the ability to reproduce. For instance, in a captive study on beluga whales and bottlenose dolphins, Romano et al. (2004) recorded a statistically elevated level of some stress-related hormones after experimental exposure to pulsed sound, though the changes were modest. Other examples of putative relationships between noise exposure and the stress response can be found in Chap. 9. Little information exists on how these changes are related to energetic and reproductive costs in marine mammals.

11.3.3.5 Temporary and Permanent Hearing Loss

While a direct link between permanent hearing loss and fitness is easily argued for marine mammals, the potential to translate TTS into fitness costs is more complicated. Any loss in sensory perception, particularly auditory perception, has the potential to impact how an animal that relies on sound to interact with its environment finds food and mates, and avoids predators. Thus, the degree to which a TTS might have an impact will be dependent on the magnitude and duration of the TTS, as well as what life history functions might overlap with it. Noise-induced hearing loss is discussed extensively in Chap. 9.

11.3.4 Population Consequences of Acoustic Disturbance

Quantifying the impact of a stressor requires an integrated framework to link disturbance-induced individual energetic imbalances to species' life histories and to estimate the proportion of the population affected by the disturbance (Pirotta et al. 2018a). The interdependency of energetics, reproduction, and survivability, and the mechanistic link between the effects of noise and its impact on population viability were formalized by the NRC in 2005. The NRC proposed a conceptual model for assessing or predicting how noise impacts on marine mammals might translate into population consequences (National Research Council 2005). The conceptual model, termed the Population Consequences of Acoustic Disturbance (PCAD), provides a framework consisting of five levels of variables connected by transfer functions (Fig. 11.1 top). The grouped variables in the model consist of (1) Sound, (2) Behavioral Change, (3) Life Functions Immediately Affected, (4) Vital Rates, and (5) Population Effects. The transfer functions describe how each group of variables affects the subsequent group of variable(s). For example, the transfer function between "Sound" and "Behavioral Change" relates how a particular sound with specified characteristics (duration, frequency, level, etc.) affects a behavioral response within a particular species. The subsequent transfer function relates the "Behavioral Change" to changes in life functions (e.g., feeding, breeding). Through this conceptual flow model, a framework for designing and conducting research to address the question of biologically significant impacts on marine mammals exposed to noise was established. Importantly, several challenging issues were noted by the NRC with respect to implementing models based on the framework (National Research Council 2005). Among these were as follows:

1. To relate changes in behavior patterns to changes in life history functions, observation

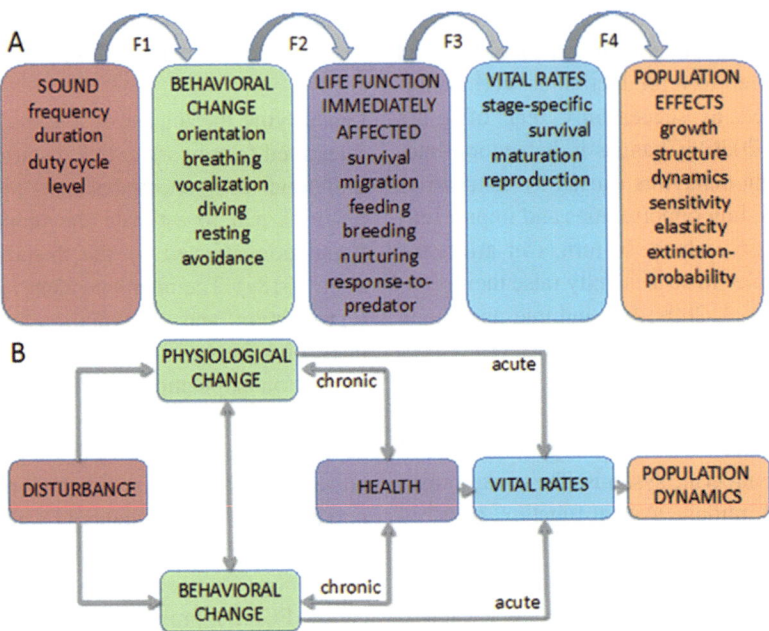

Fig. 11.1 (**a**) Conceptual layout of the Population Consequences of Acoustic Disturbance (PCAD) model. F1–F4 correspond to the transfer functions used to relate groups of variables. Adapted and reproduced with permission from the National Academy of Sciences, Courtesy of the National Academies Press, Washington, D.-C. Original image © National Research Council, 2005. Marine Mammal Populations and Ocean Noise: Determining When Noise Causes Biologically Significant Effects. https://doi.org/10.17226/11147. (**b**) The Population Consequences of Disturbance (PCoD) model shows the conceptual relationship between disturbance and its impact on health, vital rates, and ultimately population dynamics, via changes in animal physiology and behavior. Modified from New LF, Clark JS, et al. (2014) Using short-term measures of behaviour to estimate long-term fitness of southern elephant seals. Mar Ecol Prog Ser 496: 99–108; https://doi.org/10.3354/meps10547. Original image © Inter-Research, 2014. This modified version © Senigaglia et al., 2025; issued CC BY 4.0

scales need to be appropriate. To address diel changes in activity, a 24-h observation window should be made to evaluate functional short-term consequences of disturbance. Longer observation windows are required to capture functionality and variability in behavior and responsiveness that occur on seasonal scales.

2. Nearly nothing is known about how behavioral changes affect the vital rates of individuals. Information on vital rates would allow modeling population-level effects, but the lack of information on vital rates remains a critical challenge for implementing the PCAD model.

3. The measure of certain behavioral responses (e.g., respiration rate, surfacing, and avoidance), simply because they are easy to observe, will be of relatively little use in linking responses to the downstream categories of the PCAD model. Understanding the number of animals affected, the range over which they are impacted, the duration of the impact, and the cumulative impact of additional stressors will be required to implement the model for overall population impacts.

The US Office of Naval Research convened a working group in 2009 to consider the implementation of a PCAD model and how it would be parameterized (described in Harwood et al. 2014). The working group expanded the PCAD concept to include other forms of disturbance, resulting in a modified framework known as the Population Consequences of Disturbance (PCoD; Fig. 11.1 bottom). The NRC (2005) recommended that initial models (case studies) address known populations, preferably where individual animals have been tracked, but also covering diverse life history strategies. Following the recommendations, the working group began case studies using elephant seals, coastal bottlenose dolphins, beaked whales, and northern right whales. Outputs from the original case studies, which are diverse in their modeling approach to implementing the PCoD framework, are now available for elephant seals (New et al. 2014), beaked whales (New et al. 2013b), northern right whales (Schick et al. 2013), and coastal bottlenose dolphins (New et al. 2013a). Beyond the initial case studies, other models that follow the conceptual PCoD framework have been developed to assess the impact of disturbance (acoustic and otherwise) to harbor porpoises (Nabe-Nielsen et al. 2014), bottlenose dolphins (Pirotta et al. 2015), gray whales (Villegas-Amtmann et al. 2015; Villegas-Amtmann 2017), minke whales (Christiansen and Lusseau 2015), and pinniped capital and income breeders (McHuron et al. 2017).

11.3.4.1 Implementation Challenges

Parametrizing the complex PCAD/PCoD models has proven to be a significant challenge. Measuring the effect of noise disturbance on behavior is commonly achieved using visual observation or satellite tags, while physiological responses to disturbance are analyzed from tissue samples or feces from wild animals (Pirotta et al. 2018a; Rolland et al. 2012). Unfortunately, empirical data on noise disturbance effects are limited, and a critical knowledge gap in the various PCAD/PCoD models remains the mechanistic linkage of behavioral and/or physiological responses to vital rates (McHuron et al. 2022).

Accurate long-term life history data are scarce for most marine mammal populations due to the logistical difficulties of conducting research (e.g., studies in remote locations, elusive animals, or lack of regular funding) and the long lifespan of many species (McHuron et al. 2022). This is exacerbated by the fact that behavioral and physiological responses may not be necessarily linked. It is possible for animals to have physiological responses to a disturbance with no visible behavioral response. Conversely, it is possible that a behavioral response is not accompanied by a physiological response or that it will produce a delayed physiological response (e.g., chronic disruption of foraging leads to a delayed physiological change associated with fasting or reduced energy acquisition).

Even when physiological responses are noted, the consequence of the responses can be difficult to ascertain. There are efforts underway to understand how hormones (e.g., corticosteroids, catecholamines, and thyroid hormones) change in animals that are acoustically disturbed. However, understanding how hormones change is ultimately uninformative unless there is also an understanding of the physiological consequences of those changes. Hormones act as messengers that trigger and modulate metabolic, immunologic, and reproductive physiology, often interacting with one another to regulate complex processes. Yet, there is little information in marine mammals on how the magnitude and duration of shifts in hormone levels impact energy acquisition and utilization, reproductive effort/success, and health. Such information is critical to informing linkages between responses to noise and impact to vital rates. Therefore, it is imperative not only that baseline information on the natural variation in hormones be obtained to understand when shifts occur due to a disturbance but also that the physiological consequences of those changes be determined.

A better understanding of marine mammal energetics is also required to inform the costs associated with behavioral responses to disturbance. Energetically costly responses can occur through disrupted foraging opportunities, displaced prey, displacement from optimal foraging habitat, or costs associated with fleeing. Basic information on energetics (e.g., resting metabolic rate) continues to be critical for accurate modeling of the energetic consequences of disturbance. This is particularly difficult to determine on wild marine mammals, but it is necessary to understand the relative magnitude of the energetic cost due to the behavioral response.

Regardless of the advances made in monitoring noise exposures and animal behavior, knowledge of the physiological consequences of disturbance continues to be a critical gap in establishing mechanistic linkages that enable extrapolation to population consequences. To overcome such limitations, when parameters cannot be quantified empirically, they can be modeled based on other species or through expert elicitation (e.g., assessments of disturbance on harbor seals, (Thompson et al. 2013b), harbor porpoises (King et al. 2015), northern right whales (Oedekoven et al. 2015), and belugas (Tollit et al. 2016)). While assumptions must be made (see Christiansen and Lusseau 2015), these "interim" models can be updated with data and process knowledge as they become available (King et al. 2015).

11.3.4.2 Spatial and Temporal Dimension

It can be difficult for models to accurately account for the impact of a disturbance, depending upon its temporal and spatial extent and its overlap with the affected marine mammal populations. As previously mentioned, the impact of a disturbance on animal populations is a direct function of response severity (i.e., how strongly an animal responds to the stressor), species context (e.g., capital vs. noncapital breeder), and the proportion of the population affected by the disturbance. Spatial and temporal variations in habitat characteristics (e.g., bathymetry and depth, water temperature, water stratification or mixing, salinity) can affect noise propagation and local ambient noise levels. These same factors can affect the presence of prey or other factors impacting marine mammal use of the local habitat, thus affecting the spatial and temporal overlap of marine mammals in the exposure zone. If a disturbance event affects only a limited portion of an individual/population home range, the disturbance might have limited impact at the population level (Nabe-Nielsen et al. 2014). Agent-based models allow for the incorporation of spatial variation in disturbance (McHuron et al. 2022) and may translate noise-induced behavioral changes into bioenergetic costs using dynamic energy budget analysis (Mortensen et al. 2021). Continuous-time life history models (De Roos 2008) can also assess population effects of disturbance by forecasting population growth under increasing human disturbance (Manlik et al. 2022).

11.3.4.3 The Importance of including Effects on Sociality in PCoD

Many marine mammals are social, and differences in fitness across social units can arise from intra-population behavioral differences (Lahdenperä et al. 2016). Marine tourism has been shown to differentially impact social dynamics by decreasing the time spent socializing, promoting solitary foraging techniques, and disrupting animal social networks (Powell and Wells 2011; Williams and Lusseau 2006). Recorded effects of disturbance on vital rates via social modification span across a variety of species, including killer whales (*Orcinus orca*), sperm whales (*Physeter macrocephalus*), and Indo-Pacific bottlenose dolphins (*Tursiops aduncus*) (Ellis et al. 2017; Frère et al. 2010; Kopps and Sherwin 2012; Senigaglia et al. 2022; Wild et al. 2019). Cultural traditions and sociality provide "residence knowledge" and can act as a buffer to environmental changes or, alternatively, might lead to stasis due to a lack of flexibility that ultimately stymies population resilience (Brakes et al. 2021).

11.4 Cumulative Impacts

11.4.1 Multiple Sources of Disturbance and the Birth of PCoMS

11.4.1.1 Population Viability under Multiple Stressors

Marine mammals can be subjected to a multitude of anthropogenic stressors in addition to noise (e.g., climate change, fisheries bycatch, habitat loss, etc.), all of which can impact population dynamics (Lacy et al. 2017). To develop effective management strategies, it is critical to explore the cumulative impact of the multiple stressors affecting an individual and to assess how these subsequently contribute to impacts on the population. However, different life histories and a plethora of confounding factors (e.g., context, previous experience with similar stressors, temporal and spatial scale of exposure, etc.) make modeling the population viability of species under multiple stressors extremely challenging. In fact, there is a general lack of knowledge on how different stressors interact and whether these interactions are additive, synergistic, or antagonistic (Gillingham et al. 2016; Pirotta et al. 2019, 2022). Despite the high degree of uncertainty and complexity of judging multiple stressors, wildlife management requires an assessment of the potential adverse impact of human activity on long-term population viability considering multiple stressors and cumulative disturbances (Tyack et al. 2022). For instance, the 1989 Exxon Valdez oil spill in Alaska impacted the resilience of both resident and transient killer whale pods (Matkin et al. 2008), and the 2010 Deepwater Horizon leak in the Gulf of Mexico augmented the severity of the unusual mortality event caused by red tide (Carmichael et al. 2012).

11.4.1.2 Population Consequences of Multiple Stressors (PCoMS)

Efforts to accurately model concurrent multiple stressors and disturbances are in their infancy, and the impact of cumulative disturbance remains largely unexplored (Murray et al. 2019; Willsteed et al. 2023). Pirotta et al. (2022) reviewed the trade-offs between the different ways the impact from multiple stressors could be modeled. They suggested that management goals and data availability should guide model choice along the "assumption spectrum": from a full empirical approach, where trends can be inferred from data, to a more mechanistic approach, where the underlying processes and pathways linking stressors to impacts guide the investigation using Bayesian computation (Pirotta et al. 2022). A common process-driven framework was developed by the NRC to quantify the effects of multiple, overlapping stressors. Building on prior NRC reports (National Research Council 2003, 2005), they convened a committee to explore methods of evaluating cumulative effects (National Academies of Sciences and Medicine 2017). The committee report was intended to assist with creating approaches that help identify when cumulative impacts present a risk to populations or ecosystems and with developing resource management methods by which stressors can be identified and reduced to lessen population risk. Two notable recommendations were published by the committee. Firstly, a framework was presented to assess the population consequences of multiple stressors based on the prior PCAD and PCoD frameworks. The recommendation promoted the use of health indices to integrate stressors at timescales larger than observed during acute exposure experiments, but less than timescales over which vital rates are affected (e.g., as performed for the elephant seal in New et al. (2014)). Secondly, the committee proposed "interaction webs" to investigate how marine mammal abundance and distribution are affected by direct and indirect interactions of individuals with abiotic and biotic stressors, which are themselves driven by ecosystem-level changes. Such interaction webs can then be further investigated using network-based methods, such as structural equations and Bayes networks (Pirotta et al. 2022).

11.4.1.2.1 Optimal Health and Interaction Webs

The definition of health adopted by the PCoMS committee was "the ability to adapt and self-manage" and refers to the ability of an animal to maintain homeostasis under varying environmental conditions, which is effectively "allostasis" (Huber et al. 2011). Health can be assessed by body condition, markers of immune function, levels of hormones associated with stress, and other factors. Health indices, if they can be monitored and quantified, can be used as proxies for an animal's allostatic load (National Academies of Sciences and Medicine 2017), with the accumulation of stressors assessed by linkages to the health index. As has been performed with PCAD case studies, translating relationships between health indices and other compartments of the model will need to be mathematically defined so that quantitative predictions of cumulative impacts can be made. The proposal to utilize interaction webs addresses issues related to large-scale ecological drivers, which can exert population-level influences on health indices by affecting a large number of extrinsic factors to which a marine mammal is exposed (National Academies of Sciences and Medicine 2017). Interaction webs are inherently complex because they strive to characterize the way species in an ecosystem interact with each other and other components of the abiotic/biotic environment—relationships that are often nonlinear. Interaction webs are appealing due to their ability to conceptually grasp the vastness of the cumulative impacts problem. PCoMs can also be coupled with agent-based modeling where both anthropogenic disturbance and marine mammals are considered agents, and behavioral patterns can be translated into population consequences through energy balance dependent changes in vital rates (Mortensen et al. 2021).

11.4.1.2.2 Spatial and Temporal Scale of Disturbance

The information required to inform interaction webs requires sampling across temporal and spatial scales that are typically prohibitive; however, complementary analyses (e.g., inferential approaches, such as weight-of-evidence analysis) can be used to partially address this shortcoming. Marcotte et al. (2015) assessed the cumulative effect of coastal development on the Indo-Pacific humpback dolphin by characterizing the spatial and temporal pattern of individual development projects, their operations, and decommission. They combined geographic information system approaches with expert opinion on likely impacts resulting from reclamation projects, pile driving, dredging, shipping traffic, and high-speed ferry traffic. The authors concluded that the driving factor in a decline of a regional population of Indo-Pacific humpback dolphins was high-speed ferry traffic. The authors could not exhaustively evaluate all stressors, but development of the model was an important step toward understanding how to address the multivariate nature of cumulative impacts.

Modeling cumulative effects when ocean noise is an effector will be challenged by the entanglement of direct effects due to noise, the presence of noise sources (e.g., ships), and indirect effects of noise-producing activities (e.g., dredging). It will be further complicated by the large spatial scales over which underwater noise can travel and the uncertainty of anthropogenic noise presence when noise monitoring is spatially and temporally inadequate. Nevertheless, emerging technologies for assessing indices of animal health (historically and current) and efforts to characterize the spatial and temporal occurrence of other stressors continue in parallel to address the issue of cumulative impacts. Pirotta et al. (2019) used dynamic state variable models to assess the effects of environmental changes and anthropogenic disturbance on North Pacific blue whale vital rates. By modeling spatially explicit behavior and physiological and reproductive status across habitat and migratory range, the authors suggested that animals may be able to sustain high levels of human disturbance but were becoming increasingly vulnerable to frequent environmental perturbations. Anthropogenic disturbance coupled with significant environmental change (including climate change) leads to stronger effects. The impact of these effects will be greatest for adolescent and juvenile individuals (Pirotta et al. 2020). Under anthropogenic disturbance scenarios, the greatest effect was not on adult survival but on reproductive rates and reproductive success, suggesting that calving rates and intervals should be monitored as an early indicator for significant population-level effects (Pirotta et al. 2020).

A recent study from Pirotta et al. (2023) employed Bayesian state-space models in an attempt to quantify and compare the impact of multiple stressors (i.e., vessel collisions, entanglements, and resource availability) on vital rates of the critically endangered North Atlantic right whales. The comprehensive dataset used in this study, comprising over 50 years of life history and exposure data, allowed the authors to explore the mechanistic links between stressors and vital rates based on changes in individuals' health (Pirotta et al. 2023). Similarly, Murray et al. (2019) capitalized on 36 years of data to model the population trajectories of northern and southern resident killer whales exposed to multiple disturbances, including prey depletion and exposure to polychlorinated biphenyl contaminants. When data are unavailable or are subject to great uncertainty, this can be explicitly modeled using state-space models or Bayesian statistics. Unfortunately, given the challenges in obtaining baseline data and the considerable modeling effort involved, the application of cumulative effects assessment remains limited and is seldom required by environmental legislation (Willsteed et al. 2023).

11.4.2 Integrated Assessment: Evolution toward Socioecological Modeling

An approach that combines qualitative and quantitative data can allow for a more holistic view of the cumulative effects of multiple disturbances despite data paucity (Williams et al. 2016). Techniques used to merge quantitative and qualitative data with various degrees of uncertainty include "fuzzy logic" (Giabbanelli et al. 2017), Bayesian Belief Networks (BBN; Newton et al. 2007), and hybrid models (Parrott 2011). Socioecological models (SES; Desharnais et al. 2005) can be used to assess the economic and societal importance of human activities and develop a triage approach based on species' vulnerability to noise and other disturbances (Farrier et al. 2007; Morrison et al. 2016). For instance, the temporal and spatial scale of multiple anthropogenic stressors can vary, and populations subjected to multiple stressors but with limited home range will be more critically impacted by anthropogenic disturbance and environmental stochasticity (Lacy 2019; Lacy et al. 2017).

With the overarching aim of improving the sustainable management of human activities, BBN, SES, decision trees, and simulation-based analyses can be expanded to include societal interests in noise-generating activities. By integrating socioeconomic values, socioecological models can become important decision-making tools for regulatory agencies (Lacy et al. 2017). For instance, the installation, operation, and decommissioning of infrastructures for marine renewable energy generate acoustic disturbance and can potentially impact marine mammal populations, but the benefit gained by addressing climate change might offer a compelling argument to proceed despite the potential impact to local wildlife. In such instances, careful consideration should be given to the location of the development, the status of marine mammals in overlapping habitat, and the potential for population-level impacts. Thus, the evaluation of the societal value of marine mammals should be examined in the context of the trade-off between animal welfare, population impacts, and the socioeconomic benefits generated by the disturbance.

11.4.2.1 Implementation Challenges

Socioecological models allow for an exploration of marine mammal resilience to stochastic and human-induced environmental changes while providing a complementary perspective toward regulation and management. The application of socioecological modeling remains scarce due to the challenges associated with model parametrization and the definition of the mechanistic relationships between socioeconomic and biophysical elements (Rounsevell et al. 2021). Moreover, where SES are simulated and later empirically tested, differences in predicted impact and recorded environmental status can be contradictory, highlighting how simulation can overestimate or underestimate impact and lead to ineffective management actions (Stockbridge et al. 2021). Nevertheless, summary models can help triage conservation efforts and select stressors with the most severe impacts to mitigate (Pirotta et al. 2023; Willsteed et al. 2023).

11.5 Conclusion

This chapter described the importance of quantifying/qualifying the biological significance of noise disturbance and its impact on individual and population vital rates and viability, respectively. Understanding the impact of multiple and recurrent exposure to acoustic disturbance and other stressors is imperative to determining the cumulative impact of disturbance on population viability. However, differentiating between behavioral and physiological acute and chronic noise effects is challenging. A common difficulty in quantifying the biological impact of noise on marine mammals is a lack of long-term monitoring studies, including baseline information on individual and population demographics. Few marine mammal populations have been studied for sufficiently long periods to reliably quantify vital rates and population viability. This is exacerbated by an inadequate robustness in characterizing behavior and unestablished linkages between altered behavior and energetic or health consequences. Given the high intra- and interspecific variability in potential impacts, long-term studies are needed in systems in which disturbances can be adequately characterized to address population- and location-specific management issues. Collecting long-term datasets at the individual and population level is logistically and economically challenging. Only a few marine mammal species (e.g., northern elephant seals and Sarasota Bay bottlenose dolphins) have been the subject of intensive multiyear studies, and these have been utilized in model case studies that attempt to establish links between short-term responses and long-term consequences. Even in these instances, however, the models represent only a small proportion of the broad variability in species' behavior, physiology, and ecology that can be exhibited across the larger population.

Given the difficulties in obtaining an empirical evaluation of impact spanning multiple generations, modeling approaches based on recorded vital rates are being developed to predict population impacts resulting from acoustic disturbance. Recent model frameworks include the PCAD and PCoD models, which have been generalized to include diverse and multiple disturbances. While a growing scientific literature continues to highlight the importance of cumulative and integrated impact assessments, empirical applications continue to encounter multiple challenges primarily driven by data inadequacy. Consequently, examples of management plans and legislations stemming from these studies are rare. Model parametrization remains exceedingly difficult due to a scarcity of data, leaving substantial gaps in understanding and stymieing the environmental assessment of different noise-generating marine activities. Nevertheless, to avoid the paradigm of "analysis paralysis," predictive modeling of alternative management scenarios coupled with expert opinion and adaptive management can provide empirical support to proposed management actions.

References

Accardo CM, Ganley LC, Duley PA, George JC, Reeves RR, Heide-Jørgensen MP, Tynan CT, Mayo CA (2018) Sightings of a bowhead whale (*Balaena mysticetus*) in the Gulf of Maine and its interactions with other baleen whales. J Cetacean Res Manag 19(1):23–30. https://doi.org/10.47536/jcrm.v19i1.412

Andrew R, Bruce MH, James AM (2002) Ocean ambient sound: comparing the 1960s with the 1990s for a receiver off the California coast. Acoust Res Lett online 3(2):65–70. https://doi.org/10.1121/1.1461915

Armando J-L, Rojas-Bracho L, Brownell RL, Read AJ, Reeves RR, Ralls K, Taylor BL (2007) Saving the vaquita: immediate action, not more data. Conserv Biol 21(6):1653–1655. https://doi.org/10.1111/j.1523-1739.2007.00825.x

Bain DE, Smith J, Williams R, Lusseau D (2006) Effects of vessels on behavior of southern resident killer whales (*Orcinus spp.*). NMFS Contract Report No. AB133F03SE0959 and AB133F04CN0040

Baird IG, Mounsouphom B (1994) Irrawaddy dolphins (*Orcaella brevirostris*) in southern Lao PDR and the northeastern Cambodia. Nat Hist Bull Siam Soc 42:159–175

Baird RW, Martin SW, Webster DL, Southall BL (2014) Assessment of modelled received sound pressure levels and movements of satellite-tagged odontocetes exposed to mid-frequency active sonar at the Pacific Missile Range Facility: February 2011 through February 2013. NAVFAC PAC, U.S. Pacific Fleet, Report by HDR Environmental, Operations and Construction Inc

Bejder L, Samuels A (2003) Evaluating the effects of nature-based tourism on cetaceans. In: Marine mammals: fisheries, tourism and management issues, vol 1. CSIRO Publishing, pp 229–256

Bejder L, Samuels AMY, Whitehead HAL, Gales N, Mann J, Connor R, Heithaus M, Watson-Capps J, Flaherty C, Krützen M (2006) Decline in relative abundance of bottlenose dolphins exposed to long-term disturbance. Conserv Biol 20(6):1791–1798. https://doi.org/10.1111/j.1523-1739.2006.00540.x

Bejder L, Samuels A, Whitehead H, Finn H, Allen S (2009) Impact assessment research: use and misuse of habituation, sensitisation and tolerance in describing wildlife responses to anthropogenic stimuli. Mar Ecol

Prog Ser 395:177–185. https://doi.org/10.3354/meps07979

Berg Soto A, Cagnazzi D, Everingham Y, Parra GJ, Noad M, Marsh H (2013) Acoustic alarms elicit only subtle responses in the behaviour of tropical coastal dolphins in Queensland, Australia. Endanger Species Res 20(3):271–282. https://doi.org/10.3354/esr00495

Blackwell SB, Lawson JW, Williams MT (2004) Tolerance by ringed seals (*Phoca hispida*) to impact pipe-driving and construction sounds at an oil production island. J Acoust Soc Am 115(5):2346–2357. https://doi.org/10.1121/1.1701899

Bordino P, Kraus S, Albareda D, Fazio A, Palmerio A, Mendez M, Botta S (2002) Reducing incidental mortality of Franciscana dolphin *Pontoporia blainvillei* with acoustic warning devices attached to fishing nets. Mar Mamm Sci 18(4):833–842. https://doi.org/10.1111/j.1748-7692.2002.tb01076.x

Born EW, Riget FF, Dietz R (1999) Escape responses of hauled out ringed seals (*Phoca hispida*) to aircraft disturbance. Polar Biol 21(3):171–178. https://doi.org/10.1007/s003000050349

Bowles A, Stewart BS (1980) Disturbances to the pinnipeds and birds of San Miguel Island, 1979-1980. In: Jehl JR, Cooper CF (eds) Potential effects of space shuttle sonic booms on the biota and geology of the California Channel Islands: research reports. Centre for Marine Studies, San Diego State University, San Diego, pp 99–137

Brakes P, Carroll EL, Dall SR, Keith SA, McGregor PK, Mesnick SL, Noad MJ, Rendell L, Robbins MM, Rutz C (2021) A deepening understanding of animal culture suggests lessons for conservation. Proc R Soc B 288(1949):20202718. https://doi.org/10.1098/rspb.2020.2718

Brookes KL, Bailey H, Thompson PM (2013) Predictions from harbour porpoise habitat association models are confirmed by long-term passive acoustic monitoring. J Acoust Soc Am 134(3):2523–2533. https://doi.org/10.1121/1.4816577

Brotons JM, Munilla Z, Grau AM, Rendell L (2008) Do pingers reduce interactions between bottlenose dolphins and nets around the Balearic Islands? Endanger Species Res 5:1–8. https://doi.org/10.3354/esr00104

Bruno DL, Marino F (2012) A trial of acoustic harassment device efficacy on free-ranging bottlenose dolphins in Sardinia, Italy. Mar Freshw Behav Physiol 44(4):197–208. https://doi.org/10.1080/10236244.2011.618216

Bryant PJ, Lafferty CM, Lafferty SK (1984) Reoccupation of Laguna Guerrero Negro, Baja California, Mexico, by gray whales. In: Jones ML, Swartz SL, Leatherwood S (eds) The gray whale (*Eschrichtius robustus*). Academic, Orlando, pp 375–387. https://doi.org/10.1016/B978-0-08-092372-7.50021-2

Calkins DG (1979 [published 1983]) Marine mammals of Lower Cook Inlet and the potential for impact from outer continental shelf oil and gas exploration, development, and transport (NTIS: PB85-201226). National Oceanic and Atmospheric Administration, Juneau

Carmichael RH, Graham WM, Aven A, Worthy G, Howden S (2012) Were multiple stressors a 'perfect storm' for northern Gulf of Mexico bottlenose dolphins (*Tursiops truncatus*) in 2011? PLoS One 7(7):e41155. https://doi.org/10.1371/journal.pone.0041155

Carretta JV, Barlow J (2011) Long-term effectiveness, failure rates, and "dinner bell" properties of acoustic pingers in a gillnet fishery. Mar Technol Soc J 45(5):7–19. https://doi.org/10.4031/MTSJ.45.5.3

Castellote M, Clark C, Lammers M (2012) Acoustic and behavioural changes by fin whales (*Balaenoptera physalus*) in response to shipping and airgun noise. Biol Conserv 147(1):115–122. https://doi.org/10.1016/j.biocon.2011.12.021

Castellote M, Leeney RH, O'Corry-Crowe G, Lauhakangas R, Kovacs KM, Lucey W, Krasnova V, Lydersen C, Stafford KM, Belikov R (2013) Monitoring white whales (*Delphinapterus leucas*) with echolocation loggers. Polar Biol 36(4):493–509. https://doi.org/10.1007/s00300-012-1276-2

Castellote M, Brotons JM, Chicote C, Gazo M, Cerda M (2015) Long-term acoustic monitoring of bottlenose dolphins, *Tursiops truncatus*, in marine protected areas in the Spanish Mediterranean Sea. Ocean Coast Manag 113:54–66. https://doi.org/10.1016/j.ocecoaman.2015.05.017

Charif RA, Ragaman A, Muirhead CA, Pitzrick MS, Warde AM, Hall J, Pyć C, Clark CW (2013) Bowhead whale acoustic activity in the southeast Beaufort Sea during late summer 2008–2010. J Acoust Soc Am 134(6):4323–4344. https://doi.org/10.1121/1.4824679

Christiansen F, Lusseau D (2015) Linking behavior to vital rates to measure the effects of non-lethal disturbance on wildlife. Conserv Lett 8(6):424–431. https://doi.org/10.1111/conl.12166

Claridge DE (2006) Fine-scale distribution and habitat selection of beaked whales. MSc thesis, Aberdeen University

Claridge DE (2013) Population ecology of Blainville's beaked whales (*Mesoplodon densirostris*). PhD thesis, University of St Andrews

Cox TM, Read AJ, Solow A, Tregenza N (2001) Will harbour porpoises (*Phocoena phocoena*) habituate to pingers? J Cetacean Res Manag 3(1):81–86. https://doi.org/10.1111/mms.12880

Cox TM, Read AJ, Swanner D, Urian K, Waples D (2003) Behavioral responses of bottlenose dolphins, *Tursiops truncatus*, to gillnets and acoustic alarms. Biol Conserv 115:203–212. https://doi.org/10.1016/S0006-3207(03)00108-3

Cox TM, Ragen TJ, Read AJ, Vos E, Baird RW, Balcomb K, Barlow J, Caldwell J, Cranford T, Crum L, D'Amico A, D'Spain G, Fernández A, Finneran J, Gentry R, Gerth W, Gulland F, Hildebrand J, Houser D, Hullar T, Jepson PD, Ketten D, MacLeod CD, Miller P, Moore S, Mountain DC, Palka D, Ponganis P, Rommel

S, Rowles T, Taylor B, Tyack P, Wartzok D, Gisiner R, Mead J, Benner L (2005) Understanding the impacts of anthropogenic sound on beaked whales. J Cetacean Res Manage 7(3):177–187. https://doi.org/10.47536/jcrm.v7i3.729

Cranford TW, Krysl P (2015) Fin whale sound reception mechanisms: skull vibration enables low-frequency hearing. PLoS One 10(1):e0116222. https://doi.org/10.1371/journal.pone.0116222

Culik BM, Koschinski S, Tregenza N, Ellis GM (2001) Reactions of harbor porpoises *Phocoena phocoena* and herring *Clupea harengus* to acoustic alarms. Mar Ecol Prog Ser 211:255–260. https://doi.org/10.3354/meps211255

Cunningham KA, Hayes SA, Rub AMW, Reichmuth C (2014) Auditory detection of ultrasonic coded transmitters by seals and sea lions. J Acoust Soc Am 135(4):1978–1985. https://doi.org/10.1121/1.4868371

Currey RJ, Dawson SM, Slooten E, Schneider K, Lusseau D, Boisseau OJ, Haase P, Williams JA (2009) Survival rates for a declining population of bottlenose dolphins in Doubtful Sound, New Zealand: an information theoretic approach to assessing the role of human impacts. Aquat Conserv Mar Freshwat Ecosyst 19(6):658–670. https://doi.org/10.1002/aqc.1015

D'Agrosa C, Lennert-Cody CE, Vidal O (2000) Vaquita bycatch in Mexico's artisanal gillnet fisheries: driving a small population to extinction. Conserv Biol 14(4):1110–1119. https://doi.org/10.1046/j.1523-1739.2000.98191.x

D'Amico A, Gisiner R, Ketten D, Hammock J, Johnson C, Tyack P, Mead J (2009) Beaked whale strandings and naval exercises. Aquat Mamm 35(4):452–472. https://doi.org/10.1578/AM.35.4.2009.452

Danil K, St. Leger JA (2011) Seabird and dolphin mortality associated with underwater detonation exercises. Mar Technol Soc J 45(6):89–95. https://doi.org/10.4031/MTSJ.45.6.5

Dawson S, Northridge SP, Waples D, Read A (2013) To ping or not to ping: the use of active acoustic devices in mitigating interactions between small cetaceans and gillnet fisheries. Endanger Species Res 19:201–221. https://doi.org/10.3354/esr00464

De Roos AM (2008) Demographic analysis of continuous-time life-history models. Ecol Lett 11(1):1–15. https://doi.org/10.1111/j.1461-0248.2007.01121.x

Deecke VB, Slater PJB, Ford JKB (2002) Selective habituation shapes acoustic predator recognition in harbour seals. Nature 420(6912):171–173. https://doi.org/10.1038/nature01030

Demarchi MW, Holst M, Robichaud D, Waters M, MacGillivray AO (2012) Responses of Steller sea lions (*Eumetopias jubatus*) to in-air blast noise from military explosions. Aquat Mamm 38(3):279. https://doi.org/10.1578/AM.38.3.2012.279

Dendrinos P, Karamanlidis AA, Kotomatas S, Paravas V, Adamantopoulou S (2008) Report of a new Mediterranean monk seal (*Monachus monachus*) breeding colony in the Aegean Sea, Greece. Aquat Mamm 34(3):355–361. https://doi.org/10.1578/AM.34.3.2008.355

Department of the Navy (1998) Final environmental impact statement, shock testing the SEAWOLF submarine. Department of the Navy, Washington, DC, 563 pp

Department of the Navy (2001) Final overseas environmental impact statement and environmental impact statement for surveillance towed array sensor system low frequency active (SURTASS LFA) Sonar. Department of the Navy, Washington, DC, 688 pp

DeRuiter SL, Southall BL, Calambokidis J, Zimmer WM, Sadykova D, Falcone EA, Friedlaender AS, Joseph JE, Moretti D, Schorr GS (2013) First direct measurements of behavioural responses by Cuvier's beaked whales to mid-frequency active sonar. Biol Lett 9(4):20130223. https://doi.org/10.1098/rsbl.2013.0223

Desharnais F, Esseson GR, Matthews M-NR, Heard GJ, Thomson DJ, Brooke GH (2005) A generalized beamformer for localization of marine mammals. Appl Acoust 67(11-12):1213–1225. https://doi.org/10.1016/j.apacoust.2006.05.015

DiMarzio N, Moretti D, Ward JA, Morrissey R, Jarvis S, Izzi AM, Johnson M, Tyack P, Hansen A (2008) Passive acoustic measurement of dive vocal behavior and group size of Blainville's beaked whale (*Mesoplodon densirostris*) in the Tongue of the Ocean (TOTO). Can Acoust 36(1):166–173. https://jcaa.caa-aca.ca/index.php/jcaa/article/view/2007

Ducatez S, Sol D, Sayol F, Lefebvre L (2020) Behavioural plasticity is associated with reduced extinction risk in birds. Nat Ecol Evol 4(6):788–793. https://doi.org/10.1038/s41559-020-1168-8

Elliott R, Dawson S, Henderson S (2011) Acoustic monitoring of habitat use by bottlenose dolphins in Doubtful Sound, New Zealand. N Z J Mar Freshw Res 45(4):637–649. https://doi.org/10.1080/00288330.2011.570351

Ellis S, Franks DW, Nattrass S, Cant MA, Weiss MN, Giles D, Balcomb KC, Croft DP (2017) Mortality risk and social network position in resident killer whales: sex differences and the importance of resource abundance. Proc R Soc B 284(1865):20171313. https://doi.org/10.1098/rspb.2017.1313

Evans DL, England GR (2001) Joint interim report/Bahamas marine mammal stranding event of 14-16 March 2000. National Oceanic and Atmospheric Administration, Washington, DC, 61 pp

Farmer NA, Noren DP, Fougères EM, Machernis A, Baker K (2018) Resilience of the endangered sperm whale *Physeter macrocephalus* to foraging disturbance in the Gulf of Mexico, USA: a bioenergetic approach. Mar Ecol Prog Ser 589:241–261. https://doi.org/10.3354/meps12457

Farrier D, Whelan R, Mooney C (2007) Threatened species listing as a trigger for conservation action. Environ Sci Pol 10(3):219–229. https://doi.org/10.1016/j.envsci.2006.12.001

Fernández A, Edwards JF, Rodriguez F, de los Monteros AE, Herraez P, Castro P, Jaber JR, Martin V, Arbelo M (2005) "Gas and fat embolic syndrome" involving a mass stranding of beaked whales (Family Ziphiidae) exposed to anthropogenic sonar signals. Vet Pathol 42(4):446–457. https://doi.org/10.1354/vp.42-4-446

Fernández A, Sierra E, Martin V, Mendez M, Sacchinni S, Bernaldo de Quirós Y, Andrada M, Rivero M, Quesada O, Tejedor M, Arbelo M (2012) Last "atypical" beaked whales mass stranding in the Canary Islands (July, 2004). J Mar Sci Res Devel 2:107. https://doi.org/10.4172/2155-9910.1000107

Fetherston T, Turner S, Mitchell G, Guzas E (2019) Marine mammal lung dynamics when exposed to underwater explosion impulse. Anat Rec 302(5):718–734. https://doi.org/10.1002/ar.24033

Filadelfo R, Mintz J, Michlovich E, D'Amico A, Tyack PL, Ketten DR (2009) Correlating military sonar use with beaked whale mass strandings: what do the historical data show? Aquat Mamm 35(4). https://doi.org/10.1578/AM.35.4.2009.435

Fitch J, Young P (1947) Use and effect of explosives in California coastal waters. Calif Fish Game 34:53–70

Foroughirad V, Mann J (2013) Long-term impacts of fish provisioning on the behavior and survival of wild bottlenose dolphins. Biol Conserv 160:242–249. https://doi.org/10.1016/j.biocon.2013.01.001

Frantzis A (1998) The first mass stranding that was associated with the use of active sonar (Kyparissiakos Gulf, Greece, 1996). In: Evans PGH, Miller L (eds) Active sonar and cetaceans, ECS Newsletter 42, Special Supplement. 8 pp

French SS, Gonzalez-Suarez M, Young JK, Durham S, Gerber LR (2011) Human disturbance influences reproductive success and growth rate in California sea lions (*Zalophus californianus*). PLoS One 6(3):e17686. https://doi.org/10.1371/journal.pone.0017686

Frère CH, Krützen M, Mann J, Connor RC, Bejder L, Sherwin WB (2010) Social and genetic interactions drive fitness variation in a free-living dolphin population. Proc Natl Acad Sci USA 107(46):19949–19954. https://doi.org/10.1073/pnas.1007997107

Friedlaender, A. S., Hazen, E. L., Goldbogen, J. A., Stimpert, A. K., Calambokidis, J., & Southall, B. L. (2016). Prey-mediated behavioral responses of feeding blue whales in controlled sound exposure experiments. Ecol Appl 26(4):1075–1085

Gearin PJ, Gosho ME, Laake JL, Cooke L, Delong RL, Hughes KM (2000) Experimental testing of acoustic alarms (pingers) to reduce bycatch of harbour porpoise, *Phocoena phocoena*, in the state of Washington. J Cetacean Res Manag 2:1–10. https://journal.iwc.int/index.php/jcrm/article/view/483

Giabbanelli P, Gray S, Aminpour P (2017) Combining fuzzy cognitive maps with agent-based modeling: frameworks and pitfalls of a powerful hybrid modeling approach to understand human-environment interactions. Environ Model Softw 95:320–325. https://doi.org/10.1016/j.envsoft.2017.06.040

Gillingham MP, Halseth GR, Johnson CJ, Parkes MW (2016) Exploring cumulative effects and impacts through examples. In: Gillingham M, Halseth G, Johnson C, Parkes M (eds) The integration imperative. Springer, Cham. https://doi.org/10.1007/978-3-319-22123-6_6

Goldbogen JA, Southall BL, DeRuiter SL, Calambokidis J, Friedlaender AS, Hazen EL, Falcone EA, Schorr GS, Douglas A, Moretti DJ (2013) Blue whales respond to simulated mid-frequency military sonar. Proc R Soc B 280(1765):20130657. https://doi.org/10.1098/rspb.2013.0657

Harcourt R, Pirotta V, Heller G, Peddemors V, Slip D (2014) A whale alarm fails to deter migrating humpback whales: an empirical test. Endanger Species Res 25:35–42. https://doi.org/10.3354/esr00614

Harris JD (1943) Habituatory response decrement in the intact organism. Psychol Bull 40:385–422. https://doi.org/10.1037/h0053918

Harwood J, King S, Schick R, Donovan C, Booth C (2014) A protocol for implementing the interim Population Consequences of Disturbance (PCoD) approach: quantifying and assessing the effects of UK offshore renewable energy developments on marine mammal populations. Scott Mar Freshw Sci 5(2):97

Hatch L, Clark C, Van Parijs S, Frankel A, Ponirakis D (2012) Quantifying loss of acoustic communication space for right whales in and around a U.S. National Marine Sanctuary. Conserv Biol 26(6):983–994. https://doi.org/10.1111/j.1523-1739.2012.01908.x

Hazen EL, Nowacek DP, Laurent LS, Halpin PN, Moretti DJ (2011) The Relationship among oceanography, prey fields, and beaked whale foraging habitat in the Tongue of the Ocean. PLoS One 6(4):e19269. https://doi.org/10.1371/journal.pone.0019269

Hodgson AJ, Marsh H, Delean S, Marcus L (2007) Is attempting to change marine mammal behaviour a generic solution to the bycatch problem? A dugong case study. Anim Conserv 10(2):263–273. https://doi.org/10.1111/j.1469-1795.2007.00104.x

Hohn AA, Rotstein DS, Harms CA, Southall BL (2006) Report on marine mammal unusual mortality event UMESE0501Sp: multispecies mass stranding of pilot whales (*Globicephala macrorhynchus*), minke whale (*Balaenoptera acutorostrata*), and dwarf sperm whales (*Kogia sima*) in North Carolina on 15–16 January 2005. NOAA Technical Memorandum NMFS-SEFSC-537. http://hdl.handle.net/1834/19915

Holst M, Greene CR, Richardson WJ, McDonald TL, Bay K, Schwartz SJ, Smith G (2011) Responses of pinnipeds to Navy missile launches at San Nicolas Island, California. Aquat Mamm 37(2):139. https://doi.org/10.1578/AM.37.2.2011.139

Holst M, Smultea MA, Koski WR, Sayegh AJ, Pavan G, Beland J, Goldstein HH (2017) Cetacean sightings and acoustic detections during a seismic survey off Nicaragua and Costa Rica, November-December 2004. Rev Biol Trop 65(2):599–611. https://doi.org/10.15517/rbt.v65i2.25477

Houser DS, Martin SW, Finneran JJ (2013) Exposure amplitude and repetition affect bottlenose dolphin behavioural responses to simulated mid-frequency sonar signals. J Exp Mar Biol Ecol 443:123–133. https://doi.org/10.1016/j.jembe.2013.02.043

Huber M, Knottnerus JA, Green L, Horst HVD, Jadad AR, Kromhout D, Leonard B, Lorig K, Loureiro MI, Meer JWMVD, Schnabel P, Smith R, Weel CV, Smid H (2011) How should we define health? BMJ 343: d4163. https://doi.org/10.1136/bmj.d4163

Jensen FH, Bejder L, Wahlberg M, Soto NA, Johnson M, Madsen PT (2009) Vessel noise effects on delphinid communication. Mar Ecol Prog Ser 395:161–175. https://doi.org/10.3354/Meps08204

Jepson PD, Arbelo M, Deaville R, Patterson IAP, Castro P, Baker JR, Degollada E, Ross HM, Herráez P, Pocknell AM, Rodríguez F, Howie FE, Espinosa A, Reid RJ, Jaber JR, Martin V, Cunningham AA, Fernández A (2003) Gas-bubble lesions in stranded cetaceans. Nature 425:575. https://doi.org/10.1038/425575a

Johnson BW (1977) The effects of human disturbance on a population of harbor seals. Outer Continental Shelf Environmental Assessment Program (OCSEAP), Boulder, pp 422–432

Johnson SR, Greene CR, Davis RA, Richardson WJ (1989) Synthesis of information on the effects of noise and disturbance on major haulout concentrations of Bering Sea pinnipeds (NTIS PB89-191373). LGL Alaska Research Associates Inc., Anchorage, 267 pp

Johnson M, Madsen PT, Zimmer WMX, de Soto NA, Tyack PL (2004) Beaked whales echolocate on prey. Proc R Soc B 271:S383–S386. https://doi.org/10.1098/rsbl.2004.0208

Johnson M, Madsen PT, Zimmer WMX, de Soto NA, Tyack PL (2006) Foraging Blainville's beaked whales (*Mesoplodon densirostris*) produce distinct click types matched to different phases of echolocation. J Exp Biol 209(24):5038–5050. https://doi.org/10.1242/Jeb.02596

Johnson M, Hickmott LS, Aguilar Soto N, Madsen PT (2008) Echolocation behaviour adapted to prey in foraging Blainville's beaked whale (*Mesoplodon densirostris*). Proc R Soc B 275:133–139. https://doi.org/10.1098/rspb.2007.1190

Kastelein RA, Verboom WC, Jennings N, de Haan D, van der Heul S (2008) The influence of 70 and 120 kHz tonal signals on the behaviour of harbour porpoises (*Phocoena phocoena*) in a floating pen. Mar Environ Res 66(3):319–326. https://doi.org/10.1016/j.marenvres.2008.05.005

Ketten DR (1995) Estimates of blast injury and acoustic trauma zones for marine mammals from underwater explosions. In: Kastelein RA, Thomas JA, Nachtigall PE (eds) Sensory systems of aquatic mammals. De Spil Publishing, Woerden, pp 391–407

Ketten DR (2004) Experimental measures of blast and acoustic trauma in marine mammals. Annual report to the Office of Naval Research Program, FY99-00

Ketten DR, Lien J, Todd S (1993) Blast injury in humpback whale ears: evidence and implications. J Acoust Soc Am 94(3):1849–1850. https://doi.org/10.1121/1.407688

King SL, Schick RS, Donovan C, Booth CG, Burgman M, Thomas L, Harwood J (2015) An interim framework for assessing the population consequences of disturbance. Methods Ecol Evol 6(10):1150–1158. https://doi.org/10.1111/2041-210X.12411

Knudsen SK, Oen EO (2003) Blast-induced neurotrauma in whales. Neurosci Res 46:377–386. https://doi.org/10.1016/s0168-0102(03)00101-9

Kopps AM, Sherwin WB (2012) Modelling the emergence and stability of a vertically transmitted cultural trait in bottlenose dolphins. Anim Behav 84(6):1347–1362. https://doi.org/10.1016/j.anbehav.2012.08.029

Kraus SD, Read AJ, Solow A, Baldwin K, Spradlin T, Anderson E, Williamson J (1997) Acoustic alarms reduce porpoise mortality. Nature 388(6642):525. https://doi.org/10.1038/41451

Laake J, Rugh D, Baraff LS (1998) Observations of harbor porpoises in the vicinity of acoustic alarms on a set of gill net. NOAA Technical Memorandum NMFS-AFSC-84

Lacy RC (2019) Lessons from 30 years of population viability analysis of wildlife populations. Zoo Biol 38(1):67–77. https://doi.org/10.1002/zoo.21468

Lacy RC, Williams R, Ashe E, Balcomb III KC, Brent LJN, Clark CW, Croft DP, Giles DA, MacDuffee M, Paquet PC (2017) Evaluating anthropogenic threats to endangered killer whales to inform effective recovery plans. Sci Rep 7(1):14119. https://doi.org/10.1038/s41598-017-14471-0

Lahdenperä M, Mar KU, Lummaa V (2016) Nearby grandmother enhances calf survival and reproduction in Asian elephants. Sci Rep 6(1):27213. https://doi.org/10.1038/srep27213

Lafferty, K.D. and Holt, R.D. (2003), How should environmental stress affect the population dynamics of disease? Ecol Lett 6:654–664. https://doi.org/10.1046/j.1461-0248.2003.00480.x

Larsen F, Eigaard OR (2014) Acoustic alarms reduce bycatch of harbour porpoises in Danish North Sea gillnet fisheries. Fish Res 153:108–112. https://doi.org/10.1016/j.fishres.2014.01.010

Larsen F, Krog C, Eigaard OR (2013) Determining optimal pinger spacing for harbour porpoise bycatch mitigation. Endanger Species Res 20(2):147–152. https://doi.org/10.3354/esr00494

Lewis JP (1987) An evaluation of a census-related disturbance of Steller sea lions. MSc thesis, University of Alaska Fairbanks, 93 pp

Lusseau D (2003) Male and female bottlenose dolphins *Tursiops* spp. have different strategies to avoid interactions with tour boats in Doubtful Sound, New Zealand. Mar Ecol Prog Ser 257:267–274. https://doi.org/10.3354/meps257267

Lusseau D, Bejder L (2007) The long-term consequences of short-term responses to disturbance experiences from whalewatching impact assessment. Int J Comp Psychol 20(2):228–236. https://doi.org/10.46867/IJCP.2007.20.02.04

Manlik O, Lacy RC, Sherwin WB, Finn H, Loneragan NR, Allen SJ (2022) A stochastic model for estimating sustainable limits to wildlife mortality in a changing world. Conserv Biol 36(4):e13897. https://doi.org/10.1111/cobi.13897

Marcotte D, Hung S, Caquard S (2015) Mapping cumulative impacts on Hong Kong's pink dolphin population. Ocean Coast Manag 109:51–63. https://doi.org/10.1016/j.ocecoaman.2015.02.002

Marques TA, Thomas L, Ward J, DiMarzio N, Tyack PL (2009) Estimating cetacean population density using fixed passive acoustic sensors: an example with Blainville's beaked whales. J Acoust Soc Am 125(4):1982–1994. https://doi.org/10.1121/1.3089590

Marques T, Munger L, Thomas L, Wiggins S, Hildebrand J (2011) Estimating North Pacific right whale *Eubalaena japonica* density using passive acoustic cue counting. Endanger Species Res 13:163–172. https://doi.org/10.3354/esr00325

Marques T, Thomas L, Martin S, Mellinger D, Jarvis S, Morrissey R, Ciminello C-A, Dimarzio N (2012) Spatially explicit capture-recapture methods to estimate minke whale density from data collected at bottom-mounted hydrophones. J Ornithol 152(Supp 2):445–455. https://doi.org/10.1007/s10336-010-0535-7

Martin SW, Martin CR, Matsuyama BM, Henderson EE (2015) Minke whales (*Balaenoptera acutorostrata*) respond to navy training. J Acoust Soc Am 137(5):2533–2541. https://doi.org/10.1121/1.4919319

Mate BR, Harvey JT (1987) Acoustical deterrents in marine mammal conflicts with fisheries. Oregon State University Sea Grant College Program, ORESU-W-86-001, Corvallis, 116 pp. https://repository.library.noaa.gov/view/noaa/39637/noaa_39637_DS1.pdf

Matkin C, Saulitis E, Ellis G, Olesiuk P, Rice S (2008) Ongoing population-level impacts on killer whales *Orcinus orca* following the Exxon Valdez oil spill in Prince William Sound, Alaska. Mar Ecol Prog Ser 356:269–281. https://doi.org/10.3354/meps07273

Matthews L, McCordic JA, Parks SE (2014) Remote acoustic monitoring of North Atlantic right whales (*Eubalaena glacialis*) reveals seasonal and diel variations in acoustic behavior. PLoS One 9(3):e93167. https://doi.org/10.1371/journal.pone.0091367

McCarthy E, Moretti D, Thomas L, DiMarzio N, Morrissey R, Jarvis S, Ward J, Izzi A, Dilley A (2011) Changes in spatial and temporal distribution and vocal behavior of Blainville's beaked whales (*Mesoplodon densirostris*) during multiship exercises with mid-frequency sonar. Mar Mamm Sci 27(3):E206–E226. https://doi.org/10.1111/j.1748-7692.2010.00457.x

McDonald MA, Hildebrand JA, Wiggins SM (2006) Increases in deep ocean ambient noise in the Northeast Pacific west of San Nicolas Island, California. J Acoust Soc Am 120(2):711–718. https://doi.org/10.1121/1.2216565

McDonald M, Hildebrand J, Mesnick S (2009) Worldwide decline in tonal frequencies of blue whale songs. Endanger Species Res 9:13–21. https://doi.org/10.3354/esr00217

McHuron EA, Costa DP, Schwarz L, Mangel M (2017) State-dependent behavioural theory for assessing the fitness consequences of anthropogenic disturbance on capital and income breeders. Methods Ecol Evol 8:552–560. https://doi.org/10.1111/2041-210X.12701

McHuron EA, Adamczak S, Arnould JP, Ashe E, Booth C, Bowen WD, Christiansen F, Chudzinska M, Costa DP, Fahlman A (2022) Key questions in marine mammal bioenergetics. Conserv Physiol 10(1):coac055. https://doi.org/10.1093/conphys/coac055

Merchant ND, Fristrup KM, Johnson MP, Tyack PL, Witt MJ, Blondel P, Parks SE (2015) Measuring acoustic habitats. Methods Ecol Evol. https://doi.org/10.1111/2041-210X.12330

Miksis-Olds JL, Wagner T (2011) Behavioral response of manatees to variations in environmental sound levels. Mar Mamm Sci 27(1):130–148. https://doi.org/10.1111/j.1748-7692.2010.00381.x

Miksis-Olds JL, Donaghay PL, Miller JH, Tyack PL, Nystuen JA (2007) Noise level correlates with manatee use of foraging habitats. J Acoust Soc Am 121(5):3011–3020. https://doi.org/10.1121/1.2713555

Mobley JR (2005) Assessing responses of humpback whales to North Pacific Acoustic Laboratory (NPAL) transmissions: results of 2001-2003 aerial surveys north of Kauai. J Acoust Soc Am 117(3):1666–1673. https://doi.org/10.1121/1.1854475

Moore SE, Reeves RR (2018) Tracking arctic marine mammal resilience in an era of rapid ecosystem alteration. PLoS Biol 16(10):e2006708. https://doi.org/10.1371/journal.pbio.2006708

Morrison C, Wardle C, Castley JG (2016) Repeatability and reproducibility of population viability analysis (PVA) and the implications for threatened species management. Front Ecol Evol 4:98. https://doi.org/10.3389/fevo.2016.00098

Mortensen LO, Chudzinska ME, Slabbekoorn H, Thomsen F (2021) Agent-based models to investigate sound impact on marine animals: bridging the gap between effects on individual behaviour and population level consequences. Oikos 130(7):1074–1086. https://doi.org/10.1111/oik.08078

Morton AB, Symonds HK (2002) Displacement of *Orcinus orca* (L.) by high amplitude sound in British Columbia, Canada. ICES J Mar Sci 59(1):71–80. https://doi.org/10.1006/jmsc.2001.1136

Munger LM, Wiggins SM, Moore SE, Hildebrand JA (2008) North Pacific right whale (*Eubalaena japonica*) seasonal and diel calling patterns from long-term acoustic recordings in the southeastern Bering Sea, 2000-2006. Mar Mamm Sci 24(4):795–814. https://doi.org/10.1111/j.1748-7692.2008.00219.x

Murray CC, Hannah L, Doniol-Valcroze T, Wright B, Stredulinsky E, Locke A, Lacy R (2019) Cumulative effects assessment for Northern and Southern resident killer whale populations in the Northeast Pacific. Canadian Science Advisory Secretariat Research Document 2019/056

Nabe-Nielsen J, Sibly RM, Tougaard J, Teilmann J, Sveegaard S (2014) Effects of noise and by-catch on a Danish harbour porpoise population. Ecol Model 272:242–251. https://doi.org/10.1016/j.ecolmodel.2013.09.025

National Academies of Sciences, Engineering, and Medicine (2017) Approaches to understanding the cumulative effects of stressors on marine mammals. The National Academies Press, Washington, DC. https://doi.org/10.17226/23479

National Research Council (2003) Ocean noise and marine mammals. National Academies Press, New York

National Research Council (2005) Marine mammal populations and ocean noise: determining when noise causes biologically significant effects. The National Academies Press, Washington, DC. https://doi.org/10.17226/11147

Nattrass S, Lusseau D (2016) Using resilience to predict the effects of disturbance. Sci Rep 6(1):25539. https://doi.org/10.1038/srep25539

New LF, Harwood J, Thomas L, Donovan C, Clark JS, Hastie G, Thompson PM, Cheney B, Scott-Hayward L, Lusseau D (2013a) Modelling the biological significance of behavioural change in coastal bottlenose dolphins in response to disturbance. Funct Ecol 27(2):314–322. https://doi.org/10.1111/1365-2435.12052

New LF, Moretti DJ, Hooker SK, Costa DP, Simmons SE (2013b) Using energetic models to investigate the survival and reproduction of beaked whales (family Ziphiidae). PLoS One 8(7):e68725. https://doi.org/10.1371/journal.pone.0068725

New LF, Clark JS, Costa DP, Fleishman E, Hindell MA, Klanjšček T, Lusseau D, Kraus S, McMahon CR, Robinson PW, Schick RS, Schwarz LK, Simmons SE, Thomas L, Tyack P, Harwood J (2014) Using short-term measures of behaviour to estimate long-term fitness of southern elephant seals. Mar Ecol Prog Ser 496:99–108. https://doi.org/10.3354/meps10547

Newton AC, Stewart GB, Diaz A, Golicher D, Pullin AS (2007) Bayesian Belief Networks as a tool for evidence-based conservation management. J Nat Conserv 15(2):144–160. https://doi.org/10.1016/j.jnc.2007.03.001

Oedekoven, C., Fleishman, E., Hamilton, P., Clark, J. S., & Schick, R. S. (2015). Expert elicitation of seasonal abundance of North Atlantic right whales Eubalaena glacialis in the mid-Atlantic. Endangered Species Research, 29(1), 51–58. https://doi.org/10.3354/esr00699

Parks SE, Clark CW, Tyack PL (2007) Short- and long-term changes in right whale calling behavior: the potential effects of noise on acoustic communication. J Acoust Soc Am 122(6):3725–3731. https://doi.org/10.1121/1.2799904

Parrott L (2011) Hybrid modelling of complex ecological systems for decision support: recent successes and future perspectives. Ecol Inform 6(1):44–49. https://doi.org/10.1016/j.ecoinf.2010.07.001

Pellegrini A, Romeu B, Ingram S, Daura-Jorge F (2021) Boat disturbance affects the acoustic behaviour of dolphins engaged in a rare foraging cooperation with fishers. Anim Conserv 24(4):613–625. https://doi.org/10.1111/acv.12667

Pirotta E, Laesser BE, Hardaker A, Riddoch N, Marcoux M, Lusseau D (2013) Dredging displaces bottlenose dolphins from an urbanised foraging patch. Mar Pollut Bull 74(1):396–402. https://doi.org/10.1016/j.marpolbul.2013.06.020

Pirotta E, Merchant ND, Thompson PM, Barton TR, Lusseau D (2015) Quantifying the effect of boat disturbance on bottlenose dolphin foraging activity. Biol Conserv 181:82–89. https://doi.org/10.1016/j.biocon.2014.11.003

Pirotta E, Booth CG, Costa DP, Fleishman E, Kraus SD, Lusseau D, Moretti D, New LF, Schick RS, Schwarz LK (2018a) Understanding the population consequences of disturbance. Ecol Evol 8(19):9934–9946. https://doi.org/10.1002/ece3.4458

Pirotta E, Mangel M, Costa DP, Mate B, Goldbogen JA, Palacios DM, Hückstädt LA, McHuron EA, Schwarz L, New L (2018b) A dynamic state model of migratory behavior and physiology to assess the consequences of environmental variation and anthropogenic disturbance on marine vertebrates. Am Nat 191(2):E40–E56. https://doi.org/10.1086/695135

Pirotta E, Mangel M, Costa DP, Goldbogen J, Harwood J, Hin V, Irvine LM, Mate BR, McHuron EA, Palacios DM, Schwarz LK, New L (2019) Anthropogenic disturbance in a changing environment: modelling lifetime reproductive success to predict the consequences of multiple stressors on a migratory population. Oikos 128(9):1340–1357. https://doi.org/10.1111/oik.06146

Pirotta E, Hin V, Mangel M, New L, Costa DP, de Roos AM, Harwood J (2020) Propensity for risk in reproductive strategy affects susceptibility to anthropogenic disturbance. Am Nat 196(4):E71–E87. https://doi.org/10.1086/710150

Pirotta E, Thomas L, Costa DP, Hall AJ, Harris CM, Harwood J, Kraus SD, Miller PJO, Moore MJ, Photopoulou T, Rolland RM, Schwacke L, Simmons SE, Southall BL, Tyack PL (2022) Understanding the combined effects of multiple stressors: a new perspective on a longstanding challenge. Sci Total Environ 821:153322. https://doi.org/10.1016/j.scitotenv.2022.153322

Pirotta E, Schick RS, Hamilton PK, Harris CM, Hewitt J, Knowlton AR, Kraus SD, Meyer-Gutbrod E, Moore MJ, Pettis HM (2023) Estimating the effects of stressors on the health, survival and reproduction of a critically endangered, long-lived species. Oikos 2023:e09801. https://doi.org/10.1111/oik.09801

Powell JR, Wells RS (2011) Recreational fishing depredation and associated behaviors involving common bottlenose dolphins (*Tursiops truncatus*) in Sarasota Bay, Florida. Mar Mamm Sci 27(1):111–129. https://doi.org/10.1111/j.1748-7692.2010.00401.x

Rankin CH, Abrams T, Barry RJ, Bhatnagar S, Clayton DF, Colombo J, Coppola G, Geyer MA, Glanzman DL, Marsland S, McSweeney FK, Wilson DA, Wu C-F, Thompson RF (2009) Habituation revisited: an updated and revised description of the behavioral characteristics of habituation. Neurobiol Learn Mem 92(2):135–138. https://doi.org/10.1016/j.nlm.2008.09.012

Rayment W, Dawson S, Slooten L (2009) Use of T-PODs for acoustic monitoring of *Cephalorhynchus* dolphins: a case study with Hector's dolphins in a marine protected area. Endanger Species Res 10:333–339. https://doi.org/10.3354/esr00189

Read AJ, Drinker P, Northridge S (2006) Bycatch of marine mammals in U.S. and global fisheries. Conserv Biol 20(1):163–169. https://www.jstor.org/stable/3591162

Richardson WJ, Würsig B (1997) Influences of man-made noise and other human actions on cetacean behaviour. Mar Freshw Behav Physiol 29(1-4):183–209. https://doi.org/10.1080/10236249709379006

Richardson WJ, Würsig B, Greene CR Jr (1990) Reactions of bowhead whales, *Balaena mysticetus*, to drilling and dredging noise in the Canadian Beaufort Sea. Mar Environ Res 29:135–160. https://doi.org/10.1016/0141-1136(90)90032-J

Richardson WJ, Greene CR, Malme CI, Thomson DH (1995) Marine mammals and noise. Academic, San Diego. https://doi.org/10.1016/C2009-0-02253-3

Rogers TL, Ciaglia MB, Klinck H, Southwell C (2013) Density can be misleading for low-density species: benefits of passive acoustic monitoring. PLoS One 8(1):e52542. https://doi.org/10.1371/journal.pone.0052542

Rolland RM, Parks SE, Hunt KE, Castellote M, Corkeron PJ, Nowacek DP, Wasser SK, Kraus SD (2012) Evidence that ship noise increases stress in right whales. Proc R Soc B 279(1737):2363–2368. https://doi.org/10.1098/rspb.2011.2429

Romano TA, Keogh MJ, Kelly C, Feng P, Berk L, Schlundt CE, Carder DA, Finneran JJ (2004) Anthropogenic sound and marine mammal health: measures of the nervous and immune systems before and after intense sound exposure. Can J Fish Aquat Sci 61(7):1124–1134. https://doi.org/10.1139/f04-055

Rounsevell MD, Arneth A, Brown C, Cheung WW, Gimenez O, Holman I, Leadley P, Luján C, Mahevas S, Maréchaux I (2021) Identifying uncertainties in scenarios and models of socio-ecological systems in support of decision-making. One Earth 4(7):967–985. https://doi.org/10.1016/j.oneear.2021.06.003

Roy N, Simard Y, Gervaise C (2010) 3D tracking of foraging belugas from their clicks: experiment from a coastal hydrophone array. Appl Acoust 71(11):1050–1056. https://doi.org/10.1016/j.apacoust.2010.05.008

Salter RE (1979) Site utilisation, activity budgets, and disturbance responses of Atlantic walruses during terrestrial haul-out. Can J Zool 57(6):1169–1180. https://doi.org/10.1139/z79-149

Scheifele PM, Andrew S, Cooper RA, Darre M, Musiek FE, Max L (2005) Indication of a Lombard vocal response in the St. Lawrence River beluga. J Acoust Soc Am 117(3):1486–1492. https://doi.org/10.1121/1.1835508

Schick RS, Kraus SD, Rolland RM, Knowlton AR, Hamilton PK, Pettis HM, Kenney RD, Clark JS (2013) Using hierarchical Bayes to understand movement, health, and survival in the endangered North Atlantic right whale. PLoS One 8(6):e64166. https://doi.org/10.1371/journal.pone.0064166

Senigaglia V, Christiansen F, Bejder L, Gendron D, Lundquist D, Noren D, Schaffar A, Smith J, Williams R, Martinez E (2016) Meta-analyses of whale-watching impact studies: comparisons of cetacean responses to disturbance. Mar Ecol Prog Ser 542:251–263. https://doi.org/10.3354/meps11497

Senigaglia V, Christiansen F, Sprogis K, Symons J, Bejder L (2019) Food-provisioning negatively affects calf survival and female reproductive success in bottlenose dolphins. Sci Rep 9(1):8981. https://doi.org/10.1038/s41598-019-45395-6

Senigaglia V, Christiansen F, Bejder L, Sprogis K, Cantor M (2022) Human food provisioning impacts the social environment, home range and fitness of a marine top predator. Anim Behav 187:291–304. https://doi.org/10.1016/j.anbehav.2022.02.005

Siebert U, Stürznickel J, Schaffeld T, Oheim R, Rolvien T, Prenger-Berninghoff E, Wohlsein P, Lakemeyer J, Rohner S, Aroha Schick L, Gross S, Nachtsheim D, Ewers C, Becher P, Amling M, Morell M (2022) Blast injury on harbour porpoises (*Phocoena phocoena*) from the Baltic Sea after explosions of deposits of World War II ammunition. Environ Int 159:107014. https://doi.org/10.1016/j.envint.2021.107014

Silas EG, Bastian Fernando A (1985) The dugong in India—is it going the way of the dodo. In: Silas EG (ed) Symposium on endangered marine animals and marine parks, Cochin, India. Marine Biological Association of India, pp 167–176

Simonis AE, Forney KA, Rankin S, Ryan J, Zhang Y, DeVogelaere A, Joseph J, Margolina T, Krumpel A, Baumann-Pickering S (2020) Seal bomb noise as a potential threat to Monterey Bay harbor porpoise. Front Mar Sci 7(142). https://doi.org/10.3389/fmars.2020.00142

Širović A, Hildebrand JA (2011) Using passive acoustics to model blue whale habitat off the Western Antarctic Peninsula. Deep Sea Res Part II Top Stud Oceanogr 58(13-16):1719–1728. https://doi.org/10.1016/j.dsr2.2010.08.019

Širović A, Hildebrand JA, Wiggins SM, Moore SE, McDonald MA, Thiele D (2004) Seasonality of blue and fin whale calls and the influence of sea ice in the Western Antarctic Peninsula. Deep-Sea Res II 51: 2327–2344. https://doi.org/10.1016/j.dsr2.2004.08.005

Sivle LD, Kvadsheim PH, Curé C, Isojunno S, Wensveen PJ, Lam F-PA, Visser F, Kleivanec L, Tyack PL, Harris CM (2015) Severity of expert-identified behavioural responses of humpback whale, minke whale, and Northern bottlenose whale to naval sonar. Aquat Mamm 41(4):469–502. https://doi.org/10.1578/AM.41.4.2015.469

Southall BL, Braun R, Gulland FMD, Heard AD, Baird RW, Wilkin SM, Rowles TK (2006) Hawaiian melon-headed whale (*Peponocephala electra*) mass stranding event of July 3–4, 2004. NOAA Technical Memorandum NMFS-OPR-31, 73 pp

Southall BL, Nowacek DP, Bowles AE, Senigaglia V, Bejder L, Tyack PL (2021) Marine mammal noise exposure criteria: assessing the severity of marine mammal behavioral responses to human noise. Aquat Mamm 47(5):421–464. https://doi.org/10.1578/AM.47.5.2021.421

Southwell C (2005) Response behaviour of seals and penguins to helicopter surveys over the pack ice off East Antarctica. Antarct Sci 17(3):328–334. https://doi.org/10.1017/S0954102005002798

Stansbury AL, Götz T, Deecke VB, Janik VM (2014) Grey seals use anthropogenic signals from acoustic tags to locate fish: evidence from a simulated foraging task. Proc R Soc B 282(1798):20141595. https://doi.org/10.1098/rspb.2014.1595

Stewart BS (1993) Behavioral and hearing responses of pinnipeds to rocket launch noise and sonic boom. J Acoust Soc Am 94(3):1828. https://doi.org/10.1121/1.407787

Stockbridge J, Jones AR, Gaylard SG, Nelson MJ, Gillanders BM (2021) Evaluation of a popular spatial cumulative impact assessment method for marine systems: a seagrass case study. Sci Total Environ 780:146401. https://doi.org/10.1016/j.scitotenv.2021.146401

Sveegaard S, Teilmann J, Berggren P, Mouritsen KN, Gillespie D, Tougaard J (2011) Acoustic surveys confirm the high-density areas of harbour porpoises found by satellite tracking. ICES J Mar Sci 68(5):929–936. https://doi.org/10.1093/icesjms/fsr025

Thompson RF, Spencer WA (1966) Habituation: a model phenomenon for the study of neuronal substrates of behavior. Psychol Rev 73(1):16–43. https://doi.org/10.1037/h0022681

Thompson PM, Brookes KL, Graham IM, Barton TR, Needham K, Bradbury G, Merchant ND (2013a) Short-term disturbance by a commercial two-dimensional seismic survey does not lead to long-term displacement of harbour porpoises. Proc R Soc B 280(1771):20132001. https://doi.org/10.1098/rspb.2013.2001

Thompson PM, Hastie GD, Nedwell J, Barham R, Brookes KL, Cordes LS, Bailey H, McLean N (2013b) Framework for assessing impacts of pile-driving noise from offshore wind farm construction on a harbour seal population. Environ Impact Assess Rev 43:73–85. https://doi.org/10.1016/j.eiar.2013.06.005

Thompson PM, Brookes KL, Cordes LS (2015) Integrating passive acoustic and visual data to model spatial patterns of occurrence in coastal dolphins. ICES J Mar Sci 77(2):651–660. https://doi.org/10.1093/icesjms/fsu110

Tiemann CO, Porter MB, Frazer LN (2004) Localization of marine mammals near Hawaii using an acoustic propagation model. J Acoust Soc Am 115(6):2834–2843. https://doi.org/10.1121/1.1643368

Tiemann CO, Martin SW, Mobley JR (2006) Aerial and acoustic marine mammal detection and localization on navy ranges. IJOE 31(1):107–119. https://doi.org/10.1109/Joe.2006.872203

Tixier P, Gasco N, Duhamel G, Guinet C (2015) Habituation to an acoustic harassment device (AHD) by killer whales depredating demersal longlines. ICES J Mar Sci 72(5):1673–1681. https://doi.org/10.1093/icesjms/fsu166

Tollit D, Harwood J, Booth C, Thomas L, New L, Wood J (2016) Cook Inlet beluga whale PCoD expert elicitation workshop report prepared for NOAA. Fisheries. https://doi.org/10.13140/RG.2.2.28912.53763

Trasky LL (1976) Environmental impact of seismic exploration and blasting in the aquatic environment. Alaka Department of Fish & Game, Anchorage

Tyack PL, Zimmer WM, Moretti D, Southall BL, Claridge DE, Durban JW, Clark CW, D'Amico A, DiMarzio N, Jarvis S (2011) Beaked whales respond to simulated and actual navy sonar. PLoS One 6(3):e17009. https://doi.org/10.1371/journal.pone.0017009

Tyack PL, Thomas L, Costa DP, Hall AJ, Harris CM, Harwood J, Kraus SD, Miller PJO, Moore M, Photopoulou T, Pirotta E, Rolland RM, Schwacke LH, Simmons SE, Southall BL (2022) Managing the effects of multiple stressors on wildlife populations in their ecosystems: developing a cumulative risk approach. Proc Royal Soc B 289 (1987):20222058. https://doi.org/10.1098/rspb.2022.2058

Villegas-Amtmann S, Schwarz LK, Sumich JL, Costa DP (2015) A bioenergetics model to evaluate demographic consequences of disturbance in marine mammals applied to gray whales. Ecosphere 6(10):art183. https://doi.org/10.1890/ES15-00146.1

Villegas-Amtmann, S., Schwarz, L. K., Gailey, G., Sychenko, O., & Costa, D. P. (2017). East or west: the energetic cost of being a gray whale and the consequence of losing energy to disturbance. Endang Spec Res, 34:167–183. https://doi.org/10.1098/rspb.2022.2058

von Benda-Beckmann AM, Aarts G, Sertlek HO, Lucke K, Verboom WC, Kastelein RA, Ketten DR, van Bemmelen R, Lam F-PA, Kirkwood RJ, Ainsle

MA (2015) Assessing the impact of underwater clearance of unexploded ordnance on harbour porpoises (*Phocoena phocoena*) in the southern North Sea. Aquat Mamm 41(4):503–523. https://doi.org/10.1578/AM.41.4.2015.503

Ward J, Morrissey R, Moretti D, DiMarzio N, Jarvis S, Johnson M, Tyack P, White C (2008) Passive acoustic detection and localization of *Mesoplodon densirostris* (Blainville's beaked whale) vocalizations using distributed bottom-mounted hydrophones in conjunction with a digital tag (DTag) recording. Can Acoust 36(1):60–66

Wild S, Allen SJ, Krützen M, King SL, Gerber L, Hoppitt WJE (2019) Multi-network-based diffusion analysis reveals vertical cultural transmission of sponge tool use within dolphin matrilines. Biol Lett 15(7):20190227. https://doi.org/10.1098/rsbl.2019.0227

Williams R, Lusseau D (2006) A killer whale social network is vulnerable to targeted removals. Biol Lett 2(4):497–500. https://doi.org/10.1098/rsbl.2006.0510

Williams R, Lusseau D, Hammond PS (2006) Estimating relative energetic costs of human disturbance to killer whales (*Orcinus orca*). Biol Conserv 133(3):301–311. https://doi.org/10.1016/j.biocon.2006.06.010

Williams R, Clark CW, Ponirakis D, Ashe E (2013) Acoustic quality of critical habitats for three threatened whale populations. Anim Conserv 17(2):174–185. https://doi.org/10.1111/acv.12076

Williams R, Thomas L, Ashe E, Clark CW, Hammond PS (2016) Gauging allowable harm limits to cumulative, sub-lethal effects of human activities on wildlife: a case-study approach using two whale populations. Mar Policy 70:58–64. https://doi.org/10.1016/j.marpol.2016.04.023

Willsteed EA, New L, Ansong JO, Hin V, Searle KR, Cook ASCP (2023) Advances in cumulative effects assessment and application in marine and coastal management. Camb Prisms Coast Futures 1:e18. https://doi.org/10.1017/cft.2023.6

Yack TM, Barlow J, Calambokidis J, Southall B, Coates S (2013) Passive acoustic monitoring using a towed hydrophone array results in identification of a previously unknown beaked whale habitat. J Acoust Soc Am 134(3):2589–2595. https://doi.org/10.1121/1.4816585

Yamada TK, Kitamura S, Abe S, Tajima Y, Matsuda A, Mead JG, Matsuishi TF (2019) Description of a new species of beaked whale (*Berardius*) found in the North Pacific. Sci Rep 9(1):12723. https://doi.org/10.1038/s41598-019-46703-w

Yelverton JT, Richmond DR, Fletcher ER, Jones RK (1973) Safe distances from underwater explosions for mammals and birds (NTIS Report AD-766 952). Lovelace Foundation for Medical Education and Research, Albuquerque

Yurk H, Filatova O, Matkin CO, Barrett-Lennard LG, Brittain M (2010) Sequential habitat use by two resident killer whale (*Orcinus orca*) clans in Resurrection Bay, Alaska, as determined by remote acoustic monitoring. Aquat Mamm 36(1):67–78. https://doi.org/10.1578/Am.36.1.2010.67

Open Access This chapter is licensed under the terms of the Creative Commons Attribution 4.0 International License (http://creativecommons.org/licenses/by/4.0/), which permits use, sharing, adaptation, distribution and reproduction in any medium or format, as long as you give appropriate credit to the original author(s) and the source, provide a link to the Creative Commons license and indicate if changes were made.

The images or other third party material in this chapter are included in the chapter's Creative Commons license, unless indicated otherwise in a credit line to the material. If material is not included in the chapter's Creative Commons license and your intended use is not permitted by statutory regulation or exceeds the permitted use, you will need to obtain permission directly from the copyright holder.

Management of Noise

12

Capri Jolliffe, Christine Erbe, Carina Juretzek,
Jill Lewandowski, Nathan D. Merchant, Brian Miller,
Valeria Senigaglia, and Sheila J. Thornton

Contents

12.1	**Why Manage Underwater Noise?**	732
12.2	**Legislative Requirements**	733
12.3	**Best-Practice Management**	734
12.4	**Recommended Approaches for the Management of Underwater Noise**	735

C. Jolliffe (✉) · C. Erbe
Centre for Marine Science and Technology, Curtin
University, Perth, WA, Australia
e-mail: c.erbe@curtin.edu.au

C. Juretzek
Federal Maritime and Hydrographic Agency, Hamburg,
Germany
e-mail: Carina.Juretzek@bsh.de

J. Lewandowski
Center for Marine Acoustics, Bureau of Ocean Energy
Management, Sterling, VA, USA
e-mail: jill.lewandowski@boem.gov

N. D. Merchant
Centre for Environment, Fisheries and Aquaculture
Science (Cefas), Lowestoft, Suffolk, UK
e-mail: nathan.merchant@cefas.gov.uk

B. Miller
Australian Antarctic Division, Kingston, TAS, Australia
e-mail: Brian.Miller@aad.gov.au

V. Senigaglia
Securing Antarctica's Environmental Future, School of
Mathematical Sciences, Queensland University of
Technology, Brisbane, QLD, Australia
e-mail: valeria.senigaglia@qut.edu.au

S. J. Thornton
Fisheries and Oceans Canada, West Vancouver, BC,
Canada
e-mail: Sheila.thornton@dfo-mpo.gc.ca

© The Author(s) 2025
C. Erbe et al. (eds.), *Marine Mammal Acoustics in a Noisy Ocean*,
https://doi.org/10.1007/978-3-031-77022-7_12

	12.4.1	Predictions of Impact 735
	12.4.2	Management 737
	12.4.3	Ongoing Monitoring and Adaptive Management 738
12.5		**Management Challenges** 738
12.6		**Management Highlights** 741
	12.6.1	USA 741
	12.6.2	Canada 741
	12.6.3	European Union 743
	12.6.4	Germany 744
	12.6.5	Antarctica 744
	12.6.6	Australia 745
12.7		**Research Gaps** 747
	12.7.1	Limited Long-Term Monitoring Studies 747
	12.7.2	Species Coverage and Sample Size 748
	12.7.3	Understudied Noise Sources 749
	12.7.4	Understudied Impacts 750
	12.7.5	Thresholds and Significance of Auditory Injury 750
12.8		**Future Directions: A Holistic Approach to Conservation Management** 751
12.9		**Summary** 751
		References 752

12.1 Why Manage Underwater Noise?

Anthropogenic noise is recognized as a potential threat for marine mammal species that rely heavily on sound to navigate, detect predators, find food, attract mates, and maintain social cohesion (Abadi et al. 2017; Clark 1990; Darling et al. 2006; Erbe et al. 2019b; Halliday et al. 2020; Herman et al. 2013; Janik 2014). Anthropogenic noise is increasing in most parts of the world's oceans, and the overlap of noise-generating activities and biologically important areas for marine species is likely to increase as spatial and temporal ranges shift due to population recovery and climate change and areas of interest to industry expand (Erbe et al. 2018, 2019b; Guazzo et al. 2020). Anthropogenic noise impacts on marine mammals can manifest in myriad of ways, for example, as a masker of acoustic communication, echolocation, and cues for predator or prey detection (Chap. 9; Erbe et al. 2016), behavioral disturbance (Chap. 10; Southall et al. 2021a, b), changes to physiological processes (Chap. 9; Erbe et al. 2018; Rolland et al. 2012; Tal et al. 2015), and, in extreme cases, physical and auditory injury (Chap. 9; Finneran 2016; Southall et al. 2019a). The human acoustic footprint in the oceans consists of both acute and chronic noise stressors that contribute to the cumulative stress acting on existing population pressures (Chou et al. 2021).

The body of literature on the effects of anthropogenic noise on marine mammals is expanding, and there is growing evidence that noise from a variety of sources may affect marine mammals negatively. While impact thresholds and types of effects vary depending on whether a noise is continuous or impulsive, or a combination of the two, the range of possible effects is broadly the same across species. Underwater noise is a concern for environmental regulators and species conservation managers globally (International Maritime Organization 2014; European Union 2017), particularly with respect to threatened and endangered species. In recognition of the potentially significant impacts of ocean noise on living marine resources, the United Nations General Assembly adopted the Ocean Resolution focusing on ocean noise following the nineteenth meeting of the Informal Consultative Process in 2018 (United Nations 2018) (Fig. 12.1).

Fig. 12.1 The management of underwater noise varies between environmental and legislative contexts. Throughout the environmental impact assessment process, careful consideration should be given to sensitive life stages for marine mammals and the environmental values that support them. For example, many marine mammal cows and calves utilize relatively quiet and sheltered areas for resting

Regarding environmental impact assessment (EIA), increased public concern and the expansion of industrial developments into sensitive marine habitats have led to an increased focus on the potential environmental impacts of anthropogenic noise. This, coupled with advancements in the complexity and reliability of modeling techniques, has resulted in more reliable predictions of underwater noise exposure. However, challenges in interpreting impacts of noise exposure on marine mammals remain. Over the last two decades, focused and opportunistic studies have highlighted the potential for noise-generating activities to impact marine mammals. This research, in addition to previous EIAs, has presented new challenges and solutions to manage the impacts to acceptable levels. However, the combination of advanced noise modeling techniques, revised effect thresholds, and extant knowledge gaps often results in predictions of impact ranges beyond the scope of what many management regimes were designed to address.

12.2 Legislative Requirements

The level of mitigation and effort expected in the prediction and management of underwater noise can vary depending on the legislative requirements within an individual jurisdiction, expectations of regulators, the industry or entity producing the underwater noise, corporate standards and expectations within individual entities, and the influence of local stakeholders. The degree and quality of scientific data available to different jurisdictions lends to differences in the informational basis of their EIAs.

Across all jurisdictions with legislated protections or guidelines, there generally exists an expectation that underwater noise should be managed to prevent serious harm to protected fauna. For conservational issues, the level of acceptable impact is often dependent on the level of protection that is afforded to specific species or species groups. In many jurisdictions, where there is the risk of significant impact from underwater noise, common control or mitigation measures must be implemented to reduce potential impacts. Such measures are commonly adopted within an industry or other proponents and often reflect expectations of regulators when it comes to mitigating the potential impacts of underwater noise. Common mitigation measures may vary between proponents (the noise makers) and regions; for example, some jurisdictions require soft starts, pre-start observations, and implementation of exclusion zones whereby acoustic noise sources need to be shut down should marine mammals enter the zone (Lewandowski and Staaterman 2020). Many of these mitigation measures rely on conventional visual observation techniques, such as the use of marine mammal observers to watch for marine mammals from vessels or platforms during daylight hours. Some jurisdictions also require time or area closures to protect key species and habitat. Where legislation requires a specific environmental outcome to be achieved, a comprehensive EIA may be required to demonstrate that the impacts of underwater noise are appropriately managed (Lewandowski and Staaterman 2020; Chou et al. 2021).

12.3 Best-Practice Management

There is no universally accepted definition of best-practice management for underwater noise, and practices vary by country, region, species, and scenario. Commonalities among purported best practices of underwater noise can be found predominantly at the highest level: a robust EIA process. EIAs can be formulated in several different ways but in general can be viewed as an iterative multistep approach. The first step is understanding of the state of the environment or status of receptor populations. The second step is to understand activity, particularly its effects and impacts on the environment. Effects and impacts should then be predicted with an appropriate level of fidelity, such as using modeling and assessment approaches that account for the sources involved, the complexity of sound transmission in the aquatic environment, and the likelihood and severity of noise exposure on sensitive receptors. The third step is application of a mitigation hierarchy to reduce impacts and risks to a level that is acceptable, noting acceptability is influenced by various contextual factors. If possible, priority should be given to the avoidance of any impacts. A precautionary approach should be applied to mitigation, where measures to prevent harm to species at risk are not postponed due to significant scientific uncertainty in the prediction of impacts/risks. Where impacts or risks are unavoidable, then reasonable steps should be taken to mitigate and minimize them. The residual impact/risk is defined as that which is likely to eventuate after the application of all mitigation and management measures. Where substantial residual impacts or risks remain after avoidance and minimization measures are applied, then regulators must decide whether the environmental outcome is consistent with the acceptable level of impact for that species and/or environment. In some jurisdictions, compensatory mitigation, or offsets, may be considered as a suitable approach to addressing significant residual risk/impacts, which are impacts that exist after mitigations have been applied. However, the efficacy of offsets for migratory marine species in terms of meaningfully offsetting impacts is poorly studied. If an activity is approved, then the fourth step involves implementation of the activity and management plan. Rigorous monitoring for efficacy of mitigation and management programs is essential to confirming that predictions of impact are accurate and desired environmental outcomes are being achieved. The iterative part of the plan involves feedback of knowledge obtained during the prior steps (e.g., Nowacek and Southall 2016). Science, technology, and our knowledge of the natural environment are not static, but rather they tend to improve over time. Since the mitigation and management of underwater noise is dependent on these, it too should improve in lockstep with improved science, technology, and knowledge.

12.4 Recommended Approaches for the Management of Underwater Noise

12.4.1 Predictions of Impact

At its core, EIA is an orderly and systematic process for evaluating the effects of a proposed action on the environment and mitigating those effects. Best-practice management demonstrates that a certain environmental outcome at which the level of impact is acceptable can be achieved and starts at the impact prediction stage. Internationally, the field of EIA focuses on a core set of principles that govern the characterization, prediction, and evaluation of impacts. However, detailed formulae for EIA, including expectations for impact prediction and management, and the treatment of residual impacts, tend to be specific to a jurisdiction (see Chou et al. 2021).

Earlier in this book (Chap. 1), acoustic modeling approaches and limitations were discussed. Robust noise modeling leads to more reliable predictions that can better inform the design and application of mitigation measures. The prediction of impact relies not only on the modeling inputs but also on the selection of relevant effect thresholds and an assessment of the likelihood of those effects occurring. Currently, noise effect thresholds are based on a small number of studies limited in sample size in terms of species and animals. As discussed in Chaps. 9 and 10, there is likely to be significant variability in the level of noise that leads to physiological and behavioral impacts. Since these depend on the context and the species in question, it may be appropriate to apply the most conservative published thresholds that are available when predicting impacts of underwater noise on marine fauna (e.g., vulnerable, threatened, or endangered species). This should include not only those from policy papers and guidelines but also those that can be derived from peer-reviewed studies.

In an ideal world, we would measure the baseline, measure the impacts during the activity, and have real-time mitigation and post-activity monitoring to confirm the duration and severity of impacts. However, the availability of baseline information on the spatial and temporal distribution and habitat use of populations varies by species and regions. While some populations are well studied, there remains a paucity of information for others. Depending on the nature and scale of potential impacts, and the jurisdictional requirements, baseline surveys to better understand species distribution and habitat usage may be warranted (Nowacek et al. 2016; Southall et al. 2019b). Direct measurement of status and trend of managed animals and anthropogenic impacts during an activity are the gold standard for environmental management, but are not always viable or practical, particularly if the unit being managed is a widely dispersed, cryptic, and/or poorly studied population.

If monitoring of status, trend, and impacts cannot be conducted during the activity, then prediction of the consequences of impact on animals may be required. For example, modeling approaches to predict population-level impacts of anthropogenic disturbance are discussed in Chap. 11, though the veracity of their impact predictions is inherently dependent on the reliability of the input data. In such individual-based or agent-based models, impact predictions for individuals are extrapolated to impacts on other individuals and populations (local or regional), which then determines the overall level of impact to the population or management unit.

Considerations for the evaluation of predictions of underwater noise impacts to marine mammals should include, but are not limited to, the following:

- The current state of knowledge of the managed marine fauna and the level of confidence underpinning the prediction of residual impacts;
- The spatial and temporal distribution of marine mammal species, or populations, that may be affected;
- The aspects of the activity/project that may impact on, or pose a risk to, marine mammals, the vulnerability of local populations, and the

extent/severity of these potential impacts and risks;
- The regulatory requirements within the relevant jurisdiction;
- Whether legislative protections exist that set specific outcomes or levels of acceptable impact for a particular species or species group;
- The application of the mitigation hierarchy, to avoid, minimize, and, in some circumstances, offset impacts to marine fauna wherever possible;
- The management measures and approaches proposed and whether they are technically and practically feasible;
- The spatial and temporal scale of the residual impacts to marine fauna and flow-on implications for biological diversity and ecological integrity;
- The spiritual and cultural values of local communities regarding marine fauna, including specific habitats or places that may support culturally important marine fauna species and the potential impacts and risks to these values; and
- The risk posed to marine fauna should these predictions be incorrect.

The accuracy and precision of prediction of impact from underwater noise relies on appropriate sound source characterization and propagation models that consider the physical properties of the environment. Appropriately accurate predictions also rely on accurate estimation of the noise exposures likely to be experienced by relevant receptors. Typically, the prediction of impacts relies on a set of noise effect thresholds. While several European countries have specific criteria for these effects, outside of these, the thresholds that are commonly applied by action proponents, regulators, research groups, and nongovernmental organizations for temporary threshold shift (TTS) and permanent threshold shift (PTS) are those from Southall et al. (2019a). These thresholds are based on the median received level at which effects were measured in a limited number of real-world studies of a handful of species and individuals. They may not be appropriately conservative (Chap. 9), or may be overly conservative, depending on the species. Inclusion of uncertainty in the thresholds or the models can further affect interpretation of predictions (e.g., see Gedamke et al. 2011). Similarly, the thresholds commonly used to predict behavioral response from both impulsive and continuous noise may be based on the median level at which animals respond and might not consider context-dependent variability in responses or the severity of responses (Chap.10). The use of dose-response functions (e.g., Miller et al. 2014; Dunlop et al. 2018) may represent a better approach for the prediction of noise impacts than static thresholds and allow greater flexibility in the application of thresholds. Additionally, it may be appropriate for different thresholds to be applied in different environmental contexts (e.g., Southall et al. 2019c; Pirotta et al. 2021), depending on the sensitivity of the species or behaviors being undertaken in the affected areas. While such approaches may be preferable, they are often rendered nonviable due to the myriad of knowledge gaps of dose-response relationships for each context or more fundamentally challenges in measuring or predicting the context of exposures (which in turn are due to the expense and challenges involved in obtaining the necessary information).

It is important that both direct and indirect impacts be evaluated when considering potential noise impacts on marine mammals. Direct impacts are those that may arise in direct response from the underwater noise exposure, such as injury or behavioral disturbance. Indirect impacts are second-order; for example, underwater noise may directly induce a flight response or interrupt dive cycles, which could indirectly result in stranding and mortality, as observed in some beaked whale stranding events (Bernaldo de Quirós et al. 2019; Cox et al. 2006; Zimmer and Tyack 2007). Indirect impacts may also include changes in distribution or behavior of an animal because of direct impacts to their prey source. The prediction of both direct and indirect impacts should be robust and scientifically defensible, within the limits of available science. For example, it is not appropriate to assume that marine mammals will not be impacted by underwater noise because they will not remain in the area for long periods or will avoid sources of noise, unless there is evidence to support such conclusions.

12.4.2 Management

Once the presence and distribution of species and/or important habitat(s) are understood, and both direct and indirect impacts have been predicted, the mitigation hierarchy should be applied to demonstrate that impacts have been reduced to a level that is consistent with desired environmental outcomes. All attempts should be made to avoid or minimize risk or reduce the likelihood an impact will eventuate. Where avoidance of risk is not possible, mitigation and management measures informed by the impact (or predictions) should be implemented to demonstrate how environmental impacts and risks can be managed. Mitigation measures for common noise sources are discussed in Chap. 2.

Avoidance
The first step of the mitigation hierarchy is to apply those measures that would avoid impacts from the outset. Avoidance options, which may include spatial avoidance of certain habitats or temporal avoidance of seasons, are often the simplest, most effective, and cheapest way to prevent negative environmental impacts. However, to implement avoidance requires consideration at the earliest stages of planning, as it can be extremely difficult to retrospectively implement.

Minimization
Minimization measures are those that involve changing some aspect of the project or activity to reduce potential impacts. Examples include using a quieter vessel, using smaller seismic airguns or alternative sources, vibro-piling instead of impact piling, and using bubble curtains and noise attenuation systems to reduce noise levels from piling. Helpful reviews of these noise abatement options include the quieting workshops sponsored by the Global Alliance for Managing Ocean Noise (Lee et al. 2023) and the International Maritime Organization's Guidelines for Reducing Underwater Noise (International Maritime Organization 2023; Merchant 2019; and Merchant 2019). Substitutions are not always a feasible option and rarely eliminate all potential impacts; however, they can lessen the likely severity or extent of predicted impacts.

In Situ/Responsive Management Measures
Responsive management measures are measures that are implemented during the activity to reduce the predicted residual impact to a level that is consistent with the desired environmental outcome. These measures include soft starts, exclusion zones that involve monitoring an area for marine mammal presence and shutting down the acoustic source if a marine mammal moves into the zone, and adaptive management measures, such as changing the design or conduct of the activity or project in response to marine mammal observations. These types of management measures can be applied at project scales (Joint Nature Conservation Committee 2017; Nowacek et al. 2013) and regional scales (European Commission 2022).

These recommended practices for management aim to ensure that impacts and risks are reduced to the lowest possible level and provide confidence that the desired environmental outcome can be achieved. Where mitigation and management cannot reduce the impact or risk so that the desired outcome can be achieved, it may be necessary to reiterate through the steps of the EIA. This could involve reassessing whether avoidance is an option or whether some knowledge gaps could be addressed via pre-activity studies to further reduce risk. Where residual impacts are not consistent with a desired environmental outcome, then dependent on legislative requirements, offsets and/or remediation may need to be adopted. It is important to note that for many impacts and in many contexts, there may be no suitable offset or rehabilitation option. Thus, the only options may be avoidance, partial or conditioned approvals, or mitigation and management. This is particularly the case for cetaceans where there are few contexts within which offsets would be appropriate to remediate or offset impacts.

12.4.3 Ongoing Monitoring and Adaptive Management

Ideally management approaches for underwater noise would include fit-for-purpose monitoring and adaptive management programs that facilitate continual evaluation of the effectiveness of mitigation and management measures against the environmental outcome. For the purpose of this chapter, monitoring refers to the ongoing observation or surveillance for marine mammals during and after an activity as opposed to baseline surveys which are undertaken prior to an activity. Adaptive management is a process by which the mitigation and responsive management measures applied to the activity are altered to ensure desired environmental outcomes continue to be met. Even in instances where temporal and/or spatial exclusion zones have been implemented, there may be a need for adaptive processes to ensure that assumptions regarding species presence remain accurate for the project or activity duration. For example, should a higher than predicted number of animals be detected through ongoing monitoring, the spatial or temporal exclusion zone may need to be shifted or expanded to account for presence outside of the predicted periods or areas. Adaptive management measures may also include changes to monitoring and mitigation methodologies and the conduct of the activity itself. Designing appropriate monitoring and management is contingent on a thorough understanding of the presence and habitat use of marine fauna within the activity area, robust impact predictions and monitoring that are sensitive enough to inform ongoing adaptive management processes, mitigation measures that can be measured for effectiveness, and a continual demonstration that environmental outcomes are being met.

If monitoring indicates that environmental outcomes may be achieved, mitigation and management will need to be adapted to ensure that the required environmental outcome is continually met. In most instances, monitoring of the species or populations may not be feasible. In these instances, it may be appropriate to monitor aspects of habitat health or the pressure itself (e.g., monitoring relevant acoustic metrics at specific locations, such as in management zones) to provide confidence that mitigation measures are appropriate (e.g., Racca et al. 2015) (Fig. 12.2).

12.5 Management Challenges

The management of underwater noise is inherently challenging. Typical management approaches have relied on pragmatic decisions and actions that can be clearly communicated to proponents during planning and understood by all parties responsible for managing and conducting the activities. The trade-off of such pragmatism is that it can result in a "one-size-fits-all" approach that might not be as effective in achieving the required environmental outcomes or level of species protection as more complex bespoke adaptive management approaches. Underwater acoustics is a technical field, and in many jurisdictions, government officials tasked with regulating noise feel inadequately equipped to do so. Managers are often tasked with handling a suite of environmental topics making it more challenging to develop the specialized knowledge needed to effectively manage anthropogenic underwater sound. Furthermore, there may be disparities in resources between managers and proponents (e.g., the ability to procure advice from subject-matter experts) that exacerbate these issues. Baseline data on noise from the activity, extant noise sources, and protected species can also be limited, requiring additional expertise to understand the situation and to undertake surveys to close information gaps. These additional studies may need to be broadscale and long term, which may be costly. This, coupled with knowledge gaps about the biology and temporal and spatial dynamics of many animal

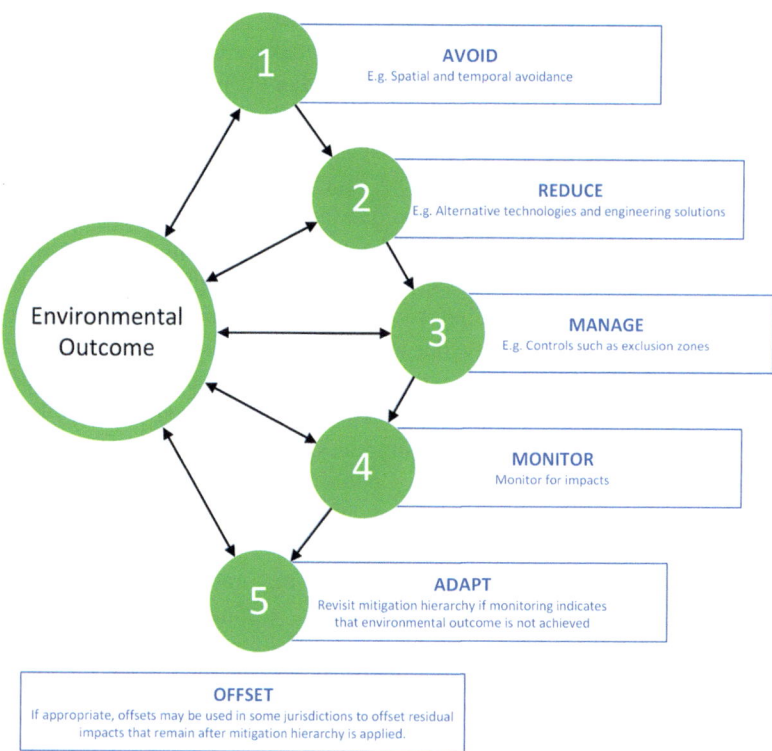

Fig. 12.2 Best-practice management approaches should be focused on environmental outcomes. Avoidance of impact should always be the primary goal; however, if that is not possible, all options to reduce and manage impacts should be explored. Management should be accompanied by monitoring programs to test whether environmental outcomes are being achieved. If environmental outcomes are not predicted to be achieved, then all stages of the mitigation hierarchy should be revisited. Offsets may be suitable in some jurisdictions to offset the residual impact after the application of the mitigation hierarchy, but should not sit within the hierarchy itself

populations, makes monitoring and management at a population level inherently challenging.

Uncertainties arising from modeling and prediction of impacts can also lead to challenges in the design and application of management measures for underwater noise. Accurate acoustic modeling depends on the availability of in situ seabed and water column parameters and the application of complex modeling approaches (Farcas et al. 2016). The range to predicted effects may exceed what can be managed using conventional visual-based observations, which are limited to good conditions and daylight hours. The effectiveness of some in situ/responsive mitigation measures relies on the detectability of marine mammals under various environmental conditions, with no solution that performs well under all conditions and for all species (e.g., Smith et al. 2020). As a result, some jurisdictions are recommending real-time passive and active acoustic monitoring, infrared imaging to improve monitoring, and tools, such as satellite imagery and geospatial modeling, to improve predictions. Like all approaches, these methods come with their own costs and limitations and are contingent on funding and regulatory expectations or requirements.

Passive acoustic monitoring can increase the detectability of certain marine mammal species, particularly those that are more easily heard than seen. In the proximity of anthropogenic noise sources, detecting marine mammal vocalizations that overlap with the noise can be challenging. Studies have shown that passive acoustic detection in these circumstances is more reliable for echolocating odontocetes (Verfuss et al. 2018). Flow noise (and vessel noise) generally impedes short towed arrays of only a few hydrophones when used to monitor for low-frequency baleen whales, though longer towed arrays have been successfully deployed in some circumstances to detect and locate some species (Abadi et al. 2017). Passive acoustic monitoring is not a reliable management tool for species that cease vocalizing in the presence of anthropogenic noise (Chap. 9; Verfuss et al. 2018). However, its effectiveness in EIA increases when used alongside other monitoring methods (Smith et al. 2020) and when used to monitor animal distribution and occupancy over large spatial and temporal scales (Van Parijs et al. 2009). For example, passive acoustic monitoring from fixed sensors can enhance baseline knowledge of the distribution and biodiversity of vocal taxa in a monitored area. It can also increase data availability on baseline environmental and anthropogenic noise (in the absence of the proposed activity) and provide ongoing monitoring of noise levels and biodiversity during and after an anthropogenic acoustic activity. Mobile platforms, such as drifters, gliders, and uncrewed vessels, represent emerging platforms from which passive acoustics could be conducted for real-time and archival monitoring of noise and vocal species (e.g., Johnson et al. 2022; Kowarski et al. 2020; Baumgartner et al. 2013; Klinck et al. 2012).

Marine underwater noise can be managed at the local, regional, state, or country level. In the European Union (EU), for example, the Marine Strategy Framework Directive (MSFD) encompasses a multitude of anthropogenic "stressors" and their potentially cumulative impacts (European Union 2017). The MSFD includes standardized procedures providing a framework to manage underwater noise, but each member state is responsible for managing noise pollution within its territorial waters and those resulting from its national activities and industries (Markus and Sánchez 2018). Consequently, the management of acoustic disturbances and potential impacts, and the expected level of mitigation, varies between countries (Erbe 2013). The difficulties of managing noise disturbance are exacerbated by the fact that noise travels under water and can cross national boundaries. In recognition of this, a growing number of international organizations, NGOs, and private initiatives directly address acoustic pollution and the need of managing it within an international context. A complete list of regulations and conventions that address noise management is beyond the scope of this chapter, and we refer to Chou et al. (2021) for a review of the subject. However, it is worth noticing the more prescriptive ones, which include the UN Convention on the Law of the Sea (UNCLOS), ASCOBANS,[1] ACCOBANS,[2] the Convention for Migratory Species of Wild Animals, the Convention on Biodiversity, and the Antarctic Treaty. The Antarctic Treaty offers a unique approach to joint environmental management of a territory with no exclusive ownership (Erbe et al. 2019a). Of particular note is the High Seas Treaty that was adopted by the United Nations in 2023 to promote and protect biodiversity in the ocean. The treaty, which is applicable to all oceans outside of other jurisdictional boundaries, sets out clear requirements for the conduct of environmental impact assessments before running activities in the high seas.

Most regulations represent "soft" laws in the form of guidelines and recommendations that are rarely enacted in national legislation. Moreover, integrated research and collaboration across governments, communities, and stakeholders is yet to be adequately achieved globally. Implementing international decisions and resolutions can be made difficult by a lack of commitment and enforcement by participating nations. A chain of custody for noise sources operating in international waters can be hard to establish and even more difficult to monitor. For these reasons, more recent approaches based on voluntary reduction of noise disturbance arose from the collective actions of the industries themselves. For instance, industrial certification programs provide guidance for reducing and mitigating adverse noise impacts. The Green Marine environmental program in North America and the Port of Vancouver EcoAction program are examples of such programs (Chou et al. 2021). Bolstering the prospects for these voluntary programs is the fact that noise from most sources (except sonar and seismic surveys) is ancillary to the activity and represents energetic waste. Under these circumstances, there could be

---

[1] The Agreement on the Conservation of Small Cetaceans of the Baltic, Northeast Atlantic, Irish, and North Seas (ASCOBANS)

[2] The Agreement on the Conservation of Cetaceans of the Black Sea, Mediterranean Sea and Contiguous Atlantic Area (ACCOBAMS)

energy savings and other benefits of switching to quieter technology. For example, using a piling technique that requires less energy may not only reduce noise but may result in a reduction in risk to health and safety of those conducting the activity.

12.6 Management Highlights

12.6.1 USA

In the USA, there is no overarching legislation or mandate addressing ocean noise. Instead, US federal government agencies use other existing laws to manage impacts of noise on species and habitat. The *Marine Mammal Protection Act 1994* (MMPA) provides the most stringent protection for marine mammals and requires that any authorized activity has no more than a negligible impact to a marine mammal *stock* (defined as below the species population level in many cases). Actions should result in "taking" only a small number of individuals from each affected stock (taking being defined as harassment or attempted harassment that has the potential to disrupt behavioral patterns or cause injury). Meeting the MMPA standards must happen before receiving any similar authority under the *Endangered Species Act 1973* (ESA).

To help guide US federal agencies in managing noise impacts on marine mammals, the National Oceanic and Atmospheric Administration (NOAA) issued noise exposure thresholds for the onset of TTS and PTS (see Chap. 9) in marine mammals exposed to anthropogenic underwater noise. NOAA also provides less formalized thresholds for the onset of behavioral harassment but may seek to adjust and formalize behavioral impact thresholds in the future. Acoustic modeling is commonly used to predict the distances at which underwater sound exposures would exceed harassment thresholds, thus enabling federal regulators to predict impacts to marine mammals, select mitigation measures, and develop protocols to monitor mitigation effectiveness. Although regulatory emphasis has historically focused on marine mammals, US federal agencies also seek to manage noise impacts on sea turtles, fishes, and invertebrates, including efforts to study hearing and develop noise impact thresholds for these groups. Many of these species are prey for marine mammals, and noise impacts on these species may indirectly impact marine mammals by affecting prey availability or quality.

US federal agencies recognize the need to look at impacts broadly and are funding efforts to measure soundscapes and conduct and archive data from long-term ecoregional monitoring programs (e.g., the Atlantic Deepwater Ecosystem Observatory Network (ADEON), NOAA's Noise Reference Stations, the US Navy's Marine Species Monitoring program, and the Bureau of Ocean Energy Management's (BOEM) regional passive acoustic monitoring networks). Agencies also support innovative risk assessment strategies, such as the population consequences of disturbance and other relativistic risk assessment frameworks (see Southall et al. 2021a, b, 2023). Finally, and most importantly, federal agencies are integrating quieting initiatives as a critical tool for reducing the levels of noise emitted into the marine environment—a key first step in transforming the management of ocean noise.

US federal agencies collaborate under multiple fora, domestically under the US National Ocean Policy's Ocean Science and Technology Subcommittee and NOAA's National Ocean Noise Strategy and through BOEM's Center for Marine Acoustics. Internationally, federal agencies coordinate through activities associated with the International Maritime Organization, United Nations Ocean Decades, and the Global Alliance for Managing Ocean Noise. These collaborations infuse learning from approaches used across different jurisdictions and continually adapt and innovate management improvements, especially at the broader temporal and spatial scales needed to effectively manage ocean noise.

12.6.2 Canada

Legislation, guidelines, and initiatives all contribute to noise management for the protection of cetaceans in Canada. Most legal protections predate our current understanding of acoustic impacts and are difficult to apply to anthropogenic noise, as they are largely discretionary (i.e., prohibitions against disturbance, harm, and harassment). The federal *Fisheries Act 1985*, where the definition of "fish" includes marine mammals, is the principal statute providing protection for fish and fish habitat. In addition to the general prohibitions against harm under the Fisheries Act, specific prohibitions against killing, fishing, or disturbance of marine mammals may be found in the *Marine Mammal Regulations 1983*. Most of Canada's cetacean species are also listed under the *Species at Risk Act 2002*, which prohibits harassment to any species listed as threatened or endangered under the Act. As with prohibitions under the Fisheries Act, enforcement is primarily corrective, rather than preventative. As underwater noise is not specifically identified in either Act, the burden of proof is left to the Crown to demonstrate whether disturbance or harm has occurred.

In recognition of the growing body of literature on vessel impacts to cetaceans, the Marine Mammal Regulations were amended in 2018 to increase restrictions on vessel approach under the prohibitions on disturbance. Vessels were provided with a schedule of approach distances that varied by location and species. These amendments moved the protections from a corrective measure with the need to demonstrate harm to a preventative action to proactively reduce the incidence of disturbance. However, the prohibition is specific to the action of approach and does not apply to vessels in transit.

Consideration of marine mammal protections is required during the assessment of coastal and oceanic development projects, as defined under the *Impact Assessment Act 2019* (formerly the *Canadian Environmental Assessment Act 2012*). In the past, scoping of major development projects (e.g., port expansion proposals, oil and gas development) did not assess the associated increase in vessel traffic, as the proponents did not have jurisdiction over vessel transits. In recent years, assessments have considered the estimated increase in vessel traffic associated with project developments. In practice, the assignment of responsibility for noise mitigation to the action proponent has been complicated by difficulties in partitioning anthropogenic noise among the different noise producers. Existing noise inputs are effectively "grandfathered" in, and new projects carry the responsibility for noise mitigation.

In 2016, the Government of Canada announced the Oceans Protection Plan (OPP), a substantial investment for the protection of Canada's coasts and waterways. In recognizing the need for, and the challenges of, mitigating anthropogenic noise, the OPP committed to better understand and address the cumulative effects of shipping on marine mammals. The OPP focused on three endangered populations of whales: the southern resident killer whale, St. Lawrence Estuary beluga, and the North Atlantic right whale (NARW). Under the OPP, the Marine Environmental Quality program increased the national research capacity, with commitments to measure current ambient underwater noise levels and to increase knowledge of noise impacts on cetaceans in support of best practices for noise mitigation. A national "three-layer" approach was developed, with a specific emphasis on (a) noise mapping of each population's habitat (e.g., Vagle et al. 2021), (b) defining each population's habitat use and function (Thornton et al. 2022a, b; Simard et al. 2023; Durette-Morin et al. 2022), and (c) analyzing the co-occurrence of ocean noise and spatial/temporal whale presence to identify key areas for mitigation (Thornton et al. 2022a, b). The research investment led to advice on priority areas for acoustic mitigation (e.g., DFO 2021), as well as a significant advancement in our understanding of noise impacts on cetaceans. The ongoing science initiatives support conservation actions by Transport Canada, such as speed-restricted zones and interim sanctuaries for killer whale populations on Canada's Pacific coast and slowdown zones and a vessel exclusion zone to protect the NARW. While measures on the Atlantic coast are focused on reducing vessel strikes, they also provide noise mitigation benefits to the endangered NARW (Marotte et al. 2022).

Various stakeholder groups play a role in noise mitigation in Canadian waters. One notable example is the Enhancing Cetacean Habitat and

Observation (ECHO) program led by the Port of Vancouver, which supports initiatives to mitigate underwater noise impacts on southern resident killer whales. ECHO program actions include voluntary slowdowns in areas of high killer whale occurrence (Burnham et al. 2021) and an effort to move vessel traffic away from whales to reduce underwater noise levels in foraging habitat (Vagle 2020). The program also supports reduced docking fees for ships that apply vessel quieting best practices and technology.

12.6.3 European Union

At the time of this writing, the EU and the UK remain unique in implementing binding legislation that explicitly addresses underwater noise pollution. Under the EU Marine Strategy Framework Directive (MSFD; European Commission 2008) and the UK Marine Strategy (UKMS), thresholds must be defined that will determine whether a given marine area has Good Environmental Status (GES) with respect to underwater noise pollution.

This top-down approach to noise management has several advantages over sector-, source-, or species-specific management paradigms. Most salient is the requirement to manage overall levels of noise pollution, since one of the primary challenges for noise management is the variety of sectors that emit noise and the fact that these sectors tend to be regulated by different departments of government. Seismic airgun surveys and marine renewable energy developments may be regulated by the energy ministry, while shipping is typically managed by the transport department, often with the involvement of coastguard agencies. The military and/or defense department usually regulates military noise sources, such as active sonar and explosives. Meanwhile, environmental targets tend to be the responsibility of the environment ministry. Without overarching legislation that compels these departments to work together to comply with an overall "noise budget" for the waters they share (Merchant et al. 2018), department-specific interests (e.g., energy security, international trade, and national security) tend to override environmental commitments. A strategic approach is therefore needed to avoid the pitfalls of this piecemeal regulatory landscape.

Implementation of the MSFD began in 2010, stimulating unprecedented levels of research, monitoring, and assessment of underwater noise pollution in European waters (Merchant et al. 2022). International cooperation has been facilitated by Regional Seas Conventions, which coordinate marine environmental policy in the Baltic Sea (HELCOM Convention), Black Sea (Bucharest Convention), Mediterranean Sea (Barcelona Convention), and Northeast Atlantic (OSPAR Convention). Joint monitoring programs for impulsive and continuous noise have now been established for the Baltic Sea and Northeast Atlantic and are under development for the Northeast Atlantic and Mediterranean Sea as part of Regional Action Plans (RAPs). These programs enable assessments to be undertaken and shared targets to be set at ecologically relevant spatial scales (e.g., in relation to the entire habitat of North Sea harbor porpoise, rather than at the level of each nation's exclusive economic zone).

The EU defined threshold values for underwater noise pollution in 2022 under the MSFD (Borsani et al. 2023; Sigray et al. 2023). These require that impulsive noise pollution does not affect more than 20% of a relevant habitat on a single day, and not more than 10% in any year, while not more than 20% of the habitat can be affected by continuous noise in any month. The areas affected are defined according to whether a specified sound level (i.e., the level of onset of biological adverse effect—LOBE) is exceeded. Much work remains to fully assess compliance with these thresholds and thereby determine whether remedial management action is required. Time will tell whether the approach established by the MSFD can bring about meaningful management of cumulative noise levels and serve as a model for other jurisdictions.

12.6.4 Germany

In Germany, the legislative framework for managing the risk of adverse environmental impact due to underwater noise by human activities is based on national law and the implementation of European directives and is complemented by administrative acts and guidelines. The federal *Nature Conservation Act 2009* (BNatSchG) is of particular importance for underwater noise management considering species conservation and the protection of important habitats for marine mammal populations. The harbor porpoise represents the key marine mammal species of concern in the German waters of the North and Baltic Seas and is strictly protected by the EU Habitats Directive (92/43/EEC) and the BNatSchG. Killing or injuring of individuals and significant disturbance of populations are not permitted, whereby TTS in harbor porpoises is considered injury.

The *Offshore Wind Energy Act 2017* (WindSeeG) and the *Offshore Installation Act 2016* (SeeAnlG) provide the legal basis for the planning approval of offshore wind and other energy installations under consideration of the *Environmental Impact Assessment Act 2022* (UVPG) and the BNatSchG. The use of low-noise installation technology is fostered by the WindSeeG. Further, the German Concept for the protection of harbor porpoises from noise emissions during the construction of offshore wind farms in the German North Sea ("Schallschutzkonzept"; BMU 2013) provides national guideline for the assessment of underwater noise exposure on marine mammals.

The Concept for the protection of harbor porpoises sets the compliance target with the prohibition of injury and killing, as required by the EU Habitats Directive and BNatSchG, to a dual noise emission threshold for the prevention of TTS due to impulsive pile driving noise (160 dB re 1 $\mu Pa^2 s$ sound exposure level and 190 dB re 1 μPa peak-to-peak sound pressure level at an effect distance of 750 m). Primary and/or secondary noise abatement measures (best offshore available technology) must be applied to prevent the threshold from being exceeded; deterrence devices must be used to displace harbor porpoises beyond the effect distance. The noise abatement technology applied to comply with the protection requirements is not specified by the approval authority, and the technical development of new techniques by the industry is welcome. The formulation of a clear target for compliance and a comprehensive monitoring program to verify compliance and effectiveness of the technical solutions were the driving force and guideline for innovation. Today, a range of commercially available noise abatement systems are successfully used in numerous windfarm projects (Bellmann et al. 2020).

The Concept for the protection of harbor porpoises defines maximum allowed thresholds of area affected by piling noise for the prevention of significant disturbance and cumulative effects. It is applicable to the entire German exclusive economic zone (EEZ) and to specific nature conservation sites. No more than 10% of the German EEZ and adjacent nature conservation sites may fall within the disturbance effect radii of simultaneous piling activities. Seasonally, no more than 1% of habitats with high species concentration or that are recognized as significant for supporting reproduction (i.e., have reproduction listed as a conservation target for that area) may fall within the disturbance effect radii. These area thresholds aim to maintain a sufficient area that is unaffected by noise to prevent significant disturbance to populations utilizing them.

12.6.5 Antarctica

The uniqueness of the Antarctic environment and its significance as habitat to many marine mammal taxa make it a unique concern for environmental managers. While many of the marine mammals that utilize the area are seasonal migrants, some populations are endemic and inhabit Antarctic waters year-round, including ice seals, Antarctic killer whales, and Antarctic minke whales.

The Antarctic represents an acoustically unique physical environment due to the confluence of a shallow sound channel of low propagation loss near the sea surface, a lack of landmasses that would be barriers to sound propagation, and seasonal ice cover that strongly influences the soundscape. These factors contribute to underwater sound potentially propagating over great distances.

The Antarctic remains relatively untouched by anthropogenic disturbance due to limited amounts of human activity that are largely constrained to the summer when sea ice is at its lowest. Human activities in the Antarctic include tourism, krill and toothfish fisheries, support activities for the Antarctic research stations, and science (Erbe et al. 2019a). The Antarctic is physically distant from most shipping lanes and other industrial noise sources and consequently represents a unique habitat that, by global standards, has low levels of anthropogenic noise.

The regulatory environment of the Antarctic is also unique, with regulation under the Antarctic Treaty System (ATS). Under the Protocol on Environmental Protection to the Antarctic Treaty (Antarctic Treaty Environmental Protocol or ATEP), procedures and regulations were established for all activities that might impact the environment of the territory covered by the treaty, as well as dependent and associated ecosystems. These procedures must be carried out or complied with before any activity in the Antarctic is granted. For instance, the preparation of environmental impact assessments according to ATEP Annex I follows one of three different paths depending on the expected level of impact determined during a mandatory preliminary assessment (Fig. 12.3). If the activity is deemed to have less than a minor and transitory impact, then it can proceed at the discretion of the nation under which the activity was proposed. If the preliminary assessment indicates the activity will cause a minor and transitory impact, then an initial environmental evaluation (IEE) is required, which is managed by the national jurisdiction and presented to the ATS Committee for Environmental Protection (CEP). If an activity is deemed to have more than a minor and transitory impact, then a comprehensive environmental evaluation (CEE) is necessary. The resulting CEE is conducted by the proponents under their national jurisdiction but *must* be reviewed by the parties of the CEP and broader Antarctic Treaty Consultative Meeting under which the CEP operates. Additionally, preliminary assessments, IEE, and CEE are all informed, in a nonbinding manner, by the Scientific Committee for Antarctic Research (SCAR). Like the CEP, the Commission for the Conservation of Antarctic Marine Living Resources (CCAMLR) is responsible for management of fisheries and the maintenance of ecological relationships over defined timescales. As top predators of species harvested by Antarctic fisheries, many marine mammal populations are nominally afforded protection under CCAMLR's ecosystem-based management program.

Underwater noise and its impacts are not explicitly mentioned under the CEP nor CCAMLR conservation measures. Each nation under the CEP retains their own discretion about impacts when conducting preliminary assessments. This can potentially lead to an inconsistent process where a noise-generating activity is deemed by one nation to have no more than a minor/transitory impact and a similar activity is deemed by another nation to require a full IEE and CEE. Finally, the same factors that contribute to low anthropogenic noise levels in the Antarctic, namely, low human presence and extreme remoteness, present challenges to enforcement and compliance of CEP and CCAMLR. Generally, it is up to each nation to demonstrate compliance with and enforcement of any conditions that are required during noise-generating activities in the Antarctic. For examples of how individual countries (e.g., Germany) operate under the Antarctic Treaty Environmental Protocol, see Erbe et al. (2019a).

12.6.6 Australia

Environmental protection is legislated under the federal *Environmental Protection and Biodiversity Conservation Act 2001* (EPBC Act) in Australia. Management of underwater noise under the EPBC Act is complemented by Policy Statements and statutory instruments including

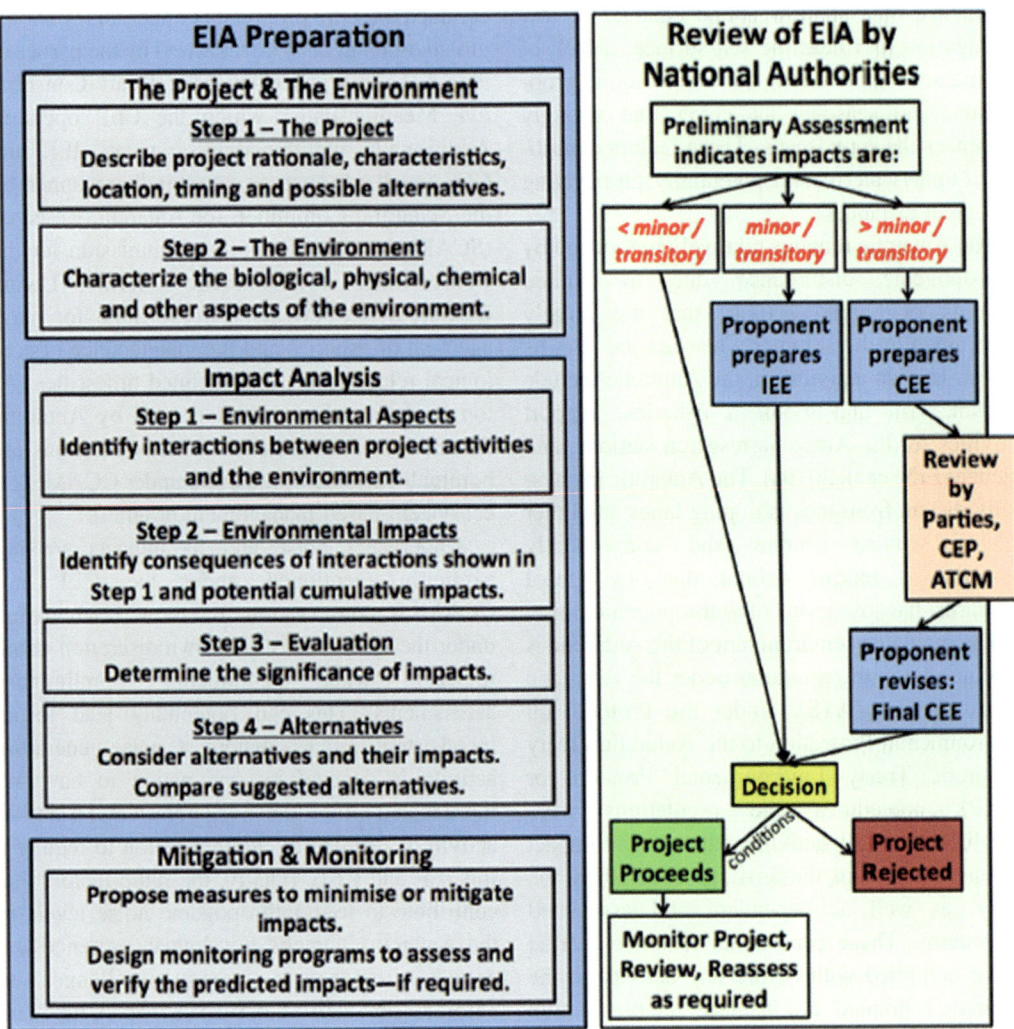

Fig. 12.3 A schematic of the EIA process for the Antarctic region. The action proponent prepares the EIA, which includes sections on the project, the environment, potential impacts of the project, and the impact mitigation, monitoring, and management that will be put in place to manage these impacts. The EIA is then reviewed by national authorities. If impacts are expected to be greater than minor or transitory, the Parties of the Antarctic Treaty, the CEP, and the Antarctic Treaty Consultative Meeting (ATCM) get to review the proposal as well. National authorities make the final decision whether a project goes ahead (potentially with imposed conditions) or not. © Erbe et al. 2019a; https://doi.org/10.3389/fmars.2019.00647. Published CC BY 4.0; https://creativecommons.org/licenses/by/4.0/

species Recovery Plans. EPBC Act Policy Statement 2.1 sets out mandatory management measures that must be applied to all seismic surveys and other sources of intense impulsive sound in Australian waters. These requirements also apply to all Australian-owned companies, citizens, and vessels in international waters. The Policy Statement includes requirements such as soft starts, the implementation of observation and shutdown zones, and low visibility and nighttime procedures whereby operations cannot continue if there has been a large number of shutdowns.

Without an appropriate permit, it is an offence to kill or injure a cetacean under the EPBC Act, and additional requirements are placed on the management of anthropogenic noise for certain species through Recovery Plans. Guidance for interpreting the Recovery Plans for blue whales and southern right whales asserts that TTS will be considered as injury and that displacement or disturbance of animals in critical foraging and resting habitats is prohibited (DAWE 2021).

These requirements present challenges for offshore industry, particularly with the overlap of important industry sites and critical foraging and resting areas for blue and southern right whales (see, e.g., Conservation Management Plan for the Blue Whale, DoE 2015; Conservation Management Plan for the Southern Right Whale, DoE 2012). While seasonal avoidance of key habitats for seismic activities has long been an expectation in Australia, industry has recently responded to regulatory and societal expectations by increasing the scope of mitigation measures for a variety of projects, particularly for threatened species with specific population management objectives (see Blue Whale Conservation Management Plan—FAQs | NOPSEMA[3]). These include the use of aerial surveys for whale presence to inform drill rig movements and power-downs, multiple vessel platforms for increased visual observation ranges in critical habitats during periods of possible whale presence (e.g., see example marine seismic survey environment plan[4]), increasingly sensitive adaptive mitigation triggers such that the sighting of one animal of a particular species can end an activity, and the maintenance of ambient noise levels at designated boundaries to critical calving habitat (see example industry environment plans at NOPSEMA[5]).

12.7 Research Gaps

Data gaps introduce uncertainty into the prediction and management of impacts. There are several critical areas of uncertainty that are fundamental to the prediction and management of impacts:

- Species presence, abundance, and trends;
- Species behavior and habitat use;
- Noise source levels, frequency content, and directionality;
- Species- and context-specific sensitivity to disturbance;
- Noise effect thresholds; and
- The potential for noise effects at the individual level to translate to population-level impacts.

While the level of uncertainty in species presence, abundance, behavior, and habitat use varies by region and species, the challenge of addressing data gaps relating to species sensitivity, noise effect thresholds, and the potential for underwater noise to have impacts at the population level are relevant at a global scale. Despite being the focus of considerable research, understanding of the effects of anthropogenic noise on marine mammal individuals and populations remains limited. It is complicated by long-term changes in environmental conditions and the challenges of observing marine mammals at sea, particularly over long periods of time.

12.7.1 Limited Long-Term Monitoring Studies

Few marine mammal populations have been studied for sufficient periods of time, with adequate robustness in characterizing behavior or in linking behavior to energetic or health consequences, to enable an accurate estimate of how a disturbance might impact an affected population. Long-term studies of baseline behavior and habitat use are needed in systems in which disturbances can be monitored and characterized. Unfortunately, collecting long-term datasets at the individual and population level is logistically

---

[3] https://www.nopsema.gov.au/blue-whale-conservation-management-plan-faqs; accessed 10 April 2024

[4] https://docs.nopsema.gov.au/A700128; accessed 10 April 2024

[5] https://info.nopsema.gov.au/home/approved_projects_and_activities; accessed 10 April 2024

and economically challenging. A few groups of marine mammals (e.g., northern elephant seal, bottlenose dolphin, and harbor porpoise) have been the subject of intensive multiyear studies (Benke et al. 2014; Gillespie et al. 2005; Harwood and Wilson 2001; New et al. 2014; Teilmann and Carstensen 2012; Wells 2014). These groups have been utilized as case studies in models that attempt to establish links between short-term responses and long-term consequences. However, it's important to recognize that the detailed behavioral data required to predict impacts of acoustic disturbance is a means to just one tool in the toolbox, and the overall goal is management of animals (or their populations). In many, if not most cases, regular surveys to estimate population abundance, distribution, and habitat use (and with repeated surveys, estimate trends over time) are likely a solid foundation for managing the effects of human activity (including noise) on most populations, though results will be with respect to the cumulative impacts that they experience, rather than particular activities.

A commitment to long-term studies of accessible marine mammal populations should be promoted both within the scientific community and within regulatory and funding agencies to facilitate the establishment of long-term, collaborative research programs. Increasing the number of species/populations for which long-term monitoring exists will ultimately provide more datasets for use as case studies. This, in turn, will better enable researchers to identify the contextual covariates most linked to biologically significant impacts across a broader range of species.

12.7.2 Species Coverage and Sample Size

Of the 137 extant marine mammal species currently recognized by the Society for Marine Mammalogy (current as of June 2023), roughly half have been studied with respect to behavioral responses to underwater noise (Chap. 10). The number and rigor of the studies varies considerably. Some species have been studied for logistical reasons, accessibility, and vulnerability or because of trends in research funding. Bottlenose dolphins are one example; the species has received significant attention due to its widespread and primarily coastal distribution, as well as their prevalence in captivity. Conversely, the various species of river dolphins (families Iniidae, Platanistidae, and Pontoporiidae) are considerably understudied despite all families containing species classified as either vulnerable or endangered. This is likely due to a combination of logistical and socioeconomic constraints, as well as the complexity of the riverine systems they inhabit.

Understanding fundamental aspects of the distribution, population dynamics, movement patterns, foraging ecology, energetics, and behavior of marine mammal species is a particular challenge in the developing world. Human and industrial development are urgent in these regions, and environmental or conservation researchers face issues related to funding, resource availability, and logistically challenging field sites (Braulik et al. 2018). Furthermore, language and funding barriers potentially hamper the dissemination of research results globally, resulting in a general lack of awareness regarding research conducted in some developing parts of the world.

Noise impact research is unevenly distributed across noise sources, partly in response to immediate industry, military, or conservation needs, but also due to feasibility (Chap. 10). For example, the responses of *Cephalorhynchus* spp. to acoustic deterrent devices (ADDs) have been investigated as a means of reducing dolphin bycatch and lowering costs to the fishing industry (e.g., through damaged nets). However, no other noise sources appear to have been explicitly studied with regard to these species, despite their primarily coastal distribution significantly overlapping with a range of anthropogenic activities. By contrast, the similarly coastal *Tursiops* spp. are featured in literature covering a variety of noise sources, including vessel traffic, construction activities, ADDs, sonar, airguns, and aircraft. Consequently, many marine mammal species remain unstudied or inadequately studied, while those that are commonly studied are often easily accessible and of limited concern with respect to species or population vitality. As human populations increase and ocean and river exploitation expands, it is imperative that the number of marine mammal species studied increases and that more targeted studies occur in regions where anthropogenic noise and other human factors are potentially impacting vulnerable and endangered species.

12.7.3 Understudied Noise Sources

Because of their military and commercial importance, certain noise sources (e.g., sonars, airguns, and ADDs) have received a significant amount of research investment, while others have not. Persistent noise sources have remained largely unstudied, particularly those that expose large areas of ocean. Chief among these are cargo/shipping vessels, which are well-known contributors to low-frequency ocean noise. With over 118,000 merchant ships from over 150 countries currently engaging in trade (see Chap. 2), many parts of the world's oceans (especially those associated with point-to-point trade routes) are persistently subjected to low-frequency shipping noise. There is a significant difference in characterizing the immediate response of a marine mammal to a passing vessel and characterizing difficult-to-determine sublethal effects that may accumulate over long time periods. Understanding the latter might require long-term monitoring, potentially over spatial scales that are logistically challenging. As with expanding the number of species that are studied, it will take concerted efforts to engage in investigations of chronic noise exposure issues.

Construction of offshore windfarms generates noise, with acute noise sources such as dynamically positioned construction vessels and piling (both hammer and vibro-piling) that operate over 3–5 years during construction of a single windfarm. While the noisiest aspect of windfarms is the construction, the operation of large-scale windfarms is associated with chronic underwater noise, albeit at a significantly lower intensity than observed during construction (see Chap. 2). Nonetheless, there is the potential for windfarms to alter underwater soundscapes and introduce infrastructure to the marine environment that may have an impact on marine mammals (see Chap. 10). As with shipping noise, understanding the impact of continuous windfarm noise would benefit from long-term studies of marine mammal habitat use in and around windfarms.

New noise sources continue to be developed (e.g., subsea oil and gas processing plants, floating liquefied natural gas facilities, newly developed sonar, and subsea mines). As new noise sources come into use, including different hammer sizes and designs for piling, attention should be given to the characteristics of the noises they emit. At a minimum, characterizing the acoustic profiles of these devices may allow the effect and impact of the noise sources to be predicted. A real difficulty with emerging technology and infrastructure is that no in situ recordings exist at the time of EIA for permit application, i.e., it has not yet been built and deployed. EIAs are therefore based on models; power upscaling from existing, older, and often significantly smaller equipment; or recordings of small prototypes or machinery parts—and perhaps not even under water, but in air (see, e.g., Williams et al. 2022 for subsea mining).

Innovations in technology mean that new approaches for the construction, operation, and decommissioning of offshore infrastructure are becoming more common. Some of these technologies reduce noise emissions at the source, or the distance noise propagates from the source, while others are designed to reduce or eliminate potentially harmful components of the noise (e.g., reduce rise time to avoid auditory system impacts). While it is important to support innovations for quieter technologies and approaches, it is also important to recognize that quieter does not mean silent. For example, down-the-hole piling techniques are often considered to be quieter, yet they still might produce noise at levels of concern for marine fauna (Guan and Reyff 2020). Similarly, marine vibroseis, a potential alternative to seismic airguns, still has the potential to cause impacts such as behavioral disturbance or masking, but to what extent requires further evaluation (Duncan et al. 2017; Matthews et al. 2021). Continuing advancements in quieting technologies hold promise, especially where technologies or methods can substantially reduce the noise introduced into the environment. However, characterizing underwater noise emissions associated with new and emerging technologies should occur prior to implementation so that the potential for environmental impact is understood before being placed in the environment.

12.7.4 Understudied Impacts

Considerable advancements have been made in our ability to model and predict population-level effects based on the population consequences of disturbance and similar models (see Chap. 11). However, these models are inherently limited by the availability of data supporting model parameterization, particularly linking functions that relate disturbance costs at the individual level to population-level effects. Data on life history parameters remain a key information gap for many species, particularly those that are migratory and/or undertake critical behaviors in areas that are difficult to access. Similarly, while considerable progress has been made, our understanding of the behavioral responses of marine mammals to a broad suite of underwater noise sources remains incomplete. Passive acoustic recorders cannot study silent animals; aerial surveys provide only a brief snapshot of an animal in time; vessel- and land-based surveys can only conduct focal follows on individuals or groups for short periods, while animal-borne tags have limited battery life and attachment times (though there has been significant improvement in attachment and tag technologies in recent years). Most behavioral response studies only monitor short-term behavioral responses, making it difficult to link responses to long-term population-level consequences of disturbance.

Predictive disturbance models (e.g., agent-based models that predict the exposure of simulated animals to sound fields based on likely movement patterns) should regularly be updated as new data become available. The predictive exposure of an animal to underwater noise and thus the potential for auditory injuries and behavioral disturbance are inherently influenced by their speed of travel, dive depth, and movement patterns. Investment in the development of these models should continue, but they should also be supported by long-term and comprehensive studies on select model species, to reduce uncertainty and raise confidence in these models as an EIA tool.

Indirect effects remain an area of research that has largely been unaddressed. For instance, many marine fishes and invertebrates are known to respond to underwater noise (Hawkins and Popper 2018; Slabbekoorn et al. 2010; Wale et al. 2013), potentially contributing to effects further up the food chain. One study involving seismic arrays suggested that they have the potential to cause significant mortality to local plankton populations (McCauley et al. 2017). The effect on marine mammals that use these plankton populations as food has received little research attention, though noise-related impacts to prey species could hypothetically reduce prey availability or quality.

12.7.5 Thresholds and Significance of Auditory Injury

Noise-induced impacts to the auditory system of marine mammals are largely rooted in experimentally determined thresholds for the onset of TTS (Southall et al. 2019a, b, c). As there are relatively few frequencies at which TTS has been measured for most species, impact thresholds used in environmental models and documents necessarily extrapolate from species and frequencies where TTS has been measured to species and frequencies where it has not been measured. Frequencies of sound where odontocetes are most susceptible to TTS do not necessarily match the frequencies where hearing is most sensitive. As a result, predictions of TTS onset at frequencies that have not been tested are likely in error, at least to some degree. Since TTS-onset data anchor the development of marine mammal auditory weighting functions (Chap. 8), future research should characterize TTS onset across the bandwidth of hearing to ensure weighting functions accurately reflect this frequency dependence. The potential for TTS to impact individual fitness is also poorly understood and requires a critical evaluation of how the magnitude and duration of a TTS translates into fitness costs (see Chap. 9 for a review of TTS in marine mammals.)

12.8 Future Directions: A Holistic Approach to Conservation Management

Marine mammals are not the sole animals impacted by noise disturbance, and the importance of an ecosystem-based management (EBM) approach to conservation cannot be overstated. Key components of EBM include multiple stressor identification, cumulative impact assessment, and the socioeconomic interests placed on the environment along the relevant spatial scale (i.e., local vs. regional and national vs. international). By clearly incorporating a human dimension to wildlife and environmental management, one can ensure early involvement of stakeholders, which, in turn, should promote a higher degree of compliance. Coupling of ecological assessment of anthropogenic impact and societal evaluation of human activities can be viewed as a requirement for sustainable management. However, the task of capturing the wide diversity of relationships and values that society places on the environment, including marine mammals, is particularly challenging. One approach that has gained increasing attention and support involves the concept of cultural ecosystem services (CES)—those benefits provided by nature to humans. There is little agreement in the scientific community over the definition of CES, but a generally accepted definition includes the immaterial benefits provided by wildlife to human well-being, such as recreational value, esthetic value, and spiritual enrichment (Hernández-Morcillo et al. 2013; Milcu et al. 2013). Marine mammals represent key contributors to CES, but as noted, they also face conservation threats from multiple stressors including fisheries interactions, vessel traffic, maritime recreational activities, and others. Delineating CES and other ecosystem services provided by marine mammals can aid the development of integrated socio-ecological models and holistic conservation plans.

12.9 Summary

Anthropogenic noise is recognized as a potential threat to marine mammal species with possible impacts including auditory masking, behavioral disturbance, stress, and hearing loss. While modeling approaches for the prediction of realistic noise exposure scenarios are becoming increasingly advanced, and modeling approaches to predict population consequences of disturbance exist, a paucity of baseline information on abundance, density, and physiological ecology exists for most species. There also remain significant gaps in our understanding of anthropogenic noise impacts to marine mammals, particularly regarding species- and context-specific responses to anthropogenic noise exposure. Further, many noise sources, particularly those that are new and emerging, remain uncharacterized. Marine mammals are complex to study, and sensitivity to underwater noise varies between species, individuals, and behavioral and environmental contexts, making it difficult to accurately predict at what level of exposure an animal may be impacted. Addressing these research gaps is important to the management of anthropogenic underwater noise, particularly from the perspective of potential population-level impacts.

The management of anthropogenic noise varies between jurisdictions, with many regulatory regimes relying on the application of commonly applied mitigation and management measures. While standardized mitigation and management approaches can reduce the potential for anthropogenic noise to impact marine mammals, there are many situations where these approaches may be insufficient to achieve the desired environmental outcome. It is important to recognize that the desired environmental outcome, and from a regulatory perspective the level of acceptable environmental impact, varies by jurisdiction. Irrespective of legislative context, a good-practice approach to the mitigation and management of environmental impacts resulting from underwater noise is one in which there is a robust application of the mitigation hierarchy to achieve the desired environmental outcome. Such an approach will often require pre-activity baseline studies, ongoing monitoring, and provisions for adaptive management to ensure that the desired environmental outcomes continue to be met over the duration of a project.

There are numerous examples from around the world where environmental regulators have pushed for higher levels of protection for specific species in particular habitats or contexts. Industry has often risen to the challenge on these occasions, providing examples of what is possible when there is joint effort among action proponents, scientists, and regulators. A management approach that encourages proponents to achieve desired outcomes has the potential to drive innovation and develop improvements for what is an "industry standard approach."

References

Abadi SH, Tolstoy M, Wilcock WSD (2017) Estimating the location of baleen whale calls using dual streamers to support mitigation procedures in seismic reflection surveys. PLoS One 12(2):e0171115. https://doi.org/10.1371/journal.pone.0171115

Baumgartner MF, Fratantoni DM, Hurst TP, Brown MW, Cole TVN, Van Parijs SM, Johnson M (2013) Real-time reporting of baleen whale passive acoustic detections from ocean gliders. J Acoust Soc Am 134(3):1814–1823. https://doi.org/10.1121/1.4816406

Bellmann MA, Brinkmann J, May A, Wendt T, Gerlach S, Remmers P (2020) Underwater noise during the impulse pile-driving procedure: influencing factors on pile-driving noise and technical possibilities to comply with noise mitigation values. Available online https://www.itap.de/media/experience_report_underwater_era-report.pdf

Benke H, Bräger S, Dähne M, Gallus A, Hansen S, Honnef CG, Jabbusch M, Koblitz JC, Krügel K, Liebschner A, Narberhaus I, Verfuß UK (2014) Baltic Sea harbour porpoise populations: status and conservation needs derived from recent survey results. Mar Ecol Prog Ser 495:275–290. https://doi.org/10.3354/meps10538

Bernaldo de Quirós Y, Fernández A, Baird RW, Brownell RL, Aguilar de Soto N, Allen D, Arbelo M, Arregui M, Costidis A, Fahlman A, Frantzis A, Gulland FMD, Iñíguez M, Johnson M, Komnenou A, Koopman H, Pabst DA, Roe WD, Sierra E, Tejedor M, Schorr G (2019) Advances in research on the impacts of anti-submarine sonar on beaked whales. Proc Biol Sci 286(1895):20182533. https://doi.org/10.1098/rspb.2018.2533

BMU (2013) Konzept für den Schutz der Schweinswale vor Schallbelastungen bei der Errichtung von Offshore-Windparks in der deutschen Nordsee (Schallschutzkonzept), Bundesministerium für Umwelt, Naturschutz und Reaktorsicherheit. Available online https://www.bfn.de/fileadmin/BfN/awz/Dokumente/schallschutzkonzept_BMU.pdf

Borsani JF, Andersson M, André M, Azzellino A, Bou M, Castellote M, Ceyrac L, Dellong D, Folegot T, Hedgeland D, Juretzek C, Klauson A, Leaper R, Le Courtois F, Liebschner A, Maglio A, Mueller A, Norro A, Novellino A, Outinen O, Popit A, Prospathopoulos A, Sigray P, Thomsen F, Tougaard J, Vukadin P, Weilgart L (2023) Setting EU threshold values for continuous underwater sound, Technical Group on Underwater Noise (TG NOISE), MSFD Common Implementation Strategy (Druon J, Hanke G, Casier M eds). Publications Office of the European Union, Luxembourg, https://doi.org/10.2760/690123.JRC133476

Braulik GT, Kasuga M, Wittich A, Kiszka JJ, MacCaulay J, Gillespie D, Gordon J, Said SS, Hammond PS (2018) Cetacean rapid assessment: an approach to fill knowledge gaps and target conservation across large data deficient areas. Aquat Cons Mar Fresh Ecol 28(1):216–230. https://doi.org/10.1002/aqc.2833

Burnham RE, Vagle S, O'Neill C, Trounce K (2021) The efficacy of management measures to reduce vessel noise in critical habitat of Southern Resident killer whales in the Salish Sea. Front Mar Sci 8:664691. https://doi.org/10.3389/fmars.2021.664691

Chou E, Southall BL, Robards M, Rosenbaum HC (2021) International policy, recommendations, actions and mitigation efforts of anthropogenic underwater noise. Ocean Coast Manage 202:105427. https://doi.org/10.1016/j.ocecoaman.2020.105427

Clark CW (1990) Acoustic behavior of mysticete whales. In: Thomas JA, Kastelein RA (eds) Sensory abilities of cetaceans, NATO ASI series, vol 196. Springer, Boston, pp 571–583. https://doi.org/10.1007/978-1-4899-0858-2_40

Cox TM, Ragen TJ, Read AJ, Vos E, Baird RW, Balcomb K, Barlow J, Caldwell J, Cranford T, Crum L, D'Amico A, D'Spain G, Fernández A, Finneran J, Gentry R, Gerth W, Gulland F, Hildebrand J, Houser D, Hullar T, Jepson PD, Ketten D, MacLeod CD, Miller P, Moore S, Mountain DC, Palka D, Ponganis

P, Rommel S, Rowles T, Taylor B, Tyack P, Wartzok D, Gisiner R, Mead J, Benner L (2006) Understanding the impacts of anthropogenic sound on beaked whales. J Cetacean Res Manage 7(3):177–187. https://journal.iwc.int/index.php/jcrm/article/view/729/453

Darling JD, Jones ME, Nicklin CP (2006) Humpback whale songs: do they organize males during the breeding season? Behaviour 143:1051–1101. https://doi.org/10.1163/156853906778607381

DAWE (2021) Guidance on key terms within the blue whale conservation management plan. Australian Government Department of Agriculture, Water and the Environment Publication, Canberra. https://www.dcceew.gov.au/sites/default/files/documents/guidance-key-terms-blue-whale-conservation-management-plan-2021.pdf

DFO (2021) Identification of areas for mitigation of vessel-related threats to survival and recovery for Southern Resident Killer Whales. DFO Can Sci Advis Sec Sci Advis Rep 2021/025

DoE (2012) Conservation Management plan for the southern right whale. Australian Government Department of Sustainability, Environment, Water, Population and Communities, Public Affairs, Canberra

DoE (2015) Conservation management plan for the blue whale. Australian Government Department of the Environment, Canberra

Duncan AJ, Weilgart LS, Leaper R, Jasny M, Livermore S (2017) A modelling comparison between received sound levels produced by a marine Vibroseis array and those from an airgun array for some typical seismic survey scenarios. Mar Pollut Bull 119(1):277–288. https://doi.org/10.1016/j.marpolbul.2017.04.001

Dunlop RA, Noad MJ, McCauley RD, Kniest E, Slade R, Paton D, Cato DH (2018) A behavioural dose-response model for migrating humpback whales and seismic air gun noise. Mar Pollut Bull 133:506–516. https://doi.org/10.1016/j.marpolbul.2018.06.009

Durette-Morin D, Evers C, Johnson HD, Kowarski K, Delarue J, Moors-Murphy H, Maxner E, Lawson JW, Davies KTA (2022) The distribution of North Atlantic right whales in Canadian waters from 2015-2017 revealed by passive acoustic monitoring. Front Mar Sci 9:976044. https://doi.org/10.3389/fmars.2022.976044

Erbe C (2013) International regulation of underwater noise. Acoust Aust 41(1):12–19

Erbe C, Reichmuth C, Cunningham KC, Lucke K, Dooling RJ (2016) Communication masking in marine mammals: a review and research strategy. Mar Pollut Bull 103:15–38. https://doi.org/10.1016/j.marpolbul.2015.12.007

Erbe C, Dunlop R, Dolman S (2018) Effects of noise on marine mammals. In: Slabbekoorn H, Dooling RJ, Popper AN, Fay RR (eds) Effects of anthropogenic noise on animals. Springer, New York, pp 277–309. https://doi.org/10.1007/978-1-4939-8574-6_10

Erbe C, Dähne M, Gordon J, Herata H, Houser DS, Koschinski S, Leaper R, McCauley R, Miller B, Müller M, Murray A, Oswald JN, Scholik-Schlomer AR, Schuster M, van Opzeeland IC, Janik VM (2019a) Managing the effects of noise from ship traffic, seismic surveying and construction on marine mammals in Antarctica. Front Mar Sci 6:647. https://doi.org/10.3389/fmars.2019.00647

Erbe C, Marley S, Schoeman R, Smith JN, Trigg L, Embling CB (2019b) The effects of ship noise on marine mammals—a review. Front Mar Sci 6:606. https://doi.org/10.3389/fmars.2019.00606

European Commission (2008) Directive 2008/56/EC of the European Parliament and of the Council of 17 June 2008, establishing a framework for community action in the field of marine environmental policy (Marine Strategy Framework Directive). Off J Eur Union L164:19–40

European Commission (2022) Zero pollution and biodiversity: first ever EU-wide limits for underwater noise. https://environment.ec.europa.eu/news/zero-pollution-and-biodiversity-first-ever-eu-wide-limits-underwater-noise-2022-11-29_en. Accessed 28 June 2023

European Union (2017) Consolidated text: Directive 2008/56/EC of the European Parliament and of the Council of 17 June 2008 establishing a framework for community action in the field of marine environmental policy (Marine Strategy Framework Directive) (Text with EEA relevance), vol OJ L 164 25.6.2008

Farcas A, Thompson PM, Merchant ND (2016) Underwater noise modelling for environmental impact assessment. Environ Impact Assess Rev 57:114–122. https://doi.org/10.1016/j.eiar.2015.11.012

Finneran JJ (2016) Auditory weighting functions and TTS/PTS exposure functions for marine mammals exposed to underwater noise. Space and Naval Warfare Systems Center Pacific, San Diego, USA

Gedamke J, Gales N, Frydman S (2011) Assessing risk of baleen whale hearing loss from seismic surveys: the effect of uncertainty and individual variation. J Acoust Soc Am 129(1):496–506. https://doi.org/10.1121/1.3493445

Gillespie D, Berggren P, Brown S, Kuklik I, Lacey C, Lewis T, Matthews J, McLanaghan R, Moscrop A, Tregenza N (2005) Relative abundance of harbour porpoises (*Phocoena phocoena*) from acoustic and visual surveys of the Baltic sea and adjacent waters during 2001 and 2002. J Cetacean Res Manag 7(1):51–57. https://journal.iwc.int/index.php/jcrm/article/view/757/480

Guan S, Reyff J (2020) Underwater sound source characteristics from down-the-hole pile driving. J Acoust Soc Am 148(4_Suppl):2627. https://doi.org/10.1121/1.5147306

Guazzo RA, Helble TA, Alongi GC, Durbach IN, Martin CR, Martin SW, Henderson EE (2020) The Lombard effect in singing humpback whales: source levels increase as ambient ocean noise levels increase. J Acoust Soc Am 148(2):542–555. https://doi.org/10.1121/10.0001669

Halliday WD, Pine MK, Insley SJ (2020) Underwater noise and Arctic marine mammals: review and policy recommendations. Environ Rev 28(4):438–448. https://doi.org/10.1139/er-2019-0033

Harwood J, Wilson B (2001) The implications of developments on the Atlantic Frontier for marine mammals. Cont Shelf Res 21(8–10):1073–1093. https://doi.org/10.1016/s0278-4343(00)00125-4

Hawkins AD, Popper AN (2018) Effects of man-made sound on fishes. In: Slabbekoorn H, Dooling RJ, Popper AN, Fay RR (eds) Effects of anthropogenic noise on animals. Springer, New York, pp 145–177. https://doi.org/10.1007/978-1-4939-8574-6_6

Herman LM, Pack AA, Spitz SS, Herman EYK, Rose K, Hakala S, Deakos MH (2013) Humpback whale song: who sings? Behav Ecol Sociobiol 67(10):1653–1663. https://doi.org/10.1007/s00265-013-1576-8

Hernández-Morcillo M, Plieninger T, Bieling C (2013) An empirical review of cultural ecosystem service indicators. Ecol Indic 29:434–444. https://doi.org/10.1016/j.ecolind.2013.01.013

IMO (2014) Guidelines for the reduction of underwater noise from commercial shipping to address adverse impacts on marine life. MEPC.1/Circ.833, vol MEPC.1/Circ.833. International Maritime Organization, United Nations

International Maritime Organization (2023) Draft revised guidelines for the reduction of underwater noise from commercial shipping to address adverse impacts on marine life. MEPC.1/Circ.906. International Maritime Organization, United Nations

Janik VM (2014) Cetacean vocal learning and communication. Curr Opin Neurobiol 28:60–65. https://doi.org/10.1016/j.conb.2014.06.010

Johnson HD, Taggart CT, Newhall AE, Lin Y-T, Baumgartner MF (2022) Acoustic detection range of right whale upcalls identified in near-real time from a moored buoy and a Slocum glider. J Acoust Soc Am 151(4):2558–2575. https://doi.org/10.1121/10.0010124

Joint Nature Conservation Committee (2017) JNCC guidelines for minimizing the risk of injury to marine mammals from geophysical surveys. Joint Nature Conservation Committee. https://data.jncc.gov.uk/data/e2a46de5-43d4-43f0-b296-c62134397ce4/jncc-guidelines-seismicsurvey-aug2017-web.pdf

Klinck H, Mellinger DK, Klinck K, Bogue NM, Luby JC, Jump WA, Shilling GB, Litchendorf T, Wood AS, Schorr GS, Baird RW (2012) Near-real-time acoustic monitoring of beaked whales and other cetaceans using a Seaglider™. PLoS One 7(5):e36128. https://doi.org/10.1371/journal.pone.0036128

Kowarski KA, Gaudet BJ, Cole AJ, Maxner EE, Turner SP, Martin SB, Johnson HD, Moloney JE (2020) Near real-time marine mammal monitoring from gliders: practical challenges, system development, and management implications. J Acoust Soc Am 148(3):1215–1230. https://doi.org/10.1121/10.0001811

Lee JA, Southall BL, Nowacek DP, Lewandowski, J (2023) Global alliance for managing ocean noise report, Practical approaches for reducing ocean noise associated with seismic exploration. 16 pp. https://docs.house.gov/meetings/II/II00/20231108/116526/HMKP-118-II00-20231108-SD011.pdf

Lewandowski J, Staaterman E (2020) International management of underwater noise: transforming conflict into effective action. J Acoust Soc Am 147:3160–3168. https://doi.org/10.1121/10.0001173

Markus T, Sánchez PPS (2018) Managing and regulating underwater noise pollution. In: Salomon M, Markus T (eds) Handbook on marine environment protection: science, impacts and sustainable management. Springer, Cham, pp 971–995. https://doi.org/10.1007/978-3-319-60156-4_52

Marotte E, Wright AJ, Breeze H, Wingfield J, Matthews LP, Risch D, Merchant ND, Barclay D, Evers C, Lawson J, Lesage V (2022) Recommended metrics for quantifying underwater noise impacts on North Atlantic right whales. Mar Pollut Bull 175:113361. https://doi.org/10.1016/j.marpolbul.2022.113361

Matthews M-NR, Ireland DS, Zeddies DG, Brune RH, Pyć CD (2021) A modeling comparison of the potential effects on marine mammals from sounds produced by marine vibroseis and air gun seismic sources. J Mar Sci Eng 9(1):12. https://doi.org/10.3390/jmse9010012

McCauley RD, Day RD, Swadling KM, Fitzgibbon QP, Watson RA, Semmens JM (2017) Widely used marine seismic survey air gun operations negatively impact zooplankton. Nat Ecol Evol 1(7):8. https://doi.org/10.1038/s41559-017-0195

Merchant ND (2019) Underwater noise abatement: economic factors and policy options. Environ Sci Pol 92:116–123. https://doi.org/10.1016/j.envsci.2018.11.014

Merchant ND, Faulkner RC, Martinez R (2018) Marine noise budgets in practice. Conserv Lett 11(3):e12420. https://doi.org/10.1111/conl.12420

Merchant ND, Putland RL, André M, Baudin E, Felli M, Slabbekoorn H, Dekeling R (2022) A decade of underwater noise research in support of the European Marine Strategy Framework Directive. Ocean Coast Manag 228:106299. https://doi.org/10.1016/j.ocecoaman.2022.106299

Milcu AI, Hanspach J, Abson D, Fischer J (2013) Cultural ecosystem services: a literature review and prospects for future research. Ecol Soc 18(3):Article No. 44. https://doi.org/10.5751/ES-05790-180344

Miller PJO, Antunes RN, Wensveen PJ, Samarra FIP, Alves AC, Tyack PL, Kvadsheim PH, Kleivane L, Lam F-PA, Ainslie MA, Thomas L (2014) Dose-response relationships for the onset of avoidance of sonar by free-ranging killer whales. J Acoust Soc Am 135(1):975. https://doi.org/10.1121/1.4861346

New LF, Clark JS, Costa DP, Fleishman E, Hindell MA, Klanjšček T, Lusseau D, Kraus S, McMahon CR, Robinson PW, Schick RS, Schwarz LK, Simmons SE, Thomas L, Tyack P, Harwood J (2014) Using short-term measures of behaviour to estimate long-term fitness of southern elephant seals. Mar Ecol Prog Ser 496:99–108. https://doi.org/10.3354/meps10547

Nowacek DP, Broker K, Donovvan G, Gailey G, Racca R, Reeves RR, Vedenev AI, Weller DW, Southall B (2013) Responsible practices for minimising and monitoring environmental impacts of marine seismic surveys with an emphasis on marine mammals. Aquat Mamm 39:356–277. https://doi.org/10.1578/AM.39.4.2013.356

Nowacek DP, Southall BL (2016) Effective planning strategies for managing environmental risk associated with geophysical and other imaging surveys. Gland, Switzerland: IUCN. 42 pp.

Nowacek DP, Christiansen F, Bejder L, Goldbogen JA, Friedlaender AS (2016) Studying cetacean behaviour: new technological approaches and conservation applications. Anim Behav 120:235–244. https://doi.org/10.1016/j.anbehav.2016.07.019

Pirotta E, Booth CG, Cade DE, Calambokidis J, Costa DP, Fahlbusch JA, Friedlaender AS, Goldbogen JA, Harwood J, Hazen EL, New L, Southall BL (2021) Context-dependent variability in the predicted daily energetic costs of disturbance for blue whales. Conserv Physiol 9(1):coaa137. https://doi.org/10.1093/conphys/coaa137

Racca R, Austin M, Rutenko A, Bröker K (2015) Monitoring the gray whale sound exposure mitigation zone and estimating acoustic transmission during a 4-D seismic survey, Sakhalin Island, Russia. Endanger Species Res 29(2):131–146. https://doi.org/10.3354/esr00703

Rolland RM, Parks SE, Hunt KE, Castellote M, Corkeron PJ, Nowacek DP, Wasser SK, Kraus SD (2012) Evidence that ship noise increases stress in right whales. Proc R Soc Lond B 279(1737):2363–2368. https://doi.org/10.1098/rspb.2011.2429

Sigray P, Andersson M, André M, Azzellino A, Borsani JF, Bou M, Castellote M, Ceyrac L, Dellong D, Folegot T, Hedgeland D, Juretzek C, Klauson A, Leaper R, Le Courtois F, Liebschner A, Maglio A, Mueller A, Norro A, Novellino A, Outinen O, Popit A, Prospathopoulos A, Thomsen F, Tougaard J, Vukadin P, Weilgart L (2023) Setting EU threshold values for impulsive underwater sound, Technical Group on Underwater Noise (TG NOISE), MSFD Common Implementation Strategy (Druon J, Hanke G, Casier M, eds). Publications Office of the European Union, Luxembourg. https://doi.org/10.2760/60215, JRC133477

Simard Y, Giard S, Roy N, Aulanier F, Lesage V (2023) Mesoscale habitat use by St. Lawrence estuary beluga over the annual cycle from an acoustic recording network. J Acoust Soc Am 154(2):635–649. https://doi.org/10.1121/10.0020534

Slabbekoorn H, Bouton N, van Opzeeland I, Coers A, ten Cate C, Popper AN (2010) A noisy spring: the impact of globally rising underwater sound levels on fish. Trends Ecol Evol 25(7):419–427. https://doi.org/10.1016/j.tree.2010.04.005

Smith HR, Zitterbart DP, Norris TF, Flau M, Ferguson EL, Jones CG, Boebel O, Moulton VD (2020) A field comparison of marine mammal detections via visual, acoustic, and infrared (IR) imaging methods offshore Atlantic Canada. Mar Pollut Bull 154:111026. https://doi.org/10.1016/j.marpolbul.2020.111026

Southall BL, Finneran JJ, Rcichmuth C, Nachtigall PE, Ketten DR, Bowles AE, Ellison WT, Nowacek DP, Tyack PL (2019a) Marine mammal noise exposure criteria: updated scientific recommendations for residual hearing effects. Aquat Mamm 45(2):125–232. https://doi.org/10.1578/am.45.2.2019.125

Southall BL, Finneran JJ, Reichmuth C, Nachtigall PE, Ketten DR, Bowles AE, Ellison WT, Nowacek DP, Tyack PL (2019b) Marine mammal noise exposure criteria: updated scientific recommendations for residual hearing effects. Aquat Mamm 45(2):125–232. https://doi.org/10.1578/AM.45.2.2019.12

Southall BL, DeRuiter SL, Friedlaender A, Stimpert AK, Goldbogen JA, Hazen E, Casey C, Fregosi S, Cade DE, Allen AN, Harris CM, Schorr G, Moretti D, Guan S, Calambokidis J (2019c) Behavioral responses of individual blue whales (*Balaenoptera musculus*) to mid-frequency military sonar. J Exp Biol 222(5):jeb190637. https://doi.org/10.1242/jeb.190637

Southall BL, Nowacek DP, Bowles AE, Senigaglia V, Bejder L, Tyack PL (2021a) Marine mammal noise exposure criteria: assessing the severity of marine mammal behavioral responses to human noise. Aquat Mamm 47(5):421–464. https://doi.org/10.1578/AM.47.5.2021.421

Southall B, Ellison W, Clark C, Tollit D, Amaral J (2021b) Marine mammal risk assessment for Gulf of Mexico G&G activities. US Department of the Interior, Bureau of Ocean Energy Management. OCS Study BOEM 2021-002

Southall BL, Dominic T, Jennifer A, Clark CW, Ellison WT (2023) Managing human activity and marine mammals: a biologically based, relativistic risk assessment framework. Front Mar Sci 10:1090132. https://doi.org/10.3389/fmars.2023.1090132

Tal D, Shachar-Bener H, Hershkovitz D, Arieli Y, Shupak A (2015) Evidence for the initiation of decompression sickness by exposure to intense underwater sound. J Neurophysiol 114(3):1521–1529. https://doi.org/10.1152/jn.00466.2015

Teilmann J, Carstensen J (2012) Negative long term effects on harbour porpoises from a large scale offshore wind farm in the Baltic—evidence of slow recovery. Environ Res Lett 7(4):045101. https://doi.org/10.1088/1748-9326/7/4/045101

Thornton SJ, Toews S, Burnham R, Konrad CM, Stredulinsky E, Gavrilchuk K, Thupaki P, Vagle S (2022a) Areas of elevated risk for vessel-related physical and acoustic impacts in Southern Resident Killer Whale (Orcinus orca) critical habitat. DFO Can Sci Advis Sec Res Doc 2022/058, vi + 47 p

Thornton SJ, Toews S, Stredulinsky E, Gavrilchuk K, Konrad C, Burnham R, Noren DP, Holt MM, Vagle

S (2022b) Southern Resident Killer Whale (Orcinus orca) summer distribution and habitat use in the southern Salish Sea and the Swiftsure Bank area (2009 to 2020). DFO Can Sci Advis Sec Res Doc 2022/037, v + 56 p

United Nations (2018) Ocean resolution adopted by the General Assembly on 11 December 2018. United Nations Law of the Sea. https://undocs.org/en/A/RES/73/124

Vagle S. Evaluation of the efficacy of the Juan de Fuca lateral displacement trial and Swiftsure Bank plus Swanson Channel interim sanctuary zones, 2019. Can Technol Rep Hydrogr Ocean Sci 2020;332:6–0

Vagle S, Burnham R, Thupaki P, Konrad C, Toews S, Thornton SJ (2021) Vessel presence and acoustic environment within Southern Resident Killer Whale (Orcinus orca) critical habitat in the Salish Sea and Swiftsure Bank area. DFO Can Sci Advis Sec Res Doc 2021/058, x + 66 p

Van Parijs SM, Clark CW, Sousa RS, Parks SE, Rankin S, Risch D, Van Opzeeland IC (2009) Management and research applications of real-time and archival passive acoustic sensors over varying temporal and spatial scales. Mar Ecol Prog Ser 395:21–36. https://doi.org/10.3354/meps08123

Verfuss UK, Gillespie D, Gordon J, Marques TA, Miller B, Plunkett R, Theriault JA, Tollit DJ, Zitterbart DP, Hubert P (2018) Comparing methods suitable for monitoring marine mammals in low visibility conditions during seismic surveys. Mar Pollut Bull 126:1–18. https://doi.org/10.1016/j.marpolbul.2017.10.034

Wale MA, Simpson SD, Radford AN (2013) Noise negatively affects foraging and antipredator behaviour in shore crabs. Anim Behav 86(1):111–118. https://doi.org/10.1016/j.anbehav.2013.05.001

Wells RS (2014) Social structure and life history of bottlenose dolphins near Sarasota Bay, Florida: insights from four decades and five generations. In: Yamagiwa J, Karczmarski L (eds) Primates and cetaceans: field research and conservation of complex mammalian societies. Springer Japan, Tokyo, pp 149–172. https://doi.org/10.1007/978-4-431-54523-1_8

Williams R, Erbe C, Duncan A, Nielsen K, Washburn T, Smith C (2022) Noise from deep-sea mining may span vast ocean areas. Science 377(6602):157–158. https://doi.org/10.1126/science.abo2804

Zimmer WMX, Tyack P (2007) Repetitive shallow dives pose decompression risk in deep-diving beaked whales. Mar Mamm Sci 23:888–925. https://doi.org/10.1111/j.1748-7692.2007.00152.x

Open Access This chapter is licensed under the terms of the Creative Commons Attribution 4.0 International License (http://creativecommons.org/licenses/by/4.0/), which permits use, sharing, adaptation, distribution and reproduction in any medium or format, as long as you give appropriate credit to the original author(s) and the source, provide a link to the Creative Commons license and indicate if changes were made.

The images or other third party material in this chapter are included in the chapter's Creative Commons license, unless indicated otherwise in a credit line to the material. If material is not included in the chapter's Creative Commons license and your intended use is not permitted by statutory regulation or exceeds the permitted use, you will need to obtain permission directly from the copyright holder.

GPSR Compliance

The European Union's (EU) General Product Safety Regulation (GPSR) is a set of rules that requires consumer products to be safe and our obligations to ensure this.

If you have any concerns about our products, you can contact us on ProductSafety@springernature.com

In case Publisher is established outside the EU, the EU authorized representative is:

Springer Nature Customer Service Center GmbH
Europaplatz 3
69115 Heidelberg, Germany

Batch number: 09745571

Printed by Printforce, the Netherlands